CAMBRIDGE

Higher

MATHEMATICS

GCSE for OCR

Student Book

Karen Morrison, Julia Smith, Pauline McLean, Rachael Horsman and Nick Asker

CAMBRIDGE
UNIVERSITY PRESS

University Printing House, Cambridge CB2 8BS, United Kingdom

Cambridge University Press is part of the University of Cambridge.

It furthers the University's mission by disseminating knowledge in the pursuit of education, learning and research at the highest international levels of excellence.

www.cambridge.org

Information on this title:
www.cambridge.org/9781107448056 (Paperback)
www.cambridge.org/9781107449909 (1 Year Online Subscription)
www.cambridge.org/9781107449855 (2 Year Online Subscription)
www.cambridge.org/9781107447967 (Paperback + Online Subscription)

First published 2015

Printed in Dubai by Oriental Press

A catalogue record for this publication is available from the British Library

ISBN 978-1-107-44805-6 Paperback
ISBN 978-1-107-44990-9 1 Year Online Subscription
ISBN 978-1-107-44985-5 2 Year Online Subscription
ISBN 978-1-107-44796-7 Paperback + Online Subscription

Additional resources for this publication at www.cambridge.org/ukschools

This resource is endorsed by OCR for use with specification J560 GCSE Mathematics.

In order to gain OCR endorsement this resource has undergone an independent quality check. OCR has not paid for the production of this resource, nor does OCR receive any royalties from its sale. For more information about the endorsement process please visit the OCR website **www.ocr.org.uk**

Contents

Find answers at: cambridge.org/ukschools/gcsemaths-studentbookanswers

Note
The colour of each chapter corresponds to the area of maths that it covers:

- Number
- Algebra
- Ratio, proportion and rates of change
- Geometry and measures
- Probability
- Statistics

Introduction

This book has been written by experienced teachers to help build your understanding and enjoyment of the maths you will meet at GCSE.

Each chapter opens with a list of skills that are covered in the chapter. The **real-life applications** section describes an example of how the maths is used in real life.

All chapters build on knowledge that you will have learned in previous years. You might need to revise some topics before starting a chapter. To check your knowledge, answer the questions in the **Before you start ...** table. You can check your answers using the free answer booklet available at **www.cambridge.org/ukschools/gcsemaths-studentbookanswers**. If you answer any questions incorrectly, you might need to revise the topic from your work in earlier years.

The chapters are divided into sections, each covering a single topic. Some chapters may cover topics that you already know and understand. You can use the **Launchpad** to identify the best section for you to start with. Answer the questions in each step. If you find a question difficult to answer correctly, the step suggests the section that you should look at.

Throughout the book, there are features to help you build knowledge and improve your skills:

This means you might need a calculator to work through a question.

This means you should work through a question without using a calculator. If this is not present, you can use a calculator if you need to.

This shows the question is from a past exam paper.

Tip

Tip boxes provide helpful hints.

Calculator tip

Calculator tips help you to use your calculator.

Learn this formula

Learn this formula boxes contain formulae that you need to know.

Key vocabulary

Important maths terms are written in **green**. You can find what they mean in **Key vocabulary** boxes and also in the **Glossary** at the back of the book.

Did you know?

Did you know? boxes contain interesting maths facts.

WORK IT OUT

Work it out boxes contain a question with several worked solutions. Some of the solutions contain common mistakes. Try to spot the correct solution and check the free answer booklet available at **www.cambridge.org/ukschools/gcsemaths-studentbookanswers** to see if you're right.

WORKED EXAMPLE

Worked examples guide you through model answers to help you understand methods of answering questions.

Some chapters contain a **Problem-solving framework**, which sets a problem and then shows how you can go about answering it.

Checklist of learning and understanding

At the end of a chapter, use the **Checklist of learning and understanding** to check whether you have covered everything you need to know.

Chapter review

You can check whether you have understood the topics using the **Chapter review,** which contains questions from the whole chapter.

A booklet containing answers to all exercises is free to download at **www.cambridge.org/ukschools/gcsemaths-studentbookanswers**.

You can find more resources, including interactive widgets, games and quizzes on **GCSE Mathematics Online**.

Acknowledgements

Questions from OCR past question papers © OCR.

These questions are indicated by .

Questions from Cambridge IGCSE® Mathematics reproduced with permission of Karen Morrison and Nick Hamshaw.

The authors would like to thank Fran Wilson for her work on GCSE Mathematics Online.

Cover © 2013 Fabian Oefner www.fabianoefner.com; p1 (top) Henry Gan/Photodisc/Thinkstock; p1 Denis Kuvaev p5 StockLite/Shutterstock; p12 (top) Yuriy S/iStock/Thinkstock; p12 Tyler Olson/Shutterstock; p20 Elnur/Shutterstock; p23 (top) agsandrew/iStock/Thinkstock; p23 wavebreakmedia/Shutterstock; p35 Andresr/Shutterstock; p38 (top) © PhotoAlto/Alamy; p38 © Monty Rakusen/Cultura/Corbis; p41 © Roger Bamber/Alamy; p50 (top) murengstockphoto/Thinkstock; p50 (bottom) Shaiith/Shutterstock; p59 (top) Monarx3d/iStock/Thinkstock; p67 kilukilu/Shutterstock; p68 (top) Huang Zheng/Shutterstock; p68 (bottom) Marafona/Shutterstock; p73 © Owen Franken/Corbis; p78 (top) Huskyomega/iStock/Thinkstock; p78 Franz Pfluegl/Shutterstock; p78/80/81 Yulia Glam/Shutterstock; p89 (left) UMB-O/Shutterstock; p89 (right) Vector House/Shutterstock; p93 Gemenacom/Shutterstock; p94 (top) Leigh Prather/iStock/Thinkstock; p94 Dariush M/Shutterstock; p117 (top) Alison Bradford Photography/iStock/Thinkstock; p117 Andrey_Popov/Shutterstock; p144 © Manuel Sulzer/cultura/Corbis; p154 (top) Mark_Stillwagon/iStock/Thinkstock; p154 Mikio Oba/Shutterstock; p159 Dmitry Kalinovsky/Shutterstock; p166 Ryan Lewandowski/Shutterstock; p174 (top) SDivin09/iStock/Thinkstock; p174 Chameleons Eye/Shutterstock; p181 Jon Milnes/Shutterstock; p186 (top) Andrey Popov/iStock/Thinkstock; p186 Leah-Anne Thompson/Shutterstock; p188 William West/Staff/Getty Images; p190 © Jumana el Heloueh/Reuters/Corbis; p198 (top) Mike Watson Images/moodboard/Thinkstock; p198 Dorling Kindersley/Getty; p205 Taiga/Shutterstock; p207 Brenda Carson/Shutterstock; p222 (top) shutter_m/iStock/Thinkstock; p228 tr3gin/Shutterstock; p237 (top) deyangeorgiev/iStock/Thinkstock; p237 Leah-Anne Thompson/Shutterstock; p245l © PCN Photography/Alamy; p245 (right) Laguna Design/Getty; p249 Suppakij1017/Shutterstock; p252 (top) David Chapman/Design pics/Valueline/Thinkstock; p252 itman__47/Shutterstock; p262 Karl Weatherly/Thinkstock; p263 iceink/Shutterstock; p270 JLRphotography/Shutterstock; p273 (top) imagean/iStock/Thinkstock; p273 Federico Rostagno/Shutterstock; p280 Stefano Spezi/Shutterstock; p292 Calvste/Shutterstock; p295 (top) Joe McDaniel/iStock/Thinkstock; p295 Dmitry Kalinovsky/Shutterstock; p266 Ilya Akinshin/Shutterstock; p305 Pixsooz/Shutterstock; p314 (top) Kerem Yucel/iStock/Thinkstock; p314 Dan Breckwoldt/Shutterstock; p340 (top) shaunnessey/iStock/Thinkstock; p340 © Jenny E. Ross/Corbis; p346 herreid/Thinkstock; p364 (top) alfimimnill/iStock/Thinkstock; p368 (left) Thomas Demarczyk/Thinkstock; p368 (right) Vitaly Edush/Thinkstock; p369 (left) Anthony Baggett/Thinkstock; p369 (right) merial/Thinkstock; p383 (top) Phil Ashley/Photodisc/Thinkstock; p383 govicinity/Shutterstock; p384 Isantilli/Shutterstock; p388 Dmitry Kalinovsky/Shutterstock; p390 Sergio Bertino/Shutterstock; p393 © Daniel Mogan/Alamy; p395 (left) Courtesy of Newcastle International Airport p395 (right) © 2ebill/Alamy; p398 Chameleons Eye/Shutterstock; p399 kravka/Shutterstock; p400 cpphotoimages/Shutterstock; p401 (top) Eugen Weide/Hemera/Thinkstock; p401 T. Fabian/Shutterstock; p403 Anteromite/Shutterstock; p405 zentilia/Shutterstock; p406 Adam Gilchrist/Shutterstock; p409 Valentyn Volkov/Shutterstock; p413 (top) Chemik11/iStock/Thinktstock; p413 racorn/Shutterstock; p415 Galina Mikhalishina/Shutterstock; p418 Lisa F. Young/Shutterstock; p423 Tamara Kulikova/Shutterstock; p430 (top) maury75/iStock/Thinkstock; p430 angellodeco/Shutterstock; p438 Monkey Business Images/Shutterstock; p452 (top) Jeffrey Collingwood/Hemera/Thinkstock; p452 Konstantin Chagin/Shutterstock; p464 Sofia Andreevna/Shutterstock; p468 (top) agsandrew/iStock/Thinkstock; p468 Sergey Kamshylin/Shutterstock; p478 Iakov Kalinin/Shutterstock; p482 (top) LongHa2006/iStock/Thinkstock; p482 Huntstock.com/Shutterstock; p498 (top) sutichak/iStock/Thinkstock; p498 Michael Winston Rosa/Shutterstock; p502 Herbert Kratky/Shutterstock; p510 (top) ninjacpb/iStock/Thinkstock; p512 Pal Teravagimov/Shutterstock; p542 (top) Comstock/Stockbyte/Thinkstock; p542 Kletr/Shutterstock; p566 (top) stevanovicigor/iStock/Thinkstock; p566 © Emma Smales/VIEW/Corbis; p570 racorn/Shutterstock; p577 Nicole Gordine/Shutterstock; p578 David Hilcher/Shutterstock; p581 (top) Digital Vision./Photodisc/Thinkstock; p581 GECO UK/Science Photo Library; p585 © Andrew Ammendolia/Alamy; p611 Svetlana Jafarova/Shutterstock; p614 (top) sdecoret/iStock/Thinkstock; p614 Marketa Jirouskova p632 (top) shutter_m/iStock/Thinkstock; p632 Pressmaster/Shutterstock; p637 vetroff/Shutterstock; p643 (top) Karol Kozlowski/iStock/Thinkstock; p643 Michael Jung/Shutterstock; p644 terekhov igor/Shutterstock; p648 Christian Delbert/Shutterstock; p658 (top) cherezoff/iStock/Thinkstock; p658 Chad McDermott/Shutterstock; p661 © Detail Nottingham/Alamy; p675 Filip Fuxa/Shutterstock; p689 (top) marekuliasz/iStock/Thinkstock; p689 Brian A. Jackson/Shutterstock; p692 Peter G./Shutterstock; p714 (top) zentilia/iStock/Thinkstock; p714 Zryzner/Shutterstock; p724 pfshots/Shutterstock; p730 (top) Tabor Gus/Corbis; p730 Vadim Ratnikov/Shutterstock; p746 (top) MarcelC/iStock/Thinkstock; p746 Balefire/Shutterstock; p749 Kristina Postnikova/Shutterstock.

1 Basic calculation skills

In this chapter you will learn how to …

- use non-calculator methods to calculate with positive and negative integers.
- perform operations in the correct order based on mathematical conventions.
- recognise inverse operations and use them to simplify and check calculations.

For more resources relating to this chapter, visit GCSE Mathematics Online.

Using mathematics: real-life applications

Everyone uses numbers on a daily basis often without really thinking about them. Shopping, cooking, working out bills, paying for transport and measuring all rely on a good understanding of numbers and calculation skills.

Tip

You probably already know most of the concepts in this chapter. They have been included so that you can revise concepts if you need to and check that you know them well.

"Number puzzles and games are very popular and there are mobile apps and games available for all age groups. Our website offers free games where you have to identify the correct order of operations to use to solve different number puzzles."

(Website designer)

Before you start …

KS3	You should be able to add, subtract, multiply and divide positive and negative numbers.	**1** Copy and complete each statement to make it true. Use only <, = or >. **a** $2 + 3 \square 4 - 7$ **b** $^{-}3 + 6 \square 4 - 7$ **c** $^{-}1 - 4 \square 20 \div {}^{-}4$ **d** $^{-}6 \times 2 \square {}^{-}7 - {}^{-}5$
KS3	You should know the rules for working when more than one operation is involved in a calculation (BIDMAS).	**2** Spot the mistake in each calculation and correct the answers. **a** $3 + 8 + 3 \times 4 = 56$ **b** $3 + 8 \times 3 + 4 = 37$ **c** $3 \times (8 + 3) \times 4 = 130$
KS3	You should understand that addition and subtraction, and multiplication and division, are inverse operations.	**3** Identify the inverse operation by choosing the correct option. **a** $14 \times 4 = 56$ A $56 \times 4 = 14$ B $14 \div 4 = 56$ C $56 \div 4 = 14$ **b** $200 \div 10 = 20$ A $200 \div 20 = 10$ B $200 = 10 \times 20$ C $10 \times 200 = 2000$ **c** $27 + 53 = 80$ A $80 = 4 \times 20$ B $80 - 27 = 53$ C $80 + 27 = 107$

Find answers at: cambridge.org/ukschools/gcsemaths-studentbookanswers

Assess your starting point using the Launchpad

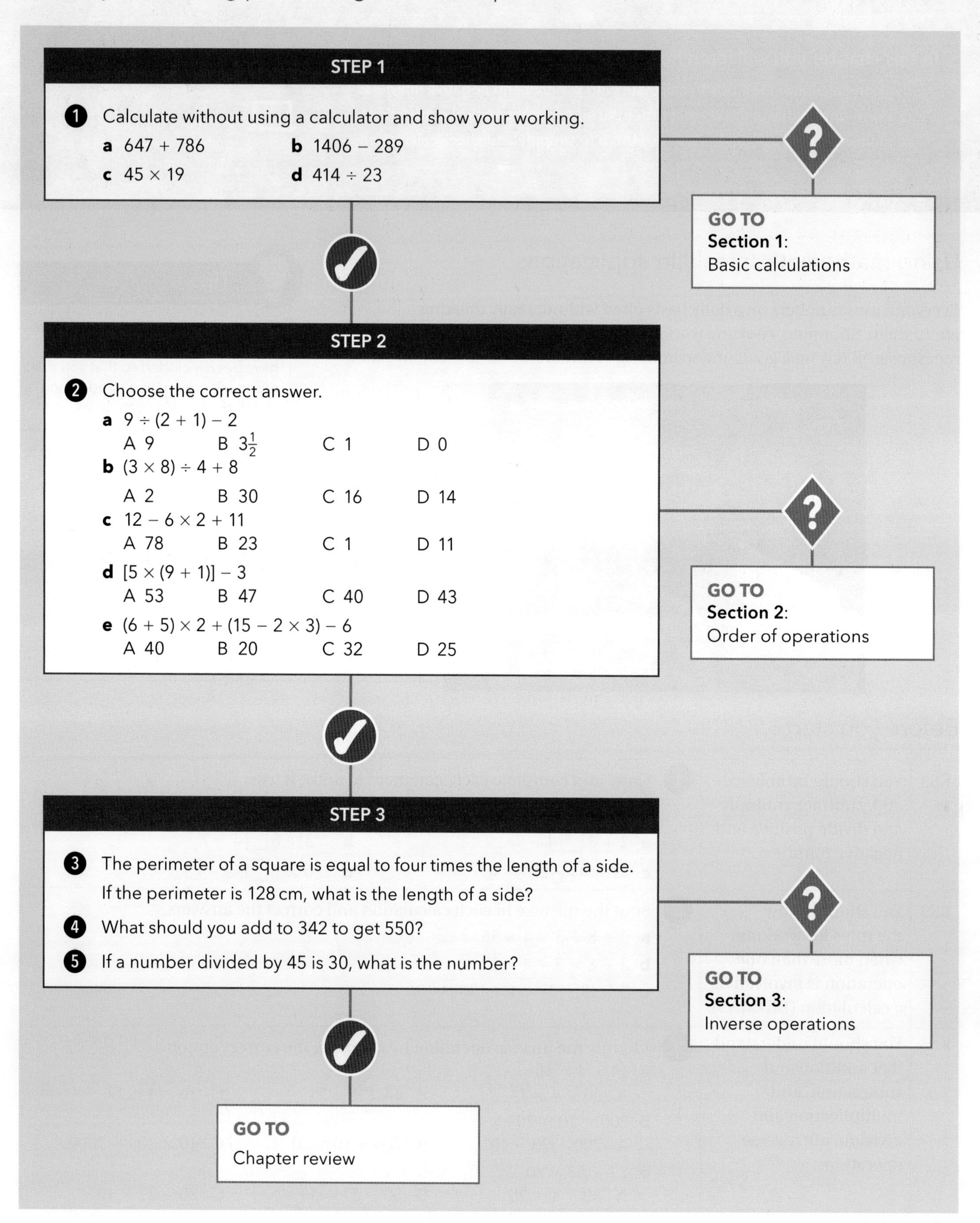

STEP 1

1. Calculate without using a calculator and show your working.
   a $647 + 786$ **b** $1406 - 289$
   c 45×19 **d** $414 \div 23$

? **GO TO Section 1:** Basic calculations

✓

STEP 2

2. Choose the correct answer.
   a $9 \div (2 + 1) - 2$
   A 9 B $3\frac{1}{2}$ C 1 D 0
   b $(3 \times 8) \div 4 + 8$
   A 2 B 30 C 16 D 14
   c $12 - 6 \times 2 + 11$
   A 78 B 23 C 1 D 11
   d $[5 \times (9 + 1)] - 3$
   A 53 B 47 C 40 D 43
   e $(6 + 5) \times 2 + (15 - 2 \times 3) - 6$
   A 40 B 20 C 32 D 25

? **GO TO Section 2:** Order of operations

✓

STEP 3

3. The perimeter of a square is equal to four times the length of a side. If the perimeter is 128 cm, what is the length of a side?
4. What should you add to 342 to get 550?
5. If a number divided by 45 is 30, what is the number?

? **GO TO Section 3:** Inverse operations

✓

GO TO Chapter review

Section 1: Basic calculations

You will not always have a calculator, so it is useful to know how to do calculations using mental and written strategies.

It is best to use a method that you are confident with and always **show your working**.

Remember that when a question asks you to find the:

- **sum** you need to add.
- **difference** you need to subtract one number from another.
- **product** you need to multiply.
- **quotient** you need to divide.

Tip

These are the kind of skills that are tested in the non-calculator examination papers.

WORK IT OUT 1.1

Look at these calculations carefully.

Discuss with a partner what methods these students have used to find the answer.

Which method would you use to do each of these calculations? Why?

① 489 + 274

400 + 200 → 600
80 + 70 → 150
9 + 4 → 13
763

② 284 − 176

2^7^8^1^4
− 176
108

③ 29 × 17

→ 30 × 17 − 17
→ 3 × 170 − 17
→ 510 − 17
→ 493

④ 15 × 62

= 30 × 31 310
= 930 310
310
930

⑤ 207 × 47

x	200	0	7
40	8000	0	280
7	1400	0	49

9400 + 0 + 329
= 9729

⑥ 2394 ÷ 42

2394
− 1680 (40)
714
− 420 (10)
294
− 210 (5)
84
− 84 (2) = 57
= 0

42 × 10 = 420
42 × 20 = 840
42 × 40 = 1680
42 × 5 = 210
42 × 2 = 84

Problem-solving strategies

Many of the written calculations that you have to do will be to solve word problems.

Here are some useful strategies and techniques to use for problem-solving.

The problem-solving framework below outlines steps to break down problems to help you solve them more easily.

Follow these steps each time you are faced with a problem to become more skilled at problem-solving and more able to self-check.

These are important skills both for your GCSE courses and for everyday life.

Problem-solving framework

Sally buys, repairs and sells used furniture at a market.

Last week, she bought a table for £32 and a bench for £18.

She spent £12 on wood, nails, varnish and glue to fix them up.

She then sold the two items on her stall for £69.

How much profit did she make on the two items?

Steps for approaching a problem-solving question	**What you would do for this example**
Step 1: Work out what you have to do. Start by reading the question carefully.	Find the profit on the two items.
Step 2: What information do you need? Have you got it all?	Cost of items = £32 + £18 Cost of repairs = £12 Selling price = £69
Step 3: Is there any information that you don't need?	In this problem you don't need to know what she spent money on for repairs. You just need to know how much she spent. Many problems contain extra information that you don't need to test your understanding.
Step 4: Decide what maths you can do.	Profit = selling price – cost You can add the costs and subtract them from the selling price.
Step 5: Set out your solution clearly. Check your working and make sure your answer is reasonable.	Cost = £32 + £18 + £12 = £62 Profit = £69 − £62 = £7 Sally made £7 profit.
Step 6: Check that you have answered the question.	Yes, you needed to find the profit and you have found it.

EXERCISE 1A

Solve these problems using written methods.

Set out your solutions clearly to show the methods you chose.

Tip

You don't always need to write anything for the first few steps in the problem-solving framework, but you should try to mentally answer the questions as you read problems to decide what to do. You should always show how you worked to solve the problem.

1. After checking the prices at three different supermarkets, Nola found out that the cheapest pack of pens was £3.90 for three. She bought 15 pens.

   How much did she pay in total and what does this work out to per pen?

   a What two things are you asked to find here?

   b How many packs of pens did she buy? Why do you need to know this?

   c How much does she pay for the 15 pens?

   d What is the cost of each pen?

2. Sandra bought a pair of jeans for £34, a scarf for £9.50 and a top for £20.

   If she had saved £100 to buy these items, how much money would she have left?

3. How many 16-page brochures can you make from 1030 pages?

4. Jason can type 48 words per minute.

   a How many words can he type in an hour and a half?

   b Approximately how long would it take him to type an article of 2000 words?

5. At the start of a year, the population of Greenside Village was 56 309.

   During the year 617 people died, 1835 babies were born, 4087 people left the village and 3099 people moved into the village.

   What was the population at the end of the year?

Did you know?

The Severn is the longest river in the UK.

6. The Amazon River is 6448 km long, the Nile River is 6670 km and the Severn is 354 km long.

   a How much longer is the Nile than the Amazon?

   b How much shorter is the Severn than the Amazon?

Tip

You will be expected to work with negative and positive values in algebra, so it is important to make sure you can do this early on in your GCSE course.

7. What is the combined sum of 132 plus 99 and the product of 36 and 127?

8. What is the result when the difference between 8765 and 3087 is added to the result of 1206 divided by 18?

Working with negative and positive integers

When doing calculations involving positive and negative **integers**, you need to remember to apply the following rules:

- Adding a negative number is the same as subtracting the number:
  $4 + {}^{-}3 = 1$
- Subtracting a negative number is the same as adding a positive number:
  $5 - {}^{-}3 = 8$
- Multiplying or dividing the same signs gives a positive answer:
  ${}^{-}4 \times {}^{-}2 = 8$ and $\frac{{}^{-}4}{{}^{-}2} = 2$
- Multiplying or dividing different signs gives a negative answer:
  $4 \times {}^{-}2 = {}^{-}8$ and $\frac{{}^{-}4}{2} = {}^{-}2$

Key vocabulary

integers: whole numbers belonging to the set {… ⁻3, ⁻2, ⁻1, 0, 1, 2, 3, …}; they are sometimes called directed numbers because they have a negative or positive sign.

EXERCISE 1B

1 What would you add to each number to get a result of 5?

a 7 **b** 3 **c** $^-1$ **d** $^-4$ **e** $^-24$

2 What would you subtract from each number to get a result of $^-8$?

a 7 **b** 3 **c** $^-1$ **d** $^-4$ **e** $^-24$

3 $^-4$ is multiplied by another number to get each result.

Work out what the other number is in each case.

a 12 **b** $^-100$ **c** $^-36$ **d** 504 **e** 0

4 By what would you divide -64 to get the following results?$^-1/2$

a 8 **b** $^-8$ **c** 2 **d** $\left(^-\frac{1}{2}\right)$ **e** $^-256$

5 Here is a set of integers: {$^-8$, $^-6$, $^-3$, 1, 3, 7}.

From the numbers in this set:

a Find two numbers with a difference of 9.

b Find three numbers with a sum of 1.

c Find two numbers whose product is $^-3$.

d Find two numbers which, when divided, will give an answer of $^-6$.

6 One more than $^-6$ is added to the product of 7 and six less than 3.

What is the result?

7 Saleem has a container of wooden dowels.

Some are 5 cm long and some are 7 cm long.

If the dowels are joined end to end, investigate what lengths between 5 cm and 150 cm **cannot** be made.

Section 2: Order of operations

Jose posted this calculation on his wall on social media.

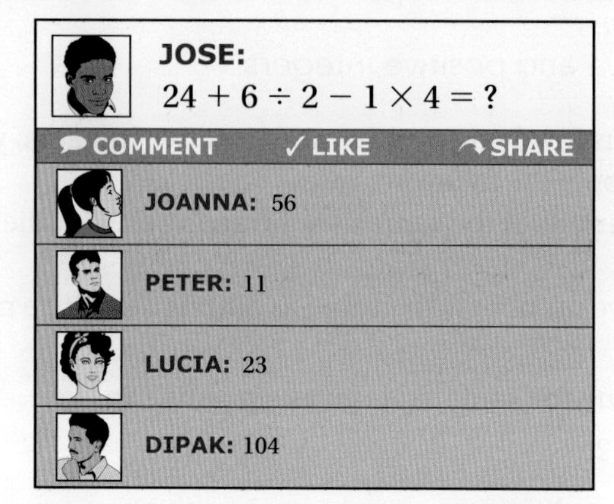

Within minutes, his friends had posted four different answers.

Which one (if any) do you think is correct? Why?

There is a set of rules that tell you the order in which you need to work when there is more than one operation.

The order of operations is:

1 Do any operations in brackets first.

2 If there are any indices in the calculation, do them next. (O in BODMAS is for 'powers of' or 'fractions of'))

3 Do division and multiplication next, working from left to right.

4 Do addition and subtraction last, working from left to right.

Tip

Many people remember these rules using the letters BIDMAS (or sometimes BODMAS).

Brackets
Indices (or **O**rders)
Divide and/or **M**ultiply
Add and/or **S**ubtract

Brackets and symbols

Brackets are used to group operations. For example:

$(3 + 7) \times (30 \div 2)$

When there is more than one set of brackets, it is best to work from the **innermost set** to the **outermost set**.

WORKED EXAMPLE 1

Solve $2((4 + 2) \times 2 - 3(1 - 3) - 10)$

$2((4 + 2) \times 2 - 3(1 - 3) - 10)$

Highlight the different pairs of brackets to help if you need to.

$2((4 + 2) \times 2 - 3(1 - 3) - 10)$
$= 2(6 \times 2 - 3(^-2) - 10)$
$= 2(6 \times 2 - 3 \times {}^-2 - 10)$

The red brackets are the innermost, so do the calculations inside these ones first. There are two lots of red brackets, so work from left to right. **Note** that you can leave $^-2$ inside brackets if you prefer because $3(^-2)$ is the same as $3 \times {}^-2$.

$= 2(12 - {}^-6 - 10)$
$= 2(8)$
$= 2 \times 8$
$= 16$

Black brackets are next. Do the multiplications first from left to right, then the subtractions from left to right.

Often a different style of bracket will be used to make it easier to identify each pair. For example, the following different types of brackets have been used below: ( ), [ ], { }.

$\{2 - [4(2 - 7) - 4(3 + 8)] - 2\} \times 8$

Other symbols can also be used to group operations. For example:

Fraction bars: $\frac{5 - 12}{3 - 8}$

Roots: $\sqrt{16 + 9}$

These symbols are treated like brackets when you do a calculation.

Calculator tip

Most modern calculators are programmed to use the correct order of operations. Check your calculator by entering $2 + 3 \times 4$. You should get 14.

If the calculation has brackets, you need to enter the brackets into the calculator to make sure it does these first.

WORK IT OUT 1.2

Which of the solutions is correct in each case?

Find the mistakes in the incorrect option.

	Option A	Option B
1	$7 \times 3 + 4$ $= 21 + 4$ $= 25$	$7 \times 3 + 4$ $= 7 \times 7$ $= 49$
2	$(10 - 4) \times (4 + 9)^2$ $= 6 \times 16 + 81$ $= 96 + 81$ $= 177$	$(10 - 4) \times (4 + 9)^2$ $= 6 \times (13)^2$ $= 6 \times 169$ $= 1014$
3	$45 - [20 \times (4 - 3)]$ $= 45 - [20 \times 1]$ $= 45 - 21$ $= 24$	$45 - [20 \times (4 - 3)]$ $= 45 - 20 \times 1$ $= 45 - 20$ $= 25$
4	$30 - 4 \div 2 + 2$ $= 26 \div 2 + 2$ $= 13 + 2$ $= 15$	$30 - 4 \div 2 + 2$ $= 30 - 2 + 2$ $= 30$
5	$\frac{18 - 4}{4 - 2}$ $= \frac{18}{2}$ $= 9$	$\frac{18 - 4}{4 - 2}$ $= \frac{14}{2}$ $= 7$
6	$\sqrt{36 \div 4} + 40 \div 4 + 1$ $= \sqrt{9} + 10 + 1$ $= 3 + 11$ $= 14$	$\sqrt{36 \div 4} + 40 \div 4 + 1$ $= \sqrt{9} + 40 \div 5$ $= 3 + 8$ $= 11$

EXERCISE 1C

1. Check whether these answers are correct.
   If not, work out the correct answer.
   - **a** $12 \times 4 + 76 = 124$
   - **b** $8 + 75 \times 8 = 698$
   - **c** $12 \times 18 - 4 \times 23 = 124$
   - **d** $(16 \div 4) \times (7 + 3 \times 4) = 76$
   - **e** $(82 - 36) \times (2 + 6) = 16$
   - **f** $(3 \times 7 - 4) - (4 + 6 \div 2) = 12$

2. Use the numbers listed to make each number sentence true.
   - **a** $\square - \square \div \square = \square$ — 9, 11, 13, 18
   - **b** $\square \div (\square - \square) - \square = \square$ — 1, 3, 8, 14, 16
   - **c** $(\square + \square) - (\square - \square) = \square$ — 4, 5, 6, 9, 12

3 Insert brackets into each calculation to make it true.

a $3 \times 4 + 6 = 30$ **b** $25 - 15 \times 9 = 90$ **c** $40 - 10 \times 3 = 90$
d $14 - 9 \times 2 = 10$ **e** $12 + 3 \div 5 = 3$ **f** $19 - 9 \times 15 = 150$
g $10 + 10 \div 6 - 2 = 5$ **h** $3 + 8 \times 15 - 9 = 66$ **i** $9 - 4 \times 7 + 2 = 45$
j $10 - 4 \times 5 = 30$ **k** $6 \div 3 + 3 \times 5 = 5$ **l** $15 - 6 \div 2 = 12$
m $1 + 4 \times 20 \div 5 = 20$ **n** $8 + 5 - 3 \times 2 = 20$ **o** $36 \div 3 \times 3 - 3 = 6$
p $3 \times 4 - 2 \div 6 = 1$ **q** $40 \div 4 + 1 = 11$ **r** $6 + 2 \times 8 + 2 = 24$

4 Each ○ represents an operation.
Fill in the missing operations to make these statements true.

a 12 ○ (28 ○ 24) = 3 **b** 88 ○ 10 ○ 8 = 8
c 40 ○ 5 ○ (7 ○ 5) = 4 **d** 9 ○ 15 ○ (3 ○ 2) = 12

5 Calculate.

a $\dfrac{7 \times \sqrt{16}}{2^3 + 7^2 - 1}$ **b** $\dfrac{5^2 \times \sqrt{4}}{1 + 6^2 - 12}$ **c** $\dfrac{2 + 3^2}{5^2 + 4 \times 10 - \sqrt{25}}$

d $\dfrac{6^2 - 11}{2(17 + 2 \times 4)}$ **e** $\dfrac{3^2 - 3}{2 \times \sqrt{81}}$ **f** $\dfrac{3^2 - 5 + 6}{\sqrt{4} \times 5}$

g $\dfrac{36 - 3 \times \sqrt{16}}{15 - 3^2 \div 3}$ **h** $\dfrac{^-30 + [18 \div (3 - 12) + 24]}{5 - 8 - 3^2}$

6 Work with a partner.

a Find a quick method for adding a set of consecutive whole numbers.
b Explain why your method works.
c Test your method on a set of consecutive negative integers.
d Does it work? Explain why or why not.

Section 3: Inverse operations

\+ INVERSE −
repeated addition repeated subtraction
× INVERSE ÷

Operations are inverses of each other if one undoes (cancels out) the effect of the other.

- Adding is the inverse of subtracting, e.g. +5 is undone by $^-5$.
- Multiplying is the inverse of dividing, e.g. ×2 is undone by ÷2
- Taking a square root is the inverse of squaring a number 4^2 is undone by $\sqrt{4}$.
- Taking the cube root is the inverse of cubing a number, e.g. 2^3 is undone by $\sqrt[3]{8}$.

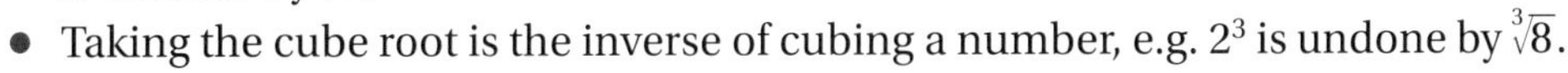

Inverse operations are useful for checking the results of your calculations, because carrying out the inverse operation gets you back to the number you started with.

For example, is $4320 - 500 = 3820$ correct?

Check by adding 500 back to the result (i.e. doing the inverse operation) to see whether it gives you 4320.

$3820 + 500 = 4320$

Tip

You will use inverse operations to solve equations and when you deal with functions, so it is important that you understand how they work.

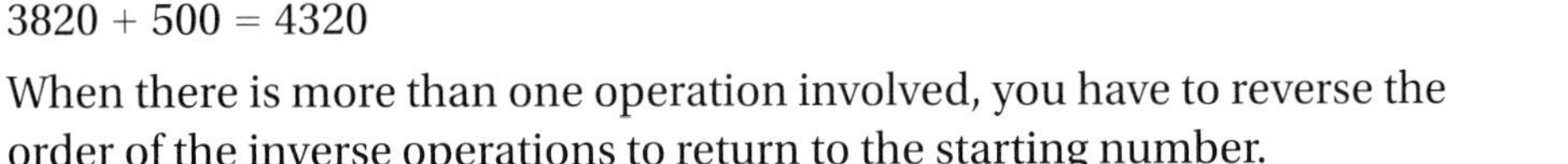

When there is more than one operation involved, you have to reverse the order of the inverse operations to return to the starting number.

For example, is $(50 + 62) \div 8 = 14$?

Check by working backwards and applying inverse operations:

$14 \times 8 - 62 = 50$

Find answers at: cambridge.org/ukschools/gcsemaths-studentbookanswers

EXERCISE 1D

1 Use inverse operations to find the missing values in each of these calculations.

a $\square + 217 = 529$ **b** $\square + 388 = 490$ **c** $\square - 218 = 182$

d $121 \times \square = {}^{-}605$ **e** ${}^{-}6 \times \square = 870$ **f** $\square \div 40 = 5400$

2 Use inverse operations to check these calculations.

a $45 \times 5 - 8 = 217$ **b** $14 + 5 \times 9 - 9 = 50$

c $(23 + 48) \times 4 = 284$ **d** $(412 - 128) \div 4 = 71$

3 The formula for finding the area of a triangle is $A = \frac{bh}{2}$.

a Find the height of a triangle with an area of $54\ \text{cm}^2$ and a base length 9 cm.

b A triangle has an area of $64\ \text{cm}^2$.
Find the height and the base length if the base is twice the height.

4 Here is an expression which includes different operations:

$$1 - \left(\frac{2}{3}(4 + 5) + 6\right) \times 7$$

a Calculate the value of the expression.

b Keep the numbers in order (from 1 to 7) but change the operations as necessary to:

i find the highest possible answer.

ii find the lowest possible answer.

c Comment on how changing the operations affected your results.

Checklist of learning and understanding

Basic calculations

- Written methods are important when you do not have a calculator.
- You can use any method as long as you show your working.
- Negative and positive numbers can be added, subtracted, multiplied and divided as long as you apply the rules to get the correct sign in the answer.

Order of operations

- In maths, there is a conventional order for working when there is more than one operation.
- Always work out brackets (or other grouping symbols) first, then powers. Multiply and/or divide next, then add and/or subtract.

Inverse operations

- An inverse operation undoes the previous operation.
- Addition is the inverse of subtraction.
- Multiplication is the inverse of division.
- Squaring is the inverse of taking the square root.
- Cubing is the inverse of taking the cube root; this applies to any power and root.

Chapter review

For additional questions on the topics in this chapter, visit GCSE Mathematics Online.

1 The cross-number puzzle on the right contains the solutions.
The clues are all calculations that involve using the correct order of operations.
Write a set of clues that would give these results.

	[1]2	[2]7						[3]3		
[4]1	4	8	[5]6			[6]1	9	7	[7]4	
[8]3	0		7						5	
	[9]4	[10]9			[11]3	2	7		[12]2	[13]4
		2			1					8
		0			[14]4	[15]2	5		[16]2	6
	[17]4	1	[18]7			1			0	
[19]2	3		[20]3	[21]2	0	4		[22]9	0	
[23]7	9	4		1			[24]7	9		

2 Use integers and operations to write ten different questions that give an answer of ⁻17.

3 **a** Work out these calculations.

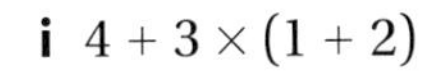

i $4 + 3 \times (1 + 2)$ *(1 mark)*

ii $\frac{4}{2} + 1 \times 3$ *(1 mark)*

iii $\frac{4 \times 3}{2 + 1}$ *(1 mark)*

b Fern is finding calculations that follow these rules

- You must use **all** the digits 1, 2, 3 and 4, but they can each be used only once
- You can add, subtract, multiply or divide as many times as you like
- You can use brackets.

For example when Fern was looking for a calculation with an **answer of 9**, she wrote down

$(4 + 3 + 2) \times 1$

Find a calculation, using her rules, which has an answer of

i 8 *(1 mark)*

ii 15 *(1 mark)*

4 On a page of a magazine there are three columns of text.
Each column contains 42 rows.
If there is an average of 32 letters per row, approximately how many letters are there on a page?

5 A stadium has seats for 32 000 people.
How many rows of 125 is this?

6 Two numbers have a sum of ⁻15 and a product of ⁻100.
What are the numbers?

7 The sum of two numbers is 1, but their product is ⁻20.
What are the numbers?

8 Josie's bank account was overdrawn.
She deposited £1000 and this brought her balance to £432.
By how much was her account overdrawn to start with?

9 We can use the formula $F = 2C + 32$ to approximately convert temperatures from Celsius to Fahrenheit.
Find the temperature in degrees Celsius when it is:

a 68 °F **b** 100 °F

2 Whole number theory

In this chapter you will learn how to ...

- identify the properties of different sets of numbers and use the correct words to talk about them.
- identify prime numbers and express any whole number as a product of its prime factors.
- find the HCF and LCM of two numbers by listing and by prime factorisation.

For more resources relating to this chapter, visit GCSE Mathematics Online.

Using mathematics: real-life applications

Numbers and basic calculations are used by people on a daily basis. A market stall holder has to quickly calculate the cost of a customer's order; a logistics manager has to order stock and divide the supplies so that they are never over or under stocked. There are many applications of basic calculation.

"Counting in multiples saves quite a bit of time. If I know that each shelf has 15 boxes and each box contains 5 reams of paper, then I know straightaway that you have 75 reams on each shelf without having to count each ream." *(Logistics manager)*

Tip

You probably already know most of the concepts in this chapter. They are included so you can revise the concepts if you need to and check that you know them well.

Before you start ...

KS3	You should be able to recognise and find the factors of a number and to list multiples of a number.	**1** If these are the factors, what is the number? **a** 1, 5, 25 **b** 1, 2, 3, 6 **c** 1, 11 **2** If these are all multiples of a number, what is the number? **a** 8, 10, 12, 14 **b** 18, 21, 27, 33 **c** 5, 20, 35, 60
KS3	You should know the first few prime numbers, square numbers and cube numbers.	**3** Which of the numbers below are: **a** prime numbers? **b** square numbers? **c** cube numbers? 0, 1, 2, 3, 4, 5, 6, 7, 8, 9, 10, 11, 12, 13, 14, 15, 16, 17, 18, 19, 20
KS3	You will need to be able to express a number as a product of its prime factors.	**4** Match each number to the product of its prime factors: **a** 450 **b** 180 **c** 120 **d** 72 A $2^3 \times 3^2$ B $2^2 \times 3^2 \times 5$ C $2^3 \times 3 \times 5$ D $2 \times 3^2 \times 5^2$

Assess your starting point using the Launchpad

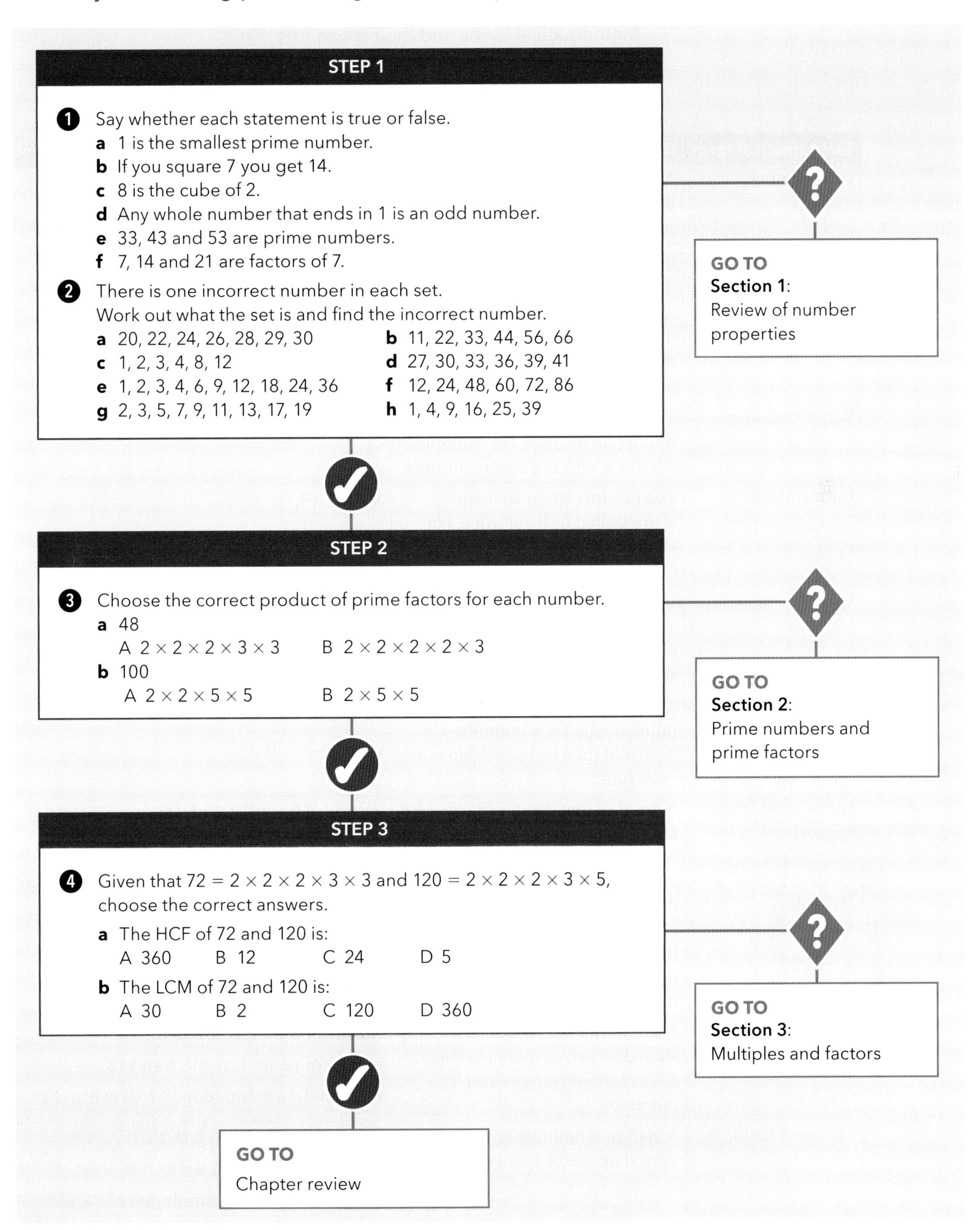

STEP 1
1 Say whether each statement is true or false.
a 1 is the smallest prime number.
b If you square 7 you get 14.
c 8 is the cube of 2.
d Any whole number that ends in 1 is an odd number.
e 33, 43 and 53 are prime numbers.
f 7, 14 and 21 are factors of 7.
2 There is one incorrect number in each set.
Work out what the set is and find the incorrect number.
a 20, 22, 24, 26, 28, 29, 30
b 11, 22, 33, 44, 56, 66
c 1, 2, 3, 4, 8, 12
d 27, 30, 33, 36, 39, 41
e 1, 2, 3, 4, 6, 9, 12, 18, 24, 36
f 12, 24, 48, 60, 72, 86
g 2, 3, 5, 7, 9, 11, 13, 17, 19
h 1, 4, 9, 16, 25, 39
?
GO TO
Section 1:
Review of number properties
STEP 2
3 Choose the correct product of prime factors for each number.
a 48
A 2 × 2 × 2 × 3 × 3
B 2 × 2 × 2 × 2 × 3
b 100
A 2 × 2 × 5 × 5
B 2 × 5 × 5
?
GO TO
Section 2:
Prime numbers and prime factors
STEP 3
4 Given that 72 = 2 × 2 × 2 × 3 × 3 and 120 = 2 × 2 × 2 × 3 × 5, choose the correct answers.
a The HCF of 72 and 120 is:
A 360
B 12
C 24
D 5
b The LCM of 72 and 120 is:
A 30
B 2
C 120
D 360
?
GO TO
Section 3:
Multiples and factors
GO TO
Chapter review

Find answers at: cambridge.org/ukschools/gcsemaths-studentbookanswers

Section 1: Review of number properties

Mathematical terms and their meanings

Make sure you remember the correct mathematical terms for the different types of numbers shown in the table.

Mathematical term	Definition	Example
Odd number	A whole number that cannot be divided exactly by 2; it has a remainder of 1.	1, 3, 5, 7, ...
Even number	A whole number that can be divided exactly by 2.	0, 2, 4, 6, 8, ...
Prime number	A whole number greater than 1 that can only be divided exactly by itself and by 1. (It only has two factors.)	2, 3, 5, 7, 11, 13, 17, 19, ...
Square number	The product when an integer is multiplied by itself. For example $2 \times 2 = 4$, so 4 is a square number.	1, 4, 9, 16, ...
Cube number	The product when an integer is multiplied by itself twice. For example $2 \times 2 \times 2 = 8$, so 8 is a cube number.	1, 8, 27, 64, ...
Root ($\sqrt{\ }$)	The number that produces a square number when it is multiplied by itself is a square root. The number that produces a cube number when it is multiplied by itself and then by itself again is a cube root.	The square root of 25 is 5. $\sqrt{25} = 5$ $(5 \times 5 = 25)$ The cube root of 8 is 2. $\sqrt[3]{8} = 2$ $(2 \times 2 \times 2 = 8)$
Factor (also called divisor)	A number that divides exactly into another number.	Factors of 6 are 1, 2, 3 and 6. Factors of 7 are 1 and 7. Factors of 25 are 1, 5 and 25.
Multiple	A multiple of a number is found when you multiply that number by a whole number. Your times tables are really just lists of multiples.	Multiples of 3 are 3, 6, 9, 12, ... Multiples of 7 are 7, 14, 21, ...
Common factor (divisor)	A common factor is a factor shared by two or more numbers. 1 is a common factor of all numbers.	Factors of 6 are 1, 2, 3, and 6. Factors of 12 are 1, 2, 3, 4, 6 and 12. 1, 2, 3 and 6 are common factors of 6 and 12.
Common multiple	A common multiple is a multiple shared by two or more numbers.	Multiples of 2 are 2, 4, 6, 8, 10, 12, ... Multiples of 3 are 3, 6, 9, 12, ... 6 and 12 are common multiples of 2 and 3.

EXERCISE 2A

1 Here is a list of numbers.

1	2	3	4	5	6	7	8	9	10	11	12	13	14	15
16	17	18	19	20	21	22	23	24	25	26	27	28	29	30

Choose and list the numbers in the box that are:

a odd.

b even.

c prime.

d square.

e cube.

f factors of 24.

g multiples of 3.

h common factors of 8 and 12.

i common multiples of 3 and 4.

2 Write down:

a the next four odd numbers after 207.

b four consecutive even numbers between 500 and 540.

c the square numbers between 20 and 70.

d the factors of 23.

e four prime numbers greater than 15.

f the first ten cube numbers.

g the first five multiples of 8.

h the factors of 36.

Key vocabulary

consecutive: following each other in order. For example, 1, 2, 3 or 35, 36, 37.

3 Say whether the results will be odd or even.

a The sum of two odd numbers.

b The sum of two even numbers.

c The difference between two even numbers.

d The square of an odd number.

e The product of an odd and an even number.

f The cube of an odd number.

Place value

Consider the number 222 222.

Each of the 2s in the number has a different place value.

The place value tells you the value of the digit; the underlined 2 in the number above has a value of 2 thousands or 2000.

Hundred thousands 100 000	Ten thousands 10 000	Thousands 1000	Hundreds 100	Tens 10	Ones/units 1
2	2	2	2	2	2

Each column in the place value table is ten times the value of the place to the right of it.

Find answers at: cambridge.org/ukschools/gcsemaths-studentbookanswers

EXERCISE 2B

1 Write each set of numbers in order from smallest to biggest.

a 432	456	348	843	654
b 606	660	607	670	706
c 123	1231	312	1321	231
d 12 700	71 200	21 700	21 007	

2 What is the value of the 5 in each of these numbers?

a 35 **b** 534 **c** 256 **d** 25 876

e 50 346 987 **f** 1 532 980 **g** 5 678 432 **h** 356 432

3 What is the biggest and smallest integer you can make with each set of digits?

Use each digit only once in each number.

a 4, 0 and 6 **b** 5, 7, 3, and 1 **c** 1, 0, 3, 4, 6, and 2

Section 2: Prime numbers and prime factors

Tip

You need to know all the prime numbers less than 20, but it will help you work faster if you can recognise them all up to 100.

Prime numbers have only two factors: 1 and the number itself.

The number 1 is not a prime number because it only has one factor.

2 is the only even prime number.

The prime numbers less than 20 are:

2, 3, 5, 7, 11, 13, 17, 19.

Prime factors

Key vocabulary

prime factor: a factor that is also a prime number.

If a factor of a number is a prime number it is called a **prime factor**.

Every integer greater than 1 can be written as a product of its prime factors.

Finding the prime numbers that multiply together to make a given number is known as **prime factorisation**.

You can find the prime factors of a number by **repeatedly dividing by prime numbers**, or by using **factor trees**.

WORKED EXAMPLE 1

Express 48 as a product of its prime factors:

a by division. **b** using a factor tree.

a

2	48
2	24
2	12
2	6
3	3

$2 \times 2 \times 2 \times 2 \times 3$

Divide by prime numbers; you should always divide to get integers. Start with the lowest divisor that is prime; always try 2 first. Divide by 2 repeatedly until you can't anymore and then try the next prime number, 3, and so on.

b

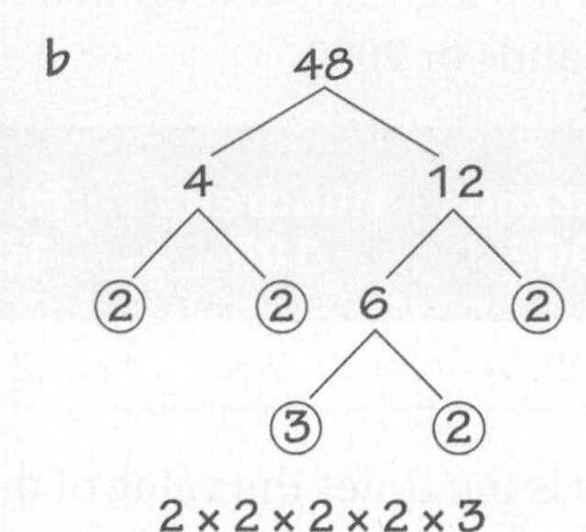

Write the number as a product of any two of its factors. Keep doing this for the factors until you cannot divide a factor anymore (i.e. until you get to a prime factor); the numbers should all be integers.

You get the same result with both methods. An integer can only be expressed in terms of its prime factors one way.

Even if you do the division in a different order and split the factors differently in the factor tree, you will always get the same result for a given number.

Here are four ways of finding the prime factors of 280.

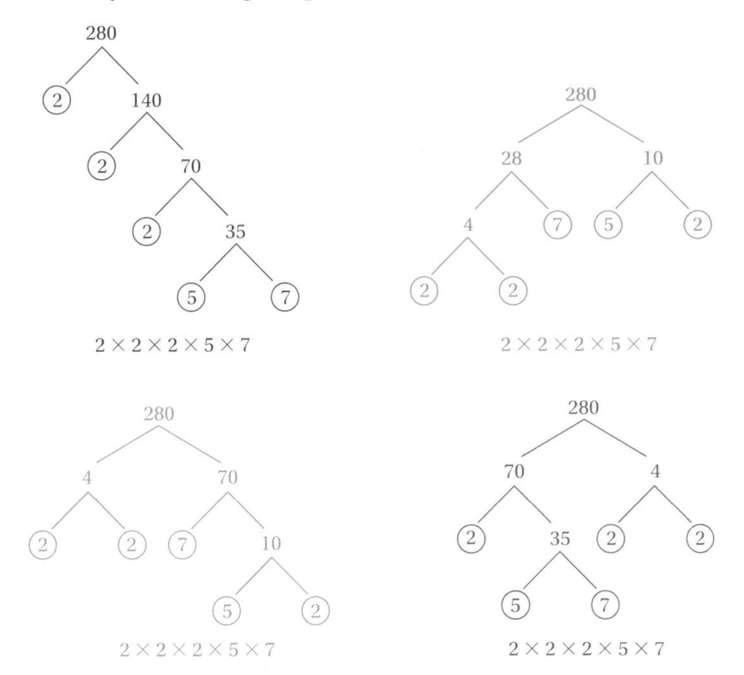

Tip

The **unique factorisation theorem** in mathematics states that each number can be written as a product of prime factors in one way only.

Different numbers cannot have the same product of prime factors.

You can use powers to write the product of prime factors in a shorter more efficient way.

Using power notation, $2 \times 2 \times 2$ is simplified to 2^3.

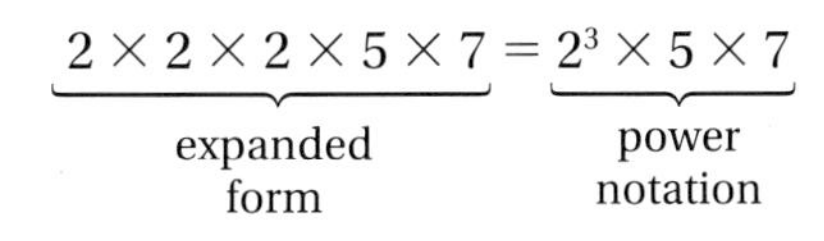

EXERCISE 2C

1 Identify the prime numbers in each set.

a 1, 2, 3, 4, 5, 6, 7, 8, 9, 10

b 50, 51, 52, 53, 54, 55, 56, 57, 58, 59, 60

c 95, 96, 97, 98, 99, 100, 101, 102, 103, 104, 105

2 Express each of the following numbers as a product of their prime factors. Use the method you prefer. Write your final answers using power notation.

a 36 **b** 65 **c** 64 **d** 84

e 80 **f** 1000 **g** 1270 **h** 1963

3 A number is expressed as $2^3 \times 3^3 \times 5$.

a What is the number? **b** Could it be any other number? Explain why.

4 Find out about the 'sieve of Eratosthenes' and show how it can be used to identify prime numbers up to 100.

Find answers at: cambridge.org/ukschools/gcsemaths-studentbookanswers

5 Over time, mathematicians have tried to find ways of identifying larger and larger prime numbers. They had difficulty testing them by dividing by all integers up to the square root of the number!

Computers have made this task easier and people are finding larger and larger prime numbers by sharing computer power.

In January 2013, a mathematician called Dr Curtis Cooper found that the number $2^{(57\,885\,161)} - 1$ was a prime number with 17 425 170 digits!

This is the 48th known 'Mersenne prime'. It has been verified.

a Find out about Marin Mersenne and why prime numbers in the form of $2^p - 1$ are known as Mersenne primes.

b The Great Internet Mersenne Prime Search (GIMPS) is a project that anyone can join.
Find out what it aims to do and whether or not any new primes have been discovered since 2013.
(www.mersenne.org is a good starting point.)

6 There are four prime numbers between 10 and 20.

a Can you find any other sets of four prime numbers between two consecutive multiples of 10?

b Explain why you cannot expect to find four prime numbers between most other consecutive multiples of 10.

Section 3: Multiples and factors

The lowest common multiple

The lowest common multiple (LCM) of two or more numbers is the smallest number that is a multiple of all the given numbers.

To find the LCM, list the multiples of the given numbers until you find the first multiple that appears in all the lists.

WORKED EXAMPLE 2

Find the LCM of 4 and 7.

4, 8, 12, 16, 20, 24, 28, 32, …	List the multiples of 4.
7, 14, 21, 28, …	List the multiples of 7.
LCM of 4 and 7 = 28	Stop listing at 28 as it appears in both lists.

The highest common factor

The highest common factor (HCF) of two or more numbers is the largest number that is a factor of all the given numbers. To find the HCF, list all the factors in each number, and pick the highest number that appears in all the lists.

WORKED EXAMPLE 3

Find the HCF of 8 and 24.	
8: $\underline{1}, \underline{2}, \underline{4}, \underline{8}$	List all the factors of 8 and of 24.
24: $\underline{1}, \underline{2}, 3, \underline{4}, 6, \underline{8}, 12, 24$	Underline the common factors.
HCF of 8 and 24 = 8	The highest underlined number is the HCF.

Tip

The LCM is used to find the lowest common denominator whenl you add or subtract fractions.
The HCF is useful for cancelling fractions. You will also use them in Chapter 3 to factorise algebraic expressions as well as in Chapter 10 on fractions.

With word problems you need to work out whether to use the LCM or HCF to find the answers.

- Problems involving the LCM usually include repeating events. You may be asked how many items you need to 'have enough' or when something will happen again at the same time.
- Problems involving the HCF usually involve splitting things into smaller pieces or arranging things in equal groups or rows.

Finding the HCF and LCM using prime factors

When you work with larger numbers you can find the HCF and LCM by writing the numbers as products of prime factors (i.e. by prime factorisation).

Once you've done that, you can use the factors to find the HCF and LCM quickly.

WORKED EXAMPLE 4

a Find the HCF of 72 and 120.

b Find the LCM of 72 and 120.

a $72 = \underline{2} \times \underline{2} \times \underline{2} \times \underline{3} \times 3$ $120 = \underline{2} \times \underline{2} \times \underline{2} \times \underline{3} \times 5$	First express each number as a product of prime factors. Underline the common factors. Here the common factors are three '2's and one '3'. 72 has an extra '3' not shared with 120, and 120 has a '5' not shared with 72.
$2 \times 2 \times 2 \times 3 = 24$. HCF of 72 and 120 = 24	Write down the common factors and multiply them out.
b $72 = \underline{2} \times \underline{2} \times \underline{2} \times \underline{3} \times \underline{3}$. $120 = 2 \times 2 \times 2 \times 3 \times \underline{5}$	First express each number as a product of prime factors. Underline the largest set of multiples of each factor across **both** lists. Here, there are three '2's in each list so underline one set; '3' appears twice in one list and once in the other, so underline the set of two '3's. The one '5' in the bottom set is the largest set of '5's so underline that one.
$2 \times 2 \times 2 \times 3 \times 3 \times 5 = 360$. LCM of 72 and 120 is 360	Write down each set of multiples and multiply them out.

Tip

Use the letters to help you remember what to do. LCM requires the **L**argest set of **M**ultiples.

EXERCISE 2D

1 Find the LCM of the given numbers.

a 9 and 18 **b** 12 and 18 **c** 15 and 18 **d** 24 and 12

e 36 and 9 **f** 4, 12 and 8 **g** 3, 9 and 24 **h** 12, 16 and 32

2 Find the HCF of the given numbers.

a 12 and 18 **b** 18 and 36 **c** 27 and 90 **d** 12 and 15

e 20 and 30 **f** 19 and 45 **g** 60 and 72 **h** 250 and 900

3 Sian has two rolls of cotton fabric. One roll has 72 metres on it, the other has 90 metres on it.

She wants to cut the fabric to make pieces of equal length without wasting any of it.

a What is the longest possible length the pieces can be?

b How many pieces of that length can be cut from each roll?

4 In a shopping centre promotion every 30th shopper gets a £10 voucher and every 120th shopper gets a free meal.

How many shoppers must enter the mall before one receives both a voucher and a free meal?

5 Amanda has 40 pieces of fruit and 100 sweets to share among the students in her class.

She is able to give each student an equal number of pieces of fruit and an equal number of sweets.

What is the largest possible number of students in her class?

6 Samir and Li start to walk in opposite directions around a walking track.

They start at the same point at the same time.

It takes Samir 5 minutes to walk round the track and it takes Li 4 minutes.

If they walk at this pace, how long will it be before they meet again at the starting point?

7 In a group of four cyclists, Lana cycles every 2nd day, Pete cycles every 3rd day, Karen cycles every 4th day and Anna cycles every 5th day.

They all cycle on 1 January this year.

a How many days will pass before they all cycle on the same day again?

b How many times a year will they all cycle on the same day?

8 Mr Abbot has three pieces of ribbon with lengths 2.4 m, 3.18 m and 4.26 m.

He wants to cut the ribbons into pieces that are all the same length.

He doesn't want any of the ribbon left over.

What is the greatest possible length for the pieces?

9 Two warning lights in a tunnel flash every 20 seconds and 30 seconds respectively.

They flashed together at 4.30 pm.

When will they next flash at the same time?

10 Francesca, Ayuba and Claire are Olympic and Paralympic contenders.

They all train on the same track.

Francesca cycles round the track and completes a lap in 20 seconds.

Ayuba runs and completes a lap in 84 seconds and Claire goes round the track in her wheelchair, taking 105 seconds.

They start at the same place at the same time.

How long will it take for all three women be at the same point again?

How many laps will they each have completed?

11 Mr Smith wants to tile a rectangular veranda with dimensions $3.3\text{ m} \times 5.7\text{ m}$ with a whole number of identical square tiles.

Mrs Smith wants the tiles to be as large as possible.

a Find the area of the largest possible tiles in cm^2.

b How many tiles will Mr Smith need to tile the veranda?

12 In an earthquake relief operation, supplies are handed out equally to all affected people at a relief centre.

On one day, 284 items of clothing, 426 food packets and 710 bottles of water are handed out.

How many people were in the relief centre on that day?

Checklist of learning and understanding

Properties of numbers

- Even numbers are multiples of 2, odd numbers are not.
- Factors are numbers that divide exactly into a number.
- Prime numbers have only two factors, 1 and the number itself.
- Square numbers are the product of a number and itself ($n \times n$).
- Cube numbers are the product of a number multiplied by itself twice ($n \times n \times n$).
- The value of a digit depends on its place in the number.

Prime numbers

- The prime numbers less than 20 are: 2, 3, 5, 7, 11, 13, 17 and 19.
- If a factor is a prime number it is called a prime factor.
- Integers can be written as the product of their prime factors.

  You find the prime factors by:
  - repeated division by prime numbers (starting from 2 and working upwards).
  - using a factor tree and breaking down factors until they are prime factors.

Multiples and factors

- The lowest common multiple (LCM) of two numbers can be found by:
  - listing the multiples of both numbers and selecting the lowest multiple that appears in both lists.
  - finding the largest set of multiples of each of the prime factors and multiplying them together.
- The highest common factor (HCF) of two numbers can be found by:
  - listing the factors of both numbers and selecting the highest factor that appears in both lists.
  - finding the common prime factors and multiplying them together.

Find answers at: cambridge.org/ukschools/gcsemaths-studentbookanswers

For additional questions on the topics in this chapter, visit GCSE Mathematics Online.

Chapter review

1 Complete the crossword puzzle on the handout provided by your teacher.

Clues

Across

1 The times tables are examples of these.

2 Whole numbers divisible by 2.

3 Another word used for factor.

4 Numbers in the sequence 1, 4, 9, 16, ...

5 An even prime number.

6 The result of a multiplication.

Down

a $n \times n \times n$ is the __ of n.

b Numbers with only two factors.

c Whole numbers that are not exactly divisible by 2.

d Number that divides into another with no remainder.

e HCF of 12 and 18.

2 Is 149 a prime number? Show how you decided.

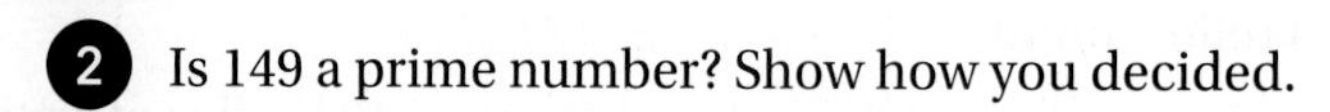

3

34	6	16	17	48	20	21

Choose from this list of numbers

a a multiple of 5, *(1 mark)*

b a factor of 24, *(1 mark)*

c a prime number, *(1 mark)*

d two numbers that add to 50, *(1 mark)*

e a square number *(1 mark)*

4 Find the HCF and the LCM of 20 and 35 by listing the factors and multiples.

5 Write 800 as a product of prime factors, giving your final answer sing powers.

6 Determine the HCF and LCM of the following by prime factorisation.

a 72 and 108 **b** 84 and 60

7 Jo, Mo and Jenny were jumping up a flight of stairs.
Jo jumped two steps at a time, Mo jumped three steps at time, while Jenny managed four steps at a time.
They started together on the first step. What is the first step they will all jump on?

8 Nick starts an exercise programme on 3rd March.
He decides to swim every third day and to cycle every fourth day.
On which dates in March will he swim and cycle on the same day?

9 A Year 10 group consists of 32 boys and 52 girls.
The teachers want to divide the students into the greatest possible number of groups that all have an equal number of boys and girls.
How many boys and girls would be in each group?

3 Algebraic expressions

In this chapter you will learn how to ...

- use algebraic notation and write algebraic expressions.
- simplify and manipulate algebraic expressions.
- use common factors to factorise expressions.
- use algebra to solve problems in different contexts.

For more resources relating to this chapter, visit GCSE Mathematics Online.

Using mathematics: real-life applications

Algebra lets you describe and represent patterns using concise mathematical language. This is useful in many different careers including accounting, navigation, building, plumbing, health, medicine, science and computing.

"You are unlikely to think about algebra when you watch cartoons or play video games, but animators use complex algebra to program the characters and make objects move." (Games designer)

Before you start ...

KS3	You need to understand the basic conventions of algebra.	**1** Choose the correct way to write each of these. **a** $n \times n$ A $2n$ B n^2 C 2^n D $2(n)$ **b** c multiplied by 3 and then added to 5 A $3c + 5$ B $3(c + 5)$ C $c + 15$ **c** n squared and then multiplied by 2 A $2n^2$ B $(2n)^2$ C $4n^2$
KS3	You should be able to substitute numbers for letters and evaluate expressions.	**2** **a** Evaluate the following expressions for $n = 5$ and $n = {}^-5$. **i** $3n + 4$ **ii** $3(n + 4)$ **b** What is the value of $\frac{(2n + 4)^2}{5}$ when $n = 3$?
Ch2	You should be able to find the highest common factor in a group of terms.	**3** Write down the HCF of: **a** $12xy$ and $18y^2$. **b** $45x$ and $50xy$.
KS3	You need to know how to apply the rules of indices to simplify expressions.	**4** Match the simplified expressions (A–D) to the mathematical statements **a–d**: **a** $a^m \times a^n$ **b** $a^m \div a^n$ **c** $(a^m)^n$ **d** a^0 A 1 B a^{m-n} C a^{m+n} D $a^{m \times n}$

Find answers at: cambridge.org/ukschools/gcsemaths-studentbookanswers

Assess your starting point using the Launchpad

STEP 1

1 Write each statement as an algebraic expression.

a Multiply n by 3 and add 4 to the result.

b Subtract 4 from n and multiply the result by 3.

c Multiply n squared by 4, add 3 and divide the result by 2.

?

GO TO
Section 1:
Algebraic notation

✓

STEP 2

2 Simplify these expressions by collecting like terms.

a $3a + 2b + 2a - b$

b $4x + 7 + 3x - 3 - x$

c $4a^2 + 8ab - 10a^2 - 5ab$

?

GO TO
Section 2:
Simplifying expressions

✓

STEP 3

3 Multiply out the brackets and simplify.

a $m(n - p)$ **b** $3(x + 5) + 4(x + 2)$ **c** $2z(z + 4) - z(z + 5)$

?

GO TO
Section 3:
Multiplying out brackets

✓

STEP 4

4 Complete the following.

a $3x + 12 = \square(x + 4)$ **b** $5x + 10y = \square(x + 2y)$

c $x^2 - 3x = \square(x - 3)$ **d** $ab - ac = a(\square - \square)$

e $-x + 7x^2 = -x(\square\square\square)$

5 Factorise each expression and write it as a product of its factors.

a $2x + 4y$ **b** $-3x - 9$ **c** $5x^2 + 5xy$

?

GO TO
Section 4:
Factorising expressions

✓

GO TO
Section 5:
Using algebra to solve problems

Section 1: Algebraic notation

In algebra, letters are used to represent unknown numbers.

For example: $x + y = 20$.

This means that two unknown numbers add up to 20.

The letters can represent many different numbers so they are called **variables**.

Letters and numbers can be combined and linked together with operation signs to form **expressions** such as $5a^3 - 2xy + 3$. This expression has three **terms**.

Terms are separated by + or − signs, and the sign belongs with the term that follows it. Terms should always be written in the shortest, simplest way:

$2 \times h$ is written as $2h$ and $x \times x \times y$ is written as x^2y.

$4x \div 3$ is written as $\frac{4x}{3}$ and $(x + 4) \div 2$ is written as $\frac{x + 4}{2}$.

The rules for the four operations are the same for letters as they are for numbers.

For example, to find a **product**, you multiply the factors together:

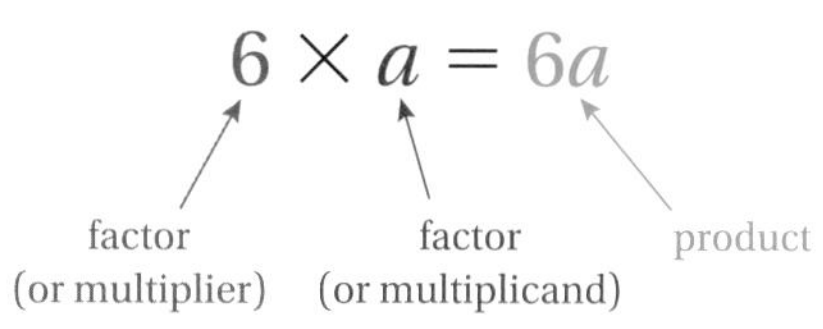

In algebraic notation, multiplication is shown by writing the factors next to each other. You write $a \times b$ as ab and $5 \times z$ as $5z$.

Key vocabulary

variable: a letter representing an unknown number.

expression: a group of numbers and letters linked by operation signs.

term: a combination of letters and/or numbers.

product: the result of multiplying numbers and/or terms together.

Equations and identities

An equation is a mathematical statement that includes an equals sign, for example $a + 3 = 27$. An identity is an equation that includes the symbol '≡', which means 'exactly the same as' or 'identical to'. This means that the left-hand side is a different way of representing the right-hand side.
For example, the following are all identities:

$a \times b \equiv ab$ $\quad 5 \times z \equiv 5z$ $\quad a \div b \equiv \frac{a}{b}$ $\quad 5 \div z \equiv \frac{5}{z}$

Identities are true for all values of the variables.

An identity is not the same as two equivalent terms. For example, you can say $2n = n + 2$ but you cannot say $2n \equiv n + 2$.

$2n = n + 2$ is an equation and has one unique answer, $n = 2$ but $2n$ and $n + 2$ are **not** identical terms. When $n \neq 2$, then $2n \neq n + 2$ so it is **not** true for all values of n, and is not an identity.

Tip

When you have numbers and letters in a term, the number is written first and letters are usually written in alphabetical order.

So, you write $5x$ not $x5$ and $3xy$ not $3yx$.

Tip

It is fine to use an = sign instead of an ≡ for example, $2 + 3 = 5$ and $2 + 3 \equiv 5$ are both accurate. But it is **not** ok to use a ≡ sign for anything other than an identity.

WORK IT OUT 3.1

Johan was asked to write three identities and to justify his choices.

Is each option correct or incorrect? Why?

Option A	Option B	Option C
$3(x - 5) \equiv 15 - 3x$ $3x - 15 \equiv 15 - 3x$ LHS is the same as the RHS, true for all values of x.	$(2x)^2 + 1 \equiv 4x^2 + 1$ LHS $= 2x \times 2x + 1$ $= 4x^2 + 1$ $=$ RHS ∴ this is an identity, true for all values of x.	$\frac{2y + 2}{3} \equiv \frac{3}{2(y + 1)}$ $2y + 2 = 2(y + 1)$ $2y + 2 = 2y + 2$ LHS is divided by 3 and on RHS 3 is divided by the expression. ∴ this is an identity, true for all values of x.

Find answers at: cambridge.org/ukschools/gcsemaths-studentbookanswers

EXERCISE 3A

1 Write the algebraic expression for:

a x multiplied by 3 and added to y multiplied by 7.

b 4 subtracted from x squared and the result multiplied by 5.

c x cubed added to y squared and the result divided by 4.

d 6 added to x, the result multiplied by 4 and then y subtracted.

e x multiplied by itself and then divided by 2.

f five less than three-fifths of a number.

2 Match the statements (**a–i**) to their correct algebraic expression from (**A–I**).

a Take a number and multiply it by 3 then add 2 to it.	**A**	$\frac{6x+2}{3}$
b Take a number and add 3 to it, then double it.	**B**	$3x + 2$
c Take a number and multiply it by itself then add three to it.	**C**	$5(x - 4)$
d Add 6 to a number then divide it by 2.	**D**	$(3x)^2$
e Take 4 away from a number all multiplied by 5.	**E**	$x^{\frac{7}{3}}$
f Square a number then multiply by 9.	**F**	$2x^2 - 3x^3$
g Square a number and multiply by 2 and subtract the number cubed multiplied by 3.	**G**	$2(x + 3)$
h Twice the sum of one third and a number.	**H**	$\frac{6+x}{2}$
i The cube root of a number multiplied by the square of a number.	**I**	$x^2 + 3$

3 Use algebra to write these as simply as possible.

a $2 \times 3a$ **b** $4b \times 5$ **c** $d \times (^-9)$

d $4a \times 3b$ **e** $5c \times 2d$ **f** $^-3m \times 4n$

g $^-2p \times (^-3q)$ **h** $a \times a$ **i** $m \times m$

j $2a \times 4a$ **k** $^-3a \times 5a$ **l** $^-2m \times (^-4m)$

m $7a \times 8ab$ **n** $^-6cd \times (^-2de)$ **o** $2a \times 2a \times 2a$

4 Write each of the following divisions as simply as possible using the conventions of algebra.

a $15x \div 5$ **b** $27y \div 3$ **c** $24a^2 \div 8$

d $7 \times 15p \div 21$ **e** $24x \div (8 \times 3)$ **f** $18y \div (6 \times 2)$

g $^-18x^2 \div 9$ **h** $^-16a^2 \div (^-4)$ **i** $15 \div (3 \times n \times n)$

Tip

Use the laws of indices for this question:
$a^m \times a^n = a^{m+n}$
$a^m \div a^n = a^{m-n}$
$(a^m)^n = a^{m \times n}$
$a^0 = 1$

5 Simplify.

a $x^6 \times x^7$ **b** $y^4 \times y^9$ **c** $3a^4 \times 5a^5$

d $2x^3 \times 5x^6$ **e** $a^2 \div a^4$ **f** $\frac{12b^7}{6b^2}$

g $\frac{18p^{10}}{9p^{11}}$ **h** $(x^4)^3$ **i** $(2a^7)^3$

j $4x^2y^3 \times 5xy^4 \div 2x^5y^3$ **k** $5x^0$ **l** $3x \times (x)^3$

6 Write expressions to represent the perimeter and area of each shape.

a

x 6

2

b

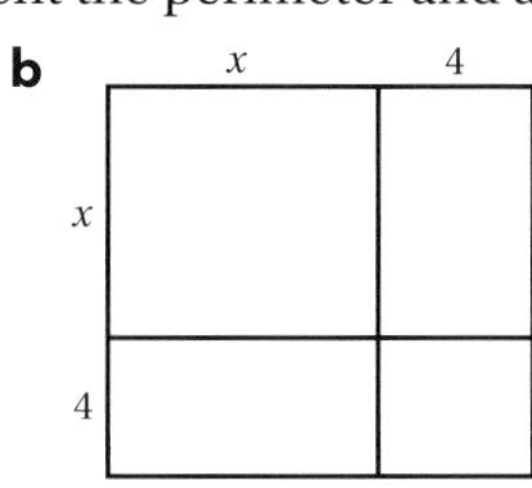

c

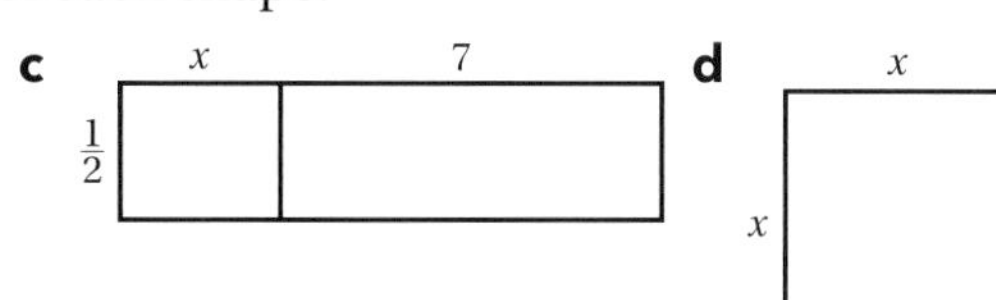

d

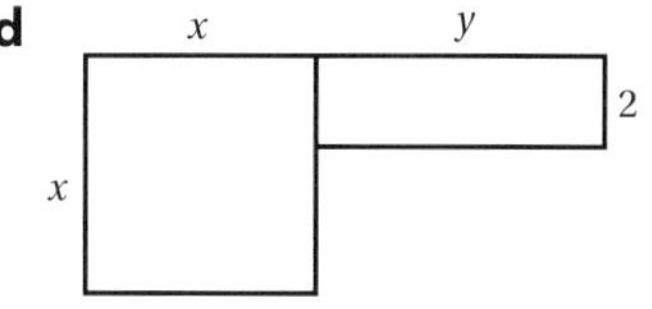

7 A man is x years old.

a How old will he be ten years from now?

b How old was he ten years ago?

c His daughter is a third of his age.
How old is his daughter?

8 A CD and a DVD together cost C pounds.

a If the CD cost £5, what did the DVD cost?

b If the DVD cost twice as much as the CD, what did the CD cost?

c If the CD cost £$(C - 15)$, what did the DVD cost?

Substitution

You can **evaluate** expressions by substituting given values for the variables into the expression.

For example, if $x = {}^-2$, you can work out that
$2x + 1 = 2 \times {}^-2 + 1 = {}^-4 + 1 = {}^-3$.

Key vocabulary

evaluate: means to find the value of, to solve.

Tip

When you substitute values into a term such as $2y$ you need to remember that $2y$ means $2 \times y$.

So, if $y = 6$, you need to write $2y$ as 2×6 and not as 26.

WORKED EXAMPLE 1

Given that $a = {}^-2$ and $b = 8$, evaluate:

a ab **b** $3b - 2a$ **c** $2a^3$ **d** $2(a + b)$

a $ab = a \times b$
$= {}^-2 \times 8$
$= {}^-16$

b $3b - 2a = 3 \times b - 2 \times a$
$= 3 \times 8 - 2 \times {}^-2$
$= 24 - ({}^-4)$
$= 28$

c $2a^3 = 2 \times a^3$
$= 2 \times ({}^-2)^3$
$= 2 \times {}^-8$
$= {}^-16$

d $2(a + b) = 2 \times (a + b)$
$= 2 \times ({}^-2 + 8)$
$= 2 \times 6$
$= 12$

Remember to do the calculation in brackets first.

Tip

Substitution is an important skill. You will need to substitute values for letters when you work with formulae for perimeter, area and volume of shapes, and when you solve problems involving Pythagoras' theorem.

EXERCISE 3B

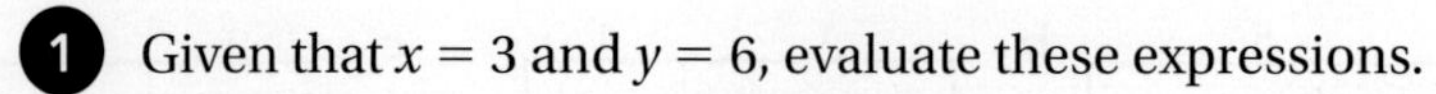

1 Given that $x = 3$ and $y = 6$, evaluate these expressions.

a $2x + 3y$ **b** $3x + 2y$ **c** $10y - 2x$

d $x + 2y$ **e** $6x + y$ **f** $5x - 5y$

g $2xy$ **h** $\frac{1}{2}xy$ **i** $2x^2 - y^2$

2 Find the value of each expression when $a = {}^{-}2$ and $b = 5$.

a ${}^{-}5ab + 10$ **b** ${}^{-}3ab - 6$ **c** $\frac{10}{b}$

d $\frac{400}{a}$ **e** $\frac{6}{a} - \frac{15}{b}$ **f** $\frac{15}{b} - \frac{24}{2a}$

g $8 - 2a + 2b$ **h** $7a - 4 + 2b$ **i** ${}^{-}3a^2$

j $4 - 3(ab)^3$ **k** $\frac{3a^2b^4}{2b^3}$ **l** ${}^{-}12a^2b \div {}^{-}2ab^2$

Section 2: Simplifying expressions

When you are asked to simplify an expression, you need to use the rules of arithmetic and algebra to write an expression as simply as possible.

Adding and subtracting like terms

Like terms have exactly the same letters or combination of letters and powers.

You can simplify expressions by adding or subtracting like terms.

$3a$ and $4a$ are **like** terms. $7xy$ and $2xy$ are **like** terms. $5x^2$ and $3x^2$ are **like** terms.

$3a + 4a = 7a$ $7xy - 2xy = 5xy$ $5x^2 - 3x^2 = 2x^2$

$5ab^2$ and $2a^2b$ are **not like** terms so $5ab^2 - 2a^2b$ cannot be simplified further.

WORK IT OUT 3.1

Here are two terms: $3x^2y$ and $2xy^2$.

Student A said that these two terms are like terms and can be added together to be written as: $5x^2y^2$

Student B said that these two terms are not like terms and can only be written added together as: $3x^2y + 2xy^2$

Which student is correct? Why?

When an expression contains many different terms you may be able to simplify it by collecting and combining like terms.

WORKED EXAMPLE 2

Simplify $2x - 4y + 3x + y$

$= 2x + 3x - 4y + y$ — Rearrange the terms so like terms are together. Keep the signs with the terms they belong to.

$= 5x - 3y$ — Combine the like terms. Remember $y = 1y$.

Multiplication and division

In Section 1 you saw how to simplify expressions by writing them without multiplication or division signs.

When you divide, you can simplify fractions by cancelling them down to lowest terms.

WORKED EXAMPLE 3

Simplify.

a $5 \times 4a$ **b** $2x \times 6y$ **c** $2a^2 \times 7ab$ **d** $12a \div {}^-4$ **e** $\frac{6x^2}{2}$ **f** $\frac{{}^-8xy}{{}^-16}$ **g** $\frac{12ab^2}{36ab}$

a $5 \times 4a = 20a$

b $2x \times 6y = 12xy$

Multiply numbers by numbers and write letters in alphabetical order.

c $2a^2 \times 7ab = 14a^3b$

$a^2 = a \times a$, so $a^2 \times a = a \times a \times a = a^3$

d $12a \div {}^-4 = \frac{12a}{{}^-4} = {}^-3a$

Write the division as a fraction and reduce it to its lowest terms by cancelling.

e $\frac{6x^2}{2} = 3x^2$

f $\frac{-8xy}{-16} = \frac{xy}{2}$

g $\frac{12ab^2}{36ab} = \frac{b}{3}$

Write the numerator as b not $1b$ by convention.

Tip

Expanding a term can help you to understand it more clearly.

$5ab^2 = 5 \times a \times b \times b$

$2a^2b = 2 \times a \times a \times b$

EXERCISE 3C

1 Decide whether each of these are like or unlike terms.

a $4a$ and $3b$ **b** $5b$ and ${}^-3b$ **c** $3b$ and $9b$

d $4p$ and $6p$ **e** $8p$ and ${}^-4q$ **f** $5a$ and $6b$

g $7mn$ and $3mn$ **h** $4ab$ and ${}^-2ab$ **i** ${}^-6xy$ and ${}^-7x$

j $9ab$ and $3a$ **k** $9x^2$ and $6x^2$ **l** $6a^2$ and ${}^-7a^2$

2 Simplify.

a $9x + 4y - 4y - 3x + 5y$ **b** $3c + 6d - 6c - 4d$

c $2xy + 3y^2 - 5xy - 4y^2$ **d** $2a^2 - ab^2 + 3ab^2 + 2ab$

e $5f - 7g - 6f + 9g$ **f** $7a^2b + 3a^2b - 4a^2b$

g $6mn^3 - 2mn^3 + 8mn^3$ **h** $3st^2 - 4s^2t + 5s^2t + 6st^2$

3 Copy and complete.

a $2a + \square = 7a$ **b** $5b - \square = 2b$

c $8mn + \square = 12mn$ **d** $11pq - \square = 6pq$

e $4x^2 + \square = 7x^2$ **f** $6m^2 - \square = m^2$

g $8ab - \square = {}^-2ab$ **h** ${}^-3st + \square = 5st$

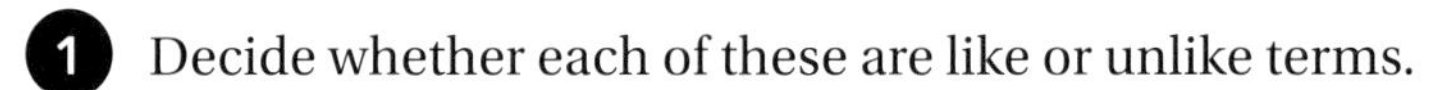

Find answers at: cambridge.org/ukschools/gcsemaths-studentbookanswers

4 Copy and complete.

a $8a \times \square = 16a$ **b** $9b \times \square = 18b$ **c** $8a \times \square = 16ab$

d $5m \times \square = 15mn$ **e** $3a \times \square = 12a^2$ **f** $6p \times \square = 30p^2$

g $^-5b \times \square = 10b^2$ **h** $4m \times \square = 12m^2n$

5 Rewrite each expression in the simplest possible form.

a $7 \times 2x \times {}^-2$ **b** $4x \times 2y \times 2z$ **c** $2a \times 5 \times a$

d $ab \times bc \times cd$ **e** $^-4x \times 2x \times {}^-3y$ **f** $\frac{1}{4x} \times 4y \times {}^-y$

g $^-9x \div 3$ **h** $^-24y \div 2x$ **i** $18x^2 \div 6$

j $^-25x^2 \div 5x$

6 Simplify.

a $\frac{4x}{8}$ **b** $\frac{3a}{9}$ **c** $\frac{^-12m}{18}$

d $\frac{15p}{21}$ **e** $\frac{22x^2}{55}$ **f** $\frac{15xy}{20}$

g $\frac{12ab}{a}$ **h** $\frac{2xy}{6xy}$ **i** $\frac{100x^2}{10xy}$

Key vocabulary

expanding: multiplying out an expression to get rid of the brackets.

Tip

Remember that $2(a + b)$ means $2 \times (a + b)$. In algebraic notation you don't write the multiplication sign.

Pay careful attention to the rules for multiplying negative and positive numbers when you multiply out.

Section 3: Multiplying out brackets

Removing brackets is called **expanding** the expression.

To expand an expression such as $2(a + b)$ you multiply each term inside the bracket by the value outside the bracket.

$$2(a + b) = 2 \times a + 2 \times b$$
$$= 2a + 2b$$

$$^-2(a + b) = {}^-2 \times a + ({}^-2 \times b)$$
$$= {}^-2a - 2b$$

WORKED EXAMPLE 4

Expand.

a $3x(y + 2z)$ **b** $^-2x(4 + y)$ **c** $^-(3x - 2)$

a $3x(y + 2z) = 3x \times y + 3x \times 2z$
$= 3xy + 6xz$

b $^-2x(4 + y) = {}^-2x \times 4 + ({}^-2x \times y)$
$= {}^-8x - 2xy$

c $^-(3x - 2) = {}^-1 \times 3x + ({}^-1 \times {}^-2)$
$= {}^-3x + 2$

Multiply each term inside the brackets by the term outside the brackets; write out each term and then simplify.

You are able to expand brackets this way because multiplication is distributive over addition and subtraction. The distributive law states that:

$a(b + c) \equiv ab + ac$

Both sides of the expression are identical, so we can use the ≡ symbol.

If two algebraic expressions are identical, the values calculated will be equal for any numbers which are substituted for the variables.

We call identical expressions identities.

You can prove this rule works by substituting in values.

WORKED EXAMPLE 5

Show that $3(2x - 4) \equiv 6x - 12$.

$x = 3, 3(2x - 4)$ $3(6 - 4) = 3(2) = 6$	Choose a value for x and substitute it into the expression on the left to solve.
$x = 3, 6x - 12$ $18 - 12 = 6$	Substitute x into the expression on the right-hand side and solve, if you get the same value, the expressions are identical.
Both expressions have a value of 6.	You can test it with other values of x to make sure it works.

WORKED EXAMPLE 6

Use substitution to determine whether $(a + b)^2 \equiv a^2 + b^2$.

Let $a = 1$ and $b = 2$	Choose small values to make your calculations as simple as possible.
$(a + b)^2 = (1 + 2)^2 = (3)^2 = 9$	
$a^2 + b^2 = 1^2 + 2^2 = 1 + 4 = 5$	
$9 \neq 5$	So the expressions are not identical.

When you have expanded an expression it may contain like terms.

Add or subtract like terms to simplify the expression further.

WORKED EXAMPLE 7

Expand and simplify. **a** $6x - 3(2x + 1)$ **b** $2x(x + y) - x(3x - 4y)$

a $6x - 3(2x + 1)$
$= 6x - 6x - 3$
$= {}^{-}3$

b $2x(x + y) - x(3x - 4y)$
$= 2x^2 + 2xy - 3x^2 + 4xy$
$= {}^{-}x^2 + 6xy$

EXERCISE 3D

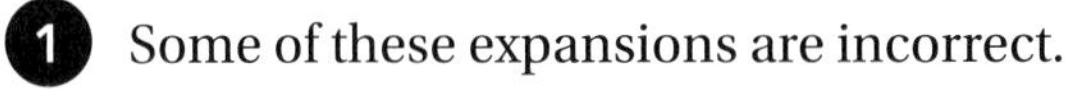

1 Some of these expansions are incorrect.

Check each one and correct those that are wrong.

a $4(a + b) = 4a + b$
b $5(a + 1) = 5a + 6$
c $8(p - 7) = 8p - 56$
d ${}^{-}3(p - 5) = {}^{-}3p - 15$
e $a(a + b) = 2a + ab$
f $2m(3m + 5) = 6m^2 + 10m$
g ${}^{-}6(x - 5) = 6x + 30$
h $3a(4a - 7) = 12a^2 - 7$
i $4a(3a + 5) = 12a^2 + 20a$
j $3x(2x - 7y) = 6x^2 - 21y$

Find answers at: cambridge.org/ukschools/gcsemaths-studentbookanswers

2 Expand and simplify.

a $2(c + 7) - 9$ **b** $(a + 2) + 7$ **c** $5(b + 3) + 10$

d $2(e - 5) + 15$ **e** $3(f - 4) - 6$ **f** $2a(4a + 3) + 7a$

g $5b(2b - 3) + 6b$ **h** $2a(4a + 3) + 7a^2$ **i** $3b(3b - 5) - 7b^2$

3 Expand and simplify.

a $2(y + 1) + 3(y + 4)$ **b** $2(3b - 2) + 5(2b - 1)$

c $3(a + 5) - 2(a + 7)$ **d** $5(b - 2) - 4(b + 3)$

e $x(x - 2) + 3(x - 2)$ **f** $2p(p + 1) - 5(p + 1)$

g $3z(z + 4) - z(3z + 2)$ **h** $3y(y - 4) + y(y - 4)$

4 The expression in each box is obtained by adding the expressions in the two boxes directly below it.

$9x + 14y$

$5x + 5y$ | $4x + 9y$

$2x + y$ | $3x + 4y$ | $x + 5y$

Complete these two pyramids.

a

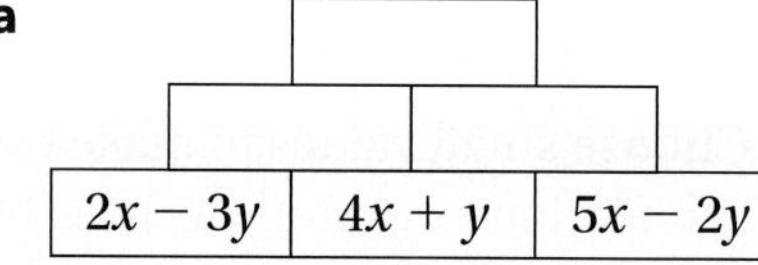

b

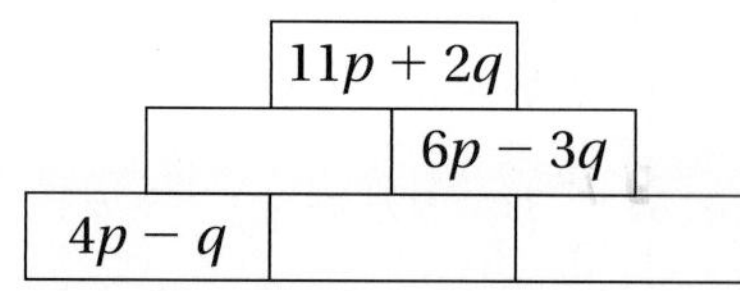

5 Use substitution to show that the following expressions are **not** identities.

a $a + a$ and a^2 **b** $3x + 4 - x + 2$ and $2x + 2$

c $(m + 2)^2$ and $m^2 + 4$ **d** $\frac{x + 3}{3}$ and $x + 1$

Key vocabulary

factorising: writing a number of expression as a product of its factors.

Section 4: Factorising expressions

Factorising is the opposite of expanding.

When you factorise an expression you use brackets to write it as a product of its factors.

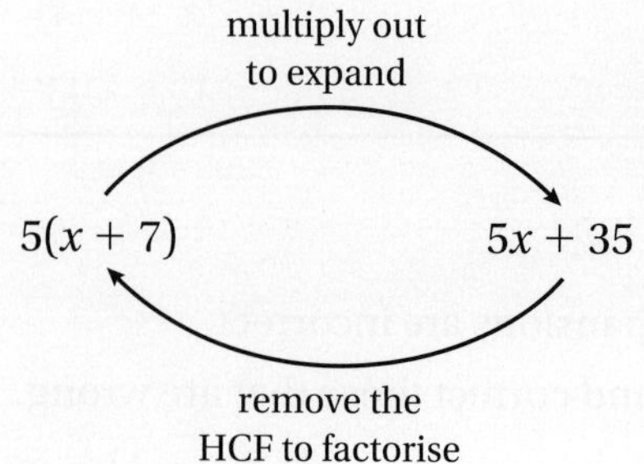

Tip

The highest common factor can be a number or a variable. It may also be a negative quantity.

If you expand $5(x + 7)$ you get $5x + 35$.

To factorise $5x + 35$ you find the highest common factor of the terms.

5 is the HCF of $5x$ and 35, so 5 is written outside the bracket and the remaining factors are written in brackets.

WORKED EXAMPLE 8

Factorise each expression. **a** $10a + 15b$ **b** $^{-}2x - 8$ **c** $3x^2 - 6xy$ **d** $3(m + 2) - n(m + 2)$

a $10a + 15b$

$10a + 15b = 5(2a + 3b)$

HCF of 10 and 15 is 5. There are no common variables.

b $^{-}2x - 8$

$^{-}2x - 8 = {}^{-}2(x + 4)$

HCF of $^{-}2$ and $^{-}8$ is $^{-}2$.

c $3x^2 - 6xy$

$3x^2 - 6xy = 3x(x - 2y)$

HCF is $3x$.

d $3(m + 2) - n(m + 2)$

$3(m + 2) - n(m + 2) = (m + 2)(3 - n)$

This looks like an expansion, but you are asked to factorise! $(m + 2)$ is common to both terms, so it is the HCF.

Combine 3 and $-n$ to make the second bracket.

EXERCISE 3E

1 Factorise each expression and write it as the product of its factors.

a $2x + 4$ **b** $12m - 18n$ **c** $3a - 3b - 6$

d $xy - xz$ **e** $5xy - 15xyz$ **f** $14ab - 21bc$

g $pq - pr$ **h** $x^2 - x$ **i** $18abc - 12ac$

j $2x^2 - 4xy$ **k** $2x^2y - 4xy^2$ **l** $^{-}6a - 12$

m $^{-}3a - 9$ **n** $^{-}xy - 5x$ **o** $^{-}x^2 + 6x$

2 Factorise.

a $7x - xy + x^2$ **b** $2xy + 4xz + 10x$ **c** $10x - 5y + 15z$

d $x(x - 2) + 5(x - 2)$ **e** $a(a - 7) - (a - 7)$ **f** $(x - 3) - 3(x - 3)$

g $3x^2y + 6xy^2$ **h** $36x^3 - \frac{1}{4}x^4$ **i** $^{-}ax^2 - ay^2$

Tip

You will learn other methods of factorising expressions in Chapter 7.

Section 5: Using algebra to solve problems

Algebra allows you to describe and make sense of patterns.

This is very useful for solving problems, particularly if they involve unknown amounts.

Working with expressions you can work out and prove rules.

WORKED EXAMPLE 9

Prove that $3n + 3$ = the sum of three consecutive numbers.

n

$n + 1$

$n + 2$

Let the first number be n. If the numbers are consecutive, you know that each number is 1 more than the previous number.

The next number must be 1 more than n, so let it be $n + 1$.

The third number is 1 more than $n + 1$, so let it be $n + 1 + 1 = n + 2$.

The three consecutive numbers are n, $n + 1$ and $n + 2$.

Continues on next page ...

$n + n + 1 + n + 2 = 3n + 3$. The sum of the three numbers is $3n + 3$, where n is the first number.	Sum the numbers to write a general rule.
$n = 205$ $3 \times 205 + 3 = 615 + 3 = 618$ Check: $205 + 206 + 207 = 618$	Test your rule using a set of consecutive numbers, for example 205, 206, 207. Substitute 205 for n in $3n + 3$.

General expressions like that in Worked example 9 are very useful for programmed operations and repeated calculations involving different starting numbers.

EXERCISE 3F

1 Are the following statements true or false?

a The expression $3z^2 + 5yx - z^2 - 6yx$ simplifies to $2z^2 - 11xy$.

b If you expand the brackets $2p(3p + q)$ you get the expression $6p^2 + 2pq$.

c This is a correct use of the identity symbol: $4(a + 1) \equiv 4a + 4$.

d $\frac{4}{x}$ always has the same value as $\frac{x}{4}$.

e x squared and added to 7 with the result divided by 3 is $\frac{x^2 + 7}{3}$.

2 In a magic square the sum of each row, column and diagonal is the same. Is this a magic square?

$m - p$	$m + p - q$	$m + q$
$m + p + q$	m	$m - p - q$
$m - q$	$m - p + q$	$m + p$

3 **a** Write an expression for each missing length in this rectangle.

b Write an expression for P, the perimeter of the rectangle.

c Given that $a = 2.1$ and $b = 4.5$, calculate the area of the rectangle. (area = length × width)

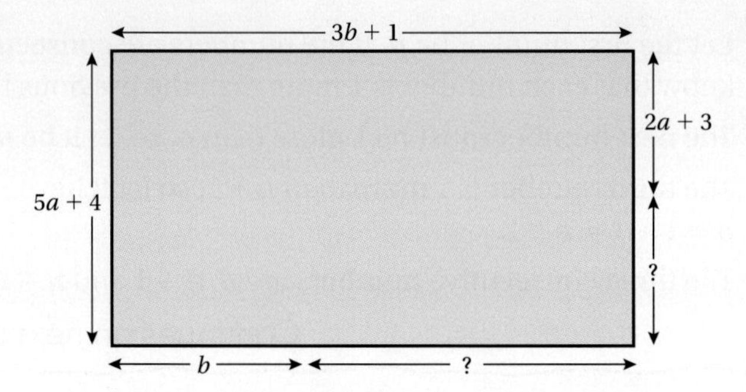

4 **a** The area of a rectangle is $2x^2 + 4x$. Suggest possible lengths for its sides.

b If the perimeter of a rectangle is $2x^2 + 4$, what could the lengths of the sides be?

5 Draw two diagrams representing areas to prove that $(3x)^2$ and $3x^2$ are different.

6 The diagram represents a room with an area of floor covered by carpet (shaded).

Divide the room into rectangles in different ways to find different, but equivalent, expressions for the floor area that is not covered by carpet.

Multiply out the expressions to confirm that they are equivalent.

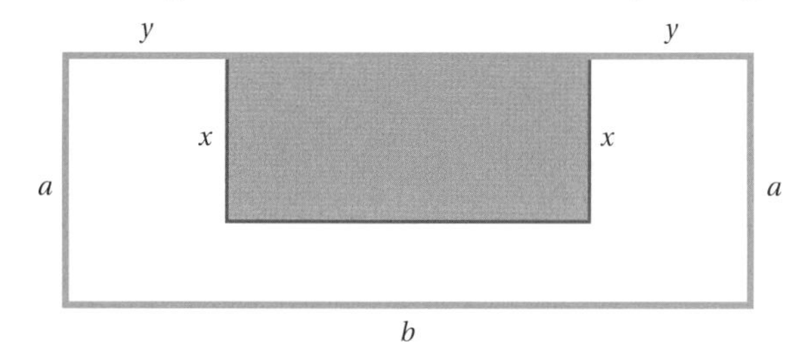

7 The number in each cell is made by adding the numbers in the two cells beneath it.

Fill in the missing expressions for each cell.

Write each expression in its simplest form.

a

	(blank)	
$2a + 4b$	(blank)	
$2a$	$4b$	$5a$

b

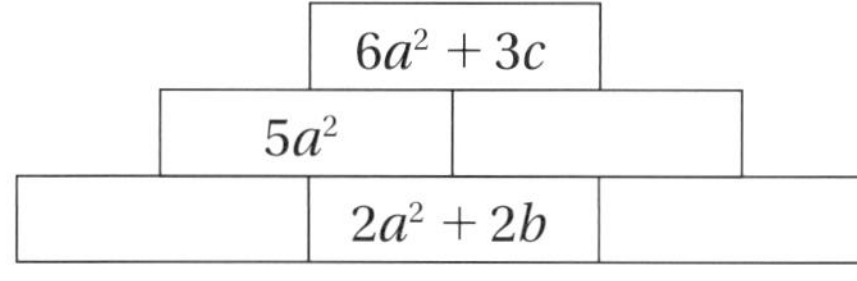

c

	$3xy - z^2$	
(blank)	$3z^2 + xy + y^2$	
$xy + 2y^2$	(blank)	(blank)

8 **a** Paul plays a 'think of a number game' with his friends and predicts what their answer will be.

These are the steps he tells his friends to follow:

- Think of a number. Double it.
- Add 6 and halve the number you have now.
- Take away the number you first thought of.

Paul then guesses that the answer is 3.

Use algebra to show why Paul will guess correctly, no matter what number his friends choose as a starting number.

b Make up a 'think of a number' problem.

Use algebra to check that it works and to see which number you end up with. Try it on a classmate to check that it works.

Find answers at: cambridge.org/ukschools/gcsemaths-studentbookanswers

9 The expression $7(x + 4) - 3(x - 2)$ simplifies to $a(2x + b)$.
Work out the values of a and b.

10 Check whether each expression has been fully simplified.
If not, simplify it further.

a	$5(g + 2) + 8g = 5g + 10 + 8g$
b	$4z(4z - 2) - z(z + 2) = 15z^2 - 10z$
c	$^-5ab \times (^-3bc) = 15ab^2c$
d	$\frac{18x^3}{3x} = \frac{6x^3}{x}$
e	$\frac{(x^5 \times x^7)}{x^4} = \frac{x^{12}}{x^4}$

11 Simplify.

a $\frac{2x}{y^3} + \frac{7x}{3y^3}$ **b** $\frac{3a}{b} + \frac{5}{2b}$ **c** $-\frac{4}{5}(25x - 100)$ **d** $-\frac{3}{5}\left(\frac{a}{3} - \frac{2}{3}\right)$

12 Simplify.

a $\frac{6x^3y^3}{xy} + \frac{2x^3y^2}{y^2} - \frac{14x^5y}{2x^2y} + \frac{12x^4y^2}{3x^2}$ **b** $\frac{4p^2q^3}{2pq} + \frac{12p^3q^3 - 8p^5q^2}{5pq^2 - pq^2}$

Checklist of learning and understanding

Algebraic notation

- You can use letters (called variables) in place of unknown quantities.
- An expression is a collection of numbers, operation signs and at least one variable.
- Each part of an expression is called a term.
- To evaluate an expression you substitute numbers in place of the variables.

Simplifying expressions

- Like terms have exactly the same variables.
- Expressions can be simplified by adding or subtracting like terms.

Multiplying out brackets

- You can multiply and divide unlike terms.
- If an expression contains brackets, you multiply them out and then add or subtract like terms to simplify it further.

Factorising

- Factorising involves putting brackets back into an expression.
- If terms have a common factor, write it in front of the bracket and write the remaining terms in the bracket as a factor. (Check by multiplying out.)

Solving problems

- Algebra allows you to make general rules that apply to any number. This is useful in problem-solving.

Chapter review

For additional questions on the topics in this chapter, visit GCSE Mathematics Online.

1 Expand and simplify.

a $3b(3b - 5) - 7b^2$

b $2x(5x + 4) - 6x(3x - 7)$

c $\frac{3}{4}(p + 2) + \frac{1}{2}(p - 1)$

d $\frac{2y}{3}(y + 5) + \frac{y}{3}(y - 4)$

2 Two of these expressions have been incorrectly simplified.
Find them and correct them.

$2x \times 6y = 12xy$ $\quad$ $2a^2 \times 7ab = 14a^3b$

$\frac{12a}{4} = 3a$ $\quad$ $\frac{6x^2}{2} = 3x^2$

$\frac{8xy}{16} = \frac{xy}{4}$ $\quad$ $\frac{12ab^2}{36ab} = \frac{b}{3}$

$\frac{15}{2x} \times \frac{2}{3x} = \frac{10}{x^2}$ $\quad$ $\frac{6a}{7b} \div \frac{2ab}{3} = \frac{9}{7b^2}$

3 Determine by substitution whether the following pairs of expressions are identities or not. For those that you think are identities, show that the left-hand side is identical to the right-hand side.

a $5(x + 3)$ and $5x + 3$

b $^{-}3(m - 2)$ and $^{-}3m - 6$

c $4(y - 3) + 2(y + 4)$ and $6y - 4$

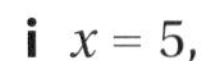

4 **a** Work out the value of $x^2 - 3x$ when

i $x = 5$, *(1 mark)*

ii $x = {}^{-}4$, *(2 marks)*

b Multiply out.

$y(y + 5)$ *(1 mark)*

c Factorise fully.

$4p^2 - 8p$ *(2 marks)*

5 The generalised rule for adding or subtracting fractions states that:

$\frac{a}{b} \pm \frac{c}{d} = \frac{(ad \pm bc)}{bd}$

A student wrote the following:

$\frac{1}{t} + \frac{1}{w} = \frac{2}{(t + w)}$

Show algebraically that this is incorrect.

6 Prove, using algebra, that the sum of two consecutive whole numbers is always an odd number.

7 Prove algebraically that the difference between the squares of any two consecutive integers is equal to the sum of these two integers.

Find answers at: cambridge.org/ukschools/gcsemaths-studentbookanswers

4 Functions and sequences

In this chapter you will learn how to ...

- generate sequences and find unknown terms in a sequence.
- interpret expressions as functions with inputs and outputs.
- work with inverse and composite functions.
- use correct notation to write rules or functions to find any term in a sequence.
- recognise and use a variety of special sequences.

For more resources relating to this chapter, visit GCSE Mathematics Online.

Using mathematics: real-life applications

Finding a pattern and working out how the parts of the pattern fit together is important in scientific discovery. Scientists use sequences to model and solve real-life problems, such as estimating how quickly diseases spread.

Tip

When you work with sequences you can draw diagrams, flow charts or tables to organise the patterns and make sense of them.

"When a new outbreak of a disease occurs I need to work out how quickly it is spreading. To do this I look at the sequence in which the numbers of victims are increasing. I use the sequence to predict how many people will become infected in a certain length of time." *(Medical researcher)*

Before you start ...

KS3 Ch 2	You need to know your multiplication tables and recognise multiples of numbers.	**1** **a** What are the first five multiples of 7? **b** Which of these are multiples of 6? 56, 66, 86, 18, 54, 36
KS3 Ch 2	You need to be able to recognise square numbers and cube numbers.	**2** **a** Which of these are square numbers? 1, 16, 66, 50, 25, 4, 6, 9, 49 **b** Which of these are **not** cube numbers? 9, 15, 27, 64, 1, 8, 125
KS3	You need to be able to spot and describe patterns.	**3** **a** Describe this pattern in words. Shape 1 Shape 2 Shape 3 **b** How many matchsticks would you need to build the sixth shape in the pattern?

Assess your starting point using the Launchpad

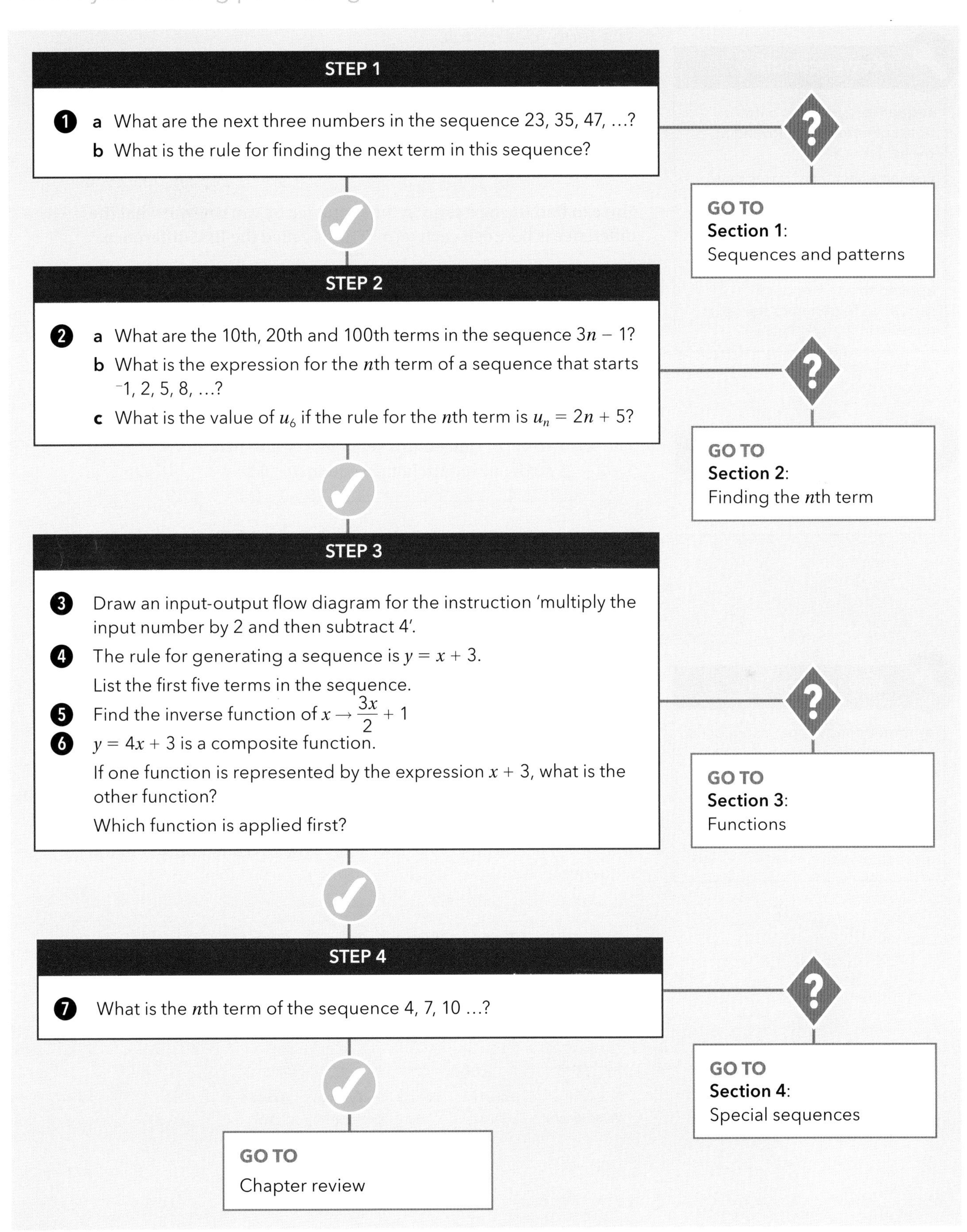

STEP 1
1 a What are the next three numbers in the sequence 23, 35, 47, …?
b What is the rule for finding the next term in this sequence?
GO TO
Section 1:
Sequences and patterns
STEP 2
2 a What are the 10th, 20th and 100th terms in the sequence $3n - 1$?
b What is the expression for the nth term of a sequence that starts $^-1$, 2, 5, 8, …?
c What is the value of u_6 if the rule for the nth term is $u_n = 2n + 5$?
GO TO
Section 2:
Finding the nth term
STEP 3
3 Draw an input-output flow diagram for the instruction 'multiply the input number by 2 and then subtract 4'.
4 The rule for generating a sequence is $y = x + 3$.
List the first five terms in the sequence.
5 Find the inverse function of $x \rightarrow \frac{3x}{2} + 1$
6 $y = 4x + 3$ is a composite function.
If one function is represented by the expression $x + 3$, what is the other function?
Which function is applied first?
GO TO
Section 3:
Functions
STEP 4
7 What is the nth term of the sequence 4, 7, 10 …?
GO TO
Section 4:
Special sequences
GO TO
Chapter review

Section 1: Sequences and patterns

The term-to-term rule

Key vocabulary

sequence: a number pattern or list of numbers following a particular order.

consecutive terms: terms that follow each other in a sequence.

first difference: the difference between one term and the next.

term-to-term rule: operations applied to any number in a sequence to generate the next number in the sequence.

A **sequence** is an ordered list or pattern.

Terms that follow each other in a sequence are called **consecutive terms**.

The first term in a sequence is called T(1), the second T(2) and so on.

You can find the next term in the sequence by working out what the difference is between each term. This is called the **first difference**.

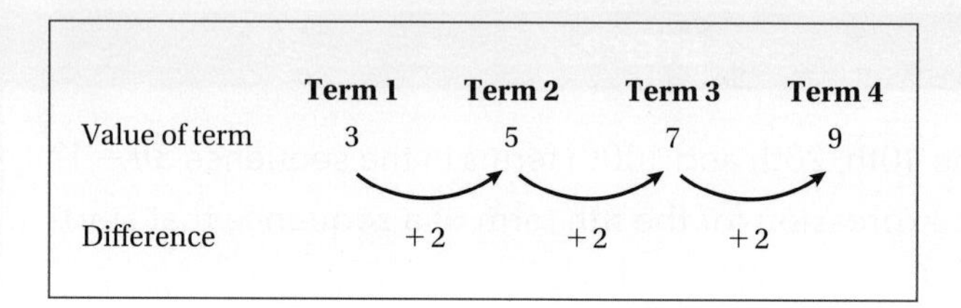

Term 1
Term 2
Term 3
Term 4
Value of term
3
5
7
9
Difference
+2
+2
+2

Tip

Each number in a sequence is called a term.

The **term-to-term rule** for this sequence is 'add two'.
$9 + 2 = 11$ so the next term in the sequence is 11.

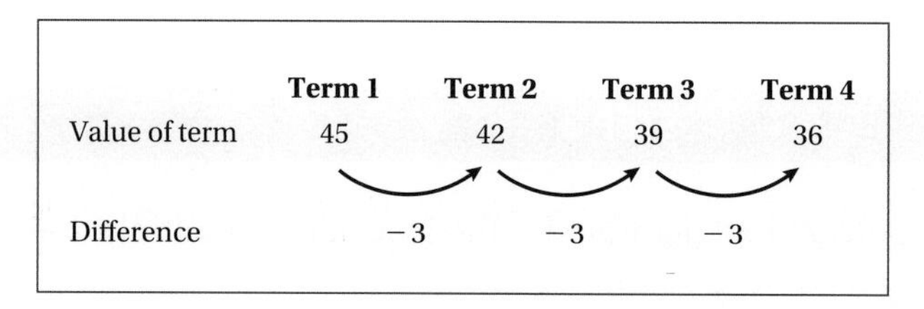

Key vocabulary

arithmetic sequence: a sequence where the difference between each term is constant.

geometric sequence: a sequence where each term is found by multiplying the previous term by a constant factor.

The term-to-term rule for this sequence is 'subtract three'.

3, 5, 7, 9, ... and 45, 42, 39, 36, ... are both **arithmetic sequences**.
In an arithmetic sequence the terms are generated by adding or subtracting a constant difference.

In a **geometric sequence**, the terms are generated by multiplying or dividing by a constant factor.

3, 6, 12, 24, ... and 1000, 500, 250, 125, ... (below) are both geometric sequences.

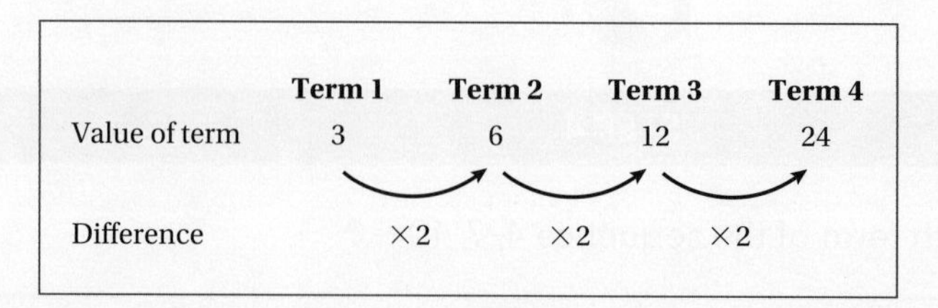

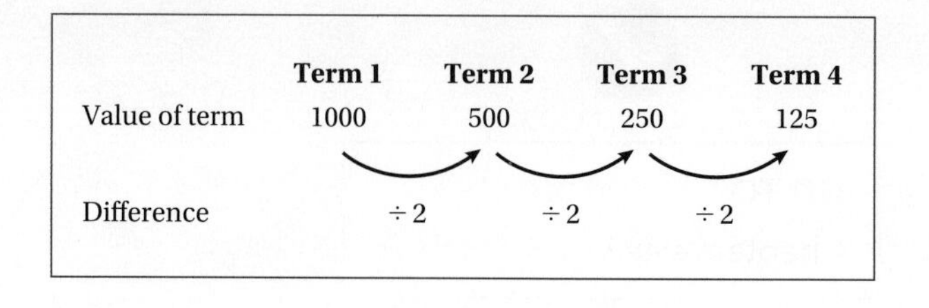

WORKED EXAMPLE 1

"I use term-to-term rules in my job. I know that each row of bricks will have three fewer bricks than the row below it, so I can work out how many bricks I need in each row."

(Bricklayer)

The first row of the wall has 57 bricks.

Use a term-to-term rule to find the number of bricks in the next three rows.

The term-to-term rule is 'subtract three'.	The number of bricks decreases by 3 for each new row.
57 bricks in the first row $57 - 3 = 54$ bricks in the second row	Subtract 3 from 57 to get the next term.
$54 - 3 = 51$ bricks in the third row $51 - 3 = 48$ bricks in the fourth row	Continue to do this for the next two terms.

EXERCISE 4A

1 Find the next three terms in each of these sequences.
Explain how you found them.

a 4, 7, 10, 13, ... **b** 38, 43, 48, 53, ...

c 27, 23, 19, ... **d** 63, 57, 51, ...

e 1, 2, 4, 8, ... **f** 64, 32, 16, ...

g 4, 12, 36, ... **h** 729, 243, 81, ...

2 Give the term-to-term rule for each of these sequences.

a 7, 14, 21, 28, ... **b** 19, 15, 11, 7, ...

c 2, 8, 32, 128, ... **d** 84, 42, 21, ...

3 Find the term-to-term rule for each of these sequences.
Use it to generate the next three terms in each sequence.

a 3.5, 5.5, 7.5, ... **b** 1.2, 2.4, 4.8, ...

c $1\frac{1}{2}$, 3, $4\frac{1}{2}$, ... **d** 8, 5, 2, ...

e 72, 36, 18, ... **f** $^{-}14$, $^{-}8$, $^{-}2$, ...

Tip

It can help to write out the sequence and label the difference between each term.

Find answers at: cambridge.org/ukschools/gcsemaths-studentbookanswers

Tip

Look for the term-to-term rule to answer this question.

4 When a ball is dropped it bounces back to half its original height.

With each new bounce it bounces back to half the height of the previous bounce.

a If a ball is dropped from 96 cm, how high will it bounce on its 4th bounce?

b How many times will it bounce before it bounces to below 1 cm?

5 T(1) of a sequence is 4 and the term-to-term rule for the sequence is 'add x'.

Find a value for x so that:

a every second term is an integer.

b every third term is a multiple of 4.

c T(2) is smaller than T(1).

Section 2: Finding the nth term

The position-to-term rule

Term-to-term rules are useful for generating the first few terms of a sequence and for finding the next term in a given sequence. They are less useful when you want to find the 50th or 100th term.

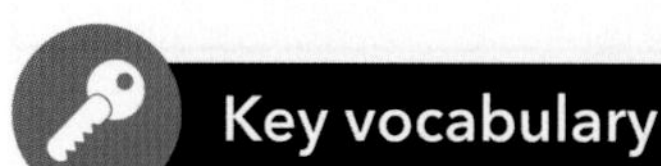

Key vocabulary

position-to-term rule: operations applied to the position number of a term in a sequence in order to generate that term.

A **position-to-term** rule allows you to work out the value of any term in a sequence if you know its position in the sequence.

The sequence 1, 3, 5, 7... can be generated using the position-to-term rule 'position number $\times 2$, subtract 1'.

To find a sequence using a position-to-term rule, you substitute the position number into the rule.

For example, the tenth term of this sequence is $(10 \times 2) - 1 = 19$.

The nth term

The notation T(n) refers to 'any term' in the sequence, where n is the position of the term. T(n) is known as the 'nth term'.

Sometimes you will be given a sequence and you will need to find a rule to find any term, T(n), in that sequence. You can usually find a rule for a sequence by looking at the difference between consecutive terms.

Remember that this is called the **first difference**.

Once you have found the first difference you can compare it with number patterns you already know to find the rule for the sequence.

Find an expression for the nth term of the sequence: 5, 8, 11, 14, 17 ...

Problem-solving framework

Steps for approaching a problem-solving question	What you would do for this example
Step 1: Identify what you have to do.	You are trying to find an expression to work out the value of any term in the sequence.

Step 2: If it is useful to have a table, draw one.

Draw a table showing the position and the term:

n	1	2	3	4	5
T(n)	5	8	11	14	17

Step 3: Start working on the problem using what you know.

Label the table with the difference between each term:

n	1	2	3	4	5
T(n)	5	8	11	14	17

+ 3 + 3 + 3 + 3

The difference in this sequence is '+ 3'.

Step 4: Connect to other sequences and compare.

Another sequence that has the same term-to-term rule of 'add 3' is the multiples of 3. The expression for the nth term of the multiples of 3 is $3n$.

Add this sequence to your table:

n	1	2	3	4	5
T(n)	5	8	11	14	17
$3n$	3	6	9	12	15

+2

Compare the multiples of 3 ($3n$) to the terms of the original sequence.

(If the difference was '+2' you would compare to $2n$; if it was '−4' you would compare it to $-4n$.)

Each term is the corresponding multiple of 3 with 2 added.

So the expression for the nth term of this sequence could be $3n + 2$.

Step 5: Check your working and that your answer is reasonable.

Test for $n = 5$:

$3n + 2 = (3 \times 5) + 2$
$= 15 + 2$
$= 17$

The fifth term is 17 so the expression is correct.

Step 6: Have you answered the question?

Yes, the expression for the nth term is $3n + 2$.

Tip

The rule for an arithmetic sequence can be calculated using the following formula: $(a + d(n - 1))$ where a = first term, and d = common difference. The **common difference** is the constant difference between terms. Try this for the example above.

Find answers at: cambridge.org/ukschools/gcsemaths-studentbookanswers

WORK IT OUT 4.1

A sequence is defined by the rule $T(n) = 3n - 2$.

What are the first five terms of the sequence?

Only one answer below is correct. Explain why the other two are wrong.

Option A	Option B	Option C
$^{-}2, 1, 4, 7, 10$	$1, 4, 7, 10, 13$	$^{-}1, 1, 3, 5, 7$

Subscript notation

The *n*th term can be written as u_n, where u represents a sequence.

This is read as 'u sub n' and is known as **subscript notation**.

u_1 is used to denote the first term, u_2 is used for the second term and so on.

You can use this notation to write term-to-term and position-to-term rules.

For example, the **position-to-term rule** of $3n - 4$ for the sequence u, would be written as: $u_n = 3n - 4$.

If you want to find the value of the 6th term, substitute 6 into the equation in place of n:

$u_6 = (3 \times 6) - 4$

$u_6 = 14$

The **term-to-term rule** '+2' would be written as: $u_{n+1} = u_n + 2$.

The 'n + 1' indicates that it is the term one more than the current term (u_n).

So, if $u_3 = 7$ then,

$u_4 = u_3 + 2$

$= 7 + 2$

$= 9$

EXERCISE 4B

1. A sequence is created using the position-to-term rule 'position ×3, subtract 1'.
   - **a** What are the first 6 terms of the sequence?
   - **b** What is the 20th term of the sequence?
   - **c** Would the 40th term of the sequence be double the value of the 20th term? Explain your answer.

2. Find the value of the following terms for each position-to-term rule.

   i 1st term **ii** 2nd term **iii** 3rd term **iv** 4th term
   v 10th term **vi** 20th term **vii** 100th term

   a $4n + 1$ **b** $4n - 5$ **c** $8n + 2$

   d $5n - \frac{1}{2}$ **e** $\frac{n}{2} + 1$ **f** $^{-}2n + 1$

3. Find the expression for the *n*th term in the sequence that begins 1, 7, 13, 19, ...

4 Find the rules for the nth term of the following sequences.

Write the rule using u_n notation.

a 3, 5, 7, 9, ... **b** 3, 7, 11, 15, ... **c** $^-1$, 4, 9, 14, ...

d 7, 12, 17, 22, ... **e** $^-3$, 0, 3, 6, ... **f** $^-1$, 6, 13, 20, ...

5 Majid conducts an experiment in science and gets the following pattern of results:

67, 73, 79, 85

Write an expression for the nth term of Majid's results.

6 Sam has planted a sunflower and notices that it measures 4.5 cm at the end of the first week, 6.7 cm at the end of the second week and 8.9 cm at the end of the third week.

a Find an expression for the nth term of this sequence, assuming the sunflower continues to grow at the same rate.

b How big will the sunflower be at the end of the 100th week?

c Explain why this is unlikely to be true.

7 Tammy has £100 saved.

She deposits £4 per week. At the end of the first week she has £104.

a If she continues to save at this rate, how much will she have after 52 weeks?

b How long will it take her to save £400?

8 A restaurant uses square tables that can seat four people.

Tables can be pushed together to create different seating arrangements.

a How many people can fit around 6 tables pushed together in this way?

b How many people can fit around 10 tables pushed together this way?

c Is it possible to seat 31 people around tables pushed together like this with no spaces?

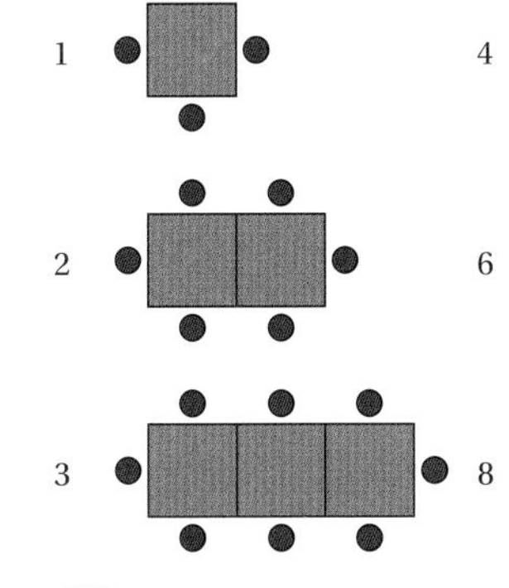

Section 3: Functions

A **function** is a rule for changing one number into another.

'Multiply by two', 'add three' and 'divide by 2 and then add 1' are examples of functions.

You can use algebra to write functions, for example 'multiply by 2' can be written as the expression '$2n$'.

Functions can be expressed in different ways:

$y = x + 3$ $\qquad$ $x \rightarrow x + 3$

These both mean the same thing: take any value of x and add 3 to it to get a result.

The steps you take to work out the value of a function can be shown as a simple flow diagram or function machine.

Key vocabulary

function: a set of instructions for changing one number (the input) into another number (the output).

Tip

You can think of a function as 'the answer you will get' if you apply this rule to a number.

Find answers at: cambridge.org/ukschools/gcsemaths-studentbookanswers

A function machine shows the input, operation and output for a given rule.

input ⟶ [rule] ⟶ output

In a function there is only **one** possible output for each input.

Generating a sequence using a function

You can generate a sequence using a function.

The table shows the outputs when you input the values 1 to 10 into the function $y = 2n + 4$

n ⟶ [$\times 2$] ⟶ [$\times 4$] ⟶ $2n + 4$

Input	Function	Output
1	$\times 2 + 4$	6
2	$\times 2 + 4$	8
3	$\times 2 + 4$	10
4	$\times 2 + 4$	12
5	$\times 2 + 4$	14
6	$\times 2 + 4$	16
7	$\times 2 + 4$	18
8	$\times 2 + 4$	20
9	$\times 2 + 4$	22
10	$\times 2 + 4$	24

So the function $y = 2n + 4$ generates the sequence
6, 8, 10, 12, 14, 16, 18 ,20, 22, 24, ...

If you know the values of a sequence, you can find the function that generates the sequence by finding the position-to-term rule.

Composite functions

Key vocabulary

composite function: a function created by combining two or more functions.

Composite functions are formed by combining two or more functions.

If you have two functions, the composite function is the result of entering the output of the first function as the input of the second function.

Using a function machine, this would look like:

input ⟶ [function 1] ⟶ output of function 1 ⟶ [function 2] ⟶ output

Or more simply:

input ⟶ [function 1] ⟶ [function 2] ⟶ output

For example, if the first function is $x + 1$, and the second function is $2x$ you would have:

x ⟶ [$+1$] ⟶ $(x + 1)$ ⟶ [$\times 2$] ⟶ $(2x + 1)$

You can also combine two functions **before** entering the input, and still get the same output as applying the functions separately.

To do this, substitute the value of x that would be the output of the first function $(x + 1)$ into the value of x in the second function: $2(x + 1)$

Then multiply out the brackets to write the rule in its simplest form

$y = 2x + 2$

Inverse functions

Every function has an **inverse function**.

You find the inverse of a function by carrying out the inverse operations in the reverse order.

For example, the operation 'times two' can be reversed by 'dividing by two', so the function $x \rightarrow 2x$ has the inverse function $x \rightarrow \frac{x}{2}$.

Key vocabulary

inverse function: a function that reverses another function.

WORKED EXAMPLE 2

Find the inverse function of $n \rightarrow 3n - 1$.

$n \rightarrow \boxed{\times 3} \rightarrow \boxed{-1} \rightarrow 3n - 1$

First write out the given function using function machines.

$\frac{(n+1)}{3} \leftarrow \boxed{\div 3} \leftarrow \boxed{+1} \leftarrow n$

To find the inverse, work in the opposite direction, replacing each rule with the inverse operation.
The inverse operation of 'subtract 1' is 'add 1'.
The inverse operation of 'multiply by 3' is 'divide by 3'.

$n \rightarrow \frac{(n+1)}{3}$

Write out the steps as a function.

EXERCISE 4C

Did you know?

$x \rightarrow x$ is called the identity function.

In this function the inverse is the same as the original function.

1. The numbers 1 to 10 are the input for the function $x \rightarrow x + 3$.
   What sequence does this create?

2. Input the numbers 1 to 10 into each function to generate a sequence.

   a $x \rightarrow x - 5$ **b** $x \rightarrow 3x$ **c** $n \rightarrow n + 7$ **d** $n \rightarrow \frac{n}{2}$

3. Input the numbers 21 to 30 into each function to generate a sequence.

   a $y = 2x$ **b** $y = x - 8$ **c** $y = \frac{x}{3}$ **d** $y = x + \frac{1}{2}$

4. Input the numbers 11 to 20 into each function to generate a sequence.

   a $n \rightarrow 3n + 5$ **b** $n \rightarrow 2n - 7$ **c** $n \rightarrow \frac{n}{2} + 4$

   d $n \rightarrow 4n + \frac{1}{2}$ **e** $n \rightarrow \frac{4}{n} + 4$

5. Write down the inverse of each of the following functions.

   a $x \rightarrow x - 7$ **b** $x \rightarrow 4x$ **c** $x \rightarrow x + 5$

   d $x \rightarrow \frac{x}{3}$ **e** $x \rightarrow 2x + 4$ **f** $x \rightarrow 4x - 5$

   g $x \rightarrow \frac{x}{5} + 3$ **h** $x \rightarrow \frac{4}{x} - 2$

Find answers at: cambridge.org/ukschools/gcsemaths-studentbookanswers

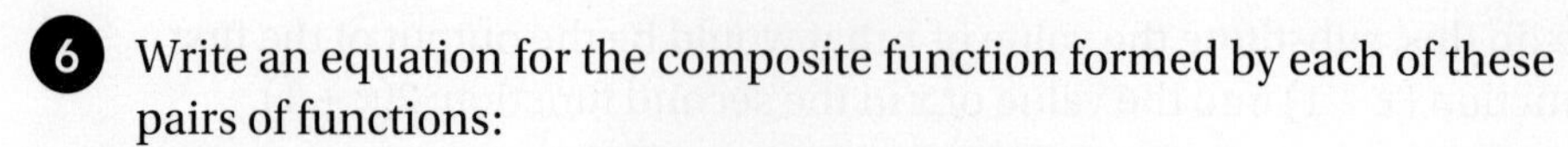

6 Write an equation for the composite function formed by each of these pairs of functions:

a function 1: $4x$ function 2: $x - 7$

b function 1: $2x$ function 2: $x + 4$

c function 1: $x - 2$ function 2: $3x$

d function 1: $x + 1$ function 2: x^2

7 **a** Write down the inverse function of $2x$.

b What is the composite function of $2x$ and its inverse function?

Section 4: Special sequences

Some patterns and sequences of numbers are well known.
You need to be able to recognise and use the following patterns.

Special sequence	Description
Simple arithmetic progression (or linear sequences)	The **difference** between consecutive terms is constant, e.g., 3, 5, 7, ... or 14, 11, 8, ...
Geometric sequences	The **ratio** between consecutive terms is constant, for example, 3, 6, 12, 24, ...
Square numbers	A square number is the product of multiplying a whole number by itself. For example, $3^2 = 3 \times 3 = 9$ Square numbers form the sequence: 1, 4, 9, 16, 25, 36, ...
Triangular numbers	These are made by arranging dots to form equilateral triangles. 1 dots 3 dots 6 dots 10 dots 15 dots The sequence is 1, 3, 6, 10, 15, ...
Quadratic sequences	These sequences are linked to square numbers. A quadratic sequence has a position-to-term rule that involves squaring one of the variables. For example, the sequence formed by the rule $n^2 + 3$ is: 4, 7, 12, 19, 28, ... The terms in a quadratic sequence do **not** increase or decrease by a constant amount. The first difference is **not** constant but the **second difference** is constant. The second difference is the difference between consecutive terms in the first difference.

Cube numbers	A cube number is the product of multiplying a whole number by itself and then by itself again. 64 s a cube number because $4^3 = (4 \times 4 \times 4) = 64$. Cube numbers form the sequence: 1, 8, 27, 64, 125, ...
Fibonacci sequences	Leonardo Fibonacci was an Italian mathematician who developed the number pattern 1, 1, 2, 3, 5, 8, 13, 21, ... while he was trying to work out how many offspring a pair of rabbits would produce over different generations. These numbers are now called Fibonacci numbers. If you start the sequence with the number 1, the term-to-term rule is 'add the previous two terms together.'

It is not possible to give an example of every possible sequence. When you are given a sequence, always look for patterns.

For example, in the sequence $\frac{1}{2}, \frac{2}{3}, \frac{3}{4}, \frac{4}{5}$ the numerator is increasing by 1 and the denominator is also increasing by 1.

Did you know?

The Fibonacci pattern is found in many natural situations.

In the Fibonacci series, the sequence of numbers is created by adding the 1st and 2nd terms together to make the 3rd term; adding the 2nd and 3rd terms together to make the 4th term and so on.

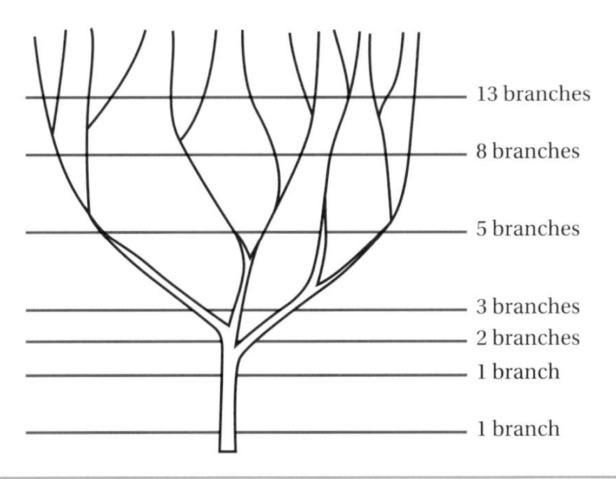

EXERCISE 4D

1 **a** Write down the sequence of the first 10 square numbers.

b How could you use this sequence to find the next 2 square numbers?

2 **a** What type of number is shown here? Explain how you know this.

b The picture represents the fifth term in the sequence.
Write down the first 10 numbers in the same sequence.

c Find the first and second differences between the terms of the sequence.

d What type of sequence is this?

3 Honeybees live in colonies known as hives.

There is one queen bee, a female, who is the only one able to lay eggs.

All other female bees are called workers, they are made when a male fertilises the queen's eggs. This means that worker bees have a mother and a father.

The males are called drones, and are made when the queen's eggs hatch without being fertilised by a male. This means that a drone has a mother but not a father.

This is a family tree of a drone; a family tree shows each generation of bee in terms of their parents.

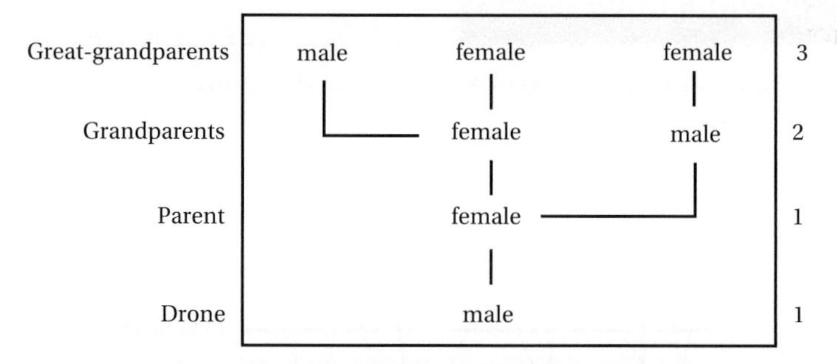

a How many great-great-grandparents does the drone have?

b The family tree shows four generations of bee.

i Copy and complete the tree so that it shows six generations.

ii Use the family tree to write a sequence for the number of bees there are in each generation.

iii Continue the sequence to find how many bees there are in the ninth generation.

iv What is the relationship between the number of bees in each generation? Do you recognise the sequence?

4 **a** Find the first 10 terms of a Fibonacci sequence that starts with the numbers 3, 4, ...

b Start a Fibonacci sequence with the numbers $^{-}2$ and 3.
Write down the first 10 terms of the sequence.

c Compare your sequence in part **b** with a Fibonacci sequence starting with 2, $^{-}3$. What do you notice?

Tip

You know the term-to-term rule of the Fibonacci sequence. How can you use this to help you?

5 The 6th and 7th terms of a Fibonacci sequence are 31 and 50.
What are the first two terms?

6 Copy and complete the table below.

Position-to-term rule	1st term	2nd term	3rd term	5th term	10th term	20th term	50th term
$n^2 + 5$							
$n^2 - 3$							
$2n^2 + 1$							
$2n^2 - 7$							

7 Write down the first six terms of the sequence formed by the rule $n^3 + 1$.

8 Write down the first 10 terms of the sequence formed by the rule $n^2 + n$.

9 Compare your answer to Question **8** with your answer for Question **2b**.

a What is the position-to-term rule for triangular numbers?

b Use your rule to find the 10th and 25th triangular numbers.

The *n*th term of a quadratic sequence

You can find the rule for the *n*th term of a quadratic sequence in a similar way to how you find the *n*th term of an arithmetic sequence.

WORKED EXAMPLE 3

Find the expression for the *n*th term in the following sequence: 4, 7, 12, 19, 28.

n	1	2	3	4	5
Tn	4	7	12	19	28

Draw a table showing the position and the term.

n	1	2	3	4	5
Tn	4	7	12	19	28

+3 +5 +7 +9

+2 +2 +2

Label the table with the difference between each term.

The first difference is not constant, it is increasing. Find the second difference by calculating the difference between the first differences.

The second difference is constant, which means this is a quadratic sequence and it will involve n^2.

n	1	2	3	4	5
Tn	4	7	12	19	28
n^2	1	4	9	16	25

+3

Square each position number and compare the resulting sequence to the original.

The square numbers are 3 fewer than the corresponding term. The expression could be $n^2 + 3$.

$Tn = n^2 + 3$

$n = 4$: $(4 \times 4) + 3 = 16 + 3 = 19$ ✓

$n = 5$: $(5 \times 5) + 3 = 25 + 3 = 28$ ✓

Test the rule works.

The general formula for finding the nth term of a quadratic equation is

$u_n = an^2 + bn + c$

where the value of a is half the second difference; the value of b depends on the first term and c is the imaginary term before the first term (T_0).

The value of a helps you determine whether the sequence involves square numbers or multiples of square numbers.

WORKED EXAMPLE 4

What is the nth term in the sequence 4, 13, 28, 49 ... ?

n	1	2	3	4
Tn	4	13	28	49

First difference: +9, +15, +21

Second difference: +6, +6

Draw a table showing the position, term and first difference. The first difference isn't constant, so find the second differences.

$u_n = an^2 + bn + c$

$a = \frac{6}{2} = 3$

Using the general formula, a is half of the second difference.

n	0	1	2	3	4
Tn	1	4	13	28	49

First difference: +9 − 6 = +3, +9, +15, +21

Second difference: +6, +6, +6

The value of c can be found by working out the value of the imaginary term before the first term (T_0).

Apply inverse operations to count back and find T_0.

$u_n = 3n^2 + bn + 1.$

Substitute $a = 3$ and $c = 1$ into the general formula.

$T_1 = 4$, so

$u_1 = 3(1)^2 + b(1) + 1 = 4$

$3 + b + 1 = 4$

$b = 4 - 3 - 1$

$b = 0$

To work out the value of b you need to substitute the value of T_1 into the formula to solve for b.

If $b = 0$, then $bn = 0$.

The formula for the nth term of the sequence 4, 13, 28, 49 ... is therefore $u_n = 3n^2 + 1$.

Substitute b into the general formula to give you the rule for the nth term.

EXERCISE 4E

1 Find a formula for the nth term for each of these sequences:

a 3, 8, 15, 24, 35

b 3, 10, 21, 36, 55

c 7, 22, 45, 76, 115

d 6, 17, 32, 51, 74

e 1, 8, 21, 40, 65

f $^-3$, 6, 23, 48, 81

g $^-2$, $^-8$, $^-18$, $^-32$, $^-50$

h 0, $^-4$, $^-12$, $^-24$, $^-40$

2 Find the next three terms in this sequence:

$1, \sqrt{2}, 2, 2\sqrt{2}, \square, \square, \square$

3 A sequence is defined as

$$x_n = \frac{n}{n+1}$$

a List the first 5 terms of this sequence.

b Determine the value of the 10th term.

4 A sequence is defined as

$$u_n = n^2 + 2n - 3$$

a List the first 5 terms of this sequence.

b Determine the value of the 15th term.

5 Look at this pattern.

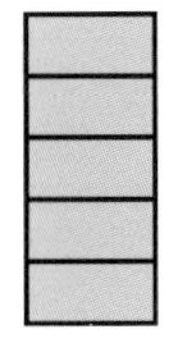 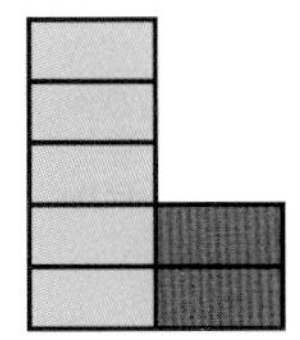 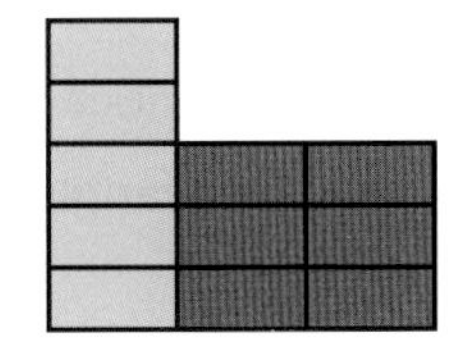 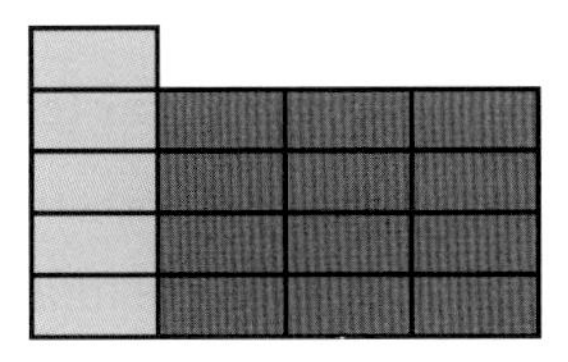 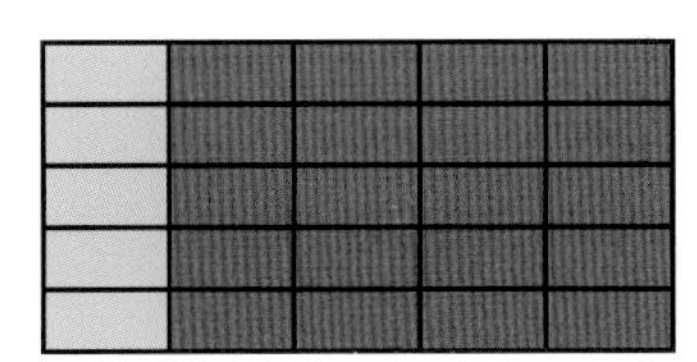

a How many tiles are there in each shape?

b How many tiles will there be in the 6th shape?

c How many tiles will there be in the nth shape?

Checklist of learning and understanding

Sequences

- Sequences can be formed using a term-to-term rule. Each term is generated by applying the same rule to the previous term.
- The position-to-term rule is used to find the value of any term in a sequence using its position in the sequence.
- Subscript notation can be used to describe position-to-term and term-to-term rules.

Functions

- A function is an expression or rule for changing one number (the input) into another number (the output).
- A sequence can be generated by inputting an ordered set of numbers into a function.
- When two or more function are applied one after the other, they make a composite function.
- A function that reverses the result of another function is called the inverse function.

Find answers at: cambridge.org/ukschools/gcsemaths-studentbookanswers

Special sequences

- It is important to be able to recognise familiar sequences such as square numbers, cube numbers, triangular numbers and Fibonacci numbers.
- In a quadratic sequence, the difference between the terms is not constant and the pattern is linked to square numbers.
- It is possible to find an expression for the nth term quadratic sequences by looking at first differences and second differences.

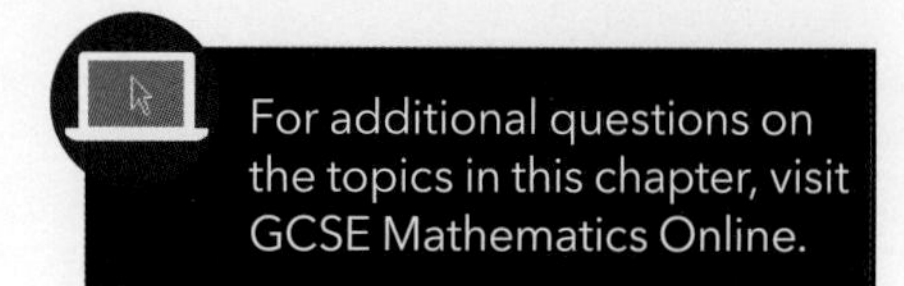

For additional questions on the topics in this chapter, visit GCSE Mathematics Online.

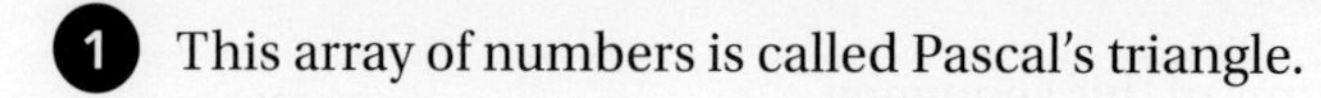

Chapter review

1 This array of numbers is called Pascal's triangle.

Each number is the sum of the two numbers above it, except for the edges which are all 1.

There are many different number patterns in the triangle.

The first row (containing the number 1) is the 0th row.

a Copy and complete Pascal's triangle to the 10th row.

b Find the total for each row.
Describe the sequence that is created by these totals.

c What sequence is represented by the diagonal series of numbers that begins in the second row 1, 3, 6, 10, ...?

d Write an expression for the nth term of the sequence 1, 3, 6, 10, ...

2 **a** Here are the first four terms of a sequence.

8 11 14 17

Write an expression for the nth term of this sequence. *(2 marks)*

b The nth term of another sequence is given by $12 - 5n$.
Write down the first three terms of this sequence. *(2 marks)*

© OCR 2012

3 A virus is infecting the population of a village.

The table shows how the number of infections increased over four days.

Day	1	2	3	4
Number of infections	8	13	18	23

a If the virus continues to infect people at the same rate, how many people will be infected on day 5?

b Assume the infection rate is constant.
Find an expression to calculate how many people will be affected on any day.

c There are 126 people in the village. How long will it be before everyone is infected?

4 The number of rabbits on an island is recorded each month.

After 1 month there are 9 rabbits.

After 2 months there are 19 rabbits.

After 3 months there are 33 rabbits.

a If the population keeps growing at the same rate, and no rabbits die, how many rabbits will there be at the end of the year?

b Explain why this sequence is not likely to reflect the real number of rabbits.

5 The number of hits on a website increases daily.

On day 1 there were 13 hits.

On day 2 there were 27 hits.

On day 3 there were 65 hits.

Assuming the number of hits continues to grow at the same rate, how many hits will there be on day 20?

6 Look at this pattern.

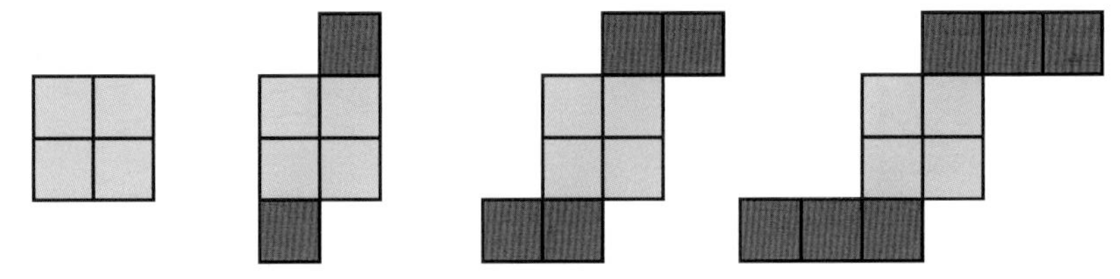

a Complete this function machine for calculating the number of tiles used to make any shape in the pattern:

input (n) → [ ] → [ ] → output

b Use your function to work out the number of tiles you would need to build the 20th, 25th and nth patterns.

7 Tamsyn is an ecologist.

She gets 28 days' holiday every year plus $\frac{1}{2}$ a day's holiday for each overnight bat survey she completes.

a Write a function for calculating how much holiday she will get if she does 10, 15 and b bat surveys in a year.

b Tamsyn's boss knows how much holiday each of her staff took last year. She wants to work out how many bat surveys they did. What function can she use to calculate this?

c Tamsyn's colleague Sofia has 30 days' holiday this year. How many surveys must she have done?

5 Properties of shapes and solids

In this chapter you will learn how to …

- use the correct geometrical terms to talk about lines, angles and shapes.
- recognise and name common 2D shapes and 3D objects.
- describe the symmetrical properties of various polygons.
- classify triangles and quadrilaterals and use their properties to identify them.

For more resources relating to this chapter, visit GCSE Mathematics Online.

Using mathematics: real-life applications

Many people use geometry in their jobs and daily lives. Artists, craftspeople, builders, designers, architects and engineers use shape and space in their jobs, but almost everyone uses lines, angles, patterns and shapes in different ways every day.

"I use a CAD package to plot lines and angles and show the direction of traffic flow when I design new road junctions."

(Civil engineer)

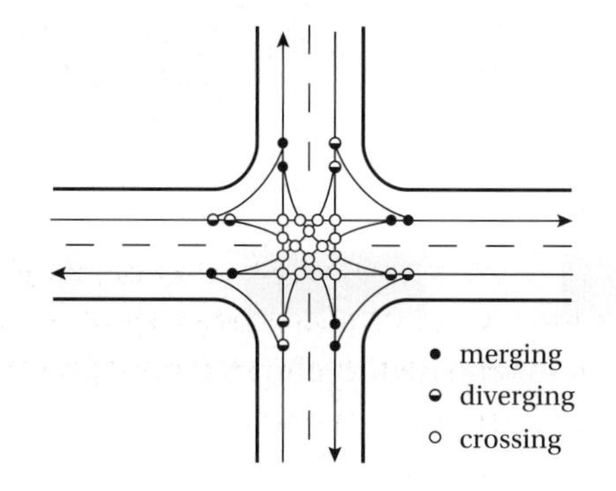

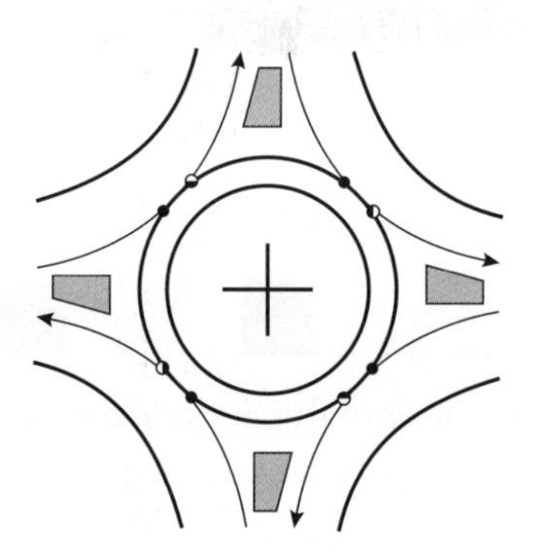

Before you start …

KS3	You should be able to use geometrical terms correctly.	**1** Choose the correct labels for each letter on the diagram. a b c d e f g h base vertex acute angle point edge right angle height face
KS3	You need to be able to recognise and name different types of shapes.	**2** **a** Identify three different shapes in this diagram and use letters to name them correctly. A B C D E **b** ABCE is one face of a solid with six faces. What type of solid could it be?

Assess your starting point using the Launchpad

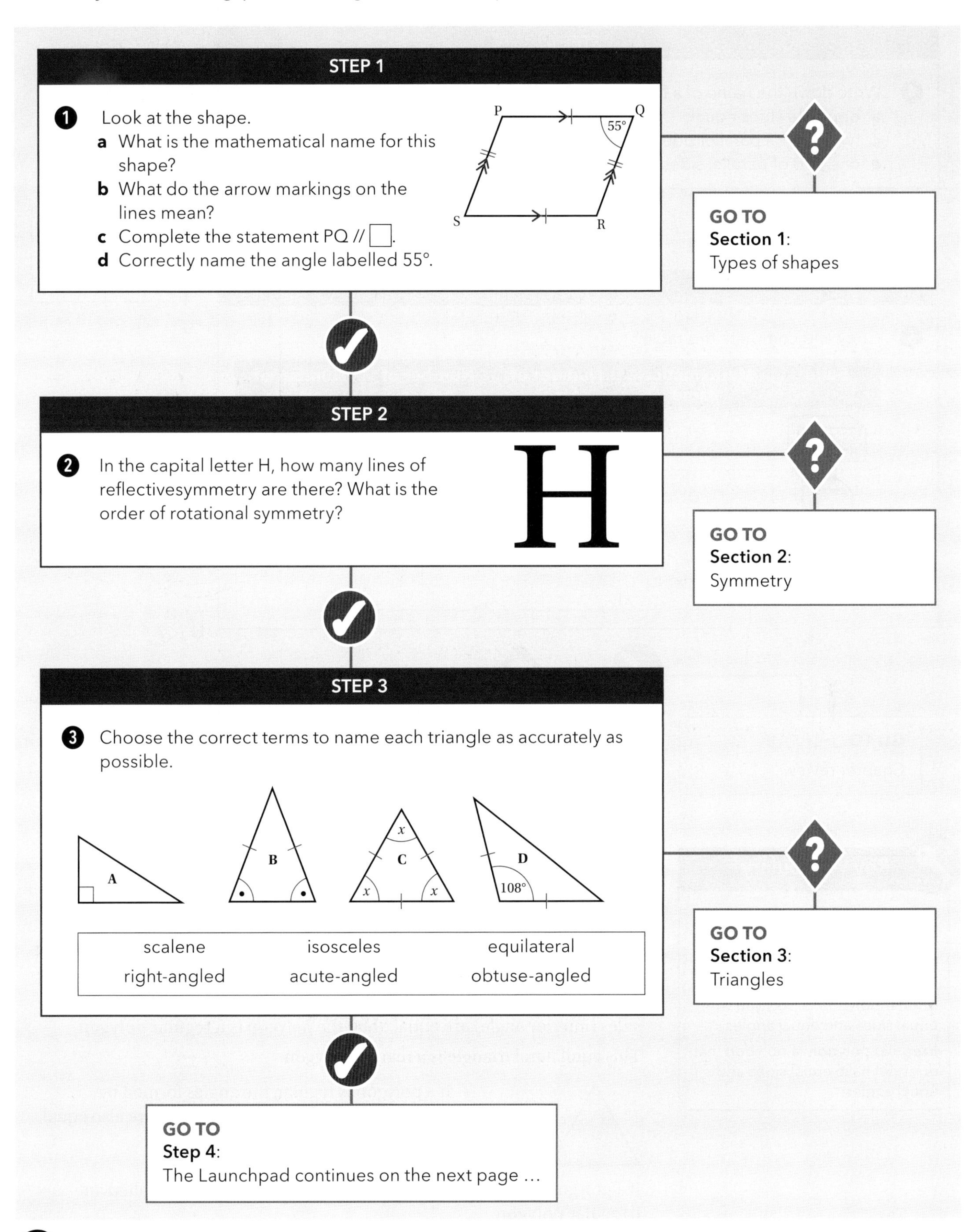

STEP 1
1 Look at the shape.
a What is the mathematical name for this shape?
b What do the arrow markings on the lines mean?
c Complete the statement PQ // ☐.
d Correctly name the angle labelled 55°.
P
Q
R
S
55°
?
GO TO
Section 1:
Types of shapes
STEP 2
2 In the capital letter H, how many lines of reflectivesymmetry are there? What is the order of rotational symmetry?
H
?
GO TO
Section 2:
Symmetry
STEP 3
3 Choose the correct terms to name each triangle as accurately as possible.
A
B
C
D
x
108°
scalene
isosceles
equilateral
right-angled
acute-angled
obtuse-angled
?
GO TO
Section 3:
Triangles
GO TO
Step 4:
The Launchpad continues on the next page …

Find answers at: cambridge.org/ukschools/gcsemaths-studentbookanswers

Launchpad continued ...

STEP 4

4 Write down the name of a four-sided shape that has:

a opposite sides equal.
b all sides equal.
c two pairs of parallel sides.
d four equal angles.
e one pair of parallel sides only.
f no parallel sides.

? **GO TO** **Section 4**: Quadrilaterals

✓

STEP 5

5 Copy and complete this table.

Solid	Mathematical name	Number of faces	Number of edges	Number of vertices
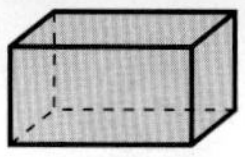				
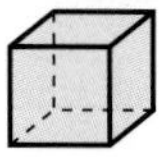				
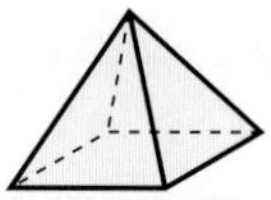				

✓ **GO TO** Chapter review

? **GO TO** **Section 5**: Properties of 3D objects

Key vocabulary

plane shape: a flat, two-dimensional shape.

polygon: a closed plane shape with three or more straight sides.

regular polygon: a polygon with equal sides and equal angles.

irregular polygon: a polygon that does not have equal sides and equal angles.

Section 1: Types of shapes

Flat shapes are called **plane shapes** or two-dimensional (2D) shapes.

A **polygon** is a closed plane shape with three or more straight sides.

Circles and ellipses (ovals) are plane shapes, but they do not have straight sides, so they are not classified as polygons.

If the sides of a polygon are all the same length and the angles between the sides (interior angle) are equal, then the polygon is a **regular polygon**.

This equilateral triangle is a regular polygon.

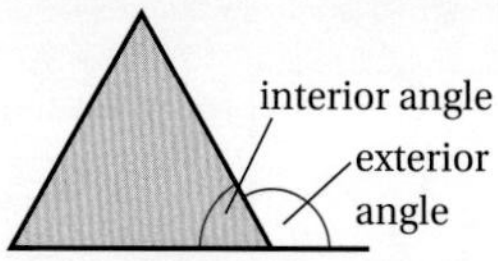

If a polygon is regular, the angles formed by extending the sides (exterior angles) are also equal.

If a polygon does not have equal sides and equal angles it is called an **irregular polygon**.

This rectangle is an irregular polygon because its sides are not all equal in length.

The fact that the angles are all equal to 90° does not make it a regular polygon.

Naming polygons

Polygons can be named according to the number of sides they have.

The table gives you the names of some polygons and shows you a regular and irregular example of each one.

Name of polygon	Number of sides	Regular polygon	Irregular polygon
Triangle	3		
Quadrilateral	4		
Pentagon	5		
Hexagon	6		
Heptagon	7		
Octagon	8		
Nonagon	9		
Decagon	10		

Key vocabulary

circumference: the distance round the outside of a circle.

diameter: a straight line from one point on the circumference to another, that passes through the centre of the circle; it is twice the length of the radius.

radius (plural **radii**): the distance of any point on the circumference from the centre of the circle.

arc of a circle: section of circumference between two points; a minor arc is the shorter distance between the two points, the major arc is the larger distance.

sector: part or slice of a circle that is enclosed by two radii and an arc; the minor sector is the smaller of the two sectors created, the major sector is the larger.

semicircle: exactly half of a circle; the diameter splits a circle into two semicircles.

chord: a straight line from one point on the circumference to another. The diameter is a chord that goes through the centre of the circle.

segment: a chord splits a circle into two segments; the smaller segment is known as the minor segment and the larger is the major segment.

tangent: a straight line that touches the circumference of a circle at only one point.

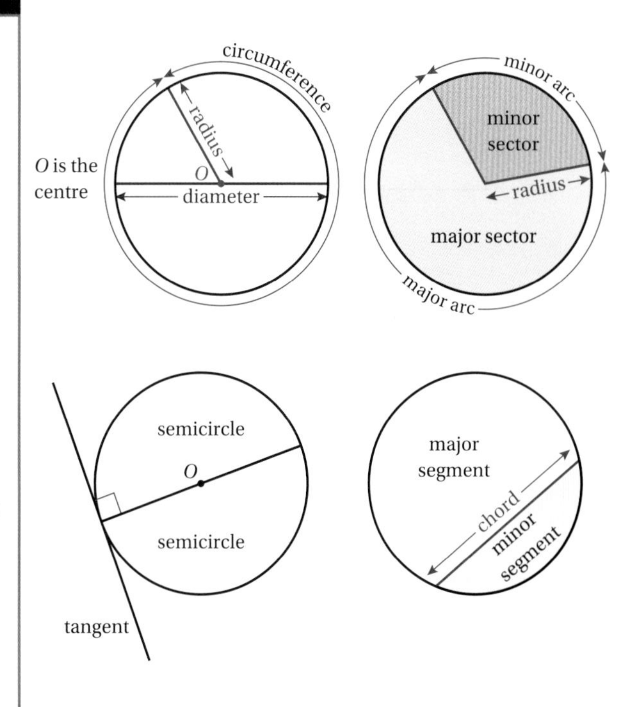

Tip

3D means an object has three dimensions or measurements: length, depth and height. It has volume and surface area.

Key vocabulary

polyhedron: a solid object with flat faces that are polygons.

Solids

Solids are three-dimensional (3D) objects.

The parts of a solid are given specific names.

Flat surfaces of a solid are called faces.

Two faces meet at the edge of a solid. Three or more faces meet at a point called a vertex. (The plural of vertex is vertices.)

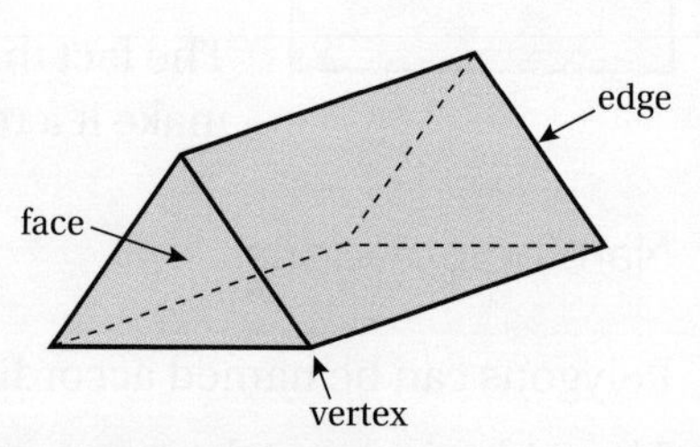

A solid with flat faces and straight edges, such as the prism above, is a **polyhedron**.

(The plural of polyhedron is polyhedra.)

Polyhedra are solid objects with flat faces that are polygons.

Cylinders, spheres and cones are not polyhedra. They are solids with a curved surface.

EXERCISE 5A

1 What is the mathematical name for each of the following shapes?

- **a** A plane shape with three equal sides.
- **b** A polygon with five equal sides.
- **c** A polygon with six vertices and six equal angles.
- **d** A plane shape with eight equal sides and eight equal internal angles.

2 Where might you find the following in real life?

- **a** A regular octagon.
- **b** A cube.
- **c** A regular quadrilateral.
- **d** An irregular pentagon.

Perpendicular and parallel lines

Perpendicular lines meet at right angles.

The symbol ⊥ means 'perpendicular to'.

In the diagram AB ⊥ CD.

The shortest distance from a point to a line is the perpendicular distance between them.

The sides of shapes are perpendicular if they form a 90° angle.

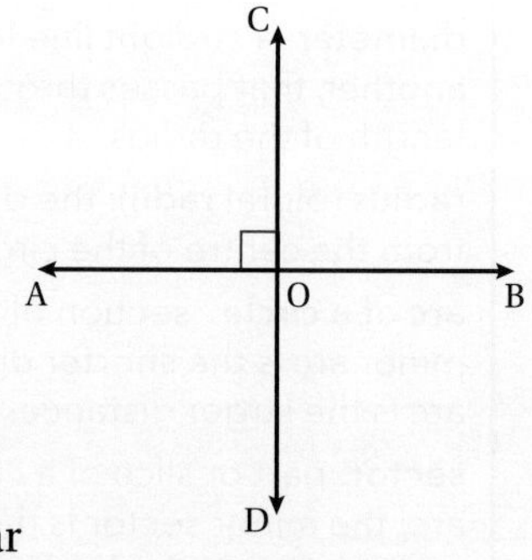

perpendicular lines

Lines are parallel if they are the same perpendicular distance apart at any point along their length.

We can say that parallel lines are **equidistant** along their length.

The symbol // means 'parallel to'. In the diagram AB // CD and MN // PQ.

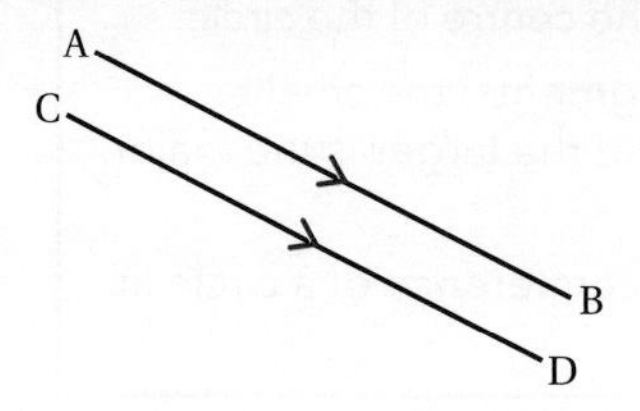

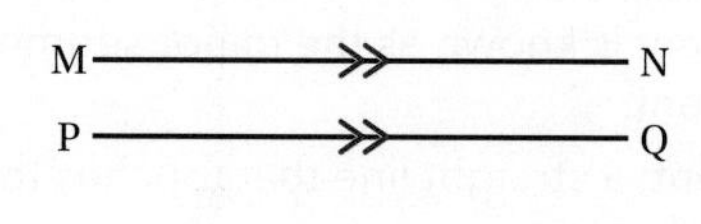

Key vocabulary

equidistant: means 'the same distance from'; if all points are equidistant they are the same distance apart.

Small arrow symbols are drawn on lines to indicate that they are parallel to each other. When there is more than one pair of parallel lines in a diagram, each pair is usually given a different set of arrow markings.

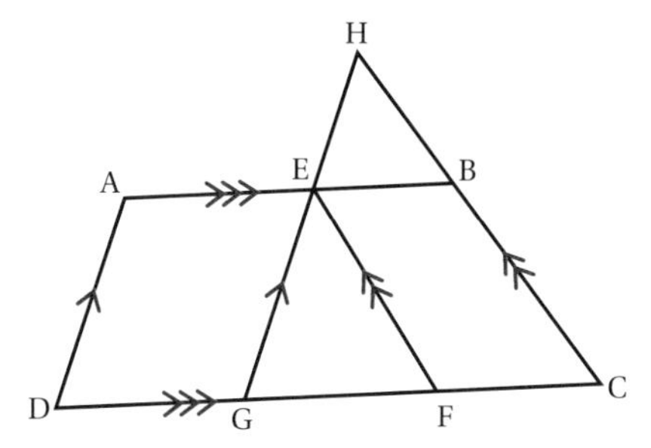

In this diagram AB // DC, AD // EG and EF // HC.

Drawing and labelling diagrams

Mathematical diagrams are drawn and labelled in particular ways so that their meaning is clear to anyone who uses them.

Shapes are labelled using capital letters on each vertex.

The letters are usually written in alphabetical order as you move round the shape. This triangle has three vertices labelled A, B and C.

The shape would be called $\triangle ABC$.

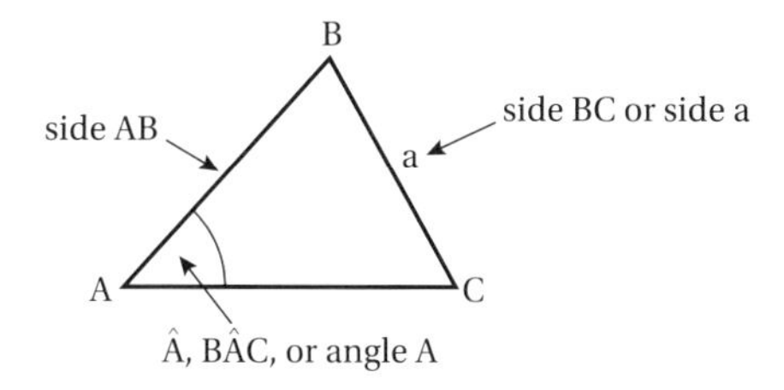

Each side of this triangle can be named using the capital letters on the vertices: AB, BC and CA (or AC).

The angles can be named in different ways.

The angle at vertex A can be named A, BAC or CAB.

Symbols can be used to label angles. For example $\angle BAC$ or $B\hat{A}C$

Sometimes single letters are used to name the sides.

In this example, side BC can also be called side a because it is opposite angle A.

This convention is often used when you work with Pythagoras' theorem and in trigonometry.

Tip

Greek letters are sometimes used to label angles. Don't be surprised to see α (alpha), β (beta), γ (gamma), δ (delta), and θ (theta) used to label angles, particularly in trigonometry.

Marking equal sides and angles

Small lines can be drawn on the sides of a shape to show whether the sides are equal or not. Sides that have the same markings are equal in length.

Curved lines and symbols such as dots or letters can be used to show whether angles are equal or not. Angles that are equal have the same marking, symbol or letter.

Find answers at: cambridge.org/ukschools/gcsemaths-studentbookanswers

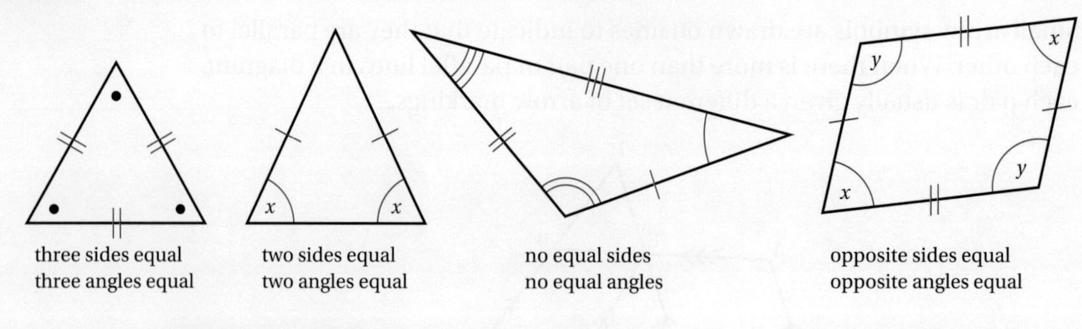

EXERCISE 5B

1 Match each description to the correct term.

a A shape that has two fewer sides than an octagon.	**i** Decagon
b A shape that has two sides more than a triangle.	**ii** Hexagon
c A shape with four sides.	**iii** Equilateral triangle
d A stop sign is an example of this shape.	**iv** Two-dimensional
e A figure that has length and height.	**v** Pentagon
f A closed plane shape with all sides x cm long and all angles the same size.	**vi** Quadrilateral
g A ten-sided figure	**vii** Square
h Another name for a regular four-sided polygon.	**viii** Regular polygon
i The more common name for a regular three-sided polygon.	**ix** Octagon

2 Look at the diagram.

Say whether the following statements are true or false.

a AF // EC.

b $\triangle$BFD is isosceles.

c CE $\perp$ BC.

d AE // BD.

e ABCE is a regular polygon.

f GB // BC.

g In $\triangle$DHJ, angle H = angle J = angle D.

h $\triangle$GHJ is a regular polygon.

3 Draw and correctly label a sketch of each of the following shapes.

a A triangle, ABC with angle B = angle C and side AB $\perp$ AC.

b A regular four-sided polygon DEFG.

c Quadrilateral PQRS such that PQ // SR but PQ $\neq$ SR and $\angle$PSR = $\angle$QRS.

4 A quadrilateral has both pairs of opposite sides parallel.

Explain mathematically why the opposite sides of the shape must also be equal.

Section 2: Symmetry

Symmetry is an important property of shapes. You can use it to identify shapes, find missing lengths and angles, and solve problems.

You need to recognise two types of symmetry in plane shapes: line symmetry and rotational symmetry.

Line symmetry

If you can fold a shape in half to create a mirror image (**reflection**) on either side of the fold the shape has line symmetry.

The fold is known as the **line of symmetry**.

Each half of the shape is a reflection of the other half so this type of symmetry is also called reflection symmetry.

Triangle A has line symmetry. The dotted line is the line of symmetry.

If you fold the shape along the line of symmetry the two parts will fit onto each other exactly.

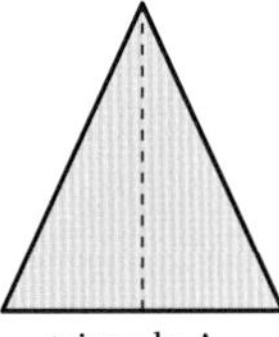
triangle A

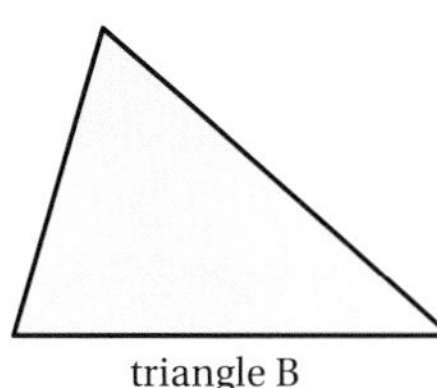
triangle B

Triangle B is not symmetrical. You cannot draw a line to divide it into two identical parts.

A shape can have more than one line of symmetry.

For example, a regular pentagon has five lines of symmetry.

Lines of symmetry can be horizontal, vertical or diagonal.

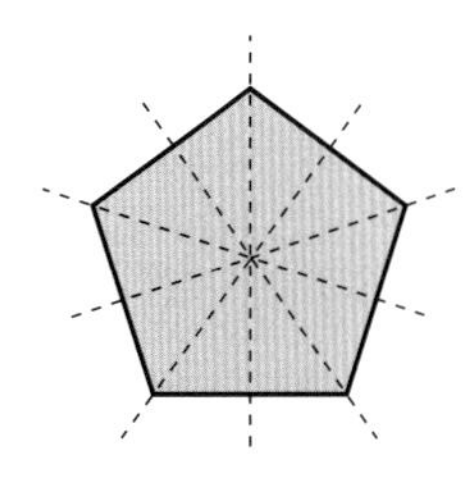

Key vocabulary

reflection: an exact image of a shape about a line of symmetry.

line (or axis) of symmetry: a line that divides a plane shape into two identical halves, each the reflection of the other.

Tip

The line of symmetry is sometimes called the mirror line. If you place a small mirror on the line of symmetry you will see the whole shape reflected in the mirror.

Rotational symmetry

A rotation is a complete turn (a movement of 360°).

A shape has **rotational symmetry** if you rotate it around a fixed point and it looks identical in different positions.

To look identical, the shape has to fit onto itself.

The **order of rotational symmetry** tells you how many times the shape will look identical before it returns to the starting point.

If you have to rotate the shape a full 360° before it appears identical again then it does **not** have rotational symmetry.

Key vocabulary

rotational symmetry: symmetry by turning a shape around a fixed point so that it looks the same from different positions.

order of rotational symmetry: how many times a shape will fit exactly onto itself when you rotate it through 360°.

Tip

You will deal with reflections in mirror lines and rotations about a fixed point again in Chapter 28 when you deal with transformations.

The order of rotational symmetry of a regular polygon depends on the number of sides it has.

A square has an order of rotational symmetry of 4 around its centre.

Find answers at: cambridge.org/ukschools/gcsemaths-studentbookanswers

You can see this in the diagram.

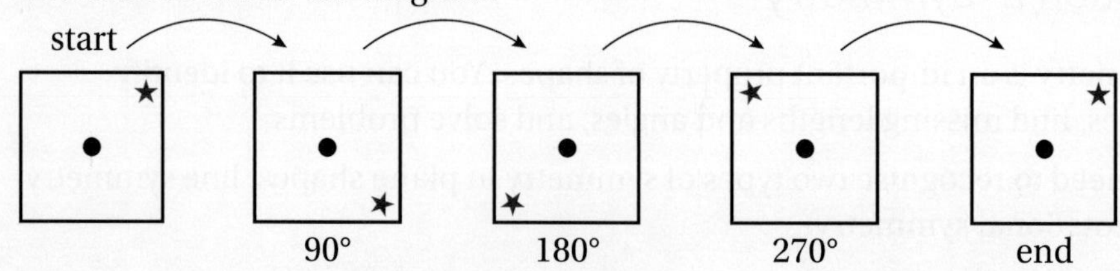

The star shows the position of one vertex of the square as it rotates.

This symbol is the national symbol for the Isle of Man.

It has an order of rotational symmetry of 3 about its centre.

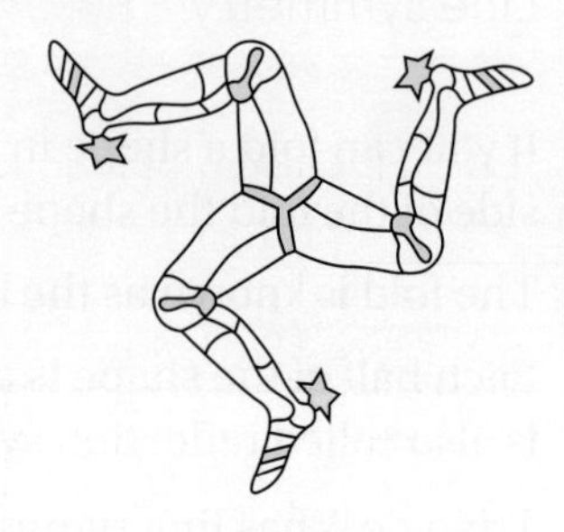

EXERCISE 5C

1 Which of the dotted lines in each figure are lines of symmetry?

a
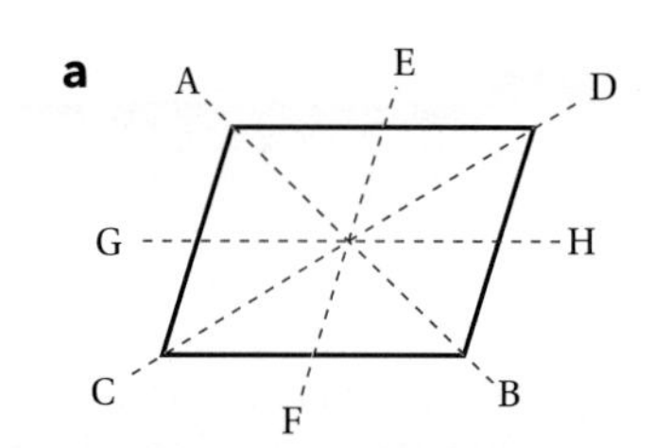

b
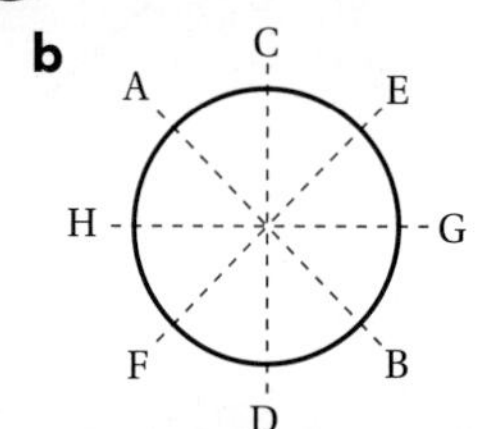

c
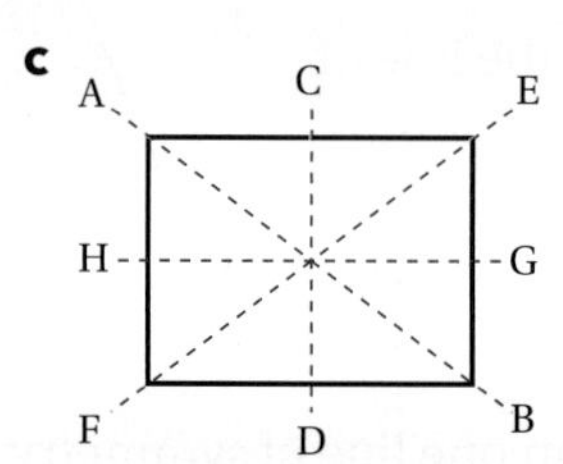

d
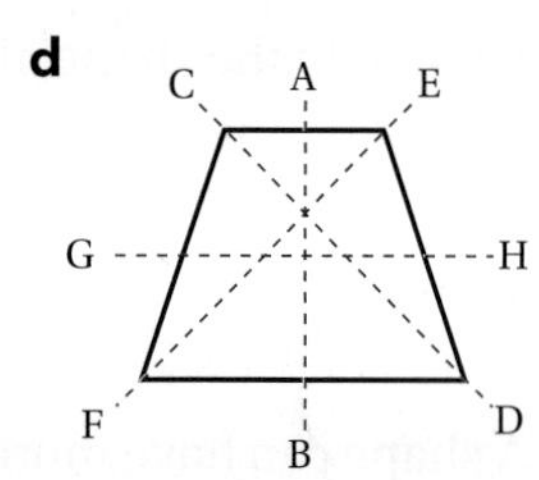

2 By sketching, investigate and work out the number of lines of symmetry and the order of rotational symmetry of each shape.

Copy and complete the table to summarise your results.

Shape	Number of lines of symmetry	Order of rotational symmetry
Square		
Rectangle		
Isosceles triangle		
Equilateral triangle		
Parallelogram		
Regular hexagon		
Regular octagon		

3 Give an example of a shape which has rotational symmetry of order 3 but which is not a triangle?

4 Which of the following letters have rotational symmetry?

C H A R

5 Look at this design. Describe its symmetrical features in as much detail as possible.

Use sketches if you need to.

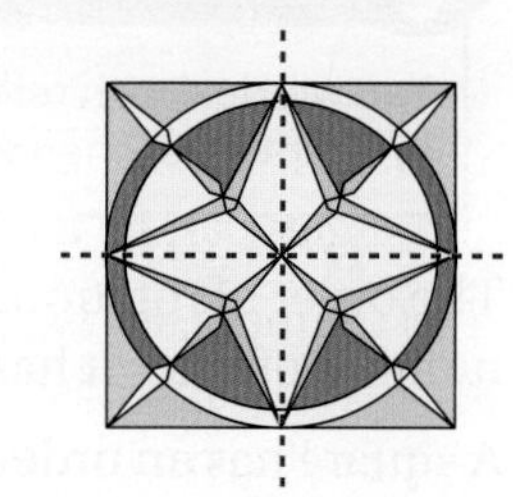

6 Metal alloy rims for tyres are very popular on modern cars.
Find and draw five alloy rim designs that you like.
For each one, state its order of rotational symmetry.

7 Sketch five different symmetrical designs or logos that you can find in your environment.
Label your sketches to indicate how the design is symmetrical.

Section 3: Triangles

Triangles are three-sided polygons that are given special names according to their properties.

Type of triangle	Properties
Scalene	• No equal sides. • No equal angles. • No line of symmetry. • No rotational symmetry.
Isosceles	• Two equal sides. • Angles at the base of the equal sides are equal. • One line of symmetry. • Line of symmetry is the perpendicular height. • No rotational symmetry.
Equilateral	• All sides equal. • Three equal angles, each is 60°. • Three lines of symmetry. • Rotational symmetry of order 3.
Acute-angled	• All angles are less than 90° (acute).
Right-angled	• One angle is a right angle (90°).
Obtuse-angled	• One angle is greater than 90° (obtuse).

Triangles can be a combination of types.

For example, an isosceles triangle could be a right-angled isosceles triangle, an acute-angled isosceles triangle or an obtuse-angled isosceles triangle depending on the size of the angles.

Find answers at: cambridge.org/ukschools/gcsemaths-studentbookanswers

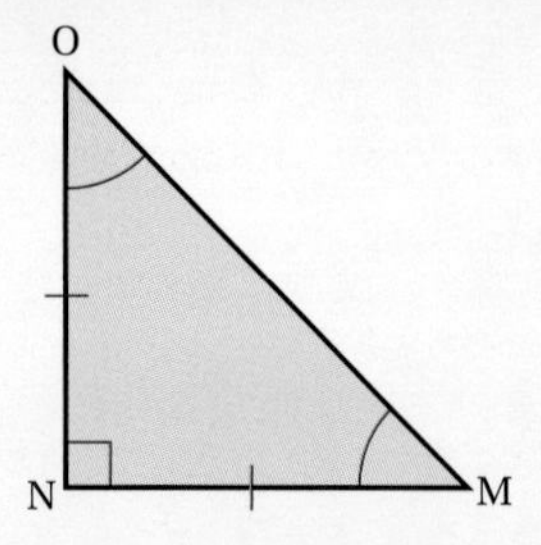

Here, triangle MNO is a right-angled isosceles triangle.

You will need to know and apply the basic properties of triangles when you work with theorems and proofs.

For example, you will use triangle properties extensively when you deal with circle theorems in Chapter 34.

Angle properties of triangles

The angles inside a shape are called interior angles.

The three interior angles of any triangle always add up to 180°.

If you extend the length of one side of a triangle you form another angle outside the triangle. Angles formed outside the triangle in this way are called exterior angles.

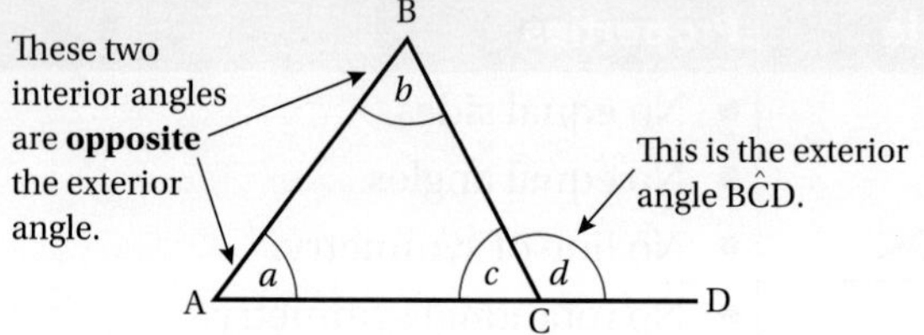

If you tear off the angles of any triangle and place them against a straight edge (180°) you can see that the interior angles add up to 180°.

You can also see that the exterior angle is equal to the sum of the two interior angles that are opposite it.

Tip

You will learn how to prove these properties using mathematical principles in Chapter 9.

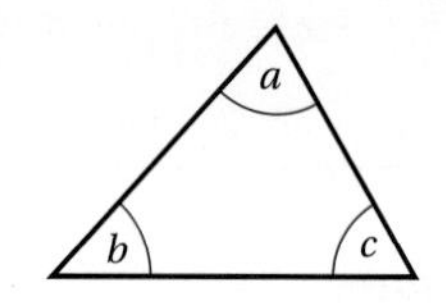

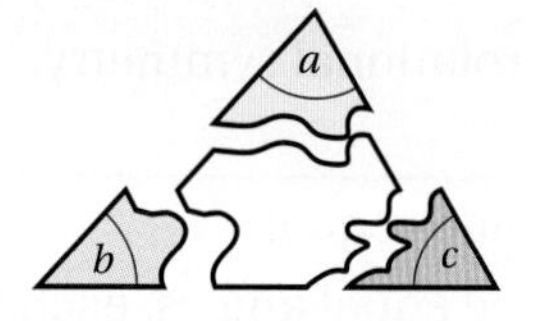

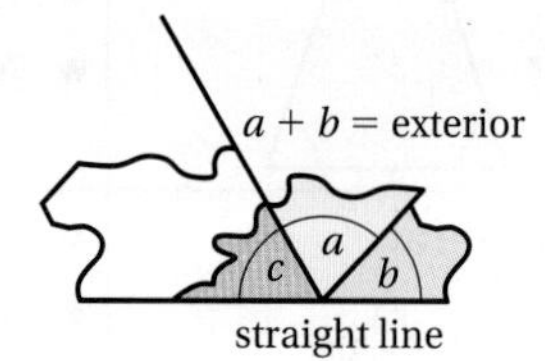

Using the properties of triangles to solve problems

You can use the properties of triangles to solve problems involving unknown angles and lengths of sides.

Problem-solving framework

Triangle ABC is isosceles with perimeter 85 mm. AB = BC and AC = 25 mm. Angle ABC = 48°.

Calculate: **a** the length of each equal side. **b** the size of each equal angle.

Steps for approaching a problem-solving question	What you would do for this example
Step 1: Work out what you have to do. Start by reading the question carefully.	You need to use the properties of isosceles triangles and the given information to find the length of two sides and the size of two angles.
Step 2: What information do you need? Have you got it all?	You should draw a labelled sketch to see whether you have the information you need. 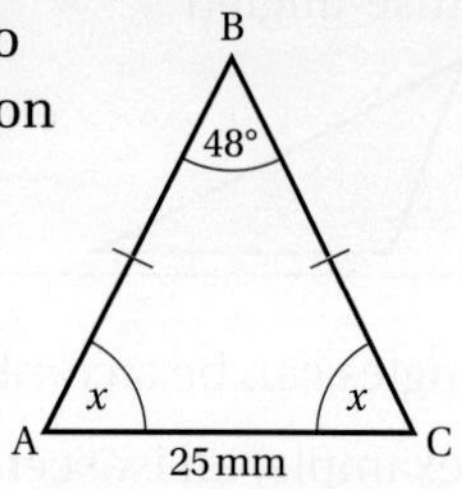 (Sketch: triangle ABC with angle B = 48°, AB and BC marked equal, base angles x and x, AC = 25 mm)

Continues on next page ...

Step 3: Decide what maths you can do.	You can use the values you already have to make equations to find the missing values.
Step 4: Set out your solution clearly. Check your working and that your answer is reasonable.	**a** Perimeter = AC + AB + BC = 85 mm So, 85 = 25 + AB + BC 85 − 25 = AB + BC 60 = AB + BC But AB = BC, so AB = BC = 30 mm Check: 30 + 30 + 25 = 85. **b** Let each equal angle be x. $48 + 2x = 180$ (angle sum of triangle) $2x = 180 - 48$ $2x = 132$ $x = 66$ Check: 66 + 66 + 48 = 180
Step 5: Check that you've answered the question.	Each equal side is 30 mm long. Each equal angle is 66°.

EXERCISE 5D

Tip

You will use properties of triangles often when you deal with trigonometry in Chapter 33. In many problems you will have to find the size of unknown angles before you can move on and solve the problem.

1 What type of triangle is this?
Explain how you decided without measuring.

2 What type of triangle is this?

Choose the correct answer.

A obtuse-angled scalene

B right-angled isosceles

C acute-angled isosceles

D obtuse-angled isosceles

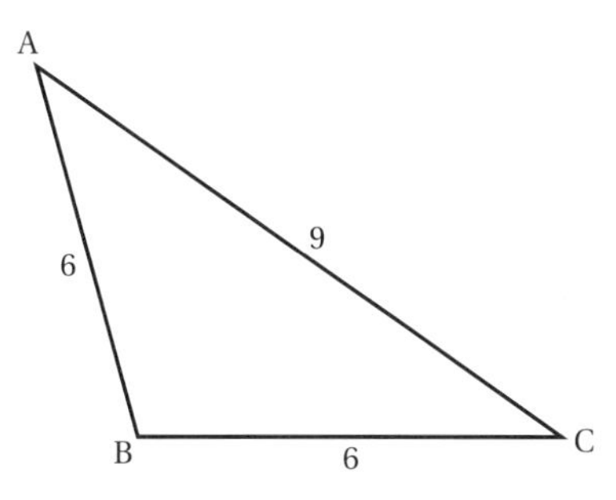

3 Which of the following triangles are not possible?

Explain why.

a An isosceles triangle with an obtuse angle.

b A scalene triangle with two angles > 90°.

c A scalene triangle with three angles 34°, 64° and 92°.

d An obtuse-equilateral triangle.

e An isosceles triangle with side lengths 6.5 cm, 7 cm and 7.5 cm.

Find answers at: cambridge.org/ukschools/gcsemaths-studentbookanswers

4 Two angles in a triangle are 38° and 104°.

a What is the size of the third angle?

b What type of triangle is this?

5 Find the size of the unknown angles in the following diagrams.

Show your working and give mathematical reasons for any deductions you make.

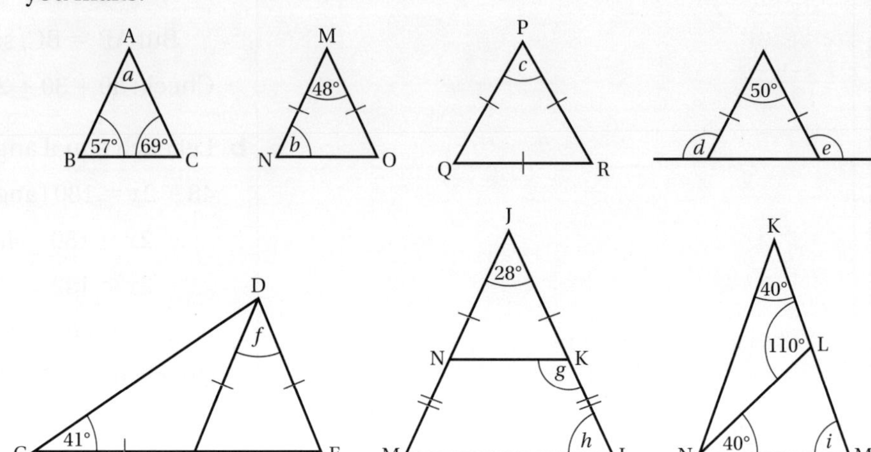

6 Isosceles triangle DEF with DE = EF has a perimeter of 50 mm.

Find the length of EF if:

a DF = 15 mm

b DF = $\sqrt{13}$

c DF = $(3x + 2)$ and DE = $(x + 4)$

Section 4: Quadrilaterals

A quadrilateral is a four-sided plane figure.

Quadrilaterals are probably the most common shape in your environment.

In the photographs below you can find rectangles, squares and trapezia.

Parallelograms and rhombuses are less common in everyday life, but the shadows of rectangles and squares often produce these shapes.

Quadrilaterals are classified and named according to their properties.

You can see from the definitions that some properties overlap.

For example, a rectangle is actually a special type of parallelogram.

The rectangle meets the definition of a parallelogram (in other words it has both pairs of opposite sides parallel), so all rectangles are parallelograms.

In this case the reverse of the statement (the converse) is not true. All parallelograms are not rectangles.

You need to know the names and basic properties of the quadrilaterals shown in the table.

Did you know?

In mathematics we call a reverse statement the converse.

Key vocabulary

adjacent: next to each other; in shapes, sides that intersect each other.

bisect: to divide exactly into two halves.

Quadrilateral	Properties
Trapezium	• One pair of opposite sides are parallel.
Kite	• Two pairs of **adjacent** sides are equal. • Diagonals are perpendicular. • One diagonal **bisects** the other. • One diagonal bisects the angles.
Parallelogram	• Both pairs of opposite sides are parallel. • Both pairs of opposite sides are equal. • Both pairs of opposite angles are equal. • Diagonals bisect each other.
Rhombus	A rhombus is a parallelogram with all sides equal. It has the same properties listed for parallelograms, but has the following special features: • All sides are equal. • Diagonals bisect at right angles. • Diagonals bisect the angles.
Rectangle	A rectangle is a parallelogram with angles of 90°. It has the same properties as listed for parallelograms, but with the following special features: • All angles are 90°. • Diagonals are equal in length.
Square	A square is a rectangle with all sides equal. It has the same properties as a rectangle, but also has these special features: • All sides are equal. • Diagonals bisect at right angles. • Diagonals bisect the angles.

Using the properties of quadrilaterals to solve problems

You can use the properties of a quadrilateral to identify and name it.

You should always state what properties you are using to justify your answer.

WORKED EXAMPLE 1

A plane shape has two diagonals. The diagonals are perpendicular.

a What shape(s) could this be? **b** The diagonals are not the same length. Which shape(s) could it **not** be?

a Two diagonals means that the shape is a quadrilateral.

Only the square, rhombus and kite have diagonals that intersect at 90°.

The shape could be a square, rhombus or kite.

Remember to include your reasoning as part of your answer.

b Of the three shapes, only the square has diagonals that are equal in length.

Therefore, it could not be a square.

The angle sum of quadrilaterals

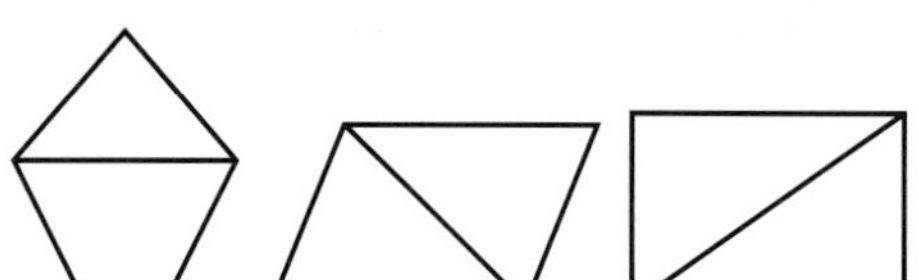

All quadrilaterals have two (and only two) diagonals. If you draw in one diagonal you divide the quadrilateral into two triangles.

You already know that the interior angles of a triangle add up to 180°.

Therefore, the interior angles of a quadrilateral are equal to $2 \times 180° = 360°$.

This is an important property of all quadrilaterals.

You can use it together with the other properties of quadrilaterals to find the size of unknown angles.

WORK IT OUT 5.1

Find the sizes of the missing angles.

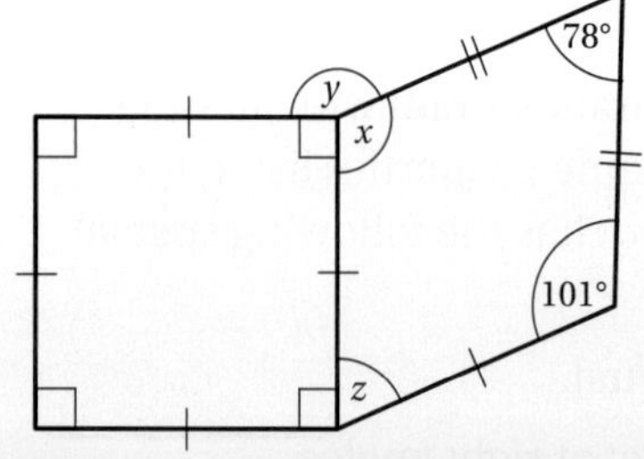

Identify the correct solution.
Find the error in the other two.

Option A	Option B	Option C
$x = 180° - 78° = 102°$ (angles on a straight line = 180°)	$x = 101°$ as the shape is a kite and opposite angles are equal	The shape is a square and a rhombus
$x + 90° + y = 360°$	If $x = 101°$ then the angles in any quadrilateral = 360°	Angle $x = 101°$
$192° + y = 360°$	$z = 360° - 78° - 101° - 101° = 80°$	Angle $z = 78°$
$y = 360° - 192° = 168°$ (angles in a quadrilateral add to 360°)	Angles round a point add to 360°	Angle $y = 360° - 101° - 90°$
$x + 78° + 101° + z = 360°$	$x + y + 90° = 360°$	$y = 360° - 191°$
$102° + 78° + 101° + z = 360°$	$360° - 90° - 101° = y$	$y = 169°$
$281° + z = 360°$	$y = 169°$	
$z = 360° - 281° = 79°$		

EXERCISE 5E

1 Identify the quadrilateral from the description.

There may be more than one correct answer.

a All angles are equal.

b Diagonals are equal in length.

c Two pairs of sides are equal and parallel.

d No sides are parallel.

e The only regular quadrilateral.

f Diagonals bisect each other.

2 You can identify a quadrilateral by considering its diagonals.

Copy and complete this table.

Shape	Diagonals are equal in length	Diagonals bisect each other	Diagonals are perpendicular
Rhombus			
Parallelogram			
Square			
Kite			
Rectangle			

3 What is the most obvious difference between a square and a rhombus?

4 Millie says that a quadrilateral has all four sides the same length.
Elizabeth says it must be a square.
Is Elizabeth correct? Give an explanation for your answer.

5 A kite has one angle of 47° and one of 133°.
What sizes are the other two angles?

6 State whether each statement is always true, sometimes true or never true.
Give a reason for your answer.

a A square is a rectangle.

b A rectangle is a square.

c A rectangle is a rhombus.

d A rhombus is a parallelogram.

e A parallelogram is a rhombus.

7 ABCD is a rhombus.
Which of the following statements are true?
Justify your answers.

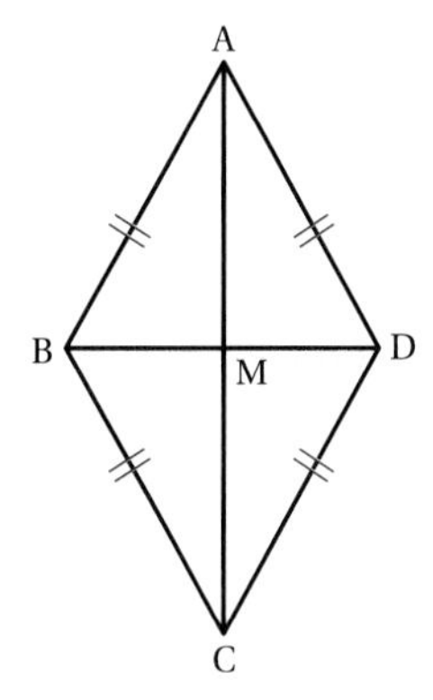

a AMB = 90°

b DB = AC

c ∠BDC + ∠ACD = 180°

d ∠ABC = ∠ADC

Find answers at: cambridge.org/ukschools/gcsemaths-studentbookanswers

8 Calculate the value of x in rhombus ABCD.

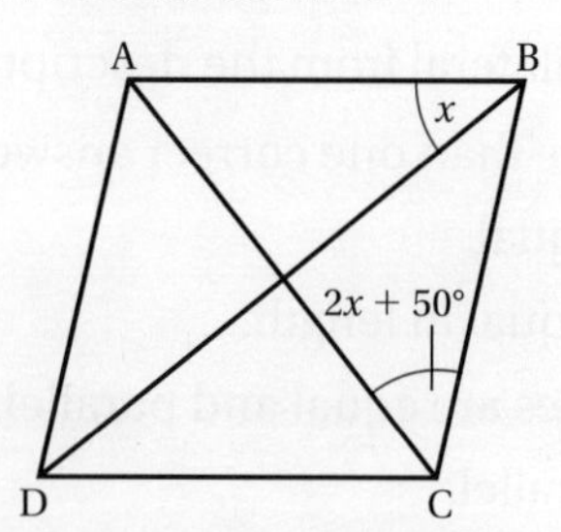

Tip

Use the problem-solving framework and think carefully about what other mathematics you can use to solve this problem.

9 In rhombus ABCD, AB = 26 mm and diagonal BD = 48 mm.
Calculate the length of diagonal AC.

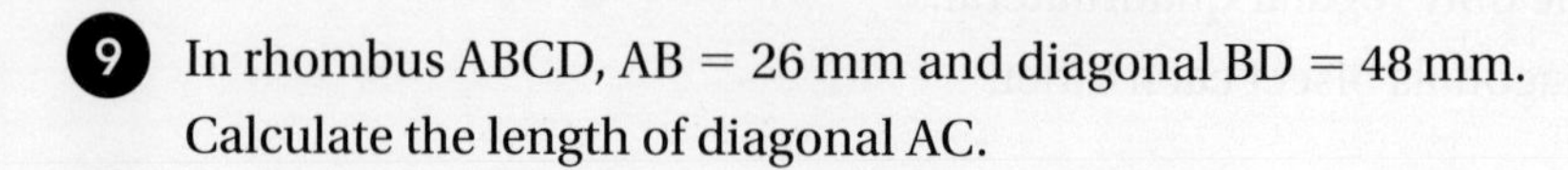

10 In this figure ABCD and ADEF are parallelograms.
Show mathematically that FECB is a parallelogram.

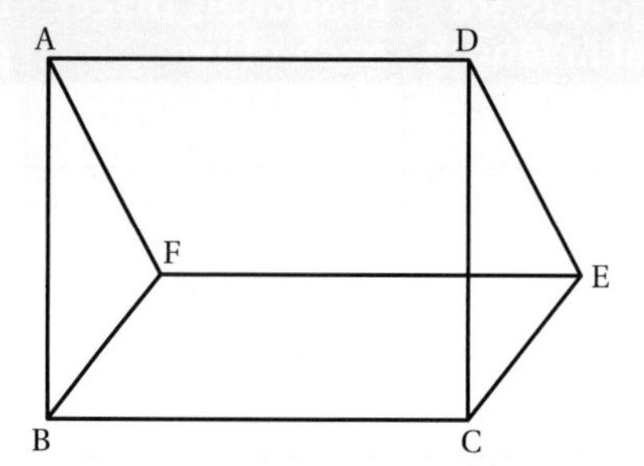

Tip

You need to know the properties of the basic polyhedra and other 3D solids.

You will use these properties to draw plans and elevations of solids in Chapter 20 and you will apply them when you solve problems relating to volume and surface area in Chapter 21.

11 Jason says that if two opposite sides of a quadrilateral are equal in length then the quadrilateral must be a parallelogram.
Is he correct?

Section 5: Properties of 3D objects

In Section 1 you learned that solids are 3D objects that have length, depth and height and that polyhedra are 3D solids with polygonal faces.

Cubes, cuboids, prisms and pyramids are all types of polyhedra.
Each type of polyhedron has some properties that are not shared by the others. For example, a cuboid has six rectangular faces. A cube also has six faces, but to be classified as a cube, the faces must all be **congruent** squares.

Key vocabulary

congruent: shapes that are identical in shape and size.

Cubes and cuboids

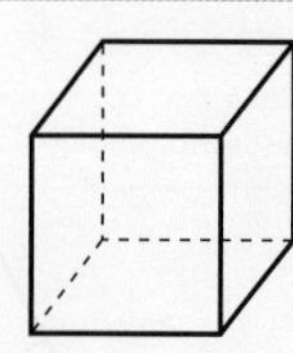

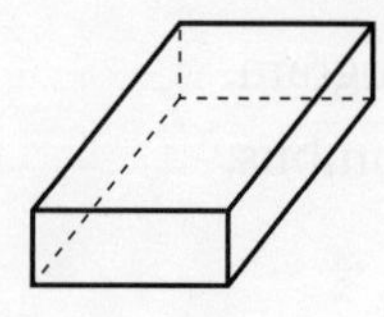

Cubes and cuboids are box-shaped polyhedra.

They have six faces, twelve edges and eight vertices.

A cube has square faces.

A cuboid has rectangular faces.

All cubes are cuboids, but not all cuboids are cubes.

Prisms

A prism is a 3D object with two congruent, parallel faces.

If the prism is sliced parallel to one of these faces the cross-section will always be the same size and shape.

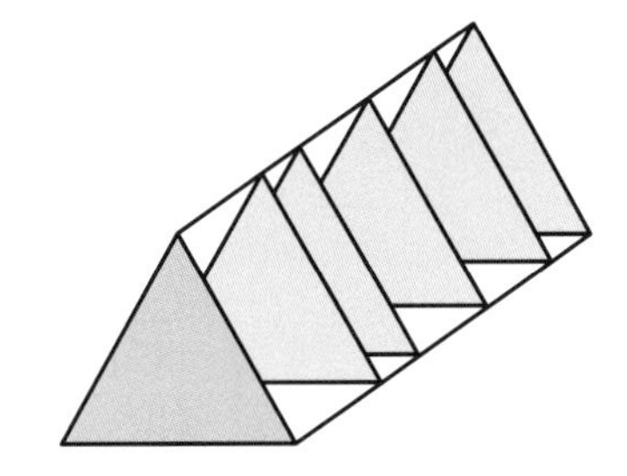

The diagram shows a prism with two triangular end faces.

You can see that slicing it anywhere along its length gives a triangular cross-section.

The parallel faces of a prism can be any shape.

If the prism is a polyhedron all of the other faces are rectangular.

Prisms are named according to the shape of their parallel faces.

pentagonal prism

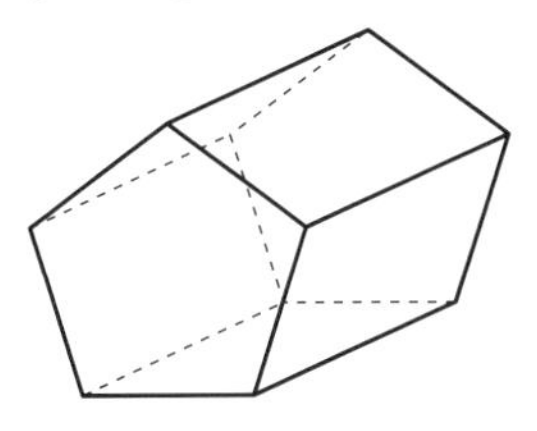

2 pentagonal faces
5 rectangular faces

hexagonal prism

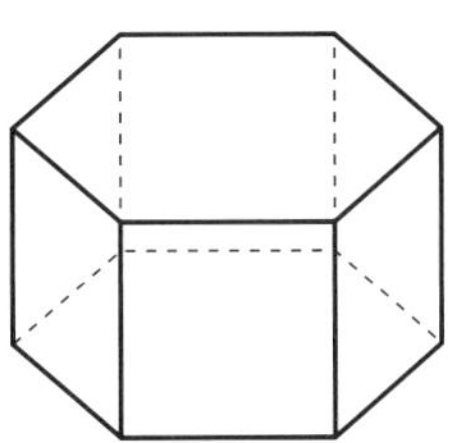

2 hexagonal faces
6 rectangular faces

octagonal prism

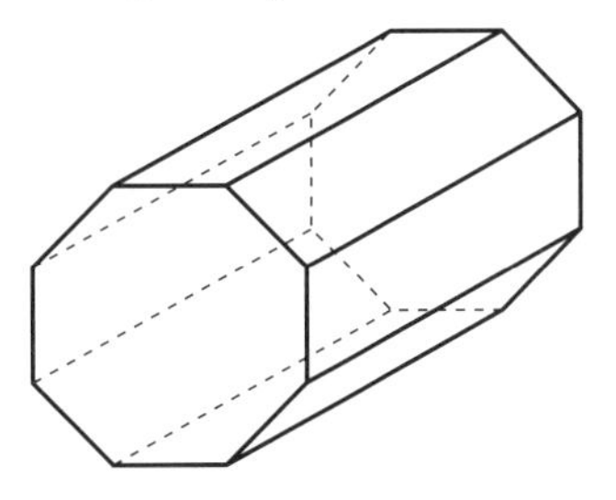

2 octagonal faces
8 rectangular faces

A cube is a square prism and a cuboid is a rectangular prism.

Pyramids

A pyramid is a polyhedron with a polygonal base and triangular faces that meet at a vertex (sometimes called the apex of the pyramid).

Pyramids are named according to the shape of their base.

triangular-based pyramid

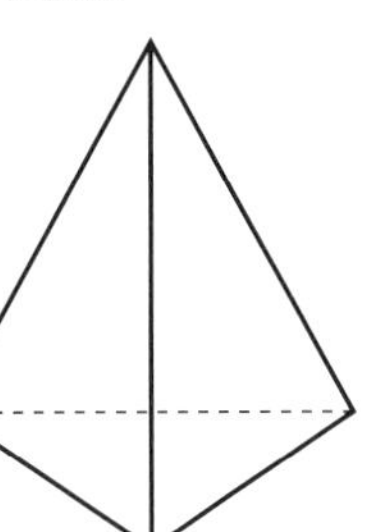

square-based pyramid

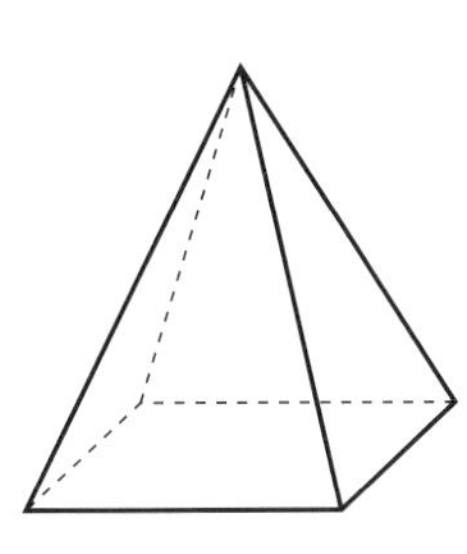

pentagonal-based pyramid

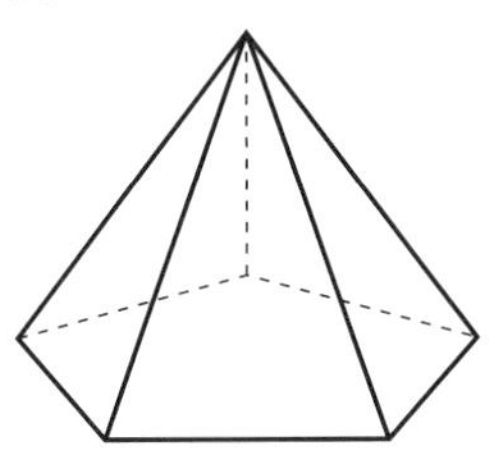

hexagonal-based pyramid

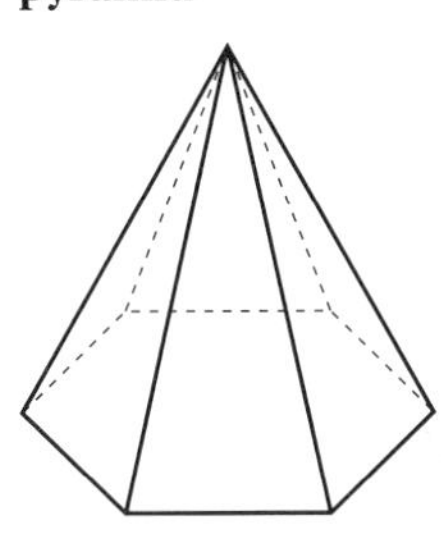

The number of sides of the base can be used to work out how many triangular faces the pyramid has. A square has four sides, so a square-based pyramid has four triangular faces.

The Louvre Museum in Paris is famous for the massive glass square pyramid at its entrance. There is another smaller, inverted square pyramid in the underground shopping mall behind the museum.

Find answers at: cambridge.org/ukschools/gcsemaths-studentbookanswers

A regular polyhedron is a solid whose faces are all congruent regular polygons.

There are only five regular polyhedra. These are shown in the table below.

Polyhedron	Vertices	Edges	Faces
Tetrahedron	4	6	4
Cube/Hexahedron	8	12	6
Octahedron	6	12	8
Dodecahedron	20	30	12
Icosahedron	12	30	20

Other solids

Cylinders, cones and spheres are also 3D objects.

They do not have straight edges or flat faces that are polygons so they are not polyhedra.

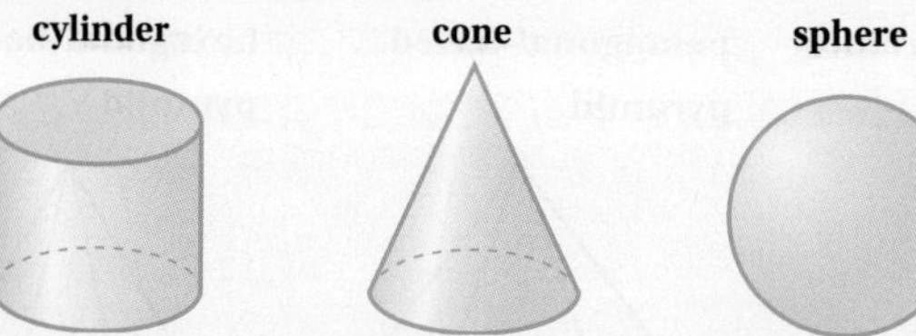

A cylinder has two circular end faces and a curved surface along its length.

A cone has a circular base and a curved surface that forms a point.

A sphere is shaped like a ball. It has only one continuous curved surface.

EXERCISE 5F

1 Sketch a possible solid you could make by combining the given solids.
For each one, state whether it is a polyhedron and how many faces, vertices and edges it would have.

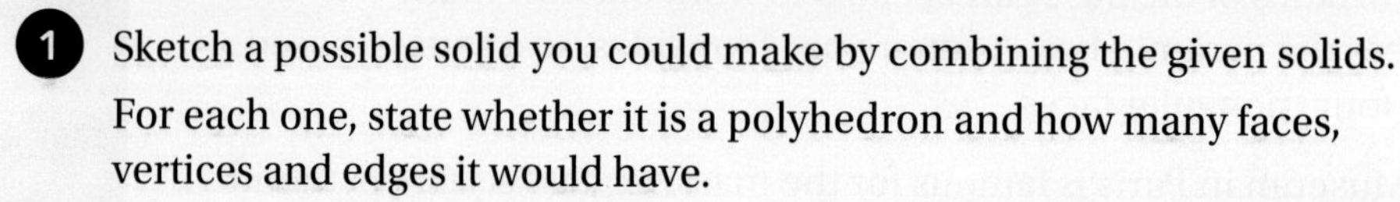

a A large and a small cylinder. **b** A cube and a square pyramid.

c Two identical pentagonal pyramids. **d** A triangular prism and a cuboid.

2 Copy and complete this table.

3D Shape	Faces (F)	Vertices (V)	Edges (E)
Cube			
Cuboid			
Triangular pyramid			
Square pyramid			
Triangular prism			
Hexagonal prism			

a Write an expression to show the relationship between F, V and E in these solids.

b Use your expression to find the number of faces in a polyhedron with 12 vertices and 30 edges.

c Show whether your expression works for each of the following solids.

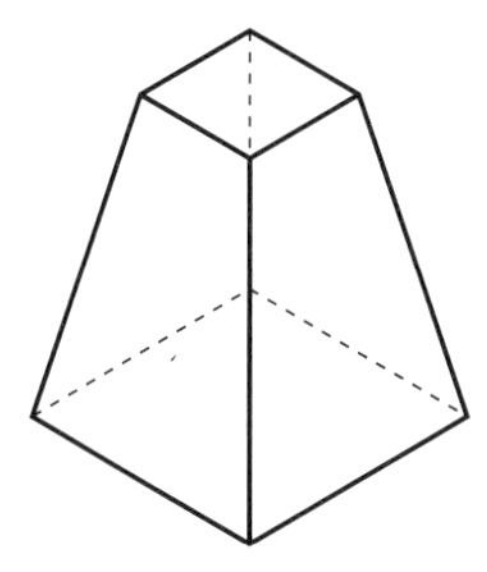

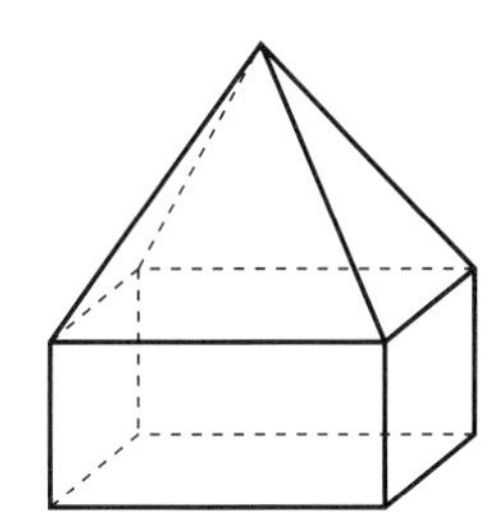

If it doesn't work, suggest a reason for this.

d Is it possible to have a polyhedron with 25 faces, 45 edges and 30 vertices?

Justify your answer.

3 Janice is building wire models of 3D objects for a school project.

How much wire would she need to build each of these shapes?

Assume there is no overlap at the vertices or joins.

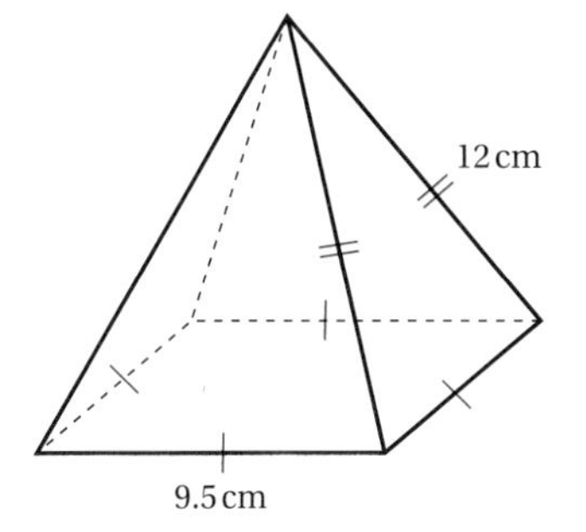

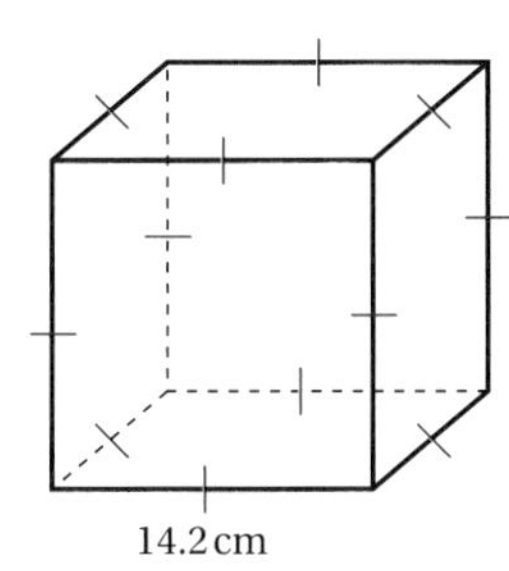

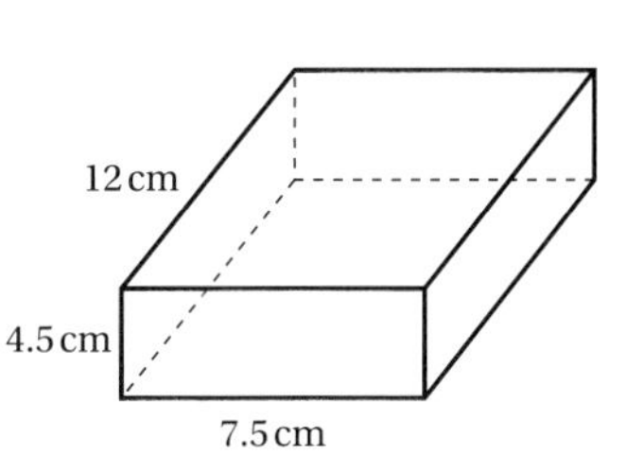

Checklist of learning and understanding

Types of shapes

- Polygons are closed plane shapes with straight sides.
- Triangles, quadrilaterals, pentagons and hexagons are all polygons.
- Circles and ovals are plane shapes, but they are not polygons.

Symmetry

- Shapes have line symmetry if they can be folded along a line of symmetry to produce two identical mirror images.
- Shapes have rotational symmetry if they fit onto themselves more than once during a 360° rotation.

Triangles

- Triangles are three-sided polygons.
- Triangles can be classified and named using their side and angle properties.
- The sum of the interior angles of a triangle is 180°.

Quadrilaterals

- Quadrilaterals are four-sided polygons.
- Quadrilaterals can be classified and named using their side, angle and diagonal properties.
- The sum of the interior angles of a quadrilateral is 360°.

Properties of 3D objects

- 3D objects are solids with length, depth and height.
- Polyhedra are solids with flat polygon faces and straight edges.
- Prisms and pyramids are polyhedra.
- Cylinders, cones and spheres are 3D objects but they are not polyhedra.

For additional questions on the topics in this chapter, visit GCSE Mathematics Online.

Chapter review

1 True or false?

- **a** A slice of pizza can be accurately described as a triangle.
- **b** A triangular pyramid has 4 vertices, 4 faces and 6 edges.
- **c** A pair of lines that are equidistant and never meet are described as being perpendicular.
- **d** A pyramid with a polygon base of n sides will have $n + 1$ vertices.
- **e** A cylinder has a uniform circular cross-section.

2 **a** Ravi has drawn these triangles on one centimetre square paper.

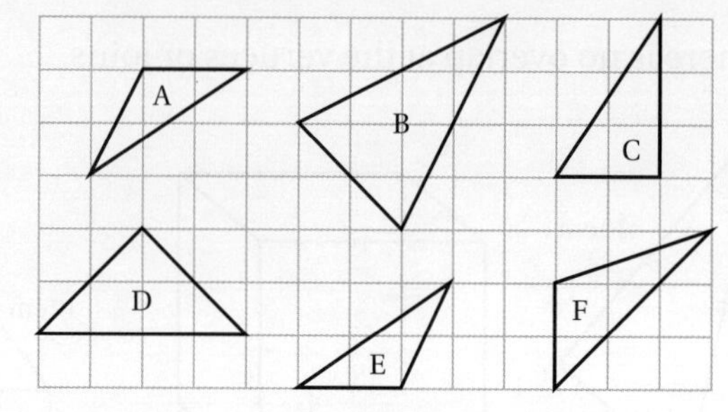

- **i** Which two triangles contain a right angle? *(1 mark)*
- **ii** Which two triangles are isosceles? *(1 mark)*
- **iii** Which two triangles are congruent? *(1 mark)*

b These are the names of some special quadrilaterals.

kite	square	parallelogram	rhombus	rectangle

Choose a quadrilateral from the list above that satisfies each set of conditions.

i
- All the sides are the same length
- All the angles are right angles
- There are four lines of symmetry *(1 mark)*

ii
- All the sides are the same length
- Opposite angles are equal
- There are only two lines of symmetry *(1 mark)*

iii
- Two pairs of adjacent sides are of equal length
- One pair of opposite angles are equal
- There is only one line of symmetry *(1 mark)*

© OCR 2013

3 Describe the symmetrical features of a regular hexagon as fully as possible.

4 Find the missing angles in this isosceles trapezium.

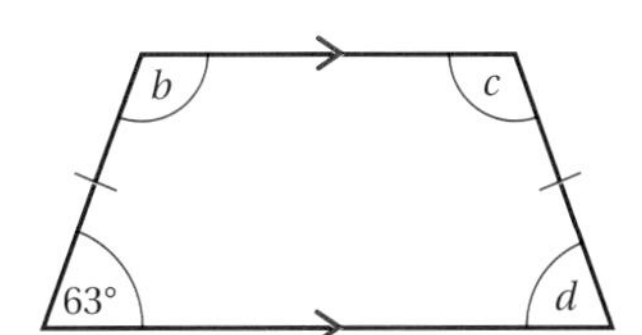

5 Is the missing angle a right angle? Explain why.

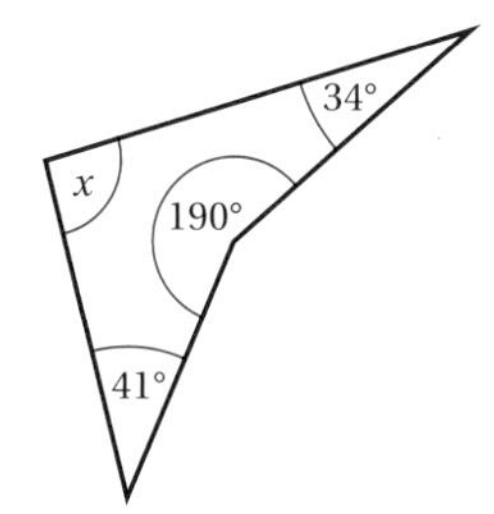

6 Find the unknown angles denoted by variables.

a

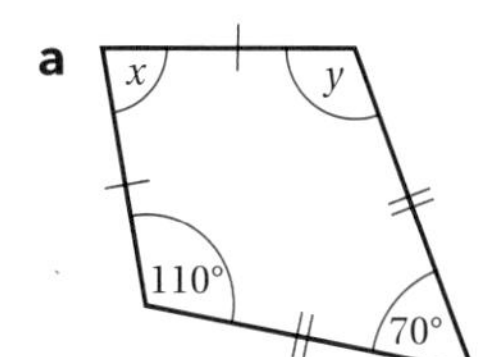

b

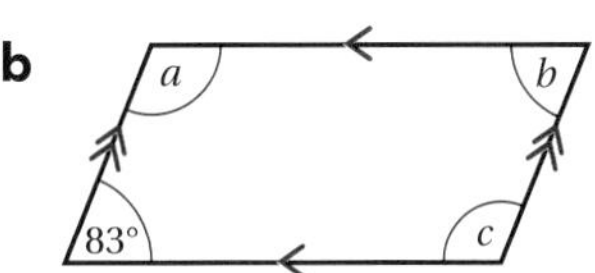

7 Is this quadrilateral a parallelogram?

Justify your answer.

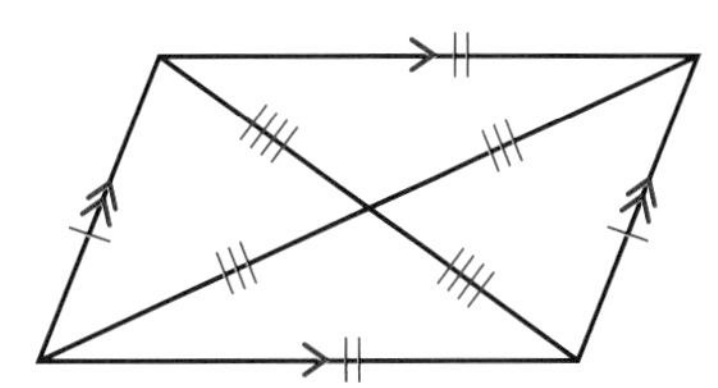

8 Niresh states that if the diagonals of a quadrilateral intersect at right angles, the quadrilateral must be a square.

Is he correct?

Justify your answer.

Find answers at: cambridge.org/ukschools/gcsemaths-studentbookanswers

6 Construction and loci

In this chapter you will learn how to ...

- use a ruler, protractor and pair of compasses effectively.
- use a ruler and a pair of compasses to bisect lines and angles and construct perpendiculars.
- use construction skills to construct geometrical figures.
- construct accurate diagrams to solve problems involving loci.

For more resources relating to this chapter, visit GCSE Mathematics Online.

Using mathematics: real-life applications

Draughtspeople and architects need to draw accurate scaled diagrams of the buildings and other structures they are working on. Although the drawings are complicated, they still use ordinary mathematical instruments like pencils, rulers and pairs of compasses to draw them.

"I prepare technical drawings and plans that are given to me by an architect. I use a CAD program, but I always start with a drawing board and plans that I draw using my ruler, set squares and pair of compasses." *(Draughtsperson)*

Before you start ...

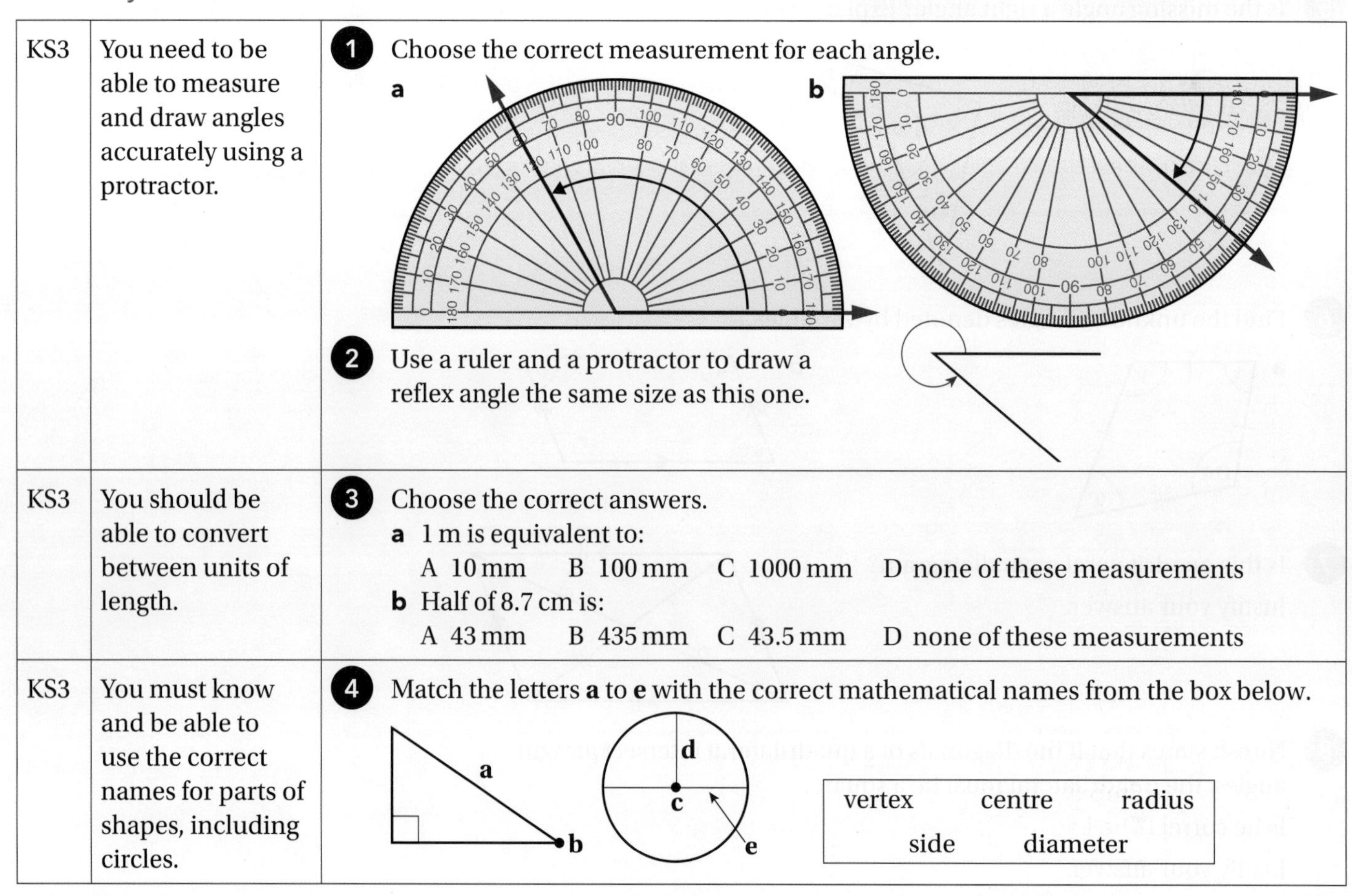

KS3	You need to be able to measure and draw angles accurately using a protractor.	**1** Choose the correct measurement for each angle. **a** **b** **2** Use a ruler and a protractor to draw a reflex angle the same size as this one.
KS3	You should be able to convert between units of length.	**3** Choose the correct answers. **a** 1 m is equivalent to: A 10 mm B 100 mm C 1000 mm D none of these measurements **b** Half of 8.7 cm is: A 43 mm B 435 mm C 43.5 mm D none of these measurements
KS3	You must know and be able to use the correct names for parts of shapes, including circles.	**4** Match the letters **a** to **e** with the correct mathematical names from the box below. **a** **b** **c** **d** **e** vertex centre radius side diameter

Assess your starting point using the Launchpad

STEP 1

1 Which of the following statements are true of this angle?

A It is an acute angle.
B It measures 120°.
C It is called QRP.
D If you extend arm QR, the size of the angle will increase.

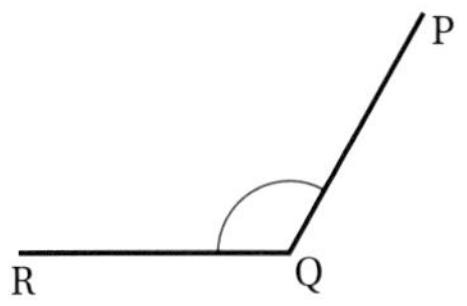

P
R
Q

2 Which of the following statements are **not** true of this circle?

A It has a radius of 5 cm.
B It has a diameter of 5 cm.
C $OC \perp AB$
D $OC = \frac{1}{2}(AB)$

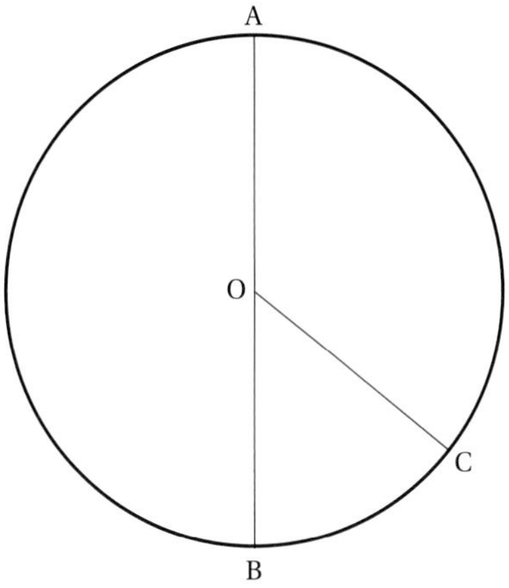

A
O
C
B

?

GO TO
Section 1:
Geometrical instruments

✓

STEP 2

3 Niresh drew the construction shown here.

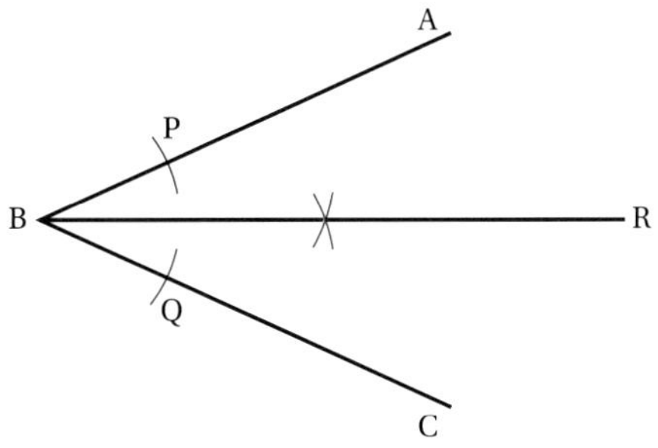

A
P
B
R
Q
C

a What do you call line BR?
b What did Niresh do to produce points P and Q?
c Given that $\angle ABC = 24°$, state the size of $\angle ABR$ without measuring it.

?

GO TO
Section 2:
Bisectors and perpendiculars

✓

GO TO
Step 3:
The Launchpad continues on the next page …

Launchpad continued ...

STEP 3

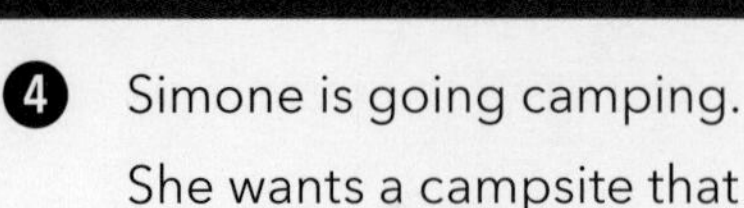

4 Simone is going camping.

She wants a campsite that is less than 50 m from the river and no further than 50 m from the showers.

She drew this diagram to help her decide where to camp.

Copy the diagram and shade the area that satisfies Simone's conditions to show where she could camp.

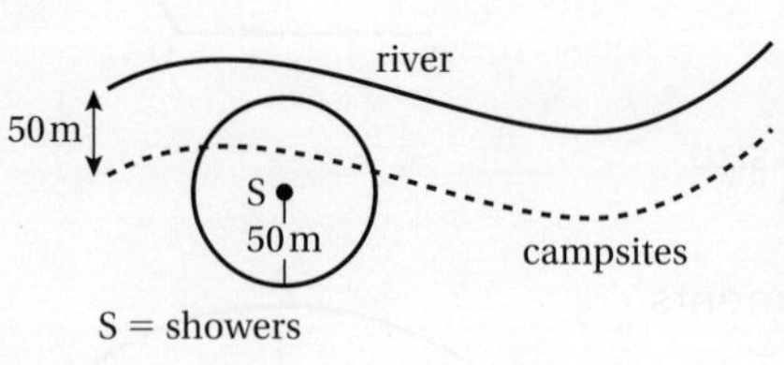

GO TO
Section 3:
Loci

✓

GO TO
Section 4:
More complex problems

Section 1: Geometrical instruments

You need to be confident in measuring and constructing lines, angles and shapes using a ruler, protractor and a pair of compasses.

Tip

Always measure and draw as accurately as you can. Use a sharp pencil. You are expected to draw lengths correct to the nearest millimetre and angles correct to the nearest degree.

Tip

You will learn more about angles in Chapter 9.

Measuring and drawing angles

Angles are created when two line segments (or rays) meet at a point. The point is called the **vertex** of the angle, and the two line segments are called its sides (sometimes referred to as the 'arms' of the angle).

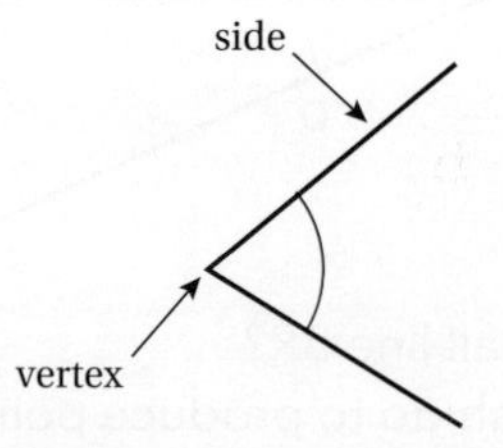

You use a protractor to measure and draw angles.

This protractor has two scales so that we can measure angles facing different directions.

To avoid measuring on the wrong scale, estimate the size of the angle before you measure. Use your knowledge of acute, right and obtuse angles to estimate as accurately as possible.

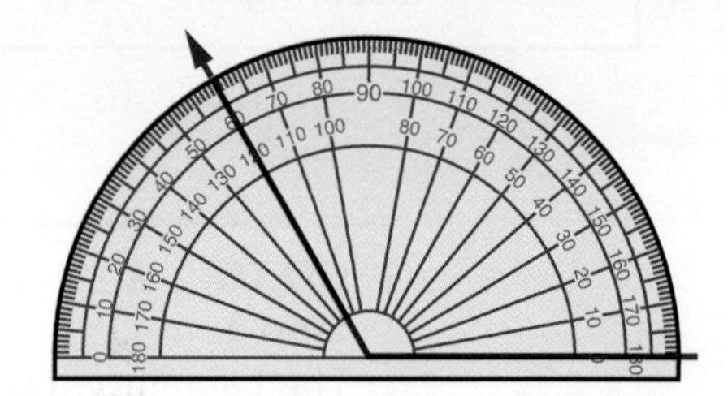

WORKED EXAMPLE 1

a Estimate and then measure the size of each red angle.

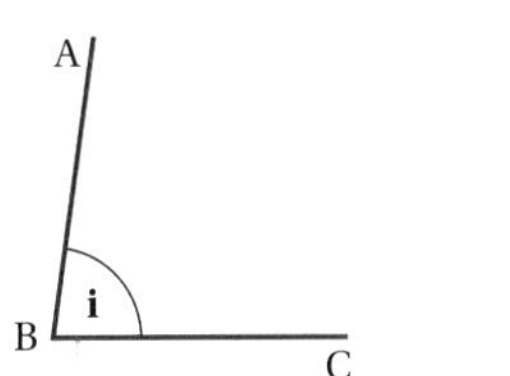

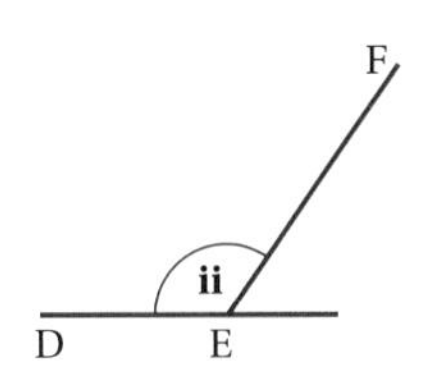

b Use your protractor to draw ∠ABC = 76°.

> **Tip**
>
> If the arms of the angle are too short to read the scale correctly, use a ruler and a pencil to extend them. This doesn't change the size of the angle, but it allows you to read the measurement more accurately.
>
> If you cannot draw on the angle (because it is in a book) you can extend the arm with the straight edge of a sheet of paper.

a i

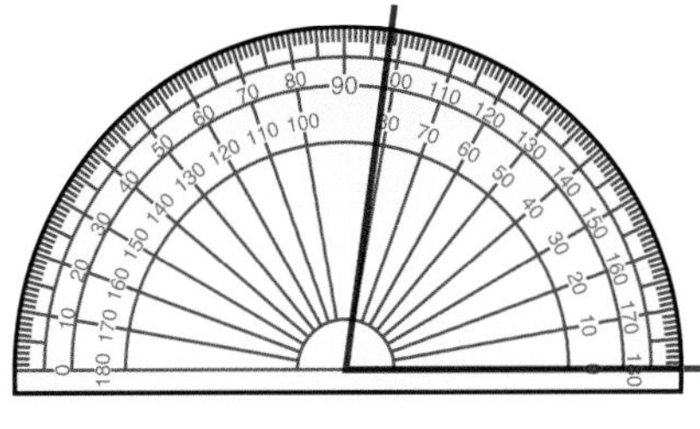

This is an acute angle but it is close to 90°. Estimate about 80°.

Use the inner scale to measure, counting from 0.

∠ABC = 82°

a ii

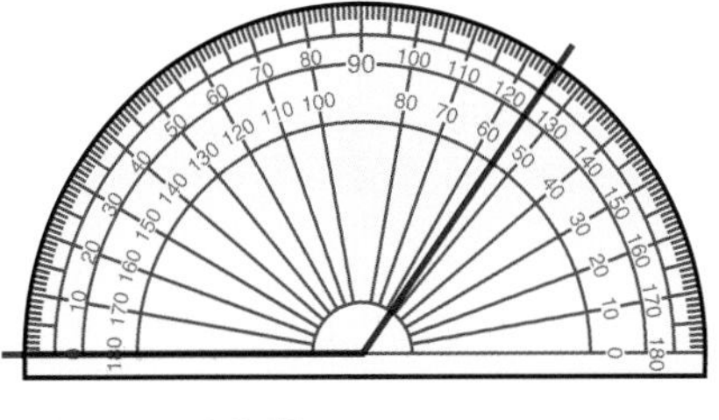

This is an obtuse angle. It is about one-third bigger than a right angle. Estimate about 130°.

Use the outer scale to measure because now this is the scale that has 0 on the arm of the angle.

∠DEF = 125°

b

B

Draw a horizontal line using your ruler. Mark B, the vertex in ∠ABC.

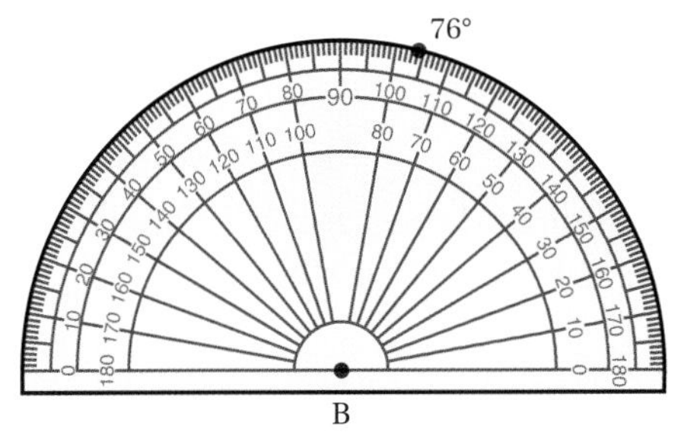

Place your protractor with its centre on B and baseline on the horizontal line. Measure and mark 76°.

A
76°
B C

Remove the protractor. Draw a line from B through the 76° marking using a ruler. Label the angle correctly.

Using a pair of compasses

A pair of compasses (sometimes just called compasses or a compass) is very useful for drawing circles and angles, and marking accurate line lengths.

Using a pair of compasses effectively takes practice.

Make sure your pencil point is sharp and that the pair of compasses are not too loose.

WORKED EXAMPLE 2

a Draw a circle with a radius of 4.5 cm.

a

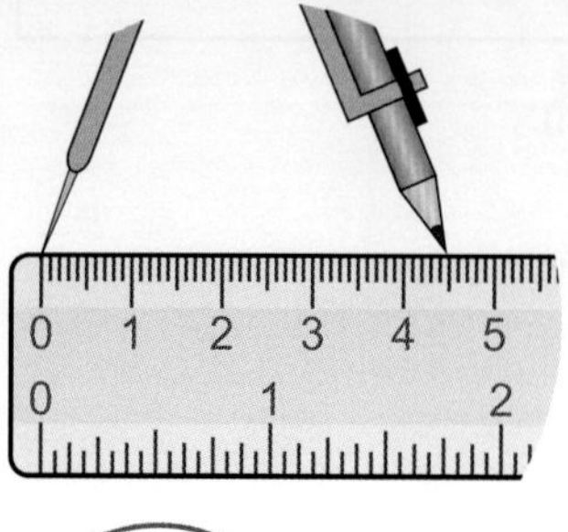

Place the pair of compasses alongside a ruler and open it to 4.5 cm.

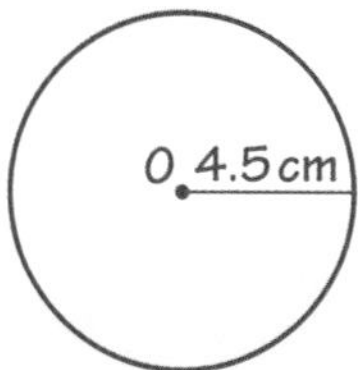

Draw a circle by pressing down on the point of the compasses, holding it at O (the centre of the circle) and sweeping the pencil round in a circle.

b Make an accurate copy of this triangle.

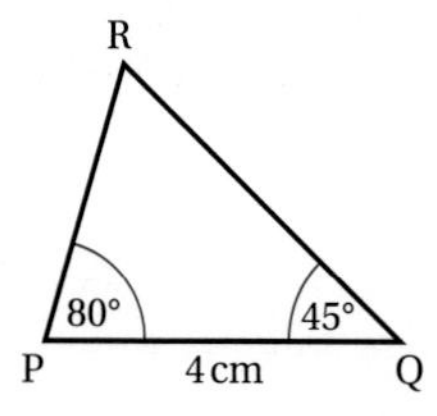

b

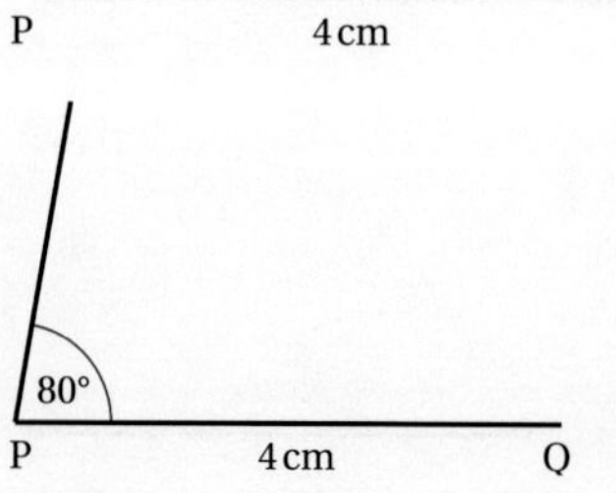

You have been given the length of one line and the size of two angles. First draw the baseline of 4 cm using a ruler, and label this PQ.

Use a protractor to measure and mark the angle 80° from point P, as per instructions in Worked example 1. Then draw a line from the mark to P.

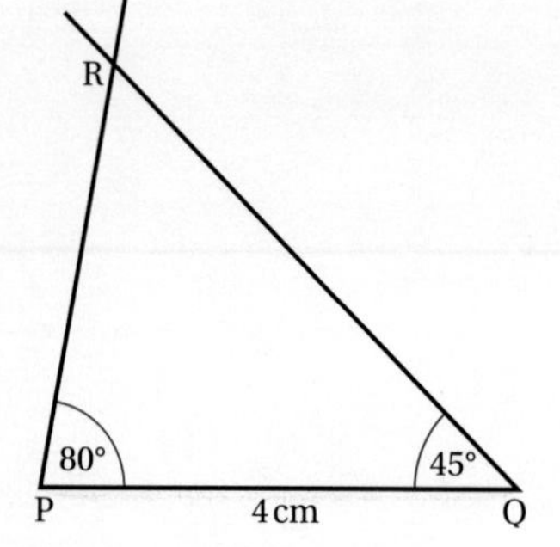

From point Q measure the angle 45° and draw a line from Q extending out so that it crosses the other line. Where the two lines cross is point R. This is the apex of the triangle.

Continues on next page ...

c Construct an equilateral triangle with side lengths 6 cm.

c A 6 cm B

Here, you have not been given the size of any angles but you have been given the length of each side. First draw the baseline of 6 cm with a ruler and label it AB.

Then set your pair of compasses to 6 cm with the point of your compasses on A. Draw an arc above the line AB, roughly where you would estimate C to be.

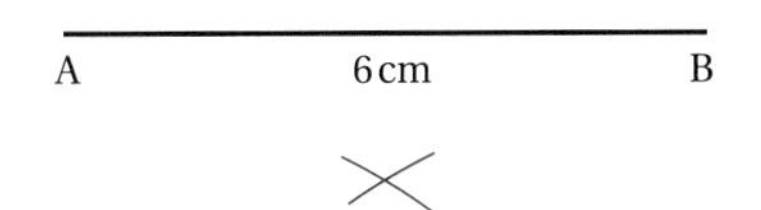

Repeat this from the other side at point B.

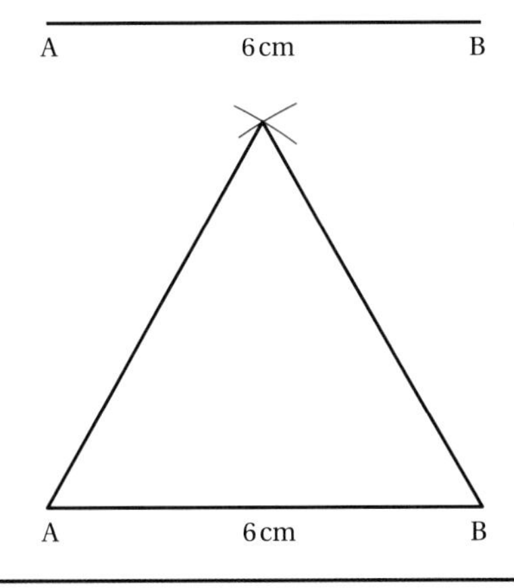

Where the two arcs cross, is the apex of the triangle, i.e., where the point C is. Use this point to complete the triangle. This method can be used for triangles with sides of different lengths by setting your pair of compasses to whatever the lengths of the sides are.

Tip

Leave the construction markings (arcs made using your pair of compasses) on your diagrams as this shows the method you used to construct them.

EXERCISE 6A

1 Use a ruler and protractor to draw and label the following angles.

a ∠PQR = 25° **b** ∠DEF = 149° **c** ∠XYZ = 90°

2 How could you use a protractor marked from 0° to 180° to measure an angle of 238°?

3 **a** Draw line MN which is 8.4 cm long.
At M, measure and draw ∠NMP = 45°.
At N, measure and draw ∠RNM = 98°.

b Explain why the lengths of MP and NR do not matter in this diagram.

4 Use a pair of compasses to construct:

a a circle of radius 4 cm. **b** a circle of diameter 12 cm.

c a circle of diameter 2 cm which shares a centre, O, with another circle of radius 5 cm.

Find answers at: cambridge.org/ukschools/gcsemaths-studentbookanswers

5 Draw a line AB which is 70 mm long.
Construct the circle for which this line is the diameter.

6 Draw a circle of radius 3.5 cm and centre O.
Use a ruler to draw any two radii of the circle. Label them OA and OB.
Join point A to point B to form triangle AOB.
Measure angles AOB, OBA and BAO. Write the measurements on your diagram.

7 Accurately copy the following diagrams using a protractor, ruler and pair of compasses.

a

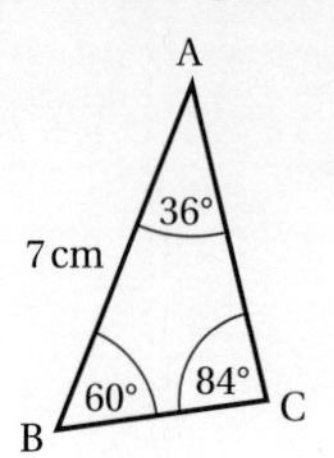

b

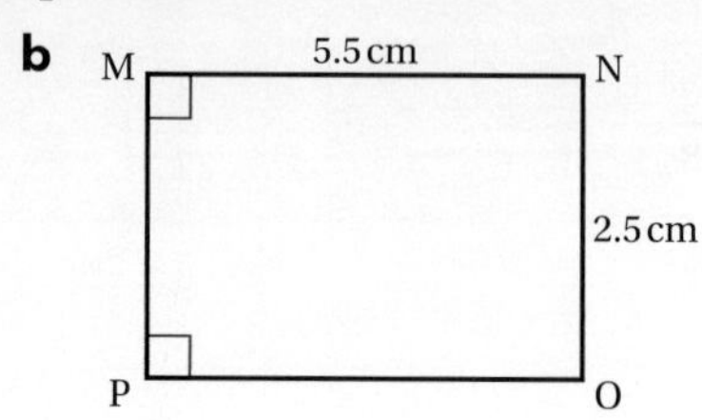

c

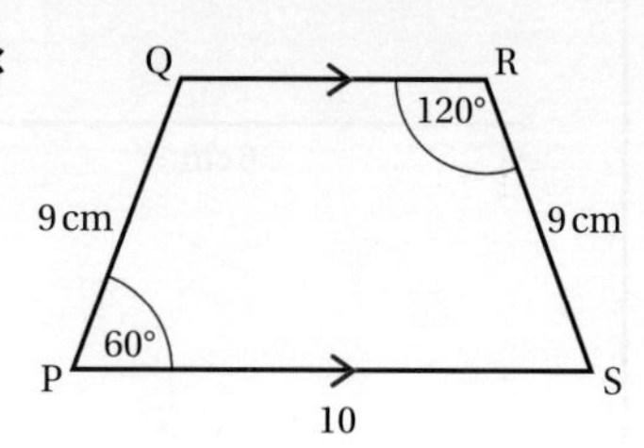

8 Prepare a step-by-step set of instructions for using only a ruler and a pair of compasses to construct:

a an equilateral triangle ABC with sides of 6.4 cm.

b a semicircle with a radius of 30 mm.

9 Use a ruler, a pair of compasses and a protractor to construct $\frac{1}{9}$th of a circle of diameter 82 mm.

Section 2: Bisectors and perpendiculars

Bisecting a line

Tip

Remember, bisect means divide exactly in two.

You can use a ruler and a pair of compasses to **bisect** any line without measuring it.

WORKED EXAMPLE 3

Bisect line AB by construction.

Place the point of a pair of compasses on A, then open them to any length that is **greater** than half way along the line AB. Draw arcs above and below the line. Keep the compasses open to the same width and place the point on B. Draw arcs above and below the line so that they cut the first set of arcs.

Use a ruler to join the points where the arcs intersect. This is the bisector of the line AB.

Key vocabulary

midpoint: the centre of a line; the point that divides the line into two equal halves.

The point where the constructed line cuts AB is called the **midpoint** of AB.
The distance from A to this point is equal to the distance from B to this point.

The constructed line is perpendicular to AB, so it is called the **perpendicular bisector** of AB.

Key vocabulary

perpendicular bisector: a line perpendicular to another that also cuts it in half.

Constructing perpendiculars

You can use your pair of compasses to construct:

- a line perpendicular to any point on a given line.
- a perpendicular line from a point above or below a given line.

Tip

Remember perpendicular means 'at right angles to'.

Construct a perpendicular at a given point on a line

Tip

Remember the symbol $\perp$ means 'perpendicular to'; see Chapter 5 if you need to.

WORKED EXAMPLE 4

Construct XY $\perp$ AB at point X. A X B

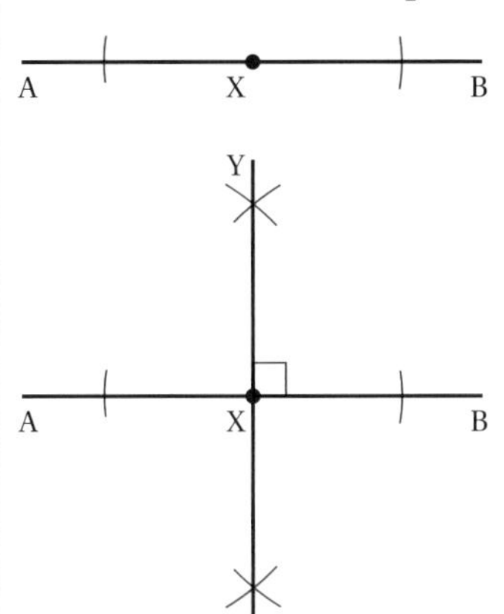

Open your pair of compasses to a width of about 4 cm. Place the point of your compasses on X. Draw two arcs to cut AB on either side of X.

Construct the perpendicular bisector of the line segment between the arcs (as per Worked example 3). Draw a line through the intersecting arcs. Label one end of it Y to produce XY.

Construct a perpendicular from a point to a line

The shortest distance from any point to a line, is the perpendicular distance from the point to the line.

WORKED EXAMPLE 5

Construct PX perpendicular to line AB from point P. •P

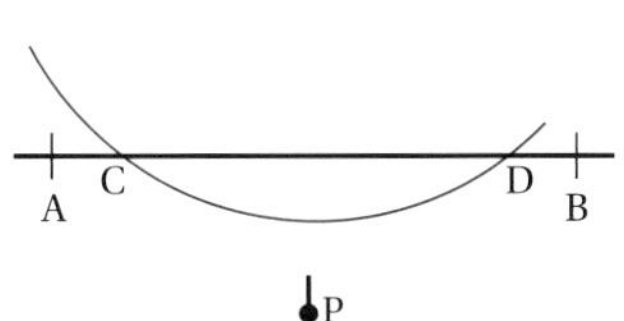

Place the point of your pair of compasses on P. Draw an arc that cuts AB in two places. Label these places C and D.

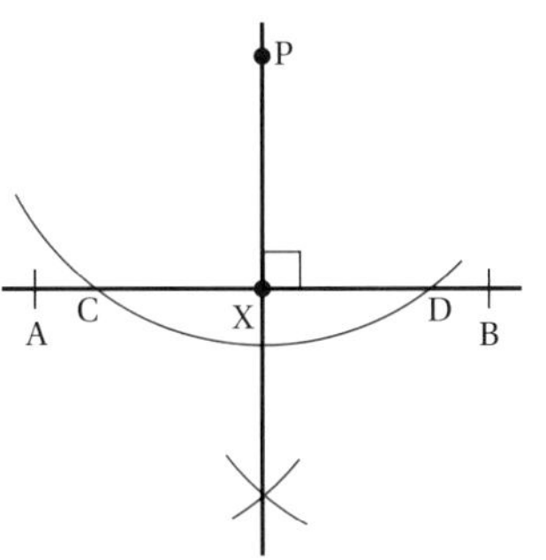

Open the pair of compasses to a width more than half the distance between C and D. Place the point on C and draw an arc on the opposite side of the line to point P. Place the point on D and draw an arc which intersects the one you just drew. Draw a line from the intersecting arcs to P. Label PX and mark the perpendicular.

Bisecting an angle

An angle bisector divides any angle into two equal halves.

WORKED EXAMPLE 6

Use a protractor to draw an angle PQR of 70°.

Construct the angle bisector of this angle without measuring.

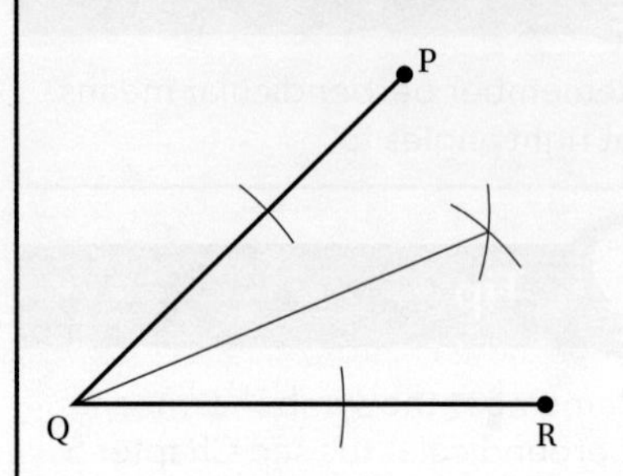

Place the pair of compasses on the vertex (Q) of the angle. Open your compasses a few centimetres and draw arcs that cut each arm of the angle. Place the point of your compasses on each arc where it cuts the arm of the angle, and keeping the width the same, draw arcs between the arms of the angle. Draw a line from the vertex of the angle through the intersection of the arcs.

This is the angle bisector.

EXERCISE 6B

1 Measure and draw the following line segments.
Find the midpoint of each by construction.

a AB = 9 cm **b** MN = 48 mm **c** PQ = 6.5 cm

2 **a** Draw any three acute angles. Bisect each angle without measuring.

b How could you check the accuracy of your constructions?

3 Draw the angles shown and then, using only a ruler and pair of compasses, bisect each angle.

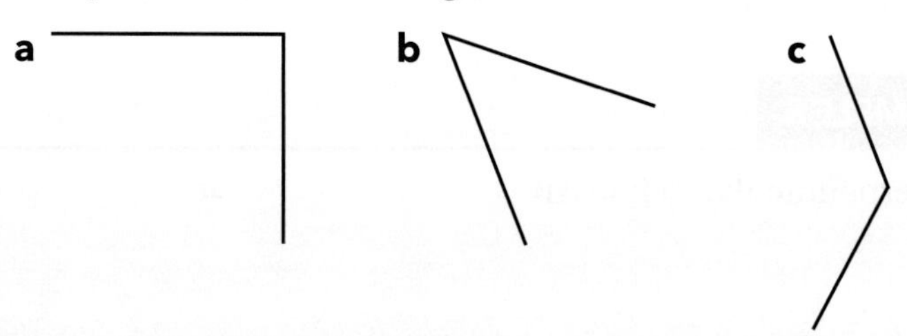

4 Draw any triangle ABC.

a Construct the perpendicular bisector of each side of the triangle.

b Use the point where the perpendicular bisectors meet as the centre, and vertex A as a radius, and draw a circle.

c What do you notice about this circle?

5 Construct equilateral triangle DEF with sides of 7 cm.

a Bisect each angle of the triangle by construction.
Label the point where the angle bisectors meet as O.

b Measure DO, EO and FO. What do you notice?

6 Draw MN = 80 mm. Insert any point A above MN.

a Construct AX ⊥ MN. **b** Draw AB // MN.

7 Draw line segments PQ and ST that intersect at point O.

- **a** Bisect SOQ.
- **b** Is the angle bisector of SOQ also the bisector of POT?

  Explain your answer.

8 Draw any two circles and use a ruler to draw in any two chords of each.

- **a** Construct the perpendicular bisector of each chord.
- **b** Describe where they meet.
- **c** What can you deduce from this?

Section 3: Loci

A **locus** is a set of points that all meet a given condition or set of conditions.

All points in the locus must meet the conditions and all points that do meet the conditions must be included in the locus.

The locus can be a single point, a line, a curve or a shaded region of points that overlap because they meet the same conditions in a particular area.

Some of the rules for loci produce shapes and lines (paths) that you are already familiar with from your work on constructions.

If you recognise the paths made by common loci you can apply what you know about those paths to solve problems.

For example, the locus of points at a given distance from a fixed point forms a circular path, i.e., a circle. You can use what you know about drawing circles to use a pair of compasses to construct this locus.

Key vocabulary

locus (plural **loci**): a set of points that satisfy the same rule.

Tip

The locus of a point at distance (r) from a fixed point (O) is a circle with centre O and radius (r).

WORKED EXAMPLE 7

A tap is located at point X.

Draw the locus of points that are exactly 50 metres from the tap.

Your diagram does not need to be to scale.

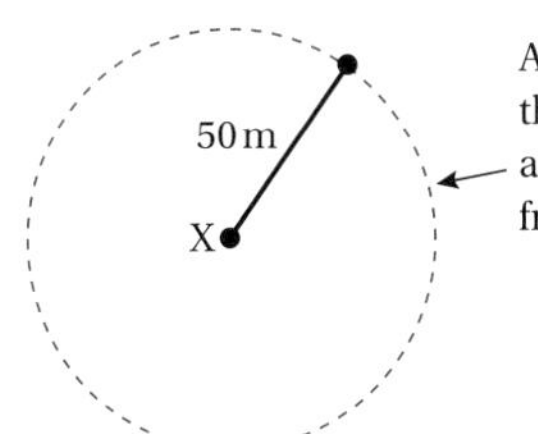
50 m
X
All of the points on the circumference are exactly 50 m from the tap.

Diagram not to scale

Use a pair of compasses to draw a circle (as per Worked example 2). Draw a point in the centre and label this 'X'. Draw in a line for the radius and label this '50 m'. Label the diagram to indicate it is not to scale. Any point on the circumference of this circle is 50 m from the tap.

In Worked example 7 the locus of points that are **less than** 50 m from the tap is the region **inside** the circle.

If you were asked to construct this locus, you would shade the interior of the circle to show that all the points inside the circumference meet the conditions of the locus.

You would also show the circumference as a **broken line** to indicate that it is **not** included in the locus.

The locus of points equidistant from two fixed points is the perpendicular bisector of the line joining the points.

Tip

If a line is included in the locus you draw it as a solid line. If the line is not included, but just shows the edge of the locus, you draw it as a broken, or dashed, line.

WORKED EXAMPLE 8

Anna lives at point A. Josie lives at point B.

They want to meet exactly midway between their homes.

Draw a diagram to show where they could meet.

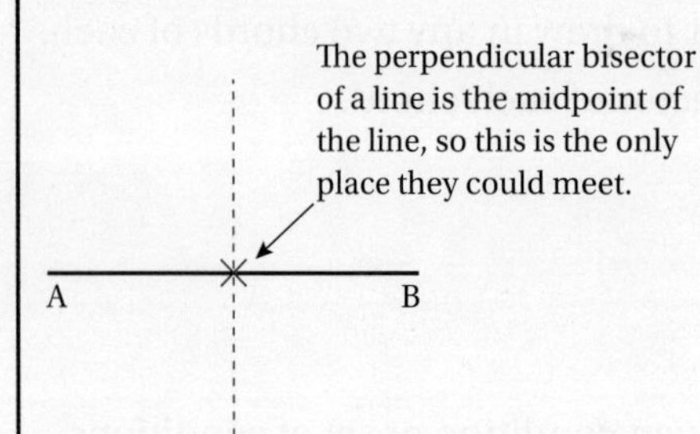

Draw a horizontal line and label the ends A, and B. Construct the perpendicular bisector of this line (as per Worked example 4).

The point where the lines cross is exactly midway between their homes.

In Worked example 8 you can take any point on the perpendicular bisector and it will be the same distance from A and B.

If Anna and Josie wanted to meet at a point that was the same distance from their homes, they could meet anywhere along the perpendicular bisector of the line AB.

However, the question asked you to find the point **exactly midway** between A and B, and the midpoint of line AB is the only point that meets that condition.

Tip

Remember a line continues to infinity in both directions, so it has no end points.

The locus of points equidistant from a line is a pair of parallel lines at the given distance. The locus lines will be on either side of the given line.

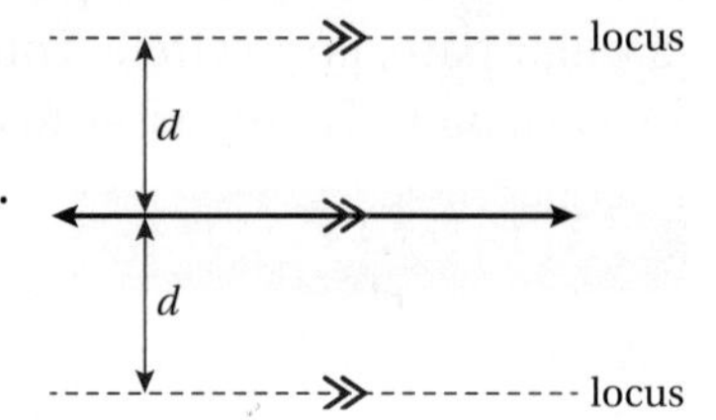

A line segment has a fixed length.

The locus of points equidistant from a line segment has to be the same distance from the line and also the same distance from its end points.

This produces an oval 'racing track' shape with all points the same distance (d) from the line segment.

Tip

The locus of points a fixed distance (d) from a line is the pair of parallel lines which are d cm away from the given line.

The locus of points at a fixed distance (d) from a line segment is a pair of parallel lines d cm away from the line segment, joined by semicircles of radius d cm at the ends of the line segment.

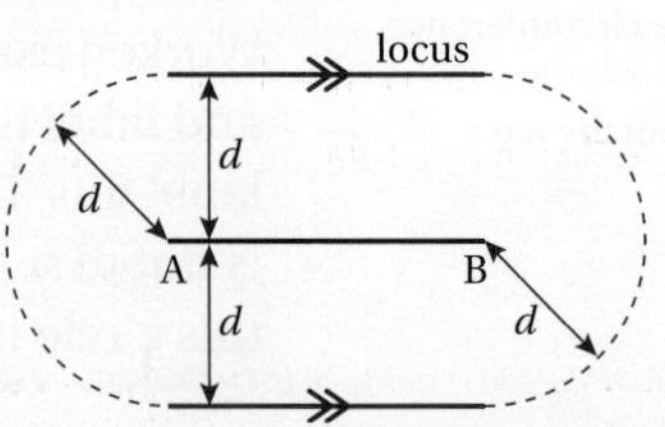

The locus of points that are equidistant from the arms of ∠ABC is shown.

If you take any point on the line BP, it will be the same distance from AB and BC.

You should recognise that BP is the angle bisector of ∠ABC.

Tip

The locus of points an equal distance from two intersecting lines is their angle bisector.

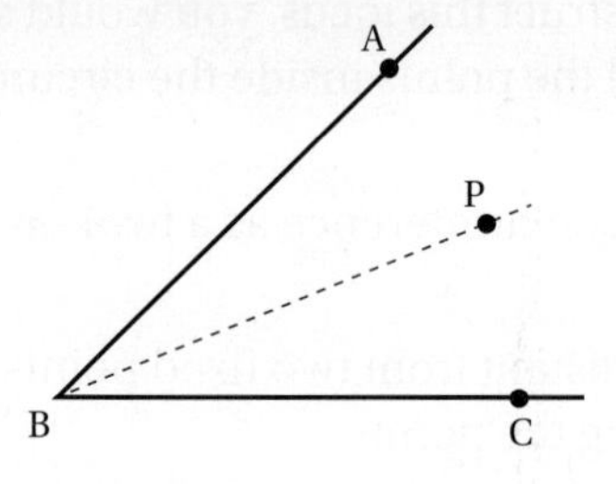

EXERCISE 6C

1 Loci are common in architecture and in line markings on sports fields. Identify and describe some of the loci in the two photographs.

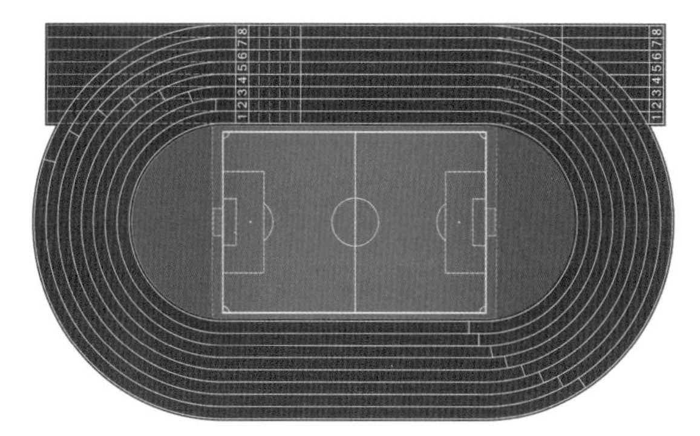

2 Without drawing them, describe the point, path or area of each locus.

a Points that are 200 km from a shop at point X.

b Points that are more than 2 km but less than 3 km from a straight fence 1 km long.

c Points that are equidistant from the two baselines of a tennis court.

d Points that are equidistant from the four corners of a soccer field.

e Points that are within 1 km of a railway line.

3 Accurately construct the locus of points 4 cm from a point D.

4 Draw $\angle MNO = 50°$.

Accurately construct the locus of points equidistant from MN and NO.

5 Draw PQ 4 cm long.

Construct the locus of points 1 cm from PQ.

6 PQ is a line segment of 5 cm.

X is a point exactly 4 cm from P and exactly 2.5 cm from Q.

Show by construction the possible locations of point X.

7 Draw a rectangle ABCD with AB = 6 cm and BC = 4 cm.

a Construct the locus of points that are equidistant from AB and BC.

b Shade the locus of points that are less than 1 cm from the centre of the rectangle.

c Construct the locus of points that are exactly 1 cm outside the perimeter of the rectangle.

8 MNOP is a square with sides of 5 cm.

Show by construction the locus of all points that are less than 1 cm from the sides of the square.

Find answers at: cambridge.org/ukschools/gcsemaths-studentbookanswers

9 Draw a diagram to show:

a the locus of the valve on the rim of a bicycle wheel as it moves along a flat road surface.

b the locus of the centre of the same wheel as it moves along the road.

Section 4: More complex problems

You need to be able to combine the construction techniques you have learned to construct accurate diagrams of shapes and to construct loci to show different situations.

In many cases you will need to decide which construction technique to use. For example, to construct a rectangle without measuring the right angles, you would need to construct the perpendiculars at two points on the base of the rectangle.

Many of the loci problems that you will have to solve could be presented in context. You might be asked to draw scaled diagrams to solve these problems. The scale might be given. For example, 1 cm : 10 km. If you are not given a scale, always state the scale that you have used.

Problem-solving framework

A, B and C represent three towns.

A mobile phone tower is to be erected in the area.

The tower is to be equidistant from towns A and B and within 30 km of town C.

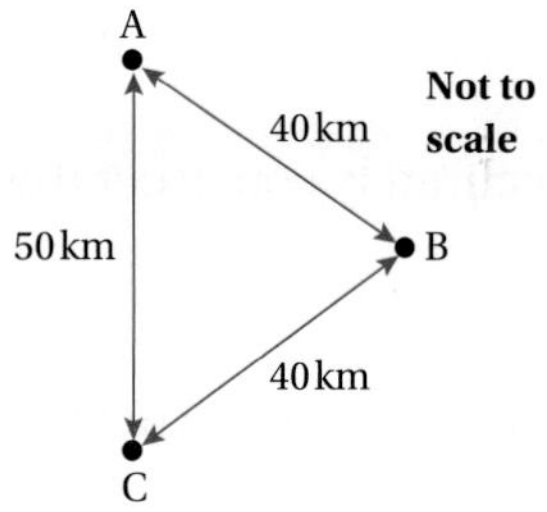

A
40 km
Not to scale
50 km
B
40 km
C

Show by accurate construction on a scale diagram all possible sites for the tower.

Use a scale of 1 cm : 10 km.

Steps for appoaching a problem-solving question	**What you would do for this example**
Step 1: Work out what you have to do. Start by reading the question carefully.	Although it doesn't say so, this is a locus problem. You have to find the locus of points equidistant from A and B and the locus of points that are less than 30 km from C. The solution is where these loci overlap.
Step 2: What information do you need? Have you got it all?	You have to draw a scale diagram. The distances and the scale are given. The conditions for the loci are given.
Step 3: Decide what maths you can do.	First work out the lengths you have to construct using the scale. The scale is 1 cm : 10 km. So: $\frac{40\text{ km}}{10} = 4\text{ cm}$ $\frac{50\text{ km}}{10} = 5\text{ cm}$ Next use these lengths to construct a triangle using your ruler and pair of compasses. Once you have the triangle you can find the loci by construction.

Continues on next page ...

Step 4: Set out your solution clearly. Check your working and that your answer is reasonable.	The tower could be built at any position along the thick red line
Step 5: Check that you've answered the question.	You have shown the overlapping loci and written a statement to answer the question.

EXERCISE 6D

1 Draw line AB = 5.2 cm.

Construct DE, the perpendicular bisector of AB.
Draw DE so that it is 44 mm long.

Mark point F on DE such that DF = FE = 22 mm.

Construct MN //AB and passing through point F.

2 Construct a parallelogram with sides of 46 mm and 28 mm and a longest diagonal of length 60 mm. Measure and write in the length of the other diagonal.

3 Accurately construct a square of side 45 mm.

4 Construct quadrilateral ABCD such that ABC = 90°, AB = DC = 2.2 cm and AD = BC = 5 cm. What kind of quadrilateral is this?

5 On a treasure map, the position of buried treasure is known to be 10 metres from the castle at point Y and 12 metres from the cave at point Z.

Y and Z are 15 metres apart.

Draw a scale diagram and mark with an X all the places where the treasure might be buried.

Use a scale of 1 cm to 2 m.

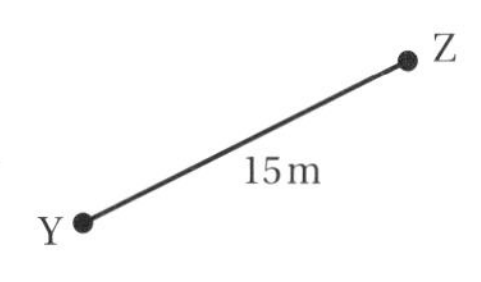

6 A monkey is in a rectangular enclosure which is 10 m by 17.5 m.

The monkey is able to stretch through the fence around its enclosure and reach a distance of 25 cm.

a Draw a scale diagram to show the locus of points that the monkey can reach outside its enclosure.

b Show on your diagram where you would place a safety barrier to make sure that people cannot touch the monkey.
Give a reason for your choice.

Find answers at: cambridge.org/ukschools/gcsemaths-studentbookanswers

7 A garden has a semicircular lawn surrounded by fencing.

If the semicircle has a diameter of 10 m and the distance from the lawn to the fence is consistently 1 m, draw an accurate scaled diagram of the lawn and fence.

8 MNOP is a rectangular field 150 m by 400 m.

A fence is to be built across the field so that it is equidistant from points M and O.

Draw a scaled plan of the field and indicate on it the position in which the fence is to be built.

9 To prevent theft, a small museum plans to put motion sensors at points A and B as shown on the plan of the building.

Each motion sensor covers 270° and picks up any movement within a range of 8 m as long as the distance is a straight line.

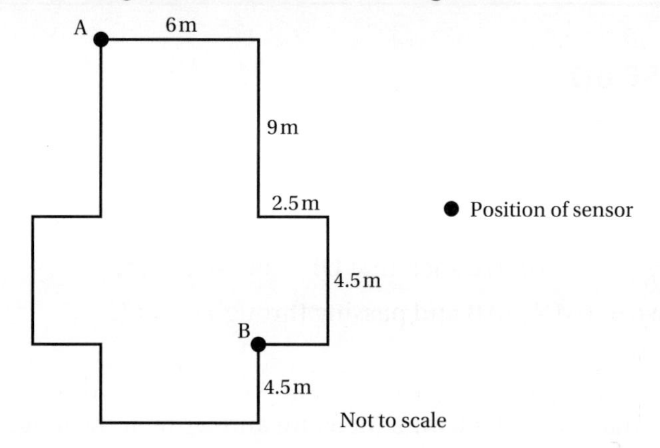

a Determine by scale drawing and construction any areas not protected by the motion sensors.

b Are these the optimal positions for the sensors?

If not, suggest where they might be located to ensure better coverage and security.

Checklist of learning and understanding

Geometry constructions

- A protractor is used to measure and draw angles.
- You can use a ruler and pair of compasses to construct perpendicular lines and to bisect lines and angles.
- The perpendicular bisector of any line cuts the line at its midpoint.
- The shortest distance from a point to a line is always the perpendicular distance.

Loci

- A locus is a set of points that meet the same conditions.
- The locus of points can be a single point, a line, a curve or a shaded area.
- Loci can be used to solve problems involving equal distances and overlapping areas.

Chapter review

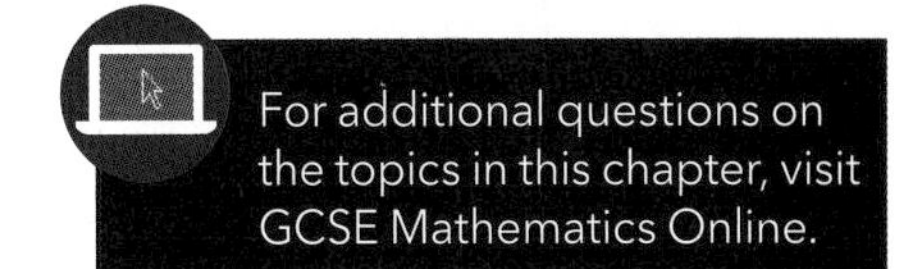

For additional questions on the topics in this chapter, visit GCSE Mathematics Online.

1 **a** Use a protractor to measure angles a and b on this clock face.

b Draw two angles which are the same size as a and b.

c Bisect the two angles you have drawn by construction.

2 Draw a line AB of length 6.5 cm and find its midpoint by construction.

Indicate the locus of points that are equidistant from A and B on your diagram.

3 Town X is due north of Town Y and they are 20 km apart.

Town Z is 25 km from Town X and 35 km from Town Y. Town Z is to the east of towns X and Y.

a Draw a scale diagram to show the location of Town Z in relation to the other two towns.

Use a scale of 1 cm : 5 km.

b A railway runs between towns X and Y such that it is equidistant from both towns.

Indicate the position of the railway on your drawing.

c The electricity supply from town X is carried on a cable that is the same distance from XZ and XY along its length.

Indicate where this cable would be.

d Salman wants to live within 10 km of town Y, but no more than 30 km from town Z.

Show by shading the area that meets these conditions.

4 Make an accurate drawing to show the loci of all points that are within 1 cm of the circumference of a circle of radius 2.5 cm.

5 Construct a parallelogram with diagonals 46 mm and 80 mm and one side of 60 mm.

Measure and fill in the sizes of the internal angles and the length of the other side.

Find answers at: cambridge.org/ukschools/gcsemaths-studentbookanswers

7 Further algebraic expressions

In this chapter you will learn how to ...

- expand the product of two or more binomial expressions.
- factorise quadratic expressions of the form $ax^2 + bx + c$.
- complete the square on a quadratic expression.
- simplify and manipulate algebraic fractions.

For more resources relating to this chapter, visit GCSE Mathematics Online.

Using mathematics: real-life applications

Situations that involve motion, including acceleration, stopping distance, velocity and distance travelled (displacement) can be modelled using quadratic expressions and formulae.

"At the site of a crash, I measure the length of the tyre skid marks and apply an equation to work out the speed at which vehicles were moving before the accident." (Police road accident investigator)

Before you start ...

Ch 3	Check you remember how to simplify expressions.	**1** **a** Simplify. $5x^2y^2 + 6xy - 11x^2y^2 - 7xy$ **b** Why are terms that are multiples of x^2y^2 and terms that are multiples of xy not 'like' terms?
Ch 3	Make sure you can multiply out brackets.	**2** Expand and simplify. $3(2x + 3y) + 2(3x - 2y)$
Ch 3	You should be able to recognise an identity.	**3** Is the identity symbol used correctly in this example? Explain why or why not. $3b(3b - 5) - 7b^2 \equiv 2b^2 - 15b$
Ch 3	You should be able to express situations using algebra.	**4** The square root of a number (x) is cubed and this is added to 36 divided by a number (y) squared. Write an expression to represent this.
Ch 1, 3	Make sure you remember the rules for multiplying and dividing in algebra.	**5** Simplify. **a** $6a \times {}^-5a$ **b** ${}^-3y \times {}^-7y^2$ **c** ${}^-2ab \div b$ **d** ${}^-6y \div {}^-5y$ **e** ${}^-2x \times {}^-5x \div {}^-4x$
KS3	Check that you remember how to apply the four operations to fractions.	**6** Match each calculation to the correct answer. **a** $\frac{1}{3} + \frac{1}{4} = ?$ **b** $\frac{2}{3} - \frac{1}{8} = ?$ **c** $\frac{5}{6} \times \frac{3}{8} = ?$ **d** $\frac{2}{7} \div \frac{1}{14} = ?$ **A** 4 **B** $\frac{7}{12}$ **C** $\frac{13}{24}$ **D** $\frac{5}{16}$

Assess your starting point using the Launchpad

STEP 1

1 Multiply out these expressions.

a $(x + 3)(x + 5)$

b $(x - 3)(x + 5)$

c $(x - 3)(x - 5)$

GO TO
Section 1:
Multiplying two binomials

STEP 2

2 Write each expression as the product of two binomials.

a $a^2 + 5a + 6$ **b** $x^2 - 3x + 2$

c $p^2 - 4p - 45$ **d** $y^2 - 16$

GO TO
Section 2:
Factorising quadratic expressions

STEP 3

3 Complete these identities by filling in the missing values:

a $x^2 + 12x - 11 \equiv (x + \square)^2 - \square$

b $x^2 + 8x + 20 \equiv (x + \square)^2 - \square$

c $x^2 - 5x - 9 \equiv \left(x - \frac{5}{2}\right)^2 - \square$

GO TO
Section 3:
Completing the square

STEP 4

4 Express as a single fraction:

a $\frac{3x}{10} + \frac{x}{10}$ **b** $\frac{3}{x + 1} + \frac{1}{x + 3}$

GO TO
Section 4:
Algebraic fractions

GO TO
Section 5:
Apply your skills

Find answers at: cambridge.org/ukschools/gcsemaths-studentbookanswers

Key vocabulary

binomial: an expression consisting of two terms.

binomial product: the product of two binomial expressions; for example, $(x + 2)(x + 3)$.

Section 1: Multiplying two binomials

A **binomial** is an expression that contains two terms.

For example:

$x + 2$ $\quad x - 4$ $\quad 3x^2 + 4$ $\quad 2x^2 - 5y^3$

A **binomial product** is the product of two binomials.

Multiplying the first two binomials above gives the binomial product $(x + 2)(x - 4)$.

You expand the product of two binomials by multiplying out the brackets.

$$(x + 2)(x - 4) = x^2 - 2x - 8$$

Tip

Writing numbers in expanded notation to do long multiplication is similar to finding a binomial product.

14×27 can be written as:

$$\begin{aligned}(10 + 4) \times (20 + 7) &= 10 \times 20 + 10 \times 7 + 4 \times 20 + 4 \times 7\\ &= 200 + 70 + 80 + 28\\ &= 378\end{aligned}$$

The area of a rectangle is useful for showing how to multiply two binomials.

Consider a large rectangle of length $(a + 2)$ metres and depth $(b + 3)$ metres.

	a	2
b	ab	$2b$
3	$3a$	6

The area of the whole rectangle must be equal to the sum of the four smaller rectangular areas shown on the diagram.

So, total area $= ab + 2b + 3a + 6$.

But the area (length $\times$ depth) of the whole rectangle is also the binomial product $(a + 2)(b + 3)$.

$$\begin{aligned}\text{Area} &= (a + 2)(b + 3)\\ &= ab + 3a + 2b + 6\end{aligned}$$

To multiply two brackets together each term in the first bracket must be multiplied by each term in the second bracket.

You can use lines to keep track of your multiplication.

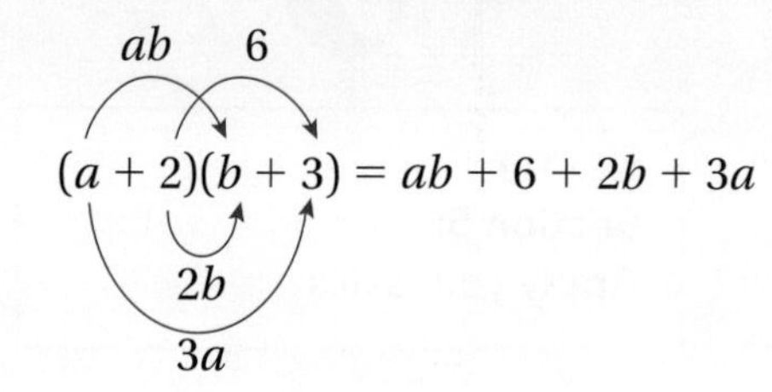

You can also use a grid to make sure you have multiplied all the terms.

$\times$	a	2
b	ab	$2b$
3	$3a$	6

The product of two binomials gives you four terms.

In the example above there are no like terms so you cannot simplify the expression any further. When the product contains like terms you add these to simplify the expression.

WORKED EXAMPLE 1

Expand and simplify, if possible.

$(x + 3)(x + 5)$.

$\times$	x	3
x	x^2	$3x$
5	$5x$	15

$x^2 + 3x + 5x + 15$

$(x + 3)(x + 5) = x^2 + 3x + 5x + 15$
$= x^2 + 8x + 15$

5x and 3x are like terms, so add them.

Key vocabulary

quadratic expression: an expression in which the highest power of x is x^2.

The expression $x^2 + 8x + 15$ is a **quadratic expression**, because the highest power of x in the expression is x squared (x^2).

The general form of a quadratic expression is $ax^2 + bx + c$ where a, b and c are real numbers and $a \neq 0$.

WORKED EXAMPLE 2

Expand and simplify, if possible.

a $(x - 2)(x + 9)$ **b** $(x - 4)(x - 7)$

a $(x - 2)(x + 9) = x^2 + 9x - 2x - 18$
$= x^2 + 7x - 18$

Notice that you get a positive and a negative like term here.

b $(x - 4)(x - 7) = x^2 - 7x - 4x + 28$
$= x^2 - 11x + 28$

Notice that the like terms are both negative here.

Perfect squares

Key vocabulary

perfect square: a binomial product of the form $(a \pm b)^2$.

A **perfect square** is the resulting product when a number, variable or expression is multiplied by itself.

When a binomial is multiplied by itself the product is a perfect square.

Find answers at: cambridge.org/ukschools/gcsemaths-studentbookanswers

For example:

$$(x + 8)^2 = (x + 8)(x + 8)$$
$$= x^2 + 8x + 8x + 64$$
$$= x^2 + 16x + 64$$

Expanding a perfect square always produces the same pattern. This pattern can be used to expand them without doing any working. It is a useful shortcut that you can use.

$(a + b)^2 = a^2 + 2ab + b^2$

- a^2 ↑ first term squared
- $2ab$ ↑ twice the product of the two terms
- b^2 ↑ second term squared

$(a - b)^2 = a^2 - 2ab + b^2$

- a^2 ↑ first term squared
- $2ab$ ↑ twice the product of the two terms
- b^2 ↑ second term squared

EXERCISE 7A

1 Expand and simplify.

a $(x + 2)(x + 5)$ **b** $(x - 2)(x - 5)$
c $(x + 2)(x - 5)$ **d** $(x - 2)(x + 5)$
e $(x + 3)(x - 4)$ **f** $(x + y)(x + y)$

2 Expand and simplify.

a $(2x + 4)(3x + 3)$ **b** $(3x + 4)(5x + 2)$ **c** $(2x - 5)(3x + 1)$
d $(4y - 3)(5y + 1)$ **e** $(3a - 5)(2a - 1)$ **f** $(2b - 5)(b - 3)$
g $(2y - 3)(3y - 5)$ **h** $(2x + 4)(2x - 6)$ **i** $(5x - 3)(4x - 1)$

3 If A = $3x + 2$ and B = $2x - 1$ determine:

a AB **b** $A^2 + B^2$ **c** $(A - B)(A + B)$

4 Determine each product.

a $\left(\frac{2}{x} + \frac{x}{2}\right)^2$ **b** $[2x - (4x + 3y)]^2$ **c** $[2x - (y + z)]^2$

5 What is the value of a if $(2x - 3)$ is a factor of $6x^2 + ax - 12$)?

The difference of two squares identity

Work through the investigation in Exercise 7B to find a shortcut for expanding binomials in the form:

$(a + b)(a - b)$

EXERCISE 7B

1 Expand each of the following binomials:

a $(x + 1)(x - 1)$ **b** $(a + 2)(a - 2)$
c $(2x - 1)(2x + 1)$ **d** $(x - 2y)(x + 2y)$

2 Write down a rule that you can use to quickly find the answer to any similar expansion.

3 In general:

$$(x + y)(x - y) \equiv x^2 - y^2$$

This is called the **difference of two squares identity**.

a How can you recognise when a binomial expansion is a difference of two squares?

b Are each of these expressions a difference of squares?

Explain why or why not.

i $(3x + 2y)(2x - 3y)$ **ii** $50a^2 - 72b^2$ **iii** $16 - (27x)^2$

Tip

You learnt about identities in Chapter 3. Remember that ≡ means that both sides are equal for all values of the variable.

The diagram illustrates the difference of two squares using areas.

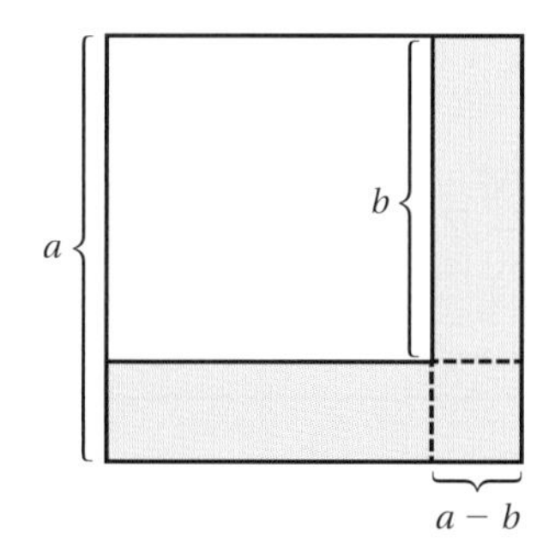

The difference in area between the two squares is the shaded part of the diagram.

The total area of the shaded part forms two rectangles and one square.

The total area of the shaded part is:

$$a^2 - 2ab + b^2 + 2ab - 2b^2 = a^2 - b^2$$

So, the difference between two squares is:

$$a^2 - b^2 \equiv (a + b)(a - b)$$

Expanding more than two factors

Consider the multiplication $3 \times 4 \times 5$.

To do this you have to multiply two numbers at a time.

For example:

$$3 \times 4 \times 5 = 12 \times 5 = 60$$

or:

$$3 \times 4 \times 5 = 3 \times 20 = 60$$

Multiplication is commutative. The order in which you multiply does not change the answer.

You can use this principle to expand three (or more) binomials.

Find answers at: cambridge.org/ukschools/gcsemaths-studentbookanswers

Tip

It is worth doing the multiplication of three or more binomials one step at a time. You are less likely to make sign errors and more likely to get the answer correct.

Tip

To work most efficiently, always check whether some factors are perfect squares or the difference of two squares and expand these first using the rules that you know.

WORKED EXAMPLE 3

Expand $(3x + 2)(2x + 1)(x - 1)$

$(3x + 2)(2x + 1)(x - 1)$

$(3x + 2)(2x + 1) = (6x^2 + 4x + 3x + 2)$ — Multiply/expand the first two binomials using either the line or the grid method.

$= (6x^2 + 7x + 2)$ — Simplify by adding like terms.

$= (6x^2 + 7x + 2)(x - 1)$ — Now multiply/expand the product with the last binomial. You can use the line method or the grid method like you used for two binomials.

×	$6x^2$	$7x$	2
x	$6x^3$	$7x^2$	$2x$
-1	$-6x^2$	$-7x$	-2

$= 6x^3 + 7x^2 + 2x - 6x^2 - 7x - 2$

$= 6x^3 + x^2 - 5x - 2$ — Collect like terms to simplify.

EXERCISE 7C

1 Expand and simplify.

a $(x + 1)(x + 2)(x + 3)$

b $(2x - 3)(x - 2)(2x - 1)$

c $(x + 1)(x + 2)(x - 2)$

d $(x - 3)(2x + 1)(3x - 2)$

2 Expand and simplify.

a $(3x - 4)^3$

b $(x + 3)(x^2 - 3x + 9)$

c $\left(\frac{1}{5x} + \frac{1}{3y}\right)\left(\frac{1}{25x^2} - \frac{1}{15xy} + \frac{1}{9y^2}\right)$

d $(x^2y^2 + x^2)(xy + x)(xy - x)$

3 The volume of a cuboid can be found using the formula LWH, where L is the length, W is the width and H is the height.

Given a cuboid of length $(2x + \frac{1}{2})$ cm, width $(x - 2)$ cm and height $(x - 2)$ cm:

a Write an expression for the volume of the cuboid in factor form.

b Expand the expression.

c Why is it sensible to expand the factors $(x - 2)(x - 2)$ first?

Section 2: Factorising quadratic expressions

You have seen that expanding a binomial such as $(x + 3)(x + 4)$ gives you a quadratic expression.

Factorising involves writing an expression as the product of its factors.

In other words, factorising means 'putting the brackets back into the expression.'

expanding

$$(x + 3)(x + 4) = x^2 + 7x + 12$$

factorising

In this section you will learn how to factorise quadratic expressions.

You need to be able to factorise quadratic expressions in the form of $x^2 + bx + c$ as well as those where the x^2 term has a coefficient that is not 0 or 1.

When the coefficient of the x^2 term is not 0 or 1, the quadratic is in the form $ax^2 + bx + c$.

Factorising a quadratic is the inverse of finding the product of (multiplying) two binomials.

In general:

$(x + a)(x + b) = x^2 + (a + b)x + ab$.

So, the middle term is the sum of a and b in the original binomials or $(a + b)$.

The third term is the product of a and b or (ab).

This pattern can be used to develop a strategy for factorising quadratics.

Work carefully through the examples to see how to do this systematically.

WORKED EXAMPLE 4

Factorise $x^2 + 7x + 12$.

$x^2 + 7x + 12$

Write down the expression you have been asked to factorise.

$= (x + \square)(x + \square)$

Start by writing brackets and insert an x in each. You can do this because to get x^2 you know you need to multiply x by x. (**Note** as both terms in the expression are positive, you can include a '+' sign in each bracket at this stage.)

Factor pairs for 12 are

1×12

2×6

3×4

Now look at the value of the constant; you know that the constants in each bracket multiplies together to get the constant of the quadratic, and that the sum of the two constants becomes the coefficient of x in the quadratic. So, what factors of 12 will add up to give 7? Write down all the factor pairs for 12.

$3 + 4 = 7$

$3 \times 4 = 12$

Identify which factor pair meets the conditions and substitute one factor into each bracket.

So, $x^2 + 7x + 12 = (x + 3)(x + 4)$

$(x + 3)(x + 4) = x^2 + 3x + 4x + 12$

$= x^2 + 7x + 12$

Expand the brackets to check your answer.

Yes, this is the expression you started with.

Quadratics that have negative terms need a bit more care.

Consider factorising $x^2 - 7x + 12$.

Find answers at: cambridge.org/ukschools/gcsemaths-studentbookanswers

You can start by writing $x^2 - 7x + 12 = (x - \square)(x - \square)$ as before.

Put negative signs in both brackets because the constant is +12 but the coefficient of x this time is a negative number, −7.

The product of two negative numbers is positive, so the two factors must both be negative:

$^-1 \times {}^-12$

$^-2 \times {}^-6$

$^-3 \times {}^-4$

$^-3 + ({}^-4) = {}^-7$, so this is the factor pair you need:

$x^2 - 7x + 12 = (x - 3)(x - 4)$

The example below shows you how to work systematically to factorise a quadratic that has negative terms.

WORKED EXAMPLE 5

Factorise $x^2 - 4x - 12$.

$x^2 - 4x - 12$

$= (x + \square)(x - \square)$

Make two sets of brackets and write an x in each as before. The constant of the quadratic is negative. To get a negative product you have to multiply a negative number by a positive number, so one bracket will have a negative sign and the other will have a positive sign. (**Note** you do not have to include a '+' or '−' sign in the brackets at this stage, you can wait until you've identified the factors.)

Factor pairs are:

1×12

2×6

3×4

Now look at the constant. It is $^-12$. Which factor pair has a difference of 4? There is a difference of 4 between 2 and 6.

$^-2 + 6 = 4$

$^-6 + 2 = {}^-4$

$x^2 - 4x - 12 = (x + 2)(x - 6)$

Which factor pair will give you $^-4$ if you add the numbers together and give you $^-12$ if you multiply them? So, the 6 goes in the bracket with the negative sign. The 2 goes in the bracket with the positive sign.

$(x + 2)(x - 6) = x^2 - 4x - 12$

Check your answer by expanding the binomial.

Yes, this is the expression you started with.

Work out the signs before you factorise by looking at the signs in the quadratic.

If the **constant is positive**, the brackets will have the **same** sign.

- If the x term is positive, both brackets will have positive signs.
- If the x term is negative, both brackets will have negative signs.

If the **constant is negative**, the brackets will have **different** signs.

- The x term is the difference between the two factors.
- The largest number in the factor pair will have the same sign as the x term in the quadratic.

When you factorise any expression, the first step should be to check for, and remove, **common factors**.

WORKED EXAMPLE 6

Factorise $4x^2 - 12x - 40$.

$4x^2 - 12x - 40$

$= 4(x^2 - 3x - 10)$ Take out the common factor of 4.

$= 4(x - 5)(x + 2)$ Factorise the quadratic.

EXERCISE 7D

1 Factorise.

a $x^2 + 7x + 6$ **b** $x^2 + 9x + 18$ **c** $x^2 + 7x + 12$

d $x^2 + 17x + 30$ **e** $x^2 + 9x + 20$ **f** $x^2 + 23x + 90$

2 Factorise.

a $x^2 - 3x + 2$ **b** $x^2 - 12x + 27$ **c** $x^2 - 11x + 30$

d $x^2 - 23x + 42$ **e** $x^2 - 15x + 54$ **f** $x^2 - 29x + 100$

3 Factorise fully.

a $2x^2 + 6x + 4$ **b** $6x^2 - 24x + 18$ **c** $5x^2 - 5x - 10$

d $2x^2 + 14x + 20$ **e** $2x^2 + 4x - 6$ **f** $3x^2 - 30x - 33$

The difference of two squares

Earlier you saw that multiplying out binomials in the form $(a + b)(a - b)$ gives a product that is the difference of two squares.

$$(a - b)(a + b) = a^2 + ab - ab - b^2$$
$$= a^2 - b^2 \text{ (simplifying)}$$

When you are asked to factorise a quadratic of the form $a^2 - b^2$, you can apply what you know about the difference of two squares to find the factors.

Tip

You can also think of factorising a difference of squares as taking the square root of each term and writing these in brackets, one with a negative sign and one with a positive sign. The order in which you write down the brackets doesn't matter.
$(a - b)(a + b) = (a + b)(a - b)$

WORKED EXAMPLE 7

Factorise. **a** $x^2 - 4$ **b** $x^2 - 36$ **c** $4a^2 - 9$ **d** $5x^2 - 45$
e $2(x - 1)^2 - 50$ **f** $x^2 - 2$

a $x^2 - 4 = x^2 - (2)^2$ Express both terms as squares.

$= (x + 2)(x - 2)$ Apply the identity: $a^2 - b^2 \equiv (a + b)(a - b)$

b $x^2 - 36 = x^2 - (6)^2$

$= (x + 6)(x - 6)$

c $4a^2 - 9 = (2a)^2 - (3)^2$

$= (2a + 3)(2a - 3)$

d $5x^2 - 45 = 5(x^2 - 9)$

$= 5(x + 3)(x - 3)$ Take out a common factor of 5.

Continues on next page ...

Tip

Some expressions are not a difference of squares until you have removed the common factor. Checking for a common factor should always be the first step when you are factorising.

e $2(x-1)^2-50$ — Take out a common factor of 2.

$= 2((x-1)^2-25))$ — $^-25 = (^+5) \times (^-5)$.

$= 2(x-1+5)(x-1-5)$

$= 2(x+4)(x-6)$ — Add like terms.

f $x^2-2 = (x+\sqrt{2})(x-\sqrt{2})$ — The square root of an imperfect square is a surd.

Mathematically a difference of squares such as $x^2 - 4$ is a special case of the quadratic expression $x^2 + bx + c$. Here, the coefficient of x is 0, so there is no x term ($x \times 0 = 0$) and the constant (c) is a negative number.

Using the difference of two squares in number problems

You can use the difference of squares to subtract square numbers such as $86^2 - 14^2$ without working out the square values which can be very large numbers.

WORKED EXAMPLE 8

Solve $86^2 - 14^2$.

$86^2 - 14^2 = (86+14)(86-14)$ — Write the subtraction as the product of its factors.

$= 100 \times 72$ — Add and subtract the values in each bracket.

$= 7200$ — Find the product.

This method can also be used to find one of the shorter sides in a right-angled triangle.

WORK IT OUT 7.1

In this right-angled triangle, the hypotenuse measures 13 cm and one of the shorter sides measures 5 cm.

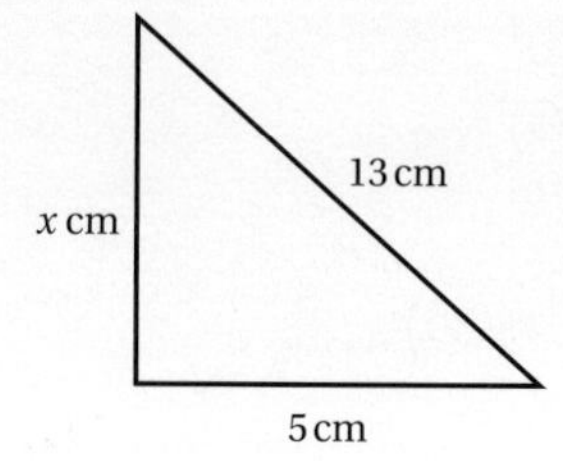

Use the difference of squares and Pythagoras' theorem to calculate the size of the unknown length.

Continues on next page ...

Which of these options is correct?

What mistakes are there in the other options?

Option A	Option B	Option C
$13^2 - x^2 = 5^2$	$x^2 = 13^2 - 5^2$	$x^2 = 13^2 - 5^2$
$(169 + x)(169 - x) = 25$	$x^2 = (13 + 5)(13 - 5)$	$x^2 = (13 - 5)(13 - 5)$
$169 - x^2 = 25$	$x^2 = 18 \times 8$	$x^2 = 8 \times 8$
$x^2 = 194$	$x^2 = 144$	$x^2 = 64$
$x = \sqrt{194}$	$x = \sqrt{144}$	$x^2 = \sqrt{64}$
$x = 13.9$ cm	$x = 12$ cm	$x = 8$ cm

Tip

You should remember Pythagoras' theorem from KS3.

The theorem states that for a right-angled triangle, the square of the length of the hypotenuse (longest side) is equal to the sum of the squares of the lengths of the other two sides.

So in the triangle shown, $a^2 = b^2 + c^2$.

You will need to know the theorem from memory.

See Chapter 32 for more information and practice.

EXERCISE 7E

1 Factorise each of the following.

a $x^2 - 36$ **b** $p^2 - 81$ **c** $w^2 - 16$

d $p^2 - 36q^2$ **e** $144s^2 - c^2$ **f** $64h^2 - 49g^2$

2 Factorise.

a $8x^2 - 2y^2$ **b** $3x^2y^2 - 12z^2$ **c** $1 - (2x - 3)^2$

d $3(x + 4)^2 - 12$ **e** $7(x - 5)^2 - 7y^2$ **f** $(x + 5)^2 - (y + 3)^2$

3 Using $(a - b)(a + b) = a^2 - b^2$, evaluate the following.

a $100^2 - 97^2$ **b** $50^2 - 48^2$ **c** $639^2 - 629^2$

d $98^2 - 45^2$ **e** $83^2 - 77^2$ **f** $1234^2 - 999^2$

4 Use the difference of two squares method to find the value of a in each triangle.

Leave the answer in square root form.

a

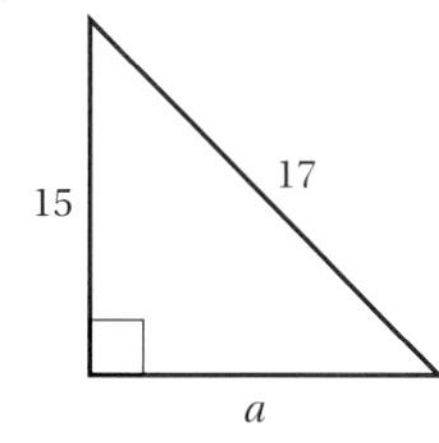

b

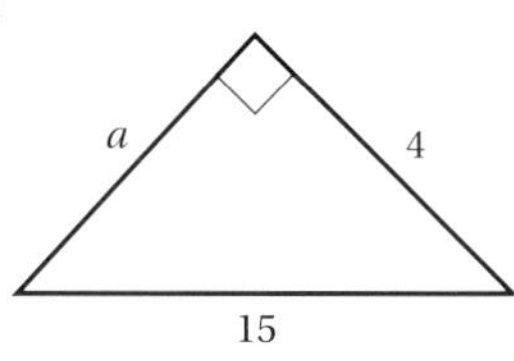

c

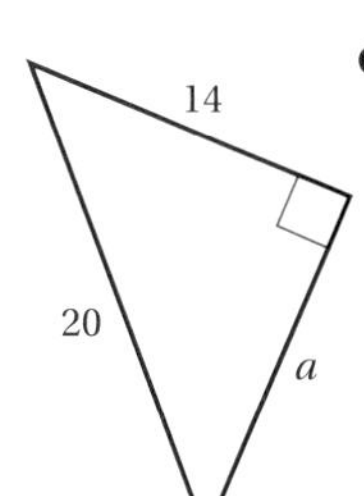

d

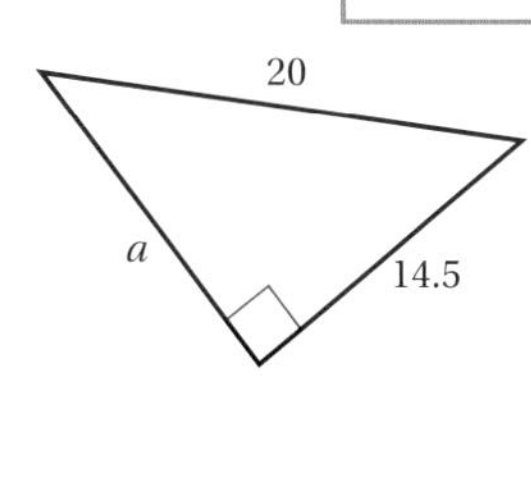

Tip

When you leave the answer in square root form you are giving the exact answer. If your answer is an irrational decimal, i.e. one that does not terminate or recur, then the square root form is known as a surd. Examples of surds are $\sqrt{2}$ and $\sqrt{3}$.

Further quadratics

The quadratic expressions you have factorised so far have had an x^2 term with a coefficient of 1. Now you will deal with examples where $a \neq 1$.

To factorise expressions in the form $ax^2 + bx + c$ (where $a \neq 1$) you need two numbers which have a sum equal to b (the coefficient of the x term). However, the two numbers must now have a product which is equal to ac (rather than just c). Finding factors that sum to b and have a product of ac is a useful shortcut to help you factorise quadratics when $a \neq 1$.

There are different methods of finding the correct factors. You may use any method. Always check your results by expanding the brackets.

Work through these examples to decide which method you prefer.

WORKED EXAMPLE 9

Factorise. **a** $2x^2 + 9x + 4$ **b** $5x^2 - 22x + 21$

a $ac =$
$2 \times 4 = 8$

In this expression $a = 2$, $b = 9$ and $c = 4$. There are no common factors. Start by working out ac.

8 and 1

Now look for factors with a sum of 9 (b) and a product of 8 (ac).

$2x^2 + 8x + x + 4$

This means that in one bracket the x term is $8x$ and in the other it is x. Rewrite the quadratic with the two x terms not simplified.

$(2x^2 + 8x) + (x + 4)$

Group the terms into two brackets, knowing that the two x terms are separated; keep the addition sign between them.

$2x(x + 4) + 1(x + 4)$

Take out any common factors for each bracket separately (write the 1 here as a reminder).

$(2x + 1)(x + 4)$

You can use the distributive law of multiplication to help take out further common factor: $a(b + c) = ab + ac$ so it follows that $2x(x + 4) + \mathbf{1}(x + 4) = (2x + 1)(x + 4)$.

b $5x^2 - 22x + 21$
$ac = 5 \times 21$
$= 105$

Start by finding ac.

$105 = 3 \times 35$
5×21
7×15

Find the factors of 105 (ac) that have a sum of $^-22$ (b).

$5x^2 - 7x - 15x + 21$

The correct pair is $^-7$ and $^-15$. So in one bracket the x term is ^-7x and in the other it is ^-15x. Rewrite the quadratic with the two x terms not simplified.

$(5x^2 - 7x) - (21 - 15x)$

Group the terms into two brackets, knowing that the two x terms are separated; keep the subtraction sign between them.

$x(5x - 7) - 3(^-7 + 5x)$

Take out any common factors for each bracket separately (write the 1 here as a reminder). Reorder the terms in the second bracket to match the first.

$(x - 3)(5x - 7)$

You can use the distributive law of multiplication to help take out a further common factor: $a(b + c) = ab + ac$ so it follows that $x(5x - 7) - 3(5x - 7) = (x - 3)(5x - 7)$.

$(5x - 7)(x - 3) = 5x^2 - 7x - 15x + 21$
$= 5x^2 - 22x + 21$

Check your solution by expanding.

EXERCISE 7F

1 Factorise each quadratic expression fully.

a $2x^2 + 7x + 5$ **b** $3x^2 + 10x + 8$ **c** $2x^2 + 8x - 90$

d $4x^2 + 16x + 15$ **e** $4x^2 - 24x + 27$ **f** $3x^2 - 6x - 105$

g $12x^2 + 12x - 9$ **h** $3x^2 + x - 10$ **i** $2x^2 - 6x - 260$

j $3x^2 - 13x - 10$

2 A rectangle has an area of $(5x^2 - 13x + 6)$ cm^2.
Find the length in terms of x if the width is $(x - 2)$ cm.

3 A triangle of base $(2x + 8)$ cm has an area of $(2x^2 + 11x + 12)$ cm^2.
Use this information to determine an expression for the perpendicular height of the triangle.

4 **a** Show how you could factorise the expression
$3(x + y)^2 + 13(x + y) + 12$
using substitution.

b Apply your method to factorise these expressions.
i $3(x - 3)^2 - (x - 3) - 24$ **ii** $8(5x + 2)^2 - 14(5x + 2) + 3$

Section 3: Completing the square

In this section you are going to learn how to **complete the square** on a quadratic expression.

This technique allows you to manipulate quadratic expressions and write them in the form $a(x \pm h)^2 + k$. This form is useful for solving quadratic equations that will not factorise and for finding the turning point of a quadratic graph. You will learn how to solve quadratics by completing the square in Chapter 8, but for now you will focus on how to manipulate the expression into this format.

Tip

You previously used the perfect square and difference of squares identities:

$(a + b)^2 \equiv a^2 + 2ab + b^2$
$(a - b)^2 \equiv a^2 - 2ab + b^2$
$a^2 - b^2 \equiv (a + b)(a - b)$

Completing the square is a combination of these two identities.

Some quadratics cannot be factorised using the methods you learnt in Section 2.

For example, in the expression $x^2 + 8x + 2$ you cannot find two numbers that have a sum of 8 and a product of 2.

You cannot just change the values in an expression, but you can add values to an expression such that those values cancel each other out (they total 0). For example, the following two expressions are equal because the +4 and −4 cancel each other out.

$$x^2 + 8x + 2 = x^2 + 8x + 4 - 4 + 2$$

You can use this principle when completing the square.

When you complete the square, you find the 'perfect square' and then add or subtract another (usually constant) term so that when the square is expanded, the additional term cancels out. So, the completed square form resembles a perfect square and another (usually constant) term. You can see this by looking again at the general expression of a completed square:

$$\underbrace{a(x \pm h)^2}_{\text{'perfect square'}} + \underbrace{k}_{\text{'additional term'}}$$

Tip

For now, you will deal with cases where $a = 1$.

Find answers at: cambridge.org/ukschools/gcsemaths-studentbookanswers

To get the 'square' part, $(x \pm h)^2$, you use the perfect square identity:

$(a + b)^2 = a^2 + 2ab + b^2$

$(x + h)^2 = x^2 + 2xh \ldots$ In this example, the term in x is $8x$.

$2xh = 8x$

$h = 4$

$(x + 4)^2$

Then, to get the 'additional term', (k), you:

Start by expanding the perfect square: $(x + 4)(x + 4) = x^2 + 8x + 16$

Our starting expression is $x^2 + 8x + 2$ but the perfect square expands to $x^2 + 8x + 16$. You need to ensure that +16 becomes +2 when the expression is expanded. Use the idea that you can add and subtract terms to cancel each other out in order to maintain the correct value of an expression.

$x^2 + 8x + 16 - 14 = x^2 + 8x + 2$

$(16 - 14) \equiv 2$

So, in order to make $(x + 4)^2$ the same as $x^2 + 8x + 2$ when expanded, we need to subtract 14.

So, the completed square form of $x^2 + 8x + 2$ is $(x + 4)^2 - 14$.

Check by expanding.

$$(x + 4)(x + 4) - 14 = x^2 + 8x + 16 - 14$$
$$= x^2 + 8x + 2$$

There is a shortcut that you can use to complete the square:

- to find the constant term (k) of the completed square form, halve the coefficient of x in the quadratic, and then square it.
- the constant value inside the square, b, is half of the coefficient of x in the quadratic.

You can see why this works by looking at the perfect square identity.

You already know that $(a - b)^2 \equiv a^2 - 2ab + b^2$.

To avoid confusion with the 'a' in the completed square form, let's replace a in the perfect square with x: $(x - b)^2 \equiv x^2 - 2xb + b^2$

The coefficient of x is $2b$.

So, $b = \frac{2b}{2}$ and b^2 is the constant term of the quadratic.

Therefore, to find the constant term of the completed square form (k), you have to halve the coefficient of x ($2b$) and then square it. The b value is simply half of the coefficient of x.

WORKED EXAMPLE 10

Complete the square for $x^2 - 8x$.

$(x \pm h)^2 + k$	Write down the general expression for the completed square form (where $a = 1$).
$h = {}^-4$	The value of h is half the coefficient of x.
$\left(\frac{-8}{2}\right)^2 = ({}^-4)^2 = 16$	The value of k is $\frac{-8}{2}$ = the same as h squared.

Continues on next page ...

$k = 16$

$(x \pm h)^2 + k$

$(x - 4)^2 + 16$

Insert the values of h and k into the general expression for the completed square form.

$(x - 4)^2 + 16$

$(x - 4)(x - 4) + 16 = x^2 - 4x - 4x + 16 + 16$

$= x^2 - 8x + 32$

Expand the expression to check that it works. It doesn't work, you are 32 over. So, k must be $^-16$ in order to cancel out the $^+16$ once the square is expanded.

$(x - 4)^2 - 16$

If the coefficient of x is an odd number, you will get a fractional value when you halve it.

When the coefficient of the x^2 term is not 1, you first need to write the coefficient of x^2 **outside** a bracket as a factor of the quadratic, and then you complete the square on the quadratic as before. The coefficient of x^2 then comes the a value in the completed square form: $a(x \pm h)^2$. In order to ensure the value of the whole expression is kept the same after you remove the coefficient of x^2 as a factor of the quadratic, you also need to divide every term inside the bracket (every term of the quadratic) by this factor.

WORKED EXAMPLE 11

Factorise $2x^2 + 3x - 2$ by completing the square.

$2\left(x^2 + \frac{3}{2}x - 1\right)$

Take the 2 outside the bracket.
Divide everything inside the bracket by 2.

$a(x \pm h)^2 + k$

$\frac{1}{2} \times \frac{3}{2} = \frac{3}{4}$

$h = \frac{3}{4}$

Write down the general expression for the completed square form, here $a = 2$.
The value of h is half of the coefficient of x which is now $\frac{3}{2}$.

$\left(\frac{3}{4}\right)^2 = \frac{9}{16}$

Find k. k is h squared.

$\left(x + \frac{3}{4}\right)^2 = \left(x + \frac{3}{4}\right)\left(x + \frac{3}{4}\right)$

$= x^2 + \frac{3}{2}x + \frac{9}{16}$

$x^2 + \frac{3}{2}x + \frac{9}{16}\left(-\frac{9}{16} - 1\right)$

Write out the perfect square that you have, then expand it to determine if you need to add or subtract k. You need to subtract k to cancel out $\frac{9}{16}$ but you need to end up with $^-1$ when you expand the brackets, so k is $\frac{25}{16}$

$2\left(\left(x + \frac{3}{4}\right)^2 - \frac{25}{16}\right)$

Now, write this in the completed square form: $a(x \pm h)^2 + k$

$2\left(\left(x + \frac{3}{4}\right)^2 - \frac{25}{16}\right) = 2\left(\left(x + \frac{3}{4}\right)\left(x + \frac{3}{4}\right) - \frac{25}{16}\right)$

$= 2\left(x^2 + \frac{3}{2}x + \frac{9}{16} - \frac{25}{16}\right)$

$= 2\left(x^2 + \frac{3}{2}x - 1\right)$

$= 2x^2 + 3x - 2$

Multiply out to check.

EXERCISE 7G

1 Find the missing terms in these quadratics to make them equivalent to the perfect square.

a $x^2 + 6x + \square = (x + 3)^2$ **b** $x^2 + 12x + \square = (x + 6)^2$

c $x^2 + 5x + \square = \left(x + \frac{5}{2}\right)^2$ **d** $x^2 + 7x + \square = \left(x + \frac{7}{2}\right)^2$

e $x^2 - 12x + \square = (x - 6)^2$ **f** $x^2 - 10x + \square = (x - 5)^2$

2 Find the missing terms in these quadratics to make them equivalent to a perfect square.

Factorise the resultant expressions by completing the square.

a $x^2 - 2x + \square$ **b** $x^2 + 2x + \square$ **c** $x^2 + 4x + \square$

d $x^2 + 6x + \square$ **e** $x^2 - \square x + \frac{1}{9}$ **f** $x^2 - \square x + 5$

g $x^2 - \square x + 25$ **h** $x^2 - \square x + 11$ **i** $x^2 + \square x + 7$

Tip

See Chapter 27 for more information on surds.

3 Factorise by completing the square.

a $x^2 + 2x - 5$ **b** $x^2 + 2x + 7$ **c** $x^2 + 4x + 1$

d $x^2 + 6x - 3$ **e** $x^2 - 6x + 6$ **f** $x^2 - 8x - 5$

g $x^2 + 8x + 25$ **h** $x^2 + 12x - 11$ **i** $x^2 + 11x + 7$

Section 4: Algebraic fractions

You can simplify algebraic fractions using the same rules that apply to arithmetic fractions.

The rules for the four operations are:

Multiplication	$\frac{a}{b} \times \frac{c}{d} = \frac{ac}{bd}$	
Division	$\frac{a}{b} \div \frac{c}{d} = \frac{ad}{bc}$	Remember $\frac{a}{b} \div \frac{c}{d} = \frac{a}{b} \times \frac{d}{c}$
Addition	$\frac{a}{b} + \frac{c}{d} = \frac{ad + bc}{bd}$	*bd* is a common denominator of the two fractions.
Subtraction	$\frac{a}{b} - \frac{c}{d} = \frac{ad - bc}{bd}$	

Tip

You will not always need to find the common denominator as shown here; sometimes you will be able to multiply just one of the fractions to get a common denominator:

e.g. $\frac{a}{b} + \frac{c}{2b} = \frac{2a}{2b} + \frac{c}{2b}$

Tip

Remember you cannot cancel out only part of a bracket.

$\frac{(x - 2)}{(x - 2)}$ can be simplified to give 1.

$\frac{(x - 2)}{2}$ cannot be simplified any further.

Multiplying and dividing algebraic fractions

You can simplify algebraic fractions by cancelling terms (using a common factor).

You may need to factorise the numerator and/or the denominator before you can cancel terms.

WORKED EXAMPLE 12

Simplify. **a** $\frac{8xy^2}{4z} \times \frac{yz}{^-2x}$ **b** $\frac{x^2 + 2x - 3}{x - 1}$ **c** $\frac{4x^2 - 9}{x + 1} \div \frac{2x + 3}{x^2 - 1}$

a $\frac{^2\cancel{8}\cancel{x}y^2}{\cancel{4}\cancel{z}} \times \frac{y\cancel{z}}{^-2\cancel{x}} = \frac{2y^3}{^-2} = {^-y^3}$

Cancel out where you can before multiplying. Multiply the numerators together and multiply the denominators together. Simplify.

b $\frac{x^2 + 2x - 3}{x - 1} = \frac{(x - 1)(x + 3)}{x - 1}$

$= x + 3$

Factorise the quadratic in the numerator then cancel the $(x - 1)$ factors.

c $\frac{4x^2 - 9}{x + 1} \div \frac{2x + 3}{x^2 - 1} = \frac{4x^2 - 9}{x + 1} \times \frac{x^2 - 1}{2x + 3}$

Rewrite the division as the equivalent multiplication by the reciprocal.

$= \frac{(2x + 3)(2x - 3)}{x + 1} \times \frac{(x + 1)(x - 1)}{2x + 3}$

Factorise using the difference of squares.

$= \frac{\cancel{(2x + 3)}(2x - 3)}{\cancel{x + 1}} \times \frac{\cancel{(x + 1)}(x - 1)}{\cancel{2x + 3}}$

Cancel common factors.

$= (2x - 3)(x - 1)$

WORK IT OUT 7.2

Zoey got these three homework questions wrong.

Find her mistakes.

Simplify each fraction correctly.

1 $\frac{3x + 2}{2x + 3} = \frac{5}{5} = 1$

2 $\frac{\cancel{x^2} - x - 6}{\cancel{x^2} + 3x + 2} = \frac{6}{5}$

3 $\frac{^x\cancel{x^2} - 1}{_1\cancel{2x} + \cancel{4}_1} \times \frac{^{2x}\cancel{4x^2} - \cancel{16}^4}{_1x + 1} = \frac{(x - 1)(2x - 4)}{4}$

Tip

You might be asked to write an expression as a single fraction. In that case, you need to find a common denominator and simplify as far as possible.

Adding and subtracting algebraic fractions

To add or subtract fractions you find a common denominator.

WORKED EXAMPLE 13

Write the following as a single fraction: **a** $\frac{2x}{5}+\frac{x}{3}$ **b** $\frac{(x+2)}{5}+\frac{(x-1)}{4}$ **c** $\frac{4}{x}+\frac{3}{2x}$ **d** $\frac{5}{2x^2}+\frac{11}{2x}$ **e** $\frac{x}{x-2}-\frac{3}{x+2}$

a $\frac{2x}{5}+\frac{x}{3}$

This is an example where the denominator is a number. Find a common denominator like you do when adding non-algebraic fractions. The lowest common multiple of 3 and 5 is 15, so use this as the common denominator.

$=\frac{6x}{15}+\frac{5x}{15}$

$=\frac{11x}{15}$

Write each fraction as an equivalent fraction with the same denominator, using the lowest common denominator, in the same way as you would for non-algebraic fractions. Add the fractions in the same way as you would for non-algebraic fractions, i.e., add the nominators. Do so by collecting like terms.

b $\frac{(x+2)}{5}+\frac{(x-1)}{4}$

$=\frac{4(x+2)}{20}+\frac{5(x-1)}{20}$

This is another example where the denominator is a number. Find a common denominator as before, and write each fraction as an equivalent fraction with that denominator.

$=\frac{4x+8+5x-5}{20}$

Expand the brackets.

$=\frac{9x+3}{20}$

Collect like terms.

$=\frac{3(3x+1)}{20}$

Take out a common factor.

c $\frac{4}{x}+\frac{3}{2x}$

This is an example of when the algebraic expression is the denominator. To find a common denominator, you multiply one of the fractions such that it is an equivalent fraction, but with the same denominator as the other fraction. Here, you multiply the numerator and the denominator of $\frac{4}{x}$ by 2, to get the common denominator of $2x$.

$=\frac{8}{2x}+\frac{3}{2x}$

$=\frac{11}{2x}$

Write each fraction as an equivalent fraction with the common denominator of $2x$. Then add the numerators as normal.

d $\frac{5}{2x^2}+\frac{11}{2x}$

The denominators are algebraic expressions, so find the common denominator by comparing the denominators. Here, the common denominator is $2x^2$.

$=\frac{5}{2x^2}+\frac{11x}{2x^2}$

To get a common denominator, multiply $\frac{11}{2x}$ by x to get a common denominator of $2x^2$.

$=\frac{(5+11x)}{2x^2}$

This expression cannot be simplified further.

Continues on next page ...

e $\frac{x}{x-2} - \frac{3}{x+2}$ $= \frac{x(x+2) - 3(x-2)}{(x-2)(x+2)}$	There is not simple factor to multiply one fraction by to get a common denominator. So, in this case, multiply the two denominators together to get a common denominator (they are both factors of their product). Don't forget to also multiply the numerators by the same amount as their respective denominators.
$= \frac{x^2 + 2x - 3x + 6}{(x-2)(x+2)}$	Expand the brackets in the numerator.
$= \frac{x^2 - x + 6}{(x-2)(x+2)}$	Simplify the numerator.
$= \frac{(x+2)(x-3)}{(x-2)(x+2)}$	The numerator is now a quadratic. Factorise it.
$= \frac{x-3}{x-2}$	Cancel to lowest terms.

EXERCISE 7H

1 Simplify.

a $\frac{8x}{10}$ **b** $\frac{9x+3}{3x+1}$ **c** $\frac{x^2-9}{x+3}$

d $\frac{4x^2-81}{2x-9}$ **e** $\frac{(x-3)(x+2)}{(x+4)(x+2)}$ **f** $\frac{(2x-5)(x+4)}{(x-4)(5-2x)}$

g $\frac{2x^2-5x+3}{2x^2-x-3}$ **h** $\frac{(125-5x^2)}{(3x^2-11x-20)}$

2 Express as single fractions.

a $\frac{4x}{5} - \frac{2x}{5}$ **b** $\frac{2x}{3} + \frac{x}{2} - \frac{3x}{4}$ **c** $\frac{7}{2x} - \frac{4}{3x}$

d $\frac{2}{x+1} + \frac{1}{x+2}$ **e** $\frac{7}{x-2} - \frac{4}{x-1}$ **f** $\frac{1}{(x-7)^2} - \frac{2}{x-7}$

3 Express as single fractions.

a $\frac{2}{x+1} + \frac{3}{x}$ **b** $\frac{x}{x-2} - \frac{3}{x+2}$ **c** $\frac{x+3}{x+2} - \frac{x+2}{x+3}$

d $\frac{1}{2x-3} + \frac{1}{2x+3}$ **e** $\frac{5}{2p+1} + \frac{1}{p}$ **f** $\frac{p+2}{p-1} + \frac{p-1}{p+2}$

g $\frac{1}{x-2} - \frac{2}{x-1} + \frac{1}{x-3}$ **h** $\frac{9}{(3x^2-3y^2)} - \frac{x}{(xy-x^2)}$

Section 5: Apply your skills

Algebraic manipulation is the basis for most mathematics at this level.

You will use the techniques you learned in this chapter regularly to solve problems in many different contexts.

For example, you will use factorising to solve and simplify equations and to sketch curves, and you will complete the square to solve some quadratic equations and find the turning point of graphs.

Find answers at: cambridge.org/ukschools/gcsemaths-studentbookanswers

EXERCISE 7I

1 Read each statement and apply what you have learned to decide whether it is true.

If not, state why not.

a $2(3b - 2) + 5(2b - 1) = 11b + 9$

b $(3x - 2)(4x + 7) = 12x^2 + 13x - 14$

c $3x^2 + 11x + 6 = (3x + 2)(3x + 3)$

d $9x^2 - 2 = (3x + \sqrt{2})(3x - \sqrt{2})$

e $x^2 + 2x - 6 = (x + 1)^2 - 7$

f $\frac{5}{x - 1} + \frac{3}{x + 2} = \frac{9x + 7}{(x - 1)(x + 2)}$

2 The area of a quadrilateral is expressed as $x^2 - 25$.

a Explain why this quadrilateral cannot be a square.

b Another quadrilateral has an area of $x^2 + 10x + 25$.
Is it a square?
How do you know?

3 Use the difference between two squares to simplify the expression $(x + 8)^2 - (x - 8)^2$.

4 Show algebraically that $(x - 1)$ is not a factor of $(^{-}2x^2 - 13x - 15)$.

5 A rectangular slot $(x - y)$ cm wide and $(x + y)$ cm long is cut out of a square metal plate with sides of $(2x - y)$ cm.

Calculate the area of metal remaining.

6 Given that $A = 3x + 2$ and $B = 2x - 1$, write each of the following expressions in terms of x in their simplest form.

a AB **b** $A^2 + B^2$ **c** $(A - B)(A + B)$

7 Evaluate the following using $(a + b)^2 \equiv a^2 + 2ab + b^2$ and $(a - b)^2 \equiv a^2 - 2ab + b^2$.

a $(1.01)^2$ **b** $(0.99)^2$

c $(4.02)^2$ **d** $(0.98)^2$

8 Use the method of completing the square to show that.

a $x^2 + 4x + 15 \geqslant 11$ **b** $x^2 + 2x + 15 \geqslant 14$

9 The three sides of a triangle are $(x + 8)$ cm, $(x + 6)$ cm and $(x - 1)$ cm.
Show algebraically that the triangle is right-angled when $x = 9$.

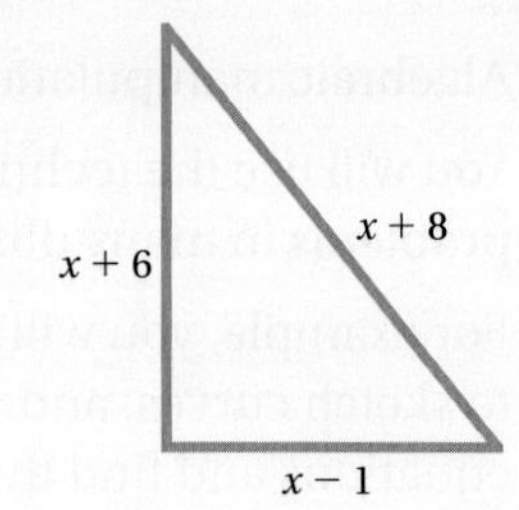

Tip

The converse of Pythagoras' theorem states that if the square of the longest side of a triangle is equal to the sum of the squares of the other two sides, then the triangle must be right-angled.

You will deal with this in more detail in Chapter 32.

10 **a** Show that the area, A cm^2, of a rectangle of perimeter 20 cm is given by the formula $A = w(10 - w)$, where w cm is the width.

b Complete the square for the quadratic expression $10w - w^2$ and hence show that $A \leqslant 25$.

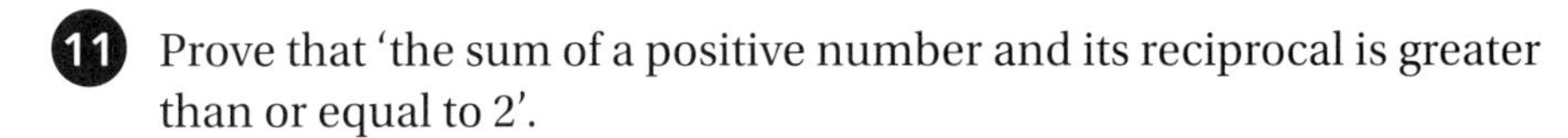

11 Prove that 'the sum of a positive number and its reciprocal is greater than or equal to 2'.

Tip

Use x for your positive integer.

12 Simplify.

a $\dfrac{(x^2 - x - 6)}{(x^2 - 8x + 15)}$

b $\dfrac{(9x^2 - 4)}{(12x - 8)}$

c $\dfrac{(a^2 + ab)}{(a^2 - ab)} \times \dfrac{(ab^2 + b^2)}{(a^3 + a^2b)}$

d $\dfrac{(3x^2 - 3)}{(2x^2 + 3x + 1)} \div \dfrac{(9 - 9x)}{(2x^2 + 7x + 3)}$

Checklist of learning and understanding

Expanding expressions

- A binomial expression is one that contains two terms.
- A binomial product is the product of two binomial expressions, for example $(x + 2)(x + 3)$.
- Expand binomial products by multiplying each term in the first bracket by each term in the second bracket.
- A binomial multiplied by itself is a perfect square. The perfect square identity is $(a \pm b)^2 \equiv a^2 \pm 2ab + b^2$.
- A binomial product in the form $(a + b)(a - b)$ gives a product that is a difference of two squares. The difference of squares identity is $(a + b)(a - b) \equiv a^2 - b^2$.

Factorising

- Factorising is the inverse of expanding.
- You can factorise a simple expression by taking out a common factor, e.g. $6ab + 3ad \equiv 3a(2b + d)$.
- You can factorise a quadratic trinomial by:
  - writing it as a product of its two binomial factors, e.g. $x^2 - x - 6 = (x + 2)(x - 3)$.
  - using the difference of two squares, e.g. $x^2 - 4 = (x + 2)(x - 2)$.
  - completing the square, e.g. $x^2 - 8x = (x - 4)^2 - 16$.
- When the terms in a bracket no longer have a factor, number or letter in common, the expression is factorised fully.

Find answers at: cambridge.org/ukschools/gcsemaths-studentbookanswers

Completing the square

- Completing the square is a technique that turns a quadratic into an equivalent expression containing a perfect square and another (usually constant) term. The completed square form is $a(x \pm h)^2 + k$.

Algebraic fractions

- Algebraic fractions can be simplified using the same rules that apply to arithmetic fractions.

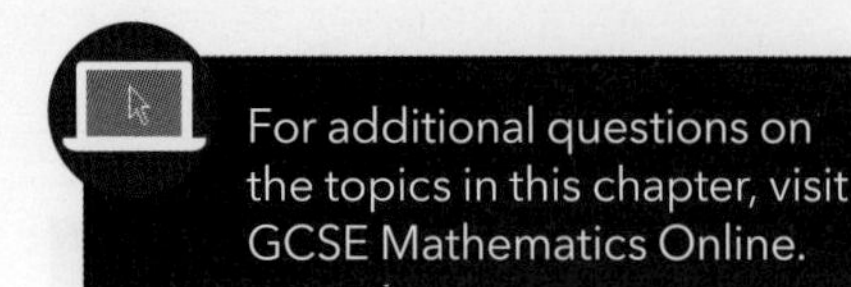

Chapter review

1 Expand and simplify.

a $(y+1)^2 + (y+2)^2 + (y+3)^2$ **b** $(x+1)(x+2)(x-3)$

2 Factorise.

a $2x^2 - 11x - 21$ **b** $^{-}6x^2 - 14x - 8$

3 Use the difference of two squares identity to calculate $1999^2 - 1998^2$.

4 The diagram shows a trapezium.

The lengths of three of the sides of the trapezium are $x - 5$, $x + 2$ and $x + 6$.

All dimensions are in centimetres.

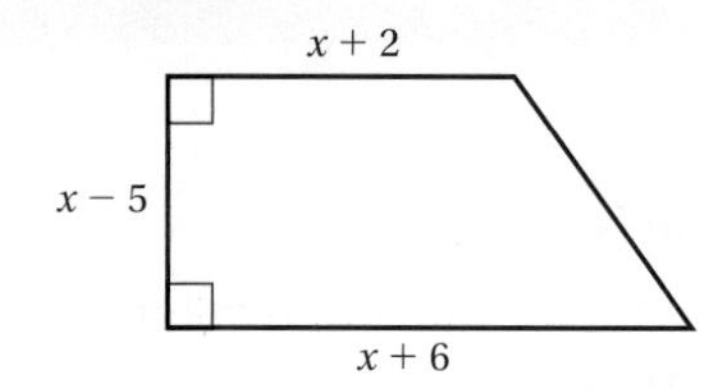

The area of the trapezium is 36 cm².

Show that $x^2 - x - 56 = 0$.

5 Write as single fractions in simplest form:

a $\frac{(x+3)}{4} + \frac{(x-5)}{3}$ **b** $\frac{1}{(x+4)} + \frac{2}{(x-4)}$

c $\frac{2x}{(3x-3)} \div \frac{4y}{(x^2-x)}$ **d** $\frac{3}{(4p+q)} + \frac{3}{(p-2q)}$

e $\frac{1}{x-2} + \frac{5}{x-2} + \frac{1}{1-3x}$ **f** $\frac{2}{x+5} + \frac{1}{x-5} + \frac{5}{x^2-25}$

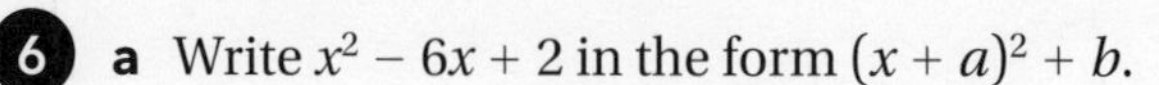

6 **a** Write $x^2 - 6x + 2$ in the form $(x+a)^2 + b$. *(3 marks)*

b Hence write down the minimum value of $x^2 - 6x + 2$. *(1 mark)*

© OCR 2012

8 Equations

In this chapter you will learn how to …

- solve linear equations and apply them in context.
- solve quadratic equations.
- set up and solve simultaneous equations.
- use graphs to find approximate solutions to equations.

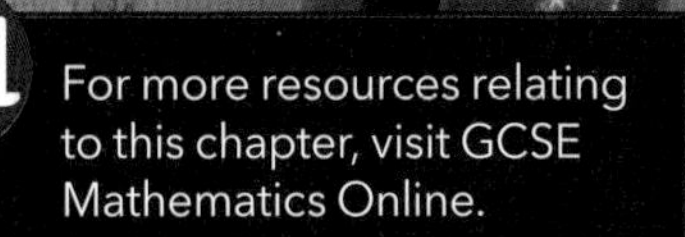

For more resources relating to this chapter, visit GCSE Mathematics Online.

Using mathematics: real-life applications

Accounting involves a great deal of mathematics. Accountants set up computer spreadsheets to calculate and analyse data. Programs such as Microsoft Excel® work by applying different equations to values in columns or cells, so you need to know what equations or formulae to use to get the results you need.

"Although the computer does the actual calculations, I have to insert different equations to tell it what operations to perform and in which order to perform them."

(Accountant)

Before you start …

Ch 3, 7	Apply your skills in using the conventions of algebraic notation to form equations.	**1** Which of the equations below correctly represent this problem? **a** y is one half the size of x **b** y is 2 more than x **c** y is the same as x multiplied by x **d** y is the square root of x A $y = x^2$ B $y - 2 = x$ C $y = \pm\sqrt{x}$ D $y = \frac{1}{2}x$
Ch 3	Check you can write an equation to represent a problem mathematically.	**2** Which of the equations below correctly represent this problem? "I think of a number, multiply it by 6 and add 1. The answer is 37. What is my number?" A $6x + 1 = 37$ B $y \times 6 = 37 + 1$ C $6a = 37$ D $(6x + 1) - 37 = 0$
Ch 1	You should be able to recognise and apply inverse operations.	**3** Complete the following statements.. **a** $7 + \square = 0$ **b** $\square - 8 = 0$ **c** $-4a + \square = 0$ **d** $5 \times \square = 1$ **e** $\frac{1}{6} \times \square = 1$ **f** $\square \times 12x = x$
Ch 7	You need to know how to factorise quadratic expressions.	**4** Match each expression to its factors. **a** $x^2 - 5x + 6$ **b** $x^2 + 3x$ **c** $x^2 - 25$ **d** $x^2 - 5$ A $x(x + 3)$ B $(x + 5)(x - 5)$ C $(x - 2)(x - 3)$ D $(x + \sqrt{5})(x - \sqrt{5})$
Ch 7	You should be able to complete the square on a quadratic expression.	**5** Complete the square on each expression:. **a** $x^2 + 4x + 10 = (x + \square)^2 + \square$ **b** $x^2 - 8x - 5 = (x - \square)^2 - \square$

Find answers at: cambridge.org/ukschools/gcsemaths-studentbookanswers

Assess your starting point using the Launchpad

STEP 1

1 Match each equation to its solution.

Equations

a $x + 7 = 19$ **b** $x - 6 = 11$ **c** $2x + 5 = 7$

d $8x = {}^-24$ **e** $2 - 3x = 8$

Solutions

A $x = 1$ B $x = 17$ C $x = {}^-2$ D $x = 12$ E $x = {}^-3$

f How can you check whether a solution is correct?

2 Solve.

a $9a - 7 = 7a + 3$ **b** $3(x + 5) = 2(x + 6)$

c $\frac{3a}{2} - 4 = \frac{1}{2}$ **d** $15(2x + 3) = 3(2x^2 - 13)$

3 Is $5x + 3 = 18$ equivalent to $6x + 3 = x + 18$?
How do you know?

4 When 16 is added to twice Jack's age, the answer is 44.
Write an equation and solve it to find Jack's age.

GO TO
Section 1:
Linear equations

STEP 2

5 If $x^2 - 2x - 3 = 0$, which pair of values is the solution?

A $x = 3$ or $x = {}^-1$ B $x = {}^-3$ or $x = 1$

6 What are the possible values of x given that $x^2 - 16 = 0$?

7 Complete the square for this expression $(x - 3)^2$ _____ $\equiv x^2 - 6x - 2$

8 Given $x^2 - 6x - 2 = 0$, identify the values of a, b and c in the equation $x = \frac{{}^-b \pm \sqrt{b^2 - 4ac}}{2a}$ then substitute these and solve for x.

GO TO
Section 2:
Quadratic equations

STEP 3

9 **a** How many positive integer solutions can you find for $x + y = 6$?

b Which of those solutions are correct if $x + y = 6$ and $x - y = 4$?

GO TO
Section 3:
Simultaneous equations

GO TO
Step 4:
The Launchpad continues on the next page …

Launchpad continued ...

STEP 4

10 A company hires out meeting rooms.

The total cost (y) can be worked out using the equation $y = 15x + 40$, where x represents the number of hours the room is hired for. The graph of this equation is the straight line shown here.

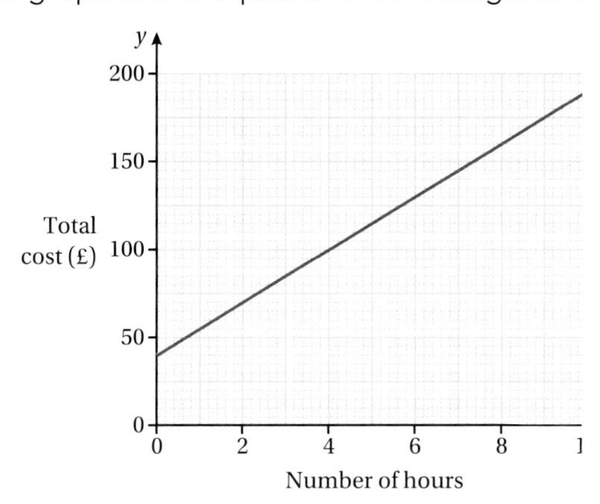

a Use the graph to find cost of hiring a meeting room for 4 hours.

b What is the value of x when y is 175?

c What is the value of y when x is 5?

GO TO
Section 4:
Using graphs to solve equations

STEP 5

11 What is the maximum number of solutions a quadratic equation can have?

Can you find an approximate solution to $x^2 - 5x + 2 = 0$ using the iteration $x_{n+1} = 5 - \frac{2}{x_n}$ with $x_1 = 4$?

GO TO
Section 5:
Finding approximate solutions by iteration

GO TO
Section 6:
Using equations and graphs to solve problems

Section 1: Linear equations

Key vocabulary

variable: a letter representing an unknown number.

linear equation: an equation where the highest power of the unknown is 1, for example $x + 3 = 7$. There are no fractional or negative powers and the resultant graph of the equation is a straight line.

An equation is a mathematical statement that contains an equal sign. For example:

$3 + 2 = 5$ $\quad$ $3 + x = 5$ $\quad$ $3 + 2 = x$ $\quad$ $2x + 3 = 6$

The unknown value is called a **variable** and can be represented by any letter but x and y are used most often.

The same letter can represent different values in different equations.

For example: in the equation $x + 1 = 4$, the value of x is 3, but in the equation $x + 2 = 3$, the value of x is 1.

When the highest power of the unknown is 1, and there are no negative or fractional powers, the equation is a **linear equation**.

Solving an equation means finding the value of the unknown letter.

In simple equations like $x + 3 = 7$ you can solve for x by inspection.

In more complex equations you can find the solution by carrying out inverse operations on both sides of the equation.

Worked example 1 shows how an equation can be changed without altering the solution (the value of x) if you carry the same operation on both sides.

WORKED EXAMPLE 1

Solve for x.

a $5x - 5 = 3x + 1$ $\quad$ **b** $2y + 17 = 5 - 6y$ $\quad$ **c** $2(3x - 1) = 2(x + 1)$

a $5x - 5 - 3x = 3x + 1 - 3x$

To find the value of x, you need to apply inverse operations to both sides of the equation so that you end up with x on its own on one side. Subtract $3x$ from each side. Simplify by collecting like terms.

$2x - 5 = 1$

$2x - 5 + 5 = 1 + 5$

Add 5 to each side. Simplify.

$2x = 6$

Divide both sides by 2.

$\frac{2x}{2} = \frac{6}{2}$

$x = 3$

b $2y + 17 + 6y = 5 - 6y + 6y$

Add $6y$ to both sides (this helps you get rid of negative signs). Add like terms.

$8y + 17 = 5$

$8y + 17 - 17 = 5 - 17$

Subtract 17 from each side.

$8y = {}^{-}12$

Simplify.

$\frac{8y}{8} = {}^{-}\left(\frac{12}{8}\right)$

Divide both sides by 8.

$y = {}^{-}\left(\frac{3}{2}\right)$

Reduce the fraction to its simplest terms.

Continues on next page ...

c $6x - 2 = 2x + 2$	Expand the brackets paying attention to the signs.
$6x - 2 - 2x = 2x + 2 - 2x$	Subtract $2x$ from each side.
$4x - 2 = 2$	Add like terms.
$4x - 2 + 2 = 2 + 2$	Add 2 to each side.
$4x = 4$	
$x = 1$	Divide both sides by 4.

EXERCISE 8A

1 Solve the following equations to find the value of x.
Check each answer by substitution.

a $3x + 10 = 5x + 3$ **b** $12x + 1 = 7x + 11$

c $5x - 2 = 3x + 6$ **d** $5x + 12 = 20 - 11x$

e $8 - 8x = 9 - 9x$ **f** $5x + 3 = 2(x + 2)$

2 Solve these equations by expanding the brackets first.

a $5(x + 3) = 3(2x + 1)$ **b** $7(x + 2) = 4(x + 5)$

c $4(x - 2) + 2(x + 5) = 14$ **d** $3(x + 1) = 2(x + 1) + 2x$

e $^{-}2(x + 2) = 4x + 9$ **f** $4 + 2(2 - x) = 3 - 2(5 - x)$

3 **a** Try to solve these two equations.

i $2(x + 8) - 3x = x + 16$

ii $4(3 + x) + 4x = 4(2x + 3)$

b If an equation is true for any value of x, it is an identity.

Which of these two equations is an identity? How do you know this?

c Do you think there is a solution for the other equation? Give a reason for your answer.

Equations containing fractions

When an equation contains fractions, it can be helpful to eliminate the denominator(s). You could do this by cross multiplying (this means that if one side of the equation has a denominator of 5, for example, you can eliminate it by multiplying each term on both sides of the equation by 5), or you could find a common denominator and add/subtract the fractions (see Chapter 7).

Find answers at: cambridge.org/ukschools/gcsemaths-studentbookanswers

WORKED EXAMPLE 2

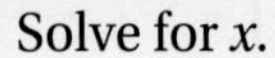

Solve for x.

$$\frac{x-1}{5}=\frac{2x+1}{4}$$

$$\frac{(x-1)}{5}=\frac{(2x+1)}{4}$$

As a first step, you might want to insert brackets around each numerator, to help you when you come to cross multiply.

$$4(x-1)=5(2x+1)$$

Cross multiply by 4 and 5 to eliminate the denominators; this means multiply both sides by 4, and then both sides by 5.

$$4x-4=10x+5$$

Expand the brackets.

$$^{-}4=6x+5$$

Subtract $4x$ from both sides.

$$^{-}9=6x$$

Subtract 5 from both sides.

$$^{-}\left(\frac{9}{6}\right)=x$$

Divide both sides by 6.

$$x=^{-}\left(\frac{3}{2}\right)$$

Write the fraction in its lowest terms.

$$\text{LHS: }\frac{(^{-}1\frac{1}{2}-1)}{5}$$

Check the answer by substitution.

$$=\frac{^{-}2\frac{1}{2}}{5}$$

$$=^{-}\left(\frac{1}{2}\right)$$

$$\text{RHS: }\frac{(2(^{-}1\frac{1}{2})+1)}{4}$$

$$=\frac{^{-}2}{4}$$

$$=^{-}\left(\frac{1}{2}\right)$$

LHS = RHS, so the solution is correct.

Tip

Cross multiplying is a way to rewrite an equation in the form $\frac{a}{b}=\frac{c}{d}$ as $ad=bc$, thus eliminating the denominators.

EXERCISE 8B

1 Solve for x.

a $\frac{2x}{3}+\frac{1}{4}=1$

b $2x-3x=4x+3$

c $3(x-5)=2(4-x)$

d $\frac{2x-3}{4}=\frac{x+1}{3}$

e $\frac{x+20}{9}+\frac{3x}{7}=6$

f $\frac{2}{x-1}=\frac{3}{x-4}$

g $\frac{x}{3}+\frac{x}{2}=10$

h $\frac{7x+3}{2}=\frac{18x-16}{8}$

i $\frac{^{-}3(4+x)}{2}=^{-}9$

j $\frac{3x-11}{5}-\frac{3x}{4}=7+4x$

k $\frac{1}{2}-\frac{(x+2)}{5}=\frac{(2x-3)}{10}$

l $\frac{(x+5)}{6}-\frac{(x+1)}{9}=\frac{(x+3)}{4}$

Forming and solving linear equations

You can use algebra to set up and then solve your own equations.

When you set up an equation you must say what the letters stand for.

Tip

Make sure you know what the unknown is and state what letter you are using to represent it. We tend to let x be the unknown, but you can choose any letter as long as you define it.

WORKED EXAMPLE 3

a The sum of two numbers is 54. If one number is 14 more than the other number, find the two numbers.

b The length of a rectangle is three times its width.
If the perimeter is 24 centimetres, find the area of the rectangle.

a Let one of the numbers be x.

Start by giving a letter to one of the unknown values.

So the other number is $(x + 14)$.
The numbers are x and $(x + 14)$.

You are told that the other number is 14 more than this.

The two numbers add up to 54.

$x + (x + 14) = 54$
$2x + 14 = 54$
$2x = 54 - 14$
$2x = 40$
$x = 20$

Use this information to form an equation and then solve for x.

$x = 20 \quad \therefore \; x + 14 = 34$

The two numbers are 20 and 34.

By solving the equation you have found one of the numbers (x). You still have to find the value of the other number.

Check that this works: $20 + 34 = 54$

Yes, the solution is correct.

b Let the width be x.
$\therefore$ the length is $3 \times x = 3x$.

Write an expression for the length and the width.

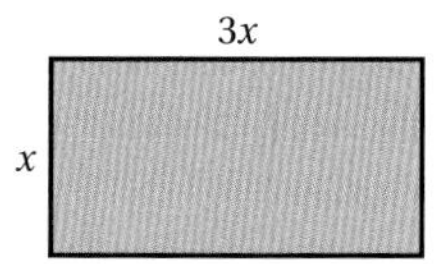

$P = 2(L + W)$

$A = L \times W$

Draw a sketch and label it.

$2(x + 3x) = 24$
$x + 3x = 12$
$4x = 12$
$x = 3$

You need to find the lengths of the sides so you can use them to work out the area.

The width is 3 cm, the length is $3 \times 3 = 9$ cm.

The area is $3 \times 9 = 27$ cm^2.

EXERCISE 8C

1 For each of the following, write an equation and solve it to find the unknown number.

a Three times a certain number is 348, what is the number?

b 7 less than a number is $^-2$, what is the number?

c 6 greater than a number is $^-4$, what is the number?

Find answers at: cambridge.org/ukschools/gcsemaths-studentbookanswers

d Two less than four times a number is 66, what is the number?

e Two consecutive numbers have a sum of 63, what are the numbers?

f Three less than twice a number is $^{-}2$. What is the number?

g When $1\frac{1}{2}$ is added to twice a certain number, the result is $4\frac{3}{4}$. What is the number?

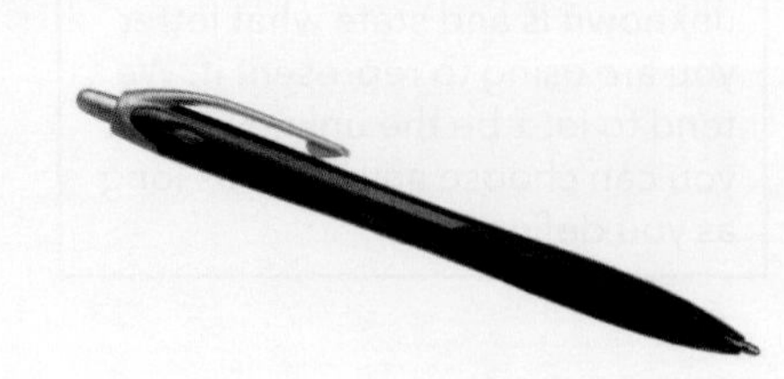

2 Form an equation and solve it to answer each question.

a When 16 is added to twice Lucy's age, the answer is 44. How old is Lucy?

b Stephen buys eight pens and receives 80p change from £20.00. How much does a pen cost, assuming each pen costs the same amount?

c Multiplying a number by 2 and then adding 5 gives the same answer as subtracting the number from 23. What is the number?

d Nick is 20 years older than his daughter. His daughter has worked out that in five years' time, she will be half her dad's age. How old is she now?

e Gina is four years older than her sister. When their ages are added together the result is 22. How old is Gina?

f A woman is three times as old as her daughter. In four years time she will be only two and a half times as old as her daughter. What is the woman's present age?

3 A square has sides of $3x$ cm. A parallelogram has sides of $2x$ cm and $(x + 9)$ cm.

a Write an expression for the perimeter of the square.

b Write an expression for the perimeter of the parallelogram.

c Given that the two quadrilaterals have the same perimeter, form an equation and solve it to find the length of one side of the square.

4 The area of this rectangle is 10 cm^2.

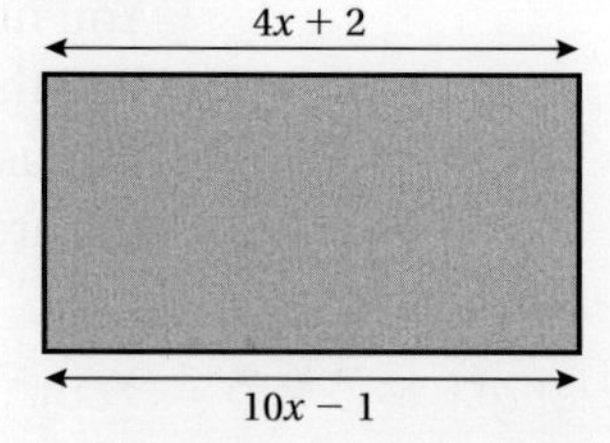

Calculate the value of x and use it to find the length and width of the rectangle.

5 In a triangle ABC, angle B is three quarters of angle A and angle C is one half of angle A. Find the size of each angle.

6 In a game of netball, the winning team won by 9 goals. In total, 83 goals were scored in the game. How many goals did each team score?

7 A fridge contains only bottles of orange, apple, blackcurrant and mango juice.

Of these, $\frac{1}{5}$ of the bottles are orange and $\frac{1}{5}$ are apple.

The fridge also contains 15 dozen bottles of blackcurrant and 30 bottles of mango juice.

How many bottles of orange and apple juice does it contain?

Section 2: Quadratic equations

A quadratic equation has at least one term with a variable that is squared (x^2) and no variables with a power higher than 2.

Consider the simple quadratic equation $x^2 = 9$. You know that $3^2 = 9$ and that $(-3)^2 = 9$. This means that the equation $x^2 = 9$ has two possible roots (or solutions). Indeed, all square numbers can have a positive or negative root.

In a quadratic equation, because you are dealing with a squared variable, you might have two **roots**.

When you are asked to solve a quadratic equation you need to give a **solution** that contains **both** roots.

So, for the quadratic equation $x^2 = 9$:

$$x = \pm\sqrt{9x} = \pm 3$$

So the two possible solutions to the quadratic equation are $x = 3$ and $x = {}^-3$.

Look at this diagram showing a graph plotted from a quadratic equation.

It shows why a quadratic can have a maximum of two solutions.

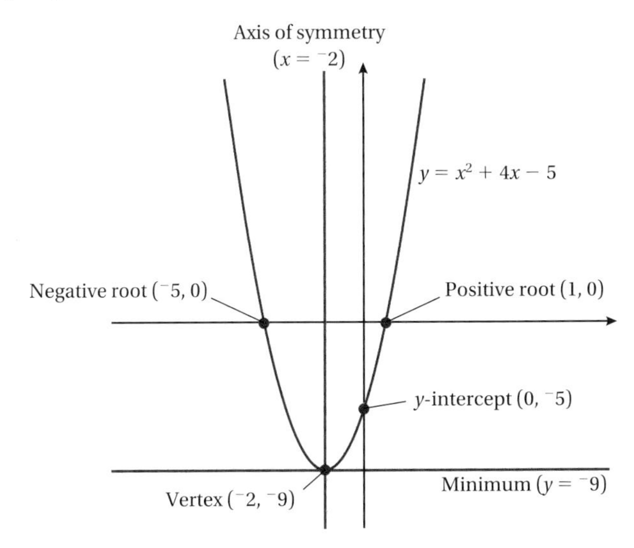

This diagram shows the quadratic function $y = x^2 + 4x - 5$.

It cuts the x-axis in two places marked as the positive root and the negative root.

These are the points at which $y = 0$, so you can solve for x by writing the equation as $0 = x^2 + 4x - 5$.

The standard form of a quadratic equation is $ax^2 + bx + c = 0$ (when the coefficient of x^2 is not 0).

Key vocabulary

roots: of an equation are the value(s) that makes the equation true; the root of a function is the value that makes the function equal to zero.

solution: all possible roots of an equation.

Tip

You learnt about functions in Chapter 4.

Tip

You will deal with graphs of quadratic equations in more detail in Chapter 19.

Find answers at: cambridge.org/ukschools/gcsemaths-studentbookanswers

Solving quadratic equations by factorising them

Before you learn how to solve quadratic equations by factorising them, you need to consider what it means if the product of two factors is 0.

If $a \times b = 0$ then either $a = 0$, or $b = 0$ or both $a = 0$ and $b = 0$.

This means that if $(x - 1)(x - 2) = 0$ then either $(x - 1) = 0$ or $(x - 2) = 0$.

If $x - 1 = 0$, then $x = 1$.
If $x - 2 = 0$, then $x = 2$.

This zero factor principle allows you to factorise the left-hand side of a quadratic equation and use the fact that one of the factors must be zero to solve it.

Tip

Make sure you remember these three methods of factorising quadratic expressions covered in Chapter 7:

- taking out a common factor, e.g. $x^2 - 6x = x(x - 6)$
- writing a quadratic as a product of binomials, e.g. $x^2 + 5x + 4 = (x + 1)(x + 4)$
- applying the difference of two squares identity, e.g. $x^2 - 100 = (x + 10)(x - 10)$

WORKED EXAMPLE 4

Solve.
a $x^2 - 3x + 2 = 0$ **b** $9x^2 - 4 = 0$ **c** $x^2 - 6x = 0$
d $3x^2 - 5x = 2$ **e** $x + 5 = \frac{14}{x}$ **f** $\frac{1}{x-1} - \frac{1}{x+3} = \frac{1}{35}$

a $x^2 - 3x + 2 = 0$

$(x - 1)(x - 2) = 0$ — Factorise the left-hand side.

Either $(x - 1) = 0$ or $(x - 2) = 0$ — Apply the zero factor principle.

$x - 1 = 0$

$\therefore \; x = 1$ — Solve both equations. Add 1 to each side.

$x - 2 = 0$

$\therefore \; x = 2$ — Add 2 to each side.

$x = 1$ or $x = 2$ — State the solution.

b $9x^2 - 4 = 0$

$(3x + 2)(3x - 2) = 0$ — Factorise using the difference of squares identity.

Either $3x + 2 = 0$ or $3x - 2 = 0$ — Apply the zero factor principle.

$3x + 2 = 0$ — Solve the equations.

$\therefore \; x = {}^{-}\left(\frac{2}{3}\right)$

$3x - 2 = 0$

$\therefore \; x = \frac{2}{3}$

$x = {}^{-}\left(\frac{2}{3}\right)$ or $x = \frac{2}{3}$ — State the solution.

Continues on next page ...

c $x^2 - 6x = 0$

$x(x - 6) = 0$ — Factorise by taking out a common factor of x.

Either $x = 0$ or $(x - 6) = 0$ — Apply the zero factor principle.

$x - 6 = 0$

$\therefore \quad x = 6$ — In this case you already have one value for x so you need only solve one equation.

$x = 0$ or $x = 6$ — State the solution.

d $3x^2 - 5x = 2$

$3x^2 - 5x - 2 = 0$ — Subtract 2 from each side to get the equation in the form $ax^2 + bx + c = 0$.

$3x(x - 2) + 1(x - 2) = 0$ — Factorise the quadratic.

$(x - 2)(3x + 1) = 0$

$x - 2 = 0$ or $3x + 1 = 0$ — Apply the zero factor principle.

$x - 2 = 0$ — Solve the equations.

$\therefore \quad x = 2$

$3x + 1 = 0$

$\therefore \quad x = {}^{-}\left(\frac{1}{3}\right)$

$x = 2$ or $x = {}^{-}\left(\frac{1}{3}\right)$ — State the solution.

e $x + 5 = \frac{14}{x}$ — This equation doesn't look like a quadratic until you rearrange it.

$x(x + 5) = 14$ — Multiply by x to get rid of the fraction.

$x^2 + 5x = 14$ — Expand the brackets – now you can see the squared variable.

$x^2 + 5x - 14 = 0$ — Subtract 14 to get the equation into standard form.

$(x + 7)(x - 2) = 0$ — Factorise.

Either $x + 7 = 0$ or $x - 2 = 0$

$\therefore x = {}^{-}7$ or $x = 2$

f $\frac{1}{x - 1} - \frac{1}{x + 3} = \frac{1}{35}$

$35(x + 3) - 35(x - 1) = (x - 1)(x + 3)$ — Multiply to get rid of the denominators.

$35x + 105 - 35x + 35 = x^2 + 2x - 3$ — Expand the brackets.

$140 = x^2 + 2x - 3$ — Collect like terms.

$x^2 + 2x - 143 = 0$ — Rewrite in the form $ax^2 + bx + c = 0$.

$(x + 13)(x - 11) = 0$ — Factorise.

Either $x = {}^{-}13$ or $x = 11$

Find answers at: cambridge.org/ukschools/gcsemaths-studentbookanswers

After working through the examples you should be able to see a clear set of steps for solving quadratic equations which can be factorised.

Step 1
If necessary, take all the terms to the left-hand side or the right-hand side to get the equation in the form $ax^2 + bx + c = 0$.

Step 2
Factorise the quadratic by:
- checking for common factors; or
- checking for difference of squares; or
- writing quadratics as a binomial product.

Step 3
Write each factor equal to 0.
Solve the equations to find the roots.

Tip

When you are dealing with a difference of two squares you can solve in the following way as well:

$4x^2 - 9 = 0 \rightarrow 4x^2 = 9 \rightarrow x^2 = \frac{9}{4} \rightarrow x = \pm\sqrt{\frac{9}{4}} \rightarrow x = \pm\frac{3}{2}$

EXERCISE 8D

1 Solve the following equations by factorising.

a $x^2 + 12x + 27 = 0$
b $x^2 - x - 30 = 0$
c $6x^2 - 7x - 10 = 0$
d $9x^2 + 4x - 5 = 0$
e $x^2 + 3x = 0$
f $5x^2 - 4x = 0$
g $x^2 - 100 = 0$
h $x^2 - 5 = 0$ (leave your answer in square root form)
i $x^2 - 6 = 0$ (leave your answer in square root form)

2 Can you solve the quadratic equation $x^2 + 4 = 0$ by factorising it? Explain.

3 Solve for x.

a $x^2 = 4(x + 8)$
b $(2x - 1)(3x + 1) = 11$
c $3x - 8 = \frac{x^2}{4}$
d $2(x + 1)^2 = (x + 1)^2 + 9$
e $\frac{x+1}{3} = \frac{10}{x}$
f $\frac{2}{2x-3} = \frac{x}{4x-6}$
g $\frac{4}{x-1} - \frac{5}{x+2} = \frac{3}{x}$
h $6(4x + 5) + \frac{7}{x}(4x + 5) = 0$

Solving quadratic equations by completing the square

Tip

Remember
$(a + b)^2 \equiv a^2 + 2ab + b^2$
$(a - b)^2 \equiv a^2 - 2ab + b^2$

In Chapter 7 you learned how to factorise quadratics by completing the square.

This method is used to solve quadratic equations if the quadratic doesn't have integral factors.

WORKED EXAMPLE 5

Solve $x^2 - 6x + 1 = 0$.

$x^2 - 6x + 1 = 0$ — The LHS cannot be factorised.

$x^2 - 6x + 9 - 9 + 1 = 0$ — Factorise by completing the square.

$x^2 - 6x + 9 - 8 = 0$

$(x - 3)^2 - 8 = 0$

$(x - 3)^2 = 8$

Solve the equation in the same way you would solve a linear equation, i.e. get x on its own on one side of the equation.

$x - 3 = \pm\sqrt{8}$ — Take the square root of both sides.

So, $x = 3 + \sqrt{8}$ or $x = 3 - \sqrt{8}$

$x = 5.83$ or $x = 0.17$

The quadratic formula

Completing the square results in a formula that can be used to solve quadratic equations.

Consider any quadratic equation $ax^2 + bx + c = 0$.

$ax^2 + bx + c = 0$	
$x^2 + \frac{b}{a}x + \frac{c}{a} = 0$	Divide all terms by a.
$x^2 + \frac{b}{a}x = {}^-\left(\frac{c}{a}\right)$	Keep only the terms in x on the LHS.
$x^2 + \frac{b}{a}x + \frac{b^2}{4a^2} = {}^-\left(\frac{c}{a}\right) + \frac{b^2}{4a^2}$	Complete the square by adding $\frac{b^2}{4a^2}$ to both sides.
$\left(x + \frac{b}{2a}\right)^2 = \frac{b^2 - 4ac}{4a^2}$	Factorise the LHS and write the RHS as a single fraction.
$x + \frac{b}{2a} = \pm\frac{\sqrt{b^2 - 4ac}}{2a}$	Take the square root of each side.
$x = {}^-\left(\frac{b}{2a}\right) \pm \frac{\sqrt{b^2 - 4ac}}{2a}$	Isolate x by subtracting $\frac{b}{2a}$.
$x = \frac{{}^-b \pm \sqrt{b^2 - 4ac}}{2a}$	Write the RHS as a single fraction.

This shows that for an equation in the standard form $ax^2 + bx + c = 0$, the two roots are:

$$x = \frac{{}^-b + \sqrt{b^2 - 4ac}}{2a} \quad \text{and} \quad x = \frac{{}^-b - \sqrt{b^2 - 4ac}}{2a}$$

Learn this formula

$$x = \frac{^-b \pm \sqrt{b^2 - 4ac}}{2a}$$

Tip

If you are asked to solve a quadratic equation to a given number of decimal places it usually means the roots are irrational and you should apply the quadratic formula to find the solution.

You need to know the quadratic formula and be able to apply it.

It will not be given to you in an exam.

Once you know the formula you can solve any quadratic equation by substituting the values of a, b and c into the formula and evaluating it.

WORKED EXAMPLE 6

Solve $3x^2 + 7x - 13 = 0$, giving your answer correct to 2 decimal places.

$$x = \frac{^-b \pm \sqrt{b^2 - 4ac}}{2a}$$

Start by identifying the values of a, b and c.

In our equation, $a = 3$, $b = 7$, $c = {^-13}$.

$$x = \frac{^-7 \pm \sqrt{7^2 - 4 \times 3 \times {^-13}}}{2 \times 3}$$

Substitute in the values for a, b and c.

$$x = \frac{^-7 \pm \sqrt{49 + 156}}{6}$$

$$x = \frac{^-7 \pm \sqrt{205}}{6}$$

$$x = \frac{^-7 \pm 14.317}{6}$$

Either $x = \frac{^-7 + 14.317}{6}$ or $x = \frac{^-7 - 14.317}{6}$

$x = 1.22$ or $x = {^-3.55}$ (correct to 2 decimal places)

Tip

When you use the quadratic formula it is best to work systematically using a step-by-step approach and paying careful attention to negative values to make sure you get an accurate solution.

EXERCISE 8E

1 Solve each equation by completing the square.

a $x^2 - x - 10 = 0$
b $x^2 + 3x - 6 = 0$
c $x(6 + x) = 1$
d $2x^2 + x = 98$
e $5x = 10 - \frac{1}{x}$
f $x - 5 = \frac{2}{x}$
g $(x - 1)(x + 2) - 1 = 0$
h $(x - 4)(x - 2) = {^-5}$
i $x^2 = x + 1$

2 Solve each of the following equations using the quadratic formula. Round the answers to 3 significant figures where necessary.

a $2x^2 - x + 6 = 4x + 5$
b $7x^2 - 3x - 6 = 3x - 7$
c $x(6x - 3) - 2 = 0$
d $0.5x^2 + 0.8x - 2 = 0$
e $(x + 7)(x + 5) = 9$
f $\frac{1}{x} + x = 7$

3 Use the quadratic formula to solve each quadratic equation.
Give your answers in simplest surd form.

a $x^2 + 5x + 5 = 0$
b $x^2 + 2x - 4 = 0$
c $x^2 + 12x + 3 = 0$
d $3x^2 + 2x - 7 = 0$
e $5x^2 + 3x - 1 = 0$
f $4x^2 - 6x + 1 = 0$

Forming and solving quadratic equations

As with linear equations, you can set up and solve quadratic equations.

Always define the letters you are using in your equation.

WORKED EXAMPLE 7

A number is added to its square and the result is 12.
What could the number be?

Let the number be x.	Define the unknown values.
$\therefore$ its square is x^2.	
$x + x^2 = 12$	Set up an equation using the information in the problem. Rearrange the equation so that it is in standard form and equal to zero.
$\therefore \quad x^2 + x - 12 = 0$	
$\therefore \ (x + 4)(x - 3) = 0$	Factorise the quadratic.
Either $x + 4 = 0$ or $x - 3 = 0$	Solve the equations.
$x + 4 = 0$	
$\therefore \quad x = {}^-4$	
$x - 3 = 0$	
$\therefore \quad x = 3$	
The number could be 3 or $^-4$.	State the solution **in the context of the problem**.

When you use quadratic equations to model real-life situations you may find that one of the solutions is not possible.

For example, if x is the length of the side of a box in metres, and you get the roots $x = 2.5$ or $x = {}^-1.75$ you can ignore the value of $^-1.75$ as this cannot be the length of an object.

You use 2.5 as the length of the side.

> **Tip**
>
> If the problem involves square units then you can probably use a quadratic equation to solve it.

WORKED EXAMPLE 8

A rectangle with an area of 28 cm^2 has one side 3 cm longer than the other.
How long are each of the shorter sides?

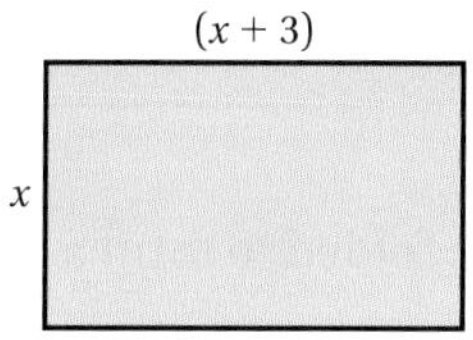

Draw a diagram and label the sides.

Let the shorter side be x.	
$\therefore$ the longer side is $x + 3$.	
$A = x(x + 3)$	Next form your equation. The area of a rectangle is length $\times$ breadth.
$28 = x(x + 3)$	But we are told the area is 28.
$x(x + 3) = 28$	Rearrange the equation to get a quadratic in the form $ax^2 + bx + c = 0$
$x^2 + 3x = 28$	
$x^2 + 3x - 28 = 0$	
$(x + 7)(x - 4) = 0$	Factorise.

Continues on next page ...

Either $(x + 7) = 0$ or $(x - 4) = 0$	Apply the zero factor principle.
$x + 7 = 0$ $\therefore \quad x = {}^{-}7$ $x - 4 = 0$ $\therefore \quad x = 4$	Solve the equations.
The shorter sides are 4 cm long.	State the answer. In this problem you can ignore the $x = {}^{-}7$ solution because ${}^{-}7$ cm is not a possible length for the side of a rectangle.

EXERCISE 8F

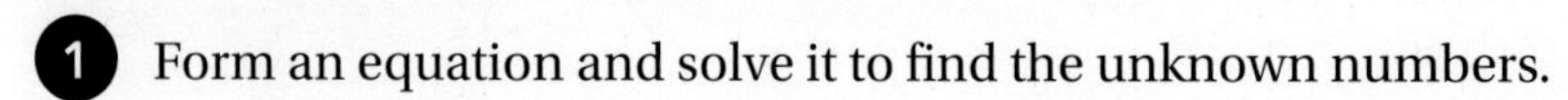

1 Form an equation and solve it to find the unknown numbers.

a The product of a certain whole number and four more than that number is 140.
What could the number be?

b The product of a certain whole number and three less than that number is 108.
What could the number be?

c When three times a number is subtracted from the square of the number the answer is 10.
What are the possible values of the number?

d The product of two consecutive positive odd numbers is 48.
What are the numbers?

2 Use Pythagoras' theorem to find the value of x in the diagram.

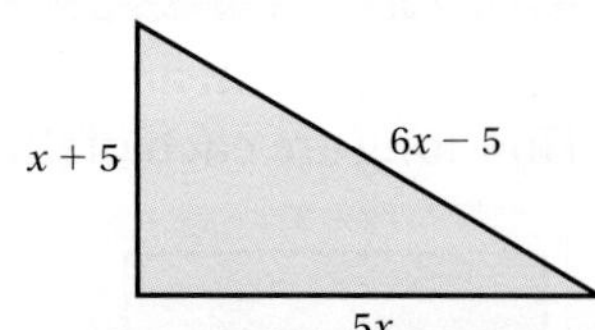

3 A metal sheet is 50 cm wide and 60 cm long.
It has squares cut out of the corners so that it can be folded to form a box with a base area of 1200 cm^2.
Find the length of the side of the squares.

4 A rectangular lawn is 18 m long and 12 m wide.
The lawn is surrounded by a path of width x m.
The area of the path is equal to the area of the lawn.
Find x.

5 The rectangles shown are equal in area.
Find the value of x and hence the dimensions of each rectangle.

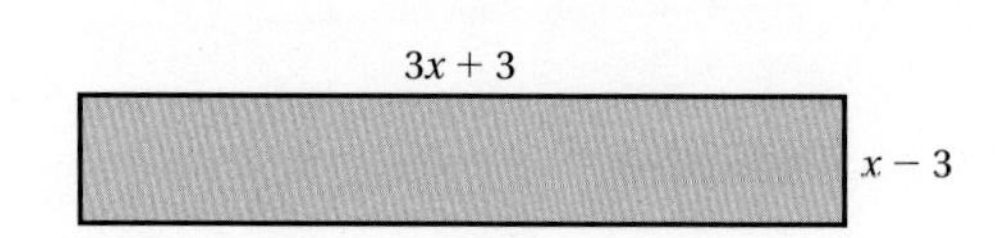

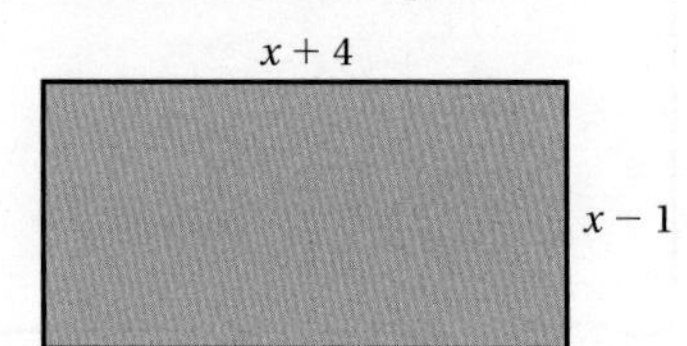

6 The perimeter of a rectangular field is 500 m and its area is 14 400 m^2.

Find the lengths of the sides.

Section 3: Simultaneous equations

Some equations contain two unknowns. For example, $x + y = 4$.

You can suggest some values for x and y, for example, $1 + 3 = 4$ or $2 + 2 = 4$.

To find the correct solution to this equation, you need more information. x and y could take many different values, including negative and fractional values.

If you are told that $x + y = 4$ and that $3x - y = 2$ you have a pair of equations in two unknowns that are true at the same time.

These are called **simultaneous equations**.

You can use the fact that both these equations are true to solve them at the same time (simultaneously).

Simultaneous equations are used to solve problems that involve two conditions that are met at the same time. The example below shows how this works.

Key vocabulary

simultaneous equations: a pair of equations with **two** unknowns that can be solved at the same time.

WORKED EXAMPLE 9

There are 30 rose bushes in a nursery. Some are red and others are white.

There are twice as many white rose bushes as red.

How many rose bushes of each colour are there?

Let w = the number of white rose bushes,
r = the number of red rose bushes

To solve this you need to set up two equations.

$r + w = 30$

There are 30 rose bushes altogether.

$w = 2r$

The number of white bushes is twice the number of red bushes.

$r + w = 30$

You can draw up a table for each equation to show some values of r and w.

r	0	10	20	30
w	30	20	10	0

$w = 2r$

r	0	5	10	15
w	0	10	20	30

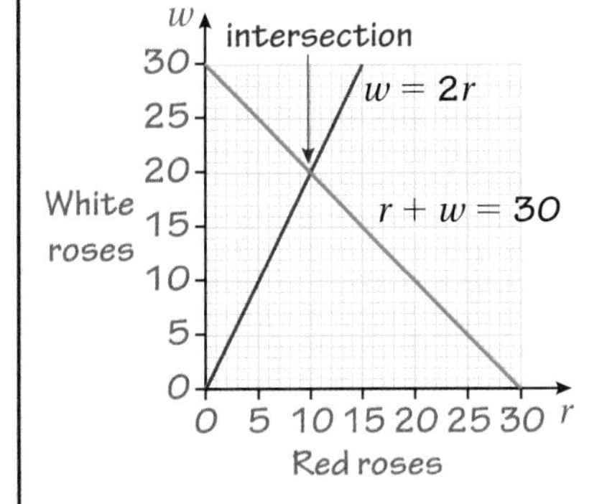

If you draw the graph of each set of values you get two lines that intersect in one place only. Where the graphs intersect, the values on the axes must be the same.

The coordinates (r, w) of the point of intersection are (10, 20), so $r = 10$ and $w = 20$.

Continues on next page ...

$r = 10$ and $w = 20$

This is the simultaneous solution of the two equations.

$r + w = 30$

$10 + 20 = 30$

Check by substitution in both equations.

$w = 2r$

$20 = 2(10) = 20$

You can draw graphs to solve simultaneous solutions but this method takes a long time and it might not be very accurate, especially if the solution values are fractions.

Tip

The substitution method is suitable when one of the equations already has x or y on its own on one side, or when you can easily rearrange one of the equations to have x or y on its own.

There are two methods of solving simultaneous equations using algebra: substitution and elimination.

The method you choose depends on how the equations are written.

Solving simultaneous equations by substitution

Problem-solving framework

Solve these simultaneous equations.

$x - 2y + 1 = 0$ and $y + 4 = 2x$

Steps for solving simultaneous equations by substitution	What you do in this example
Step 1: Number the equations (1) and (2).	$x - 2y + 1 = 0$ (1) $y + 4 = 2x$ (2)
Step 2: Rearrange one equation so x or y is on its own on the left-hand side.	Equation 2 seems simpler, so rearrange it to get: $y = 2x - 4$ (2)
Step 3: Substitute the right hand side of the rearranged equation into the other equation.	Substitute $(2x - 4)$ in place of y in (1). $x - 2(2x - 4) + 1 = 0$
Step 4: Solve the new equation (which now only has one unknown).	$x - 2(2x - 4) + 1 = 0$ $x - 4x + 8 + 1 = 0$ $^{-}3x + 9 = 0$ $^{-}3x = ^{-}9$ $x = 3$
Step 5: Substitute the solution into either of the original equations to find the other unknown.	Substitute $x = 3$ into (2) and solve. $y + 4 = 2(3)$ $y + 4 = 6$ $y = 6 - 4$ $y = 2$ Check the values for x and y satisfy both equations.

EXERCISE 8G

1 Solve the following pairs of simultaneous equations by substitution.
Check that your solutions satisfy **both** equations.

a $y = x - 2$
$y = 3x + 4$

b $y = 2x + 6$
$y = 4 - 2x$

c $y = x + 1$
$x + y = 3$

d $y = x - 2$
$3x + y = 14$

e $y = 2x + 1$
$x + 2y = 12$

f $y = 1 - 2x$
$x + y = 2$

2 Solve these simultaneous equations.

a $x + 2y = 11$
$2x + y = 10$

b $x - y = {}^{-}1$
$2x + y = 4$

c $5x - 4y = {}^{-}1$
$2x + y = 10$

d $3x - 2y = 29$
$4x + y = 24$

e $3x + y = 6$
$9x + 2y = 1$

f $3x - 2 = {}^{-}2y$
$2x - y = {}^{-}8$

Solving simultaneous equations by elimination

In this method you add or subtract the equations to get rid of one of the unknown variables.

You may need to multiply or divide one equation by a factor before you do this.

Problem-solving framework

Solve the simultaneous equations $2x + y = 8$ and $x - y = 1$.

Steps for solving simultaneous equations by elimination	What you do in this example
Step 1: Number the equations (1) and (2).	$2x + y = 8$ (1) $x - y = 1$ (2)
Step 2: Decide whether you can add or subtract to get rid of a variable.	(1) has a variable of y and (2) has a variable of ${}^{-}y$. If you add the equations, these terms will cancel out. (1) + (2) $2x + y = 8$ $x - y = 1$ $3x = 9$
Step 3: Solve the resulting combined equation (which now only has one unknown).	$3x = 9$ $x = 3$
Step 4: Substitute the solution into either of the original equations to find the other unknown.	Substitute $x = 3$ into (2). $3 - y = 1$ ${}^{-}y = 1 - 3$ ${}^{-}y = {}^{-}2$ $y = 2$

Find answers at: cambridge.org/ukschools/gcsemaths-studentbookanswers

Tip

Add if the signs are different and one variable will be eliminated.

Subtract when the signs are the same, including when they are both negative.

In some examples, you will need to form new equations before you can add or subtract to eliminate one variable.

WORKED EXAMPLE 10

Solve this pair of simultaneous equations:

$3x - 2y = 5$ (1)

$4x + 3y = 18$ (2)

Multiply (1) by 3 and (2) by 2:

$9x - 6y = 15$ (3)

$8x + 6y = 36$ (4)

Adding (3) and (4) eliminates the y terms: $^{-}6y + 6y = 0$

$17x = 51$

$x = 3$

Substituting in (2):

$4 \times 3 + 3y = 18$

$3y = 6$

$y = 2$

So $x = 3$ and $y = 2$ are the solutions that satisfy both equations.

In this example, you could could rewrite one of the equations to make either x or y the subject and use substitution, but this involves introducing fractions into the equations.

The best strategy is to create multiples of these equations, to get one pair of identical coefficients for either x or y.

Note that this created identical coefficients for y.

EXERCISE 8H

1 Solve the following pairs of simultaneous equations by elimination. Check your solutions satisfy **both** equations.

a $x - y = 2$
$3x + y = 14$

b $2x - 3y = 3$
$x + 3y = 6$

c $3x + y = 4$
$2y - 3x = {}^{-}10$

d $2x - y = 13$
$5x + y = 13$

e $x + 2y = 14$
$4x - 2y = 14$

f $^{-}x - y = 3$
$x + 5y = {}^{-}11$

g $x + y = 5$
$3x + y = 9$

h $3x + 4y = 15$
$x + 4y = 13$

i $2x - y = 7$
$4x - y = 15$

2 Solve each pair of simultaneous equations. Choose the most suitable method for doing this.

a $2x + y = 7$
$3x + 2y = 12$

b $^{-}2x + 8y = 6$
$2x = 3 - y$

c $4x + 2y = 50$
$x + 2y = 20$

d $x + y = {}^{-}7$
$x - y = {}^{-}3$

e $y = 1 - 2x$
$5x + 2y = 0$

f $y = 2x - 5$
$y = 3 - 2x$

3 Solve simultaneously.

a $2x + 5y = 10$
$x - 3y = 5$

b $x - 3y = 0$
$2x - 4y = 2$

c $-3x - 2y = 4$
$x + 7y = 5$

d $2x + 3y = 12$
$3x - 4y = {}^{-}1$

e $5x - 2y = 17$
$4x + 3y = 9$

f $2x + 3y = 1$
$5x + 4y = {}^{-}1$

Forming and solving simultaneous equations

Some problems can be described using a pair of simultaneous equations.

Once you have defined the variables and set up the equations you can use algebra to solve them.

Problem-solving framework

A field contains a number of goats and chickens.

Altogether there are 60 heads and 200 legs.

How many are there of each type of animal?

Tip

If a problem asks for two different pieces of information then it means there are two unknowns and you will need two equations to solve it.

Steps for approaching a problem-solving question	What you would do for this example
Step 1: Work out what you have to do. Start by reading the question carefully.	You have to find the number of goats (g) and the number of chickens (c)
Step 2: What information do you need? Have you got it all?	You need to know how many there are altogether. You're not told this, but you can work it out because each goat and each chicken must have one head. $g + c = 60$ Now you need another equation to link the goat and chickens. You've already used the number of heads, so it must be something to do with the legs. Goats have four legs each, so the number of goat legs is $4 \times g$. Chickens have two legs each, so the number of chicken legs is $2 \times c$. There are 200 legs in total so: $4g + 2c = 200$
Step 3: Decide what maths you can do.	There are two unknowns so you should use simultaneous equations.
Step 4: Set out your solution clearly. Check your working and that your answer is reasonable.	Let the number of goats = g and the number of chickens = c. $g + c = 60$ (1) $4g + 2c = 200$ (2) $g = 60 - c$ Make g the subject of equation (1). Substitute $(60 - c)$ for g in (2) and solve. $4(60 - c) + 2c = 200$ $240 - 4c + 2c = 200$ $240 - 2c = 200$ $240 - 200 = 2c$ $40 = 2c$ $20 = c$

Continues on next page ...

Steps for approaching a problem-solving question	What you would do for this example
	Substitute $c = 20$ into equation (1) to find g. $g + 20 = 60$ so $g = 40$ Check your result by substituting c and g into original equation (2). $4(40) + 2(20) = 200$ $160 + 40 = 200$ Yes, this works.
Step 5: Check that you've answered the question.	There are 20 chickens and 40 goats.

EXERCISE 8I

1. The sum of two numbers, a and b, is 120.
   When b is subtracted from $3a$, the result is 160.
   Find the value of a and b.

2. Two numbers have a sum of 76 and a difference of 48.
   What are the numbers?

3. A taxi company charges a flat fee plus an amount per mile.
   A journey of 10 miles costs £7 and a journey of 15 miles costs £9.
   What would you pay for a journey of 8 miles?

4. Josh and Sanjita both spent £2.20 on sweets.
   Josh bought five fizzers and four toffees.
   Sanjita bought two fizzers and six toffees.
   Work out the cost of each type of sweet.

5. Sam got eighteen 5p and 10p coins totalling £1.65 as change when he went to the shop.
   How many of each coin did he get?

6. Two children have a total of 264 stickers between them.
   One child has 6 fewer stickers than 5 times the other child's stickers.
   How many do they each have?

7. A computer store sold 4 hard drives and 10 flash drives for £200 and 6 hard drives and 14 flash drives for £290.
   Find the cost of a hard drive and the cost of a flash drive.

8. A large stadium has 21 000 seats. The seats are organised in blocks of 400 and 450 seats. There are three times more blocks of 450 seats than blocks of 400 seats.
   How many blocks of seats are there?

Simultaneous linear and quadratic equations

When the graphs of a linear equation and a quadratic equation are plotted on the same set of axes there are three possible arrangements.

Either the graphs don't intersect at all, or they intersect at one point, or they intersect at two points.

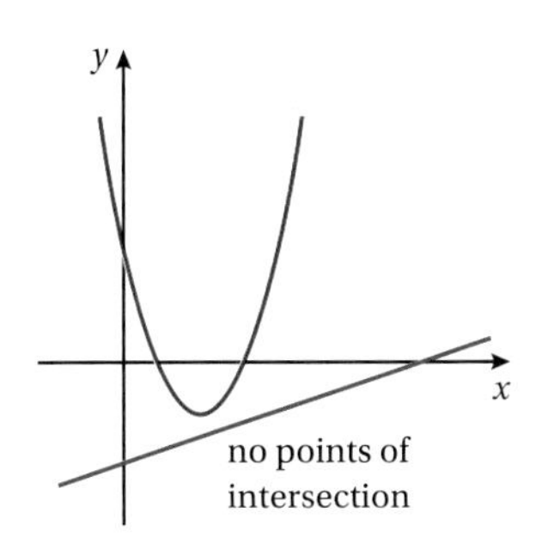

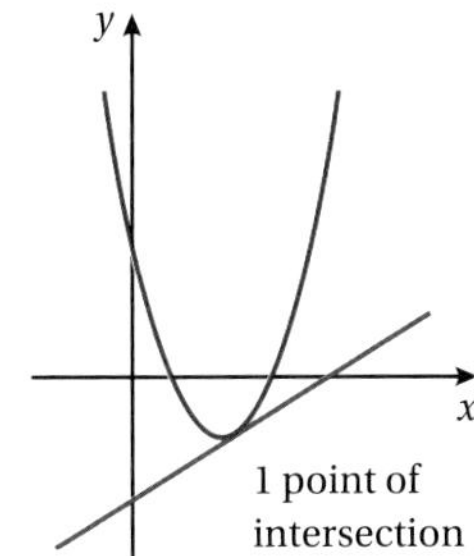

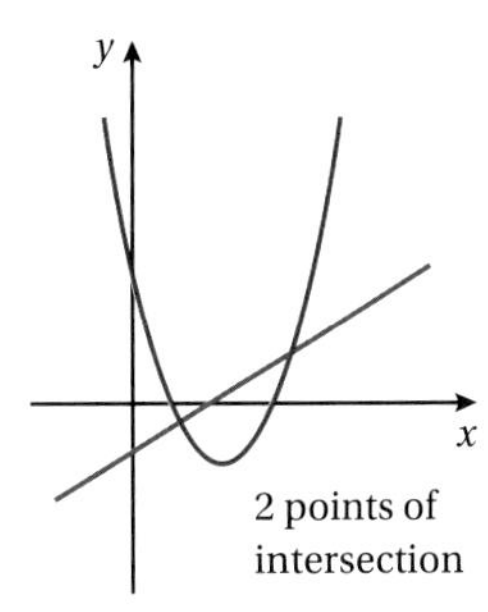

If the graphs intersect their equations can be solved simultaneously by substitution or elimination to find the point(s) of intersection.

WORKED EXAMPLE 11

Determine the point(s) of intersection for the following pairs of equations algebraically:

a $y = x^2 - 1$ (1)
$y = {}^-x + 5$ (2)

b $y = x^2 - 3x$ (1)
$y = x - 4$ (2)

Tip

You will draw and use graphs to find solutions to equations in Chapters 18 and 19. In this chapter you will focus on algebraic solutions.

a $y = x^2 - 1$ (1)
$y = {}^-x + 5$ (2)

$x^2 - 1 = {}^-x + 5$

Substituting $(x^2 - 1)$ for y.

$x^2 + x - 6 = 0$

Arrange the equation in the form $ax^2 + bx + c = 0$.

$(x + 3)(x - 2) = 0$

Factorise.

Either $x = {}^-3$ or $x = 2$

Substituting in (2): $y = {}^-({}^-3) + 5$, $y = 8$ $\quad$ $y = {}^-2 + 5$, $y = 3$

You are asked to find the points of intersection. These are given as the ordered pair (x, y) so you cannot simply give values of x and y, you need to give the ordered pairs.

$\therefore$ the two points of intersection are $({}^-3, 8)$ and $(2, 3)$.

Using $({}^-3, 8)$:

In (1), LHS = 8

RHS = $({}^-3)^2 - 1 = 9 - 1 = 8$

In (2), LHS = 8

RHS = ${}^-({}^-3) + 5 = 8$

Substitute to check that the values for x and y satisfy both equations.

Continues on next page ...

Using (2, 3):

In (1) LHS = 3

RHS = $(2)^2 - 1 = 3$

In (2) LHS = 3

RHS = $^-(2) + 5 = 3$

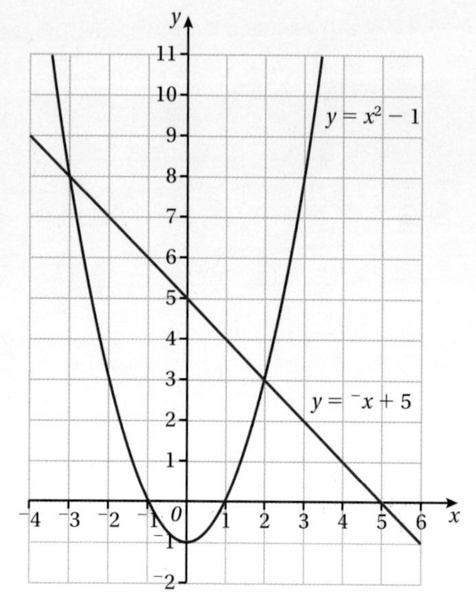

If you draw the graphs of these equations on the same set of axes, you can see that this solution works.

b $y = x^2 - 3x$ (1)

$y = x - 4$ (2)

$x - 4 = x^2 - 3x$

Substitute (2) in (1).

$x^2 - 4x + 4 = 0$

Write the equation in the form $ax^2 + bx + c = 0$.

$(x - 2)^2 = 0$

$x = 2$

Substitute x in (2).

$y = 2 - 4$

$y = ^-2$

The point of intersection is $(2, ^-2)$.

In this example there is only one solution, so the graphs will intersect at one point only.

You can always verify this graphically.

EXERCISE 8J

1 Solve each pair of simultaneous equations by substitution:

a $y = x^2$
$y = 2x - 1$

b $y = x^2$
$y = x + 2$

c $y = x^2$
$y = x - 1$

d Given that the graph of $y = x^2$ is a U-shaped parabola that goes through the origin, explain why there was no possible solution in part **c** above.

2 Find the coordinates of the point(s) of intersection of the following graphs:

a $y = x^2 + 3x + 3$ and $y = x + 2$ **b** $y = x^2 + 5x + 2$ and $y = x + 7$

c $y = x^2 + 2x + 4$ and $y = x + 6$ **d** $y = 2x^2 + 3x + 1$ and $y = 2x + 1$

e $y = 3x^2 + x + 2$ and $y = 3x + 3$ **f** $y = 6x^2 + 9x + 5$ and $y = 2x + 3$

3 The diagram shows the circular graph plotted from the equation $x^2 + y^2 = 17$ and the graph $x + y = 5$ which cuts the circle in two places.

Find the coordinates of the points of intersection of the graphs.

Tip

You will learn more about the equation of circles in Chapter 19. Here, apply what you know about the points where two graphs intersect.

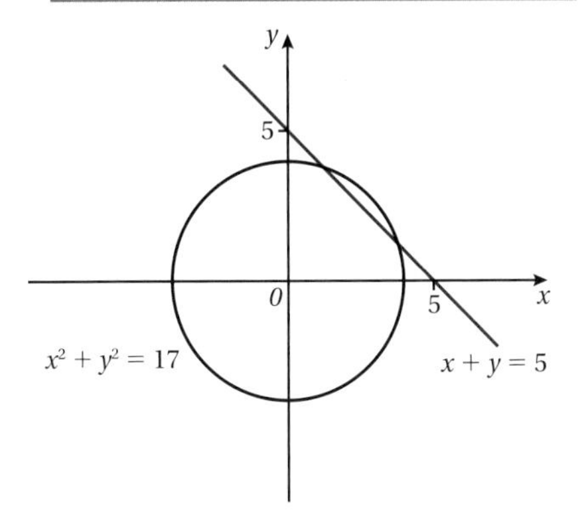

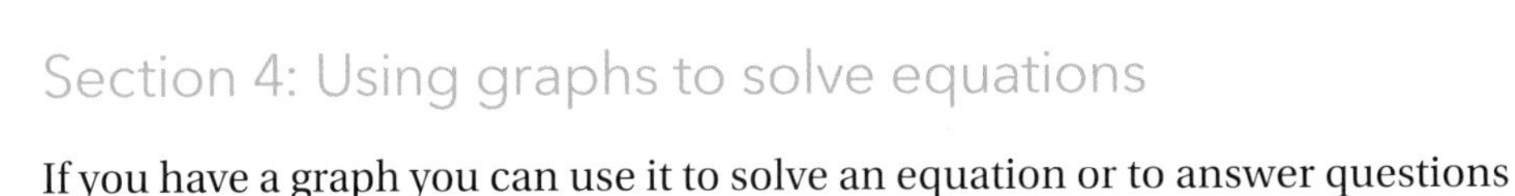

Section 4: Using graphs to solve equations

If you have a graph you can use it to solve an equation or to answer questions based on the equations.

Tip

You will work with graphs and equations again in Chapters 18 and 19.

WORKED EXAMPLE 12

This is the graph of the equation $4x + y = 2$.

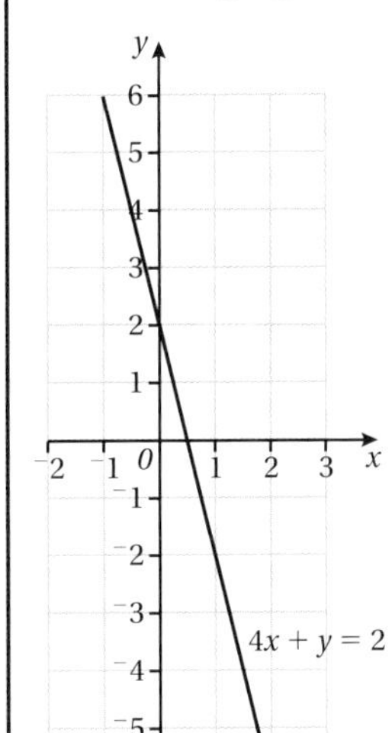

a Use the graph to estimate the value of y when:

i $x = 0$

ii $x = 1$

iii $x = 2$

b What is the value of x when $y = {}^-4$?

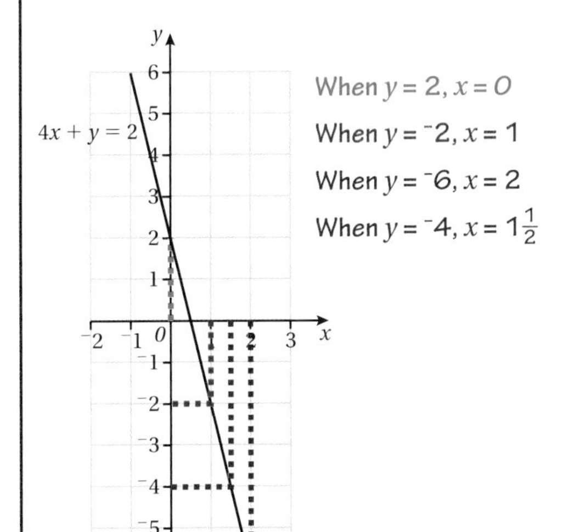

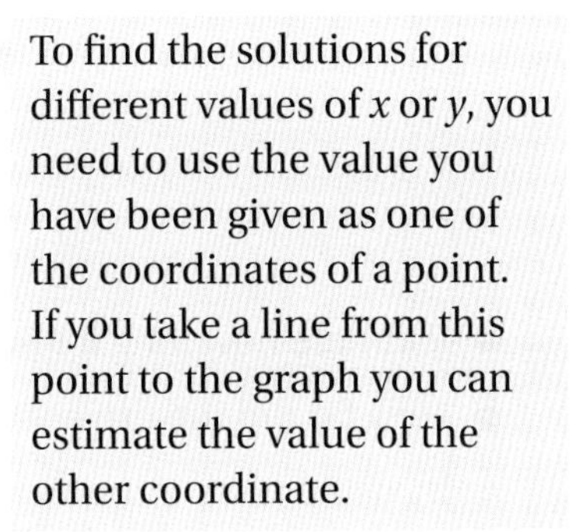

Each point on the graph represents a value of x and y that work in this equation.
To find the solutions for different values of x or y, you need to use the value you have been given as one of the coordinates of a point. If you take a line from this point to the graph you can estimate the value of the other coordinate.

Key vocabulary

parabola: the symmetrical curve produced by the graph of a quadratic function.

If there are two equations, you need two graphs. In Section 3 you used a pair of graphs to find the solution to a pair of simultaneous equations. If the equation is a quadratic, the graph will be a curve called a **parabola**. You can use the graph to find the solution of the equation; it is where $y = 0$, i.e. where the curve cuts the x-axis.

EXERCISE 8K

1 This graph represents the distance travelled by a cyclist over time.

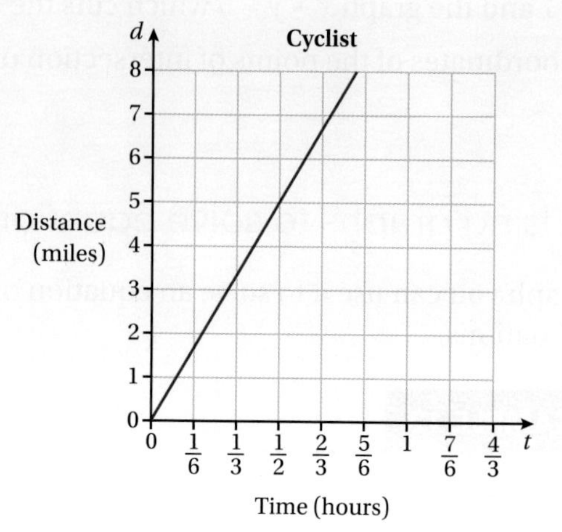

a Use the graph to estimate how far the cyclist has travelled after 30 minutes.

b How long did it take the cyclist to cover a distance of 8 miles?

c The equation $s = \frac{d}{t}$ can be used to work out the speed (s) of the cyclist.

Use values for d (distance) and t (time) from the graph to estimate the speed at which this cyclist was travelling.

2 This graph shows the distance covered over time by a motorist in a car.

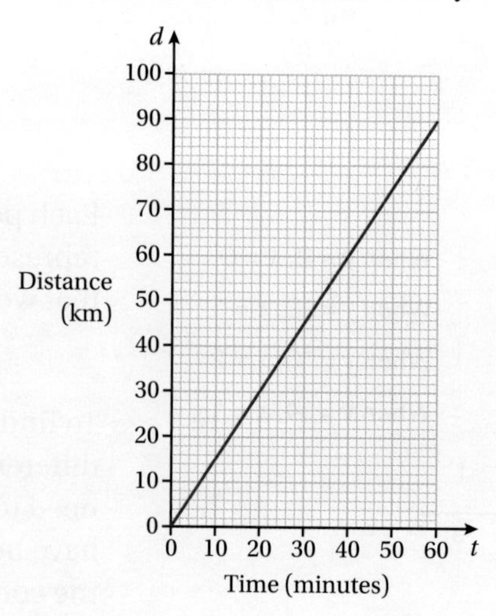

a Estimate how long it took the driver to travel 70 km.

b How far did the driver travel in the first 20 minutes?

c By using two points on the graph, estimate the speed at which the driver was travelling.

3 Use this graph of the equation $y = 3x - 2$ to estimate the value of y for the following values.

a $x = 0$ **b** $x = 1$ **c** $x = 2$

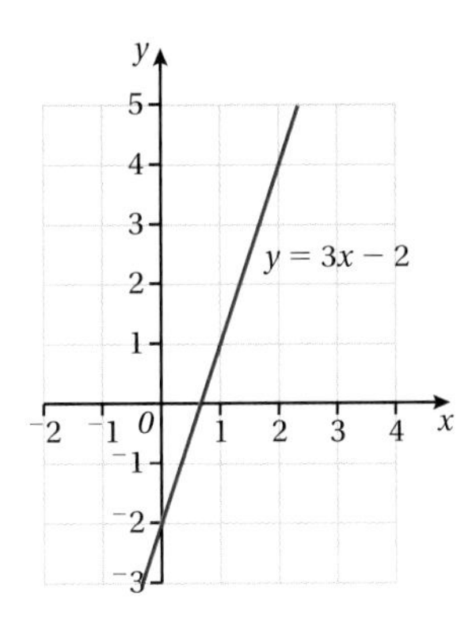

4 This graph shows how water drains from a tank at a constant rate.

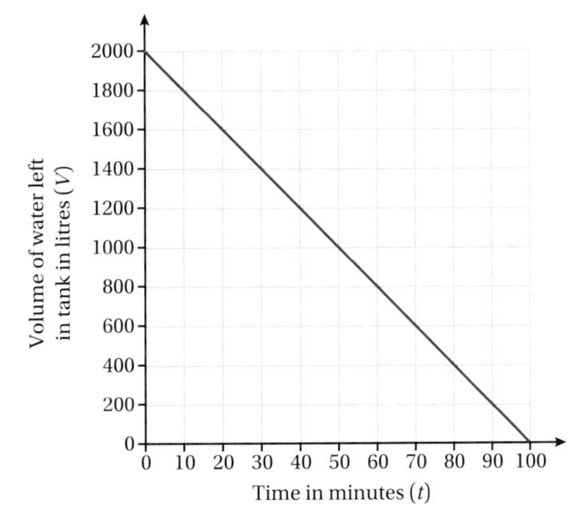

a How much water was in the tank to start with?

b How long did it take for the tank to empty?

c Zena says the equation for this graph is $y = 2000 - 20x$ and Leane says it is $y + 20x = 2000$.

Show that they are both correct, using different points from the graph.

5 This graph shows the cost of producing goods and how much money is earned from sales (revenue).

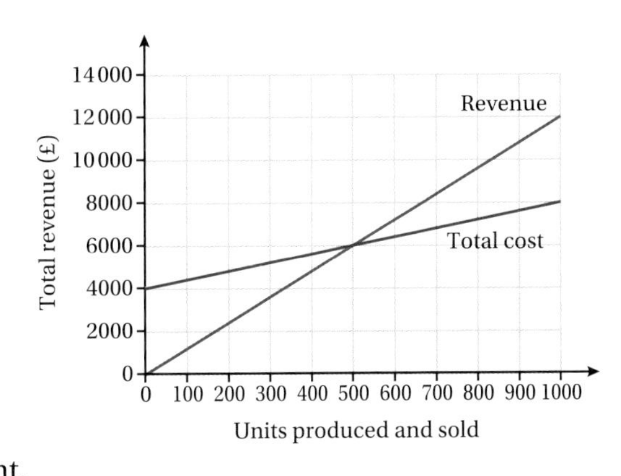

a Use the graph to estimate the point at which the costs and revenue are equal.

b The point at which costs and revenue are equal is called the break even point.

How does knowing this help a business owner?

6 This diagram show the graphs of two linear equations $y = 2x$ and $y = {}^{-}2x + 8$.

Use the graph to find the solution to the two equations.

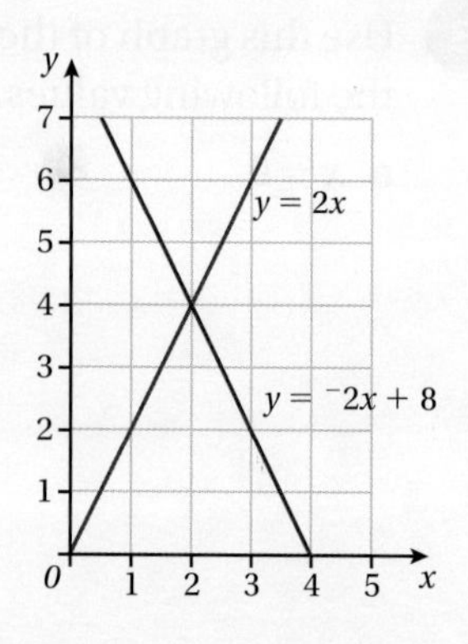

7 This graph of a quadratic equation models a stunt rider's path in the air as he does a jump.

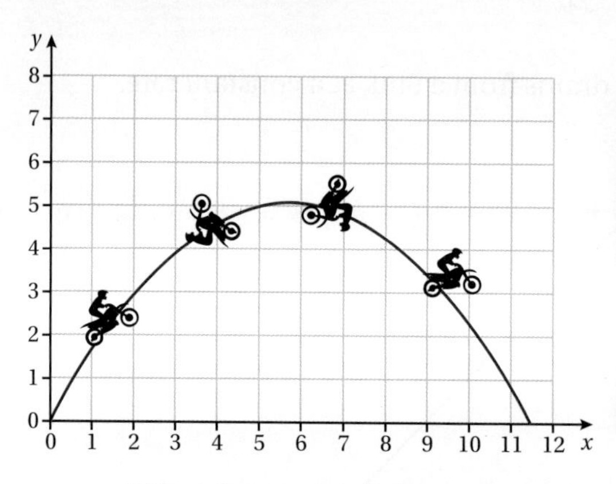

a What do you think the axes represent in this case?

b What values on the horizontal axis represent the rider taking off and landing again?

c Why are these two values useful in terms of the equation?

d Use the graph to estimate the coordinates of the maximum height reached during this jump.

8 This is the graph of a quadratic equation but the equation is not given.

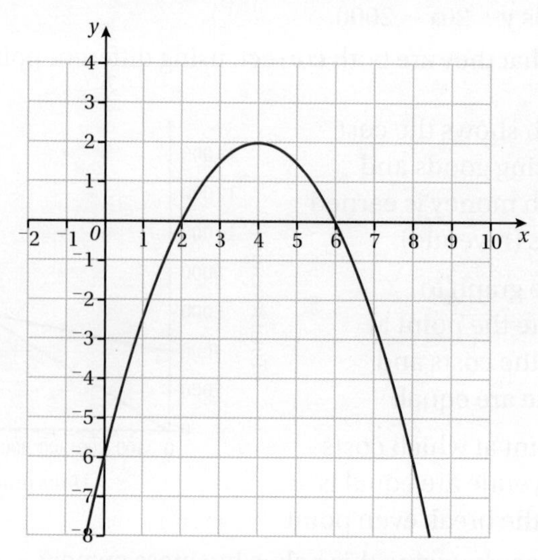

a Explain how you can use the graph to find the roots of the equation even though you don't know what the equation is.

b What are the solutions of the quadratic equation this graph represents?

c What is the quadratic equation of the graph?

d Where would the line $y = {}^{-}2x + 3$ cut the graph?
Find the solution algebraically and give your answers correct to 2 decimal places.

9 Two linear graphs are shown here.

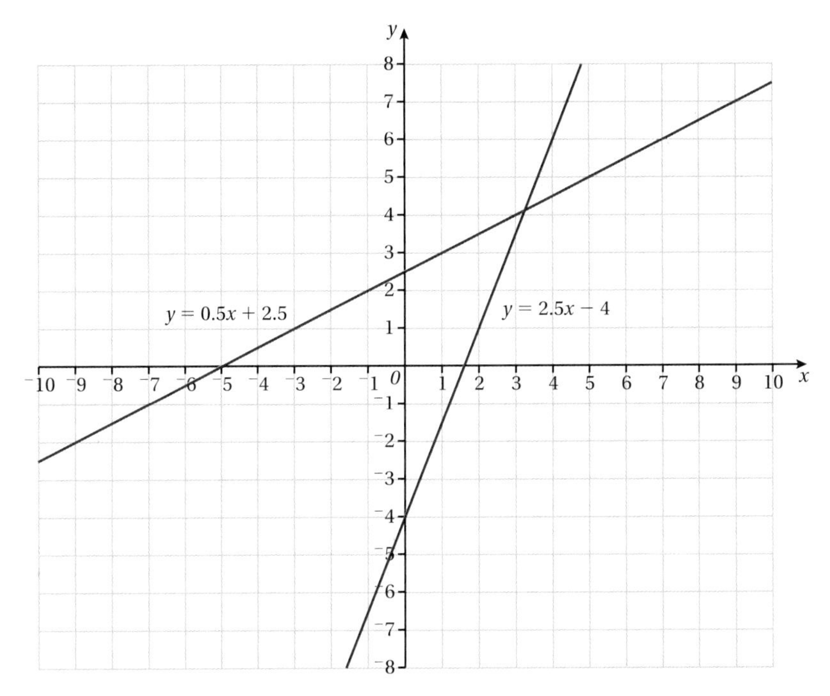

a Use the graph to estimate the values of x and y that are true for both equations.

b Find the simultaneous solution algebraically.

c What are the limitations of solving a pair of linear simultaneous equations from drawing a graph and finding the point of intersection?

10 The graph shows the bounce height of a ball dropped from different heights.

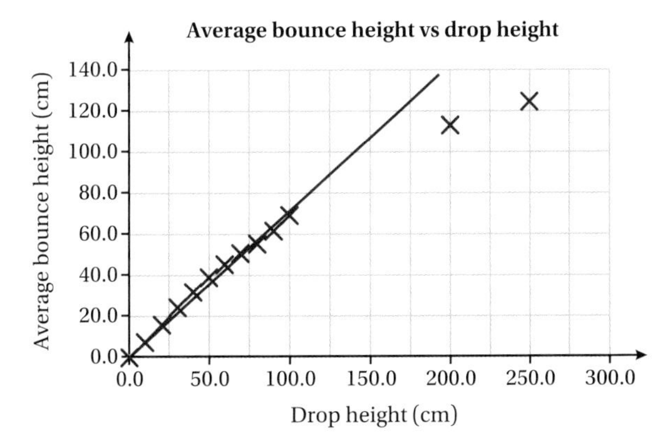

a Using the line of best fit, estimate the average bounce height for a drop of 150 cm.

b Give an alternative estimate given the results for 200 cm and 250 cm.

c Estimate the average bounce height from a drop of 300 cm.

Find answers at: cambridge.org/ukschools/gcsemaths-studentbookanswers

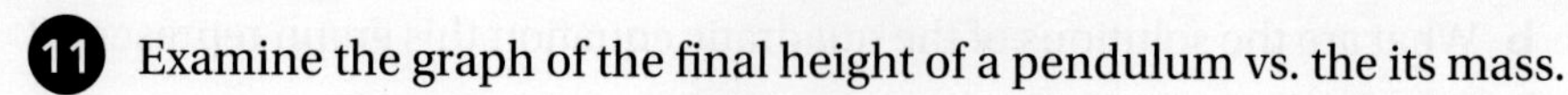

11 Examine the graph of the final height of a pendulum vs. the its mass.

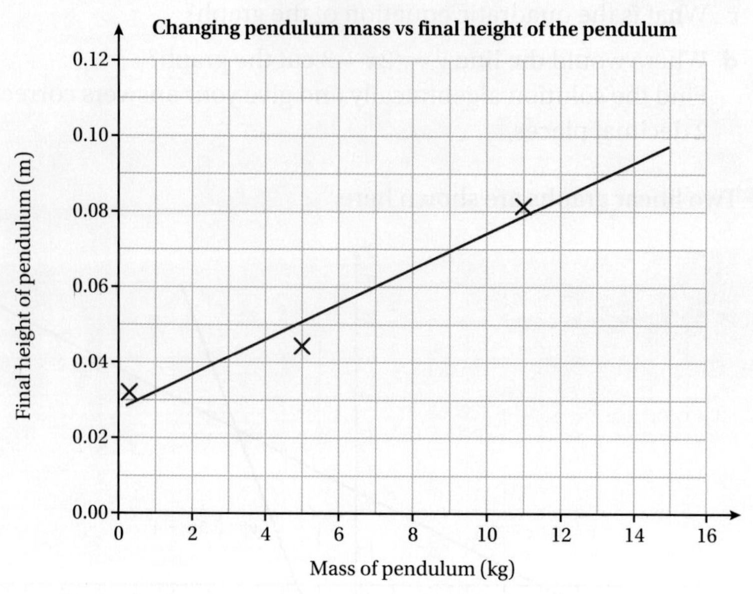

a If the height of the swing of the pendulum is 0.07 m, use the line of best fit to estimate the mass of the pendulum.

b Estimate the height of the swing for a pendulum of mass approximately 6.5 kg.

c Work out an estimate for the range of the swing achieved for a pendulum of mass 16 kg.

Section 5: Finding approximate solutions by iteration

The process of iteration is a numerical method of solving equations.

Iteration provides a method for finding successively better approximations to the roots of the equation for real values. It is useful for finding solutions to equations that cannot be solved in any other way.

Tip

You will learn about cubic equations and their graphs in Chapter 19.

It can be particularly useful for equations that contain powers of x greater than two, such as cubic graphs where the highest power of x is x^3.

The three methods discussed in this section are:

- the iteration formula
- the decimal method
- the bisection method

and they should only be used when factorising methods cannot be applied or to work with higher-order equations.

In each case, you can use your calculator carefully or set up a table of values in a spreadsheet to help you calculate values and take the next steps to get closer to approximate solutions.

Using an iterative formula

Iterate means 'repeat' or 'perform again'.

Mathematically, iteration means repeating a series of steps using the 'answer' from the previous step to get an improved 'answer' on the next step.

The starting value of x, x_1, is used to calculate x_2, which in turn is used to calculate x_3, and so on.

Did you know?

An iterative formula is created by rearranging the original equation you want to solve, to make x the subject. Subscript notation is used to indicate that you use the previous approximation of x to generate the next approximation of x.

WORKED EXAMPLE 13

Find the positive root of the quadratic equation $x^2 - 7x - 3 = 0$ to 3 decimal places given the iterative formula $x_{n+1} = 7 + \frac{3}{x_n}$ where $x_1 = 7$.

By inspection:
$x = 7$: $49 - 49 - 3 = {}^-3$
$x = 8$: $64 - 56 - 3 = 5$
0 lies between $^-3$ and 5, so x lies between 7 and 8.

First, use inspection to estimate the range in which the solution of the quadratic equation will lie. This is where the graph crosses the x axis where $y = 0$.

$x_1 = 7$
$x_2 = 7 + \frac{3}{7}$

Substitute $x_1 = 7$ into the iterative formula provided.

$x_3 = 7 + \frac{3}{x_2}$

Store the answer on your calculator (or spreadsheet) and substitute it (x_2) back into the iterative formula to get x_3.

The values for each step:
$x_1 = 7$ (first estimate)
$x_2 = 7.428571429...$
$x_3 = 7.403846154...$

Repeat the process until the values can be estimated with confidence to the required number of decimal places.

$x_4 = \mathbf{7.405}194805...$
$x_5 = \mathbf{7.405}12101...$
$x_6 = \mathbf{7.405}125047...$
$x_7 = \mathbf{7.405}124826...$

You can see the value of x converging, and after 5 calculated values you can see what the value of x is to 3 decimal places.

$x = 7.405$ (3 decimal places)

$(7.405)^2 - 7 \times 7.405 - 3 \approx 0$

Check solution by substitution.

Tip

In mathematics, **convergence** is when the terms within a sequence get increasingly close to a value L, which is called the limit of the sequence.

EXERCISE 8L

1. Find an approximate solution to $x^2 - 5x + 2 = 0$ using the iteration $x_{n+1} = 5 - \frac{2}{x_n}$ with $x_1 = 4$, correct to 3 decimal places.

2. Find an approximate solution to the square root of 18 using the iteration $x_{n+1} = \frac{1}{2}\left(x_n + \frac{18}{x_n}\right)$ with $x_1 = 4$, correct to 4 significant figures.

3. Find an answer correct to 2 decimal places for x using the iteration $x_{n+1} = \frac{4}{x_n} + 1$ with $x_1 = 2$.

Find answers at: cambridge.org/ukschools/gcsemaths-studentbookanswers

4 $x^3 - 3x + 1 = 0$

Use the iteration $x_{n+1} = \frac{-1}{x_n - 3}$ with $x_1 = 0.5$ to approximate a solution for x correct to 4 significant figures.

5 Use the iteration formula $x_{n+1} = \frac{2x_n}{3} + \frac{4}{x_n^2}$, starting with $x_1 = 2$, to find one of the roots of the cubic equation $x^3 - 12 = 0$.

Give your answer correct to 2 decimal places.

Looking for the change of sign

In terms of the graph of an equation, the change of sign indicates that the y-value has gone from positive to negative, or negative to positive.

This tells you that the graph has crossed the x-axis, i.e. this is a root of the equation.

Decimal search

You can use your calculator or a spreadsheet to help you calculate the values.

WORKED EXAMPLE 14

Here is the graph for the cubic equation $y = x^3 - 3x + 1$.

From the graph you can see there are three roots, and that the equation does not have integral solutions.

Find the third root of the function $y = x^3 - 3x + 1$ to 2 decimal places, using the decimal method.

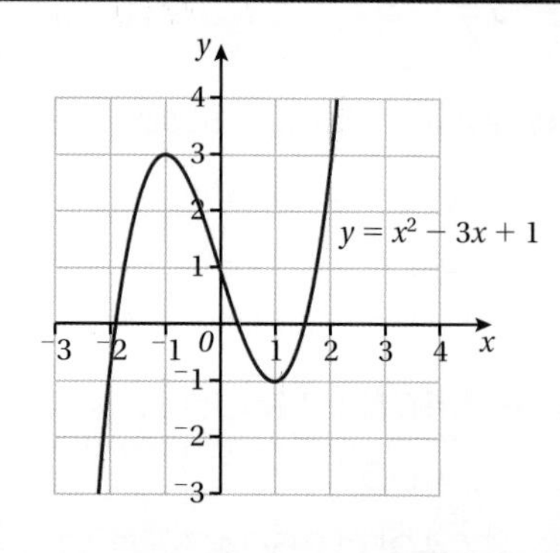

$y = x^3 - 3x + 1$

From the graph, the third root lies between (1, 0) and (2, 0). Take increments of 0.1 from $x = 1$ until you get a change of sign. Record your results in a table, or an ordered list.

x	f(x)
1	-1
1.1	-0.969
1.2	-0.872
1.3	-0.703
1.4	-0.456
1.5	-0.125
1.6	0.96

y has gone from a negative value to a positive value. This indicates that the graph has crossed the x-axis. So, the root lies between $x = 1.5$ and $x = 1.6$.

Calculate a second table of values with increments of 0.01, starting from $x = 1.5$.

x	f(x)
1.5	-0125
1.51	-0.087
1.52	-0.0482
1.53	-0.0084
1.54	0.03226

y has gone from a negative value to a positive value. So, the root lies between $x = 1.53$ and $x = 1.54$.

Continues on next page ...

x	$f(x)$
1.53	$^-0.0084$
1.531	$^-0.0044$
1.532	$^-0.0004$
1.533	0.00369

Calculate a third table of values with increments of 0.001.

The solution is between 1.532 and 1.533 (the sign changes so the curve crosses the x-axis at the value at which $y = 0$).

$x = 1.53$ (2 decimal places)

If more decimal places are required then more tables of values could be calculated.

Bisection method

In this method we look at the values calculated at **half**-interval sections.

From the graph of $y = x^3 - 3x + 1$, you can see there is a solution between $(^-1, 0)$ and $(^-2, 0)$.

WORKED EXAMPLE 15

Find the root between $x = {}^-1$ and $x = {}^-2$ to 1 decimal place for the cubic function $y = x^3 - 3x + 1$.

For $y = x^3 - 3x + 1$, when

$x = {}^-1, y = 3$

$x = {}^-2, y = {}^-1$

x	$f(x)$
$^-1.5$	2.125
$^-1.75$	0.8063

Take the half interval between $x = {}^-1$ and $x = {}^-2$ as the value of x: $^-1.5$. Substitute this value of x into the cubic function.

Substitute in the value of x that is the half interval between $x = {}^-1.5$ and $x = {}^-2$: $^-1.75$.

$^-1.875$	0.332

Substitute in the value of x that is the half interval between $x = {}^-1.75$ and $x = {}^-2$: $^-1.875$.

$^-1.9375$	$^-0.4607$

Substitute in the value of x that is the half interval between $x = {}^-1.875$ and $x = {}^-2$.

$x = {}^-1.9$ (1 decimal place)

y has gone from a negative value to a positive value.
This indicates that the graph has crossed the x-axis.
So, the root lies between $x = {}^-1.875$ and $x = {}^-1.9375$.
The method could continue to produce better accuracy.

The advantage of these methods is that the solution bounds indicate the interval in which the root lies.

The disadvantage is that an initial search may miss one or more of the roots; for example, when the x-axis is a tangent to the curve or when several roots are very close together.

Tip

You will learn about tangents to the curve in Chapter 19.

Find answers at: cambridge.org/ukschools/gcsemaths-studentbookanswers

EXERCISE 8M

1 Find an approximate solution to $x^3 + 2x - 1 = 0$ using the iteration $x_{n+1} = \frac{1}{x_n^2 + 2}$ with $x_1 = 0.5$.

Give the final answer correct to 2 decimal places and check by substitution in the original equation.

2 $y = x^3 - 6x - 4$

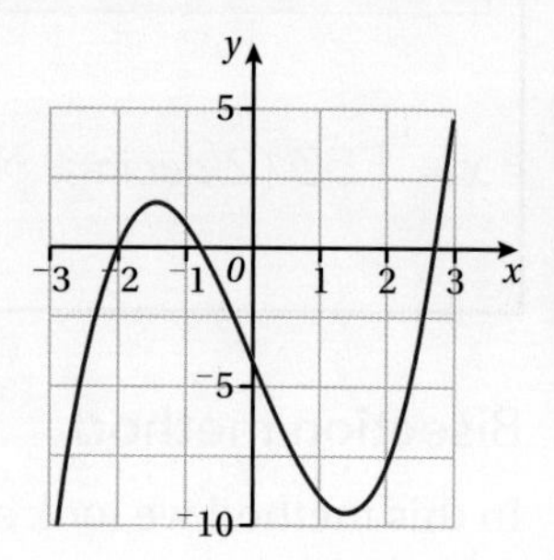

a Use the decimal search method to find the root of the cubic function $f(x) = x^3 - 6x - 4$ which lies between (2, 0) and (3, 0) correct to 2 decimal places.

b Use the interval bisection method to find the root of the cubic function $f(x) = x^3 - 6x - 4$ which lies between (0, 0) and ($^-1$, 0) correct to 1 decimal place.

Section 5: Using equations and graphs to solve problems

Work through this mixed exercise to practise and apply your skills.

EXERCISE 8N

1 Read each statement and decide whether it is true or false.
If it is false, explain why.

a $\frac{1}{2}(x - 3) = \frac{1}{3}(2x + 1)$
$x = {}^-11$ is the solution to this equation.

b $10x^2 - 25x + 10 = 0$
$(10x - 5)(x - 2) = 0 \rightarrow x = \frac{1}{2}$ and $x = {}^-2$

c When completing the square, $30\frac{1}{4}$ must be added to $x^2 - 11x$ to make a perfect square.

d $x = 3 - 4y$ (1)
$7y - 3x = 21$ (2)
Substituting equation (1) in equation (2) $\rightarrow 19y - 9 = 29 \rightarrow y = 2, x = {}^-5$.

e The diagram shows a pair of simultaneous equations, one linear and one quadratic.
The simultaneous solutions for x are both positive.

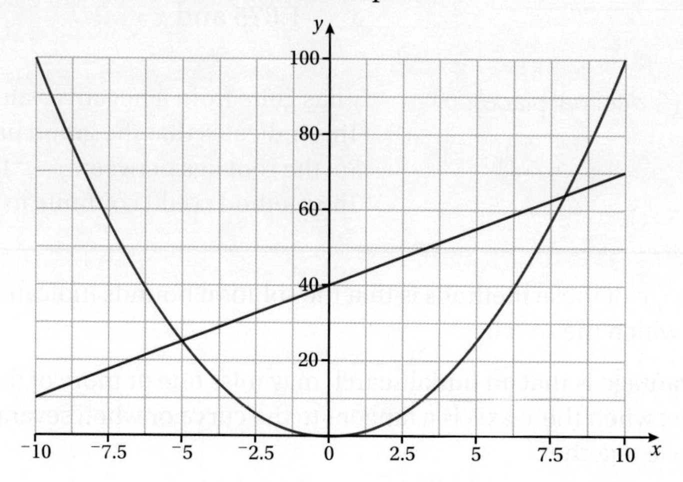

2 Solve.

a $4(x-3)=3x+4$

b $3(2x-5)=2(4x+\frac{3}{2})$

3 A gardener has 60 m of garden edging, which she uses to set out a rectangular garden with its width 5 m less than its length.

If the length of the garden is x metres:

a Find the width of the garden in terms of x.

b Hence, form an equation and solve it to find the length and width of the garden.

4 A father is 28 years older than his daughter.

In six years time, he will be three times her age.

Find their present ages.

5 Find a number such that if 5, 15 and 35 are added separately to it, the product of the first and third results is equal to the square of the second.

6 This diagram consists of two rectangles with dimensions in centimetres.

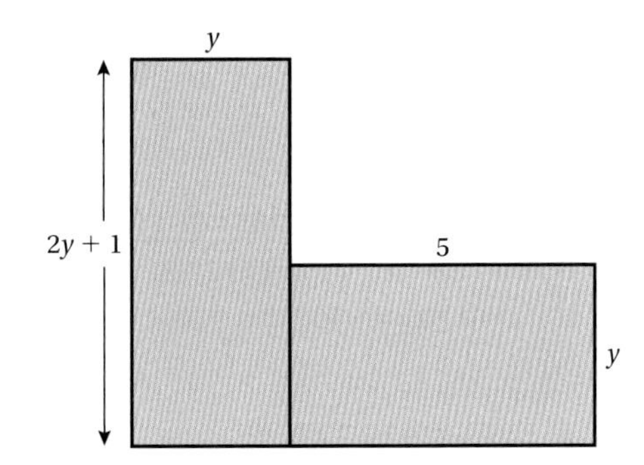

The total area is 95 cm^2.

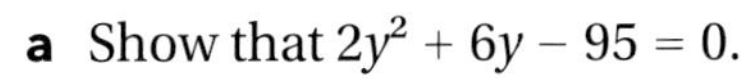

a Show that $2y^2+6y-95=0$.

b Solve the equation $2y^2+6y-95=0$ correct to 3 significant figures.

7 Solve the following equations by completing the square:

a $x^2-4x-2=0$

b $n^2=5n+4$

8 **a** Simplify the equation

$$\frac{5}{x+2}=\frac{4-3x}{(x-1)}$$

to give $3x^2+7x-13=0$.

b Solve the equation $3x^2+7x-13=0$ correct to 2 decimal places.

9 The base and height of a triangle are $x+3$ and $2x-5$.

If the area of the triangle is 20, find x.

10 For the quadratic equation $ax^2-4x+3=0$, find the values of a for which the equation has:

a one solution.

b two solutions.

c no solutions.

11 The sum of two numbers is 112 and their difference is 22.

Write a pair of simultaneous equations and solve to find the two numbers.

Find answers at: cambridge.org/ukschools/gcsemaths-studentbookanswers

Checklist of learning and understanding

Linear and quadratic equations

- A linear equation has one unknown and will have one unique value as a solution.
- A quadratic equation has a square as the highest power for the variable.
- Quadratic equations can be solved by factorising, completing the square or using the formula $x = \frac{-b \pm \sqrt{b^2 - 4ac}}{2a}$.
- Quadratic equations have a maximum of two roots. Sometimes there is only one because the values are the same, and sometimes there is only one because one value doesn't work in a particular context.

Simultaneous equations

- Simultaneous equations are a pair of equations with two unknowns that have solutions that satisfy both equations.
- Simultaneous equations can be solved graphically using the point of intersection, or algebraically by substitution or by elimination.

Iteration

- To solve higher-order equations using the process of iteration whereby numerical methods are used to find successively better approximations to the real values for the roots of an equation. The three methods discussed were:
  - iteration formulae
  - decimal search
  - bisection.

Graphs and problems

- Equations are useful for setting up problems mathematically.
- You can find or estimate the solutions of equations using graphs.
- You need two graphs to solve simultaneous equations. The solution is the point of intersection of the graphs.
- The solution of a quadratic equation is where the curve cuts the x-axis, i.e. where $y = 0$.

For additional questions on the topics in this chapter, visit GCSE Mathematics Online.

Chapter review

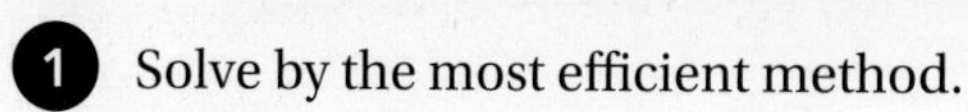

1 Solve by the most efficient method.

Leave your answers in square root form when necessary.

a $x^2 - 4x = 21$ **b** $x^2 + 10x = 2x - 12$

c $9 + x^2 + 11x = 2x - 11$ **d** $x^2 + 10x = 5$

e $2x^2 - 4x - 1 = 0$ **f** $3x^2 + 2x = 3$

2 An object is thrown upwards so that its height (h) in metres after a certain time (t) in seconds can be described using the formula $h = 20t - 4t^2$.

a How long does it take the object to first reach a height of 24 m?

b At what time does it come down to reach this height again?

3

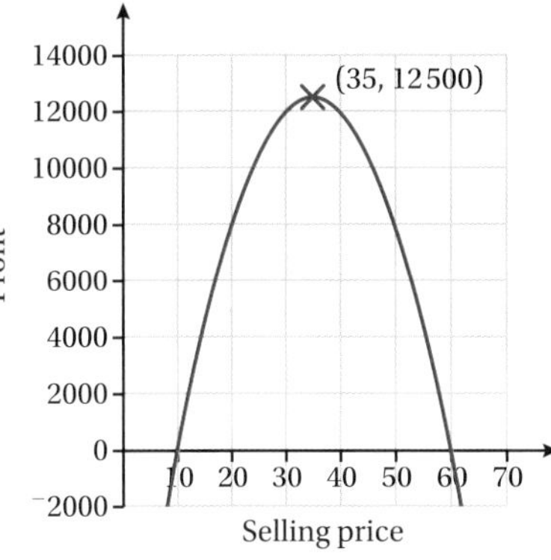

a What are the roots of the quadratic equation modelled by this parabola?

b Write the equation represented by the graph (think carefully about the shape of the parabola).

c What does the graph tell you about the selling price?

4 Are these two equations a pair of simultaneous equations?

$y = x + 2$, $5y = 5x + 10$

5 Solve the simultaneous equations $x^2 + y^2 = 25$ and $y = 3x + 1$.

6 Solve algebraically these simultaneous equations.

$y = 4x^2 - 9x - 1$

$y = 5 - 4x$ *(6 marks)*

© OCR 2013

7 Prove algebraically that the difference between the squares of any two consecutive integers is equal to the sum of these two integers.

8 The product of a number and four less than the number is equal to 16 more than twice the number.

Find the number.

9 **a** Show, by completing the square, that $y = 3x^2 + 6x - 7$ can be written in the form $y = 3(x + 1)^2 - 10$.

b Solve the equation $3(x + 1)^2 - 10 = 0$, giving your answer correct to 3 significant figures.

10 **a** Find an answer correct to 3 decimal places for x using the iteration $x_{n+1} = \frac{1}{x_n} + 2$ with a starting value of $x_1 = 2$.

b Verify that this value of x is the positive solution to the quadratic equation $x^2 - 2x - 1 = 0$.

Find answers at: cambridge.org/ukschools/gcsemaths-studentbookanswers

9 Angles

In this chapter you will learn how to ...

- apply basic angle facts to find unknown angles.
- use the angles associated with parallel lines to find unknown angles in a range of figures.
- prove that the sum of angles in a triangle is 180°.
- use known angle facts to derive the sum of exterior and interior angles of polygons.
- use angle facts and properties of shapes to justify and prove results.

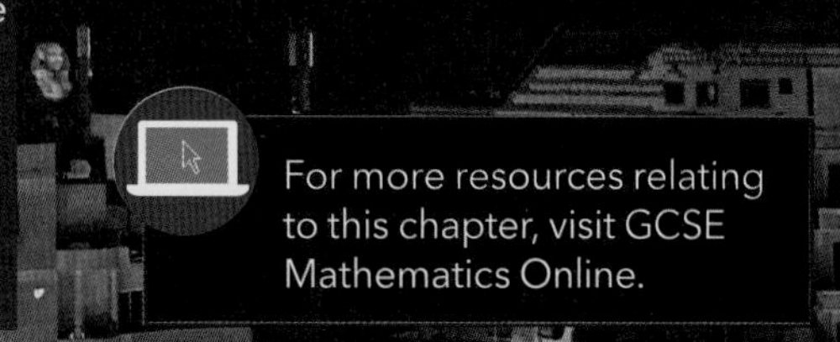

For more resources relating to this chapter, visit GCSE Mathematics Online.

Using mathematics: real-life applications

People who work in many varied and unrelated jobs rely on an understanding of angles and spatial relationships in their daily work. These include designers, architects, opticians and tree surgeons among others.

"I had to work quite carefully with the 360 degrees around the centre to place each of the 32 pods correctly on the London Eye." *(Structural engineer)*

Before you start ...

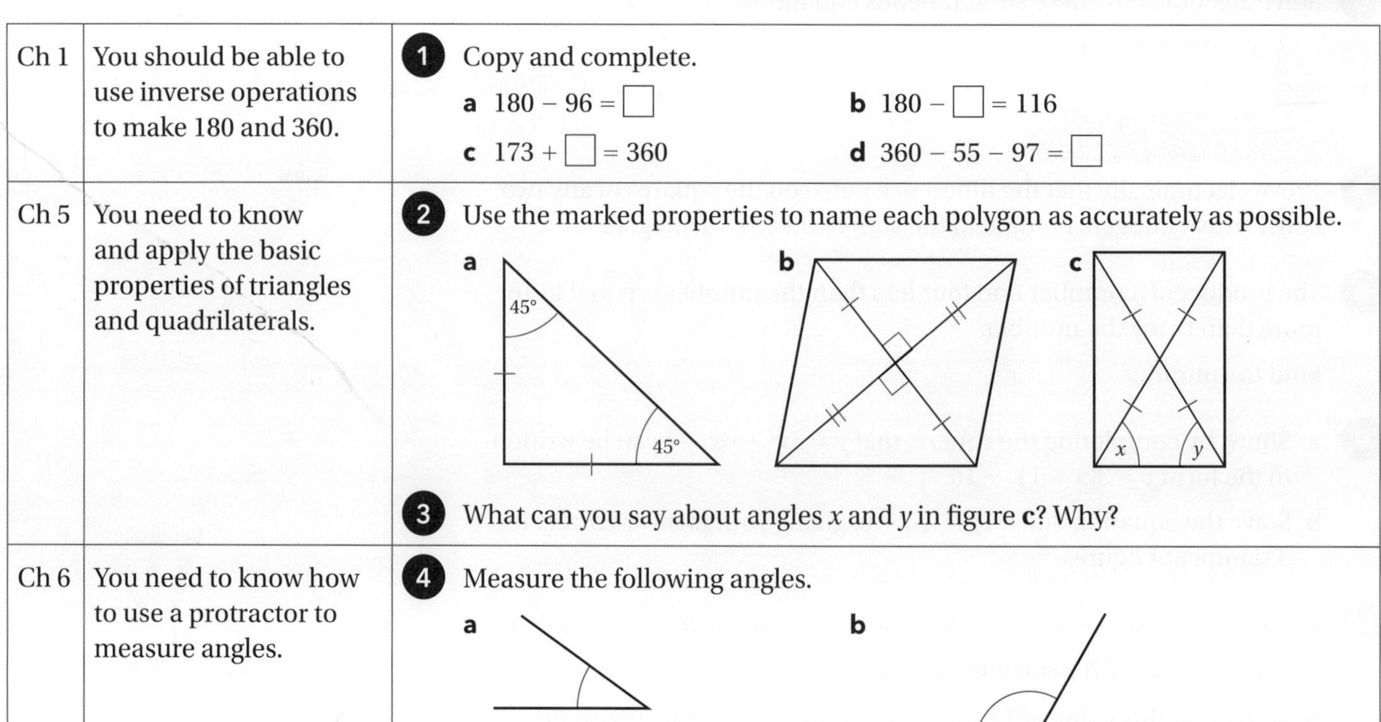

Ch 1	You should be able to use inverse operations to make 180 and 360.	**1** Copy and complete. **a** $180 - 96 = \square$ **b** $180 - \square = 116$ **c** $173 + \square = 360$ **d** $360 - 55 - 97 = \square$
Ch 5	You need to know and apply the basic properties of triangles and quadrilaterals.	**2** Use the marked properties to name each polygon as accurately as possible. **a** 45° 45° **b** **c** x y **3** What can you say about angles x and y in figure **c**? Why?
Ch 6	You need to know how to use a protractor to measure angles.	**4** Measure the following angles. **a** **b**

Assess your starting point using the Launchpad

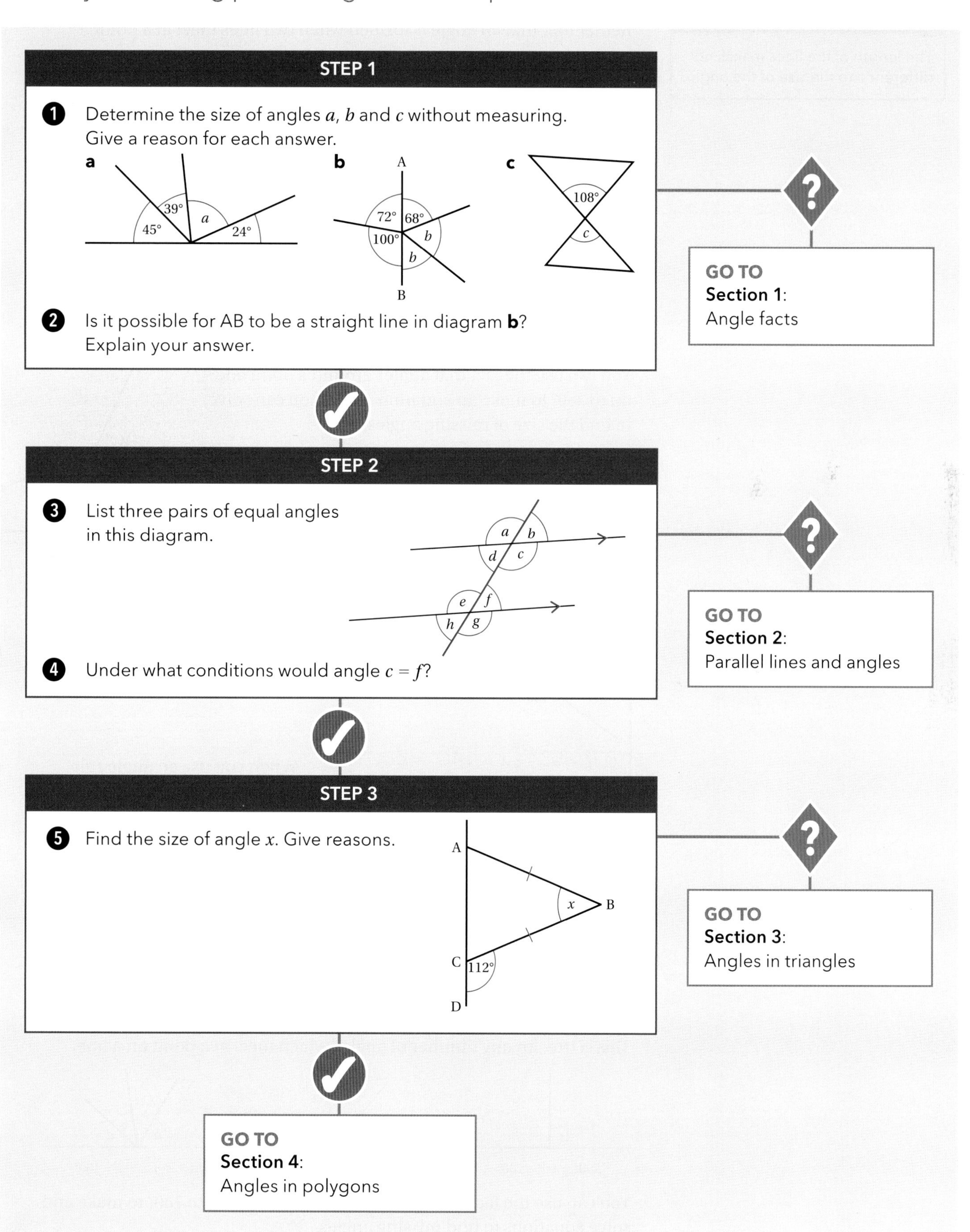

STEP 1
1 Determine the size of angles a, b and c without measuring.
Give a reason for each answer.
a
45°
39°
a
24°
b
A
72°
68°
100°
b
b
B
c
108°
c
2 Is it possible for AB to be a straight line in diagram **b**?
Explain your answer.
?
GO TO
Section 1:
Angle facts
STEP 2
3 List three pairs of equal angles
in this diagram.
a
b
d
c
e
f
h
g
4 Under what conditions would angle $c = f$?
?
GO TO
Section 2:
Parallel lines and angles
STEP 3
5 Find the size of angle x. Give reasons.
A
B
x
C
112°
D
?
GO TO
Section 3:
Angles in triangles
GO TO
Section 4:
Angles in polygons

Find answers at: cambridge.org/ukschools/gcsemaths-studentbookanswers

Tip

The length of the lines makes no difference to the size of the angle.

Section 1: Angle facts

Remember that an angle is formed when two lines meet at a point.

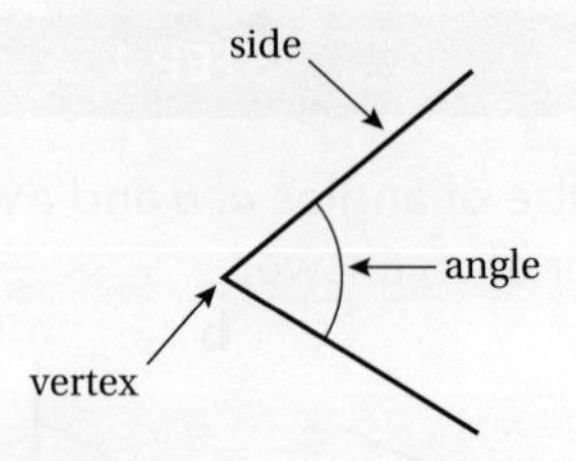

Angles around a point

The sum of angles around a point is 360°.

A 90° angle is one quarter turn, two 90° angles are half a turn and so on.

There are four quarter turns around a point: $4 \times 90° = 360°$.

You can use the fact that angles around a point add up to 360° to make an equation which you can solve to find the size of missing angles.

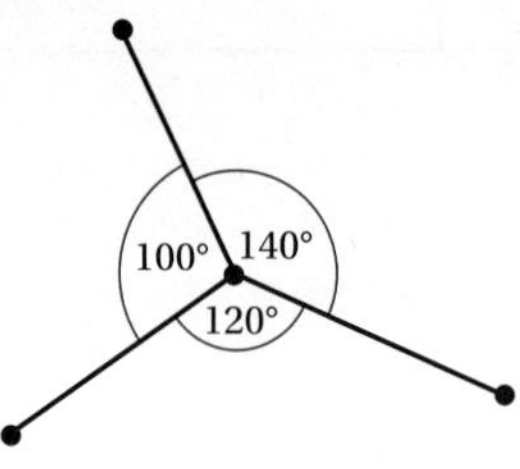

WORKED EXAMPLE 1

Find the size of x.

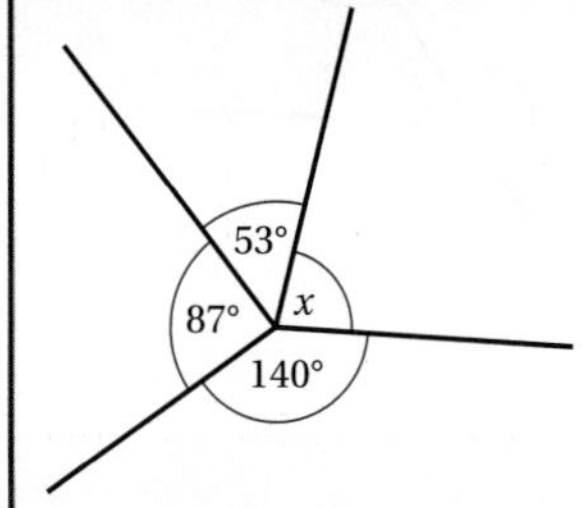

$87° + 53° + x + 140° = 360°$

$x = 360° - 140° - 53° - 87°$ (angles round a point)

$= 360° - 280°$

$= 80°$

When you use an angle rule to calculate a value, you **must** state the rule you have used alongside the equation in which you have used it.

Angles on a straight line

Angles on a straight line add to 180°.

This is true for any number of angles which meet at a point on a line.

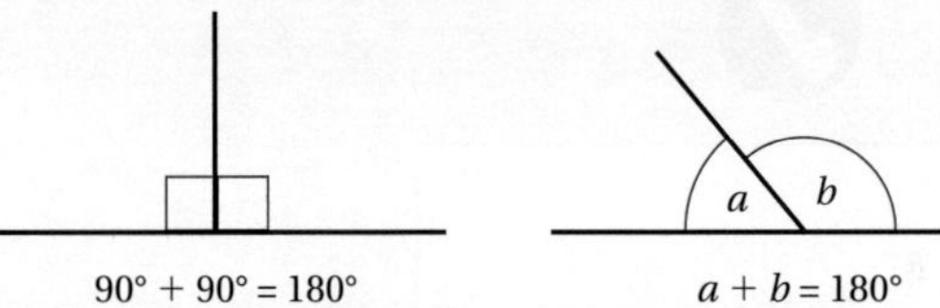

$90° + 90° = 180°$ $a + b = 180°$ $a + b + c + d = 180°$

You can use the fact that the angles on a line add up to 180° to make and solve equations to find missing angles.

WORKED EXAMPLE 2

Determine the size of x.

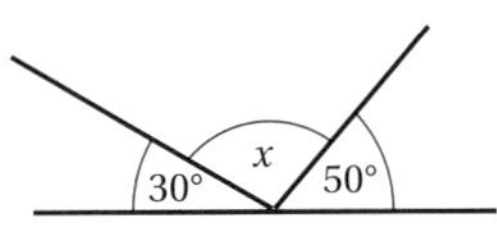

$x = 180° - 30° - 50°$ (angles on a line.)

$= 100°$

Vertically opposite angles

When two lines cross, or intersect, they form four angles.

The angles a and b are vertically opposite each other and the angles x and y are vertically opposite each other.

Vertically opposite angles are equal.

$a = b$

$x = y$

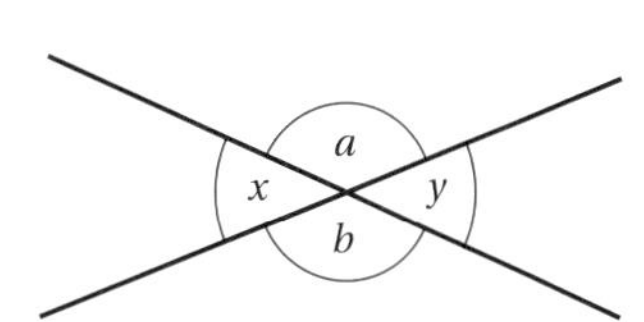

Key vocabulary

vertically opposite angles: angles that are opposite one another at an intersection of two lines. Vertical here means 'of the same vertex or point', not up and down.

You need to know these angle facts:

- Angles around a point add up to 360°.
- Angles on a straight line add up to 180°.
- Vertically opposite angles are equal.

EXERCISE 9A

1 Calculate the size of the missing angles.

a

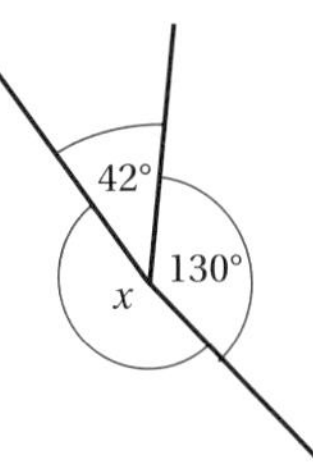

b

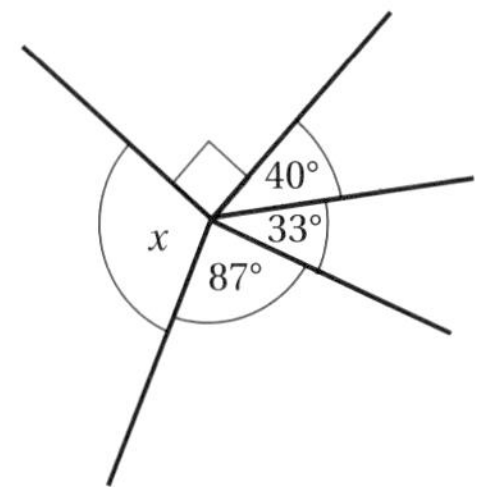

c

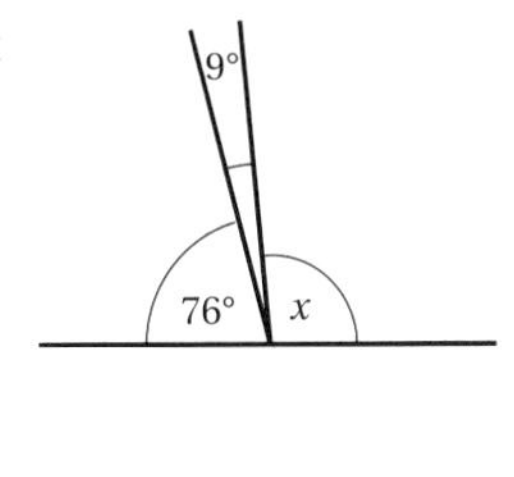

d What type of angle is x in each case?

Tip

Angles can be acute (less than 90°), obtuse (between 90° and 180°) or reflex (between 180° and 360°). A right angle is exactly 90°. An angle of 180° is a straight line.

2 Find the marked angles in each diagram. Give reasons for any deductions you make.

a Find x and y. **b** Find x. **c** Find p.

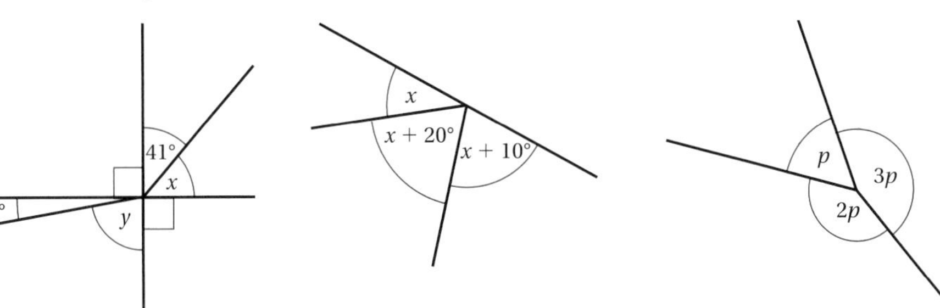

3 Explain why AE cannot be a straight line.

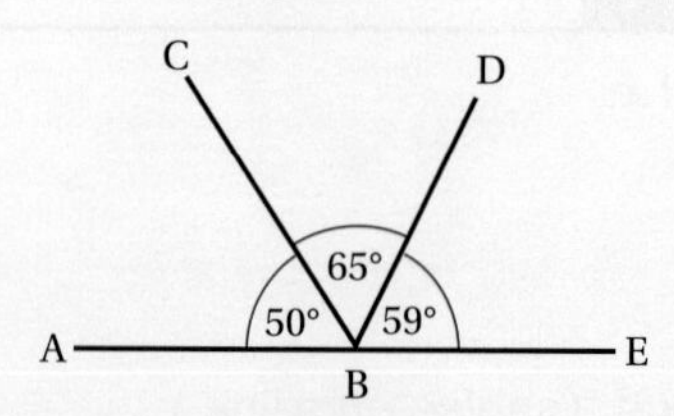

4 Calculate the size of the marked angles in each diagram.

The lines are straight lines but the diagrams are not to scale.

Show your working and give reasons for any deductions you make.

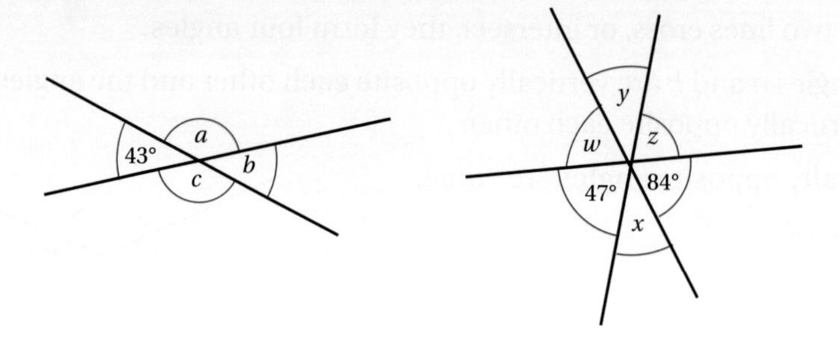

5 Given that $x = 50°$ find the size of angle z.

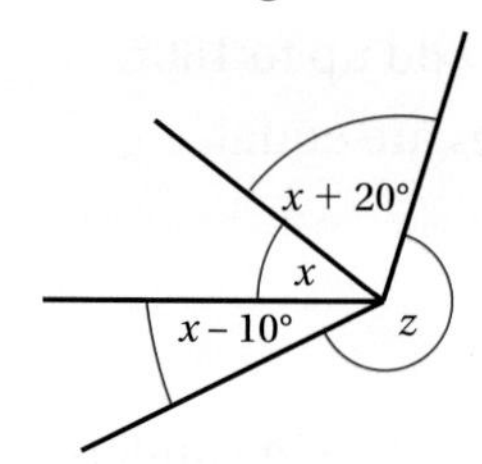

6 Use the diagram to calculate the size of each angle given the following conditions.

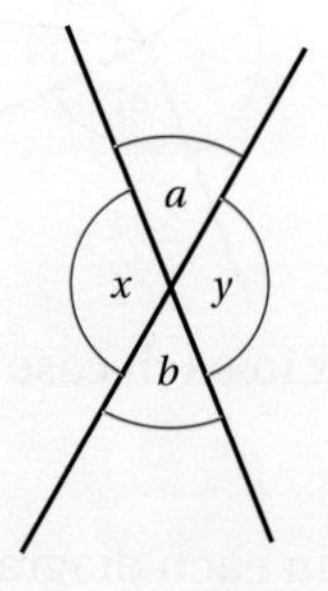

a $x = 68°$ **b** $x = a$ **c** $x - b = 112°$ **d** $a = 2x$

7 Find the value of the variables in each figure. Give reasons for your answers.

a

b

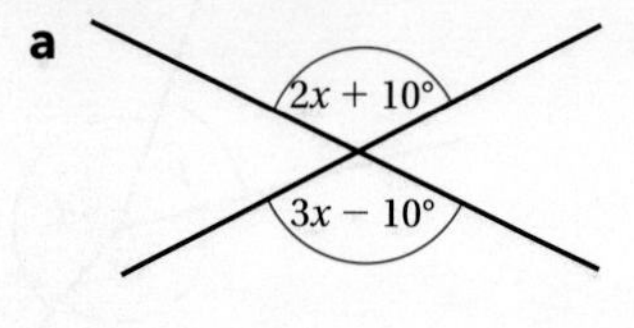

c

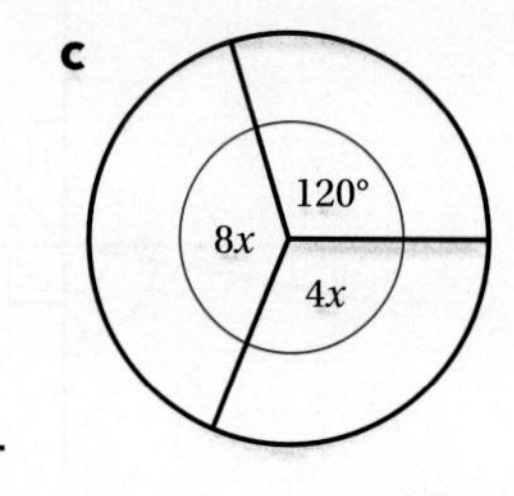

"When I'm designing and making clothes I need to be able to cut on the bias (at a given angle) and also bisect angles to add darts and fit sleeves."

(Fashion designer)

Section 2: Parallel lines and angles

Key vocabulary

transversal: a straight line that crosses a pair of parallel lines.

A line intersecting two or more parallel lines is called a **transversal**.

When a transversal intersects with parallel lines, it creates pairs of angles.

These angle pairs have particular properties.

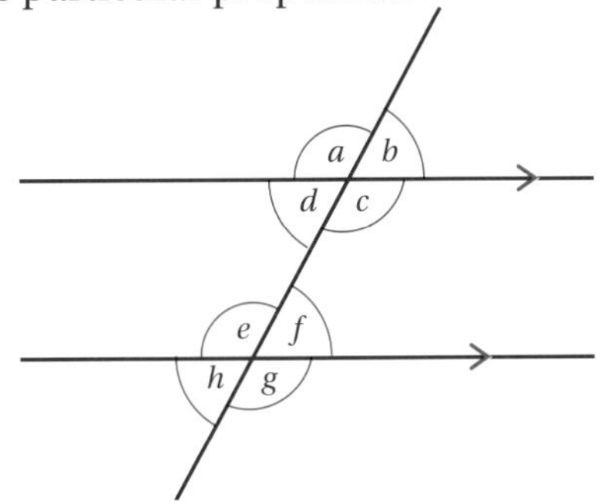

Vertically opposite angles are equal, so

$a = c$ $\quad b = d$ $\quad e = g$ $\quad f = h$

Angles on a straight line add to 180° so

$a + d = 180°$ $\quad b + c = 180°$ $\quad a + b = 180°$ $\quad c + d = 180°$

Corresponding angles

Key vocabulary

corresponding angles: angles that are created at the same point of the intersection when a transversal crosses a pair of parallel lines.

Corresponding angles formed between a transversal and each parallel line are equal.

When a transversal crosses two parallel lines, there are four pairs of corresponding angles.

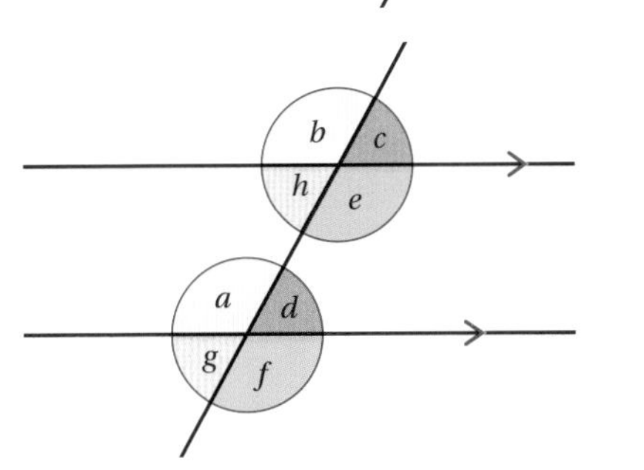

Find answers at: cambridge.org/ukschools/gcsemaths-studentbookanswers

Tip

The F-shape that you can see is what helps you identify the type of angle.

F-shape = corresponding angles

The F can be facing back-to-front as well.

Key vocabulary

alternate angles: the angles on parallel lines on opposite sides of a transversal.

The corresponding angle pairs are:

$a = b$ $c = d$ $e = f$ $g = h$

The pairs of corresponding angles form an F-shape because they are on the same side of the transversal and on the same side of the parallel lines (either both above, or both below).

Alternate angles

Alternate angles on opposite sides of the transversal on parallel lines are equal.

They can be seen as a Z-shape.

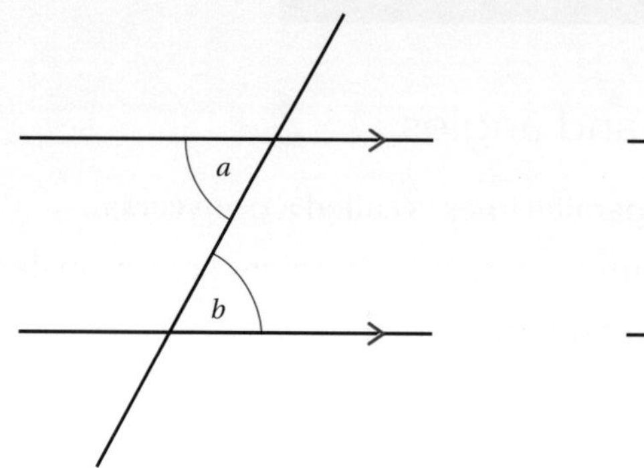

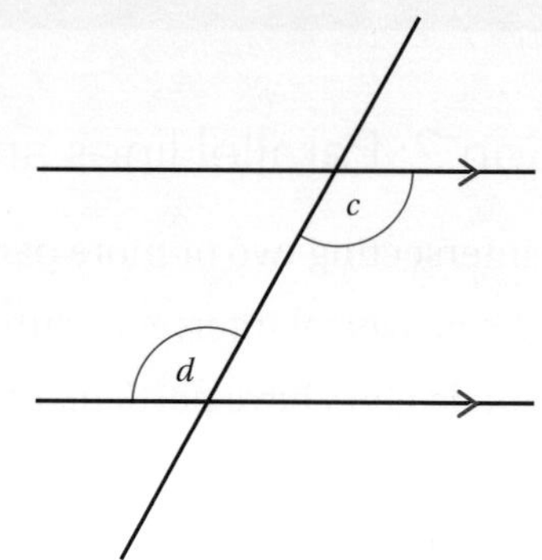

When a transversal crosses two parallel lines, four pairs of alternate angles are formed.

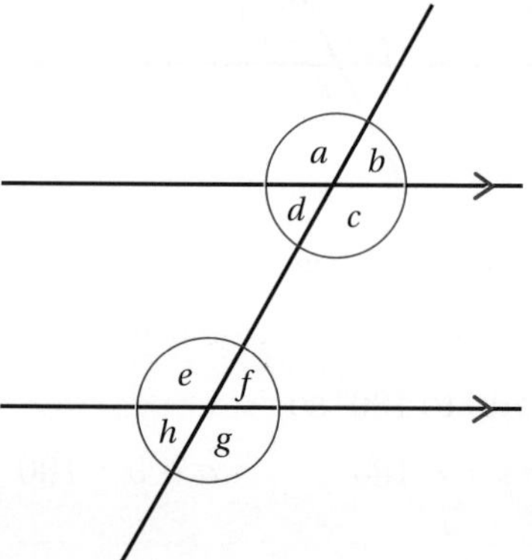

Tip

The Z-shape you can see helps you to identify the pairs of alternate angles.

Think **A** to **Z**: **A**lternate angles are **Z**-shaped.

The **Z** can be facing back-to-front as well.

The alternate angle pairs are:

$a = g$ $b = h$ $c = e$ $d = f$

Co-interior angles

Key vocabulary

co-interior angles: the angles within the parallel lines on the same side of the transversal.

supplementary angles: two angles are supplementary angles if they add up to 180°.

The angles inside the parallel lines and on the same side of the transversal are called **co-interior angles**.

Co-interior angles are **supplementary**. Read through this proof to see why.

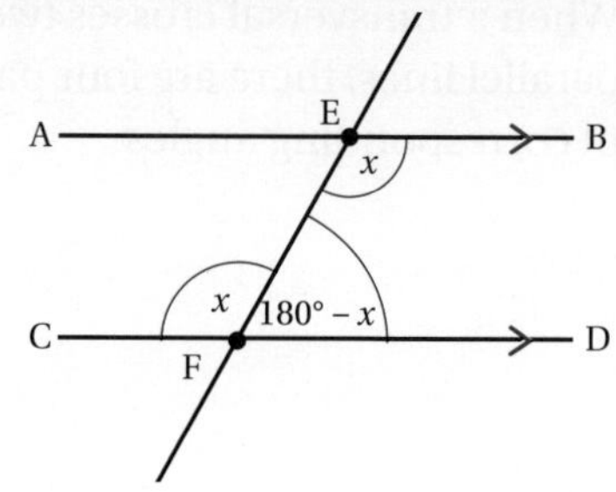

Prove that $\angle BEF + \angle EFD = 180°$

Let $\angle BEF = x$

$\therefore \angle EFC = x$ (Alternate angles are equal.)

$\angle EFC + \angle EFD = 180°$ (Angles on a line sum to 180°.)

$\therefore \angle EFD = 180° - x$

$\angle BEF + \angle EFD = x + 180° - x = 180°$

So, $\angle BEF + \angle EFD = 180°$

Summary of angle facts

This table summarises the facts you need to know about the angles associated with parallel lines.

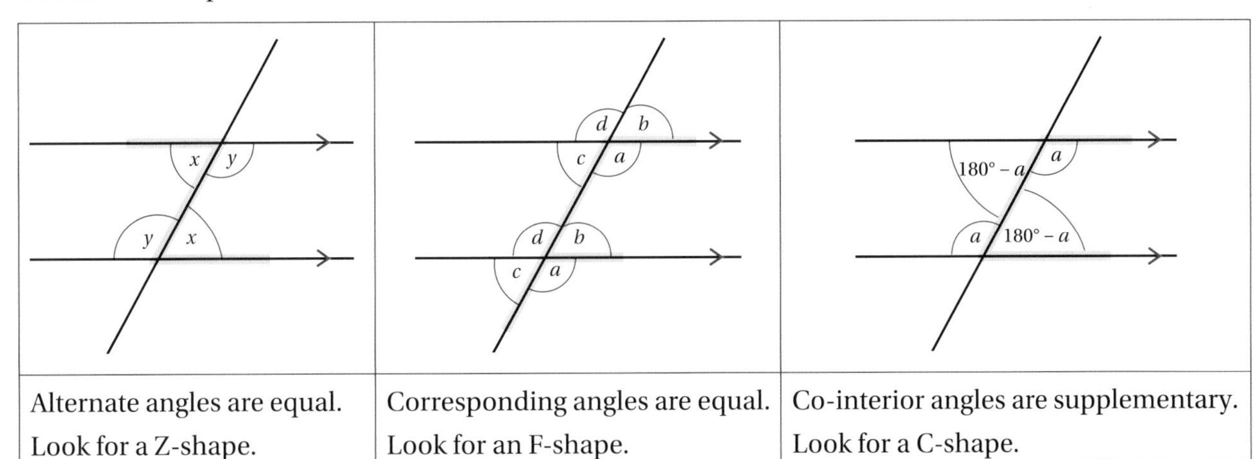

Alternate angles are equal. Look for a Z-shape.	Corresponding angles are equal. Look for an F-shape.	Co-interior angles are supplementary. Look for a C-shape.

Tip

Remember, the **Z**, **F** and **C** shapes can be facing back-to-front as well.

EXERCISE 9B

1 Find the size of the missing angles a, b, c and d.

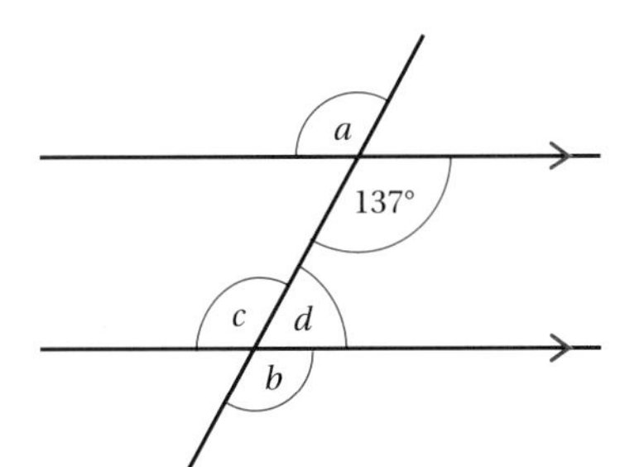

2 Given that the two poles are parallel to each other, find the angle x that the second pole makes with the incline upwards.

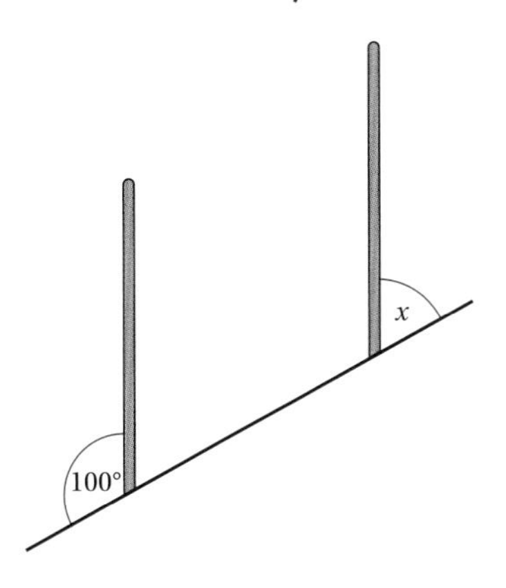

Find answers at: cambridge.org/ukschools/gcsemaths-studentbookanswers

3 Find the size of the missing angles a, b and c.

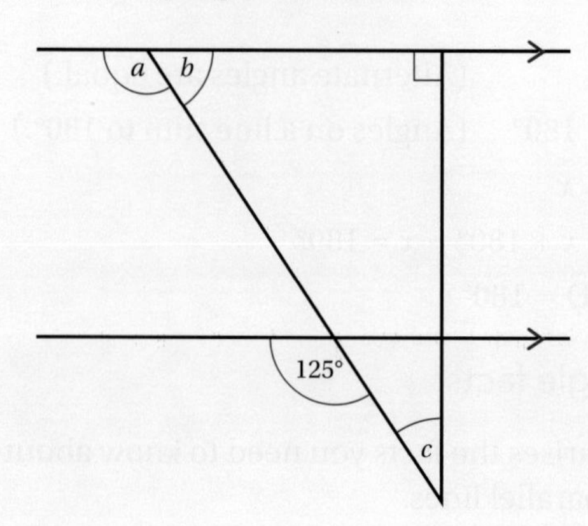

4 Find the size of angles x and y.

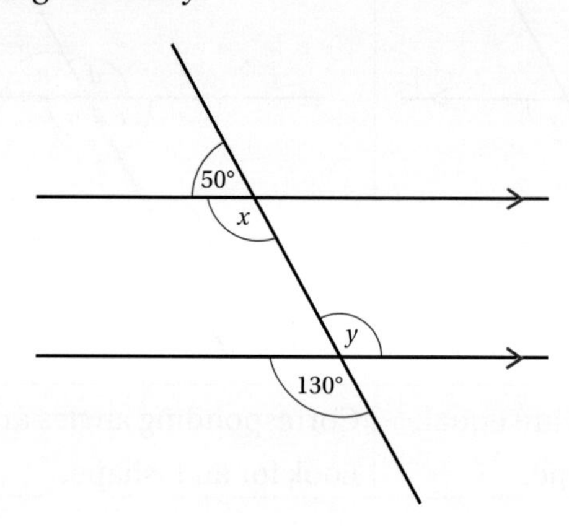

5 Find the size of the missing angles.

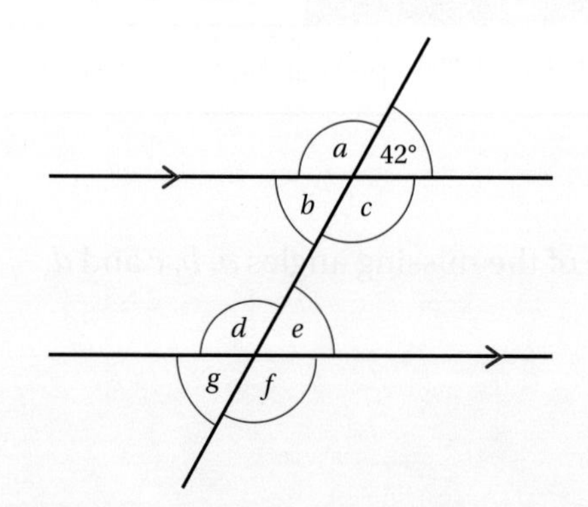

6 Find the size of $\angle$CEG.

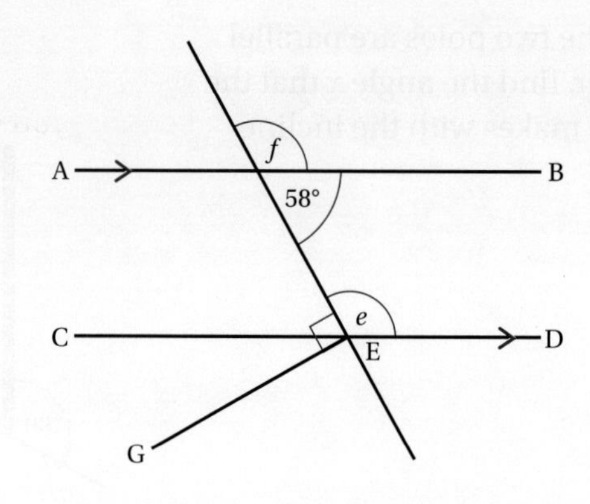

7 Find the size of ∠DCF.

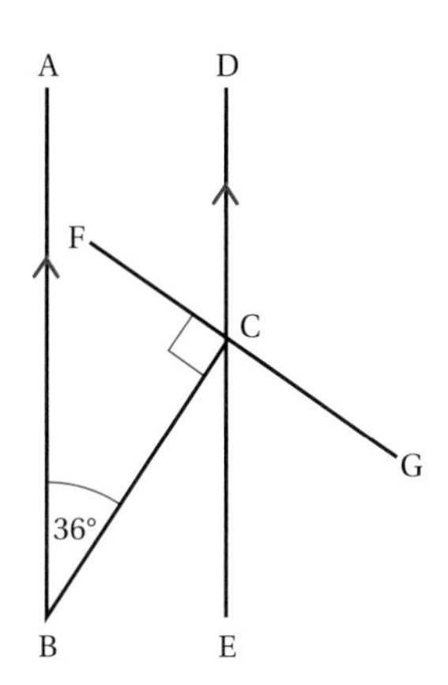

8 Find the value of x.

a

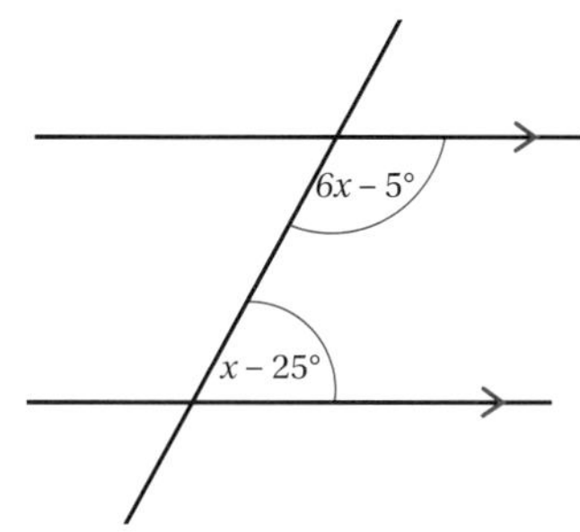

b

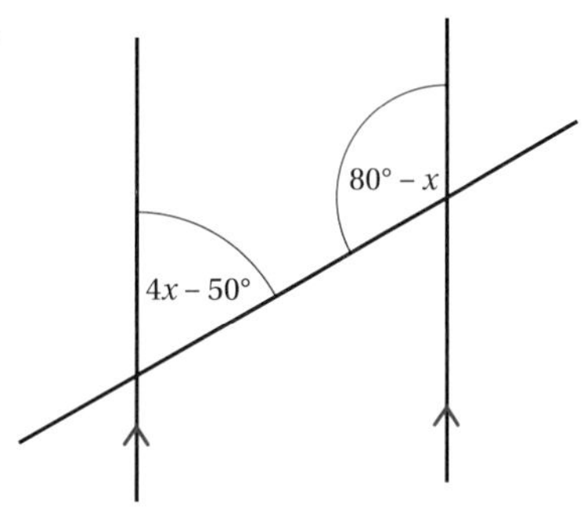

9 Decide whether AB // DC in each of the following.
Give a reason for your answer.

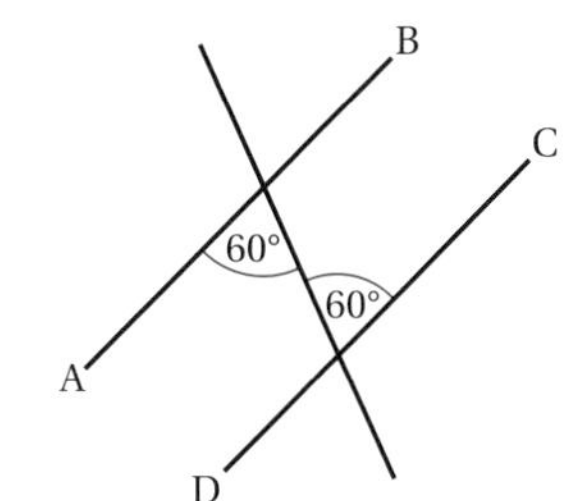

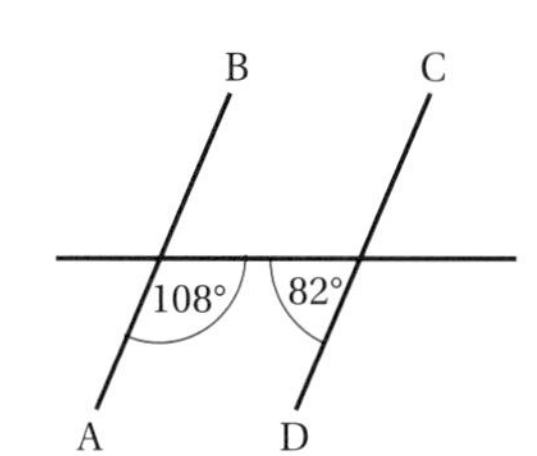

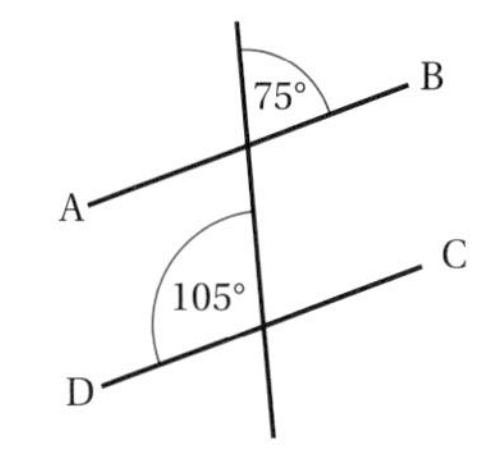

Section 3: Angles in triangles

In Chapter 5, you saw that when you tear off the interior angles of any triangle and place them on a straight edge, they will form an angle of 180°.

Tearing off the angles shows that the interior angles of a triangle add up to 180° but it is **not** a mathematical proof.

Tip

Make sure you know the differences between an equilateral triangle, a right-angled triangle, an isosceles triangle and a scalene triangle. Look back at Chapter 5 if you need to revise the properties of these triangles.

Angle sum of a triangle

To prove that the angles in a triangle add up to 180° you can construct a line parallel to one side of the triangle like this:

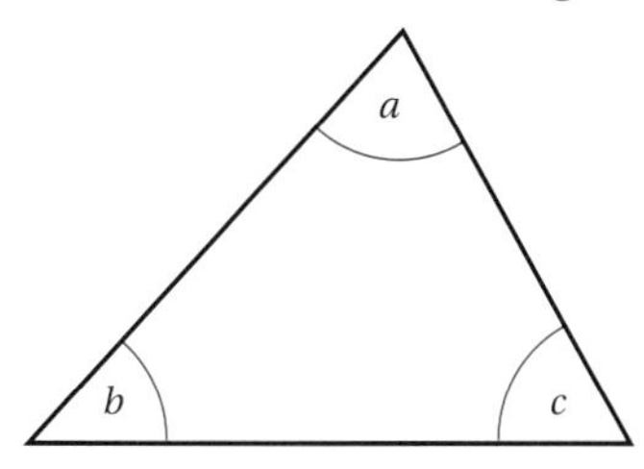

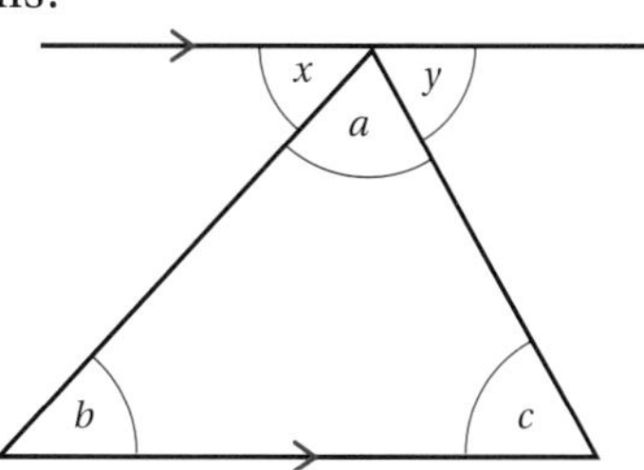

Tip

Some other important angle facts about triangles that you should learn are:

- all angles in an equilateral triangle are 60° (all angles are equal and 180° ÷ 3 = 60°).
- the base angles in an isosceles triangle are equal; this means if you know one of the base angles, or the third non-base angle, you can calculate the other angles inside the triangle.

Once you have done that, you can prove that $a + b + c = 180°$ using mathematical principles.

$x + a + y = 180°$ (Angles on a line sum to 180°.)

But: $x = b$ and $y = c$ (Alternate angles are equal.)

Substitute b for x and c for y and you prove that $a + b + c = 180°$.

Therefore the three angles of a triangle always add up to 180°.

The exterior angle is equal to the sum of the opposite interior angles

In Chapter 5 you also saw that the exterior angle of a triangle is equal to the sum of the two opposite interior angles.

There are different ways to prove this using mathematical principles.

Read through the example to see how you can prove it using angles on a line and the angle sum of triangles.

WORKED EXAMPLE 3

Show that $a + b = x$, and therefore prove that the exterior angle of any triangle is equal to the sum of the opposite interior angles.

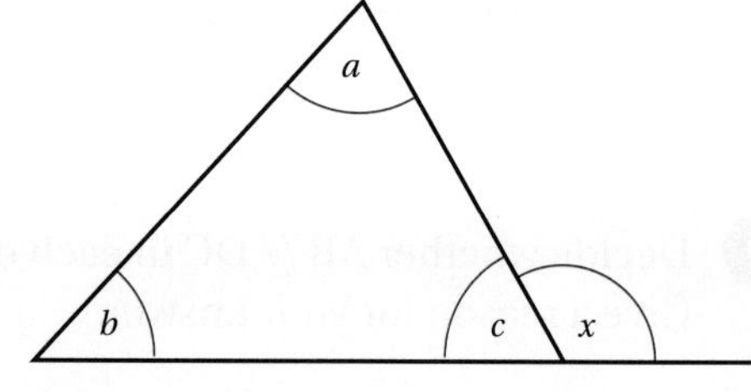

$c + x = 180°$ (angles on a line sum to 180°)

$\therefore c = 180° - x$

$a + b + c = 180°$ (angle sum of triangle)

$\therefore c = 180° - (a + b)$

But, $c = (180° - x)$ (proven above)

So, $180° - (a + b) = 180° - x$

$\therefore a + b = x$

You can now combine basic angle facts, angle facts related to parallel lines and angle facts about triangles to find unknown angles in different figures.

Always give reasons for any statements you make based on known facts.

EXERCISE 9C

1 Calculate the size of the missing angles.

a

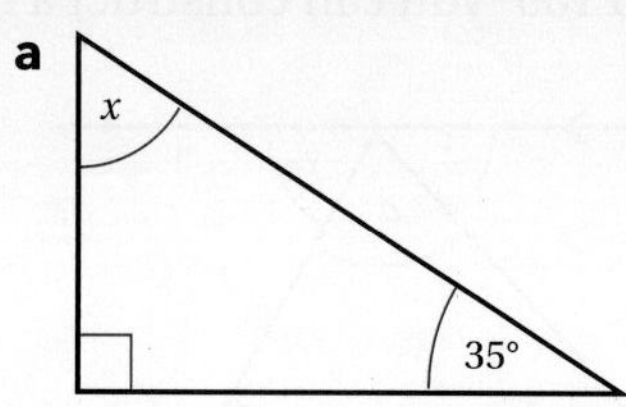

b

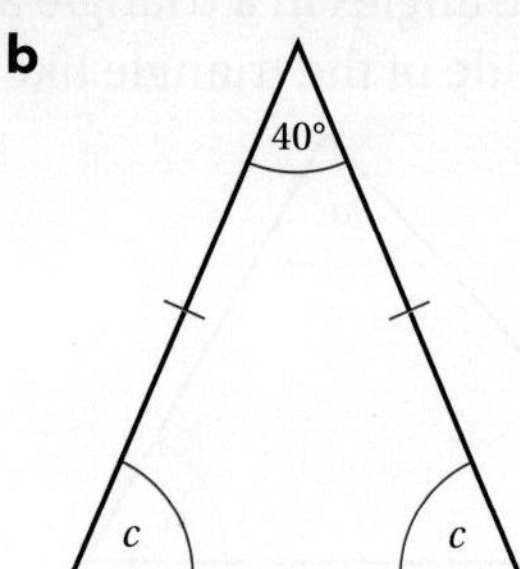

2 Find x and y.

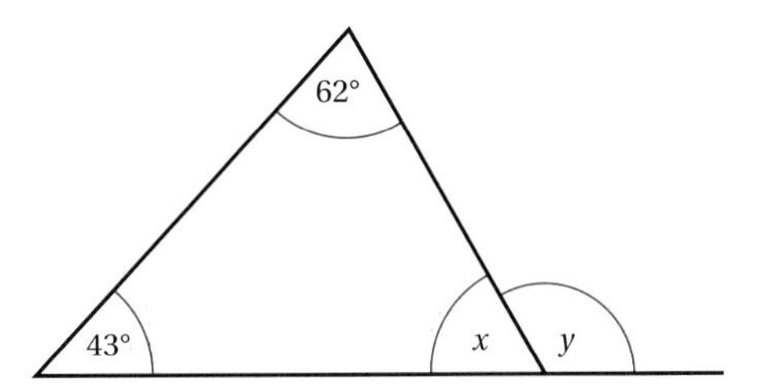

3 Find a and b.

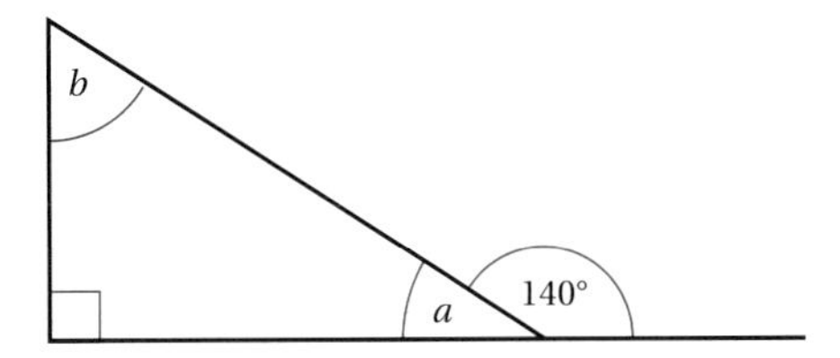

4 Find x, y and z.

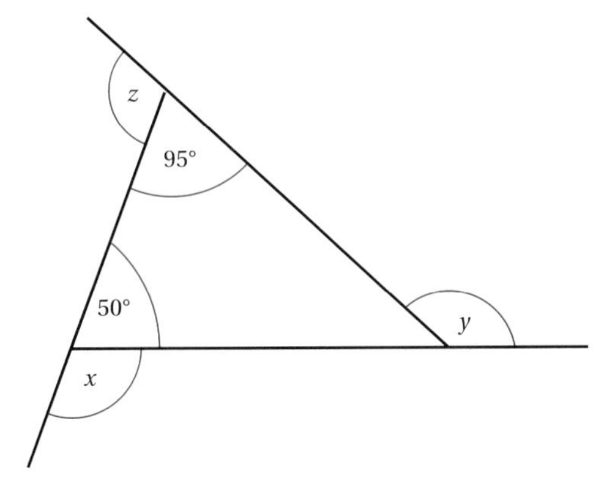

5 Work out the sizes of the angles marked a, b and c.
Give reasons to justify your answers.

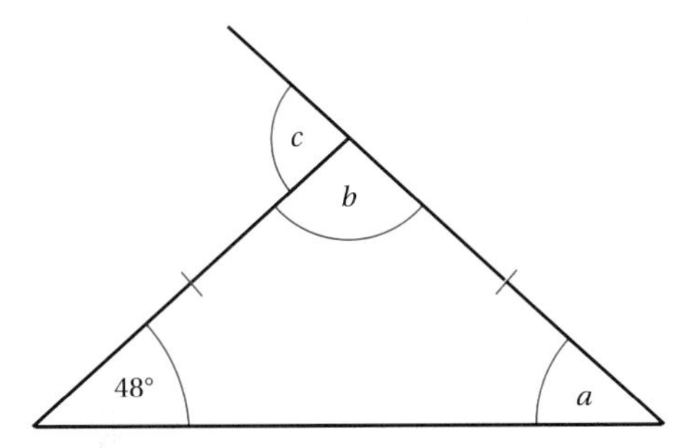

6 Calculate x.

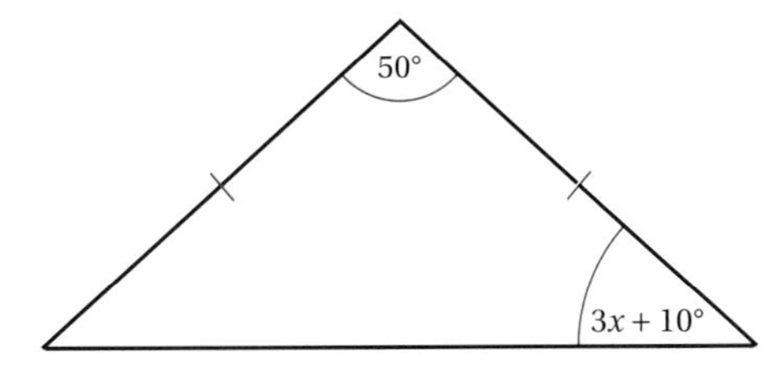

7 With reference to the figure below, show how you can prove that the exterior angle of a triangle is equal to the sum of the opposite interior angles by construction of CE // AB.

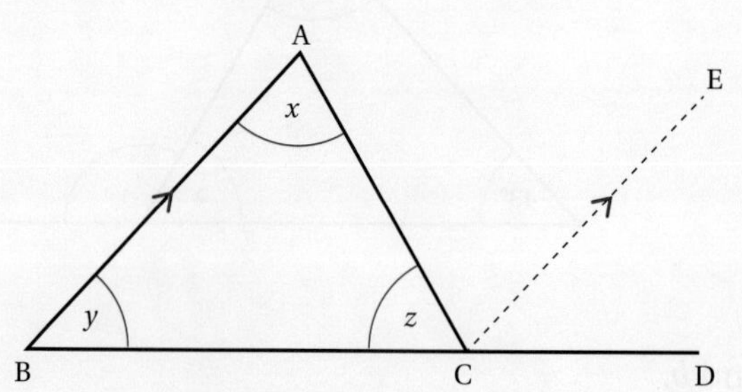

8 Find the value of the angles marked with variables in each diagram. Give reasons.

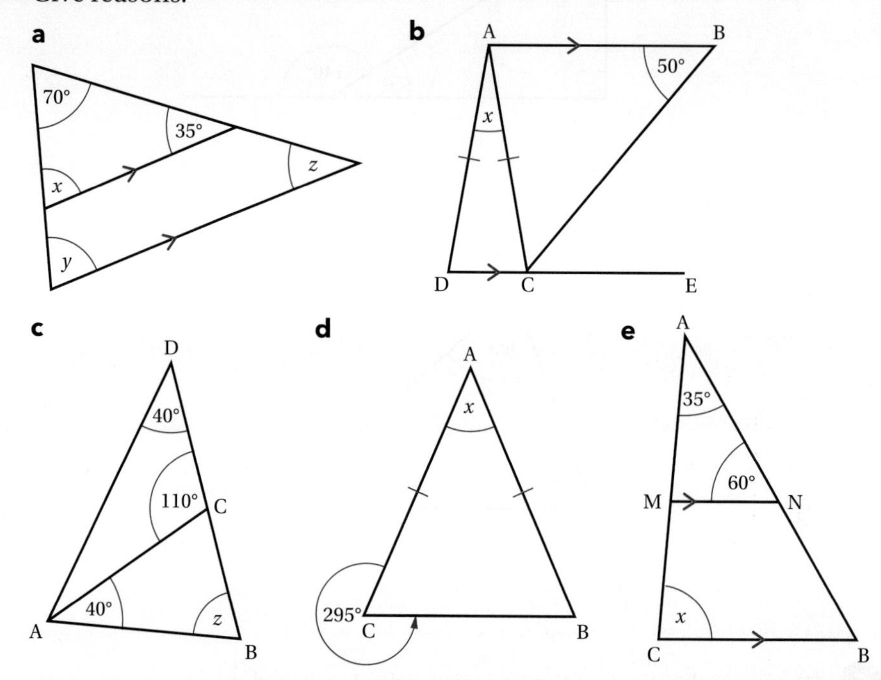

Section 4: Angles in polygons

You learnt about polygons in Chapter 5. Remember:

- A polygon is a plane shape with three or more straight sides.
- Polygons are regular if all their sides and angles are equal.
- Polygons are named according to how many sides they have: triangle (3), quadrilateral (4), pentagon (5), hexagon (6), heptagon (7) and octagon (8).

These basalt columns are formed naturally when lava cools.

The end faces are almost perfectly hexagonal.

The angle sum of a polygon

You can divide any polygon into triangles by drawing in the diagonals from one vertex.

This allows you to use the angle sum of triangles to work out the sum of the interior angles in a polygon.

This is a regular hexagon.

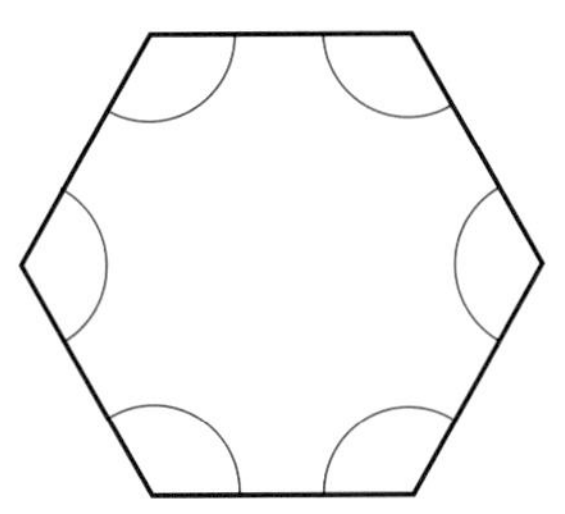

Key vocabulary

interior angles: angles inside a two-dimensional shape at the vertices or corners.

It has six **interior angles** which are all equal.

The diagonals divide the hexagon into four triangles.

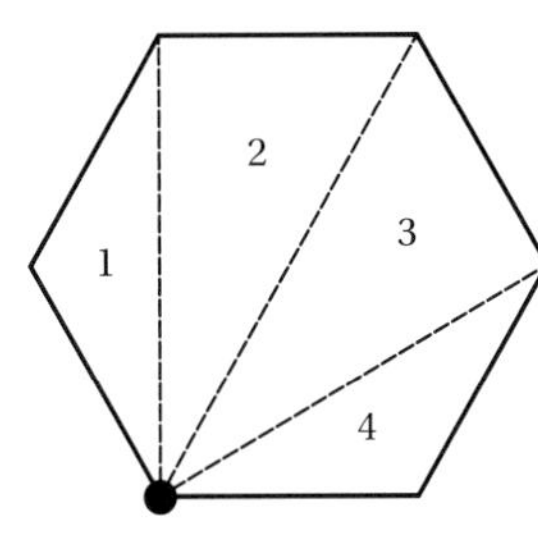

So the sum of the interior angles of a hexagon is $4 \times 180° = 720°$.

The hexagon is regular, so the six interior angles are equal in size.

$720° \div 6 = 120°$. So, each interior angle is 120°.

Work through the investigation in Exercise 9D to develop a rule for finding the angle sum of any polygon.

WORK IT OUT 9.1

Three students attempted to calculate the size of interior angles in a regular pentagon.

Which is the correct solution?

What mistakes have been made by the other students?

Option A	Option B	Option C
There are three triangles within the pentagon. $3 \times 180° = 540°$ Five angles in a pentagon. $540° \div 5 = 108°$ Each interior angle = 108°.	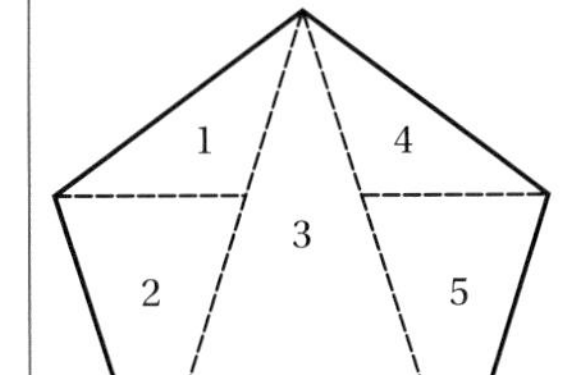 There are five triangles in a pentagon. $5 \times 180° = 900°$ $900° \div 5 = 180°$ Interior angle = 180°.	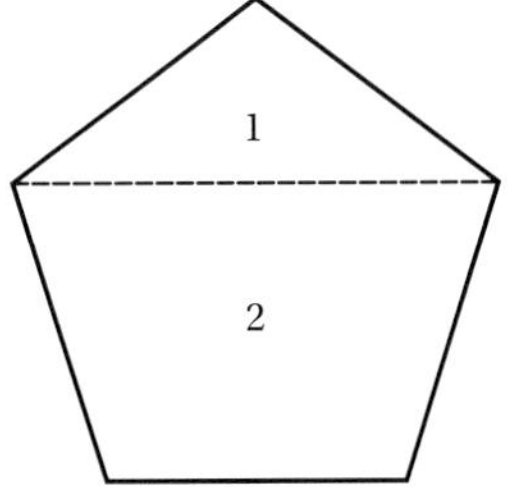 There is a trapezium and a triangle inside the pentagon. $360° + 180° = 540°$ There are five angles inside the pentagon including the central 360° which gives a total of 900°. Interior angle = $900° \div 5 = 180°$.

EXERCISE 9D

1 Draw the following polygons and divide them into triangles by drawing diagonals from one vertex as with the hexagon above.

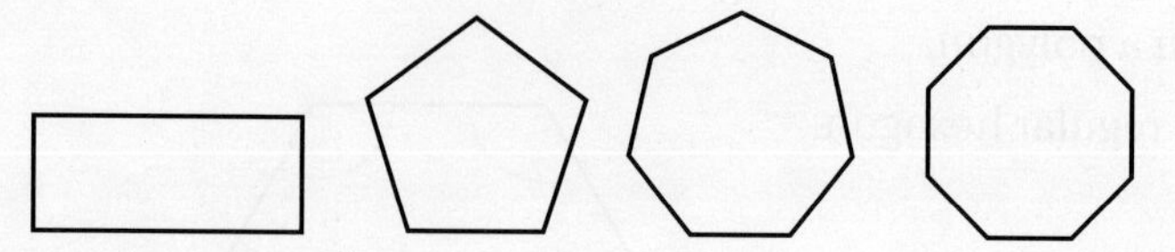

2 Predict how many triangles you could form if you did the same for a 10-sided and 20-sided polygon.

3 Copy and complete this table using your results from Questions **1** and **2**.

Number of sides in polygon	3	4	5	6	7	8	10	20
Number of triangles	1			4				
Angle sum of interior angles	180°			720°				

4 What is the relationship between the number of sides in a polygon and the number of triangles you can form in this way?

5 If a polygon has n sides, how many triangles can you form in this way?

6 Write a rule for finding the angle sum of a polygon:

a in words. **b** in general algebraic terms for a polygon of n sides.

7 Use your rule to find the angle sum of a polygon with 12 sides.

8 How could you find the size of each angle of a regular 12-sided polygon?

The sum of exterior angles of a polygon

Key vocabulary

exterior angles: angles produced by extending the sides of a polygon.

Look at the regular hexagon again.

You know that each interior angle is 120°.

By extending each side, we can form six **exterior angles**.

Like this:

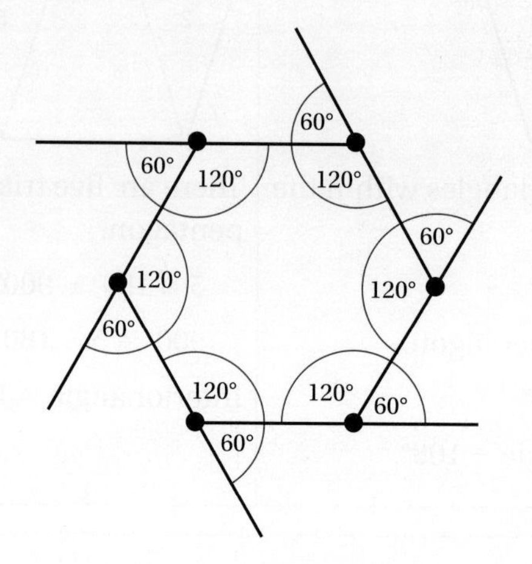

There are six sides, so this produces six pairs of angles (one interior and one exterior per pair) on straight lines.

The sum of angles on a straight line is 180°.

So, the sum of these six angle pairs is $6 \times 180° = 1080°$.

But, you already know that the sum of interior angles is $180°(n - 2) = 180° \times 4 = 720°$.

The sum of the interior and exterior angles is 1080°.

The sum of the interior angles is 720°.

So, the sum of the exterior angles of the hexagon is $1080° - 720° = 360°$.

Now consider any polygon with n sides.

The sum of interior plus exterior angles can be found using $180n$, where n is the number of sides.

The sum of the interior angles can be found using $180(n - 2)$ where n is the number of sides.

So, for a pentagon, the sum of the exterior and interior angles would be $180° \times 5 = 900°$ because there are five sides.

The sum of interior angles is $180(5 - 2) = 180° \times 3 = 540°$.

The sum of the exterior angles = the sum of the interior and exterior angles – the sum of the interior angles

$= 900° - 540°$

$= 360°$

In general terms:

Let I be the sum of interior angles.

Let E be the sum of exterior angles.

$I = 180(n - 2)$

$I + E = 180n$

$E = 180n - 180(n - 2)$

$= 180n - 180n + 360$

$= 360$

You can use these rules to find the angle sum of any polygon.

If the polygon is regular, you can also calculate the size of each interior and exterior angle.

Tip

Exterior angles of any polygon $= 360°$ or one complete turn.
For a regular polygon, the size of one exterior angle $= \frac{360}{n}$ where n is the number of sides.

WORKED EXAMPLE 4

1 For a regular 10-sided polygon, find:

a the sum of the interior angles.

Angle sum $= 180(n - 2) = 180(8) = 1440°$

In a ten-sided figure $n = 10$.

b the size of each interior angle.

There are 10 interior angles, so one angle $= \frac{1440}{10} = 144°$

c the size of each exterior angle.

$180° - 144° = 36°$

Continues on next page ...

2 A polygon has an angle sum of 2340°. How many sides does it have?

$2340 = 180(n - 2)$ — Using the angle sum rule.

$\frac{2340}{180} = n - 2$

$13 = n - 2$

$15 = n$

The polygon has 15 sides.

3 A regular polygon has an exterior angle of 18°. How many sides does it have?

Sum of exterior angles = 360

Number of angles = $\frac{360}{18} = 20$

$\therefore$ Number of sides = 20 — The number of sides equals the number of angles.

EXERCISE 9E

1 Calculate the sum of interior angles of a polygon with:

a 9 sides. **b** 17 sides. **c** 25 sides.

2 A regular polygon has 15 sides. Find:

a the sum of the interior angles. **b** the sum of the exterior angles.

c the size of an interior angle. **d** the size of an exterior angle.

3 A regular polygon has an interior angle that is three times the size of the exterior angle.

a What is the size of each exterior angle?

b What is the size of each interior angle?

c What is the name of the regular polygon?

4 Find the size of the missing interior and exterior angles x, y and z in this irregular pentagon.

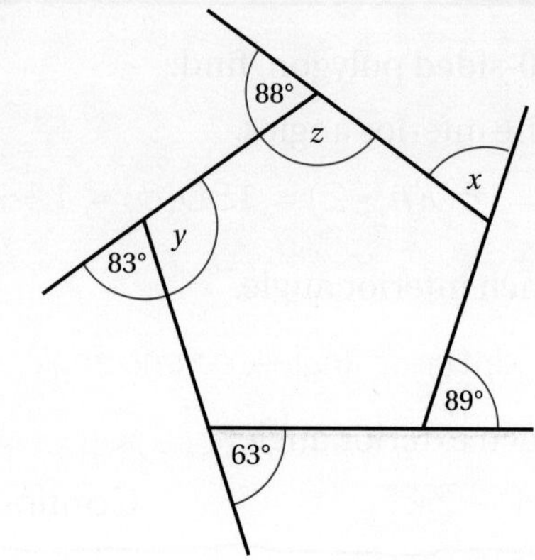

5 A pentagon has three angles that add up to 266° and two other angles that are equal in size. Determine the size of each of the other two angles.

6 A hexagon has four angles with a sum of 555° and two unknown angles.

Given that one of the unknown angles is twice the size of the other, what are their sizes?

7 Could a regular polygon have interior angles of 125°?
Justify your answer.

8 How many sides does a polygon have if the sum of interior angles is:

a 1620°. **b** 3060°.

9 The exterior angle of a regular polygon is 40°. What is the sum of its interior angles?

Checklist of learning and understanding

Basic angle facts

- Angles around a point sum to 360°.
- Angles on a straight line sum to 180°.
- Vertically opposite angles are equal.

Angles associated with parallel lines

- Corresponding angles are equal.
- Alternate angles are equal.
- Co-interior angles sum to 180°.

Geometric proofs

- Using properties of alternate and corresponding angles, you can prove that the three interior angles of any triangle sum to 180°.
- Using the angle sum of triangles and properties of angles at a line and at a point, you can find the interior and exterior angles of any polygon.

Chapter review

For additional questions on the topics in this chapter, visit GCSE Mathematics Online.

1 Work out the size of the marked angle in each diagram.

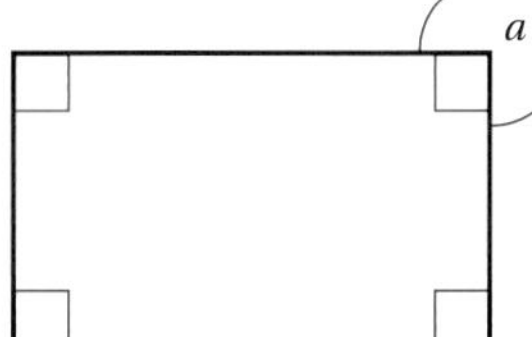

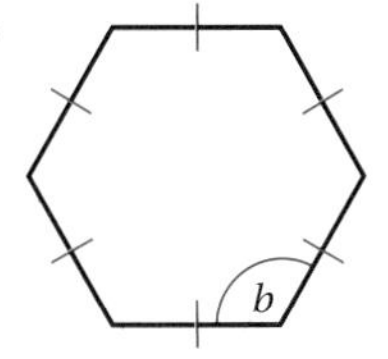

c

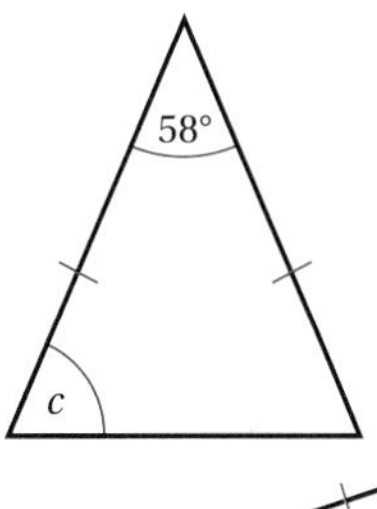

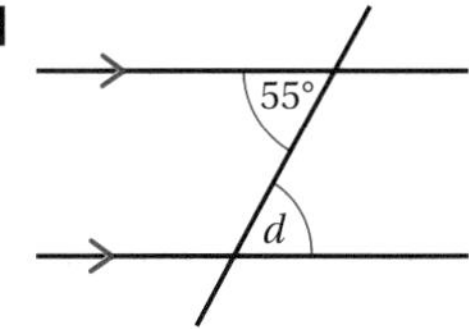

e

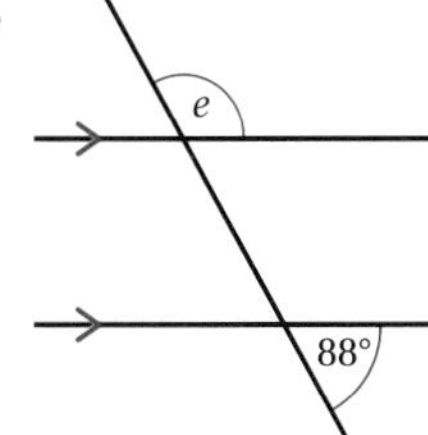

f

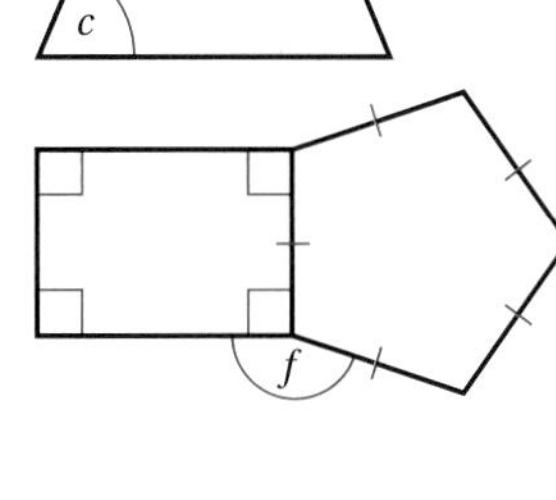

2 If the interior angle of a regular polygon is 108° what type of polygon is it?

3 Calculate $p + q + r + s + t$.

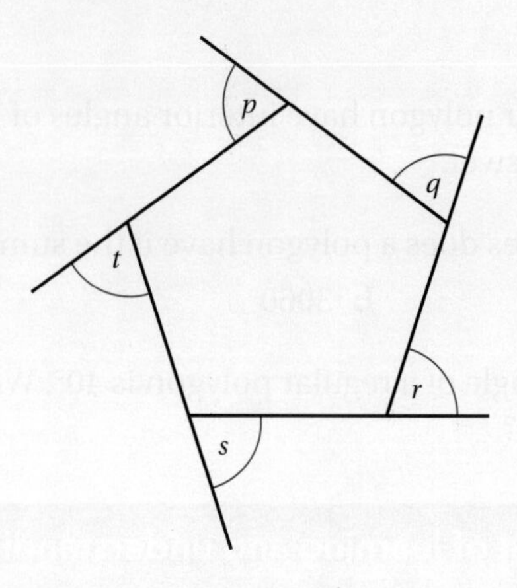

4 In this irregular hexagon, calculate the value of z.

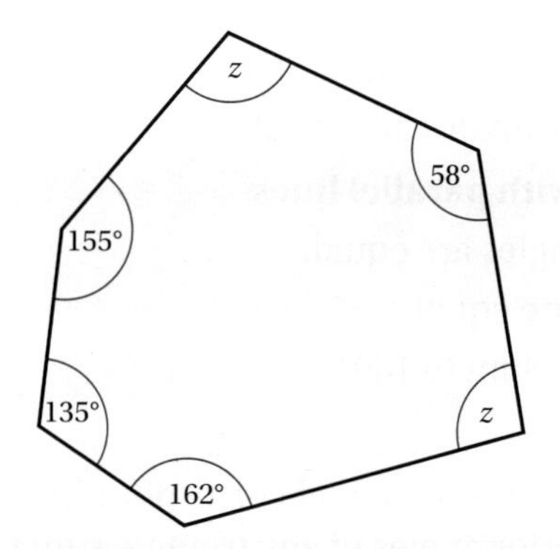

5 This diagram shows part of a regular polygon.

The interior angle is 144°.

Calculate the number of sides of the polygon.

What is the name given to this polygon?

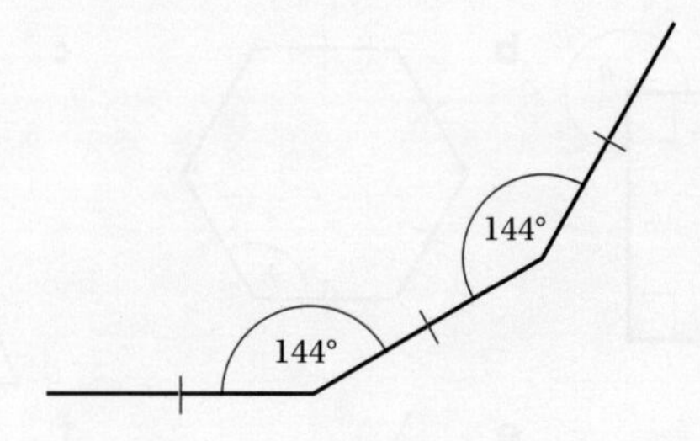

6 Calculate the exterior angle of a regular 10-sided polygon.

7 Show mathematically why the sum of the exterior angles of any polygon is 360°.

8 Find the value of x and y.

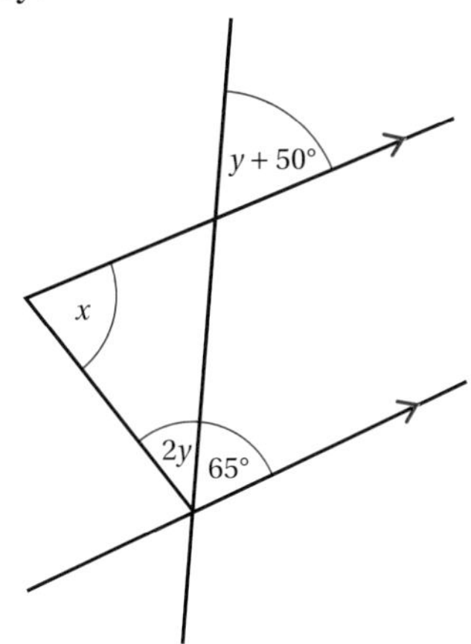

9 In the diagram ADE is a triangle.
BC is parallel to DE and DBA is parallel to EF.

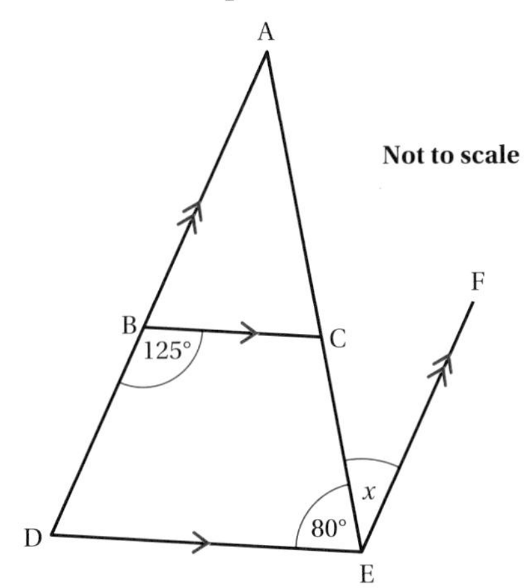

Work out angle x.
Give a reason for each step of your working. *(5 marks)*

© OCR 2013

10 Prove that quadrilateral ABCD is a trapezium.

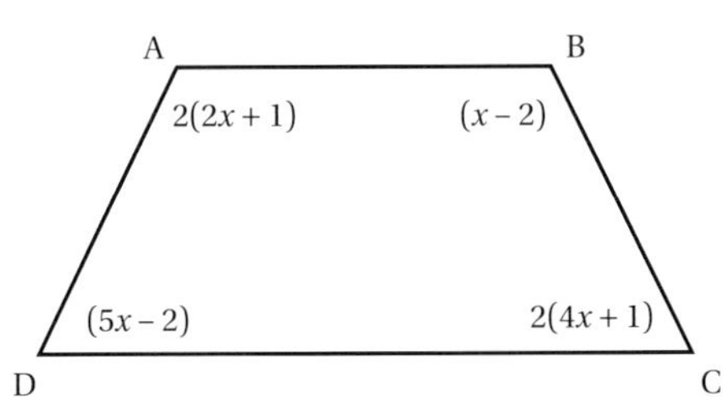

11 The sum of interior angles of an irregular polygon is 3600°.

a Determine the number of sides.

b Determine x, if half the angles in the polygon are size x and the other half are size $2x$.

Find answers at: cambridge.org/ukschools/gcsemaths-studentbookanswers

10 Fractions

In this chapter you will learn how to ...

- recognise equivalence between fractions and mixed numbers.
- carry out the four basic operations on fractions and mixed numbers.
- work out fractions of an amount.

For more resources relating to this chapter, visit GCSE Mathematics Online.

Using mathematics: real-life applications

Nurses and other medical support staff work with fractions, decimals, percentages, rates and ratios every day. They calculate medicine doses, convert between different systems of measurement and set the patients' drips to supply the correct amount of fluid per hour.

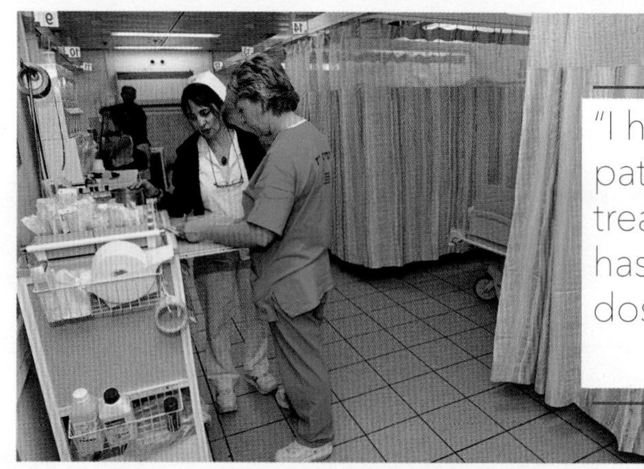

"I have to work out a treatment plan for patients who are going to have radiation treatment for various cancers. The patient has to receive a certain fraction of the total dose of radiation at each treatment session."

(Oncologist)

Before you start ...

Ch 2	Check that you can find common factors of sets of numbers.	18 24 27 28 30 32 36 **1** From this set of numbers, choose numbers that have: **a** a common factor of 9. **b** common factors 2 and 3. **c** common factors 3, 4, and 12. **d** common factors 4 and 8.
Ch 2	Find the lowest common multiple of sets of numbers.	**2** Choose the lowest common multiple of each set of numbers. **a** 5 and 10 — A 15 B 50 C 10 D 20 **b** 8 and 12 — A 12 B 96 C 36 D 24 **c** 2, 3 and 5 — A 1 B 30 C 10 D 6
Ch 1	Know the correct order for performing operations (BIDMAS).	**3** Which calculation is correct in each pair? Why? (see table below)

	Student A	Student B
a	$^-3 - 2 \times {}^-6 - 4 = 5$	$^-3 - 2 \times {}^-6 - 4 = 50$
b	$^-60 \div 5 + 3 \times {}^-4 - 8 = 28$	$^-60 \div 5 + 3 \times {}^-4 - 8 = {}^-32$
c	$13 - 2 \times {}^-6 - 5 \times 4 = 80$	$13 - 2 \times {}^-6 - 5 \times 4 = 5$

Assess your starting point using the Launchpad

STEP 1

1 Which fraction does not belong in each set?

a $\frac{3}{15}, \frac{1}{5}, \frac{6}{30}, \frac{5}{35}, \frac{4}{20}$

b $\frac{4}{7}, \frac{8}{14}, \frac{12}{21}, \frac{9}{16}, \frac{52}{91}$

c $\frac{22}{10}, \frac{11}{4}, 2\frac{3}{4}, \frac{33}{12}, 2\frac{18}{24}$

GO TO
Section 1:
Equivalent fractions

STEP 2

2 Each calculation contains a mistake. Find the mistake and write the correct answer.

a $\frac{2}{3} + \frac{3}{4} = \frac{5}{7}$

b $\frac{4}{5} - \frac{9}{10} = \frac{1}{10}$

c $\frac{2}{7} \times \frac{4}{5} = \frac{6}{35}$

d $30 \div \frac{1}{2} = 15$

GO TO
Section 2:
Operations with fractions

STEP 3

3 Which is greater in each pair?

a $\frac{5}{8}$ of 40 or $\frac{3}{5}$ of 60

b $\frac{3}{4}$ of 240 or $\frac{7}{10}$ of 300

c $\frac{1}{4}$ of $\frac{1}{2}$ or $\frac{1}{2}$ of $\frac{3}{4}$

4 If you have read 45 pages of a 240 page book, what fraction of the book remains unread?

5 What fraction of 30 minutes is 45 seconds?

GO TO
Section 3:
Finding fractions of a quantity

GO TO
Chapter review

Find answers at: cambridge.org/ukschools/gcsemaths-studentbookanswers

Tip

Fractions:
- belong to the set of real numbers (numbers that can be found on a number line).
- are written in the form $\frac{a}{b}$ where b is not equal to zero.
- are rational numbers (they can also be written as terminating or recurring decimals).
- a proper fraction is one where the numerator is smaller than the denominator.
- an improper fraction is one where the numerator is larger than the denominator; these can also be written as mixed numbers.
- a mixed number is written as an integer and a fraction, e.g. $2\frac{1}{2}$ is a mixed number.

Key vocabulary

common denominator: a number into which all the denominators of a set of fractions divide exactly.

Section 1: Equivalent fractions

Equivalent fractions represent the same value.

For example, $\frac{1}{4}$, $\frac{2}{8}$ and $\frac{16}{64}$ are equivalent.

You can find equivalent fractions by multiplying the numerator and denominator by the same number.

You can also find equivalent fractions by dividing the numerator and denominator by the same number (cancelling).

This is known as simplifying, or reducing the fraction to its simplest terms.

When you give an answer in the form of a fraction, you usually give it in its simplest form.

$\frac{6}{18} = \frac{1}{3}$ $\quad \frac{12}{27} = \frac{4}{9}$ $\quad 3\frac{5}{30} = 3\frac{1}{6}$

When you are asked to compare fractions that look different, you can write them both with a **common denominator** so that you can compare them by size, or you can cross multiply to find out whether they are equivalent or not.

WORKED EXAMPLE 1

Are the following pairs of fractions equivalent?

a $\frac{5}{6}$ and $\frac{7}{8}$ **b** $3\frac{3}{4}$ and $\frac{45}{12}$

Method 1: Using common denominators

a $\frac{5}{6} = \frac{20}{24}$ and $\frac{7}{8} = \frac{21}{24}$

Write both fractions with the same denominator by finding the lowest common multiple, and then multiplying each numerator accordingly, i.e., $6 \times 4 = 24$ so multiply 5 by 4 as well; $8 \times 3 = 24$, so multiply 7 by 3 as well.

$\frac{5}{6} \neq \frac{7}{8}$

When the fractions have the same denominator it is easy to see whether they are equivalent. It is also easy to compare them by size. Write the fractions back in their original form when writing your answer.

Method 2: By cross multiplying

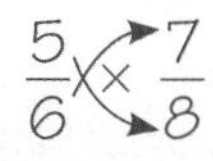

$5 \times 8 = 40$
$6 \times 7 = 42$

Cross multiplying a fraction converts it from the form $\frac{a}{b} = \frac{c}{d}$ to the form $ad = bc$. If the resulting equation is not equal then the two fractions are not equivalent.

$40 \neq 42$

$\therefore$ the fractions are not equivalent.

40 is smaller than 42, so $\frac{5}{6} < \frac{7}{8}$.

This is a useful strategy for comparing the size of fractions.

Continues on next page ...

Method 1: Using common denominators		**Method 2: By cross multiplying**	
b $3\frac{3}{4} = \frac{15}{4}$	Write the mixed number as an improper fraction. Remember, to convert a mixed number to an improper fraction, you multiply the integer part by the denominator and then add the numerator.	$3\frac{3}{4} = \frac{15}{4}$	Write the mixed number as an improper fraction.
$\frac{15}{4} = \frac{45}{12}$	Here, you can write $\frac{15}{4}$ with a denominator of 12, or write $\frac{45}{12}$ with a denominator of 4 to find a common denominator, rather than finding the LCM.	$\frac{15}{4} \times \frac{45}{12}$	$15 \times 12 = 120 + 60 = 180$ $4 \times 45 = 2 \times 90 = 180$
		$180 = 180$	$\therefore$ the fractions are equivalent.

EXERCISE 10A

1 Determine whether the following pairs of fractions are equivalent (=) or not (≠).

a $\frac{2}{5}$ and $\frac{3}{4}$ **b** $\frac{2}{3}$ and $\frac{3}{4}$ **c** $\frac{3}{8}$ and $\frac{5}{12}$ **d** $\frac{2}{11}$ and $\frac{1}{10}$

e $\frac{3}{5}$ and $\frac{9}{15}$ **f** $\frac{10}{25}$ and $\frac{4}{10}$ **g** $\frac{6}{24}$ and $\frac{5}{20}$ **h** $\frac{11}{9}$ and $\frac{121}{99}$

Tip

You can use the LCM of the denominators to find a common denominator, but any common denominator works (not just the lowest).

2 Find the equivalent fractions of $\frac{1}{4}$ with:

a denominator 32 **b** numerator 48

c numerator 27 **d** denominator 52

3 How could you use cross multiplication to find the missing values in examples like these?

a $\frac{3}{5} = \frac{18}{\square}$ **b** $\frac{\square}{51} = \frac{2}{17}$

4 Reduce the following fractions to their simplest form.

a $\frac{3}{15}$ **b** $\frac{4}{6}$ **c** $\frac{25}{100}$ **d** $\frac{5}{10}$

e $\frac{4}{12}$ **f** $\frac{-7}{21}$ **g** $\frac{36}{-24}$ **h** $\frac{60}{100}$

i $\frac{-14}{-21}$ **j** $\frac{18}{27}$ **k** $\frac{15}{21}$ **l** $\frac{-18}{-42}$

5 Write each set of fractions in ascending order.

a $\frac{3}{5}, \frac{1}{4}, \frac{9}{4}, 1\frac{3}{4}, \frac{4}{7}$ **b** $\frac{5}{6}, \frac{3}{4}, \frac{11}{3}, \frac{19}{24}, 2\frac{2}{3}$ **c** $2\frac{3}{7}, \frac{1}{7}, \frac{7}{7}, \frac{8}{14}, \frac{10}{21}, \frac{13}{7}$

6 A mediant fraction is a fraction that lies between two other fractions. They follow the general rule:

If $\frac{a}{b}$ and $\frac{c}{d}$ are two fractions, then the fraction $\frac{a+c}{b+d}$ lies between them such that $\frac{a}{b} < \frac{a+c}{b+d} < \frac{c}{d}$.

Tip

Mediant fractions should not be confused with the median value in a set of data. These concepts are not related.

Find answers at: cambridge.org/ukschools/gcsemaths-studentbookanswers

a Use this general rule to find a fraction between:

i $\frac{1}{4}$ and $\frac{3}{5}$ **ii** $\frac{4}{5}$ and $\frac{9}{11}$

b Show how you could apply the rule to find three fractions between $\frac{1}{3}$ and $\frac{3}{4}$.

c Test your results to show that the answers are correct.

d How does this work?

Find out what you can and try to explain the general rule in simple terms.

Section 2: Operations with fractions

You have already learnt in earlier school years how to add, subtract, multiply and divide fractions and mixed numbers.

Read through the examples to remind you of the key rules for operations on fractions.

Multiplying fractions

To multiply fractions multiply the numerators and then multiply the denominators. If possible, cancel before you multiply to make the calculations easier.

Tip

Cancelling in the first step means you don't have to simplify the fraction to get an answer.

$\frac{3 \times \cancel{2}^1}{{}^2\cancel{4} \times 7}$

$= 3 \times \frac{1}{2} \times 7 = \frac{3}{14}$

WORKED EXAMPLE 2

a $\frac{3}{4} \times \frac{2}{7}$

$= \frac{3 \times 2}{4 \times 7}$

$= \frac{6}{28}$ — Multiply numerators by numerators and denominators by denominators.

$= \frac{3}{14}$ — Give the answer in its simplest form.

b $\frac{5}{7} \times 3$

$= \frac{5 \times 3}{7 \times 1}$ — Think of a whole number as a fraction with a denominator of 1.

$= \frac{15}{7}$ — $\frac{15}{7}$ cannot be simplified further but it can be written as a mixed number.

$= 2\frac{1}{7}$

c $\frac{3}{8} \times 4\frac{1}{2}$

$\frac{3}{8} \times \frac{9}{2}$ — Rewrite the mixed number as an improper fraction.

$= \frac{27}{16}$ — $\frac{27}{16}$ cannot be simplified but it can be written as a mixed number.

$= 1\frac{11}{16}$

Adding and subtracting fractions

To add or subtract fractions they must have the same denominators. Find a common denominator and then find the equivalent fractions before you add or subtract the numerators.

WORKED EXAMPLE 3

a $\frac{1}{2} + \frac{1}{4}$

$= \frac{2}{4} + \frac{1}{4}$ Use 4 as a common denominator. Write $\frac{1}{2}$ as its equivalent $\frac{2}{4}$.

$= \frac{3}{4}$ Add the numerators.

b $2\frac{1}{2} + \frac{5}{6}$

$= \frac{5}{2} + \frac{5}{6}$ Rewrite mixed numbers as improper fractions.

$= \frac{15}{6} + \frac{5}{6}$ Find a common denominator.

$= \frac{20}{6}$ Add the numerators.

$= \frac{10}{3}$ or $3\frac{1}{3}$ Simplify the answer.

c $2\frac{3}{4} - 1\frac{5}{7}$

$= \frac{11}{4} - \frac{12}{7}$ Rewrite mixed numbers as improper fractions.

$= \frac{77}{28} - \frac{48}{28}$ Find a common denominator.

$= \frac{29}{28}$ or $1\frac{1}{28}$ Subtract the numerators. Simplify and write the number as mixed number.

Tip

Think of $\frac{2}{4}$ as 2 lots of 4ths, and $\frac{1}{4}$ as 1 lot of 4ths. If you combine them, you have 3 lots of 4ths, or $\frac{3}{4}$. You never add the denominators.

Dividing fractions

To divide one fraction by another fraction you multiply the first fraction by the **reciprocal** of the second fraction.

To find the reciprocal of a fraction you invert it.

So, the reciprocal of $\frac{3}{4}$ is $\frac{4}{3}$ and the reciprocal of $\frac{7}{4}$ is $\frac{4}{7}$.

The reciprocal of a whole number is a unit fraction.

For example, the reciprocal of 3 is $\frac{1}{3}$ and the reciprocal of 12 is $\frac{1}{12}$.

Key vocabulary

reciprocal: the reciprocal of a number, x, is 1 divided by x, i.e. $\frac{1}{x}$. Any number multiplied by its reciprocal is 1. For a fraction $\frac{a}{b}$, the reciprocal is $\frac{b}{a}$.

Find answers at: cambridge.org/ukschools/gcsemaths-studentbookanswers

WORKED EXAMPLE 4

a $\frac{3}{4} \div \frac{1}{2}$

$= \frac{3}{4} \times \frac{2}{1}$ — Multiply by the reciprocal of $\frac{1}{2}$, i.e. invert the fraction to $\frac{2}{1}$ and multiply.

$= \frac{6}{4}$

$= \frac{3}{2}$ or $1\frac{1}{2}$

b $1\frac{3}{4} \div 2\frac{1}{3}$

$= \frac{7}{4} \div \frac{7}{3}$ — Convert mixed numbers to improper fractions.

$= \frac{7}{4} \times \frac{3}{7}$ — Multiply by the reciprocal of $\frac{7}{3}$. Cancel the 7s.

$= \frac{3}{4}$

c $\frac{6}{7} \div 3$

$= \frac{6}{7} \times \frac{1}{3}$ — Multiply by the reciprocal of 3.

$= \frac{6}{21}$

$= \frac{2}{7}$

The rules for order of operations and negative and positive signs also apply to calculations with fractions.

EXERCISE 10B

1 Simplify.

a $\frac{3}{4} \times \frac{2}{5}$ **b** $\frac{1}{5} \times \frac{1}{9}$ **c** $\frac{5}{7} \times \frac{1}{5}$ **d** $\frac{7}{10} \times \frac{2}{3}$

e $\frac{1}{5} \times \frac{3}{8} \times \frac{-5}{9}$ **f** $\frac{2}{3} \times \frac{^-3}{4} \times \frac{^-4}{5}$ **g** $\frac{1}{2} \times \frac{2}{3} \times \frac{4}{11}$ **h** $\frac{5}{8} \times \frac{3}{7} \times \frac{2}{3}$

i $\frac{4}{25} \times \frac{^-3}{5} \times \frac{^-7}{8}$ **j** $\frac{9}{20} \times \frac{10}{11} \times \frac{1}{12}$ **k** $1\frac{2}{9} \times 1\frac{5}{22} \times 1\frac{1}{6}$ **l** $\frac{2}{7} + \frac{1}{2}$

m $\frac{1}{2} + \frac{1}{4}$ **n** $\frac{5}{8} - \frac{1}{4}$ **o** $\frac{7}{9} - \frac{1}{3}$ **p** $4\frac{3}{4} + \frac{15}{6}$

q $8\frac{2}{5} - 3\frac{1}{2}$ **r** $7\frac{1}{4} - 2\frac{9}{10}$ **s** $9\frac{3}{7} - 2\frac{4}{5}$ **t** $\frac{1}{4} \div \frac{1}{4}$

u $\frac{1}{8} \div \frac{7}{9}$ **v** $\frac{2}{11} \div \frac{-3}{5}$ **w** $3\frac{1}{5} \div 2\frac{1}{2}$ **x** $1\frac{7}{8} \div 2\frac{3}{4}$

Tip

Remember that addition and subtraction are inverse operations.

2 What should be added to $4\frac{3}{5}$ to get $9\frac{7}{20}$?

3 What should be subtracted from $13\frac{3}{4}$ to get $5\frac{1}{3}$?

4 Subtract the product of $\frac{1}{6}$ and $20\frac{4}{7}$ from the sum of $4\frac{7}{9}$ and $5\frac{5}{18}$.

5 Simplify:

a $4 + \frac{2}{3} \times \frac{1}{3}$ **b** $2\frac{1}{8} - (2\frac{1}{5} - \frac{7}{8})$ **c** $\frac{3}{7} \times \left(\frac{2}{3} + 6 \div \frac{2}{3}\right) + 5 \times \frac{2}{7}$

d $2\frac{7}{8} + (8\frac{1}{4} - 6\frac{3}{8})$ **e** $\frac{5}{6} \times \frac{1}{4} + \frac{5}{8} \times \frac{1}{3}$ **f** $\left(5 \div \frac{3}{11} - \frac{5}{12}\right) \times \frac{1}{6}$

g $\left(\frac{5}{8} \div \frac{15}{4}\right) - \left(\frac{5}{6} \times \frac{1}{5}\right)$ **h** $\left(2\frac{2}{3} \div 4 - \frac{3}{10}\right) \times \frac{3}{17}$ **i** $\left(7 \div \frac{2}{9} - \frac{1}{3}\right) \times \frac{2}{3}$

j $\left(\frac{5}{9} \times \frac{27}{35}\right) \div \frac{6}{9} + \frac{2}{3}$ **k** $1\frac{2}{3} \div \frac{5}{6} \times \frac{24}{35} \times \frac{1}{3}$

l $2\frac{1}{9} + \left[3\frac{20}{27} + \left\{\frac{5}{14} - \left(\frac{3}{35} - \frac{3}{7}\right) - \frac{1}{5}\right\} \times \frac{7}{9}\right]$

Tip

Remember the rules for order of operations apply to fractions as well.

Simplify brackets first, then powers, then multiplication and/or division, then addition and/or subtraction. When there is more than one set of brackets, work from the inner ones to the outer ones.

6 Kevin is a professional deep-sea diver.

He needs to keep track of how much time he spends underwater to make sure he has enough air left in his tank.

If he spends $9\frac{3}{4}$ minutes swimming to a wreck, $12\frac{5}{6}$ minutes exploring the wreck and $3\frac{5}{6}$ minutes examining corals, how much time has he spent underwater in total?

7 In a public park, $\frac{2}{3}$ of the area is grassed, $\frac{1}{5}$ is taken up by flower beds and the rest is paved.

How much of the park area is paved?

8 The perimeter of a quadrilateral is $18\frac{23}{60}$ m.

If the lengths of three sides are $6\frac{1}{6}$ m, $7\frac{2}{3}$ m and $1\frac{2}{15}$ m, find the length of the other side.

9 A tank contains $\frac{7}{30}$ of a litre of water.

The water has to be removed by a piece of equipment that draws out the water at a rate of $\frac{3}{5}$ of a litre per minute.

How long will it take to remove all the water?

10 There are $1\frac{3}{4}$ cakes left over after a party.

These are shared out equally among six people.

What fraction does each person get?

11 If I have $5\frac{2}{3}$ litres of juice, how many cups containing $\frac{2}{15}$ of a litre can I pour?

12 Nico buys six trays of chicken pieces for his restaurant.

Each tray contains $2\frac{1}{2}$ kg of chicken.

Each chicken meal served uses $\frac{3}{8}$ kg of chicken.

How many meals can he serve?

13 A mountaineer inserts a bolt into the rock face every $5\frac{3}{4}$ metres she climbs.

If she uses 32 bolts, what is the maximum height she has climbed?

Find answers at: cambridge.org/ukschools/gcsemaths-studentbookanswers

14 Ted has marked out three lengths of wood.

Length C is $\frac{2}{3}$ of the length of B, and length B is $1\frac{1}{3}$ times as long as A.

What is the length of piece C if A is $\frac{97}{3}$ m long?

15 Triangle ABC is isosceles with a perimeter of $12\frac{3}{4}$ cm.

Find the length of sides AB and AC given that BC is $2\frac{11}{12}$ cm.

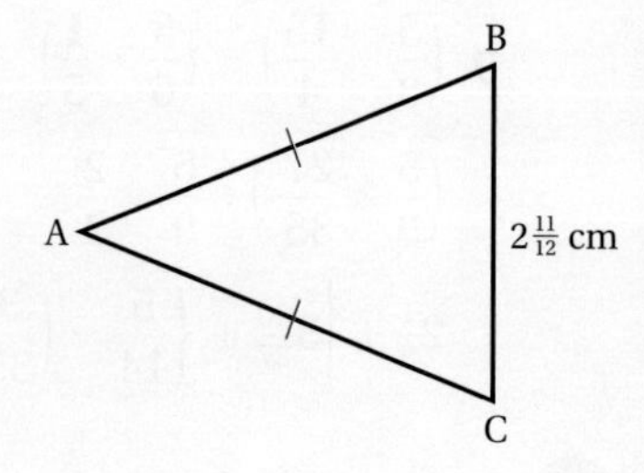

16 What is the length of the side of a square of perimeter $9\frac{3}{7}$ m?

17 On a holiday weekend Salma read $\frac{2}{9}$ of her book on Saturday, $\frac{1}{6}$ on Sunday and $\frac{5}{12}$ on Monday.

a What fraction of the book does she still have to read?

b If there are still 49 pages left for her to read, how many pages were in the book?

Section 3: Finding fractions of a quantity

Expressing one quantity as a fraction of another

It is easy to express one quantity as a fraction of another if you remember that the numerator in a fraction tells you how many parts of the whole quantity you are dealing with.

The denominator represents the whole quantity.

So, a fraction of $\frac{3}{5}$ means you are dealing with 3 of the 5 parts that make up the whole.

WORKED EXAMPLE 5

a What fraction is 20 minutes of 1 hour?

b Express 35 centimetres as a fraction of a metre.

a 20 minutes is part of 60 minutes.

$\frac{20}{60} = \frac{2}{6} = \frac{1}{3}$ of an hour.

20 minutes is the part of the whole, so it is the numerator.

The hour is the whole, so it is the denominator.

You cannot form a fraction using one unit for the numerator and another for the denominator, so you need to convert the hour to minutes.

b 35 cm is part of 100 cm.

$\frac{35}{100} = \frac{7}{20}$ of a metre.

35 cm is the part of the whole, so it is the numerator.

The metre is the whole, so it is the denominator.

Again, you cannot use centimetres and metres in the same fraction, so you convert 1 m to 100 cm.

To write a quantity as a fraction of another quantity make sure the two quantities are in the same units and then write them as a fraction and simplify.

EXERCISE 10C

1 Calculate.

a $\frac{3}{4}$ of 12 **b** $\frac{1}{3}$ of 45 **c** $\frac{2}{9}$ of 36

d $\frac{3}{8}$ of 144 **e** $\frac{4}{5}$ of 180 **f** $\frac{1}{3}$ of 96

g $\frac{1}{2}$ of $\frac{3}{4}$ **h** $\frac{1}{3}$ of $\frac{3}{10}$ **i** $\frac{4}{9}$ of $\frac{3}{14}$

j $\frac{1}{4}$ of $2\frac{1}{2}$ **k** $\frac{3}{4}$ of $2\frac{1}{3}$ **l** $\frac{5}{6}$ of $3\frac{1}{2}$

2 Calculate the following quantities.

a $\frac{3}{4}$ of £28 **b** $\frac{3}{5}$ of £210 **c** $\frac{2}{5}$ of £30

d $\frac{2}{3}$ of £18 **e** $\frac{1}{2}$ of 3 cups of sugar **f** $\frac{1}{2}$ of 5 cups of flour

g $\frac{1}{2}$ of $1\frac{1}{2}$ cups of sugar **h** $\frac{3}{4}$ of $2\frac{1}{3}$ cups of flour **i** $\frac{2}{3}$ of $1\frac{1}{2}$ cups of sugar

j $\frac{2}{3}$ of 4 hours **k** $\frac{1}{3}$ of $2\frac{1}{2}$ hours **l** $\frac{3}{4}$ of 5 hours

m $\frac{1}{3}$ of $\frac{3}{4}$ of an hour **n** $\frac{2}{3}$ of $3\frac{1}{2}$ minutes **o** $\frac{3}{15}$ of a minute

3 Express the first quantity as a fraction of the second.

a 12p of every £1 **b** 35 cm of a 2 m length

c 12 mm of 30 cm **d** 45 minutes per 8 hour shift

e 5 minutes per hour **f** 150 m of a kilometre

g 45 seconds of 30 minutes **h** 575 ml of 4 litres

4 Nick earns £18 000 per year. His friend Samir earns £24 000 per year.

What fraction of Samir's salary does Nick earn?

5 The floor area of a room is 12 m^2. Pete buys a rug that is 110 cm wide and 160 cm long.

What fraction of the floor area will be covered by this rug?

6 A technical college has 8400 books in their library.

Of these, $\frac{1}{7}$ are general reference books, $\frac{3}{7}$ are technology related, $\frac{4}{35}$ are engineering related and the rest are computer related.

Find the total number of books in each subject category.

7 A section of road $1\frac{1}{2}$ km long is to be tarred.

$\frac{1}{6}$ is tarred in week 1, $\frac{3}{5}$ in week 2 and the rest is to be completed in week 3.

Calculate the length of road tarred each week.

8 Of 60 000 people passing through a major airport in one week, $\frac{1}{6}$ are travelling first class, $\frac{1}{4}$ are travelling business class, $\frac{3}{8}$ are travelling in economy class and the rest are using no-frills, low-cost tickets.

Work out the number of people travelling in each class.

Find answers at: cambridge.org/ukschools/gcsemaths-studentbookanswers

EXERCISE 10D

The Ancient Egyptians believed that anything other than a unit fraction was unacceptable (a unit fraction has a numerator of 1).

So they wrote all fractions as the sum or difference of unit fractions.

The sum or difference always started with the largest possible unit fraction and they did not allow repetition.

So, for example, $\frac{2}{3}$ would be written as $\frac{1}{2}+\frac{1}{6}$ and not as $\frac{1}{3}+\frac{1}{3}$.

1 Try to write each of the following as the sum or difference of unit fractions:

a $\frac{5}{8}$ **b** $\frac{3}{5}$ **c** $\frac{2}{7}$ **d** $\frac{2}{9}$ **e** $\frac{3}{10}$

2 Find three unit fractions that have a sum of $\frac{2}{5}$ when added together.

3 Find a unit fraction greater than $\frac{1}{9}$ with a denominator that is a multiple of 4 and which has three additional factors.

4 Can all fractions be written as the sum or difference of unit fractions? Justify your answer.

Checklist of learning and understanding

Equivalent fractions

- Fractions that represent the same amount are called equivalent fractions.
- You can change fractions to their equivalents by multiplying the numerator and denominator by the same value or by dividing the numerator and denominator by the same value (simplifying).

Operations on fractions

- To add or subtract fractions, find equivalent fractions with the same denominator, then add or subtract the numerators; the denominators do not get added or subtracted.
- To multiply fractions, multiply numerators by numerators and denominators by denominators.
- To divide fractions, multiply the fraction being divided by the reciprocal of the divisor.

Fractions of a quantity

- The word 'of' means multiply; so to find a fraction of an amount, you multiply that amount by the fraction.
- A quantity can be written as a fraction of another as long as they are in the same units. Write one quantity as the numerator and the other as the denominator and simplify.

Chapter review

For additional questions on the topics in this chapter, visit GCSE Mathematics Online.

1 Simplify.

a $\frac{15}{90}$ **b** $\frac{195}{230}$ **c** $4\frac{18}{48}$

2 Write each set of fractions in ascending order.

a $\frac{8}{9}, \frac{4}{5}, \frac{5}{6}, \frac{3}{7}$ **b** $2\frac{2}{5}, \frac{23}{7}, 1\frac{3}{5}, \frac{16}{9}$

3 Evaluate.

a $\frac{7}{5} + \frac{3}{8} - \frac{1}{2}$ **b** $\frac{7}{5} \times \frac{3}{8} + \frac{1}{2}$ **c** $\frac{7}{5} \div \frac{3}{8} \times 3$

d $3\frac{1}{7} + 2\frac{2}{5}$ **e** $3\frac{1}{15} - 1\frac{3}{5}$ **f** $\frac{1}{7}$ of $3\frac{2}{11} + \frac{3}{4}$

g $\frac{2}{7} \times \frac{8}{18} \div 3$ **h** $28 \div \frac{3}{4} - \frac{5}{7}$ **i** $\frac{2}{9}$ of $\frac{3}{4} - \frac{1}{8}$

4 Simplify.

a $\left(\frac{3}{8} \div \frac{13}{4}\right) + \left(\frac{5}{9} \times \frac{3}{5}\right)$ **b** $2\frac{2}{3} \times \left(8 \div \frac{4}{7} + \frac{7}{8}\right)$

c $4\frac{2}{5} + 3\frac{1}{2} + 5\frac{5}{6} - 4\frac{11}{12} + 2\frac{7}{9}$ **d** $\left(7 \div \frac{2}{9} - \frac{1}{3}\right) \times \frac{2}{3}$

5 **a** Express 425 g as a fraction of $2\frac{1}{2}$ kg.

b Express 12 000 g as a fraction of 40 kg.

6 Sandy has $12\frac{1}{2}$ litres of water.

How many bottles containing $\frac{3}{4}$ litre can she fill?

7 A surveyor has to divide a 15 km^2 area of land into equal plots each measuring $\frac{1}{2}$ km^2.

How many plots can she make?

8 A vertical ladder $7\frac{2}{10}$ m long is lowered into a manhole.

If $\frac{15}{32}$ parts of the ladder are outside the manhole when the bottom of the ladder touches the ground, how deep is the manhole?

Find answers at: cambridge.org/ukschools/gcsemaths-studentbookanswers

11 Decimals

In this chapter you will learn how to ...

- express decimals as fractions and fractions as decimals.
- convert decimals to fractions and fractions to decimals.
- order fractions and decimals.
- carry out the four basic operations on decimals without using a calculator.
- solve problems involving decimal quantities.

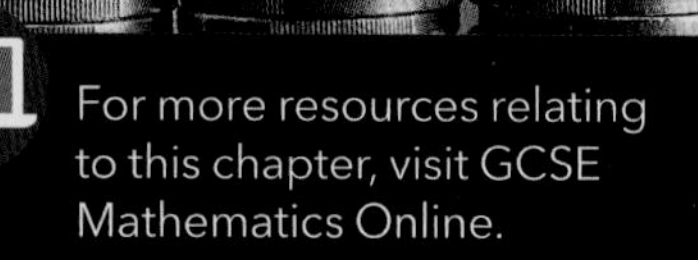

For more resources relating to this chapter, visit GCSE Mathematics Online.

Using mathematics: real-life applications

Food technologists analyse the contents of different raw and prepared foods to work out what they contain and how much there is of each ingredient. For example, how much water, protein and fat there is in a cut of meat. They use decimal fractions to give the quantities correct to tenths, hundredths or even smaller parts of a gram.

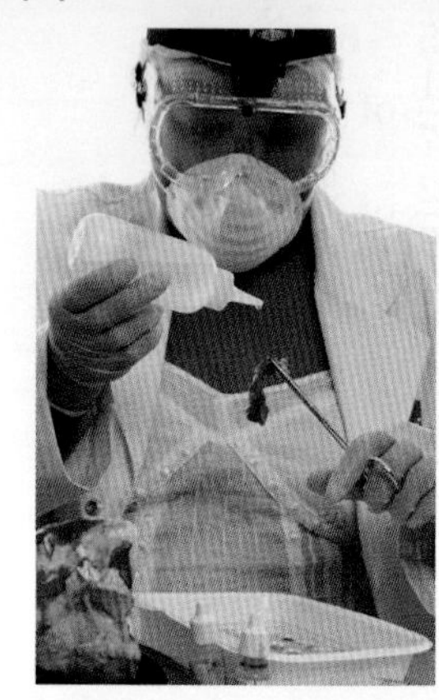

"Think about a product like a vitamin pill. We may have to list the mass of 30 different ingredients in fractions of milligrams. A milligram (mg) is 0.001 grams, a microgram (μg) is 0.001 mg and a nanogram (ng) is 0.000 000 001 g, so decimal fractions (and negative indices) are very important in my work." *(Food technologist)*

Before you start ...

KS3	You need to be able to work confidently with place value.	31.098 0.0398 300.098 0.98308 19.308 **1** Choose the number from the set that has a 3 in the: **a** hundreds position. **b** hundredths position. **c** tenths position. **d** thousandths position. **e** tens position.
KS3	Check that you can compare decimal fractions and order them by size.	**2** Fill in $<$, $=$ or $>$ between each pair of decimal fractions. **a** 0.65 [ ] 0.7 **b** 0.08 [ ] 0.01 **c** 0.8 [ ] 0.85 **d** 2.87 [ ] 0.99 **e** 4.230 [ ] 4.23
KS3, Ch 10	You need to know the fractional equivalents of some common decimals	**3** Make equivalent pairs by matching the decimals in the left-hand box with the fractions in the right. **a** 0.25 — $\frac{3}{4}$ **b** 0.375 — $\frac{3}{8}$ **c** 0.4 — $\frac{45}{90}$ **d** 0.75 — $\frac{2}{5}$ **e** 0.5 — $\frac{1}{40}$ **f** 0.025 — $\frac{4}{16}$
KS3	You need to round numbers to estimate answers.	**4** Find the approximate answer to each calculation. **a** 0.4×0.6 **b** $5.40 + 3.52 + 8.99$ **c** $99 \div 24$

Assess your starting point using the Launchpad

STEP 1

1 Write a decimal that is between:

a 2.15 and 2.16 **b** 2.155 and 2.156

c 0.6753 and 0.6754

2 Write the red digit in each number as a fraction with a denominator of 10, 100 or 1000.

a 3.0987 **b** 12.342 **c** 0.8865

3 Which is greater in each pair?

a 3.14 or $3\frac{1}{4}$ **b** 0.78 or $\frac{8}{9}$ **c** $\frac{10}{11}$ or 0.99

GO TO
Section 1:
Revision of decimals and fractions

STEP 2

4 Choose the correct answer for each calculation.

	Calculation	Possible answers		
a	24 − 2.35	A 2.165	B 216.5	C 21.65
b	19.5 − 3.45	A 16.5	B 1.605	C 16.05
c	2.25 × 3	A 675	B 67.5	C 6.75
d	18.32 × 4	A 732.8	B 73.28	C 7.328
e	7.488 ÷ 6	A 1.248	B 12.48	C 124.8
f	58.35 ÷ 3	A 0.1945	B 1.945	C 19.45

GO TO
Section 2:
Calculating with decimals

STEP 3

5 Convert each of these recurring decimals to an exact fraction in simplest terms.

a $0.\dot{2}$ **b** $0.\dot{1}\dot{8}$ **c** $1.\dot{2}\dot{1}$

GO TO
Section 3:
Converting recurring decimals to exact fractions

GO TO
Chapter review

Find answers at: cambridge.org/ukschools/gcsemaths-studentbookanswers

Section 1: Revision of decimals and fractions

Tip

This section has been included so that you can revise concepts if you need to, and check that you know them well.

Comparing decimals

The table shows the results of the men's 4×100 m relay final at the 2010 Commonwealth Games in New Delhi.

Five teams completed the race.

Team	Time (seconds)
Australia	39.14
Bahamas	39.27
England	38.74
India	38.89
Jamaica	38.79

The winning times are given as decimals. The digits that follow the decimal point indicate parts of a second (decimal fractions). There are two digits after the decimal point so the times are given correct to a hundredth of a second.

To write the times in order from fastest to slowest, compare the whole number parts of each time first. If those are the same, compare the decimal parts.

You can see that England, India and Jamaica ran the relay in less time than Australia and the Bahamas just by looking at the whole number part.

To decide who came first, second and third between England, India and Jamaica, you need to look at the decimal parts.

Tip

Remember you are comparing winning times, so you are looking for the smallest fraction of a second as this is the fastest time. The greater the fraction, the slower the team ran.

Here are the times written in a place value table.

	Tens	Ones/units	.	Tenths	Hundredths
England	3	8	.	7	4
India	3	8	.	8	9
Jamaica	3	8	.	7	9

Start by looking at the tenths.

$8 > 7$ so India came third.

England and Jamaica both have 7 in the tenths place value so compare the hundredths.

$4 < 9$ so England was faster.

The places were: England (1st), Jamaica (2nd) and India (3rd), followed by Australia then the Bahamas.

Converting decimals to fractions

Decimals can be written as fractions using place value.

Look at the time England ran the relay.

Tens	Ones/units	.	Tenths	Hundredths
3	8	.	7	4

You can see that the team ran the relay in 38 seconds and $\frac{74}{100}$ of a second.

The fraction can be simplified further: $\frac{74}{100} = \frac{37}{50}$. (÷2 top and bottom)

Any decimal can be converted to a fraction in this way. For example:

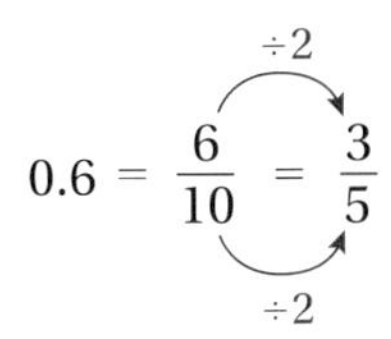

$0.6 = \frac{6}{10} = \frac{3}{5}$ (÷2) $0.25 = \frac{25}{100} = \frac{1}{4}$ (÷25) $0.375 = \frac{375}{1000} = \frac{3}{8}$ (÷125)

Converting fractions to decimals

Fractions can be converted to decimals. There are different methods of doing this and you should choose the method that is easiest for the fraction involved.

Method 1: Equivalent fractions with denominators that are powers of 10, e.g. 10, 100, 1000 and so on.	**Method 2:** Pen and paper division. Using the 'traditional' written method of division.	**Method 3:** Calculator division.
Example: express $\frac{61}{125}$ as a decimal. Multiply the numerator and denominator by 2 repeatedly until you reach a denominator that is a power of 10. $\frac{61}{125} = \frac{122}{250} = \frac{244}{500} = \frac{488}{1000} = 0.488$ $\frac{61}{125} = 0.488$ This method works well if the denominator is a factor of 10, 100 or 1000.	Example: express $\frac{5}{8}$ as a decimal. Work out 5 ÷ 8 using division. Set up a line with 5.0 below it, and write in digits of the answer above the line as you go. insert decimal point so that it aligns with the decimal point in the number you are dividing 0 . 6 2 5 8 ⟌ 5 . 0 4 . 8 0.6 × 8 0 . 2 0 0 . 1 6 0.02 × 8 0 . 0 4 0 0 . 0 4 0 0.005 × 8 0 . 0 0 $\frac{5}{8} = 0.625$	Example: express $\frac{2}{3}$ as a decimal. Input 2 ÷ 3 on your calculator.  0.666666666 The 6s continue forever (they recur). Show this by writing the answer as $0.\dot{6}$.  **Tip** Remember you write a dot above the first and last digits of the recurring number if more than one digit recurs, so 5.134134134... would be written $5.\dot{1}3\dot{4}$.

EXERCISE 11A

1 Write each of the following decimals as a fraction in its simplest form.

a 0.6 **b** 0.84 **c** 1.64 **d** 0.385 **e** 0.125
f 1.08 **g** 0.875 **h** 0.008 **i** 3.064 **j** 0.333

Find answers at: cambridge.org/ukschools/gcsemaths-studentbookanswers

Tip

When you have to compare and order ordinary fractions you can convert them all to decimals and compare them easily using place value. This is often quicker than changing them all into equivalent fractions with a common denominator.

2 Convert the following fractions to decimals without using a calculator.

a $\frac{3}{5}$ **b** $\frac{3}{4}$ **c** $\frac{18}{25}$ **d** $\frac{19}{20}$ **e** $\frac{34}{50}$

f $\frac{110}{250}$ **g** $\frac{89}{200}$ **h** $\frac{76}{500}$ **i** $\frac{185}{20}$ **j** $\frac{145}{50}$

k $\frac{11}{6}$ **l** $\frac{3}{8}$ **m** $\frac{9}{4}$ **n** $\frac{8}{9}$ **o** $\frac{19}{8}$

3 Use a calculator to convert the fractions from $\frac{1}{9}$ to $\frac{8}{9}$ into decimals.

a What pattern do you notice?

b What do you call decimals of this nature?

c Repeat this for the fractions from $\frac{1}{6}$ to $\frac{5}{6}$.
Write each decimal correctly using the notation for recurring decimals as necessary.

d Convert $\frac{1}{11}$ and $\frac{2}{11}$ to decimals.
Write the decimals correctly using notation for recurring decimals.

e Predict what $\frac{3}{11}$ and $\frac{4}{11}$ will be if you convert them to decimals.
Check your prediction using a calculator.

4 Arrange the following in descending order.

a 5.2, 5.29, 8.62, 4.92, 4.09 **b** 7.42, 0.76, 0.742, 0.421, 3.219

c 14.3, 14.72, 14.07, 14.89, 14.009 **d** 0.23, 0.26, 0.273, 0.287, 0.206

e 0.403, $\frac{1}{2}$, $\frac{2}{3}$, 0.68, 0.45, $\frac{5}{11}$ **f** $\frac{7}{9}$, $\frac{3}{8}$, 0.625, 0.88, 0.718

5 Fill in the boxes using $<$, $=$ or $>$ to make each statement true.

a 13.098 ☐ 13.099 **b** 0.312 ☐ 0.322 **c** $\frac{5}{6}$ ☐ 0.84

d 0.375 ☐ $\frac{3}{8}$ **e** 2.05 ☐ $\frac{205}{1000}$ **f** $\frac{3}{5}$ ☐ 0.7

g $\frac{2}{5}$ ☐ 0.35 **h** $\frac{18}{25}$ ☐ 0.67 **i** $\frac{1}{3}$ ☐ 0.37

6 Write a decimal fraction that is between each pair of decimals.

a 3.135 and 3.136 **b** 0.6645 and 06646 **c** 4.998 and 4.999

7 The lengths of some of the world's longest roller coaster rides are given in the table.

Roller coaster	Country	Length of ride (km)
The Beast	USA	2.243
California Screaming	USA	1.851
Formula Rossa	United Arab Emirates	2.0
Fujiyama	Japan	2.045
Steel Dragon	Japan	2.479
The Ultimate	UK	2.268

a Which is the longest roller coaster?

b Which is the shortest?

c Is the Steel Dragon longer or shorter than $2\frac{1}{2}$ km?

d Which roller coasters are longer than $2\frac{1}{4}$ km?

e Write the lengths in order from longest to shortest.

Section 2: Calculating with decimals

You need to be able to add, subtract, multiply and divide decimals without using a calculator. Working through the following activities will help you to remember how to do this.

Adding and subtracting decimals

Add or subtract decimals in columns by lining up the places and the decimal points.

WORKED EXAMPLE 1

Calculate.

a 12.7 + 18.34 + 3.087 **b** 399.65 − 245.175

```
   12.7          399.650
   18.34       − 245.175
 +  3.087        154.475
   34.127
```

Write a 0 as a place holder here.

Multiplying and dividing decimals

Tip

This is really a place value movement by the digits as you multiply or divide by a power of 10, but visually, it looks like the decimal point is moving.

To **multiply or divide by a power of 10** (10, 100, 1000, etc.):

- The digits move as many places to the left as the number of zeros when multiplying.
- The digits move as many places to the right as the number of zeros when dividing.

To **multiply decimals by decimals**:

- Ignore the decimal points and multiply the numbers.
- Place the decimal point in the answer so it has the same number of digits after the decimal point as there were altogether in the multiplication problem.

To **divide by a decimal**:

- Make the divisor a whole number by multiplying the divisor and the dividend by the same power of 10.
- Then divide as normal, keeping the decimal point in the answer directly above the decimal point in the number you are dividing; be careful to insert digits according to their correct place value.

To **divide a decimal by a whole number**:

- Use the 'traditional' written method of division but make sure you align the decimal point in the answer with the decimal point in the dividend, and keep the decimals aligned in all your working; be careful to insert digits according to their correct place value.

WORK IT OUT 11.1

Work with a partner.
Read through the notes about how to multiply and divide with decimals.
Provide two numerical examples to show how to do each of the following.

a Multiply by a power of 10.
b Divide by a power of 10.
c Multiply a decimal fraction by a decimal fraction.
d Divide by a decimal.
e Divide a decimal by a whole number.

EXERCISE 11B

1 Estimate first then calculate.

a $0.8 + 0.78$ **b** $12.8 - 11.13$ **c** $0.8 + 0.9$
d $15.31 - 1.96$ **e** 2.77×8.2 **f** 9.81×3.5

2 Evaluate without a calculator.

a $12.7 + 18.34 + 35.01$ **b** $12.35 + 8.5 + 2.91$ **c** $6.89 - 3.28$
d $34.45 - 12.02$ **e** $345.297 - 12.39$ **f** $56 + 8.345 - 34.65$
g $27.4 + 9.01 - 12.451$ **h** 0.786×100 **i** 54.76×2000
j 1.234×0.65 **k** 87.87×2.34 **l** $1.83 \div 61$
m $0.358 \div 4$ **n** $5.053 \div 0.62$ **o** $31.72 \div 0.04$

3 The world record for the men's 4×100 m relay is 36.84 seconds (Jamaica, 2008) and the Commonwealth Games record is 38.20 seconds (England, 1998).

a What is the time difference between the World Record and the Commonwealth Games Record?

b In the 2010 Commonwealth Games, the English team ran the relay in 38.74 seconds.
How much slower is this than their record in 1998?

c Each of the four runners in a relay runs 100 m.
Calculate the average time taken for 100 m during the World Record winning race.

d Do you think each runner takes the same amount of time?
Explain your reasoning.

4 Nadia wants to make a dish that requires 1.5 litres of cream.

She has four 0.385 litre cartons of cream. Does she have enough?

5 Josh takes a multivitamin every morning.

He calculates that if he takes one tablet every day for a week he will take in 1166.69 mg of vitamin C, 54.6 mg of boron and 257.95 mg of calcium.

Work out how much of each ingredient there is in a tablet.

6 The Chetty household uses about 25.75 kilowatt hours of electricity per day.

Calculate how much they will use in:

a one week.

b one (non-leap) year.

7 Sandra has £87.50 in her purse. She buys two sweaters that cost £32.99 each.

How much money does she have left?

8 George travels from York to Oxford by car.

His odometer reads 123 456.8 km when he leaves York and 123 642.7 km when he arrives in Oxford.

How far did he travel?

9 If I have 5.67 litres of juice, how many cups containing $\frac{2}{15}$ of a litre can I pour?

10 Find 0.75 of 2400.

11 Sheldon placed fence posts 0.75 m apart all along his boundary fence.

If the total fence is 51 metres long, how many posts will there be?

12 June bought 6.65 litres of petrol at £1.29 per litre. How much did she pay?

13 Toni earns £28 650 per year.

a There are 365.25 days per calendar year. How much does Toni earn each day?

b Toni works 5 days per week. How much does she get paid for 5 days' work?

c There are 52 weeks in a year. Toni is paid the same amount every week. How much does she get paid for one week's work?

d Explain the difference between your answers to **b** and **c**.

Section 3: Converting recurring decimals to exact fractions

There are three types of decimal fractions:

- exact or terminating decimals such as 0.25 and 0.375.
- non-terminating and recurring decimals in which a set of digits repeats to infinity.
- other decimals which are non-terminating and non-recurring (irrational numbers) such as π (3.141592654...) and $\sqrt{2}$ (1.414213562...).

Find answers at: cambridge.org/ukschools/gcsemaths-studentbookanswers

EXERCISE 11C

How do you know whether a fraction will give you a terminating or recurring decimal?

1. Nazeem says you can work this out by writing the denominator of the fraction as the product of its prime factors.

   Any denominator that has only 2s and/or 5s as prime factors will produce a terminating decimal.
   All the others will produce recurring decimals.

   Do you agree?
   Test Nazeem's hypothesis and state whether it is correct or not.

2. Andy says Nazeem is wrong and he uses sixths to show this:
   $6 = 2 \times 3$, so it should produce recurring decimals according to Nazeem.
   But $\frac{3}{6}$ gives an exact decimal of 0.5, so Nazeem is wrong.
   Consider Andy's argument and use it to adapt Nazeem's hypothesis so that it holds for fractions such as $\frac{3}{6}$ as well.

Writing decimals as fractions in the form of $\frac{a}{b}$

Exact and recurring decimals can be written as fractions in the form of $\frac{a}{b}$, so that they are rational numbers.

You already know how to convert exact decimals to fractions by writing them with a denominator that is a power of 10.

For example, $0.235 = \frac{235}{1000} = \frac{47}{200}$

This method doesn't work for a non-terminating decimal because you cannot give it a denominator with an infinite number of zeros.

To convert a recurring decimal to an exact fraction you can set up a pair of equations using the decimal and then use algebra to solve for x.

The answer should always be given in its simplest form.
When you convert recurring decimals, do the following:

- If the length of the recurring decimal is n, then you multiply by 10^n to find the second equation.
- Always give your answer in its simplest form.

Calculator tip

It is important to understand how your calculator deals with recurring decimals. Some calculators will round them off and other calculators will just cut them off (truncate them) at the end of the display area.

To find out what your calculator does, enter $2 \div 3$ on your calculator.

If you get 0.666 666 666 then your calculator truncates the recurring decimal.

If you get 0.666 666 667 then your calculator rounds the last available decimal place.

WORKED EXAMPLE 2

Change to a fraction.

a $0.\dot{4}$ **b** $0.\dot{2}\dot{4}$ **c** $0.2\dot{3}$

a $0.\dot{4}$

Let $x = 0.444...$ (1)

Then $10x = 4.444...$ (2)

Subtract (1) from (2).

$10x - x = 4.444 - 0.444$ — This gets rid of the recurring digits.

$9x = 4$

Divide both sides by 9 to get x.

So, $x = \frac{4}{9}$

b $0.\dot{2}\dot{4}$

Let $x = 0.\dot{2}\dot{4}$ (1)

Then $100x = 24.\dot{2}\dot{4}$ (2)

Subtract (1) from (2).

$99x = 24.\dot{2}\dot{4} - 0.\dot{2}\dot{4}$

$99x = 24$

So, $x = \frac{24}{99} = \frac{8}{33}$

c $0.2\dot{3}$

Let $x = 0.2\dot{3}$ (1) — Only one digit of the number recurs, so x is multiplied by 10.

Then $10x = 2.\dot{3}$ (2)

And $100x = 23.\dot{3}$ (3)

Subtract (2) from (3).

$100x - 10x = 23.\dot{3} - 2.\dot{3}$

$90x = 21$

So, $x = \frac{21}{90} = \frac{7}{30}$ — Note that you need three equations here to get the recurring part on its own after the decimal point.

EXERCISE 11D

1 Express each rational number as a decimal:

a $\frac{3}{8}$ **b** $\frac{5}{16}$ **c** $\frac{5}{11}$ **d** $\frac{4}{9}$

e $\frac{18}{7}$ **f** $\frac{7}{15}$ **g** $\frac{23}{7}$ **h** $\frac{8}{7}$

2 Express each decimal as a rational number:

a 0.888... **b** 2.777... **c** 8.202020... **d** 3.181818...

e 6.833333... **f** 0.6565... **g** 0.277 777... **h** 1.727272...

3 Write the following as exact fractions in simplest form:

a $0.\dot{4}$ **b** $0.\dot{7}\dot{4}$ **c** $0.8\dot{7}$ **d** $0.11\dot{4}$

e $0.\dot{9}4\dot{3}$ **f** $0.1\dot{8}5\dot{7}$ **g** $4.\dot{5}6\dot{7}$ **h** $0.11\dot{3}$

Find answers at: cambridge.org/ukschools/gcsemaths-studentbookanswers

Calculator tip

Most calculators cannot deal with recurring fractions such as $0.\dot{1}$ (although they can easily deal with 0.1), so you have to know how to convert these to exact fractions using algebra.

4 What is the exact length of line AC in this diagram?

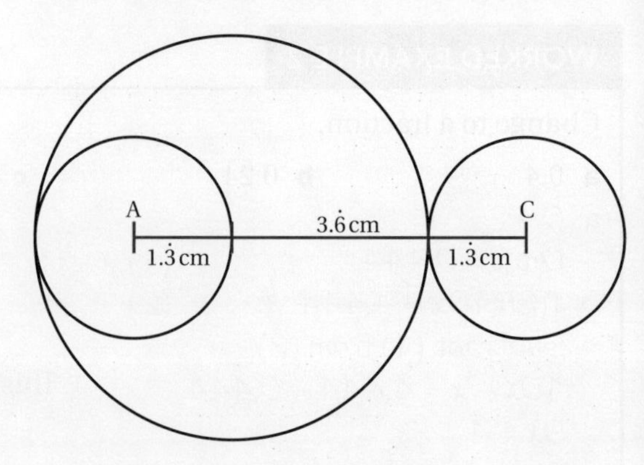

5 Write the following pairs of decimals as pairs of fractions in the form of $\frac{a}{b}$.

a 0.3 and $0.\dot{3}$ **b** 0.17 and $0.\dot{1}\dot{7}$ **c** 0.173 and $0.\dot{1}7\dot{3}$

6 Use your results to come up with a theory about writing recurring decimals as fractions.

7 Test your theory using a few examples of your own. Comment on what you find out.

Checklist of learning and understanding

Decimals and fractions

- You can express decimals as fractions by writing them with a denominator that is a power of ten and then simplifying them. For example, $0.4 = \frac{4}{10} = \frac{2}{5}$.
- You can change fractions to decimals by dividing the numerator by the denominator. The bar in a fraction stands for division. For example, $\frac{3}{4} = 3 \div 4 = 0.75$.
- Changing ordinary fractions to decimals makes it easier to compare their sizes using place value.

Calculations with decimals

- Pen and paper methods are important in a non-calculator exam paper; you can use any method as long as you show your working.
- Decimals can be added and subtracted by lining up the places and the decimal points.
- Decimals can be multiplied like whole numbers as long as you insert the decimal point so there are the same number of decimal places in the answer as there were altogether in the numbers being multiplied.
- Decimals can be divided by making the divisor a whole number (multiply both numbers by a power of ten to do this). Then divide normally and align the decimal point in the answer with the decimal point in the number being divided; be careful to insert digits according to their correct place value.
- When dividing a decimal by a whole number, use the 'traditional' written method of division but make sure you align the decimal point in the answer with the decimal point in the dividend, and keep the decimals aligned in all your working; be careful to insert digits according to their correct place value.

Recurring decimals

- Recurring decimals can be written as fractions in the form of $\frac{a}{b}$. To do this, set up a pair of equations and subtract to get rid of the recurring digits, then simplify to get an answer.

Chapter review

For additional questions on the topics in this chapter, visit GCSE Mathematics Online.

1 Arrange each set of numbers in ascending order.

a 4.2, 4.8, 4.22, 4.97, 4.08 **b** 2.96, 2.955, $2\frac{46}{50}$, $2\frac{9}{25}$, 2.12

c $\frac{3}{4}$, 0.86, $\frac{4}{5}$, 0.78, $\frac{5}{6}$, 0.91

2 Convert to decimals and insert <, = or > to compare the fractions.

a $\frac{3}{5}$ □ $\frac{12}{20}$ **b** $\frac{5}{6}$ □ $\frac{7}{11}$ **c** $\frac{2}{9}$ □ $\frac{1}{7}$

3 Write each as a fraction in its simplest terms.

a 0.88 **b** 2.75 **c** 0.008

4 **a** Increase $\frac{2}{5}$ by 2.75. **b** Reduce 91.07 by $\frac{1}{2}$ of 42.8.

c Divide 4 by 0.125. **d** Multiply 0.4 by 0.8.

5 **a** Add 4.726 and 3.09. **b** Subtract 2.916 from 4.008.

c Multiply 8.76 by 100. **d** Divide 18.07 by 1000.

e Multiply 4.12 by 0.7. **f** Simplify $\frac{32.64}{2.4}$.

6 Jarryd and Kate have £16 each. Jarryd spends 0.416 of his money and Kate spends $\frac{4}{15}$ of hers.
Who has more money left? How much more?

7 **a** Write $\frac{4}{9}$ as a recurring decimal. *(1 mark)*

b Tick the appropriate box to indicate whether each fraction can be represented by a recurring or a terminating decimal.
You do not have to find the decimal values.
The first two are done for you. *(2 marks)*

Fraction	Recurring decimal	Terminating decimal
$\frac{1}{2}$		✓
$\frac{4}{9}$	✓	
$\frac{3}{20}$		
$\frac{17}{60}$		
$\frac{73}{400}$		

c Express $0.\dot{2}\dot{7}$ as a fraction in its lowest terms. *(3 marks)*

© OCR 2013

8 Convert each recurring decimal to a fraction in its simplest form:

a $0.\dot{2}$ **b** $0.\dot{5}\dot{4}$ **c** $0.8\dot{5}$ **d** $2.4\dot{3}\dot{8}$

Find answers at: cambridge.org/ukschools/gcsemaths-studentbookanswers

12 Units and measurement

In this chapter you will learn how to ...

- work with and convert standard units of measurement.
- use and convert compound units of measurement.
- work with map scales and bearings.
- construct and use scale diagrams to solve problems.

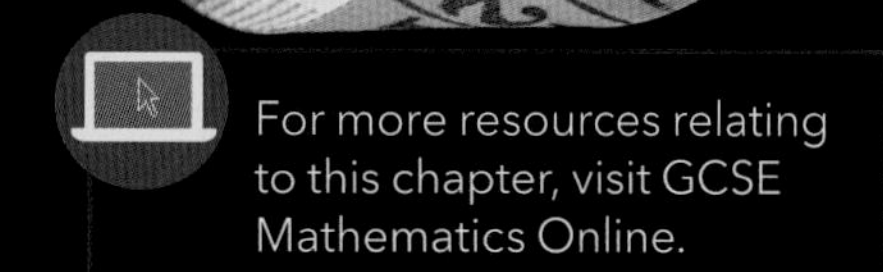

For more resources relating to this chapter, visit GCSE Mathematics Online.

Using mathematics: real-life applications

Measurement has practical applications in many different jobs, but it is also important in everyday activities. Being able to read and work with measurements is important when you make or alter clothes, work out what materials you need to build things, and weigh ingredients to make a recipe.

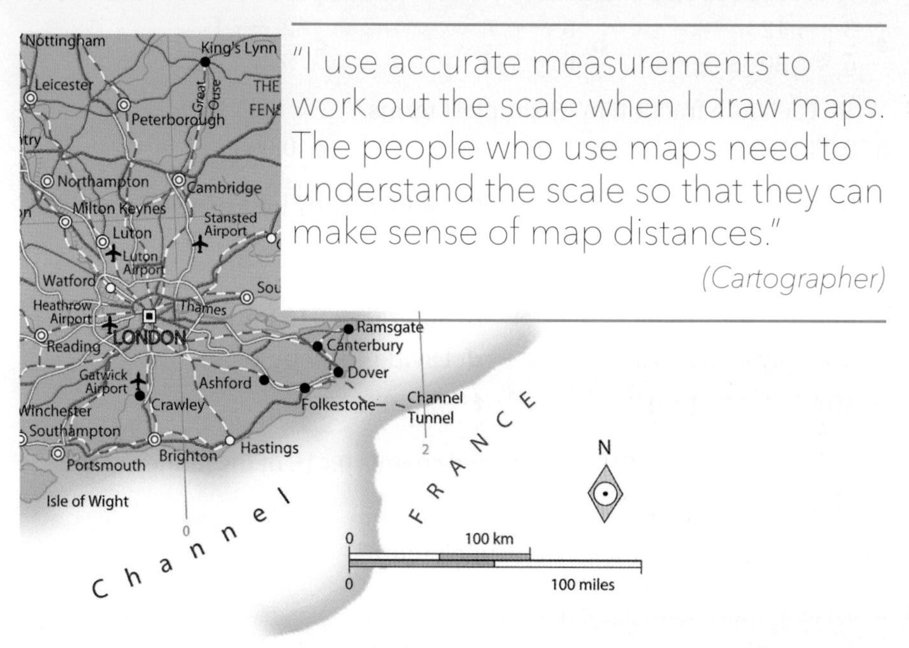

"I use accurate measurements to work out the scale when I draw maps. The people who use maps need to understand the scale so that they can make sense of map distances."

(Cartographer)

Before you start ...

KS3	You must be able to multiply and divide using multiples of 10.	**1** Work out. **a** 1000×10 **b** $10 \div 1000$ **c** $100 \div 1000$
Ch 3	You should be able to substitute numbers into a simple formula.	**2** Use the formula to work out the pay of each person: pay = hours worked × rate of pay **a** Amelia: 20 hours worked at £7 per hour. **b** Billy: 15 hours worked at £6 per hour. **c** Catrin: 40 hours worked at £5.50 per hour.
KS3	You should be able to solve problems involving direct proportion.	**3** Six pencils cost 90p. Work out the cost of: **a** 12 pencils. **b** 1 pencil. **c** 4 pencils.

Assess your starting point using the Launchpad

STEP 1

1 Convert.

a 11 569 grams into kilograms

b $4\frac{1}{2}$ hours into seconds

c 123 456 pence into pounds (£)

d 5 cm^2 into m^2

GO TO
Section 1:
Standard units of measurement

STEP 2

2 A car travels 16 kilometres in 20 minutes.

a What is the average speed of the car in kilometres per hour?

b Express this speed in m/s.

3 Annie is stuck in traffic.
She works out that her taxi is travelling at an average speed of $6\frac{2}{3}$ m/s.

a Express that speed in kilometres per hour.

b How long will it take her to cover a distance of 600 m at this speed?

c The traffic clears slightly and the average speed increases to 30 km/h. How far will she travel in 20 seconds at this speed?

GO TO
Section 2:
Compound units of measurement

STEP 3

4 A helicopter is drawn using a scale of 1 : 100.
On the scale drawing the length of the helicopter blade is 8 cm.
How long is the actual blade?

5 The helicopter takes off from a point X and flies due north for 30 km to reach point Y.

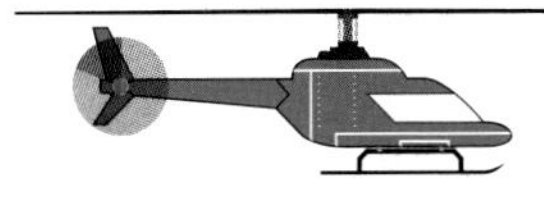

It then flies on a bearing of 150° for 15 km to reach point Z.

a Use a scale of 1 cm to represent 10 km to make a scale drawing showing this journey.

b Use your diagram, to find the bearing from X to Z.

c By measuring your diagram, find the actual distance directly between X and Z.

GO TO
Section 3:
Maps, scale drawings and bearings

GO TO
Chapter review

Find answers at: cambridge.org/ukschools/gcsemaths-studentbookanswers

Did you know?

The USA, Liberia and Myanmar are the only three countries in the world that do not officially use the metric system of measurement. All other countries have adopted the metric system.

Section 1: Standard units of measurement

You already know that we use standard metric units of measurement for recording length, area, volume, capacity, mass and money.

In the metric system, units of measurement are divided into sub-units with prefixes such as milli-, centi-, deci-, deca-, hecto- and kilo-.

The same prefixes are used for length, mass and capacity.

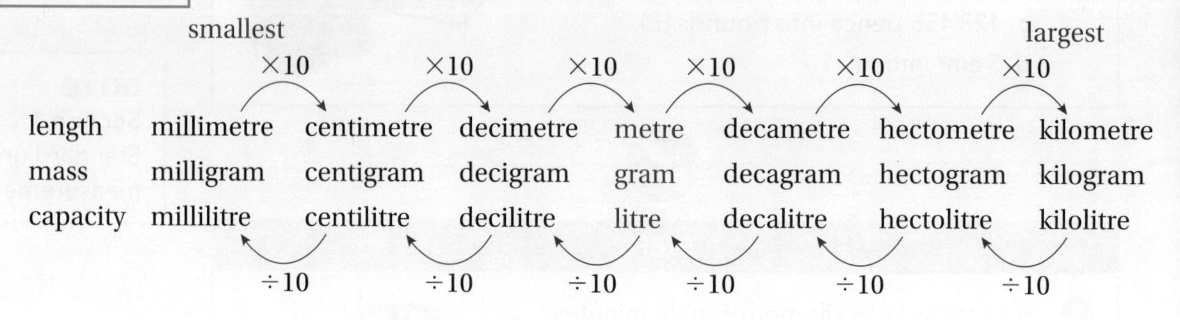

Each sub-unit is 10 times bigger than the one before it.

Centimetres are 10 times bigger than millimetres, decimetres are 10 times bigger than centimetres and so on.

Converting between units

To convert between units in the metric system you need to multiply or divide by powers of 10.

This diagram shows how to convert between centimetres and metres.

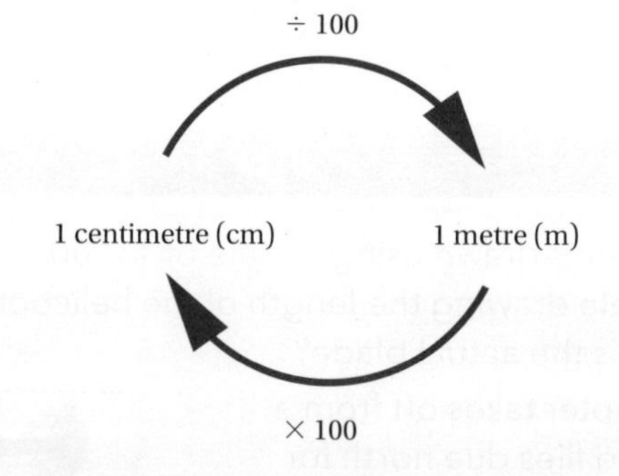

Tip

When you convert from a smaller to a larger unit there will be fewer of the larger units, so you **divide by a power of 10**.

When you convert from a larger to a smaller unit there will be more of the smaller units, so you **multiply by a power of 10**.

The **conversion factor** is 100 because you are changing across **two** sub-units.

Each sub-unit is 10 times greater or smaller than the one next to it, so here you have to multiply or divide by $10^2 = 100$.

Key vocabulary

conversion factor: the number that you multiply or divide by to convert one measure into another smaller or larger unit.

You will use the following conversions often, so it is useful to remember them.

1 centimetre (cm) = 10 millimetres (mm)

1 metre (m) = 100 centimetres (cm)

1 kilometre (km) = 1000 metres (m)

1 kilogram (kg) = 1000 grams (g)

1 tonne (t) = 1000 kilograms (kg)

1 litre (l) = 1000 millilitres (ml)

1 litre (l) = 1000 cubic centimetres (cm^3)

1 cubic centimetre (cm^3) = 1 millilitre (ml)

WORK IT OUT 12.1

In a sponsored swim the total number of lengths swum is 94.

Each length is 25 metres.

How many kilometres were swum in total?

Which of the answers below is correct?

Option A	Option B	Option C
Total number of metres $= 25 \times 94 = 2350$ metres	Total number of metres $= 25 \times 94 = 2350$ metres	Total number of metres $= 25 \times 94 = 2350$ metres
Conversion: 100 metres = 1 km	Conversion: 100 metres = 1 km	Conversion: 1000 metres = 1 km
$2350 \div 100 = 23.5$ km swum in total	$2350 \times 100 = 235\,000$ km swum in total	$2350 \div 1000 = 2.35$ km swum in total

Converting areas and volume

Area is measured in square units such as mm^2 (square millimetres), cm^2, m^2 or km^2 so any conversion factor also has to be squared.

For example, to convert cm to mm you would multiply by 10, But to convert cm^2 to mm^2 you would need to multiply by 10^2 which is 100.

The two blue rectangles below have the same area.

The conversion factor from m^2 to cm^2, and vice versa, is 10 000.

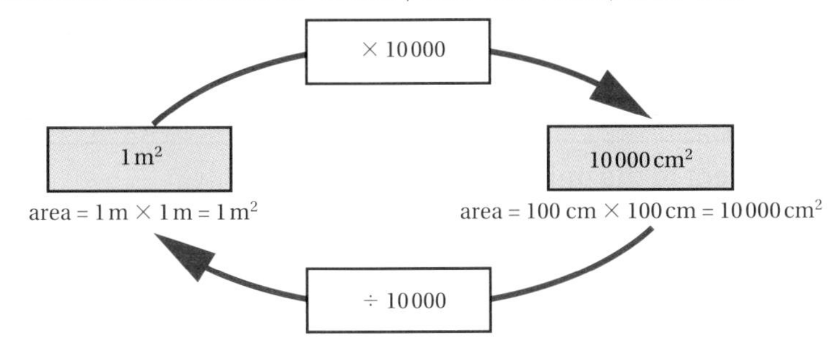

Tip

$1\ cm^2 = 100\ mm^2$ (Conversion factor = 100)
$1\ m^2 = 10\,000\ cm^2$ (Conversion factor = 10 000)
$1\ km^2 = 1\,000\,000\ m^2$ (Conversion factor = 1 000 000)

WORKED EXAMPLE 1

Convert each measure to the units given.

a $10\ m^2$ to cm^2 **b** $8.6\ km^2$ to m^2 **c** $3500\ mm^2$ to cm^2

a $10\ m^2$ to cm^2
$= 10 \times 10\,000 = 100\,000\ cm^2$ — Conversion factor = 10 000

b $8.6\ km^2$ to m^2
$= 8.6 \times 1\,000\,000 = 8\,600\,000\ m^2$ — Conversion factor = 1 000 000

c $3500\ mm^2$ to cm^2
$= 3500 \div 100 = 35\ cm^2$ — Conversion factor = 100

Find answers at: cambridge.org/ukschools/gcsemaths-studentbookanswers

Volume is measured in cubic units such as mm^3 (cubic millimetres), cm^3 or m^3 so any conversion factor also has to be cubed.

Again, to convert cm to mm you would multiply by 10, but to convert cm^3 to mm^3 you would need to multiply by 10^3 which is 1000.

The two cuboids below have the same volume.

The conversion factor from m^3 to cm^3, and vice versa, is 1 000 000.

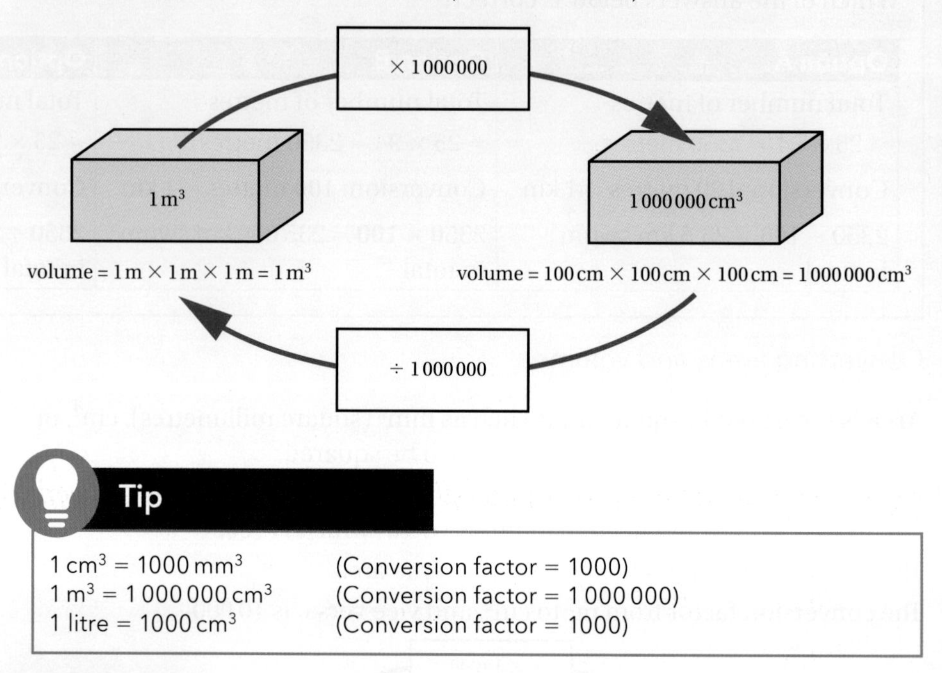

Tip

$1\text{ cm}^3 = 1000\text{ mm}^3$ (Conversion factor = 1000)
$1\text{ m}^3 = 1\,000\,000\text{ cm}^3$ (Conversion factor = 1 000 000)
$1\text{ litre} = 1000\text{ cm}^3$ (Conversion factor = 1000)

WORKED EXAMPLE 2

Convert each measurement to the unit given.

a 6.3 m^3 to cm^3 **b** $96\,500\,000\text{ cm}^3$ to m^3 **c** 750 cm^3 to mm^3

a 6.3 m^3 to cm^3
$= 6.3 \times 1\,000\,000 = 6\,300\,000\text{ cm}^3$ — Conversion factor = 1 000 000

b $96\,500\,000\text{ cm}^3$ to m^3
$= 96\,500\,000 \div 1\,000\,000 = 96.5\text{ m}^3$ — Conversion factor = 1 000 000

c 750 cm^3 to mm^3
$= 750 \times 1000 = 750\,000\text{ mm}^3$ — Conversion factor = 1000

Look at this plan of a room.

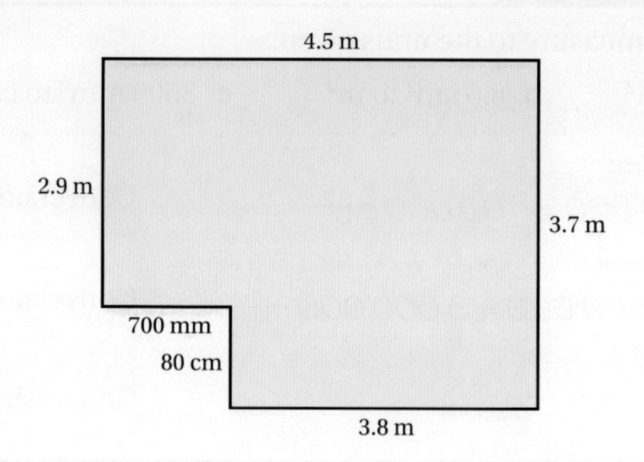

The real length of each section of wall is given but the units of measurement are different.

It is easy to make mistakes if you try to calculate with measurements in different units, so it makes sense to convert them all to the same unit before doing any calculations.

WORKED EXAMPLE 3

A builder is to put masking tape all along the line where the walls meet the ceiling of the room drawn above. How many metres of masking tape does the builder need?

80 cm ÷ 100 = 0.8 m

700 mm ÷ 1000 = 0.7 m

In this example, it makes sense to work in metres, as the answer needs to be given in metres.

Total distance around the room:

2.9 + 4.5 + 3.7 + 3.8 + 0.8 + 0.7 = 16.4 m

The builder needs 16.4 m of masking tape.

EXERCISE 12A

Tip

You will use measurement conversions when you deal with scale drawings and maps in Section 3 and when you deal with perimeter, area and volume in Chapters 15, 16 and 21.

1 Match the measurement on the left to the equivalent measurement on the right.

10 000 mm	10 l
10 000 ml	10 g
10 kg	10 m
0.01 kg	1 mm
0.1 cm	10 000 g

2 Convert the following lengths and masses into the given units to complete the following.

a 2.5 km = ☐ m **b** 85 cm = ☐ mm **c** 34 m = ☐ mm

d 1.55 m = ☐ mm **e** 0.07 m = ☐ cm **f** 5.4 kg = ☐ g

g 0.9 kg = ☐ g **h** 102 g = ☐ kg **i** 14.5 g = ☐ kg

3 Add the following capacities. Give your answers in the units indicated in brackets.

a 3.5 l + 5 l (ml) **b** 2.3 l + 450 ml (l) **c** 20 l + 4.5 l + 652 ml (l)

4 Mandy wants to use square concrete slabs to form a border round a rectangular garden.
The garden is 450 cm wide and 5.5 metres long.

a Draw a diagram to represent the garden.

b Calculate the perimeter of the garden in metres.

c If the concrete slabs are 120 cm long, how many will Mandy need to form the border?

d The slabs cost £4.55 each. Work out how much it will cost Mandy to buy the slabs she needs.

e What is the cost per metre for the concrete border?

Find answers at: cambridge.org/ukschools/gcsemaths-studentbookanswers

5 Convert each of the following into the required units.

a Total weight in kg of 3 bags of flour, each of mass 1200 g.

b The length in cm of a 7.763 m long whale.

c 3567 kg of lead into tonnes. **d** Area of 5 m^2 into mm^2.

e 96.35 m^3 of sand into cm^3. **f** 345 cm^3 of water into litres.

6 Write =, < or > between the two measurements.
Then calculate the difference between the greater and smaller measurement.
Give your answer in the most appropriate units.

a 5.7 cm □ 560 mm **b** 8 kg □ 7900 g **c** 590 l □ 59 015 cl

d 19.3 cm □ 189 mm **e** 101 cl □ 0.99 l **f** 198 cm □ 0.001 99 km

g 145 g □ 1.45 kg **h** 1987 t □ 10 987 kg **i** 2 m^3 □ 2 000 000 cm^3

7 The actual dimensions of a new kitchen sink are given on the diagram in millimetres.

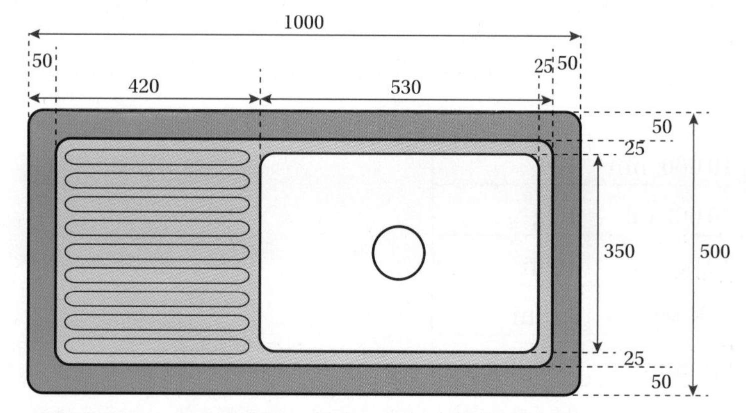

a Calculate the total perimeter of the unit in centimetres.

b What is the area of the draining board section in square millimetres?

c Convert the area of the draining board to square metres.

d The sink is 170 mm deep.
What is the approximate capacity of the sink in:

i cubic millimetres? **ii** litres?

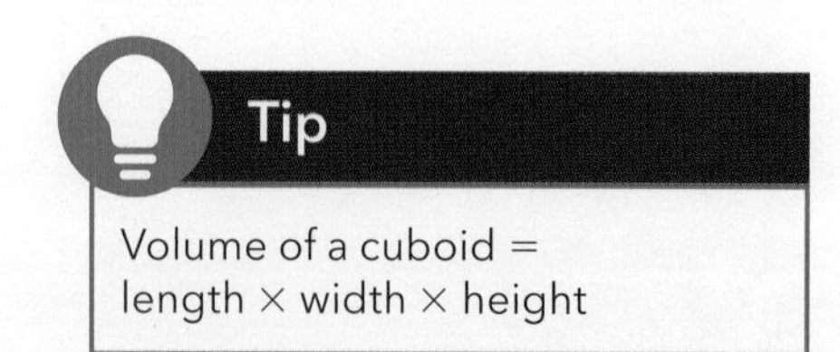

Tip

Volume of a cuboid = length × width × height

Time

The units of time we use on a daily basis are not decimal units.

To convert units of time you have to work out how many sub-units there are in the units you are working with.

For example, to convert from weeks to days, you would need to multiply by 7 as there are seven days in a week.

To convert from seconds to hours, you would need to divide by 60 to get minutes and then by 60 again to get hours.
You can also do this in one step by dividing by 3600.

Time is sometimes given in decimal form, for example 4.8 hours.

Tip

1 year = 365 days (366 in a leap year)
1 day = 24 hours = 1440 minutes
1 hour = 60 minutes = 3600 seconds
1 minute = 60 seconds

You can convert these times back to ordinary units in different ways.

There are 60 minutes in an hour, so

4.8×60 minutes = 288 minutes = 4 hours 48 minutes.

Or, you can think of this as 4 hours and 0.8 hours.

$0.8 \times 60 = 48$, so the time is 4 hours and 48 minutes.

In athletics and other timed sporting events the times are often given using decimal fractions of a second.

For example, in August 2009, Usain Bolt ran the 100 m in the World Record time of 9.580 seconds.

In the metric system, 1 second = 1000 milliseconds.

9.580 is time recorded exact to $\frac{1}{1000}$ of a second.

This is 9 seconds and 580 milliseconds. You cannot convert it in any other way.

Calculator tip

Modern scientific calculators have a mode that you can use to do sexadecimal calculation (hours, minutes and seconds). Different models work in different ways, so read the manual or check online to see how your calculator works.

12-hour and 24-hour times

The 12-hour time system uses **am** to show times from midnight to noon and **pm** for time from noon till midnight.

The 24-hour time system shows the times 00:00 to 23:59. Midnight is 00:00.

Tip

Add 12 to write a pm time using the 24-hour clock, for example 10.35pm + 12 = 22:35.
Subtract 12 to write a time between 13:00 and 23:59 using the 12-hour clock, for example 15:40 − 12 = 3.40pm.

Problem-solving framework

Mr Smith is in Moscow.
He needs to return to London for an urgent meeting in the morning.

The flight from Moscow to London takes 3.6 hours.
The local time in Moscow is 3 hours ahead of the UK.
The flight is scheduled for take-off at 18:55 local time, and on arrival it will take 45 minutes to pass through customs and exit the airport.

Mr Smith will stop to buy a coffee, sandwich and newspaper for the train.

Trains for central London leave at 5, 27 and 46 minutes past the hour, and the journey will take 29 minutes.

There is a 7 minute walk from the train station to his hotel.

What is the earliest time that Mr Smith can expect to arrive at his hotel? Give your answer using the 12-hour system of time.

Tip

When you work with time, treat hours and minutes separately. If you carry over from hours to minutes, remember you are carrying 60 minutes.

Continues on next page …

Steps for approaching a problem-solving question	What you would do for this example
Step 1: What have you got to do?	First work out the time of arrival in London and then work out the how long it takes from there to the hotel.
Step 2: What information do you need?	Flight departure times: 18:55 (local time) Time difference between London and Moscow: 3 hours Flight time: 3.6 hours Time to pass through customs: 45 minutes Train departure times: 5 past, 27 minutes past, 46 minutes past the hour Length of train journey: 29 minutes Walk time: 7 minutes
Step 3: What information don't you need?	Assume time to buy a coffee, sandwich and newspaper is negligible
Step 4: What maths can you do?	18:55 minus 3 hours = 15:55 (London time) Convert 3.6 from a decimal to time in hours and minutes: 3 hours and $0.6 \times 60 = 3$ hours 36 mins Arrival time in London: 15:55 + 3 hours 36 mins = 19:31 Add on time in customs: 19:31 + 45 mins = 20:16 Next possible train is 20:27 Time at end of train journey: 20:27 + 29 mins = 20:56 Arrival time at venue following walk: 20:56 + 7 mins = 21:03 21:03 − 12 = 9.03pm or three minutes past nine in the evening
Step 5: Have you used all the information? At this point you should check to make sure you have calculated what was asked of you.	Flight departure time ✓ Time difference ✓ Flight time ✓ Time through customs ✓ Train departure times ✓ Length of train journey ✓ Walk time ✓
Step 6: Is it correct?	Estimate to check $3\frac{1}{2}$ hours flying + 45 mins at airport + 10 mins wait + 30 mins train + 7 mins walk = about 5 hours Take off the time difference leaves 2 hours Leave 7 pm + 2 hours = 9 pm

Key vocabulary

exchange rate: the value of one currency used to convert that currency to an equivalent value in another currency.

Money

You already know how to convert between pence and pounds.

£1 = 100p   So, $£x = 100x$ pence and x pence $= £\frac{x}{100}$

The rate at which one currency is converted to another is called an **exchange rate**.

For example £1 = €1.26 and €1 = £0.794 or 79.4p (at 2014 rates).

When the exchange rate is given as 1 unit of A = x units of B, you can convert A to B by multiplying by x.

For example £1 = €1.26

So, £400 = 400 × 1.26 = €504

Currency B can be converted to currency A by dividing by x.

For example £1 = €1.26

So, €400 = $\frac{400}{1.26}$ = £317.46

EXERCISE 12B

1 The starting pistol for a road race is fired at 12:15:30.
The first runner crosses the finishing line at 14:07:22.
What was the winning time for the race?

2 Sandra is exactly 16 years old. Calculate her exact age in:

a days. **b** weeks. **c** hours. **d** seconds.

3 A boat leaves port at 14:35 and arrives at its destination $6\frac{1}{2}$ hours later.
At what time does the boat arrive?

4 An area of 250 000 cm^2 needs to be painted. A pot of paint can cover an area of 10 m^2.
How many pots of paint are needed?

5 The table gives the value of the pound against four other currencies in mid-2014.

British pound (£)	Euro (€)	US dollar ($)	Australian dollar (AS$)	Indian Rupee (INR)
1	1.26	1.70	1.80	102.28

a Convert each of the currencies into pounds using the rates in the table.

b Convert £125 to US dollars.

c How many Indian rupees would you get if you converted £45 at this rate?

d Dilshaad has 8000 Indian rupees. What is this worth in pounds at this rate?

e Explain why tourists from the UK may find India a cheap place to visit.

f How can a weaker exchange rate affect the economy of a country?

Consider the cost of imports and the value of exports in your answer.

Section 2: Compound units of measurement

Compound measures involve more than one unit of measurement.

For example:

- Rate of pay (such as pounds per hour) involves units of money and time.
- Unit pricing (such as pence per gram) involves units of money and mass (or capacity or volume).

Find answers at: cambridge.org/ukschools/gcsemaths-studentbookanswers

Tip

A forward slash symbol / is often used instead of the word 'per'. So £8.20/hour means the same as £8.20 per hour.

Compound measures usually express how many of the first unit correspond with 1 of the second unit. For example, a rate of pay of £8.20/hour shows how many pounds you earn for 1 hour of work.

You can simplify compound measures by multiplying or dividing both units by the same factor.

WORKED EXAMPLE 4

40 litres of petrol cost £52.

a What is the cost in £/litre?

a $£\frac{52}{40} = £1.30$

The compound measure £/litre tells you that pounds are the first unit.

The cost is £1.30/litre

b Convert the cost in £/litre to pence per millilitre.

b £1.30 = 130p

1 l = 1000 ml

130p per 1000 ml

We want a compound measure comparing pence and millilitres.
Convert pounds to pence.
Convert litres to millilitres.
Compare the two quantities.

$130 \div 1000 = 0.13$
$1000 \div 1000 = 1$

You want a rate per **one** millilitre, so divide both quantities by 1000.

So 130p per 1000 ml = 0.13p/millilitre

Did you know?

Speed isn't just a measure of how fast something is travelling. Run rates per over in cricket and the number of words you can type per minute are both examples of speed.

Speed

Speed compares the distance travelled to the time taken.
The units of speed depend on the situation.
A car or train's speed is often given in km/h or mph (kilometres or miles per hour).
An athlete's running speed may be given in m/s (metres per second).
You need to know the formula for calculating speed:

Learn this formula

$$\text{Average speed} = \frac{\text{distance travelled}}{\text{time taken}}$$

The speed is an average speed because a journey may involve faster and slower speeds over the given period. You can see the different speeds at different points in a car journey by looking at the speedometer. The speed shown on the speedometer is the speed at that particular time.

The triangle shows the relationship between speed, distance and time.

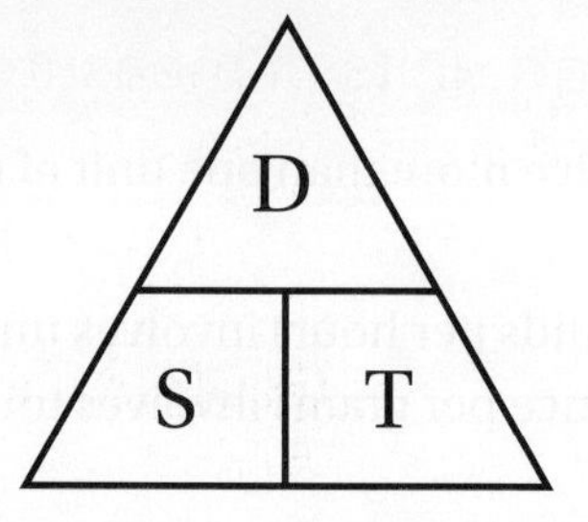

Tip

Using the triangle:

- To find distance: cover D with your finger. The position of S next to T tells you to multiply speed by time.
  Distance = Speed × Time
- To find time: cover T with your finger. The position of D over S tells you to divide distance by speed.
  Time = Distance ÷ Speed
- To find speed: cover S with your finger. The position of D over T tells you to divide distance by time.
  Speed = Distance ÷ Time

The units of speed given in a problem usually let you know what units to use. For example, if the problem talks about km/h, then express distances in kilometres and time in hours to calculate the speed.

WORK IT OUT 12.2

A car travels 330 miles in $5\frac{1}{2}$ hours.

What is the average speed of the car in miles per hour (mph)?
Which of the answers below is correct?

Option A	Option B	Option C
Speed = Distance ÷ Time	Speed = Distance × Time	Speed = Distance − Time
D = 330 miles	D = 330 miles	D = 330 miles
T = $5\frac{1}{2}$ hours = 5.5 hours	T = $5\frac{1}{2}$ hours = 5.5 hours	T = $5\frac{1}{2}$ hours = 5.5 hours
S = 330 ÷ 5.5 = 60 mph	S = 330 × 5.5 = 1815 mph	S = 330 − 5.5 = 324.5 mph

To convert speeds from one set of units to another, you need to work systematically and take care with the units.

WORKED EXAMPLE 5

An athlete runs at an average speed of 10.16 m/s. Is this faster or slower than an average speed of 40 km per hour?

$10.16 \times 60 = 609.6$ m/min
$609.6 \times 60 = 36\,576$ m/hour

Convert 10.16 m/s so that both speeds are in km per hour.
Convert seconds to hours:
× 60 to get metres per minute
× 60 to get metres per hour

$\frac{36\,576}{1000} = 36.576$ km/h

Convert metres to kilometres:
÷ 1000 to get kilometres per hour

So the athlete's speed is slower than 40 km per hour.

Tip

You will work with speed again in the context of kinematics when you deal with time-distance graphs and rates of change in Chapter 34.

EXERCISE 12C

1. Joe is 18 years old and works for a minimum wage of £5.03 per hour.
   If he works for 14 hours, how much will he earn?

2. Henry earns £8.75 per hour.
   One week he worked 36.5 hours.
   Sian earned £202.40 for working 22 hours.
   How much more than Henry does Sian earn per hour?

3. A bricklayer lays 680 bricks in 4 hours.
   How many does she lay per minute, to the nearest brick?

4. Bernie cycles 168 km in 8 hours. What is his average speed?

Find answers at: cambridge.org/ukschools/gcsemaths-studentbookanswers

5 A car travels 528 km at an average speed of 88 km/h. Work out the time taken.

6 Usain Bolt set an Olympic Record over 100 m at the London Olympics in 2012 with a time of 9.63 seconds.

a Express this speed in m/s.

b How fast is this in kilometres per hour?

c Usain Bolt is also the world record holder for the 100 m event. He set a world record of 9.58 s in August 2009. How much faster is the world record speed than the Olympic record speed? Give your answer in m/s.

7 How long would it take a zebra running at 42 km/h to cover a distance of 6.3 km?

8 A train leaves Liverpool Street at 9:37 am and arrives at Norwich at 11:43 am, a distance of 189.9 km. What is the average speed of the train?

9 A marathon runner starts a 42 km race at 05:54:10 and finishes at 08:12:37. What was her average speed for the race? Give the answer in km/h and m/s.

10 A car travelled for 20 minutes at 85 km/h and then for another 25 minutes at 100 km/h. What was its average speed over the journey?

Learn this formula

$$\text{Density} = \frac{\text{Mass}}{\text{Volume}}$$

Density and pressure

Density is the ratio between the mass and the volume of an object.

You need to know the formula for calculating the density of an object.

The triangle opposite helps you see the relationship between density, mass and volume.

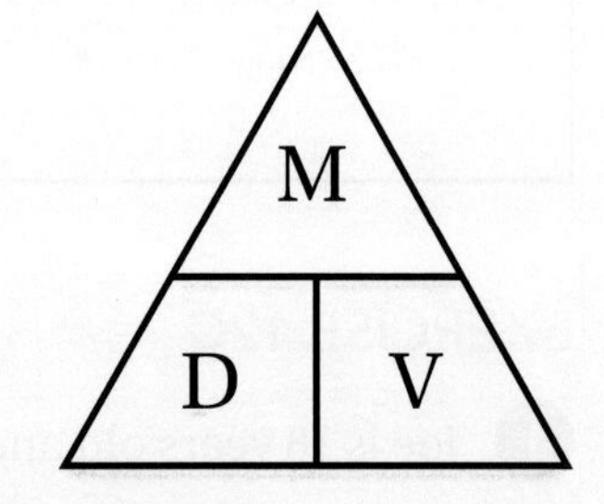

Mass = Density × Volume

$$\text{Volume} = \frac{\text{Mass}}{\text{Density}} \qquad \text{Density} = \frac{\text{Mass}}{\text{Volume}}$$

The units of density are grams per cubic centimetre (g/cm^3) or kilograms per cubic metre (kg/m^3).

Express the mass and the volume in the units given when you solve problems involving density.

WORKED EXAMPLE 6

A gold bar has a volume of 725 cm^3 and a mass of 14.5 kg.

What is the density of the gold bar in g/cm^3?

$M = 14.5\text{ kg} = 14.5 \times 1000 = 14\,500\text{ g}$
$V = 725\text{ cm}^3$

$\text{Density} = \frac{\text{Mass}}{\text{Volume}}$

Units of density are g/cm^3 so mass must be in g.

$\text{Density} = 14\,500 \div 725 = 20\text{ g/cm}^3$

Pressure is defined by the formula:

$$\text{Pressure} = \frac{\text{Force}}{\text{Area}}$$

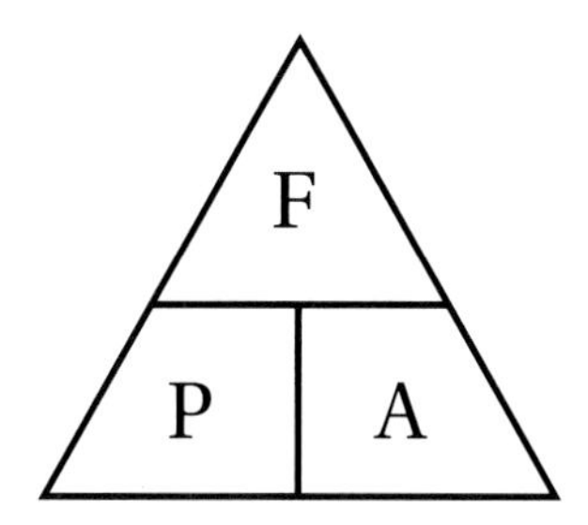

The units given for force and area give you compound units for the pressure.

For example, if force is measured in Newtons and area in mm^2, the compound unit of pressure would be N/mm^2 (newtons per mm^2).

WORKED EXAMPLE 7

A brick exerts a force of 5N.

a Calculate the pressure exerted on the ground when the brick is in each of the following positions.

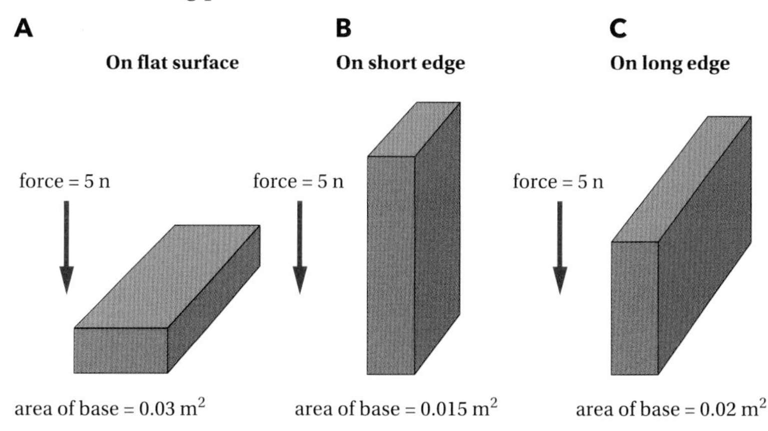

A $\text{Pressure} = \frac{5\text{ N}}{0.03\text{ m}^2} = 167\text{ N/m}^2$

B $\text{Pressure} = \frac{5\text{ N}}{0.015\text{ m}^2} = 333\text{ N/m}^2$

C $\text{Pressure} = \frac{5\text{ N}}{0.02\text{ m}^2} = 250\text{ N/m}^2$

b Which position exerts the strongest pressure?

Why?

The middle position (B) exerts the strongest pressure.

The force of 5 N is pushing down on a smaller area, so the pressure is greater.

EXERCISE 12D

1 The mass of 1 cm^3 of different substances is shown in the diagram.

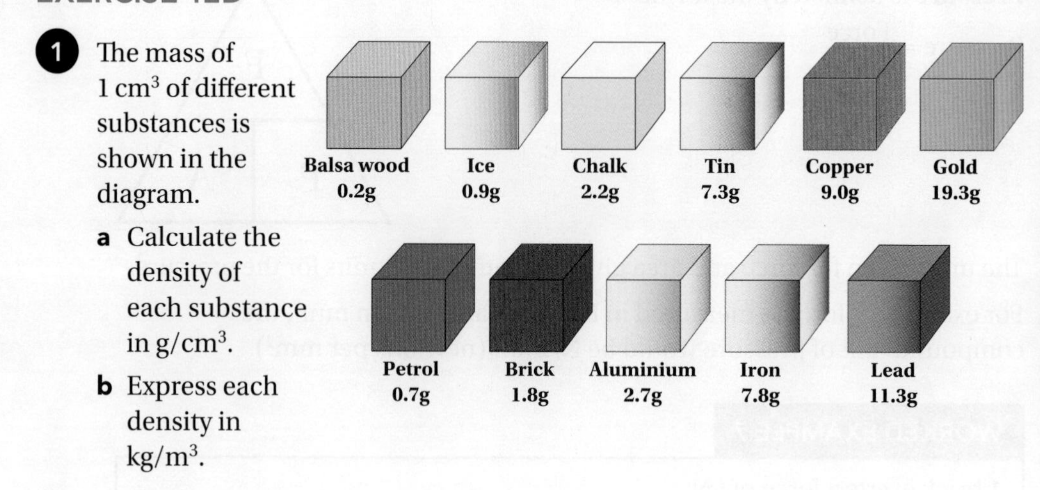

a Calculate the density of each substance in g/cm^3.

b Express each density in kg/m^3.

2 A cube of material with side length 30 mm has a mass of 0.0642 kg.
Calculate the density of the material in g/cm^3.

3 Calculate the volume of a piece of wood with a mass of 0.1 kg and a density of 0.8 g/cm^3.

4 Two metal blocks both exert a force of 18 N.
Block A is a cube with sides 100 cm long. Block B is a cuboid with a base of area 6 m^2 in contact with the floor.
Calculate the pressure exerted by each block in N/m^2.

5 A car exerts a force of 6000 N on the road.
Each of the four wheels has an area of 0.025 m^2 in contact with the road.
What pressure does the car exert on the road?

Section 3: Maps, scale drawings and bearings

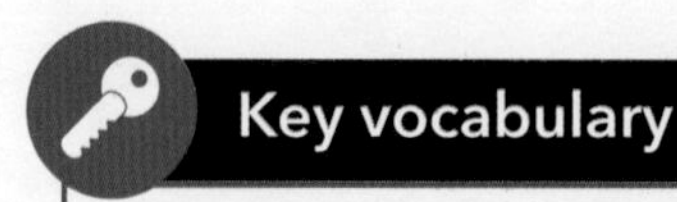

scale factor: a number that scales a quantity up or down.

A **scale drawing** is a diagram in which measurements are either reduced or enlarged by a **scale factor**.

The scale tells you by how much the dimensions were reduced or enlarged.

Maps are scaled representations of areas of the real world.

The scale of a map describes the relationship between lengths in real life and lengths on the map.

The scale allows you to convert distances that you measure on the map to real-life distances.

Map scales are shown in different ways:

- **Bar or line scales**. The divided line or bar shows you distances on the map but the number on the bar tells you what the real distances are (usually in km).
- **Ratio scales**. You will often see scale given as a ratio, for example: 1 : 25 000. This means that one unit measured on the map is equivalent to 25 000 of the same units in real life. So, 1 cm on the map represents 25 000 cm = 0.25 km in real life.

Using the map scale

Bar or line scales are useful for finding small distances. You measure the distance and then compare it with the line scale.

On the line scale below, each block is 1 cm long, so 1 cm on the map represents 1 km in real life.

To find a distance in real life:

- Measure the map distance using a piece of paper.
- Make pencil marks on the paper to record the distance.
- Compare your marked distance with the line scale.
- Read off the real distance.

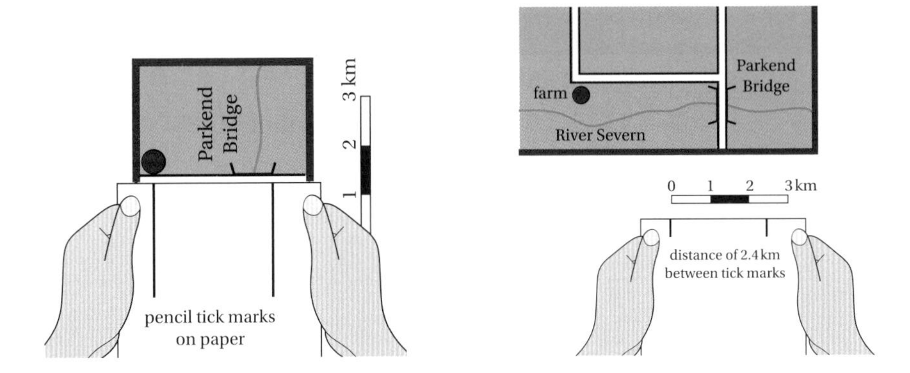

For bigger distances it is more efficient to use the ratio scale. You can convert any distance on a map to a real distance using the following formula.

Distance on the map × scale factor = distance on the ground

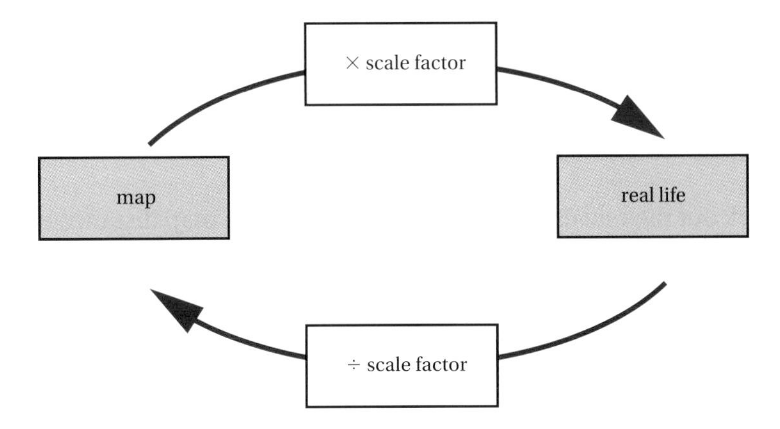

WORKED EXAMPLE 8

A scale drawing has a scale of 1 : 10.

a Calculate the real-life length when a length on the drawing is 5 cm.

5 cm on the drawing = 5 × 10 = 50 cm in real life

b Calculate the length on the drawing when the real-life length is 30 cm.

30 cm in real life = 30 ÷ 10 = 3 cm on the drawing.

Take care with the units.

When you multiply by the scale factor your answer will be in the same units that you used to measure on the map.

If the question asks for different units, you will need to convert the measurement to get the units you need.

WORK IT OUT 12.3

The distance between two towns on a 1 : 25 000 map is 3.4 cm.
How many kilometres apart are these towns in reality?

Which student has worked out the correct answer?
What have the other two done incorrectly?

Student A	Student B	Student C
$\frac{1}{25\,000} \times \frac{3.4}{1}$ = 0.000 136 cm = 1.36 km	Map distance = 34 mm Scale = 1 : 25 000 Real distance 34 × 25 000 = 850 000 mm = 8.5 km	3.4 × 25 000 = 85 000 The distance is 85 000 cm ÷ 100 = 850 m ÷ 1000 = 0.85 km

EXERCISE 12E

1. Here are three line scales. Work out what distance is represented by 1 cm in each case.

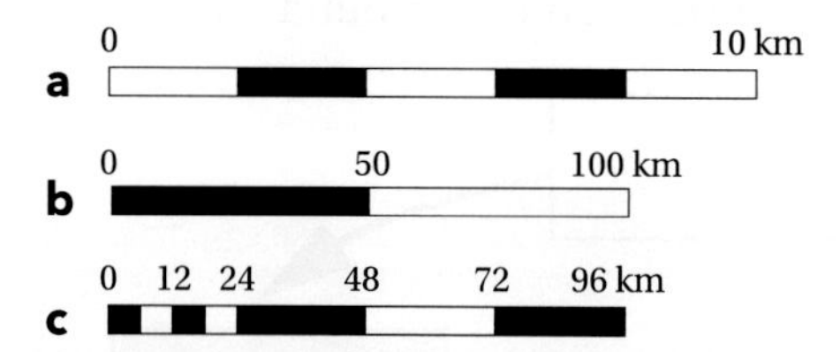

2. Work out the real distance (in kilometres) that a map distance of 45 mm would represent at each scale.

   a 1 : 120 **b** 1 : 1200 **c** 1 : 12 000
   d 1 : 120 000 **e** 1 : 1 200 000 **f** 1 : 12 000 000

3. Andrew says that a map drawn to a scale of 1 : 15 000 is a larger scale map than one drawn to 1 : 150 000.
   Is he correct? How do you know?

4. The map opposite shows three towns.

   The map has a scale of 1 : 75 000.

   Work out the actual distance directly from Coltown to Bracwich.

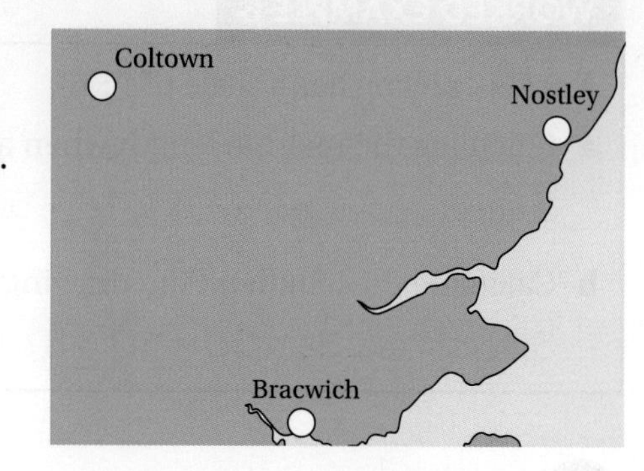

5 The red line on the map shows the flight path of a plane flying from Edinburgh to London. The flight took 55 minutes.

a Calculate the distance flown in kilometres.

b What was the plane's average speed on this flight?

6 A set of toy furniture is manufactured using a scale of 1 : 50.

Work out:

a the height of a cupboard if the model is 5 cm high.

b the width of the toy bed if the actual bed is 1.5 m wide.

c the length of a table if the model is 2.7 cm long.

7 A model of an F15 fighter jet has a scale of 1 : 32.

The real aircraft is 12.5 m long.

What is the length of the model?

8 A map of Scotland on an A4 sheet of paper has a scale of 1 : 2 000 000.

a The map distance from Inverness to Glasgow is 90 mm.
What is the real distance between these places?

b The actual distance by road from Aberdeen to Dundee is 96.5 km.
How long would this road be on the map?

9 The distance between places is often printed on road maps.

Pete has a road map that shows the driving distance from Birmingham to London as 192 km.

He estimates that $2\frac{1}{2}$ cm on the map represents a distance of 50 km.

a Work out the scale of the map.

b How long would the roads shown on the map be for Pete's journey at this scale?

Constructing scale drawings

Tip

You will use scale factors again in Chapter 30 when you deal with similar triangles and also in Chapter 28 when you deal with enlargements of shapes.

To make a scaled drawing or simple map you need to:

- Find out or measure the real lengths involved.
- Decide what size your drawing is going to be so you can work out a scale.
- Choose an appropriate scale (if you are not given one to use). Use ratio to work this out:
  scale = length on drawing : length in real life.
- Use the scale to convert the real lengths to ones you need for the scaled drawing.

WORKED EXAMPLE 9

Draw a scale plan of a rectangular park that is 115 m long and 85 m wide.
Your plan must fit into a space 7.5 cm long and 5 cm wide.

Scale = length on drawing : length in real life

Step 1: The real measurements are 115 m and 85 m.

Step 2: A scale drawing 6 cm long will fit into the given space.

= 6 cm : 115 m

= 6 cm : 11 500 cm

Convert the metres to centimetres.

= 1 : 1917

Step 3: Work out the scale.

Divide both sides of the ratio by 6 to get 1 on the left.

1 : 1917 is a clumsy scale.

Most scales are rounded.

So try a scale of 1 : 2000.

Scaled length = 115 m ÷ 2000 = 0.0575 m = 5.75 cm

Scaled width = 85 m ÷ 2000 = 0.0425 m = 4.25 cm

Step 4: Use the scale to convert the real distances.

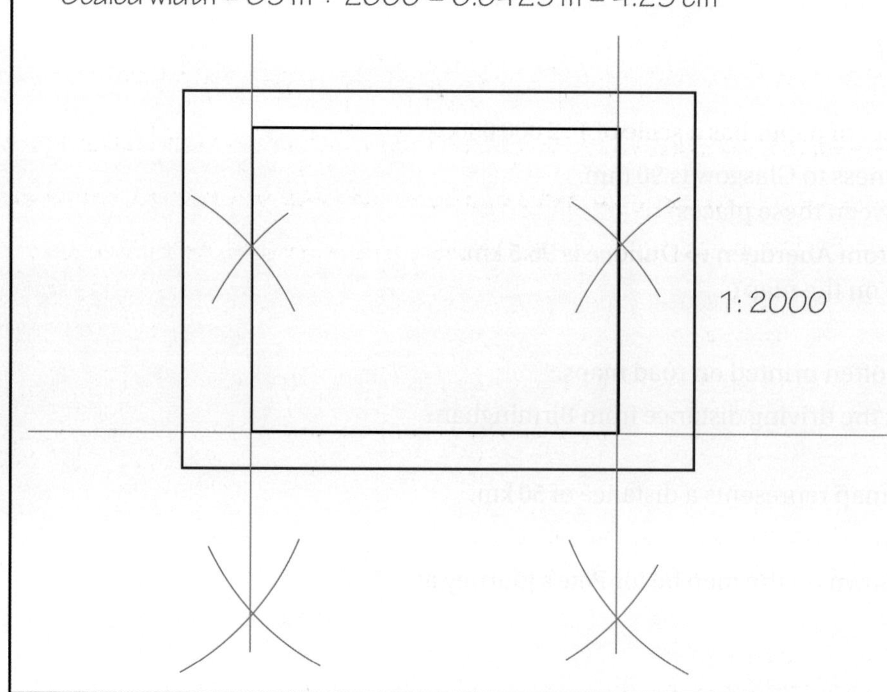

Draw a 5 cm by 7.5 cm frame.

Use your construction skills to draw an accurate rectangle 4.25 cm by 5.75 cm.

Remember to write the scale you used on the diagram.

EXERCISE 12F

1. The floor of a school hall is 40 m long and 20 m wide.
   Draw scaled diagrams to show what it would look like at each of these scales.
   a 1 : 250 **b** 1 : 500 **c** 1 : 1000

2. Measure the dimensions of your desk in centimetres.
   Work out a suitable scale and draw a scaled diagram of your desk, including anything on it.

3. Jules drew this rough plan of a classroom block at her school.
   She wrote the actual measurements on the plan.
   Use the dimensions on the plan to draw a scaled diagram of this classroom block that fits into the width of your exercise book.
   Indicate windows and doors as shown on the plan.

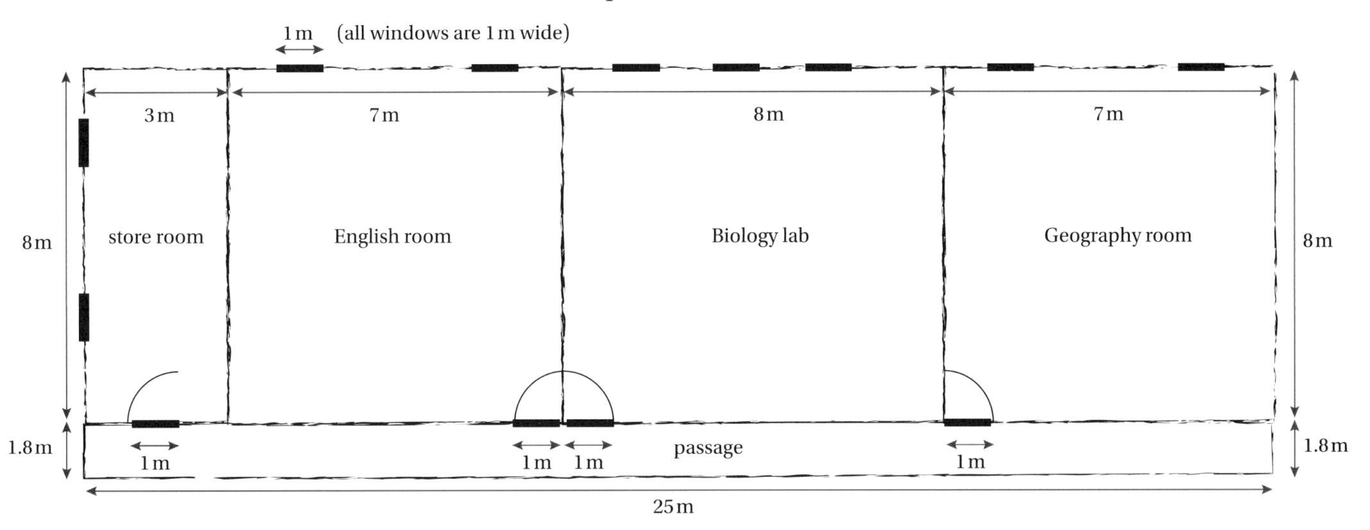

4. A plan of a house is to be drawn at a scale of 1 : 80.
   a What should the scaled dimensions of the kitchen be if the real dimensions are 4900 mm by 3800 mm?
   b Calculate the scaled length of the sink unit if it is 1.2 metres long in reality.

Bearings

Compass directions can be given using cardinal points as shown on the compass rose.

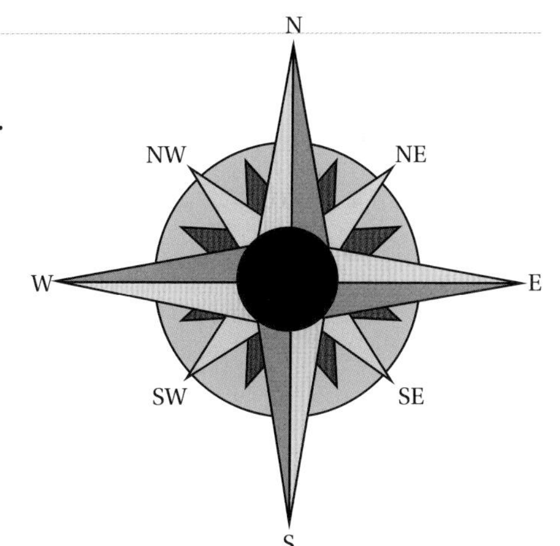

More accurate directions can be given using degrees or bearings.

Bearings are measured in degrees from 0° (North) around in a clockwise direction to 360° (which is back at North).

To measure bearings, you must place the baseline of your protractor in line with the compass direction north. Then you measure the angle from there to the given point.

The bearing is normally written as a three-figure number, so a bearing of 90° would be written as 090° (this bearing corresponds with the direction East).

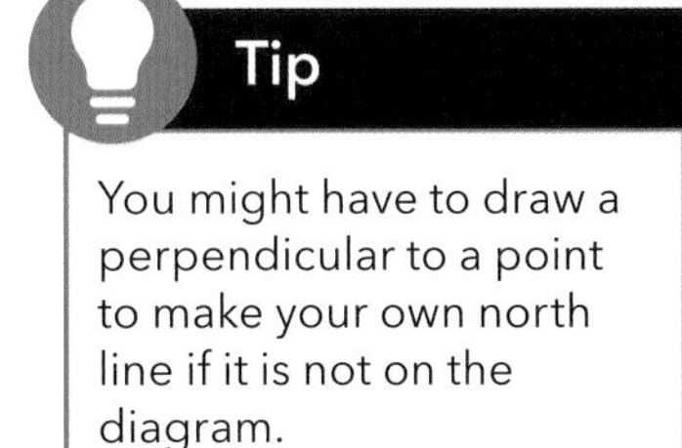

Tip

You might have to draw a perpendicular to a point to make your own north line if it is not on the diagram.

Find answers at: cambridge.org/ukschools/gcsemaths-studentbookanswers

WORKED EXAMPLE 10

Find the bearings from:

a A to B. **b** B to A.

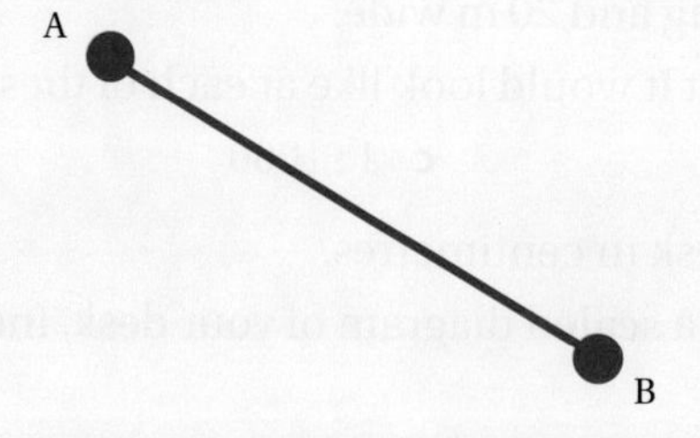

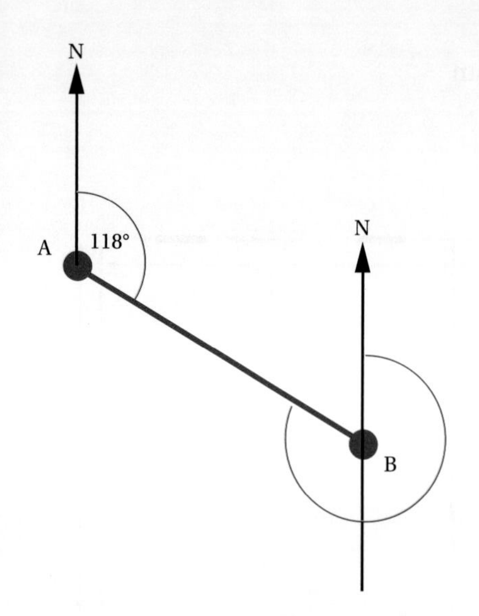

There are no north lines on the diagram so you have to draw them on before you can measure the bearings.

Draw in two perpendicular lines at A and B.

Measure the angles.

$180° + 118° = 298°$

a Bearing from A to B is 118°

b Bearing from B to A is 298°

EXERCISE 12G

1 Write the three-figure bearing that corresponds with each direction.

a Due south **b** North-east **c** West

2 Use a protractor to measure each of the following bearings on the diagram.

a A to B **b** B to A

c A to C **d** B to D

e D to A **f** D to B

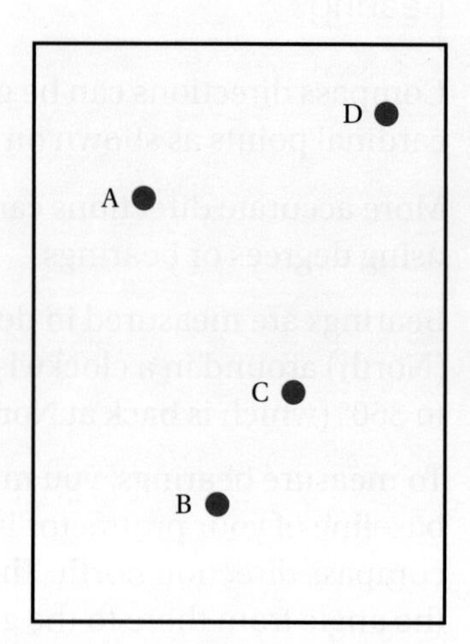

3 Beville is 140 km west and 45 km north of Lake Salina.

Draw a scale drawing with a scale of 1 cm to 20 km and use it to find:

a the bearing from Lake Salina to Beville.

b the bearing from Beville to Lake Salina.

c the shortest distance between the two places in kilometres.

4 Find the bearing from B to A, if the bearing from A to B is:

a 120° **b** 045° **c** 210°

5 Village Q is 7 km from Village P on a bearing of 060°.

Village R is 5 km from Village P on bearing of 315°.

Using a scale of 1 : 100 000, draw a diagram and use it to find:

a the direct distance from Village Q to Village R.

b the bearing of Village Q from Village R.

Checklist of learning and understanding

Standard units of measurement

- You can convert between metric units of length, mass and capacity by multiplying or dividing by powers of ten.
- To convert squared units, you need to square the conversion factors.
- To convert cubed units, you need to cube the conversion factors.
- Units of time are not metric, so you need to use the number of parts in the sub-units when you convert units of time.

Compound units of measurement

- Compound units of measurement involve more than one unit.
- Rates such as £/hour or cost per kilogram are compound units.
- Speed $= \frac{\text{distance}}{\text{time}}$
- Density $= \frac{\text{mass}}{\text{volume}}$
- Pressure $= \frac{\text{force}}{\text{area}}$

Maps, scales and bearings

- The scale of a map or diagram describes how much smaller (or bigger) the lengths on the diagram are compared to the original lengths.
- Real length = map length × scale factor
- The scale can be written as: length on diagram : real length.
- Bearings are accurate directions given in degrees from 0° to 360°. 0° corresponds with north.
- Bearings are measured clockwise from 0° and written using three figures.

Find answers at: cambridge.org/ukschools/gcsemaths-studentbookanswers

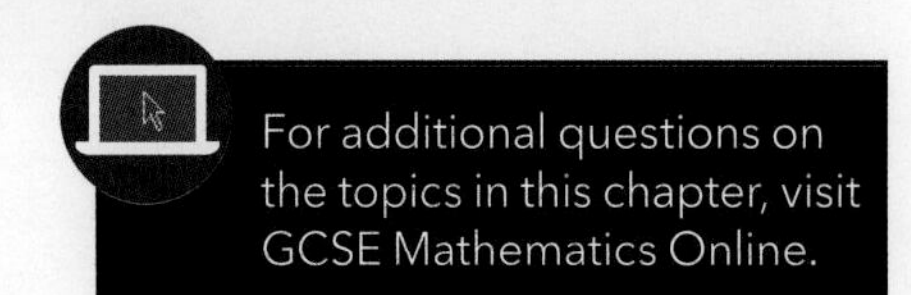

For additional questions on the topics in this chapter, visit GCSE Mathematics Online.

Chapter review

1 Match each statement to the correct number in the box below.

0.05	475	182.5	259 200

a The number of seconds in 3 days.

b The number of kilometres travelled in $2\frac{1}{2}$ hours by a car travelling at 73 km/h.

c The distance in km in real life of a length of 5 cm on a map with a scale of 1 : 1000.

d The number of litres in 475 000 millilitres.

2 True or false?

a Tony's fish tank contains 72 litres of water.
He says this is 72 000 millilitres.

b The school is 15 km from the bus stop. The bus travels at 40 km/h.
Molly says it will take her half an hour to get to school.

c The distance from Liverpool to Manchester is about 55 km.
The scale of a map is 1 : 250 000.
The distance on the map would be 5.2 cm.

3 Convert 60 000 cm^2 into m^2.

4 How many mm^2 are there in 2 m^2?

5 A car is travelling at an average speed of 80 km/h for one hour on a bearing of 120°.

a Use a scale of 1 cm to 20 km to show this journey.

b After 45 km, the driver stopped for petrol. Mark this spot on the diagram. If this was after 40 minutes, calculate his speed for that part of the journey.

6 The density of an object is 8 kg/m^3.

Work out the mass of 25 m^3 of the object.

7 A cyclist travels due east from point A for 10 km to reach point B.
She then travels 6 km on a bearing of 125° from B to reach point C.

a Use a scale of 1 cm to 2 km to represent her journey on a scale diagram.

b Find the bearing from C to A.

c Find the direct distance from C to A in kilometres.

d If it takes the cyclist $1\frac{1}{2}$ hours to cycle back using the direct route from C to A, find her average speed:

i in km/h. **ii** in m/s.

8 The scale drawing shows a coastline with two ports, A and B, and a lighthouse, L.

The scale is 2 cm represents 1 km.

A boat travels on a bearing of 128° from port A.

A plane flies out to sea so that its distance from A is always the same as its distance from B. At one point, the plane is directly over the boat.

At this point, what is the distance of the boat from the lighthouse, L?

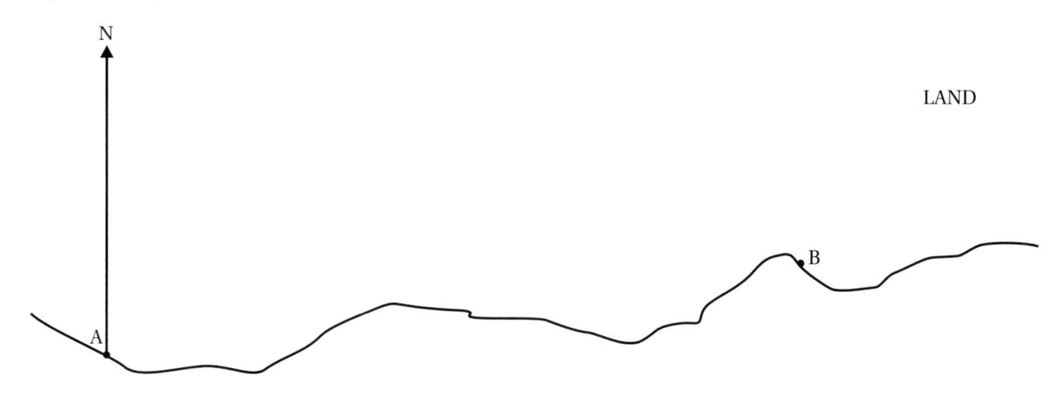

SEA

Scale: 2 cm represents 1 km

(5 marks)

9 Two motorcyclists set off from point A at the same time.
One travels at 60 km/h on a bearing of 045°.
The other travels at 55 km/h on a bearing of 072°.
Using a scale of 1 cm to 10 km, make a scaled diagram to show their journey and work out how far apart they are after $1\frac{1}{2}$ hours.

13 Percentages

In this chapter you will learn how to …

- change between fractions, decimals and percentages.
- calculate a percentage of an amount.
- express a quantity as a percentage of another.
- increase and decrease amounts by a given percentage.
- solve problems involving percentage change.

For more resources relating to this chapter, visit GCSE Mathematics Online.

Using mathematics: real-life applications

Percentages are often used in daily life to express fractions. For example, you may see adverts claiming that 76% of pets prefer a particular brand of food or that 90% of dentists recommend a particular type of toothpaste. Sale price reductions, discounts and interest rates are usually given as percentages.

Tip

You will learn about percentage change in the context of growth and decay in Chapter 35.

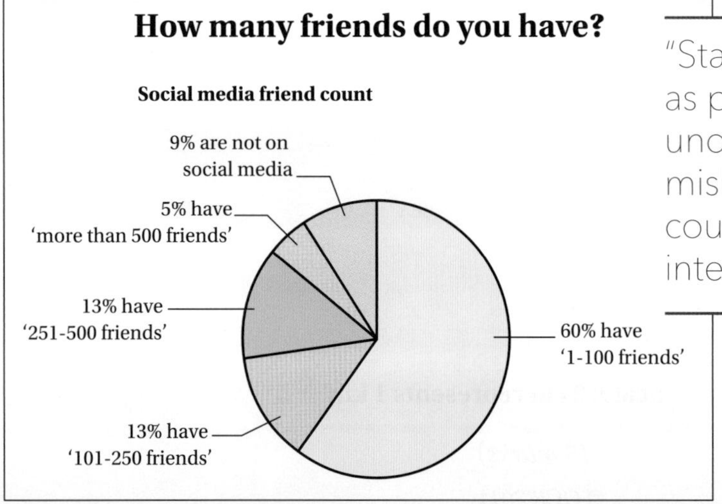

"Statistics in the media are often reported as percentages. This makes it easier to understand, but percentages can also be misleading – 60% sounds like a lot, but it could just mean three out of five people interviewed." *(Statistician)*

Before you start …

Ch 11	You need to be able to confidently multiply and divide by 100.	**1** Where should the decimal place go in each answer? **a** $210 \div 100 = 21$ **b** $21 \div 100 = 21$ **c** $0.24 \times 100 = 24$ **d** $0.024 \times 100 = 24$
Ch 10	You need to be able to express fractions in simplest terms by cancelling.	**2** Match the fractions in box A to their equivalent fraction in box B. Box A: $\frac{16}{36}$ $\frac{15}{35}$ $\frac{30}{36}$ $\frac{9}{36}$ $\frac{39}{52}$ $\frac{13}{39}$ Box B: $\frac{1}{4}$ $\frac{3}{4}$ $\frac{1}{3}$ $\frac{5}{6}$ $\frac{3}{7}$ $\frac{4}{9}$
Ch 11	You should be able to express any percentage as a decimal.	**3** Are the following statements true or false? **a** $20\% = 0.02$ **b** $25\% = 1.4$ **c** $3\% = 0.3$ **d** $12.5\% = 0.125$ **e** $1.25\% = 0.125$

Assess your starting point using the Launchpad

STEP 1

1. Write the following percentages as fractions.
   a 34%
   b 115%
2. Write these in order from smallest to largest.
   a 12%, 0.125, $\frac{7}{50}$, $\frac{5}{12}$, 19%
   b $2\frac{3}{4}$, 200%, 2.5%, 12.5%, 1.08, 1.25

? **GO TO** **Section 1**: Review of percentages

✓

STEP 2

3. What is 19 out of 25 marks as a percentage?
4. What is 50% of 128?
5. In a population of 12 500 000 people of working age, 3 400 000 are unemployed.
   What is the unemployment rate as a percentage?
6. Express 25p as a percentage of £7.55.

? **GO TO** **Section 2**: Percentage calculations

✓

STEP 3

7. Increase £20 by 9.5%.
8. Pete wants to buy a second hand car marked at £2800.
   The dealer offers him a 7.5% discount if he pays cash. What will the cash price be?
9. Mandy bought a book in a 25% off sale for £2.55. What was the original price of the book?

? **GO TO** **Section 3**: Percentage change

✓

GO TO Chapter review

Find answers at: cambridge.org/ukschools/gcsemaths-studentbookanswers

Section 1: Review of percentages

A percentage shows the number of parts per hundred.

During 2014, the maker of a mobile phone case did a survey of 872 people.

The survey results showed that 92% of people said they would feel stressed if their phone battery ran out, and 81% of people said that running out of power on their phones had led to them having a bad experience.

- 92% means 92 out of every 100.
- 81% means 81 out of every 100.

Tip

Remember that a percentage shows the number of parts per hundred.

Percentages, fractions and decimals

Percentages can be changed to fractions by writing the fraction with a denominator of 100 and simplifying.

$92\% = \frac{92}{100} = \frac{23}{25}$

$81\% = \frac{81}{100}$

To change a percentage to a decimal, write it as fraction with a denominator of 100 and then convert it to a decimal.

$92\% = \frac{92}{100} = 0.92$

$81\% = \frac{81}{100} = 0.81$

To change a fraction to a percentage, you can write it as an equivalent fraction with a denominator of 100, then multiply by 100 and add the percentage symbol.

$\frac{1}{2} = \frac{50}{100} = 50\%$ Do **not** simplify, write the fraction of 100 as a percentage.

You can also use a calculator.

To convert $\frac{2}{3}$ to a percentage, enter [2] [÷] [3] [×] [1] [0] [0]

Your display will show: 66.666666667

This is the percentage. Write it as 66.67% (correct to two decimal places).

Tip

When you use a calculator to convert a fraction to a percentage, you are actually first changing $\frac{2}{3}$ to a decimal ($2 \div 3 = 0.6666666667$) and then converting the decimal to a percentage. You do not enter the percentage sign in the calculation because the values you are entering are not percentages. The percentage is the answer you get.

To change a decimal to a percentage, write it as a fraction with a denominator of 100 and multiply it by 100, or use your calculator to multiply it by 100 directly, and write the percentage symbol.

$0.3 = \frac{3}{10} = \frac{30}{100} = 30\%$ $\quad 0.3 \times 100 = 30\%$

$0.025 = \frac{0.25}{10} = \frac{2.5}{100} = 2.5\%$ $\quad 0.025 \times 100 = 2.5\%$

$3.75 = \frac{37.5}{10} = \frac{375}{100} = 375\%$ $\quad 3.75 \times 100 = 375\%$

Comparing percentages, fractions and decimals

When you have to compare a mixed set of percentages, fractions and decimals you can compare them by changing them all to percentages.

Tip

You can also change all the values to decimals or equivalent fractions to compare them if that is easier.

Tip

Ascending order means from the smallest to largest; descending order means from the largest to smallest.

WORKED EXAMPLE 1

Write the following in ascending order: 35%, $\frac{1}{3}$, 0.38, $\frac{2}{5}$, $\frac{2}{7}$

$\frac{1}{3} \times 100 = 33.33\%$ — Convert all the fractions and decimals to percentages.

$0.38 \times 100 = 38\%$

$\frac{2}{5} \times 100 = 40\%$

$\frac{2}{7} \times 100 = 28.57\%$

$\frac{2}{7}, \frac{1}{3}, 35\%, 0.38, \frac{2}{5}$

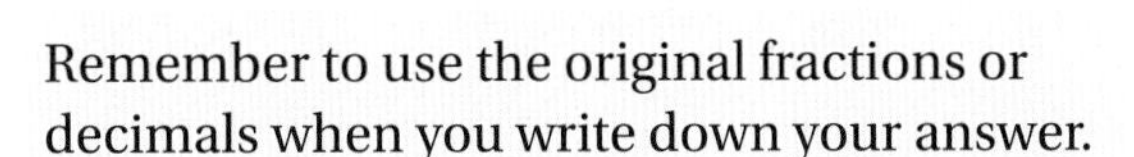

Remember to use the original fractions or decimals when you write down your answer.

EXERCISE 13A

1 Express the following as percentages.

Use fractional ($32\frac{1}{2}\%$) or decimal (2.5%) percentages where you need to.

a $\frac{5}{100}$ **b** $\frac{27}{50}$ **c** $\frac{11}{25}$ **d** $\frac{17}{20}$

e $\frac{1}{2}$ **f** $\frac{2}{3}$ **g** $\frac{5}{8}$ **h** $\frac{92}{50}$

i 0.3 **j** 0.04 **k** 0.47 **l** 1.12

m 2.07 **n** 2.25 **o** 0.035 **p** 0.007

2 Write each of the following percentages as a fraction in its simplest terms.

a 25% **b** 80% **c** 90% **d** 12.5%

e 50% **f** 98% **g** 60% **h** 22%

3 Write the decimal equivalent of each percentage.

a 82% **b** 97% **c** 45%

d 28% **e** 0.05% **f** 0.08%

g 0.006% **h** 0.0007% **i** 125%

j 300% **k** 7.28% **l** 9.007%

4 State whether the following are true or false.

a $\frac{3}{5} > 70\%$ **b** $\frac{7}{9} < 83\%$ **c** $\frac{1}{3} = 30\%$

d $67\% > 0.666$ **e** $\frac{3}{4} \neq 75\%$ **f** $0.55 \neq \frac{1}{2}$

Find answers at: cambridge.org/ukschools/gcsemaths-studentbookanswers

5 **a** If 93.5% of the students in a school have wifi at home, what percentage do not?

b If $\frac{2}{3}$ of all the sim cards sold in a mobile phone shop are pre-paid, what percentage are not pre-paid?

c 0.325 of computer users back up their work every day. What percentage do not do this?

6 Zack spends 24.7% of a day playing computer games, 0.138 of the day doing homework and $\frac{3}{8}$ of the day playing sport.

What percentage of the day is spent doing other things?

7 Write the following in ascending order.

a $\frac{1}{20}$, 30%, 0.1, $\frac{3}{5}$, 0.8%

b 0.75, 57%, 0.88, $\frac{1}{4}$, 0.15

c $\frac{2}{3}$, 0.75, 60%, $\frac{9}{10}$, 0.25

d $\frac{3}{7}$, 0.43, 45%, 0.395, $\frac{4}{9}$

e $\frac{5}{6}$, 80%, $\frac{19}{25}$, 55%, 49.3%

8 A media company states that 83.5% of its customers read the news online every day.

What fraction of the customers is this?

9 Anna pays 0.06 of her salary into her credit card account.

What percentage of her salary is this?

10 During one shift at work, Sandy spent $\frac{9}{20}$ of her time texting on her phone.

What percentage of the shift was she not texting?

Tip

Remember that:

mean = $\frac{\text{sum of values}}{\text{number of values}}$

You will cover statistics in more detail in Chapter 38.

11 Angie gets the following marks for three maths assignments:

$\frac{31}{40}$, $\frac{27}{30}$ and $\frac{13}{15}$.

a In which test did she get the best marks?

b What is her mean result for the three assignments as a percentage?

Section 2: Percentage calculations

The word 'of' means multiply.

To find a percentage of an amount you have to multiply the amount by the percentage.

You need to write the percentage as a fraction with a denominator of 100, or as a decimal, unless you are using a calculator.

WORK IT OUT 13.1

Which of the following methods was used to calculate that $9\frac{1}{2}\%$ of 400 is 38? Explain why the other methods won't work.

Method A	Method B	Method C	Method D	Method E
$\frac{9}{200} \times 400$	$\frac{19}{200} \times \frac{400}{1}$	$\frac{19}{2} \times 400$	9.5×400	$\frac{9.5}{400} \times 100$

WORKED EXAMPLE 2

What is 12% of 700?

Calculate the answer using:

a fractions. **b** decimals. **c** a calculator.

a $\frac{12}{1\not{0}\not{0}} \times 7\not{0}\not{0} = 84$

Simplify by cancelling if you can, then do the calculation. Divide by 100 to simplify the calculation to 12×7.

b 0.12×700

$= 0.12 \times 100 \times 7$

$= 12 \times 7$

$= 84$

Write 12% as the decimal 0.12. Then use what you know about multiplying decimals to do the calculation (refer back to Chapter 11 if you need help).

c 

84

Enter the calculation on your calculator as shown. **Note** that the sequence of buttons might differ on different calculators so check your manual!

Calculator tip

Make sure you know how to use the % button on your calculator.

You may need to enter '12% × 700' or '700 × 12%' (some calculators will work both ways).

On some calculators, you need to press the = but on others you may not have to.

Check how your calculator works by finding 12% of 350. The answer should be 42.

You **do** enter the percentage sign in these calculations because one of the values you are working with is a percentage.

EXERCISE 13B

1 Calculate.

a 5% of 250 **b** 9% of 400 **c** 20% of 120
d 65% of 4500 **e** 12% of 75 **f** 75% of 360
g 32% of 50 **h** 110% of 60 **i** 150% of 90

2 Calculate, give answers as mixed numbers or decimals as necessary.

a 19% of £50 **b** 60% of 70 kg **c** 45% of 35 cm
d 90% of 29 kg **e** $3\frac{1}{2}\%$ of £400 **f** 2.6% of 80 minutes
g 7.4% of £1000 **h** 3.8% of 180 m **i** $9\frac{2}{3}\%$ of 600 litres

Tip

Remember, your answer will have a unit not a percentage sign. You are not working out a percentage here, you are working out what a given percentage of a quantity is.

Find answers at: cambridge.org/ukschools/gcsemaths-studentbookanswers

3. Annie got 85% for a test that was out of 80 marks.

   What was her mark out of 80?

4. A salesperson at a mobile phone shop estimates that about 3% of phones come back for some sort of repair in the first week.

   If the shop sells 180 phones, how many can they expect to come back for repairs in the first week?

5. 46% of residents in an area throw out the local free newspaper without even looking at it, the rest read some or all of it.

   If there are 2450 residents, how many people:

   a don't look at the paper? **b** read some or all of it?

6. Of 240 trains arriving at King's Cross, 2.5% arrived early and 13.8% arrived late.

   How many trains were on time?

7. A tablet computer is advertised for £899 excluding VAT.

   VAT is 20% of the sale price.

   Nisha wanted to buy it when VAT was 17.5% of the sale price but she didn't get round to it.

   How much would she have saved if she had bought it when VAT was 17.5%?

8. 7.5% of a 620 m^2 market garden is set aside for growing tulips and the rest is used to grow vegetables.

   How many square metres of land is used to grow:

   a tulips? **b** vegetables?

9. The population of a town in Cornwall increases by about 24.8% each summer.

   If the population of the town is 12 760 before summer, approximately how many people are likely to move in during the summer?

10. Pure gold contains 24 parts (called carats) of gold to every 24 parts. $\frac{24}{24} = 100\%$ gold.

    18 carat gold contains 18 parts pure gold per 24 parts, and 9 carat gold contains 9 parts pure gold per 24 parts.

    a Work out the percentage of pure gold in

    i 9 carat gold. **ii** 18 carat gold.

    b If Naz buys an 18 carat gold ring that weighs 7.3 grams, how much pure gold does it contain?

    c If Vishnu buys a 9 carat gold pendant that has a mass of 16.3 grams, how much pure gold does it contain?

    d Do you think it is accurate to label 9 carat gold as gold? Explain your answer.

Expressing one quantity as a percentage of another

You can write one quantity as a percentage of another quantity by writing the first quantity as a fraction of the other, and then multiplying by 100 to get a percentage. The two quantities must be in the same units before you write them as a fraction.

Tip

You expressed one quantity as a fraction of another in Chapter 10. Read through that work again if you cannot remember how to do this.

WORK IT OUT 13.2

Brian has run 1500 m of a 5 km race when he gets a cramp in his foot.

What percentage of the race has he completed at this stage?

Which of these two students has got the correct answer? Why is the other one wrong?

Student A	Student B
$\frac{1500}{5} \times 100$	$\frac{1500}{5000} \times 100$
$= 300 \times 100$	$= \frac{3}{10} \times 100$
$= 300\%$	$= 30\%$

Tip

When you convert quantities to get them to the same unit, you can avoid decimal values by choosing the smaller units (for example, making both units metres in this case, rather than making them both kilometres).

Tip

Remember, the steps in the problem solving framework.

Sometimes you will need to find intermediate values before you can work out the desired percentage. Make sure you always read the question carefully so that you know what it is that you are trying to calculate.

WORKED EXAMPLE 3

Andrew has 1 kilogram of apples.

He gives 300 g to his neighbour, 400 g to his mother and keeps the rest.

What percentage of the apples does he keep for himself?

$300 + 400 = 700$

$1000 - 700 = 300$

You need to calculate what mass of apples he kept before you can calculate the percentage.

$\frac{300}{1000} = 0.3 \times 100 = 30\%$

Work out the number of grams he kept as a percentage of the total number of grams. (You could have converted the quantities to kilograms instead; it would make no difference to the final percentage.)

EXERCISE 13C

Try to answer some of the questions without using your calculator; use the non-calculator techniques you have learnt for decimals and fractions.

1 Express the first amount as a percentage of the second.

Give your answer correct to no more than two decimal places (if necessary).

a 400 m of 5 km
b 45 m of 3 km
c 150 m of 1 km
d 8 cm of 2 m
e 14 mm of 4 cm
f 19 cm of 3 m
g 25p of £4
h 66p of £3.50

Find answers at: cambridge.org/ukschools/gcsemaths-studentbookanswers

i 20 seconds of a minute

j 25 seconds of 1.5 minutes

k 750 g of 23 kg

l 800 g of 1.5 kg

m 4 days of a week

n 3 days of 6 weeks

o 800 kg of 3 tonnes

p 8.4 tonnes of 50 000 kg

q 500 mm of 2 m

r 90 mm of 14 cm

s 350 ml of 2 litres

t 5 ml of 0.5 litres

2 Sandra got 19 out of 24 for an assignment and Nina got 23 out of 30.

Which girl got the higher percentage mark?

3 In a local election there were 5400 registered voters. Of these, 3240 voted.

What percentage of registered voters voted?

Tip

Think about what quantity you are trying to express as a percentage of another quantity.

4 Mel improved his running time for the 400 m race by 3 seconds.

If his previous running time was 50 seconds, what is his percentage improvement?

5 Kenny had a box of 40 chocolates. He ate 32 of them.

What percentage of the chocolates remain?

6 Sylvia keeps a record of how many sets she wins when she plays tennis against her sister.

In the past month she won 19 out of 27 sets.

What percentage of the sets did she lose?

7 The longest kiss lasted 58 hours, 35 minutes and 58 seconds and was achieved by Ekkachai Tiranarat and Laksana Tiranarat at an event organised by 'Ripley's Believe It or Not! Pattaya', in Pattaya, Thailand, on 12–14 February 2013.

What percentage of the three-day event was this?

8 Read the label and answer the questions.

Muesli
Nutritional values
(Per 30 g serving)

Carbohydrates	19 g
(of which sugars)	6.2 g
Fat	3.8 g
Sodium	93 mg

a Calculate the combined percentage of fat and sugar in a serving of muesli.

b What percentage of a serving is sodium?

9 If x is 40% of y, what percentage of y is x?

10 $2n$ less x% is equivalent to n plus x%.

What is x%?

Section 3: Percentage change

You will often see increases or decreases in amounts expressed as percentages.

For example, you might read that the price of petrol is going to increase by 5.5%, or that the cost of mobile broadband has decreased by 15% over the past year.

Tip

You will learn about percentage change in the context of growth and decay in Chapter 35.

Increasing or decreasing an amount by a percentage

WORK IT OUT 13.3

Last September, the enrolment of students to Scaltback School was 650.

This September, the student enrolment increased by 12%.

At the same time, the registration fee of £120 decreased by 15%.

Work out:

a the new student enrolment. **b** the new registration fee.

Students A and B solved this problem using different methods. **Both** methods are correct.

Their workings are shown below.

Which method do you prefer?

Could you use your calculator to do these calculations? How?

Student A

a 650 increased by 12%

$12\% \text{ of } 650 = \frac{12}{100} \times 650$

$= 78$

$650 + 78 = 728$

728 students enrolled this September.

b 120 decreased by 15%

$15\% \text{ of } 120 = \frac{15}{100} \times 120$

$= 18$

£120 − £18 = £102

The new registration fee is £102.

Student B

a 650 increased by 12%

Old population = 100%
New population = old + increase = 100% + 12% = 112%

$112\% = \frac{112}{100} = 1.12$

$650 \times 1.12 = 728$

Tip

Multiplying by 1.12 is the same as multiplying by $\frac{112}{100}$.

728 students enrolled this September.

b £120 decreased by 15%

% (of £120)	Amount (£)
10	12
5	6
15	18

£120 − £18 = £102

The new registration fee is £102.

The new registration fee could also have been calculated as follows:

£120 decreased by 15%

Original value = 100%

New value = 100% − 15% = 85%

Find answers at: cambridge.org/ukschools/gcsemaths-studentbookanswers

So the new value is 85% of the original value.

$\frac{85}{100} \times 120 = £102$

Tip

You can express any percentage increase or decrease as a decimal or fractional multiplier.

To increase a number by x%, multiply the original value by $\left(1 + \frac{x}{100}\right)$.

To increase by 30% multiply by 1.3. To increase by 3% multiply by 1.03.

To decrease a number by x%, multiply the original value by $\left(1 - \frac{x}{100}\right)$.

To decrease by 16% multiply by 0.84. To decrease by 6% multiply by 0.94.

EXERCISE 13D

1 Increase each amount by the given percentage.

a £48 increased by 14%
b £700 increased by 35%
c £30 increased by 7.6%
d £40 000 increased by 0.59%
e £90 increased by 9.5%
f £80 increased by 24.6%

2 Decrease each amount by the given percentage.

a £68 decreased by 14%
b £800 decreased by 35%
c £90 decreased by 7.6%
d £20 000 decreased by 0.59%
e £85 decreased by 9.5%
f £60 decreased by 24.6%

3 A building is worth £125 000.

If its value increases by $3\frac{1}{2}$%, what will it be worth?

4 Josh currently earns £3125 per month.

If he receives an increase of 3.8% per month, what will his new monthly earnings be, correct to the nearest pound?

5 Sally earns £25 per shift. Her boss says she can either have £7 more per shift or a 20% increase.

Which is the better offer?

6 The membership of a sports club increased by 26% one year.

If they had 284 members the previous year, how many do they have now?

7 Sammy bought £2500 worth of shares.

At the end of the first month their value had decreased by 4.25%.

At the end of the second month Sammy checked the value again and found it had gone up 1.5% from the previous month.

Work out the value of the shares at the end of each month.

8 Amira earns £25 000 per year plus 12% commission on any sales she generates.

Calculate her annual earnings if she sold £145 250 worth of goods.

9 Crime statistics are often given as percentages, but these can be misleading.

a Use the following data to show how expressing values as a percentage increase can be misleading.

Location	Number of violent crimes in Year 1	Number of violent crimes in Year 2	% change in crime rate
Village	12	18	50%
Town	87	98	12.6%
City	1234	1230	⁻0.3%

b Which place looks like it has a high crime rate using percentages? Why?

c Which place do you think is really the most risky in terms of crime? Explain your answer.

10 This summer, an amusement park increased its entry prices by 25% to £15.00.

This summer, the number of people entering the park dropped 8% from the previous summer to 25 520.

a What was the entry price the previous summer?

b How many visitors were there the previous summer?

c If the running costs of the amusement park remained the same as the previous summer and they made a 30% profit on the entry fees in this summer, how much was their profit amount in pounds?

11 The news media reports that the winter of 2014 was 24.5% wetter than the average winter.

Explain what this means and what data you would need to work out how much more rain fell in 2014.

12 A journalist is investigating how the price of a Eurostar train ticket varies depending on whether you buy it in London or Brussels (as a result of exchange rates).

The same ticket costs €240 in London and €225 in Brussels.

Use this information to complete the statement: 'Tickets bought in London are ____% more expensive than those bought in Brussels.'

Finding original values

If you know the percentage by which an amount has increased or decreased, you can use it to find the original amount. Problems involving original values are often called reverse or inverse percentages. When you work with these problems, you need to remember that you are dealing with percentages of the **original** values.

Find answers at: cambridge.org/ukschools/gcsemaths-studentbookanswers

WORKED EXAMPLE 4

A shop is offering a 10% discount on all sale goods.
Jessie bought a bike in the sale and paid £108.
What was the original price of the bike?
Let x represent the original price of the bike.

90% of $x = £108$

If the cost is reduced by 10% then you are actually paying 90% of x, the original price.

$$\frac{90}{100}x = 108$$
$$90x = 100 \times 108$$
$$90x = 10\,800$$
$$x = \frac{10\,800}{90}$$
$$x = 120$$

You can write an equation and solve it to find the value of x.

The original price was £120.

Check this answer by decreasing £120 by 10%.

Tip

Undoing a 10% decrease is not the same as just increasing the sale price by 10%. If you add 10% to the sale price of £108, you will be adding 10% of £108 (to get £118.80), which is **NOT** the right answer.

WORKED EXAMPLE 5

Sameen sells some shares she has in a company for £3450. She makes a profit of 15%.
What did she pay for the shares originally?

Let the cost price be x.

Let x be the cost of the shares originally.
Write an equation and solve for x.

$$1.15x = 3450$$
$$x = \frac{3450}{1.15}$$
$$x = 3000$$

Profit = selling price − cost price
So here, '15% profit' means her profit is 15% of the amount she paid for the shares (the cost). So, 3450 is a 15% increase on the cost value.

She paid £3000 for the shares originally.

$3000 \times 1.15 = 3450$ ✓

Check this by increasing 3000 by 15%

EXERCISE 13E

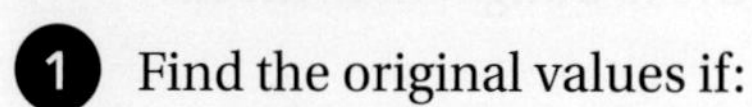

1 Find the original values if:

a 25% is £30. **b** 8% is 120 grams.
c 120% is 800 kg. **d** 115% is £2000.

2 VAT of 20% is added to most goods before they are sold.
Tourists to the UK can claim back the VAT when they leave the country.
Work out the price of each of these items without VAT to see what a tourist would pay for them.
The prices given here include VAT.

a Necklace £1200 **b** Camera £145.50
c Painting £865 **d** Boots £54.99

3 Misha paid £40 for a DVD box set in a 20% off sale.
What was the original price of the DVD set?

4 In a large school, 240 students are in Year 10. This is 20% of the school population.
a How many students are there in total in the school?
b How many students are in the other years at this school?

5 Susie was told that her pay had increased by 15%. Her new pay is £172.50.
What was her pay before the increase?

6 9 carat gold is 37.5% pure gold.
A piece of 9 carat gold jewellery is tested and found to contain 97.5 grams of pure gold.
What did the piece of jewellery weigh?

7 Julia is training for a marathon and she reduces her weight by 5% over a three-month period.
If she weighs 58 kg at the end of the period, what did she weigh at the start?

8 In a particularly hard ultramarathon, only 310 runners completed the course within the cut-off time.
If this represents 62% of the runners, how many runners started the race?

Checklist of learning and understanding

Review of percentages

- Percentage means 'parts per hundred'. Remember that percentages, fractions and decimals are all expressions of part of a whole and you can convert between them all.
- To convert a percentage to a fraction, write the percentage with a denominator of 100 and simplify.
- To change a percentage to a decimal, write it as a fraction with a denominator of 100 and then convert it to a decimal.
- To change a fraction to a percentage, find the equivalent fraction with a denominator of 100, then multiply by 100 and add the percentage sign, or use your calculator to divide the numerator by the denominator and multiply by 100.
- To change a decimal to a percentage, write it as a fraction with a denominator of 100 and multiply it by 100, or use your calculator to multiply it by 100 directly, and write the percentage symbol.
- To order a mixture of fractions, decimals and percentages, change them all to percentages or decimals first.

Percentage calculations

- To find a percentage of an amount, express the percentage as a fraction over 100 and then multiply the amount by the fraction.
- To express one quantity (quantity A) as a percentage of another quantity (quantity B), make sure the units are the same and then calculate

$$\frac{\text{quantity A}}{\text{quantity B}} \times 100.$$

Find answers at: cambridge.org/ukschools/gcsemaths-studentbookanswers

Percentage change

- To increase or decrease an amount by a percentage:
  - find the percentage amount and add or subtract it from the original amount
  - or, use a fractional or decimal multiplier:
    - to increase an amount by x%, the multiplier is $\left(1 + \frac{x}{100}\right)$.
    - to decrease an amount by x%, the multiplier is $\left(1 - \frac{x}{100}\right)$.
- To find an original value when you know the percentage increase or decrease and the new amount, write an equation and use reverse percentages to solve for x.

For additional questions on the topics in this chapter, visit GCSE Mathematics Online.

Chapter review

1 Write as fractions in their simplest form.

a 25% **b** 30% **c** 3.5%

2 Express each of these as a percentage.

a $\frac{1}{20}$ **b** $\frac{1}{8}$ **c** $\frac{8}{15}$ **d** 0.5 **e** 1.25 **f** 0.005

3 Put these values in ascending order.

15.24 $15\frac{23}{86}$ (15×1.015) $15\frac{56}{45}$

4 The value of an investment increased from £120 000 to £124 800.
What percentage increase is this?

5 The population of New Orleans was 484 674 before Hurricane Katrina.
Afterwards, the population was found to have decreased by 53.9%.
What was the population afterwards?

6 Mark has a voucher that gives him 22% off the prices at *Cordula's Hardware Store.*
Estimate how much he will pay for an electric drill that normally costs £87.99. *(3 marks)*

7 Shaz works 30 hours per week. She wants to increase this by 12%.
How many hours will she then work per week?

8 Express: **a** 3 hours as a percentage of a day.
b 750 metres as a percentage of 2 km.

9 The price of a plane ticket was reduced by 8% to £423.20.
What was the original price of the ticket?

10 A shop normally makes a profit of 32% on computer sales.
During a promotion, the marked selling prices of computers are reduced by 15%.

What is the cost price of a computer that sells for £980 during the promotion?

14 Algebraic formulae

In this chapter you will learn how to ...

- use formulae to express and solve problems.
- change the subject of a formula.
- substitute numbers into formulae to find the value of the subject.
- understand and use a range of formulae, including kinematics formulae.

For more resources relating to this chapter, visit GCSE Mathematics Online.

Using mathematics: real-life applications

Formulae are used by engineers, scientists, pharmacists, veterinarians and many other professionals. Vets use formulae to make sure they give animals the correct dosage of medicines for their age and weight. A poodle weighing 6 kg needs a far smaller dose of medicine than a 35 kg retriever.

Did you know?

Einstein's famous formula $E = mc^2$ looks deceptively simple. It describes the relationship between energy, mass and the speed of light (c). Find out more about this formula and why it is linked to the theory of relativity.

"I need to make sure I give the animals I treat the correct amount of medicine. I do this by using formulae that take into account their age, weight and the ratio between any other prescribed medicines." *(Veterinarian)*

Before you start ...

Ch 1, 3	You need to be able to substitute values into expressions.	**1** Evaluate $\frac{x + 2y}{z}$ when: **a** $x = 7, y = 4$ and $z = 2$ **b** $x = 7, y = {}^-2$ and $z = 2$ **c** $x = {}^-7, y = 4$ and $z = 4$ **d** $x = {}^-7, y = {}^-2$ and $z = 2$
Ch 8	You should be able to solve simple equations.	**2** Solve. **a** $6x = x + 35$ **b** $5x = 64 - 3x$ **c** $2(2x + 3) = x + 7$ **d** $5x - 8 = 3x + 12$
Ch 3, 5, 7, 8, 12	You should be familiar with some formulae already, and be able to identify the subject, variable(s) and any constants.	**3** Look at these two formulae for finding the area of shapes. $A = \frac{1}{2}bh$ $A = \pi r^2$ **a** What do the variables represent? **b** What is the constant in each formula? **c** What is the subject of each formula? **d** What tells you that the second formula applies to circles?

Find answers at: cambridge.org/ukschools/gcsemaths-studentbookanswers

Assess your starting point using the Launchpad

STEP 1

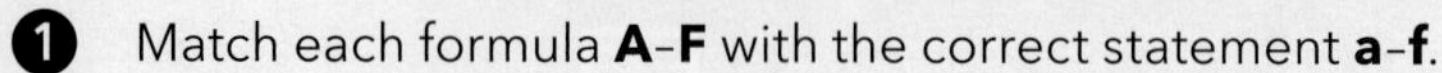

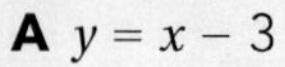

1 Match each formula **A**–**F** with the correct statement **a**–**f**.

A $y = x - 3$	**a** y is eight times the square root of one-fifth of x.
B $y = x^2 + 4$	**b** y is four more than the square of x.
C $y = 8\sqrt{\frac{x}{5}}$	**c** A car travelled 80 km in x hours at an average speed of y km/h.
D $y = 180 - x$	**d** x and y are supplementary angles.
E $y = \frac{80}{x}$	**e** A car used x litres of petrol on a trip of 80 km and the fuel consumption was y litres/100 km.
F $y = \frac{5x}{4}$	**f** x is three less than y.

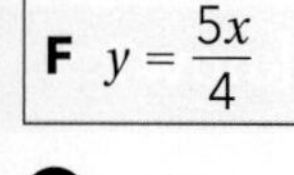

2 The cooking instructions to cook a leg of lamb are as follows:
'Preheat oven to 220°C and cook for 45 min per kg, plus an additional 20 min.'
Write a formula relating the cooking time T minutes and weight w kg.

GO TO
Section 1:
Writing formulae

STEP 2

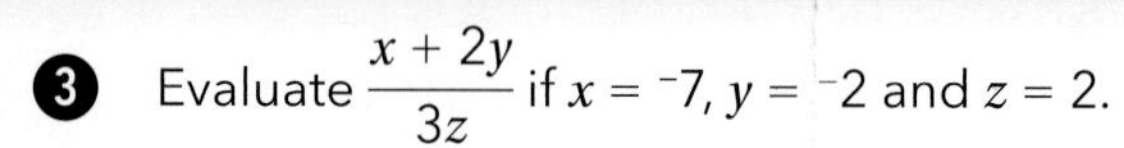

3 Evaluate $\frac{x + 2y}{3z}$ if $x = {}^-7$, $y = {}^-2$ and $z = 2$.

4 One of the kinematic equations that describes motion is $s = \frac{t(u + v)}{2}$

a How many variables does this equation contain?

b How many do you need to know to use the formula to evaluate?

c If $u = 3.5$, $v = 6.1$ and $t = 9$, what is the value of s?

GO TO
Section 2:
Substituting values into formulae

STEP 3

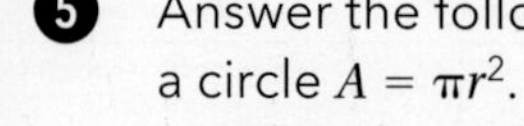

5 Answer the following questions on the formula for the area of a circle $A = \pi r^2$.

a What do each of the letters represent?

b Rewrite the formula and make r the subject.

c Can the value of r ever be a negative value?

GO TO
Section 3:
Changing the subject of a formula

GO TO
Section 4:
Working with formulae

Section 1: Writing formulae

Formulae

A **formula** is a special type of equation that shows the relationship between two or more unknown quantities.

For example, the rule for working out cooking time for a roast chicken might be cook for 40 minutes per kilogram plus an extra 20 minutes.

To make sense of a formula you need to know what the variables represent.

In the formula to calculate the area of a triangle, $A = \frac{1}{2}bh$:

A is the area in square units.

b is the length of the base of the triangle.

h is the perpendicular height of the triangle.

A, b and h are variables. The letters can be replaced by many different values.

$\frac{1}{2}$ is a constant. No matter what value you use for b or h, you have to multiply by $\frac{1}{2}$ in the formula to find the area of a triangle (Chapter 16).

A single variable on one side of the formula is called the **subject** of the formula. It is the variable (or quantity) that is being **expressed in terms of** the other variables in the formula.

In $A = \frac{1}{2}bh$, A is the subject of the formula.

Key vocabulary

formula (plural **formulae**): a general rule written as an equation showing the relationship between unknown quantities.

Tip

Formulae often use letters for the variable that relate to the value they represent. For example, in formulae for area, A is often used to represent the area and h to represent height.

Key vocabulary

subject: the variable which is expressed in terms of other variables; it is the variable on its own on one side of the equals sign.

In the formula $s = \frac{d}{t}$, s is the subject.

Writing formulae to represent real-life contexts

In Chapter 8, you formulated equations to represent information given in problems and then solved the equations to calculate the value of the unknowns. You use the same procedures to write formulae.

1 List the quantities involved. Work out whether they represent the subject, a variable, a constant or a coefficient.

2 Establish the relationship between each quantity. What is the subject being expressed in terms of?

3 Write the formula as concisely as possible using algebraic conventions.

Tip

For a reminder about variables, constants and coefficients, see Chapter 3.

WORKED EXAMPLE 1

Mary and Peter are sharing half a circular pizza.

Peter cuts himself a slice that makes an angle, a, with the straight edge of the pizza half.

He tells Mary that he will cut himself another slice the same size, and then she can have what is left.

Write a formula for the size of Mary's share of the pizza.

Subject = size of Mary's share

Variable = Peter's share

Constant = size of starting slice

List the quantities involved and determine if they are the subject, a variable, a constant or a coefficient.

Continues on next page ...

Peter's share = $2a$ Angles on a straight line equal 180°. $180° - 2a$ = size of Mary's share	Establish the relationship between each quantity. You know that Peter takes two slices with an angle of a, so you can write an expression for the total size of Peter's share. You know that angles on a straight line add up to 180° (see Chapter 9). The size of Mary's share is expressed in terms of the size of the starting slice (half a pizza) and the size of Peter's slice.
Let the size of Mary's share = M $M = 180° - 2a$	Write the formula using the language of algebra. It might be useful to draw a diagram to work out this problem.

EXERCISE 14A

1 Two friends Simon and Lucy are going to Europe on holiday. Simon has x euros and Lucy has y euros.

Write an equation for each of these statements:

a Simon and Lucy have a total of 2000 euros.

b Lucy has four times as many euros as Simon.

c If Lucy spent 400 euros, she would then have three times as many euros as Simon.

d If Lucy gave 600 euros to Simon, they would both have the same number of euros.

e Half of Simon's euros are equivalent to two-fifths of Lucy's.

2 When y is the subject and x is an unknown value, write a formula to determine y when y is:

a three more than x.

b six less than x.

c ten times x.

d the sum of $^{-}8$ and x.

e the sum of x and the square of x.

f twice x more than x plus 1.

g double x divided by the sum of x and $^{-}2$.

h half the product of π and the cube root of x divided by 3.

3 In general, temperature decreases with height above sea level.

This formula shows how temperature and height above sea level are related:

$$T = \frac{h}{200}$$

where T is the temperature decrease in °C and h is the height increase in metres.

a If the temperature at a height of 500 m is 23°C, what will it be when you climb to 1300 m?

b What increase in height would result in a 5°C decrease in temperature?

4 **a** The population of a town decreases by 2% each year. The population was initially P, and is Q after n years.

What is the formula relating Q, P and n?

b The population of a town decreases by 5% each year. The percentage decrease over a period of n years is a%.

What is the formula relating a and n?

Section 2: Substituting values into formulae

To find the value of the subject (or any variable) in a formula you need to know the value of all the other variables. The known values are substituted into the formula to work out the unknown value.

Substitute means replace the letters with the numbers you have been given.

Evaluate means calculate the numerical value.

Sometimes a formula will contain only two variables, the subject and one other.

Tip

You learnt about substitution and evaluation in Chapter 3.

WORKED EXAMPLE 2

The perimeter of a square can be found using the formula $P = 4s$, where s is the length of a side.

What is the perimeter of a square with sides of:

a 10 cm? **b** 2.5 mm?

a $P = 4s, s = 10$

$P = 4 \times 10$

$P = 40$ cm

You know that $s = 10$ cm and you need to find P. Substitute the value of s into the formula. Remember to include units in the answer.

b $P = 4s, s = 2.5$

$P = 4 \times 2.5$

$P = 10$ mm

Again, substitute the value of s into the formula and remember to include units in your answer.

WORK IT OUT 14.1

The surface area of a sphere given by the formula $A = 4\pi r^2$ where r is the radius of the sphere.

Calculate the surface area of a sphere of radius 8 cm in exact form.

Which is the correct answer?

What mistakes have been made in the incorrect answers?

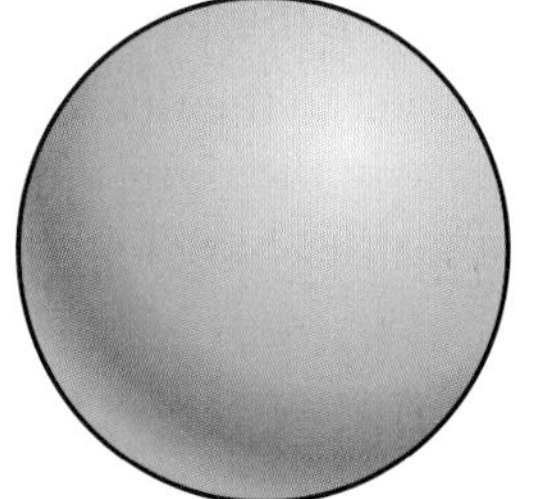

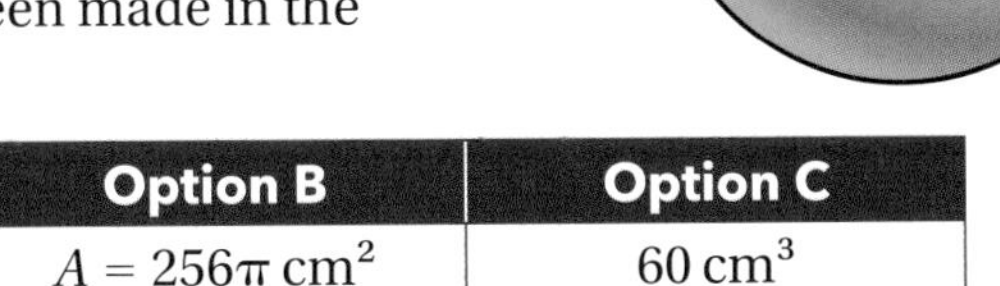

Option A	Option B	Option C
$A = 4 \times \pi \times 8^2$	$A = 256\pi$ cm^2	60 cm^3

Tip

Make sure you include units in an answer when appropriate.

Find answers at: cambridge.org/ukschools/gcsemaths-studentbookanswers

EXERCISE 14B

Tip

Formulae provide a general rule, which we apply for particular given values. It is important to be able to substitute values correctly into expressions that make up a formula and evaluate accurately.

1 Evaluate these expressions:

a $2a(a - 3b)$ when

i $a = 2$ and $b = {}^-5$ **ii** $a = {}^-3$ and $b = {}^-2$ **iii** $a = \frac{1}{3}$ and $b = \frac{1}{2}$

b $x^2 - 2y$ when

i $x = {}^-7$ and $y = 2$ **ii** $x = {}^-\left(\frac{1}{3}\right)$ and $y = \frac{5}{6}$

2 For $x = {}^-\left(\frac{a}{2}\right) - \frac{b}{c}$, find the exact value of x if:

a $a = 2, b = {}^-2, c = {}^-1$ **b** $a = 7, b = 3, c = {}^-4$

c $a = {}^-10, b = \sqrt{3}, c = \sqrt{3}$ **d** $a = \sqrt{2}, b = \sqrt{2}, c = 4$

e $a = 0, b = 0, c = 5$ **f** $a = \pi, b = \pi, c = 7$

3 For the formula $v = u + at$, find v if $u = 6$, $a = 3$ and $t = 5$.

4 Given $v^2 = u^2 + 2ax$ and $v > 0$, find the value of v (correct to 1 decimal place) when:

a $u = 0, a = 5$ and $x = 10$ **b** $u = 6, a = 4$ and $x = 15$

c $u = 2, a = 9.8$ and $x = 22$ **d** $u = 2.3, a = 4.9$ and $x = 10.6$

5 Given $\frac{1}{f} = \frac{1}{u} + \frac{1}{v}$, find the value of f when:

a $u = 2$ and $v = 4$ **b** $u = 2$ and $v = 15$

6 When a stone is thrown upward at 25 m/s, the height h metres it reaches after t seconds is given by the formula $h = 25t - 4.9t^2$.

Find the height of the stone after:

a 1 second. **b** 2 seconds.

7 $V = \pi l\{r^2 - (r - t)^2\}$ is the volume of metal in a tube where l is the length of the tube, r is the radius of the outside surface and t is the thickness of the material.

Find V when $l = 40$, $r = 5$ and $t = 0.5$ (Leave π in your answer).

8 Find the value of h, correct to one decimal place, if $h = \frac{9gRs}{2v}$ when $g = 9.8$, $R = 2.5$, $s = 3$ and $v = 7.4$.

9 The length of the hypotenuse c cm in a right-angled triangle is given by $c^2 = a^2 + b^2$ where a cm and b cm are the lengths of the perpendicular sides, as in the diagram.

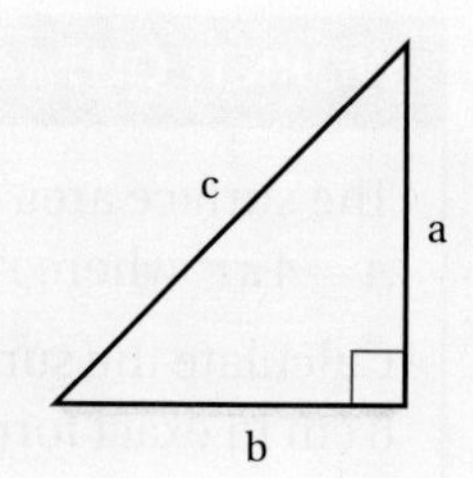

Calculate, correct to the nearest 0.1 cm, the length of the hypotenuse in the right-angled triangle whose perpendicular sides have lengths 14 cm and 25 cm.

10 The time T (in seconds) taken for one complete swing of a pendulum is given to the nearest centimetre by $T = 2\pi\sqrt{\frac{L}{g}}$ where $g = 9.8$ m/s² and L is length (in metres) of the pendulum.

Find, correct to the nearest centimetre, the length of the pendulum that takes 2 seconds to complete one swing.

Section 3: Changing the subject of a formula

You can calculate the value of **any** variable in a formula provided all the other variables are known.

When the value you want to find is not the subject of the formula you are given, you can rearrange the formula to make the desired value the subject; then you evaluate the formula as normal.

Since a formula is a type of equation, you can use inverse operations to rearrange the formula just as you would an equation; remember to do the same to **both** sides of the formula.

Be careful, sometimes the value you need to make the subject of the formula appears twice in the formula, is multiplied by a power or is reciprocal form.

Tip

See Chapter 3 for a reminder about rearranging equations.

Tip

You learnt about reciprocals in Chapter 2.

WORKED EXAMPLE 3

a Given the kinematic formula for acceleration $v^2 = u^2 + 2as$:

  i evaluate for s when $u = 8$, $v = 10$ and $a = 3$.

  ii make u the subject of the formula.

b Given the formula $P = \pi r + 2r + 2a$, make r the subject.

c Given the formula $y = \dfrac{2pt}{p - t}$, make t the subject.

a i $v^2 = u^2 + 2as$

s is not the subject of the formula, so you need to rearrange the formula in order to get s on one side on its own.

$v^2 - u^2 = 2as$

$\dfrac{v^2 - u^2}{2a} = s$

Rearrange the formula to make s the subject using inverse operations: subtract u^2 from both sides, then divide by $2a$.

$s = \dfrac{v^2 - u^2}{2a}$

Writing the subject on the left is the usual convention.

$u = 8$, $v = 10$ and $a = 3$

$s = \dfrac{10^2 - 8^2}{2 \times 3}$

$s = \dfrac{100 - 64}{6}$

$s = \dfrac{36}{6}$

$s = 6$

Substitute in the values you know.
You could alternatively substitute the numbers **before** rearranging.
Writing:

$$\begin{aligned} v^2 &= u^2 + 2as \\ 10^2 &= 8^2 + 2 \times 3s \\ 100 &= 64 + 6s \\ 100 - 64 &= 6s \\ 36 &= 6s \\ 36 \div 6 &= s \\ 6 &= s \\ s &= 6 \end{aligned}$$

Tip

Working quickly and skipping steps can lead to unnecessary mistakes. It is always a good idea to demonstrate your reasoning, especially in tests and exams.

Continues on next page ...

a ii $v^2 = u^2 + 2as$

$v^2 - 2as = u^2$

Rearrange the formula to make u the subject using inverse operations: subtract $2as$ from each side.

$u = \sqrt{v^2 - 2as}$

You want u not u^2 so find the square root of each side; write the subject on the left.

b $P = \pi r + 2r + 2a$

Rearrange the formula to make r the subject using inverse operations: subtract $2a$ from both sides.

$P - 2a = \pi r + 2r$

$P - 2a = r(\pi + 2)$

You need to get r on its own. r is a common factor of both terms $2r$ and πr, so you can factorise for r.

$\frac{(P - 2a)}{(\pi + 2)} = r$

You can now divide both sides by $(\pi + 2)$ so that you have r on its own.

$r = \frac{P - 2a}{\pi + 2}$

c $y = \frac{2pt}{p - t}$

The variable you want to make the subject is in reciprocal form, so multiply both sides by $(p - t)$ so that it is no longer a reciprocal.

$y(p - t) = 2pt$

$yp - yt = 2pt$

Multiply out the brackets.

$yp = 2pt + yt$

Add yt to both sides of the formula so that t is only on one side.

$yp = t(2p + y)$

Factorise for t.

$t = \frac{yp}{2p + y}$

Divide both sides by $(2p + y)$.

EXERCISE 14C

1 Rearrange each of these formulae to make the letter in the bracket the subject.

a $a(q - c) = d$ (q)

b $4(p - 2q) = 3p + 2$ (p)

c $5(x - 3) = y(4 - 3x)$ (x)

d $d = \sqrt{\frac{3h}{2}}$ (h)

e $y = \frac{2pt}{p - t}$ (t)

f $a = \frac{2 - 7b}{b - 5}$ (b)

g $\frac{x}{x + c} = \frac{p}{q}$ (x)

2 The formula for the sum S of the interior angles in a convex n-sided polygon is:

$S = 180(n - 2)$

Rearrange the formula to make n the subject and use this to find the number of sides in the polygon if the sum of the interior angles is:

a 1080° **b** 1800° **c** 3240°

3 The kinetic energy E joules of a moving object is given by $E = \frac{1}{2}mv^2$ where m kg is the mass of the object and v m/s is its speed.

Rearrange the formula to make m the subject and use this to find the mass of the object when its energy and speed are, respectively:

a 400 joules, 10 m/s **b** 28 joules, 4 m/s **c** 57.6 joules, 2.4 m/s

4 The formula for finding the number of degrees Fahrenheit (F) for a temperature given in degrees Celsius (C) is $F = \frac{9}{5}C + 32$

Rearrange the formula to make C the subject. Use this formula to convert these Fahrenheit temperatures to Celsius temperatures:

a 68 °F **b** 23 °F **c** 212 °F

5 The formula for the perimeter P of a rectangle l by w is $P = 2(l + w)$

If $P = 20$ cm and $l = 7$ cm, what is the length of w?

6 The area A cm² enclosed by an ellipse is given by $A = \pi ab$

Calculate to 1 decimal place the length a cm if $b = 3.2$ and $A = 25$.

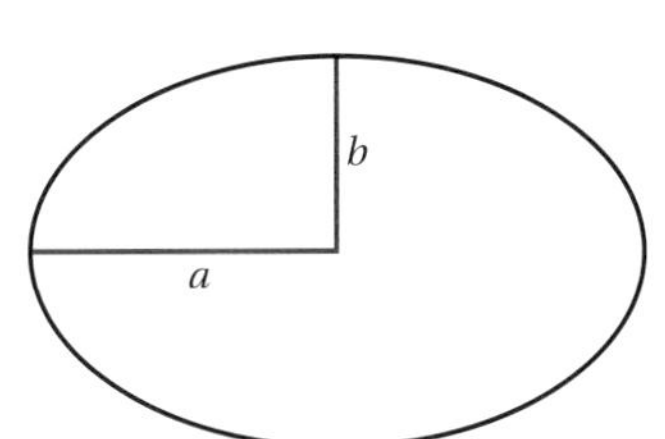

7 When an object is fired into the air at a speed of u metres per second, its height above the ground h metres and time of flight t seconds are related (ignoring air resistance) by $h = ut - 4.9t^2$

Find the speed at which an object was fired if it reached a height of 30 metres after 5 seconds.

8 For the formula $I = (180n - 360)/n$ find n when $I = 108$.

9 Given the formula $t = 2\pi\sqrt{\frac{L}{g}}$ find L if $t = \pi$ and $g = 9.8$.

Tip

Questions 6–10 are the type of question where you may decide to rearrange first and then substitute the values given to solve, or substitute the values first and then evaluate.

Did you know?

Formulae and equations are powerful tools to explore and explain the world.

This is an equation from higher-level mathematics:

$$Z_{n+1} = Z_n^2 + C$$

It is used to explore the amazing graphics called fractals.

Section 4: Working with formulae

You will use formulae as part of your work in other topics in mathematics, as well as other subjects.

Things to remember when working with formulae:

- Make sure you know what each term represents: the subject, a variable, the constant or a coefficient.
- Make sure you understand the relationship between each value.
- Substitute in the correct values.
- Include appropriate units in your answer.
- When constructing a formula, try some values to make sure it works.
- You can rearrange a formula to change the subject, but it is not always necessary when you are substituting known values.

Writing a formula can be particularly helpful when you are given lots of information and asked to find one value.

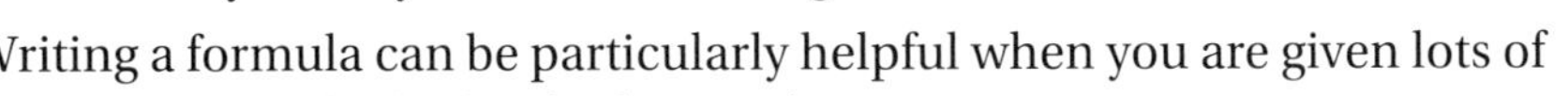

Find answers at: cambridge.org/ukschools/gcsemaths-studentbookanswers

Problem-solving framework

A group of sixth form students are planning to run a day conference.
The local university offers conference rooms for hire at a daily rate.
The students think they will have a maximum of 60 delegates and want to offer refreshments costing £4 per delegate.
The largest room they can hire costs £160 for the day.
They want to charge each delegate enough to cover the costs of running the conference.
How much should they charge each delegate if there will be 60 delegates attending the conference?

Steps for approaching a problem-solving question	**What you would do for this example**
Step 1: Identify what you have to do.	Calculate how much to charge each delegate at the conference.
Step 2: Test the problem with what you already know.	You know a formula is a general rule showing the relationship between quantities; writing a formula will help you to calculate the required value.
Step 3: What maths can you do?	**1** List the quantities involved and establish if they are the subject, a variable, a constant or a coefficient; use the language of algebra: Number of delegates (d) = 60 (variable, as this can change) Cost of refreshments = £4 per delegate (coefficient, as this value is multiplied by how many delegates there are) Cost of the room = £160 (constant, this is fixed by the university) Cost to charge each delegate = C (subject, this is what you want to find out). **2** Establish what the relationship is between each quantity. The cost to charge each delegate is the same as the cost of having each delegate at the conference (they want to break even, not make a profit). So, this is the total cost divided by the number of delegates. The total cost is equal to the cost of refreshments for each delegate and the cost of the room hire. Now put this together using algebra: Cost of conference = $4d + 160$ Cost of one delegate is this value divided by the number of delegates (d). So the formula to calculate how much to charge each delegate is: $C = \frac{4d + 160}{d}$ If there are 60 delegates: $d = 60$ $C = \frac{4d + 160}{d}$ $C = \frac{4 \times 60 + 160}{60}$ $60C = 240 + 160$ $60C = 400$ $C = 6.66666666667$ They should charge each delegate £6.67.
Step 4: Does your answer seem reasonable? Have you answered the question?	If they charged £6.67 per delegate and there were 60 delegates then they would get £400.20 which would just cover their costs. The answer is reasonable and you have answered the question as you have given a price per delegate.

Kinematics formulae

Kinematics is the study of how objects move; it is studied by mathematicians and physicists. For example, the average speed of an object can be found using the formula,

$$\text{speed} = \frac{\text{distance travelled}}{\text{time taken}}$$

In kinematics there are three important terms:

- **Velocity** – a measure of how fast an object is moving in a direction (in everyday terms this is speed).
- **Acceleration** – the rate at which the velocity of an object changes (in everyday terms this is increasing speed).
- **Displacement** – a change in the position of an object from when it started moving to where it stopped moving (in everyday terms this is like distance travelled).

In science, these terms have specific meanings; velocity, acceleration and displacement are vector quantities because they have both size and direction.

Tip

You will learn more about vectors in Chapter 28.

You will see questions that involve the following kinematics formulae:

$$v = u + at \qquad s = ut + \frac{1}{2}at^2 \qquad v^2 = u^2 + 2as$$

where:

v = final velocity — velocity is measured in units of distance and time (m/s – metres per second)

u = initial velocity

a = acceleration — acceleration in units of distance and time (m/s/s or m/s^2 – metres per second per second)

s = displacement (distance travelled) — displacement (distance travelled) in units of length (m)

t = time taken

Tip

These kinematics equations apply only to an object moving in a straight line at a constant acceleration.

WORK IT OUT 14.2

Two students studying physics were given the equation $v = u + at$ and asked to calculate the value of v when u = 12 metres per second, a = 10 metres per second per second, t = 3 seconds.

Which student has calculated v correctly, and used the correct units?

Student A	Student B
$v = 12 + 10 \times 3$	$v = 12 + 10 \times 3$
$v = 66$ m	$v = 42$ m/s

Common formulae to learn

There are some standard formulae that you will be expected to know and use. Some of these you will already have met in other chapters, others you will come across later. Make sure you know all of these formulae.

Find answers at: cambridge.org/ukschools/gcsemaths-studentbookanswers

Circumference of a circle	$C = \pi d$ or $2\pi r$ d = diameter, r = radius
Area of a circle	$A = \pi r^2$ r = radius
Pythagoras' theorem	$a^2 + b^2 = c^2$ c = hypotenuse of a right-angled triangle a and b = two other sides of a right-angled triangle
Trigonometry ratios	$\sin\theta = \frac{o}{h}$ $\cos\theta = \frac{a}{h}$ $\tan\theta = \frac{o}{a}$ Sine rule $\frac{a}{\sin A} = \frac{b}{\sin B} = \frac{c}{\sin C}$ Cosine rule $a^2 = b^2 + c^2 - 2bc\cos A$
Area of a triangle	$\frac{1}{2}ab\sin C$
The quadratic formula	$x = \frac{-b \pm \sqrt{b^2 - 4ac}}{2a}$

Tip

You will learn about the trigonometry formulae and this formula for the area of a triangle in Chapter 33.

EXERCISE 14D

1 Decide whether each of the following statements is true or false. Correct any false statements to make them true.

a Using the formula $s = ut + \frac{1}{2}at^2$ to find the value of s when $u = 4.6$, $a = 9.8$ and $t = 4$, the answer is $s = 96.8$.

b The formula $A = \pi r(r + l)$ rewritten to make l the subject becomes $l = \frac{A - \pi r}{r}$

c A formula to calculate the number n half way between two numbers x and y can be written as $n = \frac{x + y}{2}$

d In the formula to calculate the area of a circle, π is a constant and A and r are the variables.

e This formula calculates the volume of a triangular prism: $V = \pi r^2 h$

2 The Greek mathematician Hero showed that the area of a triangle with sides a, b and c is given by the formula $A = \sqrt{s(s - a)(s - b)(s - c)}$ where $s = \frac{1}{2}(a + b + c)$.

Use Hero's formula to find the area of this triangle.

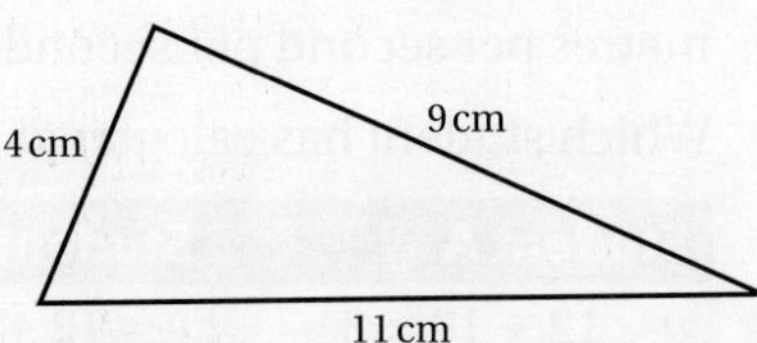

Tip

You will learn about simple and compound interest in Chapter 35.

3 Compound interest is calculated by the interest being added to a principal amount at the end of each year.

If P is the principal amount, r is the interest rate over a given period and n is number of years that the interest is compounded:

Total accrued $= P\left(1 + \frac{r}{100}\right)^n$

Find to the nearest whole £ the total accrued when £6000 is borrowed for three years at a rate of 4%. How much interest was paid?

4 What does this formula calculate?

$$x = \frac{^-b \pm \sqrt{b^2 - 4ac}}{2a}$$

Why is the sign ± included in this formula?

5 Given that the general form of a quadratic equation is $ax^2 + bx + c = 0$ use the method of completing the square to prove the formula in question 5.

> **Learn this formula**
>
> You are expected to know the formula $x = \frac{^-b \pm \sqrt{b^2 - 4ac}}{2a}$ from memory. See Chapter 8 if you need a reminder.

6 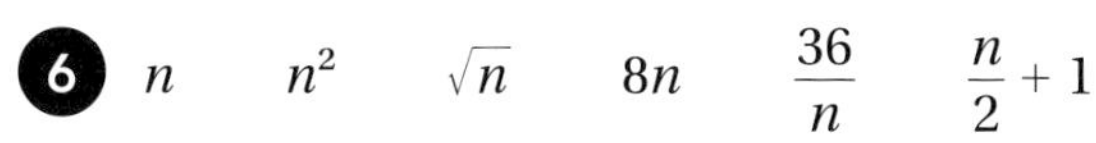

$n \quad n^2 \quad \sqrt{n} \quad 8n \quad \frac{36}{n} \quad \frac{n}{2} + 1$

a If $n = 4$ arrange in ascending numerical order.

b Arrange in ascending numerical order if $n = \frac{1}{4}$.

7 Conversion of a given temperature from the Fahrenheit scale to the Celsius scale is done by means of the formula $C = \frac{5(F - 32)}{9}$

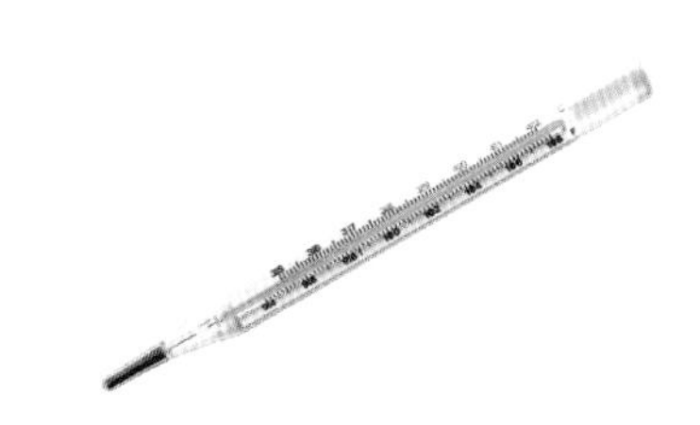

a Gallium is a soft, silvery metal that melts at temperatures above 85.57°F.
Given that your normal body temperature is about 37°C, would gallium be likely to melt if you held it in your hand? Explain your answer.

b Cast iron becomes molten at 1204°C.
Determine the Fahrenheit equivalent.

8 A formula used in life insurance is $Q = \frac{2m}{2 + m}$

a Calculate Q if $m = {}^-0.7$ (correct to 4 significant figures).

b Calculate m if $Q = 3$.

9 For a rectangle of length l cm and width w cm, the perimeter P cm is given by $P = 2(l + w)$.
Use this formula to calculate the length of a rectangle which has width 15 cm and perimeter 57 cm.

10 The distance d metres Jim's car takes to stop once the brakes are applied is given by the formula $d = 0.2v + 0.005v^2$ where v km/h is the speed of the car when the brakes are applied.
Find the distance the car takes to stop if the brakes are applied when it is travelling at each of the speeds given below. Calculate your answers correct to 3 decimal places where appropriate.

a 60 km/h **b** 65 km/h **c** 70 km/h

d Comment on how stopping distances vary with increasing speed and what this means for road safety.

11 A rocket scientist is trying to calculate how long a Lunar Explorer Vehicle will take to descend towards the surface of the moon.
He knows that if u = initial speed and v = speed at time t seconds, then:

$$v = u + at$$

where a is the acceleration and t is the time that has passed.

Rearrange the formula to make a the subject.

Find answers at: cambridge.org/ukschools/gcsemaths-studentbookanswers

12 The headmaster of a local school has to report on how well the school is performing. He does this by comparing the test scores of pupils across an entire school. He has worked out the mean but also wants to know the spread about the mean to see whether it is representative of the whole school. He uses this formula for the upper bound b of a class mean:

$$b = a + \frac{3s}{\sqrt{n}}$$

where s = sample spread about the mean, n = the sample size, a = the school mean and b = the mean maximum value.

Rearrange the formula to make s the subject of the formula.

13 When an object is fired into the air at a speed of u metres per second, its height above the ground, h metres, and time of flight, t seconds, are related (ignoring air resistance) by $h = ut - 4.9t^2$.

Find the speed at which an object was fired if it reached a height of 27.5 metres after 5 seconds.

Tip

If you need a reminder of solving quadratic equations see Chapter 8.

14 The general form of a quadratic equation is $ax^2 + bx + c = 0$.

Solve this equation with the formula $x = \frac{-b \pm \sqrt{b^2 - 4ac}}{2a}$, given $a = 1$, $b = 9$ and $c = 20$.

Checklist of learning and understanding

Writing formulae

- A formula is a general rule or equation showing the relationship between quantities.
- You can use formulae to represent real-life problems as long as you define the subject, any variable(s) you are using, and any constants or coefficients.

Substituting values into formulae

- To evaluate a formula you need to know the value of all but one of its variables.
- Substitute known values into a formula to find the unknown variable.

Changing the subject of a formula

- The subject of a formula is the variable (or quantity) that is being expressed in terms of the other variables in the formula.
- In any formula you can 'change the subject' by rearranging the formula in the same way as you rearrange equations, using inverse operations and performing the same operations to both sides.

Working with formulae

- Formulae are used across many topics in mathematics and other subjects, so make sure you are comfortable using them.
- You will come across kinematic formulae in some questions in science.

- There are some common formulae that you need to learn:
  - circumference of a circle
  - area of a circle
  - Pythagoras' theorem
  - trigonometry ratios
  - quadratic formula
  - sine and cosine rule
  - area of a triangle (using sin).

Chapter review

For additional questions on the topics in this chapter, visit GCSE Mathematics Online.

Tip

Many of the formulae you learn and prove in mathematics are for finding areas and volumes of standard 3D shapes. Research any formulae that you will need for the questions and then do the calculations.

Tip

Eliminating means you need to write a formula that does not include those variables.

1 **a** Calculate the length of the edge of a cube with a volume of 125 cm^3.

b Calculate the length of a cuboid with a total surface area of 157.36 m^2, height 6.5 m and breadth 2.2 m.

c The formula $S = 2\pi r(r + h)$ represents the total surface area of a cone. Rewrite this formula to make h the subject.

d The volume of a sphere $= \frac{4}{3}\pi r^3$

Find the volume of a sphere if the radius is $\sqrt{3}$. (Leave π and any surd values in your answer.)

e The curved surface area of a hemisphere is $2\pi r^2$. Explain why the total surface area is $3\pi r^2$.

2 Given the following two formulae

$$A = \pi r(r + 2h) \text{ and } C = r + h$$

create a formula:

a with the subject A, eliminating h.

b with the subject A, eliminating r.

3 A runner completed a 26 km race in 4 hours. After running 15 km, his average speed decreased by 2 km/hr.

$$t = \frac{d}{s}$$

where t is time in hr, d is distance covered in km and s is average speed in km/hr.

Work out the runner's two speeds for this race.

4 A cyclic quadrilateral has all its vertices on the circumference of a circle. Its area A is given by Brahmagupta's formula

$$A^2 = (s - a)(s - b)(s - c)(s - d)$$

where a, b, c and d are the side lengths of the quadrilateral and $s = \frac{1}{2}(a + b + c + d)$ is the 'semi-perimeter'.

Find the area of a cyclic quadrilateral with side lengths 8 cm, 9 cm, 10 cm and 13 cm.

15 Perimeter

In this chapter you will learn how to ...

- calculate the perimeter of simple shapes such as rectangles and triangles.
- calculate the circumference of a circle.
- calculate the perimeter of composite shapes, including circles or parts of circles.

For more resources relating to this chapter, visit GCSE Mathematics Online.

Using mathematics: real-life applications

Working out the amount of fencing needed for a field, the number of tiles needed to edge a swimming pool, or the number of perimeter security cameras needed to secure an area all require the calculation of a perimeter.

"Security cameras in a car park are only effective if they can see all round the perimeter of the car park, so I need to know the dimensions of the car park in order to determine how many cameras are needed and where to put them."

(Security camera technician)

Before you start ...

Ch 5	You must be able to recognise and name some common polygons.	**1** Match each name to the correct shape. octagon pentagon hexagon **a** **b** **c**
Ch 12	You must be able to convert between basic metric units.	**2** Convert. **a** 5 km into m **b** 12 km into cm **c** 8500 mm into m **d** 4.8 m to mm
Ch 5	You need to know and use the correct names of parts of a circle.	**3** True or false? **a** The diameter is twice the length of the radius. **b** The diameter is always shorter than the radius. **c** The angles at the centre of a circle add up to 180°.
Ch 14	You should be able to write simple formulae, substitute values into formulae and change the subject of a formula.	**4** If the side of a square is x cm, write an expression for the sum of all its sides. **5** Make l the subject of the formula $P = 2(l + w)$. **6** If $P = \pi r + 2r$, calculate r to two decimal places when $P = 10.28$ (2 dp)

Assess your starting point using the Launchpad

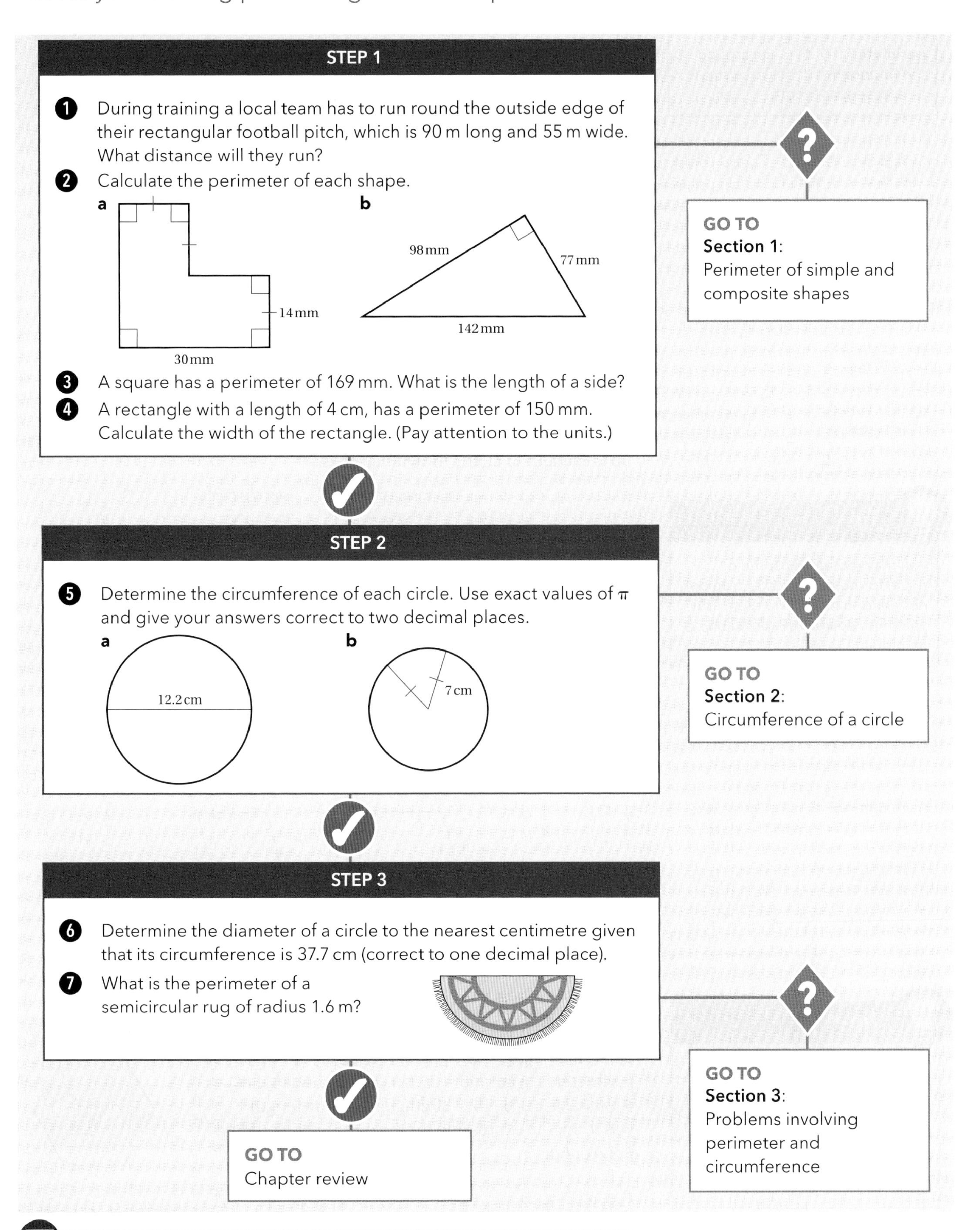

STEP 1

1. During training a local team has to run round the outside edge of their rectangular football pitch, which is 90 m long and 55 m wide. What distance will they run?
2. Calculate the perimeter of each shape.

   a

   b

3. A square has a perimeter of 169 mm. What is the length of a side?
4. A rectangle with a length of 4 cm, has a perimeter of 150 mm. Calculate the width of the rectangle. (Pay attention to the units.)

?

GO TO
Section 1:
Perimeter of simple and composite shapes

✓

STEP 2

5. Determine the circumference of each circle. Use exact values of π and give your answers correct to two decimal places.

   a

   b

?

GO TO
Section 2:
Circumference of a circle

✓

STEP 3

6. Determine the diameter of a circle to the nearest centimetre given that its circumference is 37.7 cm (correct to one decimal place).
7. What is the perimeter of a semicircular rug of radius 1.6 m?

?

GO TO
Section 3:
Problems involving perimeter and circumference

✓

GO TO
Chapter review

Find answers at: cambridge.org/ukschools/gcsemaths-studentbookanswers

Key vocabulary

perimeter: the distance around the boundaries (sides) of a shape. It represents a length.

Section 1: Perimeter of simple and composite shapes

The **perimeter** of a shape is the total distance around the boundaries of the shape. To calculate the perimeter of a shape:

- determine the lengths of all the sides, making sure they are in the same units.
- add the lengths together.

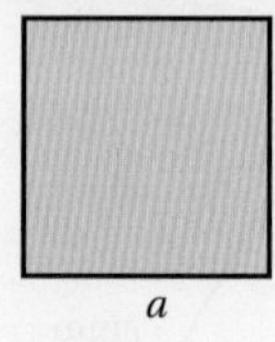
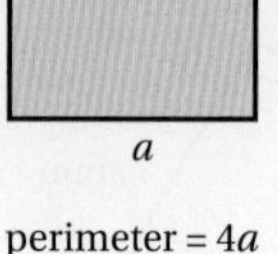

perimeter = $4a$

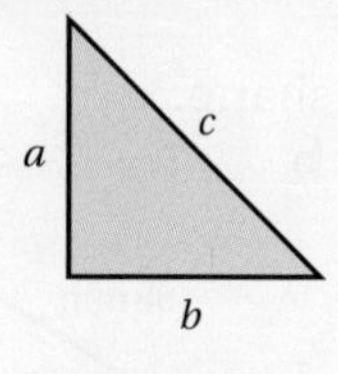

perimeter = $a + b + c$

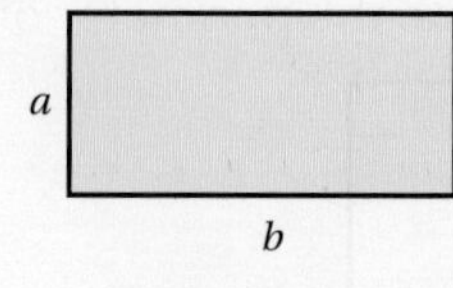

perimeter = $2(a + b)$

Using formulae to find the perimeter

The properties of different polygons can be used to derive formulae for calculating the perimeter.

This means that you can substitute in known values without having adding to up the length of all the individual sides.

Tip

You may remember some of these formulae from KS3. You do not need to memorise them, but you must be able to apply and manipulate them.

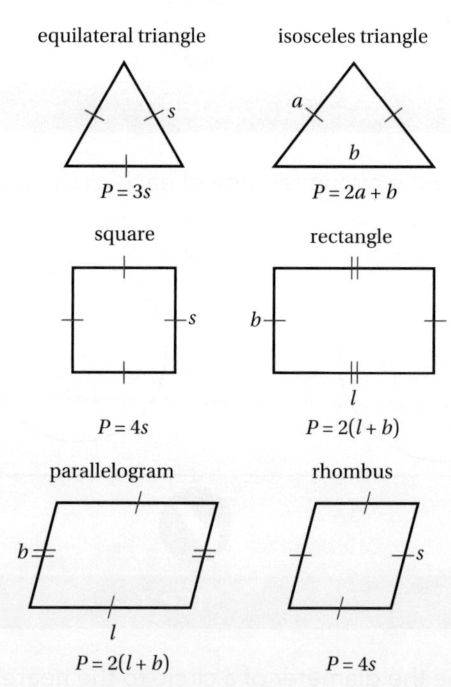

The general expression for finding the perimeter of a **regular** polygon is:

perimeter of regular polygon = side length × number of sides

You only need the length of one side to find the perimeter:

Each side length of this regular hexagon is 6 cm so the perimeter is: $6\text{ cm} \times 6 = 36\text{ cm}$. This is the same as $6 + 6 + 6 + 6 + 6 + 6 = 36\text{ cm}$. If each side length was a cm, the expression for the perimeter would be $6 \times a$ or $6a$.

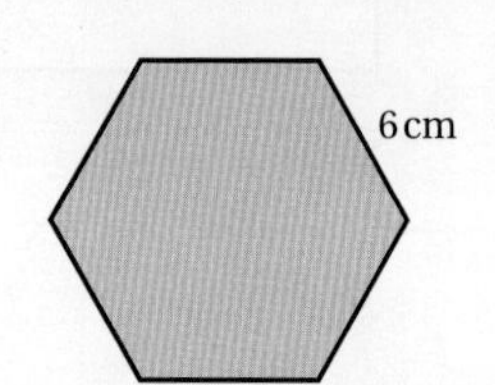

Tip

Writing a general expression is a way of showing the quantity of something when the actual numbers are not known. Numbers can be substituted into the expression to find the quantity required (see Chapter 14).

WORKED EXAMPLE 1

Calculate the perimeter of this shape if:

a $x = 5$ cm **b** $x = 4.5$ cm

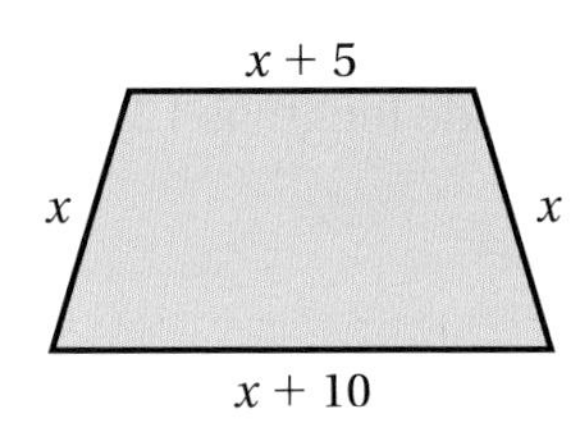

Perimeter $= x + x + 5 + x + x + 10$
$= 4x + 15$

First, write an expression for the perimeter in terms of x. Write out the perimeter as a sum of side lengths, and simplify.

a $x = 5$
perimeter $= 4x + 15$
$= 4 \times 5 + 15$
$= 35$ cm

Substitute the value of x into the formula you wrote for the perimeter.

b $x = 4.5$
perimeter $= 4x + 15$
$= 4 \times 4.5 + 15$
$= 33$ cm

Substitute the value of x into the formula you wrote for the perimeter.

Tip

Simplifying expressions by collecting like terms was covered in Chapter 3.

EXERCISE 15A

1. What is the perimeter of an equilateral triangle of side length 10 cm?

2. A yard is fully enclosed by a fence of lengths 12 m, 4.7 m, 354 cm and 972 cm.

   What is the perimeter of the yard?

3. Write expressions for the perimeter of these shapes.

   a An equilateral triangle, side length a.

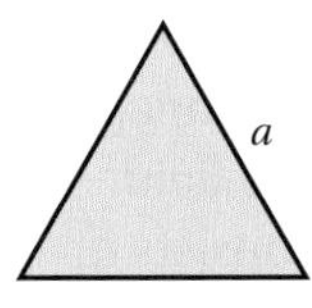

   b A rectangle, width x and length y.

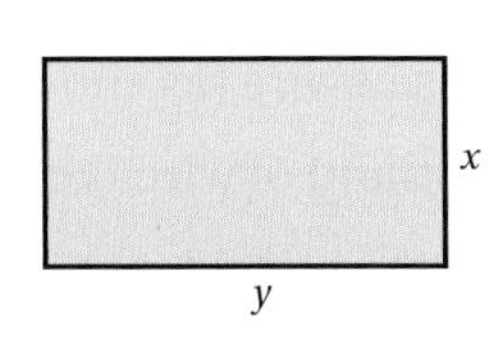

   c A regular octagon, side length z.

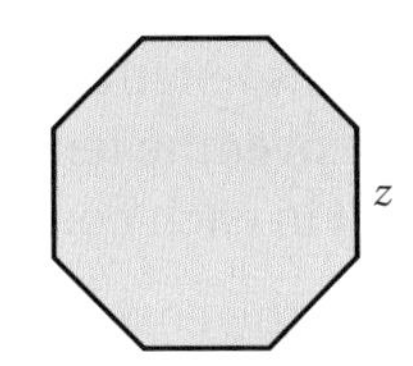

4 Work out the perimeter of this shape when

a each side length is 10 cm.

b each side length is 12 cm.

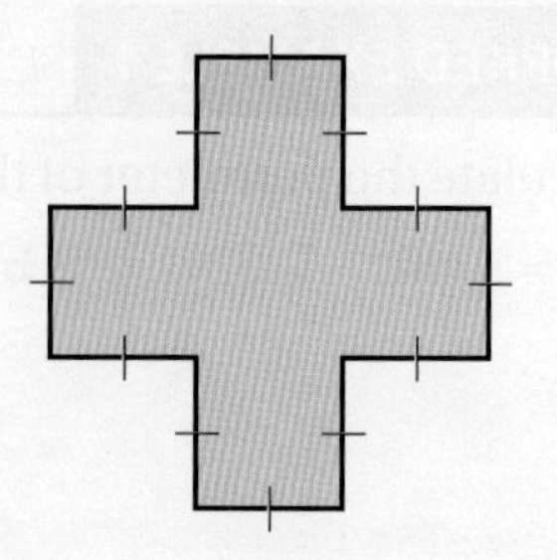

5 A tessellated hexagon creates a pattern for a patchwork quilt.

If the side length of each hexagon is 8 cm, what is the perimeter of this shape?

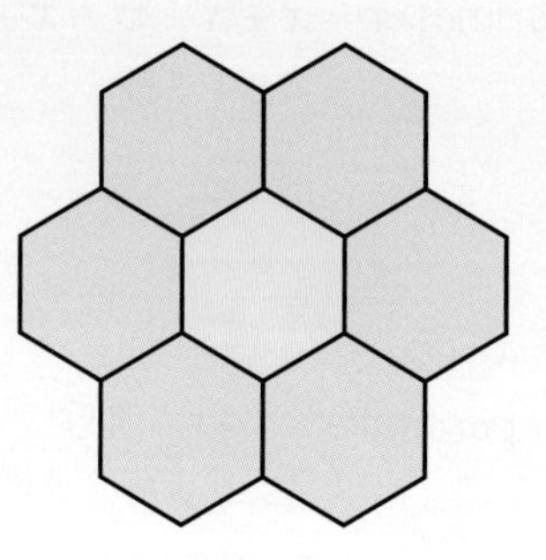

6 Each side length of this tiling pattern measures 15 cm.

Work out its perimeter.

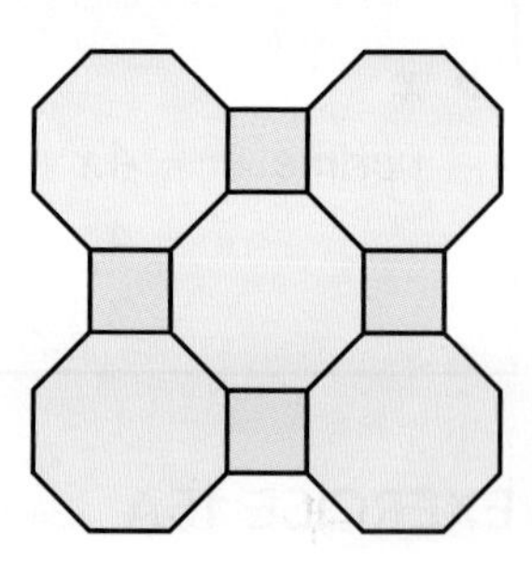

7 **a** Why are the formulae the same for calculating the perimeter of:

i a rectangle and a parallelogram?

ii a square and a rhombus?

b Why is there no formula for finding the perimeter of a trapezium?

8 A field with dimensions shown on the diagram is to be fenced to protect some sheep grazing in the field. The fence consists of upright posts and four strands of wire (as shown).

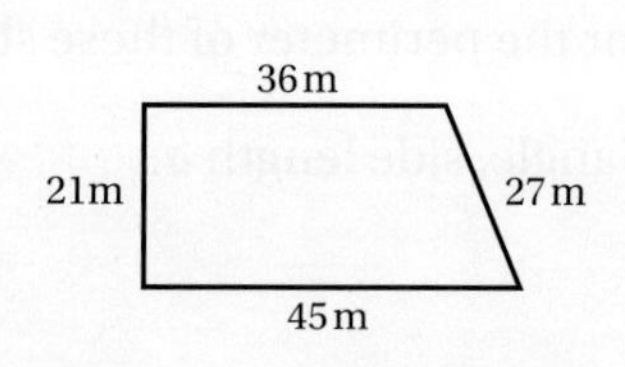

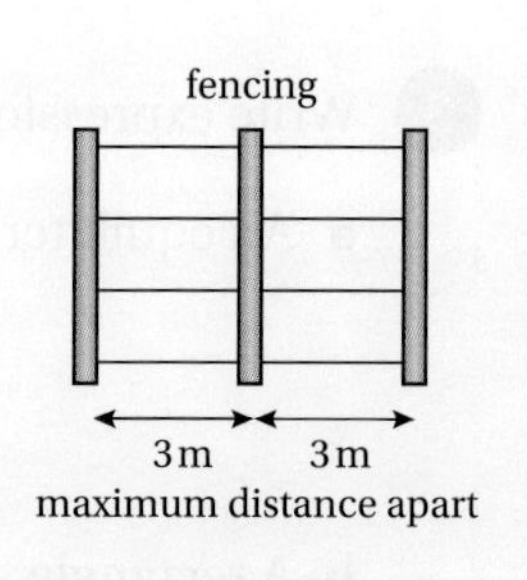

Each strand of wire is continuous and is tacked to a post where it passes one.

a Find the total length of wire needed for the fencing.

b To be structurally sound, the wire is tacked to a post every 3 m along its continuous length. Assuming there is no gate to the field (a stile is used to climb over the fence), how many posts would the fence need? There is one post in each corner of the field.

c Calculate the cost of fencing if each post costs £2.39 and the wire costs £1.78 per metre.

9 An international football field is 64 m by 75 m (the minimum size allowed).

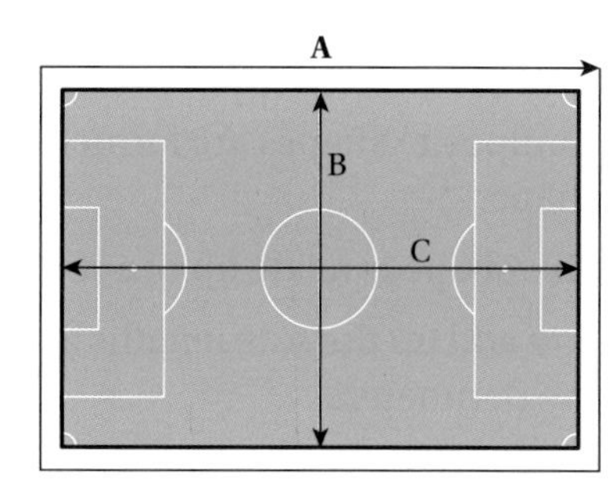

Three teams are training on the field.

Team A warms up by running 10 times around the field.

Team B runs round half the field 15 times cutting across the half-way line.

Team C also runs round half the field 15 times but they cut through the penalty spots and kickoff spot.

The routes followed by each team are shown on the diagram.

a Calculate the total distance run by each team.

b If it takes Team C, $\frac{1}{4}$ of an hour to complete their warm up laps, work out their average running speed in km/hr.

Finding lengths when the perimeter is known

You can calculate unknown lengths in a shape if you know the perimeter and have enough information about the shape, by rearranging the formula for the perimeter of that shape.

WORKED EXAMPLE 2

a What is the length of a rectangle of perimeter 20 cm and width 5.5 cm?

b Calculate the length of each side of a rhombus of perimeter 1.8 m.

a $P = 2(l + w)$ $\quad P = 20, w = 5.5$	Think about the information you have. You know the formula for the perimeter of a rectangle is $P = 2(l + w)$.
$20 = 2l + 2(5.5)$ $20 = 2l + 11$ $20 - 11 = 2l$ $9 = 2l$ $l = 4.5$ cm	Substitute the known values into the formula and solve for l. **Alternatively**, you could have rearranged the formula to make l the subject then substituted in the values.
b $P = 4s$	Write down the formula for the perimeter of a rhombus.
$s = \frac{P}{4}$ $s = \frac{1.8}{4}$ $s = 0.45$ m	Change the subject of the formula and substitute in the values you know to find s. **Alternatively**, you could have substituted the known values in first, and then rearranged to find s.

Find answers at: cambridge.org/ukschools/gcsemaths-studentbookanswers

Perimeter of composite shapes

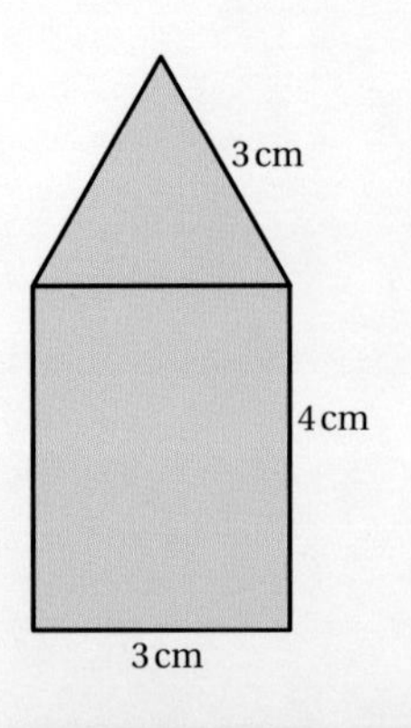

Composite shapes are formed by combining shapes or by removing parts of a shape.

This shape is made up of a rectangle and a triangle.

You add up the side lengths around the outside boundary of the shape to find the perimeter:

4 + 3 + 3 + 4 + 3 = 17 cm

You can use what you know about the properties of shapes to deduce any missing lengths.

WORKED EXAMPLE 3

Find the perimeter (P) of this composite shape.

All angles are right angles and all dimensions are in centimetres.

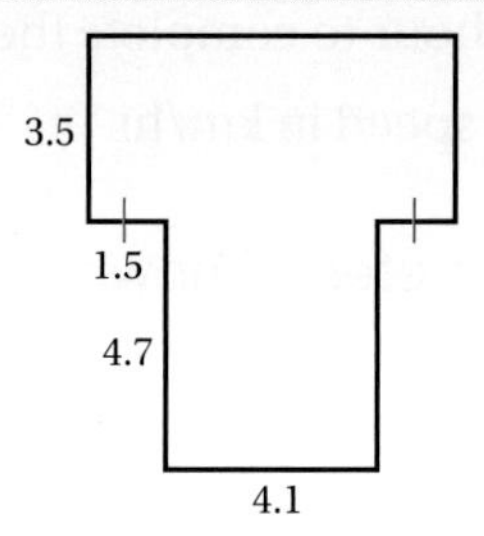

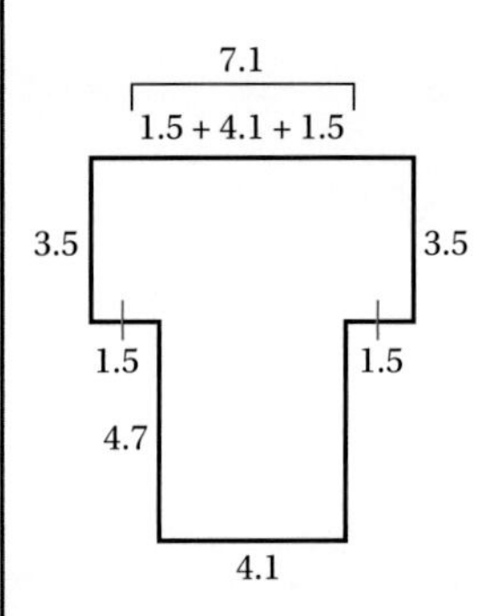

The shape is made of two rectangles. Start by working out the lengths of the missing sides. The top side is made up of the lengths 1.5, 4.1 and 1.5; you know the last bit is 1.5 because it has a mark to indicate it is the same length as the side labelled 1.5. You know this top shape is a rectangle as it has four right angles (if it was a square all four lengths would be the same). So, you know that opposite sides are equal in length, which gives you the missing length, 3.5. Similarly, the bottom shape is also a rectangle, and so the missing length is 4.7.

$P = 7.1 + 3.5 + 1.5 + 4.7 + 4.1 + 4.7 + 1.5 + 3.5$
$= 30.6$ cm

Add up the side lengths; don't forget to include the correct units in your answer.

Alternative solution

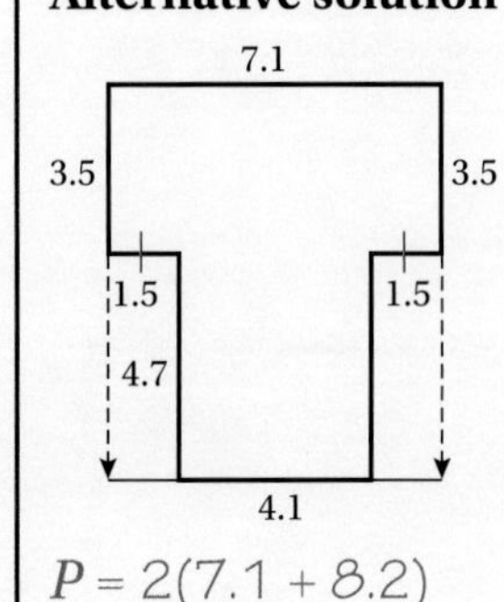

You might have spotted that the composite shape would still have the same perimeter if its two bottom corners were flipped to make a rectangle.
So, instead of adding up all the individual lengths, you could use the formula for calculating the perimeter of a rectangle instead.

$P = 2(7.1 + 8.2)$
$= 30.6$ cm

The perimeter must include **all** of the shape's boundaries; so be careful when working with composite shapes to include any **inner** boundaries in the perimeter calculation.

WORKED EXAMPLE 4

Calculate the perimeter of this shape.

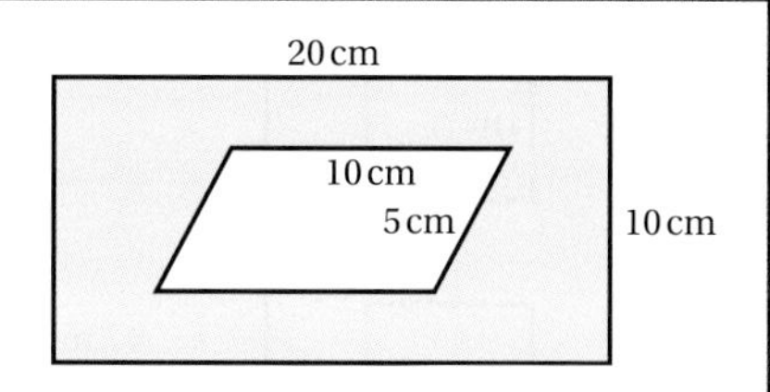

$P = 2(20 + 10) + 2(10 + 5)$
$= 2(30) + 2(15)$
$= 60 + 30$
$= 90$ cm

The shape is a rectangle with a parallelogram cut out of the middle. The perimeter of the shape is equal to the perimeter of the rectangle **and** the perimeter of the parallelogram inside. Always think carefully about what lengths are included in the perimeter of a shape.

WORK IT OUT 15.1

This shape was made by combining five identical squares with sides of 6.5 cm with four identical equilateral triangles.

Which calculation will result in the correct perimeter?

Why?

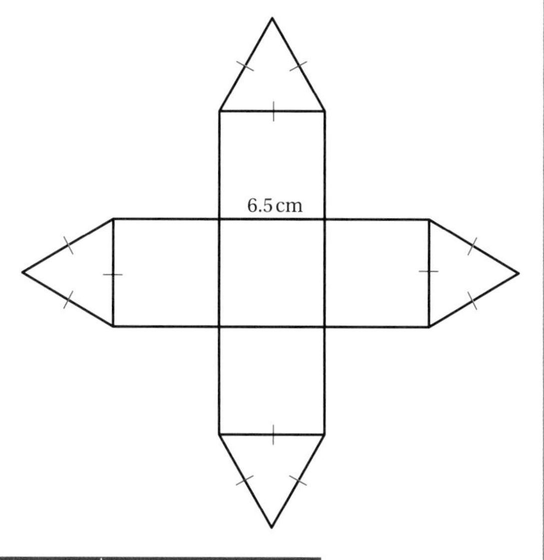

Option A	Option B	Option C
$P = 16 \times 6.5$ $= 104$ cm	$P = 20 \times 6.5$ $= 130$ cm	$P = 24 \times 6.5$ $= 156$ cm

EXERCISE 15B

1 Find the missing values.

Shape	Perimeter	Length	Width
Rectangle ABCD	242 mm	77 mm	
Parallelogram MNOP	200 mm		55 mm
Rhombus CDEF	12.25 cm		
Square PQRS	47.28 cm		

Find answers at: cambridge.org/ukschools/gcsemaths-studentbookanswers

2 Calculate the perimeter of each shape. **Diagrams not drawn to scale**

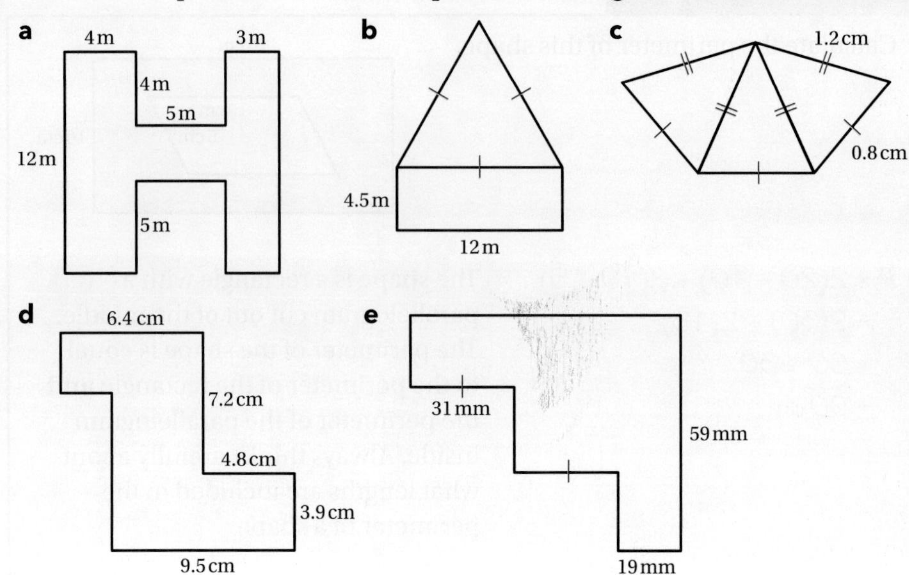

3 Petra is a mosaic artist. She has the following mosaic tiles:

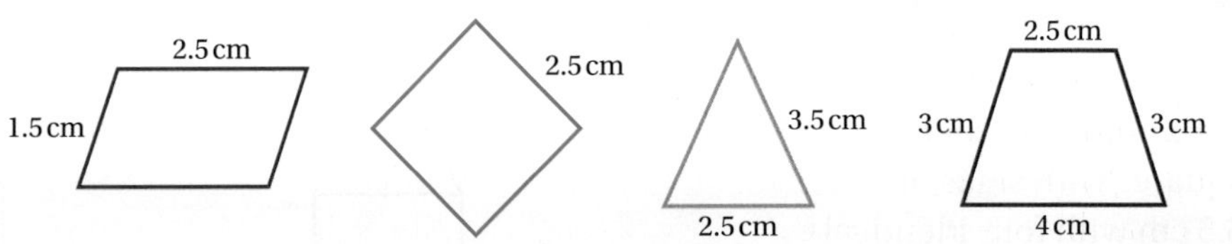

She arranges the tiles to make these shapes.

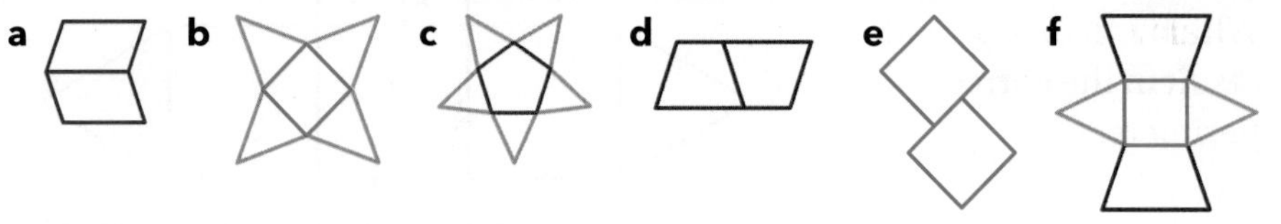

Use the dimensions above to find the perimeter of each shape.

4 The end of a maze consists of a regular pentagon with side length 5 m. If a child runs round this three times, how far will they run in total?

5 A rectangular allotment has a perimeter of 25 metres. If the length of one side is 6.8 metres, how wide is the allotment?

Tip

If there is no diagram, it is a good idea to draw a diagram to help you think about the problem and the perimeter you need to find.

Section 2: Circumference of a circle

The perimeter of a circle is called its **circumference (*C*)**.

The radius (r) is the distance from the centre to the circumference.

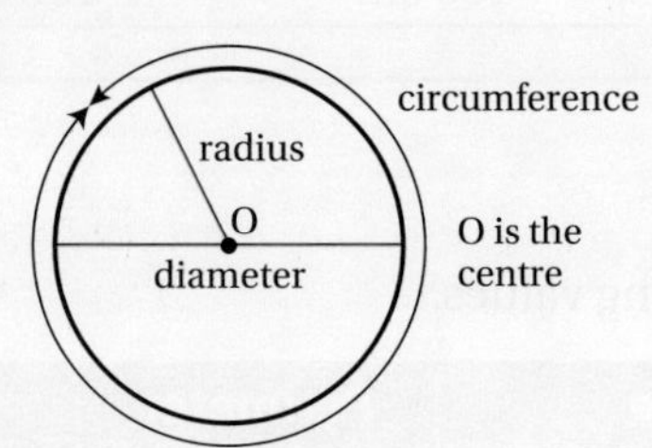

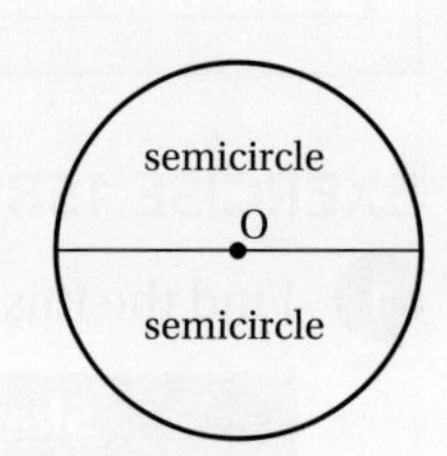

The diameter (d) is a line through the centre of a circle. The diameter is twice the radius ($2r$).

The Ancient Greeks discovered that when you measured the circumference of a circle and divided it by its diameter you got a constant ratio of approximately 3.142.

This ratio is called pi and the symbol π is used to represent it.

$$\pi = \frac{C}{d}$$

This ratio can be rearranged to give two formulae for finding the circumference of any circle.

Learn this formula

$C = \pi d$

where C = circumference and d = diameter.

Since the diameter is twice the length of the radius, the formula can also be written in terms of r.

$C = 2\pi r$

where r = radius.

Pi has no exact decimal or fractional value (it is an irrational number). So when you do calculations involving pi you might be asked to give rounded or approximate answers.

If you are asked to give an **exact** answer, this means your answer should be a multiple of π, and the π symbol should be in your answer.

Your calculator can work with exact values of pi but sometimes you will be given an approximate value of pi and asked to use that. **Read the question carefully**.

WORKED EXAMPLE 5

a Calculate the circumference of a circle with diameter 7 cm.
Give your answer to the nearest cm.

b Calculate the circumference of a circle with radius 4.5 m.
Leave your answer as a multiple of π.

c Take the value of π to be 3.142.
What is the circumference of a circle with diameter 20 m?
Give your answer to 2 decimal places.

a $C = \pi d$
$= \pi \times 7$
$= 21.991\,149 \approx 22$ cm

Use the π button on your calculator. Check the manual on your calculator to see how to use the π button; for example you might press:

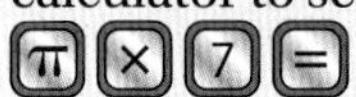

$C = 22$ cm (to the nearest cm)

Round your answer to the requested degree of accuracy; always include the degree of accuracy in your answer when using the '=' sign.

b $C = 2\pi r$
$= 2 \times \pi \times 4.5$
$= 9\pi$ m

Simplify the expression as much as possible. The question asks for the answer to be given as a multiple of π, this is the same as being asked to provide an **exact** answer. By leaving π in the answer, you are not giving an approximation of the answer.

c $C = \pi d$
$= 3.142 \times 20$
$= 62.84$ m

Use the value for π given to you in the question.

Sometimes questions that require you to calculate the circumference of a circle might not mention it directly. You need to be able to recognise what the question is asking of you.

Find answers at: cambridge.org/ukschools/gcsemaths-studentbookanswers

Problem-solving framework

Racing wheelchairs can travel at speeds of up to 45 mph and can cost up to £20 000.

The rear wheels of a chair have a diameter of 70 cm.

The front wheels have a radius of 20 cm.

Find the number of revolutions that the rear wheels will make over a 100 m race.

Steps for approaching a problem-solving question	What you would do for this example
Step 1: What have you got to do?	Find the number of revolutions (turns) that the rear wheels will make over 100 m. I need to calculate the circumference of the rear wheels so that I can see how many times the wheel turns in 100 metres.
Step 2: What information do you need?	Diameter (d) of rear wheel = 70 cm Circumference of a circle formula: $C = \pi d$ or $2\pi r$ Length of race = 100 m
Step 3: Is there any information you don't need?	The radius of the front wheel, speed of the wheelchair and its cost are all irrelevant.
Step 4: What maths can you do?	Circumference of rear wheel = $\pi d = \pi \times 70 = 219.9$ cm (to 1 dp) Change the units of the distance, so that they are the same as the circumference: 100 m = 100×100 cm = 10 000 cm Number of revolutions made by rear wheel = $10\,000 \div 219.9 = 45.5$ (to 1 dp)
Step 5: Have you used all the information? At this point you should check to make sure you have calculated what was asked of you.	All relevant information used ✓ Answered the question ✓
Step 6: Is your answer correct?	Used diameter for rear wheel ✓ Converted cm into m correctly ✓

Knowing the circumference of a circle means that you can calculate the diameter and/or the radius by rearranging the formula for the circumference.

WORKED EXAMPLE 6

A circle has a circumference of 200 cm.

Calculate the diameter of the circle to 1 decimal place.

$C = \pi d$	Write down the formula for calculating the circumference of a circle.
$200 = \pi \times d$	Substitute in the known values, then solve the equation for the unknown value (d). You could also have started by rearranging the formula to make d the subject and then substituting in the known values.
$200 \div \pi = d$	
$d = 63.7$ cm (1 dp)	

Tip

If you are asked to find r and use $C = \pi d$, remember to divide d by 2 to get r.

EXERCISE 15C

1 Use the π key on your calculator to calculate the circumference of each circle.

Give your answer correct to 2 decimal places if necessary.

a

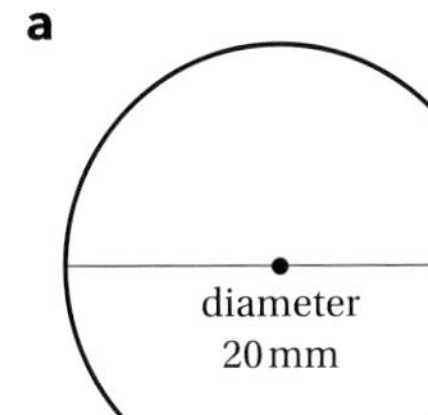

b

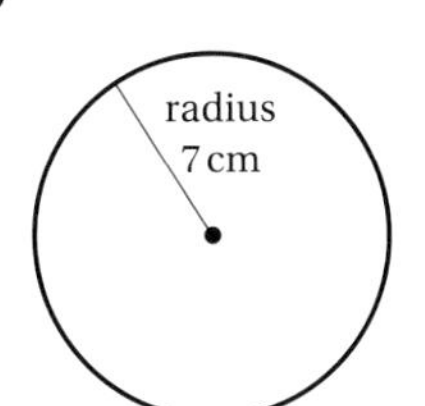

c

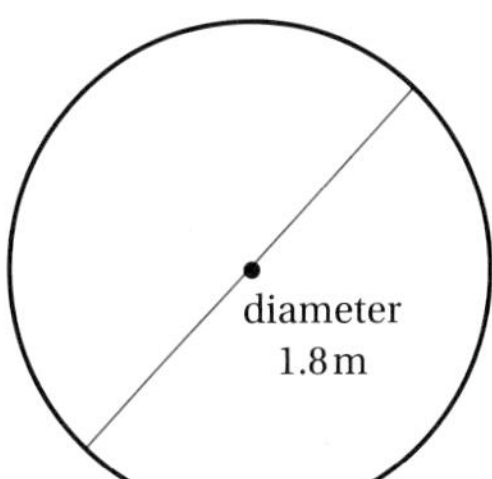

d

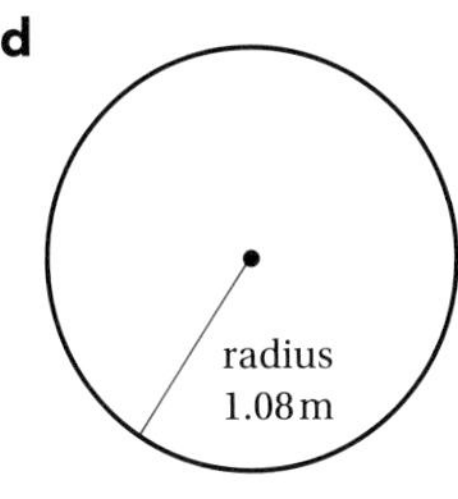

e

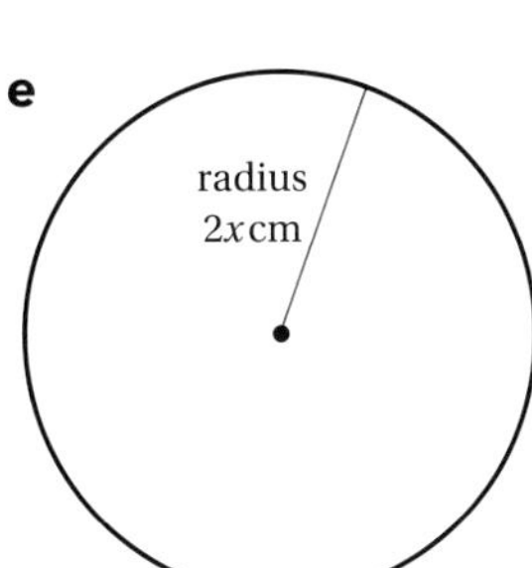

f

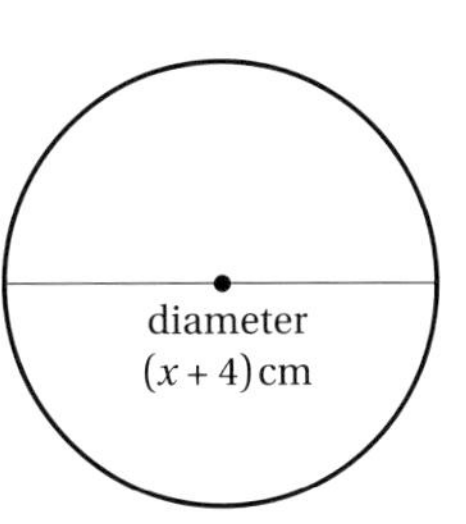

2 The alloy wheel rims of a car have a diameter of 42 cm.

Calculate the circumference of each rim.

3 A plastic slinky spring consists of 36 coils of plastic.

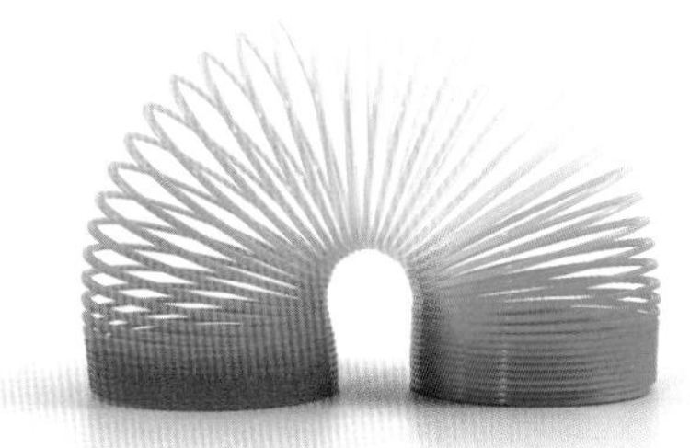

If the diameter of one coil is 55 mm, what length of plastic is needed to make the spring?

Find answers at: cambridge.org/ukschools/gcsemaths-studentbookanswers

4 Nate has a square piece of metal with sides of 8.5 cm.

He wants to cut out a round disc with a radius of at least 4 cm from the square.

a Draw a rough sketch to show this situation.

b Calculate the circumference of the disc if the radius is 4 cm.

c When he cuts the disc out, Nate finds the diameter is actually 8.3 cm.
What is the circumference of this disc?

d What is the perimeter of the piece of metal left after cutting out a disc of:

i radius 4 cm? **ii** diameter 8.3 cm?

Tip

Remember that the perimeter includes both the outer and inner boundaries.

5 Find the diameter, correct to 2 decimal places, of a circle of circumference:

a 20 mm. **b** 15.2 cm.

6 Find the radius of a round CD to the nearest mm, if its circumference is 36.33 cm.

7 What is the smallest possible square plate that you can use for a round cake of circumference 77 cm?

8 The minute hand of a clock is 75 mm long.

How far will the tip of the hand travel in one hour?

Give your answer correct to the nearest centimetre.

Tip

You learnt about the parts of a circle in Chapter 5.

Sectors of a circle

Key vocabulary

subtended: a subtended angle is one whose sides pass through the ends of an arc (or other curved line).

A **sector** is a 'slice' of a circle.

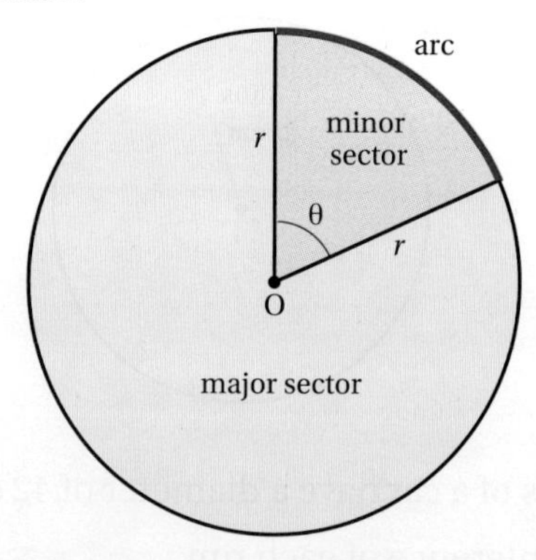

Tip

The term circumference is only used to describe the distance around a whole circle. When you deal with distance around parts of a circle you talk about the **perimeter**.

The angle θ is said to be **subtended** at the centre.

The perimeter of a sector is formed by two radii and a section of the circumference called an **arc**.

Perimeter of a sector = radius + radius + arc length.

To work out the length of an arc, you need to use what you know about angles and circles.

Angles around a point add up to 360°; this is true for the angles around the centre of a circle.

So, the subtended angle in the sector is therefore a fraction of 360°.

You can express this as $\frac{x}{360}$.

Tip

You learnt about the properties of circles in Chapter 5, and about angles in Chapter 9.

Similarly, the arc length is a fraction of the circumference.

So, to find the length of the arc you multiply the circumference of the circle by the fraction of the full circle that the sector represents:

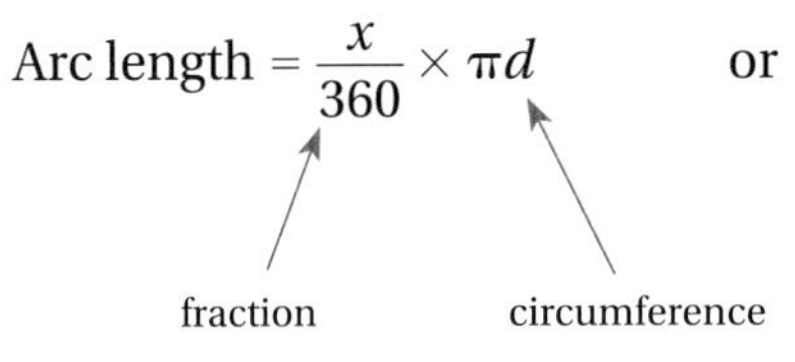

or

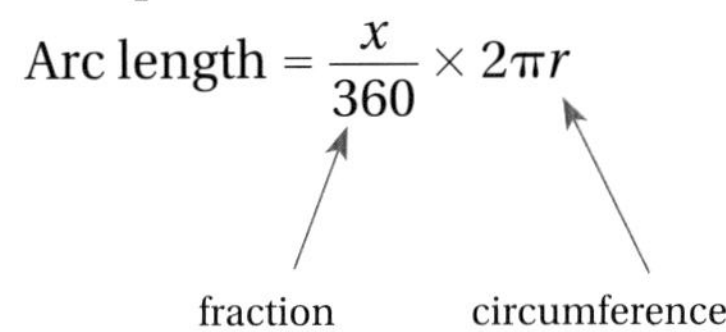

WORKED EXAMPLE 7

Find the length of the arc in each of these sectors.

a

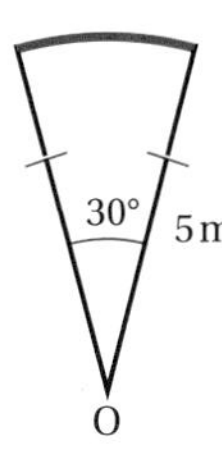

b

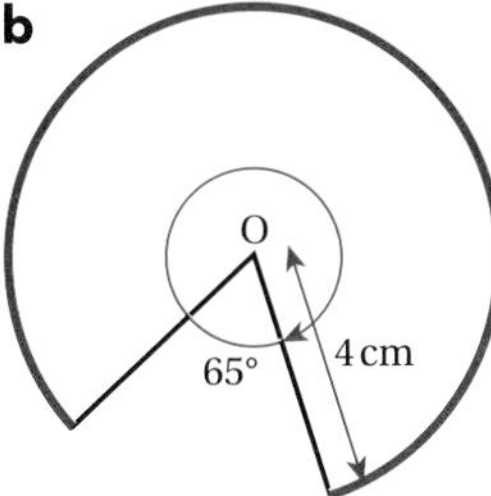

a Arc length $= \frac{x}{360} \times 2\pi r$

$= \frac{30}{360} \times 2\pi r$

$= \frac{1}{12} \times 2 \times \pi \times 5$

$= 2.62$ m (2 dp)

The subtended angle is 30°, substitute this into the formula for arc length. Use $2\pi r$ here as you have been given the radius.

If you round your answer, make sure you write what degree of accuracy you used.

b $360° - 65° = 295°$

Arc length $= \frac{295}{360} \times 2 \times \pi \times 4$

$= 20.59$ cm (2 dp)

The subtended angle is not given.
Work it out using the angle fact that angles round a point add up to 360°.

Substitute the value of the subtended angle into the formula.

The semicircle and quarter-circle (quadrant) are special examples of a sector.

- In a semicircle, the subtended angle is 180° and the arc length is half the circumference.
- In a quarter-circle, the subtended angle is 90° and the arc length is a quarter of the circumference.

EXERCISE 15D

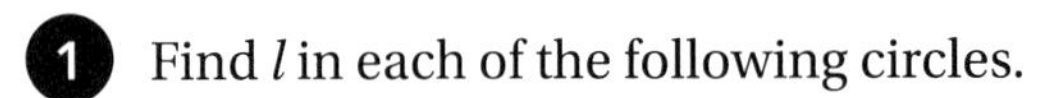

1 Find l in each of the following circles.

a

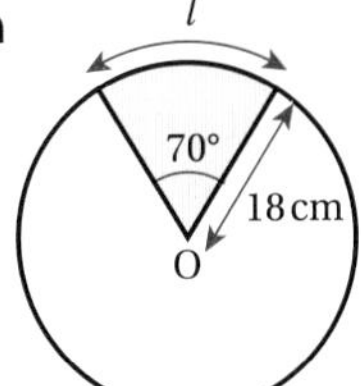

b

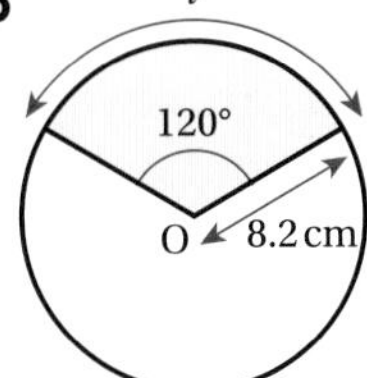

c

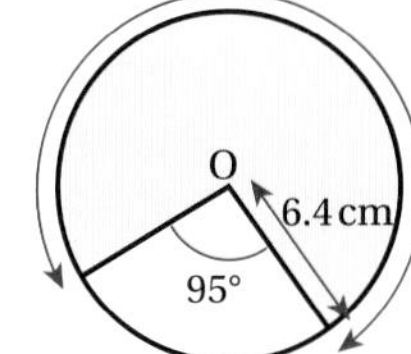

d

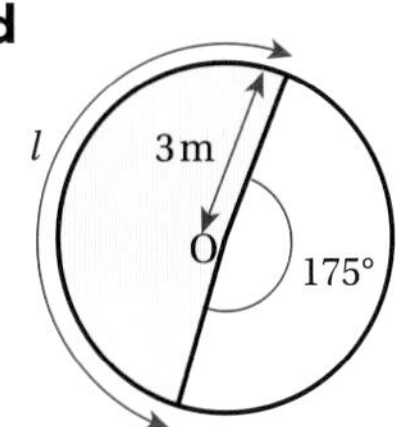

Find answers at: cambridge.org/ukschools/gcsemaths-studentbookanswers

2 Find the perimeter of each shape.

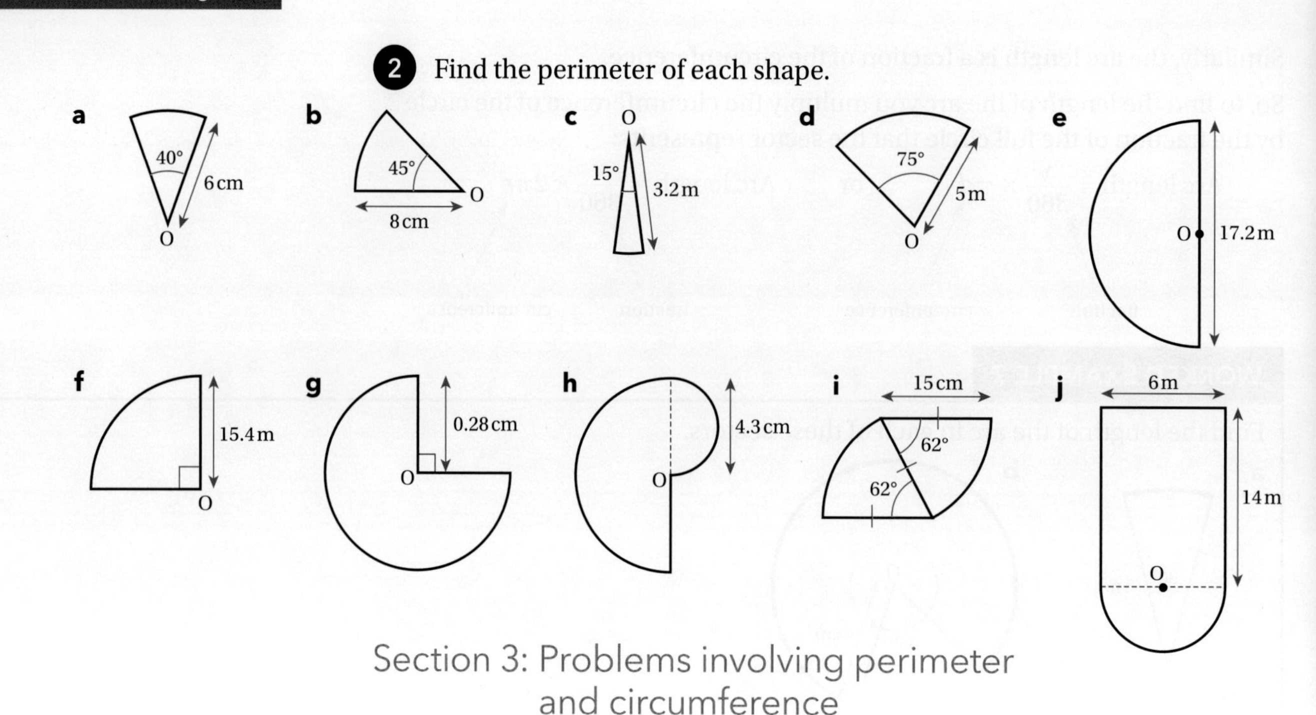

Section 3: Problems involving perimeter and circumference

You can combine your knowledge of perimeter and circumference to help you calculate the perimeter of irregular and composite shapes.

Remember to divide composite shapes into other shapes whose perimeters can be more easily calculated.

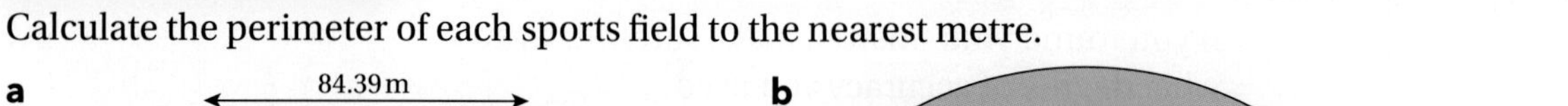

WORKED EXAMPLE 8

Calculate the perimeter of each sports field to the nearest metre.

a

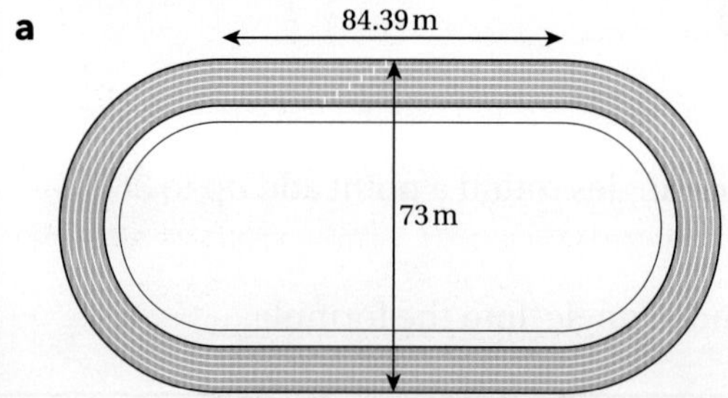

b

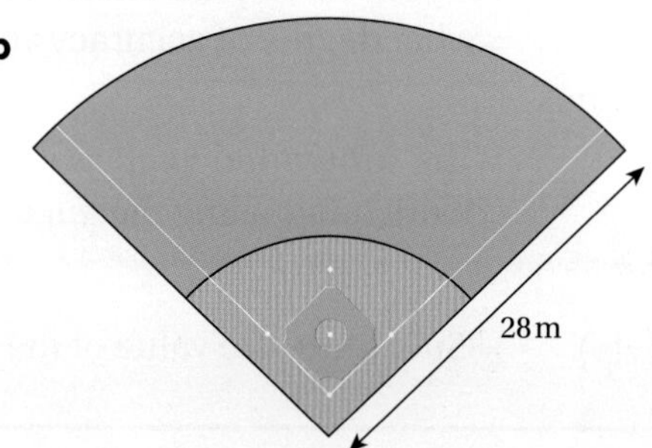

A baseball pitch is $\frac{1}{4}$ of a circle.

a Perimeter = circumference of circle + lengths of straight sides of rectangle.

The athletics track can be split into a rectangle and two semicircles. Two semicircles make one circle, so you can use this formula to calculate the perimeter.

Circumference of a circle $= \pi d$

$= \pi \times 73$

$= 229.336 \ldots$ m

Substitute the known values into each formula.

Perimeter of field $= 229.336 \ldots + (2 \times 84.39)$

$= 229.336 \ldots + 168.78$

$= 398$ m (to the nearest metre)

The three dots here indicate that the number continues; if you can, always use the exact number in the calculation and only round at the end. Some calculators allow you to store intermediate values.

Continues on next page ...

b Perimeter of a sector = $2r$ + arc length

Circumference of circle = $2\pi r = 2 \times \pi \times 28$.

Arc length of $\frac{1}{4}$ circle $= \frac{C}{4}$

$= \frac{2 \times \pi \times 28}{4}$

$= 43.98$ m

Perimeter = $2r$ + arc length

$= 2 \times 28 + 43.98$

$= 56 + 43.98$

$= 99.98$ m

$P = 100$ m correct to the nearest metre

The baseball pitch is a sector of a circle; so you can write down the formula for calculating the perimeter of a sector.

You don't have the angle inside the sector, but you do know that the pitch is $\frac{1}{4}$ of a circle; so you can calculate arc length by dividing the circumference by 4.

Substitute in the known values.

WORK IT OUT 15.2

This is a plan of a children's play area.
Tiling is to be placed around the curved edges of the sandpits.
Using $\pi = 3.142$, what is the total length of tiling required?
Which of the answers below is the correct answer?
What mistakes have been made in the other workings?

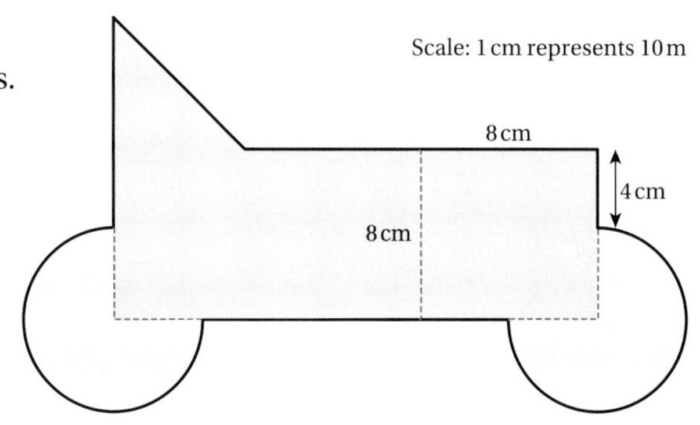

Tip

Perimeter can be worked out by measuring lengths, or by using lengths from a scale diagram. You learnt about scale drawings in Chapter 12.

Option A	Option B	Option C
Circumference of a circle = $\pi d = 2\pi r$ $r = 4$ cm	Circumference of a circle = $\pi d = 2\pi r$ $r = 4$ cm	Circumference of a circle = $2\pi d = \pi r$ $r = 4$ cm
$C = 2 \times 3.142 \times 4$ $= 3.142 \times 8$ $= 25.14$ cm	$C = 2 \times 3.142 \times 4$ $= 3.142 \times 8$ $= 25.14$ cm	$C = 2 \times 3.142 \times 8$ $= 50.27$ cm
Scale 1 : 25 So, actual circumference of the circle $= 25.14 \times 25$ $= 628.5$ cm $= 6.285$ m	Only $\frac{3}{4}$ of the sandpit needs tiling: $\frac{3}{4} \times 25.14$ cm $= 18.84$ cm	Only $\frac{3}{4}$ of the sandpit needs tiling: $\frac{3}{4} \times 50.27$ cm $= 37.7$ cm
Two sandpits, so total tiling required $= 6.285 \times 2 = 12.57$ m	Two sandpits: 18.85 cm $\times$ 2 = 37.7 cm of tiling Scale = 1 : 25 So, actual amount of edging required: 37.7 cm $\times$ 25 = 942.5 cm = 94.25 m	Two sandpits: 37.7 cm $\times$ 2 = 75.4 cm of tiling Scale = 1 : 25 So, actual amount of edging required: 75.4 cm $\times$ 25 = 1885 cm = 18.85 m

EXERCISE 15E

The diagram shows the shape and dimensions of different throwing event field areas in international competitions; the circles represent the area in which the participating athlete would stand.

Use the information in these diagrams to answer Questions **1** to **4**.

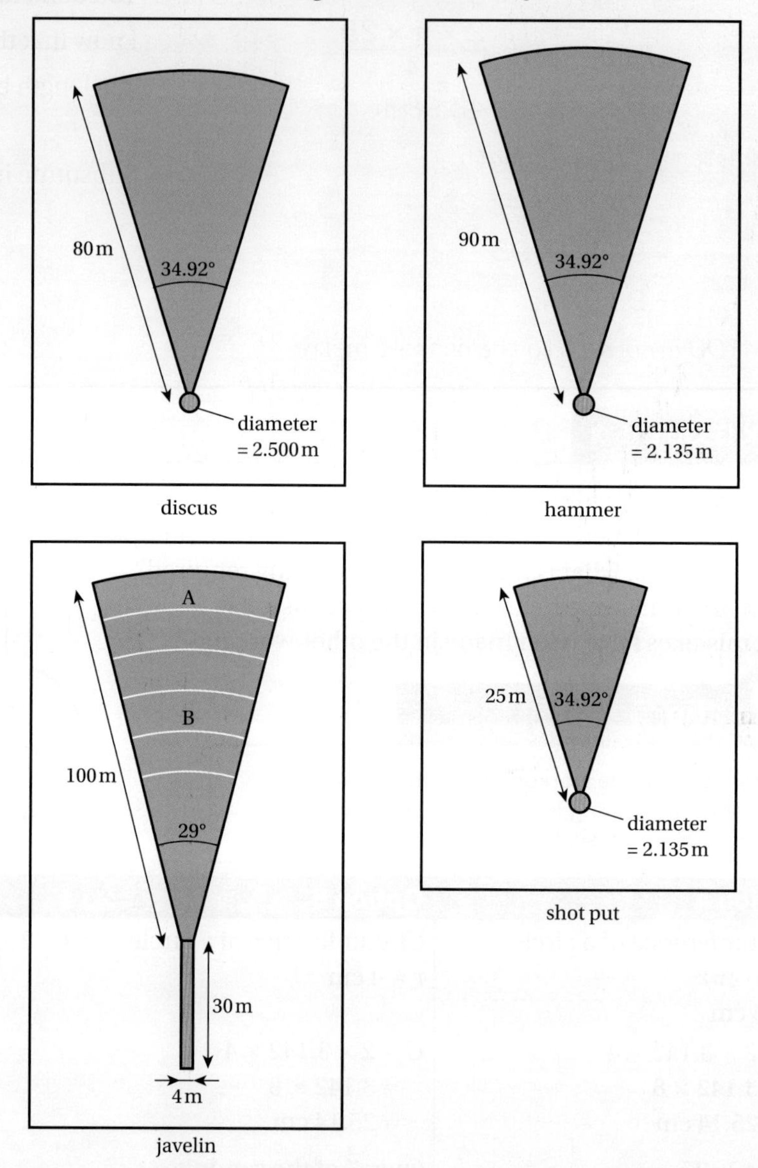

1 Calculate the length of the line painted around the outside of:

a the discus area.

b the hammer throw area.

2 Competitors in the discus, shot put and hammer throw have to remain inside a marked circle while the equipment is in their hands (before they throw it).

a Which sport has the largest marked circle?

b What is the circumference of the standing circle in the discus area?

c In shot put and hammer throw, a raised edge is built around the circumference of the starting circle. If this edge is 10 cm wide, calculate its inner and outer circumference.

3. Calculate the perimeter of the event space for javelin.

4. The curved measurement lines on each event space are 10 m apart.

   Using the javelin field, calculate the length of the lines marked A and B.

5. The position of a winning discus throw is shown on the field by a red dot.

   The angle between its path and the edge of the marked area closest to it is 10°.

   Work out how far it is from each edge of the area.

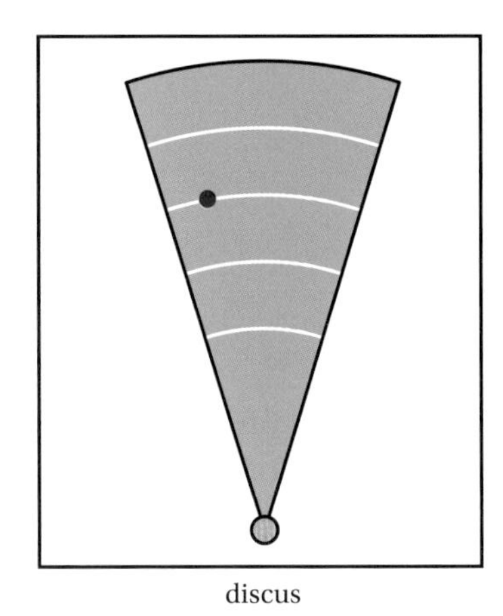

discus

6. The radius of the Earth is approximately 6378.1 km at the Equator.

   Calculate the approximate distance around the Equator. Give your answer correct to 2 decimal places.

7. A large pizza has a circumference of 94 cm.

   What is the side length of the smallest square cardboard box it will fit into?

8. Find the perimeter of this symmetrical logo.

   Use a ruler and protractor to measure and find the dimensions you need.

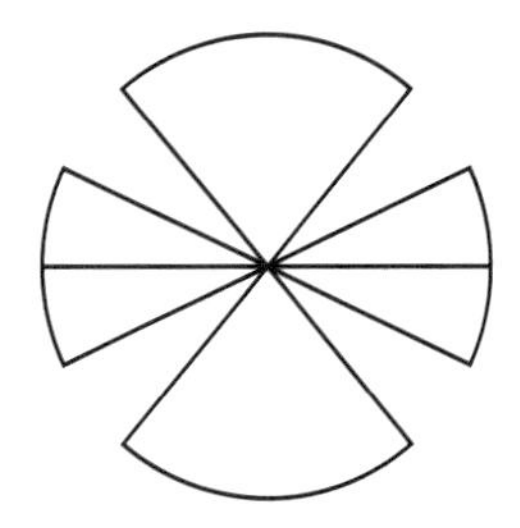

9. Garry is a carpenter who specialises in wooden abstract art.

   He creates a rectangular piece with a sector cut out of the middle.

   The sector is part of a circle of radius 5 cm, whose centre is also the centre of the rectangle.

   The rectangle's length is 6 times the radius of the circle.

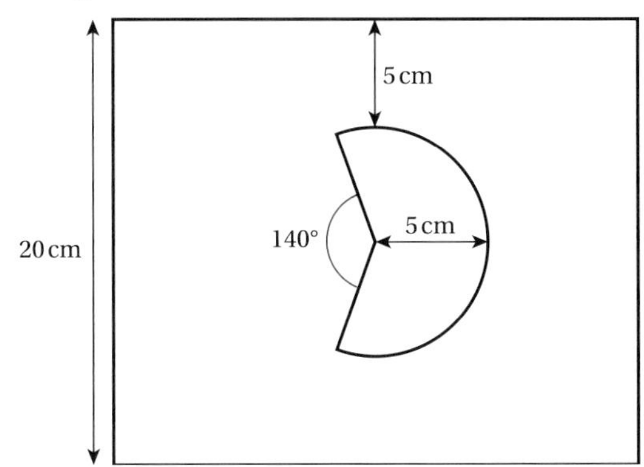

   What is the perimeter of Garry's piece of art?

Find answers at: cambridge.org/ukschools/gcsemaths-studentbookanswers

10 The diagram shows the design of a small stained glass window.

What is the perimeter of the window?

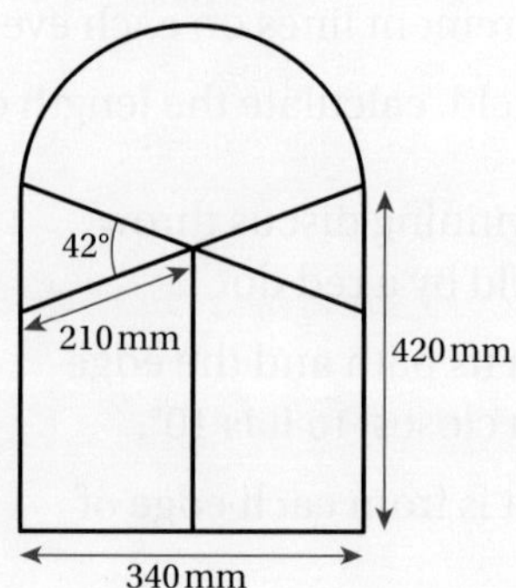

11 The London Eye is a good example of a structure involving angles and sectors of circles.

a The London eye has a diameter of 122 m. What is its circumference?

b There are 32 equally spaced pods on the outside of the wheel. Calculate the curved distance between them.

c The curved boarding platform at the base of the eye is 58 m long. How far does a capsule travel from where it leaves this platform to where it meets it again after a revolution?

d One capsule takes 30 minutes to make a complete revolution. What is its speed in metres per second?

e The London Eye rotates 7668 times per year. How many kilometres does one capsule travel in a year?

f The Eye opened in 1999. By 2013, 50 million people had ridden the Eye.

i Calculate the mean number of visitors per annum based on these figures.

ii In 2014, the mean ticket price was £26.55. If there is a 10% increase in visitors in 2014 from the mean number of visitors, work out the approximate value of ticket sales in 2014.

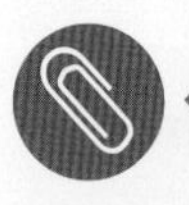

Checklist of learning and understanding

Perimeter

- Perimeter is the total distance around the boundaries of a shape; this includes both inner and outer boundaries.
- You can calculate perimeter by adding the lengths of the sides or by applying a formula based on the properties of the shape.

Circumference

- The perimeter of a circle is called its circumference.
- The formula for calculating the circumference is as follows:

  $C = \pi d$ or $C = 2\pi r$, where d = diameter and r = radius.
- A sector is a part of a circle between two radii and the circumference. You can find the arc length of a sector by working out what fraction of a circle the sector represents, and multiplying this by the circumference.

Chapter review

For additional questions on the topics in this chapter, visit GCSE Mathematics Online.

1 The perimeters of the two shapes below are equal.

What is the side length of the square?

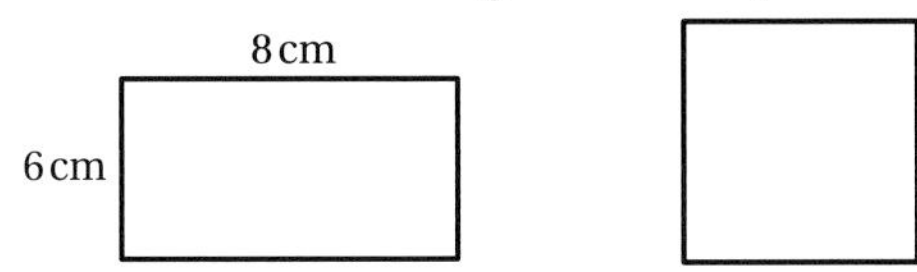

2 The perimeter of a regular pentagon is 90 cm.

Work out the length of each side.

3 A rectangular vegetable plot has a perimeter of 50 metres.

If the width is 6.5 m, what is the length of the plot?

4 An irrigator in a field can water a circular area of radius 14.5 m.

What is the circumference of the area that can be irrigated?
Use $\pi = 3.142$.

5 A pizza has a circumference of 88 cm. It is placed in a cardboard box.

Calculate the perimeter of the smallest box that it will fit into.

6 Calculate the perimeter of the shape shown.

Give your answer correct to 1 decimal place.

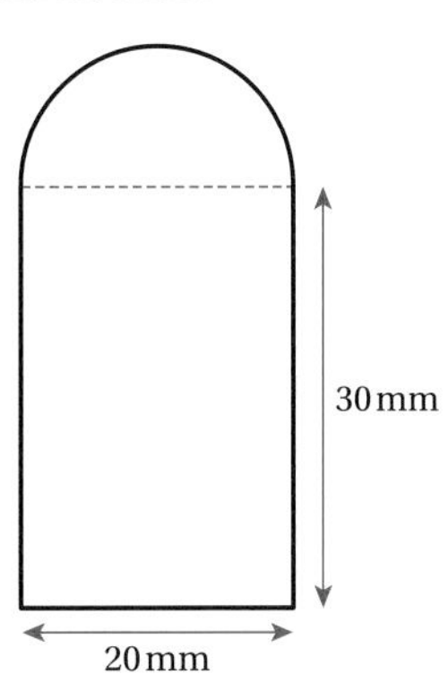

7 The diagram shows some staging for an outdoor concert.

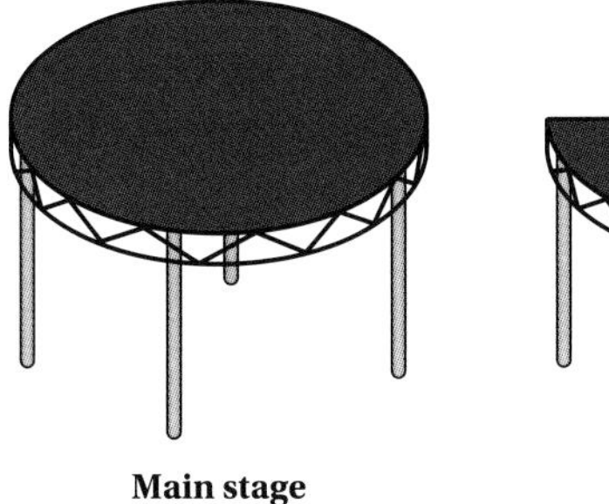

Main stage

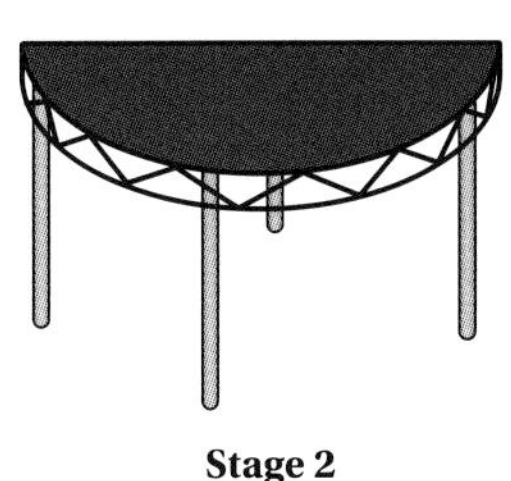

Stage 2

Stage 3

The main stage is a circle of diameter 6 m.

There are two smaller stages: stage 2 is half the size of the main stage, and stage 3 is a quarter of the main stage.

Each stage has a patterned strip along its curved edge.

a Work out the length of the patterned edge on each stage.

b A safety rubber strip is to be applied along all the edges of each stage. Work out the length of strip required for each stage. What is the total amount of rubber strip needed?

Find answers at: cambridge.org/ukschools/gcsemaths-studentbookanswers

8 Natalia is making a pendant out of polymer clay.

The design is a repeating pattern of a square that shares two sides with two radii of a circle.

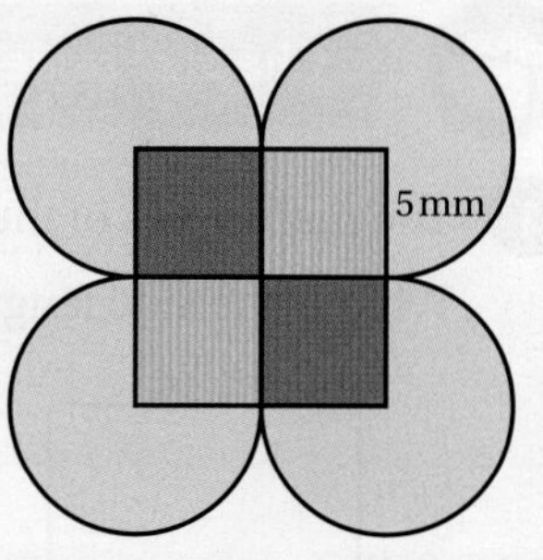

Width of one square = radius of circle

a What is the perimeter of the pattern?

b What is the perimeter of the pendant?

Natalia decides to make a pair of matching earrings but the earrings will **not** contain the central squares.

c What is the perimeter of each earring?

16 Area

In this chapter you will learn how to ...

- use formulae to find the area of different shapes, including circles and parts of circles.
- use appropriate formulae to calculate the area of composite shapes.

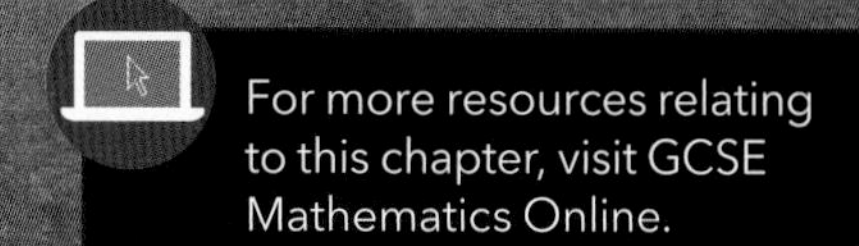
For more resources relating to this chapter, visit GCSE Mathematics Online.

Using mathematics: real-life applications

Ordering the right quantity of turf for a sports field, preparing detailed floor plans, and determining how much fertiliser is needed to treat a field crop all require knowledge and calculation of area.

"Fertiliser application rates are normally given in kilograms per hectare. One hectare is an area of 100 m × 100 m or 10 000 m^2. Applying too much or too little fertiliser to an area can have disastrous results on the crops." *(Farmer)*

Before you start ...

Ch 5	You should know the properties of quadrilaterals.	**1** Use the marked properties to name the quadrilaterals correctly. **a** **b** **c**
Ch 2	You should be familiar with square numbers and square roots.	**2** Calculate. **a** 5^2 **b** 2×10^2 **c** $3^2 + 4^2$ **3** Find the number which is squared to give each of these numbers. **a** 144 **b** 10 000 **c** 0.25
Ch 12	You need to be able to convert between square units of measurement.	**4** Complete these. **a** $5\text{ m}^2 = \square\text{ cm}^2$ **b** $\square\text{ cm}^2 = 87\,000\text{ mm}^2$ **c** $4\text{ km}^2 = \square\text{ m}^2$
Ch 14	You should be able to write simple formulae, substitute values into formulae and change the subject of a formula.	**5** Calculate A if $A = l \times w$, when $l = 2$ cm and $w = 4$ cm. **6** Make r the subject of the formula, $A = 2\pi r$.

Find answers at: cambridge.org/ukschools/gcsemaths-studentbookanswers

Assess your starting point using the Launchpad

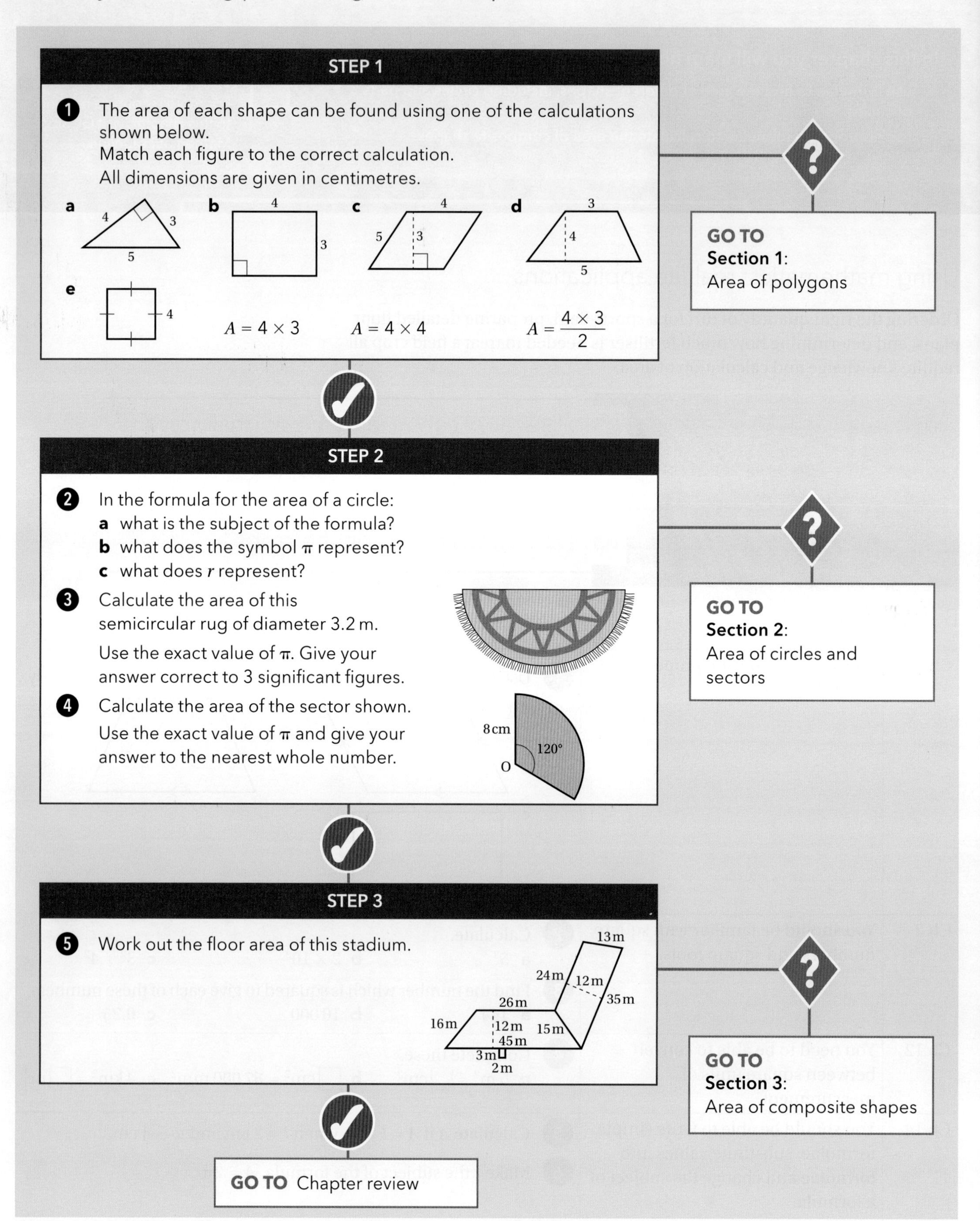

STEP 1
1 The area of each shape can be found using one of the calculations shown below.
Match each figure to the correct calculation.
All dimensions are given in centimetres.
a
4
3
5
b
4
3
c
4
5
3
d
3
4
5
e
4
$A = 4 \times 3$
$A = 4 \times 4$
$A = \frac{4 \times 3}{2}$
?
GO TO
Section 1:
Area of polygons
STEP 2
2 In the formula for the area of a circle:
a what is the subject of the formula?
b what does the symbol π represent?
c what does r represent?
3 Calculate the area of this semicircular rug of diameter 3.2 m.
Use the exact value of π. Give your answer correct to 3 significant figures.
4 Calculate the area of the sector shown.
Use the exact value of π and give your answer to the nearest whole number.
8 cm
120°
O
?
GO TO
Section 2:
Area of circles and sectors
STEP 3
5 Work out the floor area of this stadium.
13 m
24 m
12 m
35 m
26 m
16 m
12 m
15 m
45 m
3 m
2 m
?
GO TO
Section 3:
Area of composite shapes
GO TO Chapter review

Section 1: Area of polygons

The **area** of a plane shape is the amount of space it takes up. You can think of area as the number of squares (square units) that will fit inside the boundary of a shape.

Tip

You were introduced to plane shapes in Chapter 5; they are flat 2D shapes.

Area is always given in square units, for example mm^2 (square millimetres), cm^2 (square centimetres), m^2 and km^2.

Make sure you know the formulae for finding the area of common polygons.

Area of rectangles and squares

The formula for finding the area of a rectangle is

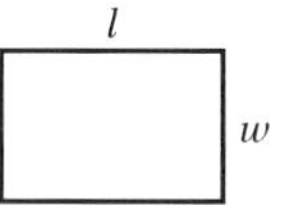

Area = length × width

$A = lw$

In a square, the length and width are equal, so the formula for area is

Area of the square $= l \times l = l^2$

$A = l^2$

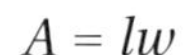

Sometimes the side length is written as s (for side) instead of l. In this case the formula for the area is written as $A = s^2$.

Area of a triangle

The formula for the area of any triangle is

Learn this formula

Area of a triangle (A) $= \frac{1}{2} \times$ base × perpendicular height

$$A = \frac{1}{2} \times b \times h$$

Tip

You were reminded about the meaning of perpendicular in Chapter 5.

You can use any side of a triangle as the base. The **perpendicular height** of the triangle **must** be perpendicular to the base. The perpendicular height can be inside the triangle, one of the sides of the triangle or outside the triangle.

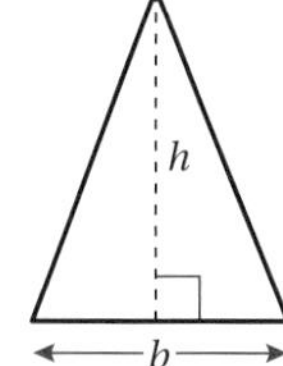

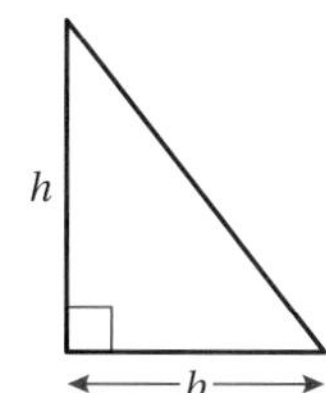

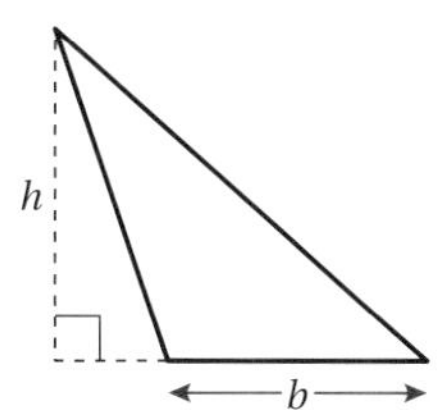

You can prove the formula for the area of any triangle works by showing that the area of a triangle is half the area of a rectangle drawn from the base of the triangle.

Find answers at: cambridge.org/ukschools/gcsemaths-studentbookanswers

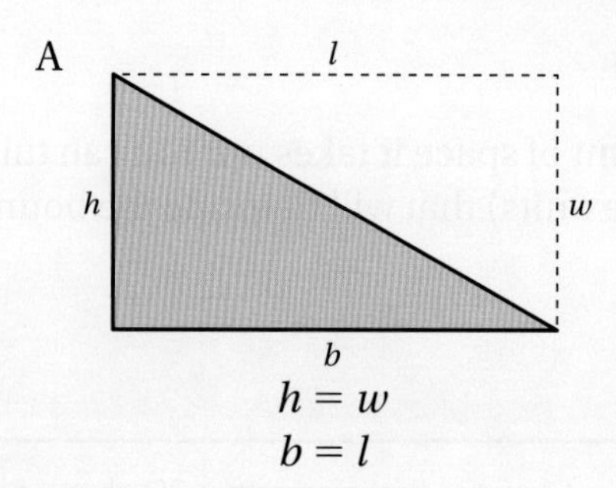

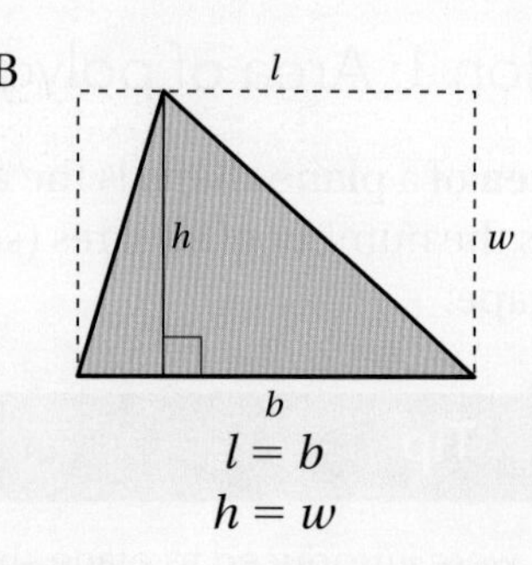

Figure A shows a right-angled triangle. Figure B shows a scalene triangle.

In both diagrams the area of the shaded part is equal to the area of the unshaded part.

Area of the rectangle = $l \times w$ so it follows that the shaded and unshaded areas are equal to $\frac{1}{2} l \times w$. You can see that l is equal to the base of the triangle and w is equal to the perpendicular height of the triangle, so the area of any triangle = $\frac{1}{2}$ × length of its base × its perpendicular height.

WORKED EXAMPLE 1

Calculate the area of each triangle.

a

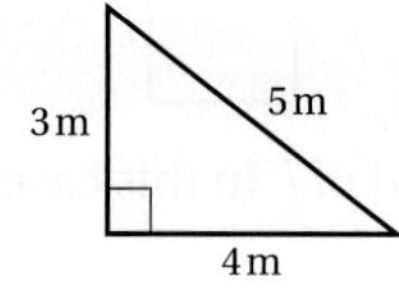

b

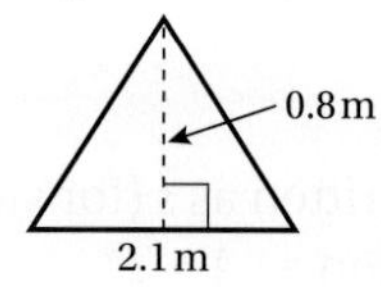

c

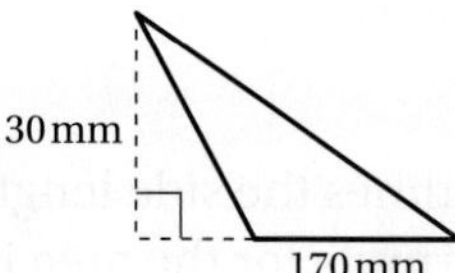

a Area $= \frac{1}{2} \times b \times h$

$= \frac{1}{2} \times 4 \times 3$

$= \frac{1}{2} \times 12$

$= 6\,\text{m}^2$

Substitute the values into the formula for the area of a triangle. In a right-angled triangle, either of the two shorter (perpendicular) sides can be used as the base and height.

Remember to include the units in the answer.

b Area $= \frac{1}{2} \times b \times h$

$= \frac{1}{2} \times 2.1 \times 0.8$

$= \frac{1}{2} \times 1.68$

$= 0.84\,\text{m}^2$

Use the side marked 2.1 as the base because the height is perpendicular to it.

c Area $= \frac{bh}{2}$

$= \frac{(170 \times 30)}{2}$

$= \frac{5100}{2}$

$= 2550\,\text{mm}^2$

This is the same formula expressed differently. Finding a half is the same as dividing by 2. The dashed line shows you the height perpendicular to the base.

Tip

The order of the multiplication is not important, so either side can be the base or the height:

$\frac{1}{2} \times 4 \times 3 = \frac{1}{2} \times 12 = 6$

$\frac{1}{2} \times 4 \times 3 = 2 \times 3 = 6$

Did you know?

Two triangles that look completely different may still have the same area.

These two triangles are equal in area.

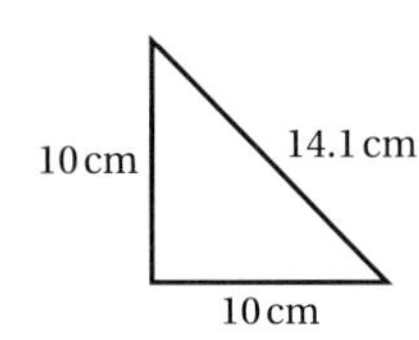

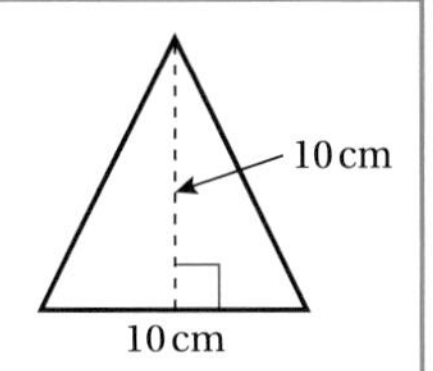

EXERCISE 16A

1 Calculate the area of each triangle.

a

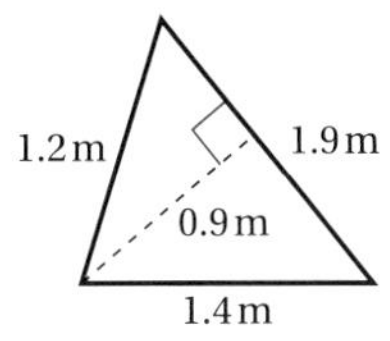

b

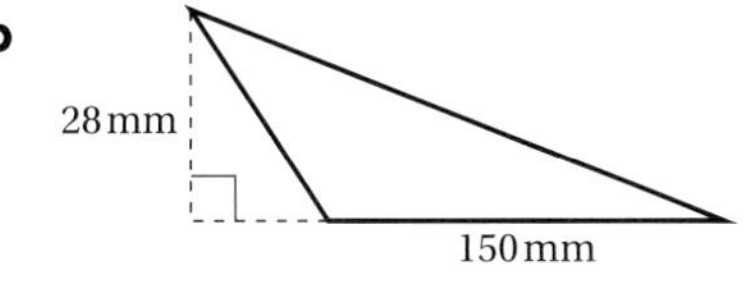

2 The area of a triangle is 36 m^2 and its perpendicular height is 6 m.

Work out the length of its base.

3 Work out the total area of this kite.

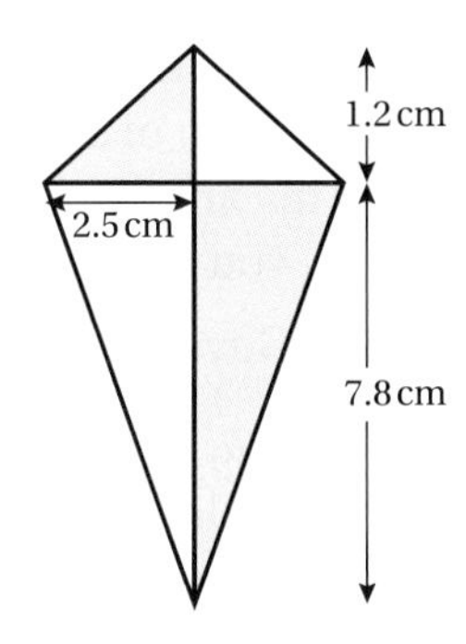

4 A triangle has a base of 3.6 m and a height of 50 cm.
What is its area in metres squared?

5 A triangle of area 0.125 m^2 has a base 25 cm long.
What is its height in centimetres?

6 Calculate the total area of the sails on this small boat.

The mast is 2.7 m tall.

The smaller sail extends $\frac{2}{3}$ of the way up the mast.

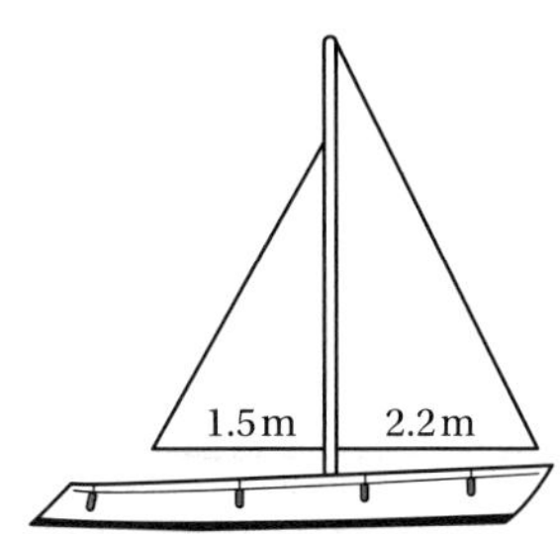

Tip

Use what you know about fractions to help you. See Chapter 10 if you need to.

7 These flags were seen at an international convention centre.

Eritrea

Nepal

Guyana

Trinidad and Tobago

Antigua and Barbuda

Tip

A ratio of 3 : 4 means 3 parts to 4 parts, so the dimensions of the flag as a whole are based on 7 parts. What fraction of the whole does the top triangle represent? Use what you know about fractions to help you; see Chapter 10 if you need to.

a Nepal is the only nation in the world that doesn't have a quadrilateral flag.

The flag consists of two overlapping triangles.

$\frac{1}{4}$ of the area of the top triangle overlaps the bottom triangle.

The dimensions of the top triangle to the bottom one are in the ratio 3 : 4.

If the top triangle is 84 cm high and 63 cm wide at its base, what is the area of the whole flag?

b The flag of Eritrea contains three triangles. If the flag dimensions are 1.2 m by 60 cm, calculate the area of each triangle on the flag.

c If the flag of Trinidad and Tobago is 1.3 m by 87 cm and the width of the black and white flash is 32.5 cm along the edges of the flag, calculate the area of each red triangle.

d A small decorative flag of Antigua and Barbuda is 210 mm × 297 mm.

i What is the area of each red triangle?

ii Estimate the area of the white triangle based on these dimensions. Show how you made your estimate.

Tip

Use what you know about fractions to help you. Look back at Chapter 10 if you need to.

e On the flag of Guyana, the point of the black line around the red triangle extends $\frac{4}{9}$ of the length of the flag. Assuming a flag is 270 mm by 450 mm, calculate:

i the area covered by the red and black triangle.

ii the area of each green triangle.

iii the area of the 'golden arrow' made by the yellow and white colour on the flag.

f Investigate the dimensions of the Union Jack. What percentage of the flag area is each colour?

Area of a parallelogram

The formula for the area of a parallelogram is:

Learn this formula

Area of a parallelogram (A) = base × perpendicular height

$$A = b \times h$$

The base of this parallelogram is b and the **perpendicular height** is h.

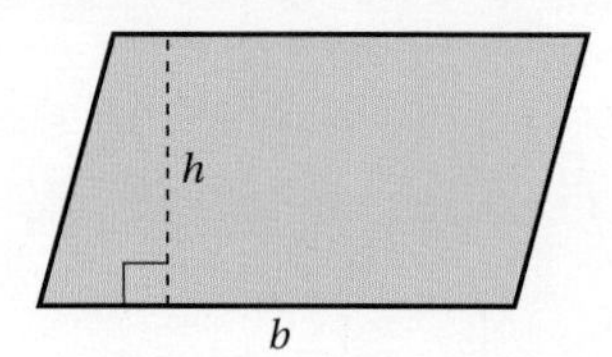

Tip

You must use the perpendicular height and **not** the slant height.

You can see how this formula works by removing a right-angled triangle from one end of the parallelogram, and joining it to the other end to form a rectangle. The parallelogram and rectangle have the same area.

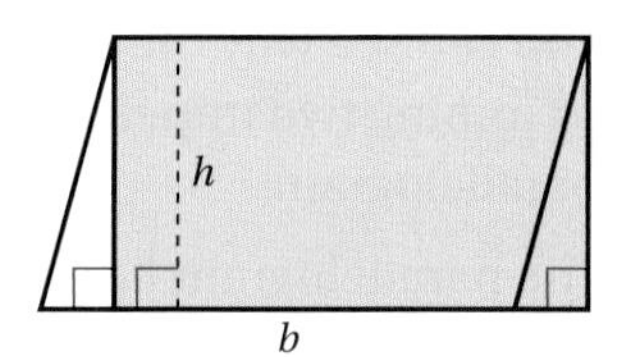

WORKED EXAMPLE 2

Calculate the area of the parallelogram.

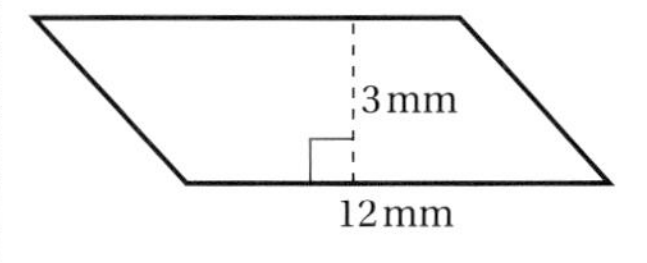

Area $= b \times h$	Substitute the known values into the formula.
$= 12 \times 3 = 36\text{ mm}^2$	Remember to use the correct units!

You can use the formula for the area of a triangle to derive the formula for the area of a parallelogram.

A parallelogram can be divided into two triangles.

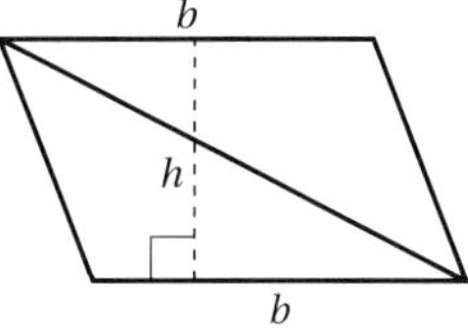

So, area of parallelogram $= \left(\frac{1}{2} \times b \times h\right) + \left(\frac{1}{2} \times b \times h\right) = b \times h$

It might not be immediately obvious what the important values are in a diagram; think carefully about what information you need.

WORK IT OUT 16.1

A parallelogram is made by combining a rectangle and two right-angled triangles.

Which of these options will give the correct area?

Which dimensions are incorrect in the other two? Why?

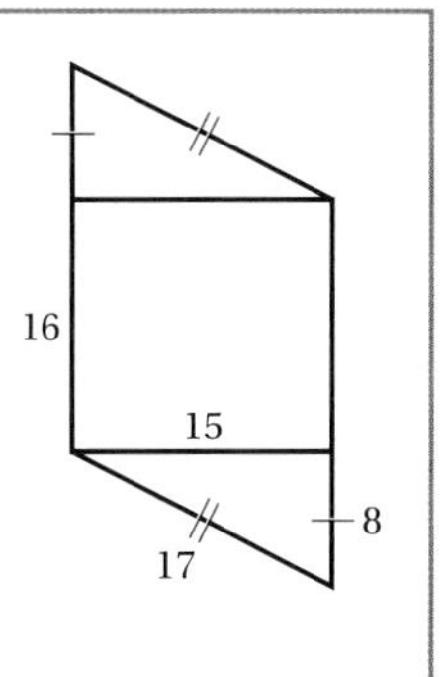

Option A	Option B	Option C
$A = bh$	$A = bh$	$A = bh$
$= 16 \times 17$	$= 24 \times 17$	$= 24 \times 15$

Area of a trapezium

The formula for the area of a trapezium is,

Area of a trapezium $(A) = \frac{1}{2} \times (a + b) \times h$, where a and b are the lengths of the parallel sides.

A trapezium has parallel sides a and b, and **perpendicular height** h.

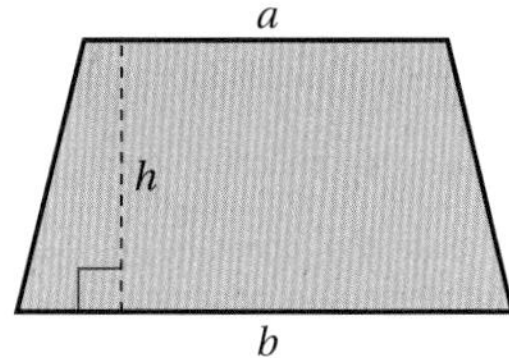

Find answers at: cambridge.org/ukschools/gcsemaths-studentbookanswers

You can see why the formula works by joining two trapezia to form a parallelogram.

The parallelogram has a base of length $a + b$, and perpendicular height h.

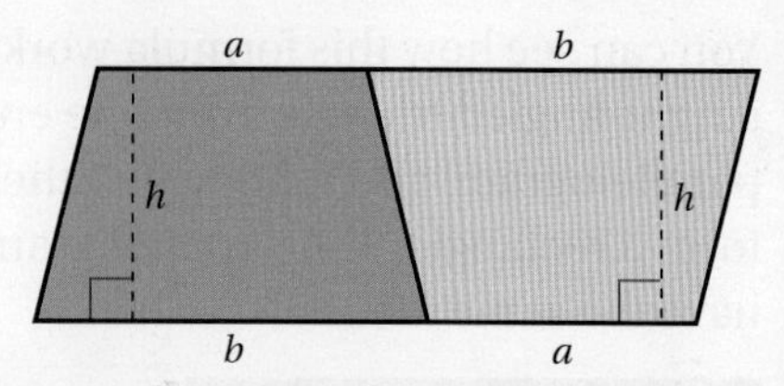

Area of a parallelogram (A) = base × perpendicular height

$$A = (a + b) \times h$$

Each trapezium is half the area of the parallelogram.

WORKED EXAMPLE 3

Calculate the area of the trapezium.

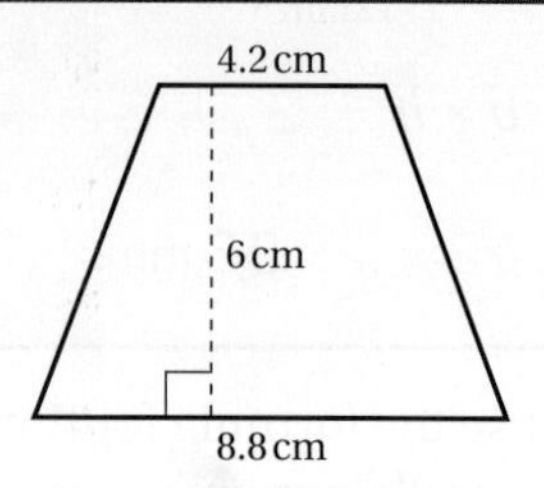

$\text{Area} = \frac{1}{2} \times (a + b) \times h$ — Write down the formula for area of a trapezium.

$= \frac{1}{2} \times (4.2 + 8.8) \times 6$ — Substitute in the known values, make sure you have the perpendicular height.

$= \frac{1}{2} \times 13 \times 6$

$= 39\text{ cm}^2$ — Don't forget to use the correct units.

When calculating area, make sure you know which shape you are dealing with and be careful to use the correct formula.

WORK IT OUT 16.2

A solar farm is being built in a field. Each solar panel measures 98 cm by 150 cm. If 500 panels can fit onto the field, what is the area of panels being used, in m^2?

Which of the options below is correct?

What errors were made in each of the others?

Option A	Option B	Option C
Area of one panel:	Area of one panel:	Area of one panel:
$98 \times 150 = 14\,700\text{ cm}^2$	$\frac{1}{2} \times (98 \times 150) = 7350\text{ cm}^2$	$98 \times 150 = 14\,700\text{ cm}^2$
$1\text{ m}^2 = 10\,000\text{ cm}^2$	$1\text{ m}^2 = 10\,000\text{ cm}^2$	$1\text{ m}^2 = 100\text{ cm}^2$
$1\text{ panel} = 14\,700 \div 10\,000 = 1.47\text{ m}^2$	$1\text{ panel} = 7350 \div 10\,000 = 0.735\text{ m}^2$	$1\text{ panel} = 14\,700 \div 100 = 147\text{ m}^2$
$500\text{ panels} = 1.47\text{ m}^2 \times 500 = 735\text{ m}^2$	$500\text{ panels} = 0.735\text{ m}^2 \times 500 = 367.5\text{ m}^2$	$500\text{ panels} = 147\text{ m}^2 \times 500 = 73\,500\text{ m}^2$

EXERCISE 16B

1 Use the formula $A = bh$ to calculate the area of each parallelogram.

a

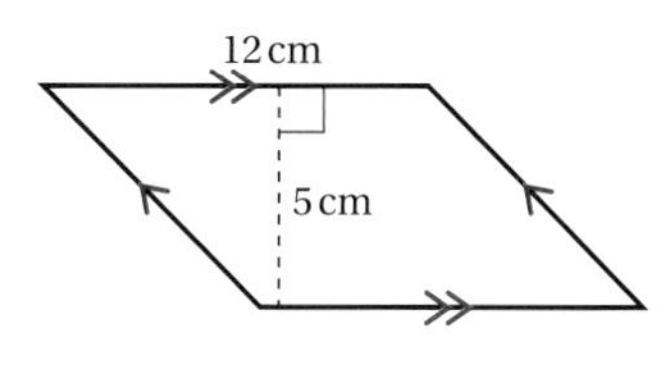

b

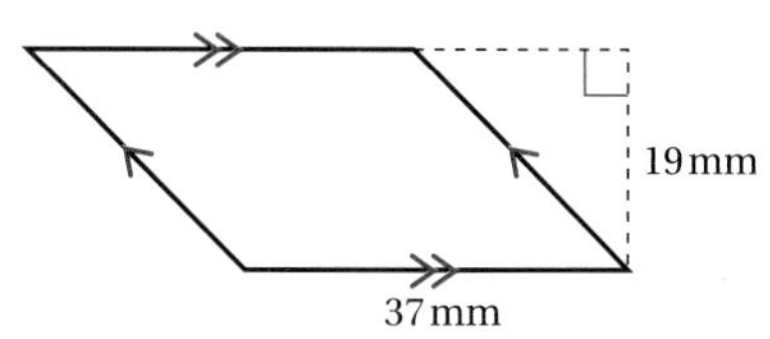

c

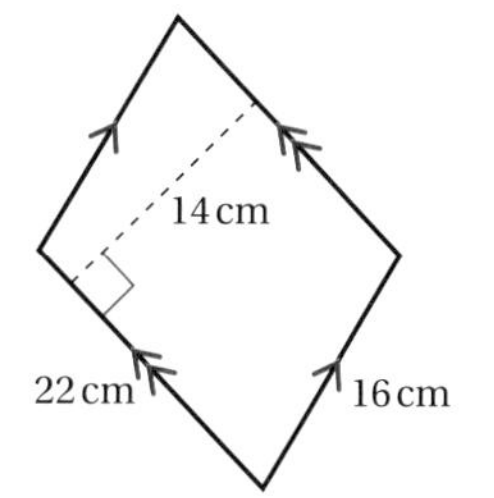

d

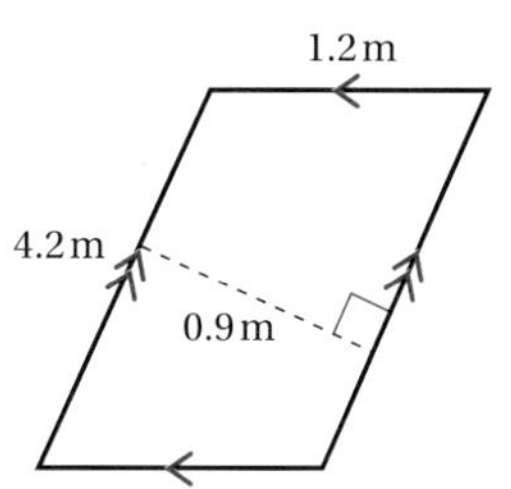

2 Use the formula $A = \frac{1}{2}(a + b)h$ to calculate the area of each trapezium.

a

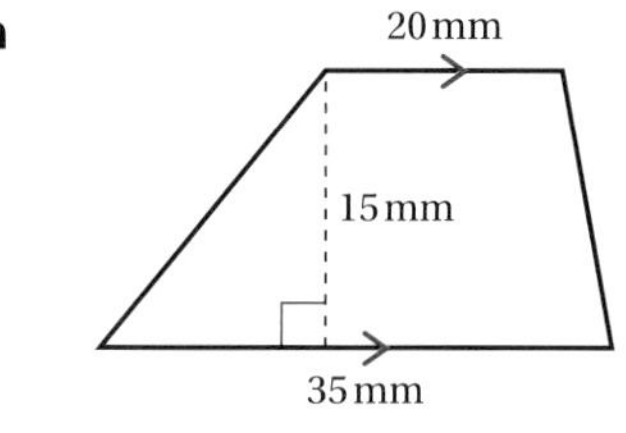

b

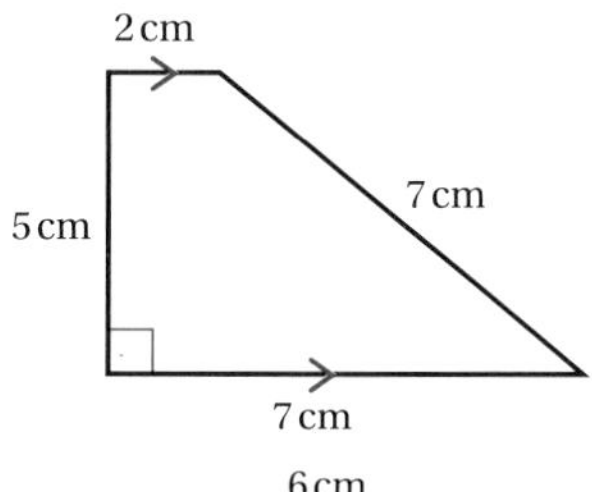

c

16 cm
3 cm
5 cm
19 cm

d

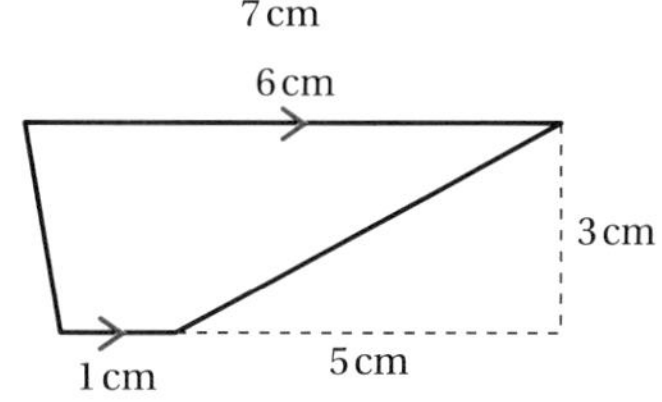

3 The area of this shape is 96 cm^2.
What is its length?

8 cm
l

4 The area of a parallelogram is 40 m^2.
If the perpendicular height is 10 cm, what is the length of the base?

5 The area and one other measurement is given for each shape.
Find the unknown length in each figure.

a

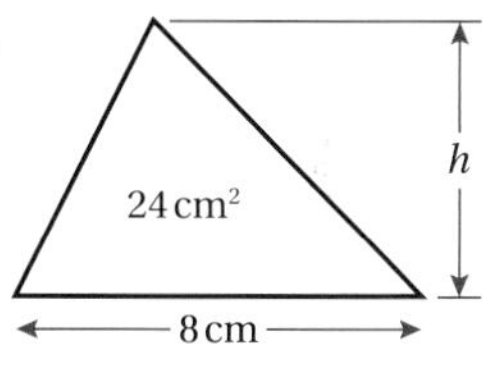

b

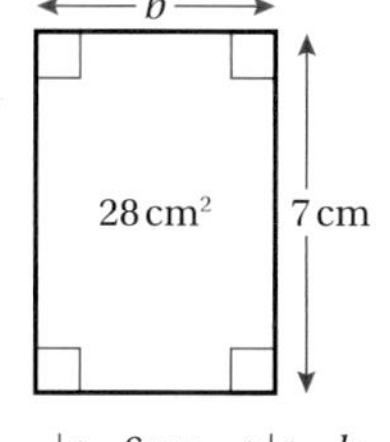

c

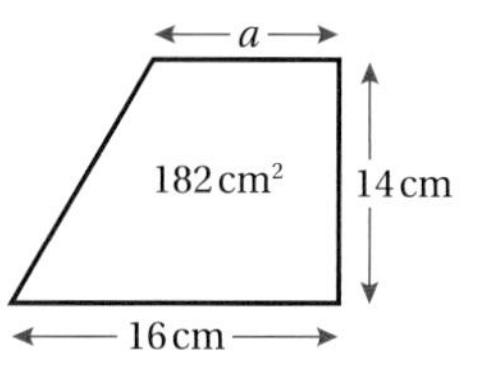

d

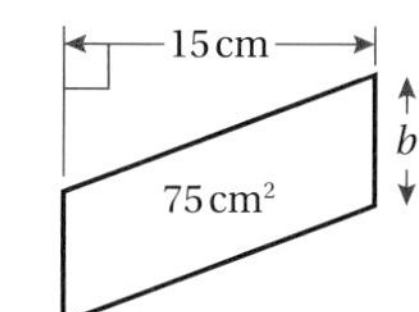

e

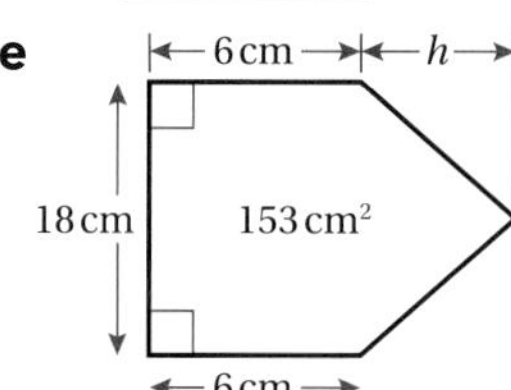

6 Amira works in a community development programme that helps people to grow organic vegetables.
The area of land available in one community is shown on the plan.

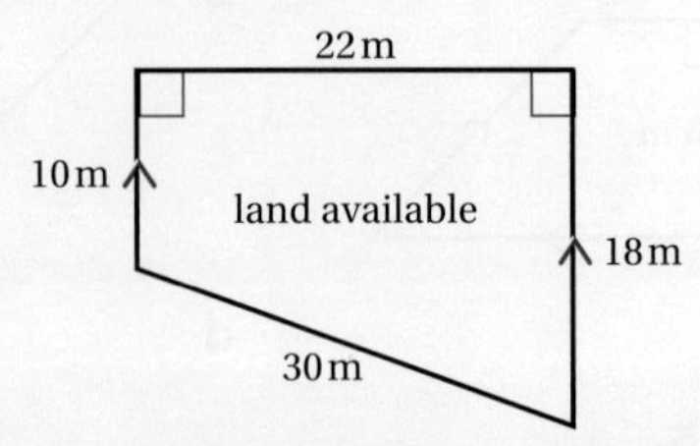

a Calculate the area of the available land.

b To prepare the soil, the community has to lay down 25 kg of soil and 10 kg of compost per square metre of land. Work out how much soil and compost they will need.

c There is a gate 2 m wide along the 22 m boundary. The rest of the land needs to be fenced. Work out the total amount of fencing needed.

Tip

Remember that the perimeter of a shape is its boundary. You learnt about perimeter in Chapter 15.

d Once the first crop is planted, an organic fertiliser is mixed with water and applied to the area. The instructions for mixing the powdered fertiliser with water are:

'Mix 43 g/litre and the application rate is 125 litres per hectare.'

i Work out how many litres are needed for this area. Remember one hectare is 10 000 m^2.

ii The powder comes in $\frac{1}{2}$ kg tubs.

How many applications can the community get from one tub?

7 Write an expression in simplest terms for the area of each shape.

a

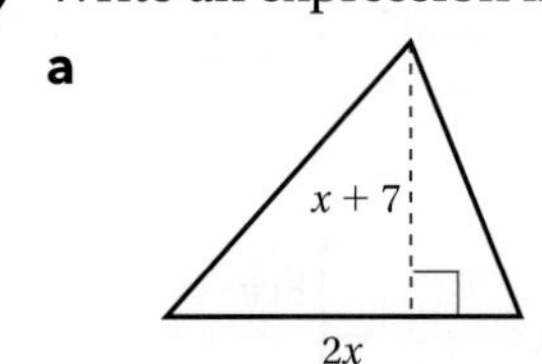

b $4x + 6$, $2x - 1$

c

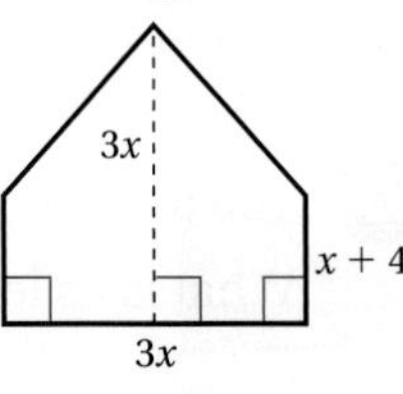

d $\frac{7x}{y}$, $\frac{4y}{x}$

e

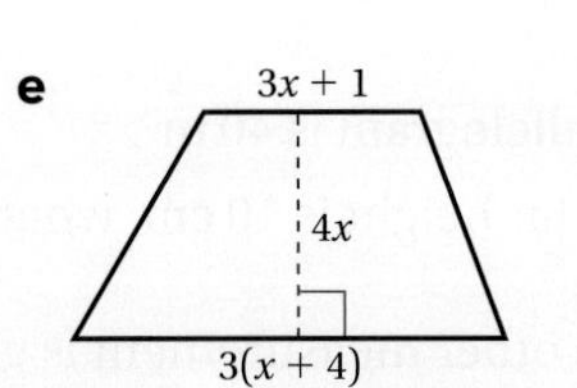

f

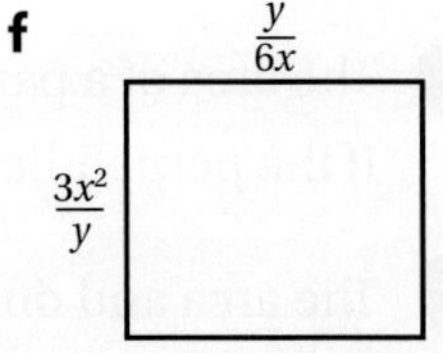

Section 2: Area of circles and sectors

In Chapter 15 you worked with the diameter, radius and circumference of circles and sectors of circles. You also used approximate and exact values of pi in circle calculations. You will meet these terms again in this section.

Tip

Use the π key of your calculator to find the area and circumference of circles, unless you are given an approximate value to use. Leave the value you get on the display for the next step, and only round off to the required number of places when you have a final value.

The area of a circle is calculated using the formula

Area of a circle $(A) = \pi r^2$

where r = radius

If you are given the diameter, d, you can still use this formula by remembering that:

$r = \frac{1}{2}d$

WORKED EXAMPLE 4

Calculate the area of this circle correct to 3 significant figures.

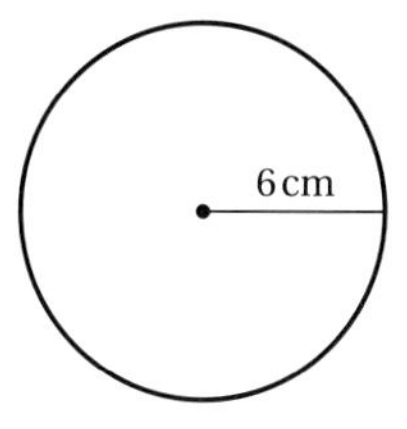

$A = \pi r^2$	Write down the formula for the area of a circle. You will need to know this formula from memory.
$= \pi \times 6 \times 6 = 113.0973355\ \text{cm}^2$ $= 113\ \text{cm}^2$ (to 3 sf)	Substitute in the known values. Remember to use the π key on your calculator, and to store the values so you only have to round the final answer.

If you know the area of a circle, you can find the length of a radius (or the diameter) by rearranging the formula for the area.

WORKED EXAMPLE 5

Calculate the radius of a circle with area 50 cm², correct to 2 decimal places.

$A = \pi r^2$	Write down the formula for the area of a circle.
$50 = \pi \times r^2$	Substitute in the known values.
$r^2 = \frac{50}{\pi}$ $r = \sqrt{\frac{50}{\pi}} = 3.989422804$	Rearrange the formula to make r the subject using inverse operations and applying the same operation to both sides. (**Alternatively**, you could first rearrange the formula to make r the subject and *then* substitute in the known values.)
$r = 3.99$ cm (to 2 dp)	Use the π key on your calculator, and round your final answer to 2 decimal places.

Find answers at: cambridge.org/ukschools/gcsemaths-studentbookanswers

Area of a sector

Remember that a sector is a slice of a circle.

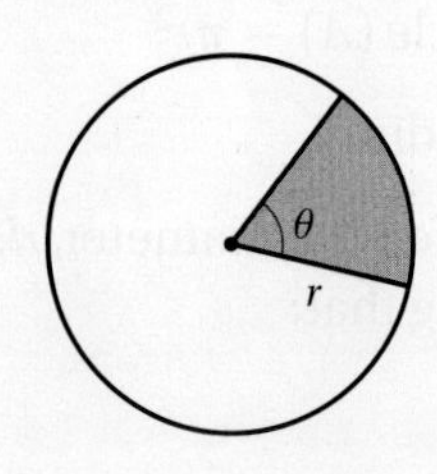

The area of a sector can be calculated using the formula,

Area of sector (A) $= \frac{\theta}{360} \times$ area of circle

$$= \frac{\theta}{360} \times \pi r^2$$

This formula is based on similar principles to those that you used to find the arc length in Chapter 15.

If a sector can be represented as $\frac{\theta}{360}$ of a whole circle, and the area of a whole circle $= \pi r^2$ then it follows that the area of sector $= \frac{\theta}{360} \times$ area of circle.

Remember the special cases:

- A semicircle is half a circle, so its area is half the area of a circle $\left(\frac{\pi r^2}{2}\right)$.
- A quarter-circle is one quarter of a circle, so its area is one quarter of the area of a circle $\left(\frac{\pi r^2}{4}\right)$.

WORKED EXAMPLE 6

Calculate the area of the sector shown.

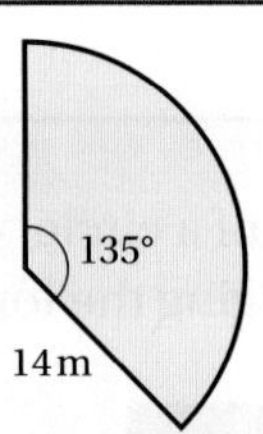

$A = \frac{\theta}{360} \times \pi r^2$ — Write down the area of a sector.

$= \frac{135}{360} \times \pi \times r^2$

$= 0.375 \times \pi \times 14^2$ — Substitute in the known values and calculate. Use the π key on your calculator, and only round your answer.

$= 230.90706$

$= 230.91\ \text{m}^2$ (to 2 dp) — Make sure you use the correct units and state the degree of accuracy to which you have rounded.

EXERCISE 16C

1 Find the area of each circle.

Use the exact value of pi and give your answers correct to 1 decimal place.

a

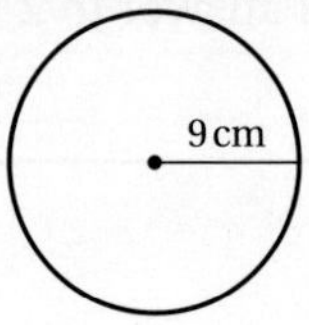

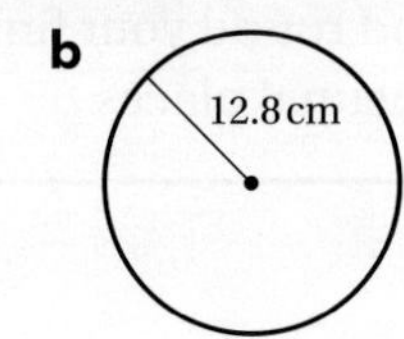

b

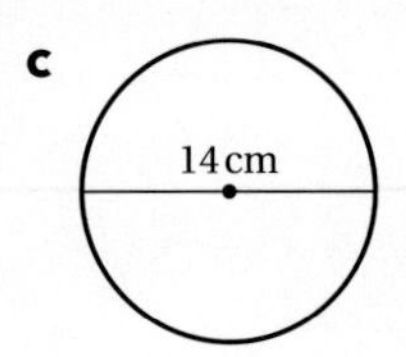

c

d

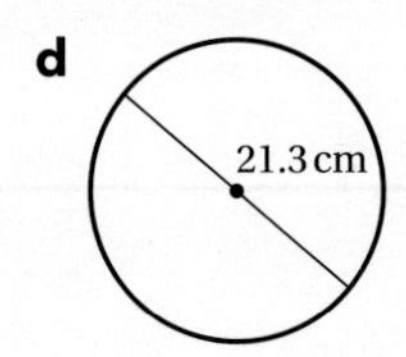

2 Use the formula $A = \frac{\theta}{360} \times \pi r^2$ to find the area of each sector.

a

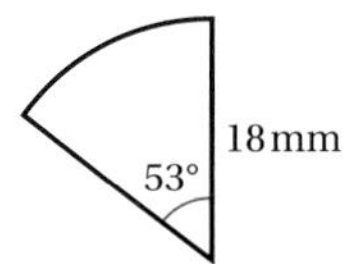

b

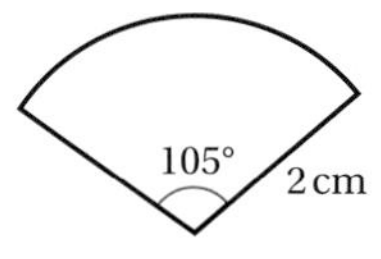

c

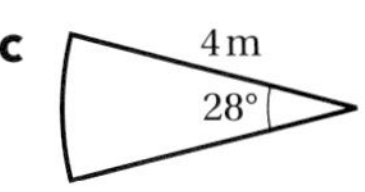

d

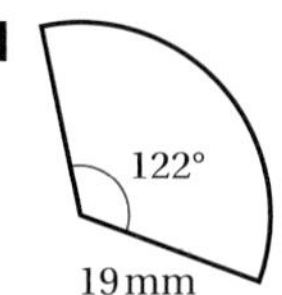

3 What is the difference in area between these two sectors?

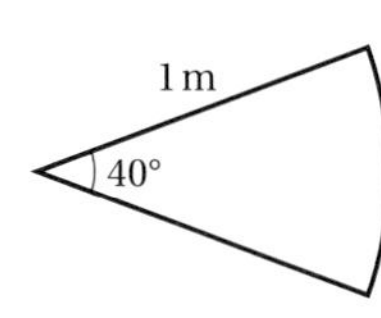

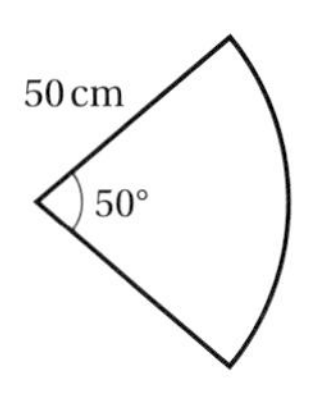

4 A pizza has a diameter of 14 cm.

a Calculate the area of the pizza.

b Estimate the area of a round plate that the pizza can fit onto with about 1 cm space around the edge.

5 A pair of sunglasses has circular lenses, each 8.4 cm in diameter.

a What is the total area of the tinted surface of the lenses?

b What is the smallest round frame size needed for each lens?

6 The area for discus at an international event has these dimensions.

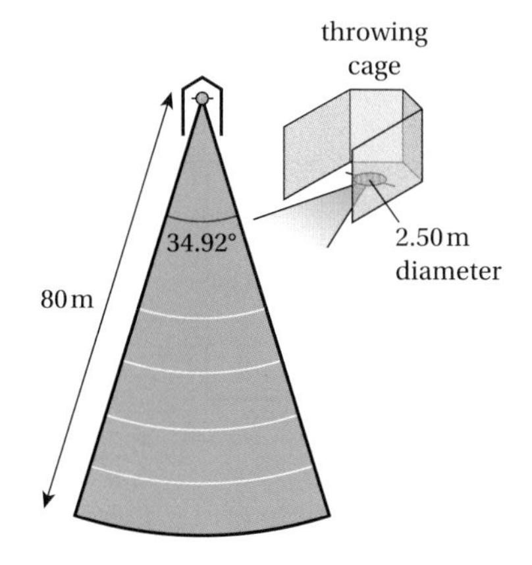

a Calculate the area of the grass in the landing zone.

b What is the area of the starting circle in the throwing cage?

7 A circular disc has a circumference of 75.398 mm.

Show clearly how you could use this information to find the area of one side of the disc.

Find answers at: cambridge.org/ukschools/gcsemaths-studentbookanswers

8 A jeweller is making a tapered tube to set a diamond.

She works out the dimension of the curved piece of metal she needs based on the diameter and depth of the stone and draws a rough sketch like this.

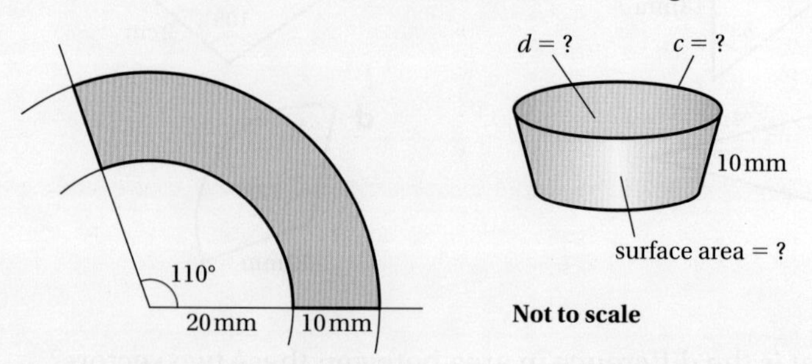

Apply what you know about the area and circumference of circles to work out:

a the area of the flat curved section shown on the diagram.

b the diameter and circumference of the top of the finished tube.

c the diameter and circumference of the bottom of the finished tube.

Section 3: Area of composite shapes

Remember that composite shapes are formed by combining shapes or by removing parts of a shape.

Tip

You saw how to calculate the perimeter of composite shapes in Chapter 15.

You can find the area of composite shapes in different ways.

Addition of parts

- Divide the figure into smaller known shapes whose area can be found directly.
- Calculate the area of each part separately.
- Add together the areas of all the parts to find the total area.

For example:

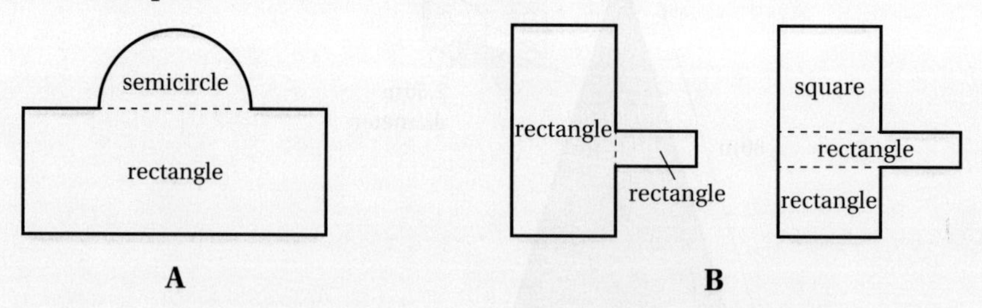

Figure A can be divided into a rectangle and a semicircle.

Figure B can be divided in different ways. The first way requires fewer calculations.

Subtraction of parts

When one figure is 'cut out' of another, you have to find the area of the larger figure and subtract the area of the 'cut out' figure.

For example:

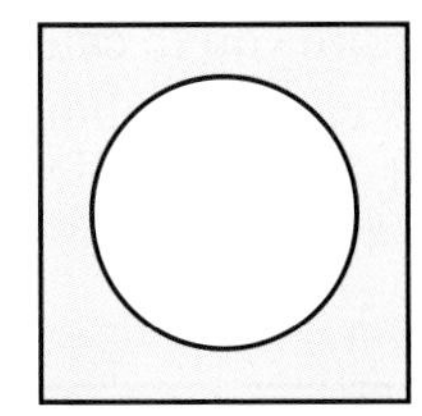

the shaded area = area of square − area of circle

Sometimes you need to visualise what the larger shape would look like *before* the smaller shape is removed from it.

For example:

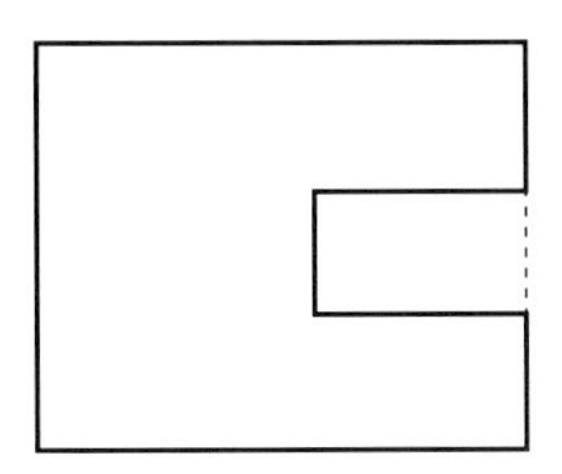

Tip

Some problems can be solved using either the 'addition of parts' or the 'subtraction of parts' method. Think carefully about which method will be more efficient before you decide which one to use.

Drawing in a broken line helps to visualise the shape as a large rectangle with a smaller rectangle cut out of it. Then,
the area of the shape = Area of larger rectangle − area of smaller rectangle.

For example, you could have also found the area of the shape above by splitting the shape into three rectangles and adding each area, but this would require more calculations.

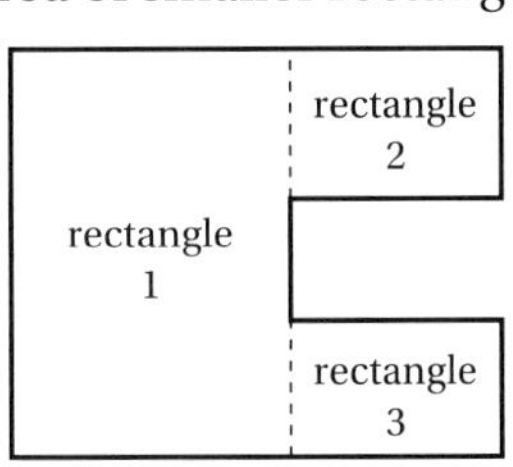

Tip

Copy diagrams into your exercise book so you can draw on them and mark dimensions. This helps you keep track of your working.

WORKED EXAMPLE 7

Calculate the area of this shape.

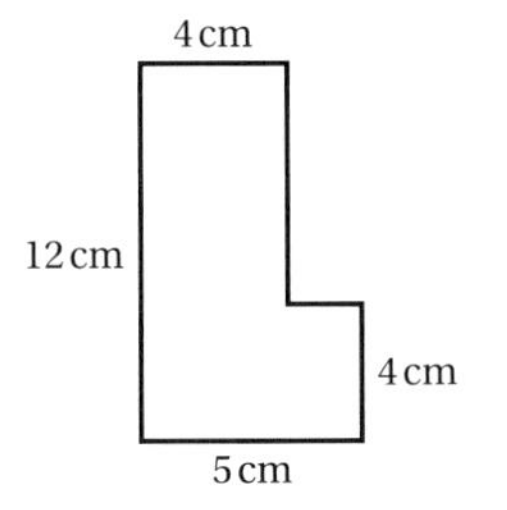

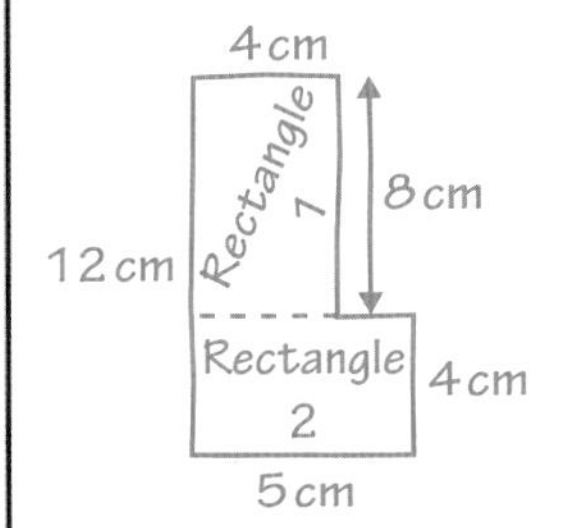

Use the 'addition of parts' method and split the shape into Rectangle 1 (4 cm × 8 cm) and Rectangle 2 (4 cm × 5 cm).

Area of Rectangle 1 = $4 \times 8 = 32$ cm^2

Area of Rectangle 2 = $4 \times 5 = 20$ cm^2

Calculate the area of each rectangle.

Area of shape = $32 + 20 = 52$ cm^2

Add the areas together to find the total area of the original shape.

Tip

This shape can also be split into:
Rectangle 1: 12 cm × 4 cm
Rectangle 2: 4 cm × 1 cm

Find answers at: cambridge.org/ukschools/gcsemaths-studentbookanswers

Sometimes questions on area might also require you to calculate the perimeter or circumference of a shape. Think carefully about what is being asked of you and what information you need.

WORKED EXAMPLE 8

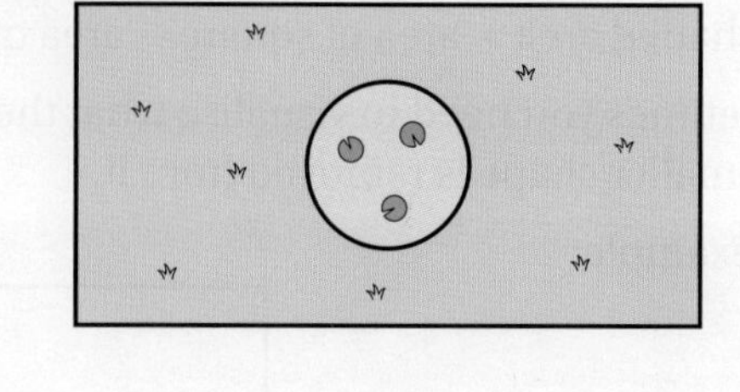

Gwen has a rectangular garden of width 15 m and length 20 m.
She plans for her garden to be all grass except for a circular pond in the middle.

The pond has a radius of 5 m to its inside edge.

a The pond needs a flat layer of netting across its surface.
The netting will be fixed to a circular wire that is pinned to the inside rim of the pond.
How much netting and wire will Gwen need?
Take the value of π to be 3.142.

b Gwen has 15 kg of grass seed.
What area must she sprinkle the grass seed over?

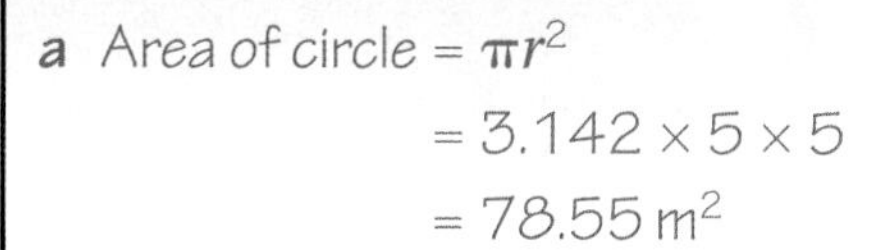

a Area of circle $= \pi r^2$
$= 3.142 \times 5 \times 5$
$= 78.55\ m^2$

Circumference of circle $= 2\pi r$
$= 2 \times 3.142 \times 5$
$= 31.42\ m$

Gwen needs $78.55\ m^2$ of netting and 31.42 m of wire.

> You are told the netting sits flat on the pond's surface and is fixed to the inside rim, so you can assume that the netting is a circle with radius of 5 m. You are also told that the wire is circular and is pinned to the inside rim of the pond; ignoring the width of the wire which you are not given, you can assume that the length of the wire is the circumference of a circle with radius 5 m. Use the appropriate formulae for the area and circumference of a circle, and substitute in the value given for the radius.

b Area to be seeded =
area of garden − area of pond's surface

> You are told that all of Gwen's garden, except for the pond, will be grass. So, the area you need to calculate is the area of the rectangular garden minus the area of the pond's surface.

Area of rectangle = length × width
$= 15 \times 20 = 300\ m^2$

> Use the formula for the area of a rectangle and substitute in the known values. Then substitute this value and the answers to part **a** into your formula for calculating the area to be seeded.

Area of pond $= 78.55\ m^2$

Area to be seeded $= 300 - 78.55$
$= 221.45\ m^2$

Sometimes questions that require area calculations might not ask directly for an area, but instead require you to apply the knowledge you have to find a different value. You need to recognise what information is being asked for and what knowledge you need to apply.

Problem-solving framework

Jose wants to paint the side of his house.

He chooses paint that costs £25.99 for a 2.5 litre tin.

If each tin of paint can cover an area of 12 m^2, how many tins of paint will Jose need?

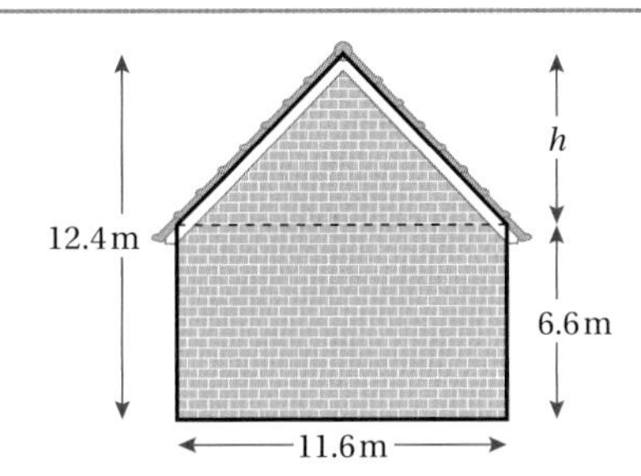

Steps for approaching a problem-solving question	What you would do for this example
Step 1: What have you got to do?	You need to find out how many tins of paint are needed.
Step 2: What information do you need?	Number of tins of paint = area to be painted ÷ area each tin of paint can cover. So you need to calculate the area of the side of the house. Use the 'addition of parts' method and split the side of the house into a rectangle and triangle. Area of rectangle $= l \times w$ $l = 12\text{ m}$, $w = 7.5\text{ m}$ Area of triangle $= \frac{1}{2} \times b \times h$ $b = 12\text{ m}$, $h =$ perpendicular height Each tin of paint covers an area of 12 m^2.
Step 3: What information don't you need?	The price and volume of each tin of paint is not relevant.
Step 4: What maths can you do?	Area of rectangle $= l \times w$ $= 12 \times 7.5 = 90\text{ m}^2$ Area of triangle $= \frac{1}{2} \times b \times h$ Perpendicular height, h, of triangle $= 15 - 7.5 = 7.5\text{ m}$ Area of triangle $= \frac{1}{2} \times 12 \times 7.5$ $= 45\text{ m}^2$ Area to be painted $= 90 + 45$ $= 135\text{ m}^2$ Number of tins of paint $= 135 \div 12 = 11.25$ Jose needs 12 tins of paint.
Step 5: Have you used all the relevant information? Have you answered the question?	All relevant information used ✓ The question asked for the number of tins of paint needed ✓
Step 6: Does your answer seem reasonable?	Check answer: $12 \times 12 = 144\text{ m}^2$ is enough paint to cover the side of the house; but $11 \times 12 = 132\text{ m}^2$ would not be enough paint. The question asked for the number of tins, not the volume of paint used. As you can't have 11.25 tins, Jose would need to round up to 12 tins in order to have enough paint. The answer of 12 tins is reasonable.

EXERCISE 16D

1 Find the total area of each of these shapes. Show all your working.

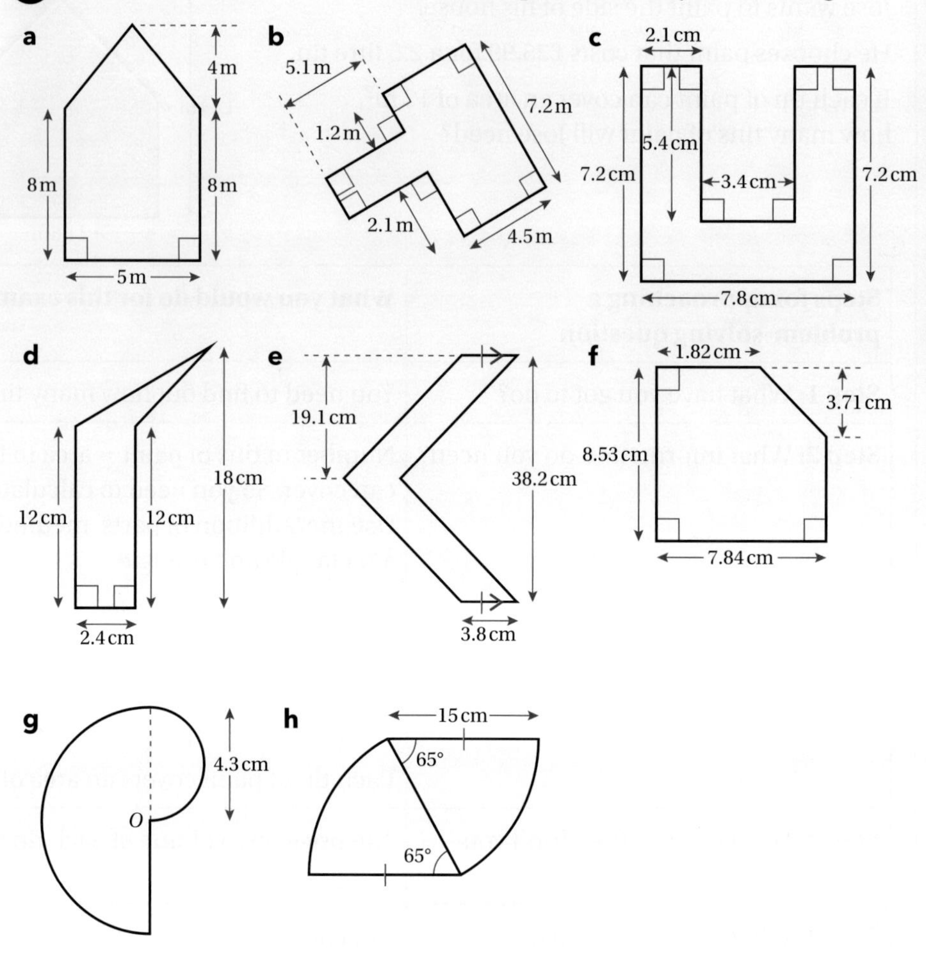

2 Find the area of the shaded part of each figure.

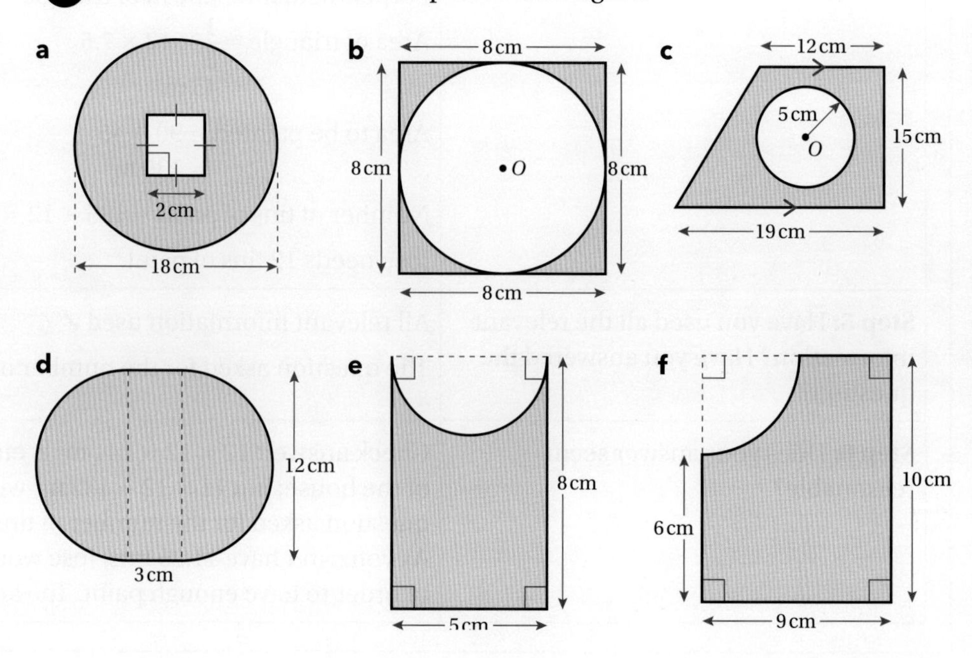

g **h**

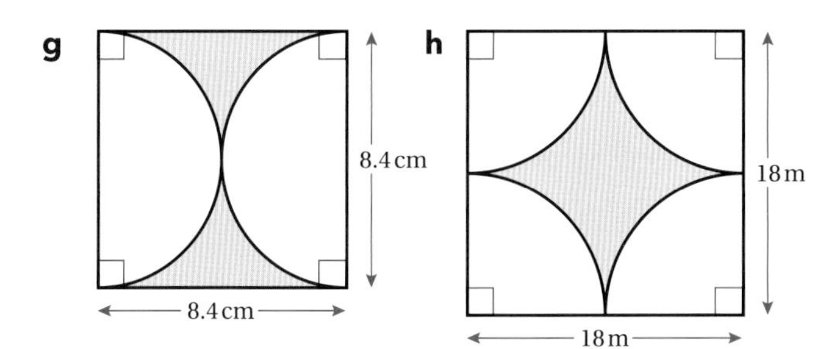

3 Calculate the perimeter and area of the shaded region in each figure.

a

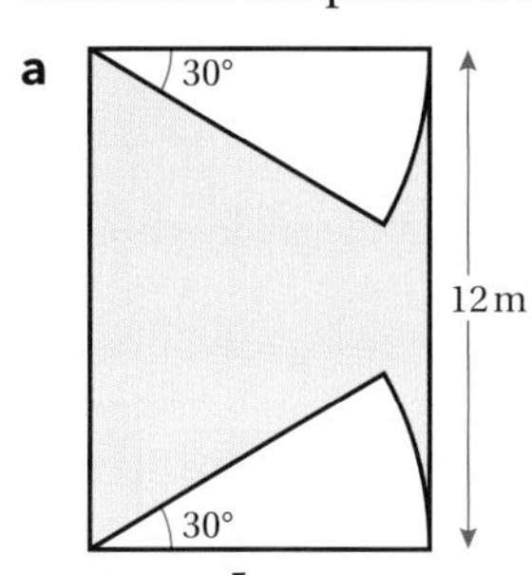

b

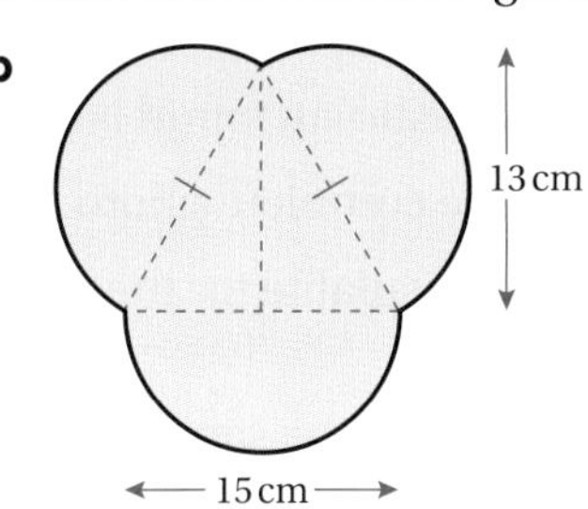

c

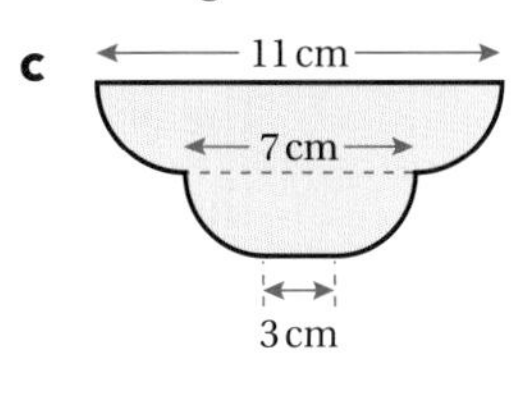

4 How many rectangular tiles 20 cm by 30 cm would you need to tile the area shown?

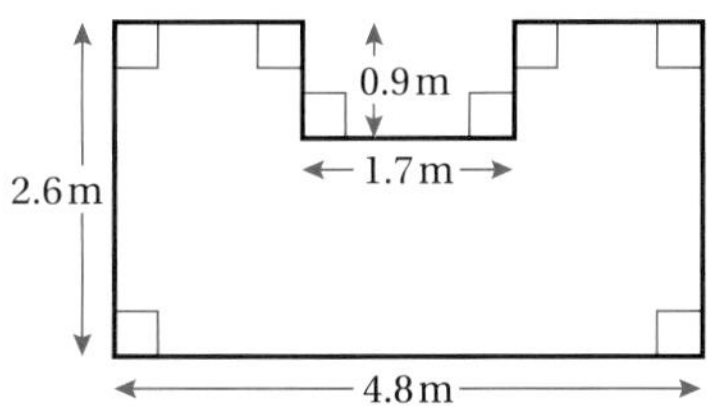

A net is a 2D shape that can be folded to create a 3D solid.

5 The net of a cylinder is shown.

The rectangle has dimensions 10 cm × 16 cm, and each circle has radius 2.55 cm.

Using 3.142 as an approximate value of π, calculate the surface area of the cylinder.

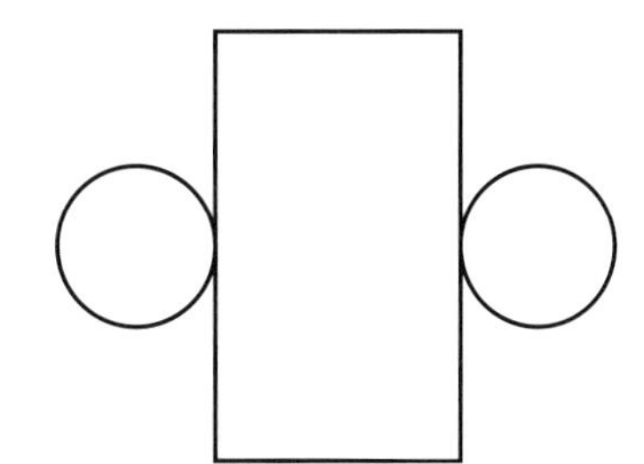

Remember that the surface area of a solid is the total area of each of its faces.

6 The diagram shows a circular mirror with a decorative frame.

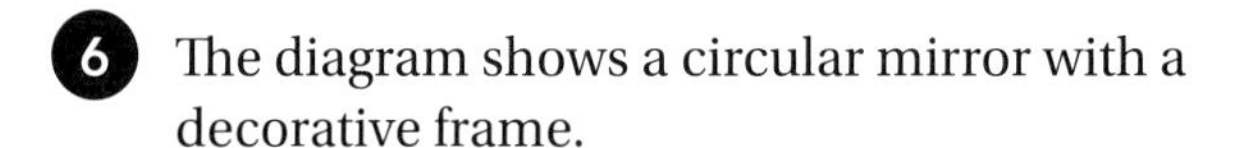

The frame is 12 cm wide all the way round.

The frame width is $\frac{4}{5}$ of the radius of the **whole** piece.

a Calculate the area of the whole piece. Use 3.142 as an approximate value of π.

b Calculate the area of the frame surrounding the mirror.

If 12 cm is $\frac{4}{5}$ of the radius, then what is the radius? Refer back to Chapter 10 on fractions if you need to.

7 A piece of fondant icing is rolled out into the shape of a square.

A circle with radius 11 cm is cut out from the square.

Given that the largest circle possible is cut out, find the difference between the area of the circular icing and the area of the square.

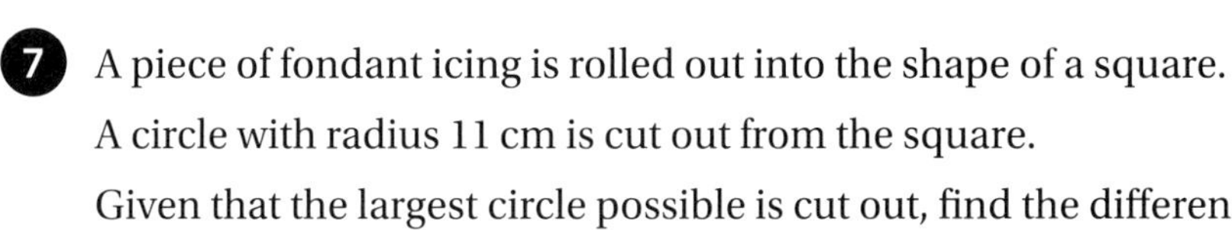

Find answers at: cambridge.org/ukschools/gcsemaths-studentbookanswers

8 A semicircular sandpit sits at the end of a lawn.

A cover is to be placed over the sandpit.

What is the area that needs to be covered?

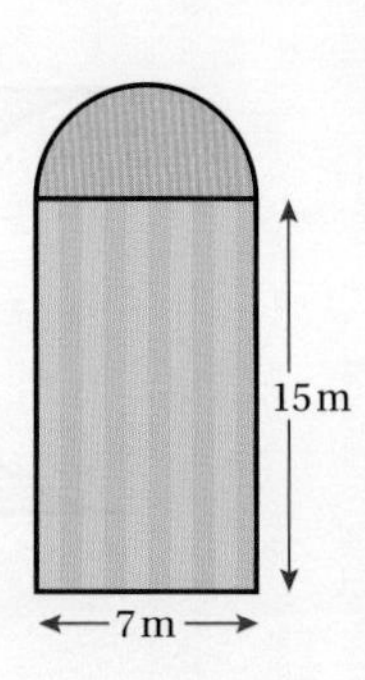

9 A circular photo frame has a plastic surround.

The width of the plastic surround is 8 cm.

The diameter of the complete photo frame is 28 cm.

Calculate the area available for the photo.

10 The shapes below have the same area.

The second shape is a square.

What is the side length of the square?

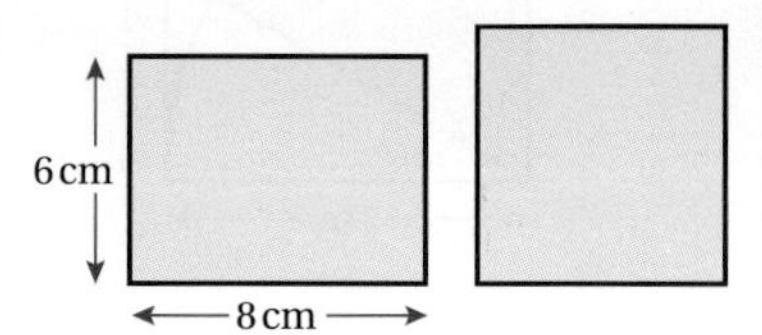

11 Car parking bays are either based on rectangles or parallelograms as shown on this model.

a The Department of the Environment's Planning Service specifies the following minimum standing space dimensions for cars and light vans:

Cars 2.4×4.8 m

Vans 2.4×5.5 m

These dimensions do not take into account space between vehicles and space for access and/or unloading.

Based on this, suggest some suitable dimensions for both rectangular and parallel parking bays.

b Is it feasible for both shapes to have the same area? Give a reason for your answer.

c What are the advantages and disadvantages of using each type of parking bay?

d Given a rectangular area 120 m by 200 m, with access possible from all sides, sketch the parking arrangement you would recommend and justify your choices.

Checklist of learning and understanding

Area of polygons

- Area is the amount of surface covered by a plane shape. Area is always given in square units.
- Area can be calculated using formulae. Some common formulae are:

Rectangle	Square	Triangle	Parallelogram	Trapezium
$A = lw$	$A = l^2$	$A = \frac{1}{2}bh$	$A = bh$	$A = \frac{1}{2}(a + b)h$

Area of circles

- Area of a circle $= \pi r^2$.
- Area of a sector is a fraction of the area of the whole circle.
- Area of a sector $= \frac{\theta}{360} \times \pi r^2$.

Composite shapes

- The area of composite shapes can be found by splitting them into known shapes.
- You can use the 'addition of parts' method to calculate the areas of smaller shapes separately and then add them together to find the total area.
- You can use the 'subtraction of parts' method to calculate the area of a shape with a 'cut out' section by finding the area of the larger shape and subtracting the area of the 'cut out' shape. Sometimes you need to visualise what the larger shape would look like *before* the smaller shape is cut out.
- The area of some composite shapes can be calculated using either the 'addition of parts' or the 'subtraction of parts' method; use whichever one leads to fewer calculations.

Chapter review

For additional questions on the topics in this chapter, visit GCSE Mathematics Online.

1 A rectangular vegetable plot has an area of $100\ m^2$.
If the width is 6.5 m, what is the length of the plot?

2 The net of a square-based pyramid is shown.
The square has sides of length 5 cm.
The triangles have a perpendicular height of 4.3 cm.

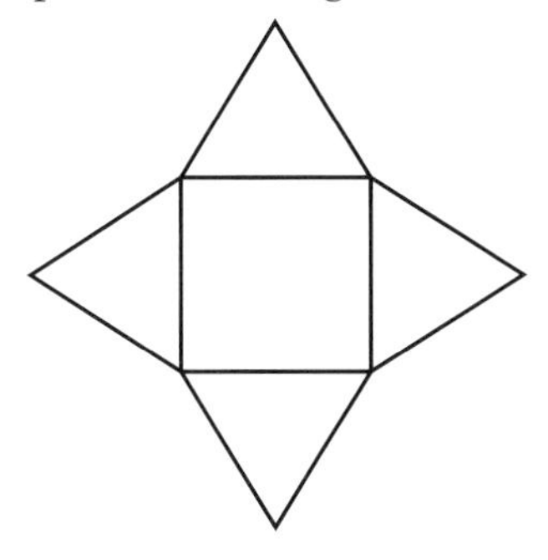

Calculate the surface area of the square-based pyramid.

Find answers at: cambridge.org/ukschools/gcsemaths-studentbookanswers

3 A square flower bed of side length 3 m sits exactly in the middle of a rectangular lawned area of side length 5 m and width 450 cm.

Calculate the area of grass.

4 An irrigator in a field can water a circular crop within a radius of 14.5 m. What is the total area of field that can be irrigated?

5 Calculate the area of the shape shown.

Give your answer in cm^2.

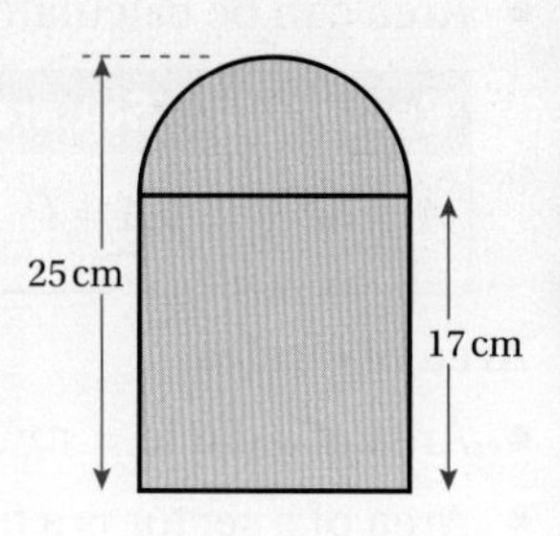

6 A kite has dimensions as shown in the diagram.

Not to scale

0.3 m

0.5 m

0.7 m

a Work out the area of the kite. *(4 marks)*

b Convert your answer in part **a** to cm^2. *(1 mark)*

© OCR 2013

7 Calculate the area of the shape below.

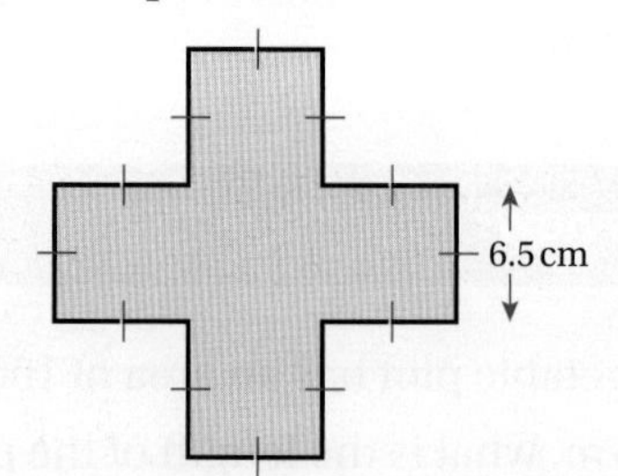

8 Guy is paid £0.15 for every square metre of grass he cuts.

How much would he be paid for cutting the grass in this garden?

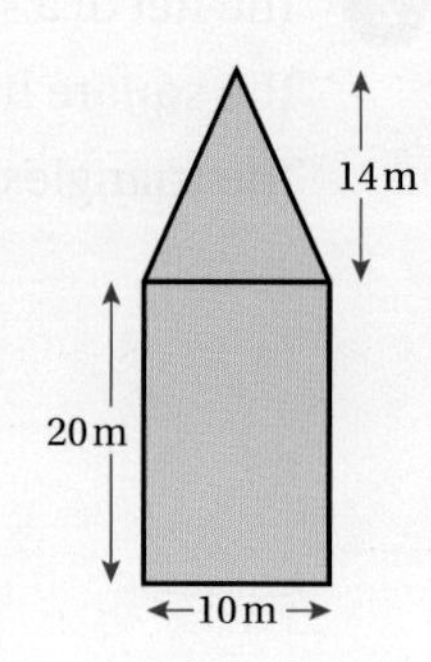

9 A stained glass window is a semicircle with radius 30 cm. Calculate:

a the perimeter of the window.

b the area of glass.

17 Approximation and estimation

In this chapter you will learn how to ...

- approximate values by rounding them to different degrees of accuracy or truncating.
- use approximations to estimate and check the results of calculations.
- understand and apply limits of accuracy in numbers and measurements.
- calculate the upper and lower bounds of a calculation (for discrete and continuous quantities).

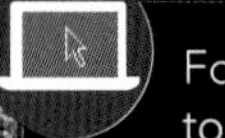

For more resources relating to this chapter, visit GCSE Mathematics Online.

Using mathematics: real-life applications

When you estimate what you spent over the weekend, or look at an object and guess it is about $2\frac{1}{2}$ m long, or say things like "I live about 15 kilometres from school" you are estimating and using approximate values.

"I round off the prices to the nearest pound and keep a mental running total of the costs of things I put in my trolley so I know that I am not over-spending." *(Consumer)*

Before you start ...

KS3	You should be able to use rounding to quickly estimate the answers to calculations.	**1** Use rounded values to estimate and decide if each answer is correct. **Do not do the calculation.** **a** $312 - 56 = 256$ **b** $479 \times 17 = 3142$ **c** $350 + 351 - 96 = 798$
Ch 11	You should be able to calculate with decimals and estimate to decide whether an answer is reasonable.	**2** Say whether each statement is true or false. **a** $5.8 \times 6.72 \approx 42$ **b** $3.789 + 234.6 \approx 4 + 230$ **c** $0.00432 + 3.55 \approx 4$ **d** $4 \times \pi \approx 12$
Ch 11	You need to be able to work confidently with decimals and place value.	**3** Write the number halfway between: **a** 3.0 and 5.0 **b** 3.5 and 3.6 **c** 0.02 and 0.07

Find answers at: cambridge.org/ukschools/gcsemaths-studentbookanswers

Assess your starting point using the Launchpad

STEP 1

1 Round each value to the degree of accuracy specified.

a 86 to the nearest 10.
b 1565 to the nearest 1000.
c 134.1234 to 2 decimal places.
d 19.999 to 1 decimal place.
e 1235.26 to 1 significant figure.
f 234 650 034 to 3 significant figures.

2 The length of a metal component is found to be 0.937 cm. What is its length to the nearest millimetre?

3 The cost of an international call on an itemised phone bill is given as £5.159 32.

a What is this amount truncated to the nearest penny?
b What is this amount rounded to two decimal places?

GO TO
Section 1:
Rounding

STEP 2

4 Estimate the cost of 12 packets of seeds at £1.36 each.

5 Approximately how many litres of petrol can you get for £20 if each litre costs £1.89?

6 Use rounding to find an approximate answer to each calculation.

a $\dfrac{784 + 572}{109}$ **b** $(2.099)^2$ **c** $\sqrt{\dfrac{3.803 + 7.52}{3.29}}$

GO TO
Section 2:
Approximation and estimation

STEP 3

7 A piece of wire is 10 m long, correct to the nearest metre. If the actual length of the wire is x, complete the following statement to show the longest and shortest possible lengths that this wire could be.

$\square \leqslant x < \square$

8 If $a = 3.6$ (correct to 1 decimal place) and $b = 14$ (correct to the nearest whole number), find the upper and lower bounds of:

a $a + b$ **b** ab **c** $\dfrac{a + b}{a}$

GO TO
Section 3:
Limits of accuracy

GO TO
Chapter review

Section 1: Rounding

Approximation allows you to give numbers in a more convenient form by writing them in a simpler, but less accurate way. For example, you might tell your friends that you pay £10 a month for your mobile phone contract when really you pay £10.32.

Approximate values are created by **rounding**.

Rounding a number means writing it with fewer non-zero digits to make it less precise.

Numbers can be rounded to different **degrees of accuracy**, such as 'to the nearest whole number' or to 'three significant figures'.

To round a number to a given degree of accuracy, find the digit in the specified place and look at the next digit to the right. It helps to think of its position on a number line.

- If it is **less than 5**, you keep the digit in the specified place the same; this results in the original number being rounded **down**.
- If it is 5 or greater than 5, you increase the digit in the specified place by 1; this results in the original number being rounded **up**.

The 'specified place' will vary depending on the degree of accuracy used.

Key vocabulary

rounding: writing a number with fewer non-zero digits by replacing some digits with zeros.

degree of accuracy: the number of places to which you round a number, for example to the nearest whole number, 2 decimal places, 3 significant figures.

Rounding to the nearest …

You might be asked to round a number to the nearest ten, hundred or thousand and so on.

If the number you are rounding is:

- an **integer**, after rounding you need to replace each digit to the right of the specified place with a zero.
- a **decimal number**, after rounding you replace each digit to the right of the specified place with a zero up to the decimal point, but you **do not** include the decimal point or any digits after the decimal point; the resulting approximate value will be an integer.

Tip

The zeros act as placeholders to make sure that the approximate value is the appropriate size in terms of place value. For example, ensuring that 232 is rounded to 200, not 2! Think of a number line and where the number lies. 232 is between 200 and 300.

Rounding 456 189 to the nearest ten:

456 1<u>8</u>9	Find the specified place, this is the digit in the 'tens' place.
456 190	Look at the digit to its right. 9 > 5 (over half way) so 8 is increased by 1. To keep the approximate value a six-digit number, the digit to the right of the specified value is replaced with a placeholder zero. 456 189 has been **rounded up**.

Rounding 456 189 to the nearest hundred:

456 <u>1</u>89	Find the digit in the 'hundreds' place and look at the digit to its right. 8 > 5, so the 1 is increased by 1 and the next two digits to the right are replaced by zeros.
456 200	456 189 has been **rounded up**.

Rounding 456 189 to the nearest thousand:

45<u>6</u> 189	Find the digit in the 'thousands' place and look at the digit to its right. 1 < 5 (less than half way), so the 6 stays the same and the next three digits to the right are replaced by zeros.
456 000	456 189 has been **rounded down**.

Find answers at: cambridge.org/ukschools/gcsemaths-studentbookanswers

If you are asked to round a decimal number to the **nearest whole number**, the specified value is the digit in the units/ones place and the digit to the right will be the first digit after the decimal place.

WORKED EXAMPLE 1

a Round these numbers to the nearest whole number.

i 0.5 **ii** 23.1 **iii** 0.034 235 **iv** 2 583 943.34

b Round 415.75 to

i the nearest ten. **ii** the nearest hundred.

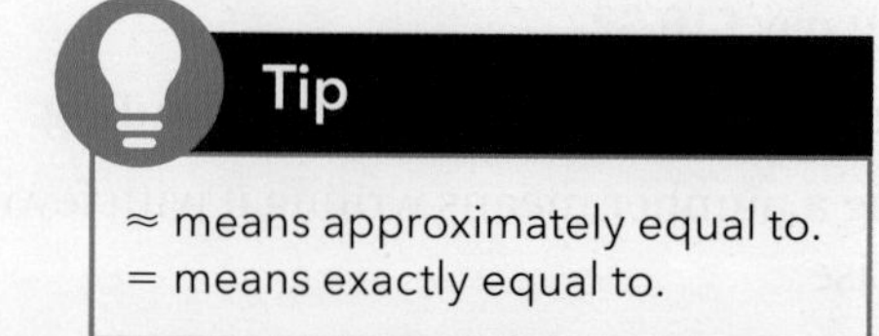

$\approx$ means approximately equal to.
= means exactly equal to.

a **i** 0.5

The specified place is in the units/ones place. Look at the digit to the right (i.e., after the decimal place). It is 5, so the digit in the specified place increases by 1.

$0.5 \approx 1$ (to the nearest whole number)

Because you are rounding to the nearest **whole number**, you **do not** include the decimal point or replace the digits after it with zeros. State the degree of accuracy used in your answer.

ii $23.1 \approx 23$ (to the nearest whole number)

You use the same approach as for part **a i**, except that $1 < 5$ so the digit in the specified place stays the same.

iii $0.034235 \approx 0$ (to the nearest whole number)

Same approach as parts **a i** and **ii**.

iv $2583943.34 \approx 2583943$ (to the nearest whole number)

Same approach as parts **a i** and **ii**.

b **i** 415.75

$415.75 \approx 420$ (to the nearest ten)

Same approach as for part **a**, except the specified place is in the tens place. Replace all digits to the right of the specified place up to the decimal point with zeros but **do not** include the decimal point or digits after it.

ii $415.75 \approx 400$ (to the nearest hundred)

Same approach as for part **b i**, except the specified place is in the hundreds place.

EXERCISE 17A

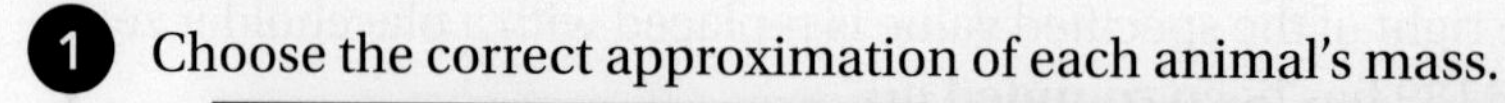

1 Choose the correct approximation of each animal's mass.

	Mass of animal		Rounded to	Identify the correct one	
a	Cow	635 kg	Nearest 10 kg	A 630 kg	B 640 kg
b	Horse	526 kg	Nearest 100 kg	A 500 kg	B 600 kg
c	Sheep	96 kg	Nearest 10 kg	A 90 kg	B 100 kg
d	Dog	32 kg	Nearest 10 kg	A 30 kg	B 40 kg
e	Cat	5.2 kg	Nearest 10 kg	A 0 kg	B 10 kg

2 **a** Round each value to the nearest whole number.

i 54.8 **ii** 10.6 **iii** 9.4 **iv** 12.3

b Round each value to the nearest 10.

i 26 **ii** 57.5 **iii** 111.1 **iv** 35 814

c Round each value to the nearest 100.

i 458 ii 5732 iii 2389 iv 35 814

d Round each value to the nearest 1000.

i 2590 ii 176 iii 35 814 iv 66 876

e Round the following to the nearest hundred thousand.

i 123 456 ii 1 234 567 iii 12 354 642 iv 123 456 789

f Round the following to the nearest million.

i 545 000 ii 555 000 iii 14 354 642 iv 546 267 789

3 a A food bill is £27.60. How much is this to the nearest pound?

b There are 27 students in a class. What is this to the nearest ten students?

c I have £175 saved up. What is this to the nearest £100?

d You need 167 cm of material to make a kite. Approximately how many metres do you need?

e Sue says that the population of the United Kingdom is 63 793 234 which is 63.7 million to the nearest hundred thousand people. Is she correct?

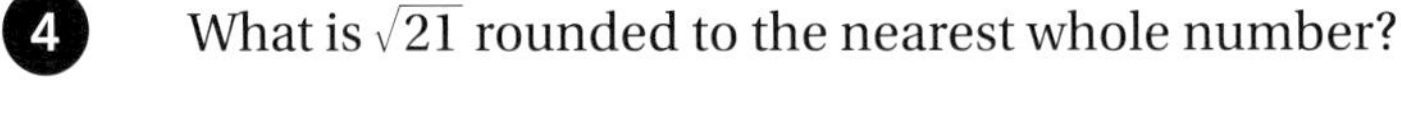

4 What is $\sqrt{21}$ rounded to the nearest whole number?

Rounding to a decimal place

Key vocabulary

decimal place (dp): place value position of a digit to the right of the decimal point.

A **decimal place (dp)** is the position of a digit to the right of the decimal point.

The number of decimal places relates to place value:

Place value after decimal point	.	tenths	hundredths	thousandths	...
Number of decimal places	.	1	2	3	...

When rounding to a specified decimal place you must remember to:

- include the decimal point in your answer.
- include any digits after the decimal point up to and including the specified decimal place.
- **not** include digits to the right of the specified place.

WORKED EXAMPLE 2

Round:

a 54.149 to 1 decimal place.

b 0.8751 to 2 decimal places.

c 0.100 24 to 3 decimal places.

a 54.1̲49
54.1

The first decimal place is the first digit after the decimal point. Look at the digit to its right; $4 < 5$ so the specified value stays the same. Delete the digits to the right of the specified place.

b 0.87̲51
0.88

The second decimal place is two digits after the decimal point, round as normal. Remember to include the decimal point and any digits after the decimal point up to and including the specified place, and take off the digits to the right.

c 0.100̲ 24
0.100

Round as in part **b**, except the specified place is the third digit after the decimal. Although 0.100 is the same as 0.1, you include the two placeholder zeros to show that the number is rounded to 'or accurate to' 3 decimal places.

Find answers at: cambridge.org/ukschools/gcsemaths-studentbookanswers

Suitable levels of accuracy

Tip

In some situations, you have to round up or down to answer a particular question. In this case, you use common sense rather than the rules of rounding.

Different situations require different levels of accuracy. The level of accuracy that is suitable for rounding depends on the situation. If you are not given a degree of accuracy to work to, you need to decide whether it is sensible to use a rounded value or whether it needs to be an exact value. You also need to consider whether a rough whole number is good enough or whether you should use decimals.

Here are some example answers from calculations, the appropriate degree of accuracy chosen for the answer, and the reasoning for choosing that accuracy.

Answer		
Before rounding	**After rounding**	**Reasoning**
Number of tins of paint needed = 9.15	10 tins	You can only buy whole numbers of tins so you round to the nearest whole number. Following the rules of rounding, you would round down to 9 tins. **However,** if you bought 9 tins you wouldn't have enough paint, so actually it is more appropriate here to round up.
Cost of repairs = £4.568 96	£4.57	The smallest denomination of pounds sterling is 1p (1p = £0.01), so round to two decimal places. Using the rules of rounding, you would round up.
Driving distance to Kent = 133.736 486 4 km	134 km	It's not reasonable to give the answer to 1 or 2 decimal places because that's more accurate than anyone would need to know (2 dp would give the distance to the nearest 10 m). As driving distances are generally recorded in kilometres, the nearest kilometre seems the most reasonable degree of accuracy. Using the rules of rounding, you would round up.
Temperature for weather forecast = 36.895 79°	37°	Most people want to have a general idea of the weather: is it hot, cold, warm, raining, and so on, so rounding to the nearest whole number is appropriate. Using the rules of rounding, you would round up.

When you calculate with decimals or significant figures and there isn't an obvious degree of accuracy to work to, then you normally round to no more than the degree of accuracy given in the original values.

When providing a rounded answer, **always state the degree of accuracy used.**

WORKED EXAMPLE 3

Calculate 4.13×2.07.

Key the calculation into your calculator.

8.5491

Given the size of the values in the question, the answer seems too precise.

8.55 (2 dp)

As the question uses values to 2 decimal places, round your answer to 2 decimal places. State the degree of accuracy you have used.

EXERCISE 17B

1 Round each number to:

i 1 dp. **ii** 2 dp. **iii** 3 dp.

a 4.526 38 **b** 25.256 37 **c** 125.617 38
d 0.537 921 **e** 32.3972 **f** 0.8993

2 Write each value correct to two decimal places.

a 19.869 03 **b** 302.0428 **c** 0.292
d 0.205 28 **e** 21 245.8449 **f** 0.0039
g 0.0972 **h** 0.999 999 9 **i** 99.997

3 Round each value to a suitable level of accuracy. Explain your decisions.

a A large dog weighs 24.4872 kg.
b To calculate a circumference, I use the value pi = 3.141 592 653 589 793 238 46...
c Dan's car can travel 13.7895 km per 1.000 098 7 litres of petrol.
d My share of a phone bill is £14.098 76.

4 What is $\sqrt{21}$ to 1 dp?

5 The department of the environment stated that "The overall total extent of land and sea protected in England through national and international protected areas increased from 1 million to 2 million hectares between 2000 and 2013."

What level of accuracy do you think was used in each of these figures?

Significant figures

The first **significant figure (sf)** in a number is the first non-zero digit when you read the number from left to right. It is considered the most **significant** figure because it tells you the number's overall size in terms of place value, i.e., if it is in the millions, thousands, hundreds or tenths and so on.

All digits that follow the first significant figure are significant, including any zeros. For example:

First significant figure	120 000 000	208.130	1.000 87	0.000 560 3
Second significant figure	120 000 000	208.130	1.000 87	0.000 560 3
Third significant figure	120 000 000	208.130	1.000 87	0.000 560 3

When rounding to a given number of significant figures,

- replace each digit to the right of the specified place with a zero (using zero as a placeholder).
- if the original number is a **decimal**, and the specified place is:
  - **before** the decimal point, do not include the decimal point or any digits after it.
  - **after** the decimal point, include the decimal point and all digits (zero or non-zero) between the decimal point and the specified place but delete any digits to the right of the specified place.

Key vocabulary

significant figure (**sf**): the most significant figure (digit) in a number is the first non-zero digit when reading the number from left to right.

Tip

One significant figure does not mean that you will have only one digit in the answer. 18 756 is 20 000 correct to 1 significant figure, not 2. Always think of the size of the original number you are rounding. Think of its position on a number line.

Find answers at: cambridge.org/ukschools/gcsemaths-studentbookanswers

WORKED EXAMPLE 4

Write each figure correct to the given degree of accuracy.

a 308 000 000 (2 sf) **b** 476.372 (4 sf)

c 2531.8 (2 sf) **d** 0.004 36 (1 sf)

a

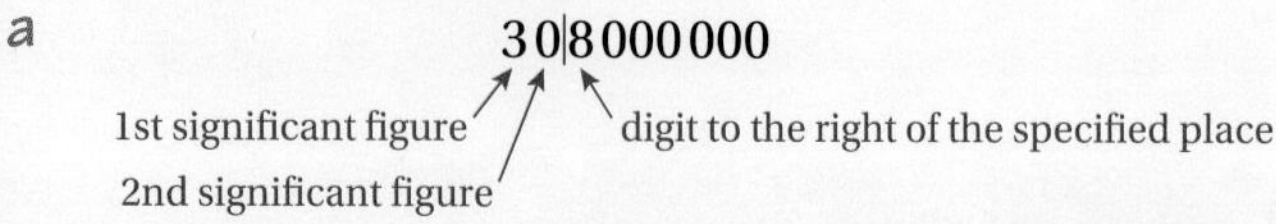

Read the number from left to right to find the first non-zero digit; this is the first significant figure. You need the second significant figure, so count on one digit. '0' is the second significant figure. Look at the digit to the right of the specified place; it is 8.

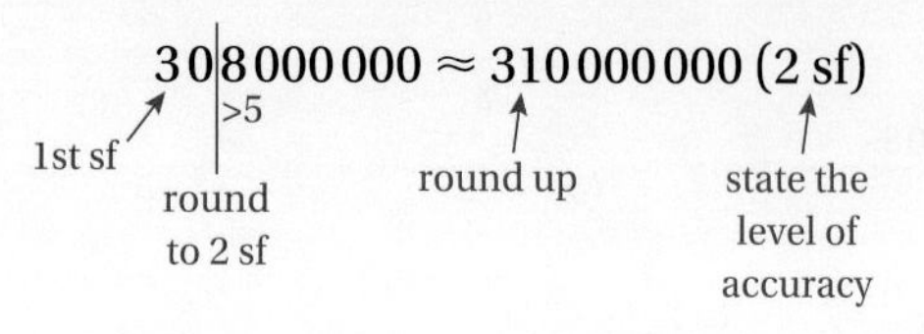

include the zeros: 31 is not the same as 310 000 000.

$308\,000\,000 \approx 310\,000\,000$ (2 sf)

$8 > 5$ so you increase the digit in the specified place by 1. Remember to replace all digits to the right with zero as placeholders (31 is not the same as 310 000 000!). Don't forget to include the degree of accuracy in your answer.

b $476.\underline{3}72 \approx 476.4$ (4 sf)

The fourth significant figure is '3'. As the specified place is after the decimal point, you **do not** include any digits after the specified place.

c $2\underline{5}31.8 \approx 2500$ (2 sf)

The second significant figure is '5'. Replace digits to the right with placeholder zeros. As the specified place is before the decimal point, do not include the decimal point or any digits to the right of it.

d $0.00\underline{4}36 \approx 0.004$ (1 sf)

The first significant figure is '4'. As the specified place is after the decimal place, you include the decimal point and any digits between it and the specified place but **do not** include any digits after the specified place.

Tip

Rounding numbers to a given number of decimal places means that you start the rounding at the specified place after the decimal point. Rounding numbers to significant figures means you start at the specified significant figure, which can be before or after the decimal point.

Significant figures are particularly useful for scientific or medical calculations where you deal with either very small or very large values.

Imagine that having 0.0001 ml or more of substance x in the bloodstream was dangerous. If the machine that measured the volume of the substance in the blood only measured to 2 decimal places, the following result would be negative: $0.000\,345$ ml ≈ 0.00 (2 dp). However, if the machine rounded to 2 sf, you would get a positive result: $0.000\,345$ ml $\approx 0.000\,35$ ml (2 sf).

The moon is about 384 400 km from the Earth. Say the precise measurement was 384 400.434563 km, rounding to a decimal place would give a much more detailed value than would make sense for such a large distance. Rounding to a significant figure would mean you could use a value that is accurate to the nearest km, or hundred km or thousand km, which is more appropriate for such a large distance.

EXERCISE 17C

1 **a** Round each value to 1 sf.

i 789 **ii** 3874 **iii** 69 356 **iv** 0.0456

b Write correct to 2 sf.

i 789 **ii** 3145 **iii** 0.003 325 **iv** 0.000 749 9

c Express each number correct to 3 sf.

i 789 **ii** 46 712 **iii** 0.004 214 **iv** 753 413

d Round each value to 2 sf.

i 37.673 **ii** $^{-}4127$ **iii** 3.0392 **iv** 1 999 000

e Write correct to 3 sf.

i 37.673 **ii** $^{-}4127$ **iii** 3.0392 **iv** 1 999 000

2 Explain why it is more useful to round a value such as 0.000 134 567 to 2 significant figures than to 2 decimal places.

3 **a** Pi $\approx$ 3.141 592 6. What is this to 3 sf?

b The density of a gas is 1.234 kg/m^3. What is this to 2 sf?

c The speed of light is 299 792 458 m/s. What is this to 2 sf?

d The strength of gravity at the Earth's surface is 9.806 65 m/s^2. What is this to 3 sf?

4 **a** What is $\sqrt{21}$ to 2 significant figures?

b How might it affect the accuracy of your results if you rounded $\sqrt{21}$ to 2 decimal places in one part of the calculation and to 2 significant figures in another?

Truncation

Truncation is when all the digits past a given point are cut off **without** rounding.

You need to be aware that your calculator display might give you a truncated value rather than a rounded value.

Test your calculator by entering [2] [÷] [3], the precise answer is the recurring decimal $0.\dot{6}$.

If your calculator display shows:

- 0.6666666666 your calculator has **truncated** the value to 10 decimal places.
- 0.6666666667 your calculator has **rounded** the value to 10 decimal places.

Key vocabulary

truncation: cutting off all digits after a certain point without rounding.

Tip

$2 \div 3 = \frac{2}{3} = 0.\dot{6}$; you learnt about recurring decimals in Chapter 11.

Tip

When you use rounded or truncated values during calculations, your results will not be completely accurate. You will learn more about this in Section 3.

Find answers at: cambridge.org/ukschools/gcsemaths-studentbookanswers

EXERCISE 17D

1. Truncate each number to 2 decimal places.
   a 37.673 **b** $^{-}$4.1275 **c** 3.0392 **d** 0.997

2. Truncate each number after the third significant figure.
   a 4.526 38 **b** 25.256 37 **c** 125.617 38
   d 0.537 921 **e** 32.397 **f** 200.6127

3. Consider splitting a bill of £20 equally between three people.
   What would each person pay?
   What method of approximation is most useful for deciding?

4. Computers can work out values to millions of decimal places.
   This level of accuracy is not always required, so computer programmers often work with truncated values to reduce the number of decimal places in calculations.
   What will each number below be if a programmer truncated it to 6 decimal places?
   a 1254543.35769504895849... **b** 0.00034789349030490...
   c 2.457578453954784...

Tip

The use of three dots at the end of a number indicates that the number continues on for more digits than are shown.

Did you know?

Truncation is used in statistics. A truncated mean is an average worked out by discarding (cutting off) very high or very low values in the data.

Key vocabulary

estimate: an approximate value or calculation.

Section 2: Approximation and estimation

Approximate values are useful for checking the results of calculations are sensible.

If your **estimate** and your actual answer are not similar, then you may have made a mistake in your calculation.

When estimating an approximate answer, you can generally round to 1 sf.

WORKED EXAMPLE 5

Estimate then calculate the value of $\frac{8.3 \times 536}{2.254 \times 9.612}$.

$\frac{8.3 \times 536}{2.254 \times 9.612} \approx \frac{8 \times 500}{2 \times 10}$

$\approx \frac{4000}{20}$

≈ 200

Round each number to 1 sf, then do the calculation.

$\frac{8.3 \times 536}{2.254 \times 9.612} = 205.340\,780\,4$

Key the calculation into your calculator to get the precise answer. Comparing this with your estimate tells you that your calculated answer is reasonable. If your calculator had given you an answer of 2053.407 ... for example, then you would know that you had probably made an error when keying in the calculation and would need to do it again.

Estimating is also useful for giving you an idea of what to expect.

"I estimate measurements and prices to quote an approximate cost so that customers have an idea of what a building job will cost them." *(Builder)*

WORK IT OUT 17.1

A group of four friends are travelling to a festival.

They intend to split the cost of everything between them.

The costs are: tent hire, £86.50; travel, 140 km round trip with petrol costing roughly 20p per kilometre, and camping entry tickets at £44 per night for three nights.

They decide to estimate how much they will each have to pay.

Which is the best estimate?
Why is it better than the others?

Estimate A	Estimate B	Estimate C
The cost of tent hire is roughly £88, which is £22 per person split between four. Petrol costs roughly £28 (140×0.2) which is £7 per person. The camping ticket costs are £44 × three nights, which is roughly £120 which is £30 per person. So total cost per person is approximately £22 + £7 + £30 = £59.	The tent hire is roughly £100, which is £25 per person. The petrol is roughly £150 × 0.2, which is £30, so split four ways is £7.50. The camping costs about £50 × 3 = £150, which is about £40 each. Total cost per person is approximately £25 + £7.50 + £40 = £72.50.	The tent hire is roughly £80, which is £20 each. Petrol is about £100 × 0.2 = £20 so £5 per person. Camping cost is about £40 × 3 = £120 for three nights, so about £30 each. Total cost per person is approximately £20 + £5 + £30 = £55.

EXERCISE 17E

1 Estimate the following by rounding each number to 1 sf.

a 111.11×3.6 **b** 378×1.07

c 0.99×16.7 **d** 13.6×0.48

e $\pi \times (5.3)^2$ **f** 4.8×12.5

g $\frac{192}{17.2}$ **h** $\frac{58.38}{0.5185}$

2 Which calculation would provide the best estimate (A, B or C)?

a 186×9.832 A 200×10 B 190×9 C 190×10

b $15.76 \div 7.6$ A $15 \div 7$ B $16 \div 8$ C $16 \div 7$

Find answers at: cambridge.org/ukschools/gcsemaths-studentbookanswers

3 Estimate the following by rounding each number to 1 sf.

a $\frac{82.65 \times 0.4654}{42.4 \times 2.43}$ **b** $\frac{16.96 + 3.123}{16.96 - 6.432}$

c $\frac{979 \div 43.6}{2.36 \times 0.23}$ **d** $\frac{979 \div 492.9}{21.6 \div 43.87}$

4 Estimate the following.

a $\sqrt{\frac{3.2 \times 4.05}{0.39 \times 0.29}}$ **b** $\sqrt{\frac{4.1 \times 11.9}{7.9 \times 0.25}}$

5 Shafiek runs a cross country race at an average speed of 6.25 m/s.

a Estimate how far he will have run after 6 minutes.

b Estimate how long it takes him to cover 1467 m.

6 Look at the calculator display answers for each calculation. Use your estimation skills to say whether the answer is sensible or not, without actually doing the calculation.

	Calculation	Display
a	$3 \times \pi \times 5^2$	125.6637061
b	5×8.9	445
c	50×8.9	445
d	3×192.5	57.75
e	$\frac{\sqrt{86}}{2.8 \times 16.18}$	0.204697565
f	$0.0253 \div 0.45$	56.22222222

7 A parallelogram has an area of 54.67 cm². The base of the parallelogram is 7.9 cm long.

a When a student tries to find the height on his calculator, he gets a result of 69 202 531.

This is clearly wrong.

Give the height correct to 2 significant figures.

b Estimate the perimeter of the parallelogram to the nearest cm.

8 Estimate the square root of 50 if the square root of 49 is 7 and the square root of 64 is 8.

9 Given that $\frac{0.514 \times 76.3}{2.4^2} = 6.8087$ to 4 dp, estimate the value of $\frac{51.4 \times 7.63}{24^2}$.

10 Find an approximate value of $\frac{2876}{31 \times 33}$.

Section 3: Limits of accuracy

All measurements of mass, height, length, area and capacity are given to a certain level of accuracy. Even with very accurate measuring instruments, these quantities cannot be measured exactly because they are **continuous variables**.

When you are given a measurement you assume it is accurate except for the last digit as this has been rounded. The rules of rounding mean the measurement has to fall within certain limits. The limits are determined by the degree of accuracy used to round.

The smallest value a measurement can take is called the lower bound of the measurement.

The largest value it can take is called the upper bound.

The difference between the upper and lower bounds is called the error interval. This indicates the range of values in which the precise value could fall.

For example, a piece of wood is measured and its length recorded as 47 cm **to the nearest cm**.

Use what you know about rounding to determine the upper and lower bounds:

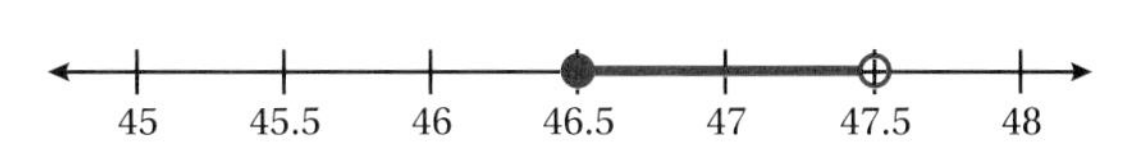

If the length was less than 46.5 cm, rounding to the nearest cm would give 46 cm.

If the length was 47.5 cm or more, rounding to the nearest cm would give 48 cm.

If we let l represent the length of the piece of wood, the error interval can be expressed as:

$46.5 \text{ cm} \leqslant l < 47.5 \text{ cm}$

This is called **inequality notation** and it means the length of the wood is greater than or equal to 46.5 cm and less than 47.5 cm.

For any measurement correct to a given level of accuracy, the exact values lie in a range half a unit below and half a unit above the measurement.

Look at the following examples and note the rules that apply for decimals and significant figures:

Example	Lower and upper bounds	Error interval
		(let the value be x)
0.5 rounded to 1 dp	0.45 and 0.55	$0.45 \leqslant x < 0.55$
0.65 rounded to 2 dp	0.645 and 0.655	$0.645 \leqslant x < 0.655$
0.7663 rounded to 4 dp	0.76625 and 0.76635	$0.76625 \leqslant x < 0.76635$
15 rounded to 2 sf	14.5 and 15.5	$14.5 \leqslant x < 15.5$
320 to 2 sf	315 and 325	$315 \leqslant x < 325$
2.32 rounded to 2 dp	2.315 and 2.325	$2.315 \leqslant x < 2.325$

For a **truncated** value, the upper and lower bounds are **not** ± half a unit because the value is cut off not rounded; this means that the digit to the right of the specified place is irrelevant, and it is only the digit in the specified place that is considered.

For example, if 6.6 is a value truncated to 1 dp, then the error interval is:

$6.6 \leqslant x < 6.7$

Key vocabulary

continuous variable: data that can take any numerical value within a range; it can be measured.
lower bound: the smallest value that a number (given to a specified accuracy) can be.
upper bound: the largest value that a number (given to a specified accuracy) can be.
error interval: the difference between the upper and lower bounds.

Tip

You will learn more about continuous and discrete data in Chapter 35.

Tip

It is helpful to draw a number line to work out the upper and lower bounds of a value. The closed circle means the value **is** included, the open circle means it is **not** included.

Tip

$\leqslant$ means less than or equal to
$\geqslant$ means greater than or equal to
$<$ means less than
$>$ means greater than

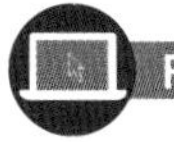

Find answers at: cambridge.org/ukschools/gcsemaths-studentbookanswers

WORKED EXAMPLE 6

Use inequality notation to write down the error interval for:

a 10 cm correct to the nearest cm.

b 22.5 kg to 1 decimal place.

c 128 000 correct to 3 sf.

a 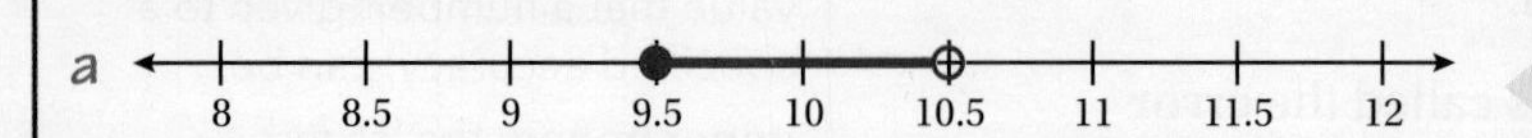

The degree of accuracy is 'to the nearest cm' so half a unit is 0.5cm. The error interval is half a unit above and below, so it is 10 cm ± 0.5 cm. 9.5 cm would be rounded to 10 cm, and 10.5 cm would be rounded to 11 cm. Check using a number line; use a solid circle for the lower bound and an open circle for the upper bound.

Let the length be x.

$9.5 \text{ cm} \leqslant x < 10.5 \text{ cm}$

Write the error interval using inequality notation, don't forget to include the correct symbols: 'greater than or equal to' and 'less than'.

b 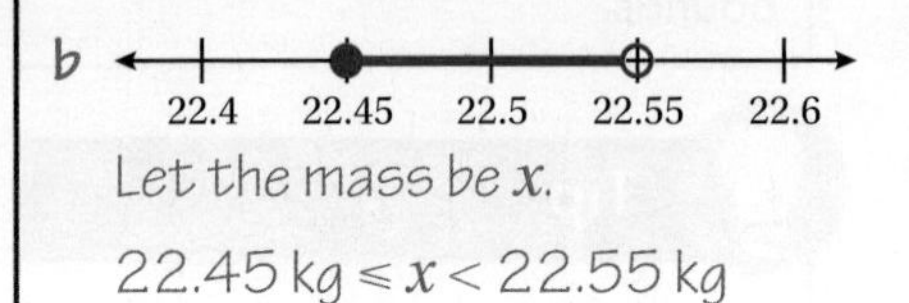

Let the mass be x.

$22.45 \text{ kg} \leqslant x < 22.55 \text{ kg}$

The degree of accuracy is 'to 1 decimal place'; so half a unit is 0.05 kg. The error interval will be 22.5 kg ± 0.05 kg. Draw a number line to check.

c 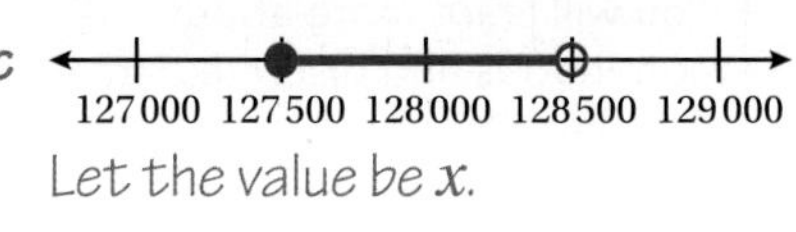

Let the value be x.

$127\,500 \leqslant x < 128\,500$

The degree of accuracy is '3 sf', so half a unit is 500. The error interval is 128 000 ± 500. Draw a number line to check.

Sometimes you will need to be able to identify when a question is asking for upper and lower bounds.

WORKED EXAMPLE 7

The healthy mass of a male labrador is between 27 kg and 34 kg to the nearest kilogram.

Sophia's dog has a mass of 34.34 kg. In terms of his mass, is he healthy?

Let m be the mass in kg.

$26.5 \leqslant m < 34.5$

Yes, the dog is healthy.

The question requires you to calculate the upper bound. Half a unit of 1 kg is 0.5 kg. Write the error interval. The mass is less than the upper bound.

EXERCISE 17F

1 Find the lower and upper bound of each value.

a 96 rounded to 2 sf

b 96.0 rounded to 3 sf

c 96.00 rounded to 4 sf

d 0.6 rounded to 1 dp

e 0.06 rounded to 1 dp

f 0.60 rounded to 2 dp

g 3.142 rounded to 3 dp

h 9.9 rounded to 2 sf

2 The following lengths were measured to the nearest millimetre.

Write down an error interval for each one using inequality notation.

Let the length be L in each case.

a 4.9 cm **b** 12.520 m

c 43.0 cm **d** 29 mm

3 **a** There are 36 litres of petrol in a car's tank, to the nearest litre.

What is the least possible volume of petrol in the tank?

b A length of wood is 1.4 m to the nearest cm.

Is it possible for the wood to be 137 cm long?

c The weight of a stone is 43.4 kg to the nearest tenth of a kg.

What is the least and greatest weight it could be?

4 A calculator has truncated the following number to 4 decimal places.

34.5638

Write down the error interval for this number.

5 Hilal ran 100 m in 15.3 seconds.

The distance is correct to the nearest metre and the time is correct to 1 decimal place.

Write down the upper and lower bounds of:

a the distance he ran.

b the actual time taken.

6 The length of a rope is 4.5 metres to the nearest 10 cm.

The actual length of the rope is L metres.

Find the range of possible values for L, giving the answer as an inequality.

The upper and lower bounds of a calculation

When you use approximate values in calculations, the error interval of each value is often compounded so that the answer has a larger interval of error than any of the individual values used to calculate it. Worked example 8 helps to demonstrate this.

WORKED EXAMPLE 8

A rectangle has sides of 10 cm and 6 cm to the nearest centimetre.

a Calculate the limits of accuracy of the area of the rectangle.

b What is the difference between the minimum and maximum value for

i each length? **ii** the area?

possible lengths
$9.5 \leqslant l < 10.5$

10 cm

6 cm

possible widths
$5.5 \leqslant w < 6.5$

Draw a sketch. Find the error interval of each measurement.

a Smallest possible area $= 9.5 \times 5.5$
$= 52.25\ \text{cm}^2$

From this you can see that the lowest bound of the length is 9.5 cm and the lowest bound of the width is 5.5 cm. Use these values to calculate the smallest possible area.

Largest possible area $= 10.5 \times 6.5$
$= 68.25\ \text{cm}^2$

The upper bound of the length is 10.5 cm and the upper bound of the width is 6.5 cm. Use these values to calculate the largest possible area.

Let a be the area of the rectangle.

The limits of accuracy for the area are

$52.25\ \text{cm}^2 \leqslant a < 68.25\ \text{cm}^2$.

Write these values as an error interval for the area, a, of the rectangle.

b **i** The difference between the minimum and maximum value for each of the lengths is 1 cm for each length.

ii The difference between the minimum and maximum value for the area is $16\ \text{cm}^2$.

Here, the interval of error for each value in the calculation has compounded, so that there is a much larger interval of error in the answer.

This is why you are often encouraged **not** to use rounded values during a calculation, and to round only the final answer.

Tip

In real terms this means that your area calculation could be out by $16\ \text{cm}^2$. If you were coating the area with platinum, this could make quite a difference to the amount you needed and the cost.

Key vocabulary

discrete values: countable values using whole numbers.

Discrete and continuous quantities

When working with upper and lower bounds, it is important to understand the difference between **discrete values** and continuous quantities.

You already know that measurements are continuous quantities. Length, mass, capacity and other measures can take any value between the given points (within a range).

Quantities that can be counted, such as the number of people, number of cars or number of buildings are all discrete values. They can only take certain values; you can't have half a person for example!

The upper and lower limits still lie within half a unit of the degree of accuracy used when rounding, but the nature of discrete values means that the limit of accuracy for the same value can be expressed in different ways.

WORKED EXAMPLE 9

The population of a village is 400 to the nearest 100.

If n represents the number of residents in the village, what is the error interval?

$350 \leqslant n < 450$

or

Calculate the error interval just as you would for a continuous variable. Half a unit is 50 people, so the error interval is 400 ± 50. So the lower bound is 350 as this rounds up to 400, and the upper bound is 450 as this rounds up to 500.

$350 \leqslant n \leqslant 449$

However, because you can't have a fraction of a person you can also write the upper bound as $\leqslant 449$ because there is no possible number the value can take between 449 and 450. (You **can't** do this with a continuous variable because there are an infinite number of values it could take between two integers.)

EXERCISE 17G

1 12 kg of sugar are removed from a sack containing 50 kg.

Each measurement is correct to the nearest kilogram.

Find the upper and lower bounds of the mass of the sugar left in the container.

2 The dimensions of a rectangle are 3.61 cm and 2.57 cm, each correct to 3 sf.

a Write down the upper and lower bounds of each measurement.

b Find the upper and lower bounds of the area of the rectangle.

c Write the upper and lower bounds of the area correct to 3 significant figures.

3 In the diagram, the measurements of a piece of land are given correct to 1 decimal place.

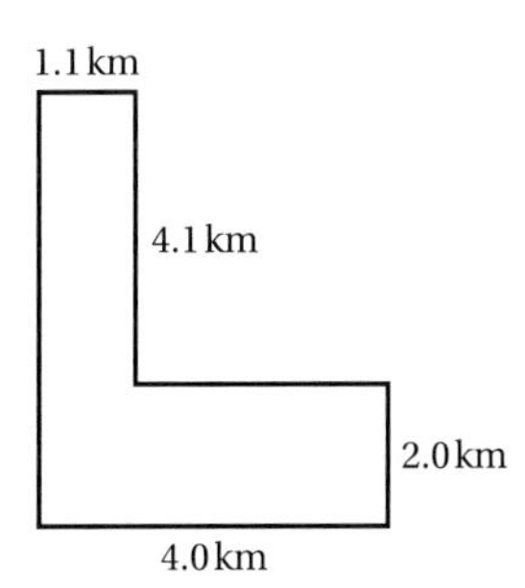

a Calculate the limits of accuracy of the area of the land.

b A surveyor calculates the area to be 12.51 km^2. What is the greatest possible error in taking that value?

4 Henry has £100 to the nearest £1 and he spends £30 to the nearest £1.

What is the least amount of money he can have left?

5 **a** A rectangle is drawn to the nearest cm, with width 5 cm and length 6 cm. What is the:

i greatest length of perimeter? **ii** smallest length of a diagonal?

b Another rectangle measures 3 cm by 7 cm.

If this is also accurate to the nearest cm, what is its maximum possible area?

6 In order to try to calculate pi (π) John measures a circle.
The circumference (C) is 40 cm to the nearest cm and the diameter (d) is 12 cm to the nearest cm.
Given the formula for the circumference of a circle, what are the highest and lowest values for John's calculation of π?

7 Mishka runs 100 m in 12.32 seconds to 2 dp.
Write down the range of values for her average speed.

8 A chocolate manufacturer recently reduced the weight of its standard chocolate bar from 49 g to 45 g, to the nearest gram.
What is the maximum weight of chocolate that could have been removed?

9 In an experiment, the velocity of a model racing car is worked out by timing how long it takes to cover a marked length on a track.
The velocity is found using the formula $v = \frac{d}{t_2 - t_1}$, where d is the distance covered in metres, t_1 is the starting time and t_2 is the finishing time on a stopwatch, measured in seconds.
In a particular trial, the car travels 1.000 m (to the nearest mm), starting at 0.2 s and ending at 1.4 s (both correct to the nearest 0.1 second).
Write an inequality to express the values between which v falls in this trial.

Checklist of learning and understanding

Approximate values

- Approximate values are created by rounding a value to a given degree of accuracy.
- When you round to a specified place, if the digit to the right is 5 or more (half way or more), then the original number is rounded up. If it is less than 5 (less than half way), then the original number is rounded down.
- Rounding to a given number of decimal places means that you start the rounding at the specified place after the decimal point. Rounding to significant figures means you start rounding at the specified significant figure, which can be before or after the decimal point.
- Truncating a number means removing all digits after a specified place without rounding.

Estimation

- Complex calculations can be estimated without using a calculator by using approximations of each term in the calculation to make a simple calculation.
- Estimating the answer first can help you to spot when the answer to a calculation is incorrect.

Level of accuracy

- Measurement of continuous variables is really an approximation of the value within an upper and lower bound. Therefore, the measurement of a continuous variable can be expressed as an error interval using inequality notation ($\leqslant$, $<$).

Chapter review

For additional questions on the topics in this chapter, visit GCSE Mathematics Online.

1 Are the following true or false?

	Original number	Approximation	Answer
a	123.456	Rounded to 2 dp (2 decimal places)	123.456
b	123.456	Rounded to 1 sf (1 significant figure)	120.000
c	123.456	Rounded to 1 dp (1 decimal place)	123.5
d	123.456	Rounded to 4 sf (4 significant figures)	123.5
e	123 456.789	Truncated to 1 dp (1 decimal place)	123 456.8

2 Estimate the value of $\frac{6.68^2}{4.76 \times 21.2}$.

3 A market trader sells 470 apples for £73. Roughly how much does an apple cost?

4 If $521 \times 32 = 16\,672$, estimate the value of $16\,672 \div 3.2$.

5 Estimate the value of $\frac{39.9 + \sqrt{0.934}}{(15.4 - 4.3)^2}$.

6 Chocolate bars are supposed to have a mass of 50 g with a tolerance of 5%.

If they are weighed to the nearest 10 grams will they be within this tolerance?

7 The voltage V of an electronic circuit is given by Ohm's law which is the formula

$V = IR$ where I is the current in amps and R is the resistance in Ohms.

Given that $V = 316$ correct to 3 significant figures and $R = 19.2$ correct to 3 significant figures, calculate the lower bound of I.

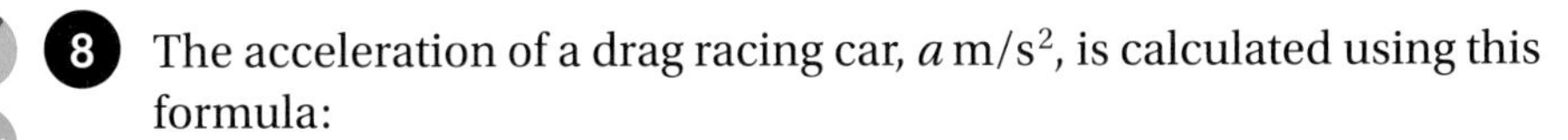

8 The acceleration of a drag racing car, a m/s^2, is calculated using this formula:

$$a = \frac{v - u}{T - t}$$

u m/s is the velocity after time t seconds and v m/s is the velocity after time T seconds.

A drag racing car accelerates along a straight course.
After 1.3 seconds its velocity is 49.5 m/s; after 3.26 seconds its velocity is 124.1 m/s.

Work out an **estimate** of the acceleration of the drag racing car. *(3 marks)*

© OCR 2011

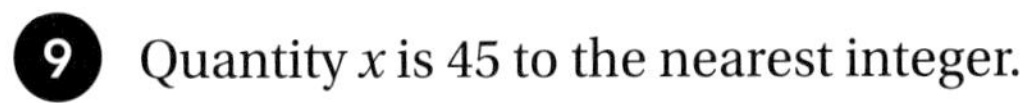

9 Quantity x is 45 to the nearest integer.

Quantity y is 98 to the nearest integer.

Calculate upper and lower bounds for x as a percentage of y to 1 decimal place.

Find answers at: cambridge.org/ukschools/gcsemaths-studentbookanswers

18 Straight-line graphs

In this chapter you will learn how to …

- use a table of values to plot graphs of linear functions.
- identify the main features of straight-line graphs and use them to sketch graphs with equations in the form $y = mx + c$.
- find the equation of a straight line using the gradient and points on the line.
- identify parallel lines from the equation of the line in the form $y = mx + c$.
- identify perpendicular lines from the equation of the line in the form $y = mx + c$.
- find the equation of a tangent to a circle with centre (0, 0).

For more resources relating to this chapter, visit GCSE Mathematics Online.

Using mathematics: real-life applications

This is a building in London nicknamed *The Gherkin*. The curves and lines of the building were designed using complex equations and their graphs. Architecture is just one of many professions in which people plot and use graphs in their work.

"When designing a new building, I use graphs to help identify and describe the structural properties the building needs to have." *(Architect)*

Before you start …

<table>
<tr><td>Ch 4</td><td>You should remember how to generate terms in a sequence using a rule.</td><td>1 Use the rule $T(n) = 3n - 2$ to complete this table.
<table><tr><th>n</th><td>1</td><td>3</td><td>5</td><td>10</td></tr><tr><th>$T(n)$</th><td></td><td></td><td></td><td></td></tr></table></td></tr>
<tr><td>KS3</td><td>You should be able to give the coordinates of points on a grid.</td><td>2 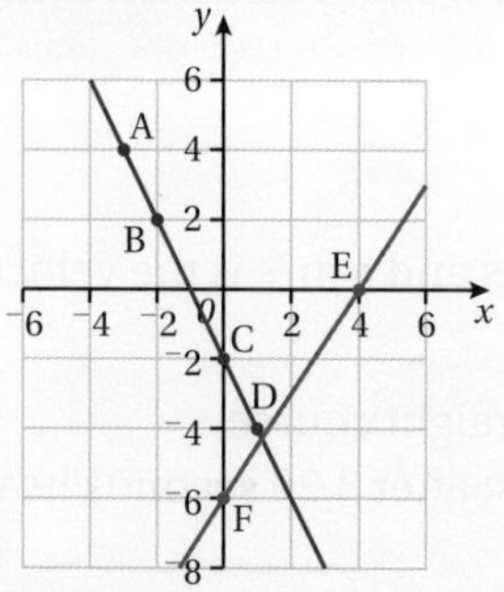

a Write down the coordinates of points A, D and E.
b What point has the following coordinates?
i $(^-2, 2)$ ii $(0, ^-6)$
c What is the name given to the point (0, 0)?</td></tr>
<tr><td>Ch 8</td><td>You must be able to rearrange and solve equations.</td><td>3 Solve for x.
a $4 - 3x = 13$ b $\frac{x}{7} = 6$ c $^-3(5x + 2) = 0$
4 If $y = 2x + 5$:
a find y when $x = {}^-2$.   b find x when $y = 8$.</td></tr>
<tr><td>Ch 14</td><td>You should remember how to change the subject of a formula.</td><td>5 Make y the subject of each equation.
a $^-2x - y + 1 = 0$ b $2x + 3y = 6$ c $x - 2y = {}^-2$</td></tr>
</table>

Assess your starting point using the Launchpad

STEP 1

1 Copy and complete the table of values for each function.

a $x - y = 2$

x	$^{-}2$	$^{-}1$	0	1
y				

b $x + y = 4$

x	$^{-}2$	$^{-}1$	0	1
y				

c $2x + y + 2 = 0$

x	$^{-}2$	$^{-}1$	0	1
y				

d $x - 2y + 2 = 0$

x	$^{-}2$	$^{-}1$	0	1
y				

GO TO
Section 1:
Plotting graphs

2 The graphs of two of the functions from Question **1** are shown here.

Match each graph to its equation.

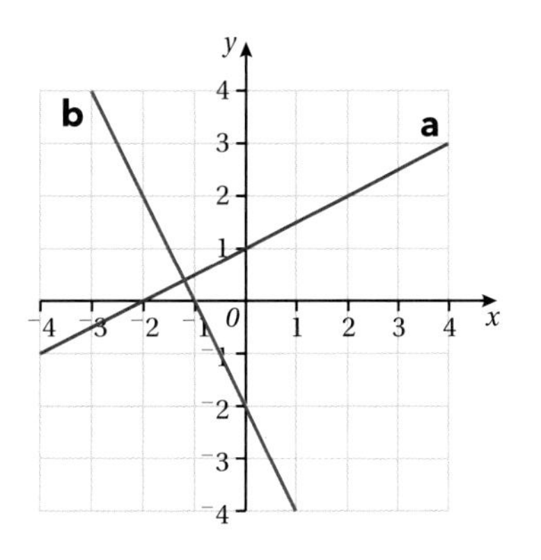

STEP 2

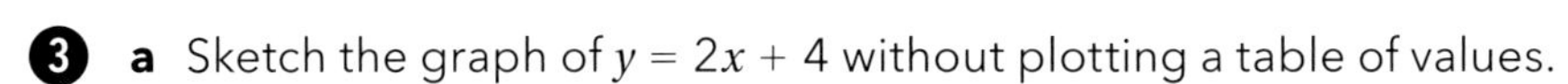

3 **a** Sketch the graph of $y = 2x + 4$ without plotting a table of values.

b Find the gradient and y-intercept of the resulting straight line.

4 Find the equation of the straight line that passes through the points (1, 4) and (3, 7).

5 A line cuts the x-axis at 4 and the y-axis at 5. What is its gradient?

GO TO
Section 2:
Using the features of straight-line graphs

GO TO
Step 3:
The Launchpad continues on the next page …

Find answers at: cambridge.org/ukschools/gcsemaths-studentbookanswers

Launchpad continued ...

STEP 3

6 Which of these lines are parallel to each other?

a $y = {}^{-}3x + 3$ **b** $y = 7 - 3x$ **c** $y = 3x + 7$

d $y = \frac{1}{3}x + 3$ **e** $y = 7 - 2x$

7 A line is parallel to the line $y = \frac{1}{2}x$ and passes through the point (2, 4). What is its equation?

8 How do you know $y = \frac{1}{2}x + 3$ crosses the line $y = 5 - 2x$ at right angles?

9 Find the tangent to the circle $x^2 + y^2 = 25$ that passes through point (${}^{-}3$, 4).

GO TO
Section 3:
Parallel lines, perpendicular lines and tangents

GO TO
Section 4:
Working with straight-line graphs

Tip

You learnt about functions in Chapter 4.

Section 1: Plotting graphs

Graphs can be used to show the relationship between two variables.

You can think of a graph as a picture of a **function**. The line of the graph shows what happens when you apply a rule to x to get a value of y.

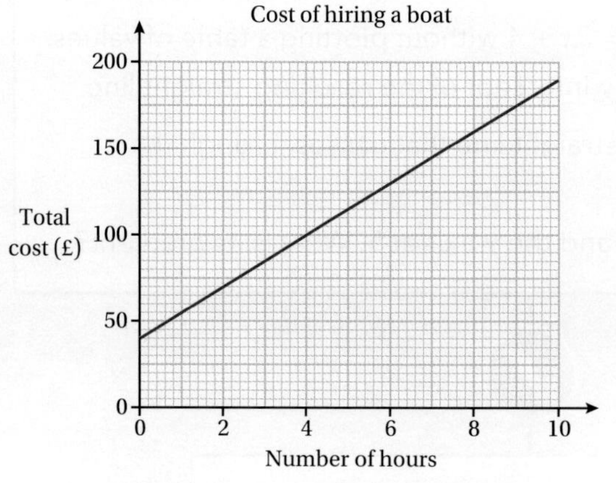

This graph shows the relationship between the cost of hiring a boat and the time the boat is hired for.

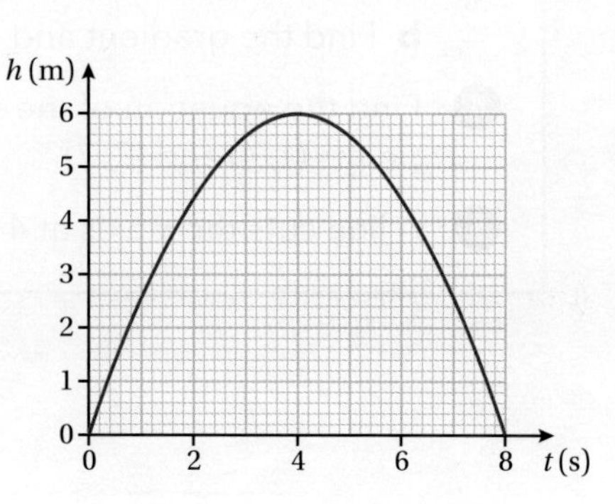

This graph shows the height of a projectile from the time it is released to the time it falls to the ground.

For every value of x on the graph there is a corresponding value of y. Each pair of x- and y-values form the **coordinates** (x, y) of a point on the line.

Functions that produce straight lines when you plot their x- and y-values are called **linear functions**. You know from Chapter 8 on equations, that a linear function or equation is one where the highest power of x is 1.

Key vocabulary

coordinates: an ordered pair (x, y) identifying position on a grid.

Plotting linear functions

The graphs of a linear function are called straight-line graphs.

You can **plot** the graph by first drawing up a table of values.

Substitute each value of x into the function to find the corresponding values of y. Then plot the (x, y) coordinates on a set of axes.

Key vocabulary

plot: drawing a graph accurately by marking points on a grid using coordinates.

WORKED EXAMPLE 1

Draw up a table of values and plot the graph of $y = 2x + 1$.

x	$^{-}1$	0	1	2
y	$^{-}1$	1	3	5

Choose some values for x. Substitute each x-value into the function to find the corresponding y-value.

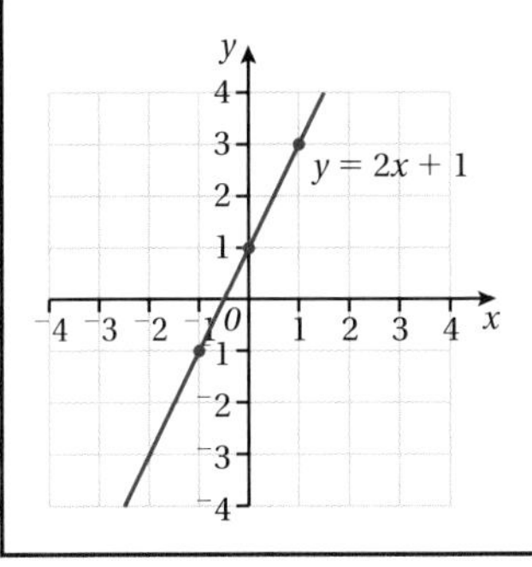

Plot at least three points using the coordinates. Points $(^{-}1, {}^{-}1)$, $(0, 1)$ and $(1, 3)$ have been plotted here.

Use a ruler to draw a straight line through the points. Label the graph with the equation.

Tip

When you draw a graph, continue the line in both directions through the points, as far as you can. Don't just join the three plotted points together: they are just three of the infinite number of points on the line.

EXERCISE 18A

1 Complete a table of values for each function. Then plot the graphs.

a $y = x$ **b** $y = x + 2$ **c** $y = 3x - 5$

d $y = 6 - x$ **e** $y = 2x + 1$ **f** $y = x - 1$

g $y = {}^{-}2x + 3$ **h** $y = 4 - x$ **i** $y = 3x - 2$

2 What is the minimum number of points you need to plot a straight line accurately? Why?

Find answers at: cambridge.org/ukschools/gcsemaths-studentbookanswers

Key vocabulary

gradient: a measure of the steepness of a line.

Gradient = $\frac{\text{change in } y}{\text{change in } x}$

***x*-intercept**: the point where a line crosses the x-axis when $y = 0$.

***y*-intercept**: the point where a line crosses the y-axis when $x = 0$.

Section 2: Using the features of straight-line graphs

The main characteristics of a straight-line graph are:

- the **gradient**, or slope of the graph.
- the ***x*-intercept** (where it crosses the x-axis).
- the ***y*-intercept** (where it crosses the y-axis).

Gradient

The gradient is a measure of how steep a line is.

It is the **vertical** distance travelled divided by the **horizontal** distance travelled as you move from left to right along the line.

$$\text{Gradient} = \frac{\text{vertical rise}}{\text{horizontal run}} = \frac{\text{difference in } y\text{-values}}{\text{difference in } x\text{-values}}$$

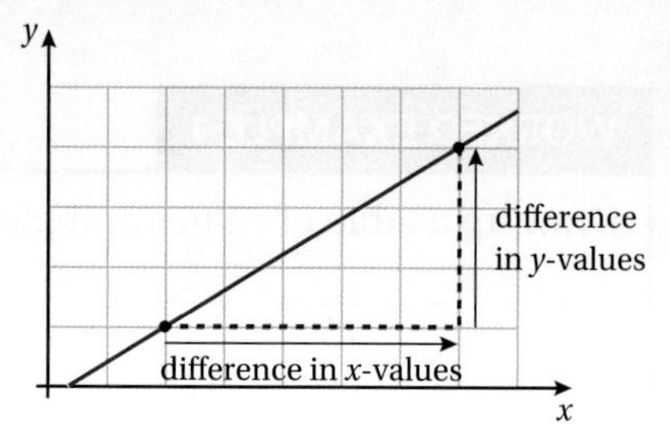

Calculating the gradient from a graph

The graph of the linear equation $y = 2x + 4$ has been plotted using the coordinates $(1, 6)$, $(0, 4)$, $(^-4, ^-4)$.

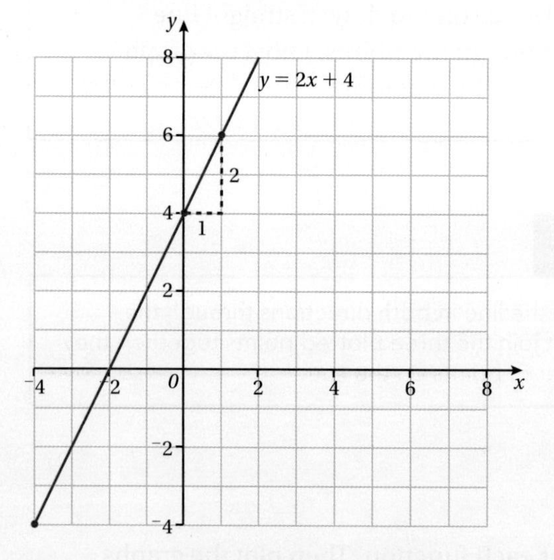

A right-angled triangle has been drawn between the points $(1, 6)$ and $(0, 4)$. The vertical height of the triangle is the same as the difference in y-values, and the horizontal length of the triangle is the same as the difference in x-values. You can use these to calculate the gradient.

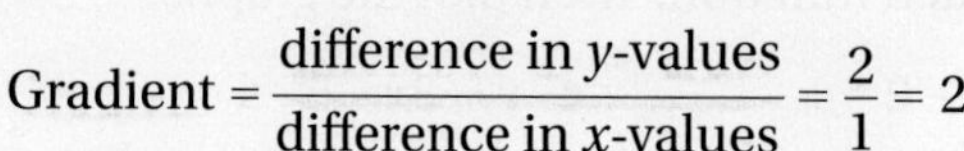

$$\text{Gradient} = \frac{\text{difference in } y\text{-values}}{\text{difference in } x\text{-values}} = \frac{2}{1} = 2$$

A gradient of 2 means that for every one unit the graph moves to the right, it moves up two units. When a graph slopes **up** as you move from left to right, it is said to have a **positive** gradient.

When a graph slopes **down** as you move from left to right, it has a **negative** gradient.

Tip

You can use **any** two points on the line.

WORKED EXAMPLE 2

Calculate the gradient of each line.

a

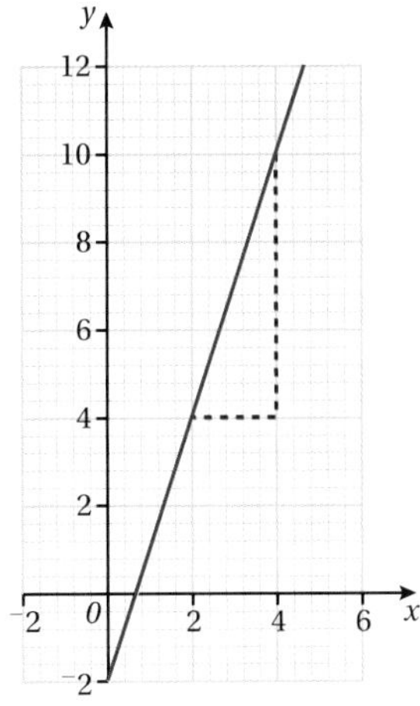

Movement up is positive

b

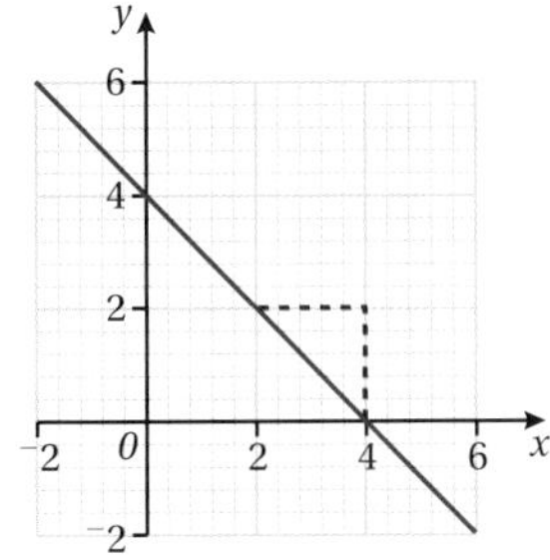

Movement down is negative

Tip

Be careful to check the scale on each axis when calculating the difference in x- and y-values. The vertical rise might be two squares but if each square represents two units, then the rise is four units, not two.

a Gradient $= \dfrac{\text{difference in } y\text{-values}}{\text{difference in } x\text{-values}}$

$= \dfrac{6}{2}$

$= 3$

Using the scale on the y-axis, work out the length of the vertical line of the right-angled triangle: this is the difference in y-values. Using the scale on the x-axis, work out the horizontal length of the triangle: this is the difference in x-values. As you move from the first point to the second, the graph is moving up so the gradient is positive.

b Gradient $= \dfrac{\text{difference in } y\text{-values}}{\text{difference in } x\text{-values}}$

$= \dfrac{-2}{2}$

$= -1$

Use the right-angled triangle to work out the differences as before. As you move from the first point to the second point, the graph is moving down, so the gradient is negative. Show this by making the difference in y-values negative.

Calculating the gradient using two points

You can represent any two points on the line algebraically as (x_1, y_1) and (x_2, y_2).

If you look at the graph, you can see:

difference in y-values $= y_2 - y_1$

difference in x-values $= x_2 - x_1$

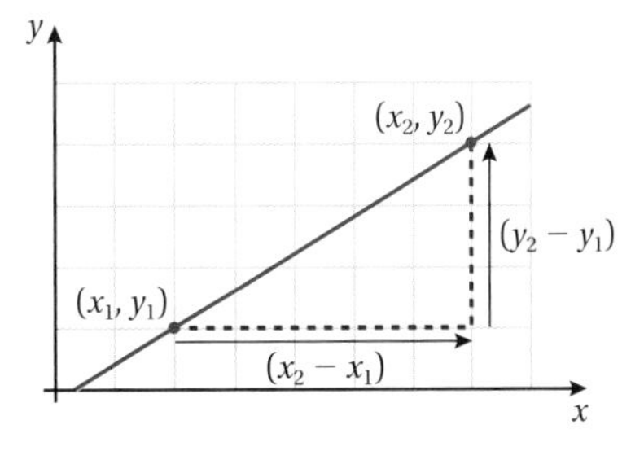

This means the gradient can be calculated as follows:

$$\text{Gradient} = \frac{\text{difference in } y\text{-values}}{\text{difference in } x\text{-values}} = \frac{(y_2 - y_1)}{(x_2 - x_1)}$$

In turn, this means that if you know the coordinates of two points on a straight-line graph, you can use them to find the gradient without drawing the graph.

WORKED EXAMPLE 3

Find the gradient of the straight line that passes through the points (1, 4) and (3, 8).

$$\text{gradient} = \frac{(y_2 - y_1)}{(x_2 - x_1)}$$

Substitute the *x*- and *y*-values from the coordinates into the formula.

$$= \frac{(8 - 4)}{(3 - 1)} = \frac{4}{2}$$

$$= 2$$

The gradient of the line that passes through the points (1, 4) and (3, 8) is 2.

EXERCISE 18B

1 Calculate the gradient of each line.

a

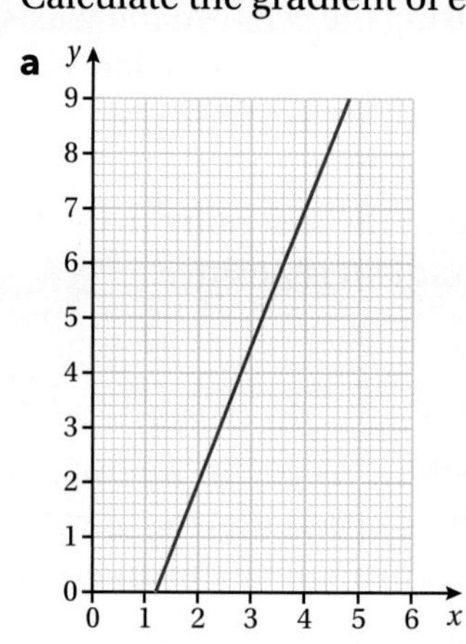

b

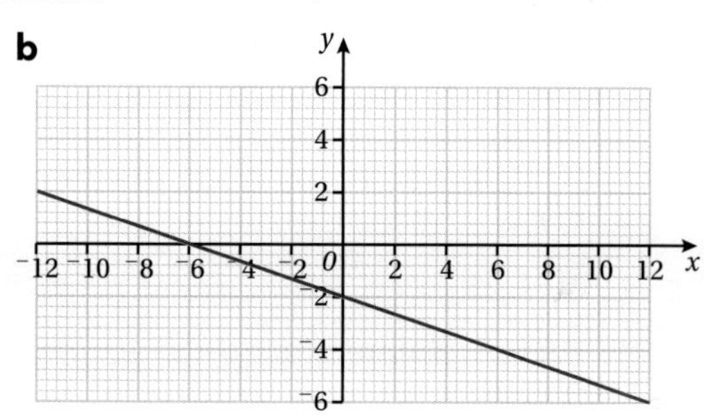

c

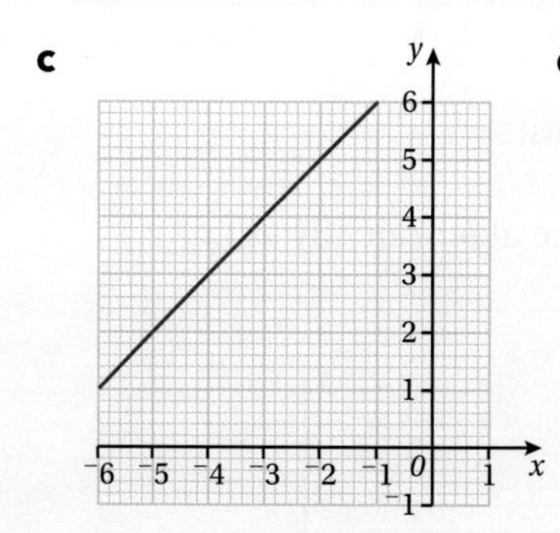

d

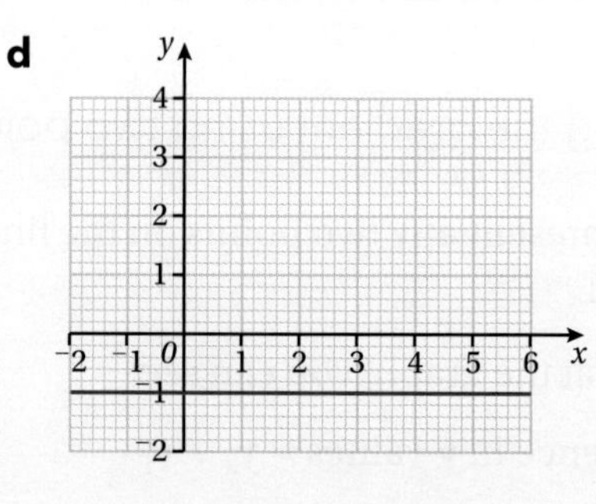

2 Plot the graph of $y = 2x + 3$ and calculate its gradient.

3 Calculate the gradient of each line.

Leave your answer as a fraction in lowest terms if necessary.

a

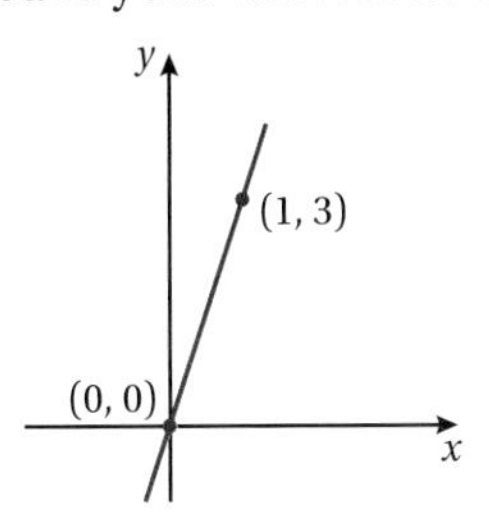

b

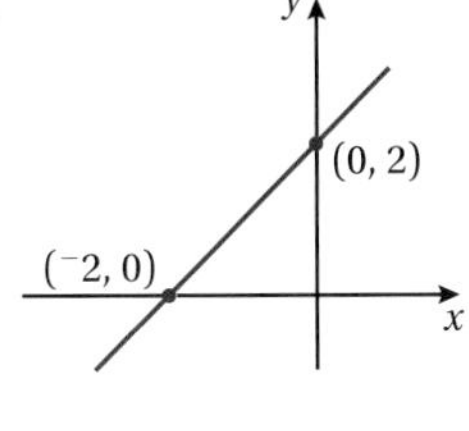

c

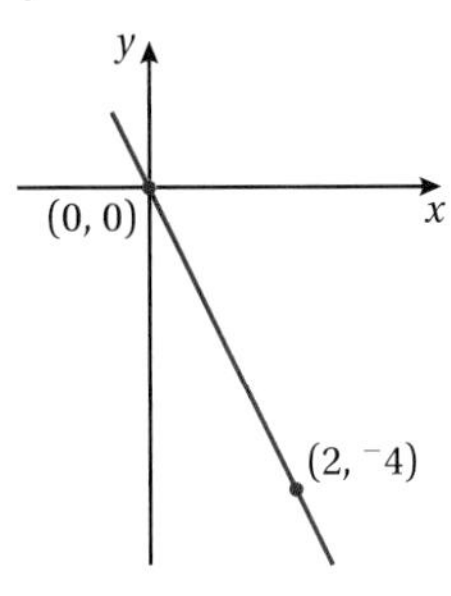

d

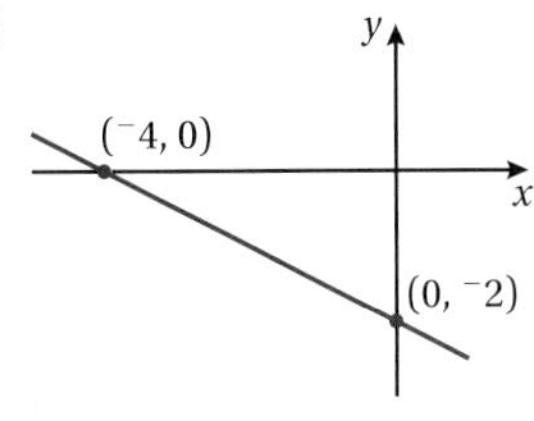

e

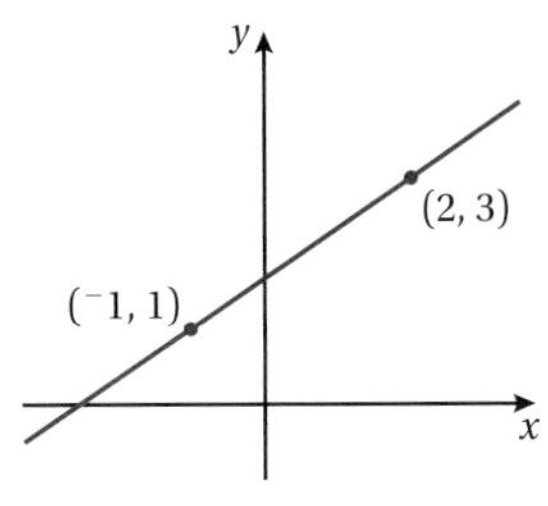

f

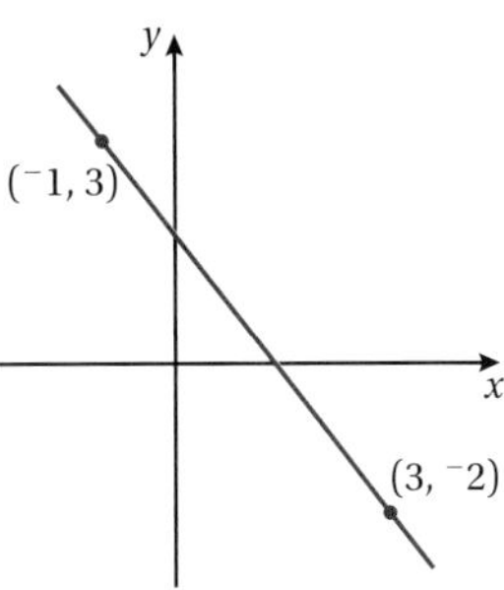

4 Find the gradient of the line that passes through points A and B in each case.

a A(1, 2) and B(3, 8)　　**b** A(0, 6) and B(3, 9)

c A(-1, -4) and B(-3, 2)　　**d** A(3, 5) and B(7, 12)

The x-intercept and y-intercept

The **x-intercept** is where the line crosses the x-axis, and the **y-intercept** is where the graph crosses the y-axis.

Look at the coordinates of these points.

All points on the x-axis have a y-value of 0. All points on the y-axis have an x-value of 0.

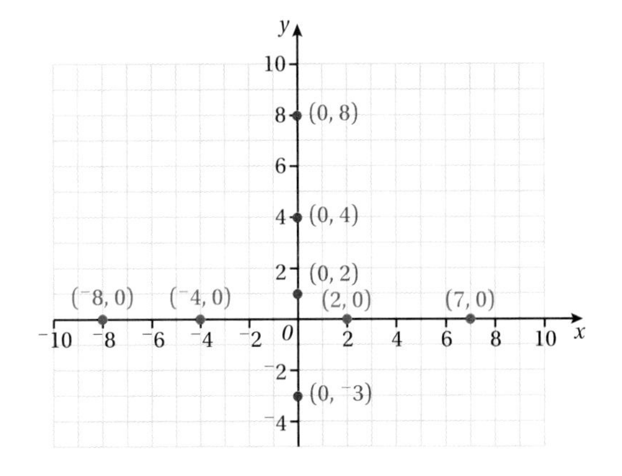

Now consider these two lines.

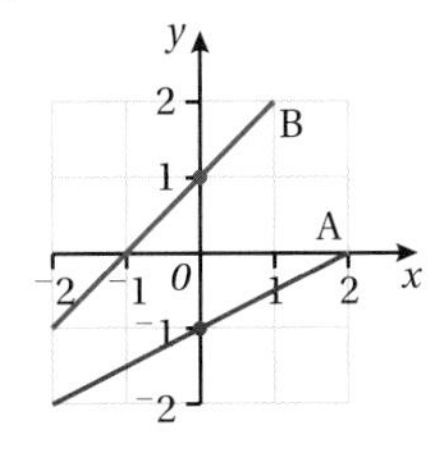

Line A has an x-intercept at (2, 0) and a y-intercept at (0, -1).

Line B has an x-intercept at (-1, 0) and a y-intercept at (0, 1).

This shows that you can identify an x- or y-intercept from its (x, y) coordinates:

- when the x-value is zero, the corresponding y-value is the y-intercept.
- when the y-value is zero, the corresponding x-value is the x-intercept.

These points are important and you will use them to sketch graphs and to find the equation of a graph.

Find answers at: cambridge.org/ukschools/gcsemaths-studentbookanswers

The general form of a linear equation

When a linear equation is in the general form $y = mx + c$, it is known as the **gradient-intercept** form.

Tip

If a linear equation is not written in the gradient-intercept form you can rearrange it so that it is; see Chapter 8 if you need a reminder on how to rearrange equations.

Tip

You learnt about coefficients and constants in Chapter 7.

This form tells us the value of the:

- **gradient** – the **coefficient** m is the gradient of the graph.
- ***y*-intercept** – the **constant** c is the y-intercept.

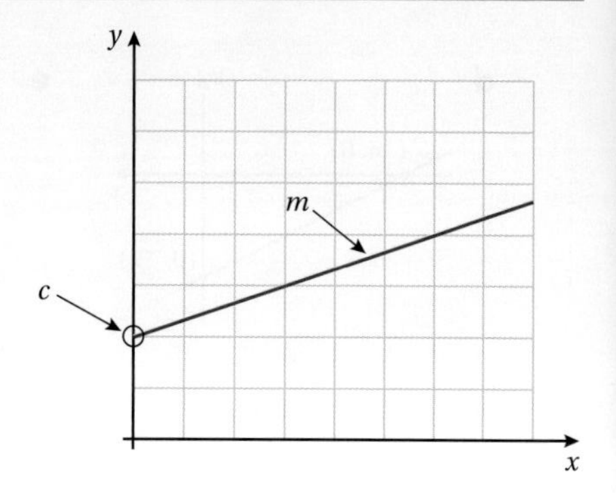

To demonstrate, look at the graph of $y = -x + 2$.

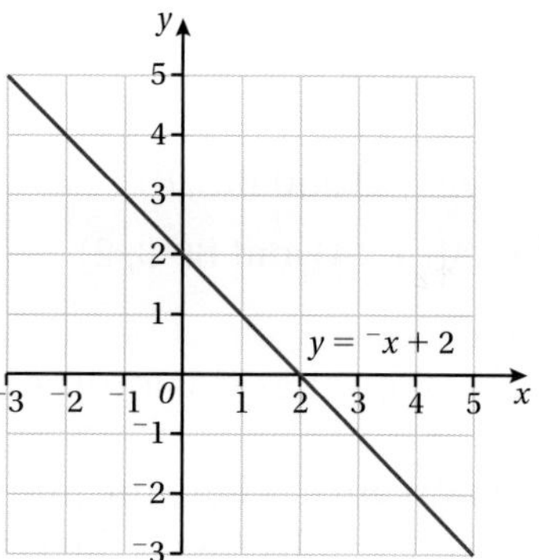

Tip

A gradient of $^-1$ means that for every one unit the graph moves to the right it moves one unit down.

From the graph, the gradient of the line is $\frac{1}{^-1} = {}^-1$.

The y-intercept is 2.

In the equation $y = {}^-x + 2$, the coefficient of x is $^-1$ and the constant is 2.

This means that for any straight-line graph, you can write its equation in the form $y = mx + c$ if you know the gradient and the y-intercept.

WORKED EXAMPLE 4

Write the equation of each line.

a

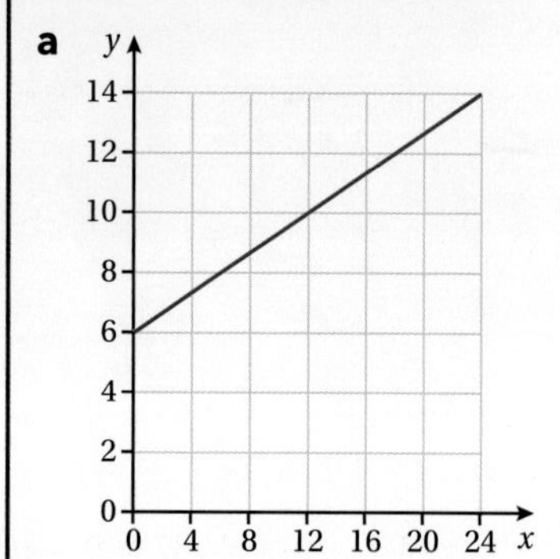

b A straight line with a gradient of $^-2$ and a y-intercept of 3.

Continues on next page ...

a

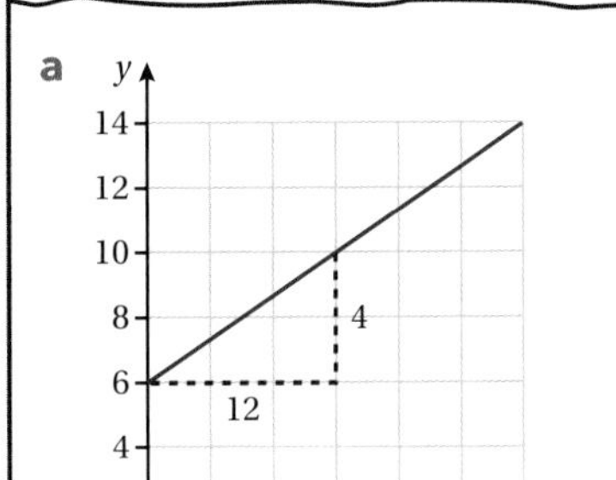

Draw in the right-angled triangle to work out the difference in y-values and the difference in x-values.

$\text{Gradient} = \dfrac{\text{difference in } y\text{-values}}{\text{difference in } x\text{-values}}$

$= \dfrac{4}{12} = \dfrac{1}{3}$

$c = 6$

y-intercept $= (0, 6)$ so $c = 6$

The equation of the line is $y = \dfrac{1}{3}x + 6$

Calculate the gradient (m) and read off the y-intercept (c).

Substitute the values into the gradient-intercept form of the equation: $y = mx + c$

$x = 24$

$y = \dfrac{1}{3}x + 6$

$= 8 + 6$

$= 14$ ✓

You can use some points on the line to check your equation works. Substitute the value of x from the coordinate of the point into the equation and see if the resulting y-value matches the coordinate of the point.

b $y = mx + c$

$m = {}^{-}2, c = 3$

$y = {}^{-}2x + 3$

Substitute the given values of m and c into the gradient-intercept form of the equation.

The gradient-intercept form also enables you to identify the gradient and y-intercept without having to plot the graph.

WORK IT OUT 18.1

Find the gradient and y-intercept of the linear function $y = {}^{-}2x + 4$.

Do not plot a graph.

Think about which direction the line moves across the page.

Which of the answers below is correct?

Option A	Option B	Option C
$y = {}^{-}2x + 4$	$y = {}^{-}2x + 4$	$y = {}^{-}2x + 4$
2 is the coefficient of x.	$^{-}2$ is the coefficient of x.	$^{-}2$ is the coefficient of x.
4 is the constant.	4 is the constant.	4 is the constant.
So, the gradient is 2 and the y-intercept is 4.	So, the gradient is $^{-}2$ and the y-intercept is 4.	So, the gradient is $^{-}2$ and the y-intercept is 4.
The graph goes up to the right.	The graph goes down to the right.	The graph goes down to the left.

Find answers at: cambridge.org/ukschools/gcsemaths-studentbookanswers

Finding the equation of the line from points on the line

If you know the gradient and at least one point on the line, you can find the y-intercept of a straight-line graph and thus write its equation. Substitute the gradient and the x- and y-values of the point into the gradient-intercept form and solve for c.

This also means that you can write the equation if you only know two points on the line. You can use those points to calculate the gradient of the line, and then use the gradient and one of the points to solve for c as before.

Tip

These are algebraic methods.

WORKED EXAMPLE 5

a Find the equation of the line passing through points (3, 11) and (6, 7).

b Find the equation of a line that has the same gradient as the line $y = \frac{1}{2}x - 3$ and passes through the point (⁻1, 2).

a Gradient $= \frac{y_2 - y_1}{x_2 - x_1}$

$= \frac{7 - 11}{6 - 3} = \frac{-4}{3}$

So, $y = \frac{-4}{3}x + c$. — Substitute the calculated gradient into the gradient-intercept form.

(6, 7) is a point on the line. — Pick one of the points and substitute the values of x and y into the equation. Solve for c.

So, $7 = \frac{-4}{3} \times 6 + c$

$7 = {}^{-}8 + c$

$15 = c$

The equation is $y = \frac{-4}{3}x + 15$. — Write the equation of the line.

b $y = mx + c$ — You are told that the line has the same gradient as the line $y = \frac{1}{2}x - 3$, so you know that the equation you are trying to find will contain $m = \frac{1}{2}$. Substitute the values you know into the equation to find the value of c.

$y = \frac{1}{2}x + c$

$2 = (\frac{1}{2} \times {}^{-}1) + c$

$c = \frac{5}{2}$

The equation is $y = \frac{1}{2}x + \frac{5}{2}$. — Write the equation of the line.

Key vocabulary

sketch: a basic graph showing the direction, gradient and y-intercept; it is not drawn by plotting a table of values.

Using the gradient-intercept form to sketch a graph

You can use the gradient-intercept form of a linear equation to **sketch** its graph without plotting values from a table.

WORKED EXAMPLE 6

Sketch the graph of $y = 4x + 4$.

$y = 4x + 4$

y-intercept $= 4$

$m = 4 = \frac{4}{1}$

From the equation you know that the gradient, m, is 4 and the y-intercept is 4.

You know that the gradient $= \frac{\text{vertical rise}}{\text{horizontal run}}$ so you can use m to work out the rise and run.

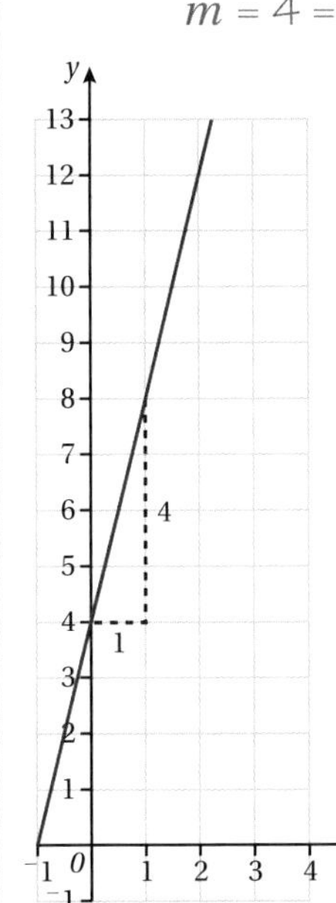

Draw a pair of axes and plot the y-intercept at (0, 4). Then use the rise of 4 and run of 1 to plot another point at (1, 8). Join the two points and draw a line that passes through them.

EXERCISE 18C

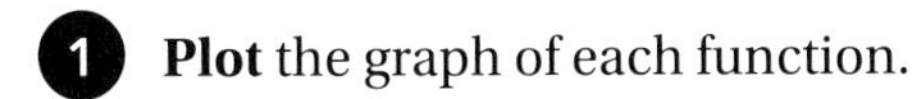

1 **Plot** the graph of each function.

Write a description of each one, giving the y-intercept and gradient, and stating if the gradient is positive or negative.

a $y = 3x - 2$

b $y = {}^{-}2x + 3$

c $y = \frac{1}{2}x - 1$

d $y = x - 1$

2 Rearrange each equation so it is in the gradient-intercept form ($y = mx + c$).

Sketch each of the lines.

a $3x - 2y = 6$

b $6x + 2y + 10 = 0$

c $3y - 6x + 12 = 0$

d $2y - x + 18 = 0$

e $6y - 2x + 18 = 0$

f $2x - 3y + 12 = 0$

Find answers at: cambridge.org/ukschools/gcsemaths-studentbookanswers

3 Match each graph to the correct linear equation.

a $y = x + 1$ **b** $y = 3 - x$ **c** $y = 9 - 3x$ **d** $y = x + 4$ **e** $y = {}^-2x + 20$

A

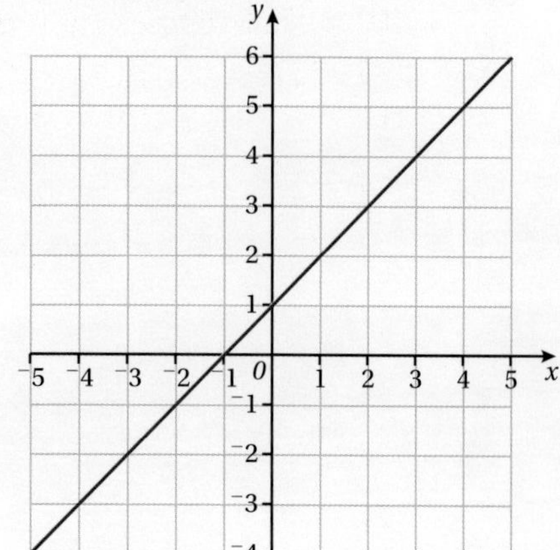

B

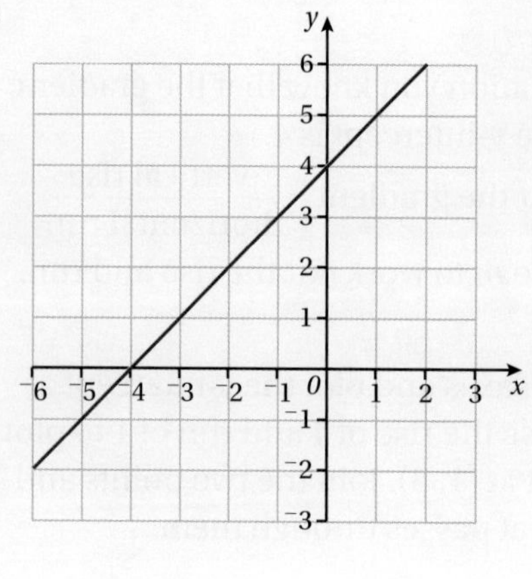

C

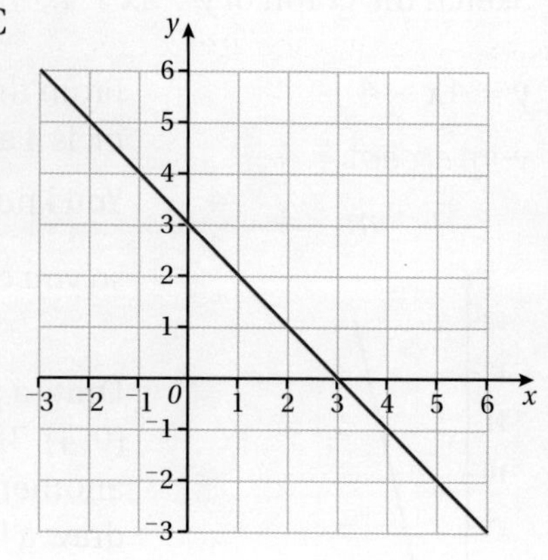

D

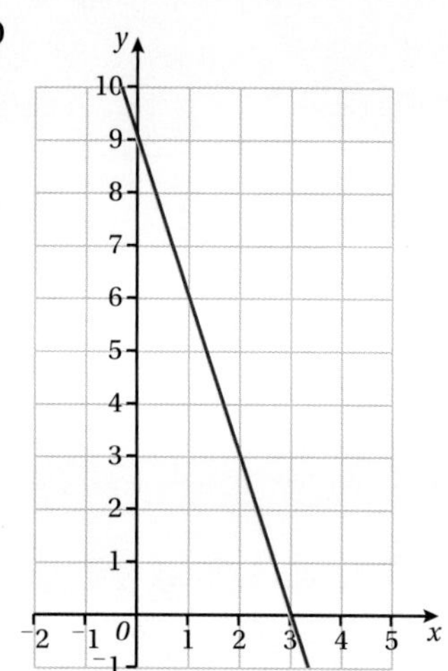

E

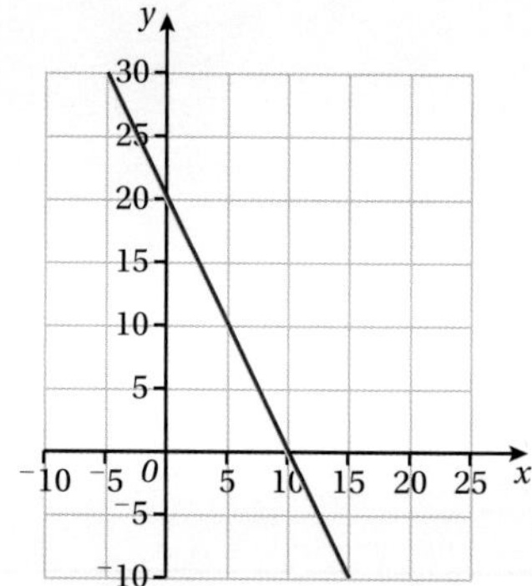

4 Sketch the graph of each line by calculating the coordinates of the x- and y-intercepts.

Write down the gradient of each graph.

a $2x + y = 4$ **b** $3x + 4y = 12$ **c** $x + 2y = 1$ **d** $3x + y = 2$

e $x - y = 4$ **f** $x - y = 1$ **g** $4x - 2y = 8$ **h** $3x - 4y = 12$

5 For each equation, find c if the given point is on the line.

a $y = 3x + c$ $(1, 5)$ **b** $y = 6x + c$ $(1, 2)$

c $y = {}^-2x + c$ $({}^-3, {}^-3)$ **d** $y = \frac{3}{4}x + c$ $(4, -5)$

6 Find the equation of the line passing through each pair of points.

a $(0, 0)$ and $(6, {}^-2)$ **b** $(0, 0)$ and $({}^-2, {}^-3)$

c $({}^-2, {}^-5)$ and $({}^-4, {}^-1)$ **d** $({}^-2, 9)$ and $(3, {}^-1)$

Tip

Find the equation of the line before you try to find the coordinates of points on it.

7 **a** A line passes through the point $(2, 4)$ and has gradient 2. Find the y-coordinate of the point on the line when $x = 3$.

b A line passes through the point $(4, 8)$ and has gradient $\frac{1}{2}$. Find the y-coordinate of the point on the line when $x = 8$.

c A line passes through the point $({}^-1, 6)$ and has gradient ${}^-1$. Find the y-coordinate of the point on the line when $x = 4$.

Section 3: Parallel lines, perpendicular lines and tangents

Parallel lines

Tip

You learnt about parallel and perpendicular lines in Chapter 5.

These three graphs are parallel to each other.

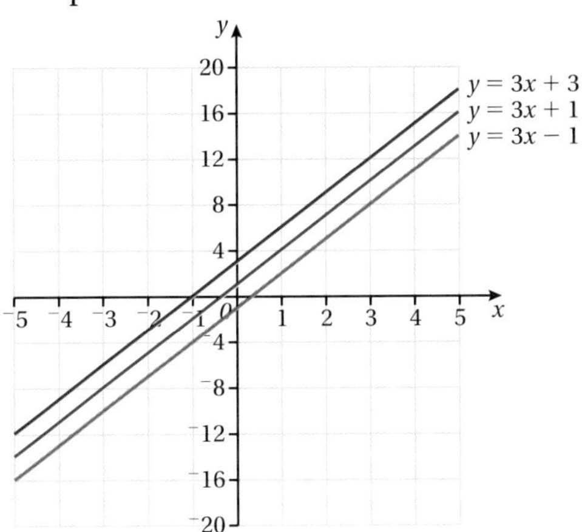

If you look at the equations of each line you can see that they all have the same gradient.

Lines with equal gradients are parallel to each other.

These graphs are members of a family of parallel lines with a gradient of positive 3. Each line in the family can be defined by the equation $y = 3x + c$.

If equations written in the form $y = mx + c$ have identical values for m, they have the same gradient and therefore they are parallel lines.

Problem-solving framework

Lucia is helping her classmates by checking their work.

Who has correctly grouped together two sets of parallel lines and who hasn't?

Tip

In mathematics, one way to prove a statement is true is to follow a series of statements that end in a valid conclusion. You can use this method to prove that lines are parallel.

Carlos	**Jonathan**	**Monika**
$y = 4 - 2x$	$y = \frac{1}{3}x + 1$	$y = x - 1$
$y + 2x = 5$	$3y + x = 1$	$y + x = 1$
$y = -2x + 1$	$y = 3x + 1$	$y = 1 - x$
$y = 3x + 1$	$2y = x + 1$	$x = y + 1$
$y - 3x = -1$	$2y - x = 3$	$x - y = 1$
$y = 2 + 3x$	$y = \frac{1}{2}x - 1$	$x = 1 - y$

Steps for approaching a problem-solving question	What you would do for this example
Step 1: Draw a diagram?	A diagram might take too long, especially if you are working with many different lines.
Step 2: Identify what you have to do.	Determine who has and hasn't correctly grouped parallel lines.

Continues on next page ...

Step		In $y = mx + c$ form	Gradient
Step 3: Test the problem with what you know.	If lines are parallel they will have an identical value for the gradient. The gradient is the coefficient of x in the equation (the value of m).		
Step 4: What maths can you do?	Rearrange the equations so they are all in the form $y = mx + c$ to compare the gradients (m). If there are identical values for m then Lucia's classmates have correctly shown/proved the lines are parallel.		
Step 5: Set your working and solutions out clearly. Check your working and make sure your answer is reasonable.	**Carlos**	In $y = mx + c$ form	Gradient
	$y = 4 - 2x$		$^{-}2$
	$y + 2x = 5$	$y = 5 - 2x$	$^{-}2$
	$y = {}^{-}2x + 1$		$^{-}2$
	Gradient of each line is $^{-}2$.		
	$y = 3x + 1$		3
	$y - 3x = {}^{-}1$	$y = 3x - 1$	3
	$y = 2 + 3x$		3
	Gradient of each line is 3.		
	Carlos has correctly grouped parallel lines as each group has the same gradient.		
	Jonathan	In $y = mx + c$ form	Gradient
	$y = \frac{1}{3}x + 1$		$\frac{1}{3}$
	$3y + x = 1$	$y = \frac{1}{3} - \frac{1}{3}x$	$^{-}\left(\frac{1}{3}\right)$
	$y = 3x + 1$		3
	The lines do not all have the same gradient.		
			Gradient
	$2y = x + 1$	$y = \frac{1}{2}x + \frac{1}{2}$	$\frac{1}{2}$
	$2y - x = 3$	$y = \frac{3}{2} + \frac{1}{2}x$	$\frac{1}{2}$
	$y = \frac{1}{2}x - 1$		$\frac{1}{2}$
	The lines do have the same gradient of $\frac{1}{2}$.		
	Jonathan correctly grouped one group but not the other.		
	Monika	In $y = mx + c$ form	Gradient
	$y = x - 1$		1
	$y + x = 1$	$y = 1 - x$	$^{-}1$
	$y = 1 - x$		$^{-}1$
	$x = y + 1$	$y = x - 1$	1
	$x - y = 1$	$y = x - 1$	1
	$x = 1 - y$	$y = 1 - x$	$^{-}1$
	Monika did not correctly group either group.		
	Checked answers are accurate ✓		
	They are reasonable because it is easy to see where her classmates could have gone wrong, mistakenly thinking that a positive and negative gradient of the same number is the same gradient.		
Step 6: Check that you have answered the question.	Yes. You state who has and hasn't grouped the lines correctly and have proved your answer.		

EXERCISE 18D

1 **a** If $y = (2a - 3)x + 1$ is parallel to $y = 3x - 4$, find the value of a.

b If $y = (3a + 2)x - 1$ is parallel to $y = ax - 4$, find the value of a.

2 Find the equation of the blue line.

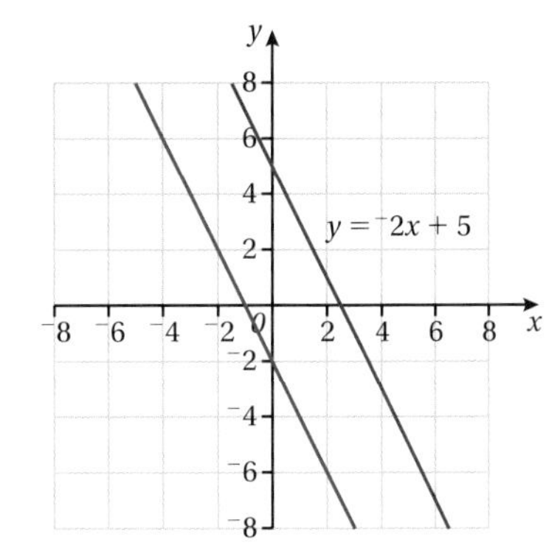

3 The vertices (corners) of a quadrilateral have coordinates A(1, 6), B(3, 14), C(15, 16) and D(13, 8).

a Find the gradient of the line AB.

b Find the equation of the line AB.

c Prove that ABCD is a parallelogram.

Tip

You learnt about the properties of quadrilaterals in Chapter 5.

4 Investigate lines that are parallel to the axes.

How are the equations for these graphs different to those for slanted graphs? Why?

Perpendicular lines

Tip

You learnt about perpendicular lines in Chapter 5; you were reminded in Chapter 1 that a 'product' is the result of a multiplication.

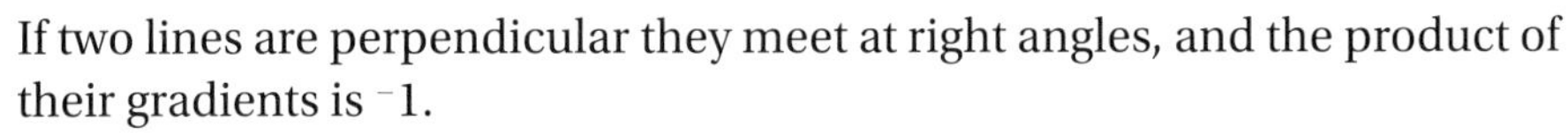

If two lines are perpendicular they meet at right angles, and the product of their gradients is $^-1$.

$m_1 \times m_2 = {}^-1$, where m is the gradient of the line.

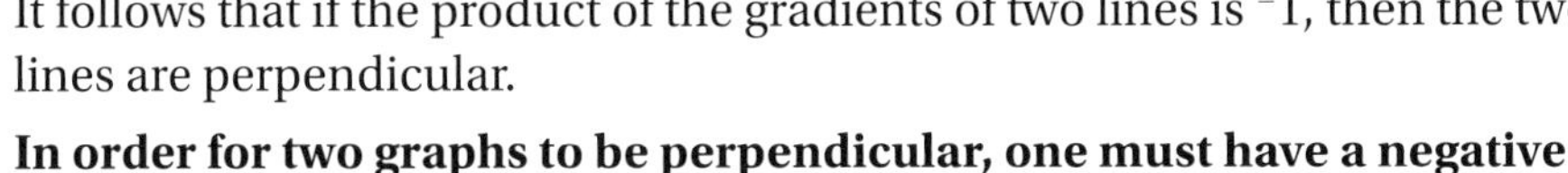

It follows that if the product of the gradients of two lines is $^-1$, then the two lines are perpendicular.

In order for two graphs to be perpendicular, one must have a negative gradient and the other must have a positive gradient.

The graphs of $y = \frac{^-1}{3}x + 2$ and $y = 3x - 4$ are perpendicular to each other.

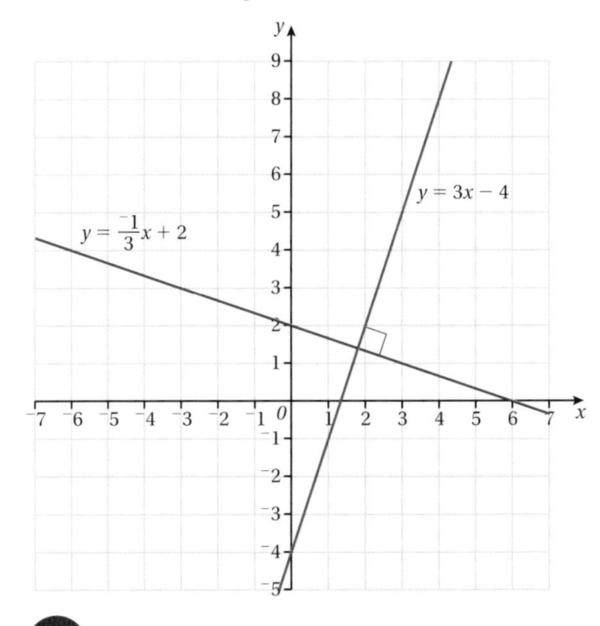

Graph $y = \frac{^-1}{3}x + 2$ has negative gradient $m = \frac{^-1}{3}$.

Graph $y = 3x - 4$ has positive gradient $m = 3$.

The product of the gradients $\frac{^-1}{3} \times 3 = {}^-1$.

Tip

You saw in Chapter 1 that when you multiply any number by its inverse the product is always 1.

The inverse of any integer, x, is $\frac{1}{x}$.

So for any value, $x \times \frac{1}{x} = \frac{x}{x} = 1$.

When one value of x is negative, the product will also be negative.

Find answers at: cambridge.org/ukschools/gcsemaths-studentbookanswers

WORKED EXAMPLE 7

What is the gradient of a linear graph that is perpendicular to $y = \frac{-4}{5}x - 2$?

$m = \frac{-4}{5}$

$m_1 \times m_2 = -1$

$x \times \frac{1}{x} = \frac{x}{x} = 1$

The gradient of the first line is the value of m in $y = mx + c$. We know that the product of multiplying this gradient by the unknown gradient will be -1 for any line perpendicular to it. We also know that multiplying a number by its reciprocal gives a product of 1.

$\frac{-4}{5} \times m_2 = -1$

$\frac{-4}{5} \times \frac{5}{4} = \frac{-20}{20} = -1.$

The reciprocal of $\frac{-4}{5}$ is $-\frac{5}{4}$

But if both sides are negative, we will get a result of positive 1. Therefore the gradient of the other line must be positive.

So the gradient of a graph perpendicular to the line given is $\frac{5}{4}$.

You can write the equation of a perpendicular line if you know its gradient and a point on the line.

WORKED EXAMPLE 8

Given that $y = \frac{2}{3}x + 2$, determine the equation of the straight line that is:

a perpendicular to this line and which passes through the origin.

b perpendicular to this line and which passes though the point $(-3, 1)$.

a $y = mx + c$

$m = \frac{-3}{2}$

$c = 0$

The equation of the line is $y = \frac{-3}{2}x$.

Perpendicular gradients have a product of -1, so the gradient of the line is the negative reciprocal of $\frac{2}{3}$. If the line passes through the origin, (0, 0) then the y-intercept is at $y = 0$.

b Using $m = \frac{-3}{2}$ from part **a**.

$y = \frac{-3}{2}x + c$

$x = -3$ and $y = 1$

$1 = \frac{-3}{2}(-3) + c$

$1 = \frac{9}{2} + c$

$c = -3\frac{1}{2}$

$y = \frac{-3}{2}x - 3\frac{1}{2}$.

As the line is also perpendicular to the given equation, you know that its gradient will be the same as your answer to part **a**. Substitute the gradient, and the values of x and y from the given point, into the equation and solve for c.

When an equation has more than one fraction, it can be easier to read and neater to express, if the fractions were eliminated. So, you could also express this as $2y = -3x - 7$ by multiplying both sides by 2.

EXERCISE 18E

1 Which of the lines $y = 4x$, $y = 3x + 2$ and $y = x$ is perpendicular to the line $4y + x = {}^{-}2$?

2 Show that the line through the points A(6, 0) and B(0, 12) is:

a perpendicular to the line through P(8, 10) and Q(4, 8).

b perpendicular to the line through M($^{-}4$, $^{-}8$) and N($^{-}1$, $^{-}6\frac{1}{2}$).

3 Given the graph $2x + 5y = 20$:

a Write the equation of a parallel graph that crosses the y-axis at $^{-}2$.

b Give the equation of a graph that intersects the first graph at the x-axis and which passes through point (6, $^{-}4$).

4 Write the equations of these two lines from information in the graph and prove that the lines are perpendicular.

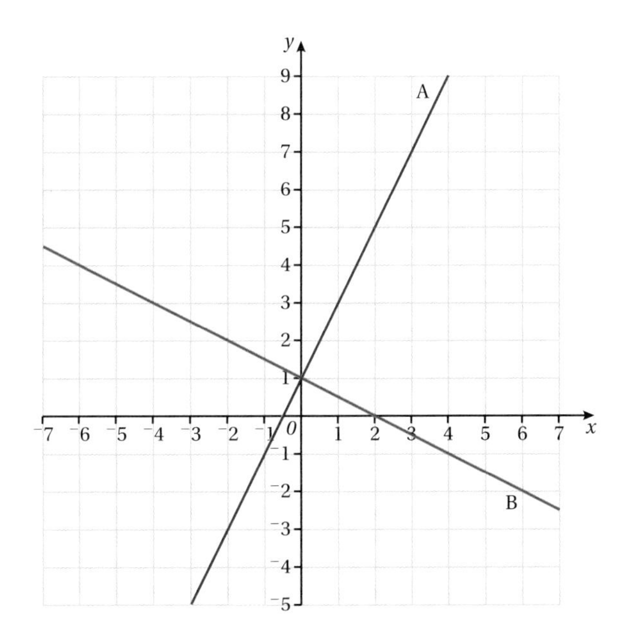

5 Sketch each of the following graphs.

a $y = 3x + 2$ **b** $y = {}^{-}2x - 1$ **c** $2y = x + 8$

d $x - y = {}^{-}3$ **e** $y + 4 = x$ **f** $3x + 4y = 12$

6 Is the triangle formed by joining P($^{-}1$, 3), Q(5, 1) and R($^{-}2$, 0) right-angled?

Finding the equation of a tangent to a circle

You saw in Chapter 5 that a **tangent** to a circle is a straight line that touches the circumference at a single point but doesn't cross it.

The radius meets the tangent at a right angle where it touches the circumference. So, the radius of a circle is perpendicular to the tangent at the point of tangency.

Find answers at: cambridge.org/ukschools/gcsemaths-studentbookanswers

You know that the product of the gradients of perpendicular lines is $^-1$, and you know that the tangent is perpendicular to the radius. So,
$m_{\text{tangent}} \times m_{\text{radius}} = {}^-1$.

If you know the gradient of the radius of a circle at a given point, you can find the equation of the tangent to the circle at that point.

WORKED EXAMPLE 9

Find the equation of the tangent to a circle with centre (0, 0) at the point (4, 3).

Tip

Circles can be plotted using an equation. You will learn about the equation of a circle in Chapter 19.

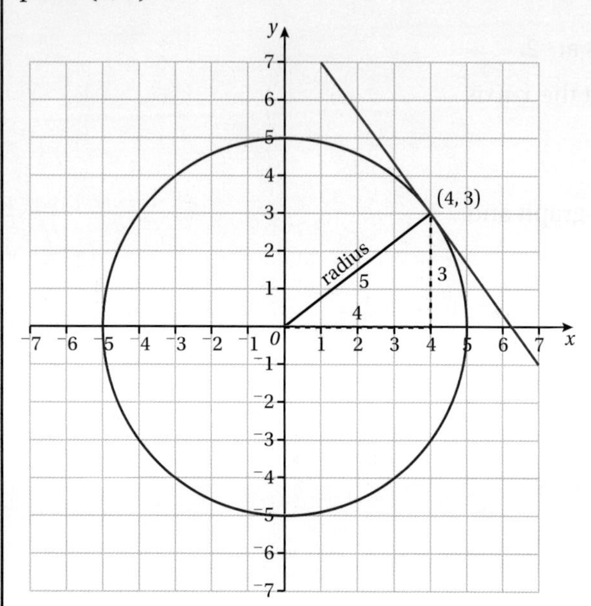

$m_{\text{radius}} = \frac{3}{4}$ — Calculate the gradient of the radius from the graph.

$m_{\text{tangent}} \times \frac{3}{4} = {}^-1$ — Calculate the gradient of the tangent: you know the radius is perpendicular to the tangent so the product of their gradients is $^-1$.

$m_{\text{tangent}} = \frac{^-4}{3}$ — You know that the product of multiplying a fraction by its reciprocal is 1, so the gradient of the tangent is the negative of the reciprocal of the radius' gradient.

$y = 3$ when $x = 4$

$3 = 4 \times \frac{^-4}{3} + c$ — Calculate the y-intercept algebraically by substituting gradient, and the values of x and y from the given point, into the general form of the equation, $y = mx + c$.

$c = 3 + \frac{16}{3} = \frac{25}{3}$

$y = \frac{^-4}{3}x + \frac{25}{3}$ — Multiply both sides by 3 to remove the fractions.

The equation of the tangent to the circle at (4, 3) is $3y = {}^-4x + 25$

EXERCISE 18F

1 This diagram shows a circle with centre (0, 0) and radius 5.

The tangent touches the circle at (3, $^{-}4$).

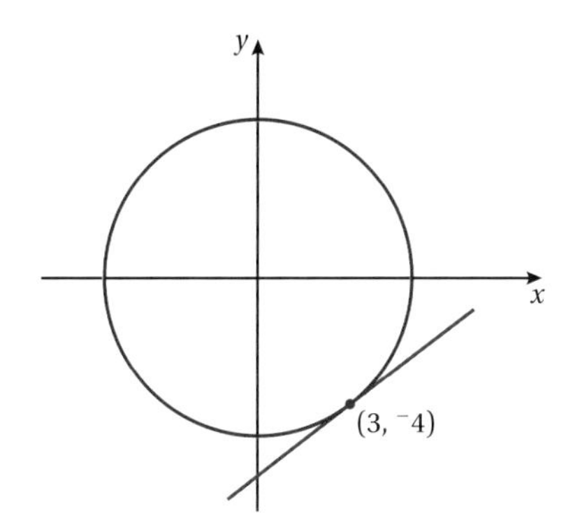

a Copy the diagram. Draw in a radius from (0, 0) to the point (3, $^{-}4$).

b What is the gradient of the line of this radius?

c What is the gradient of the tangent to the circle at point (3, $^{-}4$)?

d Find the equation of this tangent.

Tip

The easiest way to draw this circle is to place the point of your pair of compasses at (0, 0), and the pencil at the point (5, 0) and then draw the circle; the circumference of this circle should pass through the point (3, $^{-}4$).

2 Determine the equation of the tangent at the point given, to a circle with centre (0, 0).

a radius = $\sqrt{5}$ point of tangency = (2, 1)

b radius = $\sqrt{80}$ point of tangency = ($^{-}4$, 8)

c radius = $\sqrt{90}$ point of tangency = (3, 9)

3 Determine the equations of the tangents at the points (0, 9) and (9, 0), to a circle with centre at the origin and radius 9.

Section 4: Working with straight-line graphs

You need to be able to interpret straight-line graphs.

Interpreting graphs means you can:

- determine the equation of the graph.
- calculate the gradient of a line using given information.
- use graphs to model and solve problems, including solving simultaneous equations. This means that you are able to answer questions by reading the correct information off a graph and know what it means.

Tip

You learnt how to solve simultaneous equations in Chapter 8. Remember that the point of intersection between two graphs is the simultaneous solution to the equation of each line.

Tip

You will learn more about interpreting graphs in Chapter 36 in the context of direct and indirect proportion.

WORKED EXAMPLE 10

a Sketch the graphs of $x + 3y = 6$ and $y = 2x - 5$.

b What are the coordinates of the point of intersection of the two graphs?

c Show by substitution that these values are the simultaneous solution to the equations $x + 3y = 6$ and $y = 2x - 5$.

a $x + 3y = 6$

$y = 2 - \frac{1}{3}x$

Let $x = 0$

$y = 2 - \left(\frac{1}{3} \times 0\right)$

$y = 2$

y-intercept $(0, 2)$

Let $y = 0$

$0 = 2 - \frac{1}{3}x$

$\frac{1}{3}x = 2$

$x = 6$

x-intercept $(6, 0)$

Sketch the graphs using your preferred method.

Here, the first graph has been sketched by plotting the intercepts and drawing a straight line through them. Rearrange the first equation into the form $y = mx + c$. For both equations:
substitute $x = 0$ into the equation to find the y-intercept; then
substitute $y = 0$ into the equation to find the x-intercept.

$y = 2x - 5$

y-intercept $= ^-5$

gradient $= 2$

Here, the graph has been sketched using the gradient and the y-intercept.

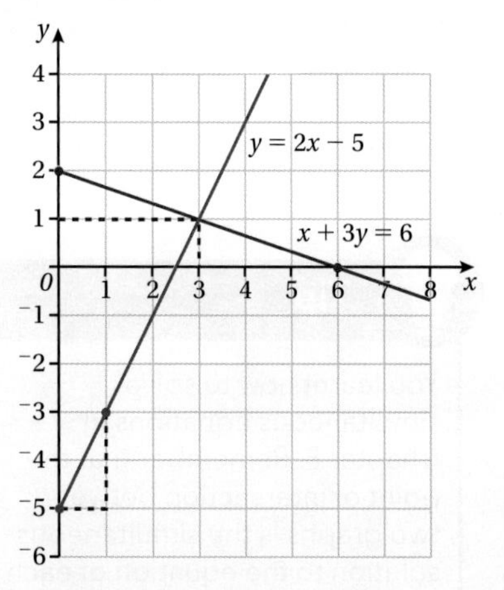

b The point of intersection is $(3, 1)$.

Read this from the graph; the point of intersection is the point at which the two graphs cross each other.

c Let $x = 3$ and $y = 1$

Substitute in $x + 3y = 6$:

$3 + 3(1) = 6$.

Substitute in $y = 2x - 5$:

$1 = 2(3) - 5$

$1 = 6 - 5$

$1 = 1$

Use the substitution method to solve the simultaneous equation. See Chapter 8 for a reminder if you need to.

The solutions work for both equations.

EXERCISE 18G

1 Each table of values has been generated from a function.

i

x	$^-2$	$^-1$	0	1	2	3
y	$^-4$	$^-3$	$^-2$	$^-1$	0	1

ii

x	-2	$^-1$	0	1	2	3
y	4	3	2	1	0	$^-1$

a Which of these functions would produce each set of values?
$y = {}^-x + 2$, $y = 2x - 1$, $y = {}^-2x + 4$, $y = x - 2$

b Plot the graphs of each function.

c Onto the same grids, draw a line parallel to each graph which crosses the y-axis at (0, 3) and write its equation.

d Determine the gradient of lines perpendicular to each pair of parallel lines.

2 Write down the equation of the line that has:

a gradient 3 and y-intercept 5.

b gradient $^-1$ and y-intercept 4.

c gradient $\frac{3}{4}$ and y-intercept $^-2$.

d gradient $^-\left(\frac{1}{7}\right)$ and y-intercept 0.

3 Find and write in gradient-intercept form ($y = mx + c$) equations which satisfy the following statements.

a A linear equation that does not pass through the first quadrant.

b Two lines whose gradients differ by 2.

c An equation of a straight line that passes through (2, 3) and has a gradient of 3.

d An equation of a vertical line and a horizontal line.

4 The equations of two lines are $y = x + 3$ and $y = mx - 2$.

a For what values of m will the two lines not intersect?

b For what values of m will the two lines intersect?

c Given that the lines intersect at (5, 8), find m.

5 Calculate the gradient of the line shown in the diagram and write down the equation of the line.

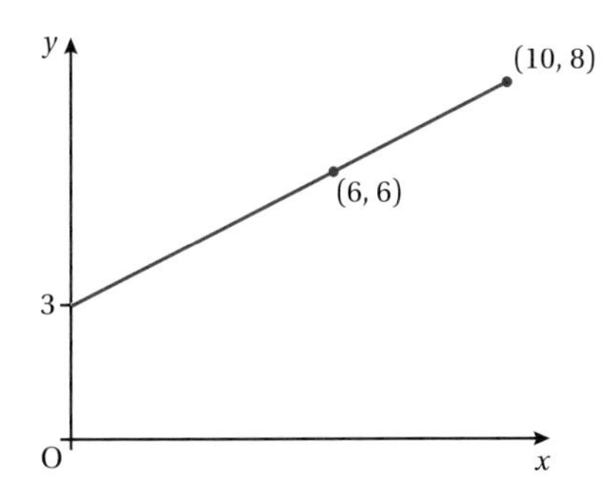

6 Sketch the graph of each equation.

a $y = 2x - 3$ **b** $y = 3x - 2$ **c** $x + 2y = 4$

d $2x - y = 1$ **e** $y - 3x = 6$

7 Find the equation of the line that passes through each pair of points.

a $(5, 6)$ and $(^{-}4, 10)$ **b** $(3, 4)$ and $(^{-}2, 8)$ **c** $(^{-}2, 6)$ and $(1, 10)$

8 **a** Find the equation of the line with gradient $^{-}4$ that passes through the point $(0, ^{-}6)$.

b Find the equation of the line with gradient $^{-}4$ that passes through the point $(3, 8)$.

c Find the equation of the line that passes through the points $(^{-}4, 8)$ and $(^{-}6, ^{-}2)$.

9 The line passing through the points $(^{-}1, 6)$ and $(4, b)$ has gradient $^{-}2$. Find the value of b.

10 **a** Is the gradient of this straight line 2 or $^{-}2$? Write down the equation of the line.

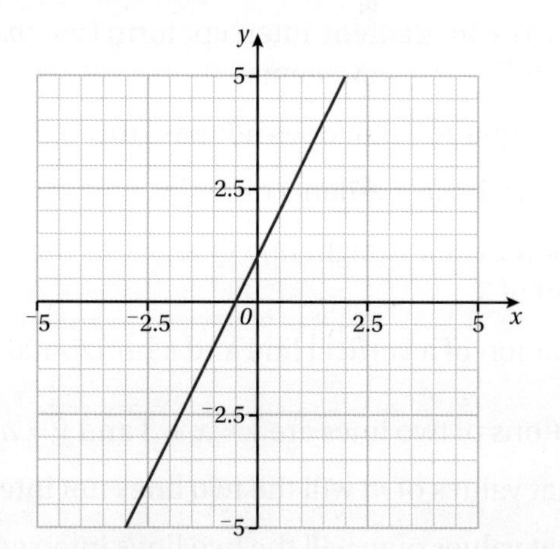

b Write the equation of this line.

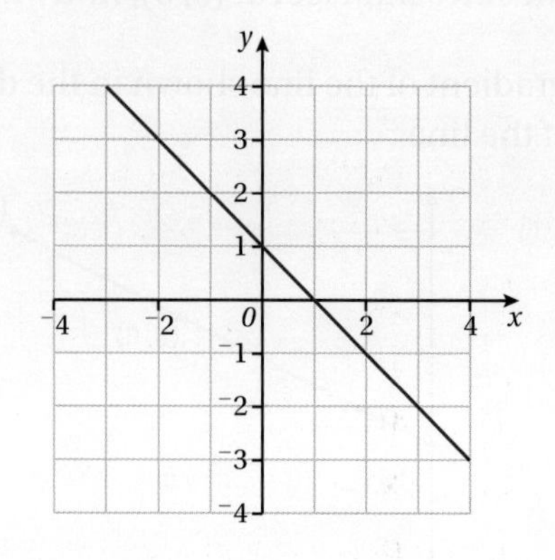

11 Find the equations of the four straight lines that would intersect to make this rhombus.

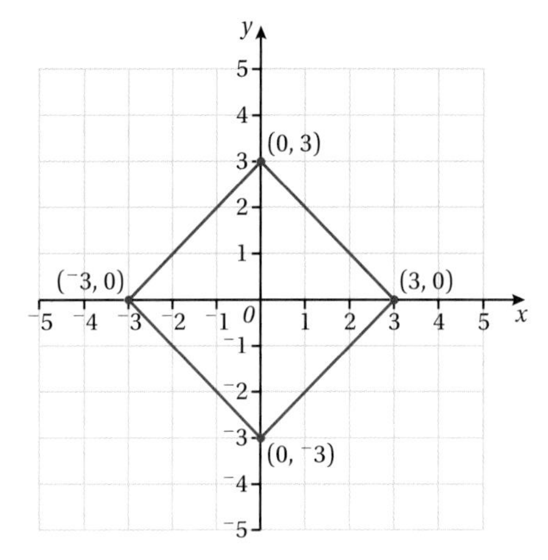

12 Show that the points A(1, 1), B(3, 11) and C(⁻2, ⁻14) all lie on the same line (are collinear) and find the equation of this line.

13 Given the circle with centre (0, 0) and radius 5, determine the equation of the tangent that touches the circle at $(^-1, ^-2\sqrt{6})$.

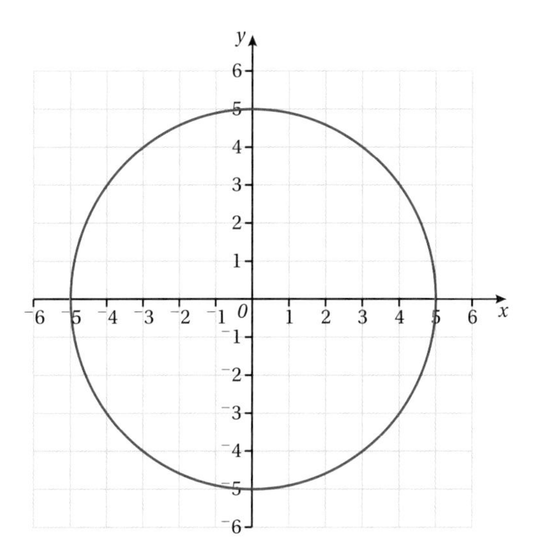

Checklist of learning and understanding

Plotting linear functions

- The graph of a linear function is called a straight-line graph.
- You can use the equation of a line to generate a table of x- and y-values.
- Choose any three (x, y) values, plot them and join the points to draw the graph.

Features of straight-line graphs

- When a linear equation is in the general form $y = mx + c$, it is known as the **gradient-intercept** form, where m is the gradient and c is the point where the line cuts the y-axis.
- You can find the equation of a straight line if you have one of the following:
  - the gradient and the y-intercept.
  - two points on the line.
  - one point and the gradient.
- You can sketch graphs using the gradient and y-intercept or using the x- and y-intercepts.

Find answers at: cambridge.org/ukschools/gcsemaths-studentbookanswers

Parallel and perpendicular lines, and tangents to a circle

- Parallel lines have the same gradient so the value of m is equal when their equation is written in the form $y = mx + c$.
- The product of the gradients of two perpendicular lines is $^-1$:
  $m_1 \times m_2 = {}^-1$
- A tangent to a circle touches it at one point only. The gradient of a tangent is $m_{\text{tangent}} \times m_{\text{radius}} = {}^-1$. You can use the gradient of the tangent to find the equation of the tangent.

For additional questions on the topics in this chapter, visit GCSE Mathematics Online.

Chapter review

1 Draw these lines on the same grid.

a $y = x + 1$ **b** $y = 2x + 5$ **c** $y + 2 = 4x$

2 **i** Write the equation of each line.

a

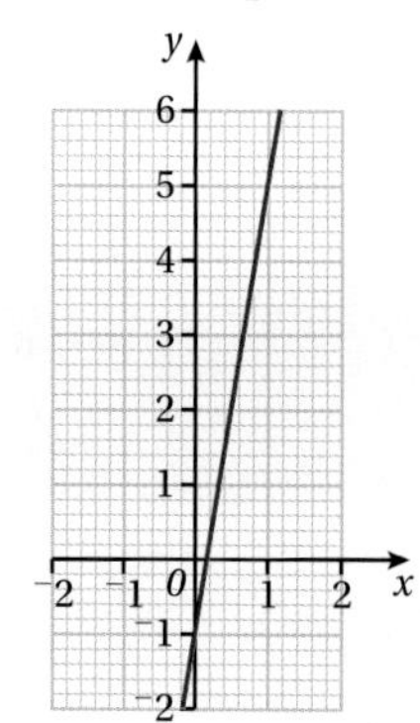

b

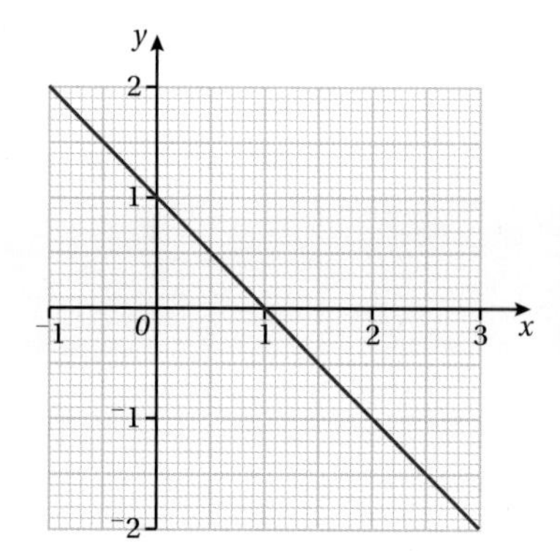

c

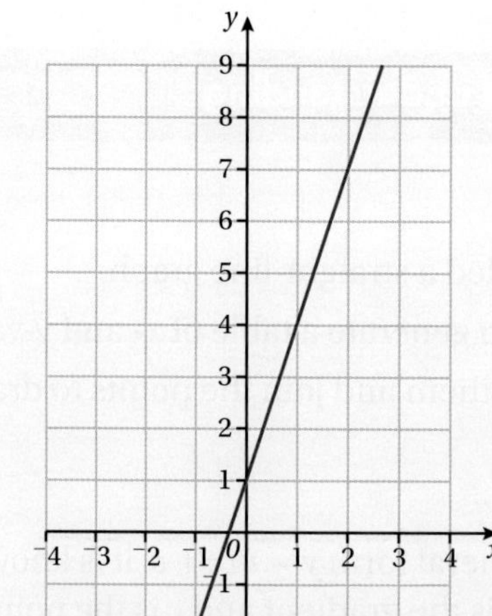

d

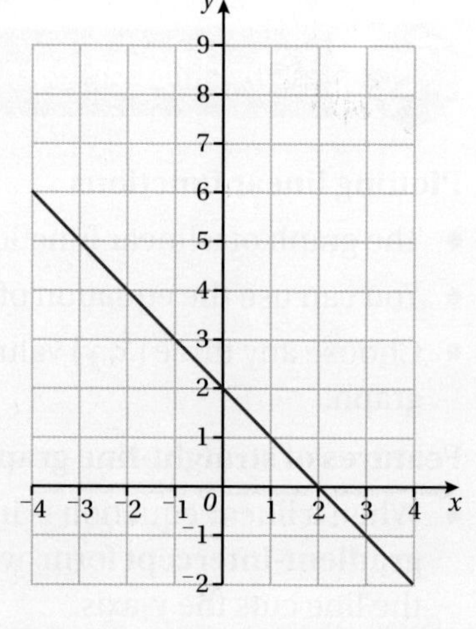

ii Determine the equation of the line parallel to line b passing through point $(0, {}^-1)$.

iii Is the line $2y + 6 = 2x$ perpendicular to graph **d**?

3 Find the equation of the line that passes through the two points $(2, 4)$ and $(6, {}^-12)$.

4 Here are six equations of straight lines, each labelled with a letter.

A	B	C
$y = 4x - 7$	$y = 3x + 14$	$y = 2x + 5$

D	E	F
$y = {}^{-}3x + 1$	$y = 14x - 7$	$y = 4x + 3$

Choose the correct letters to make each statement true.

Line ________ is the steepest line.

Lines ________ and ________ are parallel.

Lines ________ and ________ meet on the y-axis. *(3 marks)*

© OCR 2013

5 **a** Draw a graph of the two lines $y = 3x - 2$ and $y + 2x = 3$ and find their point of intersection.

b Show by substitution that this is the simultaneous solution to the two equations.

6

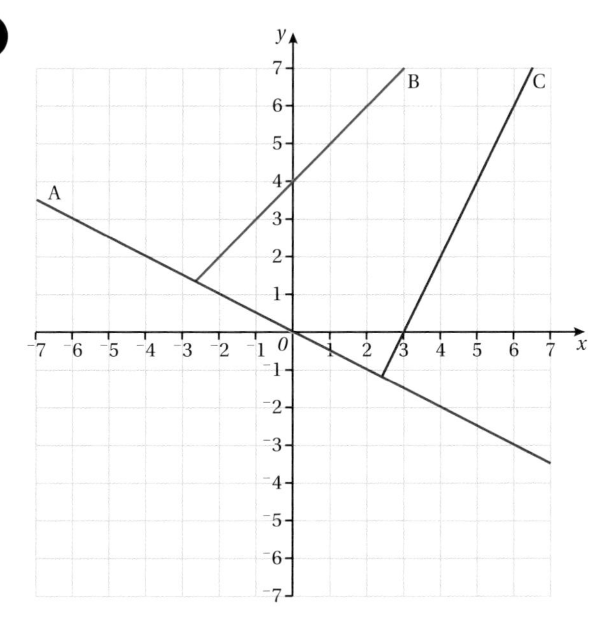

a Find the equation of each line A, B and C.

b Show that A $\perp$ C.

c Show that B and C are not parallel.

d Determine the x-intercept of B algebraically.

e Determine the y-intercept of C algebraically.

f Determine the equation of a line passing through (2, 3) such that the four lines on the graph form a trapezium with its base passing through the origin.

Tip

The symbol $\perp$ means 'perpendicular to'; see Chapter 5 if you need to.

19 Graphs of equations and functions

In this chapter you will learn how to …

- plot and sketch graphs of quadratic functions.
- identify the main features of graphs of quadratic functions and equations.
- plot and sketch other polynomials and reciprocal functions.
- recognise and sketch graphs of exponential functions.
- recognise and use the equation of a circle with centre at the origin.

For more resources relating to this chapter, visit GCSE Mathematics Online.

Using mathematics: real-life applications

Graphs are used to process information, make predictions and generalise patterns from sets of data. The nature of the data and the relationship between values determines the shape and form of the graph.

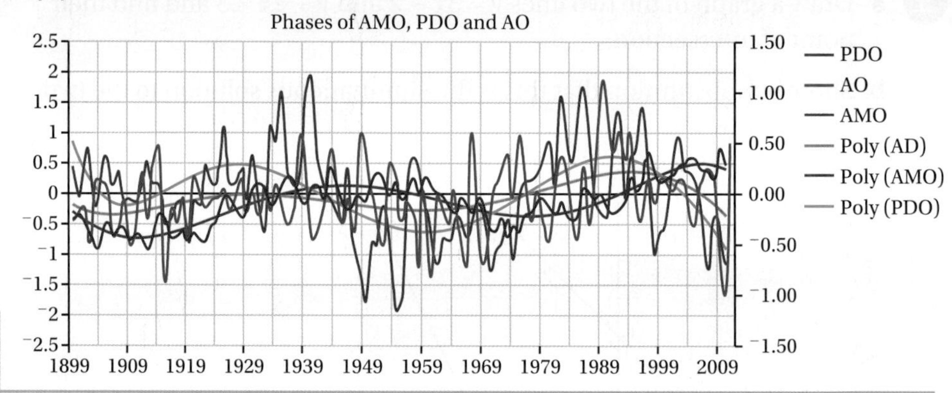

"I study the Earth using gravity, magnetic, electrical, and seismic methods. I used this graph in a study of the Pacific and Atlantic Oceans. I need to be able to understand equations and recognise the features of graphs to understand and interpret it."

(Geophysicist)

Before you start …

Ch 18	You should be able to interpret equations of linear graphs.	**1** For the graph $y = 3x + 1$: **a** identify the gradient of the graph. **b** give the coordinates of the y-intercept. **c** find the value of x when $y = {}^{-}14$. **d** show that it is parallel to the graph $2y - 6x = {}^{-}4$.
Ch 4, 18	You must be able to generate a table of values from a function.	**2** Given $y = 3x^2 + 1$, complete the table of values. x: ⁻2, ⁻1, 0, 1, 2 y: (blank)
Ch 8	You need to be able to find the roots of a quadratic equation algebraically.	**3** What are the roots of: **a** $x^2 + 2x - 8 = 0$? **b** $x^2 + 5x = {}^{-}4$? **4** Complete the square to find the solution of $x^2 + 4x - 6 = 0$.

Assess your starting point using the Launchpad

STEP 1

1. How many points do you need to plot the graph of a linear equation?
2. Sketch the graph of $y = 2x + 1$.

?

GO TO
Section 1:
Review of linear graphs

✓

STEP 2

3. **a** Is this the graph of $y = x^2 + 1$ or $y = {}^-x^2 + 1$?
   b How can you tell?
   c Is the turning point a maximum or a minimum?
   d What are the coordinates of the vertex?
   e For what values of x is $y = 0$?

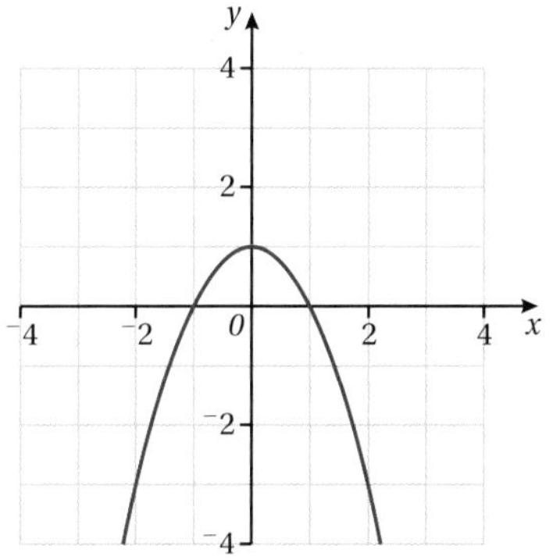

?

GO TO
Section 2:
Graphs of quadratic functions

✓

STEP 3

4. **a** What type of equation is $y = x^3$?
   b How many points do you need to plot the graph of $y = x^3$?
5. Given $y = \frac{1}{x}$:
   a explain what happens when $x = 0$.
   b what happens to the value of y as the value of x increases?
   c what is the value of y when $x = 60$?

?

GO TO
Section 3:
Graphs of other polynomials and reciprocals

✓

GO TO
Step 4:
The Launchpad continues on the next page …

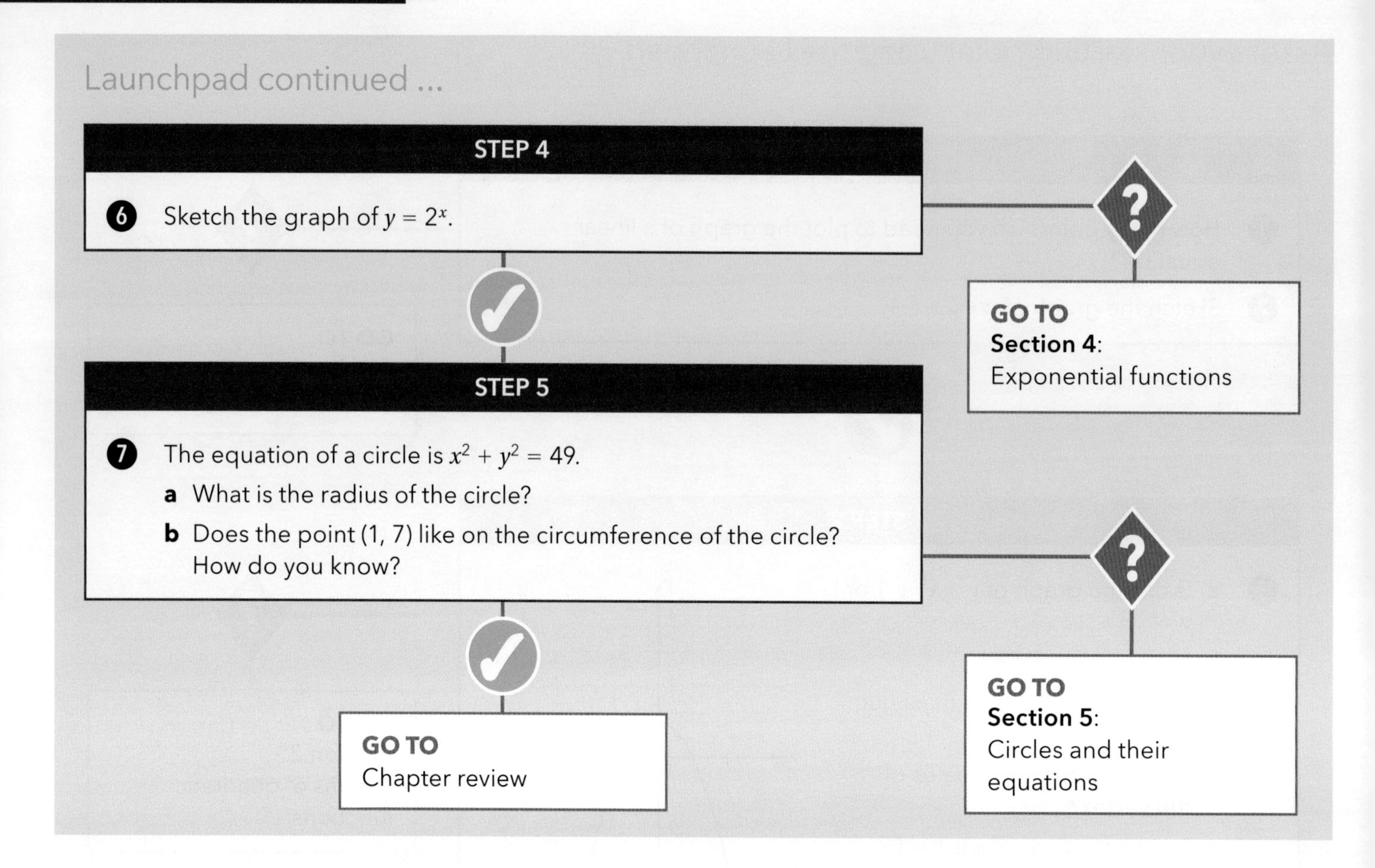

Section 1: Review of linear graphs

In Chapter 18 you learnt how to plot, sketch and interpret linear graphs. You also learnt that the general form of a linear function is $y = mx + c$, where m is the gradient of the graph and c is the y-intercept.

In this section you are going to review some of that work and learn to recognise line graphs when the general equation looks slightly different from the form $y = mx + c$.

Graphs in the form of $y = mx$

In the equation $y = mx + c$, the value of c tells you where the graph cuts the y-axis. When there is no value of c in the equation you get a graph in the form of $y = mx$.

WORK IT OUT 19.1

These are three linear functions.

$y = 3x \qquad y = {}^{-}3x \qquad y = \frac{1}{3}x$

There is a common point which all the graphs pass through. What is that point? How do you know?

Option A	Option B	Option C
(0, 3)	(⁻3, 0)	(0, 0)

Any linear equation of the form $y = mx$ passes through the origin with a gradient of m.

The graphs of $y = mx$ and $y = {}^{-}mx$ are shown here.

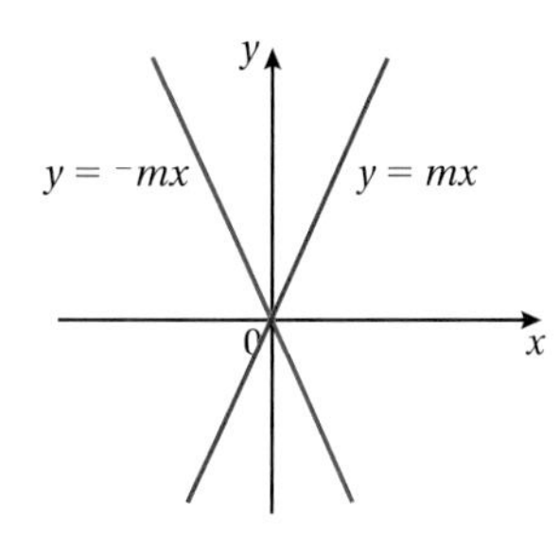

If $m = 1$, the equation is written as $y = x$.
If $m = {}^{-}1$, the equation is written as $y = {}^{-}x$.

$y = x$ is the line that passes through the origin going up from left to right at an angle of 45°.

$y = {}^{-}x$ is the line that passes through the origin going down from left to right making an angle of 45°.

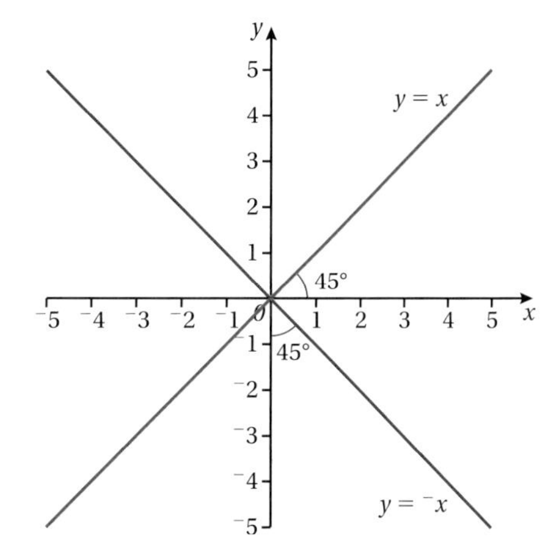

Sketching $y = mx$ ...	Examples	Notes
If m is greater than 1.	$y = 3x$, $y = 7x$	The line still passes through the origin but is steeper than $y = x$.
If m is a value between 0 and 1.	$y = \frac{1}{2}x$, $y = \frac{1}{5}x$	The line still passes through the origin but is less steep than $y = x$.
If m is a negative value.	$y = {}^{-}3x$, $y = \frac{{}^{-}1}{2}x$ $y = {}^{-}3x$, $y = \frac{{}^{-}1}{2}x$	The line still passes through the origin, but will go down from left to right like $y = {}^{-}x$.

Vertical and horizontal lines

The equations of some lines are in the form of $x = a$ or $y = b$. (Where a and b are constant values.) Equations in this form tell you that there is only one value of x or y for each graph.

Consider the equation $x = 7$.

The graph of this equation passes through the point (7, 0). It also passes through all points on the grid with an x-coordinate of 7. For example: (7, ⁻4), (7, ⁻1), (7, 3), (7, 50).

Now consider the equation $y = 7$.

This graph would need to pass through the point (0, 7) and all other points with a y-coordinate of 7. For example: (3, 7), ($^-2$, 7), (7, 7) and (50, 7).

The graphs of $x = 7$ and $y = 7$ are shown here.

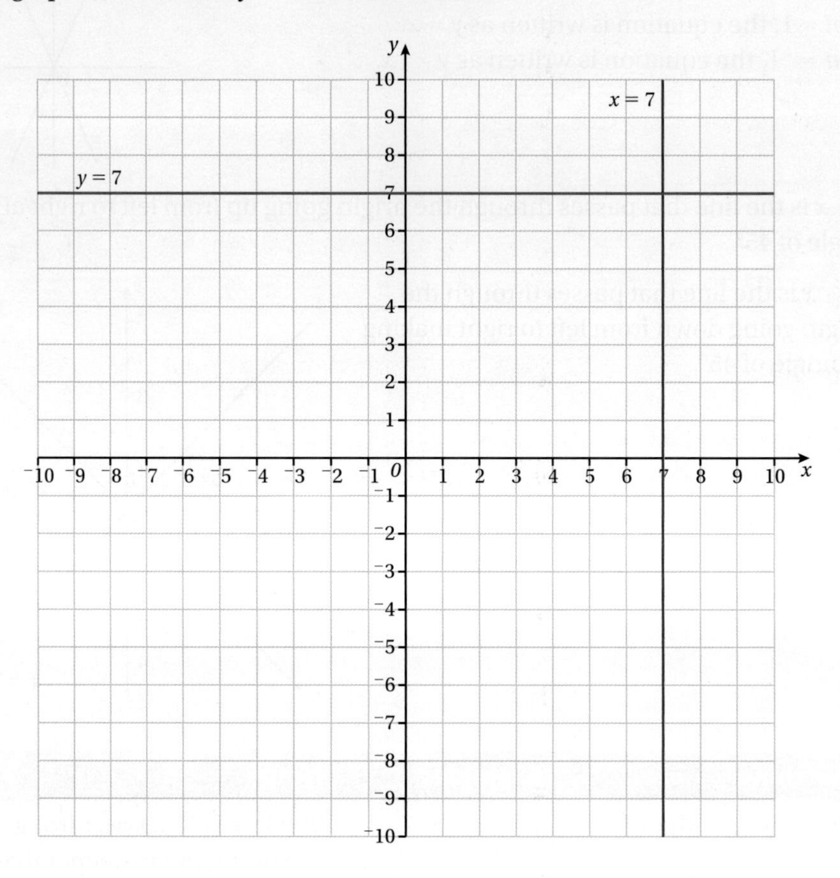

In general:

- Any graph in the form of $x = a$ is parallel to the y-axis and passes through a on the x-axis.
- Any graph in the form of $y = b$ is parallel to the x-axis and passes through b on the y-axis.
- The values of a and b can be positive or negative.

Tip

You should be able to recognise and sketch the graphs of any line in the form $x = a$ or $y = b$.

The axes themselves can be described using equations.

The x-axis is the line $y = 0$ and the y-axis is the line $x = 0$.

EXERCISE 19A

1 Write down which of the graphs on the grid on the right can be described in each of the following ways. There may be more than one correct answer.

a The x-coordinate of each point is equal to the y-coordinate.

b The gradient is negative.

c The general form of the graph is $y = mx$.

d The y-coordinate is 6 times the x-coordinate.

e The y-coordinate is 6 less than the x-coordinate.

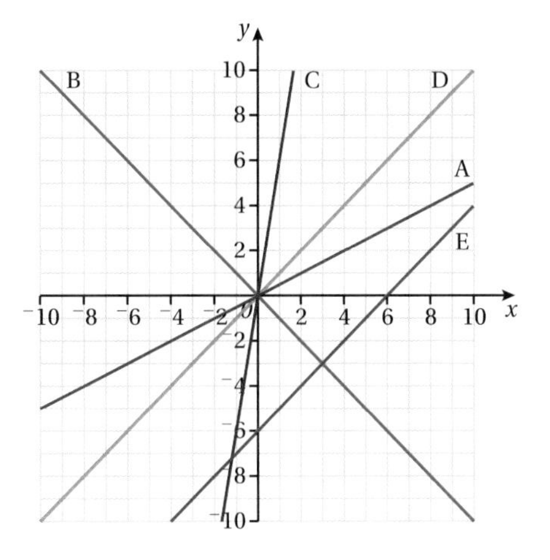

2 Give the equation of each line on the grid in Question **1**.

3 **a** Write down the equation of each line A to F.

b Determine the equation of the line parallel to D which passes through point (0, 2).

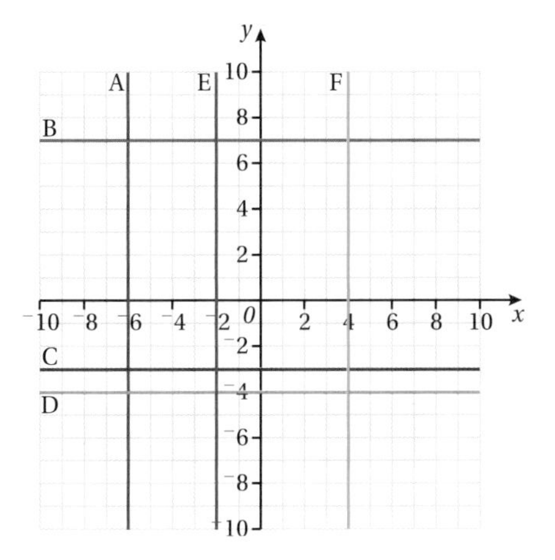

4 Draw the following graphs on a grid numbered from $^-6$ to 6 on each axis.

a $x = {}^-3$ **b** $y = 5$ **c** $y = {}^-3$ **d** $x = 5$

5 What type of quadrilateral is enclosed by the four lines you drew in Question **4**? Give reasons for your answer.

6 **a** Write the equations of two lines that would divide the quadrilateral formed in Question **4** into two identical rectangles.

b What is the mathematical name for lines that divide shapes into two identical halves?

c It is possible to draw two other lines that divide the quadrilateral into two equal halves.

i Draw these lines on your diagram.

ii Determine the equation of each line.

7 Plot the graphs of:

a $y + 2x = 6$ **b** $3x - 9y = 21$ **c** $x - y = {}^-4$

8 Sketch the graph of each linear equation.

a $y = 2x$ **b** $y = {}^-8x$ **c** $y = {}^-\left(\frac{1}{4}x\right)$

d $y = x + 7$ **e** $y = {}^-2x - 1$

Tip

Equations in the form $ax + by = c$ can be rearranged into the general form $y = mx + c$ to help you calculate the (x, y) points and plot the linear graph.

Find answers at: cambridge.org/ukschools/gcsemaths-studentbookanswers

Key vocabulary

parabola: the symmetrical curve produced by the graph of a quadratic function.

Did you know?

The path of moving objects, such as this basketball, can be modelled by a parabola.

Section 2: Graphs of quadratic functions

You saw in Chapters 7 and 8 that a quadratic expression has the form $ax^2 + bx + c$, where $a \neq 0$.

The graph of a quadratic function is a curve called a **parabola**.

The simplest equation of a parabola is $y = x^2$.

This is the graph of $y = x^2$.

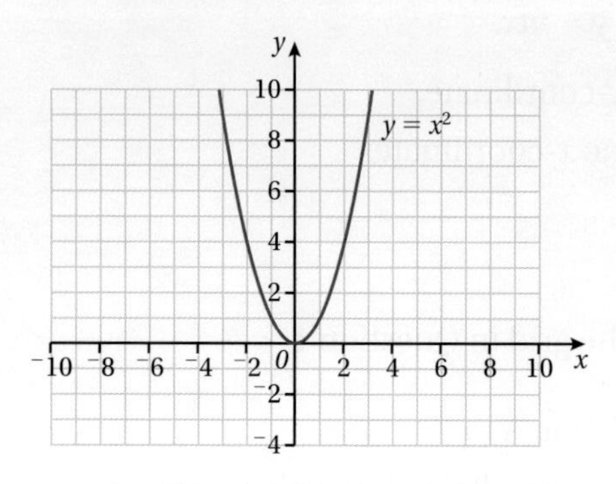

You can plot the graph of a quadratic function by drawing up a table of values.

WORKED EXAMPLE 1

Plot the graph of $y = x^2 - 2x - 8$.

x	⁻3	⁻2	⁻1	0	1	2	3	4	5
y	7	0	⁻5	⁻8	⁻9	⁻8	⁻5	0	7

Build up a table of points that satisfy the equation $y = x^2 - 2x - 8$. To make the graph as accurate as possible, include at least five points (the more the better), and both positive and negative values of x.

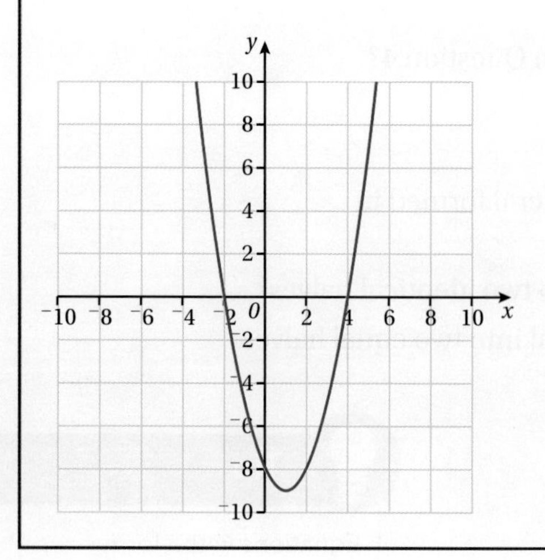

Plot all these points to draw the graph and produce a smooth curve drawing through and beyond the points calculated.

EXERCISE 19B

1 Plot the graph of each of these equations for $^-3 \leqslant x \leqslant 3$ on the same grid.

a $y = x^2$ **b** $y = {}^-x^2$ **c** $y = x^2 + 1$ **d** $y = x^2 - 4$

2 Plot the graph of each quadratic equation for whole number values of x in the given range:

a $y = x^2 + 2x - 3$ $\quad {}^-4 \leqslant x \leqslant 2$

b $y = x^2 + x - 2$ $\quad {}^-3 \leqslant x \leqslant 2$

c $y = x^2 + 3x$ $\quad {}^-4 \leqslant x \leqslant 1$

3 You may need to plot some more graphs to answer these questions.

a If the coefficient of x^2 is greater than 1 what impact does it have on the shape of the parabola?

b If the coefficient of x^2 is a value between 1 and 0, what impact does it have on the shape of the parabola?

c What happens to the graph of the parabola if a constant value is added; consider the difference between the graph of an equation such as $y = ax^2$ and $y = ax^2 + c$.

d How does the graph of the parabola differ from the basic $y = x^2$ graph when the coefficient of x^2 is negative?

4 Plot the graph of $y = x^2 - x - 2$ for ${}^-2 \leqslant x \leqslant 3$.

a Solve the equation $x^2 - x - 2 = 0$.

b How would you read the solution to the equation $x^2 - x - 2 = 0$ from the graph?

Features of parabolas

Quadratic graphs have characteristics that you can use to sketch and interpret them.

The main features of a parabola with a general equation of $y = ax^2 + bx + c$ are:

- the turning point or vertex of the graph – this is the point at which the graph changes direction; this has the same x-value as the axis of symmetry.
- the axis of symmetry – a line which divides the parabola into two symmetrical halves; this line passes through the turning point.
- the y-intercept – a parabola can only have **one** y-intercept.
- the x-intercepts – a parabola can have 0, 1 or 2 x-intercepts depending on the position of the graph.

The graph either has a minimum turning point or a maximum turning point.

If a is positive, the graph will go down to a minimum turning point (this is the lowest point of the graph and is called the minimum).

If a is negative, graph will go up to a maximum turning point (this is the highest point of the graph and is called the maximum).

Tip

The value of x^2 will always be positive, so it is the sign of the coefficient a that determines if the turning point is a maximum or minimum.

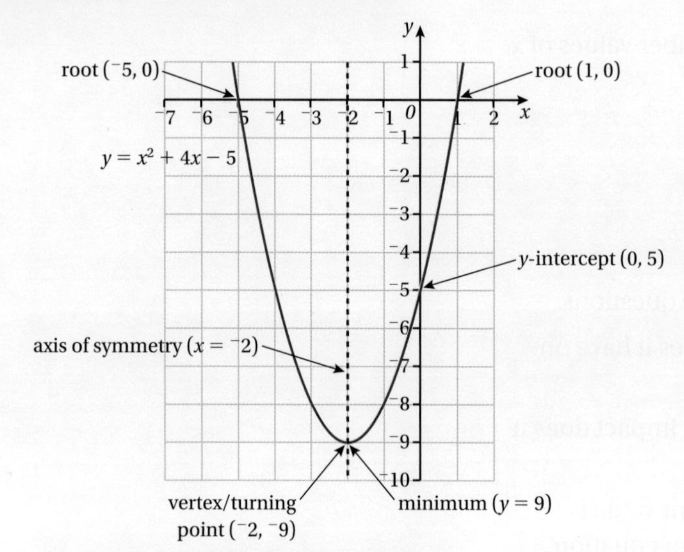

The diagram shows the main features of a parabola drawn from the equation $y = x^2 + 4x - 5$.

Note that the axis of symmetry is $x = {}^-2$, which is also the x-value of the vertex.

The x-intercepts of a parabola are the roots (or solution) of the quadratic equation that defines the graph. You can find the roots graphically by reading their value off the graph.

You can also solve the quadratic equation algebraically to find its roots and therefore work out the points at which its curve crosses the x-axis. This is useful when you have to sketch the graph.

In the example on the left, the roots of the equation are $x = {}^-5$ and $x = 1$.

The y-intercept can also be read off the graph or found algebraically:

$$x = 0$$

$$y = x^2 + 4x - 5$$

$$= 0^2 + (4 \times 0) - 5 = {}^-5$$

So the y-intercept is the point $(0, {}^-5)$.

Tip

In Chapter 8, you learnt how to solve quadratic equations algebraically. Revise that chapter if you have forgotten how to do this.

Identifying the turning point

You can find the turning point algebraically by completing the square to give you the equation in the **vertex** form $y = a(x \pm h)^2 + k$. In this form, the vertex is at the point (h, k).

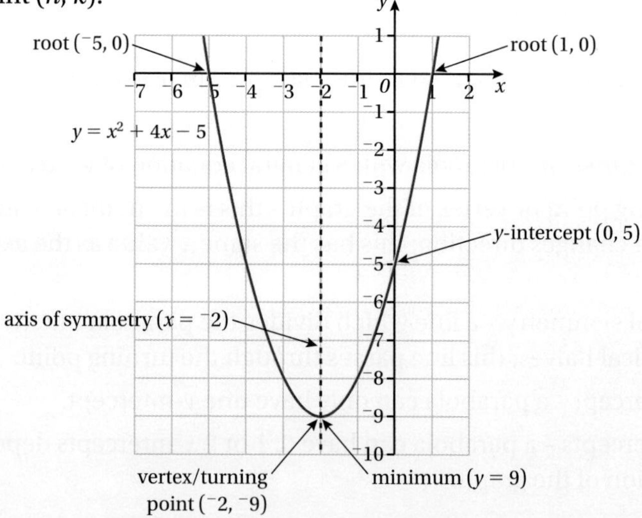

This is because $(x \pm h)^2 \geqslant 0$, so the minimum value of y when a is positive, or the maximum value of y when a is negative, is when $(x \pm h)^2 = 0$.

For example, returning to the graph of $y = x^2 + 4x - 5$, completing the square gives: $(x + 2)^2 - 9$

$(x + 2)^2 \geqslant 0$ when $x = {}^-2$, and $y = {}^-9$. So the turning point, is the point $({}^-2, {}^-9)$.

The axis of symmetry goes through the turning point, so the axis of symmetry of this graph is the line $x = {}^-2$.

All parabolas are symmetrical and will have an axis of symmetry. Not all parabolas will cross the x-axis and have real roots.

WORK IT OUT 19.2

What are the main features of this parabola?
Choose the correct option.

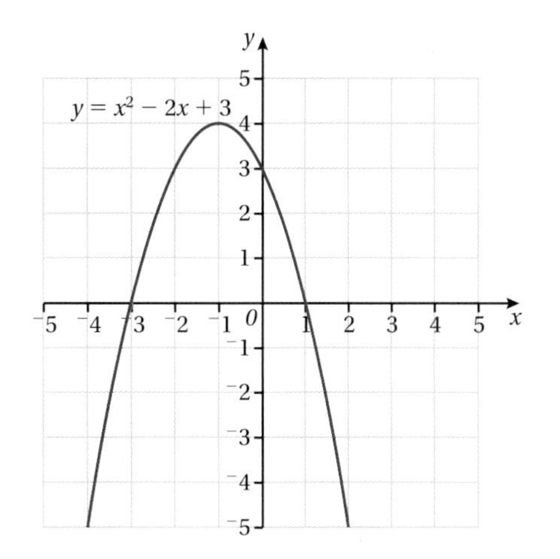

Option A	Option B	Option C
y-intercept (0, 3) vertex ($^-1$, $^-4$) is a minimum turning point. Axis of symmetry $x = 1$; roots ($^-3$, 0) and (1, 0)	y-intercept (0, 3) vertex ($^-1$, 4) is a maximum turning point. Axis of symmetry $x = 1$; roots ($^-3$, 1) and (1, 1)	y-intercept (0, 3) vertex ($^-1$, 4) is a maximum turning point. Axis of symmetry $x = {^-1}$; roots ($^-3$, 0) and (1, 0)

x-intercepts and roots of a quadratic equation

The x-intercepts of a parabola are the roots (or solution) of the quadratic equation that defines the graph. You can find the roots graphically by reading their value off the graph.

You can also solve the quadratic equation algebraically to find its roots and therefore work out the points at which its curve crosses the x-axis. This is useful when you have to sketch the graph.

A parabola will have no intercepts if the quadratic equation cannot be solved.

EXERCISE 19C

1 For each parabola determine:

i the turning point and whether it is a minimum or maximum.
ii the axis of symmetry.
iii the y-intercept.
iv the x-intercepts.
v the roots of the equation used to generate the graph.

a

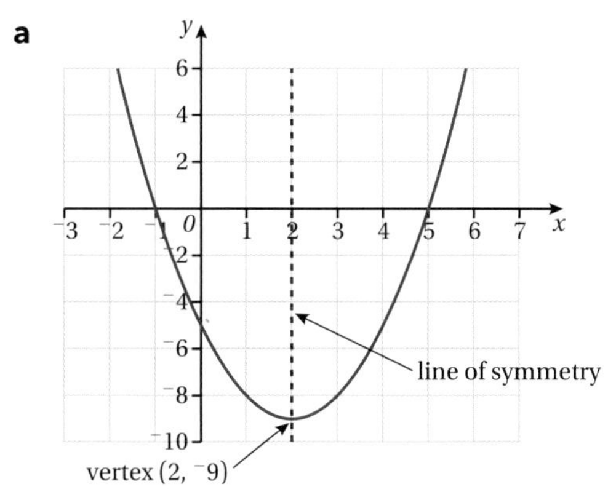

b

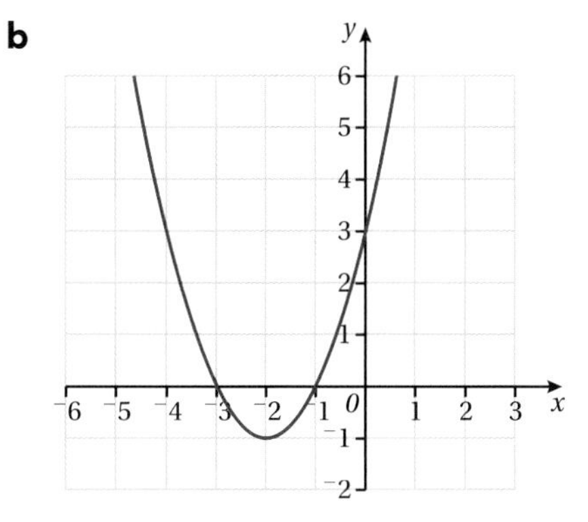

Find answers at: cambridge.org/ukschools/gcsemaths-studentbookanswers

c

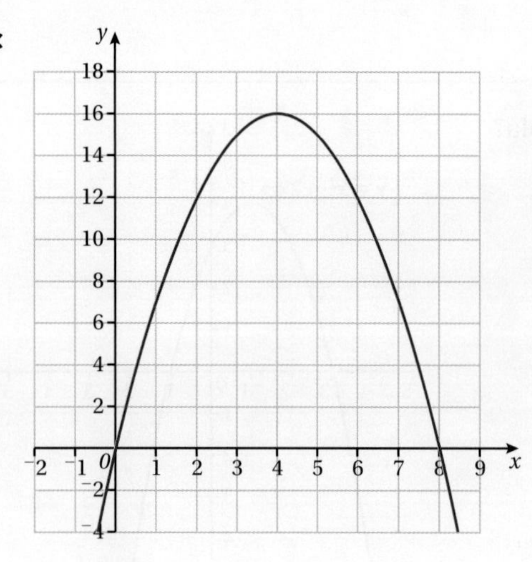

d

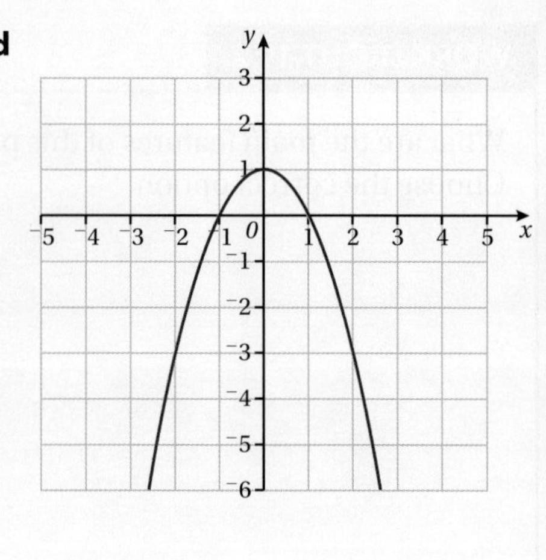

2 Rewrite the equation $y = 3x^2 + 6x + 3$ in the form $y = a(x - h)^2 + k$.

a Determine the y-intercept.

b Find the axis of symmetry and the vertex.

c Determine the x-intercepts.

d Sketch the graph of the equation labelling the main features.

Sketching quadratic graphs

You can use the characteristics of a parabola to sketch graphs without drawing up a table of values:

- Write the equation in the general form $y = ax^2 + c$.
- Use the sign of a to determine if the graph goes up to a maximum turning point or down to a minimum turning point.
- Work out the y-intercept (this is given by c in the equation).
- Calculate the x-intercepts by substituting $y = 0$ and solving for x.
  If the x-intercept is at the origin ($x = 0$) or there are no x-intercepts (no solution), find the coordinates of one point on the graph.
- Mark the y-intercept and x-intercepts (or single point) and use the shape and symmetry of the graph as a guide to draw a smooth curve.
- Label your graph.

Tip

Draw a smooth curve to join the points and try to make your graph as symmetrical as possible.

WORKED EXAMPLE 2

Sketch the graph of $y = 3x^2$.

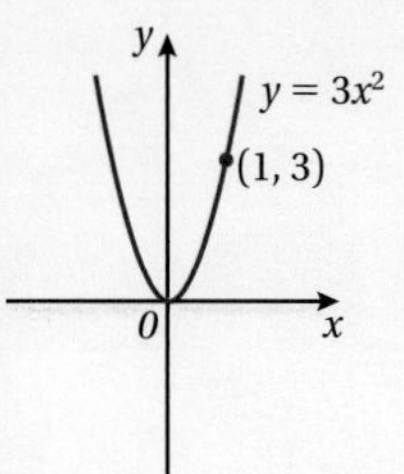

$a = 3$ so the graph goes down to a minimum turning point.

There is no constant, so $c = 0$ and the y-intercept is at the point $(0, 0)$.

When $y = 0$ then $x = 0$, so the x-intercept is at $(0, 0)$; find one point on the curve.

When $x = 1$

$y = 3(1)^2 = 3$

So, $(1, 3)$ is a point on the curve.

Sketch the graph by marking the y-intercept and the point $(1, 3)$. As the graph is in the form $y = ax^2 + c$, you know that the turning point has the same coordinate at the y-intercept and the axis of symmetry is the line $x = 0$, use this and the point $(1, 3)$ to create a curve.

Label the graph.

WORKED EXAMPLE 3

Sketch the graph of $y = {}^{-}x^2 + 4$.

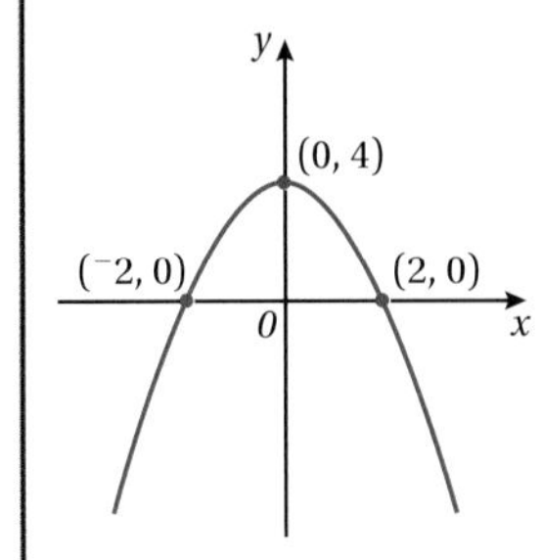

$a = {}^{-}1$, so the graph goes up to a maximum turning point. $c = 4$, so y-intercept is $(0, 4)$.

x-intercepts when $y = 0$

$0 = {}^{-}x^2 + 4$

$\therefore x^2 - 4 = 0$

This is a difference of squares

$(x + 2)(x - 2) = 0$

$x + 2 = 0$ or $x - 2 = 0$

$x = {}^{-}2$ or $x = 2$

So, intercepts are $({}^{-}2, 0)$ and $(2, 0)$.

Sketch: mark the intercepts; the turning point is the same as the y-intercept; the axis of symmetry is $x = 0$. Add labels.

EXERCISE 19D

1 Noor sketched these graphs but she didn't write the equations on them. Use the features of each graph to work out what the correct equations are.

a

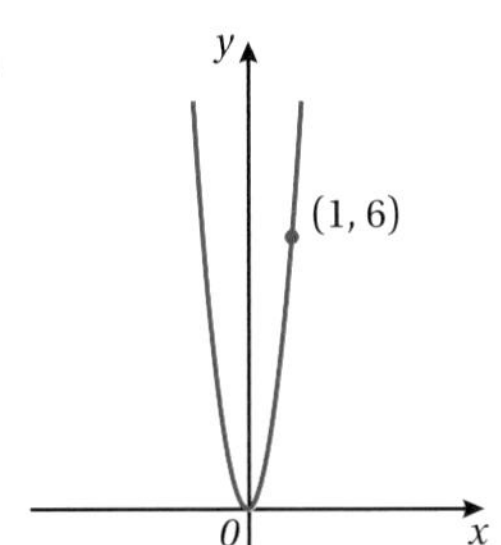

b

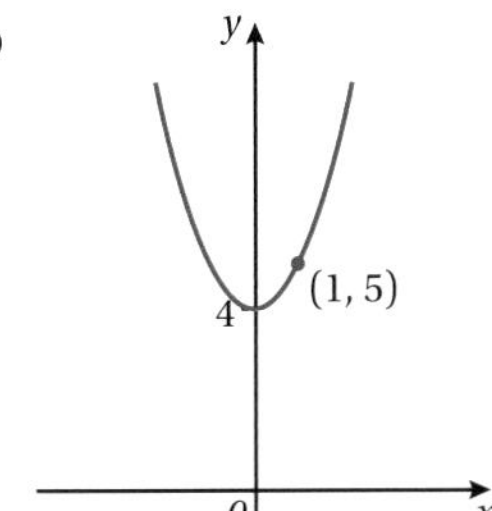

c

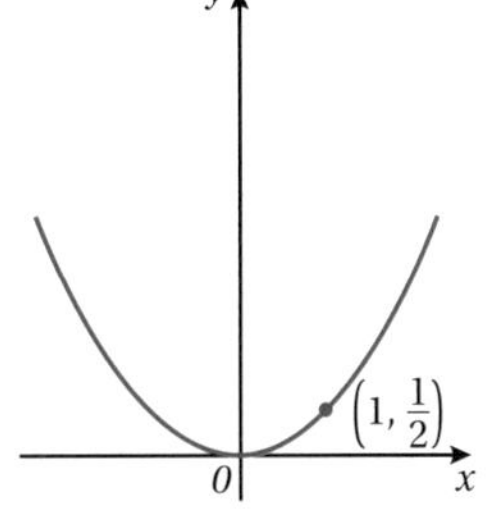

d

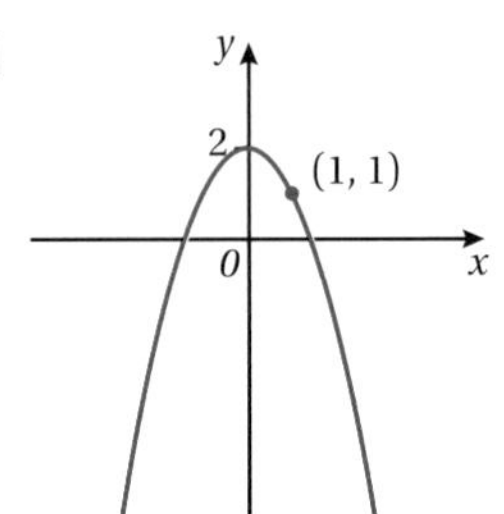

e

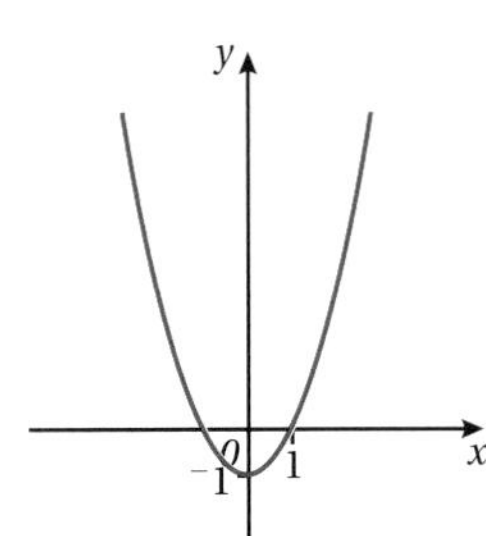

f

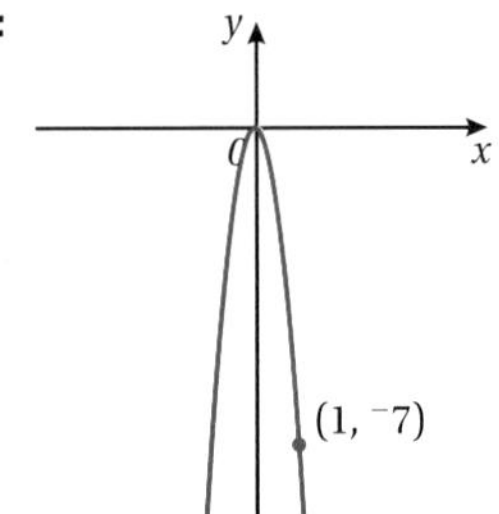

2 Sketch the graph of each of these quadratic equations on the same grid.

$y = x^2$ $\quad y = x^2 + 2$ $\quad y = 3x^2$ $\quad y = \frac{1}{2}x^2 - 2$ $\quad y = {}^{-}x^2 + 2$

3 For each equation, determine:

i the y-intercept. **ii** the x-intercept(s).
iii the axis of symmetry. **iv** the turning point.

Use the results to sketch and label each graph.

a $y = x^2 + 2x - 3$ **b** $y = 2x^2 + 4x + 3$ **c** $y = 4x - x^2$
d $y = x^2 + 2x - 8$ **e** $y = x^2 - 8x + 12$ **f** $y = {}^{-}x^2 - 6x - 10$
g $y = 2(x - 3)(x + 5)$ **h** $y = 4x^2 + 16x + 7$ **i** $x^2 + 3x - 6 = y$
j $2x^2 + x = 8 + y$

Find answers at: cambridge.org/ukschools/gcsemaths-studentbookanswers

Key vocabulary

polynomial: an expression made up of many unlike terms with positive powers for the variables.

Tip

You saw in Chapter 7 that a **binomial** is an expression with **two** unlike terms.

Tip

In both graphs, $c = 0$ and so has not been included in the written equation.

Section 3: Graphs of other polynomials and reciprocals

A **polynomial** is an expression with many unlike terms; all the variables have positive powers.

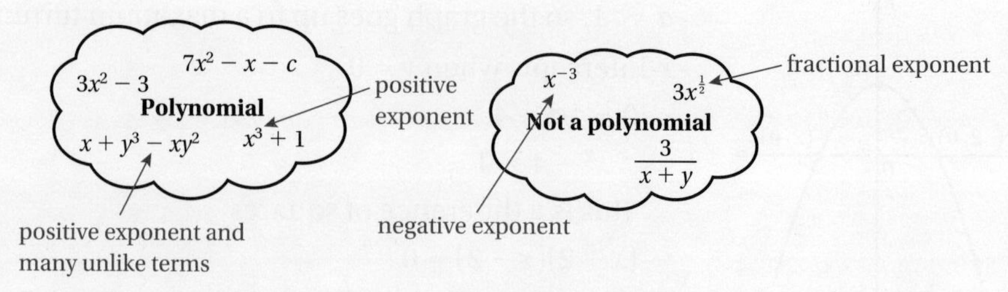

If the highest power of x is 3, the expression is called a cubic expression.

For example, $2x^3$ and $2x^3 + x^2 + 3$ are both cubics.

The simplest equation of a cubic graph is $y = x^3$.

All basic cubic graphs have a similar shape. The diagram shows the basic shape of cubic graphs in the form of $y = ax^3 + c$.

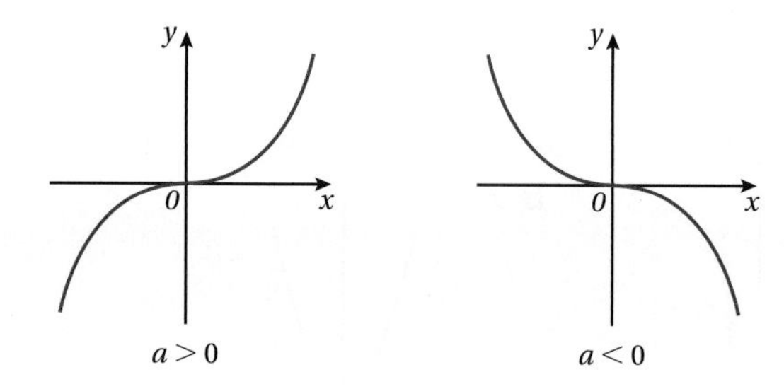

The shape on the left occurs when a is positive. We call this an increasing curve.

The shape on the right occurs when a is negative. We call this a decreasing curve.

The larger the value of a, the steeper the curve.

The constant, c, is the y-intercept.

To draw an accurate graph of a cubic equation, you plot at least five points including both positive and negative values of x, then draw a smooth curve through and beyond them.

This is a table of values for $y = x^3$.

x	$^-3$	$^-2$	$^-1$	0	1	2	3
y	$^-27$	$^-8$	$^-1$	0	1	8	27

Tip

Remember that when you cube a negative number you will get a negative result. For example: $(^-1)^3 = {^-1} \times {^-1} \times {^-1} = {^-1}$

Note that the curve passes through the origin but it is not symmetrical.

For more complicated cubic equations with two or more terms, you should work out whole number and half number values to make sure you plot the graph as accurately as possible.

You might find it easier to evaluate each term separately in the table and add them to find y-values.

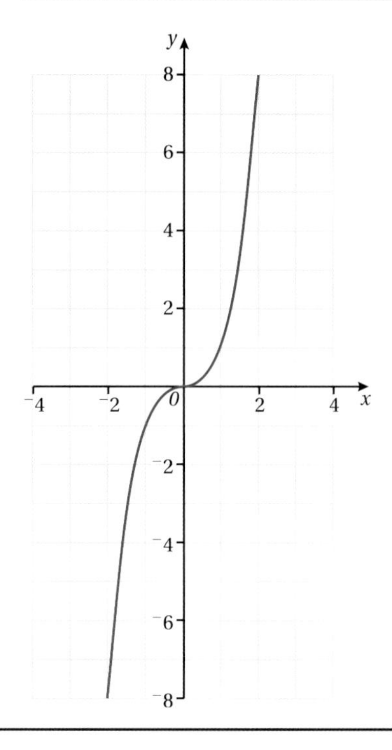

WORKED EXAMPLE 4

Draw the graph of the equation $y = x^3 - 6x$ for $^-3 \leqslant x \leqslant 3$

x	−3	−2	−1	0	1	2	3
x^3	−27	−8	−1	0	1	8	27
$-6x$	18	12	6	0	−6	−12	−18
y	−9	4	5	0	−5	−4	9

Determine whole number values first.

x	−2.5	−1.5	−0.5	−0.5	−1.5	−2.5
x^3	−15.625	−3.375	−0.125	0.125	3.375	15.625
$-6x$	15	9	3	−3	−9	−15
y	−0.625	5.625	2.875	−2.875	−5.625	0.625

Construct a separate table for in-between values of x.

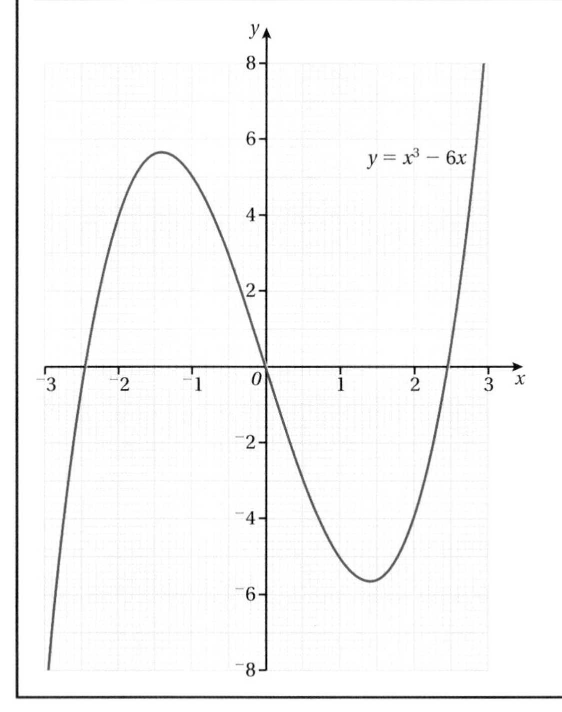

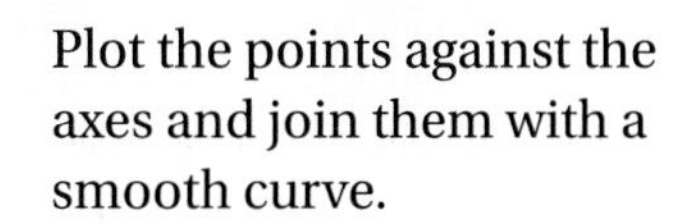

Plot the points against the axes and join them with a smooth curve.

EXERCISE 19E

1. Complete a table of values for whole number values of x from $^-3$ to 3 for $y = {^-x^3}$.

   How does this graph differ from the graph of $y = x^3$?

2. Use a table of values to sketch the following pairs of cubic graphs.

   Plot each pair on the same grid, but use a separate grid for each pair.

   a $y = {^-2x^3}$ and $y = 2x^3$

   b $y = \frac{1}{2}x^3$ and $y = {^-\left(\frac{1}{2}x^3\right)}$

3. Work with a partner to compare the pairs of graphs you drew in Question **2**.

   Discuss how you could sketch the graph of $y = {^-4x^3}$ if you were given the graph of $y = 4x^3$.

4. Complete a table of values for whole number values of x from $^-3$ to 3 for these equations and draw a graph of each curve.

   a $y = x^3 + 1$ **b** $y = x^3 - 2$

5. Look at the graphs and their equations in Question **4**.

   What is the y-intercept in each graph?

6. The red line is the graph $y = x^3$.

   What are the equations of graphs A and B?

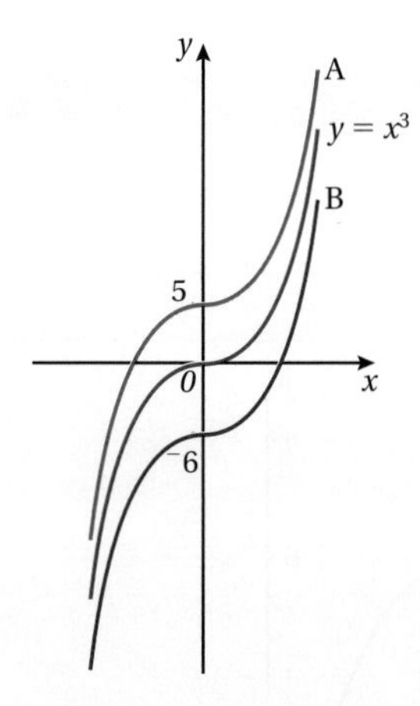

7. Plot the graph of each cubic equation for the given values of x.

   a $y = x^3 + 3x^2$ $\quad ^-3 \leqslant x \leqslant 3$

   b $y = x^3 - 3x + 1$ $\quad ^-3 \leqslant x \leqslant 4$

Reciprocal functions

Every number has a **reciprocal** except for 0, as $\frac{1}{0}$ cannot be defined.

A **reciprocal function** relates x to its reciprocal, and has the general equation $y = \frac{a}{x}$, where a is a constant value. You might also see this expressed as $xy = a$.

The graphs of reciprocal functions have a characteristic shape called a **hyperbola**. Each graph is made of two non-connected curves which are mirror images in opposite quadrants of the grid.

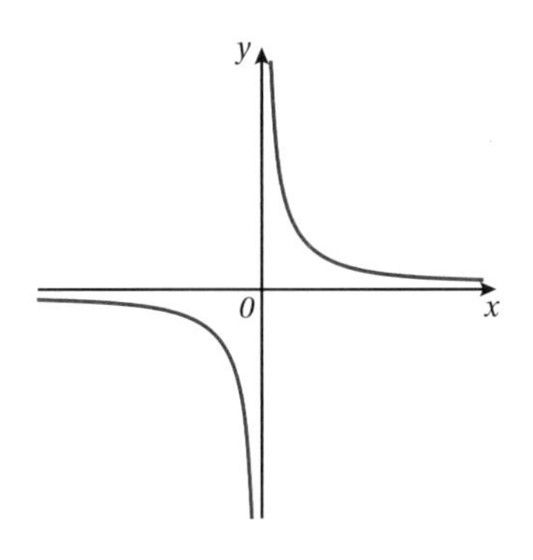

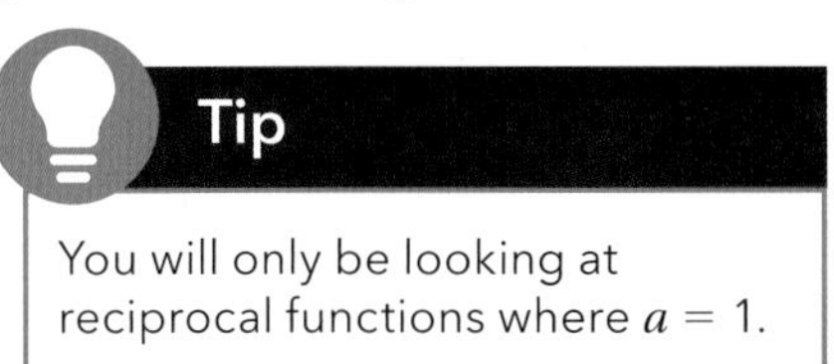

Tip

You will only be looking at reciprocal functions where $a = 1$.

Key vocabulary

hyperbola: the curved graph(s) formed by a reciprocal function; the curve of $y = \frac{1}{x}$ gets increasingly close to the x-axis and y-axis but never touches them.

This is the table of values for the equation $y = \frac{1}{x}$.

x	$^-3$	$^-2$	$^-1$	$\frac{^-1}{2}$	$\frac{^-1}{3}$	0	$\frac{1}{2}$	$\frac{1}{3}$	1	2	3
y	$\frac{^-1}{3}$	$\frac{^-1}{2}$	$^-1$	$^-2$	$^-3$	N/A	2	3	1	$\frac{1}{2}$	$\frac{1}{3}$

In order to draw a reciprocal graph accurately you need many points that include integer and some non-integer values of x.

Note that there is no y-value when $x = 0$ or x-value when $y = 0$ because division by 0 is undefined.

To draw the graph:

- Plot the (x, y) values from the table.
- Join the points with a smooth curve.
- Write the equation on both parts of the graph.

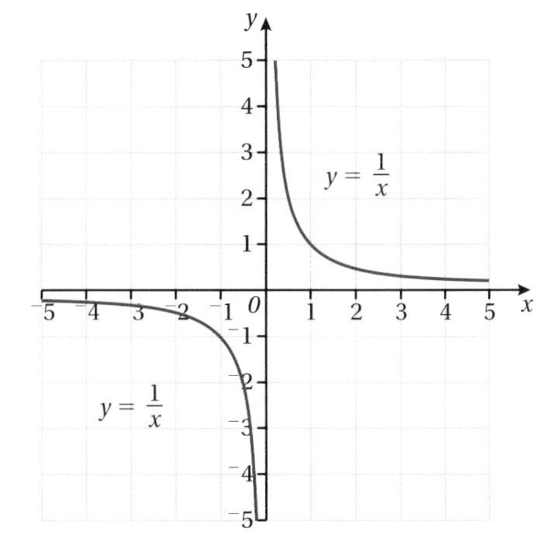

Its curves are in quadrants 1 and 3.

The graph of $y = {}^-\left(\frac{1}{x}\right)$ has its curves in quadrants 2 and 4.

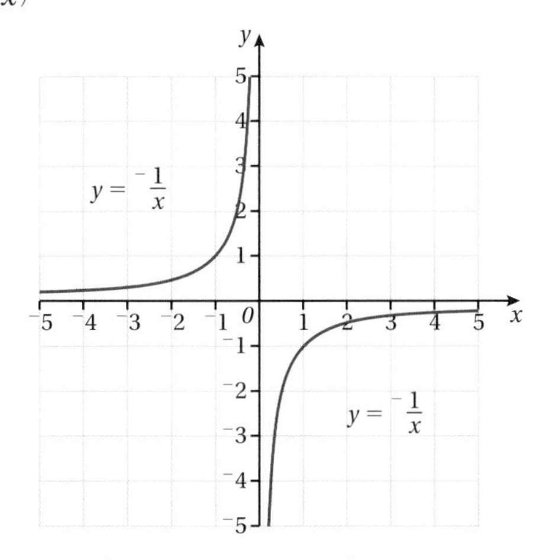

Note that as the digit for x (ignoring the sign) gets bigger the value for y gets closer and closer to 0 but the graph never actually meets the x-axis.

Find answers at: cambridge.org/ukschools/gcsemaths-studentbookanswers

EXERCISE 19F

1 Plot each of the following graphs on the same grid using x-values from $^-5$ to 5.

a $y = \frac{1}{x}$ **b** $y = \frac{1}{x} + 1$ **c** $y = \frac{1}{x} + 3$ **d** $xy = 1$

2 Use your graphs from Question **1** to describe how the constant c in the equation $y = \frac{1}{x} + c$ changes the reciprocal graph of $y = \frac{1}{x}$.

3 Neo says that the line $y = x$ is the line of symmetry of the graph $y = \frac{1}{x}$. Is he correct? Explain your answer.

4 **a** Copy and complete each table for the given values of x. Plot the graphs on the same grid.

i $y = \frac{2}{x}$

x	$^-4$	$^-2$	$^-1$	1	2	4
y						

ii $y = \frac{6}{x}$

x	$^-6$	$^-3$	$^-1$	1	3	6
y						

iii $xy = {^-12}$

x	$^-10$	$^-8$	$^-6$	$^-4$	$^-2$	2	4	6	8
y									

iv $y = \frac{8}{x}$

x	$^-8$	$^-6$	$^-4$	$^-2$	1	2	4	6	8
y									

b Compare the graphs that you have drawn in part **a**. How does the value of the constant in the equation affect the graph?

5 Here are four reciprocal graphs.

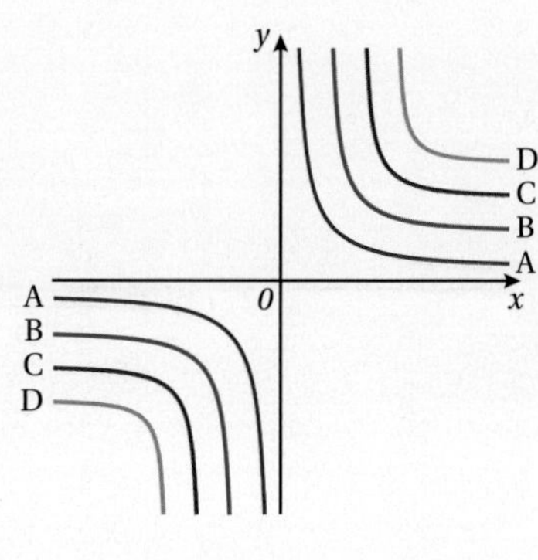

a Without doing any calculation, match each of these equations to a graph.

$y = \frac{2}{x}$ $y = \frac{4}{x}$ $y = \frac{8}{x}$ $y = \frac{10}{x}$

b Explain how you worked out which equation belonged with each graph.

6 Plot the graph for each of the reciprocal equations for the given values of x.

a $y = \frac{5}{x}$ $\quad$ $^-5 \leqslant x \leqslant 4$

b $y = \frac{3}{x+2}$ $\quad$ $x = {}^-16, {}^-12, {}^-8, {}^-4, 0, 4, 8, 12, 16$

c $y = \frac{1}{x-2}$ $\quad$ $^-4 \leqslant x \leqslant 6$

7 Use what you now know about the shape of the graphs of the basic cubic and reciprocal functions from Questions **1** and **2** to sketch diagrams of these equations:

a $y = x^3 + 2$ $\quad$ **b** $y = {}^-x^3$ $\quad$ **c** $y = \frac{2}{x}$

d $y = \frac{1}{x} - 1$ $\quad$ **e** $y = \frac{^-1}{x}$ $\quad$ **f** $y = \frac{1}{x} + 2$

Use ICT to check your answers and try some more versions of a basic cubic and a reciprocal by changing the value of the constant numbers in the equations.

8 The equation of graph **a** is $y = \frac{1}{x}$. What is the equation of graph **b**?

a

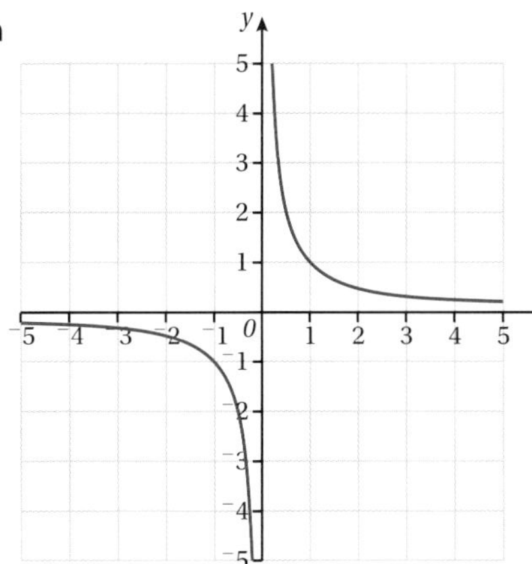

b

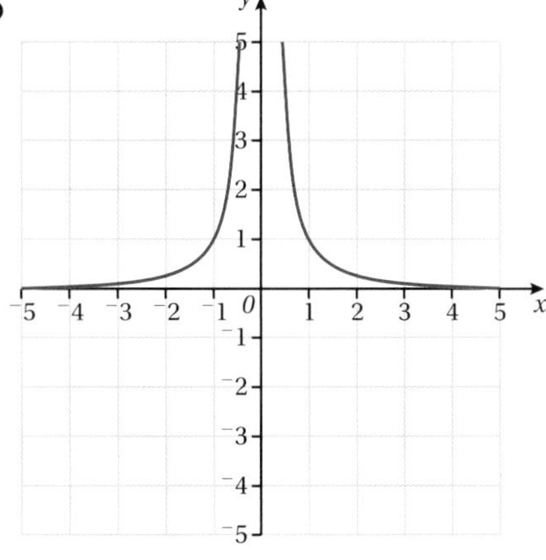

Section 4: Exponential functions

An **exponential function** has a number as the base and a variable as the **exponent**. For example $y = 2^x$.

The general form of the exponential function $y = ak^x$ (where k is positive) produces a graph called an exponential curve. When $a = 1$, you simply write the equation as $y = k^x$

Key vocabulary

exponential function: a function of the form $y = ak^x$, where $k > 1$.

exponent: the value of a power.

WORKED EXAMPLE 5

Complete a table of values for $y = 2^x$ for $^-4 \leqslant x \leqslant 4$ and draw the graph.

x	$^-4$	$^-3$	$^-2$	$^-1$	0	1	2	3	4
y	$\frac{1}{16}$	$\frac{1}{8}$	$\frac{1}{4}$	$\frac{1}{2}$	1	2	4	8	16

Calculate at least five points by substituting different values of x into the equation as normal. Use some negative values of x as well.

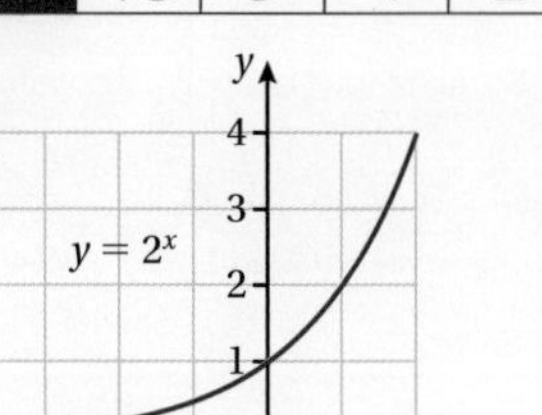

Plot all of the calculated points and draw a smooth curve through and beyond the points plotted.

You can see from the example that the curve rises rapidly towards the right. This is the typical shape of graphs of **exponential growth**.

As the values of x decrease the curve gets closer and closer to the x-axis without ever touching it. This is because there is no defined value for x when $y = 0$. Exponential growth happens when $k > 1$ and x is positive. i.e., when $y = k^x$.

Tip

No matter how the negative value of x decreases there will still be a value for y. See Chapter 25.

Because k^x is always positive, the curve will never extend below the x-axis. k cannot be 1 because $1^x = 1$ and so the equation $y = 1$ is a straight line. Similarly, k cannot be a negative value because the y-value might not be defined.

Exponential curves do not pass through the origin and they are not symmetrical.

The value of k affects the steepness of the curve. The graphs of $y = 2^x$, $y = 3^x$ and $y = 5^x$ are shown here on the same grid. The graph of $y = 5^x$ is much steeper than $y = 2^x$ and $y = 3^x$ because powers of 5 (5, 25, 125, 625) increase in value faster than powers of 2 and 3. The higher the value of k the steeper the curve. Note that they all have a y-intercept at (0, 1) and they all approach the x-axis to the left.

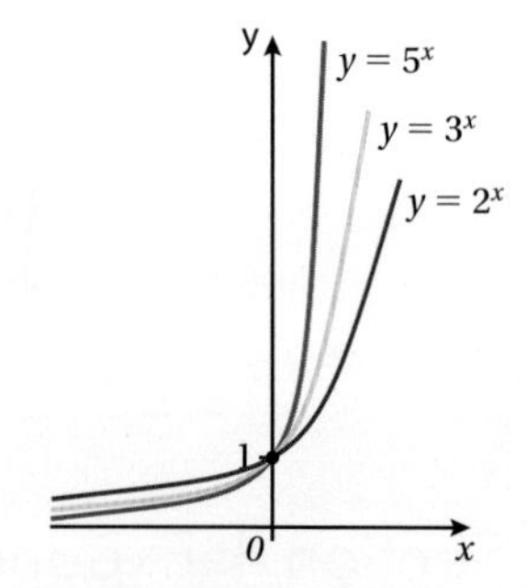

Tip

Remember that any value raised to a power of 0 is equal to 1. $k^0 = 1$.

When $x = 0$ and $a = 1$, $y = k^0 = 1$, so all graphs in the form $y = k^x$ will pass through the point (0, 1).

When the graph slopes down to the right (decreases) it is known as **exponential decay.** This happens when $0 < k < 1$, or x is negative.

Tip

You will learn about negative and fractional exponents in Chapter 25.

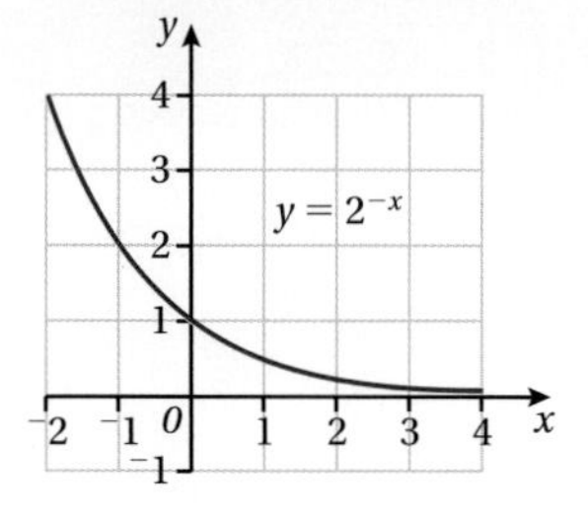

The function $y = k^{-x}$ is the reciprocal of $y = k^x$, its graph is a reflection of $y = k^x$ with the y-axis as the line of symmetry.

Graphing $y = 2^{-x}$ produces a reflection of the graph $y = 2^x$ about the y-axis.

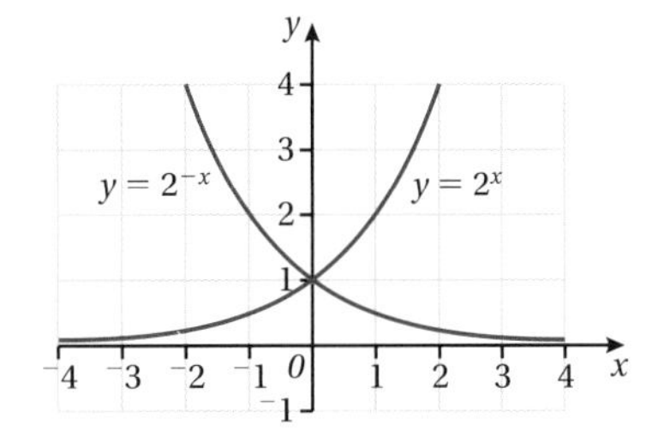

Tip

You will learn about transforming curves in Chapter 41.

Modelling using exponential functions

Exponential functions are used to model situations involving rapid growth or decay, for example population growth and compound interest.

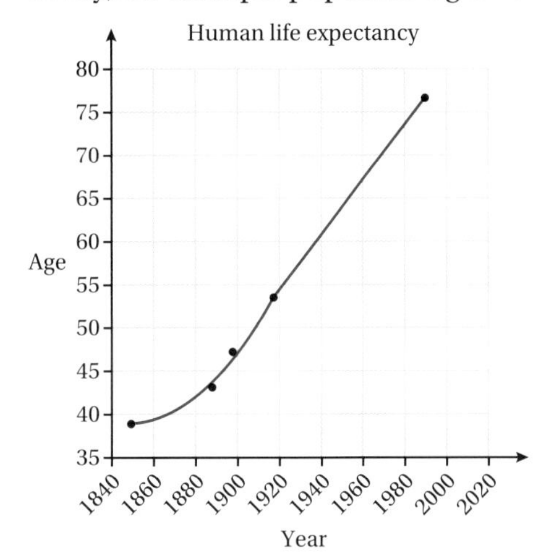

Tip

You will learn more about exponential growth and decay in Chapter 35.

In many real life applications, exponential functions are multiplied by a fixed value. For example, radioactivity levels in a radioactive material decrease exponentially over time at a rate of $R = 80 \times 0.55^t$, where R is the level of radioactivity and t is the time in hours.

This is where the a value in the general form of exponential functions comes into play.

In $y = ak^x$, $a > 0$ and is a constant value.

The curve of $y = ak^x$ has all the same features as the graph of $y = k^x$ except that the y-intercept is at $(0, a)$. The value of a will change the value of k, so will consequently impact on the steepness of the graph.

WORKED EXAMPLE 6

The temperature of two different metals heated in a furnace can be described using the rule

temperature = initial temperature $\times\ 3^{\text{time}}$.

Compare the increase in temperatures over four minutes for two samples of the same metal with initial temperatures of $\frac{1}{2}$ °C and 2°C. What do the graphs indicate?

x	0	1	2	3	4
$\frac{1}{2} \times 3^x$	$\frac{1}{2}$	$\frac{3}{2}$	$\frac{9}{2}$	$\frac{27}{2}$	$\frac{81}{2}$
2×3^x	2	6	18	54	162

Draw up a table of values for the given information. Use only positive values as the time is given in minutes and this cannot be negative.

Continues on next page ...

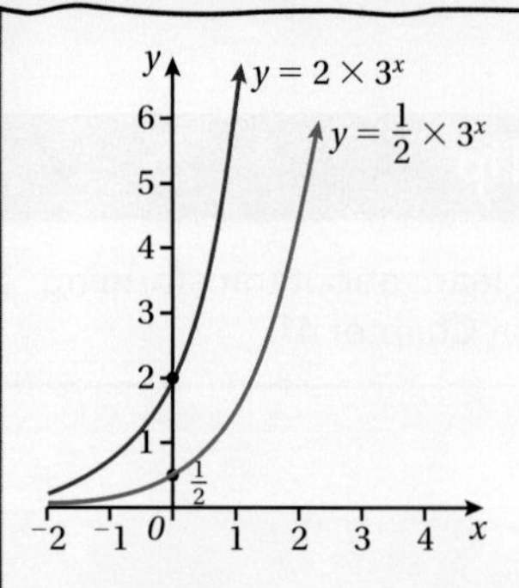

The first few values are sufficient to compare the graphs.

The graph of $y = 2 \times 3^x$ is steeper than the graph of $y = \frac{1}{2} \times 3^x$, indicating that metal at a higher initial temperature will increase in heat much faster than the same metal at a lower initial temperature.

EXERCISE 19G

1 Produce a table of values and draw the graph of each equation for the given values of x.

Draw each set of graphs on the same axes.

a **i** $y = 3^x$ **ii** $y = 1.1 \times 3^x$ **iii** $y = 2.5 \times 3^x$ values: $^-2 \leqslant x \leqslant 3$

b **i** $y = 5^x$ **ii** $y = 2 \times 5^x$ **iii** $y = \frac{1}{2} \times 5^x$ values: $^-1 \leqslant x \leqslant 2$

2 **a** Plot the graph of the equation $y = 2^x$ on a grid. (Use the table of values from Worked Example 5 to do this.)

b Sketch the graphs of $y = 4^x$ and $y = 2^{-x}$ in relation to $y = 2^x$.

3 Complete a table of values, plot the points and draw the graph for $y = \left(\frac{1}{4}\right)^x$ for $^-3 \leqslant x \leqslant 3$.

4 Draw a sketch graph of the following equations.

a $y = 3^x$ **b** $y = 1^x$

5 Consider the equation $P = 5 \times (0.85)^t$

a Will this equation result in an increasing or decreasing curve? Give a reason for your answer.

b Where will this graph cut the y-axis?

c Use the information to draw a sketch of this graph.

Section 5: Circles and their equations

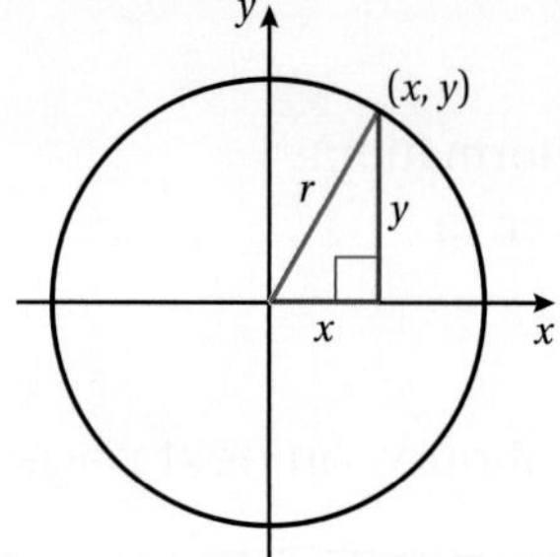

The equation of a circle is based on the fact that every point on the circumference is the same distance from the centre.

Any circle that has its centre on the origin and a radius of r can be defined by the equation:

$$x^2 + y^2 = r^2$$

WORK IT OUT 19.3

What is the equation of a circle centre the origin with a radius of 4 units?

Option A	Option B	Option C
$(x + y)^2 = 4$	$x^2 + y^2 = 16$	$x^2 + y^2 = 4$

EXERCISE 19H

1 $x^2 + y^2 = 25$ is a circle centre the origin.

a What is the value of the radius?

b Verify that the following points lie on the circle: (3, 4), ($^-3$, 4).

c List the coordinates of four other points that would also lie on this circle.

2 Sketch the graph of each circle, marking the intercepts on each of the axes.

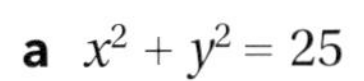

a $x^2 + y^2 = 25$ **b** $x^2 + y^2 = 1$ **c** $x^2 + y^2 = 2$

d $x^2 + y^2 = \frac{9}{4}$ **e** $x^2 = 5 - y^2$

> **Tip**
>
> To sketch the graph, set your pair of compasses to (0, r) on the x-axis and draw a circle with the origin as the centre.

3 **a** Which of these points lie on the circle $x^2 + y^2 = 100$?

(6, 8) (10, 10) (20, 80) ($^-6$, 8) $(5\sqrt{2}, 5\sqrt{2})$ (10, 0)

b Which of these points lie on the circle $x^2 + y^2 = 169$?

(5, 12) (100, 69) ($^-5$, $^-12$) ($^-5$, 12) $(^-13\sqrt{2}, 13\sqrt{2})$ (0, 13)

4 Write the equation of each of the circles in this diagram.

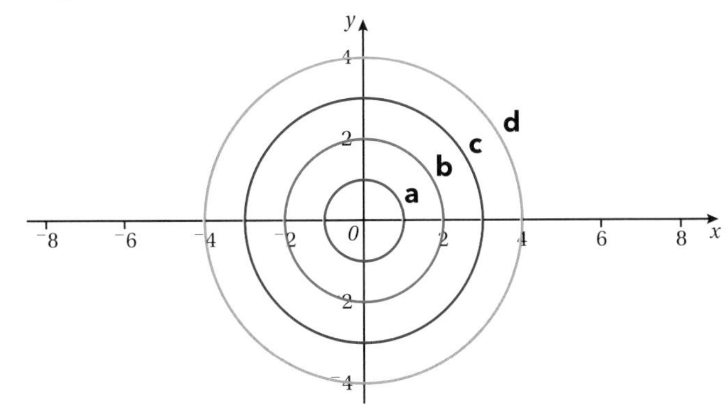

Checklist of learning and understanding

Linear functions

- Linear functions produce straight-line graphs.
- The general form of the linear function is $y = mx + c$.
- Graphs $x = a$ are vertical lines parallel to the y-axis.
- Graphs $y = a$ are horizontal lines parallel to the x-axis.
- Lines of the form $y = mx$ go through the origin.

Find answers at: cambridge.org/ukschools/gcsemaths-studentbookanswers

Quadratic functions

- The graphs of quadratic equations such as $y = ax^2$ and $y = ax^2 + c$ are called parabolas.
- When a is positive the graph goes down to a minimum point. When a is negative the graph goes up to a maximum point. The y-intercept is given by c.
- Parabolas have a turning point which can be a minimum or maximum depending on the shape of the graph. The axis of symmetry has the same x-value as the turning point $x = 0$.

Polynomials and reciprocals

- To draw graphs of polynomials first calculate a table of values that satisfy the equation for a range of values for x.
- A cubic function is a curve defined by $y = ax^3 + c$.
- A reciprocal function is a graph made of two non-connected curves in opposite quadrants defined by $y = \frac{a}{x}$ or $xy = $ a.

Other curved graphs

- An exponential function is a steeply increasing or decreasing curve defined by $y = ak^x$.
- Any circle with a centre of (0, 0) and radius r, can be defined by $x^2 + y^2 = r^2$.

For additional questions on the topics in this chapter, visit GCSE Mathematics Online.

Chapter review

1 Draw the graphs of the straight lines $y = 2x - 5$ and $2y - x = 5$.
What is the point of intersection of these two lines?

2 **a** What are the roots of the quadratic equation represented by this graph?

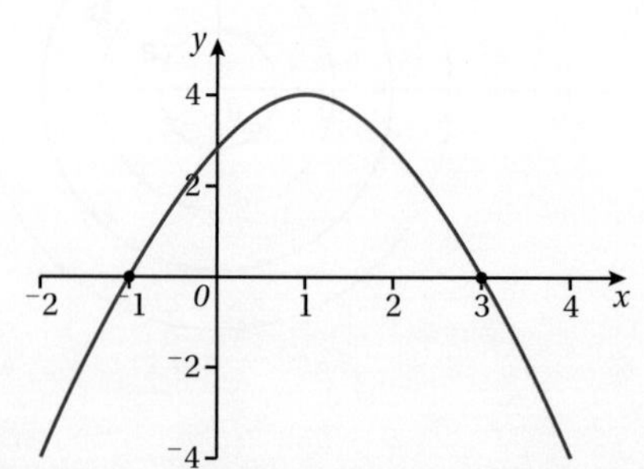

b Show that the equation of the graph is $y = {}^{-}x^2 + 2x + 3$.

c Complete the square for $x^2 - 2x - 3 = 0$ to find the coordinates of the turning point and the axis of symmetry for the graph of $y = {}^{-}x^2 + 2x + 3$.

3 Each statement below is sometimes true and sometimes false.

For each statement sketch a graph to show an example where it is true and an example where it is false.

The first one has been done for you. *(3 marks)*

Statement	True	False
A straight line graph goes through the origin.		
The gradient of a straight line graph is positive.		
A quadratic equation $ax^2 + bx + c = 0$ has two positive solutions.		

© OCR 2013

4 Draw a sketch diagram of the circle $x^2 + y^2 = 25$.

On the same diagram sketch the two linear functions represented by the equations $y = 2x - 2$ and $y = {}^{-}\left(\frac{3}{4}x\right) + 6\frac{1}{4}$.

Verify that one of the lines will intersect with the circle at two points and the other line will be a tangent to the circle.

What are the coordinates of the point of contact of the tangent with the circle?

5 Determine the equation of each graph using what you know about the features of different graphs.

a

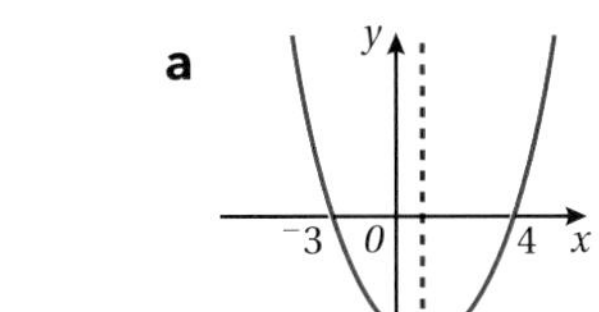

b

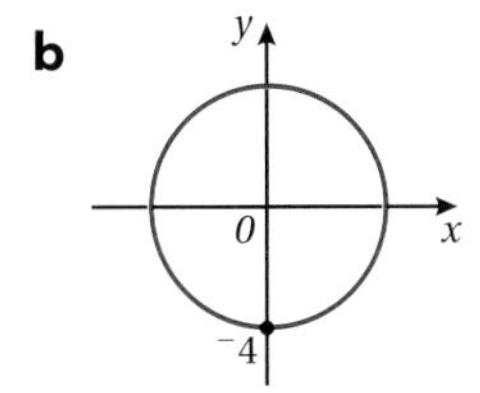

c

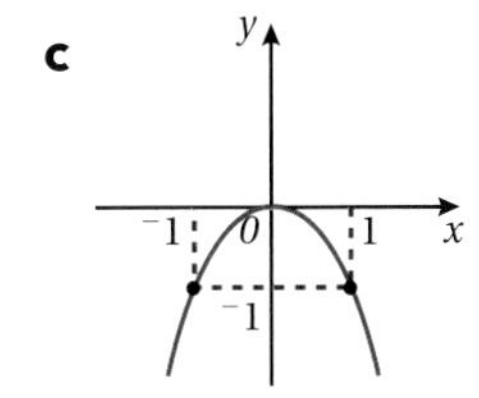

d

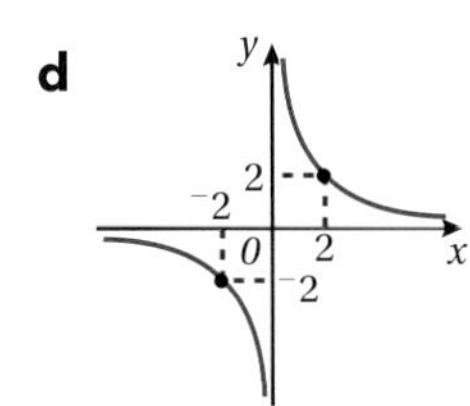

e

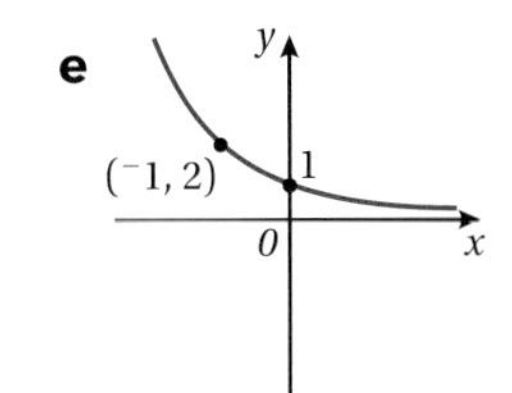

f

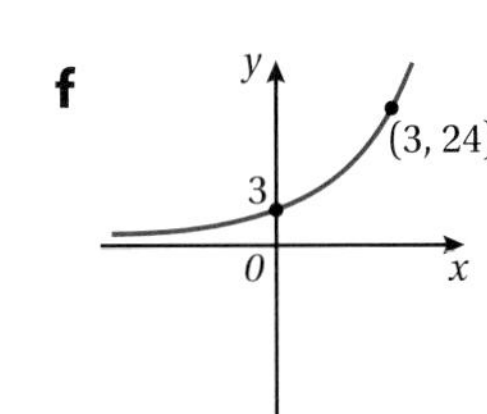

Find answers at: cambridge.org/ukschools/gcsemaths-studentbookanswers

20 Three-dimensional shapes

In this chapter you will learn how to ...

- work with 2D representations of 3D objects.
- construct and interpret plans and elevations of 3D objects.

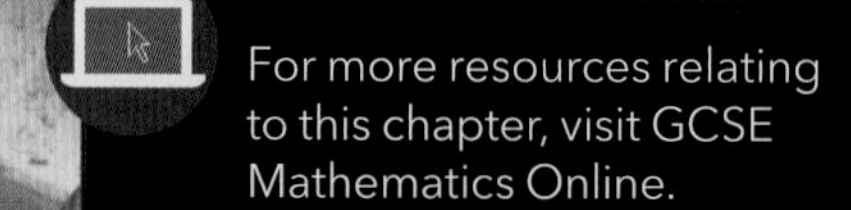

For more resources relating to this chapter, visit GCSE Mathematics Online.

Using mathematics: real-life applications

Buildings, engine parts, vehicles and packaging are all carefully planned and designed before they are built. Most design work starts on paper or screen using two-dimensional (2D) images to represent the final three-dimensional (3D) objects.

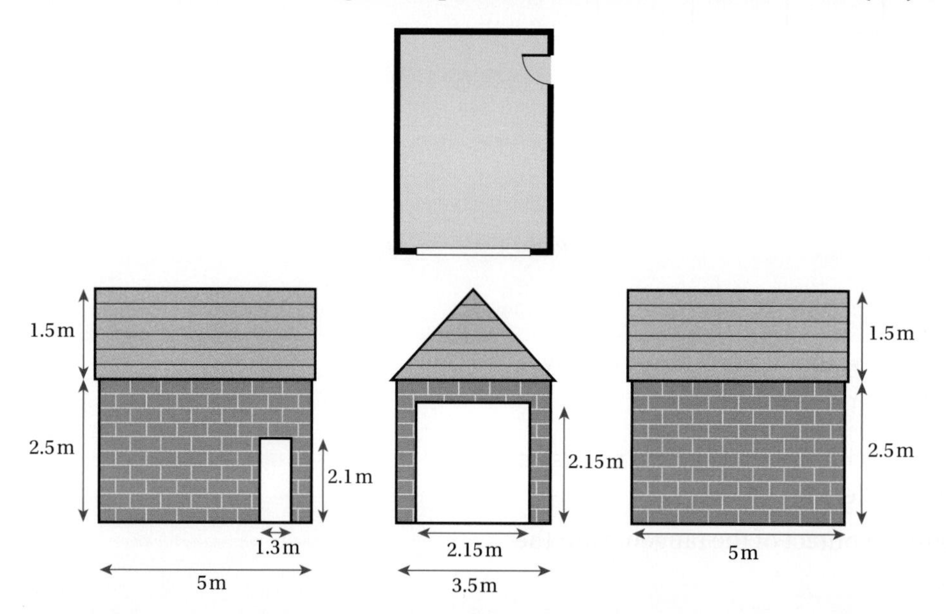

"No one will buy an apartment that isn't built yet if they don't know what it is going to look like! When we sell a development we show people floor plans as well as elevations from all four sides. Sometimes we also have a 3D scale model of the development." *(Estate agent)*

Before you start ...

Ch 5	You must be able to identify and name some common 3D solids.	**1** Name each of these 3D solids. **a** **b** **c** **d**
Ch 5	You should know the basic properties of polygons and other 3D solids.	**2** True or false? Correct the false statements. **a** A cube has 4 faces. **b** A cube has 12 edges. **c** A cuboid has 8 vertices.
Ch 6	You must be able to accurately construct lines and angles using your ruler and a pair of compasses.	**3** Construct and bisect a right angle ABC. **4** Use a pair of compasses to draw a circle of radius 5 cm. Construct a hexagon inside the circle.

Assess your starting point using the Launchpad

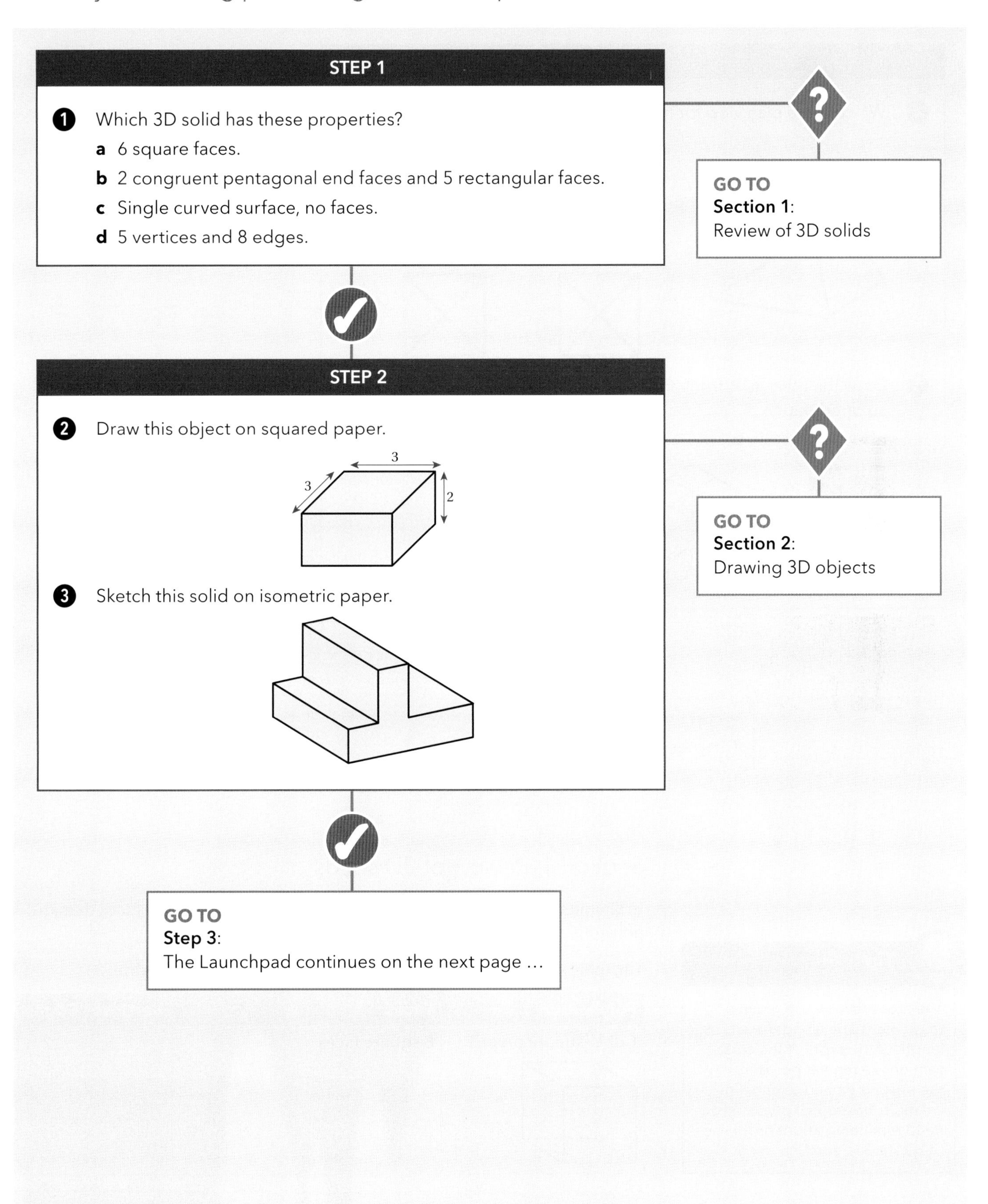

STEP 1

1. Which 3D solid has these properties?
   - **a** 6 square faces.
   - **b** 2 congruent pentagonal end faces and 5 rectangular faces.
   - **c** Single curved surface, no faces.
   - **d** 5 vertices and 8 edges.

?

GO TO
Section 1:
Review of 3D solids

✓

STEP 2

2. Draw this object on squared paper.

3. Sketch this solid on isometric paper.

?

GO TO
Section 2:
Drawing 3D objects

✓

GO TO
Step 3:
The Launchpad continues on the next page …

Find answers at: cambridge.org/ukschools/gcsemaths-studentbookanswers

Launchpad continued ...

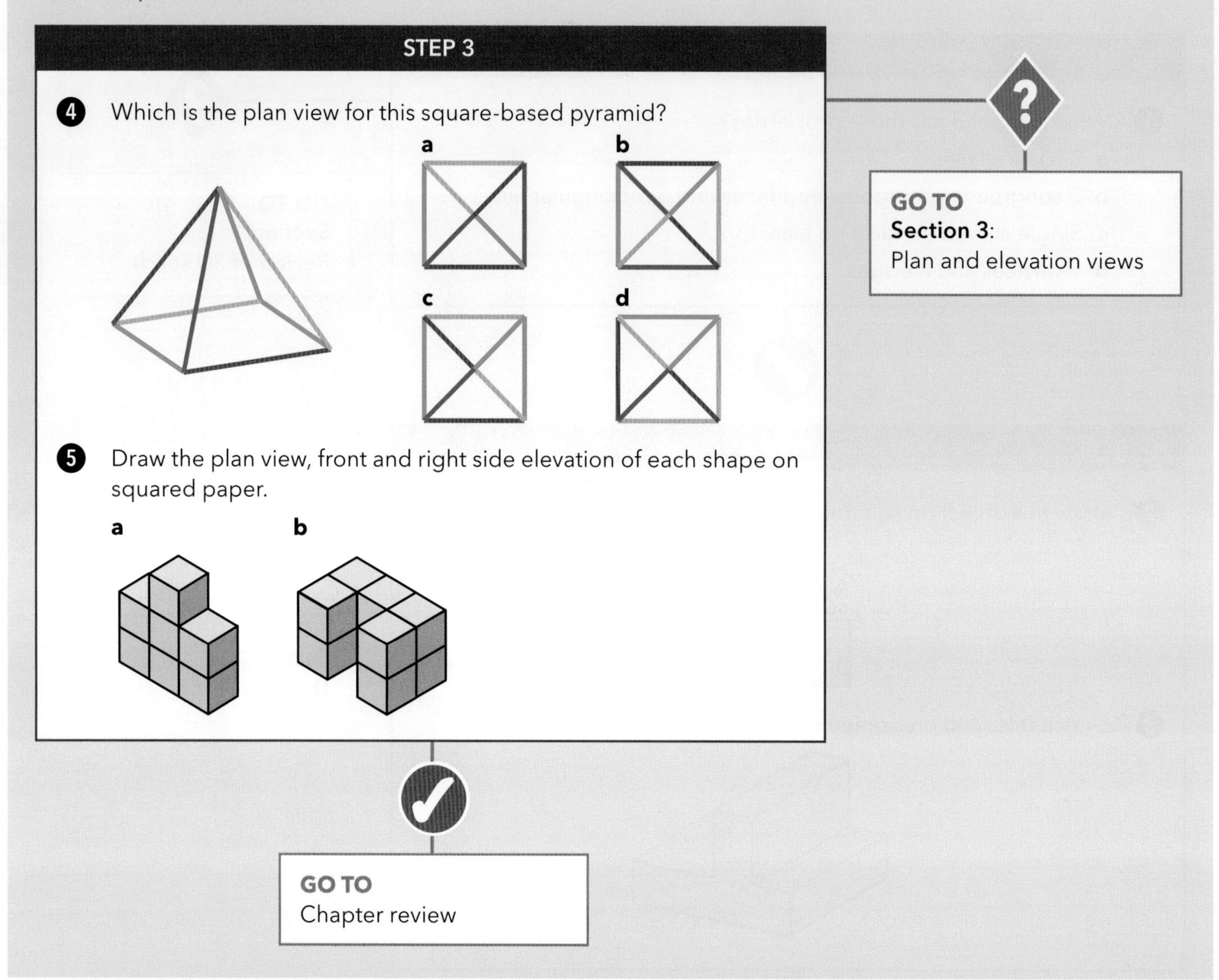

STEP 3
4 Which is the plan view for this square-based pyramid?
a
b
c
d
5 Draw the plan view, front and right side elevation of each shape on squared paper.
a
b
?
GO TO
Section 3:
Plan and elevation views
GO TO
Chapter review

Section 1: Review of 3D solids

The cross section of a prism is the same shape along the entire length of the prism. If the cross section is a regular polygon, the prism is named after it. For example, hexagonal prism. If you need reminding about properties of 2D shapes, see Chapter 5.

In Chapter 5 you saw that polyhedra are solid shapes with flat faces that are polygons.

The table summarises the main properties of different polyhedra.

Polyhedron	Faces	Vertices	Edges
Cube (square prism)	6 square faces	8	12

Cuboid (rectangular prism)	2 congruent rectangular end faces 4 rectangular faces	8	12
Triangular prism	2 congruent triangular end faces 3 rectangular faces	6	9
Pentagonal prism	2 congruent pentagonal end faces 5 rectangular faces	10	15
Triangular pyramid	1 triangular base 3 triangular faces that meet at an apex	4	6
Square-based pyramid	1 square base 4 triangular faces that meet at an apex	5	8

Cylinders, cones and spheres are also 3D shapes but they are **not** polyhedra; they do not have flat faces that are polygons.

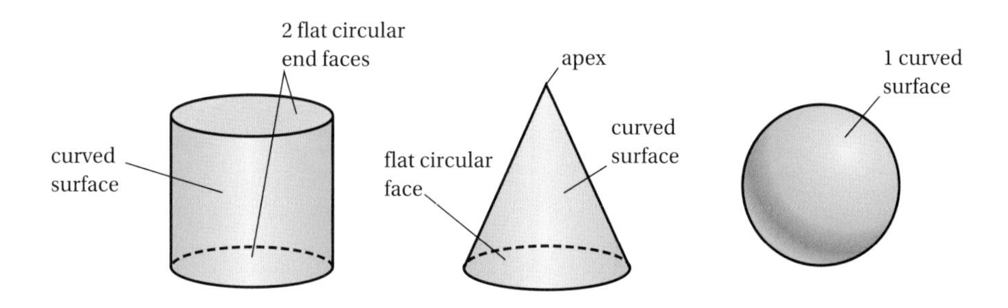

Did you know?

Leonhard Euler was a Swiss mathematician and physicist who noticed a pattern between the number of faces, vertices and edges in polyhedra. Euler's theorem states that $F + V = E + 2$ for any convex polyhedron, where F = number of faces, V = number of vertices, E = number of edges.

WORKED EXAMPLE 1

Describe each object fully by referring to its properties.

State if it is a polyhedron or not, giving a reason for your decision.

a 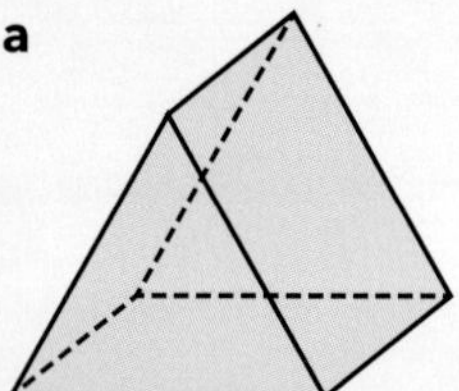 **b** 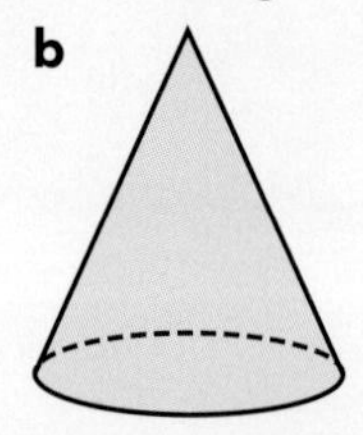 **c** 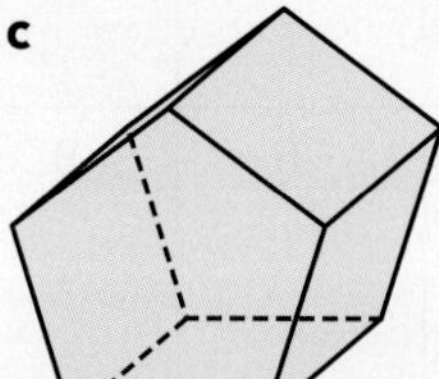 **d** 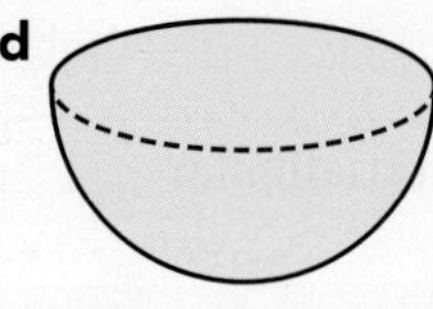

a Two end faces are congruent triangles. There are also three rectangular faces. It has 6 vertices and 9 edges. Its cross-section is a regular polygon. It is a polyhedron because all of its faces are polygons.

b The shape has one end face that is a circle. It has one vertex and a single curved face. It is not a polyhedron because not all of its faces are polygons.

c It has two congruent pentagon faces and 5 rectangular faces. It has 15 edges and 10 vertices. It has a regular cross-section. It is a polyhedron because all its faces are polygons.

d The shape has a flat circle face and a single curved face. It has no vertices. It is not a polyhedron because not all of its faces are polygons.

> Make sure you use the correct vocabulary. Don't forget to include how many vertices and edges it has.

EXERCISE 20A

1 What is the mathematical name for each of the following shapes?

a A solid with 6 faces; the 2 end faces are congruent rectangles.

b There are 4 vertices and 6 edges. At one of the vertices, 3 triangular faces meet. The base of the shape is a triangle.

c A solid with 6 vertices and 2 congruent triangular end faces.

d A regular solid with 6 identical faces, 8 vertices and 12 edges.

2 Which solid or solids, does each photo remind you of?

 **a** 

 **b** 

c

d

3 Where might you find the following in real life?

a A sphere.

b A cube.

c A cone.

d A cylinder.

Section 2: Drawing 3D Objects

You need to be able to make drawings of 3D objects, and to interpret drawings of 3D objects from different perspectives.

It can be challenging to draw a 3D object because you are trying to show three dimensions on a two-dimensional plane (your paper or computer screen).

There are a number of ways of drawing 3D objects to show their features in 2D. As you read through each method, try it out on rough paper.

Prisms and cylinders using end faces

You can make fairly realistic drawings of prisms and cylinders by visualising the position of their end faces.

Cuboid

a

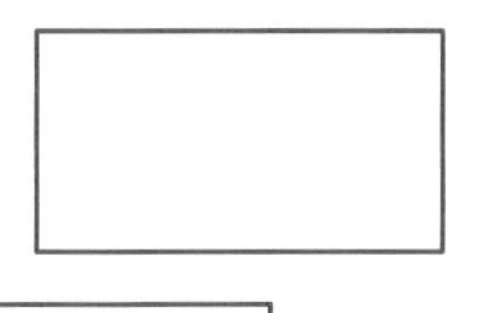

A cuboid is a rectangular prism. Visualise where the rectangular end faces would be and draw those first. Draw two rectangles, positioning one above the other, with one of them further over to one side.

b

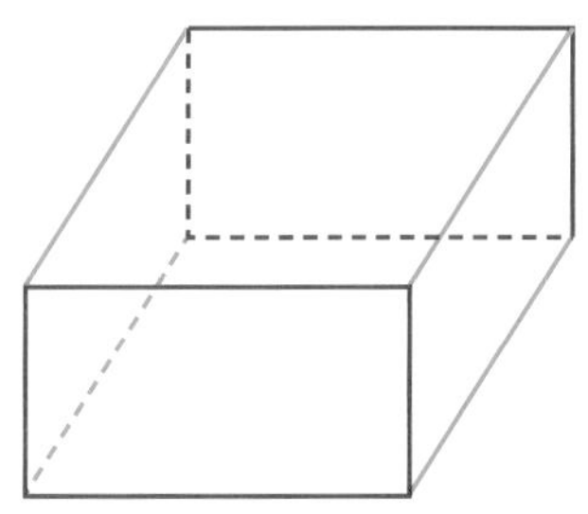

Then join the two rectangles together by drawing a line from each vertex on one rectangle to the corresponding vertex on the other rectangle. For any edges that you wouldn't be able to see, use a broken line.

Cylinder

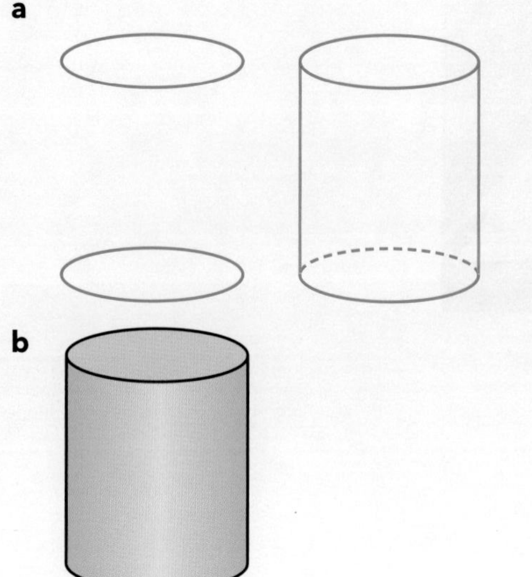

First draw the two circular end faces. Draw them as o[...] to get a more realistic drawing. Then draw in two line[...] join the end faces.

Shading can make a drawing look more realistic.

Triangular prism

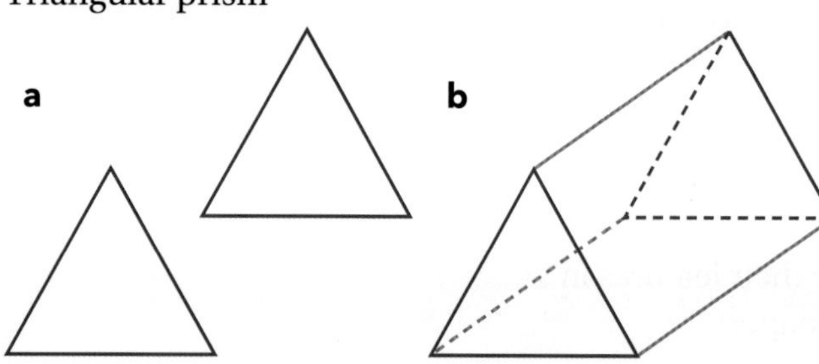

First draw the two triangular end faces. Then draw in lines to join the vertices. For any edges that you woul[...] be able to see, use a broken line.

The same process can be used for any shaped prism, just make sure you rememb[...]

- start with the two end faces
- join all corresponding vertices
- shading can help make the shape look more realistic.

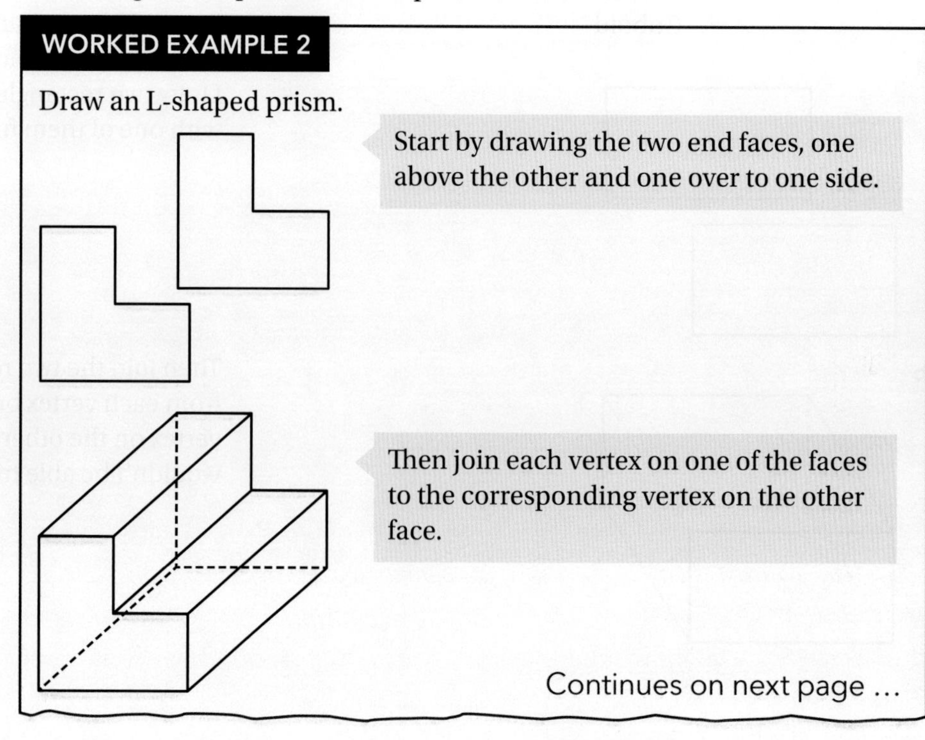

WORKED EXAMPLE 2
Draw an L-shaped prism.
Start by drawing the two end faces, one above the other and one over to one side.
Then join each vertex on one of the faces to the corresponding vertex on the other face.

Continues on next page ...

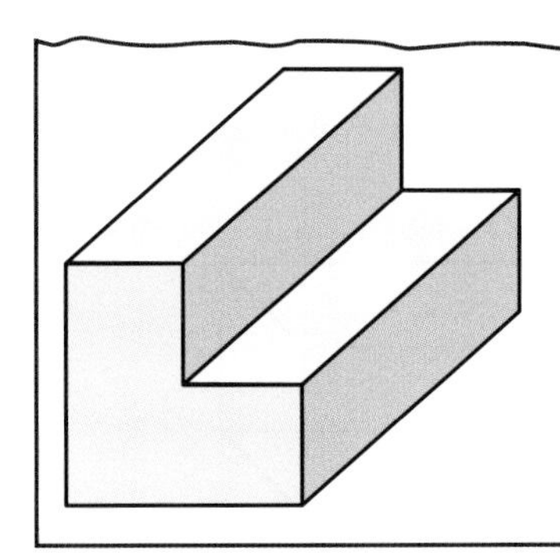

Use shading to make the shape look more realistic.

Prisms and pyramids from parallel lines

Tip

Parallel lines were covered in Chapter 5.

You can draw square and rectangular prisms and square-based pyramids using two pairs of parallel lines as a starting point.

Prism

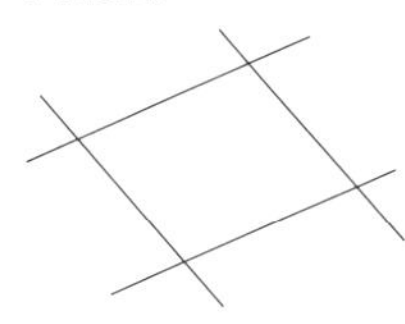

Begin by drawing two pairs of parallel lines that intersect (cross each other).

Then draw vertical lines of equal length down (or up) from each intersection. Complete the shape by joining the ends to make a prism.

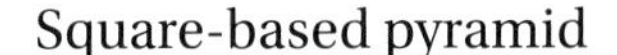

Square-based pyramid

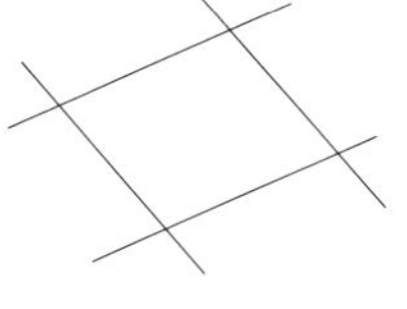

Start by drawing two pairs of parallel lines that intersect.

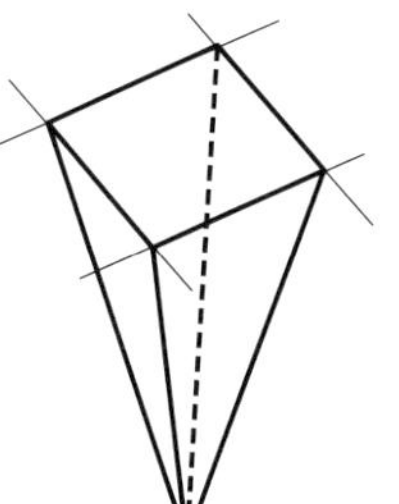

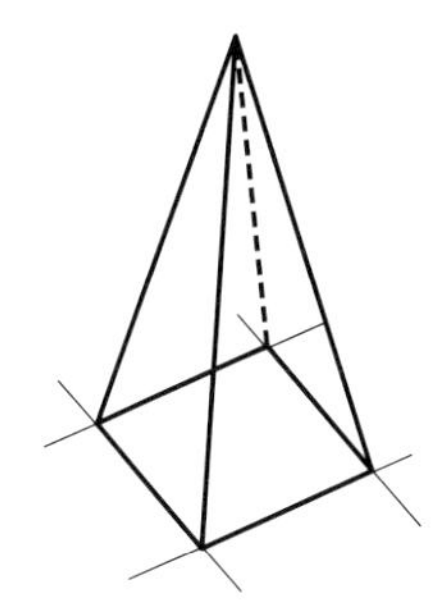

Mark a point above, below or to the side, of the parallel lines. Draw lines from the intersections of the parallel lines to the point.

Drawing shapes on squared or isometric grids

Tip

'Viewpoints' means that you could look at a shape from the front, the back, the side, or from one of its edges. The viewpoint from which you look at a shape will determine which of its faces/edges/vertices you can see and which would be hidden.

3D objects can be drawn on squared or isometric grids to show the object from different viewpoints.

The vertical lines on each type of grid are used to represent the vertical edges of the 3D object. You draw along the lines at an angle on the paper to represent the horizontal edges of the 3D object.

Find answers at: cambridge.org/ukschools/gcsemaths-studentbookanswers

The diagram below shows a cube and a cuboid drawn on a square grid. The shapes are drawn as if viewed 'face-on'.

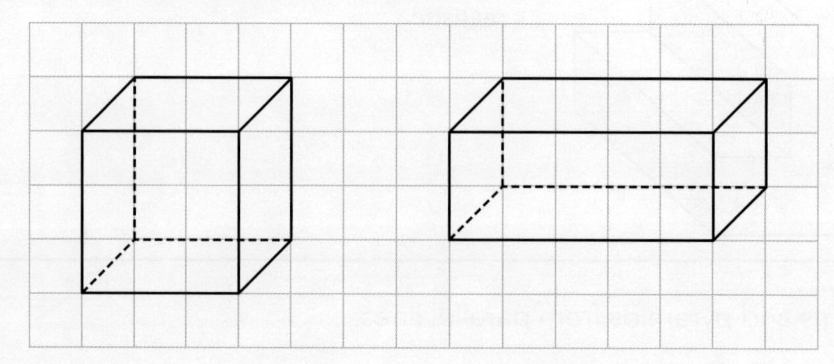

Key vocabulary

isometric grid: special drawing paper based on an arrangement of triangles.

This diagram shows the same objects drawn on an **isometric grid**. The shapes are drawn as if viewed from one of their edges.

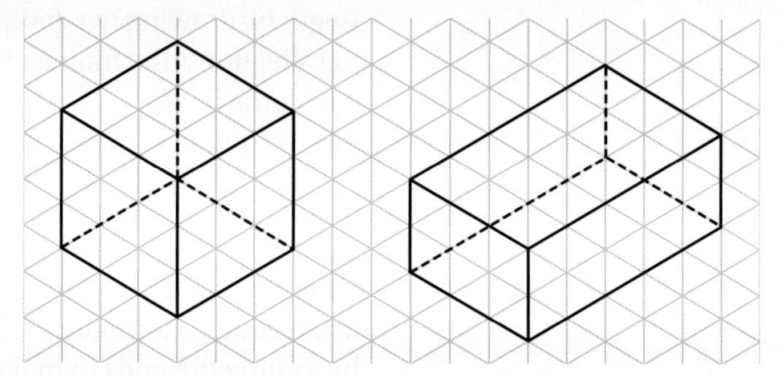

Remember that when you draw 3D shapes you use broken lines to show the edges that would **not** be seen from that viewpoint.

Isometric drawings

Isometric drawings are used to visually represent 3D objects in 2D in technical and engineering drawings.

This diagram shows the design of an engineering component on isometric paper.

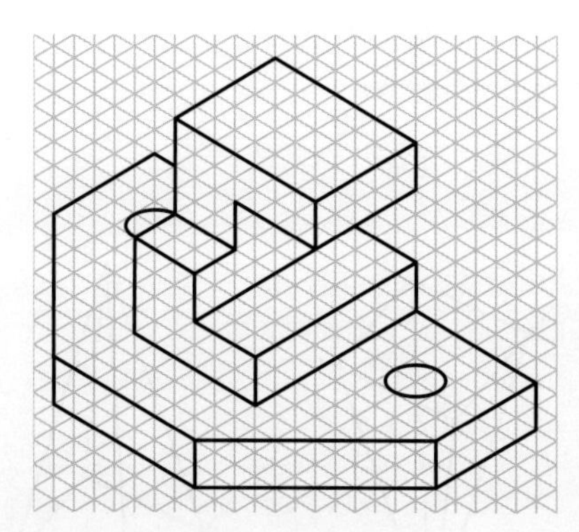

Isometric paper is very useful for drawing solids built from cubes.

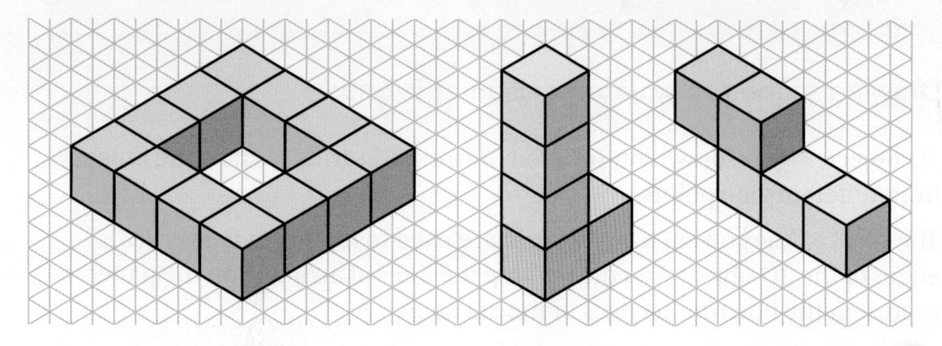

WORKED EXAMPLE 3

Draw this shape on an isometric grid.

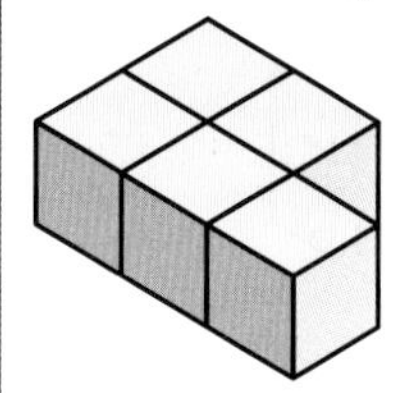

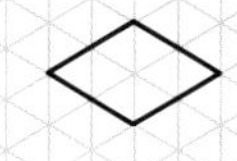

Use the lines on the grid to help you draw the horizontal face of one of the cubes; here, the top-most yellow cube is the starting point

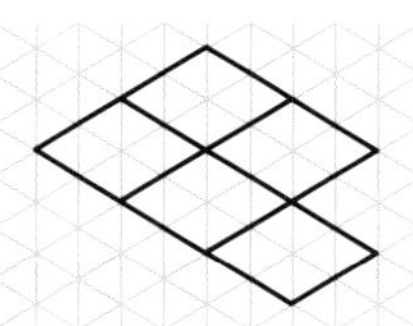

Use that face to draw in the other horizontal faces.

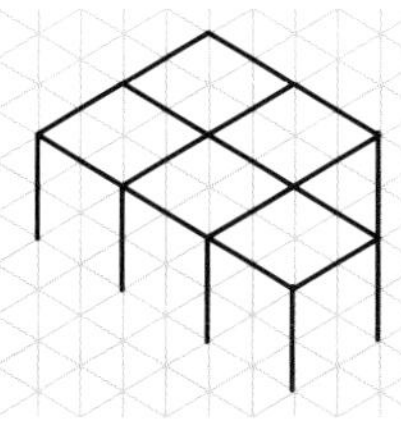

Use the vertical lines on the grid to draw in the vertical edges.

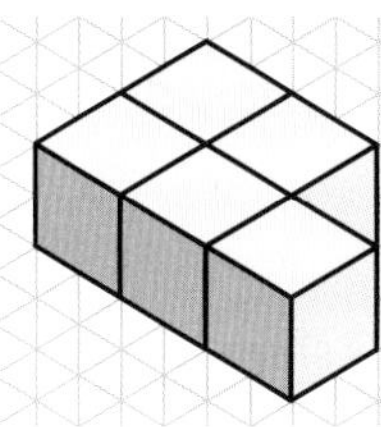

Use the angled lines to draw in the bottom horizontal edges.

You can use the 'end faces' or 'parallel lines' methods you saw earlier in the chapter on square and isometric grids. The grid makes it easier for you to make sure that the end faces are the same size.

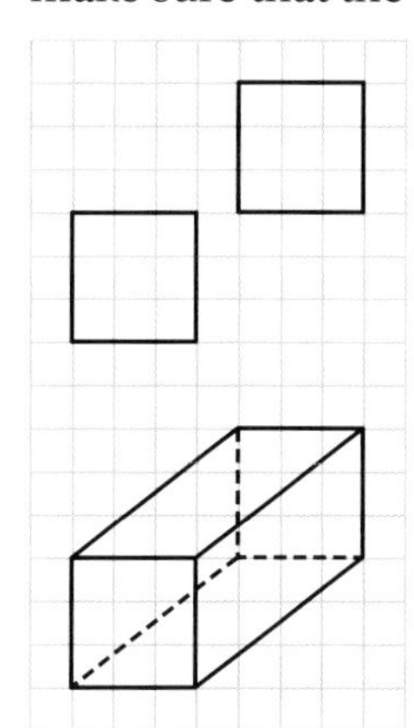

A cuboid drawn on a square grid using the 'end faces' method.

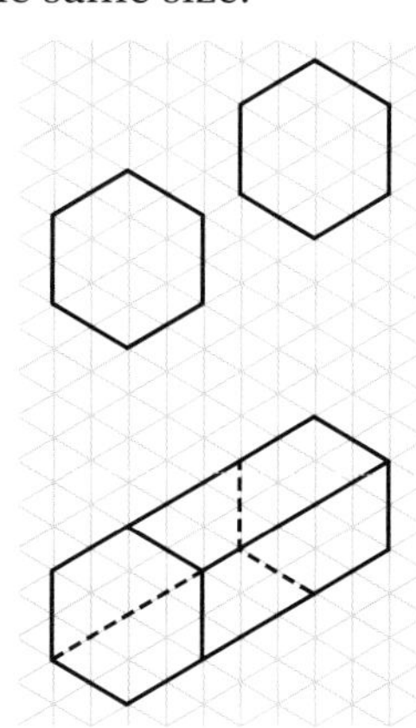

A hexagonal prism drawn on an isometric grid using the 'end faces' method.

Tip

Be very careful to join up corresponding vertices. Remember to use a sharp pencil and a ruler to make your diagrams look professional.

WORK IT OUT 20.1

Students were asked to draw this view of a shape on an isometric grid.

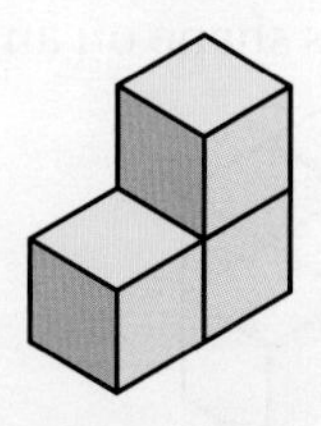

This is how they started their sketches.

Student A	Student B	Student C

Which student is likely to end up with the correct view of the shape?

What are the others doing incorrectly?

EXERCISE 20B

1 Draw the following objects without using a grid.

a 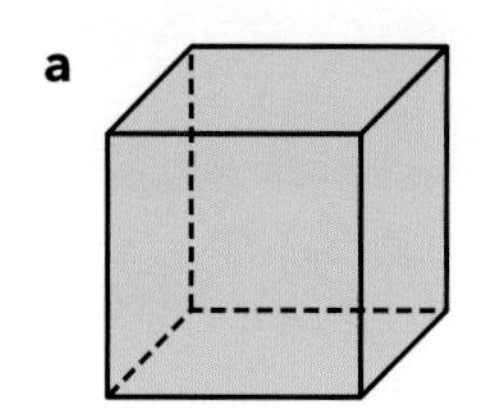

b 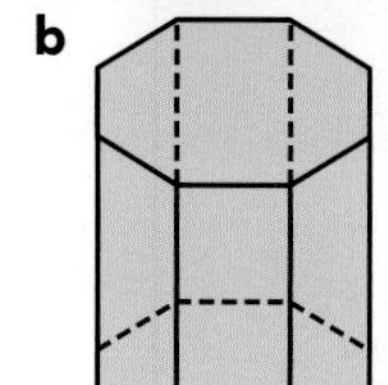

c 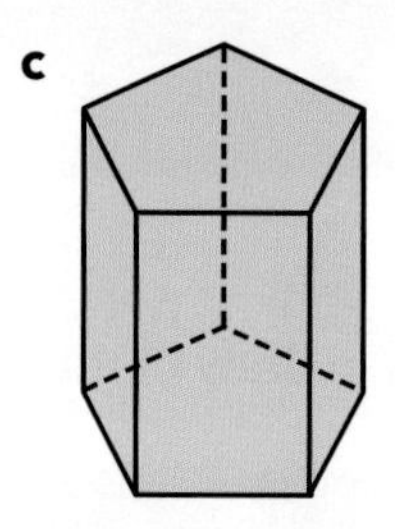

d 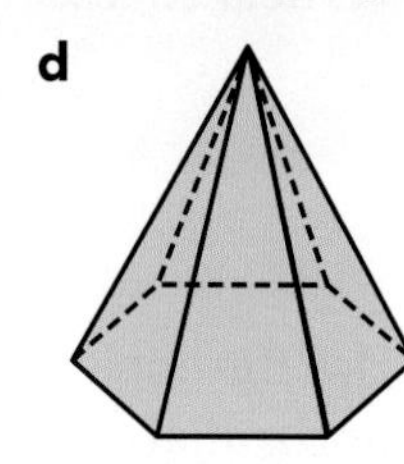

e 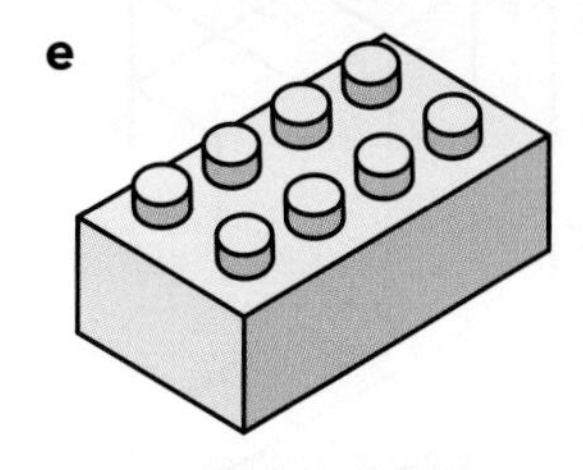

2 One of the parallel end faces of three different prisms is shown here.

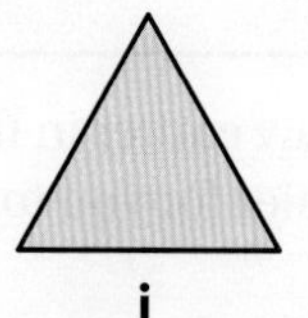
i

ii

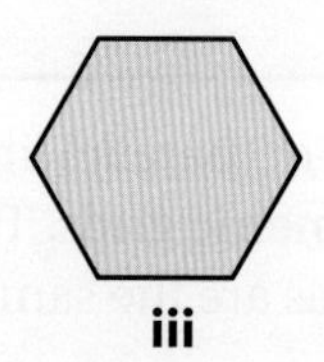
iii

a Sketch each prism on squared paper.

b Sketch each prism on isometric paper.

c Compare the two drawings. How does the grid affect what your drawing looks like?

3 Draw the following shapes on an isometric grid.

a 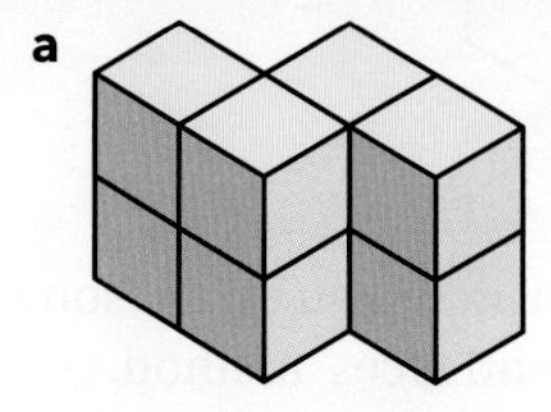

b 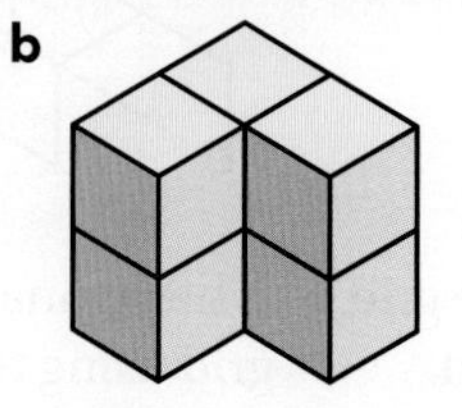

c 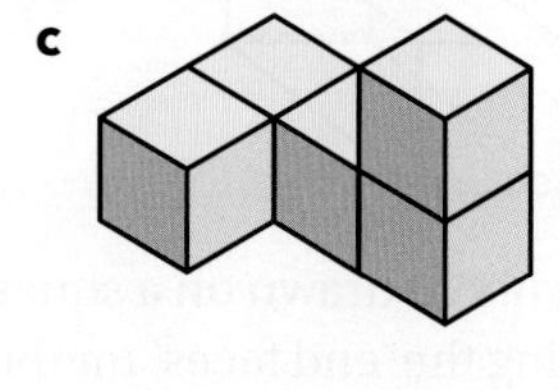

4 The diagrams show different shapes made from cubes.
If there are no cubes missing from the layers you cannot see, how many cubes would you need to build each shape?

a

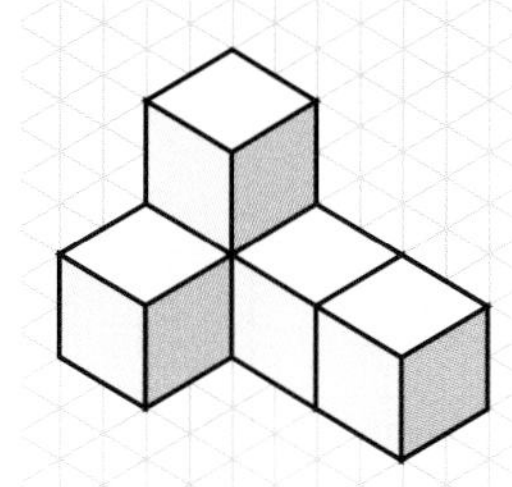

b

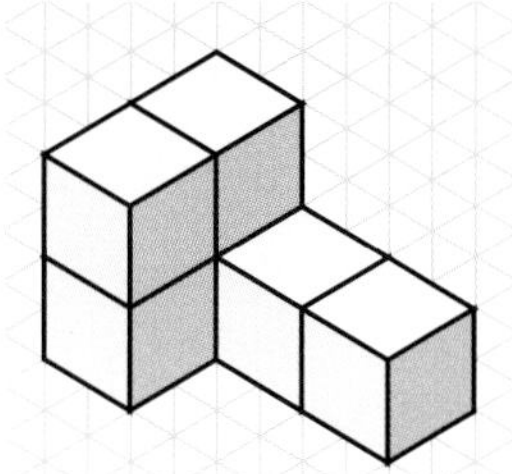

c

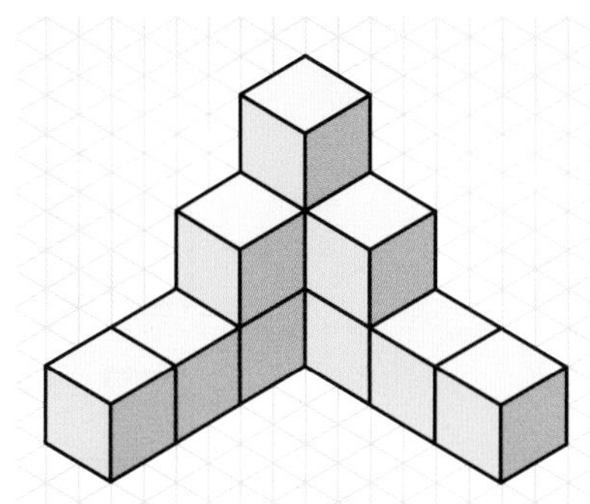

d

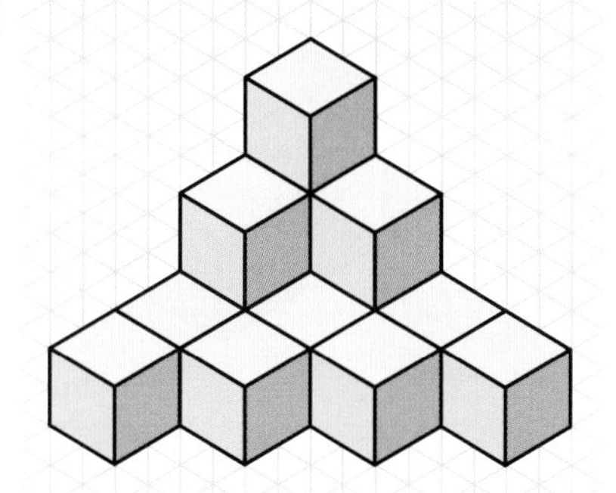

5 Shading 3D drawings can create optical illusions.

a Are there six or seven cubes in this diagram?

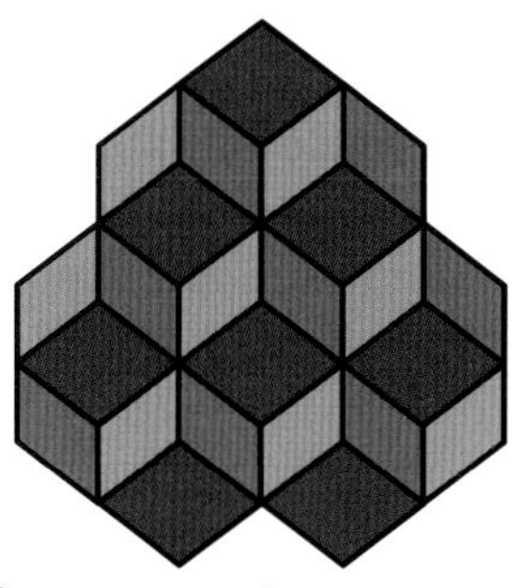

b Is this a large cube with a small cube cut out of one corner?
Or a small cube inside the corner of a large open shape with three sides?

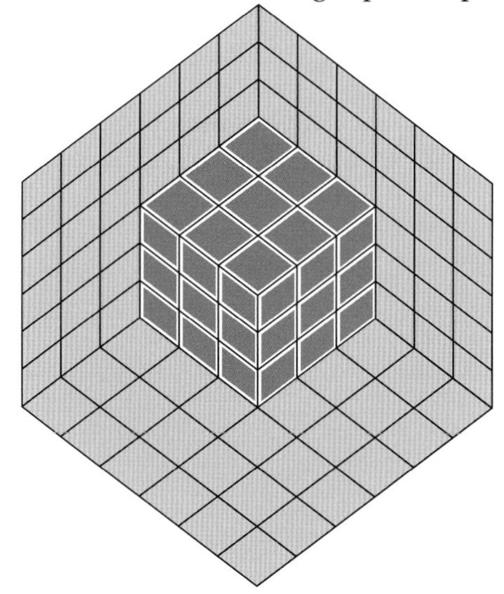

c Use an isometric grid to reconstruct these diagrams.
Shade your diagrams to form the optical illusion.

Key vocabulary

plan view: the view of an object from directly above.

elevation: a view of an object from the front, side or back.

Section 3: Plan and elevation views

A **plan** view shows how a 3D object would look if you viewed it from directly above.

The front **elevation** is the view of a 3D object from its front.

The side elevation is the view from the side.

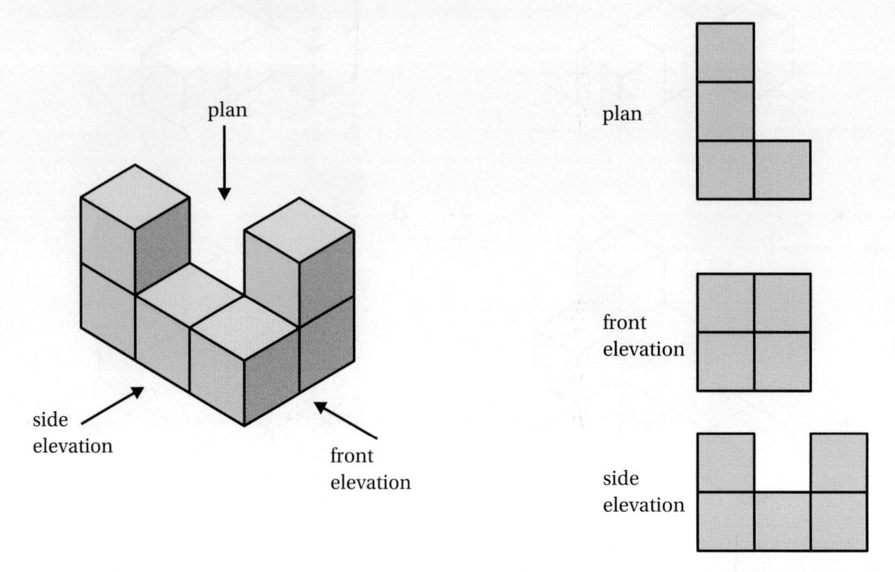

When you draw plans or elevations you show any immediate changes in height as solid lines. You use broken lines to indicate any hidden edges.

Think carefully about what the parts of the shape you cannot see will look like. For example, the shape below, must have a fourth (hidden) cube in order to support the top cube; you would need to include this cube when drawing the back and side elevations.

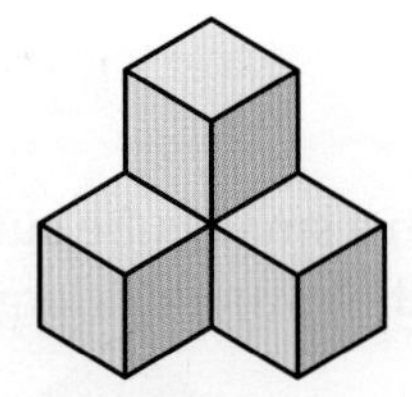

Look at the shape below. You can think of it as a prism with trapezium-shaped end faces. The front is higher than the back.

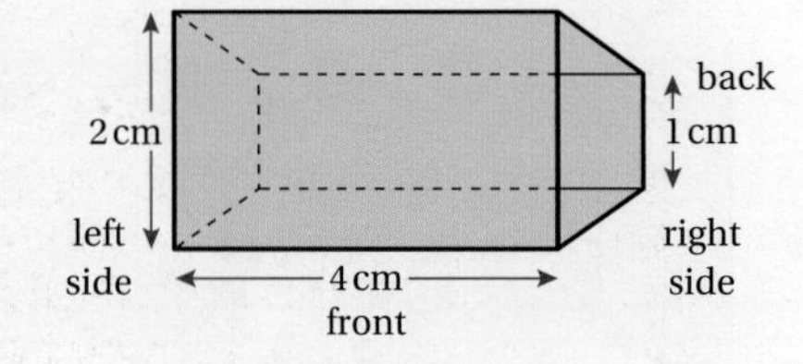

Plan

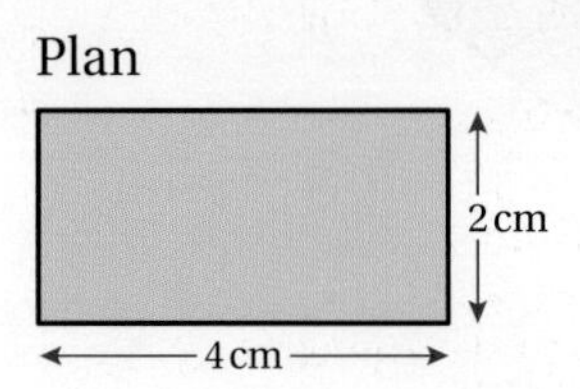

The plan view is a rectangle. Even though the top of the object slopes down to the back, if you look at it from above, it will look like a rectangle. The plan view is normally drawn above the front view because the two views will be the same width.

Elevations

The left and right elevations are reflections of each other. They are drawn on the left and right side of the front elevation.

The front and back elevations look the same, except that on the back elevation there is a solid line to show that the height of the back face is shorter than the front. You can see the same line in the front elevation but here it is broken to indicate that it is a hidden height. From these elevations, you would not be able to tell that the change in height from the front and back is not immediate. It is sloping; you need the side views to know this.

You get different information from different views because each one shows two of the three dimensions of the solid:

- The plan view shows the length and width of the solid.
- The front view shows the length and height of the solid.
- The side views show the width and height of the solid.

WORKED EXAMPLE 4

Draw the plan, front and side elevations of this solid.

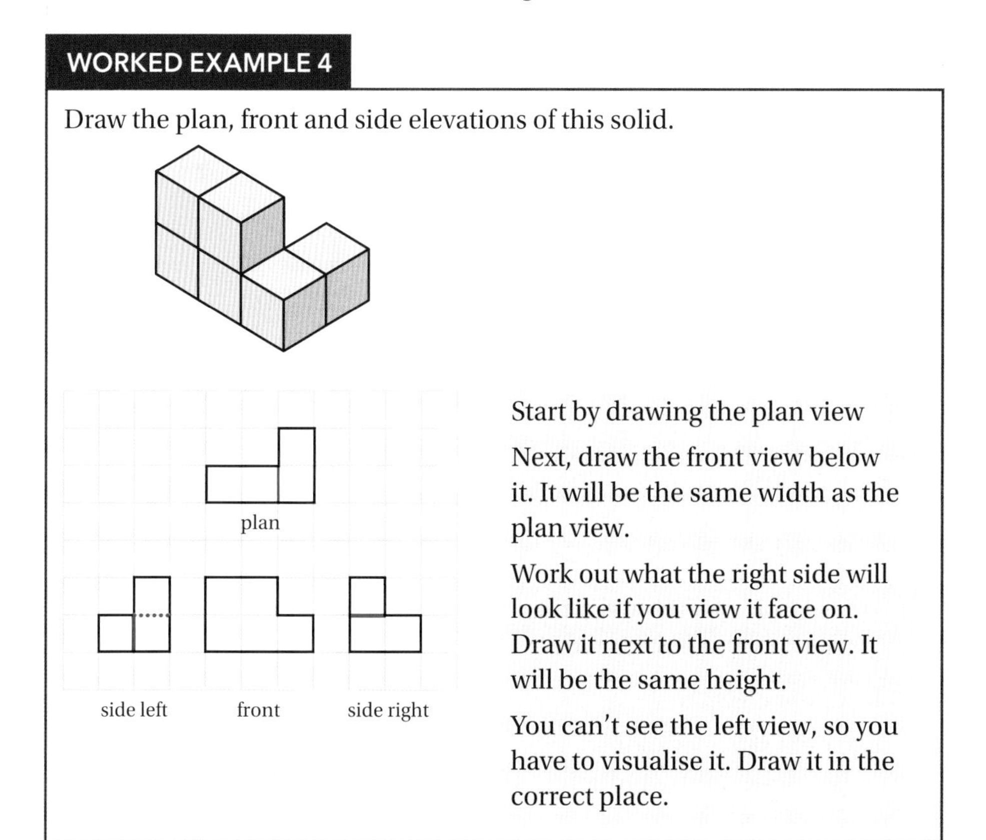

Start by drawing the plan view

Next, draw the front view below it. It will be the same width as the plan view.

Work out what the right side will look like if you view it face on. Draw it next to the front view. It will be the same height.

You can't see the left view, so you have to visualise it. Draw it in the correct place.

EXERCISE 20C

1 Select the correct plan view of each object.

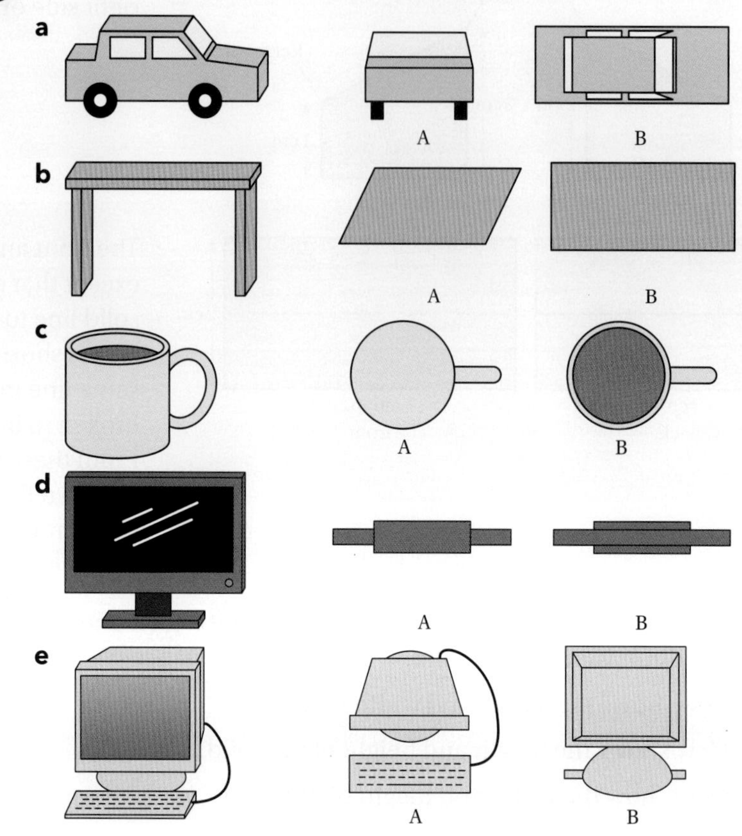

2 **a** Match each shape to its plan and elevation.

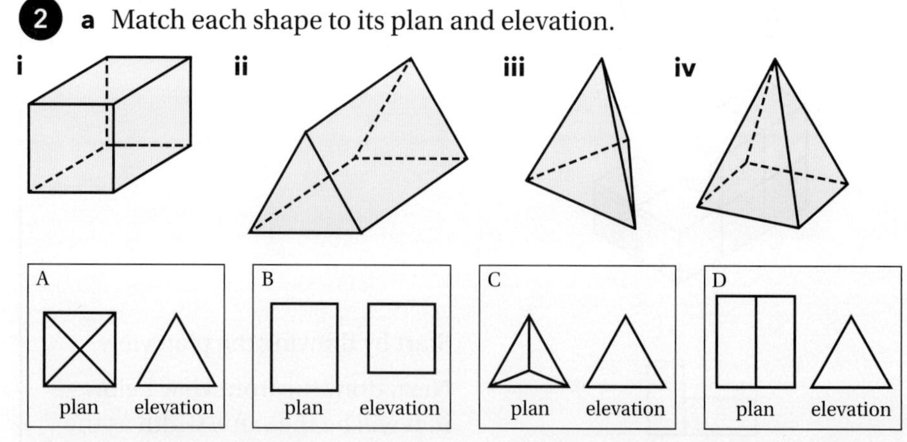

b Sketch and label the elevations that are not shown for each shape.

3 For each set of differently coloured cubes, draw:

a a plan.

b a front elevation.

c a right side elevation (from the right-hand side).

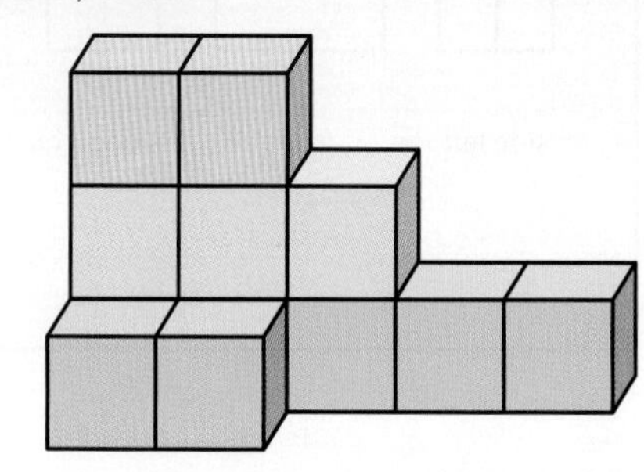

4 Draw the plan, the front elevation and the side elevation of the shape below.

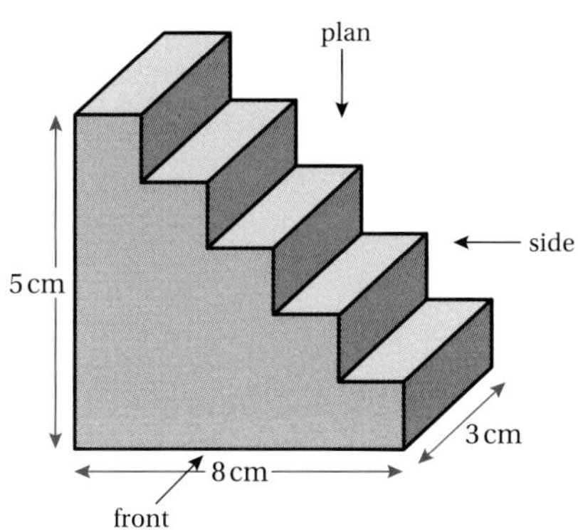

5 The plan view and elevations of different solids are shown.

Use these views to work out what each solid looks like and draw it on an isometric grid.

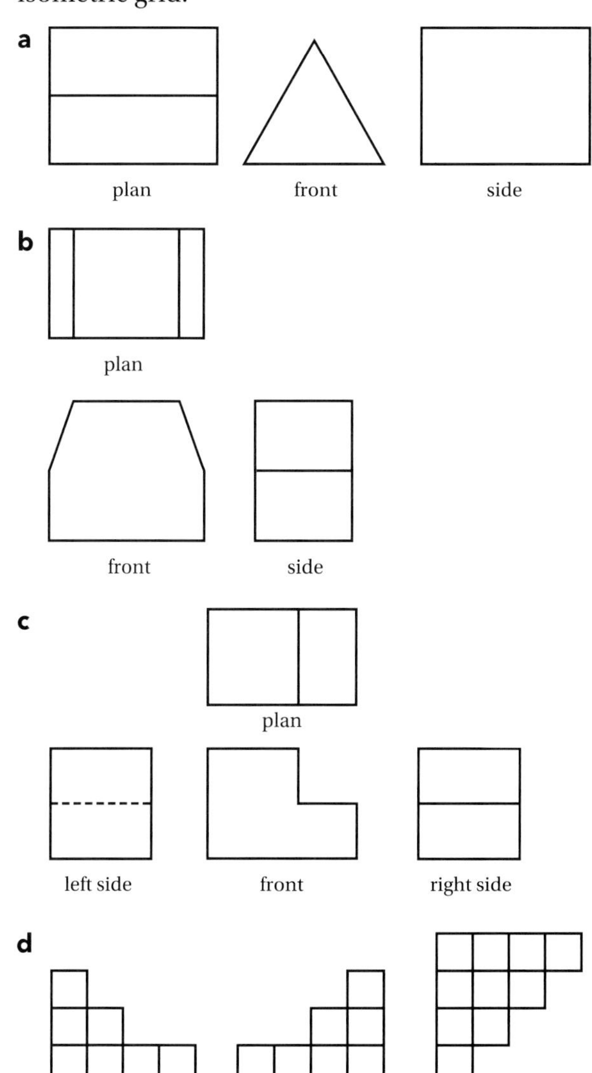

Checklist of learning and understanding

Review of 3D solids

- The number and shape of the faces, and the number of edges and vertices can be used to identify and name 3D solids.
- Prisms are shapes with two end faces that are congruent polygons, and they have a regular cross-section.
- Pyramids have a polygon base, and triangular sides that meet at an apex.
- Cylinders, cones and spheres are 3D shapes, but they are not polyhedra (polyhedra have faces that are polygons and no curved sides).

2D representations of 3D shapes

- 3D shapes can be represented by 2D drawings on squared or isometric grids.
- Hidden edges are shown as broken lines.

Plans and elevations

- A plan is a view from above a shape.
- An elevation is a view from the front, sides or back of a shape.
- Immediate changes in height are shown using a solid line; hidden edges are shown using a broken line.

For additional questions on the topics in this chapter, visit GCSE Mathematics Online.

Chapter review

1 **a** Which solid is being described?

i It has a square cross-section and 12 edges.

ii A solid with two identical circular bases.

iii 12 vertices, two congruent hexagon faces.

iv Circular flat face and one vertex.

v Three triangular faces meet at an apex.

b Draw each solid.

2 **a** Match each block of cubes to the correct plan and elevation.

b Identify any missing elevations and draw them for each shape.

i

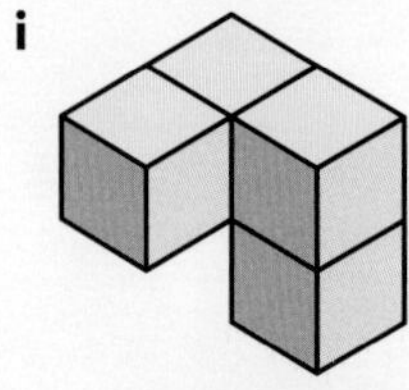

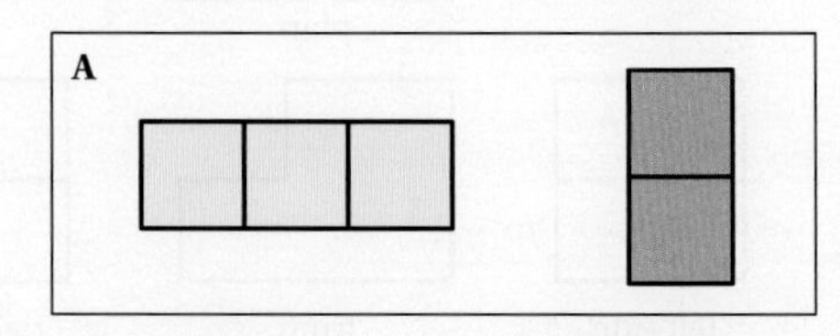

ii

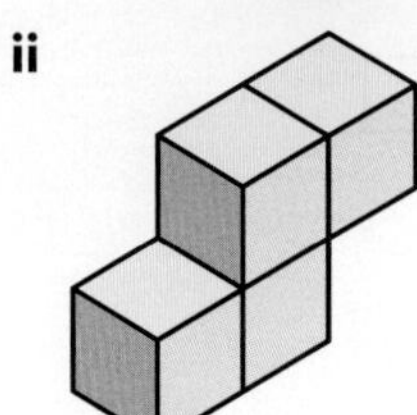

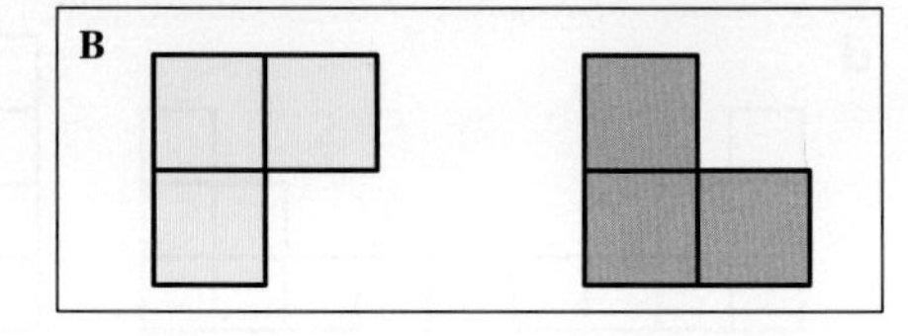

iii 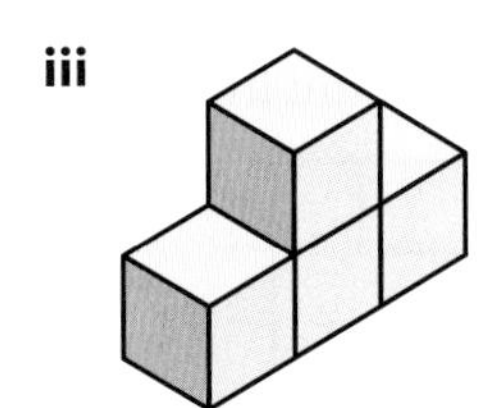

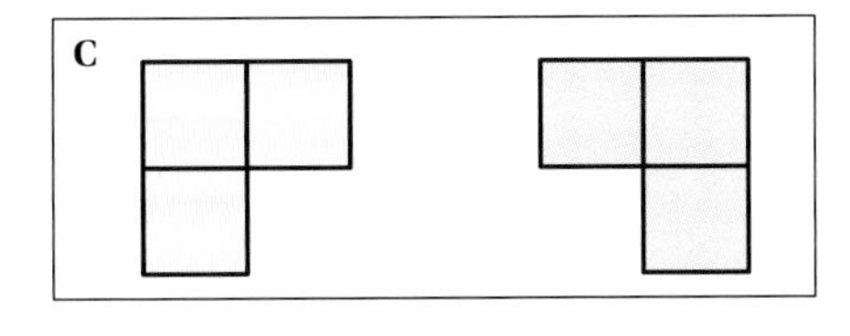

iv

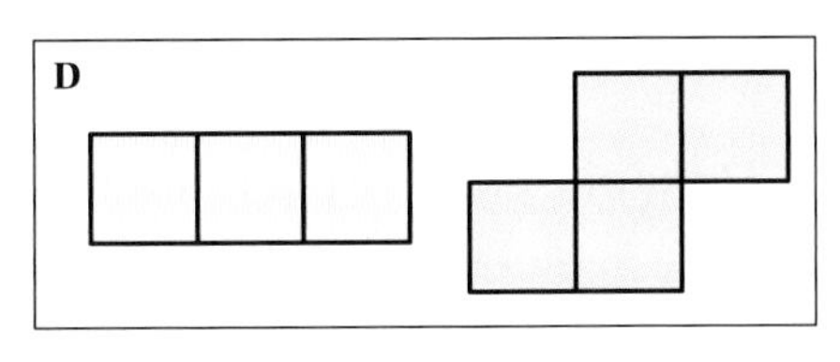

3 The plan and elevation of a solid built from cubes is shown here.

Work out what the solid looks like and sketch it accurately on an isometric grid.

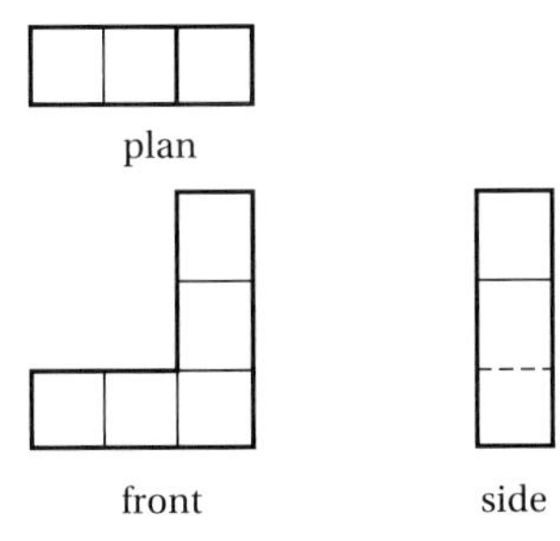

4 Draw this shape:

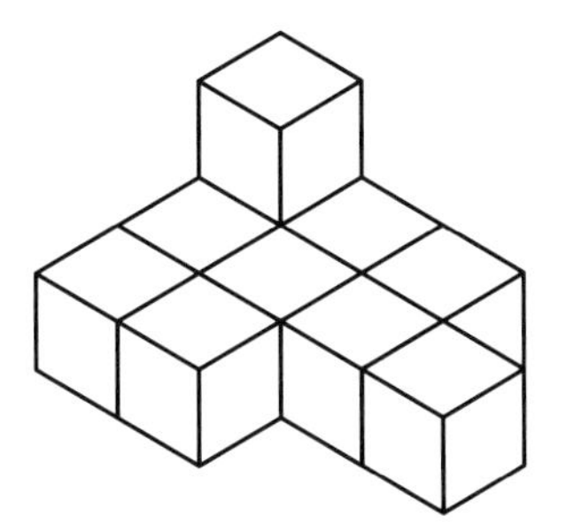

a on square paper.

b on isometric paper.

5 Draw each of the following onto isometric paper:

a a cube

b a 3D shape with 4 rectangular faces

c a triangular-based pyramid.

6 This is the front and right side elevation of a solid built from cubes.

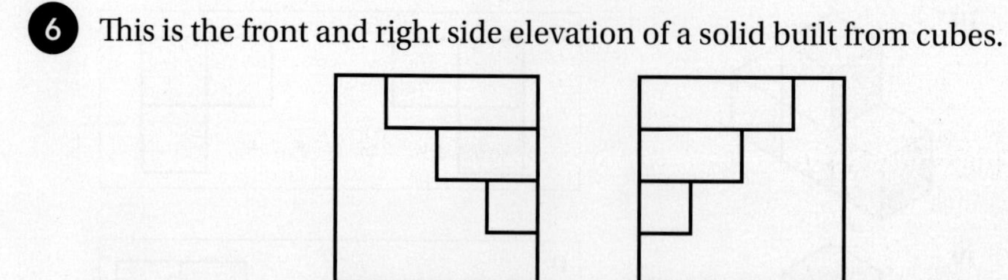

a Draw one possible plan view of this shape.

b How many cubes would be needed to build this shape?

7 This is the plan view and front elevation of a computer generated solid built using 32 cubes.

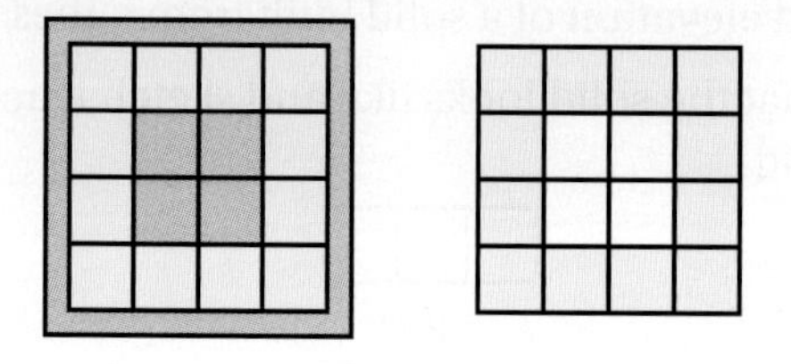

Work out what the shape might look like and draw an accurate diagram of the 3D shape on an isometric grid.

21 Volume and surface area

In this chapter you will learn how to ...

- calculate the volume and surface area of cuboids and other prisms.
- calculate the volume and surface area of cylinders.
- solve volume and surface area problems involving composite shapes.

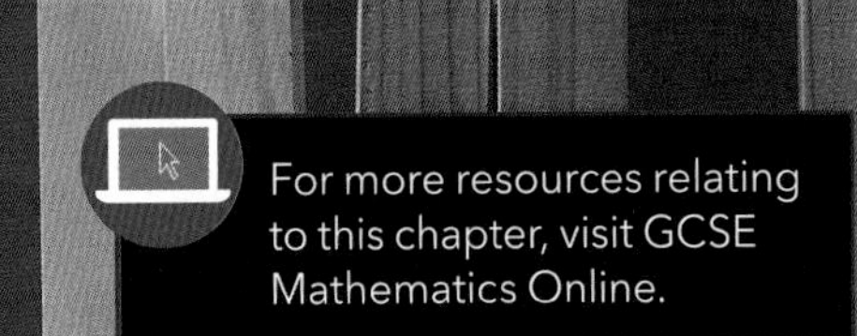

For more resources relating to this chapter, visit GCSE Mathematics Online.

Using mathematics: real-life applications

Freight costs are dependent upon the volume of material being transported. Freight rates are calculated using the container volume measured against the length of the container. The longer the container, the higher the freight cost.

"To transport a container full of apples from Felixstowe, England to Le Havre in France I have to let the freight operator know the volume of apples I have to transport as well as the dimensions of the crates. I am then quoted a transport cost." *(Apple farmer)*

Before you start ...

Ch 5, 20	You need to be able to recognise and identify solid objects, and understand and use their properties	**1** Name each object as accurately as possible from the description. **a** A 3D object with 6 identical square surfaces. **b** A 3D solid with 2 parallel circular faces. **c** An object with a square base and triangular side faces that meet at an apex. **d** A 3D object with a circular base and 1 vertex. **e** A 3D object with many flat surfaces that are polygons. **f** A polyhedron with 2 triangular and 3 rectangular faces. **2** A shape has 6 faces, 8 vertices and 12 edges. **a** What could it be? **b** What additional information do you need to name the shape more accurately?
Ch 16	You must be able to calculate the area of plane shapes.	**3** What is the formula for the area of a circle? **4** What is the area of a right-angled triangle with sides of 3 cm, 4 cm and 5 cm?

Find answers at: cambridge.org/ukschools/gcsemaths-studentbookanswers

Assess your starting point using the Launchpad

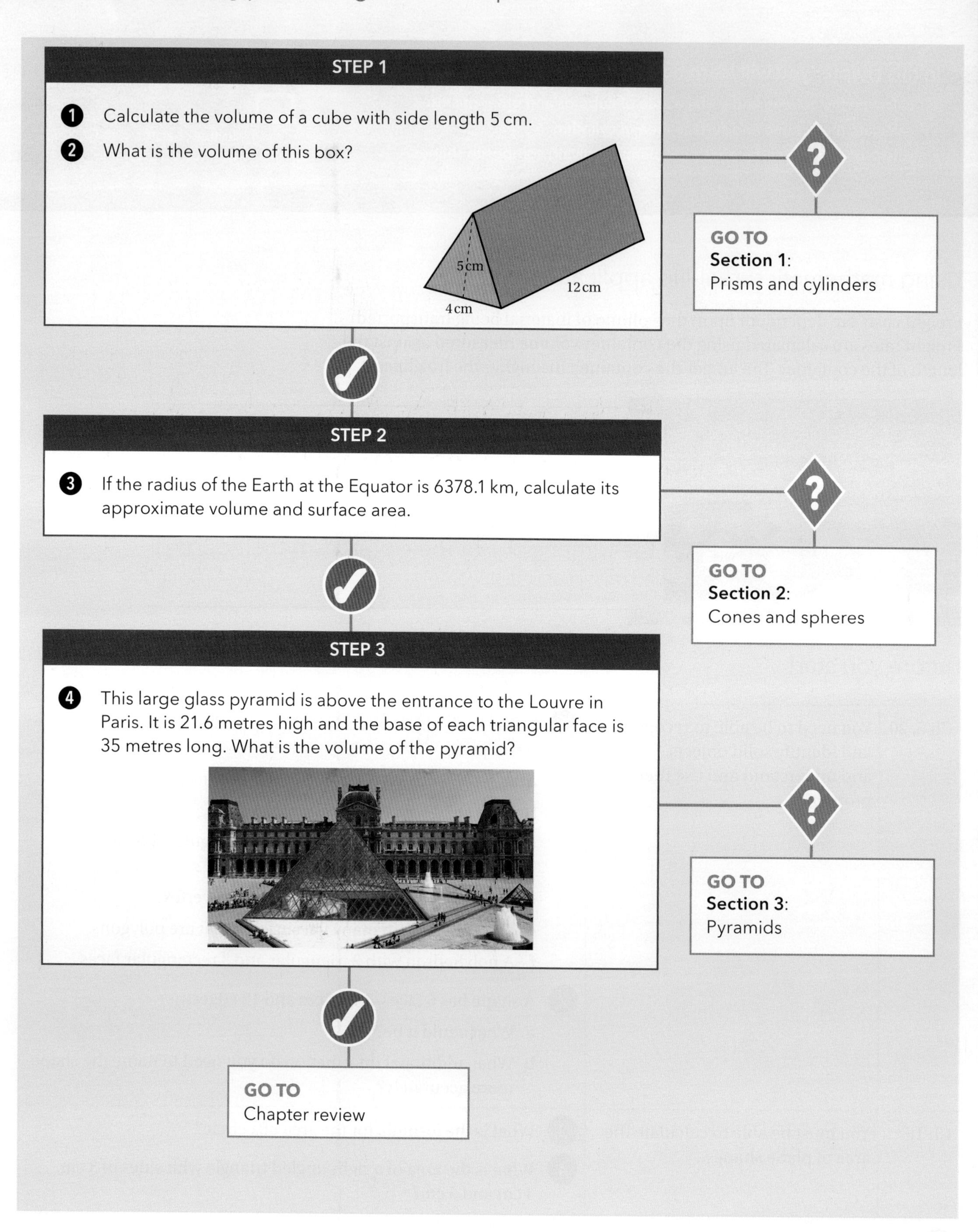

STEP 1
1 Calculate the volume of a cube with side length 5 cm.
2 What is the volume of this box?
5 cm
12 cm
4 cm
GO TO
Section 1:
Prisms and cylinders
STEP 2
3 If the radius of the Earth at the Equator is 6378.1 km, calculate its approximate volume and surface area.
GO TO
Section 2:
Cones and spheres
STEP 3
4 This large glass pyramid is above the entrance to the Louvre in Paris. It is 21.6 metres high and the base of each triangular face is 35 metres long. What is the volume of the pyramid?
GO TO
Section 3:
Pyramids
GO TO
Chapter review

Section 1: Prisms and cylinders

In Chapter 5, you learnt that a prism is a 3D object with two parallel and congruent polygonal end faces and a uniform cross-section along its length.

You also learnt that a cylinder is not a prism because the circular end faces of a cylinder are not polygons. However, it is very similar: a cylinder also has a uniform cross-section and two parallel and congruent end faces.

This shape sorter is really a prism sorter. You should be able to name the shapes sticking out of it.

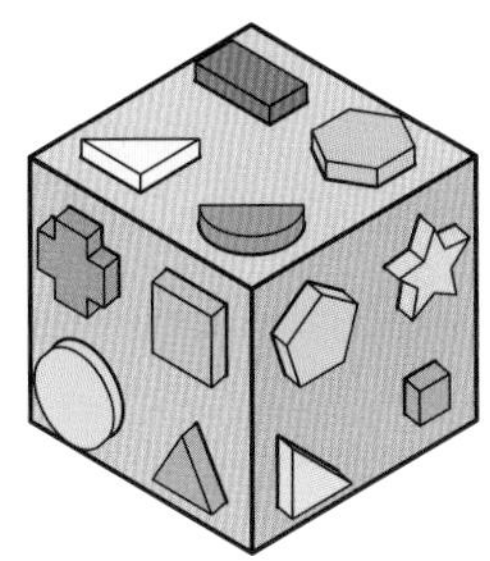

Key vocabulary

right prism: a prism with sides perpendicular to the end faces (base).

The diagram below shows examples of **right prisms**. One of the end faces is known as the base of the object. The other faces are rectangles, perpendicular to the base.

cube (square prism)

rectangular prism

triangular prism

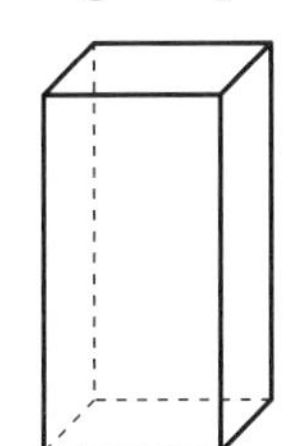

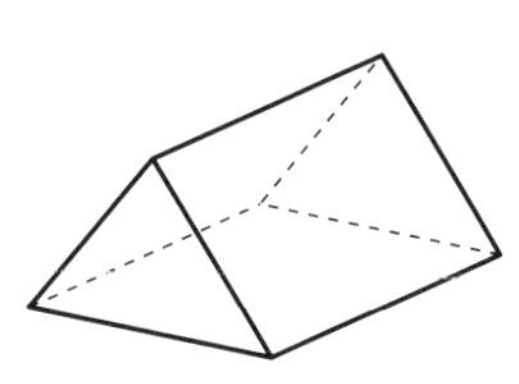

Volume

The volume of an object is the three-dimensional space that it occupies. Volume is given in cubic units, such as mm^3, cm^3 and m^3 (for solids).

You can find the volume of any right prism by finding the area of one end face (base) and multiplying this by its length.

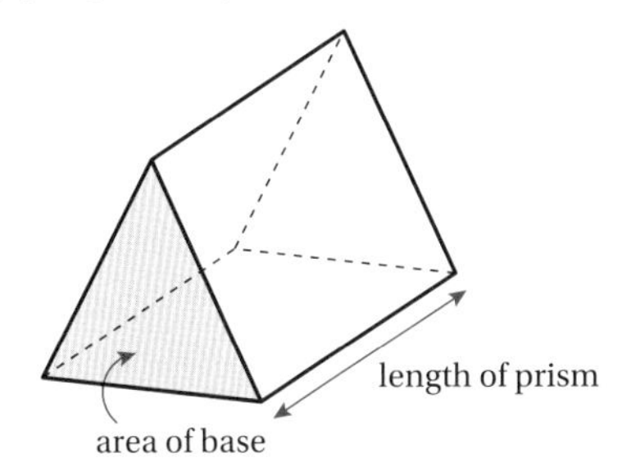

Tip

You learnt about nets of solids in earlier school years; a net is a 2D shape that can be folded to make a 3D solid.

Surface area

Surface area is the total area of the faces of a three-dimensional object.

Sketching a rough net of the object can help you to see what faces to include when you calculate the surface area.

Find answers at: cambridge.org/ukschools/gcsemaths-studentbookanswers

The net of a cuboid shows that the surface area includes the area of six faces.

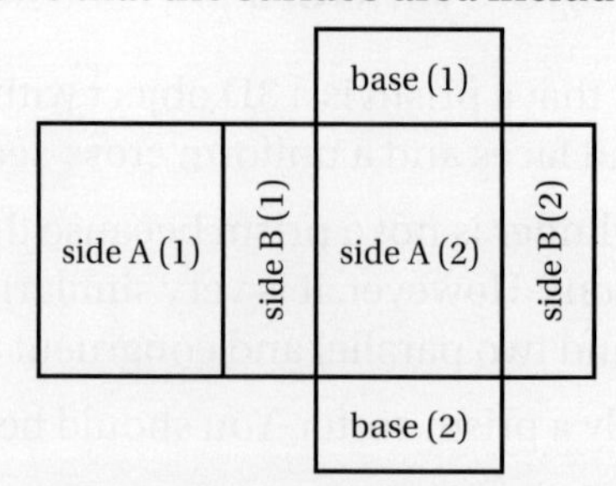

The surface area of the cuboid is calculated by adding the area of each of its faces. The opposite faces match, so

Surface area = 2(area of side A) + 2(area of side B) + 2(area of base).

A cube has six identical square faces. You can use the following formulae for volume and surface area:

Volume of a cube = x^3

Surface area of a cube = $6x^2$

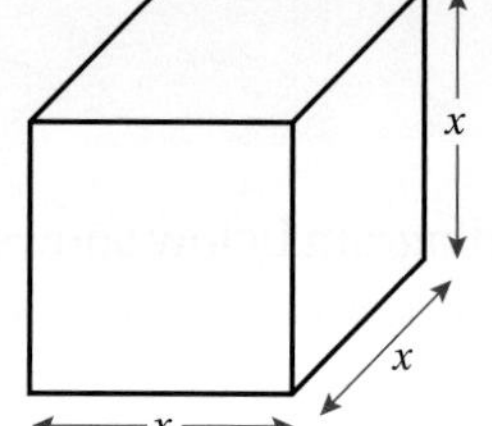

WORKED EXAMPLE 1

Calculate the volume and surface area of this cuboid.

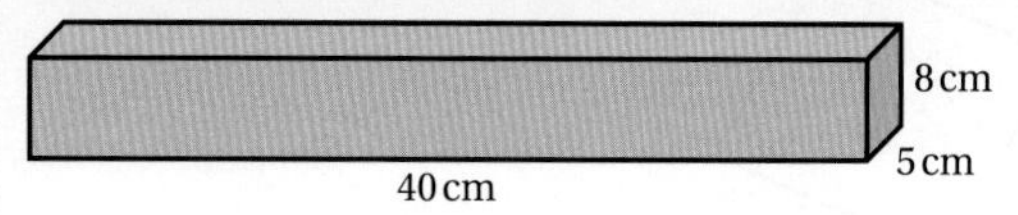

Volume = area of base × length

= 40 × 5 × 8 = 1600 cm^3

Use the formula for volume of a cuboid and substitute in the known values. Use the correct units for volume.

Surface area = 2 × (40 × 5) + 2 × (40 × 8)

= 400 + 640 + 50

= 1090 cm^2

The two end faces are the same, and the four other sides are the same. Write a formula and substitute in the values. Use the correct units for area.

Prisms with bases of other shapes

The same general formula is used to find the volume and the surface area of any prism.

This is a triangular prism. The base is a triangle and the side faces are rectangles

Area of a triangular base $= \frac{1}{2}bh$

Volume of triangular prism $= \frac{1}{2} \times b \times h \times l$

Surface area of prism = 2(area of triangular base) + area of three side faces

$= 2\left(\frac{1}{2} \times b \times h\right) + (a \times l) + (c \times l) + (b \times l)$

Tip

Remember that the triangular base could be isosceles, equilateral or scalene; this will determine if the rectangular faces have the same or different area.

WORKED EXAMPLE 2

Find the volume and surface area of this triangular prism.

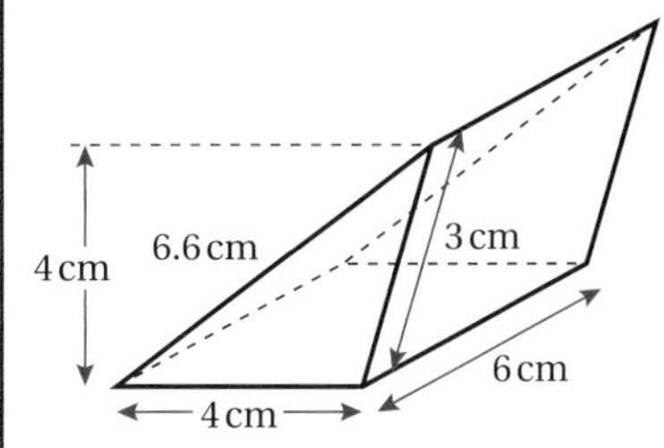

Area of the base $= \frac{1}{2} \times b \times h$

$= \frac{1}{2} \times 4 \times 4 = 8\text{ cm}^2$

Remember to use the perpendicular height of the triangle to calculate its area.

Volume of prism = area of base × length

$= 8 \times 6 = 48\text{ cm}^3$

Remember to use the correct units for volume.

Surface area of base triangles $= 8 + 8 = 16\text{ cm}^2$

Surface area of side face 1 $= 3 \times 6 = 18\text{ cm}^2$

Surface area of side face 2 $= 4 \times 6 = 24\text{ cm}^2$

Surface area of side face 3 $= 6.6 \times 6 = 39.6\text{ cm}^2$

Total surface area $= 97.6\text{ cm}^2$

Notice that the triangular base is a scalene triangle. This means that each rectangular face will have a different area. Remember to use the correct units for area.

Rearranging the formula

You can change the subject of the formula to find the length of a prism if you know the volume and area of the base. You can also find the area of the base if you know the volume and length.

For example, the volume of a triangular prism is 100 cm³ and the area of the end face is 25 cm². How long is the prism?

volume = area of the triangle × length $v = al$ so $v \div a = l$

$100 \div 25 = l = 4\text{ cm}$

Other prisms

Prisms with bases that are trapeziums are quite common. Rubbish skips, wheelbarrows, planters and many other containers take this shape.

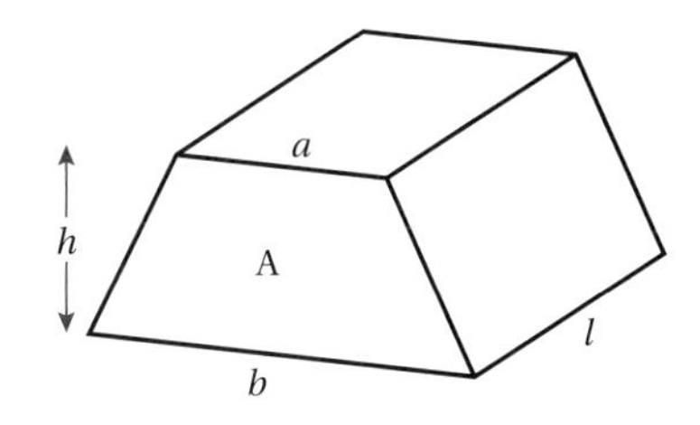

Area of the trapezium base $= \frac{1}{2}(a + b) \times h$

Volume of prism = area of the trapezium × length

WORK IT OUT 21.1

What is the volume of soil that can be contained in this skip?

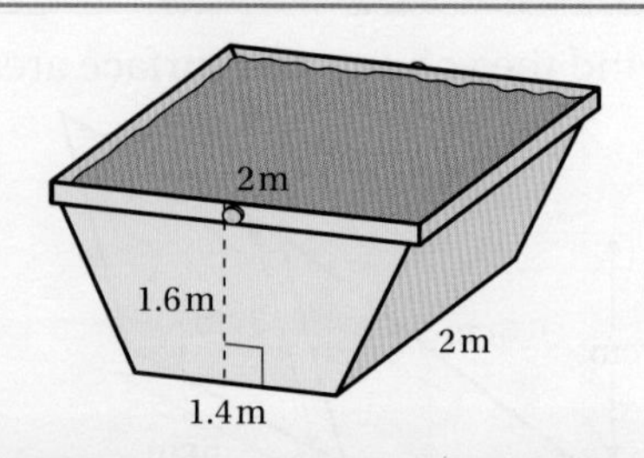

Which of the following is the correct calculation?

Calculation A	Calculation B	Calculation C
Area of the trapezium	Area of the trapezium	Area of the trapezium
Area = $\frac{1}{2}(a + b) \times h$	Area = $\frac{1}{2}(a - b) \times h$	Area = $\frac{1}{2} \times b \times h$
Area = $\frac{1}{2} \times 3.4 \times 1.6 = 2.72\text{ m}^2$	Area = $\frac{1}{2} \times 0.6 \times 1.6 = 0.48\text{ m}^2$	Area = $\frac{1}{2} \times 1.4 \times 1.6 = 1.12\text{ m}^2$
Volume = area of the trapezium × length	Volume = area of the trapezium × length	Volume = area of the trapezium × length
Volume = $2.72 \times 2 = 5.44\text{ m}^3$	Volume = $0.48 \times 2 = 0.96\text{ m}^3$	Volume = $1.12 \times 2 = 2.24\text{ m}^3$

Cylinders

Cylinders are not prisms, but you can find their volume and surface area in the same way as you do with prisms.

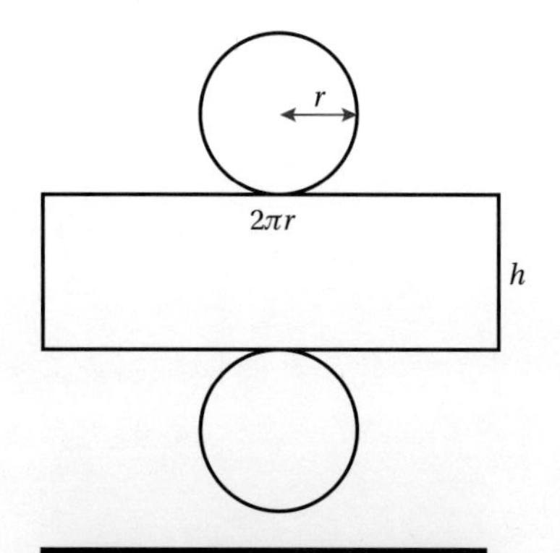

The net of a cylinder shows that the curved surface forms a rectangle when it is flattened out. The length of the rectangle is equivalent to the circumference of the circular base.

The surface area of a cylinder is calculated using the formula:

Surface area = area of curved surface + 2(area of circular base)

$$S = 2\pi rh + 2\pi r^2$$

WORKED EXAMPLE 3

A road roller has a roller on the front which is filled with water to make it heavy. The tank of water in the roller has a radius of 0.95 m and a width of 2.4 m.

What volume of water is needed to fill the tank?

Volume of water = area of the circle × length

$= \pi r^2 \times l$

$= 3.14 \times 0.95 \times 0.95 \times 2.4$

$= 6.801\,24\text{ m}^3$

$= 6.8\text{ m}^3$ (1 dp)

Round your answer to a suitable degree of accuracy. (See Chapter 17 if you need a reminder.)

Problem-solving framework

You are painting the four walls of a room and it needs two coats of paint. The room is 10 m long, 7 m wide and 3 m high. There is a door which is 2 m high and 1.5 m wide, and a window that is 2.6 m high and 2.2 m wide. You want to colour the room blue and it should take five days to paint it.

Paint covers 5 m^2 per litre and you can buy it in 5-litre tins. Each pot costs £14.99. How many 5-litre paint tins will you need to buy?

You should add 10% into your calculations for special circumstances.

Steps for approaching a problem-solving question	**What you would do for this example**
Step 1: What have you got to do?	Paint a room with two coats of paint. Work out how many 5-litre tins of paint are needed.
Step 3: What information do you need?	Room dimensions are needed for surface area calculations. Door and window dimensions need to be taken away from the coverage area. It is useful to estimate before you start. The area of the four walls is roughly 100 m^2 and the door and window are roughly 10 m^2. So coverage is $2 \times 90\ m^2 = 180\ m^2$. Plus 10% takes it roughly to 200 m^2. This means $200 \div 5 = 40$ litres. Eight 5-litre tins are needed.
Step 4: What information don't you need?	The colour of paint, the length of time taken and the cost of the paint are not needed.
Step 5: What maths can you do?	Calculate the surface area of the room: Walls 1 & 3: $10 \times 3 = 30\ m^2$ Walls 2 & 4: $7 \times 3 = 21\ m^2$ Total surface area $= 30 + 30 + 21 + 21 = 102\ m^2$ Door area $= 2 \times 1.5 = 3\ m^2$ Window area $= 2.6 \times 2.2 = 5.72\ m^2$ Total surface area for painting: $102 - 3 - 5.72 = 93.28\ m^2$ Two coats required $= 93.28 \times 2 = 186.56\ m^2$ $10\% = 18.656\ m^2$ Total paint coverage required: $18.656 + 186.56 = 205.216\ m^2$ 1 litre = 5 m^2 coverage of paint $205.216 \div 5 = 41.0432$ litres of paint are required 5-litre tins of paint can be bought: $41.0432 \div 5 = 8.208\,64$ You will need to buy 9 tins.
Step 6: Have you done it all?	Room size less the door and window sizes ✓ Two coats of paint + 10% ✓ Litres divided by 5 for the number of 5-litre tins required; total divided by the coverage of 1 tin of paint ✓
Step 7: Is it correct?	Checked against estimate ✓

EXERCISE 21A

1 Calculate the volume and surface area of each object. (Each object is a closed object.)

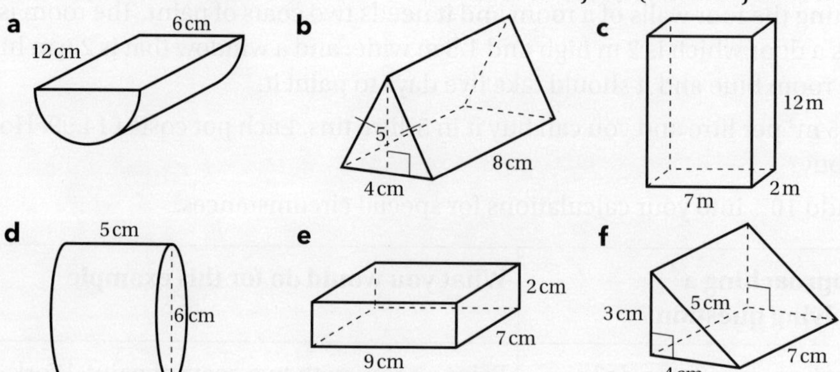

2 If 1 litre = 1000 cm³ what is the volume, in litres of the aquarium below?

3 The volume of a cube is 144 m³. What is the length of each side?

4 An Olympic-sized swimming pool is 50 m long, 25 m wide, and the water is 2 m deep. What is the volume of water in an Olympic-sized swimming pool?

5 What is the volume of this triangular prism? Give your answer correct to 2 dp.

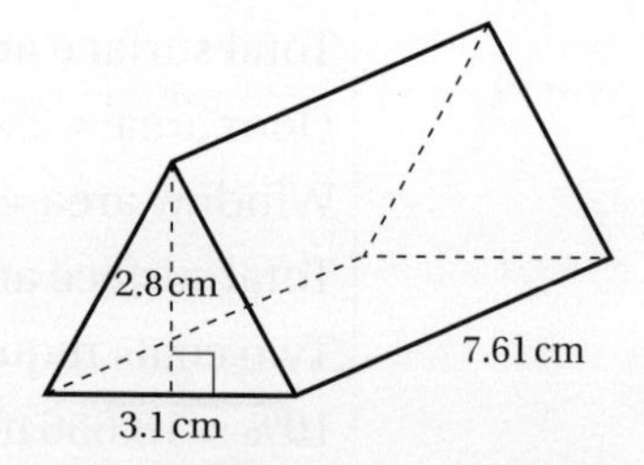

6 What is the surface area of one side of this roof, for the solar panel calculations?

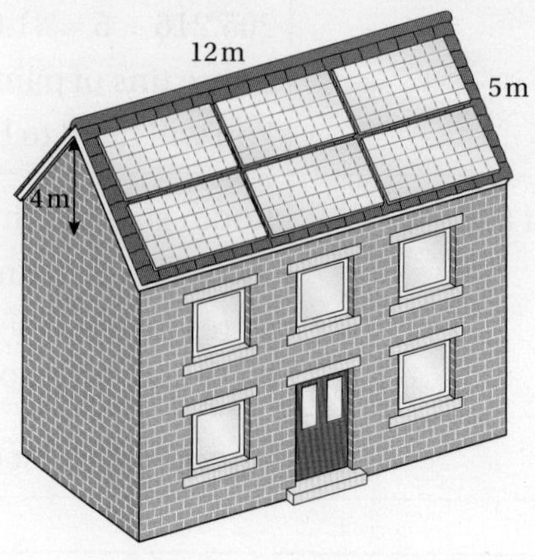

7 A tin can contains oil up to $\frac{3}{4}$ of its full height.

Show that the volume of oil will be $\frac{3}{4}$ of the volume of the whole tin.

Height of the tin = 45 cm; width = 15 cm; depth = 8 cm.

8 A cylindrical water tank with a diameter of 1.2 m and height of 1.6 m needs painting around the outside of its curved surface. What area needs to be painted?

9 You can make two types of candle – a short fat one with a radius 4 cm and height 5 cm or a tall thin one with a radius of 2 cm and height of 20 cm. Which one will require the most candle wax?

10 **a** What is the volume of the metal in a length of a pipe with a hollow radius of 10 cm, an outer radius of 12 cm and a length of 20 cm?

b What is the volume of the hollow centre of the pipe?

11 A cube with side of x cm has a surface area of 150 cm^2. Calculate x.

12 Calculate the volume of the object below.

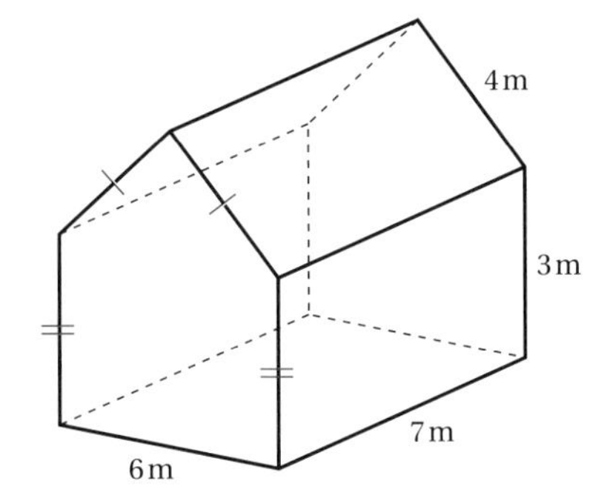

13 Calculate the volume of this solid piece of wood with a cylindrical hole drilled through the middle.

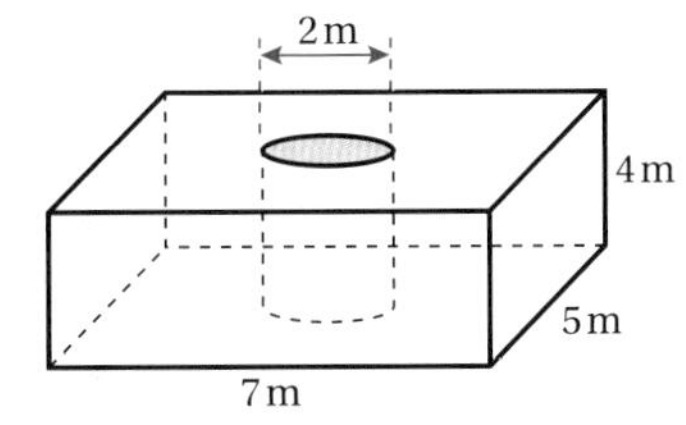

Tip

Think back to how you calculated the area of composite 2D shapes in Chapter 16; subtract the volume of the 'cut out' shape from the volume of the larger shape.

14 Find in terms of x the volume and surface area of a cuboid with sides of length x, $(x + 2)$ and $(x + 3)$.

15 Write a formula for the volume of a cube with a side equal to $a + b$.

Find answers at: cambridge.org/ukschools/gcsemaths-studentbookanswers

Tip

You don't need to learn these formulae. The formulae will be given if you need to use them in a test or exam. Investigate on the internet if you want to find proofs of these formulae.

Tip

In some problems involving cones you may need to use Pythagoras' theorem to find the perpendicular height using the radius and the slant height.

Section 2: Cones and spheres

Cones

The formula for the volume of a cone is

$$\frac{1}{3} \times \text{area of circular base} \times h$$

where h is the **perpendicular height** from the base to the apex of the cone.

The area of the base can be found using the formula for the area of a circle, πr^2.

So, the volume of cone $= \frac{1}{3}\pi r^2 h$.

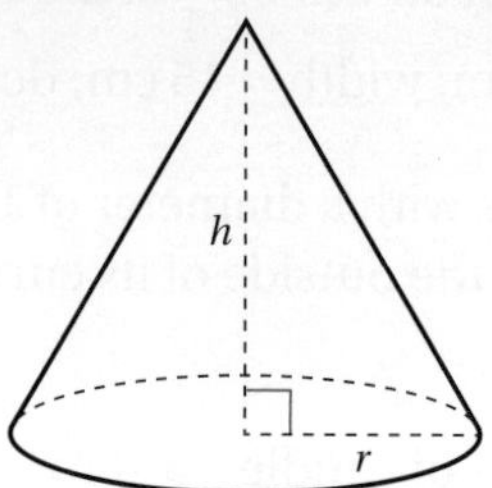

WORKED EXAMPLE 4

Find the volume of a cone of radius 12 cm with a perpendicular height of 14 cm.

Volume $= \frac{1}{3}(\pi r^2)h = \frac{1}{3}(3.142 \times 12 \times 12) \times 14$
$= 2111.424 \text{ cm}^3$

You are given all the values you need. Substitute them into the formula.

The area of the curved surface of a cone is πrs, where r is the radius of the base, and s is the **slant height** of the cone. To find the total surface area of a solid cone you need to include the curved surface area as well as the circular base.

Therefore, the total surface area (S) of the cone is:

S = area of curved surface + area of base

$= \pi rs + \pi r^2$

The slant height of a cone can be calculated using Pythagoras' Theorem if the dimensions of the height and base are given. The curved surface area of the cone can also be given by $\pi r\sqrt{h^2 + r^2}$.

Problem-solving framework

You are selling ice creams with two sizes of cone. One cone has a radius of 3 cm and is 5 cm long; the other has a radius of 4.5 cm and is 6.5 cm long.

You are going to sell strawberry, vanilla and chocolate flavours.

If the price of the first cone is £1.50, prove through a comparison of the volume of ice cream that you cannot simply charge double that price for the second cone. Can you recommend a suitable price to charge based upon your findings?

Continues on next page ...

Steps for solving problems	What you would do for this example
Step 1: What have you got to do?	Compare the volume of the two cones and recommend a suitable price for the second cone based on the comparison.
Step 2: What information do you need?	The radius and length of the small cone; the radius and length of the large cone; the formula for the volume of a cone $\frac{1}{3} \times \pi r^2 \times h$.
Step 3: What information don't you need?	The flavours are irrelevant.
Step 4: What maths can you do?	Volume of the small cone $= \frac{1}{3} \times \pi \times 3 \times 3 \times 5 = 47\text{ cm}^3$ (2 sf) Volume of the large cone $= \frac{1}{3} \times \pi \times 4.5 \times 4.5 \times 6.5 = 138\text{ cm}^3$ (2 sf) About three times as much ice cream will fit inside, so charging double would mean you would be selling the ice cream too cheaply. As the second cone contains three times as much ice cream, a reasonable price would be three times as much as the price of the first cone: $1.50 \times 3 = £4.50$.
Step 5: Have you done it all?	Yes, I have shown double the price isn't right and recommended a new price. ✓
Step 6: Is it correct?	Yes - double-check made. ✓

Spheres

A sphere is any perfectly round object.

The volume of a sphere is equal to $\frac{4}{3}\pi r^3$ where r is the radius of the sphere.

The surface area of a sphere is equal to $4\pi r^2$, where r is the radius of the sphere.

Many objects include spheres or parts of spheres in their structure.

Tip

You do not need to remember these formulae as you will be given them in the exam.

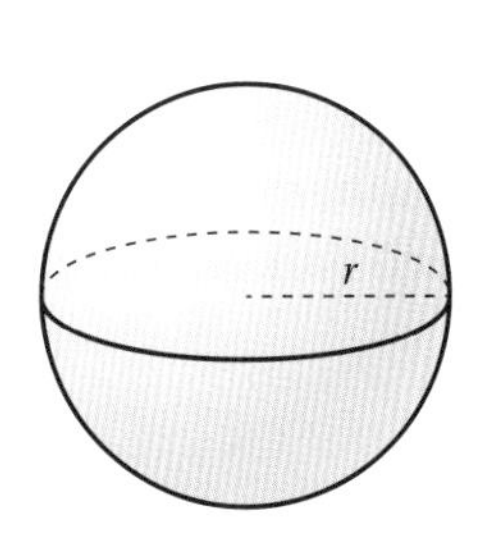

WORKED EXAMPLE 5

Find the surface area and volume of a sphere of radius 3 cm.
Use 3.142 as an approximation of pi.

Surface area $= 4 \times 3.142 \times 3 \times 3 = 113.112\text{ cm}^2$

Volume $= \frac{4}{3} \times 3.142 \times 3 \times 3 \times 3 = 113.112\text{ cm}^3$

Substitute known values into the relevant formula. Don't forget to use the correct units!

WORKED EXAMPLE 6

The radius of the Earth is approximately 6378.1 km.

a Find the approximate volume and surface area of the Earth. Use pi = 3.142.

b 70% of the surface of the Earth is covered with water. What is the surface area of land?

a $V = \frac{4}{3} \times 3.142 \times 6378.1^3 \approx 1\,086\,832\,412\,000\,\text{km}^3$

Surface area $= 4 \times 3.142 \times 6378.1^2 \approx 511\,201\,962.3\,\text{km}^2$

b 30% of the Earth's surface area is land.

$0.3 \times 511\,201\,962.3 = 153\,360\,588.7\,\text{km}^2$

If 70% is water then 30% is land. Refer back to Chapter 13 for a reminder on percentages if you need to.

Tip

In calculations with such large values, you would normally give your answers in standard form. You will deal with this in Chapter 26.

EXERCISE 21B

1 Calculate the volume and surface area of each object.

a

3 cm
11.6 cm
12 cm

b

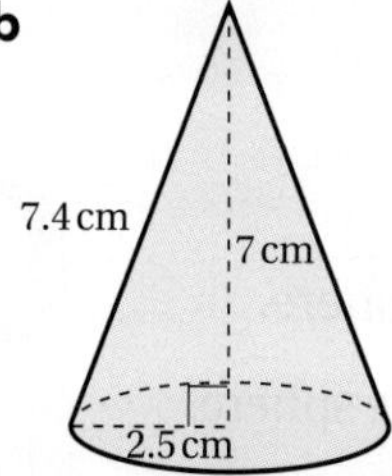

c

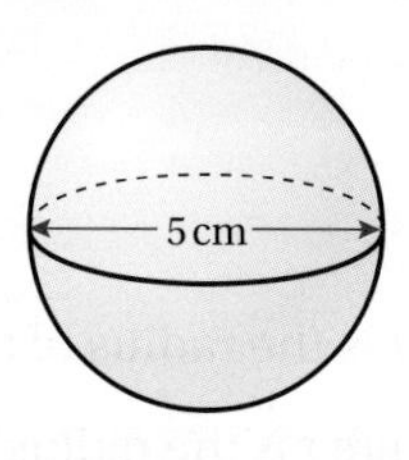

d

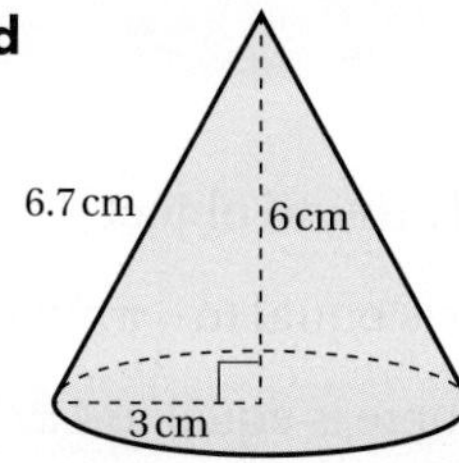

e

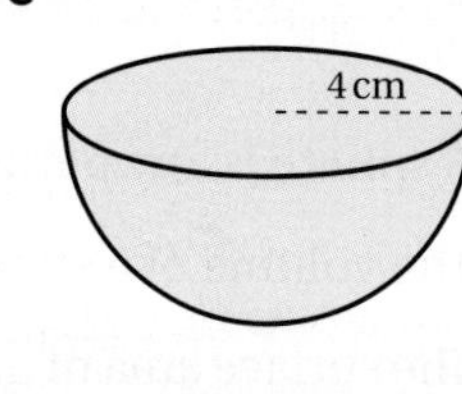

2 Earth's moon has a mean radius of 1738 km. Use the exact value of pi to find its approximate volume.

3 The table below gives some standard diameters of spherical balls used in different sports. Calculate the surface area of each ball. Assume they are round and ignore any dimples on the surface.

	Sport	Standard diameter
a	snooker	52.5 mm
b	tennis	6.35 cm
c	football	15 cm
d	golf	42.7 mm
e	bowling	21.6 cm
f	basketball	25.4 cm
g	hockey	3 cm
h	baseball	74 mm
i	cricket	7 cm

4 A factory needs to calculate the volume and surface area of plastic cones which have the dimensions given in centimetres in the table. Calculate each volume and surface area.

	Radius (r)	Slant height (s)
a	5	10
b	18	34
c	7	21
d	16	22
e	60	64
f	9	26
g	30	52

5 A conical tent has a circular base with a diameter of 3 m, and a slant height of 3 m.

What is its volume?

Composite solids

Objects in real life are very rarely composed of just one kind of geometric object. Most buildings involve a combination of solid shapes, and modern buildings often incorporate unusual shapes into their designs.

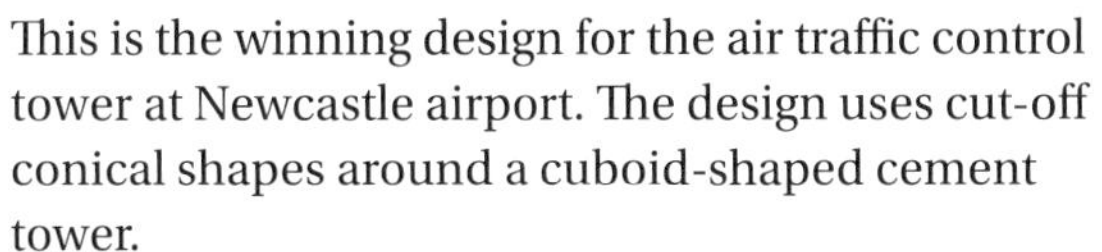

This is the winning design for the air traffic control tower at Newcastle airport. The design uses cut-off conical shapes around a cuboid-shaped cement tower.

This is the North Gate Bus Station in Northampton. Many different solids have been used in the design.

When you worked with area in Chapter 16 you split composite shapes into known shapes and found the area of each shape separately. You can use the same technique to find the volume of composite solids.

To find the total surface area of a composite solid you need to find the area of each section separately. However, you cannot just automatically use the formula for this because the area of some faces will overlap and not form part of the 'outside' or surface area of the solid.

Tip

It is useful to develop a system for checking that you have included all the surfaces, when you are finding the surface area of a composite shape.

Find answers at: cambridge.org/ukschools/gcsemaths-studentbookanswers

WORKED EXAMPLE 7

Calculate the total volume and surface area of the object shown below.
Use 3.142 as the value of π in your calculations and give final answers correct to 2 decimal places.

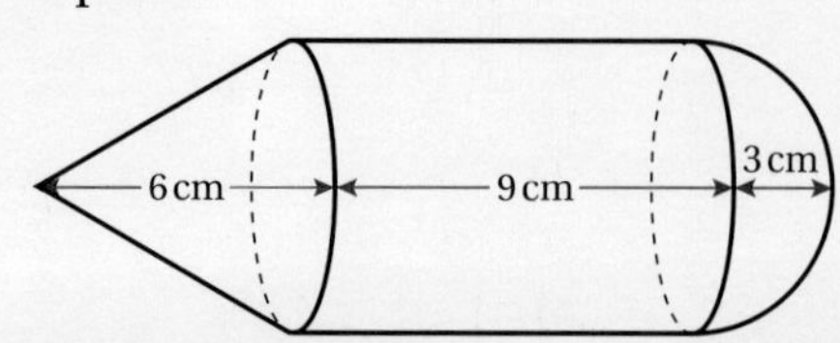

The object consists of a cone, cylinder and half a sphere.
Surface area of the cone = area of lateral curved surface without the area of base (as this overlaps with the end of the cylinder and is **inside** the shape). The slant height of the cone is 6.7 cm and the radius is 3 cm.

Volume	Surface area
Volume of cone $= \frac{1}{3}(\pi r^2)h$ $V = \frac{1}{3} \times 3.142 \times 3^2 \times 6$ $V = 56.556\ \text{cm}^3$	Cone: $S = \pi rl$ (where l is the slant height) $= 3.142 \times 3 \times 6.7$ $= 63.1542\ \text{cm}^2$
Volume of cylinder $= \pi r^2 h$ $V = 3.142 \times 3^2 \times 9$ $V = 254.502\ \text{cm}^3$	Cylinder (without top and base): $S = 2\pi rh$ $= 2 \times 3.142 \times 3 \times 9$ $= 169.668\ \text{cm}^2$
Volume of half sphere $= \frac{2}{3}\pi r^3$ $V = \frac{2}{3} \times 3.142 \times 3^3$ $V = 2.0946666 \times 27$ $V = 56.556\ \text{cm}^3$	Half sphere: $S = \frac{1}{2} \times 4\pi r^2$ $= \frac{1}{2} \times 4 \times 3.142 \times 3^2$ $= \frac{1}{2} \times 113.112\ \text{cm}^2$ $= 56.556\ \text{cm}^2$
Total volume = 56.556 + 254.502 + 56.556 $= 367.61\ \text{cm}^3$ (2 dp)	Total surface area = 63.1542 + 169.668 + 56.556 $= 289.38\ \text{cm}^2$ (2 dp)

The top and base are inside the shape so are not included in the surface area of the composite shape.

Volume of a sphere $= \frac{4}{3}\pi r^3$.
The volume of half a sphere is therefore half of this:
$\frac{1}{2} \times \frac{4}{3} = \frac{2}{3}$

When the composite shape is made of a shape with another shape cut out of it, you subtract the volume of the 'cut out' shape from the volume of the larger shape.

EXERCISE 21C

1 Find the surface area of each solid. Give your answers correct to the nearest cm^2 or mm^2.

a

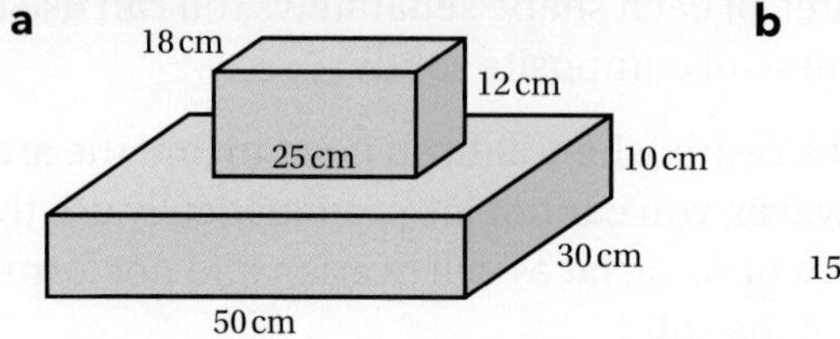

b

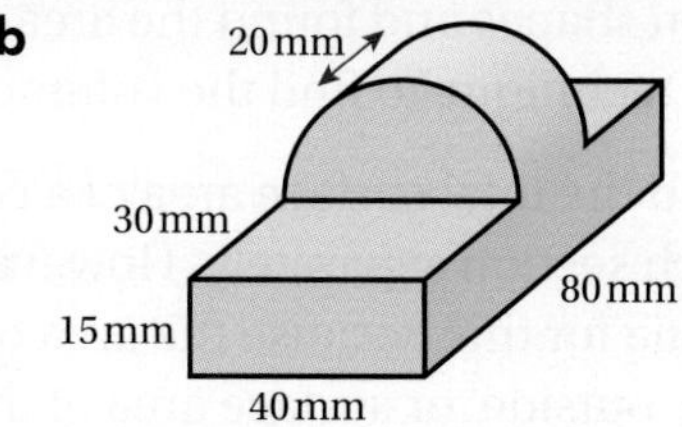

2 Calculate the volume of this capsule.

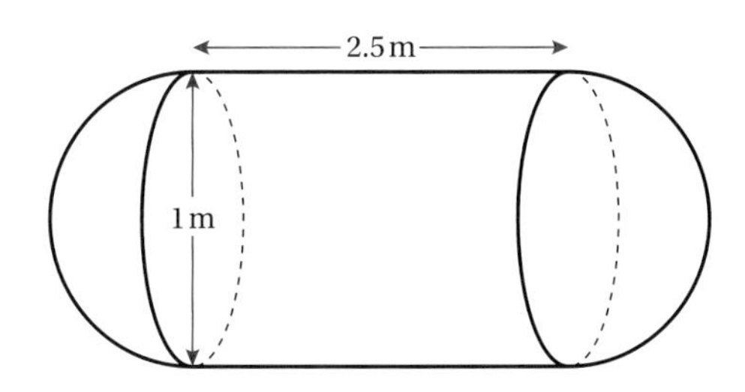

3 Calculate the volume and external surface area of this water tank.

Assume the bottom section is half a cylinder and the top is a right rectangular prism.

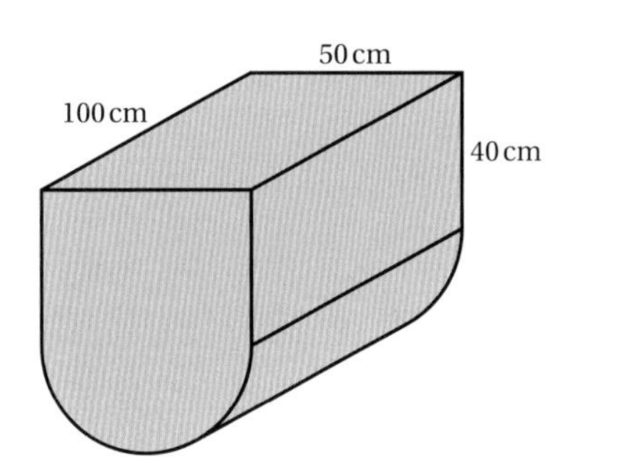

4 These metal blocks have areas cut out of them. For each block, calculate:

a the volume of metal.

b the total surface area to be coated with rust inhibitor.

i

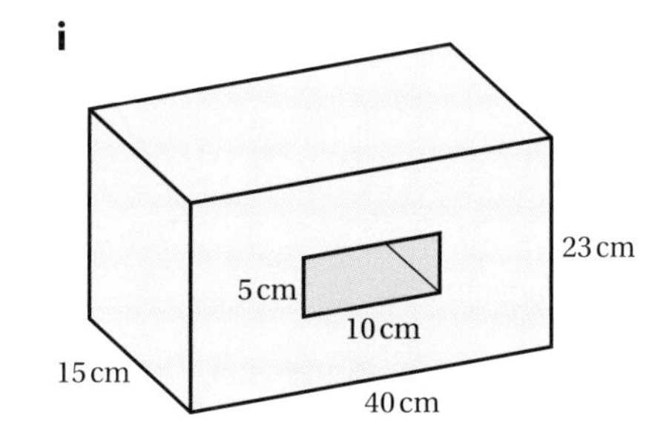

ii

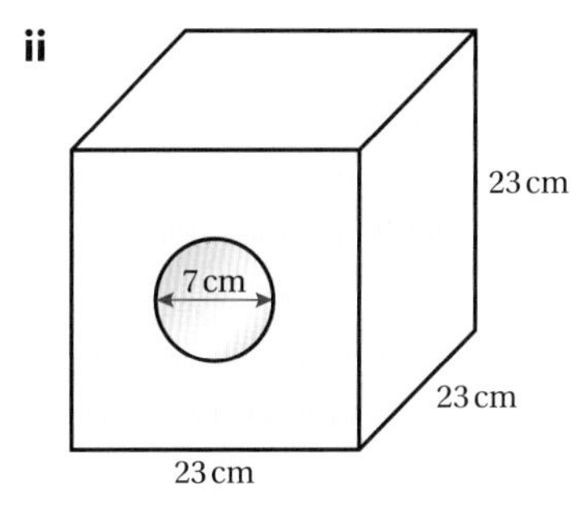

5 Determine the volume of water in a swimming pool that is 6 m wide and 30 m long. The shallow end is 2 m deep and the deep end is 3.5 m deep.

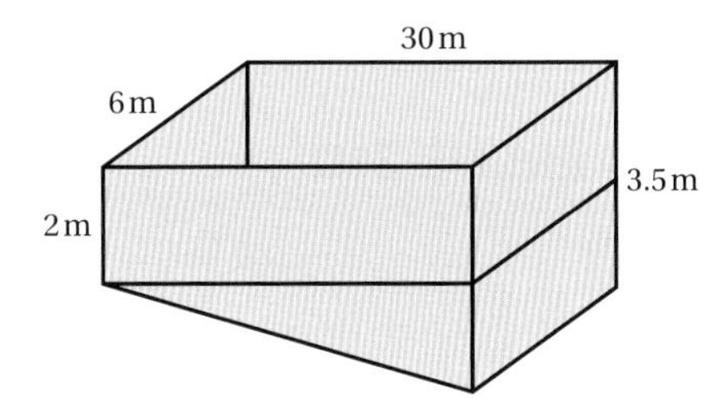

6 What is the volume of a fish tank in the shape of a regular hexagonal prism if the hexagon has equal sides of 10 cm and the height of the tank is 30 cm?

7 Calculate the volume of the following prism. All measurements are in metres. Calculate your answer correct to one decimal place, if necessary.

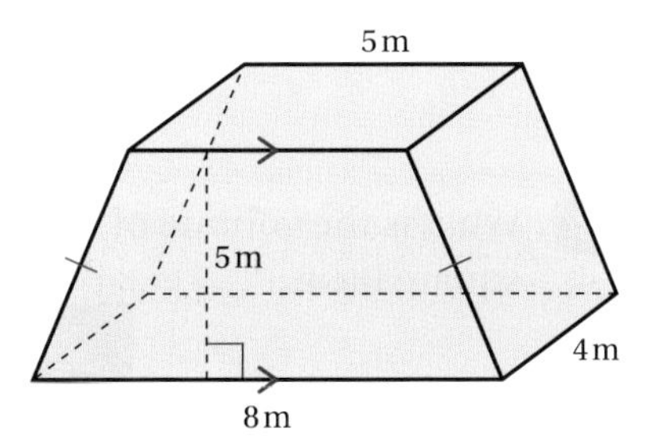

8 A container must have a volume of between 800 ml and 1 litre and the height must be no more than 15 cm. Draw up a table to show possible dimensions of a cylinder and cone that would meet these requirements.

Tip

Think back to the work you did on upper and lower bounds in Chapter 17.

Find answers at: cambridge.org/ukschools/gcsemaths-studentbookanswers

Section 3: Pyramids

Pyramids are named according to the shape of their base.

The volume of a pyramid is $\frac{1}{3}$ of the volume of a prism with the same base area and height.

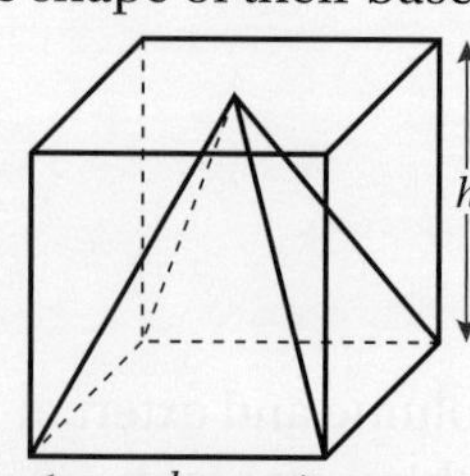

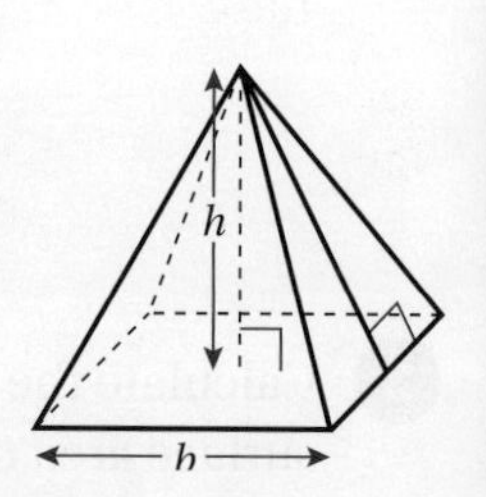

Volume of a pyramid $= \frac{1}{3}$ area of base × perpendicular height.

The surface area of a pyramid is the total area of the base plus the area of each triangular side.

WORKED EXAMPLE 8

Calculate the volume and the surface area of the square-based pyramid opposite.

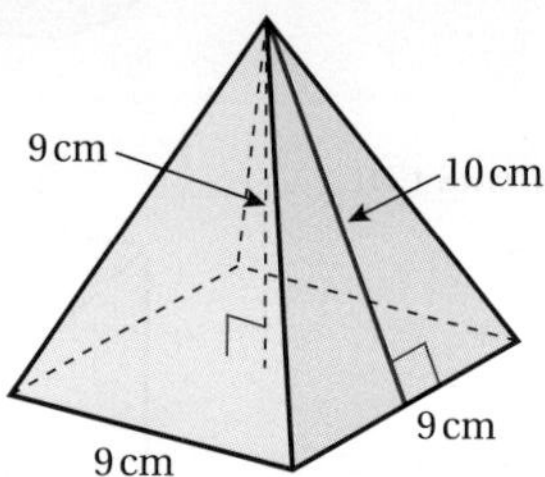

$\text{Volume} = \frac{1}{3} \times \text{area of base} \times h$

$= \frac{1}{3} \times 9 \times 9 \times 9 = 243 \text{ cm}^3$

$\text{Surface area} = b \times b + 4 \times \left(\frac{1}{2} \times \text{slant height} \times b\right)$

$= (9 \times 9) + 4 \times \left(\frac{1}{2} \times 10 \times 9\right)$

$= 81 + 180 = 261 \text{ cm}^2$

Surface area = area of square base + 4 × area of triangular sides. The height of the triangular side is the slant height.

EXERCISE 21D

1 The six pyramids below have either square or triangular bases.

Calculate the volume and the surface area of each one.

a

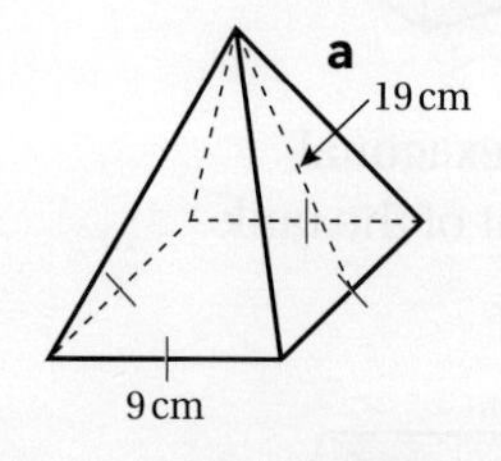

b

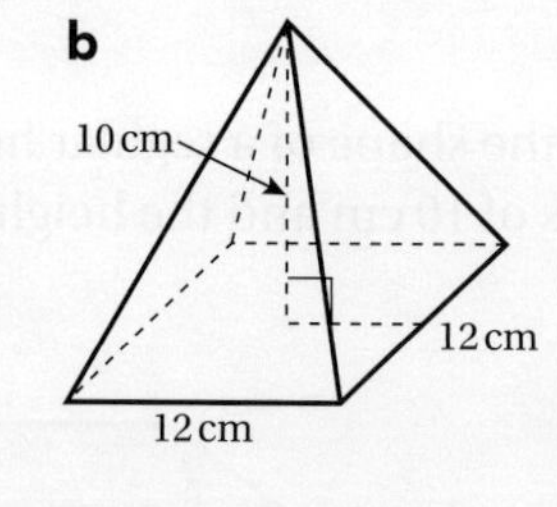

c

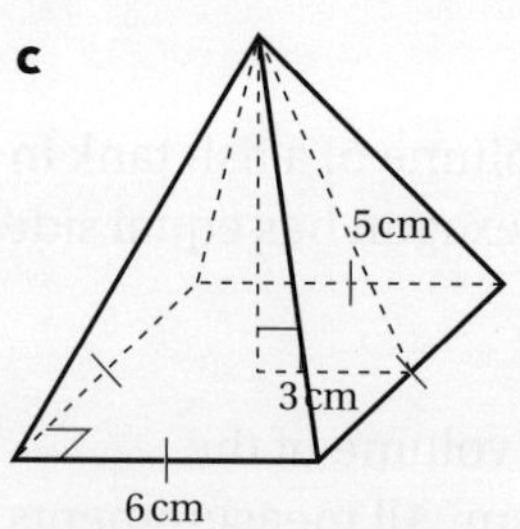

d

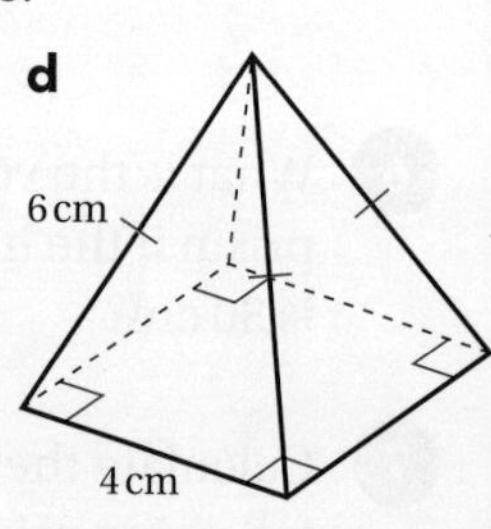

2 What is the volume of the Great Pyramid in the photograph? It has a square base.

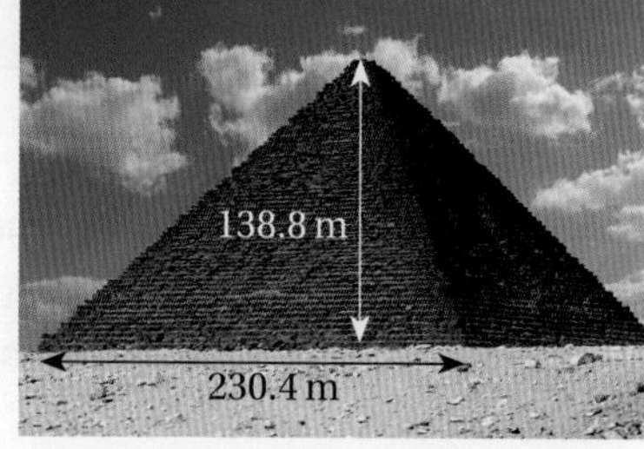

3 What is the difference in the volumes of a pyramid with a square base of side 6 m and a pyramid with an equilateral triangle with side 6 m as a base, if both have a perpendicular height of 8 m?

4 A square-based pyramid has base sides of $6x$ and a perpendicular height of $4x$. Find, in terms of x, the volume and the surface area of the pyramid.

5 The wooden sculpture on the right is a triangular-based pyramid.

The base is an equilateral triangle with sides of 1 m. The height of the sculpture is 2 m.

Calculate the volume of wood used to make the sculpture.

6 A regular tetrahedron is a pyramid with 4 faces that are equilateral triangles.

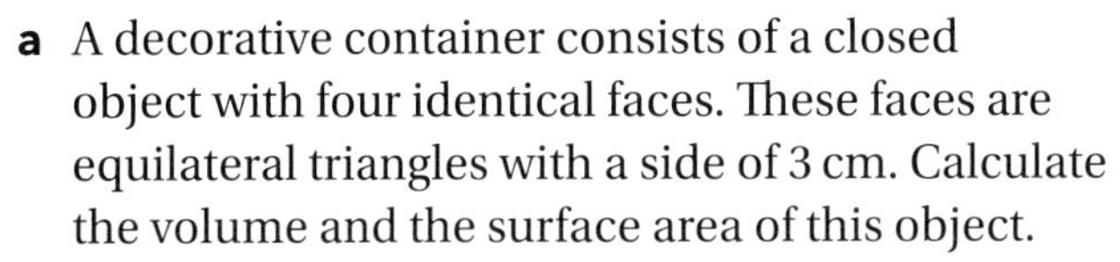

a A decorative container consists of a closed object with four identical faces. These faces are equilateral triangles with a side of 3 cm. Calculate the volume and the surface area of this object.

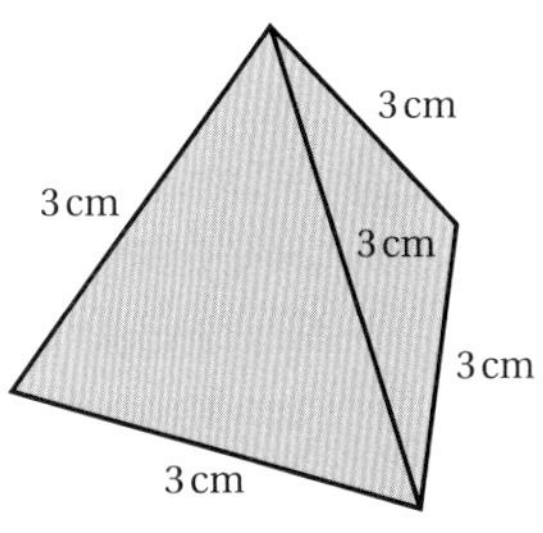

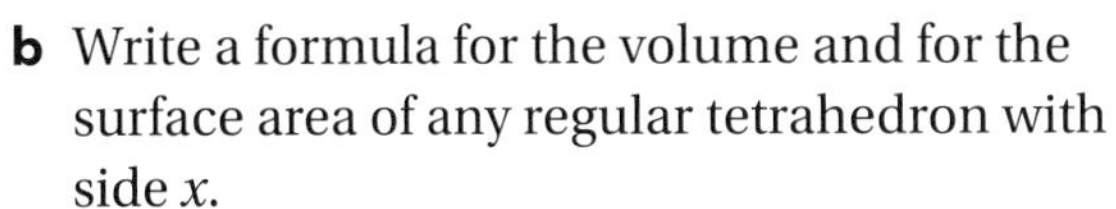

b Write a formula for the volume and for the surface area of any regular tetrahedron with side x.

7 Ancient Egyptians used objects called obelisks in their architecture. They consisted of a square-based column with a pyramidal structure on the top.

Calculate the volume and surface area of the obelisk of Queen Hatshepsut in the photograph. It is 30 m high, the square base has an area of 5 m^2 and the pyramid itself is 1.5 m high.

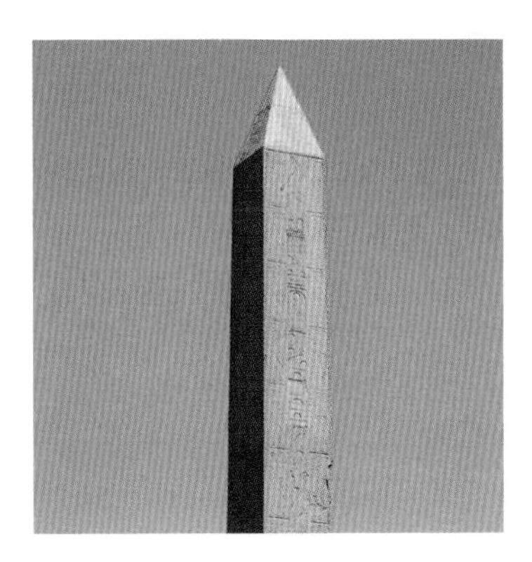

Checklist of learning and understanding

Volume

- Volume is the amount of space a 3D object occupies.
- Volume is calculated in cubic units.
- The volume of a prism and a cylinder = area of base × length.
- The volume of a cone = $\frac{1}{3}$ × area of base × height.
- The volume of a sphere = $\frac{4}{3}\pi r^3$.
- The volume of a pyramid = $\frac{1}{3}$ × area of base × height.

Surface area

- The surface area of a solid is the combined areas of all the external faces.
- You will need to apply what you know about calculating the area of 2D shapes to find the surface area of a solid.

Tip

Remember length, area and volume represent very different measurements: 1 cm, 1 cm^2 and 1 cm^3. What do these look like?

Find answers at: cambridge.org/ukschools/gcsemaths-studentbookanswers

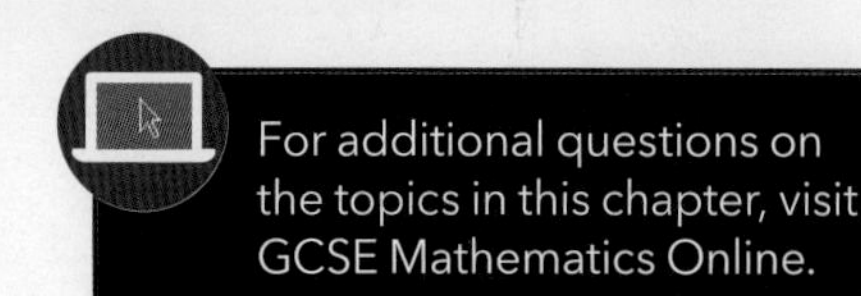
For additional questions on the topics in this chapter, visit GCSE Mathematics Online.

Chapter review

1 How much canvas is in this tent, including the ground sheet on the floor? (Assume the shape is a triangular prism.)

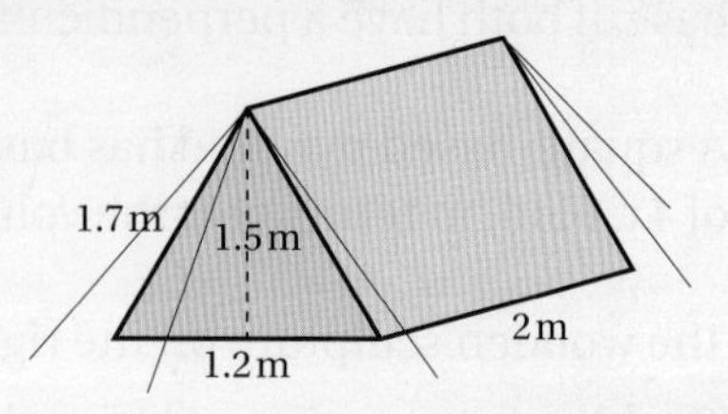

2 What is the volume of this model house?

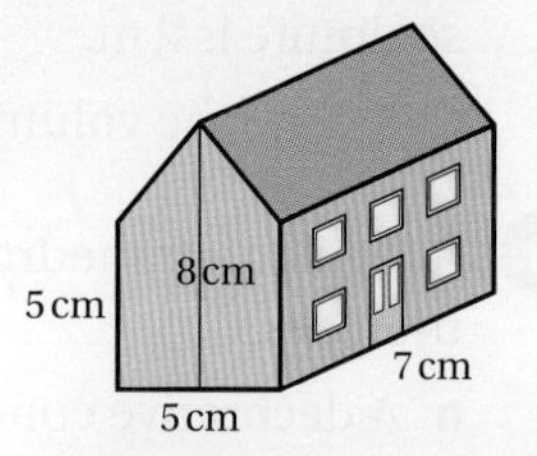

3 What is the volume of this piece of art sculpture?

It is made of a cube with a cylinder cut out through the middle of it.

4 A cheese is a cylinder of radius 7 cm and depth 5 cm.

The cheese is totally covered with a thin coating of wax.

A slice of the cheese is cut so that the top is the sector of the circle of angle 34°. *(6 marks)*

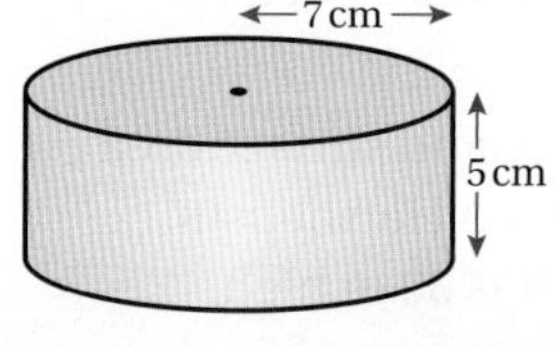

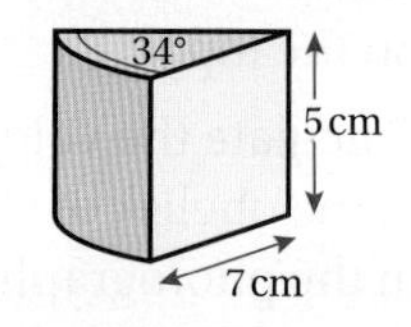

Work out the area of the wax coating on this slice of cheese. *© OCR 2013*

5 The volume of a room is needed in order to work out which air conditioning unit is required. Calculate the volume of a room measuring $23\text{ m} \times 14\text{ m} \times 13\text{ m}$.

6 The dimensions of a cube are whole numbers. If the volume of this cube is 64 cm^3 which of the following whole numbers could be a side length?

A 4 B 10 C 8 D 16 E 5

7 Calculate:

a the volume of a tin of dog food.

b the surface area of the printed label.

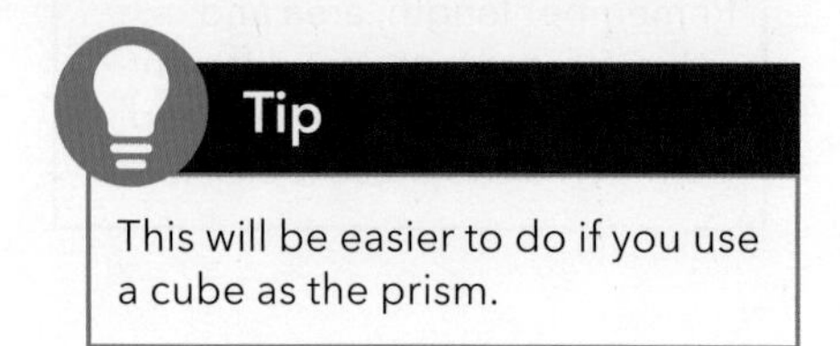

Tip

This will be easier to do if you use a cube as the prism.

8 How could you prove geometrically that the volume of a pyramid is $\frac{1}{3}$ of the volume of a prism with the same base area and height?

22 Calculations with ratio

In this chapter you will learn how to ...

- work with equivalent ratios.
- divide quantities in a given ratio.
- identify and work with fractions in ratio problems.
- apply ratio to real contexts and problems, such as those involving conversion, comparison, scaling, mixing and concentrations.

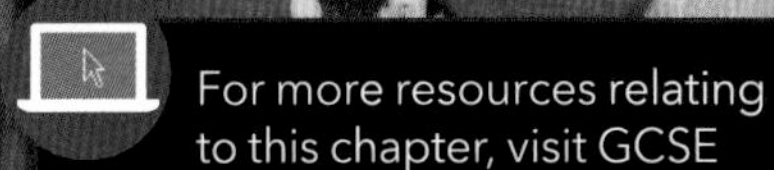

For more resources relating to this chapter, visit GCSE Mathematics Online.

Using mathematics: real-life applications

Converting between different currencies, working out which packet of crisps is the best value for money, mixing large quantities of cement and scaling up a recipe to cater for more people all involve reasoning using ratios.

"Every day customers bring me paints to match. I have to understand how changing the ratio of base colours affects the colour of the paint and how to scale the quantities up and down for larger or smaller amounts of paint. If I get it wrong, customers will have patches of different colours and their walls would look quite strange." *(Paint technician)*

Before you start ...

Ch 10	You need to be able to identify and simplify fractions.	**1** **a** In a class of 35 pupils 21 are boys. What fraction of the class are girls? **b** What fraction of this shape is shaded? Write your answer in its simplest form.
Ch 10	You need to be able to find a fraction of a quantity.	**2** Find $\frac{2}{3}$ of 42.
Ch 10	You need to be able to find an original amount given a fraction.	**3** There are 51 parents of students in the audience at a school play. These parents make up $\frac{3}{4}$ of the audience. How many people are in the audience?

Find answers at: cambridge.org/ukschools/gcsemaths-studentbookanswers

Assess your starting point using the Launchpad

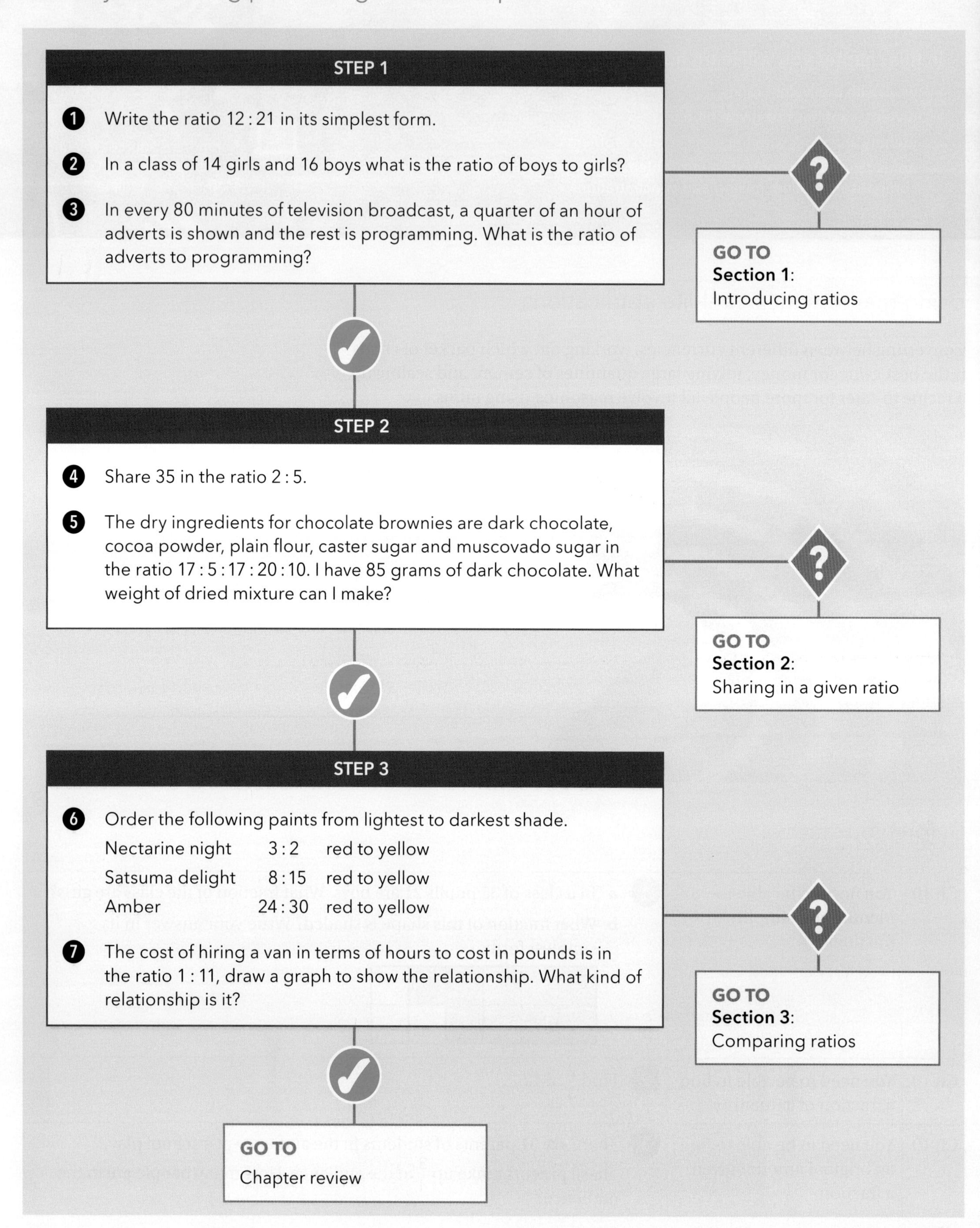

STEP 1

1. Write the ratio 12 : 21 in its simplest form.
2. In a class of 14 girls and 16 boys what is the ratio of boys to girls?
3. In every 80 minutes of television broadcast, a quarter of an hour of adverts is shown and the rest is programming. What is the ratio of adverts to programming?

? **GO TO Section 1:** Introducing ratios

✓

STEP 2

4. Share 35 in the ratio 2 : 5.
5. The dry ingredients for chocolate brownies are dark chocolate, cocoa powder, plain flour, caster sugar and muscovado sugar in the ratio 17 : 5 : 17 : 20 : 10. I have 85 grams of dark chocolate. What weight of dried mixture can I make?

? **GO TO Section 2:** Sharing in a given ratio

✓

STEP 3

6. Order the following paints from lightest to darkest shade.

Nectarine night	3 : 2	red to yellow
Satsuma delight	8 : 15	red to yellow
Amber	24 : 30	red to yellow

7. The cost of hiring a van in terms of hours to cost in pounds is in the ratio 1 : 11, draw a graph to show the relationship. What kind of relationship is it?

? **GO TO Section 3:** Comparing ratios

✓

GO TO Chapter review

Section 1: Introducing ratios

Many colours of paint can be mixed from the four base colours: blue, yellow, red and white.

Think about mixing green paint. You would need to know which base colours to mix. You would also need to know how much of each colour to mix to get dark green or light green. The amount of each colour is important for getting the same shade of green each time you make it.

Paint technicians can mix the same shade of green over and over by mixing yellow and blue paints in a particular **ratio**.

Artists and designers use a special chart with thousands of numbered shades of colours to make sure they get the exact shade they want. The number allows the colour to be mixed using the correct ratio of base colours.

Ratio describes how parts of equal size relate to each other. A ratio of yellow to blue paint of 1 : 3 means one unit of yellow for every three units of blue. This gives a very dark green.The order in which a ratio is written is important. A ratio of 2 : 5 means 2 parts to 5 parts. Each part is equal in size.

A ratio of yellow to blue paint of 5 : 1 means five units of yellow for every one unit of blue. This would give a much lighter green.

The diagram shows a ratio of yellow to blue of 3 : 9.

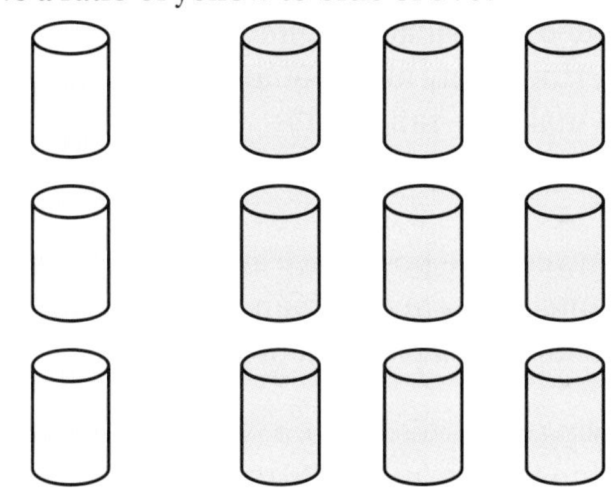

Dividing by three simplifies the ratio of 3 : 9 to give 1 : 3.

$$3 : 9 \xrightarrow{\div 3} 1 : 3$$

Mixing paint in the ratio 3 : 9 would give the same colour as mixing it in the ratio 1 : 3 because the colours are mixed in the same ratio. The yellow paint makes up the same **proportion** of the mix in both cases.

The ratios 3 : 9 and 1 : 3 are **equivalent** ratios.

The difference between ratio and proportion

A ratio compares two or more quantities with each other. A proportion compares a quantity to the 'whole' of which it is a part.

For example, in the dark green paint mixture, the ratio of yellow paint to blue paint is 3 : 9 or 1 : 3. The proportion of yellow paint in the dark green paint mixture is $\frac{3}{12}$, $\frac{1}{4}$ or 25%.

Key vocabulary

ratio: the comparison between two or more amounts in relation to each other.

Tip

With many ratio questions, drawing a picture of the situation can help you work it out.

Key vocabulary

proportion: a comparison of a part, or amount, to the whole; often expressed as a fraction, percentage or ratio.

equivalent: having the same value; two ratios or fractions are equivalent if one is a multiple of the other because they will cancel to the same simplest term.

Tip

The 'whole' of a ratio 3 : 9 is 3 + 9 = 12; the 'whole' of the ratio 1 : 3 is 1 + 3 = 4.

Find answers at: cambridge.org/ukschools/gcsemaths-studentbookanswers

EXERCISE 22A

1 36 girls, 45 boys and 9 teachers went on a school trip.

a What is the ratio of boys to girls?

b What is the ratio of pupils to teachers?

c What is the ratio of pupils to people on the trip?

d The school policy is that each teacher can be responsible for no more than 10 pupils. Does this trip meet this requirement?

2 Look at each diagram. What is the ratio of shaded squares to unshaded squares in each? Write the answers in simplest form.

a 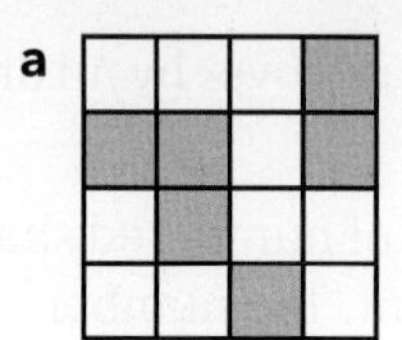

b 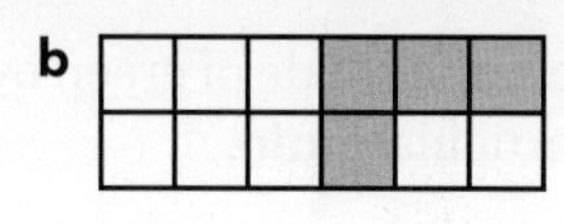

c 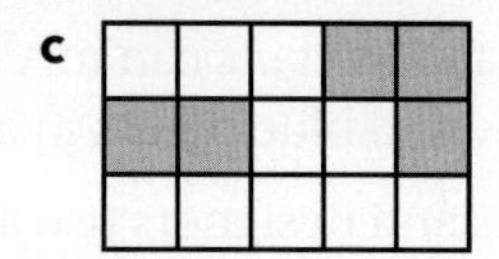

3 Look at each diagram. What is the ratio of shaded squares to total squares in each? Write the answers in simplest form.

a 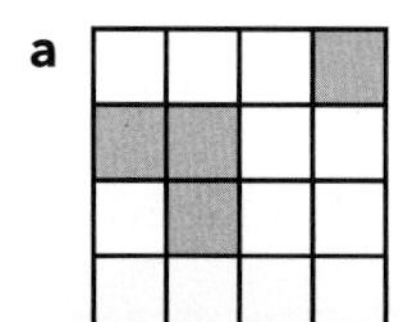

b 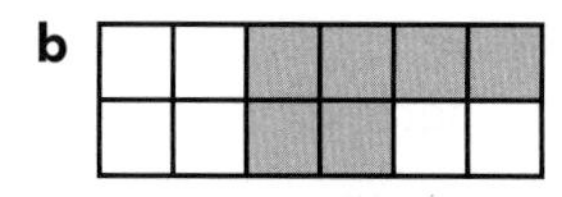

c 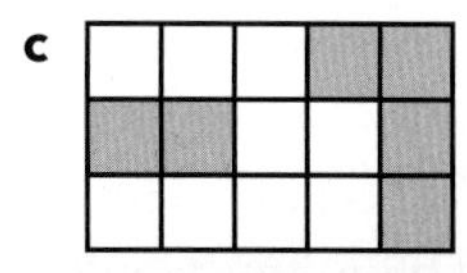

Tip

Think about what the ratio would have been **before** it was simplified to 1 : 3, and how many parts there are in the whole.

4 The ratio of shaded to unshaded squares in this diagram is 1 : 3. How many more squares need to be shaded to make the ratio 2 : 3?

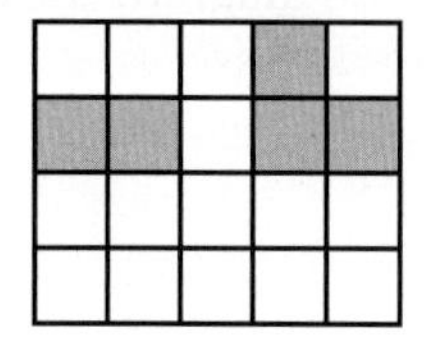

Tip

Ratios do not include units. To compare measured amounts you need to make sure they are written in the same units.

5 The distance between the post office and the bank on the local high street is represented as 5 cm on a map. In real life this distance is 20 m. What is the scale of the map (as a ratio)?

6 On a scale drawing of a cruise ship, a cabin is 8 cm from the restaurant. On the actual ship the distance is 76 m. Express the distances as a ratio.

7 A natural history programme lasts 90 minutes. The crew recorded 60 hours of footage. What is the ratio of footage used to footage recorded?

8 **a** Use the diagram to find the ratio of:

i side AB to side AC.

ii side EB to side DC.

iii side AE to side AD.

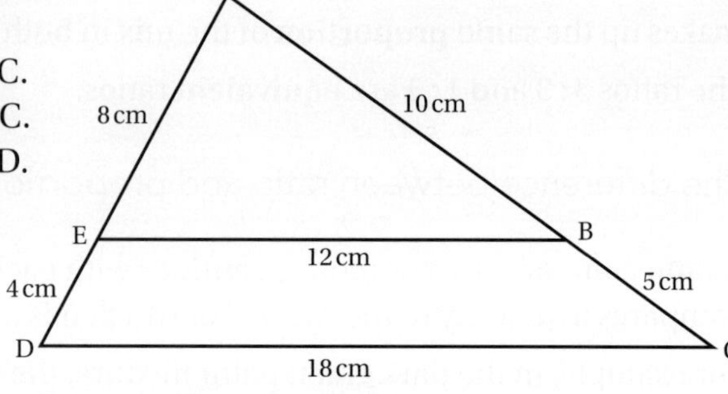

b What does this tell you about triangles ABE and ACD?

c On this basis, what is the ratio of the angle AEB to angle EDC?

Tip

You will learn more about similarity in Chapter 29, but for now think back to work you have done in earlier school years.

9 A jam recipe uses 55 g of fruit for every 100 g of jam. The rest is sugar. What is the ratio of fruit to sugar?

10 An adult ticket for the cinema is one and a half times the price for a child's ticket. What is the ratio of the price of an adult ticket to the price of a child's ticket? Write the ratio in its simplest form.

Tip

Ratios can include decimal numbers, but not when they are written in their simplest form; the simplest form always uses integers.

11 According to recent statistics $\frac{3}{5}$ of 16 year olds have a mobile phone. What is the ratio of 16 year olds with mobiles to those without?

12 After an increase of 20% in the number of boys in a school, the ratio of boys to girls is 3 : 4. If there are now 630 pupils in the school, how many boys were there originally?

13 If $\frac{1}{5}$ of chocolates in a box are dark chocolate, $\frac{1}{2}$ are milk and the rest are white, what is the ratio of dark : milk : white chocolate?

Section 2: Sharing in a given ratio

Often you will be given a ratio and asked to share an amount using that ratio.

For example, a group of three office workers form a lottery syndicate. Together they buy eight lottery tickets a week.

Simon pays £1 a week, Oliver £3 and Lucy £4. They win £32 000.

Should they each get an equal share of the winnings? If not, how should they share their winnings? What is the fairest way?

The fairest way would be for each member to receive winnings in the same ratio as they bought tickets.

The winnings should therefore be distributed between Simon, Oliver and Lucy in the ratio 1 : 3 : 4. This can be represented using a diagram, where each box represents the number of parts of the whole each individual should receive.

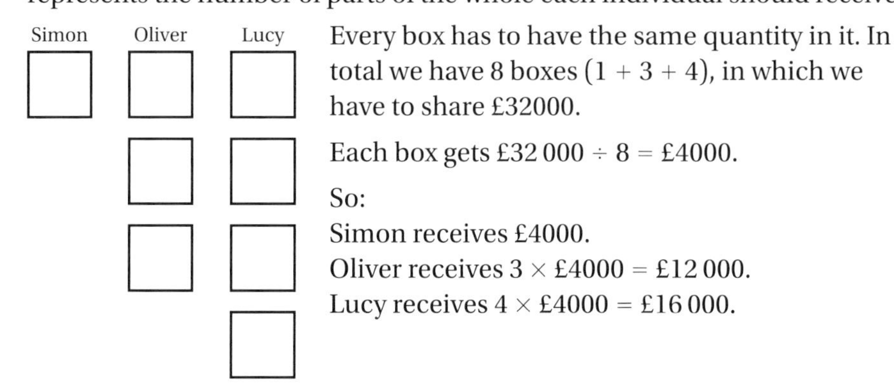

Every box has to have the same quantity in it. In total we have 8 boxes $(1 + 3 + 4)$, in which we have to share £32000.

Each box gets £32 000 ÷ 8 = £4000.

So:

Simon receives £4000.

Oliver receives 3 × £4000 = £12 000.

Lucy receives 4 × £4000 = £16 000.

Tip

The box method shown in the example is useful for working out shares in a given ratio problems.

You can also think in terms of fractions and use what you know about finding fractions of a quantity.

Tip

You learnt how to find a fraction of a quantity in Chapter 10.

1 : 3 : 4 gives 8 parts, so each person gets the following:

Simon receives $\frac{1}{8} \times 32\,000 = £4000$.

Oliver receives $\frac{3}{8} \times 32\,000 = £12\,000$.

Lucy receives $\frac{4}{8} \times 32\,000 = \frac{1}{2} \times 32\,000 = £16\,000$.

The final step is to double check that the shared quantities sum to the original amount:

£4000 + £12 000 + £16 000 = £32 000

Find answers at: cambridge.org/ukschools/gcsemaths-studentbookanswers

EXERCISE 22B

1. Share 144 in each of the given ratios.

   a 1 : 3   b 4 : 5   c 11 : 1
   d 2 : 3 : 1   e 1 : 2 : 5   f 2 : 7 : 5 : 4

2. To make mortar you mix sand and cement in the ratio of 4 : 1.

   a How much sand is needed to make 25 kilograms of mortar?

   b What fraction of the mix is cement?

3. The first bi-colour £2 coin was issued in 1998. The inner circle is made of cupronickel. This is copper and nickel in the ratio 3 : 1. The inner circle weighs 6 grams. How much copper is used to make the centres of ten £2 coins?

4. Flaky pastry is made by mixing flour, margarine and lard in the ratio 8 : 3 : 3 and then adding a drizzle of cold water.

   a How much of each ingredient is needed to make 350 g of pastry?

   b What fraction of the pastry does the margarine and lard make together?

5. The sides of a rectangle are in the ratio of 2 : 5. Its perimeter is 112 cm.

   a What are the dimensions of the rectangle?

   b Use these dimensions to calculate its area.

6. Orange squash is made by mixing one part cordial to five parts of water. How much squash can you make with 750 ml of cordial?

7. Two-stroke fuel is used to power small engines. It is produced by mixing oil and petrol in the ratio of 1 : 20. How much oil needs to be mixed with 10 litres of petrol to make two-stroke fuel?

8. Tiffin is a sweet made by crushing biscuits and mixing them with dried fruit, butter and cocoa powder. The ratio of biscuit to dried fruit to butter to cocoa powder is 5 : 6 : 2 : 2. How much of each ingredient is needed to make 600 g of tiffin?

9. In a music college, the ratio of flute to oboe to string to percussion players is 7 : 2 : 15 : 1. If the college has 175 students. How many play an oboe?

10. The ratio of red to green to blue to black to white pairs of socks in a drawer is 2 : 3 : 7 : 1 : 4. If there are 8 pairs of white socks, how many pairs are there altogether?

11. Potting compost is made by mixing loam, peat and sand in the ratio of 7 : 3 : 2. If a gardener has 4.5 kg of peat and plenty of loam and sand, how much potting compost can she make?

Section 3: Comparing ratios

It is often useful to write ratios in the form $1:n$ (where n represents a number) so that they are in the same form and you can compare them directly by size.

Did you know?

The scale of maps is given as a ratio in the form of $1:n$.
For example, 1 : 25 000.

WORKED EXAMPLE 1

Red and white paints can be mixed to make pink paint.

Which of the mixes below will give the lightest shade of pink?

A

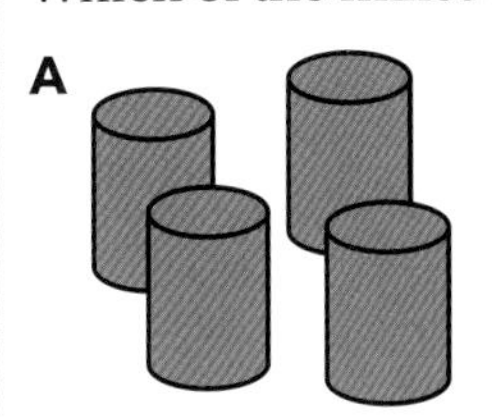

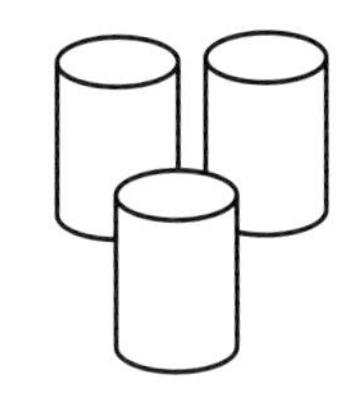

B

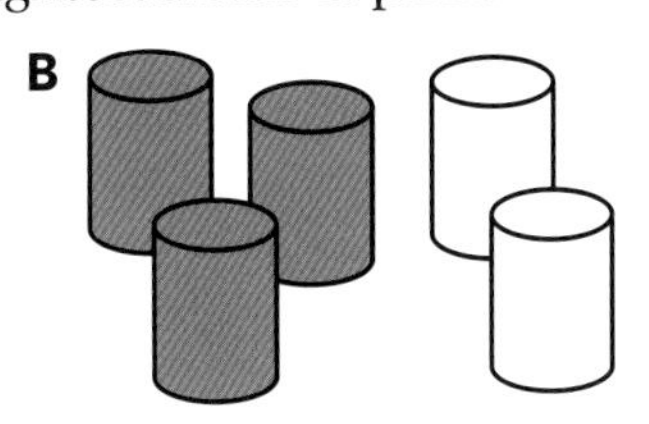

C

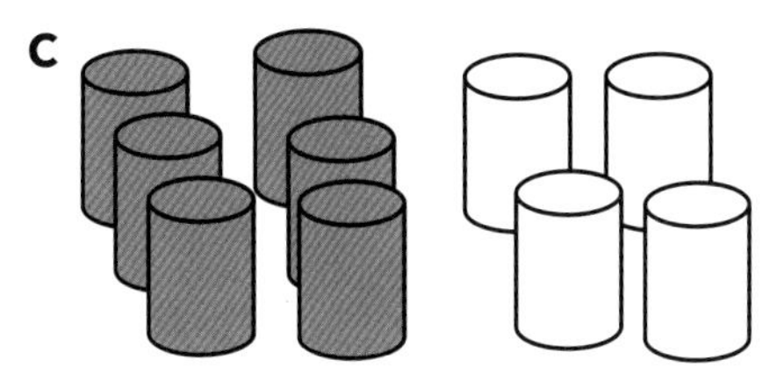

The ratios of red to white paint are:

A 4 : 3 B 3 : 2 C 6 : 4

First, work out the ratio of red to write paint in each diagram.

A $\frac{4}{4}:\frac{3}{4} = 1:0.75$ B $\frac{3}{3}:\frac{2}{3} = 1:0.67$ C $\frac{6}{6}:\frac{4}{6} = 1:0.67$.

Paint A has the greatest amount of white paint per unit of red paint, i.e. 0.75 tins of white for 1 tin of red, so this will be the lightest shade of pink.

Change these to form $1:n$. For each diagram, divide both parts of the ratio by the first part. Give the answers as decimals to make the comparison simpler.

Ratios in the form of $1:n$ are also useful for converting from one unit to another. For example, the ratio of inches to centimetres is 1 : 2.54.

This means that 1 inch is equivalent to 2.54 cm.

So, 2 inches = 2 × 2.54 cm and 12 inches = 12 × 2.54 cm.

Tip

Ratios in the form of $1:n$ can also be written as $n:1$.

This is a **linear** relationship and it can be shown as a straight-line graph. The equation of the line is $y = 2.54x$. Notice that the ratio of inches to centimetres (the ratio of $x:y$) is 1: 2.54. The equation $y = 2.54x$ and the ratio 1: 2.54 both state that for every value of x, y is 2.54 times larger.

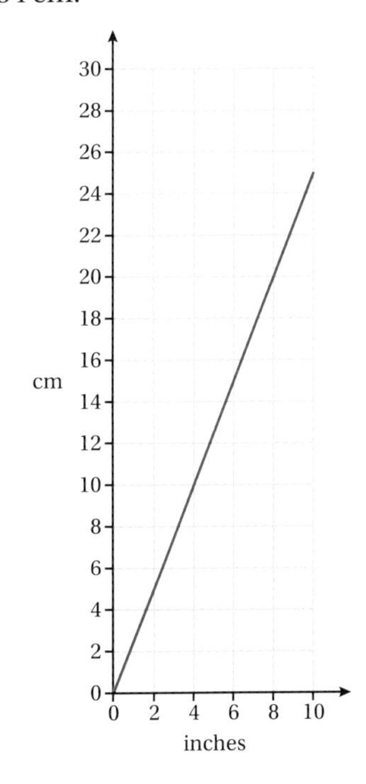

Tip

You learnt about linear graphs in Chapter 18.

Golden ratio

The golden ratio, $1:\frac{1+\sqrt{5}}{2} = 1:1.618$ has been studied and used for centuries. Artists, including Leonardo Da Vinci and Salvador Dali often produced work using this ratio. The ratio can also be seen in buildings, such as the Acropolis in Athens. The golden ratio is said to be the most attractive way to space out facial features.

The diagram shows how the golden ratio can be worked out using the dimensions of a 'golden' rectangle. The large rectangle ACDF is similar to BCDE. Hence the ratio of $a : a + b$ is equivalent to $b : a$.

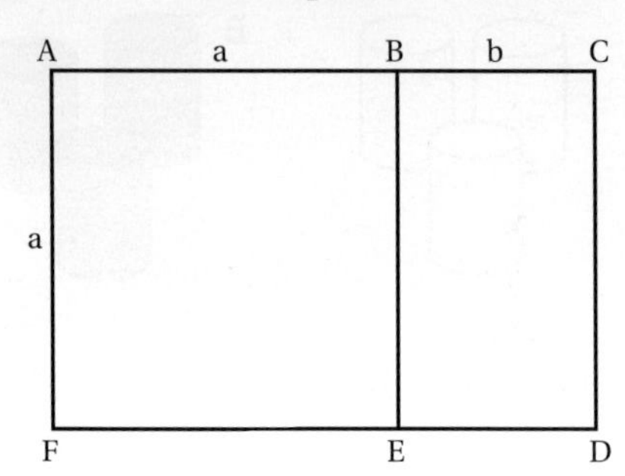

An approximate numerical value for this ratio can be found by measuring.

WORKED EXAMPLE 2

How golden are your hands?

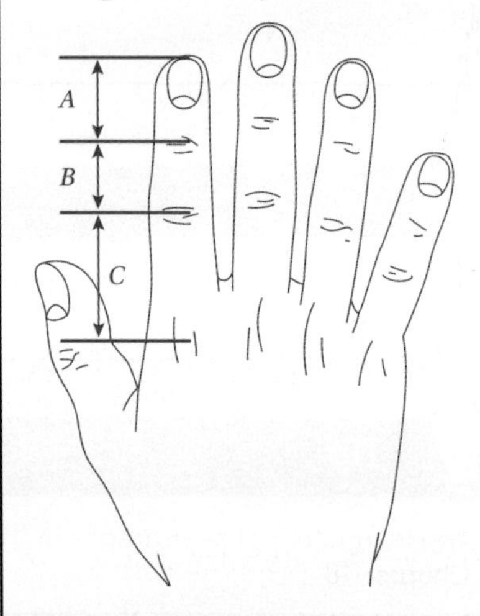

Measure the distances A, B and C on your hand.

Distance B = 22 mm

Distance C = 40 mm

Distance A = 13 mm

Length of hand = 170 mm

Wrist to elbow = 240 mm

B : C = 22 : 40 = 1 : 1.81

A : B = 13 : 22 = 1 : 1.69

Hand : wrist-elbow = 170 : 240 = 1 : 1.41

All the ratios are close to 1 : 1.618 which is the golden ratio.

Now calculate these ratios and write them in the form $1 : n$.

Distance B : Distance C

Distance A : Distance B

Length of your hand : Distance from your wrist to your elbow

Can you see anything special about these ratios?

The closer your results are to 1.1618 the more golden your hand!

You worked with the sequence of Fibonacci numbers in Chapter 4. This sequence follows the golden ratio. If you calculate the ratio of consecutive numbers in the Fibonacci sequence you will find that the ratio gets closer and closer to the actual golden ratio $1 : \frac{1+\sqrt{5}}{2}$ as you get further along the sequence.

EXERCISE 22C

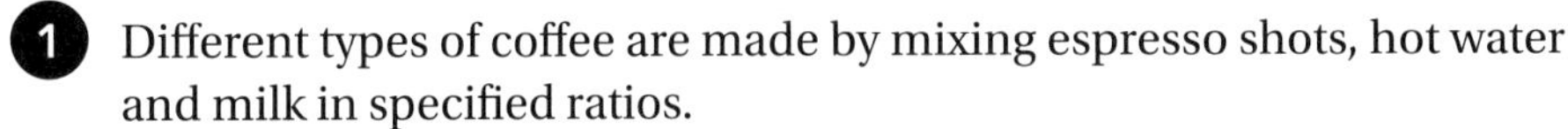

1 Different types of coffee are made by mixing espresso shots, hot water and milk in specified ratios.

Espresso	1 : 0 : 0
Double espresso	2 : 0 : 0
Flat white	1 : 2 : 1
Cappuccino	1 : 0 : 2
Latte	1 : 0 : 4

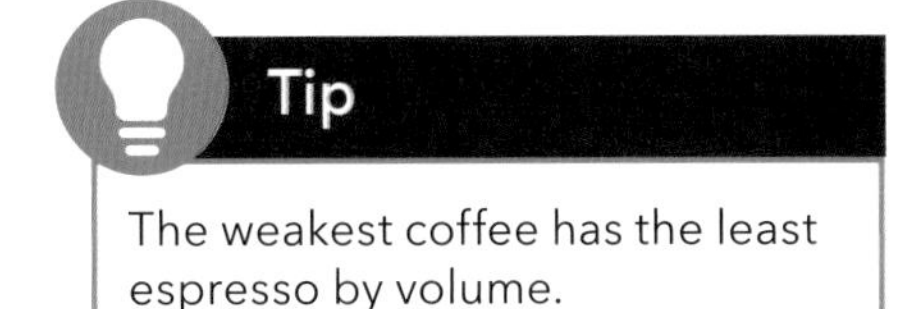

Tip

The weakest coffee has the least espresso by volume.

Put the drinks in order of strength, weakest first.

2 When Jon was going on holiday he used this graph to convert between pounds and euros.

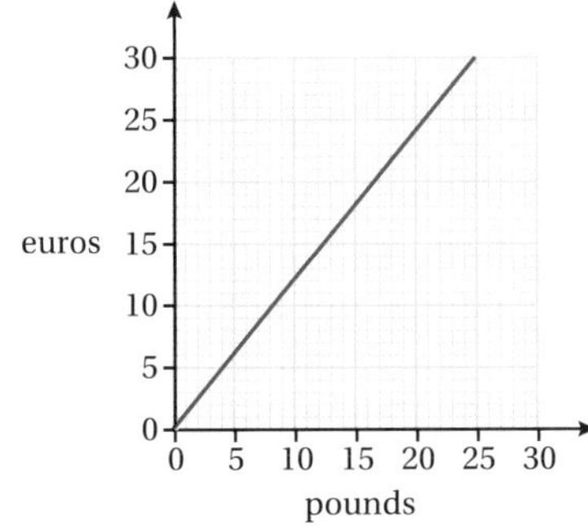

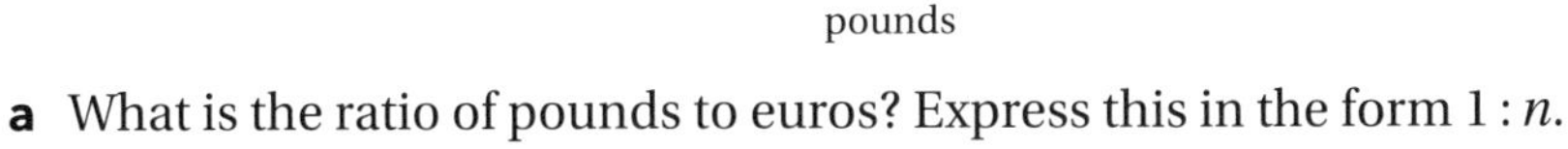

a What is the ratio of pounds to euros? Express this in the form $1 : n$.

b What is the ratio of euros to pounds? Express this in the form $1 : n$.

3 This graph shows the relationship between ounces and grams.

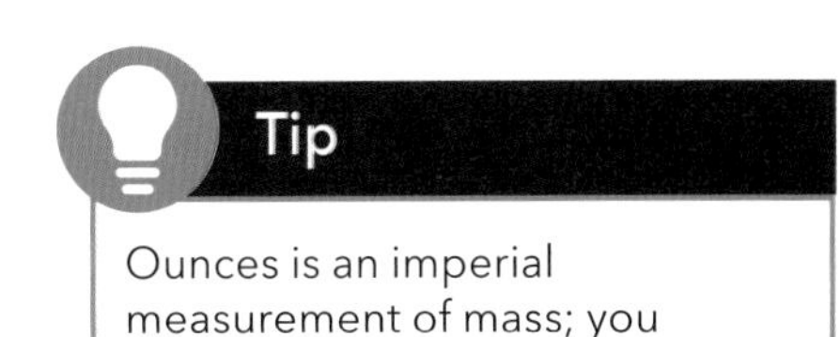

Tip

Ounces is an imperial measurement of mass; you worked on metric units in Chapter 12.

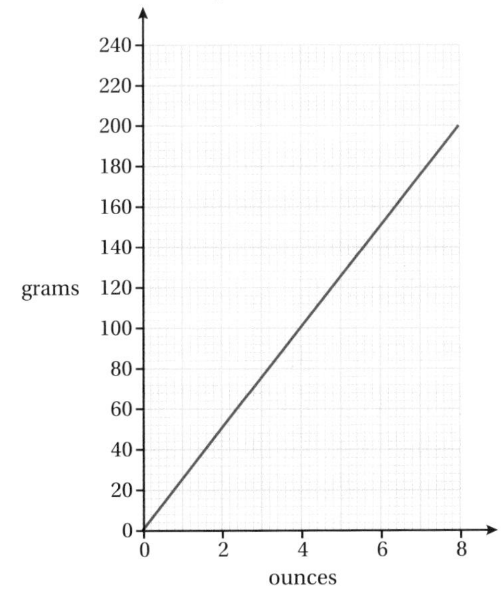

a What is the ratio of ounces to grams? Express this in the form $1 : n$.

b What is the ratio of grams to ounces? Express this in the form $1 : n$.

Find answers at: cambridge.org/ukschools/gcsemaths-studentbookanswers

4 The ratio of litres to gallons is 1 : 1.8.

a Draw a conversion graph to show this relationship.

b What is the ratio of gallons to litres in the form 1 : n?

5 Three siblings; Daisy, aged 5, Patrick, 8, and Imogen, 12, share sweets in the same ratio as their ages. Imogen gets 21 more sweets than Daisy.

a How many sweets were there to begin with?

b What fraction of the sweets did Patrick get?

6 The ratio of kilometres to miles is approximately 8 : 5. A car travels at 45 miles per hour for 20 minutes. Approximately how many kilometres does it travel?

7 This recipe for sausage casserole serves 6.

Find the quantities of ingredients needed to serve 4 people. Show all the steps in your answer.

Sausage casserole (serves 6)

- 12 sausages
- 3 tins of tomatoes
- 450 g potatoes
- 9 tsp mixed herbs
- 600 ml vegetable stock

8 Gill and her sister Bell share a box of chocolate. Bell gets $\frac{1}{3}$ of the box. Gill shares her chocolates with her best friend Katy in the ratio 4 : 3. Katy gets 12 chocolates. How many chocolates were there in the box?

9 A quarter of a box of chocolates are white chocolate. The ratio of dark to milk chocolates is 2 : 5. There are 7 white chocolates. How many more milk chocolates than dark chocolates are there?

10 A, B and C are three pulley wheels. For every 3 turns A makes, B makes 4 turns. For every 2 turns B makes C makes 3 turns.

a What is the ratio of the turns A makes to the turns B makes?

b What is the ratio of the turns C makes to the turns B makes?

c What is the ratio of the turns A makes to the turns C makes?

d Pulley wheel A makes 24 turns. How many turns does C make?

e Pulley wheel C makes 36 turns. How many turns does A make?

11 A 210 cm ribbon is cut into two sections. The longer piece is 2 and a half times the length of the shorter piece.

a What is the ratio of the longer piece to the shorter piece of ribbon?

b How long is each piece?

12 Each month a sunflower's height increases by 40%. What was the ratio of the height of the sunflower on 1st May to its height on 1st August?

13 Which is a better deal, 325 g jar of chocolate spread for 66p or a 1 kg tub for £1.99?

14 The ratio of Moisha's height at age 3, to her height at age 4 is 15 : 16.

a What percentage increase is this?

b During this time Moisha grew 7 cm. If Moisha keeps growing at the same rate how tall will she be when she is 8?

Checklist of learning and understanding

Notation

- A ratio compares two or more quantities with each other.
- The order in which a ratio is written is important. A ratio of 2 : 5 means 2 parts to 5 parts. Each part is equal in size.
- A proportion compares a quantity to the 'whole' of which it is a part; this can be written as a fraction, percentage or ratio.

Simplifying ratios

- Ratios can be simplified by dividing both parts of the ratio by a common factor.
- Two ratios are equivalent if one is a multiple of another, i.e. if they can be cancelled to the same simplest terms.
- Expressing ratios in the form $1 : n$ makes it easy to compare ratios. If three different ratios of $a : b$ were converted to $1 : n$ and compared, then the ratio with the highest value of n has the greatest amount of b per amount of a.
- Ratios can be useful for converting from one unit to another where the relationship is **linear**. The relationship can be represented by a straight-line graph.

Sharing in a given ratio

- The box method can be used to tackle problems that involve sharing a quantity in a given ratio. To share quantity Q in the ratio $a : b : c$, divide the quantity evenly into $a + b + c$ boxes.
- You can also work out the fraction each part is of the whole, and multiply the quantity by the fraction.

Chapter review

For additional questions on the topics in this chapter, visit GCSE Mathematics Online.

1 What is the ratio of the diameter of a circle to its circumference in the form $1 : n$?

2 The three angles of a triangle are in the ratio 3 : 3 : 4. What information can you give about the triangle?

3 The ratio of the five angles in a pentagon are 1 : 1 : 1 : 1 : 1, what information does this tell you about the pentagon? How do you know this?

4 The ratio of an exterior angle to an interior angle of a regular polygon is 1 : 3. How many sides does the polygon have?

5 The ratio of the angles in a triangle is 1 : 2 : 1. What information can you give about the triangle? If the longest side is 10 cm, how long are its other two sides?

Tip

To answer the questions about circles, triangles, pentagons and rectangles you may need to look again at Chapter 5. For the questions on area and volume you may need to look again at Chapters 16 and 21.

Find answers at: cambridge.org/ukschools/gcsemaths-studentbookanswers

6 **a** Joe and Pam planted crocus bulbs in their gardens.

They shared a bag of 250 crocus bulbs.

The table shows the colour of the flower from each bulb.

	Yellow	Purple	White	Totals
Joe	64	40		125
Pam	56		32	125
Totals	120			250

i Complete the table. *(3 marks)*

ii Write the ratio 64 : 56 as simply as possible. *(1 mark)*

b Sumita bought a pack of 60 crocus bulbs which produced Yellow, Purple or White flowers. The ratio Yellow : Purple : White was 7 : 5 : 3.

How many of the 60 bulbs produced White flowers? *(3 marks)*

© OCR 2012

7 The ratio of the sides of a rectangle is 3 : 4. After an enlargement by a scale factor of 5 what is the ratio of the same two sides?

8 Gareth and John share a box of chocolates. Gareth gets $\frac{3}{5}$ of the box. The ratio of white to milk to dark chocolates in John's share is 1 : 2 : 1. John gets 4 white and dark chocolates in total. Gareth gets twice as many white chocolates as John, and has an equal number of dark and milk. How many types of each type of chocolate were in the box?

9 The ratio of the sides of two squares is 3 : 4. What is the ratio of their areas?

10 The ratio of the edge of two cubes is 5 : 2.

a What is the ratio of their surface areas?

b What is the ratio of their volumes?

11 What is the ratio of vowels to consonants in the English alphabet?

12 What is the ratio of prime numbers to square numbers between (and including) 1 and 20 in its simplest form?

13 Share 360 in the ratio 3 : 5 : 1.

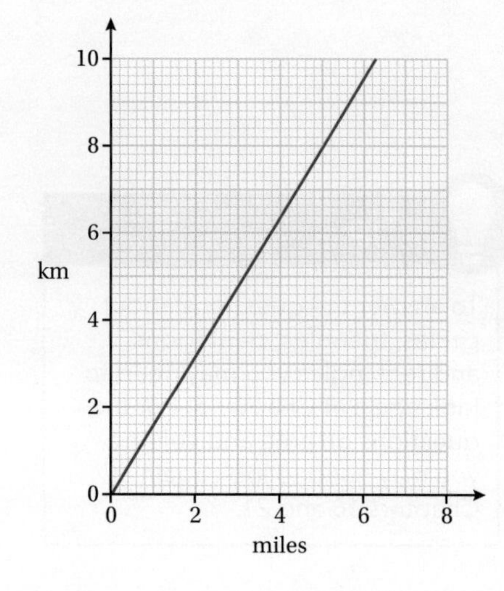

14 Using the graph to the left express the ratio of miles to kilometres in the form 1 : *n*.

15 In a car park, three-quarters of the cars are not silver, but are blue, red, black or yellow. The proportion of blue to red to black to yellow cars is 6 : 2 : 3 : 1. There are six more black cars than yellow cars. How many cars of each colour are in the car park?

Tip

Don't forget about the silver cars!

23 Basic probability and experiments

In this chapter you will learn how to …

- represent and analyse outcomes of probability experiments.
- relate relative frequency to theoretical probability.
- calculate probabilities in different contexts

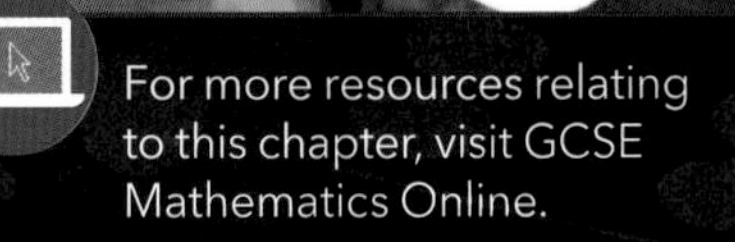

For more resources relating to this chapter, visit GCSE Mathematics Online.

Using mathematics: real-life applications

Software developers use probability when they build applications. Apps such as speech recognition, speech synthesis, key-word spotting and predictive text all rely on probability. In speech recognition for example, the software analyses the audio input and finds the most likely stream of text based on the audio. So, when you say a name into your phone instead of dialling, the software chooses the most likely name from your contact list.

"Knowing how to use and apply probability was one of the requirements when I was interviewed for this programming job." *(Computer programmer)*

Before you start …

Ch 10, 11	You need to know how to calculate with fractions, decimals and percentages	**1** Choose the correct answer without doing the calculations. **a** 0.13×0.24 A 0.312 B 0.0312 C 0.00312 **b** $0.82 + 0.18$ A 0.1 B 100 C 1 **c** 0.08% of 50 A 4 B 0.4 C 0.04
Ch 17	You need to be able to round decimals to 1, 2 or 3 places.	**2** Which answers are rounded incorrectly? Why? **a** $0.705\,882\,352 \approx 0.706$ **b** $0.316\,666\,6 \approx 0.316$ **c** $0.989\,087 \approx 1.09$
Ch 10, 11, 13	You should be able to find equivalent fractions, decimals and percentages.	**3** Match the equivalent pairs. $\frac{39}{52}$ 0.25 75% 50% $\frac{26}{52}$ $\frac{13}{52}$ 0.077 $\frac{4}{52}$

Calculator tip

Make sure you know how to enter common fractions such as $\frac{17}{23}$ into your calculator and how to convert between fractions and decimals; see Chapters 10 and 11 if you need to.

Find answers at: cambridge.org/ukschools/gcsemaths-studentbookanswers

Assess your starting point using the Launchpad

STEP 1

1 Maria rolls a normal six-sided dice 300 times.

a Based on the theoretical probability of rolling a six, how many sixes would you expect her to roll?

b In her experiment she obtains a six 113 times. What is the relative frequency of rolling a six?

c If she rolled the same dice 650 times, how may sixes could she expect? Why?

? **GO TO Section 1**: Review of probability concepts

✓

STEP 2

2 Two normal six-sided dice are rolled and the sum of the numbers on their faces is recorded.

a Calculate P(12).

b Which sum has the greatest probability? What is the probability of rolling this sum?

c What is P(*not* even)?

d What is P(sum < 5)?

? **GO TO Section 2**: Further probability

✓

GO TO Section 3: Working with probability

Section 1: Review of probability concepts

Expressions of probability

Each of the following statements indicates the likelihood or probability of something happening.

- I definitely didn't pass that test because I couldn't answer a single question.
- It is unlikely to rain today.
- I'm sure Sarah will be elected captain. Everyone says they will vote for her.

The likelihood, or chance, of an **event** can be given more mathematically using the probability scale.

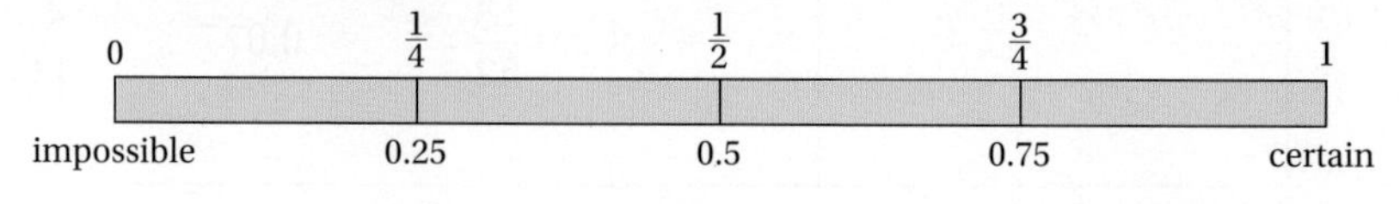

Key vocabulary

event: a set of possible outcomes of an experiment or situation, to which you give a probability.

The smaller the fraction, the less likely it is that the favourable outcome will happen.

An outcome is the mathematical word for the result of an experiment. The favourable outcome is the particular outcome you are calculating the probability for.

- An impossible outcome (such as passing a test if you didn't answer any questions) is given a probability of 0.
- A certain outcome (such as winning an election if you get all the votes) is given a probability of 1. All other events are given a probability between 0 and 1.

The sum of the probabilities of all the possible outcomes of an event is 1.

If the probability of a given outcome is P(E), then the probability of that outcome **not** happening is 1 – P(E). Probabilities can be expressed as fractions, decimals or percentages. In different situations, probabilities can be easier to visualise when given in a particular form.

Fraction	Being told the probability of winning the jackpot on a fruit machine is $\frac{1}{266\,144}$ (just over 1 in quarter of a million) is more meaningful than being told the probability is 0.00037%
Decimal	You can quickly compare decimal values on the probability scale because you don't have to work with different denominators.
Percentage	Percentages are often used in media report about probability. For example, 'a probability of $\frac{28}{37}$ is not as meaningful to the general public as 'a 76% chance'.

There are two ways of assigning probability:

- Experimental probability – where you do an experiment, observe the **outcome** and record the results.
- Theoretical probability – where you calculate a probability based on fairness and symmetrical properties of results.

Key vocabulary

outcome: a single result of an experiment or situation.

Tip

A favourable outcome can also be referred to as a successful outcome.

Tip

In real life, people often say things like "I have a 1 in 5 chance of getting the job". You should not give mathematical probabilities in the form of a 1 in 5 chance, 1 to 5 or 1 : 5 in your answers.

Experimental probability

You can do experiments to determine the probability that something will happen.

A class did an experiment to test whether toast always lands butter-side down. They got 30 volunteers to throw toast into the air and they recorded how many times it landed butter-side facing down. They recorded it to be 16 out of 30 times.

The number of trials in this experiment is 30.

There are two possible outcomes for each trial: butter-side down or butter-side up.

Tip

At GCSE the terms 'relative frequency' and 'experimental probability' get used interchangeably. So if a question asks for the 'relative frequency' and another asks for 'experimental probability', both questions are asking for the same thing.

The frequency of each outcome is how many times it occurred. The frequency of butter-side down is 16 and the frequency of butter-side up is 14 (because $30 - 16 = 14$).

$$\text{Relative frequency} = \frac{\text{number of favourable outcomes recorded}}{\text{total number of trials}}$$

The relative frequency of butter side down is $\frac{16}{30} \approx 0.533$.

The relative frequency of butter side up is $\frac{14}{30} \approx 0.467$.

This means that the experimental probability of toast landing butter-side down is only slightly higher than it landing butter-side up.

Did you know?

The dice that are used for gambling in casinos are carefully made and checked to make sure that they are fair and unbiased. Biased dice (or coins, or spinners) will give some results more often than others. For example, confidence tricksters might use a coin that has been weighted to land on heads every time it is flipped, or a dice that has been manipulated to land on a six. When you are working with probability problems, assume the equipment is fair and unbiased unless you are told differently.

Tip

In probability questions, a 'normal dice' is a six-sided dice numbered from 1 to 6. Unless specifically stated, assume this is the case. Similarly, assume coins are fair and that there is an equal chance of having a girl or boy at birth.

Tip

Here two events are happening at the same time: 'rolling dice A' and 'rolling dice B'; this is a combined event. You will learn more about combined events in Chapter 24.

EXERCISE 23A

1 Work with a partner. You will need two normal dice.

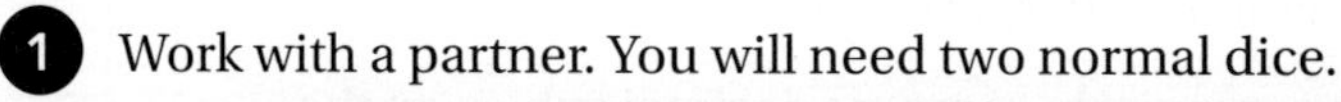

a Use words from the box to describe the probability of each of these outcomes when you roll two dice at the same time and add the total on the two faces.

	Outcome	Predicted probability
i	A total $\leqslant 12$	
ii	An even number	
iii	An odd number	
iv	A total of 1	
v	Exactly 12	
vi	A total > 4	

Impossible   Highly unlikely   Unlikely   Evens
Likely   Highly likely   Certain

b Carry out an experiment in which you roll the two dice 50 times and record the frequency of each of the outcomes listed in the table. Bear in mind that one trial may meet more than one outcome.

c Compare your results with your predictions. How well did you predict the outcomes?

2 Find the experimental probability of:

a getting heads with one flip of a coin if the coin landed heads up 96 times in 180 trials.

b rolling a 6 with a dice given that when the dice was rolled 300 times, the frequency of rolling a 6 was 54.

c getting an even number on a dice if an odd number was rolled 33 times in 60 trials.

3 Two dice were rolled 80 times and the total shown on the faces was recorded. This table gives the frequency of each outcome.

Total	2	3	4	5	6	7	8	9	10	11	12
Frequency	5	2	8	6	12	14	11	8	7	3	4

a Calculate the relative frequency of getting a total of 7 as a fraction and a decimal.

b What is the experimental probability of not getting a total of 7?

c What is the experimental probability of rolling two sixes in this experiment?

d What is the experimental probability of getting a total less than 6?

4 A market-research company did a survey to find out what brand of shampoo people bought most often. The results are given in the table.

Brand	**Frequency**	**Relative frequency**
Silk-e-shine	123	
Get knotted	105	
Goldilocks	83	
Bubbly stuff	92	
Total		

a What was the sample size for the survey?

b Calculate the relative frequency of buying each brand.

c Use the results of this survey to estimate the probability that a person chosen at random is likely to buy Silk-e-shine shampoo.

5 Mira calls customers who have had their car serviced to check whether they are happy with the service they received. She kept this record of what happened for 200 calls made one month.

Result	**Frequency**
Spoke to customer	122
Phone not answered	44
Left message on answering machine	22
Phone engaged or out of order	10
Wrong number	2

a Calculate the relative frequency of each event as a decimal fraction.

b Is highly likely, likely or unlikely that the following events will occur when Mira makes a call?

i The call will be answered by the customer.

ii The call will be answered by a machine.

iii She will dial the wrong number.

Find answers at: cambridge.org/ukschools/gcsemaths-studentbookanswers

6 The results of an on-campus student council election are shown below. A total of 4000 students voted in the election.

Candidate	Votes
Alexia Adams	1445
Zunaid Darcey	1593
Amitab Smith	483
Nicky Chin	

a How many votes did Nicky Chin get?

b What is the probability that a randomly selected student voted for Zunaid Darcey?

c What is the probability that a randomly selected student did not vote for Alexia Adams?

7 A catering company has yellow, red and black candles which it chooses at random to put on tables. The probability of the chosen candle being yellow is 0.083. A candle is three times as likely to be red as it is to be yellow. Calculate the probability of the candle being black.

8 A local government agency carried out a census of 500 000 people working in the city. They collected the following data.

Qualifications	Frequency	Language abilities	Frequency
Postgraduate diploma/degree	74 500	English only	123 000
First degree/diploma	92 350	English and one other language	209 500
No post-school qualifications	333 150	Multilingual (English plus at least two other languages)	167 500

If a person included in this census is selected at random, what is the probability that the person:

a has a first degree or a diploma?

b is able to speak English only?

c has some post-school qualification?

d is able to speak a language other than English?

Theoretical probability

Key vocabulary

equally likely: having the same probability of happening.

random: not predetermined.

When you flip a fair coin there are two possible outcomes: head and tails. You have the same chance of getting heads as you have of getting tails, so the outcomes are **equally likely**. This does not mean that if you flip a coin six times in a row that you will get three heads, and three tails. Although the outcomes are equally likely, they are also **random**. You could get six heads in a row or six tails in a row. However, the more often you flip the coin, the closer you will get to an equal number of heads and tails.

For equally likely outcomes you can calculate the probability using a formula.

$$\text{Probability of an event} = \frac{\text{number of favourable outcomes}}{\text{total number of outcomes}}$$

WORKED EXAMPLE 1

A bag of clean laundry contains 5 blue shirts, 6 red shirts, 7 green shirts and 7 white shirts. A student grabs one shirt at random from the bag.

What is the probability that it is green?

$5 + 6 + 7 + 7 = 25.$ Calculate the total number outcomes.

$P(\text{green}) = \frac{7}{25}$

$= 0.28$

7 green shirts means 7 favourable outcomes. Put the values into the formula and calculate the probability.

EXERCISE 23B

1 There are 19 girls and 17 boys in a classroom. The teacher puts their names into a bag and draws one at random.

What is the probability that the outcome will be a boy?

2 Camilla rolls an unbiased six-sided dice.

What is the probability that she rolls an even number?

3 Nick and Vijay are playing a game in which they take turns to roll two unbiased dice with the numbers 1 to 6 on them. They find the product of the two dice. If the product is odd, Nick gets a point, if the product is even, Vijay gets a point. The first person to 20 points wins.

Predict which student is most likely to win and justify your answer.

4 Is the reasoning in each of these statements correct? Explain why or why not.

a Since there are 26 letters in the alphabet, the probability that a name will start with X is $\frac{1}{26}$.

b My first three children were boys, so the next one must be a girl.

c There are ten teams in the tournament, so the probability of any team winning is $\frac{1}{10}$.

d The probability that a family will go on holiday in August is $\frac{1}{12}$.

e This team has won the last four matches, so they are certain to win the next one too.

Section 2: Further probability

Probability is a value that we use to predict what we expect to happen. In reality, there is no guarantee that any particular outcome will occur. You know that the probability of getting heads when you flip a coin is $\frac{1}{2}$ and you may expect to get heads about half of the time, but in reality you could flip a coin 20 times and get 18 heads.

Similarly, an insurance company may use statistical data to work out that drivers between the ages of 17 and 23 are more likely to have an accident than older drivers. This does not mean that a 19-year-old driver will definitely have an accident.

A good understanding of probability and how it works will help you make sense of chance and risk in daily life.

The probability of an event not happening

The probability of Amanda scoring a goal in a netball match is 0.6. In this situation there are two possible outcomes: either Amanda scores a goal, or she does not. All probability situations can be reduced to two possible outcomes. For example, win or not win, heads or not heads, rolling a six or not rolling a six. When you express the outcomes in this way we say they are **complementary**.

When you add the probability of an event and its complement you get 1.

$P(\text{heads}) + P(\text{not heads}) = \frac{1}{2} + \frac{1}{2} = 1$ (Not heads is the same as tails.)

$P(\text{six}) + P(\text{not six}) = \frac{1}{6} + \frac{5}{6} = 1$ (Not six is the same as rolling 1, 2, 3, 4, or 5.)

$P(\text{Amanda scoring}) + P(\text{Amanda not scoring}) = 0.6 + 0.4 = 1$

In general terms we can say that
$P(\text{event occurring}) + P(\text{event not occurring}) = 1$.

If you rearrange this equation you get $P(\text{event not occurring}) = 1 - P(\text{event occurring})$.

If $P(E)$ is the probability of an event (E) happening, and $P(E')$ is the probability of that event not happening, then $P(E') = 1 - P(E)$.

WORK IT OUT 23.1

A laboratory tested 500 batches of tablets and found 4 to be contaminated. What is the probability that a batch of tablets produced in this laboratory would be:

a contaminated? **b** not contaminated?

Which of these answers is the correct answer to the following question? Why is the other one wrong?

Option A	Option B
a $P(\text{contaminated}) = \frac{4}{500} = 0.8\%$	**a** $P(\text{contaminated}) = \frac{4}{500} = 0.008$
b $P(\text{not contaminated}) = \frac{496}{500} = 92\%$	**b** $P(\text{not contaminated}) = 1 - 0.008 = 0.992$

Key vocabulary

mutually exclusive: events that cannot happen at the same time.

Mutually exclusive events

Mutually exclusive events cannot happen at the same time. For example, you cannot get an even number and a three at the same time when you roll a dice, and you cannot pick a vowel and a consonant if you choose one letter of the alphabet at random.

Imagine you have a bag with 3 red, 2 yellow and 5 green sweets in it and you are allowed to choose one sweet at random.

- The probability of you choosing a red sweet is $\frac{3}{10}$.
- The probability of you choosing a yellow sweet is $\frac{2}{10}$ or $\frac{1}{5}$.
- The probability of you choosing a green sweet is $\frac{5}{10}$ or $\frac{1}{2}$.
- The sum of probabilities of the three events is 1.

You cannot pick a red sweet and a yellow sweet at the same time, so the events P(red) and P(yellow) are mutually exclusive.

You can, however, work out the probability of choosing either a red or a yellow sweet by adding their individual probabilities.

There are 3 red and 2 yellow so $\frac{5}{10}$ of the sweets are either red or yellow.

EXERCISE 23C

1 Michelle catches the C125 bus to work. Over a period of 227 working days she did not manage to get a seat on the bus 58 times.
Calculate the experimental probability of her getting a seat on the bus.

2 What is the probability that the number on a raffle ticket drawn at random from a set of tickets numbered 0 to 99 (inclusive) will be:

a divisible by 2?

b not a multiple 10?

c a multiple of 8?

d not a multiple of 8?

3 The probability of a basketball player missing a goal is given as 0.432. What is the probability that the player will not miss?

4 A packet holds 300 sweets in five different flavours. The probability of choosing a particular flavour is given in the table.

Flavour	Strawberry	Lime	Lemon	Blackberry	Apple
P(flavour)	0.21	0.22	0.18	0.23	

a Calculate P(apple).

b What is P(not apple)?

c Calculate the probability of choosing P(not lemon or lime).

d Calculate the number of sweets of each flavour in the packet.

5 Students in a school have five extra-curricular clubs to choose from. The probability that a student will choose each club is given in the table.

Club	Computers	Sewing	Woodwork	Choir	Chess
P(club)	0.57	0.2	0.2	0.02	0.01

a Calculate P(sewing or woodwork).

b Calculate P(not chess or choir).

c If 55 students have to choose a club, how many would you expect to choose sewing?

d If 4 students chose choir, calculate how many students chose computers. (Assume the probabilities are correct.)

Find answers at: cambridge.org/ukschools/gcsemaths-studentbookanswers

Organising outcomes – tables and frequency trees

With probabilities it is important to find ways of recording and organising the information to show how many favourable outcomes there are, how many outcomes there are in total, and to find the figures you need to make decisions and calculate probabilities. Tables and simple diagrams called frequency trees are used to do this.

A doctor was interested in whether patients knew the difference between having a cold and having the flu. He collected data one winter and kept track of 42 patients. He organised his data as follows.

Two-way table

	Actual diagnosis	
Self-diagnosis	**Cold**	**Flu**
Cold	7	4
Flu	12	19

Frequency tree

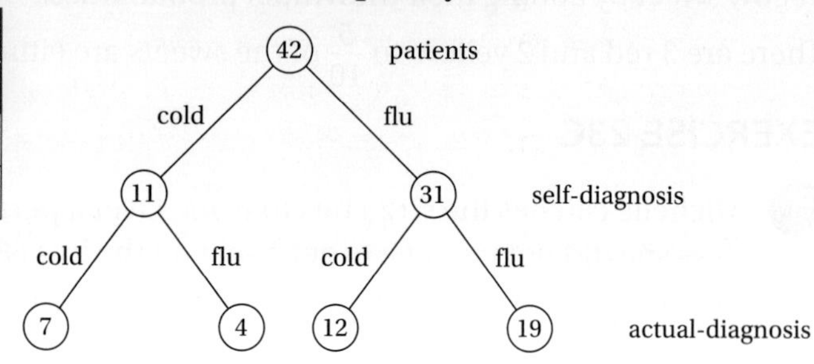

Frequency trees

A frequency tree shows the actual frequency of different outcomes.

The branches of the tree show the paths or decisions and the 'leaves' show the actual data for each path. Both the two-way table and the frequency tree show the same information, but the frequency tree is clearer because it shows how many patients thought they had a cold or flu without you having to add the data in the table.

Frequency trees allow you to understand and make sense of complicated probabilities. For example, if a woman is screened for breast cancer and her test results come back positive, she is likely to assume that she has breast cancer. But research has shown that there are many false-positive results.

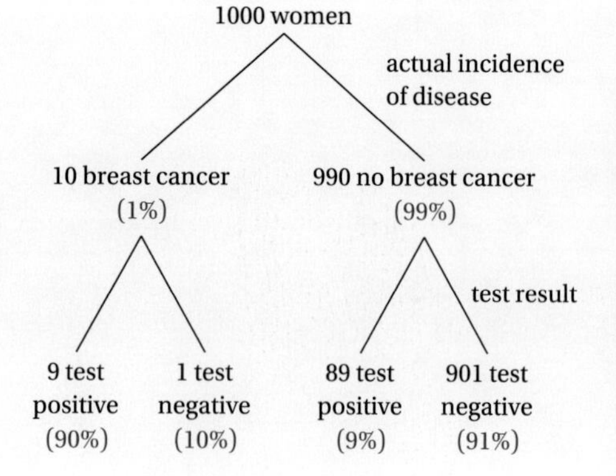

This frequency tree shows how many women who test positive for breast cancer will really have cancer.

The frequency tree shows that 98 women test positive on the screening test.

Of these, only 9 actually have breast cancer.

$\frac{9}{98} = 0.09 \ldots = 9\% \text{ (1 sf)}$

So only 9 out of every 98 women (around 1 out of every 10) who get a positive screening test for breast cancer will actually have cancer.

Tip

Frequency trees are organisational tools and they are often used in computer programming (they are sometimes called binary trees). They are not the same as probability tree diagrams which you will deal with in Chapter 24.

EXERCISE 23D

1 A hotel chain keeps track of which customers use of its spa. Here are its results.

Gender	Use the spa	Don't use the spa
Female	780	232
Male	348	640

guests
gender
spa use
2000
F
M
use
don't use
use
don't use

a Complete this frequency tree to show this data.

b Are male or female guests more likely to use the spa?

2 Of 60 patients visiting a doctor's rooms, 42 think they will need prescription medication, the others think they probably won't need a prescription. Of those who think they will need a prescription, 13 do not get one. Altogether 36 patients do need a prescription.

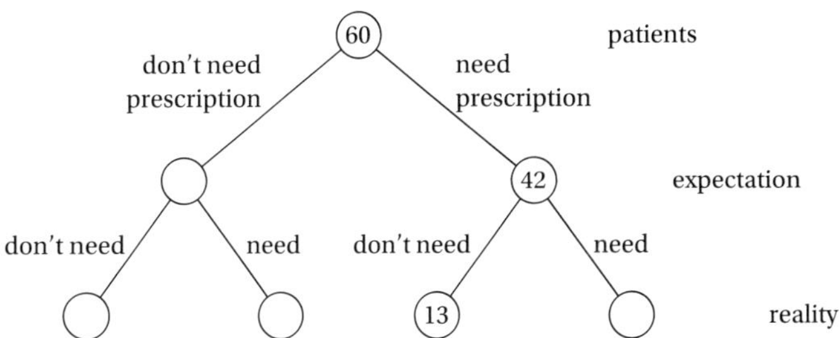

a Complete the frequency tree to show the actual numbers.

b Assuming this is a representative situation, what is the probability that a patient that thinks they need a prescription will actually need one?

c What percentage of patients who thought they would not need a prescription actually needed one?

3 After completing a multiple-choice test, Andy predicted that he got 16 of the 20 questions correct. Of the 16 he predicted that he'd answered correctly, he got 3 wrong. Altogether he got 17 out of 20.

a Draw a frequency tree to show this information.

b Comment on how well he predicted the outcomes of the test.

4 80 volunteers take a flu test to help the medical researchers work out how accurate the test is. Of the volunteers, 17 people have the flu, the others do not. The results show that 1 of the flu-positive people gets a negative result on the test and 2 of the flu-negative people get a positive result on the test.
Draw a frequency tree to show the actual results.

Section 3: Working with probability

In real life, people tend to use probability quite informally to explain things and to predict what will happen in the future. You may have heard people say things like:

- You must drive carefully on this bend because there are always accidents here.
- It is never sunny here in February.
- Most people prefer to wear sandals in summer.
- There is a very high risk of HIV among intravenous-drug users.
- If it rains on match day, the other team has a better chance of winning.
- We are only selling 10 000 tickets, so you have an excellent chance of winning the car.
- Young people who haven't had a driving licence for very long have more accidents than older drivers.

These statements are all expressions of probability that make a claim about the future based on experience, patterns or trends.

But, each statement can be untrue in individual cases. For example, Pete might say: 'You must drive carefully on this bend because there are always accidents here.' His friend Ahmed might say: 'I don't agree, I drive along that road all the time and I have never seen an accident there.'

Both statements can be true. In reality, the traffic department would want to know the relative frequency of accidents $\left(= \dfrac{\text{number of cars involved in accidents}}{\text{total number of cars using the road}}\right)$ before they make decisions about the safety of the bend.

Understanding probability allows you to think more critically about statements like the ones above and to work out more accurately what the chance is of different things happening.

WORKED EXAMPLE 2

A consumer organisation commissioned a series of tests to determine the average lifetime of a locally produced solar garden lamp. The results of the tests are shown in the table:

Lifetime of solar lamp, L (hours)	$0 \leqslant L < 1000$	$1000 \leqslant L < 2000$	$2000 \leqslant L < 3000$	$3000 \leqslant L$
Frequency	30	75	160	35

a Use the results of the tests to estimate the probability that a solar lamp will last for less than 3000 hours but at least 1000 hours.

b If the hardware depot orders 2000 solar lamps. How many of them can they expect to last for more than 3000 hours?

a $P(1000 \leqslant L < 3000 \text{ hours}) = \dfrac{75}{300} + \dfrac{160}{300}$

$= \dfrac{235}{300} = 0.783$

'At least 1000 hours' means greater than or equal to 1000 hours. Add together the frequencies for the appropriate ranges and substitute them into the formula for relative frequency/experimental probability.

b $P(\text{lasts more than } 3000 \text{ hours}) = \dfrac{35}{300}$

$\dfrac{35}{300} \times 2000 = 233.\dot{3}$

Calculate the relative frequency for a lamp lasting for more than 3000 hours. Then multiply this by the number of lamps in the order.

If these statistics are correct, 233 of the solar lamps should last more than 3000 hours

Sometimes the way a problem is worded can be confusing, but the actual calculations in probability generally use the same principles. When you have to solve word problems involving basic probability you can generally do this by organising your work and following the steps in the problem-solving framework.

Problem-solving framework

Nick is throwing a ball randomly at a wall on the side of a building. The side of the building is 2 m high and 10 m wide. There are three windows on the side of the building, each window is 2 m wide and 1 m high.

What is the probability that Nick will hit a window when he throws the ball at the wall?

Express your answer as a percentage.

Steps for approaching problem-solving questions	What you would do for this example
Step 1: What are you trying to work out?	The probability of hitting any of the windows.
Step 2: What information do you need?	The area of the wall and the area of the windows.
Step 3: What maths can you do?	Area of wall = $10\text{ m} \times 2\text{ m} = 20\text{ m}^2$ Area of windows = $3 \times (2\text{ m} \times 1\text{ m}) = 3 \times 2\text{ m}^2 = 6\text{ m}^2$ Apply this information to the formula for calculating relative probability: P(hits window) = $\frac{6}{20} = \frac{3}{10}$ (Convert to a percentage) $\frac{3}{10} \times 100 = 30\%$ There is a 30% probability that Nick will hit a window.

EXERCISE 23E

1 Nina and Maria made up a game with an eight-sided dice. The sides of the dice are labelled 6, 24, 9, 29, 15, 7, 18 and 12. The chance of rolling each number is equally likely.

They take turns to roll the dice. Nina wins the roll if the dice shows a multiple of 2. Maria wins the roll if the dice shows a multiple of 3.

a Is this a fair game? Give a reason for your answer.

b What is the theoretical probability that the dice will show a multiple of 3?

2 During a netball competition, the same coin was flipped 20 times. Busi claimed the coin was unfair because it landed on tails only 5 out of the 20 times.

She says the probability of getting tails when you flip a coin is 0.5, so if you flip the coin 20 times you should get $20 \times 0.5 = 10$ tails.

Was she correct? Explain your answer.

3 A local educational authority wants to introduce random drug testing in secondary schools. It claims the tests have a very small false-positive rate of one half of one per cent.

a Express one half of one per cent as a decimal.

b The parents at a school with 800 students object to the test. They claim that 4 students could incorrectly test positive for drug use.
Are the parents' concerns valid? Give a reason for your answer.

c There were 38 937 secondary school students under this authority in the year they wanted to do the drug testing.
If they were all tested for drug use, how many of them would you expect to be incorrectly identified as using drugs?

Tip

A false-positive in a drug test means that a person who is not using drugs tests positive for drug use.

Find answers at: cambridge.org/ukschools/gcsemaths-studentbookanswers

4 Professional athletes are routinely tested for banned performance enhancing substances. The testing authority estimates only 1% of the athletes tested are actually using banned substances. If an athlete is using banned substances, 90% of the time he or she will test positive in the test (in other words, fail the drug test). But, 10% of the athletes who are not using banned substances will also test positive (in other words, fail the drug test even though they are not using banned substances).

a Complete this table to show how many athletes will pass or fail the drug test for every 1000 athletes tested.

Status	Test positive (i.e. fail drug test)	Test negative (i.e. pass drug test)	Total
Athletes who are using banned substances			10
Athletes who are not using banned substances			990
Total			1000

b Represent the same information on a frequency tree.

c If an athlete tests positive for the banned substances, what is the chance that he or she is not actually using the substance? Give your answer as a percentage.

d If an athlete tests negative for the substances, is it certain that he or she is not using them? Explain your answer.

5 Amit designed a round spinner out of plastic to be used in a game at a school fund-raising event. The spinner was divided into quarters, coloured red, green, yellow and white. His friend Nickali said the spinner might be biased because Amit didn't sand it down smoothly. Amit disagreed, but they decided to test whether it was fair or not by spinning it 1000 times and recording the outcomes. These are the results.

Outcome	Red	Green	Yellow	White
Frequency	295	248	238	219

Does the evidence suggest the spinner is biased?

6 A shopkeeper did a survey to find out which customers buy fresh fish at his shop every week.

	Females		Males	
	Buy fish	Don't buy fish	Buy fish	Don't buy fish
20–50 years old	23	56	19	25
Over 50 years old	45	26	13	5

a How many people were surveyed?

b How many people over 50 were surveyed?

c A customer from the survey is chosen at random. What is the probability that the customer is:

i male? **ii** male or over 50? **iii** over 50 and buys fish?

d How could this information help the shopkeeper plan advertising and marketing campaigns?

7 There are four main blood groups: A, B, AB and O. The following data about blood types was collected by a blood bank based on the blood types of 500 blood donors.

Blood type	Number of donors
A	220
B	49
AB	21
O	210

a What percentage of donors belongs to group O?

b What is the probability of a donor having blood type AB?

c People with blood type O negative are often called universal donors because their blood can usually be given to people of any blood type without any bad reaction. Only about 7% of the population are O negative.

How many of these donors would you expect to have O negative blood?

d Given these statistics, what is the probability that a baby will be born with blood type AB?

e Why is theoretical probability not very useful for predicting blood type?

8 80 people are asked if they can tell the difference between butter and margarine. 37 say they can, 24 say no and 19 say they are not sure. The interviewer then carries out a blind taste test. Of those who said they could tell the difference, 14 got it wrong; of those who said no, 9 got it right; and of those who said they were not sure, 14 got it wrong.

Draw a frequency tree to show the outcomes of this experiment.

Checklist of learning and understanding

Probability

- Probability is a measure of how likely something is to happen.
- The probability scale ranges from 0 to 1. Impossible events have a probability of 0 and certain events have a probability of 1. It is not possible to have a negative probability (< 0) or a probability greater than 1.
- Probabilities between 0 and 1 can be expressed as fractions, decimals or percentages.
- A favourable outcome is the particular outcome of an event for which you are calculating the probability.
- The sum of probabilities will always total 1.
- The probability of a favourable outcome happening is equal to 1 minus the probability of it not happening. So, $P(E) = 1 - P(\text{not } E)$

Theoretical probability

- Probability of a favourable outcome $= \dfrac{\text{number of favourable outcomes}}{\text{number of possible outcomes}}$

Find answers at: cambridge.org/ukschools/gcsemaths-studentbookanswers

Experimental probability

- Relative frequency or experimental probability, tells you how often a favourable outcome occurs in an experiment as a fraction of the number of trials in that experiment.

  Relative frequency/experimental probability = $\dfrac{\text{frequency of a favourable outcome}}{\text{number of trials}}$
- Tables and frequency trees can be used to organise the outcomes of different experiments.
- Statistical data (empirical evidence) can also be used to find relative frequency of particular events.
- The relative frequency of an event can be used to predict future outcomes.

Mutually exclusive events

- Mutually exclusive events cannot happen at the same time. For example, you cannot throw a 1 and a 5 at the same time when you roll a dice. We **add** together the probabilities of mutually exclusive outcomes.

For additional questions on the topics in this chapter, visit GCSE Mathematics Online.

Chapter review

1 A coin is flipped a number of times giving the following results.

Heads: 4083

Tails: 5917

a How many times was the coin flipped?

b Calculate the relative frequency of each outcome.

c What is the probability that the next flip will result in heads?

d Jess says she thinks the results show that the coin is biased. Do you agree? Give a reason for your answer.

2 A bag contains 10 red, 8 green and 2 white counters. Each counter has an equal chance of being chosen. Calculate:

a the probability of choosing a red ball.

b the probability of choosing a red or a green ball.

c the probability of not choosing a white ball.

d P(ball is not red).

3 Research shows that the probability of a person being left-handed is 0.23.

How many left-handed people would you expect in a population of 25 000?

4 The probability of a sim card for a mobile phone being faulty is found to be 0.0265.

What percentage of sim cards are not faulty?

5 Jill interviews 64 people to get their opinions about sending texts when in company. 44 of those interviewed say it is rude to send texts in company. Jill then observes the people at a large event. Of those who said it was rude to send texts in company, 13 sent texts when at the table with others. Of those who said it was acceptable, 9 did not send texts when in company.

Complete the frequency tree to show this data.

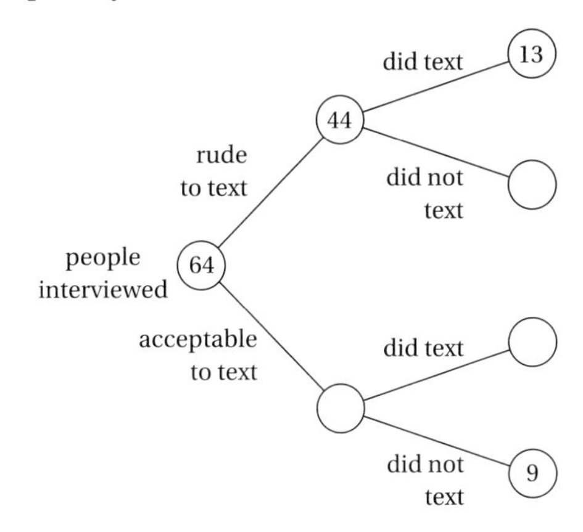

6 Mina used a computer program to simulate drawing a playing card at random from a shuffled pack. She recorded the suit of each card drawn. These are her results for 1000 trials.

Suit drawn	Hearts	Diamonds	Spades	Clubs
Frequency	238	240	264	258

a Calculate the relative frequency of each outcome.

b Do these results indicate that the simulation is fair and unbiased? Give a reason for your answer.

24 Combined events and probability diagrams

In this chapter you will learn how to ...

- use a range of sample space diagrams to list outcomes of combined events.
- apply the addition rule and use various representations to solve probability problems.
- understand conditional probability and solve problems involving conditional probability.

For more resources relating to this chapter, visit GCSE Mathematics Online.

Using mathematics: real-life applications

Medical researchers have developed a range of tests to detect drug use, blood-alcohol levels, disease markers and genetic and birth defects in unborn children. The probability that the test results are accurate is very high, but it is not often 100%.

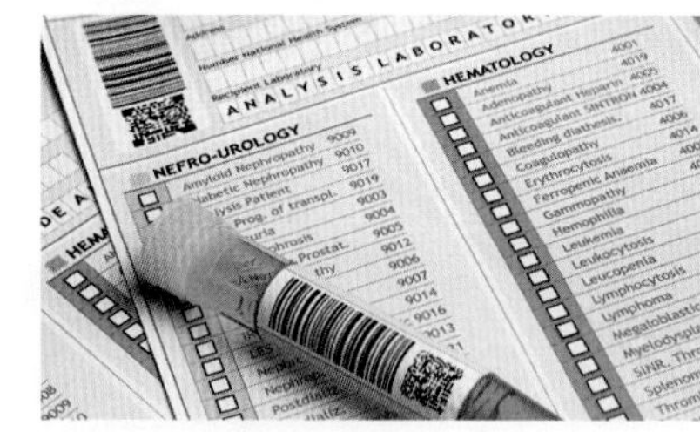

"An incorrect test result can be devastating. People can be convicted of drink-driving, be expelled from competitive sports, risk surgery or decide to terminate a pregnancy based on test results, so it is really important to understand the probability of a good test giving a bad result." *(Medical statistician)*

Before you start ...

Ch 10, 11	You'll need to be able to calculate effectively with fractions and decimals.	**1** These calculations are all incorrect. What should the answers be? **a** $\frac{1}{8}+\frac{1}{4}=\frac{1}{12}$ **b** $\frac{2}{3}+\frac{1}{5}=\frac{2}{15}$ **c** $1-\frac{3}{5}=-\frac{2}{5}$ **d** $\frac{2}{3}\times\frac{2}{5}=\frac{2}{15}$ **e** $0.3\times0.6=1.8$
Ch 23	You should be familiar with the vocabulary of basic probability.	**2** Select the correct term from the box for each definition. event outcomes random equally likely relative frequency **a** The number of times an event is recorded divided by the total number of trials conducted. **b** The set of results of an experiment. **c** Something of interest, such as getting heads when you flip a coin. **d** Things that have the same chance of happening are ... **e** Not predetermined.
Ch 23	Check that you can list all the possible outcomes of an experiment.	**3** Complete each list of possible outcomes. **a** Two students are to be chosen at random from a group of males and females: FF, FM, ... **b** Two coins are to be flipped at the same time: HH, ... **c** Two cards are selected from a set of three cards labelled A, B and C and placed next to each other in the order they are drawn: AB, AC, ...

Assess your starting point using the Launchpad

STEP 1

1 The Venn diagram shows the different sports chosen by students from one particular class. T represents students who play tennis and S represents those who take swimming.

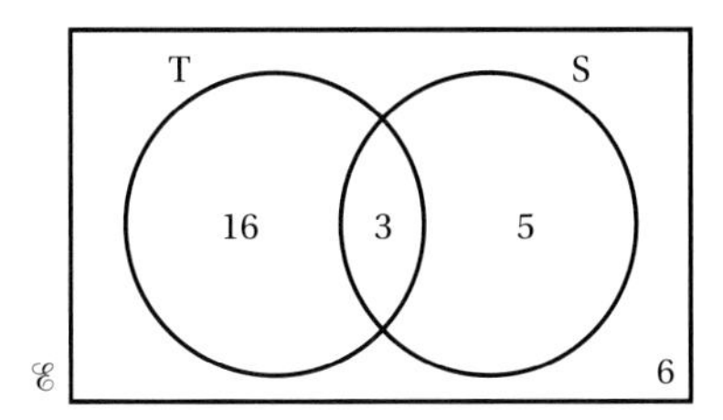

a How many students are in the class?

b How many students play tennis?

c How many students play tennis and swim?

d If a student is chosen at random from the class, what is the probability that he or she will take swimming?

2 Complete the tree diagram to show all the possible outcomes if you spin this spinner twice in a row. Add the probability of landing on each colour to the tree diagram.

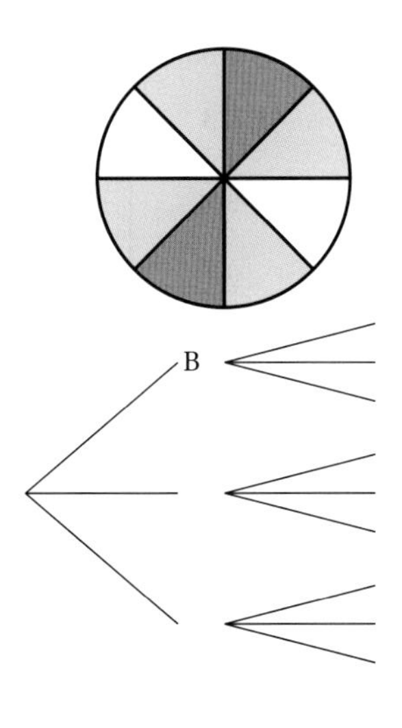

GO TO

Section 1:

Representing combined events

GO TO

Step 2:

The Launchpad continues on the next page ...

Launchpad continued ...

STEP 2

3 The diagram gives the probability of drawing hearts or twos at random from a normal pack of 52 playing cards.

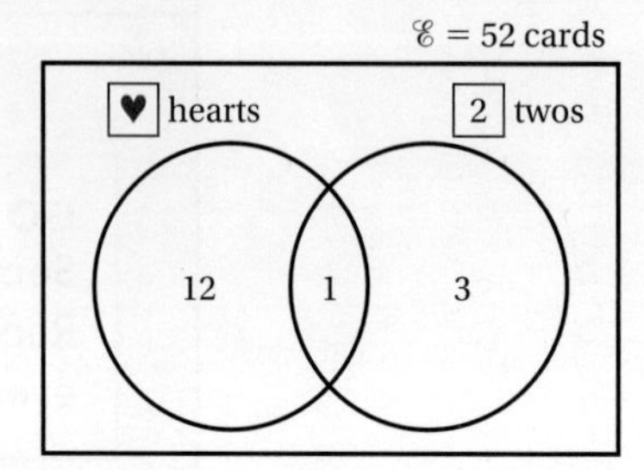

What is the probability that a card drawn at random will be:

a both a heart and a two?

b either a heart or a two?

c neither a heart nor a two?

GO TO
Section 2: Theoretical probability of combined events

STEP 3

4 A black (B) or white (W) counter is drawn at random from a box containing both black and white counters. The counter is replaced before a second counter is drawn. The possible outcomes and the probabilities of each outcome are shown on the right.

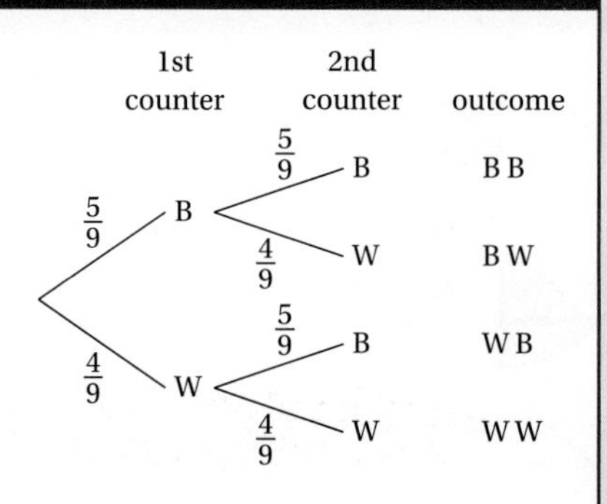

What is the probability of:

a drawing a black counter on the first draw?

b drawing two counters the same colour?

c drawing a white counter first and a black counter second?

5 **a** Draw a tree diagram to show the sample space for the genders of the children in a two-child family.

b If the older child is a girl, calculate the probability that the younger child is a boy.

GO TO
Section 3: Conditional probability

GO TO
Chapter review

Section 1: Representing combined events

When two or more events happen at the same time they are called **combined events** because there is a combination of outcomes.

Lists and tables are useful for showing the sample space (all the possible outcomes) of simple events. For example, you can list the sample space for rolling a six-sided dice (1, 2, 3, 4, 5, 6).

For the more complex sample spaces of combined events, tables are more useful than lists. In some cases you don't have to list all the possible outcomes. For example, let's say you want to know how many ways there are to get a total score of 7 when you roll two ordinary dice. You can draw up a table like this one:

Number on dice	1	2	3	4	5	6
1	2	3	4	5	6	7
2	3	4	5	6	7	
3	4	5	6	7		
4	4	6	7			
5	6	7				
6	7					

Once you get to a sum of 7 you can stop because the next sum will be greater than that.

The table shows there are six ways of getting a score of 7: (1, 6), (2, 5), (3, 4), (4, 3), (5, 2) and (6, 1).

Even though you haven't filled in the empty blocks, you can still see that there are 36 possible outcomes (it is a 6×6 table). So the probability of getting 7 is $\frac{6}{36}$ or $\frac{1}{6}$.

Key vocabulary

sample space: a list or diagram that shows all possible outcomes from two or more events.

Tables and grids

Two-way tables and grids let you work faster and see quickly whether you have left out, or repeated any outcomes.

WORKED EXAMPLE 1

Represent the sample space for flipping a coin and rolling a dice using:

a a table.

b a grid.

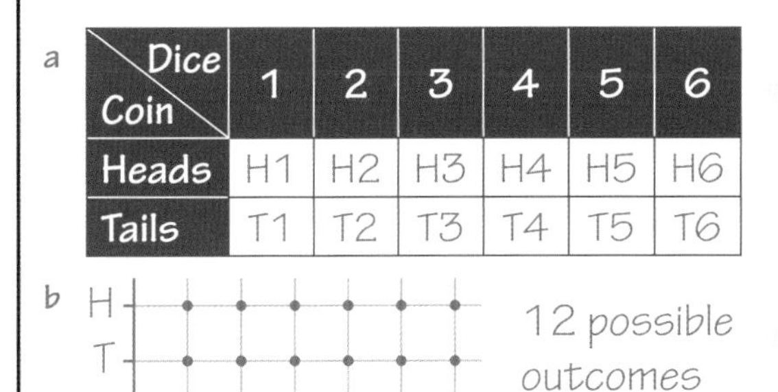

a

Dice / Coin	1	2	3	4	5	6
Heads	H1	H2	H3	H4	H5	H6
Tails	T1	T2	T3	T4	T5	T6

Make the header of each column a different score for one roll of a dice, and the header of each row one result from flipping a coin. Fill in the table to find all the possible combinations.

b

Use a similar process to creating the two-way table but here use a set of axes to list each result and match them by drawing dots.

Tip

Tree diagrams show probabilities, not actual responses. This is the main difference between them and the frequency trees you worked with in Chapter 23.

Tree diagrams

A tree diagram is a branching diagram that shows all the possible outcomes (sample space) of one or more activity.

To draw a tree diagram:

1. Make a dot to represent the first activity.
2. Draw branches from the dot to show all possible outcomes of that activity only.
3. Write the outcomes at the end of each branch.
4. Repeat the same process for the next activity, except that the starting dot is drawn at the end of **each** outcome from the previous activity.

These two diagrams both show the possible outcomes for throwing a dice and flipping a coin at the same time. Both diagrams are correct.

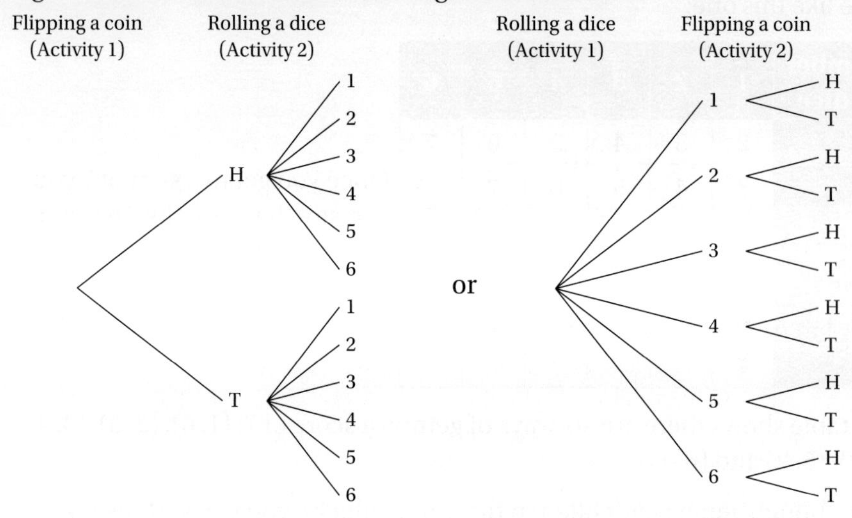

Once you've drawn a tree diagram you can list the possible outcomes by following the paths along the branches. Listing the combinations lets you work out the probability of different events.

WORKED EXAMPLE 2

a Draw a tree diagram to show that when the probability of having a boy or a girl is equal, there are eight possible combinations of boys and girls in a three-child family. Using the sample space, calculate the probability of:

i having three girls.

ii having three children of the same gender.

Tip

You should remember how to calculate theoretical probability from Chapter 23.

1st child	2nd child	3rd child	Possible combinations
B	B	B	B B B
		G	B B G
	G	B	B G B
		G	B G G
G	B	B	G B B
		G	G B G
	G	B	G G B
		G	G G G

1. Draw a dot for the first born child.
2. Draw two branches as there are only two possible outcomes.
3. Label the outcomes (one B and one G).
4. Repeat this at the end of each branch for the second and third child.

List the possible combinations.

You can see from the diagram there are 8 possible combinations of boys and girls.

Continues on next page ...

$P(3 \text{ girls}) = \frac{1}{8}$

You can use the diagram to find different probabilities. There is only one outcome that produces three girls.

$P(\text{all the same gender}) = \frac{2}{8} = \frac{1}{4}$

There are two outcomes that produce three children of the same gender.

EXERCISE 24A

Did you know?

This tree diagram assumes that a boy or a girl is equally likely for each pregnancy. In reality, the probability of having a boy or a girl varies by family and by country. Worldwide, the probability of having a boy is a little higher than having a girl. The UN for example, estimates that in 2013 there were 107 boys born for every 100 girls born.

1 Use a grid to represent the sample space for:

- **a** flipping two coins.
- **b** choosing a letter at random from the word CAT and flipping a coin.
- **c** drawing one counter each from two bags containing red, blue and yellow counters.

2 **a** Draw a table to show:

- **i** all possible combinations of scores when you roll two dice.
- **ii** the sample space for flipping a coin and spinning a spinner with sectors A, B, C and D.

b For each table above, make up five probability questions that could be answered from the tables. Exchange questions with a partner and try to answer each other's questions.

3 Draw a tree diagram to show the sample space when three coins are tossed one after the other.

4 Sandy has a bag containing a red, a blue and a green pen. Complete this tree diagram to show the sample space when she takes a pen from the bag at random, replaces it, and then takes another pen.

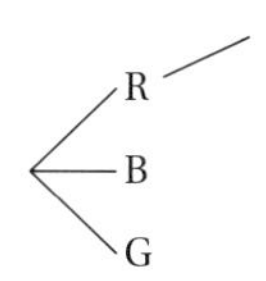

5 In a knockout quiz, the winner goes on to the next round. Naresh takes part in a four-round quiz and he estimates that he has an equal chance of winning or losing each round.

- **a** Using W to represent win and L to represent loss, draw a tree diagram to show all possible outcomes for Naresh.
- **b** How many possible outcomes are there?
- **c** What is the probability that he will win the first round, given his own estimate of his chances?

6 Here are two groups of jelly beans.

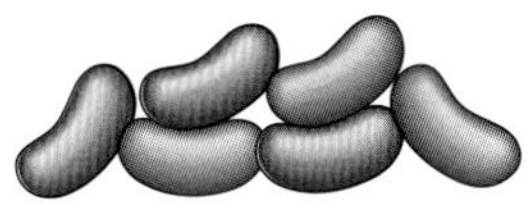

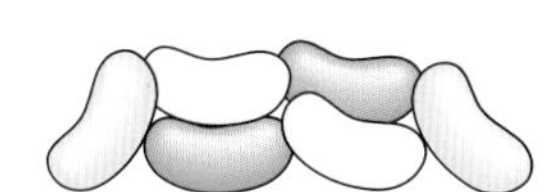

- **a** Draw a tree diagram to show the sample space for taking one jelly bean, at random, from each group of jelly beans.
- **b** Based on this, does it seem that any particular combination of colours has a higher probability than another? Is that the reality? Explain your reasoning.

Find answers at: cambridge.org/ukschools/gcsemaths-studentbookanswers

Tip

Set notation: the element symbol ∈ means 'is an element of'; the symbol ∉ means 'is not an element of'.

Venn diagrams

Venn diagrams show the mathematical relationships between sets of data. Different events (sets of outcomes) are represented by circles inside a rectangular frame. The rectangular frame represents all the possible outcomes, i.e. the sample space (this is called the universal set).

Look at the Venn diagram and read through the information to revise the main features of Venn diagrams.

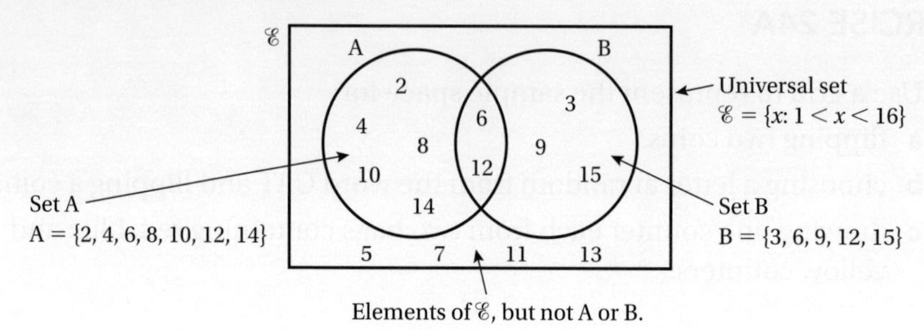

ℰ is the universal set. In this, case this is the whole numbers between 1 and 16.

ℰ = {set of numbers between 1 and 16}.

This can also be written in set notation as: ℰ = $\{x: 1 < x < 16\}$.

The circles A and B represent sets.

Set A is the set of even numbers between 1 and 16 and it can be listed as: A = {2, 4, 6, 8, 10, 12, 14}.

Set B is the set of multiples of three between 1 and 16 and it can be listed as B = {3, 6, 9, 12, 15}.

There are seven elements in Set A. This can be written as n(A) = 7, which means 'the number of elements in set A is 7'.

There are five elements in Set B, so n(B) = 5.

Set A and Set B have two elements in common. The numbers 6 and 12 are written in the overlapping section of the circles to show that they are elements of both sets.

The numbers 5, 7, 11 and 13 are not elements of A or B but they are elements of the universal set so they are written inside the rectangle, but outside the circles.

Tip

The curly brackets { } are used to show that you are describing a set. The numbers between the brackets are elements of the set.

Intersection, union and complement of sets

Venn diagrams can also represent operations between sets. The three important operations for probability work are intersection, union and complement.

The shaded area represents the **intersection** between A and B. The intersection of two sets is the elements that are common to (shared by) both sets.

$A \cap B = \{6, 12\}$

$n(A \cap B) = 2$

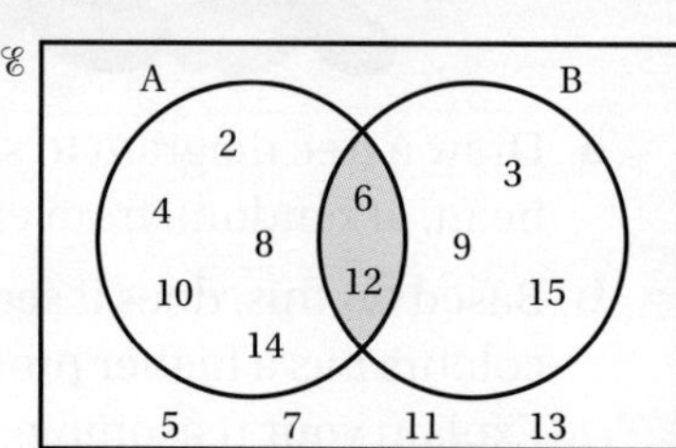

The shaded area here represents the **union** of A and B. This is the **combined elements** of both sets with no elements repeated.

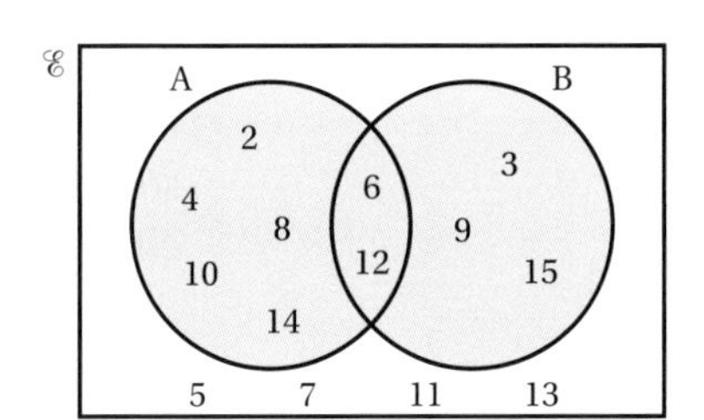

$A \cup B = \{2, 3, 4, 6, 8, 9, 10, 12, 14, 15\}$

$n(A \cup B) = 10$

The **complement** of a set refers to all the elements in the universal set other than the ones in the given set. The complement of Set A is shaded below.

$A' = \{3, 5, 7, 9, 11, 13, 15\}$

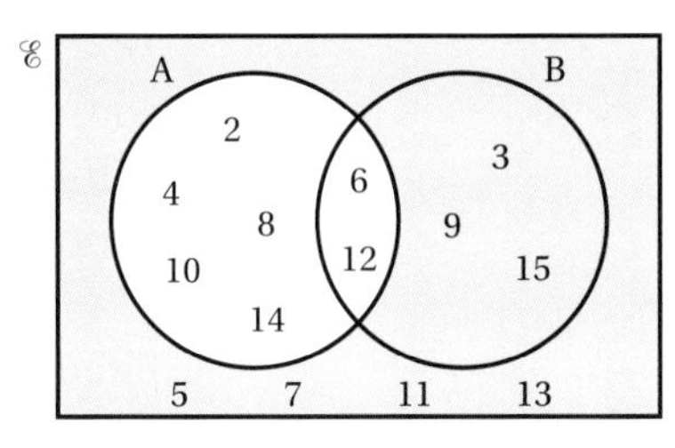

Tip

You may be given the elements of the different sets and asked to draw a Venn diagram.

You will learn more about how to solve probability problems using Venn diagrams in Section 2.

WORKED EXAMPLE 3

Given that $\mathscr{E} = \{x: x \text{ is a letter from a to h inclusive}\}$, $A = \{a, b, c, e\}$ and $B = \{c, d, e, f, g\}$, draw a Venn diagram to represent this information.

$\mathscr{E} = \{a, b, c, d, e, f, g, h\}$

$A = \{a, b, c, e\}$

$B = \{c, d, e, f, g\}$

Start by comparing the sets to find the intersection and any elements which are in the universal set but which are not in A or B (the complement of A and B or $(A \cup B)'$).

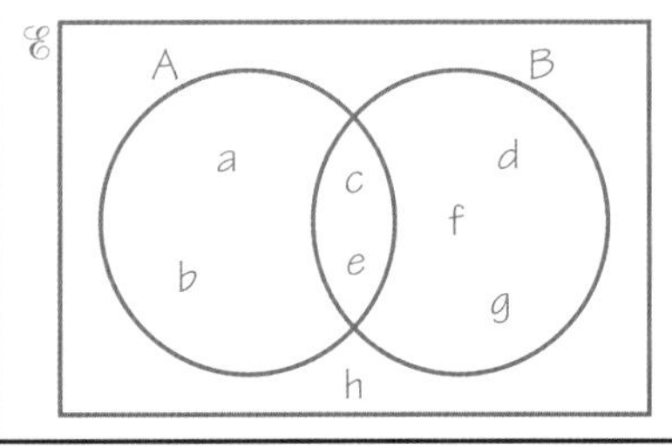

c and e are elements of A and B, so
$A \cap B = \{c, e\}$
h is not in A or B, so $(A \cup B)' = h$
(in other words, h is outside the two circles)
Draw the diagram and label it correctly.

In some problems you may be given information and have to define the sets yourself. In some cases you can't list the separate elements of the sets so you write the number of elements in each set.

WORKED EXAMPLE 4

In a survey of 25 people it was found that they all liked either chocolate or ice cream. 15 people said they liked ice cream and 18 said they liked chocolate.

Draw a Venn diagram and use it to work out the probability that a person chosen at random from this group will like both chocolate and ice cream.

Tip

You don't know the names of the people, so you can't list them in the diagram. You do know how many of each response there was, so you can just write the number of people in the diagram.

$\mathscr{E} = \{\text{number of people surveyed}\}$, so, $n(\mathscr{E}) = 25$

$C = \{\text{people who like chocolate}\}$, so, $n(C) = 18$

$I = \{\text{people who like ice cream}\}$, so, $n(I) = 15$

Start by defining the sets and writing the information in set language.

Continues on next page ...

$n(C) + n(I) = 15 + 18 = 33$

$n(C \cap I) = 8$

But there were only 25 people surveyed, so 8 people must have said they liked both chocolate and ice cream (since 33 − 25 = 8).

ℰ

C I

10 + 8 = 18 who like chocolate → 10 | 8 | 7 ← 8 + 7 = 15 who like ice-cream

Use the figures to draw your Venn diagram. There are 18 people in total who liked chocolate and 8 of these also like ice cream, so the number who just like chocolate is: 18 − 8 = 10; enter this value in the left circle of your diagram and label this circle 'C'. Similarly, of the 15 who liked ice cream, 8 also liked chocolate so ice cream only: 15 − 8 = 7.

$$P(\text{person likes both}) = \frac{\text{number of people who like both}}{\text{number of people surveyed}}$$

$$= \frac{8}{25} = 0.32$$

Finally calculate the probability

EXERCISE 24B

1 $\mathscr{E} = \{x: 1 \leqslant x \leqslant 20\}$, $A = \{x: 5 < x \leqslant 12\}$ and $B = \{x: x \text{ is a factor of } 24\}$.

a Draw a Venn diagram to show this information.

b Use your Venn diagram to find:

i $A \cap B$. **ii** $A \cup B$. **iii** $n(A)$.

iv $n(A')$. **v** $n(B')$.

2 Nadia has 20 pairs of shoes. Six pairs are sports shoes, four pairs are red and only one of the pairs of sports shoes is red. Draw a Venn diagram to show this information and work out the probability that a pair of shoes chosen at random from her shoe collection will be neither red nor sports shoes.

3 A factory employs 100 people. 47 of the employees have to work with moving machinery, so if they have long hair they have to tie it back. 35 employees have long hair, and of these, some work with moving machinery. 23 employees neither have long hair nor work with moving machinery. Draw a Venn diagram to show this information and use it to work out the probability of a random employee having to tie his or her hair back at work.

4 Of the first 20 students to walk into a classroom, 13 were wearing headphones and 15 were sending texts. Four students were not wearing headphones or sending texts. Represent this information on a Venn diagram and state how many students were wearing headphones while sending texts when they walked into class.

5 A group of 200 people at a function were questioned to find out about their food preferences. It was found that 110 people ate red meat, 135 ate chicken, and 15 ate neither. Calculate the probability that a person chosen at random will eat:

a red meat and chicken. **b** only red meat. **c** only chicken.

6 A health-policy researcher asked 700 shoppers in one morning which of three items they bought at their local supermarket.

Of these 700 shoppers, 324 bought ready-made meals, 213 shoppers bought fresh produce (fruit or vegetables) and 245 bought dairy products.

- 237 of the shoppers bought only ready-made meals.
- 43 of them bought both fresh produce and ready-made meals, but no dairy products.
- 32 of them bought only ready-made meals and dairy products.
- 122 of them bought only fresh produce.

Represent this information on a Venn diagram. Use the diagram to find out:

a how many of the shoppers bought all three of the items.

b how many bought only fresh produce and dairy products, but no ready-made meals.

c how many bought only dairy.

Section 2: Theoretical probability of combined events

You can use sample space diagrams (tables, grids, tree diagrams and Venn diagrams) to find the probability of getting a favourable outcome from probabilities of combined events.

Use the sample space diagram to calculate the number of favourable outcomes and the total number of possible outcomes, then use the formula for calculating theoretical probability that you saw in Chapter 23.

Once you have identified all the possible outcomes, you can mark the ones that are favourable and use these to find the probability of different outcomes. You can still use the following formula to work this out:

$$\text{P(event with favourable outcome)} = \frac{\text{number of favourable outcomes}}{\text{total number of possible outcomes}}$$

WORKED EXAMPLE 5

Jay has six cards with the numbers 0, 0, 2, 2, 3 and 7 on them. He picks a card, returns it and then picks another at random.

a Draw a grid to show the sample space.

b Use the grid to find the probability that Jay will pick:

i two numbers that are the same. **ii** two numbers that add up to 7.

a

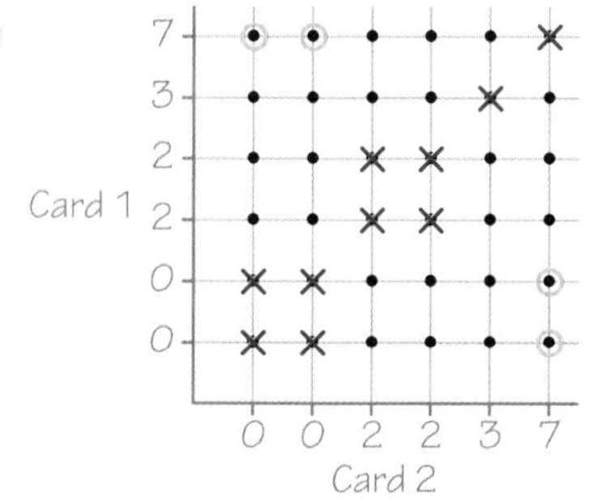

Draw the sample space. Use this to mark all the combinations where the two numbers are the same (here a cross has been used). Then mark all the combinations where the two numbers sum to 7 (here, a circle has been used).

b i P(two numbers the same) $= \frac{10}{36}$

$= \frac{5}{18}$

or 0.28 (correct to 2 dp)

The grid shows there are 36 possible outcomes. The successful outcomes are marked with a cross on the grid.

ii P(sum of 7) $= \frac{4}{36}$

$= \frac{1}{9}$

The successful outcomes are circled on the grid.

The product rule for counting

In the example above, the total number of possible outcomes is 36. You can count the dots on the grid, but you can also find the total by multiplying: $6 \times 6 = 36$.

The product rule for counting states that if there are m ways of doing one thing and n ways of doing another, then the total number of combinations of the two ways can be found by multiplying together m and n.

WORK IT OUT 24.1

A tube station has two entrance turnstiles and three exit turnstiles. How many possible options are there to leave and enter the station?

Which answer is correct? List all the possible outcomes to show this.

Option A	Option B
Two entrances = 2 ways in Three exits = 3 ways out Total number of ways in and out = 2 + 3 = 5	For each of the two ways in, there are three ways out. $2 \times 3 = 6$ possible options.

The problem on the previous page can also be solved by drawing a tree diagram or grid.

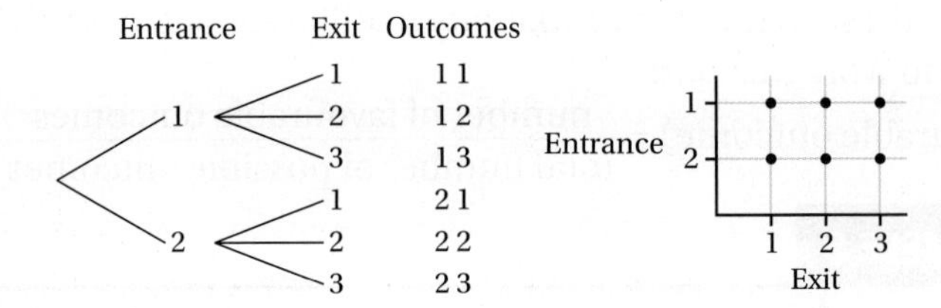

The product rule is useful for quickly working out the total number of possible outcomes without drawing the sample space.

WORKED EXAMPLE 6

How many three-digit numbers can be made from the digits 2, 3, 4 and 5:

i if each digit can be used more than once?

ii if each digit can only be used once?

[D1] [D2] [D3]

$4 \times 4 \times 4 = 64$ ways of arranging the digits into a three-digit number.

> There are 4 ways that you could have D1, there are 4 ways you can have D2 and 4 ways you have can D3.

[D1] [D2] [D3]

$4 \times 3 \times 2 = 24$ ways of arranging the digits.

> The number of options reduces with each digit; once, you've chosen the first digit, there are only 3 options for the second digit and so on. So there are 4 ways you can have D1, 3 ways you can have D2 and 2 ways you can have D3.

EXERCISE 24C

Sunil has to choose a six-character password for his phone. He decides to choose two letters and four digits, in that order, with no repetition of letter or digits. How many password options does he have?

2 In a lottery, if there are 40 numbers to choose from and you need to match six (different) numbers to win the jackpot.

How many different tickets are possible?

3 Anna wants to buy running shoes from a specialist running shop. She wants to choose:

- either neutral, cushioned or stability shoes
- from three top brands
- one of five trendy colours.

How many different shoes can she choose between?

4 How many combinations are possible for a four-digit code number made from digits 1 to 9? The digits cannot be repeated.

5 There are five questions in a multiple choice test and the answers for each are A, B, C or D. A hacker uses a computer to generate answers. How many sets of answers must she generate to make sure that one of the sets is 100% correct?

6 You can choose from 8 different fillings for a wrap. How many different combinations are there if you order a wrap with 3 fillings?

Different types of events

The type of event determines whether you add or multiply the probabilities.

Mutually exclusive events and the addition rule

Imagine you have a bag with 3 red, 2 yellow and 5 green sweets in it and you are allowed to choose one sweet at random. You cannot pick a red sweet and a yellow sweet at the same time, so the events P(red) and P(yellow) are mutually exclusive.

You can work out the probability of choosing *either* a red *or* a yellow sweet.

There are 3 red and 2 yellow sweets, so $\frac{5}{10}$ of the sweets are either red or yellow.

$$\text{P(red or yellow)} = \text{P(red)} + \text{P(yellow)} = \frac{3}{10} + \frac{2}{10} = \frac{5}{10} = \frac{1}{2}$$

We can say that P(A or B) = P(A) + P(B) where A and B are mutually exclusive events.

This is called the **addition law** for mutually exclusive events.

Tip

You should remember from Chapter 23 that mutually exclusive events cannot happen at the same time.

Tip

Questions involving 'either-or' events normally involve mutually exclusive events, so they can usually be solved by adding the probabilities.

Events that are not mutually exclusive

When you list the elements in the union of sets you do not repeat shared elements (those in the intersection).

In set language, we can write this as $n(A \cup B) = n(A) + n(B) - n(A \cap B)$. The elements in the intersection of sets are **not** mutually exclusive and this affects your probability calculations.

WORKED EXAMPLE 7

The Venn diagram shows the possible outcomes when a six-sided dice is rolled. Set A = {prime numbers} and Set B = {odd numbers}. Use the diagram to find the probability of rolling a number that is either odd or prime.

Tip

The word 'or' means it belongs in one set or the other, so you need to deal with the union of the sets.

$P(A \text{ or } B) = P(A) + P(B) - P(A \text{ and } B)$

$P(A) = \frac{3}{6}$

$P(B) = \frac{3}{6}$

$P(A \text{ and } B) = \frac{2}{6}$

The total number of outcomes is the denominator, and it's easier to add and subtract the fractions if you don't simplify the fractions first. Remember that P(A) includes the values in the intersection so there are 3 outcomes out of 6 that are in set A.

So, $P(A \text{ or } B) = \frac{3}{6} + \frac{3}{6} - \frac{2}{6} = \frac{4}{6} = \frac{2}{3}$

You can see this is true by looking at the diagram. The combined elements of A and B are 1, 2, 3 and 5, giving you $\frac{4}{6}$ numbers falling into one or the other of these sets. We don't want to add the numbers that fall into the intersecting part twice which is why we subtract $n(A \cap B)$ in the formula.

Key vocabulary

independent events: events that are not affected by what happened before.

Independent events

When the outcome of one event does not affect the outcome of the others, we say they are **independent events**.

Rolling a dice and flipping a coin are independent events. The score on the dice doesn't affect whether you get heads or tails.

Tree diagrams are useful for solving problems involving independent events if you write the probabilities of the events on the branches.

Here is the tree diagram showing possible outcomes for throwing a dice and flipping a coin at the same time (H is used for heads and T is used for tails).

This is the same diagram as in the previous section but now the probability of each outcome is written at the side of each branch.

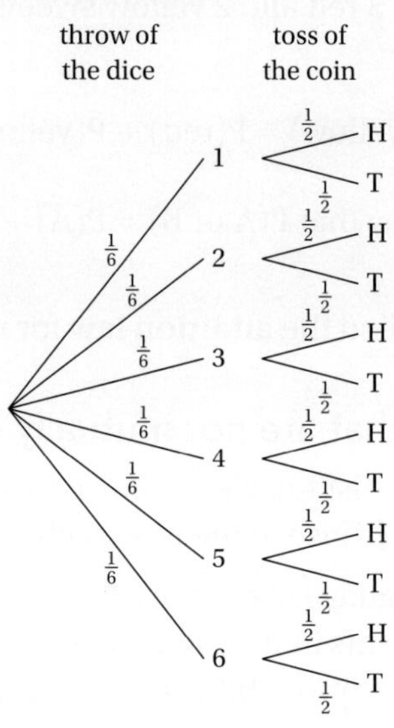

The probability of combined events on a tree diagram

To find the probability of one particular combination of outcomes multiply the probabilities on consecutive branches, for example, the probability of throwing a 5 **and** getting heads is $\frac{1}{6} \times \frac{1}{2} = \frac{1}{12}$.

This is called the **multiplication law**.

P(A **and** B) = P(A) × P(B)

Tip

It can be helpful to use a colour to mark the route along the branches to show which events you are dealing with.

Combining the laws

To find the probability when you have independent events and there is more than one favourable combination, **or** when the combinations are mutually exclusive:

- multiply the probabilities on consecutive branches for each favourable outcome
- add the probabilities.

For example,

P(rolling 1 **or** 2 and getting an H) = P(rolling a 1 and getting an H) + P(rolling a 2 and getting an H)

$$= \left(\frac{1}{6} \times \frac{1}{2}\right) + \left(\frac{1}{6} \times \frac{1}{2}\right)$$

$$= \frac{1}{12} + \frac{1}{12} = \frac{2}{12} = \frac{1}{6}.$$

WORKED EXAMPLE 8

Two coins are flipped together. Draw a tree diagram to find the probability of getting:

a two tails. **b** one head and one tail.

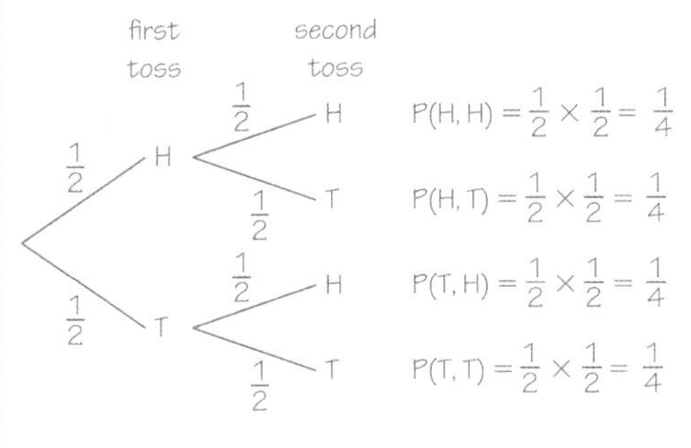

Draw a tree diagram and label each branch with the probability of that particular outcome. List all the possible outcomes.

a P(TT) = P(T on 1st flip) × P(T on 2nd flip)

$$= \frac{1}{2} \times \frac{1}{2} = \frac{1}{4}$$

Follow the branches to find the probability of getting a T **and** another T; multiply the probabilities along the branches.

b P(HT or TH) = P(HT) + P(TH)

$$= \left(\frac{1}{2} \times \frac{1}{2}\right) + \left(\frac{1}{2} \times \frac{1}{2}\right)$$

$$= \frac{1}{4} + \frac{1}{4} = \frac{1}{2}$$

There are two different ways that you could get a head and a tail: HT **or** TH. The two different combinations are mutually exclusive. Calculate the probability of each outcome first, by multiplying the probabilities along the branches, then add the two probabilities together.

Key vocabulary

dependent events: events in which the outcome is affected by what happened before.

Dependent events

When the outcome of one event affects the outcome of the other events, we say they are **dependent events.**

Here are 4 red and 2 yellow sweets.

Suppose you choose one sweet at random and eat it before you select a second sweet. What is the probability of the second sweet being red?

The answer to this depends on what colour the first sweet was. If the first sweet was red then the probability that the second sweet is red is $\frac{3}{5}$ because there are only 5 sweets left and only 3 of those are red.

If the first sweet was yellow then the probability that the second one is red is $\frac{4}{5}$. There are still only 5 sweets left to choose from, but this time 4 of them are red.

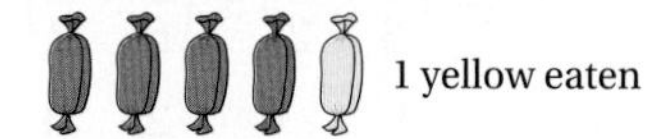

For dependent events you can find the probability by adapting the multiplication rule to accommodate the dependent event.

P(A and then B) = P(A) × P(B given that A has occurred)

You can use tree diagrams to help work this out.

WORKED EXAMPLE 9

A box contains three yellow, four red and two purple marbles. A marble is chosen at random and not replaced before choosing the next one. If three marbles are chosen (without replacement) what is the probability of choosing:

a three red marbles

b a yellow, red and purple marble in that order?

Tip

Notice in both parts only a partial tree diagram was drawn; you are only interested in certain outcomes so there is no need to draw them all.

a

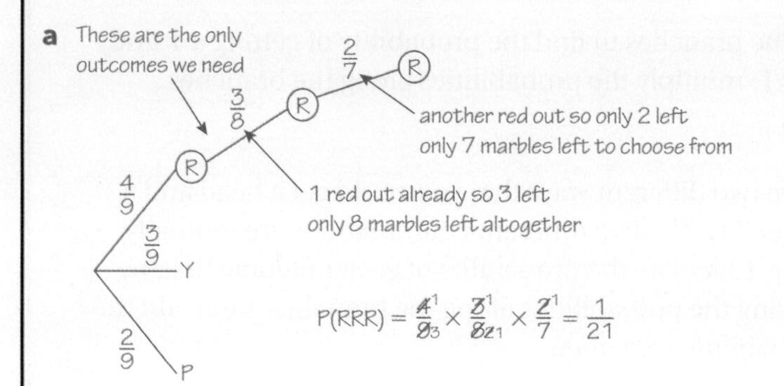

b

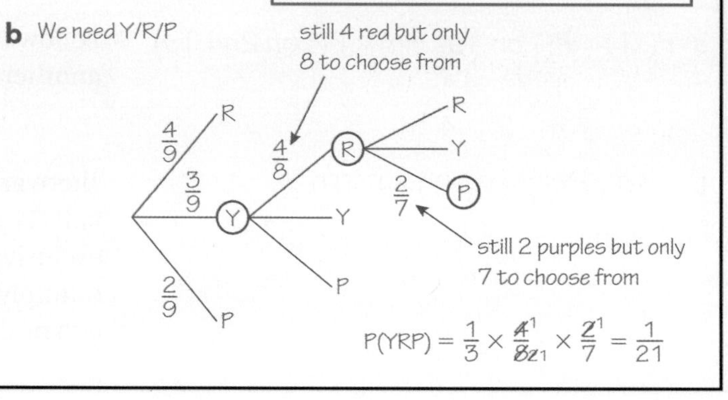

Probability using a Venn diagram

When you use Venn diagrams to solve problems it is important to choose the correct operation (union, intersection or complement) to solve the problem. The wording of the problem normally gives you clues about which operation you need.

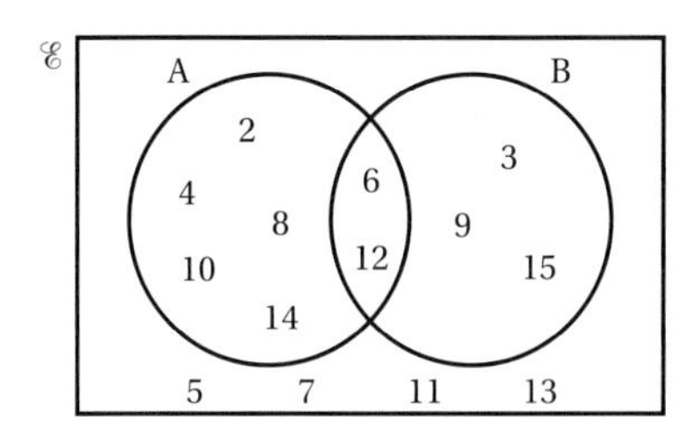

For example, if you have Set A = {x: x is an even number} and Set B = {x: x is a multiple of 3} you may be asked questions like the ones in the first column of the table.

Question: What is the probability of a number ...	Which operation is involved in finding the solution?	Why?
... being even and not a multiple of 3	No operations	The probability will involve all the elements in circle A, including those in the intersection with circle B.
... being even **and** a multiple of 3	Intersection	The word 'and' tells you are looking for numbers that are elements of both sets (i.e. those in the overlapping section).
... being even **or** a multiple of 3	Union	The word 'or' tells you it can be in either of the two sets, so you need to include all the elements of both sets – without repeating any in the intersection.
... **not** being even	Complement	All the numbers that are outside the set of even numbers must be included. This means all the numbers outside circle A: the numbers outside both circles, and the numbers inside circle B, but not in the intersection.
... being **neither** even **nor** a multiple of 3	Complement	'Neither, nor' tells you that two sets have to be excluded, you are looking for the elements outside the circles in this case.

The product rule for counting can be used to help solve some types of problems without drawing the sample space.

WORKED EXAMPLE 10

Josh has four cards labelled 1, 2, 3 and 4. He draws two cards at random to make a 2-digit number, with the first card drawn being used to make the tens digit.

a How many possible outcomes are there?

b What is the probability of making 32?

a [D1] [D2]

$4 \times 3 = 12$ possible outcomes

4 options for first digit leaves 3 for the second.

b P(3 then 2) = P(3) × P(2, given 3 has occurred)

$= \frac{1}{4} \times \frac{1}{3} = \frac{1}{12}$

Alternatively, as the digits don't repeat you should be able to tell straight away that there is only one favourable outcome out of the 12 total outcomes, so P(3 then 2) $= \frac{1}{12}$.

Tip

You need to understand the difference between independent and dependent events to make sense of conditional probability in Section 3.

EXERCISE 24D

1 Nico is on a bus. He amuses himself by choosing a consonant and a vowel at random from the names of towns on road signs. The next road sign is DUNDEE.

a Draw up a sample space diagram to list all the options that Nico has.

b Calculate P(D and E).

c Calculate P(D and E or U).

d Calculate P(N′ and U).

2 A bag contains 3 red counters, 4 green counters, 2 yellow counters and 1 white counter. Two counters are drawn from the bag one after the other, without being replaced.
Calculate:

a P(2 red counters).

b P(2 green counters).

c P(2 yellow counters).

d P(white and then red).

e P(white or yellow in any order, but not both).

f P(white or red in any order, but not both).

g P(white or yellow first and then red or green).

3 A card is randomly selected from a pack of 52 playing cards and its suit is noted. The card is not replaced. Then a second card is chosen.

a Draw a tree diagram to represent this situation.

b Use the tree diagram to find the probability that:

i both cards are hearts.

ii both cards are clubs.

iii the first card is red and the second card is black.

4 Mohammed has four Scrabble® tiles with the letters A, B, C and D on them. He draws a letter at random and places it on the table, then he draws a second letter and a third, placing them down next to the previously drawn letter.

a What is the probability that the letters he has drawn spell the words:

i cad? **ii** bad? **iii** dad?

b What is the probability that he will not draw the letter B?

c What is Mohammed's chance of drawing the letters in alphabetical order?

5 In a standard pack of cards, A = {hearts} and B = {kings}.
If a card is picked at random, determine:

a P(A). **b** P(B). **c** P(A and B). **d** P(A or B).

6 During January in Manchester it rained on 16 days and the temperature fell below 6 °C on 25 days. Draw a Venn diagram to determine the maximum and minimum possible number of days in January that were below 6 °C and rainy.

Section 3: Conditional probability

Conditional probability is used when we need to work out the probability of one event happening when we already know that another has happened. The information about the first event changes the sample space and affects the calculation. For two events A and B, the notation P(B|A) is used to refer to the conditional probability of B happening, given that A has already happened. We read B|A as 'B given A'.

If two events A and B are independent, you will find that it is always true that P(B|A) = P(B). This is because the probability of the first event does not affect the probability of the second event.

You saw earlier in the chapter that when events are dependent, you calculate the probability of a combined event using the formula
P(A and then B) = P(A) × P(B given that A has happened).

Using the notation for conditional probability, you can write this as:

P(A and B) = P(A) × P(B|A)

To find P(B|A) you can rearrange the formula:

$$P(B|A) = \frac{P(A \text{ and } B)}{P(A)}$$

Tip

When you deal with Venn diagrams, this rule can be written in set language as $P(B|A) = \frac{P(A \cap B)}{P(A)}$.

WORKED EXAMPLE 11

Jona has five sweets: one blue, one yellow, one red, one orange and one green.

a Jona picks a sweet and replaces it. What is the probability that the second sweet he picks up is yellow if his first sweet is blue?

b Jona then decides to eat two sweets. What is the probability that the second sweet he eats is yellow given that the first sweet he eats is blue?

Tip

When the problem states that the items are 'replaced' then the events are independent and the probabilities don't change. When the item is not replaced, the outcome of the second depends on the outcome of the first event, and the probabilities change.

a $P(B|A) = P(B)$

$P(B) = \frac{1}{5}$

As the first sweet is replaced, picking each sweet is an independent event, because picking a blue sweet and then replacing it does not affect the probability of the picking a yellow sweet.

b $P(B|A) = \frac{P(A \text{ and } B)}{P(A)}$ $\quad P(A) = \frac{1}{5}$ $\quad P(A \text{ and } B) = \frac{1}{5} \times \frac{1}{4}$

$P(B|A) = \frac{P(A \text{ and } B)}{P(A)} = \frac{\frac{1}{5} \times \frac{1}{4}}{\frac{1}{5}}$

$= \frac{1}{4}$

Note that the probability you are trying to find is not that Jona ate a blue **and** a yellow sweet, it is that he ate a yellow sweet given that he had already eaten a blue one. Once he has eaten a blue sweet, there are only 4 sweets left so the new total number of outcomes is 4.

1/5 → blue sweet
- 1/4 → orange sweet
- 1/4 → green sweet
- 1/4 → red sweet
- 1/4 → yellow sweet → $\frac{1}{5} \times \frac{1}{4}$

not blue sweet

You can see the formula $P(B|A) = \frac{P(A \text{ and } B)}{P(A)}$ works by drawing a tree diagram and following the branches for the dependent events of 1st sweet blue and 2nd sweet yellow. As you are only interested in the combined event where the first sweet is blue, you can draw a partial tree diagram.

For some conditional probability problems it is easier to use Venn diagrams.

WORKED EXAMPLE 12

In a group of 50 students, 36 students work on tablet computers, 20 work on laptops and 12 work on neither of these. If a student is chosen at random, what is the probability that he or she:

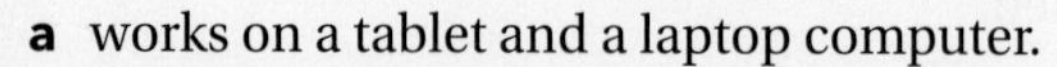

a works on a tablet and a laptop computer.

b works on at least one type of these types of computer.

c works on a tablet given that he or she works on a laptop.

d doesn't work on a laptop, given that he or she works on a tablet.

> **Tip**
>
> When a problem asks for the probability of at least one event you can either list all the possible outcomes or you can use the fact that P(at least one happens) = 1 − P(none of the events happen).

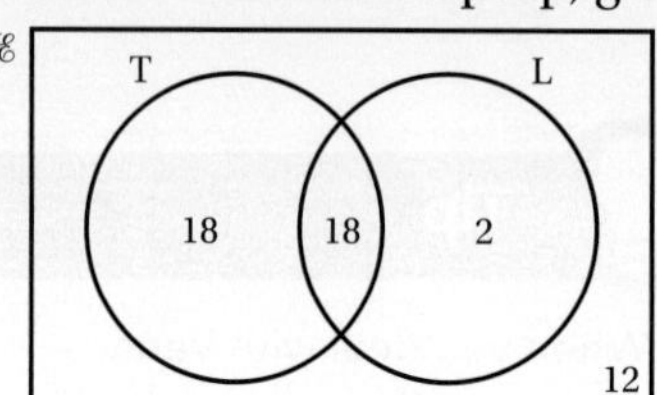

$50 - 12 = 38$
$T = 36$ students
$L = 20$ students
56
-38
18 ← work on both

a $P(\text{works on both}) = P(T \cap L) = \frac{18}{50} = \frac{9}{25}$

Using the Venn diagram, this is just the number in the intersection divided by the total number of students.

b $P(\text{works on at least one}) = 1 - P(\text{works on neither})$

$= 1 - \frac{12}{50}$

$= \frac{38}{50} = \frac{19}{25}$.

P(works on neither) is the number outside the circles.

c $P(B|A) = \frac{P(A \cap B)}{P(A)} = \frac{n(L \cap T)}{n(L)}$

$P(T|L) = \frac{18}{20} = \frac{9}{10}$

As this is a Venn diagram you use the amended formula. P(T|L) is dependent on students already using a laptop and so the probability is calculated using the total number of students who use a laptop not the total number of students. n(A and B) is the number of people in the intersection of the Venn diagram .

d $P(B|A) = \frac{P(A \text{ and } B)}{P(A)} = \frac{n(T \text{ and } L')}{n(T)}$

$P(L'|T) = \frac{18}{36} = \frac{1}{2}$

The P(L′|T) can be calculated using the number in the T circle of the Venn diagram. The total number of outcomes is no longer 50, it is 36 as this is the total number of students who use tablets.

EXERCISE 24E

1 There are 21 students in a class. 12 are boys and 9 are girls. The teacher chooses two different students at random to answer questions.

a Draw a tree diagram to represent this information.

b Find the probability that:

i both students chosen are boys. **ii** both students are girls.

iii the second student is a girl given that the first student was a boy.

c The teacher chooses a third student at random. What is the probability that:

i all three students chosen are girls?

ii at least one student is a girl?

iii the third student is a girl, if the first two students were boys?

2 A cleaner accidentally knocked the name labels off three students' lockers. The labels say Raju, Sam and Kerry. The tree diagram shows the possible ways of replacing the labels.

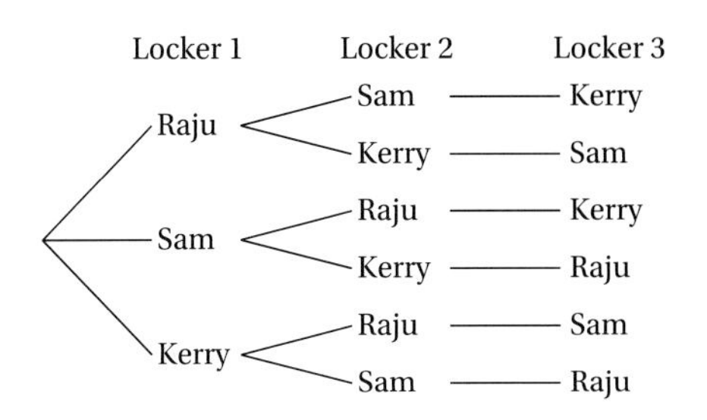

a Copy the diagram and write the probabilities next to each branch.

b Are these events dependent or independent? Why?

c How many correct ways are there to match the name labels to the lockers?

d How many possible ways are there for the cleaner to label the lockers?

e If the cleaner randomly stuck the names back onto the lockers, what is the chance of getting the names correct?

f What is the probability of getting the labels on Locker 2 and 3 correct, given that the first one is correctly labelled Kerry?

3 A climatologist reports that the probability of rain on Friday is 0.21. If it rains on Friday, there is a 0.83 chance of rain on Saturday; if it doesn't rain on Friday, the chance of rain on Saturday is only 0.3.

a Draw a tree diagram to represent this situation.

b Use your diagram to work out the probability of rain on:

i Friday and Saturday.

ii Saturday given that it was dry on Friday.

4 In a group of 25 people 15 like cappuccino (C) and 17 like latte (L) and 2 people like neither. Using an appropriate sample space diagram, calculate the probability that a person will:

a like cappuccino.

b like cappuccino given that he or she likes latte.

5 100 teenagers went on a summer camp during which 80 of them went hiking, 42 went sailing and each student did at least one of these activities.

a Draw a Venn diagram to show how many teenagers did both activities.

b If a teenager is randomly selected, find the probability that he or she:

i went hiking but not sailing.

ii went sailing given that he or she went hiking.

6 In a group of 120 students, 25 are in the sixth form, 15 attend maths tutorials and 4 of the students are sixth formers who attend maths tutorials. What is the probability that a randomly chosen student who attends maths tutorials will be a sixth former?

7 Three counters are removed one at a time from a bag containing two red and six yellow counters. Find the probability that:

a the three counters are the same colour.

b at least one counter is red.

c you get exactly one red counter.

d the second counter is red given that the first is yellow.

Find answers at: cambridge.org/ukschools/gcsemaths-studentbookanswers

8 When Nadia takes the train from Monday to Thursday the probability that she gets a seat is 95%. When she takes the train on a Friday or Saturday, the probability that she gets a seat is 70%. Assuming that she is equally likely to take the train on any day from Monday to Saturday, determine the probability that:

a she gets a seat.
b it is Saturday and she gets a seat.

9 **a** Explain how you can know whether two events A and B are dependent or independent, given the probability of A and the probability of A given B.

b The probability of drawing a red marble from a bag is $\frac{1}{8}$. The probability of drawing a blue and then a red marble from a bag is $\frac{2}{15}$. Is the marble returned to the bag after the first draw? Explain.

Checklist of learning and understanding

Representing combined events

- The sample space of an event is all the possible outcomes of the event.
- When an event has two or more stages it is called a combined event.
- Lists, tables, grids, tree diagrams and Venn diagrams can be used to represent combined events.

Calculating probabilities for combined events

- For mutually exclusive events P(A or B) = P(A) + P(B).
- For independent events P(A and then B) = P(A) × P(B).
- When the combined outcome of independent events are mutually exclusive, you need to add the probabilities after you have obtained the probability of each event using multiplication.
- For dependent events P(A and then B) = P(A) × P(B given that A has happened).
- For 'at least' problems P(at least A) = 1 − (not A).

Conditional probability

- For independent events, $P(B|A) = P(B)$.
- For dependent events, $P(B|A) = \frac{P(A \text{ and } B)}{P(A)}$

For additional questions on the topics in this chapter, visit GCSE Mathematics Online.

Chapter review

1 Choose the most appropriate method and use it to represent the sample space in each of the following.

a A coin is flipped and an octagonal dice with faces numbered 0 to 7 is rolled at the same time.
b Boxes A, B and C contain pink and yellow tickets. A box is selected at random and a ticket is drawn from it.
c The number of ways in which three letters P, A and N may be arranged to form a three letter sequence.
d In a class of 24 students, 10 take art, 12 take music and 5 take neither.

2 Two normal six-sided dice are rolled simultaneously. Draw a sample space for this information and hence calculate the probability of rolling:

a double 2. **b** at least one 4.

c a total greater than 9. **d** a total of 6 or 7.

3 The letters from the word MANCHESTER are written on cards and placed in a bag.

a What is the probability of drawing a vowel if one letter is drawn at random?

b Copy and complete this tree diagram to show all probabilities for when a letter is drawn from the bag, noted and replaced and then another letter is drawn.

1st draw — 2nd draw
Vowel — Vowel / Consonant
Consonant — Vowel / Consonant

c Use the tree diagram to determine the probability of drawing:

i two vowels. **ii** two consonants.

iii a vowel and a consonant. **iv** at least one consonant.

d Explain why drawing the letters can be considered independent events in this case.

e How could you change the experiment to make the events dependent?

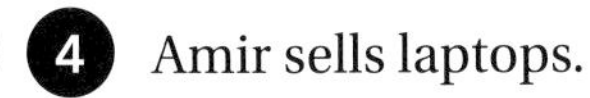

4 Amir sells laptops.

Before selling each laptop, he checks the hard drive and the screen.

The probability that the hard drive is faulty is $\frac{1}{10}$.

The probability that the screen is faulty is $\frac{1}{5}$.

These probabilities are independent.

a Complete the tree diagram to represent this information. *(2 marks)*

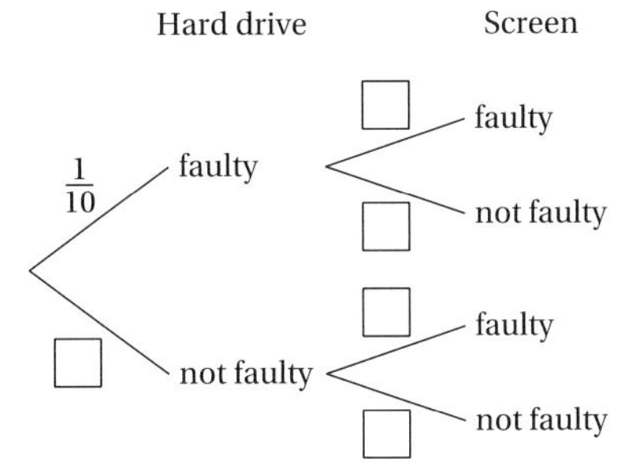

b Amir tests a laptop at random.

Find the probability that both the hard drive and the screen are **not** faulty. *(2 marks)*

5 There are 50 students in a year group. 30 have brown eyes, 9 have fair hair and 3 have both brown eyes and fair hair. Represent this information on a Venn diagram and use it to determine the probability that a student chosen at random from this group:

a has neither brown eyes nor fair hair.

b has brown eyes but not fair hair.

c has fair hair given that he or she has brown eyes.

d does not have fair hair given that he or she does not have brown eyes.

6 Find the number of different ways that all the letters in the word PROBABILITY can be arranged.

Find answers at: cambridge.org/ukschools/gcsemaths-studentbookanswers

25 Powers and roots

In this chapter you will learn how to ...

- use positive and negative integers and fractional powers to represent numbers in index notation.
- calculate with powers and roots.
- apply the rules for multiplying and dividing indices.

For more resources relating to this chapter, visit GCSE Mathematics Online.

Using mathematics: real-life applications

Powers and roots are used in many different occupations. Builders, painters and decorators need to work out areas using square units. Financial advisors and investors have to perform calculations involving powers and roots to work out the value of investments over time.

Calculator tip

Make sure you know which buttons to use to evaluate different powers and find different roots of numbers.

"I have to be able to understand, manipulate and evaluate different formulae to find the best investment. Many of these formulae involve fractional powers and different roots."

(Personal financial adviser)

Before you start ...

Ch 1	You should be able to quickly add and subtract pairs of integers mentally.	**1** Choose the correct sign: <, = or >. **a** $^{-}3 + {}^{-}3 \square 4 + 2$ **b** $6 - 7 \square 3 - 4$ **c** $4 - {}^{-}5 \square -3 + {}^{-}6$ **d** $^{-}2 + 6 \square 9 - 5$
Ch 1	You need to be able to find the squares, cubes, square roots and cube roots of numbers.	**2** Choose the correct answer. **a** The area of a square with sides of 3 cm. A 6 cm^2 B 9 cm^2 C 12 cm^2 **b** $\sqrt[3]{27}$ A 9 B 5.2 C 3 **c** $\sqrt{810000}$ A 9 B 90 C 900
Ch 10	You need to be able to find the reciprocal of a number or fraction.	**3** Find the reciprocal of each number. Choose from the values in the box. **a** $\frac{3}{4}$ **b** 12 **c** $1\frac{2}{5}$ $\frac{12}{1}$ $\frac{4}{3}$ $\frac{1}{12}$ $\frac{7}{5}$ $\frac{5}{7}$

Assess your starting point using the Launchpad

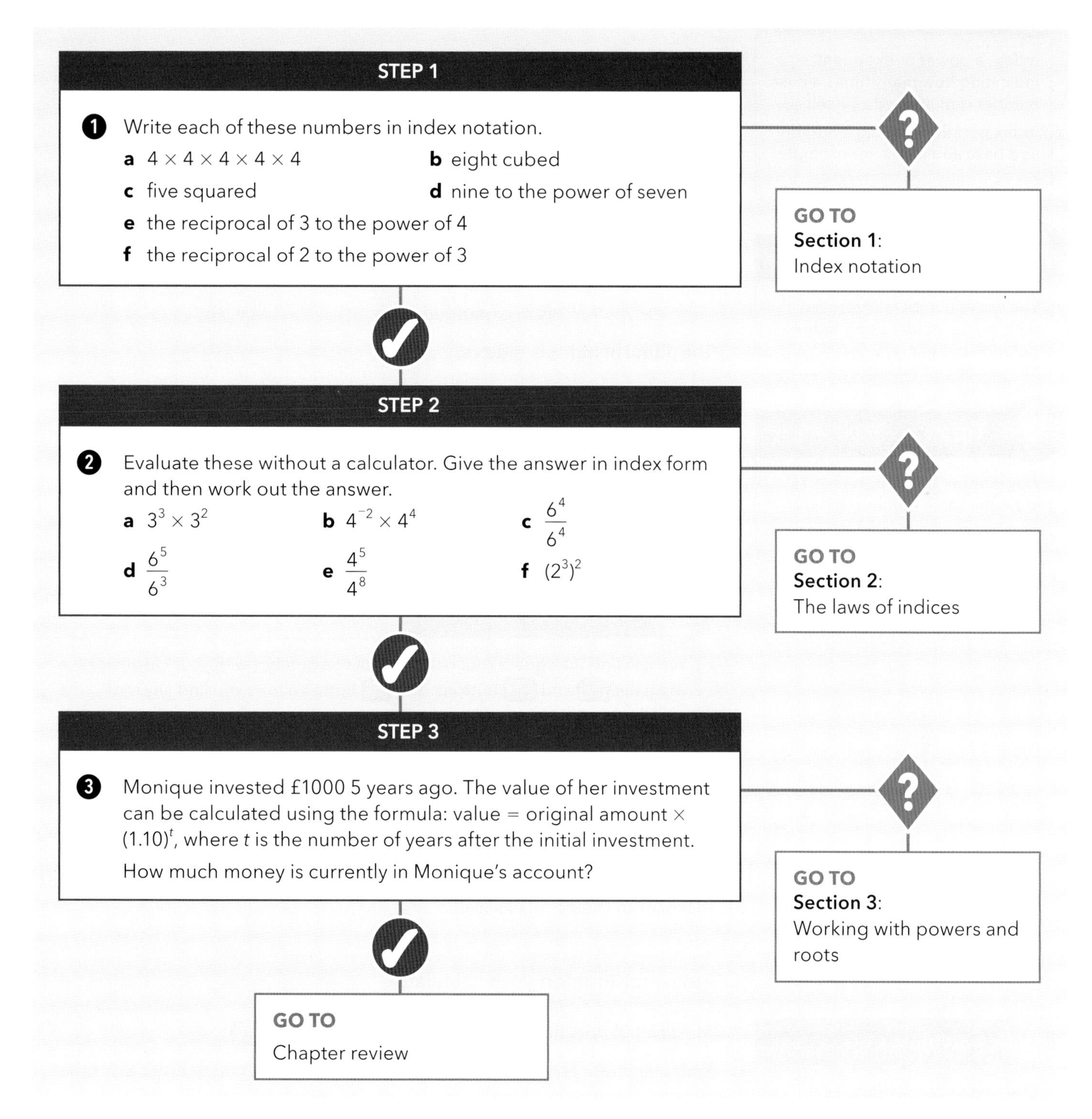

STEP 1
1 Write each of these numbers in index notation.
a $4 \times 4 \times 4 \times 4 \times 4$
b eight cubed
c five squared
d nine to the power of seven
e the reciprocal of 3 to the power of 4
f the reciprocal of 2 to the power of 3
GO TO
Section 1:
Index notation
STEP 2
2 Evaluate these without a calculator. Give the answer in index form and then work out the answer.
a $3^3 \times 3^2$
b $4^{-2} \times 4^4$
c $\frac{6^4}{6^4}$
d $\frac{6^5}{6^3}$
e $\frac{4^5}{4^8}$
f $(2^3)^2$
GO TO
Section 2:
The laws of indices
STEP 3
3 Monique invested £1000 5 years ago. The value of her investment can be calculated using the formula: value = original amount × $(1.10)^t$, where t is the number of years after the initial investment.
How much money is currently in Monique's account?
GO TO
Section 3:
Working with powers and roots
GO TO
Chapter review

Section 1: Index notation

You already know that you can use powers to write repeated multiplications in a shorter form.

For example: $6 \times 6 = 6^2$ and $5 \times 5 \times 5 = 5^3$

5^3 — This is the index. The index is also called the power or the exponent. / This is the base.

Find answers at: cambridge.org/ukschools/gcsemaths-studentbookanswers

Key vocabulary

index: a power or exponent indicating how many times a base number is multiplied by itself.

index notation: writing a number as a base and index, for example 2^3.

Tip

You wrote numbers in terms of their prime factors using index notation in Chapter 2.

Tip

Any number to the power of 1 stays the same number, so you don't write powers of 1.

The **index** tells you how many times the base number is multiplied by itself.

$$4^3 = 4 \times 4 \times 4$$

$$7^4 = 7 \times 7 \times 7 \times 7$$

$$9^{12} = 9 \times 9 \times 9 \times 9 \times 9 \times 9 \times 9 \times 9 \times 9 \times 9 \times 9 \times 9$$

When you write a number using an index you are using **index notation**.

7^4 is in index notation.

This means that numbers can be expressed as powers of their factors using index notation.

When you write the multiplication out in full you are using expanded form.

$7 \times 7 \times 7 \times 7$ is in expanded form.

The plural of index is indices.

Roots

You know that $5^2 = 25$ and $\sqrt{25} = 5$ and that $2^3 = 8$ and $\sqrt[3]{8} = 2$.

Mathematically, finding the root of a number is the inverse of working out the power of the number.

Calculator tip

You can find the square root or cube root of a number with your calculator using the [√] and [∛] buttons. The [ˣ√] button allows you find any root (fourth root, fifth root, and so on). On some calculators you would enter [1] [6] [ˣ√] [4] [=] to find $\sqrt[4]{16}$. Check how your calculator works.

Powers of 2, 3, 4 and 5

It is useful to recognise the first few powers of 2, 3, 4 and 5. This can help you to work out their roots as well.

Index notation on your calculator

Calculator tip

Check your calculator to see which button(s) it has.

Most calculators have one key to square a number: [x^2].

Your calculator is also likely to have a key that allows you to enter any other powers quickly and easily. It may be [y^x] or [x^y] or [a^b].

To enter 13^4, you press: [1] [3] [y^x] [4] [=].

You will get a result of 28 561.

EXERCISE 25A

1 Draw up a table like this one.

Base \ Index	⁻3	⁻2	⁻1	0	1	2	3	4	5
2	$2^{-3} = \frac{1}{8}$	$2^{-2} =$	$2^{-1} = \frac{1}{2}$	$2^0 = 1$	$2^1 = 2$	$2^2 = 4$	$2^3 =$	$2^4 = 16$	$2^5 =$
3									
4									
5									

a Use a calculator to work out the missing values in the table. Some powers of two have been done as an example.

b Compare the positive and negative values for the same index. What do you notice?

c Compare the powers of 2 with the powers of 4. What do you notice?

d How can you decide quickly that a number is *not* a power of 5?

2 Use the table that you completed in Question 1 to decide whether each statement is true or false.

a $2^4 = 4^2$ **b** $2^5 > 3^5$ **c** $2^0 = 5^0$

d $2^1 = 2^{-1}$ **e** $3^4 > 4^3$ **f** $4^4 < 3^4$

g $5^2 = 2^5$ **h** $3^5 > 5^3$ **i** $2^{-3} > 3^{-2}$

j $3^{-3} = \frac{1}{27}$ **k** $5^{-1} = 4^{-1}$ **l** $2(2^{-1}) = 1$

3 Use your table to work these out **without** using a calculator.

a $\sqrt{25}$ **b** $\sqrt[3]{8}$ **c** $\sqrt[4]{256}$

d $\sqrt[3]{125}$ **e** $\sqrt[5]{243}$ **f** $\sqrt[3]{64}$

g $\sqrt[3]{8} + \sqrt[4]{625}$ **h** $\sqrt{2500}$ **i** $\sqrt[5]{32} + \sqrt[4]{81}$

j $\sqrt[3]{27\,000}$ **k** $\sqrt[4]{160\,000}$ **l** $\sqrt[5]{3125} \times \sqrt[4]{625}$

4 Evaluate each expression without using a calculator.

a 2^3 **b** 6^2 **c** 1^8

d 8^3 **e** 10^4 **f** 10^6

g $2^3 - 1^5$ **h** $1^6 + 7^2$ **i** $2^4 \times 2^2$

j $2^4 + 4^2$ **k** $2^3 \times 2^4$ **l** $3^3 \times 3^3$

m $2^4 \div 2^3$ **n** $4^5 \div 4^3$ **o** $7^2 \times 10^3$

p 7×10^6 **q** $2 \times 10^2 + 3 \times 10^3$ **r** $6^2 \times 10^6$

5 Use your calculator to evaluate the following.

a 4^6 **b** 12^3 **c** 8^5 **d** 7^4

e 15^3 **f** 10^4 **g** 28^2 **h** 25^3

6 Use a calculator to find the value of each expression.

a $12^3 - 2^8$ **b** $20^4 - 15^2$ **c** $15^3 \times 15^2$

d $3^{12} + 3^4$ **e** $3^6 + 2^8$ **f** $35^3 \div 5^3$

Find answers at: cambridge.org/ukschools/gcsemaths-studentbookanswers

  Fill in $<$ or $>$ to make each statement true. You can use your calculator to help.

a $4^6 \square 6^4$ **b** $10^3 \square 3^{10}$ **c** $4^9 \square 9^4$

d $15^2 \square 2^{15}$ **e** $9^8 \square 8^9$ **f** $2^{10} \square 10^2$

Zero and negative indices

Look at this table of powers of 10.

Index notation	Expanded form	Value
10^6	$10 \times 10 \times 10 \times 10 \times 10 \times 10$	1 000 000
10^5	$10 \times 10 \times 10 \times 10 \times 10$	100 000
10^4	$10 \times 10 \times 10 \times 10$	10 000
10^3	$10 \times 10 \times 10$	1 000
10^2	10×10	100
10^1	10	10

If you look at the table you can see that each value is $\frac{1}{10}$ of the value above it. (In other words $10^6 \div 10 = 10^5$).

If you continue dividing by 10 you get this pattern for smaller and smaller indices:

Index notation	Expanded form	Value
10^6	$10 \times 10 \times 10 \times 10 \times 10 \times 10$	1 000 000
10^5	$10 \times 10 \times 10 \times 10 \times 10$	100 000
10^4	$10 \times 10 \times 10 \times 10$	10 000
10^3	$10 \times 10 \times 10$	1 000
10^2	10×10	100
10^1	10	10
10^0	$10 \div 10 = 1$	1
10^{-1}	$1 \div 10 = \frac{1}{10}$	$\frac{1}{10}$
10^{-2}	$\frac{1}{10} \div 10 = \frac{1}{100}$	$\frac{1}{100}$
10^{-3}	$\frac{1}{100} \div 10$	$\frac{1}{1000}$
10^{-4}	$\frac{1}{1000} \div 10$	$\frac{1}{10\,000}$

Tip

Remember that we use the reciprocals to change fraction divisions into multiplications.

$\frac{1}{10} \div 10 = \frac{1}{10} \times \frac{1}{10} = \frac{1}{100}$

The pattern in the table gives us two very important facts about indices.

Any number with an index of 0 is equal to 1: $a^0 = 1$ (except for 0^0 which is undefined).

So, for example: $5^0 = 1$ and $7^0 = 1$.

Any number with a negative index is equal to its reciprocal with a positive index: $a^{-m} = \frac{1}{a^m}$.

So, for example: $4^{-2} = \frac{1}{4^2}$ and $5^{-3} = \frac{1}{5^3}$.

Tip

An index can also be a fraction. You will deal with fractional index notation in Section 2 because it is easier to understand how it works when you know the laws of indices.

EXERCISE 25B

1 Write each of the following using positive indices only.

a 2^{-1} **b** 3^{-1} **c** 4^{-1} **d** 3^{-2}

e 4^{-3} **f** 3^{-5} **g** 3^{-4} **h** 6^{-6}

i 34^{-5} **j** x^{-3} **k** m^{-2} **l** $3x^{-4}$

2 Express the following with negative indices.

a $\frac{1}{3}$ **b** $\frac{1}{5}$ **c** $\frac{1}{7}$

d $\frac{1}{3^2}$ **e** $\frac{1}{4^5}$ **f** $\frac{1}{2^6}$

g $\frac{1}{7^2}$ **h** $\frac{1}{10^5}$ **i** $\frac{1}{2^2}$

j $\frac{1}{12^3}$ **k** $\frac{1}{10^4}$ **l** $\frac{1}{3(2)^2}$

m $\frac{1}{x^2}$ **n** $\frac{1}{x^3}$ **o** $\frac{4}{y^2}$

3 Fill in $=$ or $\neq$ in each of these statements.

a $10^{-1} \square \frac{1}{10}$ **b** $6^0 \square 1$ **c** $6^{-1} \square \frac{1}{6}$

d $10^{-2} \square \frac{2}{10}$ **e** $6^{-3} \square \frac{1}{6^3}$ **f** $10^0 \square 1$

g $6^{-4} \square \frac{1}{6^4}$ **h** $\frac{1}{10^4} \square 10^{-4}$ **i** $\frac{1}{6^3} \square \frac{3}{6}$

j $\frac{1}{4}(x^{-1}) \square 4x$ **k** $\frac{10}{m^5} \square {}^{-}10m^5$ **l** $0.5x^{-3} \square \frac{1}{2x^3}$

4 Write each of the following in index notation with base numeral 2.
(In other words, rewrite them in the form of 2^x.)

a 2 **b** 16 **c** 64

d $\frac{1}{8}$ **e** 0.25 **f** 1

g $\frac{1}{32}$ **h** $\sqrt{16}$ **i** ${}^{-}\sqrt{64}$

Section 2: The laws of indices

Calculators cannot deal with powers and variables in algebra, for example $x^2 \times 2x^3$, and it takes a long time to simplify expressions like these by first writing them in expanded form.

The laws of indices are a set of rules that allow you to multiply and divide powers without writing them out in expanded form.

Multiplying numbers in index notation

Look at these examples:

$3^4 \times 3^2 = (3 \times 3 \times 3 \times 3) \times (3 \times 3) = 3^6$ Can you see a short-cut?

$3^4 \times 3^2 = 3^{4+2} = 3^6$

$2^3 \times 2^5 = (2 \times 2 \times 2) \times (2 \times 2 \times 2 \times 2 \times 2) = 2^8$

$2^3 \times 2^5 = 2^{3+5} = 2^8$

Find answers at: cambridge.org/ukschools/gcsemaths-studentbookanswers

To multiply two numbers in index notation you add the indices.

$$a^m \times a^n = a^{m+n}$$

This law works for all indices, including negative indices.

For example, $2^3 \times 2^{-2} = 2^{3+(-2)} = 2^1 = 2$.

Dividing numbers in index notation

Look at these examples to find the short-cut for division.

$$2^5 \div 2^2 = \frac{2 \times 2 \times 2 \times 2 \times 2}{2 \times 2} = 2^3$$

$$3^4 \div 3^2 = \frac{3 \times 3 \times 3 \times 3}{3 \times 3} = 3^2$$

You should notice that

$$2^5 \div 2^2 = 2^{5-2} = 2^3 \text{ and } 3^4 \div 3^2 = 3^{4-2} = 3^2.$$

To divide two numbers in index notation you subtract the indices.

$$a^m \div a^n = a^{m-n}$$

This law works for all indices, including negative indices.

$$2^2 \div 2^4 = 2^{2-4} = 2^{-2}$$

You can understand how this works by looking at the expanded notation:

$$2^2 \div 2^4 = \frac{2 \times 2}{2 \times 2 \times 2 \times 2}$$ If you cancel you get $\frac{1}{2 \times 2}$.

$\frac{1}{2^2}$ is equal to 2^{-2}.

Powers of a power

$(3^2)^3$ means 3^2 to the power of 3 which is $3^2 \times 3^2 \times 3^2$.

$3^2 \times 3^2 \times 3^2 = 3^6$ (When you multiply powers, you add them.)

This means that $(3^2)^3 = 3^6$.

Tip

The laws of indices also help to show that $a^0 = 1$:
$4^3 \div 4^3 = 4^{3-3} = 4^0$
You already know that any number divided by itself is 1, so $4^3 \div 4^3 = 1$.
But this is also equal to 4^0 so 4^0 must equal 1.

To find the power of a power you multiply the indices.

$$(a^m)^n = a^{mn}$$

This law works for all indices, including negative indices.

$$(4^3)^{-4} = 4^{(3)(-4)} = 4^{-12}$$

You will use these laws of indices over and over in algebra, so it is important to know them well. Here is a summary:

Law of indices for ...	
multiplication	$a^m \times a^n = a^{m+n}$
division	$a^m \div a^n = a^{m-n}$
powers of indices	$(a^m)^n = a^{mn}$

EXERCISE 25C

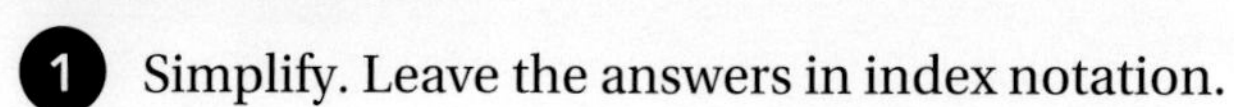

1 Simplify. Leave the answers in index notation.

a $2^4 \times 2^3$ **b** $10^2 \times 10^5$ **c** $4^3 \times 4^3$

d 5×5^6 **e** $2^4 \times 2^7$ **f** $3^2 \times 3^{-4}$

g $2^{-2} \times 2^5$ **h** $3^0 \times 3^2$ **i** $2 \times 2^3 \times 2^{-5}$

j $3^2 \times 3^2 \times 3$ **k** $10^2 \times 10^{-3} \times 10^2$ **l** $10^0 \times 10^{-2} \times 10^2$

2 Simplify. Leave the answers in index notation.

a $6^4 \div 6^2$ b $10^5 \div 10^2$ c $6^5 \div 6^3$ d $6^3 \div 6^5$

e $10^3 \div 10^5$ f $3^{10} \div 3^0$ g $3^8 \div 3$ h $10^4 \div 10^4$

i $\frac{5^4}{5^{-2}}$ j $\frac{10^6}{10^{-4}}$ k $\frac{3^{-2}}{3^{-3}}$ l $\frac{2^0}{2^3}$

3 Simplify each expression. Give the answers in index notation.

a $(2^2)^3$ b $(2^3)^3$ c $(2^4)^2$ d $(10^2)^2$

e $(10^2)^3$ f $(10^4)^2$ g $(2^4)^{-3}$ h $(10^{-2})^2$

i $(10^2)^{-3}$ j $(3^4)^{-2}$ k $(2^3)^0$ l $(2^2 \times 2^3)^2$

4 Say whether each statement is true or false. If it is false, write the correct answer.

a $3^3 \times 3^5 = 3^8$ b $3^8 \div 3^2 = 3^4$ c $10^8 \div 10^2 = 10^6$

d $(3^3)^2 = 3^6$ e $121^0 = 1$ f $4^5 \times 4^2 = 4^7$

g $3^{10} \div 3^2 = 3^5$ h $(4^2)^4 = 4^8$ i $(3^2)^0 = 1$

Fractional indices

The law of indices for multiplication can be used to explain the meaning of fractional indices.

Read through these examples carefully.

$3^{\frac{1}{2}} \times 3^{\frac{1}{2}} = 3^{\frac{1}{2}+\frac{1}{2}} = 3^1 = 3$ Add the indices using the multiplication law.

You also know that $\sqrt{3} \times \sqrt{3} = 3$

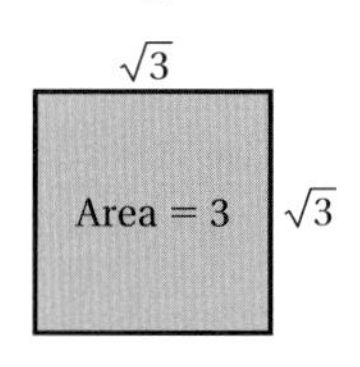

So, $3^{\frac{1}{2}} \times 3^{\frac{1}{2}} = 3$ and $\sqrt{3} \times \sqrt{3} = 3$ $3^{\frac{1}{2}}$ must equal $\sqrt{3}$ for this to be the case.

$\therefore 3^{\frac{1}{2}} = \sqrt{3}$

Similarly:

$8^{\frac{1}{3}} \times 8^{\frac{1}{3}} \times 8^{\frac{1}{3}} = 8^{\left(\frac{1}{3}+\frac{1}{3}+\frac{1}{3}\right)} = 8^1 = 8$ Add the indices using the multiplication law.

You also know that $\sqrt[3]{8} \times \sqrt[3]{8} \times \sqrt[3]{8} = 8$

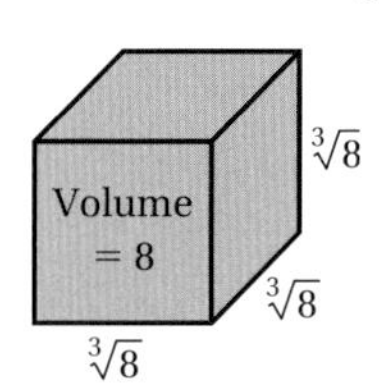

So, $8^{\frac{1}{3}} \times 8^{\frac{1}{3}} \times 8^{\frac{1}{3}} = 8$ and $\sqrt[3]{8} \times \sqrt[3]{8} \times \sqrt[3]{8} = 8$ $8^{\frac{1}{3}}$ must equal $\sqrt[3]{8}$ for this to be the case.

$\therefore 8^{\frac{1}{3}} = \sqrt[3]{8}$

From this you can see that $a^{\frac{1}{2}}$ is the square root of a and $a^{\frac{1}{3}}$ is the cube root of a.

$\sqrt{a} = a^{\frac{1}{2}}$ and $\sqrt[3]{a} = a^{\frac{1}{3}}$

This gives us a general rule:

the *n*th root of *a* is the same as *a* to the power of $\frac{1}{n}$.

So, $\sqrt[n]{a} = a^{\frac{1}{n}}$

Find answers at: cambridge.org/ukschools/gcsemaths-studentbookanswers

WORKED EXAMPLE 1

a Rewrite as roots.

i $5^{\frac{1}{2}}$ **ii** $4^{\frac{1}{5}}$

b Express each value in index notation.

i $\sqrt{90}$ **ii** $\sqrt[4]{2-x}$

a i $5^{\frac{1}{2}} = \sqrt{5}$

ii $4^{\frac{1}{5}} = \sqrt[5]{4}$

The denominator tells you what root you are dealing with.

b i $\sqrt{90} = 90^{\frac{1}{2}}$

ii $\sqrt[4]{2-x} = (2-x)^{\frac{1}{4}}$

The root tells you the denominator of the fraction.

Indices that are not unit fractions

The general law above applies to fractions where the numerator is 1 (a unit fraction). You may also get indices where the fraction does not have a numerator of 1. The index law for raising a power of a power can be used to make a general rule for these fractions.

Look at these examples to see how this works:

$4^{\frac{2}{3}} = \left(4^{\frac{1}{3}}\right)^2$ $\quad \frac{1}{3} \times 2 = \frac{2}{3}$ Use the power of indices law

$2^{\frac{3}{4}} = \left(2^{\frac{1}{4}}\right)^3$ $\quad \frac{1}{4} \times 3 = \frac{3}{4}$ Use the power of indices law

You already know that a unit fraction gives a root. So you can rewrite these expressions by writing the unit fraction inside the bracket as a root like this:

$\left(4^{\frac{1}{3}}\right)^2 = \left(\sqrt[3]{4}\right)^2$ and $\left(2^{\frac{1}{4}}\right)^3 = \left(\sqrt[4]{2}\right)^3$

So, $4^{\frac{2}{3}} = \left(\sqrt[3]{4}\right)^2$ and $2^{\frac{3}{4}} = \left(\sqrt[4]{2}\right)^3$

In general terms:

$a^{\frac{m}{n}} = \left(a^{\frac{1}{n}}\right)^m = \left(\sqrt[n]{a}\right)^m$ The nth root of a to the power of m.

Tip

It is much quicker to evaluate an expression in root form on your calculator than it is to try and enter $4^{\frac{2}{3}}$.

WORKED EXAMPLE 2

Write in index notation and then work out the value of each expression.

a $27^{\frac{2}{3}}$ **b** $5^{\frac{3}{4}}$ **c** $27^{-\frac{2}{3}}$

a $27^{\frac{2}{3}} = \left(\sqrt[3]{27}\right)^2$

Rewrite as an integer power. Remember $\frac{2}{3} = 2 \times \frac{1}{3}$.

$= (3)^2$

$= 9$

You should know that the cube root of 27 is 3, but you can work it out if you need to.

b $5^{\frac{3}{4}} = \left(\sqrt[4]{5}\right)^3$

$= 3.344$ (3 dp)

Rewrite as an integer power. If you round your answer, make sure you include the degree of accuracy to which it was rounded.

Continues on next page ...

c $27^{-\frac{2}{3}} = \frac{1}{(\sqrt[3]{27})^2} = \frac{1}{9}$

The index is negative and fractional. Start by writing the reciprocal of the positive index $\frac{1}{a^m}$ then because it is also a fractional index you know you need to include a root, $(\sqrt[n]{a})^m$. **Alternatively**, start by writing it as a root and then as a reciprocal. You are fine to leave your answer a fraction .

Tip

The index laws apply to all indices, including negative and fractional indices.

$2^3 \times 2^{-2} = 2^{3+(-2)} = 2^1 = 2$ and $3^{\frac{1}{2}} \times 3^{\frac{1}{4}} = 3^{\left(\frac{1}{2}+\frac{1}{4}\right)} = 3^{\frac{3}{4}}$

$2^2 \div 2^4 = 2^{2-4} = 2^{-2}$ and $3^{\frac{3}{4}} \div 3^{\frac{1}{2}} = 3^{\left(\frac{3}{4}-\frac{1}{2}\right)} = 3^{\frac{1}{4}}$

$(4^3)^{-4} = 4^{(3)(-4)} = 4^{-12}$ and $\left(2^{\frac{1}{4}}\right)^{\frac{1}{2}} = 2^{\left(\frac{1}{4}\times\frac{1}{2}\right)} = 2^{\frac{1}{8}}$

EXERCISE 25D

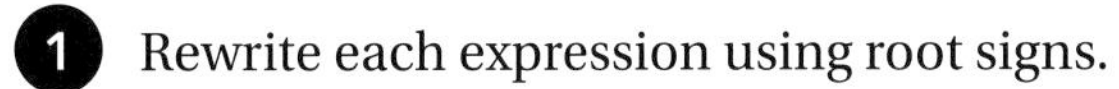

1 Rewrite each expression using root signs.

a $3^{\frac{1}{2}}$ **b** $4^{\frac{1}{3}}$ **c** $5^{\frac{1}{4}}$ **d** $6^{\frac{1}{2}}$

e $4^{\frac{1}{9}}$ **f** $5^{\frac{2}{3}}$ **g** $4^{\frac{3}{8}}$ **h** $6^{\frac{2}{9}}$

2 Write in index notation.

a $\sqrt{6}$ **b** $\sqrt[3]{4}$ **c** $\sqrt[3]{11}$ **d** $\sqrt[4]{9}$

e $(\sqrt[3]{3})^4$ **f** $\sqrt[5]{7}$ **g** $(\sqrt[3]{7})^2$ **h** $2(\sqrt[3]{3})^5$

3 Evaluate.

a $8^{\frac{1}{3}}$ **b** $32^{\frac{1}{5}}$ **c** $8^{\frac{4}{3}}$

d $216^{\frac{2}{3}}$ **e** $256^{\frac{3}{4}}$ **f** $256^{-\frac{1}{4}}$

g $125^{-\frac{4}{3}}$ **h** $\left(\frac{8}{27}\right)^{-\frac{1}{3}}$ **i** $\left(\frac{8}{18}\right)^{-\frac{1}{2}}$

Section 3: Working with powers and roots

Powers and roots are used in many different contexts. Builders, painters and decorators need to work out areas using square units (powers of 2 and square roots). Volume calculations use cube units (powers of 3 and cube roots). Calculations dealing with growth rates use different powers and roots and many science formulae rely on being able to work with powers and roots. In situations involving numbers that are not perfect squares or cubes (or other powers) an estimate of the power or root of a value may be used.

Estimating powers and roots

You can use what you already know about square and cube numbers and their roots to estimate the approximate value of other powers and roots. For example, you can estimate the square root of a number that is not a perfect square by working out where it lies on a number line in relation to the square number below it and above it.

Find answers at: cambridge.org/ukschools/gcsemaths-studentbookanswers

WORKED EXAMPLE 3

Estimate $\sqrt{98}$.

$\sqrt{81} < \sqrt{98} < \sqrt{100}$

$9 < \sqrt{98} < 10$

9 — 9.5 — 10

$\sqrt{81}$ — half way — $\sqrt{100}$

ignore values <9.5

$\sqrt{98}$ is closer to $\sqrt{100}$ so try 9.9

First find the perfect squares closest to 98. These are 81 and 100.

Now use trial and improvement to estimate more accurately.

Check: $9.9 \times 9.9 = 98.01$

As $98.01 \approx 98$, we can state $\sqrt{98} \approx 9.90$ (2 dp)

If you were asked to estimate $\sqrt{98}$ to the nearest whole number it would be 10.

Tip

Remember your estimate may not be an exact value, but it should be close. Usually a value correct to two decimal places is good enough. If you use a calculator you will see that $\sqrt{98}$ is 9.899, so 9.90 is a fairly accurate estimate.

You can use a similar method using your calculator to find the approximate value of a power by trial and improvement.

WORKED EXAMPLE 4

If $2^x = 18$, estimate the value of x.

$2^4 = 16$ $\quad 2^5 = 32$

Use what you know.

So $4 < x < 5$

16 (18) — 32

2^4 — 2^5

close to 16, so x must be closer to 4 than to 5

Try $2^{4.1} = 17.14$ Too small

$2^{4.2} = 18.37$ Too big

Now use trial and improvement.

x must be between 4.1 and 4.2

Try $2^{4.15} = 17.75$ Too small

$2^{4.16} = 17.88$ Too small

$2^{4.17} = 18.00$ Correct to 2dp

So $x \approx 4.17$ (2 dp)

If you were asked to estimate x to the nearest whole number it would be 4.

EXERCISE 25E

1 Estimate the following roots. Show your working.

a $\sqrt{72}$ **b** $\sqrt{33}$ **c** $\sqrt{6}$ **d** $\sqrt[3]{29}$ **e** $\sqrt[3]{^-200}$ **f** $\sqrt[4]{37}$

2 Use a calculator to find the roots in question 1 correct to two decimal places. How close were your estimates?

3 Find the value of x in each of the following by trial and improvement. Show your working.

a $2^x = 25$ **b** $3^x = 36$ **c** $2^x = 280$

d $x^4 = 1296$ **e** $x^3 = 12$ **f** $x^3 = 7000$

4 Find four different pairs of values for a and m which will satisfy the equation $a^m = 81$.

5 Use trial and improvement methods to find the length of sides of a square piece of plastic with an area of $90\ \text{mm}^2$.

6 A cube of sodium has a volume of $800\ \text{mm}^3$. Find the lengths of the side of the cube correct to 2 decimal places using trial and improvement methods.

Solving problems involving powers and roots

In this section you will use the rules of indices and what you already know about powers and roots to solve problems in real life contexts.

Some problems seem difficult and confusing because they are so long and wordy. A good strategy for these problems is to rewrite the problem using only the most important words and numbers.

Problem-solving framework

Steps for approaching a problem-solving question	What you would do for this example
Step 1: Read the problem carefully and highlight the important words and numbers.	Naresh wants to have £5000 saved in three years' time to buy a car. He finds an account that offers an interest rate of 2.9% compounded annually. His sister tells him that you can work out how much you need to put away now to have an amount in the future using the formula Principal amount (P) = future value (F) $\times (1.029)^{-3}$. Use the formula to calculate how much Naresh should put into the account to have £5000 in 3 years' time.
Step 2: Jot down the important words and numbers only. Write down what it is you need to calculate.	Future value: £5000 Formula: $P = F(1.029)^{-3}$ We need to find P.
Step 3: Rewrite the problem again in mathematical terms to show what you need to do.	$P = £5000 \times (1.029)^{-3}$ Now you can see that the complicated word problem is really quite a simple substitution problem that you can work out in one step with your calculator.
Step 4: Do the necessary calculation.	5 0 0 0 × 1 . 0 2 9 x^y 3 +/− = 4589.06 Naresh needs to put £4589.06 into the account now.

Calculator tip

On some calculators when you might instead have to enter

5 0 0 0 × 1 . 0 2 9 x^y − 3 =

Work out how to enter a negative index on your calculator.

EXERCISE 25F

1 Shamila receives £2500. She wants to invest it for 10 years in an account that offers 5% growth but she wants to know how much money she will have after the ten-year period. There is a formula to work this out quickly:

$$\text{Value of future investment} = \text{original amount} \times (1.05)^{10}$$

a Work out how much money Shamila will have in the investment after ten years.

b How much will she have if she decides to spend £500 and put the rest of the money into this investment?

c How could you change this formula to work out the future value of an investment if the interest rate is 3% and the period of the investment is 18 months?

d Use your formula to find the future value of £3200 invested under these conditions.

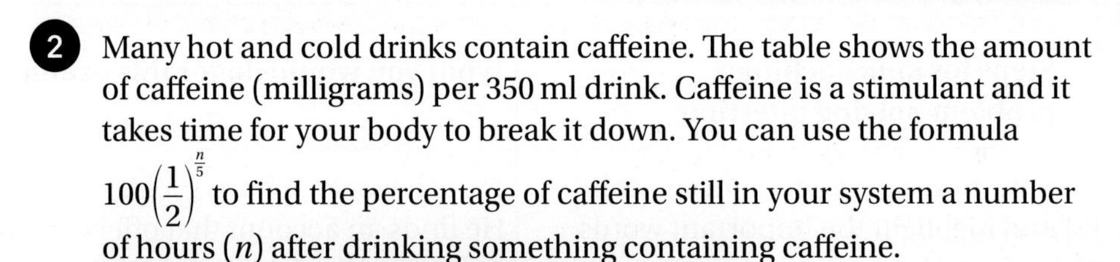

2 Many hot and cold drinks contain caffeine. The table shows the amount of caffeine (milligrams) per 350 ml drink. Caffeine is a stimulant and it takes time for your body to break it down. You can use the formula $100\left(\frac{1}{2}\right)^{\frac{n}{5}}$ to find the percentage of caffeine still in your system a number of hours (n) after drinking something containing caffeine.

Drink (350 ml)	Amount of caffeine (mg)
Americano coffee	154
Cappuccino	154
Ceylon tea	63
Iced tea	25.5
Cola A	34
Diet cola A	45
Cola zero	45
Cola B	38
Diet cola B	38
Cola one	54
Cream soda	29
Energy drink	120

Marie and Suki apply the formula to find out how many milligrams of caffeine will remain in their system 3 hours after drinking a 350 ml energy drink. This is how each student worked:

Marie's working	Suki's working
% caffeine $= 100(\frac{1}{2})^{\frac{n}{5}}$	$\% C = 100(0.5)^{\frac{3}{5}}$
$= 100(0.5)^{\frac{3}{5}}$	$= 65.98\%$
$= 2.5\%$	65.98% of 120 mg = 79.18 mg
2.5% of 120 mg = 3 mg	

a Which answer is correct? What did the other student do wrong?

b Billy has two energy drinks and a cola zero. How many milligrams of caffeine will be left in his system after 4 hours?

c Karen has three cups of Ceylon tea at breakfast time. How much caffeine will be left in her system $2\frac{1}{2}$ hours later when it is teatime at her office?

d What percentage of caffeine is left in your body $\frac{1}{2}$ hour after drinking any caffeinated drink?

3 The time (t seconds) a ball takes to hit the ground after being dropped from a height (h metres) can be found using the formula $t = \sqrt{\frac{h}{4.9}}$.

a Work out how long it will take a ball dropped from a height of 3.6 metres to hit the ground.

b Matt drops a ball from a height of 2.5 m and Nina drops a ball from a height of 3.6 m.

i Whose ball will hit the ground first?

ii How many seconds later will the second ball hit the ground?

4 Biologists have discovered that many measurements in humans and other mammals are in proportion to the mass of the body. This has allowed them to develop a number of formulae for estimating a number of different measurements. Here are four different formulae which use mass (m) in kilograms to determine other measurements.

Mass of the brain (B) in kilograms	Surface area of the skin (S) in square metres	Resting metabolic rate (C) (Calories consumed at rest)	Time (T) it takes for the blood to circulate in seconds
$B = 0.01m^{\frac{2}{3}}$	$S = 0.0096m^{\frac{7}{10}}$	$C = 70(\sqrt[4]{m})^3$	$T = 17.4(\sqrt[4]{m})$

Tip

These formulae are all based on a constant of proportionality (k). You will learn more about this in Chapter 36 when you deal with direct and inverse proportion.

a Work out all of these values based on your own mass in kilograms.

b Compare the blood circulation time for an elephant (average mass 5000 kg) and a human male (average mass 70 kg).

c What is the brain of a 4.5 kg cat likely to weigh?

d Find the surface area of the skin of a mouse of mass 0.03 kg and a cat of mass 4.5 kg.

e How many calories does a 145 kg lion consume while lying in the sun sleeping?

f Why are all the values found using these formulae only approximates?

Find answers at: cambridge.org/ukschools/gcsemaths-studentbookanswers

5 Kepler's law can be used to work out the time (T) in Earth days that it takes for a planet to complete an orbit around the Sun. The formula for this is $T = 0.2R^{\frac{3}{2}}$, where R is the mean distance from the planet to the Sun in millions of kilometres.

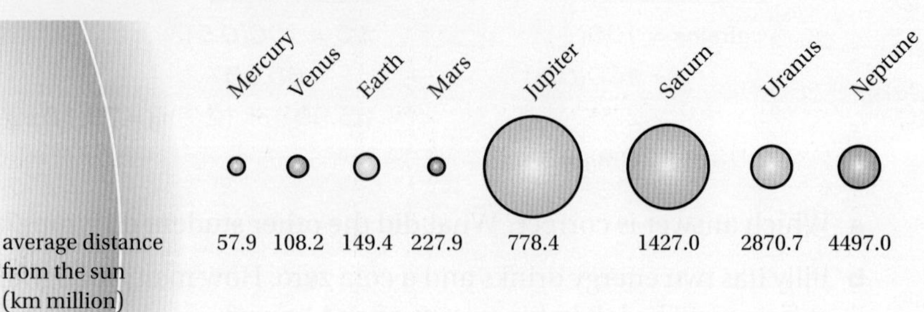

Tip

If you don't get 365 days as an answer for Question 5a you know you are doing something wrong!

a Use the formula to find the time it takes the Earth to orbit the Sun.

b Which takes longer to orbit the Sun: Jupiter or Uranus? How much longer does it take?

c Which planet orbits the Sun in the shortest time? Explain how you know this and then work out how long it takes.

Checklist of learning and understanding

Index notation

- Numbers can be expressed as powers of their factors using index notation.
  - $2 \times 2 \times 2 \times 2$ can be written in index notation as 2^4.
  - The 4 is the index (also called the power or the exponent) and it tells us how many times the base (2) must be multiplied by itself.
- Any number to the power of 0 is equal to 1: $a^0 = 1$.
- A negative index is the reciprocal of a positive index: $a^{-m} = \frac{1}{a^m}$

Laws of indices

- Multiplication: $a^m \times a^n = a^{m+n}$
- Division: $\frac{a^m}{a^n} = a^{m-n}$
- Raising a power: $(a^m)^n = a^{mn}$
- Fractional power: $a^{\frac{m}{n}} = \left(a^{\frac{1}{n}}\right)^m = (\sqrt[n]{a})^m$

Working with powers and roots

- Finding the root of a number is the inverse of raising the number to a power. For example $4^2 = 16$ and $\sqrt{16} = 4$.
- Powers and roots can be estimated using a number line to locate them in relation to known powers and roots and then finding an accurate estimate by trial and improvement.
- Many real-life calculations rely on formulae containing integer and fractional indices and roots.

For additional questions on the topics in this chapter, visit GCSE Mathematics Online.

Chapter review

1 Write each number in index form.

a $8 \times 8 \times 8 \times 8 \times 8$

b three cubed

c $\frac{1}{64}$

d $(\sqrt{5})^3$

2 Put these expressions in order from smallest to greatest.

a $3^4, \sqrt{81}, 10^2, 4^3, 2 \times \sqrt{121}$ **b** $4^5, 5^4, 10^3, 96^0, 3^2, 20^2$ **c** $\sqrt[3]{4}, 4^{\frac{3}{2}}, 4^2, \left(\frac{1}{4}\right)^{\frac{3}{2}}$

3 Write these numbers with positive indices.

a 3^{-3} **b** 2^{-10} **c** 5^{-2}

4 Use the laws of indices to simplify each expression and write it as a single power of 4.

a $4^2 \times 4^2$ **b** $4^6 \times 4^{-3}$ **c** $4^7 \div 4^3$

d $4^3 \div 4^5$ **e** $(4^3)^2$ **f** $(4^{-2})^2$

5 Evaluate. Check your answers with a calculator.

a $\sqrt{121}$ **b** $\sqrt{0.25}$ **c** $\sqrt[3]{125}$

d $\sqrt[5]{32}$ **e** $\sqrt[4]{81}$ **f** $\sqrt{\frac{1}{4}}$

g $9^{\frac{3}{2}}$ **h** $(^-27)^{\frac{2}{3}}$ **i** $8^{\frac{4}{3}}$

6 Estimate the length of each side of a cube of volume 35 cm^3. Show your working.

7 Electricians use the formula $V = \sqrt{PR}$ to work out voltage (V) when P is the power in watts and R is the resistance in ohms. Calculate the voltage when $P = 2000$ and $R = 24.2$.

8 The length of time (T seconds) it takes for a pendulum of length (L) in metres to swing through one complete movement can be calculated using the formula $T = 2\pi\left(\frac{L}{9.8}\right)^{\frac{1}{2}}$.

a Work out how long it will take for a pendulum of length 0.25 m to complete one swing.

b How does the length of the pendulum affect how long it takes to complete one swing? Justify your answer.

9 The radius of a cylindrical container can be found using the formula $r = \sqrt{\frac{V}{\pi h}}$ where V is the volume and h is the height of the container. Pete has a 12 cm tall cylindrical tin with a volume of 500 cm^3. He wants to use it to store round biscuits of diameter 7 cm. Will they fit?

10 **a** Evaluate **i** 17^0 *(1 mark)*

ii 4^{-3} *(2 marks)*

b The distance, d, in miles to the horizon is given by the formula

$$d = \left(\frac{3h}{2}\right)^{\frac{1}{2}}$$

where h is the height, in feet, of an observer's eyes above sea level.

i How far away is the horizon from a man whose eyes are 6 feet above sea level? *(2 marks)*

ii From the top of a cliff, Samira can see the horizon 12 miles away. Find the height above sea level of Samira's eyes. *(3 marks)*

Find answers at: cambridge.org/ukschools/gcsemaths-studentbookanswers

26 Standard form

In this chapter you will learn how to ...

- convert numbers to and from standard form.
- use a calculator to solve problems with numbers in standard form.
- apply the index laws to add, subtract, multiply and divide numbers in standard form with and without using a calculator.

For more resources relating to this chapter, visit GCSE Mathematics Online.

Using mathematics: real-life applications

The study of stars, moons and planets involves huge numbers. Astronomers use standard form to make it easier to write or type very large quantities, to make them easier to compare and to allow them to calculate with and without calculators. The Sun has a mass of 1.988×10^{30} kg. This is a number with 27 zeros and it would be clumsy and impractical to have to write it out each time you wanted to use it.

Calculator tip

Make sure you know how your calculator deals with exponents and that you have it in the correct mode to do calculations involving exponents.

"In astronomy we work with very large and very small numbers. There are 100 000 000 000 000 000 000 000 known stars alone! Imagine having to write this number out in full every time you wanted to use it! It is much easier to write 1×10^{23}." *(Astronomy student)*

Before you start ...

Ch 11	You should be able to calculate efficiently with decimals.	**1** Evaluate these without using a calculator. **a** $2.9 + 5.8$ **b** $12.5 - 3.8$ **c** 4.5×1.5 **d** $4.5 \div 0.3$
Ch 17	You need to be able to round numbers to a given number of significant figures.	**2** Choose the correct answer. **a** 507 000 000 rounded to 2 sf. A 50 700 B 510 000 000 **b** 1.098 rounded to 3 sf. A 1.10 B 1.09 **c** 0.006 25 rounded to 1 sf. A 0.6 B 0.006
Ch 25	You should know how to apply the laws of indices.	**3** State whether the following are true or false. **a** $3^2 \times 3^3 = 3^6$ **b** $x^5 \times x^3 = x^8$ **c** $x^{-3} \times x^4 = x$ **d** $\frac{x^4}{x^5} = x$ **e** $\frac{x^{-4}}{x^2} = x^{-6}$

Assess your starting point using the Launchpad

STEP 1

In Questions **1** to **4**, choose the correct answer.

1 2.4×10^7 expressed as an ordinary number.

A 0.000 000 24 B 240 000 000 C 0.000 000 024
D 24 000 000 E 2 400 000

2 7×10^{-3} expressed as an ordinary number.

A 0.0007 B 7000 C 0.007
D 70 000 E 0.000 07

3 23 500 written in standard form.

A 2.35×10^{-4} B 2.35×10^4 C 235×10^2
D 23.5×10^3 E 2.35×10^{-2}

4 0.000 023 1 written in standard form.

A 23.1×10^{-5} B 2.31×10^5 C 2.31×10^{-4}
D 2.31×10^4 E 2.31×10^{-5}

? **GO TO** **Section 1**: Expressing numbers in standard form

✓

STEP 2

5 Use your calculator to evaluate 2 400 000 × 23 000. Give your answer in standard form.

6 Use your calculator to calculate $\frac{4.6 \times 10^{-3}}{1.84 \times 10^4}$. Write your answer in ordinary form.

? **GO TO** **Section 2**: Calculators and standard form

✓

STEP 3

7 Calculate, without using your calculator.

a $(4 \times 10^{11}) \times (2 \times 10^{-6})$ **b** $\frac{3.2 \times 10^{16}}{7.9 \times 10^4}$

c $1.63 \times 10^7 - 8.43 \times 10^6$

? **GO TO** **Section 3**: Working in standard form

✓

GO TO Chapter review

Find answers at: cambridge.org/ukschools/gcsemaths-studentbookanswers

Section 1: Expressing numbers in standard form

Writing out very large or very small numbers takes time and it is easy to make mistakes and skip zeros when you do calculations. We use standard form to write these numbers in a simpler way using powers of 10.

A number is in standard form when it is written as a product (×) of a factor and a power of 10. The factor must be greater or equal to 1 and smaller than 10.

> Algebraically, we can say that any number (x) can be expressed in the form of:
>
> $$x \times 10^n, \text{ where } 1 \leqslant x < 10 \text{ and } n \text{ is an integer.}$$

For example, 2×10^2 and 1.2×10^{-2} are both in standard form.

Having a good understanding of place value will help you to understand standard form.

Consider $3 \times 10^4 = 30\,000$ and $3 \times 10^{-4} = 0.0003$.

The index (power of ten) gives you important information.

- Multiplying by 10^4 moves the decimal point 4 places to the right.
- Multiplying by 10^{-4} moves the decimal point 4 places to the left.

Tip

In the UK we use the term **standard form**; this notation is also called scientific notation.

Writing a number in standard form

To write a number in standard form:

- Place the decimal point after the first non-zero digit.
- Find the power of 10 needed to move the decimal point back to its original position. In other words, work out the number of places the decimal point has moved and the direction in which it has moved.
- Write the number as a decimal (between 1 and 10) multiplied by a power of 10.

WORKED EXAMPLE 1

Express these numbers in standard form.

a 416 000 **b** 0.0037

a 416 000

4.16

4.1 6 0 0 0

Find the number between 1 and 10 and insert the decimal point after the first non-zero digit.
Work out how many decimal places the point needs to move to get back to the original number.

$416\,000 = 4.16 \times 10^5$

The decimal point needs to move five places to the right (+), so the power of ten is 5.

Continues on next page ...

b 0.0037
3.7
0.0 0 3.7

Find the number between 1 and 10. Work out how many decimal places the point needs to move to get back to the original number.

$0.0037 = 3.7 \times 10^{-3}$

The decimal point needs to move three places to the left (−), so the power of 10 is ⁻3.

$0.0037 = \frac{37}{10\,000}$

Alternatively, you could convert the decimal to a fraction first.

$= \frac{3.7}{1000}$

Divide the top and bottom by 10 to get a numerator between 1 and 10.

$= 3.7 \times 10^{-3}$

Look at the number of zeros in the denominator to find the index. As the original number is less than 1, the index must be negative.

Converting from standard form to ordinary numbers

To convert numbers from standard form to ordinary numbers or decimals you need to look at the power of ten and move the decimal point this number of places to the left or right.

WORKED EXAMPLE 2

Write as ordinary numbers.

a 3.25×10^5 **b** 2.07×10^{-5}

a 3.25×10^5
3.2 5 0 0 0

The power of ten is 5. Move the decimal point 5 places to the right (positive direction).

3 2 5 0 0 0
$3.25 \times 10^5 = 325\,000$

Fill in the correct number of zeros; notice that the digits '2' and '5' count as two of the 5 places so only three zeros are needed.

b 2.07×10^{-5}
2.0 7

Move the decimal point 5 places to the left (negative direction).

.0 0 0 0 2 0 7

Fill in the correct number of zeros; notice that the '2' counts as one of the 5 places (it is the first significant figure), so only four zeros are needed before the decimal point.

$2.07 \times 10^{-5} = 0.000\,020\,7$

Remember to write the 0 before the decimal point as well.

EXERCISE 26A

1 Express each of the following in standard form.

a 321 000 **b** 1340 **c** 40 050
d 3 010 000 **e** 0.08 **f** 0.0001
g 32 000 000 **h** 910 000 **i** 0.000 031 255
j 0.000 000 241 52 **k** 0.003 05 **l** 0.201
m 34 000 **n** 0.000 34 **o** 0.009
p 2.45 **q** 0.000 426 **r** 0.426

2 Express each of the following as a basic numeral.

a 1.4×10^{2} **b** 4.8×10^{4} **c** 2.9×10^{3}
d 3.25×10^{2} **e** 3.25×10^{-1} **f** 3.67×10^{5}
g 4.5×10^{7} **h** 2.13×10^{-2} **i** 3.209×10^{4}
j 3.46×10^{-3} **k** 1.89×10^{-4} **l** 7×10^{-7}
m 1.03×10^{-2} **n** 1.025×10^{-3} **o** 2.09×10^{-5}

3 Express each of the following quantities in standard form.

a In 2011 the population of the Earth reached 7 000 000 000.
b The distance from the Earth to the Moon is approximately 240 000 miles.
c There are about 100 000 000 000 000 (a hundred trillion) cells in your body.
d Some cells are about 0.000 000 2 metres in diameter.
e The surface area of the Earth's oceans is about 140 million square miles.
f An angstrom is a unit of measure. One angstrom is equivalent to 0.000 000 000 1 metre.
g Humans blink on average about 6 250 000 times per year.
h A dust particle has a mass of about 0.000 000 000 753 kg.

4 Write each quantity out in full as an ordinary number.

a The area of the Atlantic Ocean is 3.18×10^{7} square miles.
b The space between tracks on a DVD disk is 7.4×10^{-4} mm.
c The diameter of the silk used to weave a spider's web is 1.24×10^{-6} mm.
d There are 3×10^{9} possible ways to play the first four moves in a game of chess.
e A sheet of paper is about 1.2×10^{-4} m thick.
f The distance between the Sun and Jupiter is about 7.78×10^{8} km.
g The Earth is about 1.5×10^{11} km from the Sun.
h The mass of an electron is about $9.109\,382\,2 \times 10^{-31}$ kg.

5 Order these numbers from smallest to largest.

1.75×10^{2} 3×10^{-3} 9.9×10^{1} 5.7×10^{-2} $3.654\,645 \times 10^{20}$ 1.75×10^{4}

Section 2: Calculators and standard form

You can use a scientific calculator to enter calculations in standard form. The calculator will also give you an answer in standard form if it has too many digits to display on the screen.

Entering standard form calculations

You will need to use the [$\times 10^x$], [Exp] or [EE] button on your calculator. These are known as the exponent keys and they all work in the same way, even though they may look different on different calculators.

When you are using the exponent function key of your calculator you don't enter the $\times$ 10 part of the calculation because the calculator function automatically includes that part.

Use your own calculator to work through this example. You should get the same result even if your function key is different to the one in the example.

WORKED EXAMPLE 3

Calculate:

a 2.134×10^4 **b** 3.124×10^{-6}

a 2.134×10^4

[2] [.] [1] [3] [4] [$\times 10^x$] [4] [=]

$2.134 \times 10^4 = 21\,340$

Enter the digits into your calculator, and write down the answer.

b 3.124×10^{-6}

[3] [.] [1] [2] [4] [Exp] [+/–] [6] [=]

$3.124 \times 10^{-6} = 0.000\,003\,124$

Use the correct key for your calculator to enter the negative 6. Some calculators might still display this number in standard form as it has too many digits to fit the display.

Calculator tip

Calculators work in different ways and you need to understand how your own calculator works. Make sure you know what buttons to use to enter standard form calculations, how to read and make sense of the display, and how to convert your calculator answer into decimal form.

Making sense of the calculator display

The answer your calculator displays will depend on the calculator you use. Here are two ways in which calculators display answers in standard form:

5.98 -06

This is 5.98×10^{-6}

2.56E24

This is 2.56×10^{24}

To give the answer in standard form read the display and write the answer correctly.

To give the answer as an ordinary number, apply the rules you know to convert it from standard form to ordinary form.

Find answers at: cambridge.org/ukschools/gcsemaths-studentbookanswers

EXERCISE 26B

Enter each of these numbers into your calculator using the correct function key and write down what appears on the display.

a 4.2×10^{12} **b** 1.8×10^{-5} **c** 2.7×10^{6}

d 1.34×10^{-2} **e** 1.87×10^{-9} **f** 4.23×10^{7}

g 3.102×10^{-4} **h** 3.098×10^{9} **i** 2.076×10^{-23}

Here are ten different calculator displays giving answers in exponential form. Write each answer correctly in standard form.

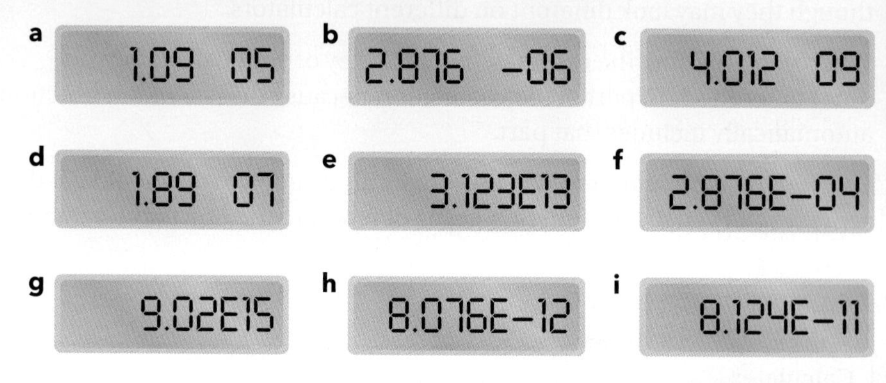

Significant figures

When you work with standard form, you will often be asked to give the answers in standard form correct to a given number of significant figures.

You already know how to round answers to a given number of significant figures (refer to Chapter 17 if you have forgotten). When you are working with decimal values, you need to remember that none of the zeros before a non-zero digit are significant.

- 0.003 is correct to 1 significant figure.
- 0.01 is correct to 1 significant figure.
- 0.10 is correct to 2 significant figures.

A zero after a non-zero digit **is** significant.

EXERCISE 26C

Use your calculator to do these calculations. Give your answers in standard form correct to 3 significant figures.

a 4216^{6} **b** $(0.000\,09)^{4}$

c $0.0002 \div 2500^{3}$ **d** $65\,000\,000 \div 0.000\,0045$

e $(0.0029)^{3} \times (0.003\,65)^{5}$ **f** $(48 \times 987)^{4}$

g $\dfrac{4525 \times 8760}{0.000\,020}$ **h** $\dfrac{9500}{0.0005^{4}}$

i $\sqrt{5.25} \times 10^{8}$ **j** $\sqrt[3]{9.1} \times 10^{-8}$

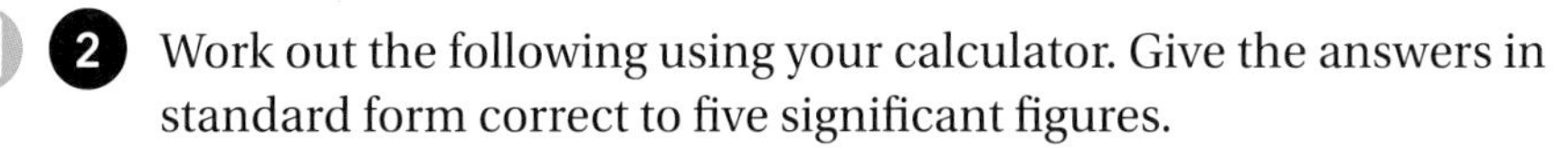

2 Work out the following using your calculator. Give the answers in standard form correct to five significant figures.

a 4234^5 **b** $0.0008 \div 9200^3$

c $(1.009)^5$ **d** $123\,000\,000 \div 0.00076$

e $(97 \times 876)^4$ **f** $(0.0098)^4 \times (0.0032)^3$

g $\dfrac{8543 \times 9210}{0.000\,034}$ **h** $\dfrac{9745}{(0.0004)^4}$

i $\sqrt[3]{4.2 \times 10^{-8}}$

Section 3: Working in standard form

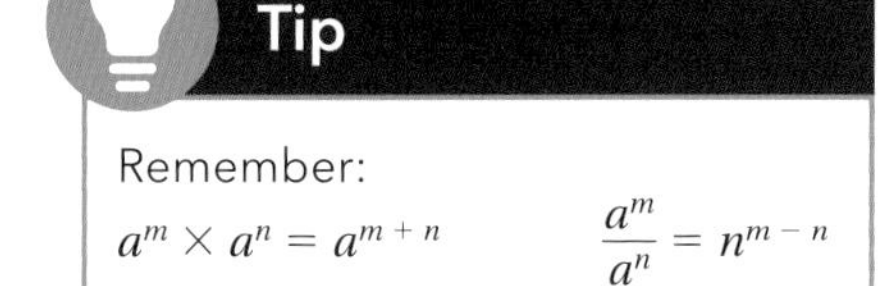

Tip

Remember:

$a^m \times a^n = a^{m+n}$ $\dfrac{a^m}{a^n} = n^{m-n}$

Writing numbers in standard form allows you to use the laws of indices to calculate quickly without using a calculator.

Multiplying and dividing numbers in standard form

When you multiply powers of ten, you add the indices.

When you divide powers of ten, you subtract the indices.

WORKED EXAMPLE 4

Do these calculations without using a calculator. Give your answers in standard form.

a $(3 \times 10^5) \times (2 \times 10^6)$ **b** $(2 \times 10^{-3}) \times (3 \times 10^{-7})$ **c** $(2 \times 10^3) \times (8 \times 10^7)$ **d** $\dfrac{2.8 \times 10^6}{1.4 \times 10^4}$ **e** $\dfrac{4 \times 10^8}{9 \times 10^5}$

a $(3 \times 10^5) \times (2 \times 10^6)$

$3 \times 2 \times 10^5 \times 10^6$

$= 6 \times 10^{5+6}$

Remove the brackets and group like terms. Using the multiplication law of indices, add the indices.

$= 6 \times 10^{11}$

Write the answer in standard form.

b $(2 \times 10^{-3}) \times (3 \times 10^{-7})$

$2 \times 3 \times 10^{-3} \times 10^{-7}$

$= 6 \times 10^{-3+-7}$

Remove the brackets and group like terms. Using the multiplication law of indices, add the indices.

$= 6 \times 10^{-10}$

Write the answer in standard form.

c $(2 \times 10^3) \times (8 \times 10^7)$

$2 \times 8 \times 10^3 \times 10^7$

$= 16 \times 10^{3+7}$

$= 16 \times 10^{10}$

$= 1.6 \times 10 \times 10^{10}$

$= 1.6 \times 10^{11}$

But 16 is greater than 10 so this is not in standard form. If you think of 16 as 1.6×10 you can change it to standard form.

d $\dfrac{2.8 \times 10^6}{1.4 \times 10^4} = \dfrac{2.8}{1.4} \times \dfrac{10^6}{10^4}$

$= 2 \times 10^{6-4}$

Subtract the indices to divide the powers.

$= 2 \times 10^2$

Continues on next page ...

e $\frac{4 \times 10^8}{9 \times 10^5} = \frac{4}{9} \times \frac{10^8}{10^5}$ — Rewrite so that like terms are grouped.

$= 0.44 \times 10^3$

$= 4.4 \times 10^{-1} \times 10^3$

$= 4.4 \times 10^2$

0.44 is smaller than 1 so this is not standard form. If you think of 0.44 as 4.4×10^{-1} you can change it to standard form.

EXERCISE 26D

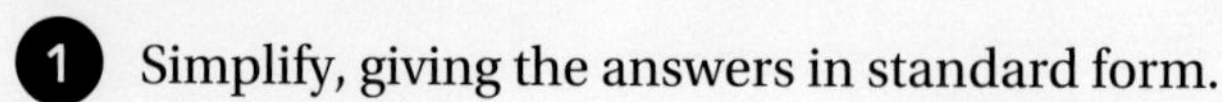

1 Simplify, giving the answers in standard form.

a $(2 \times 10^{13}) \times (4 \times 10^{17})$
b $(1.4 \times 10^{8}) \times (3 \times 10^{4})$
c $(1.5 \times 10^{13}) \times (1.5 \times 10^{13})$
d $(0.2 \times 10^{17}) \times (0.7 \times 10^{16})$
e $(9 \times 10^{17}) \div (3 \times 10^{16})$
f $(8 \times 10^{17}) \div (4 \times 10^{16})$
g $(1.5 \times 10^{8}) \div (5 \times 10^{4})$
h $(2.4 \times 10^{64}) \div (8 \times 10^{21})$

2 Simplify, giving the answers in standard form.

a $(2 \times 10^{-4}) \times (4 \times 10^{-16})$
b $(1.6 \times 10^{-8}) \times (4 \times 10^{-4})$
c $(1.5 \times 10^{-6}) \times (2.1 \times 10^{-3})$
d $(11 \times 10^{-5}) \times (3 \times 10^{2})$
e $(9 \times 10^{17}) \div (4.5 \times 10^{-16})$
f $(7 \times 10^{-21}) \div (1 \times 10^{16})$
g $(4.5 \times 10^{8}) \div (0.9 \times 10^{-4})$
h $(11 \times 10^{-5}) \times (3 \times 10^{2}) \div (2 \times 10^{-3})$

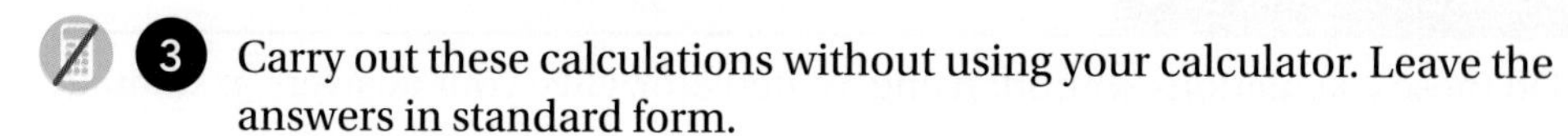

3 Carry out these calculations without using your calculator. Leave the answers in standard form.

a $(3 \times 10^{12}) \times (4 \times 10^{18})$
b $(1.5 \times 10^{6}) \times (3 \times 10^{5})$
c $(1.5 \times 10^{12})^3$
d $(1.2 \times 10^{-5}) \times (1.1 \times 10^{-6})$
e $(0.4 \times 10^{15}) \times (0.5 \times 10^{12})$
f $(8 \times 10^{17}) \div (3 \times 10^{12})$
g $(1.44 \times 10^{8}) \div (1.2 \times 10^{6})$
h $(8 \times 10^{-15}) \div (4 \times 10^{-12})$

4 The speed of light is approximately 3×10^8 metres per second. How far will the light travel in:

a 10 seconds?
b 20 seconds?
c 10^2 seconds?
d 2×10^3 seconds?

5 A human being blinks approximately 6.25×10^6 times per year.

a How often will you blink in 5 years? Give the answer in standard form and as an ordinary number.

b If there were 7.2×10^9 people on the planet, how many blinks would there be in a year?

Adding and subtracting in standard form

You already know from algebra that you can add or subtract like terms, and that you cannot add or subtract unlike terms. The same rules apply to powers of ten. You can only add or subtract powers of ten if the powers are identical. This means that you sometimes have to manipulate the expressions to get like terms that you can add or subtract.

For example, you can change 3×10^6 into $3 \times 10^4 \times 10^2$ because $10^4 \times 10^2 = 10^6$.

You can also manipulate the number by multiplying it or dividing it by ten to increase or decrease the powers. For example, you can change 1.6×10^3 into 16×10^2 or into 0.16×10^4 but you must remember to write your answer correctly in standard form.

Calculator tip

If you are using a calculator you don't need to manipulate the terms. You just need to enter them correctly.

WORKED EXAMPLE 5

Simplify and give the answers in standard form.

a $(2.5 \times 10^6) + (3.2 \times 10^8)$ **b** $6 \times 10^{-3} - 3 \times 10^{-4}$

a $(2.5 \times 10^6) + (3.2 \times 10^8)$	
$= 2.5 \times 10^6 + 3.2 \times 10^6 \times 10^2$	Rewrite the second term as its factors to get like powers.
$= 10^6(2.5 + 3.2 \times 10^2)$	Factorise by removing the common factor. $3.2 \times 10^2 = 320$.
$= 10^6(2.5 + 320)$	Add the numbers.
$= 10^6(322.5)$	$322.5 = 3.225 \times 10^2$.
$= 3.225 \times 10^8$	Rewrite the answer in standard form.
b $6 \times 10^{-3} - 3 \times 10^{-4}$	
$= 60 \times 10^{-4} - 3 \times 10^{-4}$	6×10^{-3} is the same as 60×10^{-4}.
$= 57 \times 10^{-4}$	This is not yet in standard form.
$= 5.7 \times 10^{-3}$	Rewrite the answer in standard form.

You can also convert the standard form expressions to ordinary numbers and add or subtract them. You can then write the answer back in standard form.

For example, $6 \times 10^{-3} - 3 \times 10^{-4}$

$$\begin{array}{r} 0.0060 \\ -\ 0.0003 \\ \hline 0.0057 \\ \hline \end{array}$$

$= 5.7 \times 10^{-3}$

If you do this, remember to write the numbers so the place values line up and make sure you convert your answer to standard form.

EXERCISE 26E

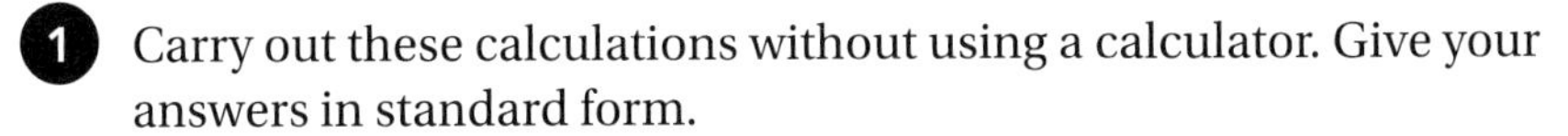

1 Carry out these calculations without using a calculator. Give your answers in standard form.

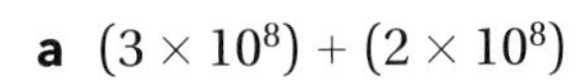

a $(3 \times 10^8) + (2 \times 10^8)$ **b** $(3 \times 10^{-3}) - (1.5 \times 10^{-3})$

c $(1.5 \times 10^5) + (3 \times 10^6)$ **d** $(6 \times 10^7) - (4 \times 10^6)$

e $(4 \times 10^{-4}) + (3 \times 10^{-3})$ **f** $(5 \times 10^{-3}) - (2.5 \times 10^{-2})$

Find answers at: cambridge.org/ukschools/gcsemaths-studentbookanswers

2 The Pacific Ocean has a surface area of approximately 1.65×10^8 km² and the Atlantic Ocean has a surface area of approximately 1.06×10^8 km².

a Which ocean has the greater surface area?

b How much larger is it?

c If the total surface area of the world's oceans is 361 000 000 km², work out the combined surface area of the other three oceans (the Indian, Southern and Arctic), giving the answer in standard form.

3 The Earth is approximately 9.3×10^7 miles from the Sun and 2.4×10^5 miles from the Moon. How much further is it from Earth to the Sun than from the Earth to the Moon?

4 A scientist studying viruses finds that Virus A has a diameter of 3×10^{-8} m and Virus B has a diameter of 3×10^{-7} m.

a Which has the smaller diameter?

b What is the difference in diameter between the two viruses?

c If the two viruses were placed alongside each other, what would the length of their combined diameters be in millimetres?

Solving problems using standard form

Standard form is very useful for solving problems involving very large or very small quantities. Problems involving modelling growth rates of bacteria in biology, population growth and decline in ecology, and future values in economics are often best solved using functions that rely on expressing values in standard form.

Computer engineering problems often involve very small values, as engineers try to fit more and more software onto smaller and smaller components.

Problem-solving framework

Yasin has worked out that he has a space 0.000 003 27 m wide, 0.000 000 2 m long and 0.000 116 m high into which he must fit a similarly shaped component with a volume of 8.034×10^{-17}. Will it fit?

Steps for approaching a problem-solving question	What you would do for this example
Step 1: Work out what you need to do.	Find the volume of the space. Compare it with the volume of the component to see which is bigger.
Step 2: Look for information that will help you.	The dimensions are for length, width and height, so shape must be a cuboid with a volume of $L \times W \times H$. The component volume is given in standard form, so you need the volume in standard form to compare it.

Continues on next page ...

Step 3: What maths can you do?	Convert the dimensions to standard form. 3.27×10^{-6} 2×10^{-7} 1.16×10^{-4} Multiply $= 3.27 \times 2 \times 1.16 \times 10^{-6} \times 10^{-7} \times 10^{-4}$ $= 7.5864 \times 10^{-17}$.
Step 4: Set out the solution clearly, making sure you have answered the original question.	Space available $7.5864 \times 10^{-17} < 8.034 \times 10^{-17}$ (Component volume) $\therefore$ The component will not fit.

EXERCISE 26F

1 Data storage in computers is measured in gigabytes. One gigabyte is 2^{30} bytes.

a Write 2^{30} in standard form correct to 3 significant figures.

b There are 1024 gigabytes in a terabyte. How many bytes is this? Give your answer in standard form correct to 3 significant figures.

2 The display on a mobile phone screen is made up of small square areas called pixels. The density of pixels per unit determines the image quality – the more pixels in an area, the clearer the picture. Most mobile phone manufacturers talk about pixels per inch (ppi). This is the number of pixels on one side of a square inch. You can see this in the images on the left.

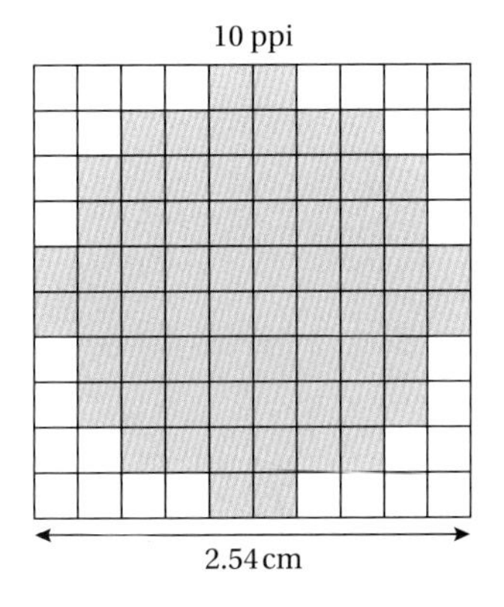

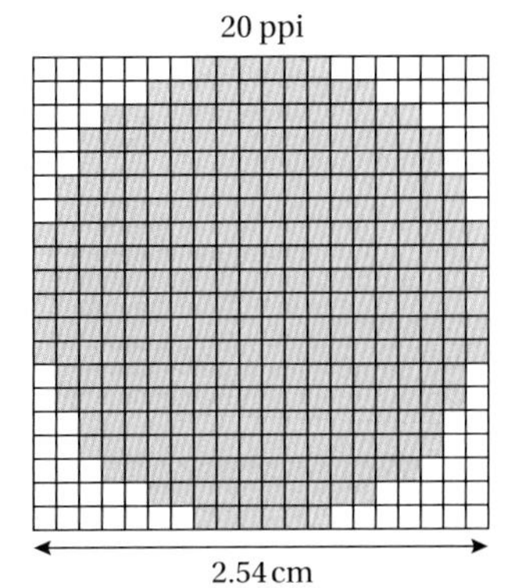

a How many pixels are there in a 2.54 cm × 2.54 cm area of the 10 ppi screen?

b Calculate the area (in cm^2) covered by one pixel at a density of 20 ppi. Give your answer in standard form correct to 3 significant places.

c At a density of 20 ppi, how many pixels would you have on a 4 inch by 3 inch rectangular screen?

d On a particular phone, one pixel is 1×10^{-3} cm wide and 4×10^{-5} cm long. What is the area of a pixel?

e What is the total area of 2.5×10^2 pixels of that size?

3 A nanometer (nm) is a very small unit of measure. One nanometer is 1.0×10^{-9} m. Express the following measurements in nanometers.

a 33 m **b** 21 mm

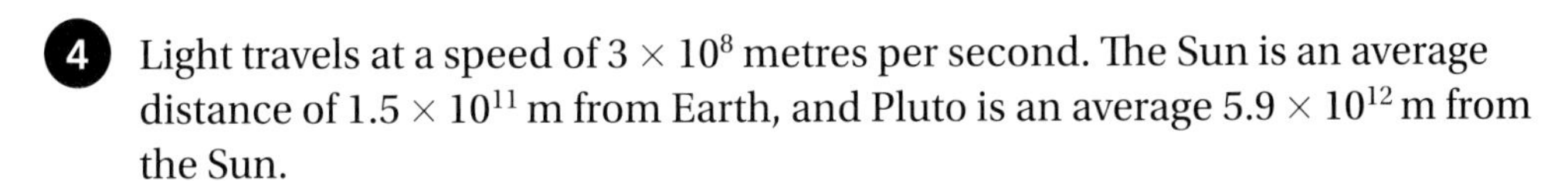

4 Light travels at a speed of 3×10^8 metres per second. The Sun is an average distance of 1.5×10^{11} m from Earth, and Pluto is an average 5.9×10^{12} m from the Sun.

a Work out how long it takes light from the Sun to reach Earth (in seconds). Give your answer in both ordinary numbers and standard form.

b How much longer does it take for the light to reach Pluto? Give your answer in both ordinary numbers and standard form correct to three significant places.

5 An immunologist cultures two sets of bacteria. Culture X contains 5.8×10^{11} bacterial cells, Culture Y contains 4.8×10^{9} bacterial cells. She combines the two cultures in one incubation flask.

a How many cells are there when she combines the two cultures?

b The bacteria numbers double every 8 hours. How many cells will there be after two days?

6 Use the dimensions given in Questions **3** and **4** of Exercise 26A to construct five mixed standard form problems. (Work out the solutions as well.) Exchange these with another student and try to solve each other's problems.

Checklist of learning and understanding

Standard form

- Very large and very small numbers can be written in standard form by expressing them as the product of a value greater than or equal to 1 and less than 10, and a power of 10.
- Any number can be expressed in the form $x \times 10^n$, where $1 \leqslant x < 10$ and n is an integer. Positive powers of ten indicate large numbers and negative powers of ten indicate small numbers.

Using a calculator

- The exponent function of the calculator allows you to enter calculations in standard form without entering the $\times$ 10 part of the calculation.
- When a number has too many digits to display, the calculator will give the answer in standard form.

Calculations in standard form

- You can multiply and divide numbers in standard form by applying the laws of indices; you can manipulate expressions to group like terms so that you can do the calculations with standard form without a calculator.
- When adding and subtracting numbers in standard form without a calculator, it can be easier to first convert them to an ordinary number to do the calculation and then convert your answer to standard form. If you are using a calculator, you can enter the calculation directly using the appropriate keys. If the powers of ten are the same, you can also use factorising.

For additional questions on the topics in this chapter, visit GCSE Mathematics Online.

Chapter review

1 Express the following numbers in standard form.

a 45 000 **b** 80 **c** 2 345 000

d 32 000 000 000 **e** 0.0065 **f** 0.009

2 Write the following as ordinary numbers.

a 2.5×10^3 **b** 3.9×10^4

c 4.265×10^5 **d** 1.045×10^{-5}

3 Use a calculator and give the answers in standard form.

a $5 \times 10^4 + 9 \times 10^6$ **b** $3.27 \times 10^{-3} \times 2.4 \times 10^2$

c $5(8.1 \times 10^9 - 2 \times 10^7)$ **d** $(3.2 \times 10^{-1}) - (2.33 \times 10^{-6})$ (to 3 sf)

e $\frac{4.22 \times 10^7 \times 3.25 \times 10^6}{4 \times 10^5}$ **f** $2.13 \times 10^6 \div (5.67 \times 10^{-5})$

4 Simplify the following without using a calculator and give the answers in standard form.

a $(1.44 \times 10^7) + (4.3 \times 10^7)$ **b** $(4.9 \times 10^5) \times (3.6 \times 10^9)$

c $(3 \times 10^4) + (4 \times 10^3)$ **d** $(4 \times 10^6) \div (3 \times 10^5)$

5 The distance between interconnecting lines on a silicon chip used in computer processing is 4×10^{-8} m. What is:

a twice this distance? **b** a quarter of this distance?

c ten times this distance?

6 The Sun has a mass of approximately 1.998×10^{27} tonnes. The planet Mercury has a mass of approximately 3.302×10^{20} tonnes.

a Which has the greater mass?

b How many times heavier is the greater than the smaller mass?

7 The table shows the areas of some countries.

Country	Area (square kilometres)
Brazil	
China	9.33×10^6
India	2.97×10^6
Mongolia	1.55×10^6
Tunisia	1.55×10^5
Turkey	7.70×10^5

a The area of Brazil is 8 459 417 square kilometres.
Complete the table to show the area of Brazil.
Give the area in standard form, correct to three significant figures.
(1 mark)

b Complete the following sentences.
The area of ________________ is about twice the area of Turkey.
The area of China is about three times the area of ______________.
(2 marks)

c The population of India in 2011 was approximately 1.19×10^9.
Estimate the population density of India in people per square kilometre. *(2 marks)*

© OCR 2013

8 The diameter of the Earth (at the Equator) is approximately 1.27×10^4 km.

a What is the approximate radius of the Earth at the Equator?
Give this answer in standard form.

b Given that $C = 2\pi r$, calculate the approximate distance around the Earth at the Equator in kilometres.

c If the volume of a sphere is $V = \frac{4}{3}\pi r^3$, where r is the radius, calculate the approximate volume of Earth correct to two decimal places.
(The Earth is not a perfect sphere, but assume that it is for your calculation.)

Find answers at: cambridge.org/ukschools/gcsemaths-studentbookanswers

27 Surds

In this chapter you will learn how to …

- calculate exactly with surds.
- simplify expressions containing surds.
- manipulate surds.

For more resources relating to this chapter, visit GCSE Mathematics Online.

Using mathematics: real-life applications

Surds are only really used when you are doing mathematical calculations that require exact answers. For all practical purposes, surds are approximated. You cannot tell a builder to cut a length of steel that is $\sqrt{2}$ metres long, because $\sqrt{2}$ is an irrational number; so you would be more likely to specify an approximate length of 1.41 metres.

"The widths and lengths of A-series rectangular paper were developed using the ratio $1:\sqrt{2}$ to mathematically construct a rectangle of area 1 m^2 (A0 size). In real life, paper is cut to approximate millimetre sizes, so A0 is 841 mm × 1189 mm rather than $841(\sqrt{2})$ which is 1189.353606 … mm."

Before you start …

Ch 2	You'll need to be able to write numbers as a product of their factors.	**1** Match the equivalent pairs. **a** $3^2 \times 5$ **b** $3 \times 2 \times 5$ **c** $2^3 \times 3$ **i** 24 **ii** 30 **iii** 45
KS3	You should be very familiar with Pythagoras' theorem.	**2** For each triangle, select the correct statement of Pythagoras' theorem. **a** 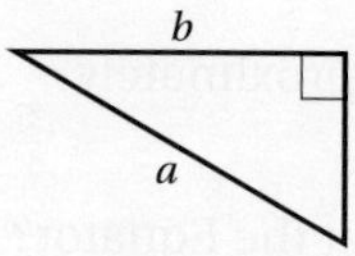 $a^2 = b^2 + c^2$ $b^2 = a^2 + c^2$ $c^2 = a^2 + b^2$ $a^2 + b^2 = c^2$ $a = b^2 + c^2$ **b** 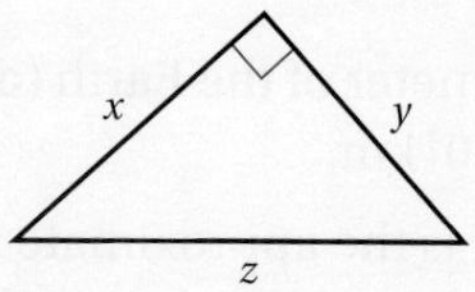 $z^2 = y^2 - x^2$ $x^2 + z^2 = y^2$ $z^2 + y^2 = x^2$ $x^2 - y^2 = z^2$ $x^2 + y^2 = z^2$
Ch 2, 25	You need know the basic laws of indices and how they apply to squares and roots.	**3** Decide whether each statement is true or false. **a** $\frac{18x^3}{6x^3} = 3$ **b** $\frac{(2y^3)^4}{(4y^6)^2} = y$
Ch 3, 7	You should be able to simplify expressions by collecting like terms.	**4** Find the error in each of these answers. **a** $5x + 2y - 3x + y = 2x + 2y$ **b** $5x^2 + 2x - 3x^2 = 4x^3$ **c** $6x - 2y + 5 + y - 2x - 7 = 4x + y - 2$

Assess your starting point using the Launchpad

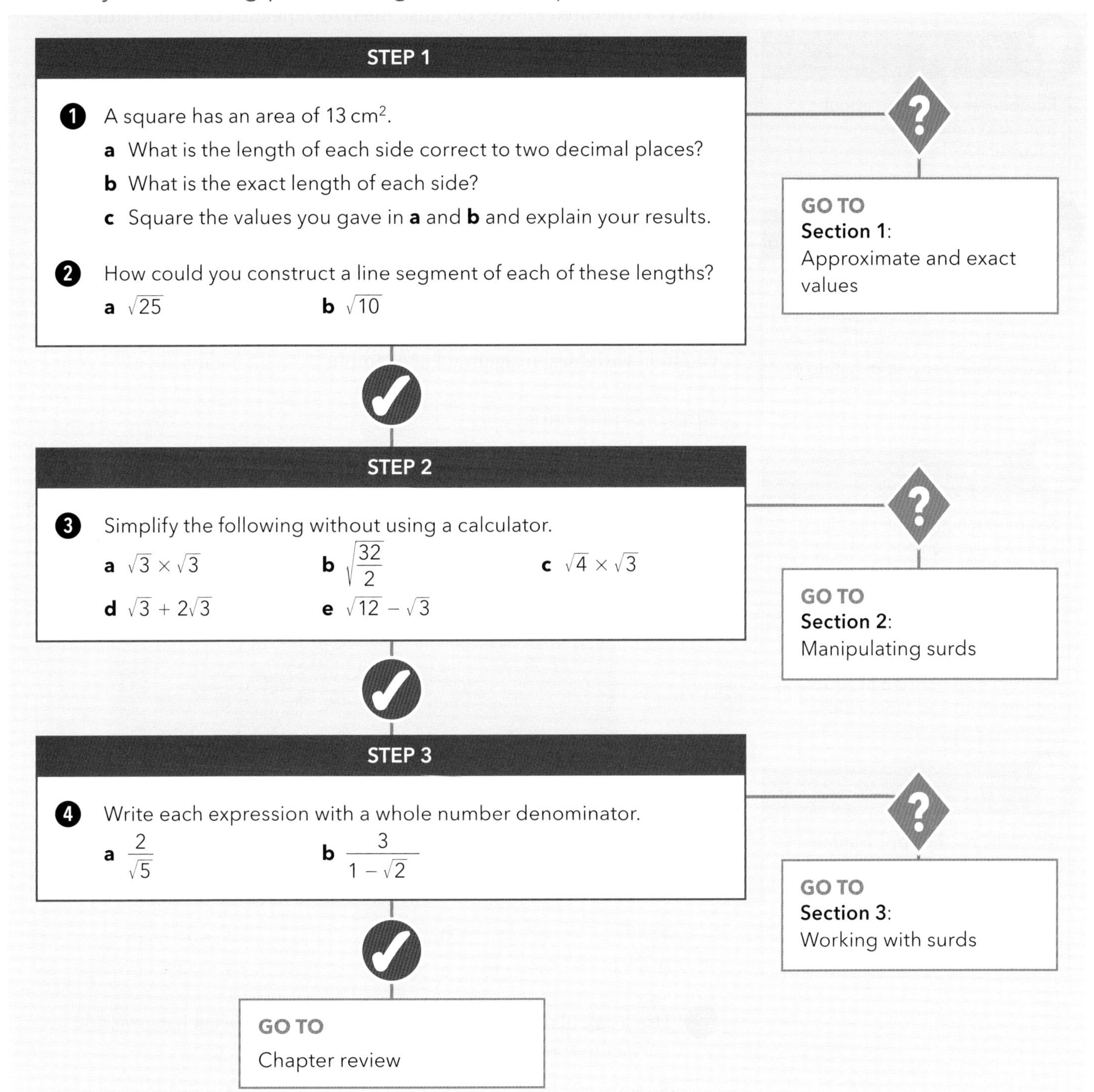

STEP 1
1 A square has an area of 13 cm^2.
a What is the length of each side correct to two decimal places?
b What is the exact length of each side?
c Square the values you gave in **a** and **b** and explain your results.
2 How could you construct a line segment of each of these lengths?
a $\sqrt{25}$ b $\sqrt{10}$
GO TO
Section 1:
Approximate and exact values
STEP 2
3 Simplify the following without using a calculator.
a $\sqrt{3} \times \sqrt{3}$ b $\sqrt{\frac{32}{2}}$ c $\sqrt{4} \times \sqrt{3}$
d $\sqrt{3} + 2\sqrt{3}$ e $\sqrt{12} - \sqrt{3}$
GO TO
Section 2:
Manipulating surds
STEP 3
4 Write each expression with a whole number denominator.
a $\frac{2}{\sqrt{5}}$ b $\frac{3}{1 - \sqrt{2}}$
GO TO
Section 3:
Working with surds
GO TO
Chapter review

Section 1: Approximate and exact values

When you work with Pythagoras' theorem you often have to find the square root of a number that is not itself a perfect square.

For example, you may need to work with **irrational numbers** such as $\sqrt{2}$, $\sqrt{5}$ and $\sqrt{11}$. These are all **surds**.

The value $\sqrt{2}$ is the number which produces 2 when it is squared ($\sqrt{2} \times \sqrt{2} = 2$).

Key vocabulary

irrational number: a number that cannot be written in the form of $\frac{a}{b}$ or as a terminating or repeating decimal.

surd: if $\sqrt[n]{a}$ is an irrational number, then $\sqrt[n]{a}$ is called a surd.

Find answers at: cambridge.org/ukschools/gcsemaths-studentbookanswers

If you try to find $\sqrt{2}$ on your calculator you will get a result of 1.414 213 562.

This is a truncated answer because the non-repeating decimal values continue to infinity, so it only represents an approximate value for $\sqrt{2}$.

If you enter 1.414 213 562 into your calculator and square it you will not get 2 as an answer; you will get the truncated value 1.999 999 999.

A surd can only be expressed as an exact value using the root sign ($\sqrt{\ }$).

The expressions $\sqrt{12}$, $\sqrt{3}$ and $\sqrt[3]{7}$ are all surds. Expressions such as $\sqrt{81}$, $\sqrt[3]{125}$ and $\sqrt[4]{1296}$ are **not** surds because they have **rational** solutions: $\sqrt{81} = 9$, $\sqrt[3]{125} = 5$ and $\sqrt[4]{1296} = 6$.

The expression $\sqrt[3]{7}$ means the cube root of 7. This is not the same as $3\sqrt{7}$, which means $3 \times \sqrt{7}$.

When you work with surds you can usually treat them in the same way that you treat variables in algebra. For example, $3\sqrt{7} + 2$ means $3 \times \sqrt{7} + 2$.

Tip

Remember you learnt about truncation and rounding in Chapter 17.

Key vocabulary

rational number: a rational number is a number that can be expressed in the form of $\frac{a}{b}$ (or as its equivalent as a terminating or repeating decimal).

Approximate values

Approximate values of surds are useful in many situations. For example, if you want to know the length of a diagonal path across a rectangular 12 m by 7 m field you can use Pythagoras' theorem and rounding to work out that the path is approximately 13.89 m in length. You will work with approximate values often in your maths course.

Tip

Keeping a number in surd form until you get to your final answer means that you don't carry through any approximation or rounding errors that could affect the final answer. This is important for the very small tolerances permitted in structural engineering and other fields.

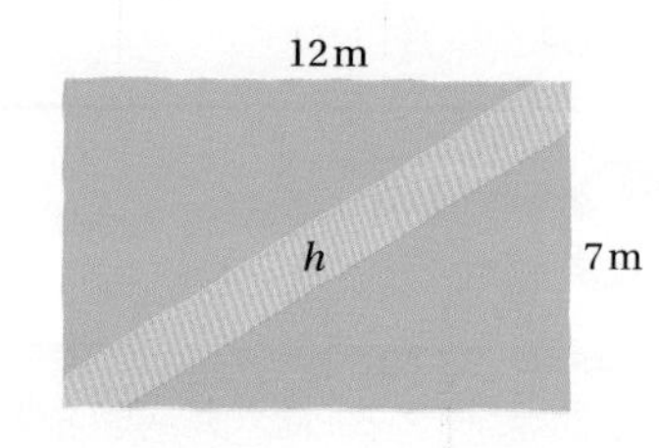

Tip

You can find the approximate value of any surd by using a calculator and your knowledge of rounding decimals. You will learn to do this without using a calculator in Section 3.

EXERCISE 27A

1 Use a calculator to find the approximate value of each of the following surds. Give your answers correct to three decimal places.

a $\sqrt{7}$ **b** $\sqrt{12}$ **c** $\sqrt{51}$

d $\sqrt{75}$ **e** $^{-}\sqrt{3}$ **f** $^{-}\sqrt{47}$

2 Use a calculator to find the approximate value of each expression correct to three decimal places.

a $2\sqrt{2}$ **b** $3\sqrt{5}$ **c** $^{-}3\sqrt{12}$

d $10\sqrt{2}$ **e** $4(2\sqrt{3})$ **f** $^{-}3(2\sqrt{18})$

3 Find the approximate value of each expression, giving your answers correct to 3 decimal places.

a $\sqrt{2} + \sqrt{3}$ **b** $\sqrt{8} - \sqrt{2}$ **c** $\sqrt{2 + 3}$

d $\sqrt{8 - 2}$ **e** $2\sqrt{3} + 3\sqrt{5}$ **f** $^{-}2\sqrt{3} + 3\sqrt{5}$

Tip

You were reminded in Chapter 1 that the root sign functions like a set of brackets in an expression, so you have to apply the correct order of operations and solve the root first.

Exact values

Values are left as surds when you need to express an answer precisely.

Exact values of surds are used in many technical calculations, such as determining resultant forces and velocities in physics, margins of error in chemistry and other sciences, and peak-to-peak voltages in electronics.

WORKED EXAMPLE 1

Use Pythagoras' theorem to calculate the following exactly.

a length AC **b** height EG

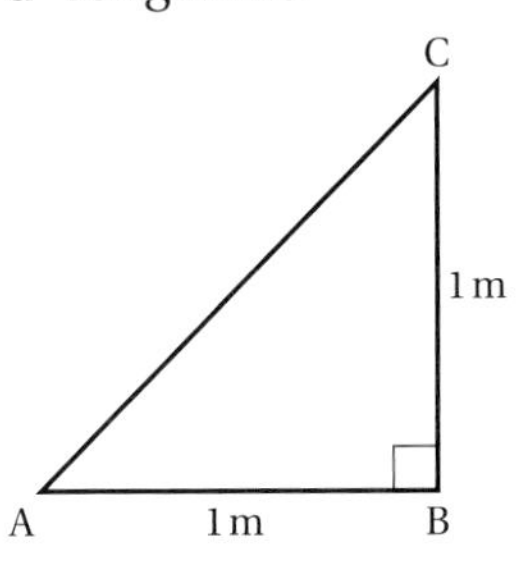

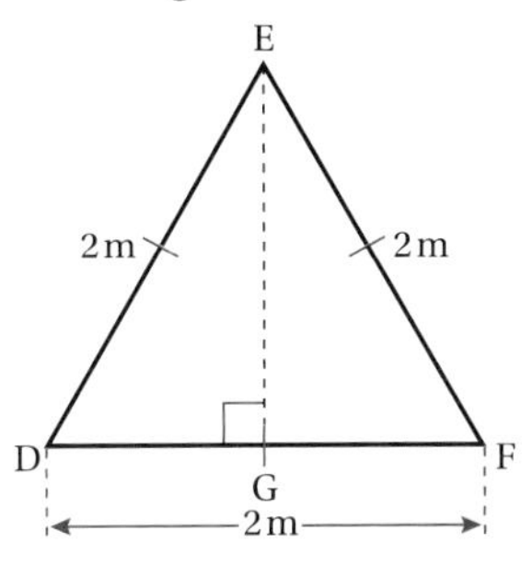

a $(AC)^2 = 1^2 + 1^2$

$(AC)^2 = 1 + 1 = 2$

$AC = \sqrt{2}$ m

You need to know Pythagoras' theorem from memory. See Chapter 32 if you need a reminder. Don't forget units!

b $(DE)^2 = (EG)^2 + 1^2$

$(EG)^2 = 4 - 1 = 3$

$EG = \sqrt{3}$m

DEF is an equilateral triangle and the line EG divides the line DF in half. So DG = 1.

The answers here are left in surd form because the question asks for an **exact** calculation.

EXERCISE 27B

1 Find the exact length of a side of a square with area:

a 14 cm^2. **b** 20 m^2. **c** 17 cm^2.

2 What is the exact circumference of a coin with a radius of $\sqrt{3}$ cm?

3 What is the exact value of $\left(\frac{9}{\sqrt{3}}\right)^2$?

4 Two identical square mosaic tiles have a combined area of 14 cm^2. What is the exact length of the sides of each square?

5 The area of a square plot of land is 50 m^2.

a What is the exact length of each side of the plot?

b What is the exact diagonal distance across the plot?

Tip

When a question asks for an answer in exact form it means you should leave any surds in root form. Do not use approximate values in either the calculations or the answer.

Find answers at: cambridge.org/ukschools/gcsemaths-studentbookanswers

6 Nico is a jeweller who makes square plates of platinum to use in various pieces of jewellery. Each square has a diagonal of 4 cm.

a What is the exact length of each side of such a square?

b Use the exact length to calculate the exact area of a square.

c Give the length of the sides correct to:

i 2 dp. **ii** 3 dp. **iii** 4 dp.

d Calculate the area of each square using the approximate values.

e Nico works out the cost of the metal by finding the area of 100 squares and then multiplying this amount by £1245. Use the exact area as well as the three approximate areas you found to show how the four different values for the area might impact on the price of the metal.

f Which value is most beneficial to Nico if he is calculating a selling price? Why?

Tip

It is common to talk about all expressions containing surds as 'surds', even though technically the expression $2\sqrt{5}$ is the product of a whole number and a surd, and $\sqrt{3} + \sqrt{5}$ is the sum of two surds.

Tip

Remember that taking a square root and squaring are **inverse operations**.

Section 2: Manipulating surds

When you do calculations with exact values using surds you can treat them a bit like variables in algebra – simplifying them and rearranging them so you can add and subtract like surds and multiply and divide easily. Generally you won't need a calculator to work with exact values of surds.

You already know some general rules for working with square roots. These rules apply to surds as well and they are useful for simplifying and manipulating surds.

If x and y represent *positive* integers then by definition:

Surd	**Example**
$(\sqrt{x})^2 = x$	$(\sqrt{25})^2 = \sqrt{25} \times \sqrt{25} = 5 \times 5 = 25.$
	$(\sqrt{3})^2 = \sqrt{3} \times \sqrt{3} = 3.$
$\sqrt{x^2} = x$	$\sqrt{5^2} = \sqrt{25} = 5.$
$\sqrt{x} \times \sqrt{y} = \sqrt{xy}$	$\sqrt{9} \times \sqrt{4} = 3 \times 2 = 6$ and $\sqrt{9 \times 4} = \sqrt{36} = 6.$
$\sqrt{x} \div \sqrt{y} = \sqrt{\frac{x}{y}}$	$\sqrt{36} \div \sqrt{9} = 6 \div 3 = 2$ and $\sqrt{\frac{36}{9}} = \sqrt{4} = 2.$

WORKED EXAMPLE 2

Is each statement true or false?

a $\sqrt{35} = \sqrt{5} \times \sqrt{7}$ **b** $(\sqrt{49})^2 = 49$ **c** $\sqrt{30} \div \sqrt{6} = 5$ **d** $3\sqrt{3} = \sqrt{9} \times \sqrt{3}$

a True — Using the rule $\sqrt{x} \times \sqrt{y} = \sqrt{xy}$

b True — Using the rule $(\sqrt{x})^2 = x$ $(\sqrt{49})^2 = (7)^2 = 49$

c False — Using the rule $\sqrt{x} \div \sqrt{y} = \sqrt{\frac{x}{y}}$ $\sqrt{30} \div \sqrt{6} = \sqrt{\frac{30}{6}} = \sqrt{5}$

d True — $3\sqrt{3} = 3 \times \sqrt{3}$ and $3 = \sqrt{9}$

Simplifying surds

A surd is in its simplest form when the number under the root sign is as small as possible. This means that it has no factors greater than 1 that are perfect squares. For example, $\sqrt{18}$ can be written as $\sqrt{9} \times \sqrt{2}$, and $\sqrt{9}$ can be written as 3, so you get the expression $3\sqrt{2}$. This is considered to be a simpler form of the surd than $\sqrt{18}$.

To simplify a surd, you can use the fact that $\sqrt{xy} = \sqrt{x} \times \sqrt{y}$ and write it as the product of two smaller square roots, one of which is a square number. You can see how to do this in the examples below.

WORKED EXAMPLE 3

Simplify the following surds.

a $\sqrt{27}$ **b** $3\sqrt{32}$ **c** $^-2\sqrt{320}$ **d** $2\sqrt{12} \times 4\sqrt{3}$

a Factors of 27: 9×3

$\sqrt{27} = \sqrt{9} \times \sqrt{3}$ — Write the surd as a product of two smaller roots.

$= 3 \times \sqrt{3}$ — Take the root of the perfect square.

$= 3\sqrt{3}$ — Write the answer using algebraic conventions.

b Factors of 32: 16×2

$3\sqrt{32} = 3 \times \sqrt{16} \times \sqrt{2}$ — Write the surd as a product of two smaller roots.

$= 3 \times 4 \times \sqrt{2}$ — Take the root of the perfect square.

$= 12 \times \sqrt{2}$ — Multiply the integers.

$= 12\sqrt{2}$ — Write the answer using algebraic conventions.

c Factors of 320: 16×20

Factors of 20: 5×4

$^-2\sqrt{320} = {}^-2 \times \sqrt{16} \times \sqrt{20}$ — Write the surd as a product of two smaller roots.

$= {}^-2 \times 4 \times \sqrt{4} \times \sqrt{5}$ — Take the root of 16 and factorise 20.

$= {}^-2 \times 4 \times 2 \times \sqrt{5}$ — Take the root of 4.

$= {}^-16\sqrt{5}$ — Multiply the integers and write the answer correctly.

d Factors of 12: 3×4 — Only 12 can be factorised: 4×3

$2\sqrt{12} \times 4\sqrt{3} = 2 \times \sqrt{4} \times \sqrt{3} \times 4\sqrt{3}$ — Write $\sqrt{12}$ as product of two smaller roots.

$= 2 \times 2 \times \sqrt{3} \times 4\sqrt{3}$ — Simplify.

$= 16 \times \sqrt{3} \times \sqrt{3}$ — Multiply integers.

$= 16 \times 3$ — Square the surds.

$= 48$

In the examples above, the surds can be simplified because the factors used included square numbers.

If you choose factors that are not square numbers, you have to do far more work to get to a simplified answer.

For example, $\sqrt{320}$ could be factorised as $\sqrt{8} \times \sqrt{40}$ instead of $\sqrt{16} \times \sqrt{20}$. In that case, you cannot find the roots as both $\sqrt{8}$ and $\sqrt{40}$ are themselves surds and you have to factorise them further before you can simplify the surd.

Tip

Check whether the number in the root sign can be divided by a perfect square (4, 16, 25, 36, 49, 64, 81, 100, ...) and if so, use the largest possible square number as one of the factors to make simplification quicker.

Find answers at: cambridge.org/ukschools/gcsemaths-studentbookanswers

EXERCISE 27C

1 What is the simplest form of each surd? Choose the correct answer.

a $\sqrt{28}$ A $4\sqrt{7}$ B $\sqrt{4} \times \sqrt{7}$ C $2\sqrt{7}$ D $2\sqrt{14}$ E $28\sqrt{1}$

b $\sqrt{12}$ A $12\sqrt{1}$ B $2\sqrt{3}$ C $3\sqrt{4}$ D $4\sqrt{3}$ E $\sqrt{3} \times \sqrt{4}$

c $\sqrt{72}$ A $2\sqrt{36}$ B $6\sqrt{2}$ C $\sqrt{36} \times \sqrt{2}$ D $\sqrt{8} \times \sqrt{9}$ E $3\sqrt{8}$

d $5\sqrt{12}$ A $5\sqrt{4} \times \sqrt{3}$ B $15\sqrt{4}$ C $10\sqrt{3}$ D $3\sqrt{10}$ E $7\sqrt{3}$

e $\sqrt{320}$ A $\sqrt{16} \times \sqrt{20}$ B $\sqrt{16 \times 4 \times 5}$ C $4\sqrt{20}$ D $32\sqrt{10}$ E $8\sqrt{5}$

2 Simplify.

a $\sqrt{8}$ **b** $\sqrt{24}$ **c** $\sqrt{28}$ **d** $\sqrt{45}$

e $\sqrt{54}$ **f** $\sqrt{68}$ **g** $\sqrt{60}$ **h** $\sqrt{126}$

i $\sqrt{90}$ **j** $\sqrt{200}$ **k** $\sqrt{117}$ **l** $\sqrt{243}$

3 Explain how you can tell at a glance that the following surds are already in their simplest form:

$\sqrt{2}$ $\sqrt{11}$ $\sqrt{13}$ $\sqrt{53}$ $\sqrt{83}$ $\sqrt{101}$

4 Write each surd in its simplest form.

a $3\sqrt{8}$ **b** $^-4\sqrt{24}$ **c** $5\sqrt{20}$

d $^-5\sqrt{60}$ **e** $3\sqrt{56}$ **f** $^-2\sqrt{128}$

g $^-4\sqrt{45}$ **h** $^-3\sqrt{68}$ **i** $7\sqrt{108}$

5 **a** Complete the following.

i $2\sqrt{7} = 2 \times \sqrt{7}$
$= \sqrt{\square} \times \sqrt{7}$
$= \sqrt{\square \times 7}$
$= \sqrt{\square}$

ii $^-3\sqrt{6} = {}^-3 \times \sqrt{6}$
$= {}^-\sqrt{\square} \times \sqrt{6}$
$= {}^-\sqrt{\square \times 6}$
$= {}^-\sqrt{\square}$

b Why do you leave the minus sign outside the root in part **ii**?

Tip

An entire surd has no rational factors or terms, only a number that is not a perfect square under the root sign, so for example $\sqrt{18}$ is an entire surd, but $3\sqrt{2}$ is not. Apply the inverse procedure to get back to an entire surd when you have a whole number and a surd in the expression.

6 Rewrite each of the following in the form $\sqrt{n}$.

a $3\sqrt{2}$ **b** $4\sqrt{3}$ **c** $3\sqrt{6}$

d $4\sqrt{11}$ **e** $^-2\sqrt{7}$ **f** $^-3\sqrt{3}$

g $^-4\sqrt{17}$ **h** $^-2\sqrt{11}$ **i** $12\sqrt{3}$

7 **a** What would you need to do to arrange these sets of surds, in order from smallest to largest (without using your calculator)?

i $4\sqrt{2}$ $2\sqrt{3}$ $3\sqrt{3}$ **ii** $8\sqrt{3}$ $6\sqrt{7}$ $5\sqrt{7}$

iii $3\sqrt{7}$ $2\sqrt{10}$ $4\sqrt{3}$ **iv** $5\sqrt{6}$ $8\sqrt{2}$ $6\sqrt{3}$

b Apply your method and arrange the surds in order.

c Compare your answers and your method with a partner. Could you work more efficiently? If so, how?

Adding and subtracting surds

You can use algebra to simplify expressions containing surds. For example, $3\sqrt{5} + 4\sqrt{5}$ can be added because $\sqrt{5}$ is common to both terms. Using factorising you can take out a common factor of $\sqrt{5}$ to get

$$\sqrt{5}\,(3 + 4) = \sqrt{5}\,(7) = 7\sqrt{5}$$

An expression such as $2\sqrt{5} + 4\sqrt{3}$ cannot be added because it has no surds that are common factors. (You can also think about $\sqrt{5}$ and $\sqrt{3}$ as unlike surds.) As with like terms in algebra, you can only add or subtract terms which have the same surds in them.

Before you add or subtract, you might need to express surds in their simplest terms.

WORKED EXAMPLE 4

Calculate the exact value of $5\sqrt{2} - 2\sqrt{8}$.

$= 5\sqrt{2} - 2 \times \sqrt{4} \times \sqrt{2}$ — Factorise $\sqrt{8}$

$= 5\sqrt{2} - 2 \times 2 \times \sqrt{2}$ — Take the root of 4.

$= 5\sqrt{2} - 4\sqrt{2}$ — Both expressions include the same surd, so you can factorise for $\sqrt{2}$

$= \sqrt{2}$

WORK IT OUT 27.1

A teacher gave her class some multiple choice questions for homework. One student's answers are given below. Three of the answers are correct and two are incorrect. What errors do you think this student made to get the incorrect answers? What are the correct answers for these questions?

Circle the correct answers.

1. $8\sqrt{4} - \sqrt{4} =$

   A $8\sqrt{4}$ (B $7\sqrt{4}$) ✓   C 8   D 0   E 7

2. $7\sqrt{3} + 3\sqrt{2} + 5\sqrt{3} =$

   A $15\sqrt{5}$ B $15\sqrt{8}$ C $12\sqrt{3} + 3\sqrt{2}$ (D $12\sqrt{6} + 3\sqrt{2}$) ✗   E $15\sqrt{3}$

3. $5\sqrt{2} - 2\sqrt{8} =$

   A 1   B 0   C $^{-}3\sqrt{2}$ (D $\sqrt{2}$) ✓   E $3\sqrt{^{-}6}$

4. $4\sqrt{5} - \sqrt{2} + 6\sqrt{5} - 3\sqrt{2} =$

   A $10\sqrt{5} - 3$ B $6\sqrt{3}$ (C $10\sqrt{5} - 4\sqrt{2}$) ✓   D $10\sqrt{10} - 4\sqrt{2}$ E $2\sqrt{5} - 2\sqrt{2}$

5. $\sqrt{27} + 2\sqrt{5} + \sqrt{20} - 2\sqrt{3} =$

   (A $2\sqrt{8} + \sqrt{47}$) ✗   B $\sqrt{3} + 4\sqrt{5}$ C $4\sqrt{3} + \sqrt{5}$ D $5\sqrt{3} - 4\sqrt{5}$ E $-\sqrt{3} + 4\sqrt{5}$

Find answers at: cambridge.org/ukschools/gcsemaths-studentbookanswers

EXERCISE 27D

1 Provide examples using square numbers to show that when x and y are positive integers:

a $\sqrt{x} + \sqrt{y} \neq \sqrt{x + y}$ b $\sqrt{x} - \sqrt{y} \neq \sqrt{x - y}$

2 Simplify by adding or subtracting.

a $2\sqrt{4} + 3\sqrt{7} + 4\sqrt{4}$ b $\sqrt{2} + 3\sqrt{2} + 2\sqrt{5}$

c $2\sqrt{5} + 3\sqrt{3} + 2\sqrt{5} + 5\sqrt{3}$ d $9\sqrt{2} + 2\sqrt{3} - 7\sqrt{2} + 3\sqrt{3}$

e $4\sqrt{5} - 2\sqrt{2} + 2\sqrt{5} + 5\sqrt{2}$ f $4\sqrt{2} + 4\sqrt{3} - 3\sqrt{2} - 6\sqrt{3}$

3 Simplify.

a $\sqrt{2} + \sqrt{8}$ b $\sqrt{28} - \sqrt{7}$

c $3\sqrt{6} + \sqrt{24}$ d $3\sqrt{5} - \sqrt{20}$

e $5\sqrt{63} - 7\sqrt{28}$ f $4\sqrt{45} - 2\sqrt{20}$

4 Simplify.

a $\sqrt{75} + \sqrt{27} - 2\sqrt{3}$

b $^{-}4\sqrt{11} + 8\sqrt{10} - 2\sqrt{11} - 2\sqrt{10}$

c $2\sqrt{75} - \sqrt{45} + 2\sqrt{20}$

d $2\sqrt{12} - \sqrt{20} - \sqrt{27} + 2\sqrt{45}$

e $3\sqrt{54} + 4\sqrt{24} - 2\sqrt{96}$

f $6\sqrt{50} - 2\sqrt{24} + 4\sqrt{32} + \sqrt{54}$

5 A rectangular component has side dimensions $(3 - \sqrt{3})$ cm by $(3 + \sqrt{48})$ cm.

Calculate the exact perimeter of the component.

Multiplying surds

You've already applied the rules $\sqrt{x} \times \sqrt{y} = \sqrt{xy}$ and $\sqrt{x} \times \sqrt{x} = x$ to simplify surds. Some expressions will contain both surds and brackets. To simplify these, you apply the basic rules you know from algebra.

WORKED EXAMPLE 5

Simplify.

a $\sqrt{5} \times \sqrt{7}$ b $2\sqrt{7} \times 3\sqrt{3}$ c $\sqrt{2}(4 - 3\sqrt{2})$ d $(3 + \sqrt{2})(3 - \sqrt{2})$ e $(\sqrt{2} + \sqrt{3})^2$

a $\sqrt{5} \times \sqrt{7} = \sqrt{35}$

Use the rule $\sqrt{x} \times \sqrt{y} = \sqrt{xy}$.

b $2\sqrt{7} \times 3\sqrt{3} = 2 \times 3 \times \sqrt{7} \times \sqrt{3}$
$= 6\sqrt{21}$

Multiplication can be done in any order, so group and multiply like terms.

Continues on next page ...

c $\sqrt{2}(4-3\sqrt{2})=\sqrt{2}\times 4-\sqrt{2}\times 3\sqrt{2}$ — Multiply both terms in the brackets by $\sqrt{2}$ paying attention to the negative sign.

$=4\sqrt{2}-3\times 2$ — There is now no surd in the second term because $\sqrt{2}\times\sqrt{2}=2$

$=4\sqrt{2}-6$

d $(3+\sqrt{2})(3-\sqrt{2})$ — Use the rules for binomial products, see Chapter 7 if you need a reminder.

$=3\times 3-3\times\sqrt{2}+\sqrt{2}\times 3-\sqrt{2}\times\sqrt{2}$

$=9-3\sqrt{2}+3\sqrt{2}-2$ — The two surd terms cancel out.

$=9-2$

$=7$

e $(\sqrt{2}+\sqrt{3})^2=(\sqrt{2}+\sqrt{3})(\sqrt{2}+\sqrt{3})$ — Expand the brackets and then simplify the answer.

$=\sqrt{2}\times\sqrt{2}+\sqrt{2}\times\sqrt{3}+\sqrt{3}\times\sqrt{2}+\sqrt{3}\times\sqrt{3}$

$=2+\sqrt{6}+\sqrt{6}+3$ — Collect like terms.

$=5+2\sqrt{6}$

Dividing surds

Surds can be divided using the rule $\sqrt{x}\div\sqrt{y}=\sqrt{\frac{x}{y}}$. You may need to simplify surds so that you can find common factors and cancel them.

WORKED EXAMPLE 6

Simplify.

a $\sqrt{104}\div\sqrt{13}$ **b** $\frac{3\sqrt{21}}{\sqrt{3}}$ **c** $\frac{6+2\sqrt{20}}{2}$

a $\sqrt{104}\div\sqrt{13}=\sqrt{\frac{104}{13}}$ — Apply the rule $\sqrt{x}\div\sqrt{y}=\sqrt{\frac{x}{y}}$

$=\sqrt{8}$

$=2\sqrt{2}$ — Simplify $\sqrt{\frac{104}{13}}$. Give your answer in simplest form.

b $\frac{3\sqrt{21}}{\sqrt{3}}=3\sqrt{\frac{21}{3}}$ — Apply the rule $\sqrt{x}\div\sqrt{y}=\sqrt{\frac{x}{y}}$

$=3\sqrt{7}$ — Simplify $\sqrt{\frac{21}{3}}$. Give your answer in simplest form.

c $\frac{6+2\sqrt{20}}{2}=\frac{6}{2}+\frac{2\sqrt{20}}{2}$ — Remember the rules for simplifying fractions.

$=3+\sqrt{20}$ — Divide both terms by 2.

$=3+2\sqrt{5}$ — Write the surd in simplest form.

Rationalising the denominator

There are many calculations in algebra (and in calculus) which are made easier if a fraction containing a surd has the surd in the numerator and an integer in the denominator.

Removing surds from the denominator of a fraction is called rationalising the denominator, and it is done using the principles you learnt when you made equivalent fractions.

$\frac{x}{\sqrt{y}}$ is equivalent to $\frac{x}{\sqrt{y}} \times \frac{\sqrt{y}}{\sqrt{y}}$ (because $\frac{\sqrt{y}}{\sqrt{y}} = 1$)

Multiplying will give you $\frac{x\sqrt{y}}{y}$ (because $\sqrt{y} \times \sqrt{y} = y$) and you have removed the surd from the denominator.

WORKED EXAMPLE 7

Rationalise the denominators.

a $\frac{5}{\sqrt{7}}$ **b** $\frac{2\sqrt{5}}{\sqrt{6}}$

a $\frac{5}{\sqrt{7}} = \frac{5}{\sqrt{7}} \times \frac{\sqrt{7}}{\sqrt{7}} = \frac{5\sqrt{7}}{7}$ — Multiply by $\frac{\sqrt{7}}{\sqrt{7}}$.

b $\frac{2\sqrt{5}}{\sqrt{6}} = \frac{2\sqrt{5}}{\sqrt{6}} \times \frac{\sqrt{6}}{\sqrt{6}}$ — Multiply by $\frac{\sqrt{6}}{\sqrt{6}}$.

$= \frac{2\sqrt{30}}{6}$ — Remember the rule $\sqrt{x} \times \sqrt{y} = \sqrt{xy}$.

$= \frac{\sqrt{30}}{3}$ — Simplify/cancel the fraction $\frac{2}{6} = \frac{1}{3}$.

Key vocabulary

conjugate: binomial expressions with the same terms but opposite signs are said to be conjugate.

Binomial denominators

$(\sqrt{x} + \sqrt{y})$ and $(\sqrt{x} - \sqrt{y})$ are called **conjugate** surds. The surds in each expression are the same, but the signs differ.

If you multiply the conjugate surds, the answer contains no surds and therefore is rational.

$(\sqrt{x} + \sqrt{y})(\sqrt{x} - \sqrt{y}) = (\sqrt{x})^2 - (\sqrt{y})^2 = x - y$

We can use this principle when the denominator is a binomial. This example shows how rationalising the denominator can make a complicated fraction into a simpler expression to work with.

Tip

You should recognise this principle as the difference of squares rule that you applied in Chapter 7.

WORKED EXAMPLE 8

Write this fraction with a rational denominator $\frac{3}{2-\sqrt{3}}$.

$\frac{3}{2-\sqrt{3}}$	The conjugate surd is $2+\sqrt{3}$.
$\frac{3}{2-\sqrt{3}} = \frac{3}{2-\sqrt{3}} \times \frac{2+\sqrt{3}}{2+\sqrt{3}}$	Using the method for rationalising denominators but using the conjugate surd: $\frac{x}{\sqrt{y}}$ is equivalent to $\frac{x}{\sqrt{y}} \times \frac{\sqrt{y}}{\sqrt{y}}$
$= \frac{3(2+\sqrt{3})}{4-3}$	$(2+\sqrt{3})(2-\sqrt{3}) = 4 - 3$ (difference of squares)
$= 6 + 3\sqrt{3}$	Multiply to remove the brackets. The rational denominator is 1; you know that convention is to not include the denominator if it is 1.

Tip

Rationalising the denominator is useful for simplifying in exact calculations. If you only need an approximate value, use your calculator and you won't need to rationalise the denominator.

EXERCISE 27E

1 Simplify.

a $\sqrt{7} \times \sqrt{3}$ **b** $\sqrt{3} \times \sqrt{5}$ **c** $\sqrt{3} \times \sqrt{12}$

d $2\sqrt{7} \times 3\sqrt{5}$ **e** $^-3\sqrt{11} \times 4\sqrt{3}$ **f** $2\sqrt{15} \times 3\sqrt{3}$

g $2\sqrt{13} \times 3\sqrt{13}$ **h** $2\sqrt{6} \times 5\sqrt{3}$ **i** $4\sqrt{2} \times 5\sqrt{3}$

j $\sqrt{27} \times \sqrt{72}$ **k** $\sqrt{48} \times \sqrt{45}$ **l** $2\sqrt{20} \times 3\sqrt{24}$

2 Simplify.

a $\sqrt{14} \div \sqrt{2}$ **b** $\sqrt{26} \div \sqrt{13}$ **c** $\frac{\sqrt{5}}{\sqrt{10}}$

d $\frac{\sqrt{2}}{\sqrt{20}}$ **e** $\frac{\sqrt{45}}{\sqrt{5}}$ **f** $\frac{3\sqrt{7}}{3}$

g $\frac{5\sqrt{6}}{10}$ **h** $4\sqrt{\frac{60}{5}}$ **i** $\frac{6\sqrt{22}}{\sqrt{2}}$

j $\frac{12\sqrt{12}}{4\sqrt{3}}$ **k** $\frac{^-3\sqrt{24}}{\sqrt{6}}$ **l** $\frac{^-2\sqrt{27}}{\sqrt{12}}$

Tip

Simplify the surds first so that you work with smaller numbers.

3 Simplify fully.

a $\frac{3\sqrt{6} \times 4\sqrt{3}}{4}$ **b** $\frac{2\sqrt{5} \times 4\sqrt{6}}{\sqrt{10}}$ **c** $\frac{3\sqrt{5} \times 2\sqrt{8}}{6\sqrt{20}}$

d $\frac{\sqrt{3} \times \sqrt{15}}{3\sqrt{5}}$ **e** $\frac{5\sqrt{6} \times -2\sqrt{5}}{2\sqrt{15}}$ **f** $\frac{^-\sqrt{12} \times \sqrt{27}}{2\sqrt{6} \times 2\sqrt{2}}$

4 Expand and simplify.

a $\sqrt{5}(\sqrt{3}+2)$ **b** $2\sqrt{3}(5-\sqrt{3})$ **c** $(\sqrt{2}+1)(\sqrt{6}-2\sqrt{3})$

d $(2\sqrt{5}+\sqrt{7})^2$ **e** $(\sqrt{2}+3)(\sqrt{3}+5)$ **f** $(2-\sqrt{5})(\sqrt{5}-1)$

g $(4\sqrt{3}-\sqrt{2})(4\sqrt{3}+\sqrt{2})$ **h** $(\sqrt{7}+\sqrt{2})^2$ **i** $(\sqrt{3}-\sqrt{5})^2$

Find answers at: cambridge.org/ukschools/gcsemaths-studentbookanswers

5 Express each of the following in simplest form with a rational denominator.

a $\frac{5}{\sqrt{3}}$ **b** $\frac{1}{\sqrt{5}}$ **c** $\frac{^-2}{\sqrt{3}}$

d $\frac{\sqrt{2}}{\sqrt{3}}$ **e** $4\sqrt{\frac{3}{2}}$ **f** $\frac{^-3}{4\sqrt{7}}$

g $\frac{2+\sqrt{3}}{2\sqrt{3}}$ **h** $\frac{2+\sqrt{5}}{\sqrt{5}}$ **i** $\frac{\sqrt{10}-\sqrt{5}}{5\sqrt{10}}$

6 Rationalise each denominator and simplify.

a $\frac{3}{3-\sqrt{2}}$ **b** $\frac{\sqrt{2}}{\sqrt{11}+3}$ **c** $\frac{\sqrt{3}}{\sqrt{5}-\sqrt{2}}$

d $\frac{3}{3\sqrt{2}+4}$ **e** $\frac{8}{\sqrt{5}+1}$ **f** $\frac{1-\sqrt{3}}{2-\sqrt{5}}$

g $\frac{\sqrt{5}+1}{7-3\sqrt{5}}$ **h** $\frac{2+\sqrt{3}}{5-\sqrt{3}}$ **i** $\frac{\sqrt{2}-3}{\sqrt{6}+2\sqrt{3}}$

Section 3: Working with surds

In this section you are going to apply the skills you've just learnt to solve problems where some or all of the values are expressed as surds. Unless you are asked to give an approximate value, leave your answers in exact form.

In maths you generally learn new skills and concepts and then you have to use them to solve problems. In the case of surds, the problems will generally be similar to those that you have already worked with in algebra or geometry. The only real difference will be that the values are given in surd form.

The following steps are useful for solving all problems, including those involving surds.

Problem-solving framework

Find length DE. Give the answer in the simplest possible exact form.

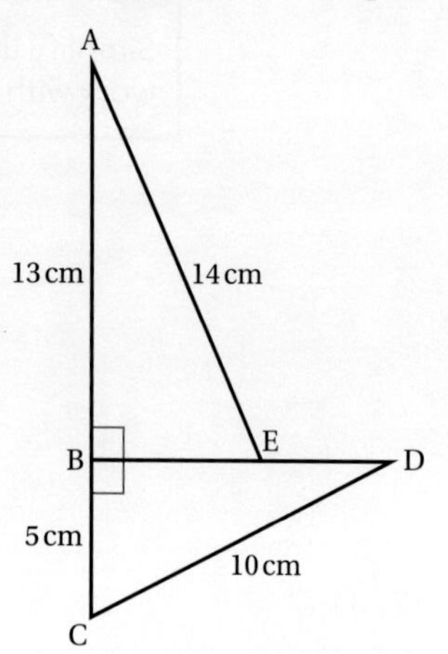

Continues on next page ...

Steps for approaching a problem-solving question	What you would do for this example
Step 1: What you have to do?	Find DE and leave the answer in exact form.
Step 2: What information do you know?	DE = BD − BE BD and BE are sides of right-angled triangles.
Step 3: What maths can you do?	The triangles are right-angled and you need to find a length, so use Pythagoras. Find BE and BD and then subtract them. Leave all values in exact form. $BE = \sqrt{(14^2 - 13^2)}$ Pythagoras $= \sqrt{196 - 169}$ $= \sqrt{27}$ Simplify the surd. $= 3\sqrt{3}$ $BD = \sqrt{10^2 - 5^2}$ Pythagoras $= \sqrt{100 - 25}$ $= \sqrt{75}$ $= 5\sqrt{3}$ Simplify the surd. The surds are the same, so you can subtract them by factorising for $\sqrt{3}$. $BD - BE = 5\sqrt{3} - 3\sqrt{3}$ $= 2\sqrt{3}$ cm
Step 4: Have you answered the questions and does it seem reasonable?	DE = $2\sqrt{3}$ cm ✓ $2\sqrt{3}$ cm ≈ 3.5 cm. This length seems reasonable compared with the other dimensions of the triangles.

EXERCISE 27F

1 Find the exact area and perimeter of each shape.

a

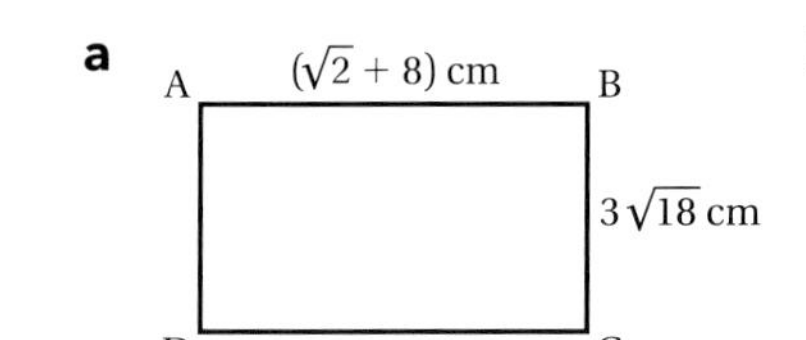

b

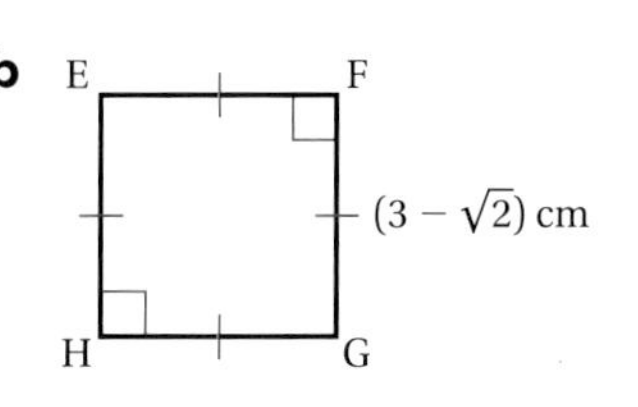

c

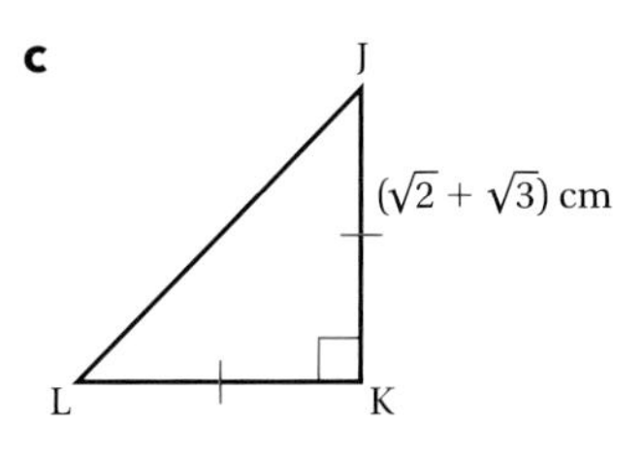

d

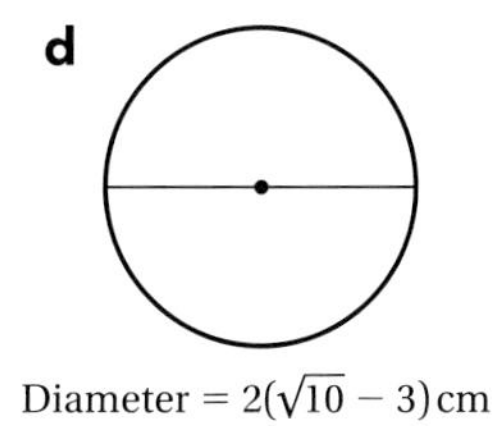

e

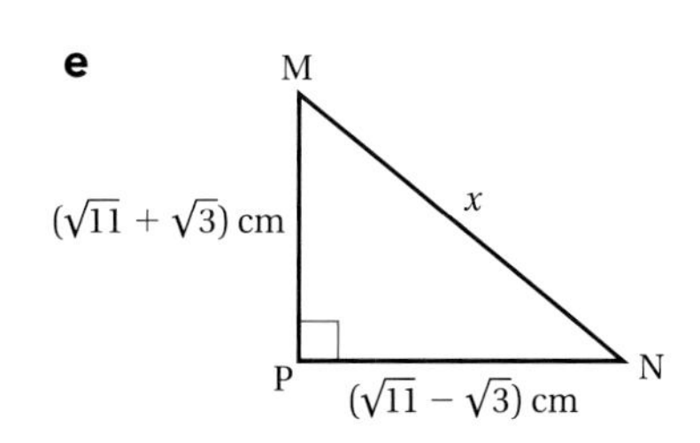

2 Calculate the exact area of a rectangle ABCD with sides of $(6 + \sqrt{5})$ cm and $(6 - \sqrt{5})$ cm.

3 If the area of the front of a circular disk is 12π cm², what is its exact radius?

4 The chart shows the relationship of paper sizes in the A-series papers. The ratio of sides of each sheet is $1 : \sqrt{2}$.

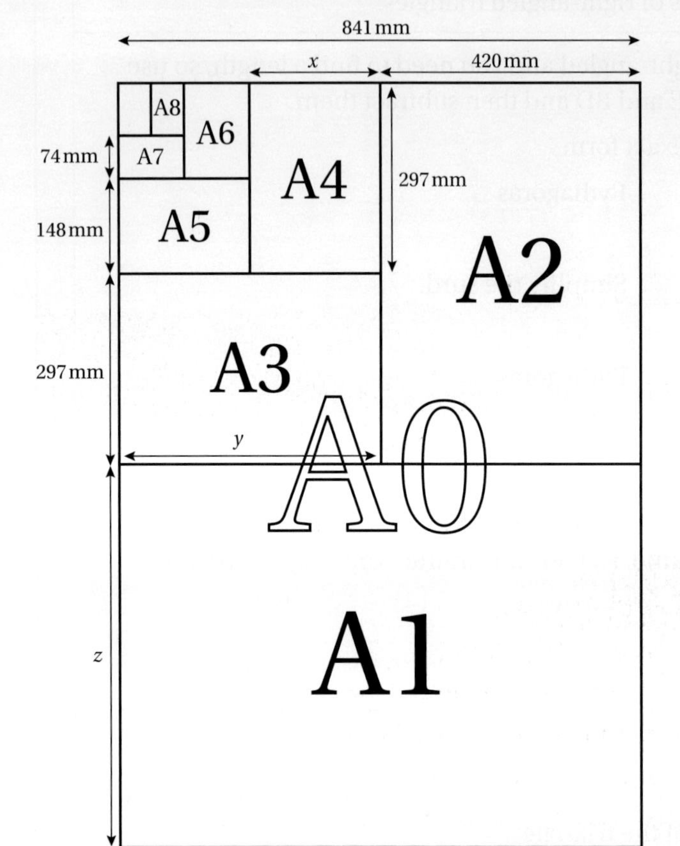

a Calculate the length of the sides marked x, y and z, giving your answers to the nearest whole millimetre.

b Size A0 paper is a rectangle of area 1 m².

i Express the area in cm².

ii Use the ratio of sides $1 : \sqrt{2}$ to determine the length of the diagonal of a sheet of A0 paper given the width is 841 mm.

c For any sheet of A-series paper with shorter side x wide, express the length of the diagonal (z) in terms of x.

5 A metal cube $(2\sqrt{3} + 4\sqrt{5})$ cm high is to be coated with a rust inhibitor. Calculate the exact area of the surface to be painted.

6 Given that ABC is a right-angled isosceles triangle with AB = BC, determine sin A, leaving your answer in surd form.

7 If x is a positive integer, find $\sqrt{x^3 + 2x^2 + x}$.

8 A square has sides of length $3\sqrt{8}$. Find:

a the area of the square.

b the length of a diagonal.

9 Find the exact perimeter of a square of area 150 cm².

Checklist of learning and understanding

Surds are exact values

- Surds can only be expressed exactly using the root sign. Using a calculator to get a decimal value for a surd gives an approximate and/or truncated value.

Surds can be simplified

- Surds can sometimes be simplified by writing them as factors and taking the roots of any factors that are perfect squares.

Adding and subtracting surds

- Expressions containing surds can be simplified by finding common factors.

Multiplying and dividing surds

- Surds can be multiplied and divided using the rules:
  - $(\sqrt{x})^2 = x$
  - $\sqrt{x^2} = x$
  - $\sqrt{x} \times \sqrt{y} = \sqrt{xy}$
  - $\sqrt{x} \div \sqrt{y} = \sqrt{\frac{x}{y}}$
- When expanding expressions with brackets, apply the rules of algebra.

Rationalising the denominator

- Fractions with a surd in the denominator can be rewritten as equivalent fractions with an integer denominator.
- When the denominator is a binomial for example $(\sqrt{x} + \sqrt{y})$ you can use the conjugate surd $(\sqrt{x} - \sqrt{y})$ to rationalise the denominator.

Chapter review

For additional questions on the topics in this chapter, visit GCSE Mathematics Online.

Two sets of answers to an exercise involving simplifying surds are given below. For each question:

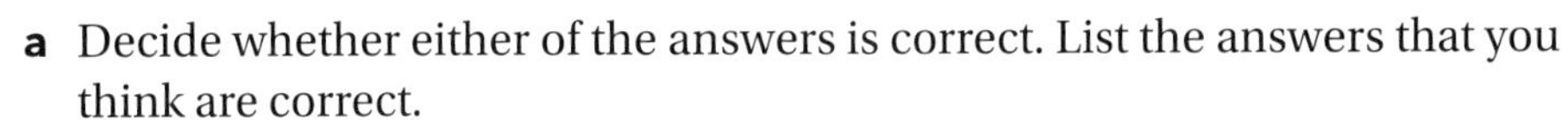

a Decide whether either of the answers is correct. List the answers that you think are correct.

b If neither answer is correct find the correct answer, showing your working.

	Question	Answer A	Answer B
1	$\sqrt{18}$	$\sqrt{6} \times \sqrt{3}$	$3\sqrt{2}$
2	$\sqrt{45}$	$3\sqrt{5}$	$5\sqrt{3}$
3	$3\sqrt{5} + \sqrt{3} + \sqrt{5} - 2\sqrt{3}$	$3\sqrt{5} - \sqrt{3}$	7.2122
4	$\sqrt{14} + \sqrt{8}$	$\sqrt{22}$	$\sqrt{112}$
5	$\sqrt{5} \times \sqrt{5} \times \sqrt{5} \times \sqrt{5}$	25	$4\sqrt{5}$
6	$\sqrt{3} \times \sqrt{3} \times \sqrt{3} + \sqrt{3}$	$4\sqrt{3}$	$3\sqrt{3}$
7	$\frac{6\sqrt{35}}{2\sqrt{5}}$	$3\sqrt{30}$	$3\sqrt{7}$
8	$\frac{1 + \sqrt{5}}{\sqrt{5}}$	2	$\frac{2}{\sqrt{5}} + 1$
9	$\frac{10 + 5\sqrt{7}}{5}$	$\sqrt{7} + 2$	$2 + 5\sqrt{7}$
10	$\frac{2}{2\sqrt{7} + 5}$	$\frac{2(\sqrt{7} - 5)}{2}$	$\frac{4\sqrt{7} - 10}{3}$

11 Simplify.

$$\frac{6 + \sqrt{2}}{\sqrt{2}}$$

Give your answer in the form $a\sqrt{2} + b$. *(3 marks)*

© OCR 2013

Find answers at: cambridge.org/ukschools/gcsemaths-studentbookanswers

28 Plane vector geometry

In this chapter you will learn how to ...

- represent vectors as a diagram or column vector.
- add and subtract vectors.
- multiply vectors by a scalar.
- use vectors to construct geometric arguments and proofs.

For more resources relating to this chapter, visit GCSE Mathematics Online.

Using mathematics: real-life applications

Vectors have huge applications in the physical world: for example, mathematical modelling of objects sliding down slopes with varying amounts of friction, working out how far objects can tilt before they tip over and making sure two ships don't crash in the night. All these problems involve the use of vectors.

"When landing at any airport I have to consider how the wind will blow me off course. Over a set amount of time I expect to travel through a particular vector but I have to add on the effect the wind has on my flight path. If I didn't do this accurately I would struggle to land the plane safely." *(Pilot)*

Before you start ...

Ch 18	You need to be able to plot coordinates in all four quadrants.	**1** Draw a set of axes going from $^-6$ to 6 in both directions. Plot the points A(2, 3), B($^-3$, 4) and C($^-2$, $^-3$).
Ch 2	You need to be able to add, subtract and multiply negative numbers.	**2** Calculate. **a** $3 - 7$ **b** $^-4 + 11$ **c** $^-5 - 18$ **d** $^-4 \times 7$ **e** $^-3 \times {}^-9$
Ch8	You need to be able to solve simple linear equations.	**3** Solve. **a** $12 = 4m - 36$ **b** $2k + 15 = 7$ **c** $^-6 + 5d = {}^-41$
Ch 8	You need to be able to solve simultaneous linear equations.	**4** Solve. $3x + 2y = 8$ and $4x - 3y = 5$

Assess your starting point using the Launchpad

STEP 1

1 Give the column vector for $\overrightarrow{HG}$.

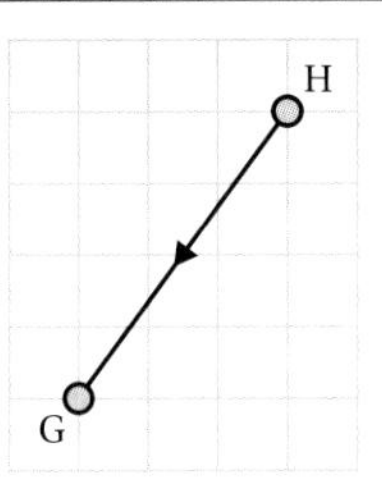

2 Draw the triangle ABC where

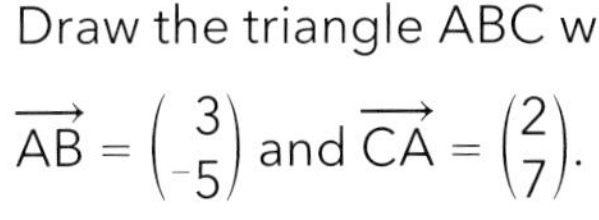

$\overrightarrow{AB} = \begin{pmatrix} 3 \\ -5 \end{pmatrix}$ and $\overrightarrow{CA} = \begin{pmatrix} 2 \\ 7 \end{pmatrix}$.

GO TO
Section 1:
Vector notation and representation

STEP 2

3 $\mathbf{j} = \begin{pmatrix} -1 \\ 3 \end{pmatrix}$ $\mathbf{k} = \begin{pmatrix} 2 \\ 1 \end{pmatrix}$ $\mathbf{l} = \begin{pmatrix} -4 \\ -2 \end{pmatrix}$

Write the following as single vectors.

a $\mathbf{j} + \mathbf{k}$ **b** $2\mathbf{k} - \mathbf{l}$

4 Find the values of f and g.

$\begin{pmatrix} 10 \\ g \end{pmatrix} - 4\begin{pmatrix} f \\ -3 \end{pmatrix} = \begin{pmatrix} -2 \\ 18 \end{pmatrix}$.

5 In the diagram

$\overrightarrow{AC} = \begin{pmatrix} 14 \\ 2 \end{pmatrix}$ and $\overrightarrow{AB} = \begin{pmatrix} 9 \\ 12 \end{pmatrix}$.

Find:

a $\overrightarrow{CA}$ **b** $\overrightarrow{CA} + \overrightarrow{AB}$

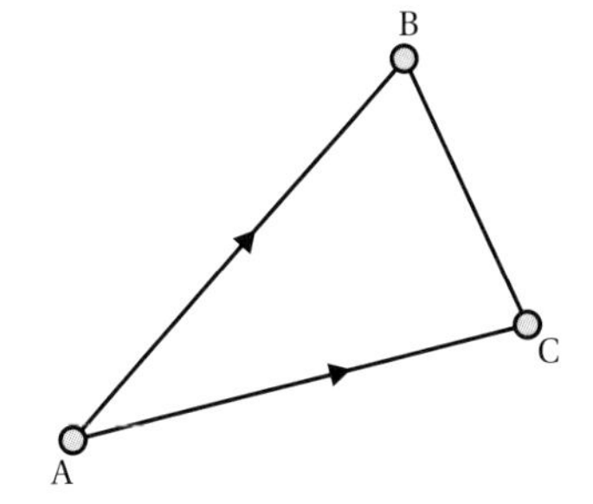

6 Which of these vectors are parallel?

$\begin{pmatrix} -3 \\ 4 \end{pmatrix}$ $\begin{pmatrix} 9 \\ 16 \end{pmatrix}$ $\begin{pmatrix} 15 \\ -20 \end{pmatrix}$ $\begin{pmatrix} -3 \\ 2 \end{pmatrix}$

GO TO
Section 2:
Vector arithmetic

STEP 3

7 ABCD is a square.

$\overrightarrow{AB} = \mathbf{x}$, $\overrightarrow{BC} = \mathbf{y}$

If the ratio of AB : AE is 1 : 2, find

a $\overrightarrow{BE}$ **b** $\overrightarrow{AF}$

M is the midpoint of EF.

c Find: $\overrightarrow{AM}$

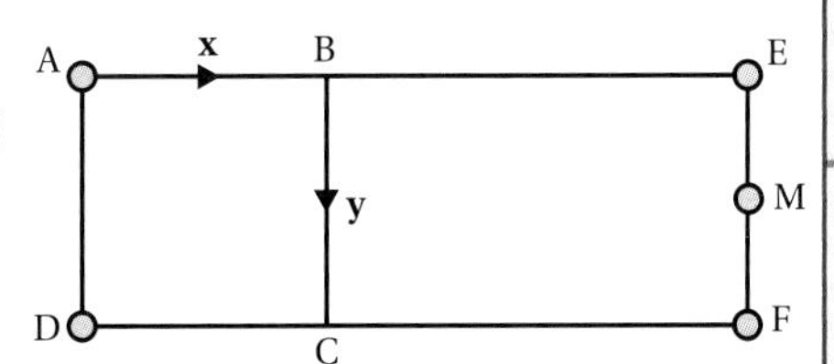

GO TO
Section 3:
Using vectors in geometric proofs

GO TO
Chapter review

Find answers at: cambridge.org/ukschools/gcsemaths-studentbookanswers

Section 1: Vector notation and representation

Key vocabulary

vector: a quantity that has both magnitude and direction.

displacement: a change in position.

A **vector** describes movement from one point to another, it has a direction and a magnitude (size).

Vectors can be used to describe many different kinds of movement. For example: **displacement** of a shape following translation, displacement of a boat during its journey, the velocity of an object, and the acceleration of an object.

A vector that describes the movement from A to B can be represented by:

- an arrow in a diagram

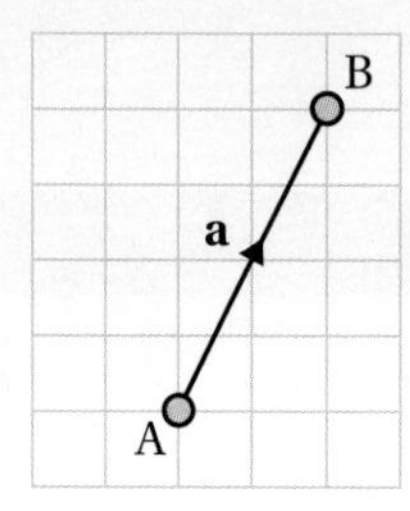

- $\overrightarrow{AB}$ (arrow indicates direction)
- **a** (if handwritten this would be underlined, $\underline{a}$)
- a column vector $\begin{pmatrix} x \\ y \end{pmatrix}$

Tip

You include the negative sign because you are now moving in the opposite direction.

If you were to travel along this vector in the opposite direction, from B to A, you would represent this vector as:

- $\overrightarrow{BA}$
- $^-$**a**
- $\begin{pmatrix} ^-x \\ ^-y \end{pmatrix}$

Column vectors

In a column vector x represents the **horizontal** movement; y represents the **vertical** movement.

	Movement	
	x	y
positive	right	up
negative	left	down

In the diagram, $\overrightarrow{AB} = \begin{pmatrix} 2 \\ 4 \end{pmatrix}$.

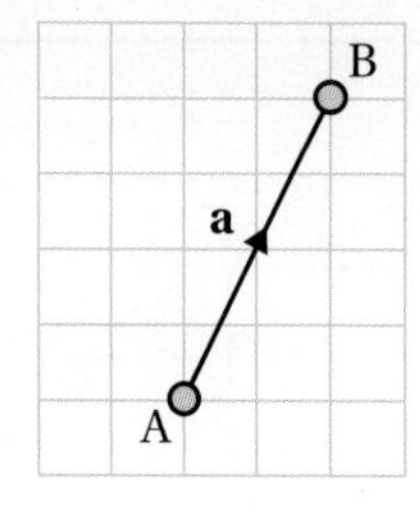

EXERCISE 28A

1 Match up equivalent representations of the vectors.

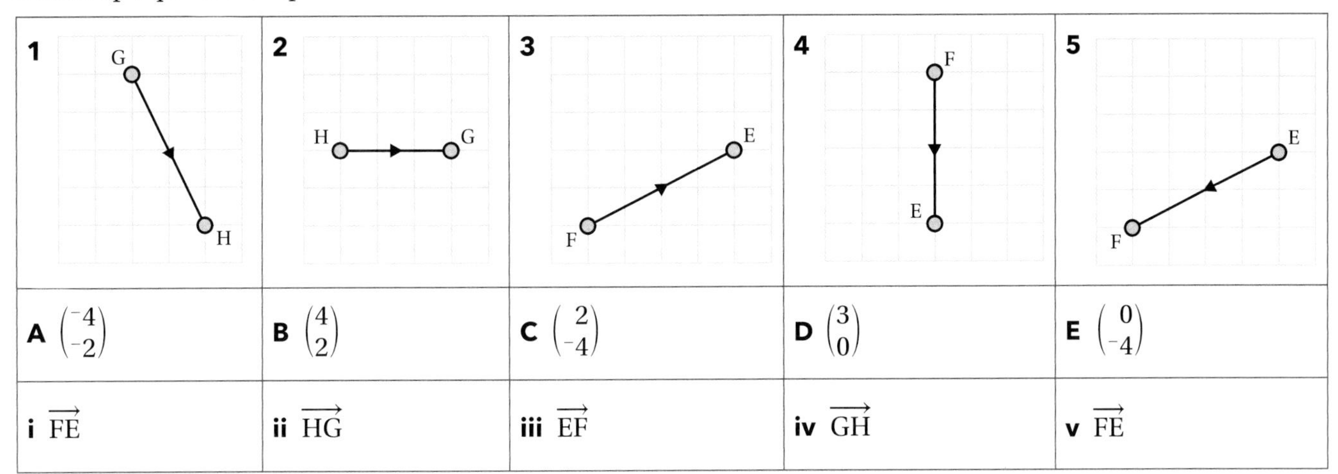

A $\begin{pmatrix}-4\\-2\end{pmatrix}$	**B** $\begin{pmatrix}4\\2\end{pmatrix}$	**C** $\begin{pmatrix}2\\-4\end{pmatrix}$	**D** $\begin{pmatrix}3\\0\end{pmatrix}$	**E** $\begin{pmatrix}0\\-4\end{pmatrix}$
i $\overrightarrow{FE}$	**ii** $\overrightarrow{HG}$	**iii** $\overrightarrow{EF}$	**iv** $\overrightarrow{GH}$	**v** $\overrightarrow{FE}$

2 Use the diagram to answer the following questions.

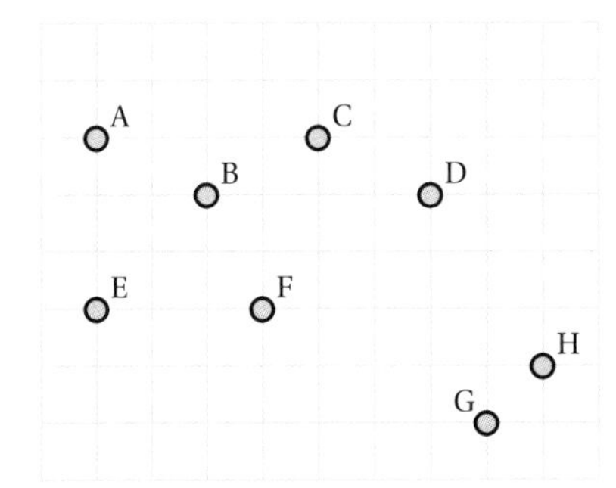

Find each of the vectors:

a $\overrightarrow{AB}$ **b** $\overrightarrow{DC}$ **c** $\overrightarrow{BC}$

d $\overrightarrow{DF}$ **e** $\overrightarrow{HF}$ **f** $\overrightarrow{BH}$

g What do you notice about $\overrightarrow{AB}$ and $\overrightarrow{DC}$?

h What do you notice about $\overrightarrow{AB}$ and $\overrightarrow{BH}$?

3 Draw a pair of axes where x and y vary from -8 to 8.

Plot the point A $(2, -1)$.

Then plot points B, C, D, E, F and G where:

$\overrightarrow{AB} = \begin{pmatrix}2\\7\end{pmatrix}$ $\overrightarrow{AC} = \begin{pmatrix}-3\\7\end{pmatrix}$ $\overrightarrow{AD} = \begin{pmatrix}-6\\3\end{pmatrix}$

$\overrightarrow{AE} = \begin{pmatrix}5\\3\end{pmatrix}$ $\overrightarrow{AF} = \begin{pmatrix}-3\\-1\end{pmatrix}$ $\overrightarrow{AG} = \begin{pmatrix}2\\-1\end{pmatrix}$

4 **a** Find the vector from point A with coordinates $(3, -4)$ to point B with coordinates $(-1, 2)$.

b Give the coordinates of two more points E and F where the vector from E to F is the same as $\overrightarrow{AB}$.

Find answers at: cambridge.org/ukschools/gcsemaths-studentbookanswers

Tip

The midpoint of the line KL is halfway along the line from K to L.

5 **a** Find the vector from point K with coordinates ($^-2$, $^-1$) to point L with coordinates ($^-8$, 9).

b Use your answer to find the coordinates of the midpoint of KL.

6 These vectors describe how to move between points A, B, C and D.

$\overrightarrow{AB} = \begin{pmatrix} 2 \\ 1 \end{pmatrix}$ $\overrightarrow{BC} = \begin{pmatrix} 1 \\ 0 \end{pmatrix}$ $\overrightarrow{DA} = \begin{pmatrix} ^-1 \\ 2 \end{pmatrix}$

Draw a diagram showing how the points are positioned to form the quadrilateral ABCD.

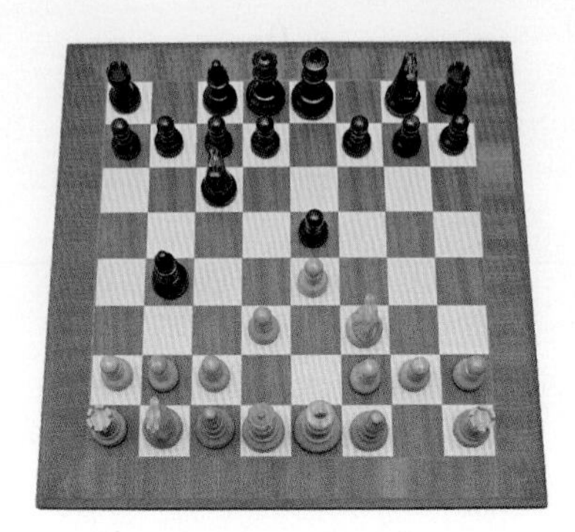

7 In a game of chess, different pieces move in different ways.

- A king can move one square in any direction (including diagonals).
- A bishop can move any number of squares diagonally.
- A knight moves two squares horizontally and one square vertically or two squares vertically and one horizontally.

A chessboard is eight squares wide and eight squares long.

What vectors can the following pieces move?

a Bishop **b** King **c** Knight

For part **a**, you will need to think algebraically.

Tip

You should remember Pythagoras' theorem and trigonometry from earlier years, but see Chapters 32 and 33 if you need a reminder.

8 How would you find the length (magnitude) of a vector?

How could you describe its direction?

Use these diagrams and your knowledge of Pythagoras and trigonometry to help design a method.

Vector $\begin{pmatrix} 2 \\ ^-4 \end{pmatrix}$

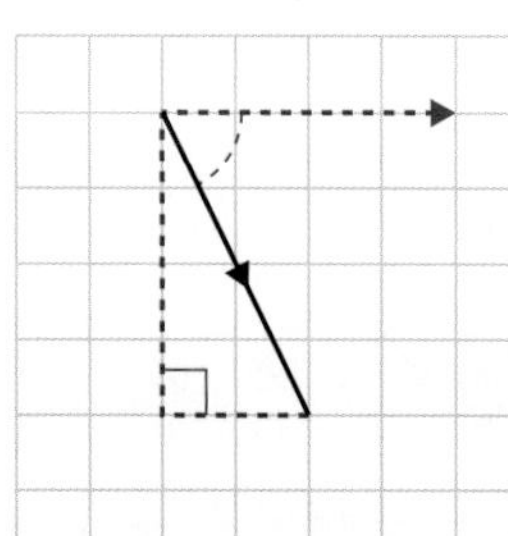

Vector $\begin{pmatrix} 3 \\ 5 \end{pmatrix}$

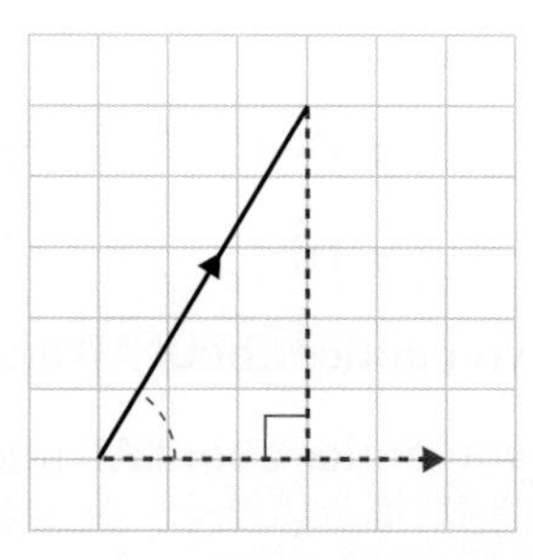

Section 2: Vector arithmetic

Addition and subtraction

The diagram shows $\overrightarrow{AB} = \begin{pmatrix} 2 \\ 4 \end{pmatrix}$, $\overrightarrow{BC} = \begin{pmatrix} 4 \\ ^-2 \end{pmatrix}$, and $\overrightarrow{AC} = \begin{pmatrix} 6 \\ 2 \end{pmatrix}$.

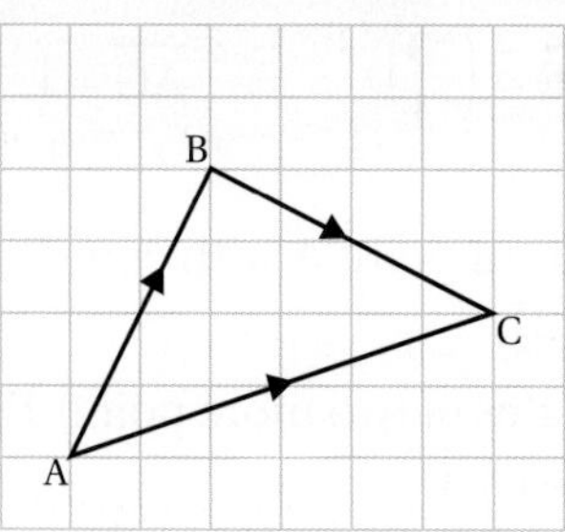

Moving from A to B and then from B to C is the same as moving directly from A to C. In other words, you can take a 'shortcut' from A to C by adding together $\overrightarrow{AB}$ and $\overrightarrow{BC}$.

$\overrightarrow{AC}$ is known as the **resultant** of $\overrightarrow{AB}$ and $\overrightarrow{BC}$.

$\overrightarrow{AB} + \overrightarrow{BC} = \overrightarrow{AC}$

$\begin{pmatrix} 2 \\ 4 \end{pmatrix} + \begin{pmatrix} 4 \\ ^-2 \end{pmatrix} = \begin{pmatrix} 6 \\ 2 \end{pmatrix}$

The diagram shows $\overrightarrow{FG}$ and $\overrightarrow{HG}$.

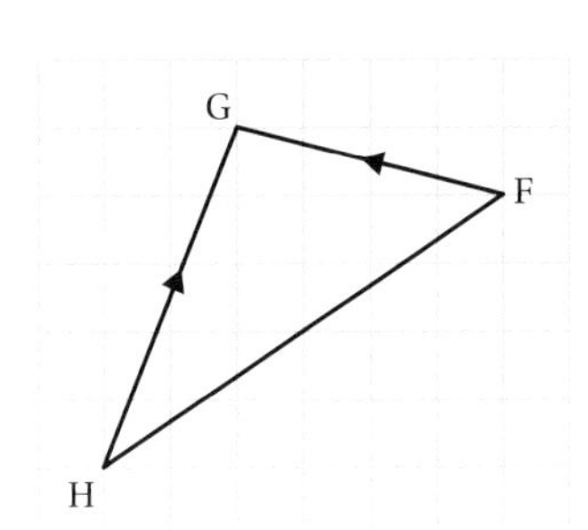

To find $\overrightarrow{FH}$ you need to travel along $\overrightarrow{FG}$, then travel along $\overrightarrow{HG}$ in the opposite direction.

So, **subtract** $\overrightarrow{HG}$.

$\overrightarrow{FG} - \overrightarrow{HG} = \overrightarrow{FH}$

$\begin{pmatrix} ^-4 \\ 1 \end{pmatrix} - \begin{pmatrix} 2 \\ 5 \end{pmatrix} = \begin{pmatrix} ^-4 - 2 \\ 1 - 5 \end{pmatrix} = \begin{pmatrix} ^-6 \\ ^-4 \end{pmatrix}$

Multiplying by a scalar

Multiplying a vector by a scalar results in repeated addition.

This is the same as multiplying the x-component by the scalar, k, and the y-component by the same scalar, k.

Key vocabulary

scalar: a numerical quantity (it has no direction).

$\overrightarrow{AB} = \begin{pmatrix} 4 \\ ^-1 \end{pmatrix}$ and $\overrightarrow{CD} = \begin{pmatrix} 12 \\ ^-3 \end{pmatrix}$.

$CD = 3\overrightarrow{AB}$

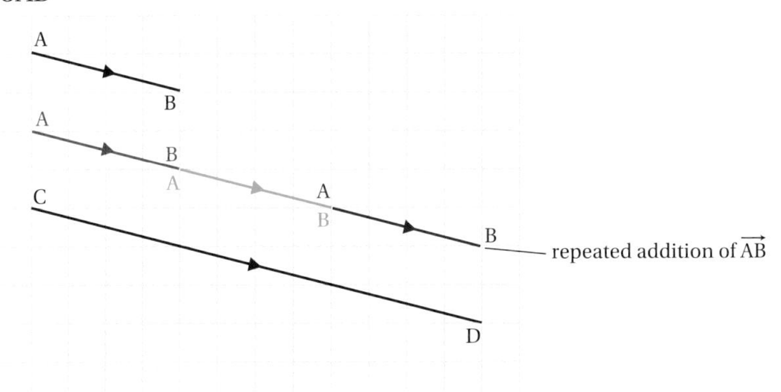

$$3 \times \begin{pmatrix} 4 \\ ^-1 \end{pmatrix} = \begin{pmatrix} 4 \\ ^-1 \end{pmatrix} + \begin{pmatrix} 4 \\ ^-1 \end{pmatrix} + \begin{pmatrix} 4 \\ ^-1 \end{pmatrix}$$ — repeated addition

$$= \begin{pmatrix} 3 \times 4 \\ 3 \times ^-1 \end{pmatrix}$$ — multiplying the x-component by the scalar 3; multiplying the y-component by the scalar 3

$$= \begin{pmatrix} 12 \\ ^-3 \end{pmatrix}$$

Multiplying a vector by a scalar k results in a parallel vector with a magnitude multiplied by k.

Key vocabulary

parallel vectors: occur when one vector is a multiple of the other.

Vectors are parallel if one is a multiple of the other.

WORK IT OUT 28.1

Which of the following vectors are parallel?

$\mathbf{a} = \begin{pmatrix} 3 \\ -1 \end{pmatrix}$ $\mathbf{b} = \begin{pmatrix} 4 \\ -3 \end{pmatrix}$ $\mathbf{c} = \begin{pmatrix} 9 \\ -3 \end{pmatrix}$ $\mathbf{d} = \begin{pmatrix} 6 \\ 2 \end{pmatrix}$ $\mathbf{e} = \begin{pmatrix} -6 \\ 2 \end{pmatrix}$

Option A	Option B	Option C
Vectors **a** and **c**	Vectors **d** and **e**	Vectors **a**, **c** and **e**

EXERCISE 28B

1 $\mathbf{p} = \begin{pmatrix} -3 \\ 2 \end{pmatrix}$ $\mathbf{q} = \begin{pmatrix} 5 \\ -1 \end{pmatrix}$ $\mathbf{r} = \begin{pmatrix} -3 \\ -2 \end{pmatrix}$ $\mathbf{s} = \begin{pmatrix} 4 \\ -7 \end{pmatrix}$

Write each of these as a single vector.

a $\mathbf{p} + \mathbf{q}$ **b** $\mathbf{s} - \mathbf{r}$ **c** $4\mathbf{p}$

d $-3\mathbf{s}$ **e** $\mathbf{p} + \mathbf{q} + \mathbf{r}$ **f** $2\mathbf{p} + \mathbf{q} - 2\mathbf{s}$

g Which of the results from parts **a** to **f** are parallel to the vector $\begin{pmatrix} 3 \\ -2 \end{pmatrix}$?

2 Give three vectors parallel to $\begin{pmatrix} 2 \\ -3 \end{pmatrix}$.

3 Find x, y, s and t in each of the following vector calculations.

a $\begin{pmatrix} x \\ 3 \end{pmatrix} + \begin{pmatrix} 5 \\ y \end{pmatrix} = \begin{pmatrix} 9 \\ 3 \end{pmatrix}$ **b** $\begin{pmatrix} 10 \\ y \end{pmatrix} - \begin{pmatrix} x \\ -3 \end{pmatrix} = \begin{pmatrix} -2 \\ 8 \end{pmatrix}$

c $\begin{pmatrix} x \\ -3 \end{pmatrix} + \begin{pmatrix} -6 \\ y \end{pmatrix} = \begin{pmatrix} 11 \\ -8 \end{pmatrix}$ **d** $s\begin{pmatrix} x \\ 12 \end{pmatrix} = \begin{pmatrix} 7 \\ -24 \end{pmatrix}$

e $s\begin{pmatrix} -12 \\ y \end{pmatrix} = \begin{pmatrix} 3 \\ -8 \end{pmatrix}$ **f** $\begin{pmatrix} 2 \\ -4 \end{pmatrix} + s\begin{pmatrix} 5 \\ y \end{pmatrix} = \begin{pmatrix} 17 \\ 14 \end{pmatrix}$

g $\begin{pmatrix} x \\ -4 \end{pmatrix} - s\begin{pmatrix} -5 \\ -3 \end{pmatrix} = \begin{pmatrix} 20 \\ 5 \end{pmatrix}$ **h** $s\begin{pmatrix} 3 \\ 4 \end{pmatrix} + t\begin{pmatrix} 2 \\ -2 \end{pmatrix} = \begin{pmatrix} 18 \\ 10 \end{pmatrix}$

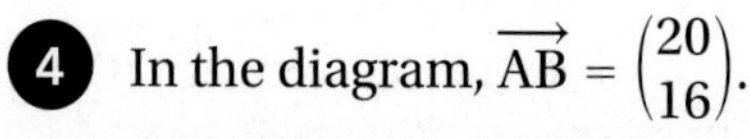

Tip

You can multiply a vector by a fractional scalar if you need to divide. If you need a reminder on fractions see Chapter 10; if you need a reminder of how to calculate ratios, see Chapter 22.

4 In the diagram, $\overrightarrow{AB} = \begin{pmatrix} 20 \\ 16 \end{pmatrix}$.

The ratio of AC : CB is 1 : 3.

a Find $\overrightarrow{AC}$.

b Find $\overrightarrow{BC}$.

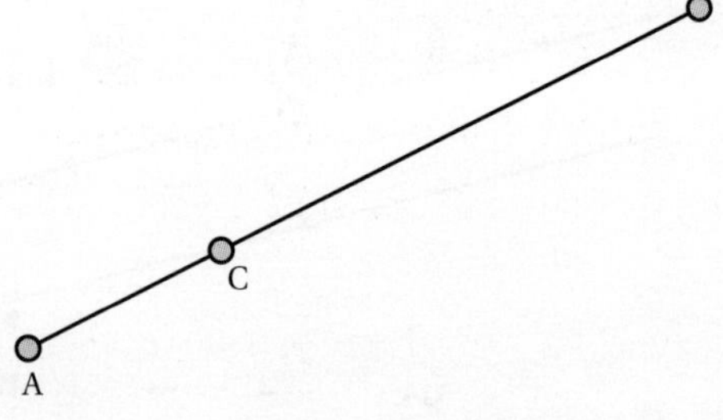

5 These vectors describe how to move between points E, F, G and H, which are four vertices of a quadrilateral.

$\overrightarrow{EF} = \begin{pmatrix} 3 \\ -1 \end{pmatrix}$ $\overrightarrow{HG} = \begin{pmatrix} 6 \\ -2 \end{pmatrix}$ $\overrightarrow{EH} = \begin{pmatrix} 0 \\ -1 \end{pmatrix}$

a What can you say about sides EF and HG?

b Predict what kind of quadrilateral EFGH is.

c Draw the quadrilateral and find $\overrightarrow{GF}$.

6 ABCD is a quadrilateral. $\overrightarrow{AB} = \overrightarrow{DC}$ and $\overrightarrow{DA} = \overrightarrow{CB}$.

What kind of quadrilateral is ABCD? How do you know this?

Section 3: Using vectors in geometric proofs

Vectors can be used to prove geometric results. You can use them to

- identify parallel lines
- find midpoints
- share lines in a given ratio.

For example, in this triangle $\overrightarrow{AB} = -\mathbf{a} + \mathbf{b}$

$= \mathbf{b} - \mathbf{a}$

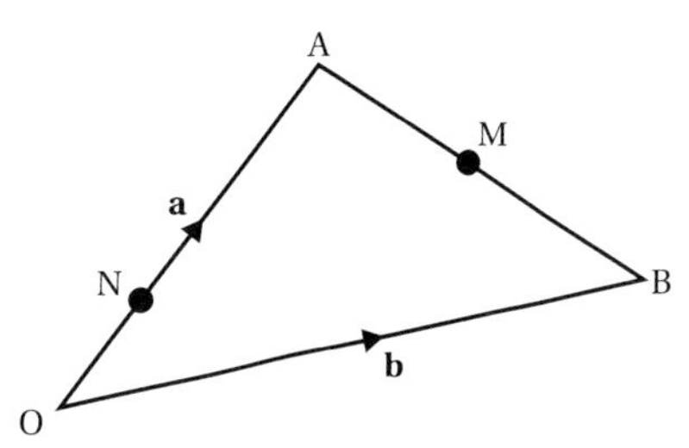

M is the midpoint of AB.

The ratio of ON : NA is 1: 2

- You can use what you know about vectors to find $\overrightarrow{AM}$.

  You know that AM is half of AB, so it follows that $\overrightarrow{AM}$ is half of the journey from A to B, so

  $\overrightarrow{AM} = \frac{1}{2}(\mathbf{b} - \mathbf{a})$

- You can use what you know about the effect of scalars on vectors to calculate $\overrightarrow{ON}$.

  You know that the ratio of ON : NA is 1 : 2.

  So, you know that the point N is such that $2\overrightarrow{ON} = \overrightarrow{NA}$, so

  $\overrightarrow{ON} = \frac{1}{3}\mathbf{a}$

WORKED EXAMPLE 1

In this parallelogram, M is the midpoint of AC, and N is the midpoint of BC.

Prove that $\overrightarrow{MN}$ is parallel to $\overrightarrow{AB}$.

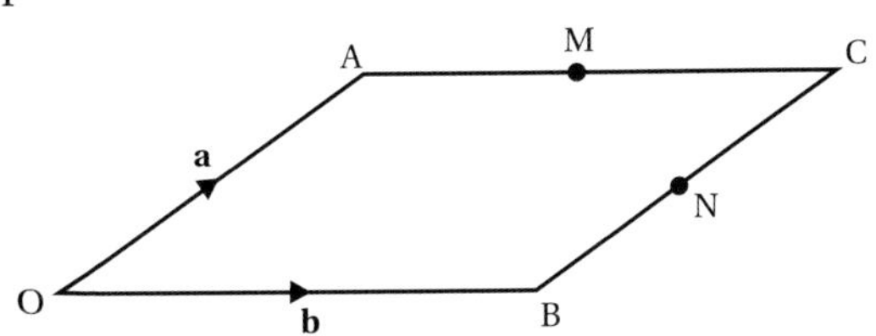

$\overrightarrow{AB} = \mathbf{b} - \mathbf{a}$

$\overrightarrow{MN} = \overrightarrow{MA} + \overrightarrow{AO} + \overrightarrow{OB} + \overrightarrow{BN}$

$= -\frac{1}{2}\mathbf{b} - \mathbf{a} + \mathbf{b} + \frac{1}{2}\mathbf{a}$

$= \frac{1}{2}(\mathbf{b} - \mathbf{a})$

Since $\overrightarrow{MN}$ is a multiple of $\overrightarrow{AB}$, they are parallel.

Tip

If you're struggling with a question, highlight sides that are labelled with vectors or that are parallel to any given vectors. Then identify the journey between your two points by using these highlighted lines.

WORKED EXAMPLE 2

In the regular hexagon shown, $\overrightarrow{EF} = \mathbf{e}$ and $\overrightarrow{JI} = \mathbf{j}$.

It is possible to move between any two vertices of the regular hexagon using combinations of vectors **e** and **j**.

Find the vector that describes each journey.

a $\overrightarrow{EG}$ **b** $\overrightarrow{HJ}$ **c** $\overrightarrow{EJ}$

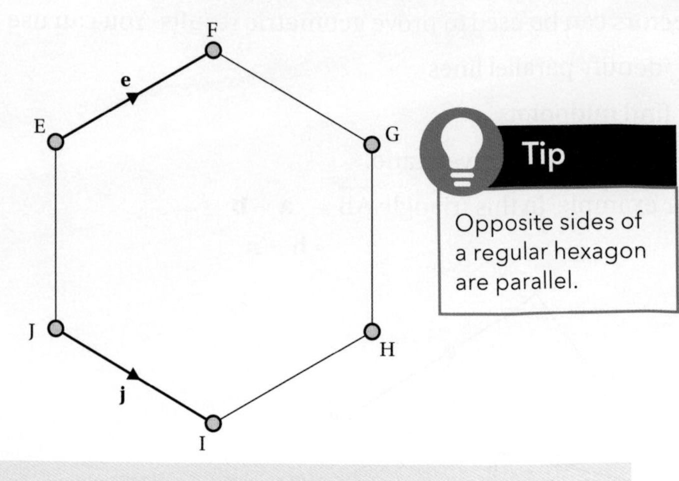

Tip

Opposite sides of a regular hexagon are parallel.

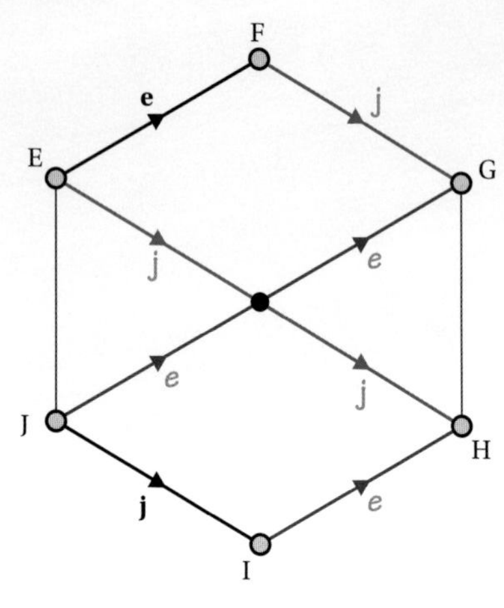

Drawing a few additional lines parallel to the vectors given can help you to see a solution.

Write these in terms of **e** and **j**.

a $\overrightarrow{EG} = \overrightarrow{EF} + \overrightarrow{FG}$

Using addition of vectors, $\overrightarrow{EG}$ is the resultant vector.

$\overrightarrow{FG} = \mathbf{j}$

$\overrightarrow{EG} = \mathbf{e} + \mathbf{j}$

Using the properties of a regular hexagon, the line FG is parallel to JI. Parallel vectors are a multiple of each other, in this case the scalar is 1.

b $\overrightarrow{HJ} = \overrightarrow{HI} + \overrightarrow{IJ}$

Using addition of vectors, $\overrightarrow{HJ}$ is the resultant vector.

$\overrightarrow{HI} = {}^{-}\mathbf{e}$

Using the properties of a regular hexagon, the line IH is parallel to EF. Parallel vectors are a multiple of each other, in this case the scalar is 1. You are travelling from H to I, in the opposite direction of **e**, so need the negative of **e**.

$\overrightarrow{HJ} = {}^{-}\mathbf{e} - \mathbf{j}$

$\overrightarrow{IJ}$ is the opposite direction of $\overrightarrow{JI}$.

c $\overrightarrow{EJ} = \overrightarrow{EH} + \overrightarrow{HI} + \overrightarrow{IJ}$

Using addition of vectors, $\overrightarrow{EJ}$ is the resultant vector.

$\overrightarrow{EH} = 2\mathbf{j}$

Using the helpful additional lines, EH is parallel to IJ and twice its length. It is moving in the same direction.

$\overrightarrow{HI} = {}^{-}\mathbf{e}$

$\overrightarrow{HI}$ is the opposite direction to $\overrightarrow{IH}$.

$\overrightarrow{IJ} = {}^{-}\mathbf{j}$

$\overrightarrow{IJ}$ is the opposite direction to $\overrightarrow{JI}$.

$\overrightarrow{EH} = 2\mathbf{j} - \mathbf{e} - \mathbf{j} = \mathbf{j} - \mathbf{e}$

EXERCISE 28C

1 In the diagram, $\overrightarrow{AC} = \begin{pmatrix} 10 \\ 2 \end{pmatrix}$ and $\overrightarrow{AB} = \begin{pmatrix} 8 \\ 14 \end{pmatrix}$.

M is the midpoint of BC.

Find:

a $\overrightarrow{CA}$

b $\overrightarrow{CA} + \overrightarrow{AB}$

c $\overrightarrow{CM}$

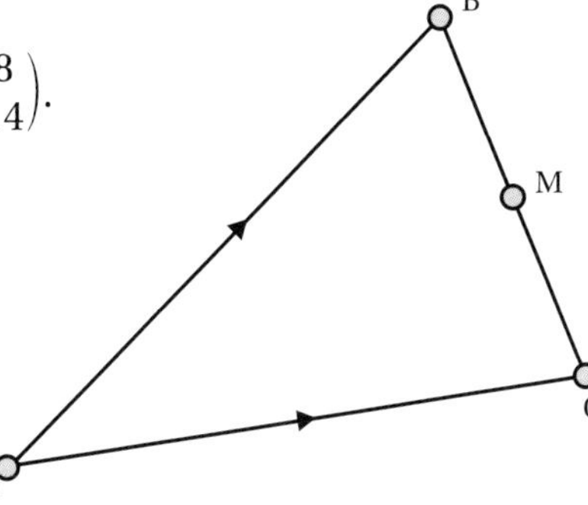

2 Two triangles have vertices ABC and DEF:

A(0,0), B(3,2), C(2, 5)

D(1,1), E(7,5), F(5,11).

Compare the vectors.

a $\overrightarrow{AB}$ and $\overrightarrow{DE}$

b $\overrightarrow{AC}$ and $\overrightarrow{DF}$

c What does this tell you about the triangles ABC and DEF?

3 MNOP is a square.

Find the vectors. Explain your answers.

a $\overrightarrow{NO}$

b $\overrightarrow{OP}$

c $\overrightarrow{MO}$

d $\overrightarrow{PN}$

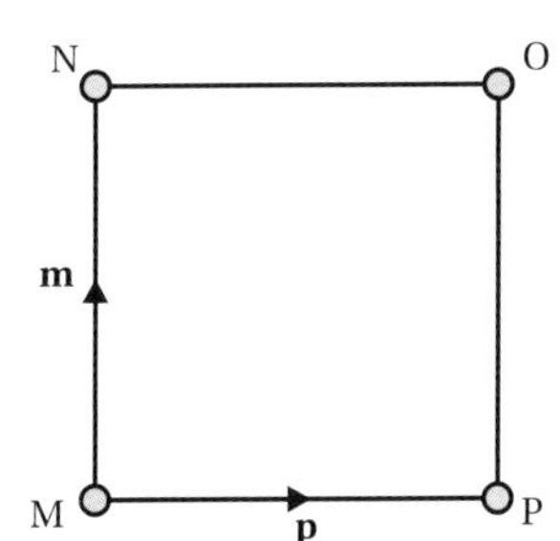

4 ABCD is a parallelogram.

M is the midpoint of side BC, N the midpoint of CD.

$\overrightarrow{AB} = \mathbf{p}$ and $\overrightarrow{BM} = \mathbf{q}$.

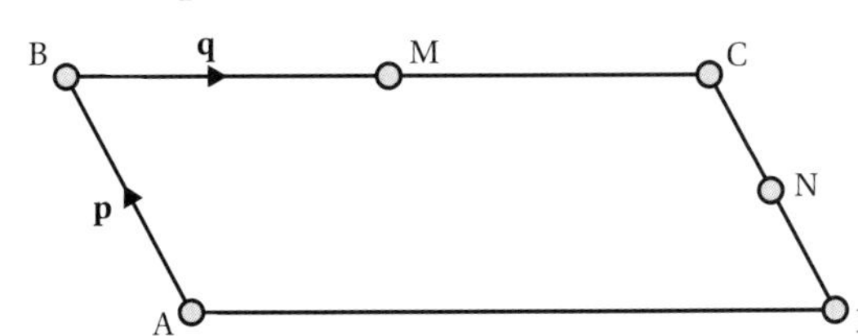

Find the vectors. Explain your answers.

a $\overrightarrow{AC}$

b $\overrightarrow{DB}$

c $\overrightarrow{MD}$

d Show that $\overrightarrow{NM}$ is parallel to $\overrightarrow{DB}$.

5 The diagram shows four congruent triangles forming a tessellating pattern, and vectors **m** and **n**.

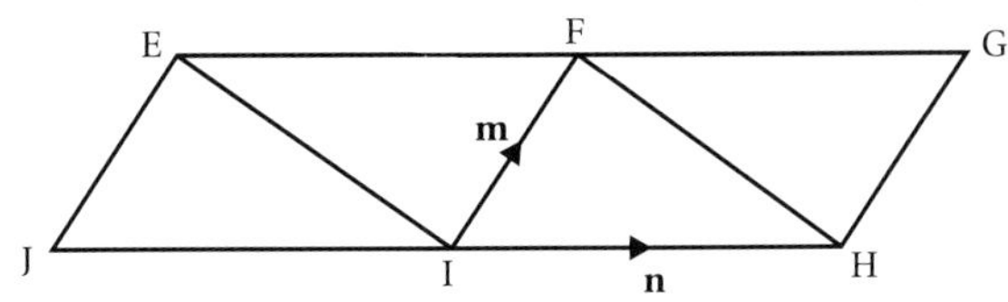

Find the vectors

a $\overrightarrow{IJ}$

b $\overrightarrow{EJ}$

c $\overrightarrow{JF}$

d $\overrightarrow{EH}$

Tip

Congruent shapes are exactly the same shape and size. You will learn more about congruency in Chapter 30.

Find answers at: cambridge.org/ukschools/gcsemaths-studentbookanswers

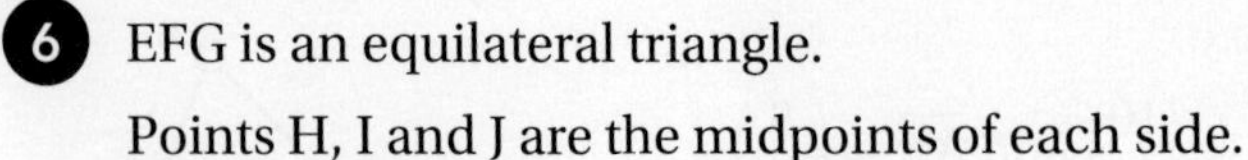

6 EFG is an equilateral triangle.

Points H, I and J are the midpoints of each side.

$\overrightarrow{EF} = \mathbf{e}$ and $\overrightarrow{EG} = \mathbf{g}$.

Find the vectors

a $\overrightarrow{EH}$ **b** $\overrightarrow{JE}$ **c** $\overrightarrow{FG}$ **d** $\overrightarrow{HI}$ **e** $\overrightarrow{IJ}$

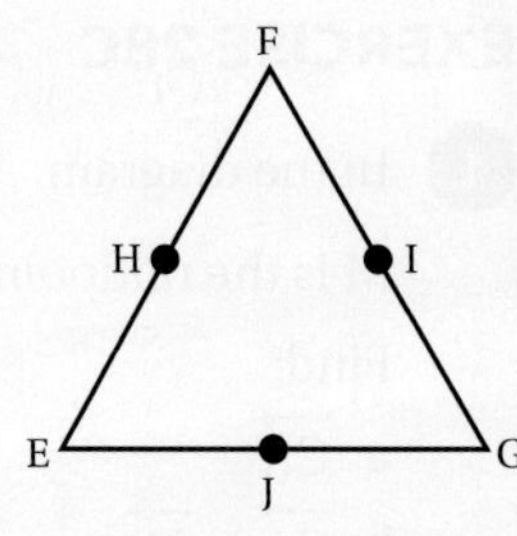

What can you say about triangle HIJ?

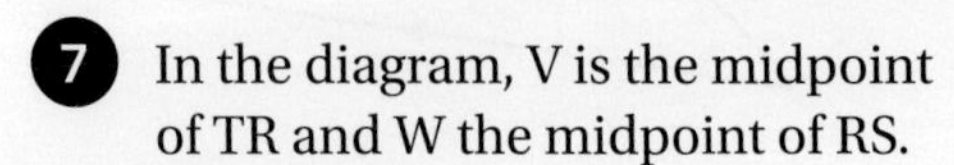

Tip

You will need to use Pythagoras' theorem and trigonometric ratios; if you can't remember these, see Chapters 32 and 33.

7 In the diagram, V is the midpoint of TR and W the midpoint of RS.

The ratio of TU : US is 1 : 4.

$\overrightarrow{SR} = \mathbf{r}$ and $\overrightarrow{TU} = \mathbf{t}$.

Find the vectors

a $\overrightarrow{TS}$ **b** $\overrightarrow{UR}$

c $\overrightarrow{VR}$ **d** $\overrightarrow{WV}$

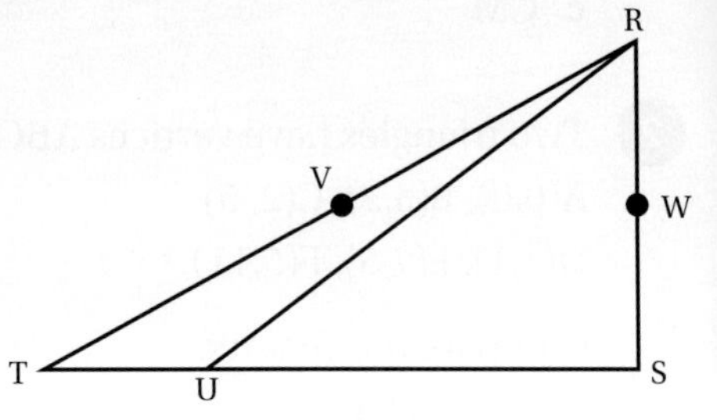

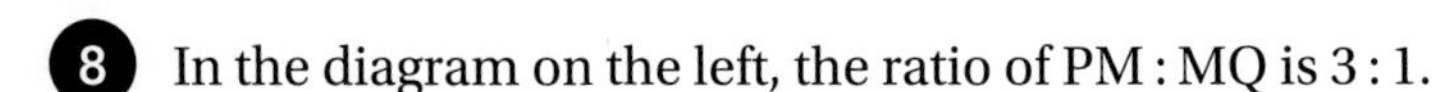

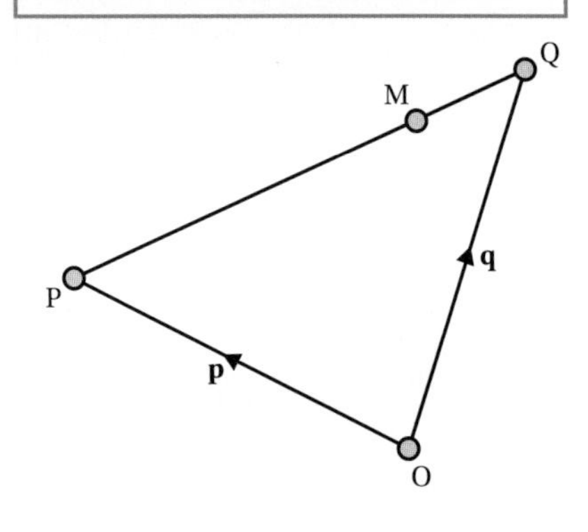

8 In the diagram on the left, the ratio of PM : MQ is 3 : 1.

a Find $\overrightarrow{PQ}$ **b** Find $\overrightarrow{PM}$ **c** Show $\overrightarrow{OM} = \frac{1}{4}(3\mathbf{q} + \mathbf{p})$

9 A river runs from east to west at 3 m every second.

It is 12 m wide.

James can swim at 1.5 m every second.

He sets off to cross the river.

How far off course is he when he reaches the other river bank?

How far does he actually swim?

Checklist of learning and understanding

Notation

- Vectors can be written in a variety of ways: $\overrightarrow{AB}$, $\mathbf{a}$, $\begin{pmatrix}1\\2\end{pmatrix}$.

Addition and subtraction

- To add or subtract vectors simply add or subtract the x- and y-components.

$\begin{pmatrix}3\\2\end{pmatrix} + \begin{pmatrix}2\\-4\end{pmatrix} = \begin{pmatrix}5\\-2\end{pmatrix}$ $\quad$ $\begin{pmatrix}-1\\4\end{pmatrix} - \begin{pmatrix}5\\-6\end{pmatrix} = \begin{pmatrix}-6\\10\end{pmatrix}$

Multiplication by a scalar

- To multiply by a scalar you can use repeated addition; or multiply the x-component by the scalar and the y-component by the scalar.

$3\begin{pmatrix}-2\\1\end{pmatrix} = \begin{pmatrix}-2\\1\end{pmatrix} + \begin{pmatrix}-2\\1\end{pmatrix} + \begin{pmatrix}-2\\1\end{pmatrix} = \begin{pmatrix}-6\\3\end{pmatrix}$ $\quad$ $3\begin{pmatrix}-2\\1\end{pmatrix} = \begin{pmatrix}-6\\3\end{pmatrix}$

- Multiplying a vector by a scalar quantity produces a parallel vector; you can identify that vectors are parallel if one vector is a multiple of the other. Parallel vectors can be part of the same line and described using a ratio.

Using vectors in geometric proofs

- You can use vectors to identify parallel lines, find midpoints and share lines in a given ratio.

Chapter review

For additional questions on the topics in this chapter, visit GCSE Mathematics Online.

1 What is the difference between coordinate $(^-2, 3)$ and vector $\begin{pmatrix} -2 \\ 3 \end{pmatrix}$?

2 Match the parallel vectors.

a $= \begin{pmatrix} -6 \\ 2 \end{pmatrix}$ **b** $= \begin{pmatrix} 1 \\ 3 \end{pmatrix}$ **c** $= \begin{pmatrix} 3 \\ -1 \end{pmatrix}$ **d** $= \begin{pmatrix} 7 \\ 21 \end{pmatrix}$

e $= \begin{pmatrix} -2 \\ 4 \end{pmatrix}$ **f** $= \begin{pmatrix} -6 \\ 12 \end{pmatrix}$ **g** $= \begin{pmatrix} -1 \\ 2 \end{pmatrix}$

3 Calculate.

a $\begin{pmatrix} 1 \\ -2 \end{pmatrix} + \begin{pmatrix} -2 \\ -1 \end{pmatrix}$ **b** $\begin{pmatrix} 0 \\ -3 \end{pmatrix} - \begin{pmatrix} -2 \\ 4 \end{pmatrix}$ **c** $-3\begin{pmatrix} 2 \\ -1 \end{pmatrix}$

4 EFG is a straight line.

EF : FG = 2 : 3

Find:

a $\overrightarrow{EG}$ **b** $\overrightarrow{FG}$

c $\overrightarrow{OF}$

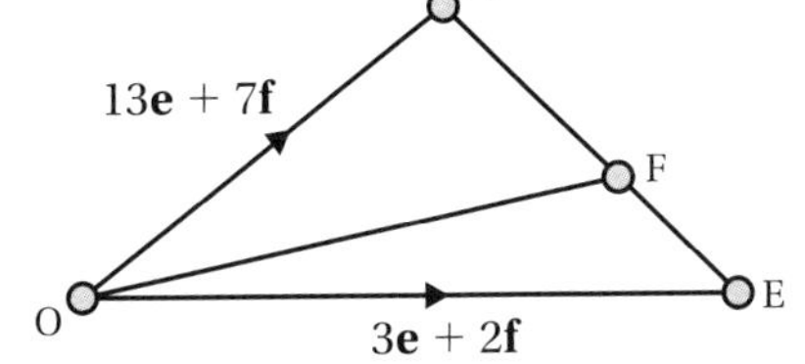

5 The points G and H lie on line EF.

$\overrightarrow{EF} = 12\mathbf{e} - 18\mathbf{f}$

The ratio of EG : GH : HF is 1 : 3 : 2.

a Draw a sketch of this situation.

b Find the vectors

i $\overrightarrow{GH}$ **ii** $\overrightarrow{HE}$

6 In triangle OAB, $\overrightarrow{OA} = 6\mathbf{a}$ and $\overrightarrow{OB} = 6\mathbf{b}$.

M is the midpoint of OB and N is the midpoint of AB.

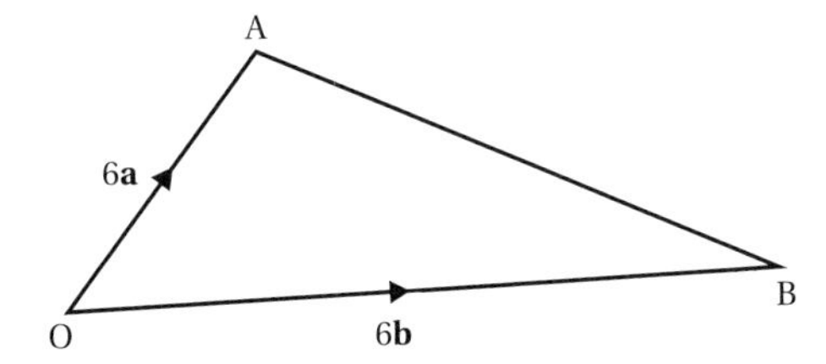

In this question give your answers in their simplest form in terms of **a** and **b**.

a Find $\overrightarrow{AB}$. *(1 mark)*

b Find $\overrightarrow{ON}$. *(2 marks)*

G is a point on AM such that $AG = \frac{2}{3}AM$.

c **i** Find $\overrightarrow{AM}$. *(1 mark)*

ii Find $\overrightarrow{OG}$. *(2 marks)*

d What do your answers tell you about the points O, G and N? *(1 mark)*

Find answers at: cambridge.org/ukschools/gcsemaths-studentbookanswers

29 Plane isometric transformations

In this chapter you will learn how to ...

- carry out rotations, reflections and translations.
- identify and describe rotations, reflections and translations.
- describe translations using column vectors.
- perform multiple transformations on a shape and describe the results.

For more resources relating to this chapter, visit GCSE Mathematics Online.

Using mathematics: real-life applications

You can see examples of reflections, rotations and translations all around you. Patterns in wallpaper and fabric are often translations, images reflected in water are reflections and the blades of a wind turbine are a good example of rotation.

Tip

Tracing paper is very useful for work with transformations. Don't be afraid to ask for it in an exam.

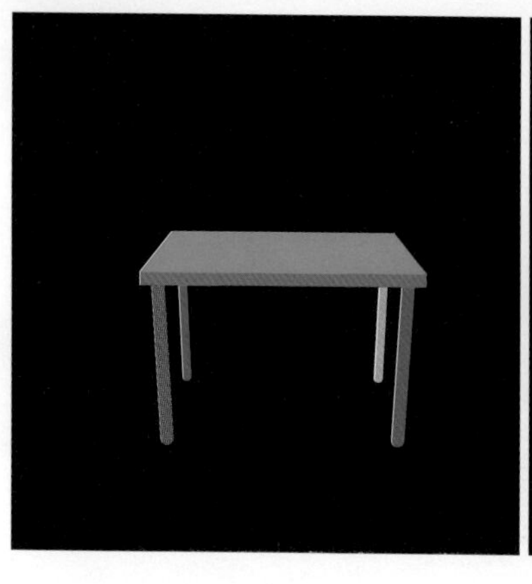

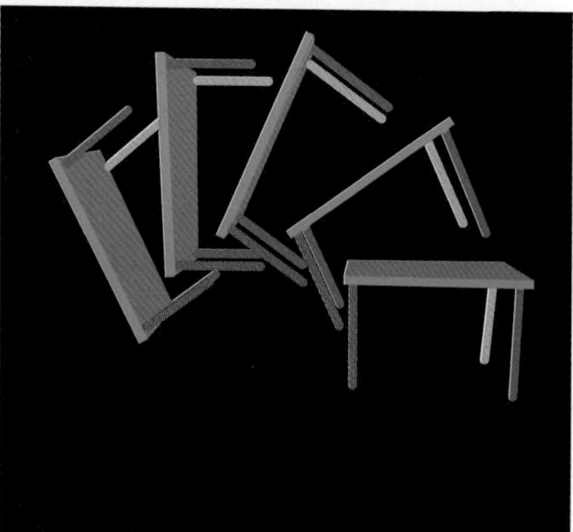

"I use transformations all the time when I program computer graphics. Transformations allow me to position objects, shape them and change the view I have of them. I can even change the type of perspective that is used to show something."

(Computer programmer)

Before you start ...

Ch 9	You need to know what angles of 90°, 180° and 270° look like and also the directions clockwise and anti-clockwise.	**1** How many degrees is each angle? State whether each arrow is showing clockwise or anti-clockwise movement. **a** **b** **c**
Ch 18, 19	You need to know how to plot straight-line graphs in the form $x = a$, $y = a$ and $y = x$.	**2** Draw the graph for each equation. **a** $y = 2$ **b** $y = x$ **c** $x = {}^{-}1$ **d** $y = {}^{-}x$
Ch 27	You need to know what a vector is and how they describe movement.	**3** **a** What is the difference between the coordinate (3, 2) and the vector $\begin{pmatrix} 3 \\ 2 \end{pmatrix}$? **b** What is the relationship between the vectors $\begin{pmatrix} ^{-}1 \\ 3 \end{pmatrix}$ and $\begin{pmatrix} 3 \\ 1 \end{pmatrix}$?

Assess your starting point using the Launchpad

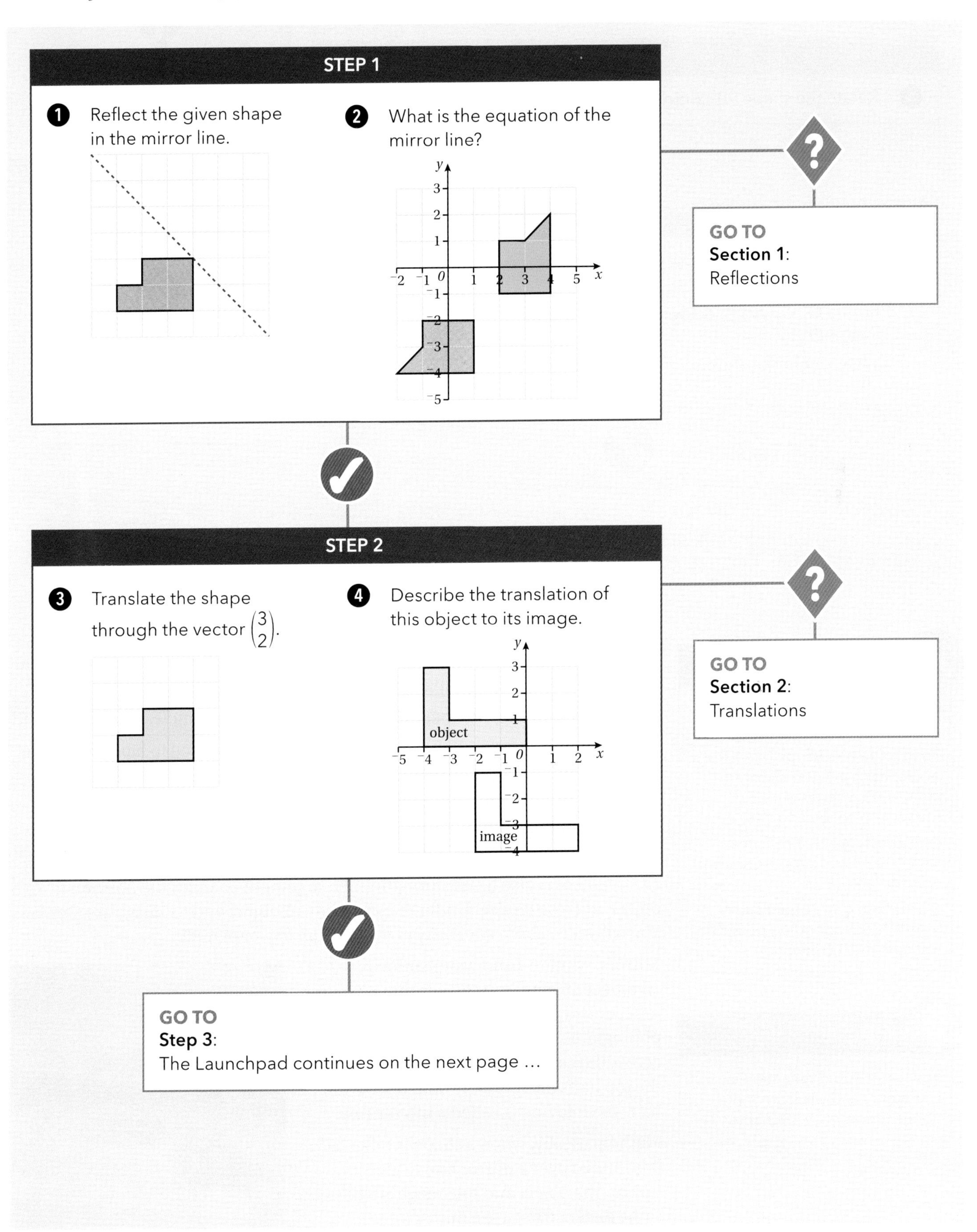

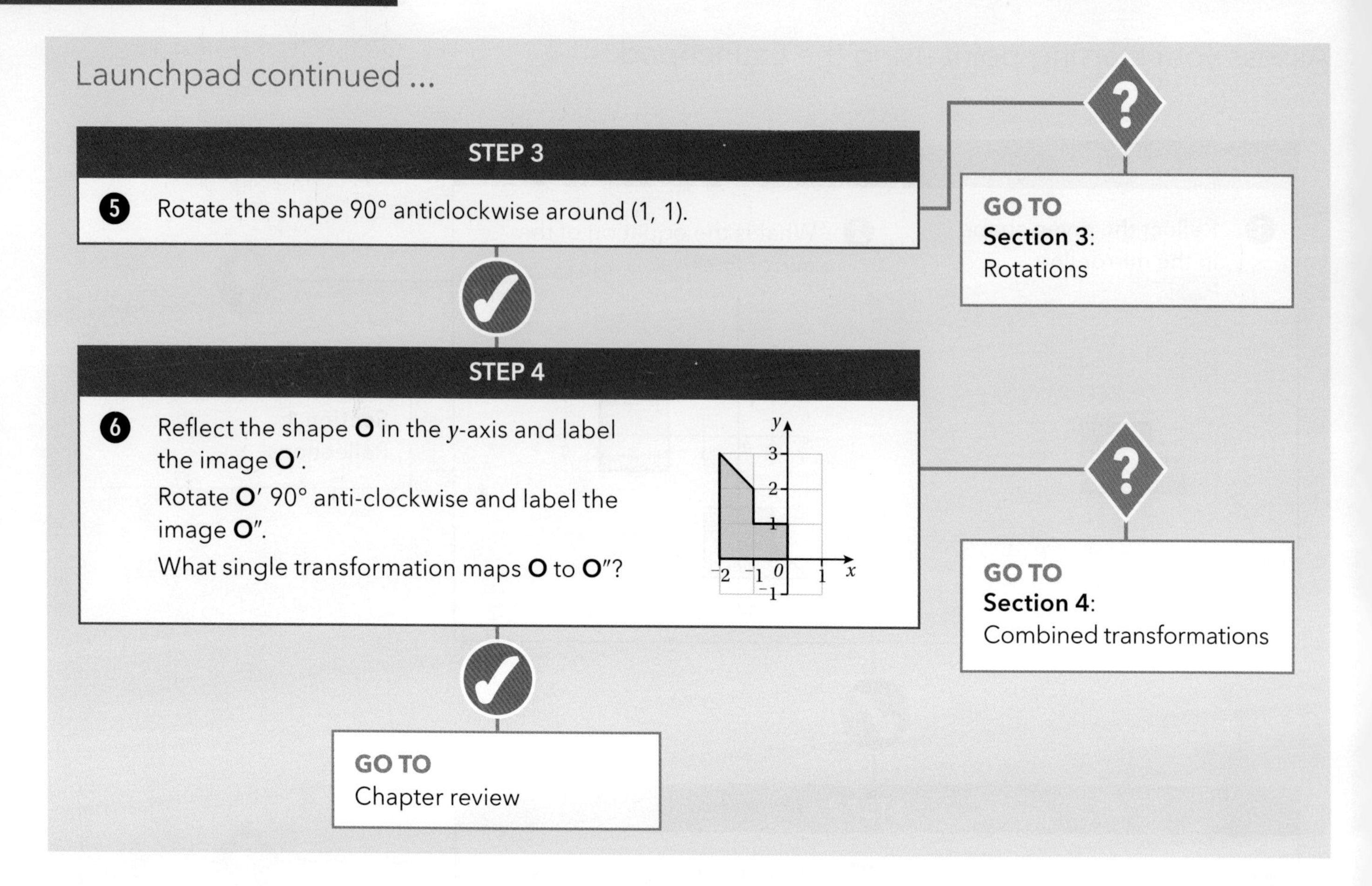

Section 1: Reflections

Key vocabulary

object: the original shape (before it has been transformed).

image: the new shape (once the object has been transformed).

congruent: shapes that are identical in shape and size.

similar: shapes that have the same shape and proportions but are a different size.

mirror line: a line equidistant from all corresponding points on a shape and its reflection.

Tip

You learnt about reflection in Chapter 5, and will learn more about congruency in Chapter 29.

A transformation is a change in the position of a point, line or shape. When you transform a shape you change its position or its size.

The original point, line or shape is called the **object**. For example, the triangle ABC.

The transformation is called the **image**. The symbol ′ is used to label the image. For example, the image of triangle ABC is A′B′C′.

Reflection, rotation and translation change the position of an object, but not its size. Under these three transformations an object and its image will be **congruent**.

Enlargement is also a transformation. Enlargement changes the position of an object and also its size. Under enlargement an object and its image are **similar**. You will learn about similarity in more detail in Chapter 30.

Mirrors, windows and water surfaces all reflect objects. You can see the reflection of clouds and trees clearly in the photograph. If you draw a line horizontally across the centre of the image and fold it, the top half will fit exactly onto the bottom half. The fold line is called a **mirror line**.

Mathematically, when a shape is reflected it is flipped over a mirror line to give its image. The object and the image are the same distance from the mirror line.

EXERCISE 29A

1 Reflect the triangle in the line $x = 1$, and then reflect the triangle and the resultant image in the line $y = {}^{-}1$.

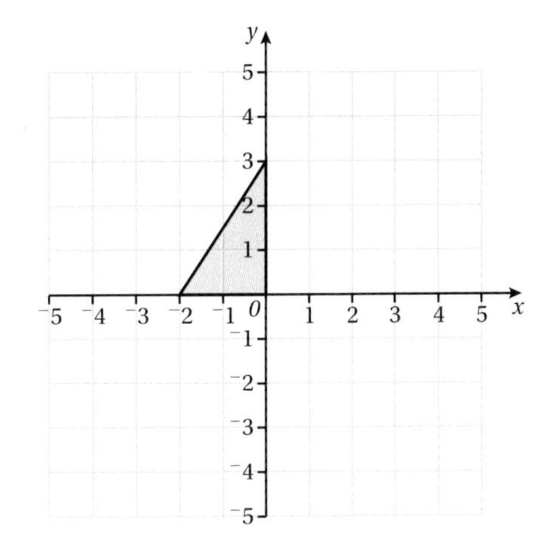

2 Reflect this shape in the line $y = x$, and then reflect the shape and the resultant image in the line $y = 1 - x$.

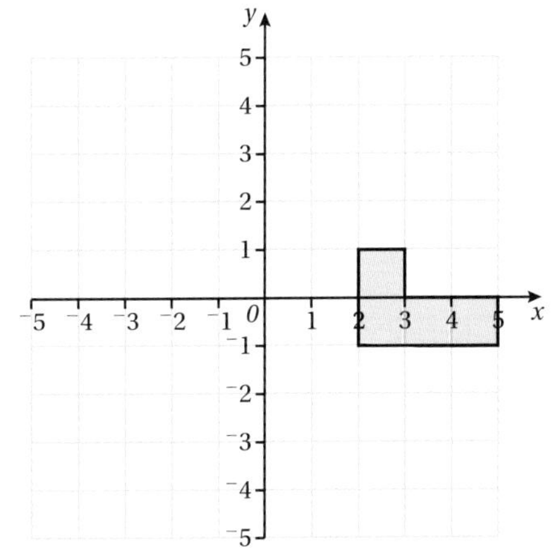

3 Carry out the ten reflections on the grid below to reveal the word.

Shape **A** in the line $y = x$

Shape **B** in the x-axis

Shape **C** in the line $y = {}^{-}2$

Shape **D** in the line $y = 4$

Shape **E** in the line $x = 7$

Shape **F** in the line $y = {}^{-}x$

Shape **G** in the line $x = {}^{-}4$

Shape **H** in the line $x = 1.5$

Shape **I** in the line $y = 6$

Shape **J** in the line $y = x$

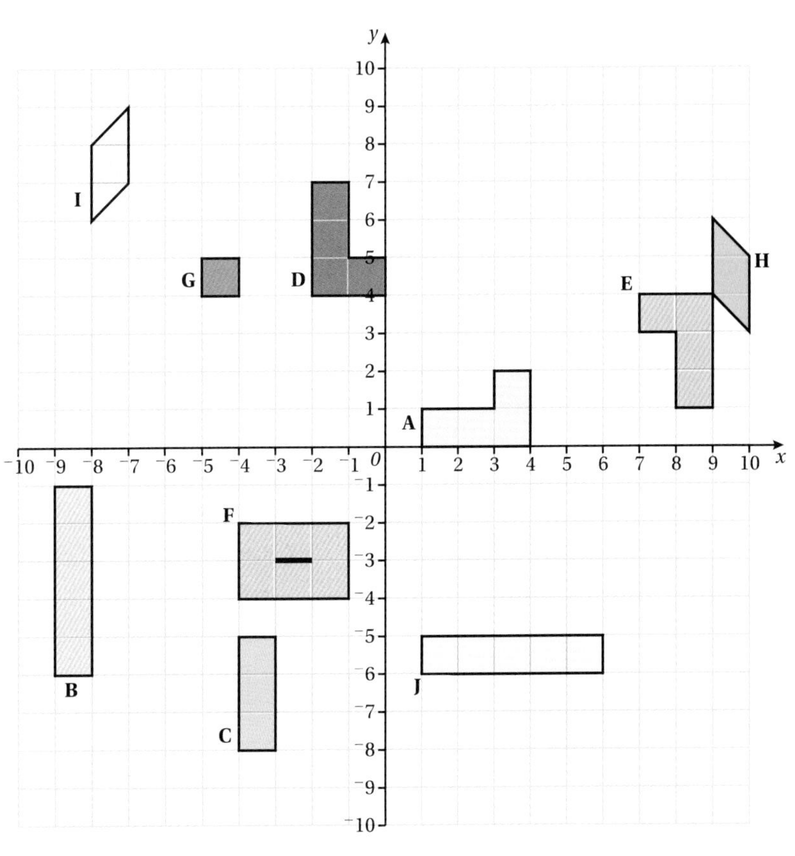

4 Draw a pair of axes going from $^{-}10$ to 10 in both directions. Draw two line segments the first with ends at (0, 0) and (1, 1) the second with ends at (1, 1) and (2, 1). Show how these lines can be made into:

a a hexagon with two successive reflections.

b an octagon with three successive reflections.

5 Make up your own reflection puzzle for another student to solve.

Describing reflections

You need to be able to draw a mirror line on a diagram. You must also be able to give the equation of the mirror line when the reflection is shown on a coordinate grid. This is known as describing the reflection.

The mirror line is the perpendicular bisector of two corresponding points in a reflection. So, if you can't 'spot' a mirror line, you can join two corresponding points and construct the perpendicular bisector to find it.

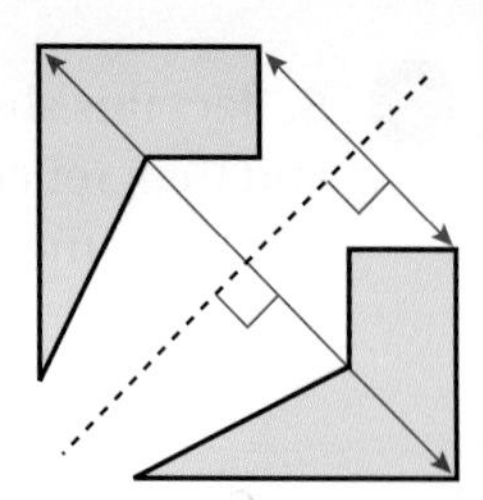

Tip

To check a reflection, trace the **object**, the **image** and the mirror line. Fold the tracing paper along the mirror line; the shapes should match up exactly.

EXERCISE 29B

1 Find the equation of the mirror line in each reflection.

a

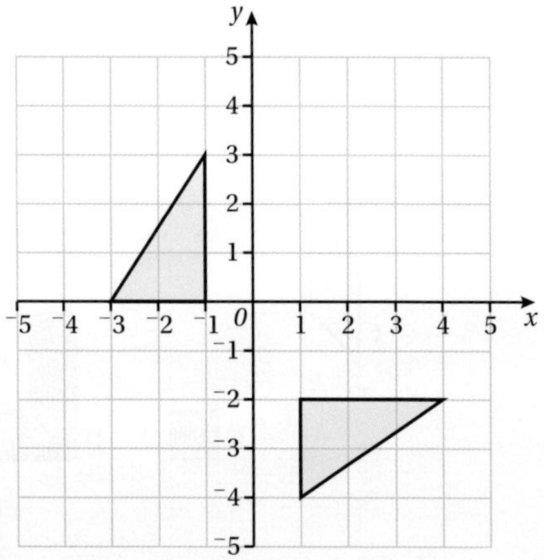

b

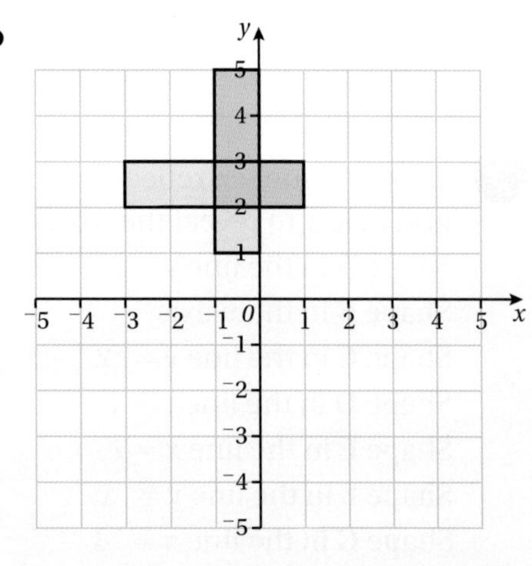

c

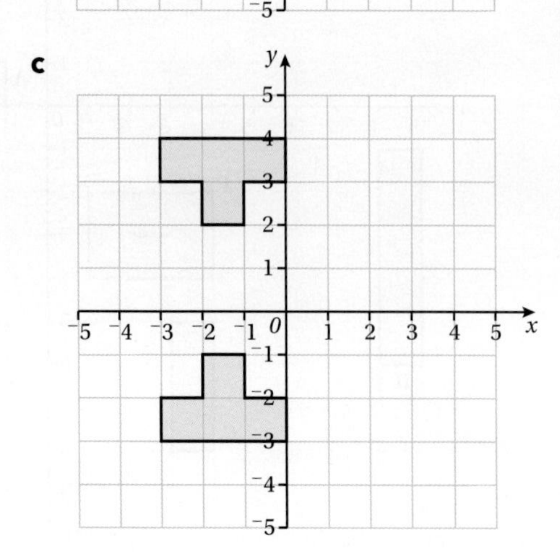

d

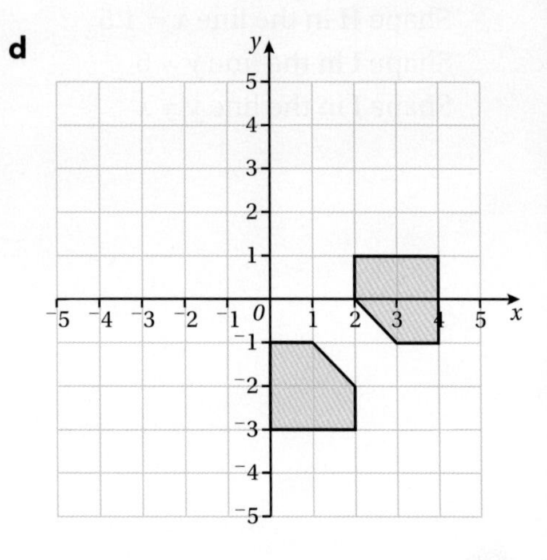

2 **a** Fully describe each of the following reflections.

i Shape **A** to shape **E**.

ii Shape **C** to shape **G**.

iii Shape **G** to shape **E**.

iv Shape **B** to shape **F**.

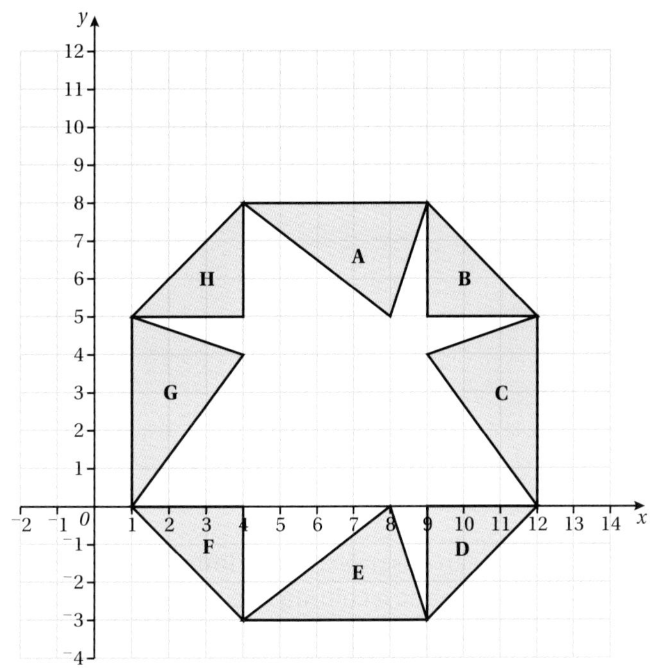

b Challenge another student to describe a reflection of two triangles you choose.

3 Trace each pair of shapes and construct the mirror line for the reflection.

a

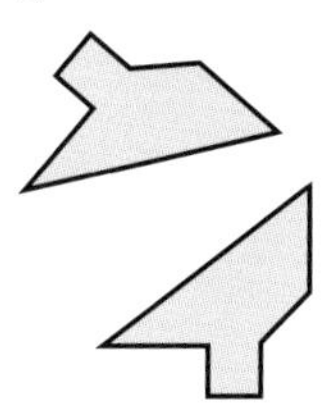

b

c

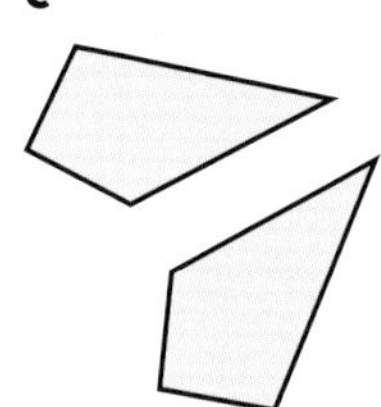

4 **a** In each diagram below, describe the multiple reflections that take the original shaded shape to its image.

i

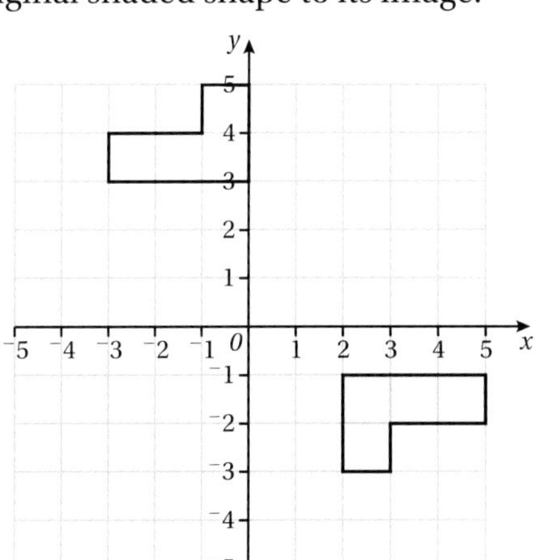

ii

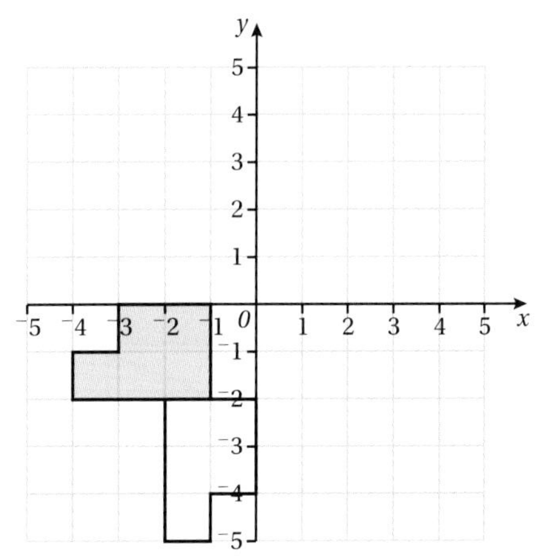

b Is there more than one answer?

c Does the order of the reflections matter?

d Is there any easy way to tell the minimum number of reflections needed?

Section 2: Translations

Key vocabulary

orientation: the position of a shape relative to the grid.

A translation is a slide along a straight line. (Think about pushing a box across a floor.) The translation can be from left to right (horizontal), up or down (vertical) or both (horizontal and vertical).

The image is in the same **orientation** as the object and every point on the shape moves exactly the same distance in exactly the same direction. Translated shapes are congruent to each other.

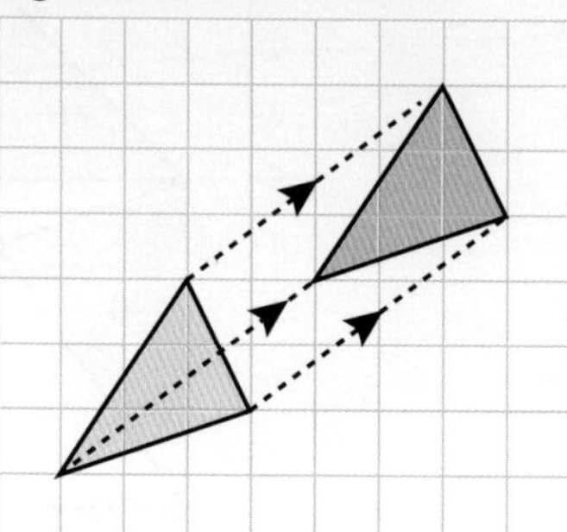

Tip

You learnt about vectors in Chapter 28.

You can describe translations on a coordinate grid using column vectors. Remember, a column vector shows horizontal displacement over vertical displacement.

Tip

Drawing on the grid to show the movements can help you to avoid unnecessary mistakes.

WORKED EXAMPLE 1

Describe the translation ABC to A′B′C′ by means of a column vector.

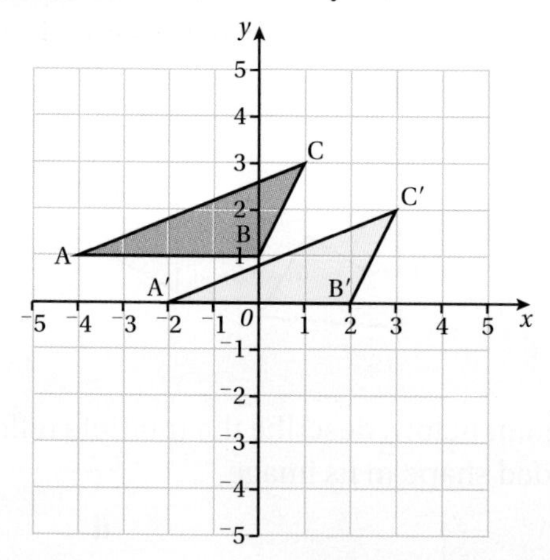

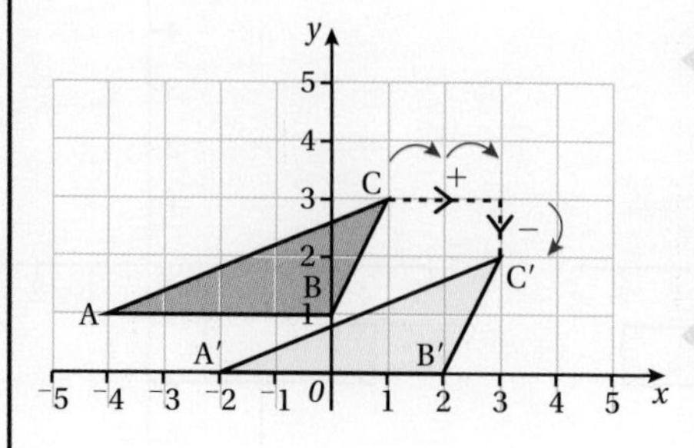

Take any point on the object and find the corresponding point on the image. Look at point C and point C′.

Work out how the point has been translated horizontally and vertically. To get from C to C′ you move:

2 units to the right = +2

1 unit down = -1

The translation is $\begin{pmatrix} 2 \\ -1 \end{pmatrix}$.

Write this as a column vector.

WORK IT OUT 29.1

This shape is translated through a vector of $\begin{pmatrix} ^{-}2 \\ 4 \end{pmatrix}$. Draw its image.

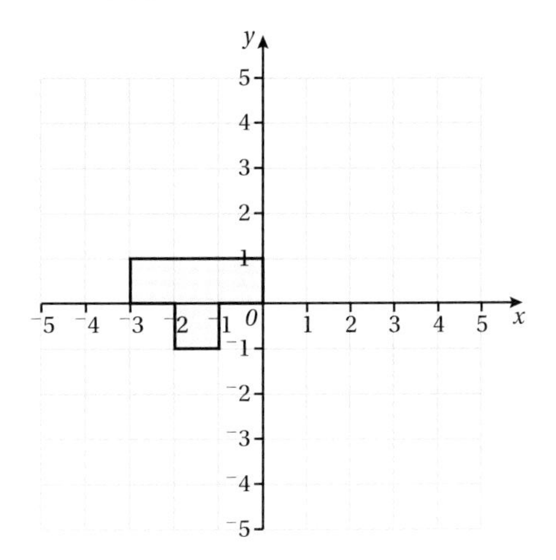

Which one of these answers is correct?
What has gone wrong in each of the others?

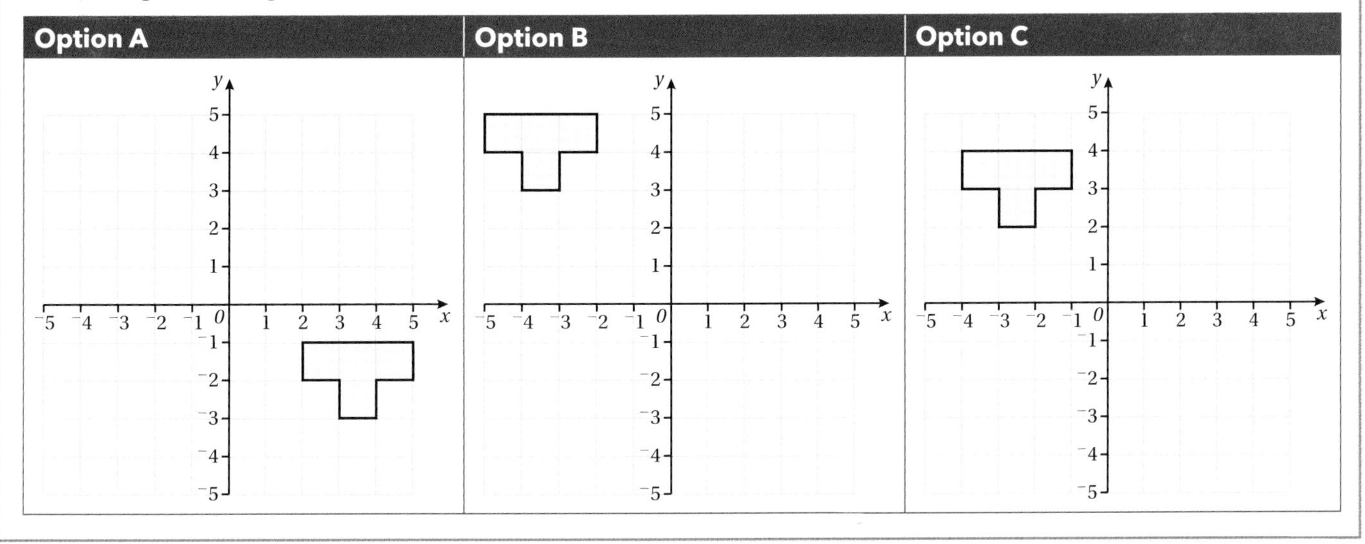

EXERCISE 29C

1 Translate each shape as directed.

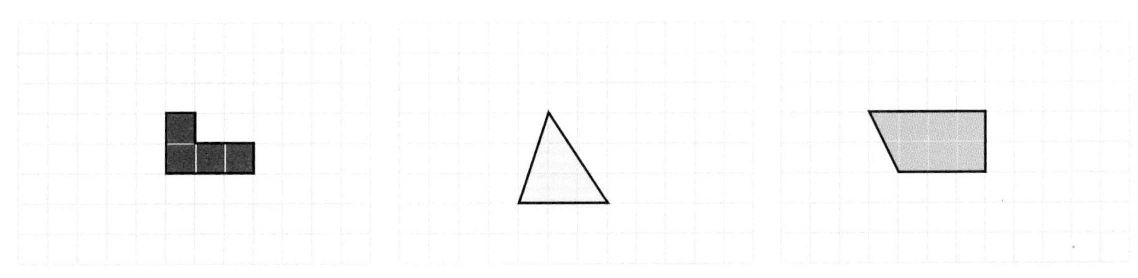

a Translate 6 right and 2 up. **b** Translate 3 right and 1 down. **c** Translate 2 down and 4 right.

2 Translate each shape using the given vector.

a $\begin{pmatrix} 3 \\ -1 \end{pmatrix}$ b $\begin{pmatrix} -1 \\ 2 \end{pmatrix}$ c $\begin{pmatrix} 0 \\ 4 \end{pmatrix}$

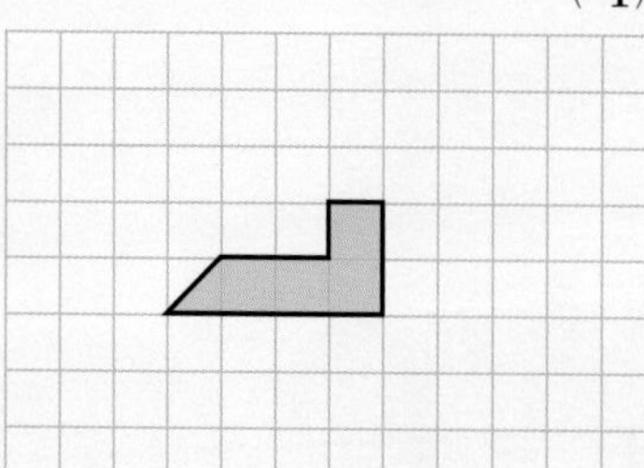

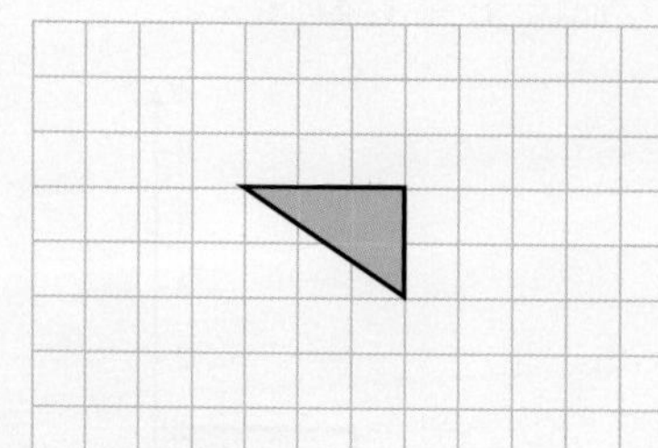

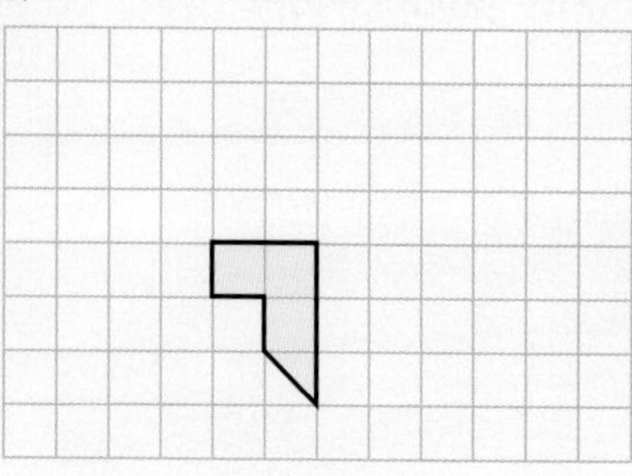

3 Translate each shape by the given vector, then give the name of the shape you have put together.

a Translate shape **A** by $\begin{pmatrix} -1 \\ -3 \end{pmatrix}$.

b Translate shape **B** by $\begin{pmatrix} 1 \\ 5 \end{pmatrix}$.

c Translate shape **C** by $\begin{pmatrix} 2 \\ -1 \end{pmatrix}$.

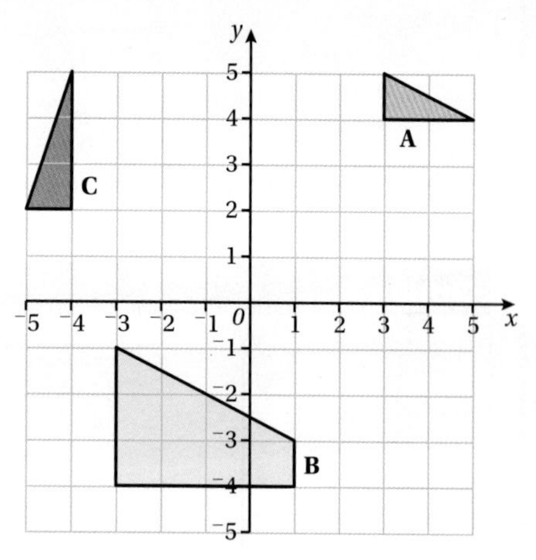

4 Translate each piece of this jigsaw using the vectors on the right.

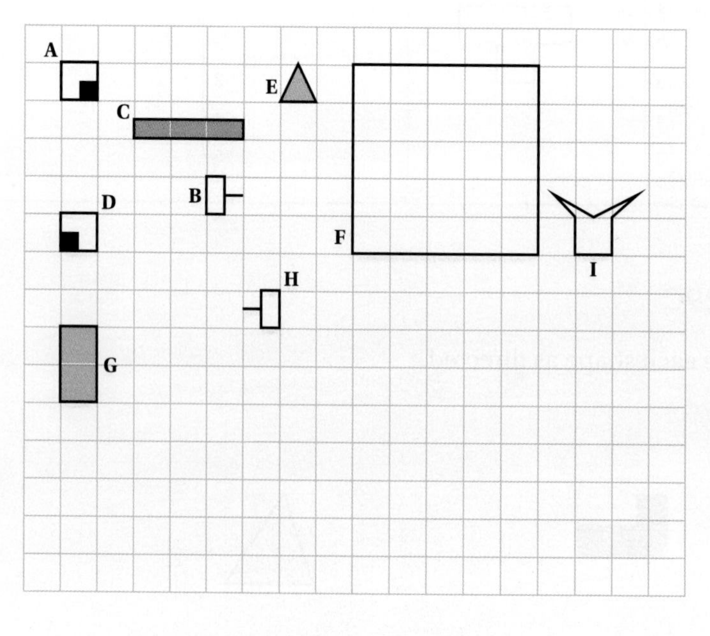

A $\begin{pmatrix} 12 \\ -8 \end{pmatrix}$

B $\begin{pmatrix} 12 \\ -5 \end{pmatrix}$

C $\begin{pmatrix} 10 \\ -9 \end{pmatrix}$

D $\begin{pmatrix} 14 \\ -4 \end{pmatrix}$

E $\begin{pmatrix} 7 \\ -9 \end{pmatrix}$

F $\begin{pmatrix} 3 \\ -7 \end{pmatrix}$

G $\begin{pmatrix} 13 \\ -5 \end{pmatrix}$

H $\begin{pmatrix} 5 \\ -2 \end{pmatrix}$

I $\begin{pmatrix} -1 \\ -2 \end{pmatrix}$

Describing translations

You should be able to use vectors to describe a translation. Remember to count between corresponding points on the two shapes.

Tip

Make sure you count from the object to the image and write this as a column vector (not a coordinate)!

WORK IT OUT 29.2

Which transformations below are reflections and which are translations? How did you make your decision?

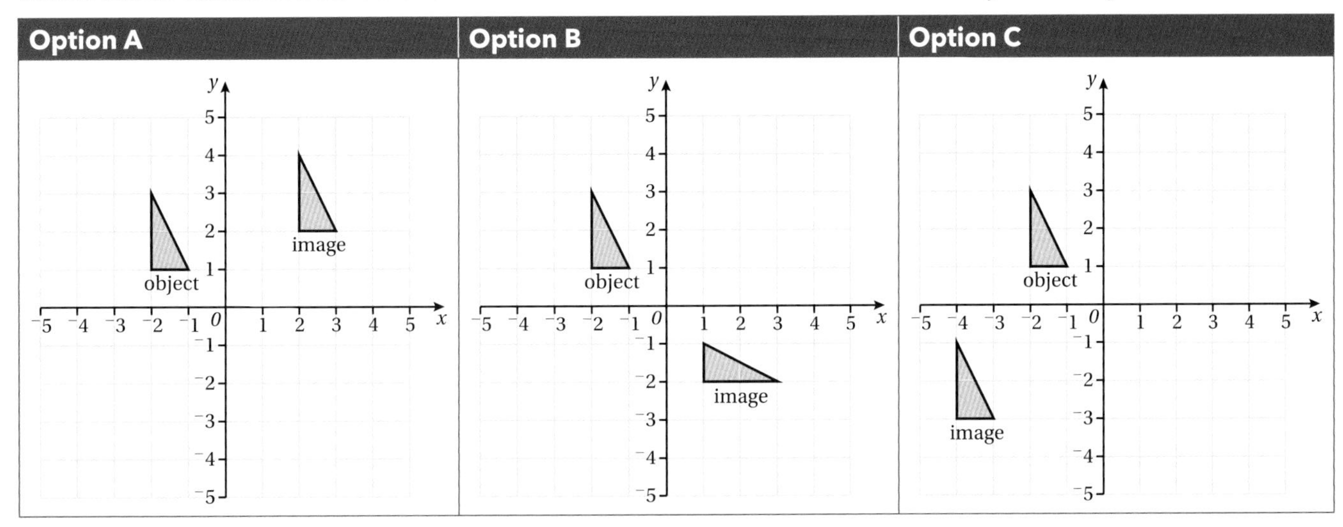

EXERCISE 29D

1 Here are some completed translations. The objects (**A** to **E**) are shown in colour and the images are in grey. Write column vectors to describe the translation from each object to its image.

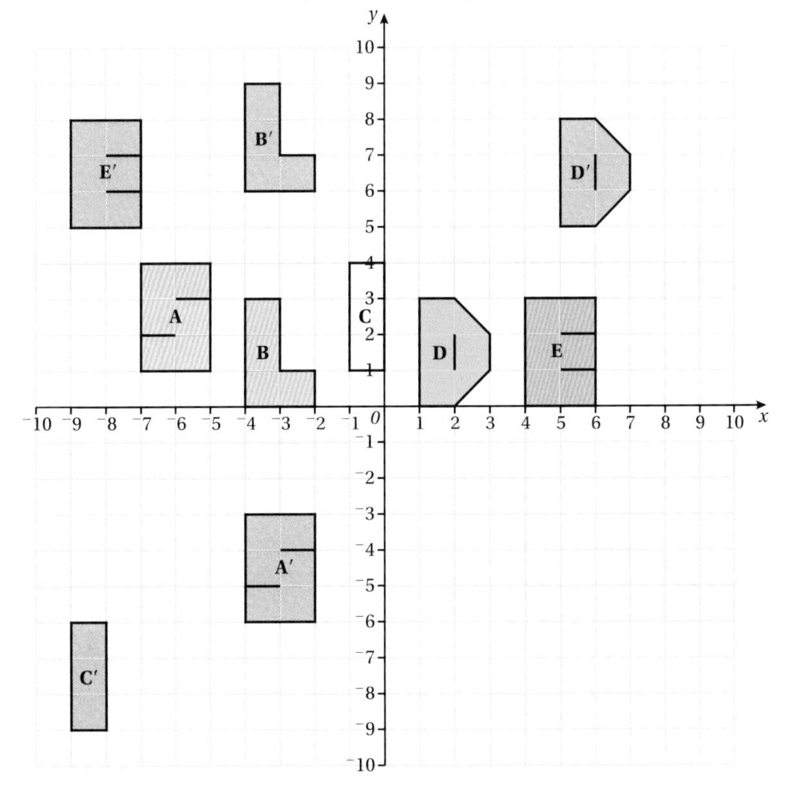

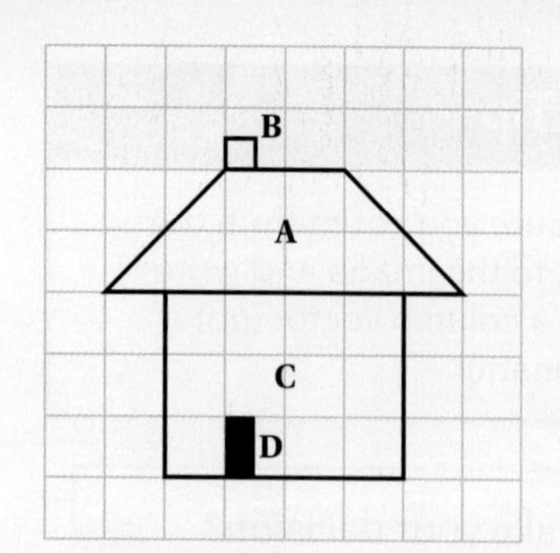

2. Work on a square grid. Draw the four objects (**A** to **D**) used to make this image in any position on the grid.
Make up translation instructions for moving the four objects to form the image.
Exchange with a partner and perform the translations to make sure their instructions are correct.

Section 3: Rotations

A rotation is a turn. An object can turn clockwise or anti-clockwise around a fixed point called the **centre of rotation**. The centre of rotation may be inside, on the edge of, or outside the object.

A rotation changes the orientation of a shape, but the object and its image remain congruent.

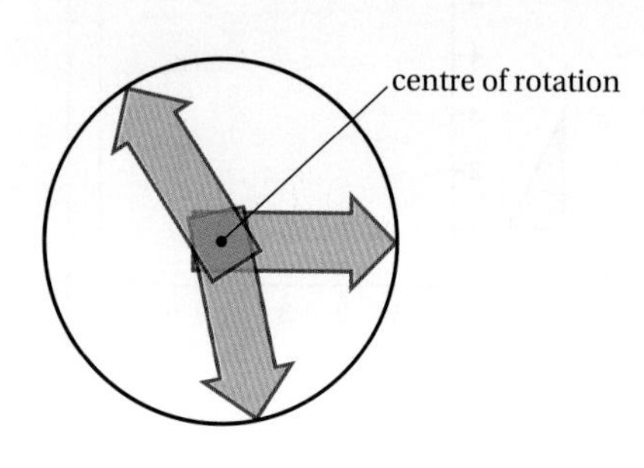

When you rotate a shape, the distance from the centre of rotation to any point on the object remains the same. Each point travels in a circle around the centre. Think about the tip of a blade on a wind turbine or a child sitting on a roundabout. When they rotate, the paths they trace out are circles.

To carry out a rotation you need to know the centre of rotation as well as the angle and direction of rotation. At this level, all rotations will be in multiples of 90°.

WORK IT OUT 29.3

This shape is rotated anti-clockwise, with centre of rotation (0, 1) through an angle of 90°. What is its image?

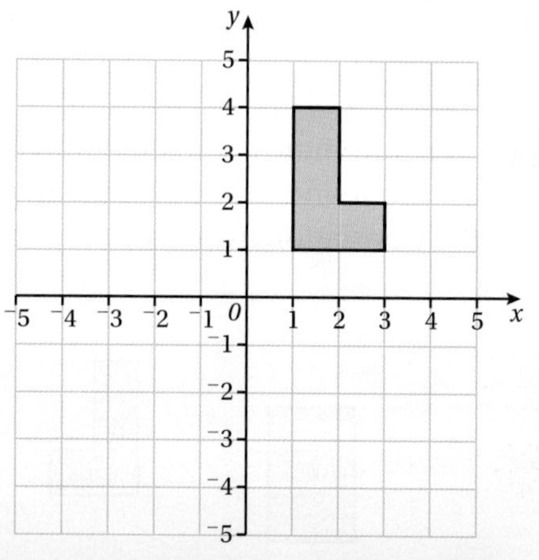

Which one of these answers is correct?
What has gone wrong in each of the others?

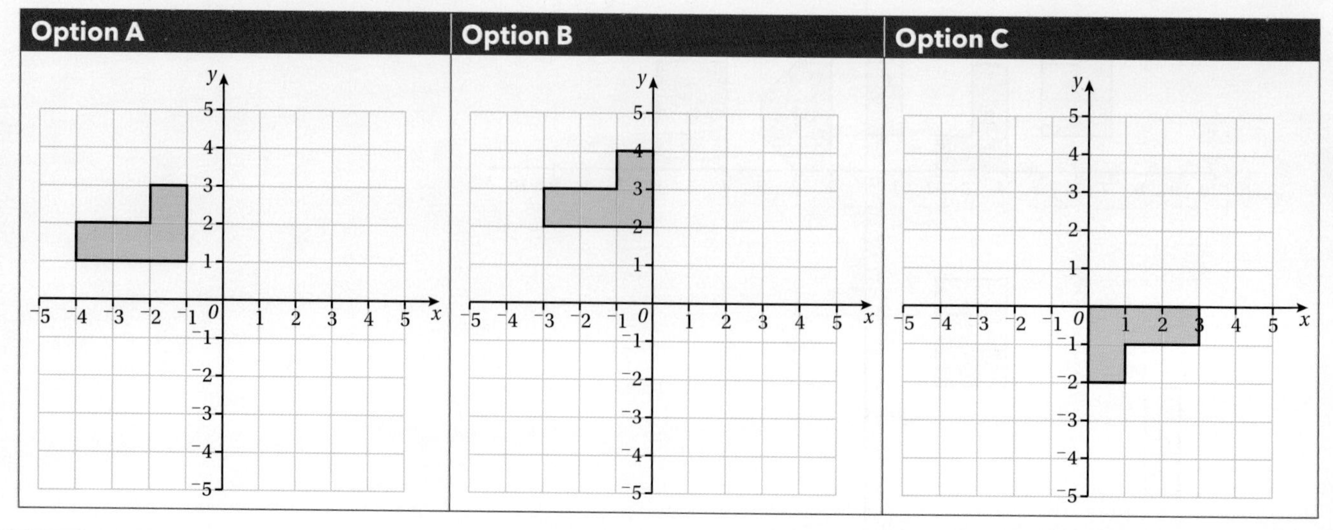

EXERCISE 29E

1 Rotate the triangle 180° about the origin.

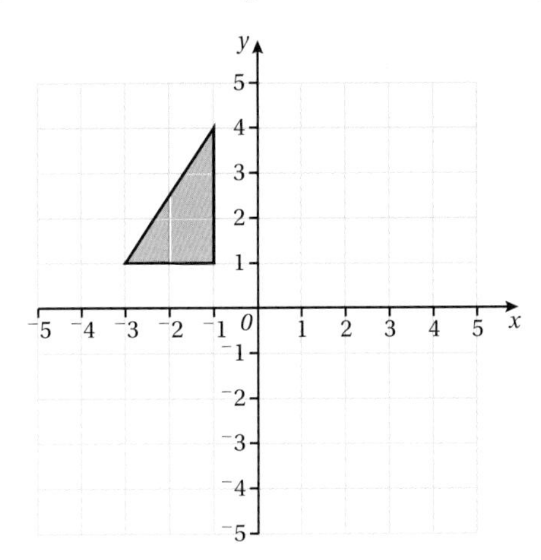

2 Rotate the shape 90° clockwise around the point (1, 1).

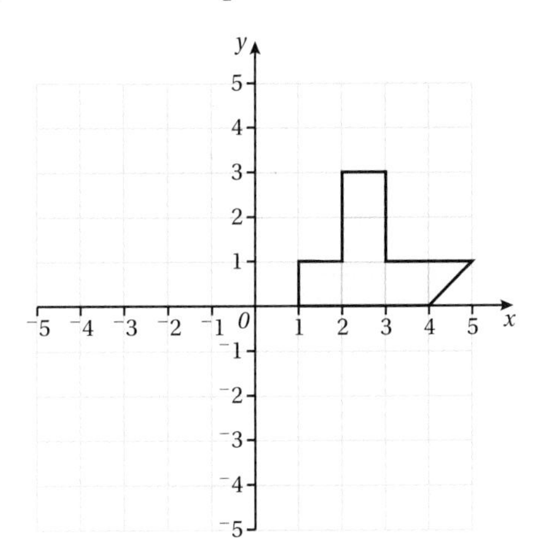

3 Rotate the shape 90° anti-clockwise around the point (⁻2, 1).

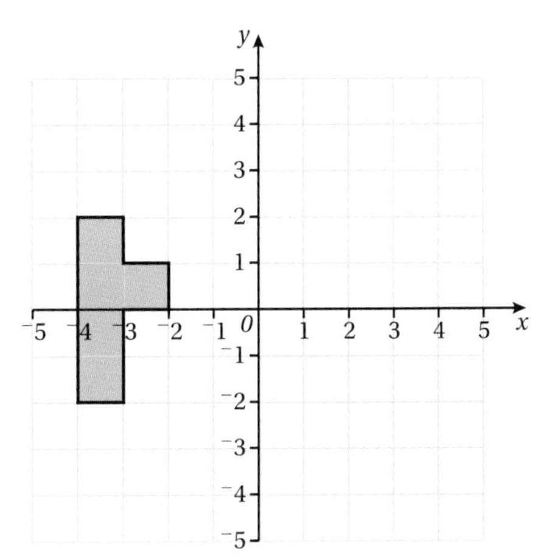

4 Rotate the shape 90° anti-clockwise around the point (2, 1).

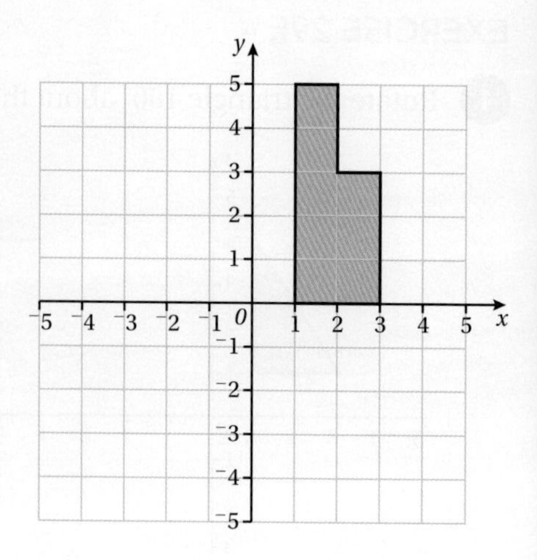

5 Rotate each shape as directed.

Shape **A**: 90° anti-clockwise around the point (⁻1, ⁻1).

Shape **B**: 180° around the point (2, 3).

Shape **C**: 90° clockwise around the point (1, 0). Label this D.

Shape **D**: 180° around the point (⁻3.5, 2).

Shape **E**: 180° around the point (3, 1).

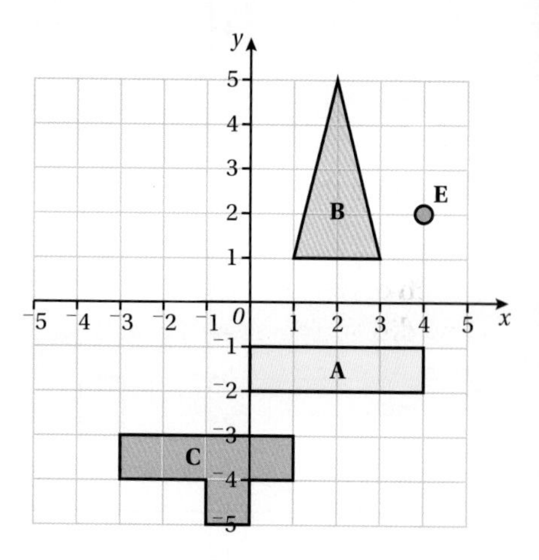

6 This image was designed by drawing a triangle and rotating this around the origin in multiples of 90°.

What do you notice about the coordinates of the vertices of the triangle? Would this work if you rotated an image around a different point? Why?

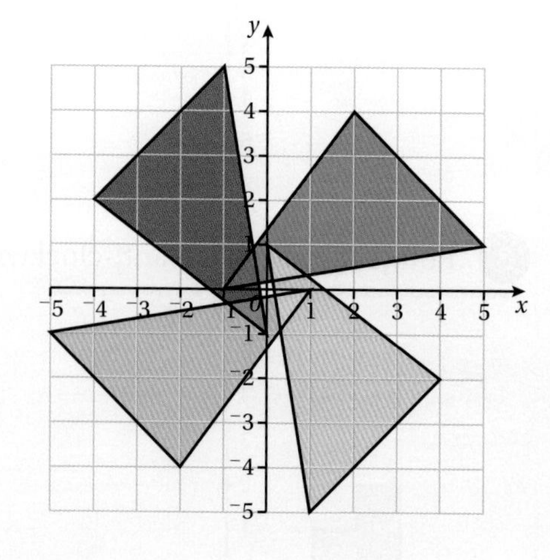

Describing rotations

Tip

Always give the direction and angle of the rotation from the object to the image.

To describe a rotation you give a centre, angle and direction. Very often you can find the centre using tracing paper and trial and error. Trace the object and rotate the tracing paper using different centres of rotation. Spotting the centres improves with practice.

You can also use construction to find the centre of rotation. Corresponding points on an object and its image under rotation lie on the circumference of a circle. If you join the corresponding points to make a chord, the perpendicular bisector of the chord will go through the centre of the circle.

So by drawing the perpendicular bisectors of **two** chords you can determine the centre.

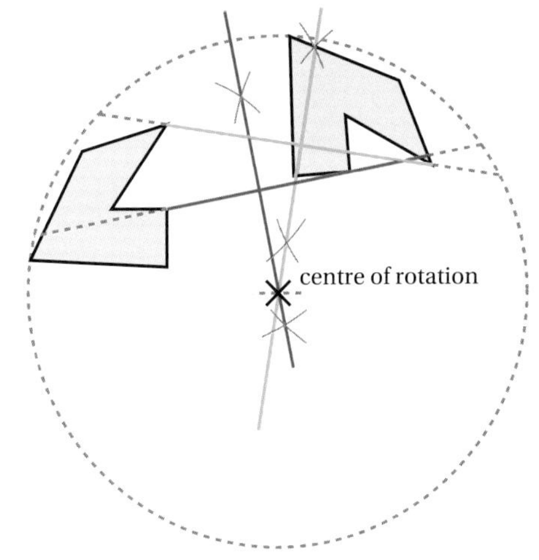

The broken lines show you the circle and chord to help demonstrate the principle; you do not need to draw these!

WORK IT OUT 29.4

Which of these transformations are reflections, which are rotations and which are translations? How did you make your decision? Were there any transformations you couldn't identify?

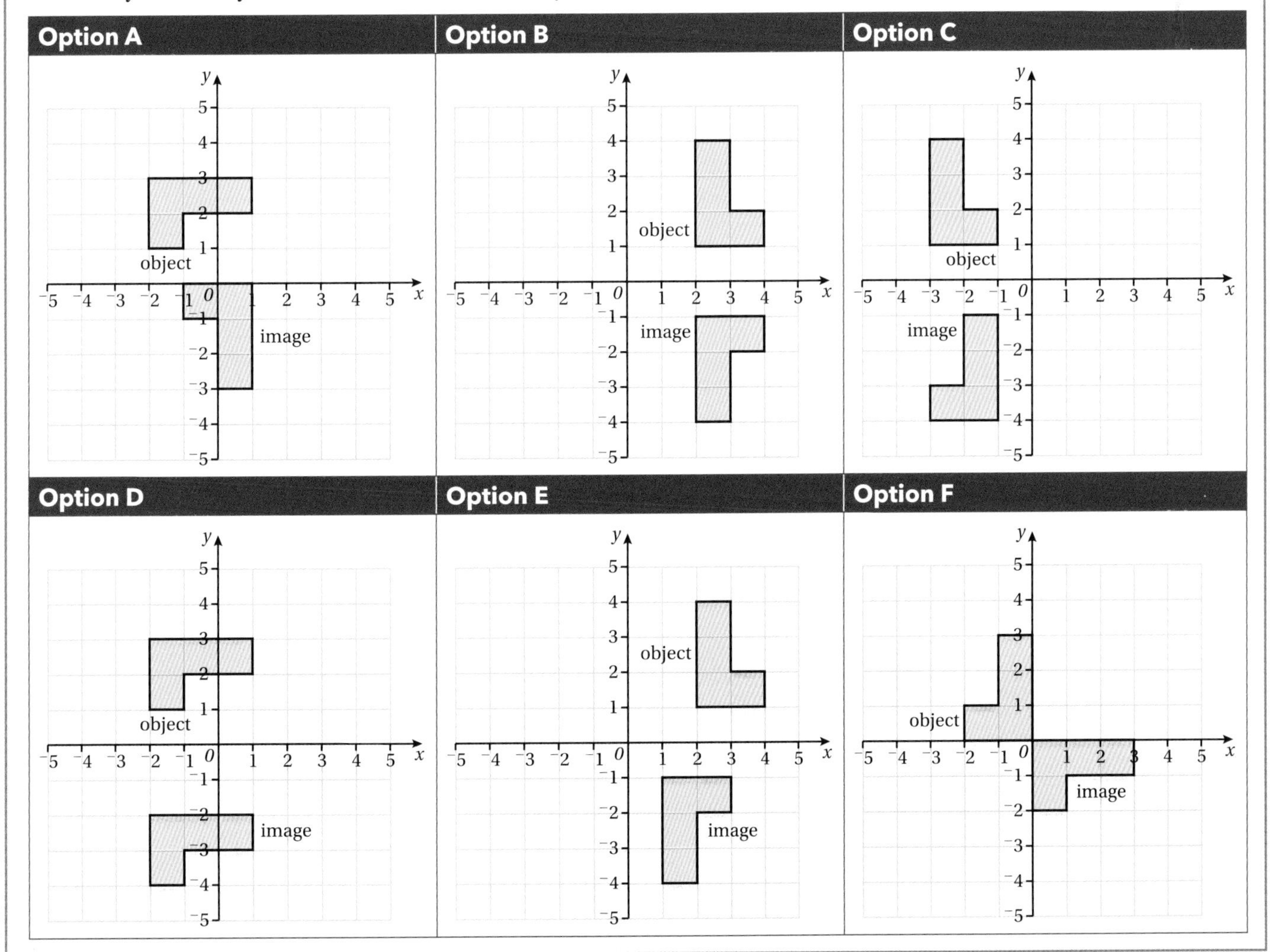

Find answers at: cambridge.org/ukschools/gcsemaths-studentbookanswers

EXERCISE 29F

1 Describe each of the following rotations.

a

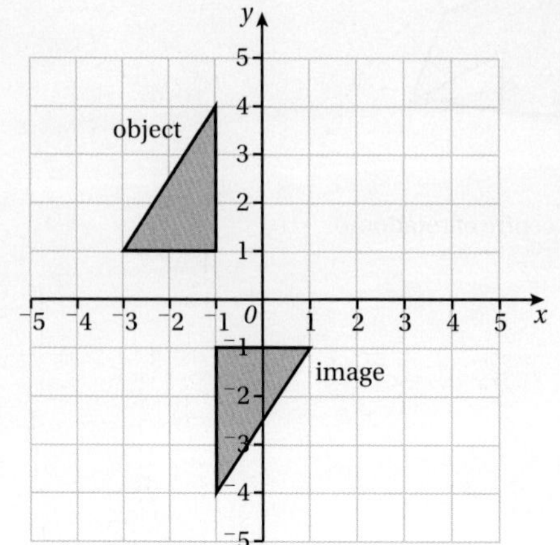

b

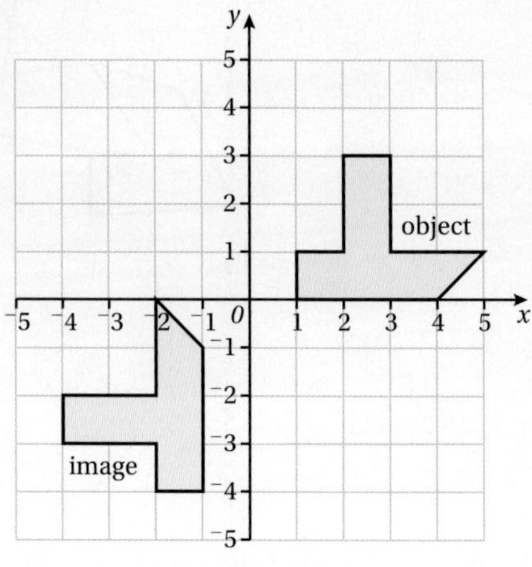

c

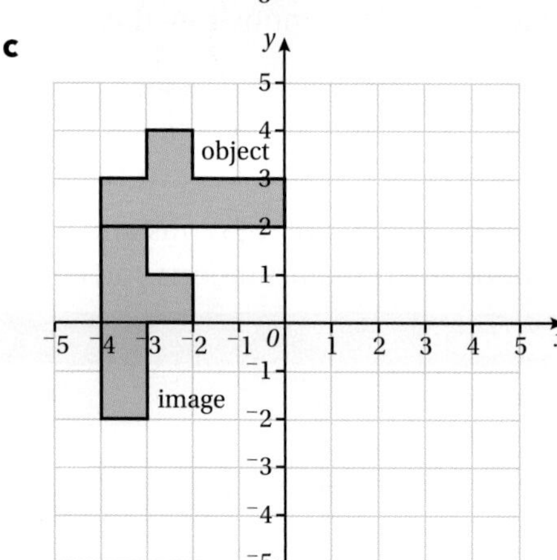

d

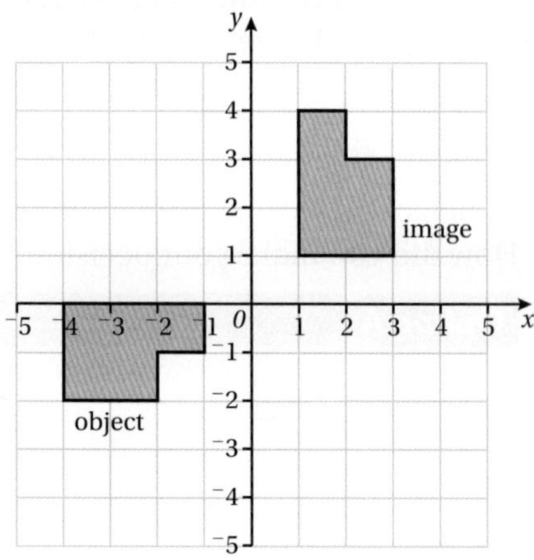

2 This section of wallpaper has been designed using rotations. A **coordinate** grid has been overlaid.

Identify as many different rotations as you can.

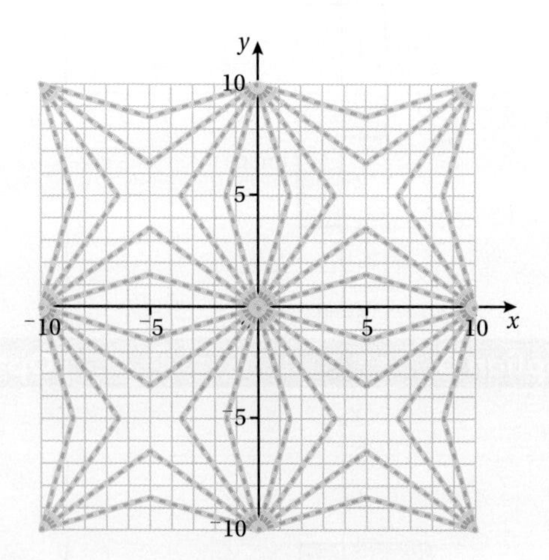

3 Use construction to locate the centres of rotation for each pair of shapes.

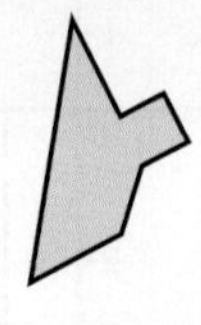

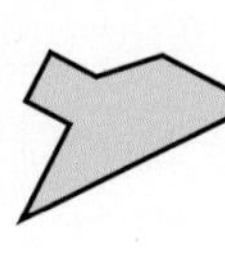

Section 4: Combined transformations

You have seen that an object can undergo a single transformation to map it onto an image. An object can undergo two (or more) transformations in a row. The transformations do not have to be of the same type. For example, an object could be reflected in the line $y = 0$ and then rotated clockwise through 90° about a vertex. Sometimes a combined transformation can be described by a single, equivalent transformation.

EXERCISE 29G

1 On your copy of this diagram, reflect the object **A** in the line $x = 2$ and label it **A′**.

Rotate shape **A′** 90° anti-clockwise around the point (2, ⁻2) label it **A″**.

Describe the single transformation that maps shape **A** to shape **A″**.

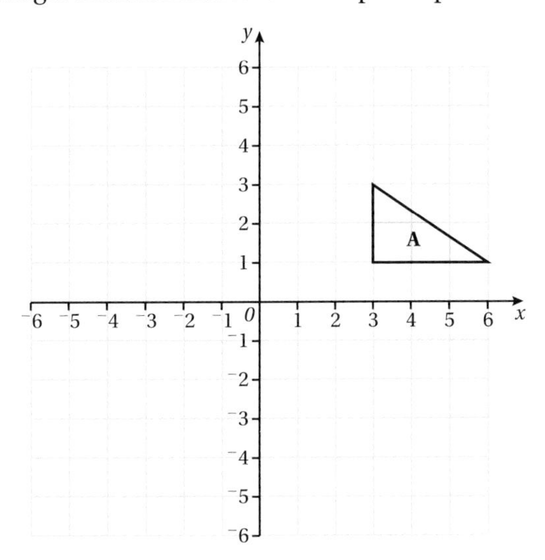

2 On your copy of this diagram, translate **F** through a vector of $\begin{pmatrix} -3 \\ 2 \end{pmatrix}$ and label it **F′**.

Rotate **F′** 180° around the point (⁻3, 0) and label if **F″**.

Describe the single transformation that maps shape F onto to shape **F″**.

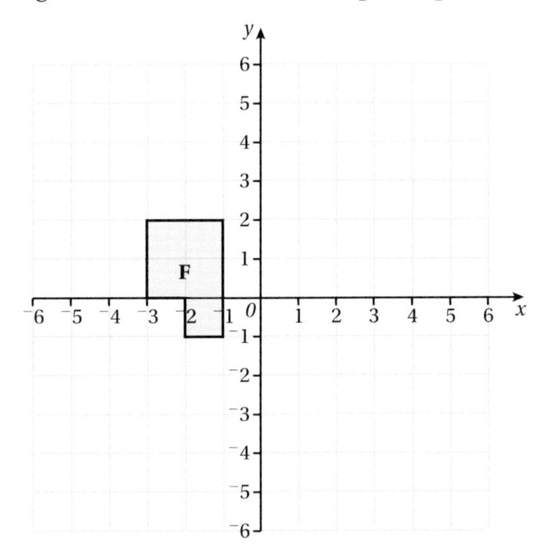

3 On your copy of this diagram, reflect **K** in the y-axis label it **K′**.

Rotate **K′** 90° clockwise around the point (2, ⁻1) and label it **K″**.

Reflect **K″** in the line $x = 0$, label it **K‴**.

Describe the single transformation that maps shape **K** onto shape **K‴**.

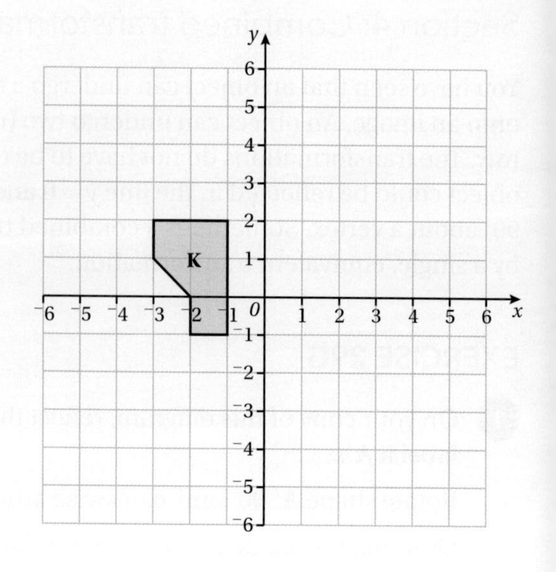

4 **a** Describe fully the combined transformation from shape **A** to **A′**.

b Does the order of the transformations matter?

c Find another solution.

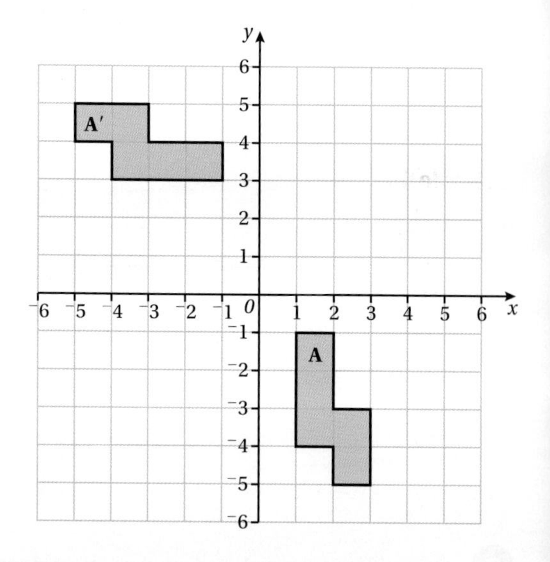

5 A shape is transformed by carrying out a reflection in the x-axis followed by a reflection in the y-axis. What single transformation has the same effect?

6 What two transformations would have the same effect as rotating the following shape 90 degrees anti-clockwise around the point (⁻2, 0)?

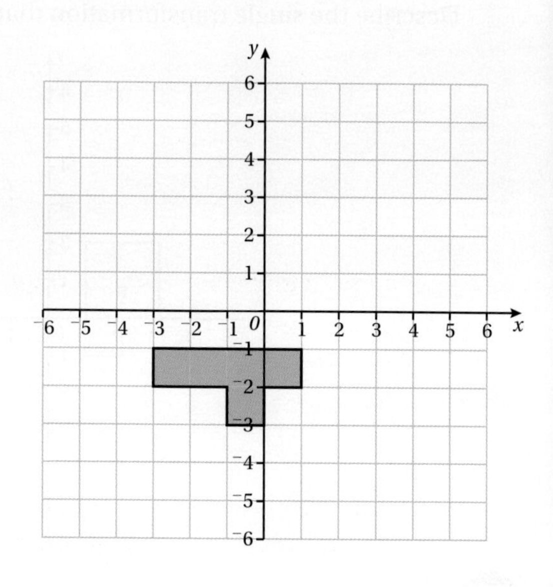

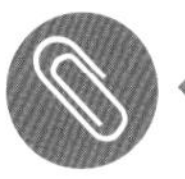

Checklist of learning and understanding

Transformations

- A transformation is a change in the position of a point, line or shape. When you transform a shape you change its position or its size.
- The original point, line or shape is called the **object**: for example, triangle ABC.
- The transformation is called the **image**. The symbol ′ is used to label the image. For example, the image of triangle ABC is A′B′C′.
- Different types of transformation include reflections, translations, enlargements and rotations.

Reflections

- Reflections change the orientation of a shape but the image remains congruent.
- To describe a reflection the equation of the mirror line needs to be given.
- The mirror line is the perpendicular bisector of the line linking any two corresponding points on the image and object.

Translations

- Translations leave the orientation of the shape unchanged but move it horizontally and/or vertically.
- Translations are described using vectors.

Rotations

- A rotation is a turn around a centre. Rotations are described by giving the coordinates of the centre of rotation, angle and direction of the rotation.

Combined transformations

- Transformations can be combined by completing one after another; the transformations do not have to be of the same type. Sometimes the result of combined transformations can be achieved with a single, equivalent transformation.

Chapter review

For additional questions on the topics in this chapter, visit GCSE Mathematics Online.

1 Which of the following statements are true? Explain your reasoning

a The images constructed by reflecting, rotating, or translating are congruent to the objects you started with.

b The images constructed by reflecting, rotating, or translating are similar to the objects you started with.

c The images constructed by reflecting, rotating, or translating are in the same orientation to the objects you started with.

d The images constructed by reflecting, rotating, or translating have the same angles as the objects you started with.

Find answers at: cambridge.org/ukschools/gcsemaths-studentbookanswers

2 Describe fully the transformation from:

a shape **A** to shape **B**.

b shape **B** to shape **C**.

c shape **C** to shape **A**.

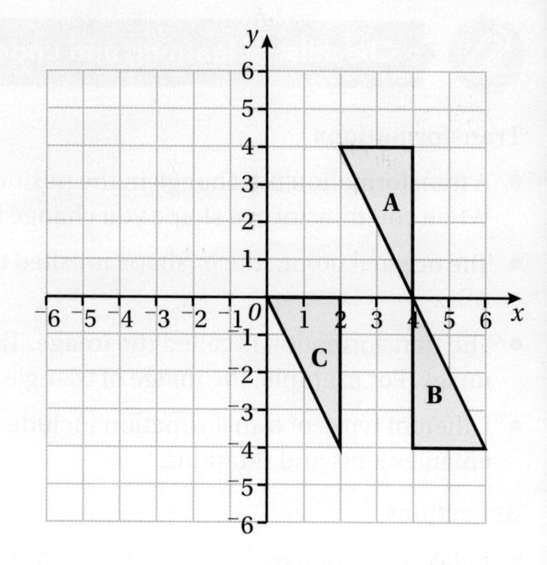

Tip

Tessellation is a pattern of shapes that fit together perfectly, with no gaps or overlaps.

3 This triangle is used to create a tessellating pattern.
This is produced using multiple translations and one rotation.
Explain how this could be done.

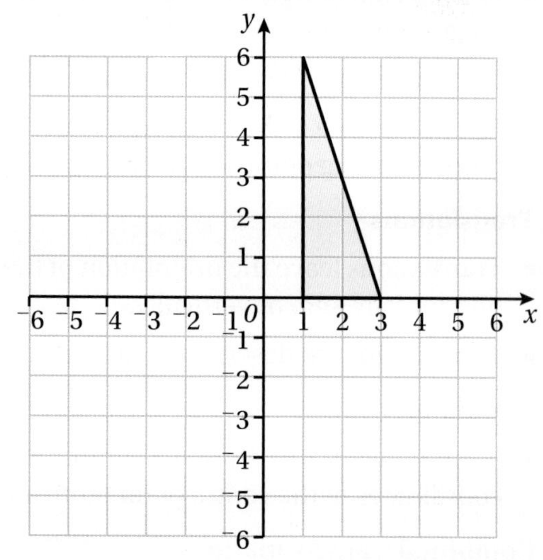

4

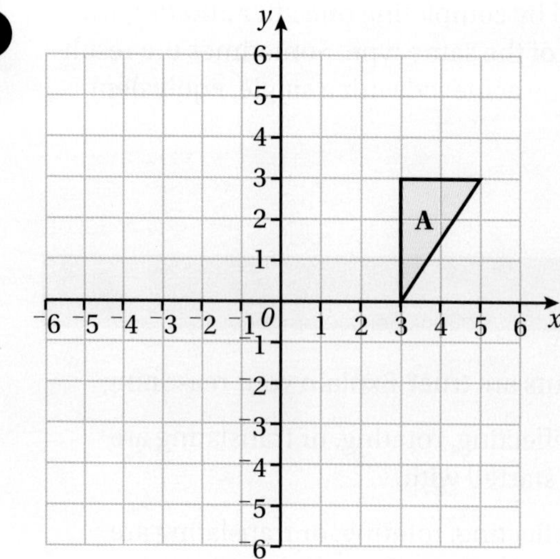

Rotate shape **A** 90° clockwise around the point (1, 2). Label it **B**.
Reflect shape **B** in the line $y = x$. Label it **C**.
Translate shape **C** through the vector $\begin{pmatrix} -1 \\ 1 \end{pmatrix}$. Label it **D**.
Describe the single transformation that maps shape **A** onto shape **D**.

5 Look at the dancing figure below. Describe how the figure can be drawn using only transformations of shapes **A**, **B**, **C** and **D**.

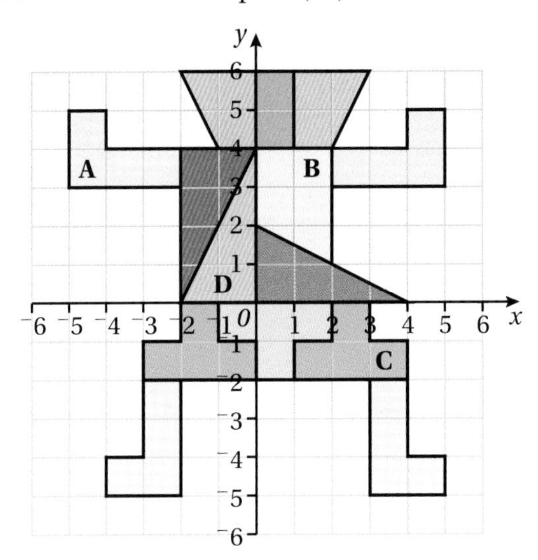

30 Congruent triangles

In this chapter you will learn how to …

- prove that two triangles are congruent using the cases SSS, ASA, SAS, RHS.
- apply congruency in calculations and simple proofs.

For more resources relating to this chapter, visit GCSE Mathematics Online.

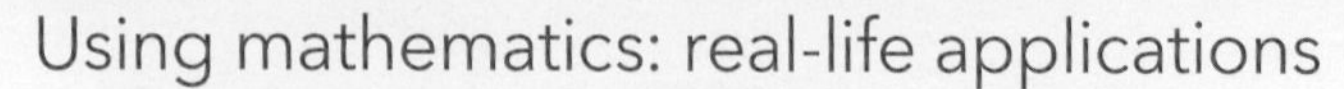

Using mathematics: real-life applications

Congruent triangles are used in construction to reinforce structures that need to be strong and stable.

"When designing any bridge I have to allow for reinforcement. This ensures that the bridge doesn't collapse under heavy traffic. Any bridge I design has many congruent triangles."

(Structural engineer)

Before you start …

Ch 5	You need to know how to label angles and shapes that are equal.	**1** Here are two identical triangles. **a** Write down a pair of sides that are equal in length. **b** What angle is equal in size to angle BAC? **c** Write down another pair of angles that are equal in size.
Ch 9	You need to know basic angle facts.	**2** Match up the correct statement with the correct diagram. **a** Vertically opposite angles are equal. **b** Alternate angles are equal. **c** Corresponding angles are equal.
Ch 5, 9	You should be able to apply angle facts to find angles in figures and to justify results in simple proofs.	**3** Decide whether each statement is true or false. **a** Angle DBE = 40° (alternate to angle ADB). **b** Angle BEC = 50° (complementary to angle ADB). **c** Angle BDE = angle BED = 70°. **d** Triangle ABD, triangle BDE and triangle BCE are congruent.
Ch 5	You need to know and be able to apply the properties of triangles and quadrilaterals.	**4** What is the value of x? Choose the correct answer. **A** 60° **B** 30° **C** 45° **D** 50°

Assess your starting point using the Launchpad

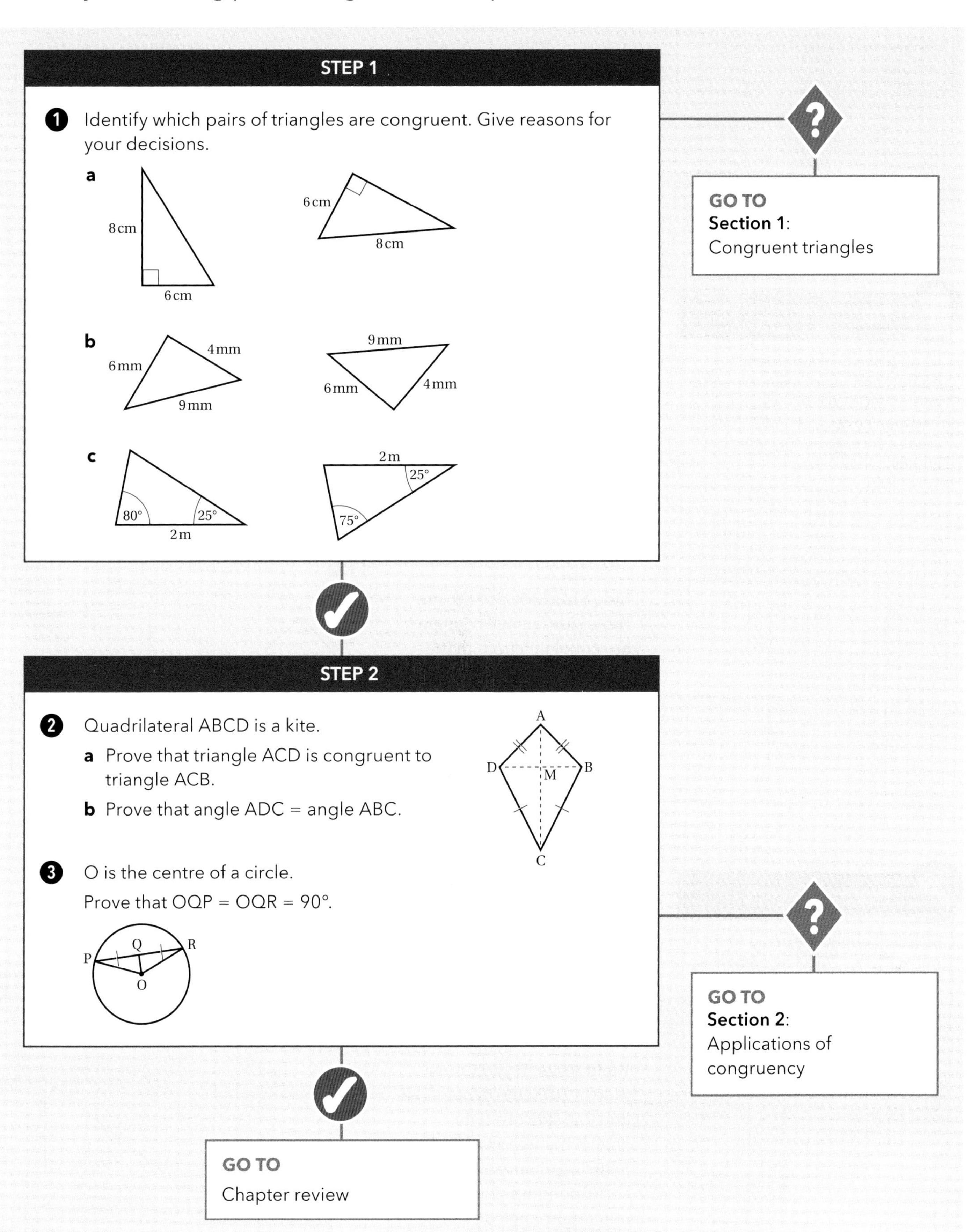
STEP 1
1 Identify which pairs of triangles are congruent. Give reasons for your decisions.
a
8 cm
6 cm
6 cm
8 cm
b
6 mm
4 mm
9 mm
9 mm
6 mm
4 mm
c
80°
25°
2 m
2 m
25°
75°
GO TO
Section 1:
Congruent triangles
STEP 2
2 Quadrilateral ABCD is a kite.
a Prove that triangle ACD is congruent to triangle ACB.
b Prove that angle ADC = angle ABC.
A
B
C
D
M
3 O is the centre of a circle.
Prove that OQP = OQR = 90°.
P
Q
R
O
GO TO
Section 2:
Applications of congruency
GO TO
Chapter review

Key vocabulary

congruent: identical in shape and size.

Tip

If you place two congruent triangles on top of each other the angles and sides will match up. The matching sides and angles are the corresponding sides or angles.

Key vocabulary

included angle: the angle between two lines that meet at a vertex.

Section 1: Congruent triangles

Congruent triangles are identical in shape and all corresponding measurements are equal.

The corresponding sides are equal in length. The corresponding angles are the same size.

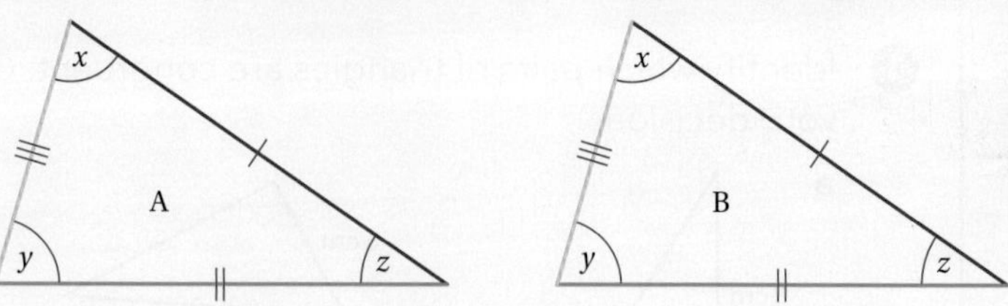

Congruent triangles can have different orientations. (You worked with congruent shapes in different orientations when you dealt with reflections and rotations in Chapter 29.) When the triangles are in different orientations you need to think carefully about the corresponding sides and angles.

This diagram shows triangle A and B from the example above in different orientations.

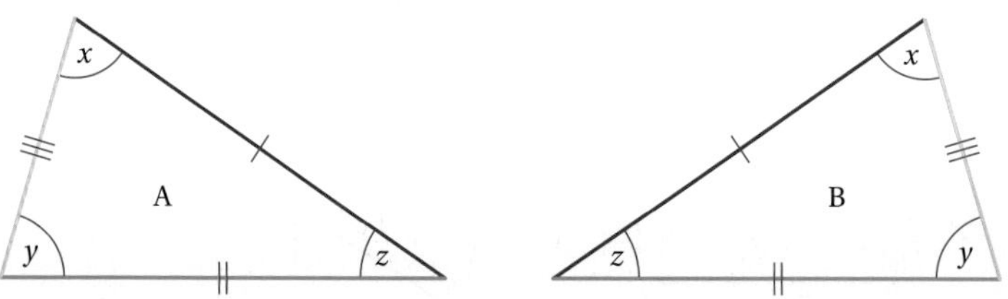

Two triangles are congruent if one of the following sets of conditions is true.

Condition	Diagram
Side Side Side or **SSS:** the three sides of one triangle are equal in length to the three sides of the other triangle.	A, B, C; D, E, F
Angle Side Angle or **ASA**: two angles and one side of one triangle are equal to the corresponding two angles and side of another triangle.	A, B, C; D, E, F
Side Angle Side or **SAS**: two sides and the **included angle** of one triangle are equal to two sides and the included angle of the other triangle.	A, B, C; D, E, F
Right Angle Hypotenuse Side or **RHS**: the hypotenuse and one side of a right-angled triangle are equal to the hypotenuse and the corresponding side of the other right-angled triangle.	A, B, C; D, E, F

Tip

It is important to write the letters of the vertices of the two triangles in the correct order. When we write that triangle ABC is congruent to triangle DEF, it means that:

$\hat{A} = \hat{D}$
$\hat{B} = \hat{E}$
$\hat{C} = \hat{F}$
and
AB = DE
AC = DF and
BC = EF.

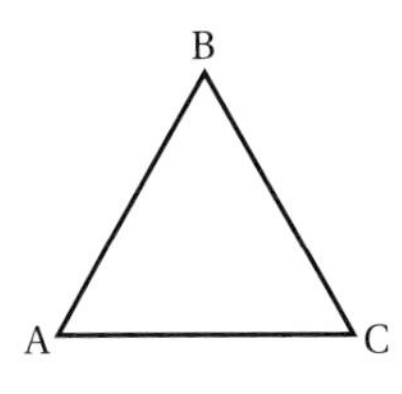

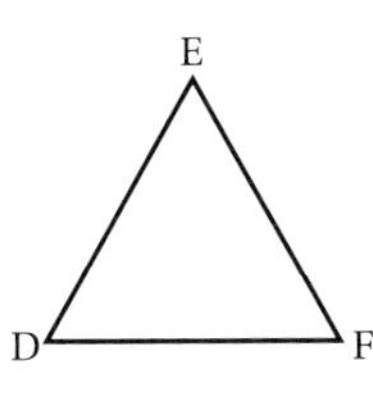

The conditions in the table are the minimum conditions for proving that triangles are congruent. No other combinations of side and angle facts are sufficient to tell us whether a triangle is congruent or not.

For example:

Two triangles with all their angles equal can still be very different sizes.

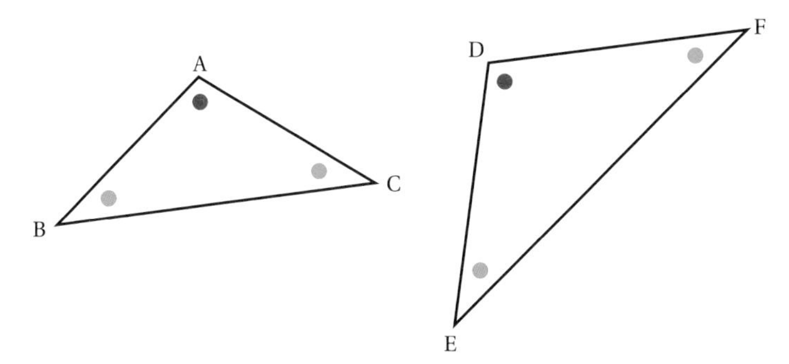

If you are given two triangles that have two equal sides and one equal angle, but where the equal angle is not included (between the two given sides), you cannot tell whether the triangles are congruent or not. The third side may have a different length in the two triangles.

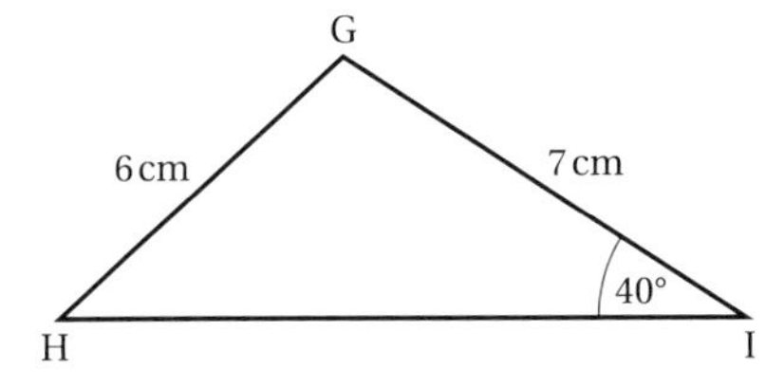

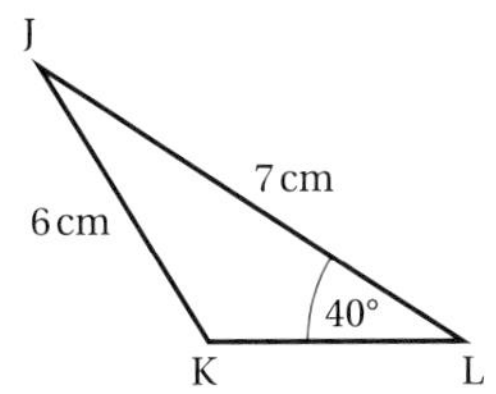

Although a pair of triangles with one of these sets of information may still be congruent, the conditions given are not sufficient to **prove** that they are.

Tip

If two congruent shapes are drawn in different orientations it is sometimes hard to see which angles and sides match each other. To help you, trace one shape onto tracing paper and label its vertices, then rotate and/or flip the paper over to help see which sides and angles match up.

WORK IT OUT 30.1

Here are three proofs for congruence for the pair of triangles.

Which one uses the correct reasoning?

Why are the others incorrect?

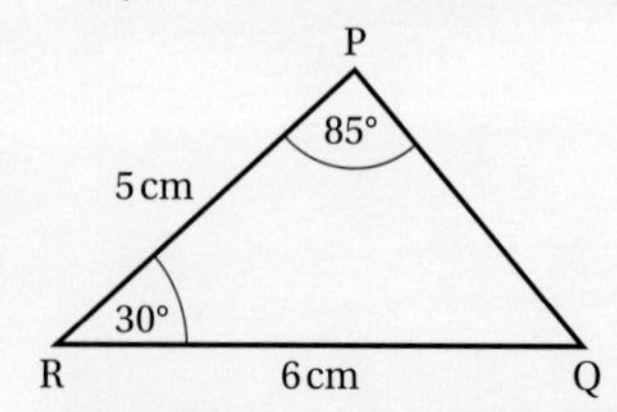

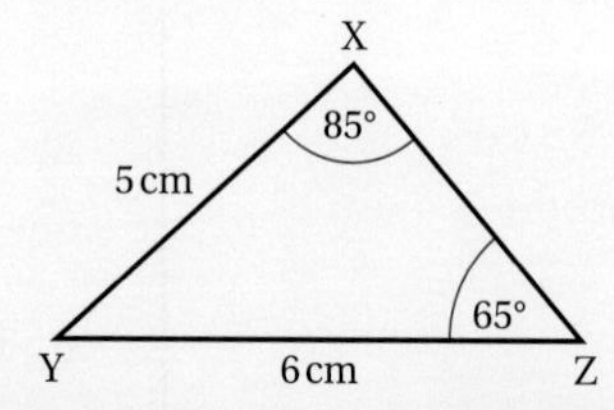

Option A	Option B	Option C
In triangle PQR and triangle XYZ: PR = XY = 5 cm angle P = angle X = 85° RQ = YZ = 6 cm so triangle PQR is congruent to triangle XYZ (SAS).	In triangle PRQ and triangle XYZ: In triangle PRQ, angle Q = 65° (sum of angles in a triangle) angle Q = angle Z = 65° RQ = YZ so the triangles are congruent.	In triangle PRQ and triangle XYZ: PR = XY = 5 cm RQ = YZ = 6 cm In triangle XYZ, angle Y = 30° (sum of angles in a triangle) so triangle PRQ is congruent to triangle XYZ (SAS).

EXERCISE 30A

1 Match up each of the congruency descriptions (SSS, ASA, SAS, RHS) with each pair of triangles below:

a

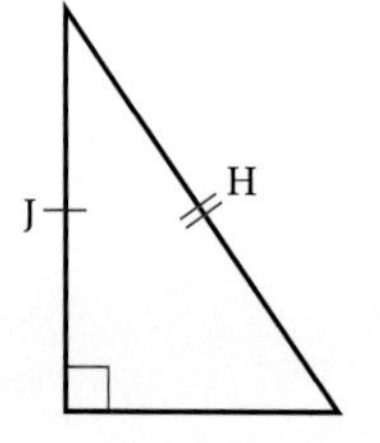

b

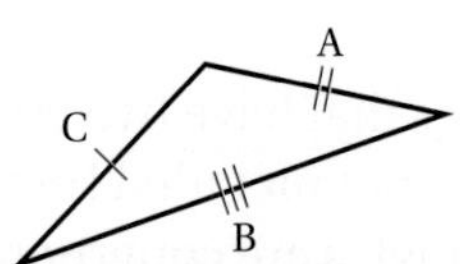

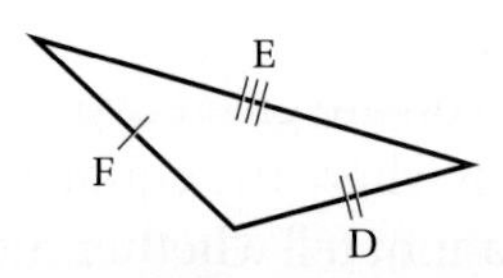

c

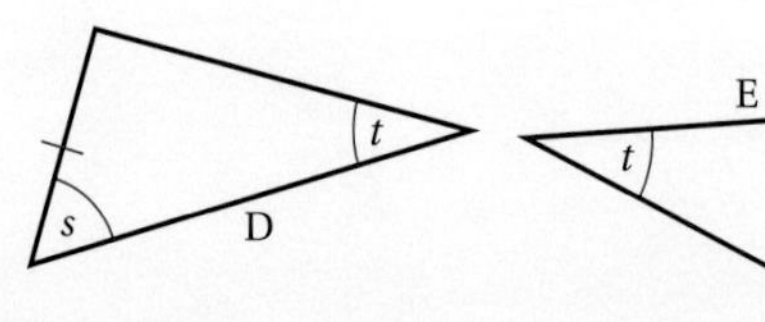

d

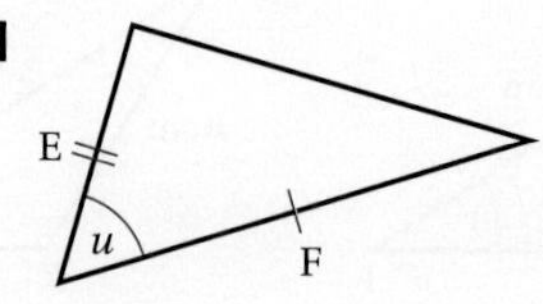

2 State whether each pair of triangles is congruent or not. For those that are, state the conditions that make them congruent.

a

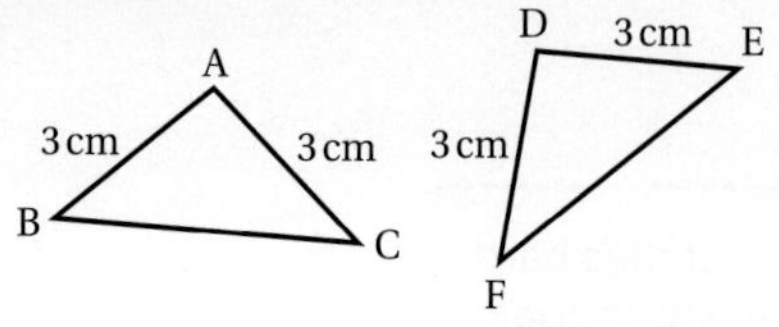

b

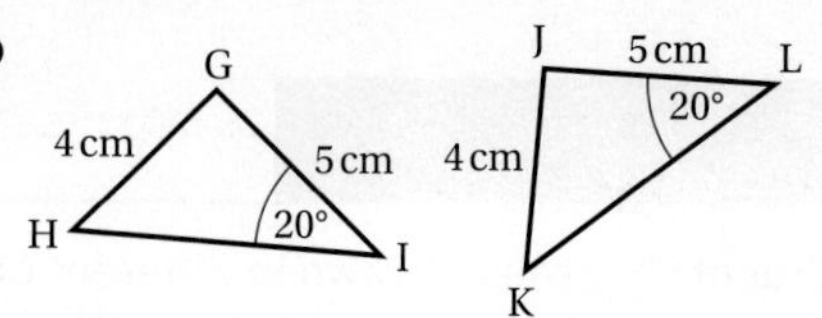

c

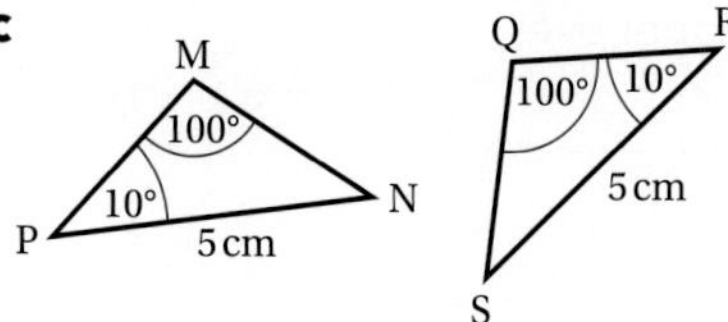

d

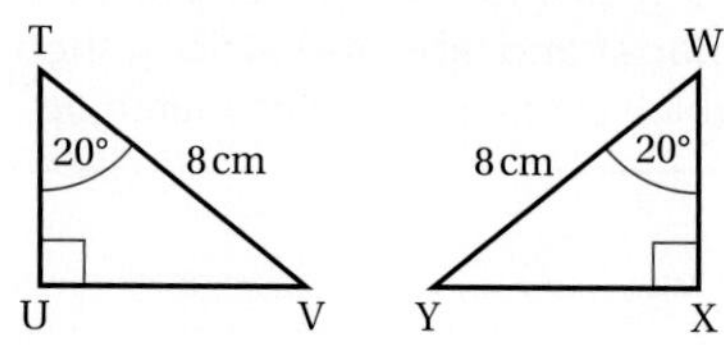

3 Prove that triangle ABC is congruent to triangle DEC.

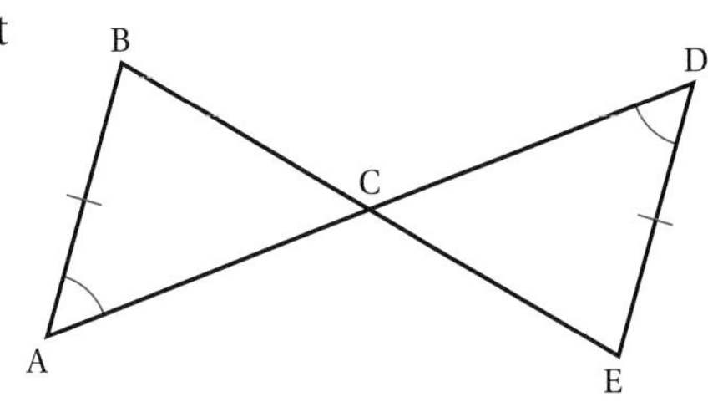

4 Write down two different proofs for congruence of triangles DEF and DGF.

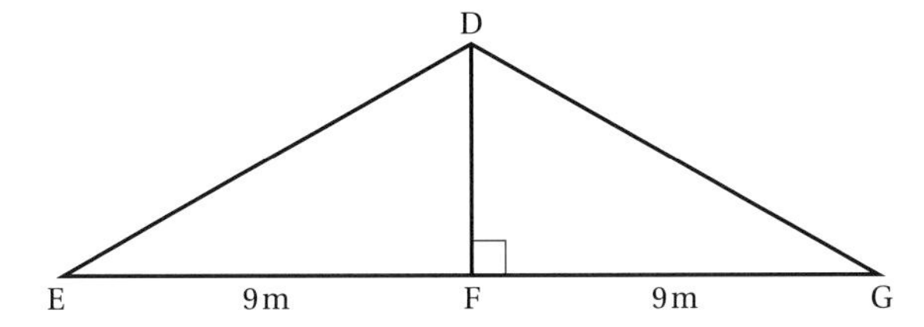

5 In the diagram, PQ is parallel to SR and QT = TR = 2 cm.

Prove that triangle PQT is congruent to triangle SRT.

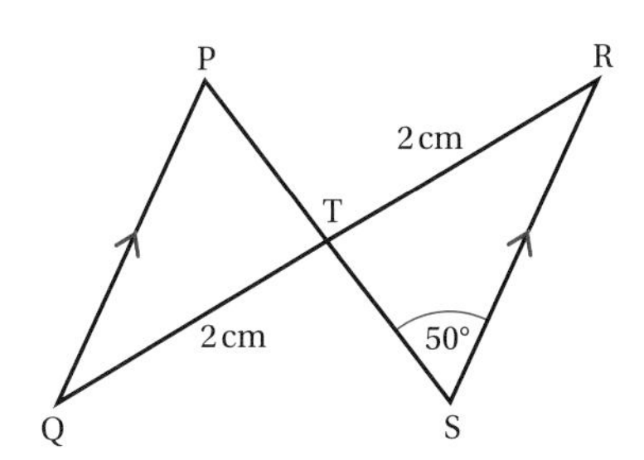

6 Prove that triangles ABE and CBD in the figure are congruent, giving full reasons.

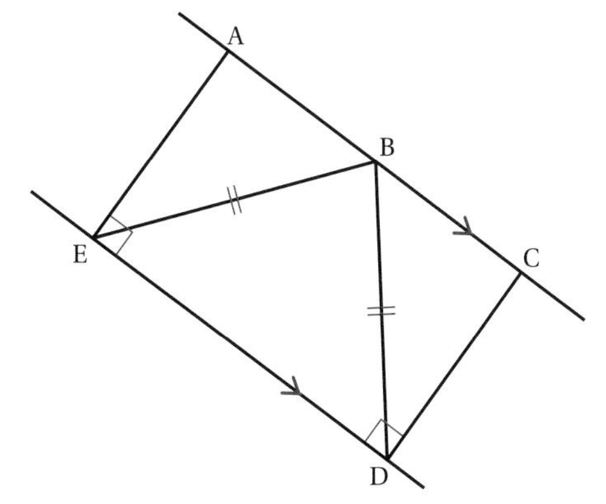

7 Triangle ABD is isosceles. AC is the perpendicular height. Prove that triangle ABC is congruent to triangle ADC.

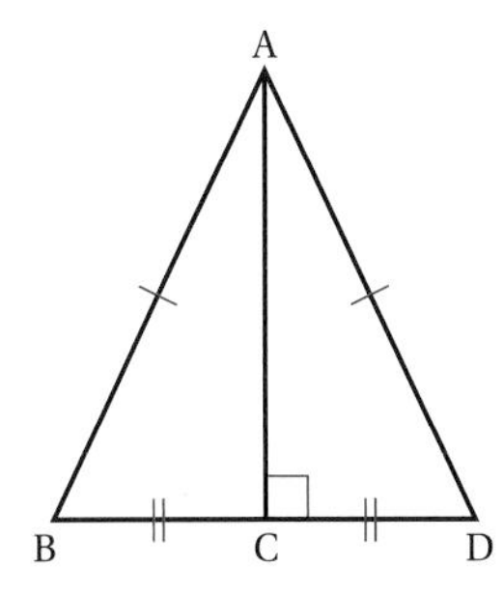

8 In the figure on the right, PR = SU and RTUQ is a kite.

Prove that triangle PQR is congruent to triangle SQU.

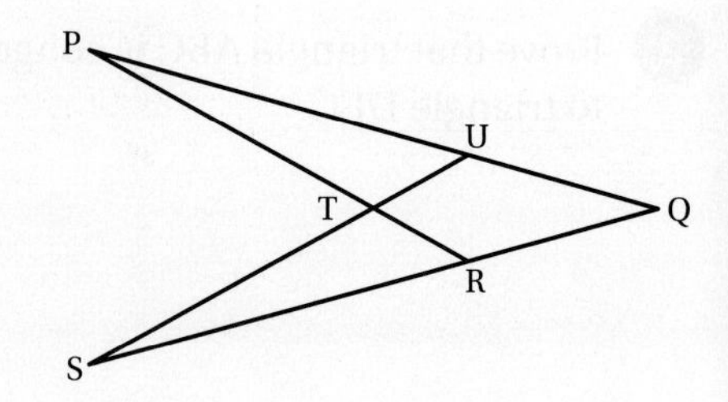

9 ABCD in the figure is a kite.
Prove that:

a triangle ADB is congruent to triangle CDB.

b triangle AED is congruent to triangle CED.

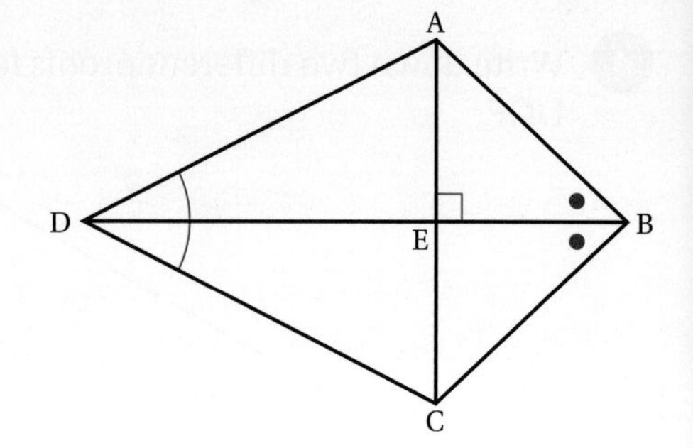

10 Quadrilateral ABCD is a rhombus.

Write down three sets of congruent triangles in this diagram and write down the cases of congruence.

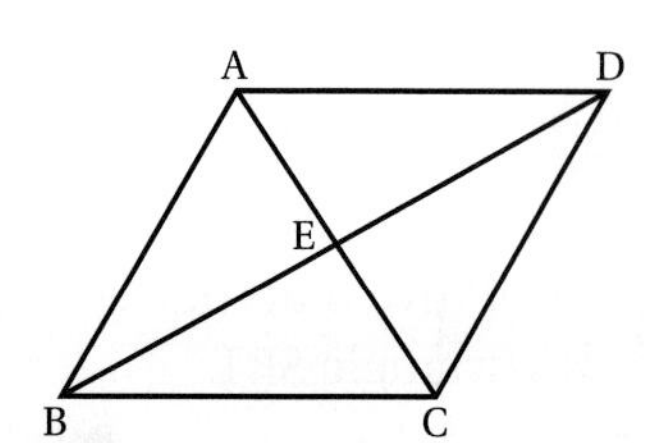

11 O is the centre of two concentric circles. Prove that triangle MPO is congruent to triangle NQO and give reasons.

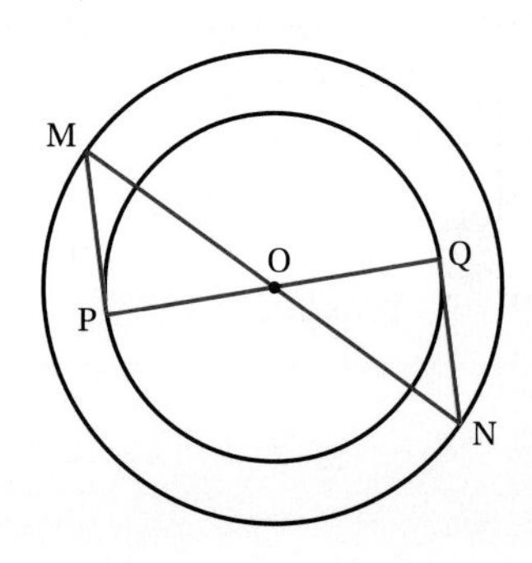

Section 2: Applying congruency

Whenever you learn new skills and concepts in geometry, you add them to your toolbox and use them in problem-solving. So you will need to combine what you've learnt previously with your new skills to solve problems.

Problem-solving framework

In the diagram, AM = BM and PM = QM.

a Prove that triangle AMP is congruent to triangle BMQ.

b Prove that AP // QB.

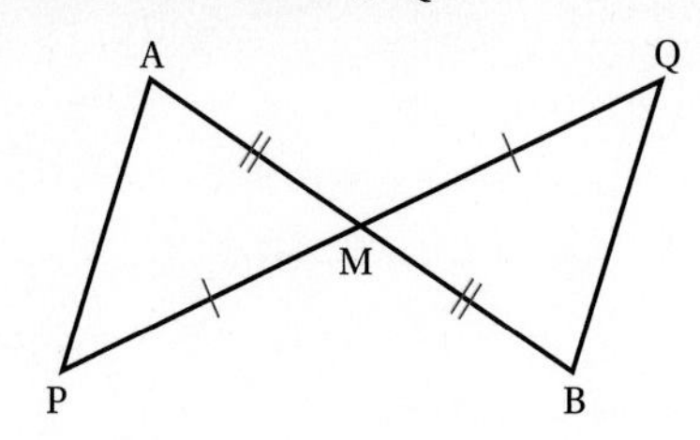

Tip

You saw in Chapter 5 that the symbol // means parallel.

Continues on next page ...

Steps for approaching a problem-solving question	What you would do for this example
Step 1: Read the question carefully to decide what you have to find.	Find mathematical evidence to show that the triangle AMP is congruent to triangle BMQ; look for SSS, SAS, RAH or ASA. Find evidence to show that lines AP and QB are parallel.
Step 2: Write down any further information that might be useful.	Line AB and PQ intersect, so the two triangles also have vertically opposite angles. Vertically opposite angles are equal (see Chapter 9, if you need to).
Step 3: Decide what method you'll use.	We are given two equal sides and we can see that the included angle is also equal, so use SAS to prove congruence.
Step 4: Set out your working clearly.	**a** In triangles AMP and BMQ: AM = BM (given). PM = QM (given). angle AMP = angle BMQ (vertically opposite angles at M). So triangle AMP is congruent to triangle BMQ (SAS). **b** Angle APM = BQM (matching angles of congruent triangles). So AP // QB (alternate angles are equal).

WORKED EXAMPLE 1

Triangle DEF is divided by GF into two smaller triangles.
Prove that FG is perpendicular to DE in the diagram.

In triangle FGD and triangle FGE:
Side FG is common to both triangles.
DG = GE Given
DF = EF Given
So the triangles are congruent (SSS).
Angle DGF = angle EGF and the two angles lie on a straight line (angles on a straight line add up to 180°).
So each angle is 90°, and FG is perpendicular to DE.

EXERCISE 30B

1 In the diagram on the right, prove that KL = ML.

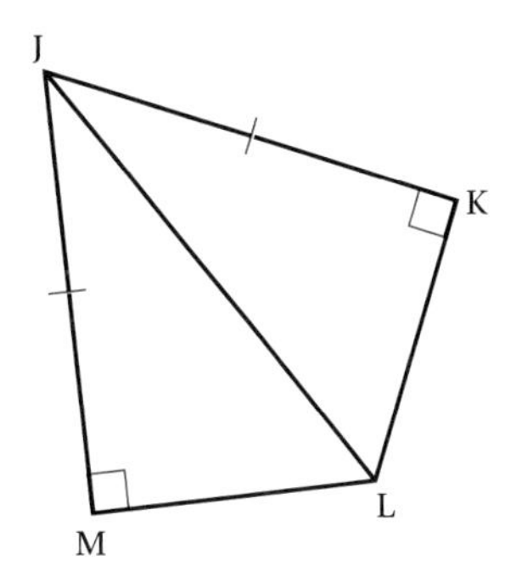

2 Use the facts given in the diagram to:

a prove that angle ABE = angle EDC.

b prove that quadrilateral ABCD is a parallelogram.

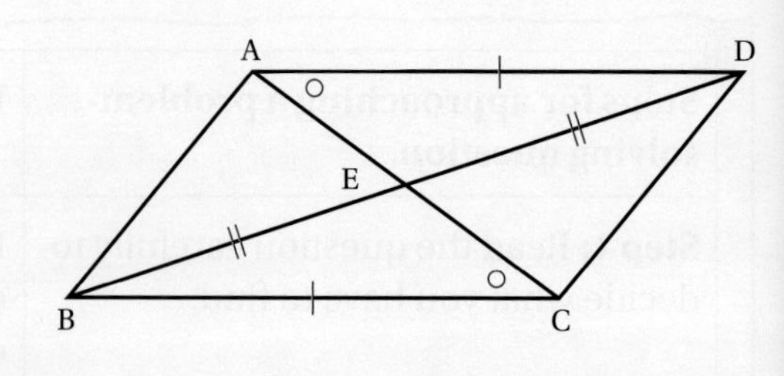

3 In the quadrilateral, SP = SR and QP // RS. Angle QRP = 56°.

a Calculate the size of angle PQR and give reasons.

b Find the size of angle PSR and give reasons.

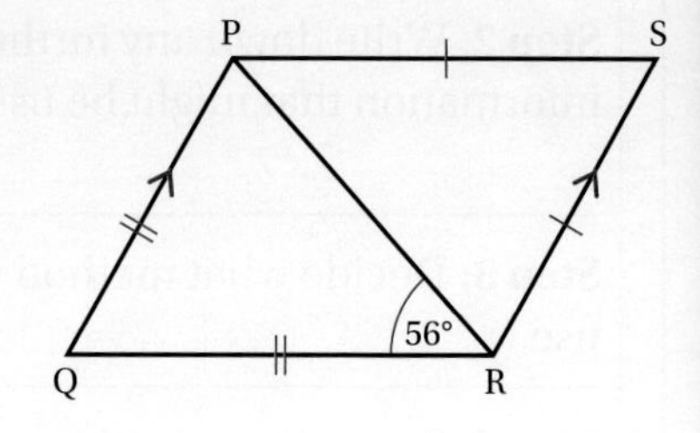

4 In the diagram, PQ = PT, QR = ST and angle PRS = angle PST.
Prove that triangle PRS is isosceles.

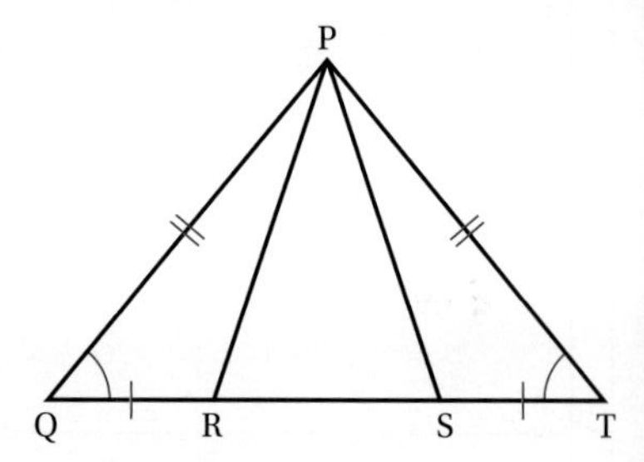

5 In the figure below, prove that angle EAD = angle ECD.

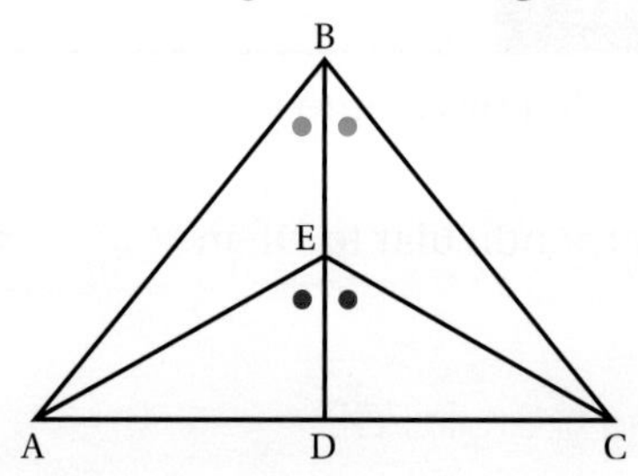

6 Prove that the diagonals of a rectangle are equal in length.
Use the diagram to help, if needed.

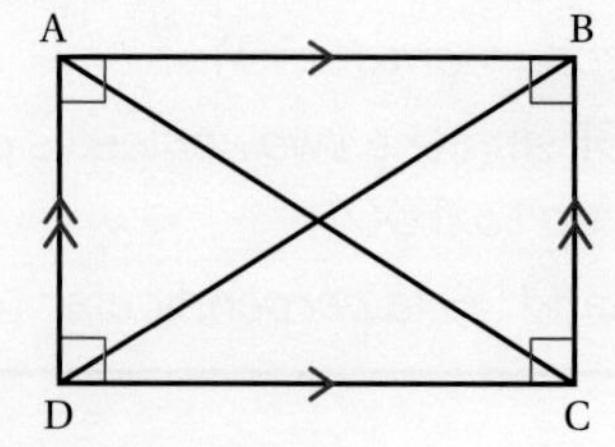

7 In the parallelogram ABCD, the point X and Y are on the diagonal such that DX = DA and BY = BC. Angle ADX = 40°.

a Find the size of angle XYC.

b Prove that AY = CX.

c Prove that triangle AYX and CXY are congruent.

d Prove that AYCX is a parallelogram.

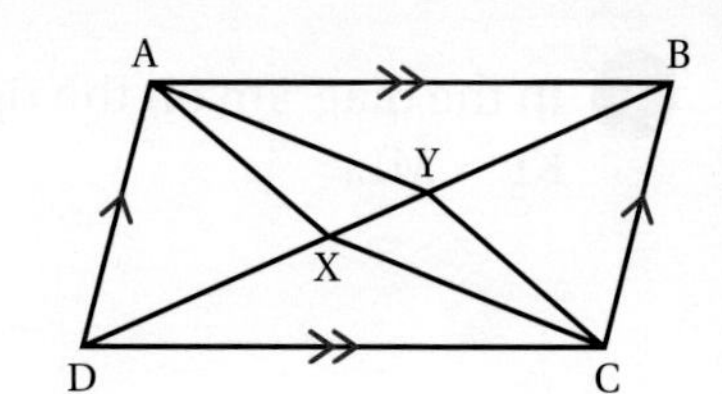

8 ABCD is a straight line such that AB = BC = CD and BCPQ is a rhombus.
The lines produced through AQ and DP meet at R.
Prove that ∠ARD is a right angle.

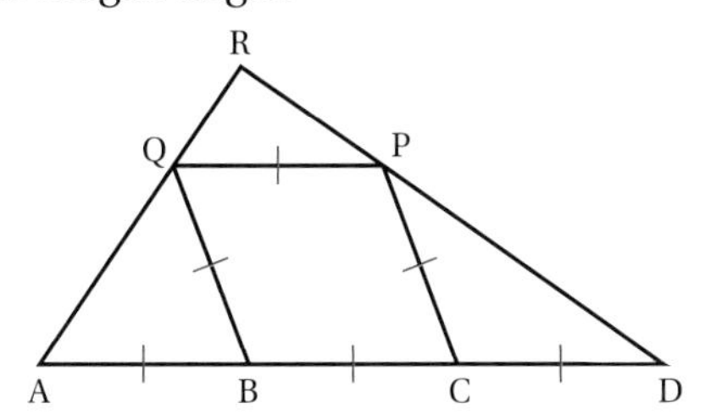

9 The trapezium ABCD has angle BCD = 70° and CDA = 50°.

Find the other two angles in the trapezium.

10 AC and BD are the diagonals of a quadrilateral ABCD.
AC and BD are perpendicular, meeting at M. Name the quadrilateral(s) if:

a M is the midpoint of AC and BD.

b M is the midpoint of AC and BD and AB = BD.

c M is the midpoint of AC.

11 **a** The points P and Q are chosen on the diagonal BD of the square ABCD so that BP = DQ.

i Prove that the triangles ABP, CBP, ADQ and CDQ are all congruent.

ii Prove that APCQ is a rhombus.

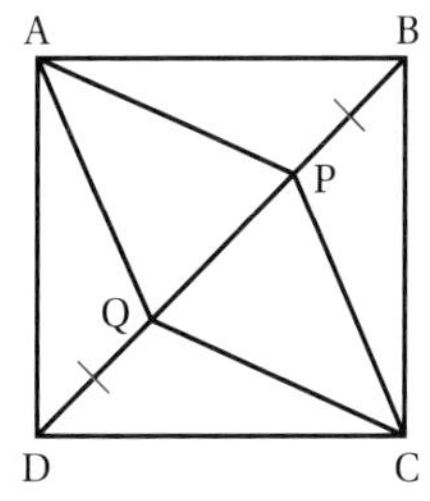

b In △ABC, MA is the bisector of ∠BAC.
The line MF is parallel to BA, and MG is parallel to CA.

Prove that AFMG is a rhombus.

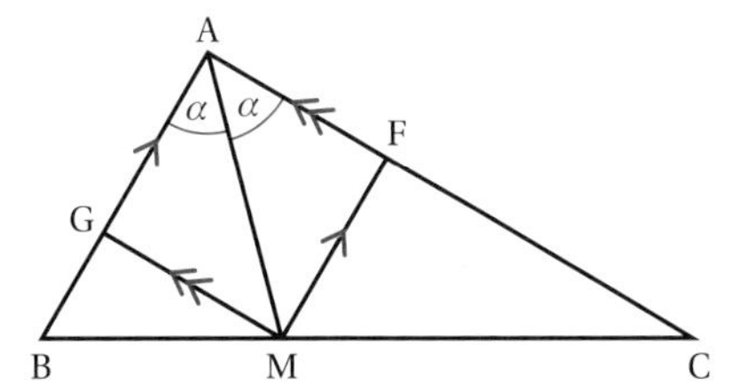

12 Prove that the quadrilateral below is not a trapezium.

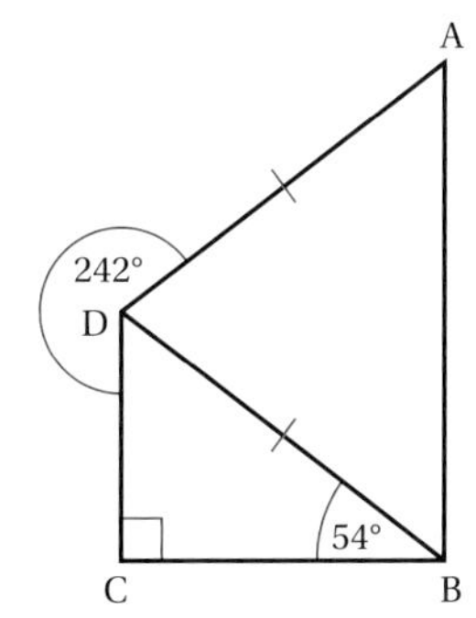

Checklist of learning and understanding

Congruent triangles

- You can prove that two triangles are congruent using one of the four cases of congruence:
  - Side Side Side or SSS: the three sides of one triangle are equal in length to the three sides of the other triangle.
  - Angle Side Angle or ASA: two angles and one side of one triangle are equal to the corresponding two angles and side of another triangle.
  - Side Angle Side or SAS: two sides and the included angle of one triangle are equal to two sides and the included angle of the other triangle.
  - Right Angle Hypotenuse Side or RHS: the hypotenuse and one side of a right-angled triangle are equal to the hypotenuse and one other side of the other right-angled triangle.
- You can apply your knowledge of congruency to help prove other geometrical features such as parallel and perpendicular lines.

For additional questions on the topics in this chapter, visit GCSE Mathematics Online.

Chapter review

1 State whether each pair of triangles is congruent or not. Give a reason for your answer and give the vertices of the triangles in the correct order.

a

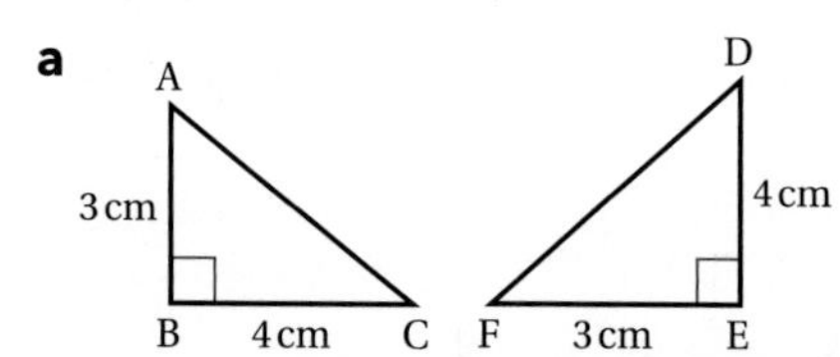

b

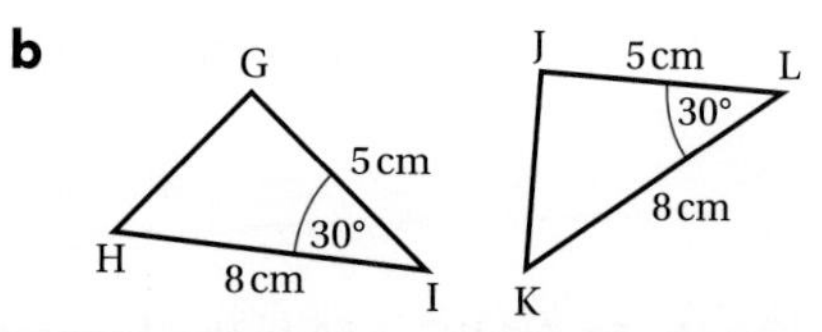

c

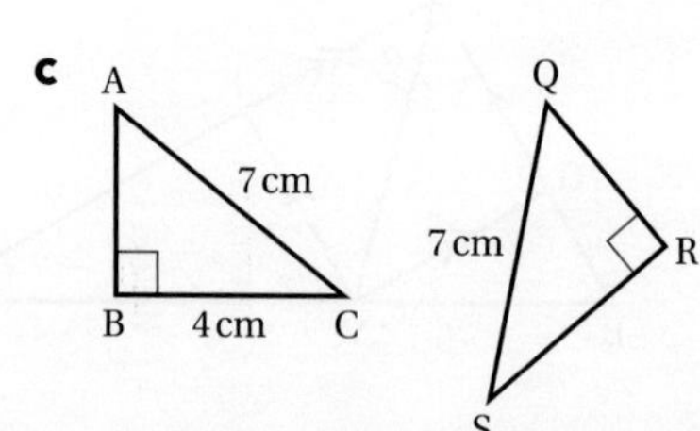

d

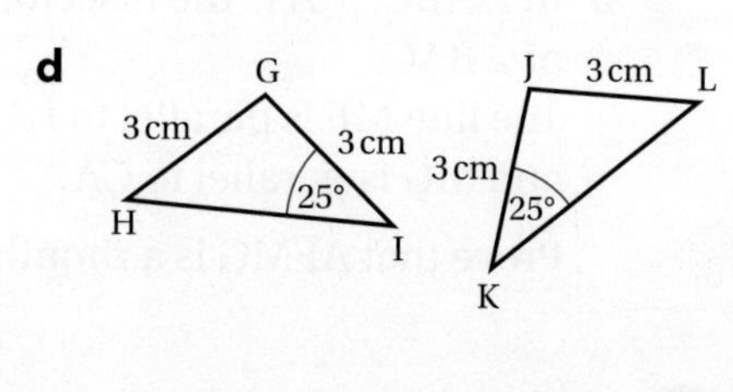

2 In the figure below, prove that angle T = angle R.

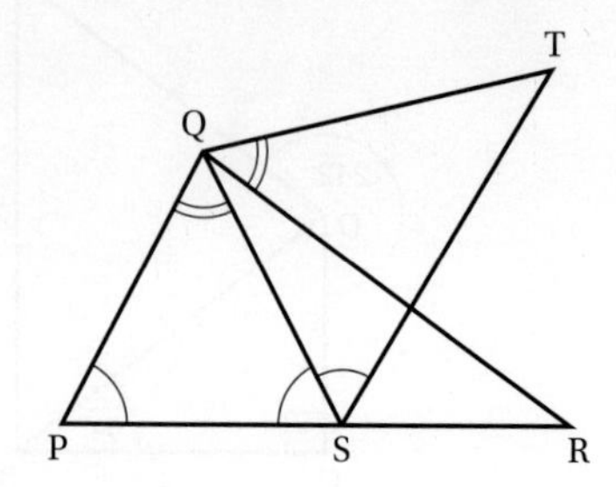

3 Use triangle congruence to prove that angle EBC = angle ECB in the figure below.

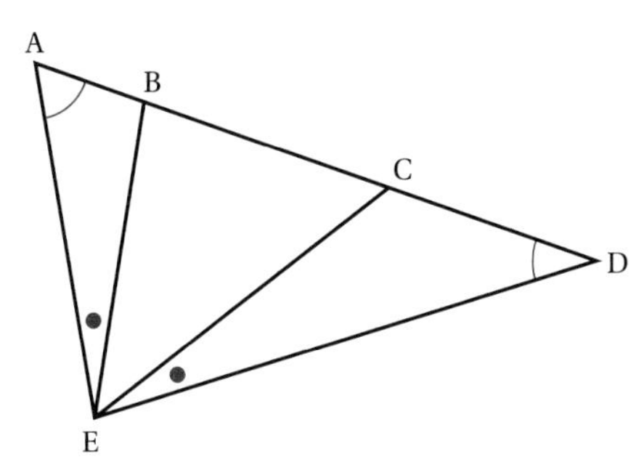

4 In the figure below, triangle ABD lies between two parallel lines. BC = CA = AD. Prove that angle EAD = 2 × angle ABC.

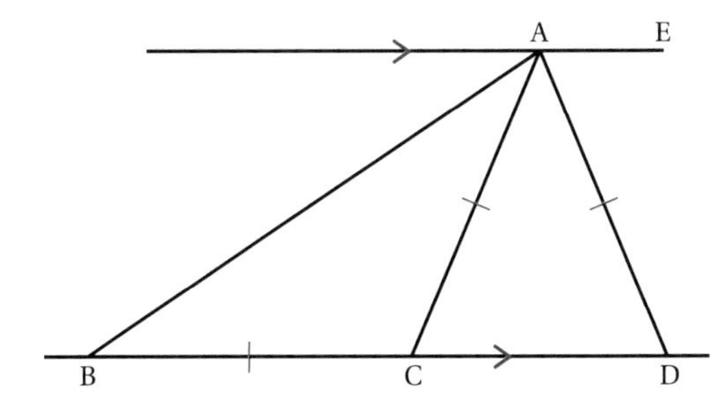

5 Triangle UVW is congruent to triangle UZY.

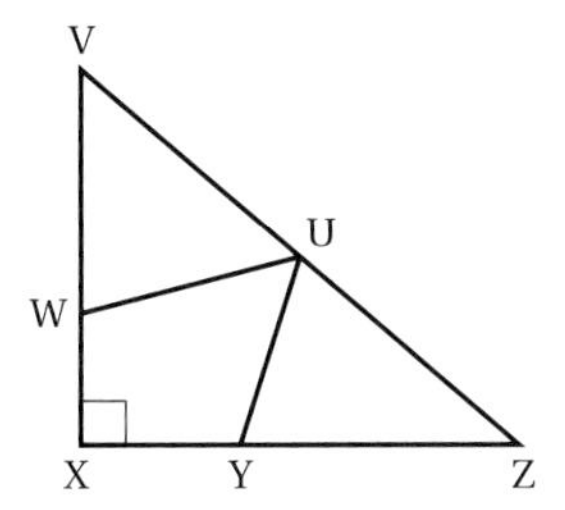

Prove that quadrilateral UWXY is a kite.

6 Given that triangle MNP is congruent to triangle NPQ, prove that quadrilateral MNPQ is a square.

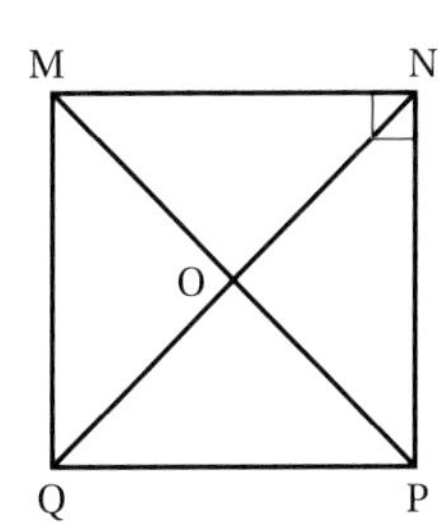

31 Similarity

In this chapter you will learn how to …

- identify similar triangles and prove that two triangles are similar.
- work with positive, fractional and negative scale factors to enlarge shapes on a grid.
- find the scale factor and centre of enlargement of a transformation.
- understand the relationship between lengths, areas and volumes of similar objects.

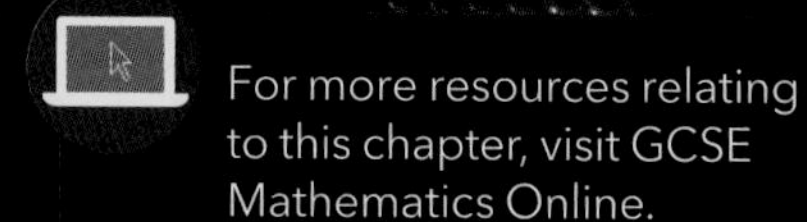

For more resources relating to this chapter, visit GCSE Mathematics Online.

Using mathematics: real-life applications

When you enlarge a photo, project an image onto a screen or make scaled models you are dealing with similarity. Many toys and other objects are scaled, similar, versions of larger objects from real life.

"I work with scale drawings and scale models all the time. The models are mathematically similar to the real planes so the clients can see what they are buying. We made these scaled models to display at an international air show." *(Aircraft designer)*

Before you start …

Ch 9	You need to be able to label angles correctly.	**1** **a** Which angle is a right angle? **b** What size is angle DOA? **c** What size is angle BOD? 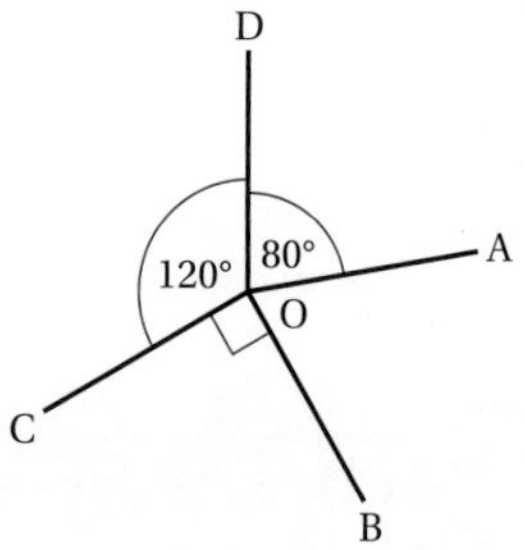 (Diagram: lines OA, OB, OC, OD from O; angle DOA = 80°, angle COD = 120°, right angle at O between OC and OB)
Ch 30	You need to be able to prove that two triangles are congruent.	**2** Prove that the triangles below are congruent, giving reasons. 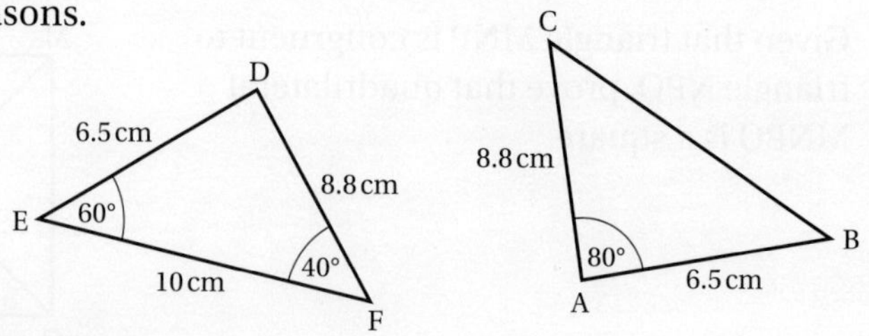 (Triangle DEF: ED = 6.5 cm, DF = 8.8 cm, EF = 10 cm, angle E = 60°, angle F = 40°. Triangle ABC: AC = 8.8 cm, AB = 6.5 cm, angle A = 80°)
Ch 8	You need to know how to solve simple equations using inverse operations.	**3** Solve **a** $3x = 24$ **b** $15 = 6h$ **c** $6.25 = 25k$
Ch 22	You need to be able to recognise numbers in equivalent ratios.	**4** Which pair of numbers is in the same ratio as 3 : 2? A 6 : 5 B 4 : 6 C 0.15 : 0.1 **5** Given that $\frac{4}{15} = \frac{x}{90}$, find x.
Ch 1, 2, 25	You need to be able to find powers of whole numbers, fractions and decimals.	**6** Calculate the following: **a** $\left(\frac{1}{2}\right)^3$ **b** 4.5^2 **c** 15^3

Assess your starting point using the Launchpad

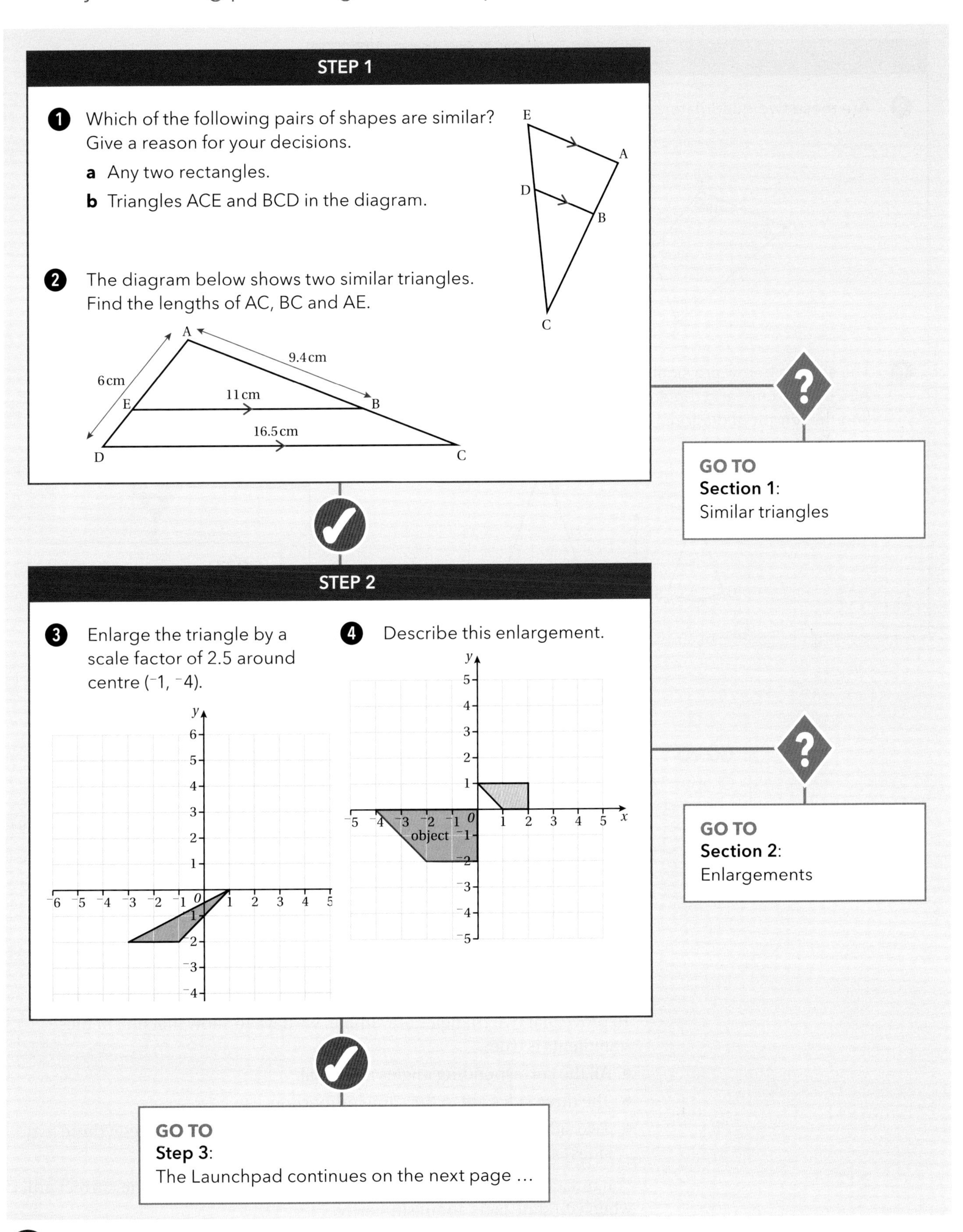

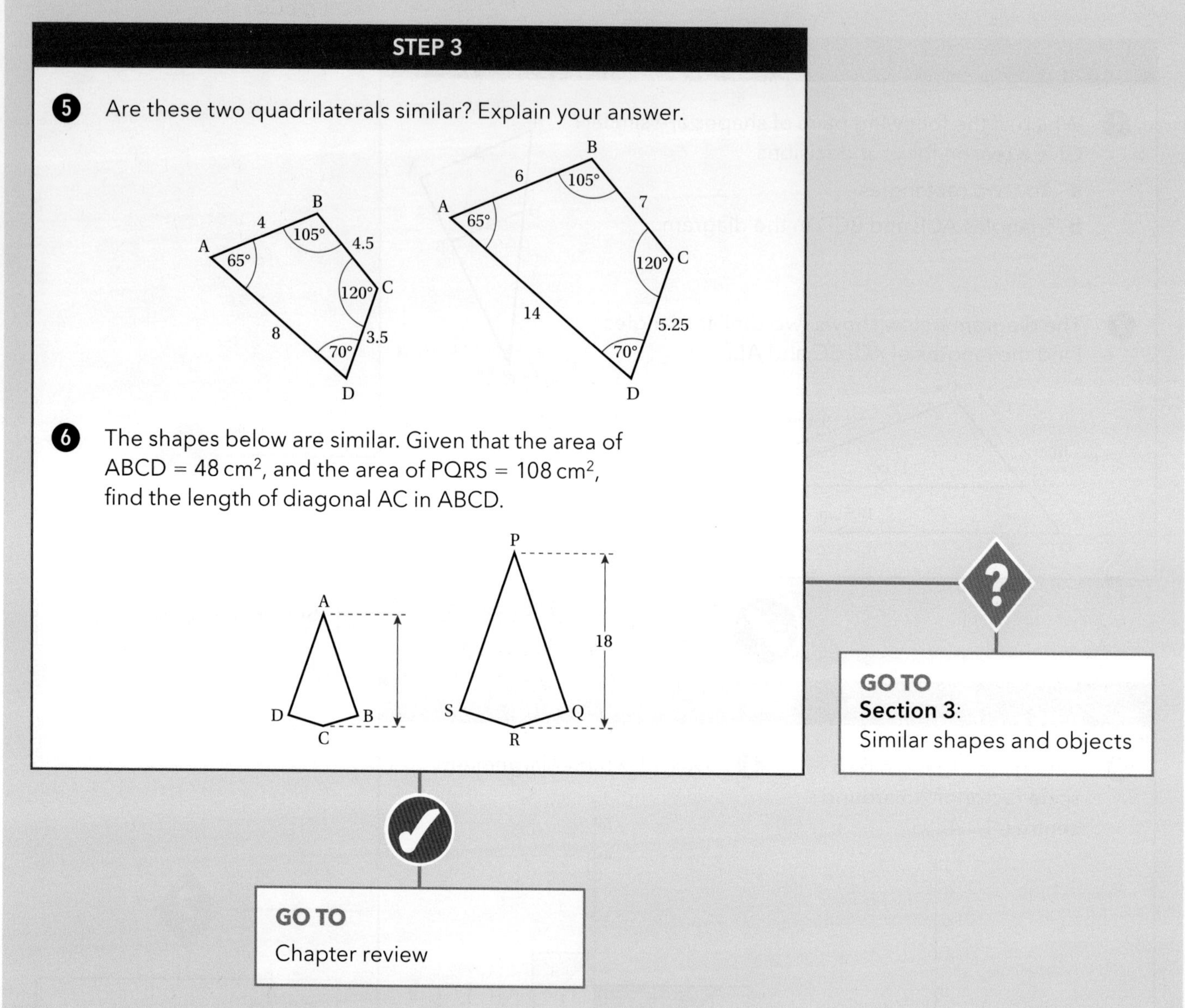
Launchpad continued ...
STEP 3
5 Are these two quadrilaterals similar? Explain your answer.
A B C D 4 105° 4.5 65° 120° 8 3.5 70°
A B C D 6 105° 7 65° 120° 14 5.25 70°
6 The shapes below are similar. Given that the area of ABCD = 48 cm², and the area of PQRS = 108 cm², find the length of diagonal AC in ABCD.
A B C D
P Q R S 18
?
GO TO
Section 3:
Similar shapes and objects
GO TO
Chapter review

Section 1: Similar triangles

Two shapes are mathematically similar if they have the same shape and proportions but are different in size.

If the corresponding angles in two triangles are equal, then the corresponding sides will be in proportion, and the triangles will be similar.

To prove that two triangles are similar, we have to show that one of these statements is true:

- All the corresponding angles are equal.
- The three sides are in the same proportion.
- Two sides are in proportion and the included angles (between these two sides) are equal.

You must name triangles with the corresponding vertices in the correct order when you state facts about similarity.

WORKED EXAMPLE 1

Prove that the triangles ABC and RTS are similar.

State which angles are equal and which sides are in proportion.

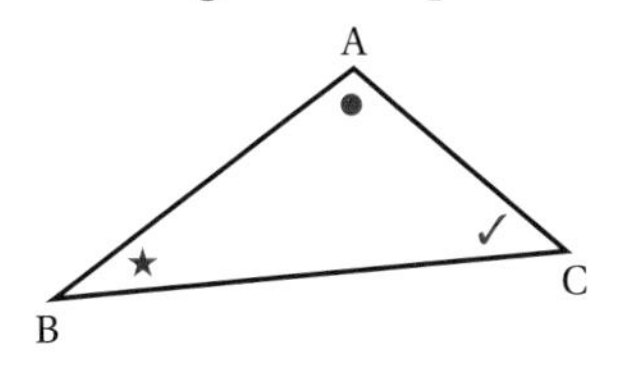

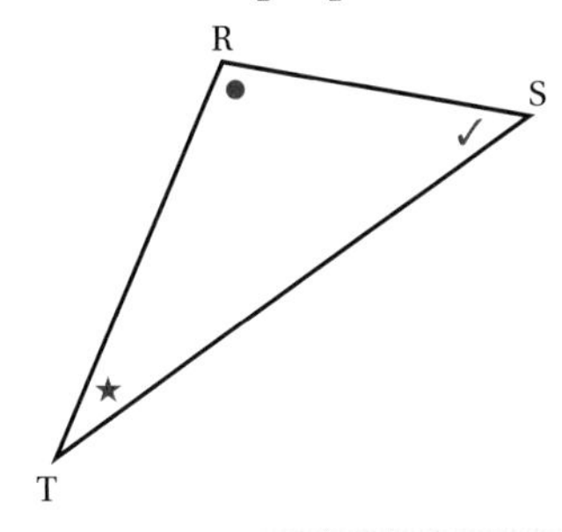

Triangle ABC is similar to triangle RTS because all the corresponding angles are equal:
$\angle A = \angle R$, $\angle B = \angle T$ and $\angle C = \angle S$.

This means that the three sides are in the same proportion, so $\frac{AB}{RT} = \frac{AC}{RS} = \frac{BC}{TS}$

Remember to write the corresponding vertices in the same order in each triangle.

WORKED EXAMPLE 2

Given that the following relationship exists between the sides of triangle PQR and triangle WYX, write down which angles are equal.

$$\frac{PQ}{WY} = \frac{PR}{WX} = \frac{QR}{YX}$$

Triangle PQR is similar to triangle WYX.

Therefore $\angle P = \angle W$; $\angle Q = \angle Y$ and $\angle R = \angle X$.

It might help you to visualise each triangle in your mind.

Finding unknown lengths using proportional sides

In similar triangles the ratio of the lengths of any pair of corresponding sides is equal. You can therefore use the ratio of corresponding sides to find the lengths of unknown sides in similar figures.

Problem-solving framework

In the figure, triangle ABC is similar to triangle QRP.
Find the length of x.

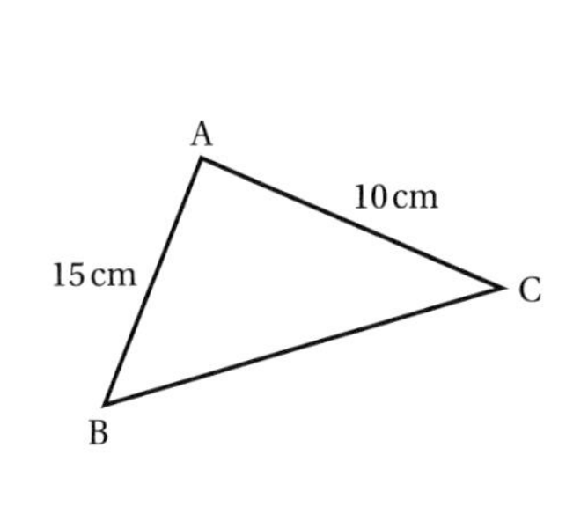

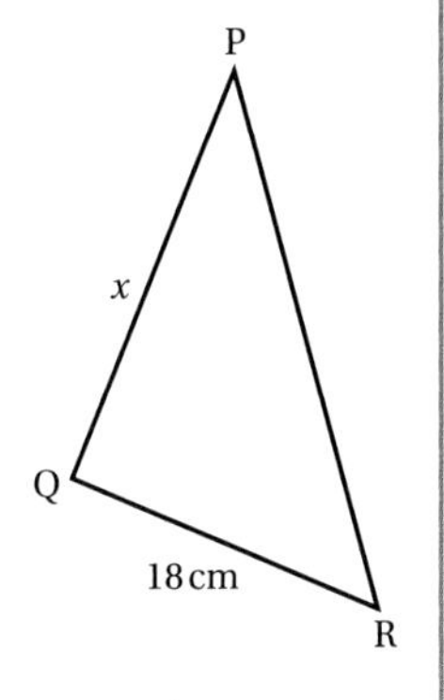

Continues on next page ...

Steps for approaching a problem-solving question	What you would do for this example
Step 1: What do you have to do?	Use the fact that two triangles are similar to calculate an unknown length in one of the triangles.
Step 2: Write down what you know.	That the triangles are similar. Check that the vertices are written in the correct order: ABC is similar to QRP. You know the length of two sides of triangle ABC and one side of triangle QRP.
Step 3: What maths can you do?	Find the ratio between the sides: AB corresponds to QR so the ratio is 15 : 18. AC corresponds to QP so $\frac{AC}{QP} = \frac{AB}{QR}$ Write a proportion with the unknown side: $\frac{10}{x} = \frac{AB}{QR}$ so $\frac{10}{x} = \frac{15}{18}$ Solve for x: $x = \frac{18 \times 10}{15} = 12$ cm $x = 12$ cm
Step 4: Have you answered the question?	Yes, you have used the principle of similarity between the two triangles to find the unknown length.

EXERCISE 31A

1 Each diagram below contains a pair of similar triangles. Identify the matching angles and the sides that are in proportion. Explain your reasoning using the correct angle vocabulary.

a

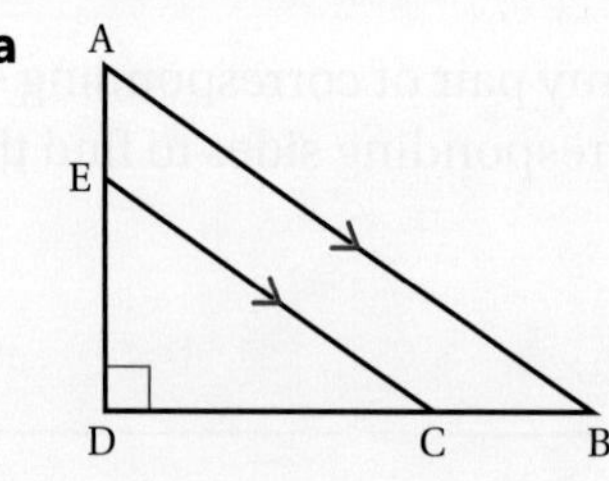

b

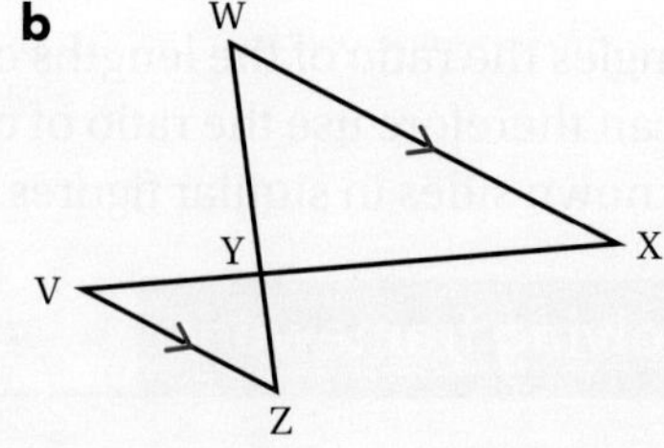

c

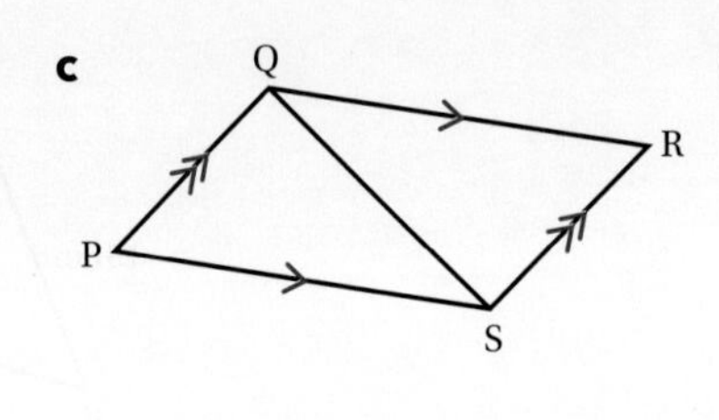

d

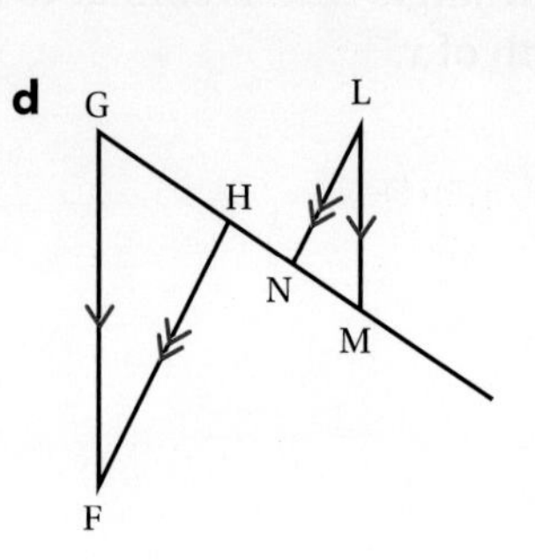

2 Are the following pairs of triangles similar? Explain your answers.

a

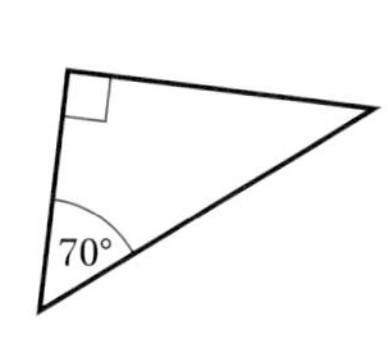

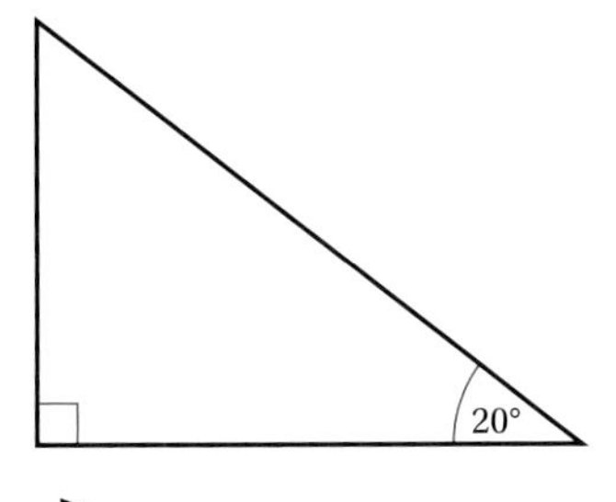

b

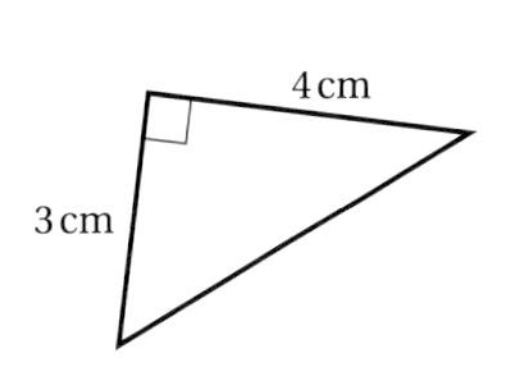

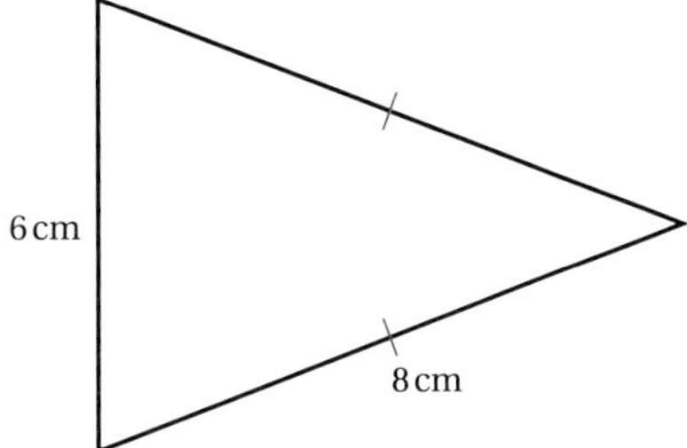

c

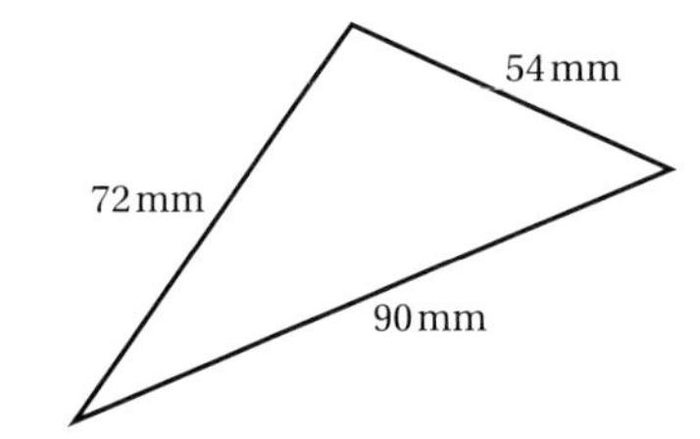

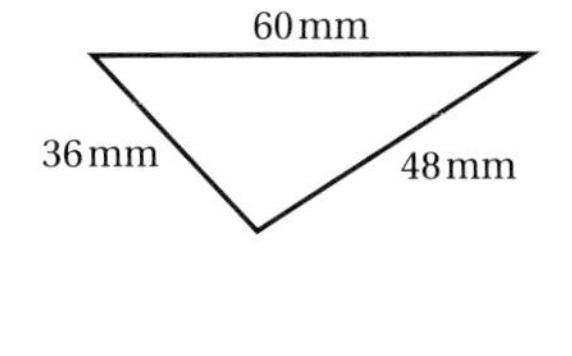

3 Are the statements below true or false? Explain your reasoning and give a counter-example for any statement you believe is false.

a All isosceles triangles are similar.
b All equilateral triangles are similar.
c All right-angled triangles are similar.
d All right-angled triangles with an angle of 30° are similar.
e All right-angled isosceles triangles are similar.
f No two scalene triangles are ever similar.

4 Each diagram below contains three similar triangles.
Identify the matching angles and sides in each group of triangles, explaining your reasoning.

a

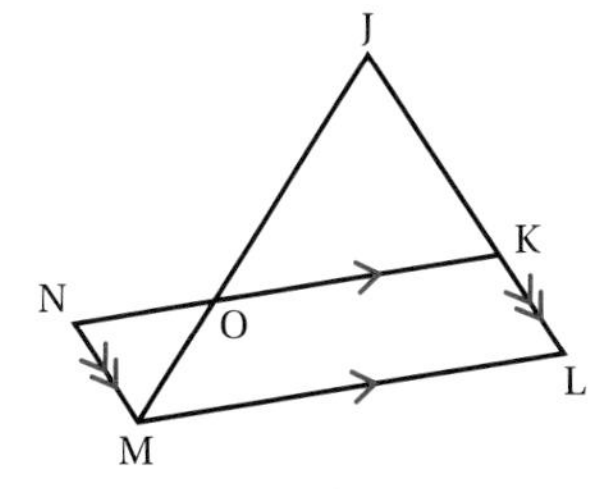

b

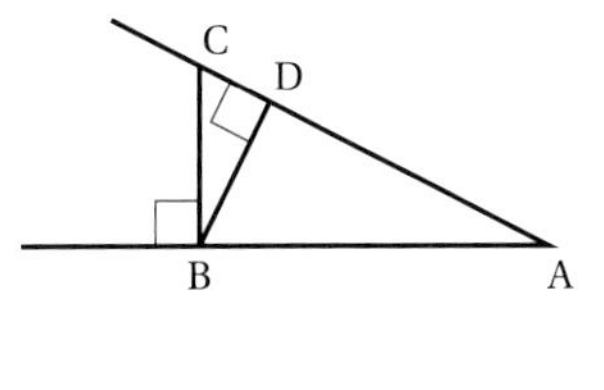

5 The two shapes below are similar.
Find the missing lengths c and d.

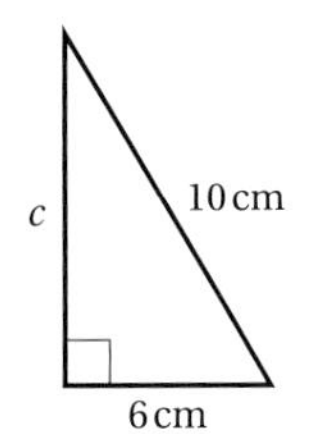

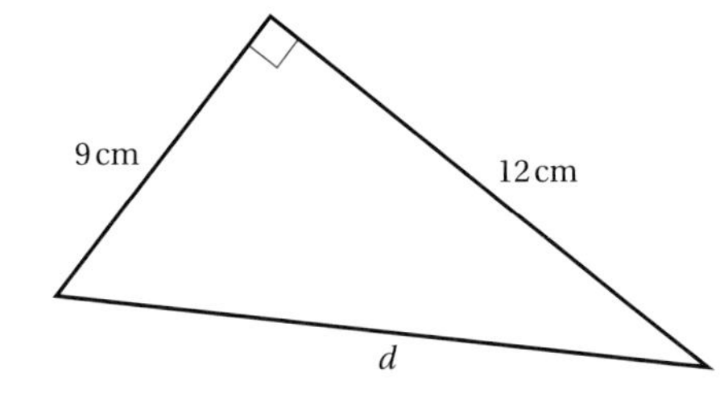

Find answers at: cambridge.org/ukschools/gcsemaths-studentbookanswers

6 The two shapes below are similar.
Find the missing lengths e and f.

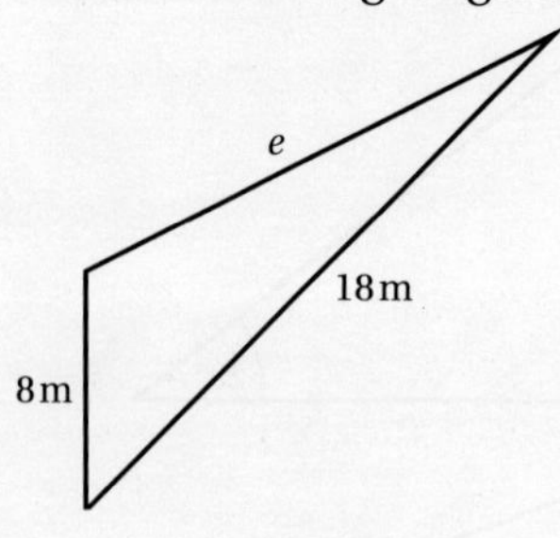

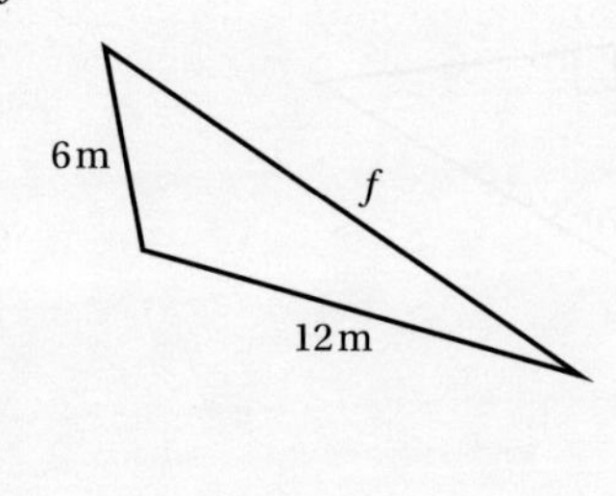

7 Find the lengths of AE, CE and AD.

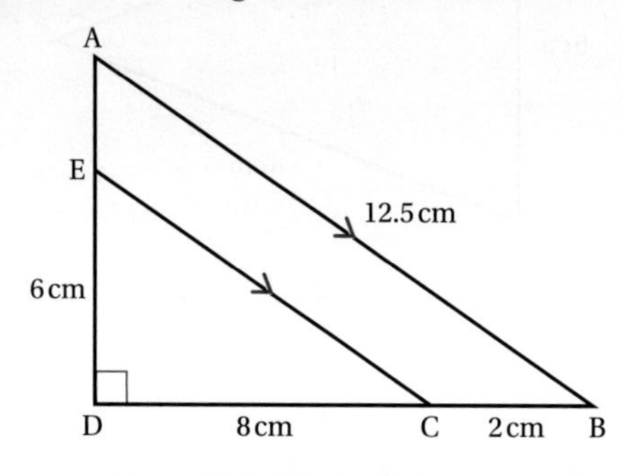

8 Find the lengths of YZ and XY.

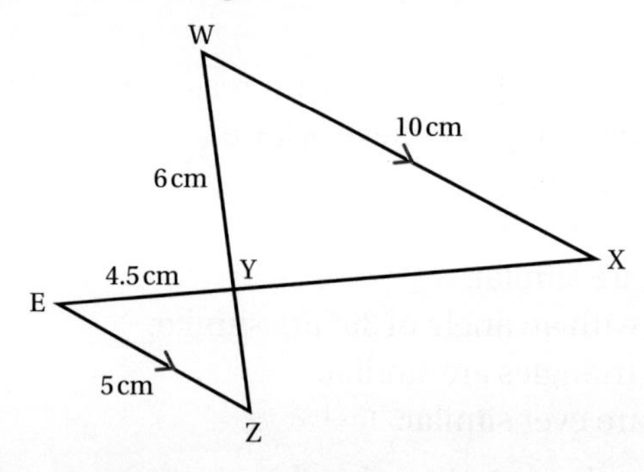

9 The diagram shows a part of a children's climbing frame.
Find the length of BC.

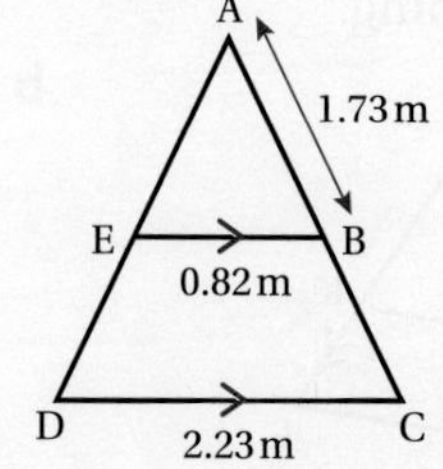

10 Swimmer A and Boat B, shown in the diagram, are 80 m apart.
Boat B is 1200 m from the lighthouse C.
The height of the boat is 12 m and the swimmer can just see the top of the lighthouse at the top of the boat's mast when her head lies at sea level.
What is the height of the lighthouse?

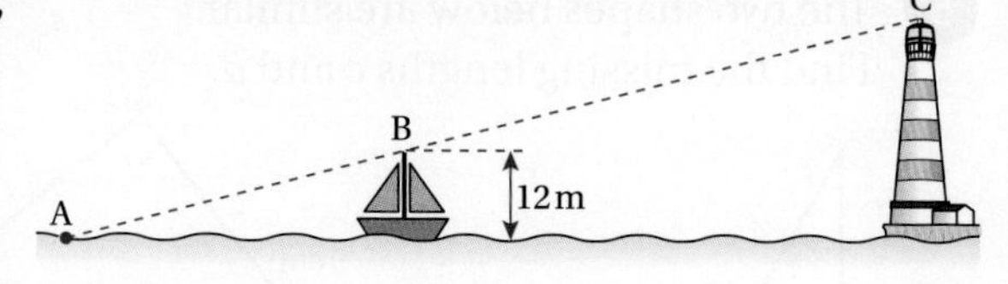

Section 2: Enlargements

An enlargement is a transformation that changes the position of a shape as well as its size. Under enlargement, an object and its image are similar shapes.

When your teacher projects an image onto a whiteboard, you see an enlargement of whatever is on the computer screen. In mathematics, we use the word enlargement for all transformations that produce similar images, even if the image is smaller than the original object. When you take a photo, the image you see on your screen is considered to be an enlargement of the scene in front of you, even though it is smaller.

Key vocabulary

scale factor: a number which scales a quantity up or down.

To construct an enlargement of a shape, you multiply the length of each side by the **scale factor**. You saw scale factors when you learnt about maps and scale drawings in Chapter 12.

Enlarged shapes are mathematically similar to each other, because their angles remain unchanged.

Before you enlarge a shape, consider its new dimensions.

Has it got bigger? Has it stayed the same size? Has it got smaller?

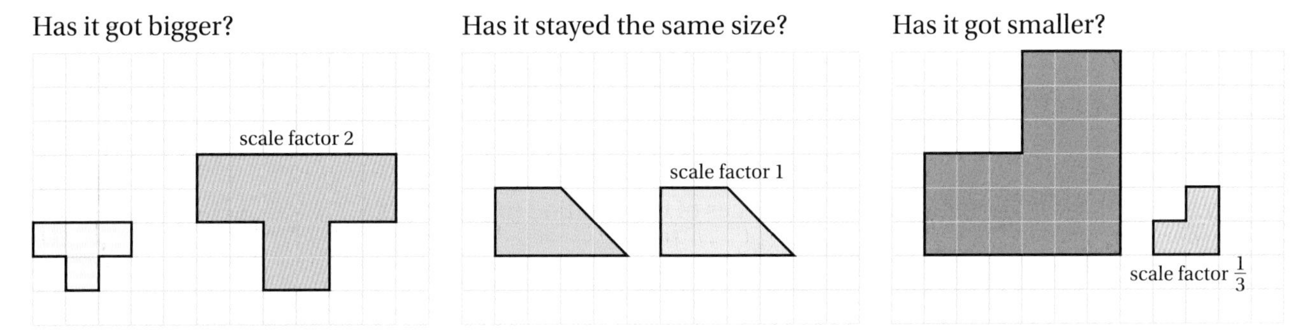

WORKED EXAMPLE 3

Draw an enlargement of triangle ABC by a scale factor of 2.

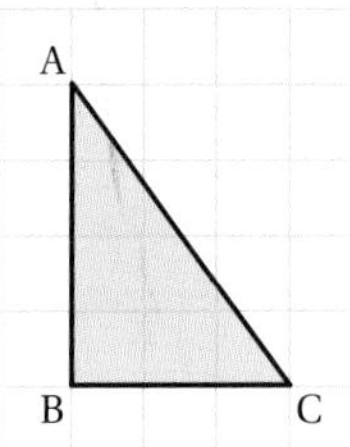

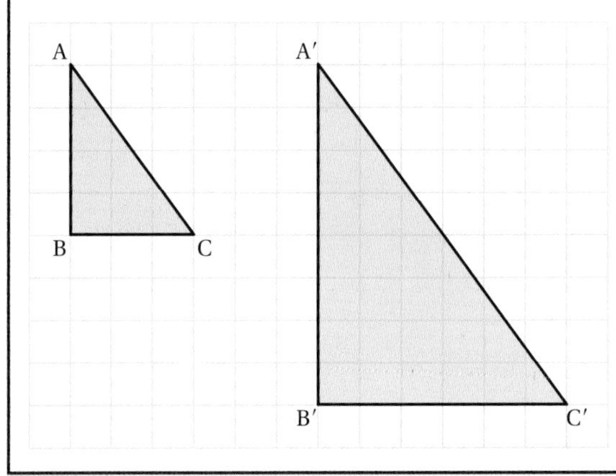

To increase triangle ABC by a scale factor of 2, increase each length by the scale factor:

$A'B' = AB \times 2 = 8$

$B'C' = BC \times 2 = 6$

$A'C' = AC \times 2 = \begin{pmatrix} 3 \\ -4 \end{pmatrix} \times 2 = \begin{pmatrix} 6 \\ -8 \end{pmatrix}$

The image of triangle ABC is triangle A′B′C′.

Find answers at: cambridge.org/ukschools/gcsemaths-studentbookanswers

Notice that triangle ABC is similar to triangle A′B′C′ and that the sides are in proportion.

EXERCISE 31B

 Enlarge each shape as directed.

- **a** Enlarge A by a scale factor of 3.
- **b** Enlarge B by a scale factor of 0.5.
- **c** Enlarge C by a scale factor of $1\frac{1}{2}$.

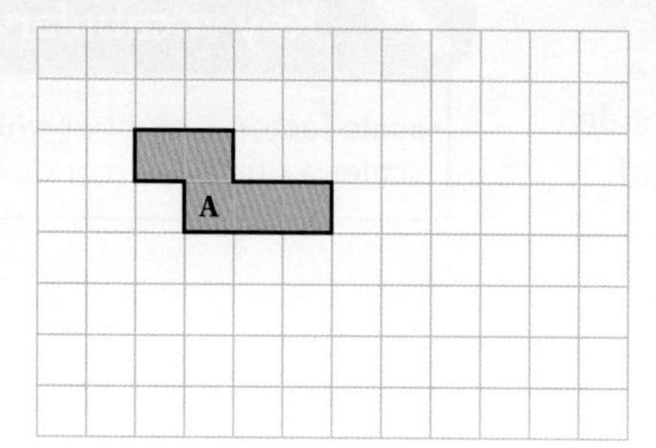

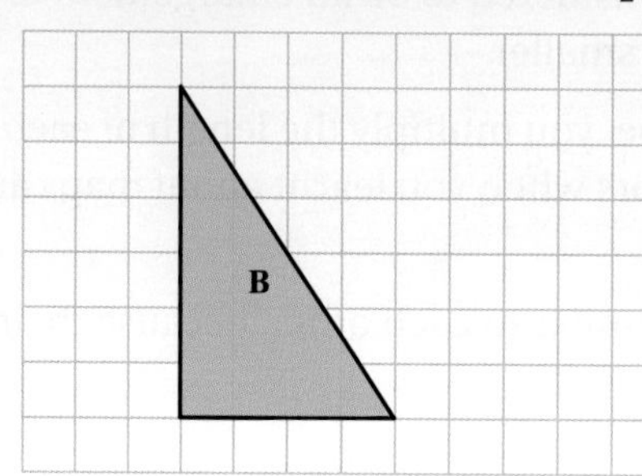

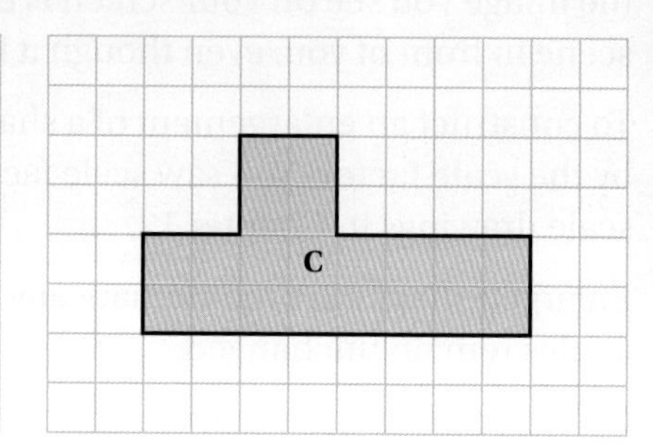

- **d** What do you notice about the enlargement when the scale factor, x, is $0 < x < 1$?

The centre of enlargement

You need two pieces of information to accurately draw an enlargement.

- The scale factor.
- The centre of enlargement.

The centre of enlargement is the point from where the enlargement is measured. In Worked example 3, the enlargements were drawn on another position on the grid. When we use a centre of enlargement, we draw the enlargement in a certain position in relation to the original object.

The table below shows how to enlarge a shape from a given centre of enlargement by a scale factor of 2.

Step 1: Find the distance from the centre of enlargement to a point on the object. You can draw a ray from the centre to the point.	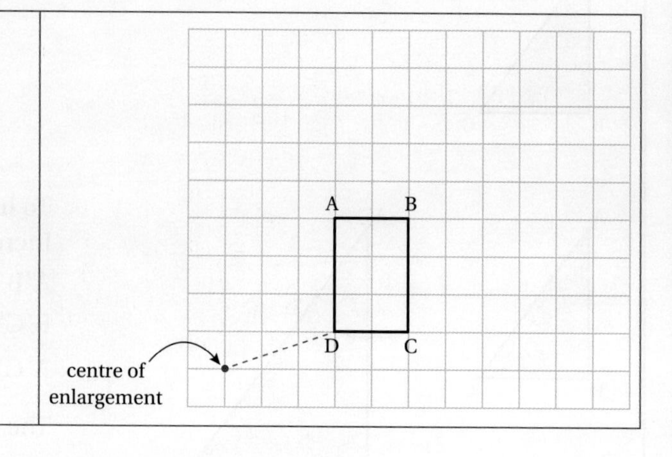

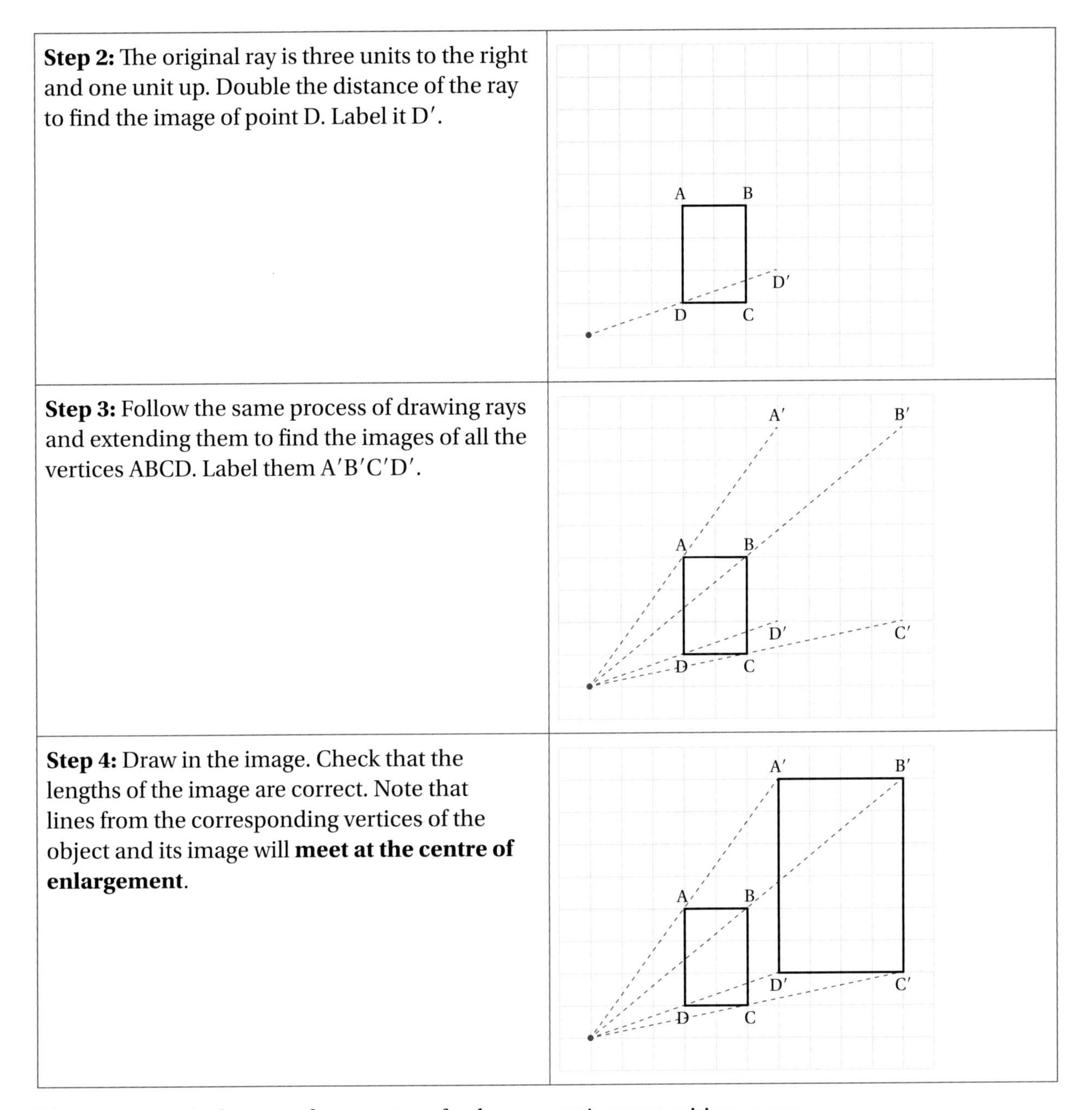

Step 2: The original ray is three units to the right and one unit up. Double the distance of the ray to find the image of point D. Label it D′.
A
B
D
C
D′
Step 3: Follow the same process of drawing rays and extending them to find the images of all the vertices ABCD. Label them A′B′C′D′.
A′
B′
A
B
D′
C′
D
C
Step 4: Draw in the image. Check that the lengths of the image are correct. Note that lines from the corresponding vertices of the object and its image will meet at the centre of enlargement.
A′
B′
A
B
D′
C′
D
C

The procedure is the same for a centre of enlargement in any position, even for a centre of enlargement inside the shape itself.

WORK IT OUT 31.1

This triangle is enlarged from centre ($^-3$, 4) by a scale factor of 2. Draw its image.

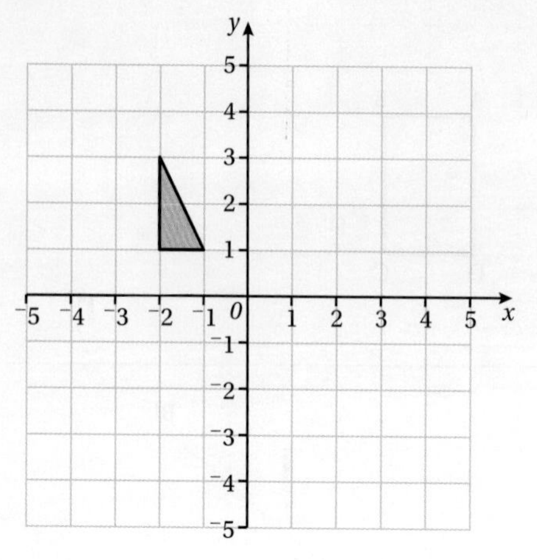

Tip

Sketch the new shape before you construct the enlargement. Draw one ray to identify the new position of the shape. After drawing the enlargement, add in additional rays to check that it is in the correct position.

Which one of these answers is correct? Why are the others wrong?

How many marks would you give the incorrect answers if you were the teacher? Why?

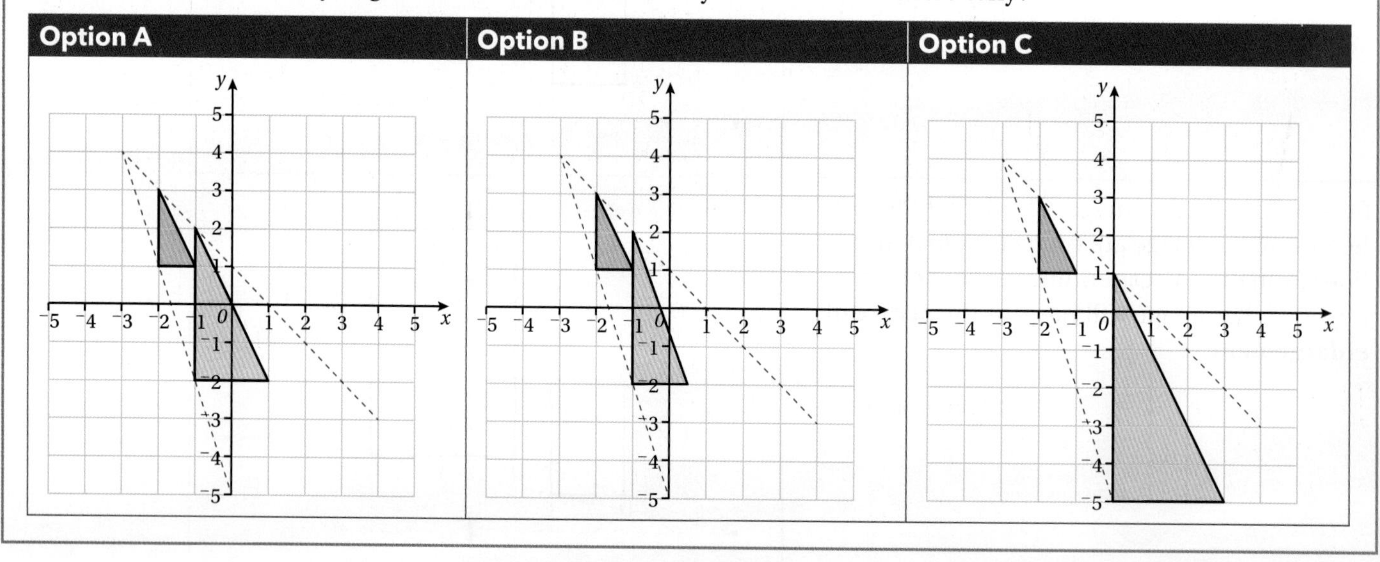

A fractional scale factor

If the scale factor is a fraction **less than 1 but greater than zero**, the image will be smaller than the object.

WORKED EXAMPLE 4

Enlarge the triangle on the grid by a scale factor of $\frac{1}{2}$ through the given centre of enlargement.

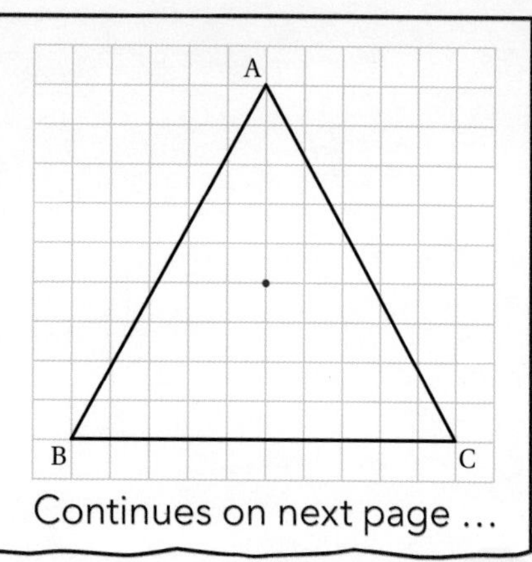

Continues on next page ...

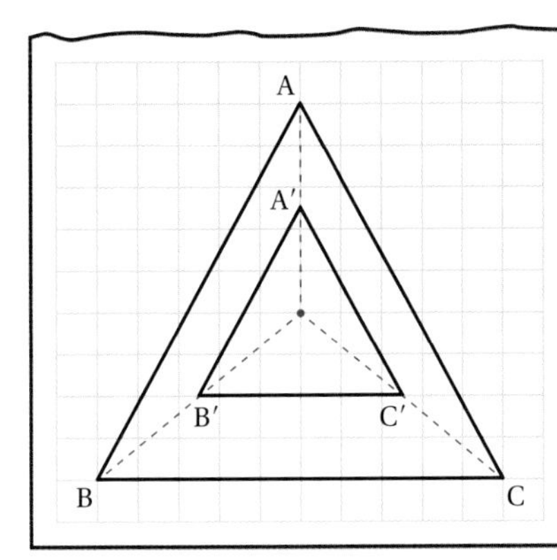

In this case, the rays from each vertex of the object are halved to find the position of the image.

EXERCISE 31C

1 Enlarge each shape as directed, using the point C as the centre of enlargement.

a Enlarge shape R by a scale factor of 3.

b Enlarge shape S by a scale factor of 2.

c Enlarge shape T by a scale factor of $\frac{1}{2}$.

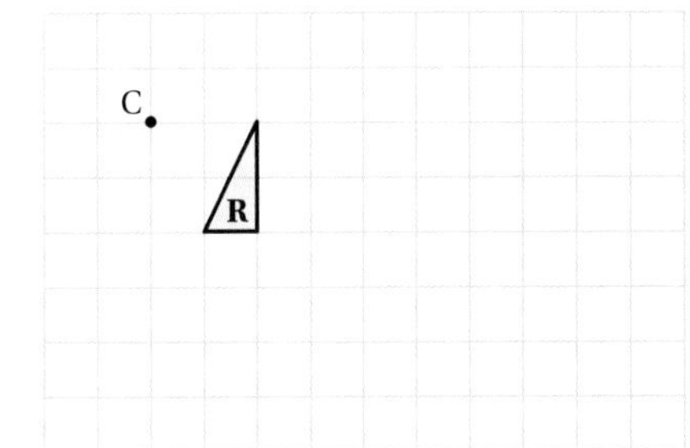

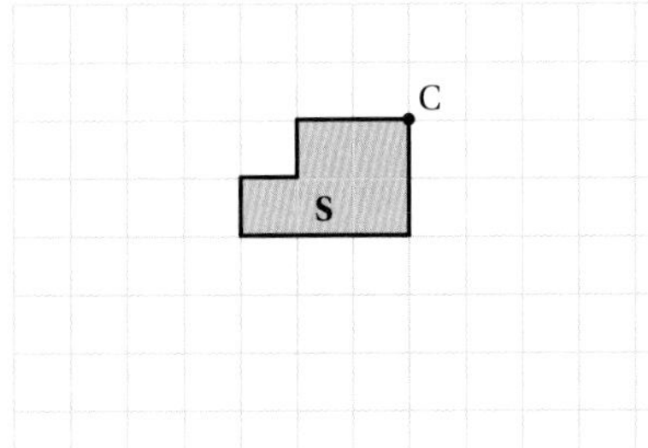

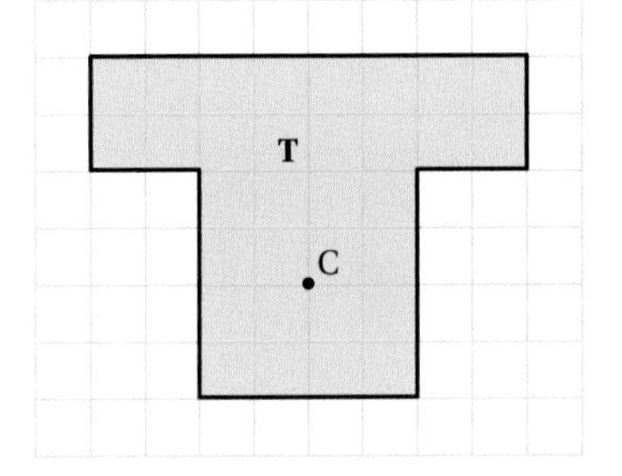

2 Enlarge the given shape by a scale factor of 3, using the origin as the centre of enlargement.

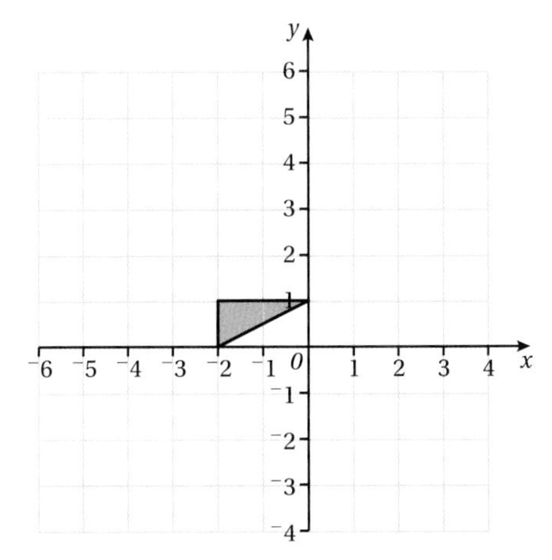

3 Enlarge the given shape by a scale factor of 2 using ($^-4$, 3) as the centre of enlargement.

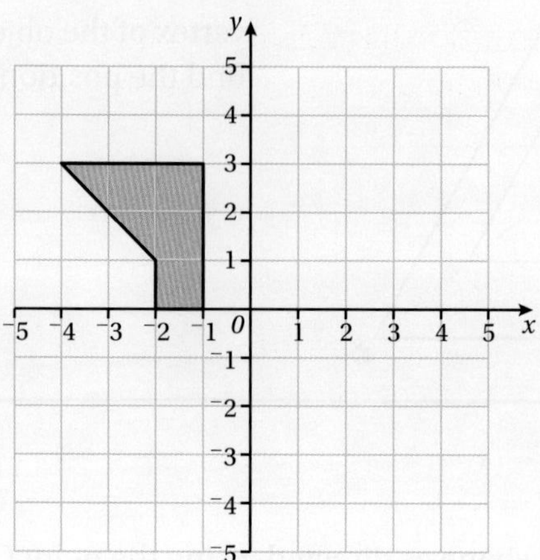

4 Enlarge the given shape by a scale factor of $\frac{1}{3}$ using ($^-5$, 2) as the centre of enlargement.

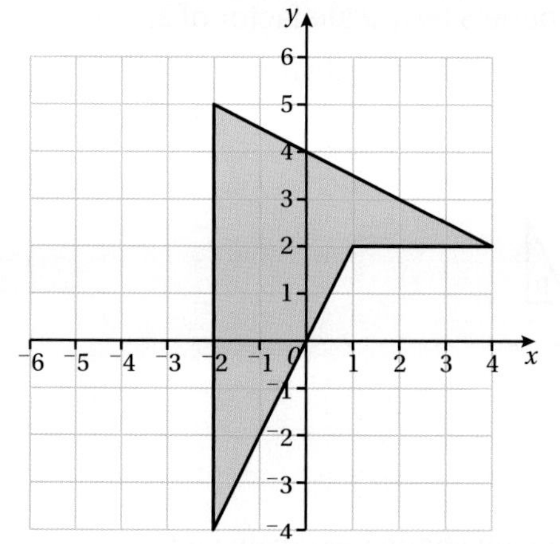

5 Enlarge the given shape by a scale factor of $1\frac{1}{2}$ using (0, 1) as the centre of enlargement.

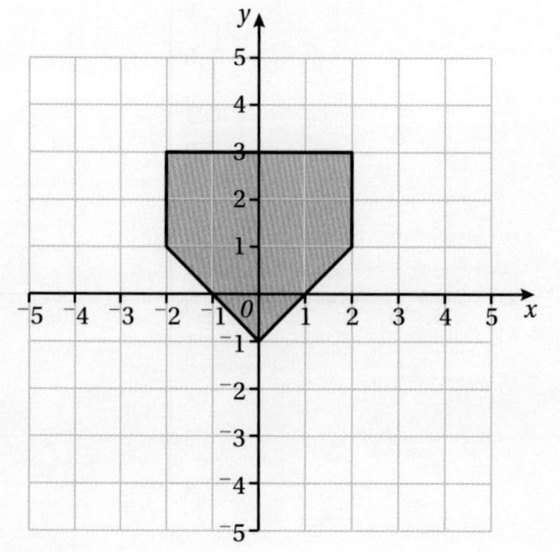

Negative scale factors

A negative scale factor changes the **orientation** of the shape.

The points of the image remain on their ray but they are located on the other side of the centre of enlargement in an inverted position.

Fractional scale factors that are $^-1 < x < 0$ will cause a reduction in the size of the image as well as a change in orientation.

Scale factors less than $^-1$ will result in an enlargement of the shape and a change in the orientation

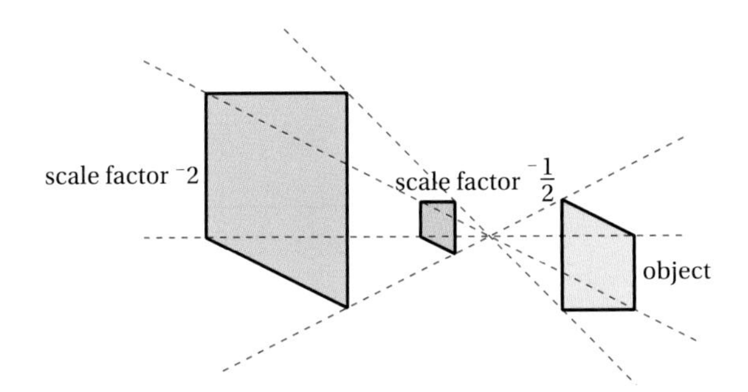

WORKED EXAMPLE 5

Enlarge the rectangle PQRS by a scale factor of $^-2$, using the origin as the centre of enlargement.

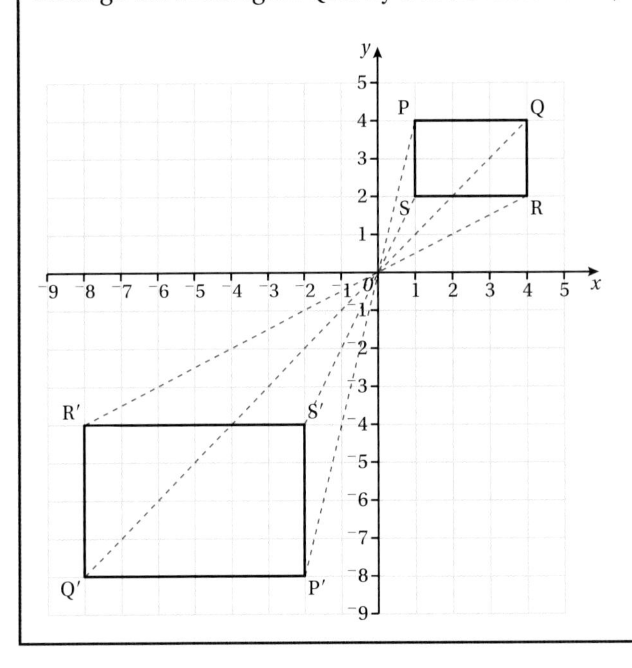

For each vertex, the coordinates are multiplied by $^-2$ to find the image of the point. For example, S has the coordinates (1, 2) and S′ has coordinates ($^-2$, $^-4$).

EXERCISE 31D

1 Enlarge the given shape by a scale factor of $^-1$, using the origin as the centre of enlargement.

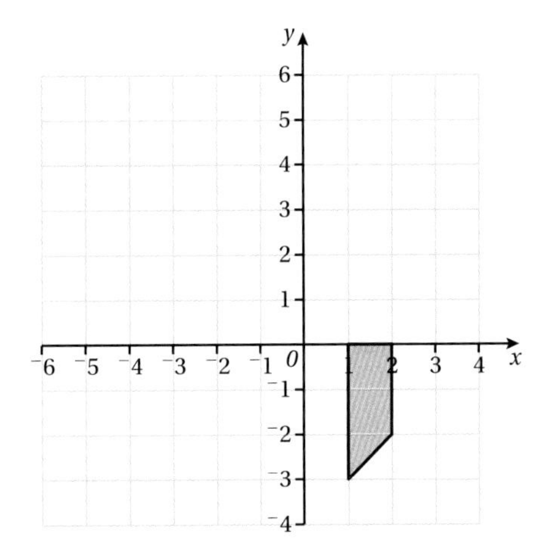

2 Enlarge the given shape by a scale factor of $^{-}2$ using $(^{-}2, 2)$ as the centre of enlargement.

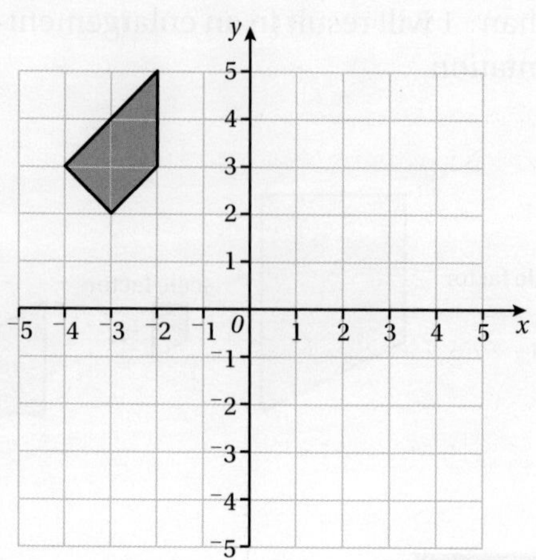

3 Enlarge the given shape by a scale factor of $\frac{^{-}1}{2}$ using $(^{-}3, 0)$ as the centre of enlargement.

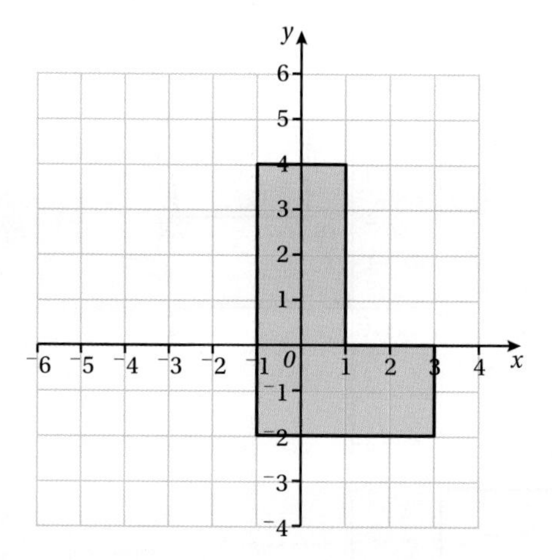

4 Enlarge the given shape by a scale factor of $^{-}1\frac{1}{2}$ using $(^{-}1, 1)$ as the centre of enlargement.

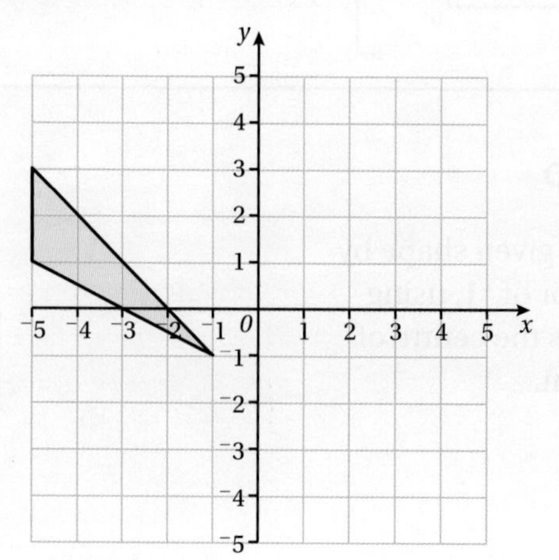

Properties of enlargements

The centre of enlargement can be anywhere: inside the object, on a vertex or side of the object or outside the object.

A scale factor greater than 1 will enlarge the object. A scale factor smaller than 1 will reduce the size of the object, but this is still called an enlargement.

The object and its image are **similar under enlargement**.

- Sides are in the ratio $1 : k$, where k is the scale factor.
- The area of an object and its image will be in the ratio $1 : k^2$, where k is the scale factor.
- An object and its image have corresponding angles the same size.
- For **positive** scale factors ($x > 0$), an object and its image have the same orientation.

Describing enlargements

To describe an enlargement you need to give:

- the scale factor
- the centre of enlargement.

The ratio of sides gives the scale factor.

To find the centre of enlargement you need to draw lines from corresponding vertices of the object and its image to find the point where they meet.

WORK IT OUT 31.2

What scale factors have been used to enlarge this shape?

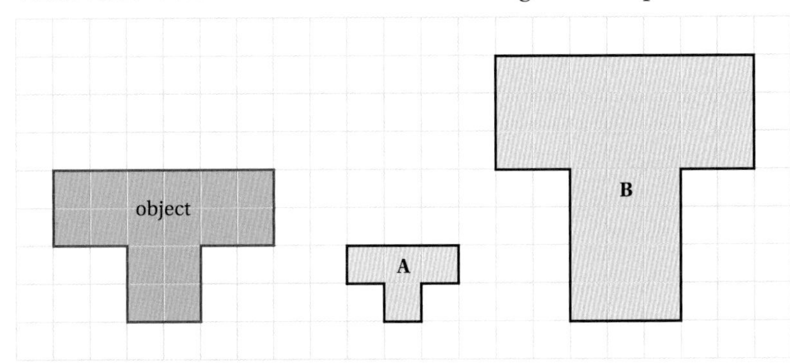

Which one of these students' answers is correct?

What feedback would you give each student to make sure they don't make the same mistakes again?

Ben	Ellie	Rosie
A scale factor of 2 has been used to produce shape **A**.	Since the sides have halved in length to get **A**, the scale factor is $\frac{1}{2}$.	To get shape **A** you have to take away one square along the bottom and along each side edge. So the scale factor is $^{-}1$.
The top of shape **A** is 3 squares across, multiply this by 2 to get 6, the length of the top of the object.	Shape **B** isn't an enlargement. Each side has been increased by different numbers of squares.	Shape **B** is a scale factor of $1\frac{1}{2}$ because 2 squares have become 3 squares.
B can't be an enlargement. Its top is 7 squares and the object is 6. You can't do that using multiplication. Maybe it's $^{-}1$?		

Find answers at: cambridge.org/ukschools/gcsemaths-studentbookanswers

EXERCISE 31E

1 Which of the following houses are enlargements of house A? For each enlargement state the scale factor.

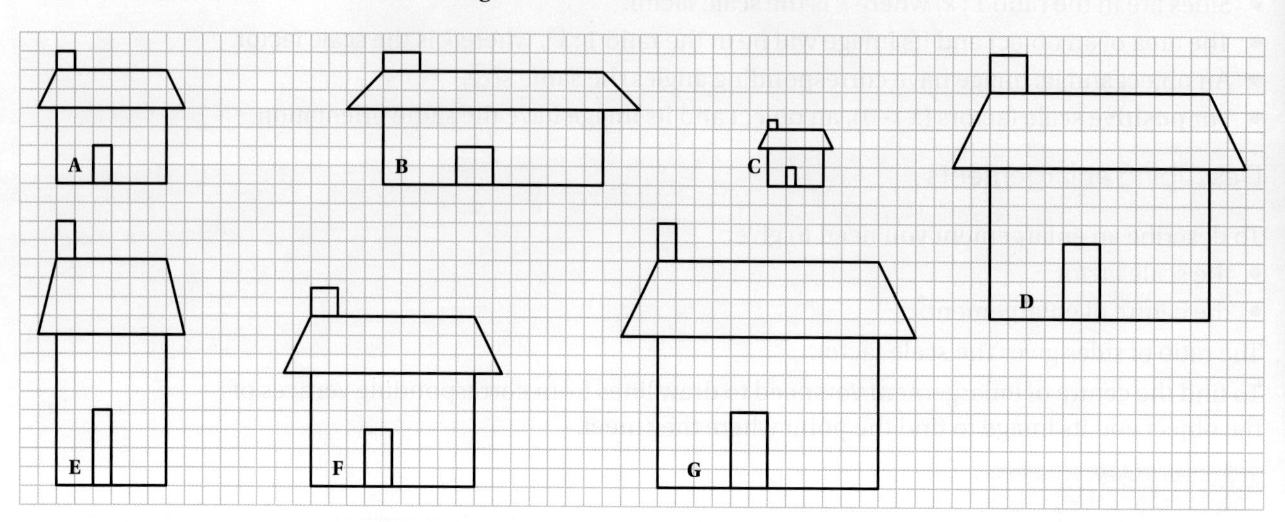

2 These diagrams each show an object and its image after an enlargement.

Describe each of these enlargements by giving both the scale factor and the coordinates of the centre of enlargement. In each case the object is labelled.

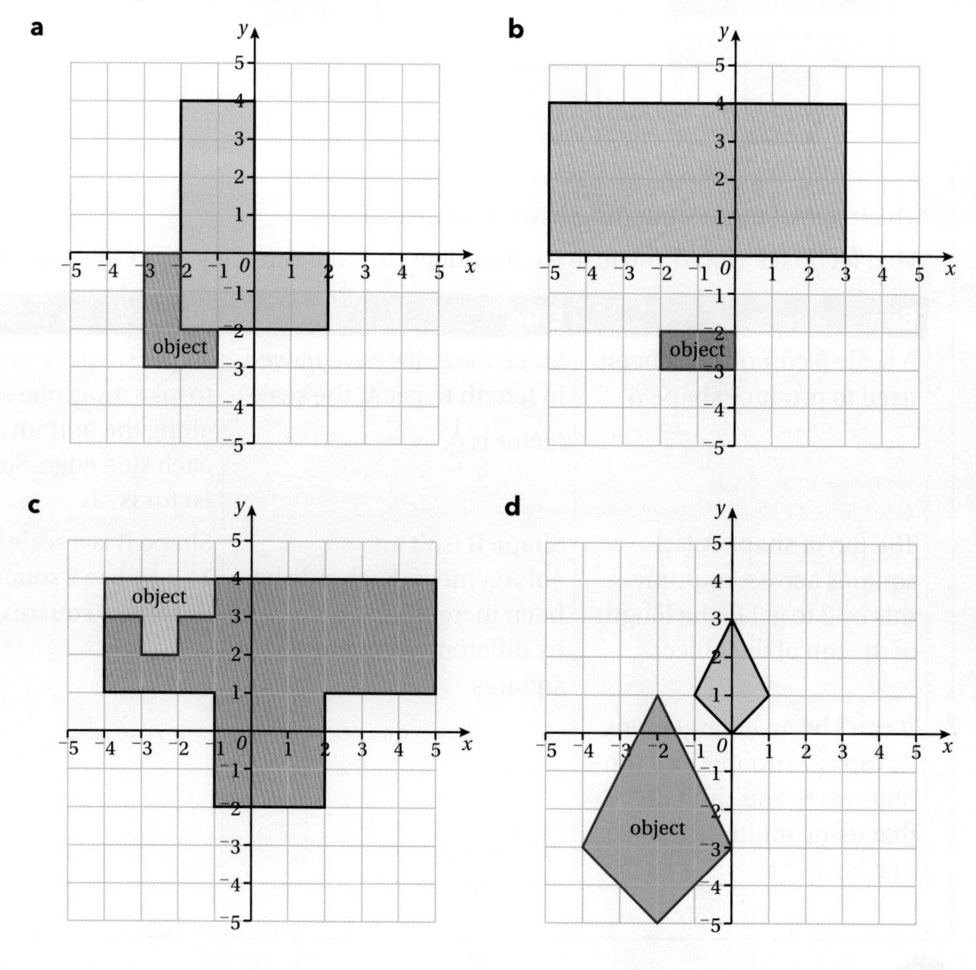

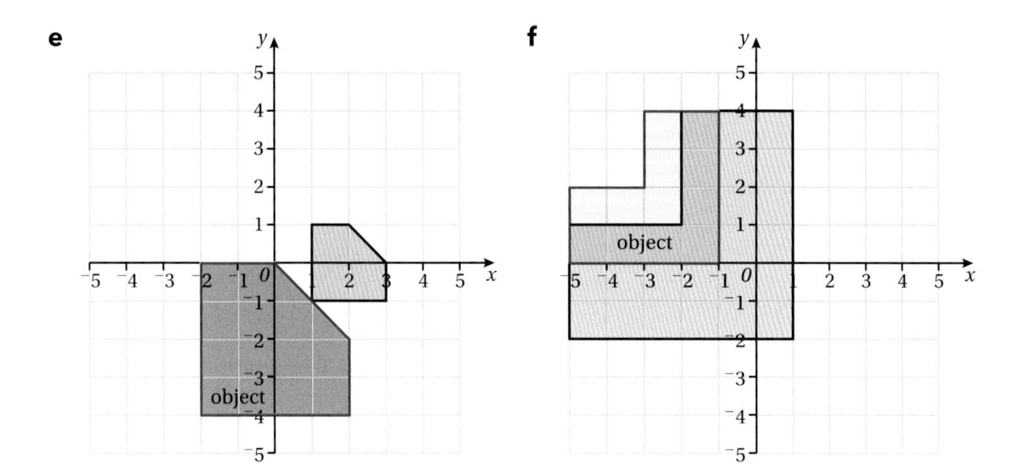

3 Describe each of these enlargements. In each case the object is labelled.

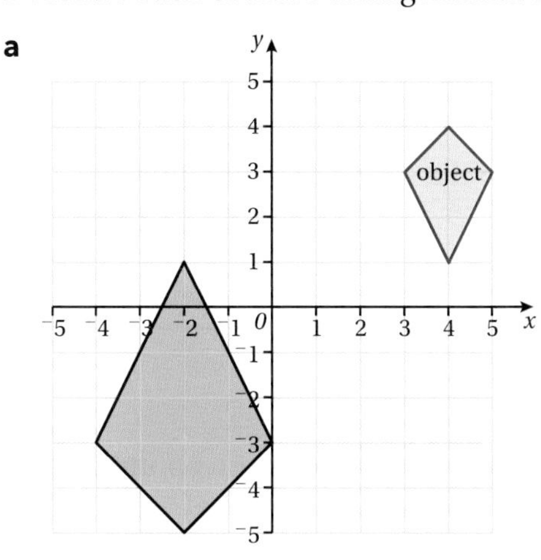

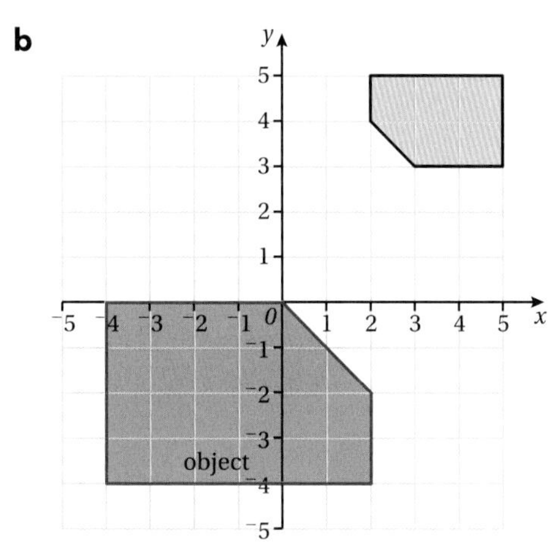

Section 3: Similar shapes and objects

A polygon is similar to another polygon if it is an enlargement of the original polygon. So two polygons will be similar if the angles in one polygon are equal to the corresponding angles in the other polygon, **and** the ratio of the corresponding sides from the one polygon to the other are kept the same.

For polygons other than triangles, equal angles alone are not sufficient to prove similarity.

Find answers at: cambridge.org/ukschools/gcsemaths-studentbookanswers

WORKED EXAMPLE 6

Compare each of the quadrilaterals B to D below to the first quadrilateral, A.

Which of them are similar to A?

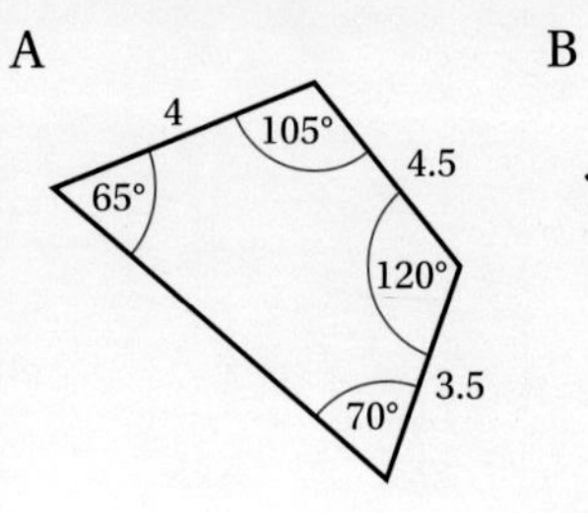

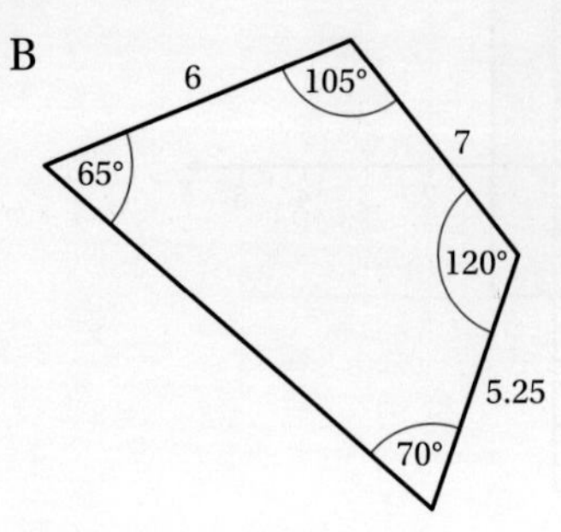

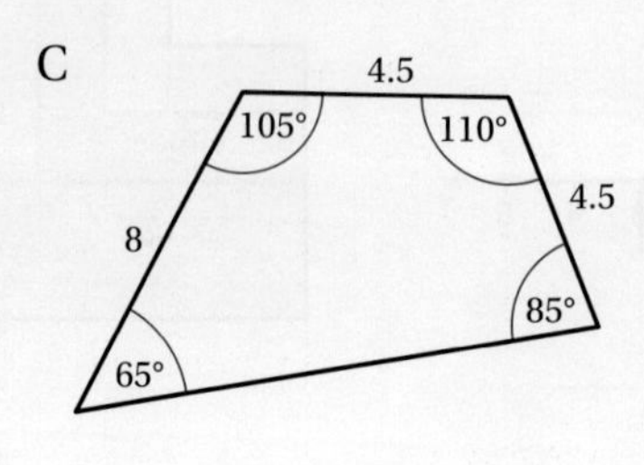

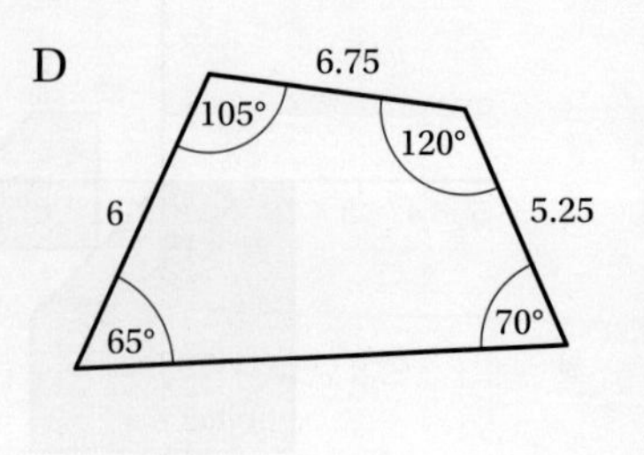

$\frac{4}{6} \neq \frac{4.5}{7}$. So B is not similar to A.

The angles in C are different to the angles in A, so C is not similar to A.

B has the same size angles as A, but the sides are not in proportion.

D has corresponding angles equal to those in A.

$\frac{4}{6} = \frac{4.5}{6.75} = \frac{3.5}{5.25}$. D is similar to A.

D is an enlargement of A with a scale factor of 1.5.

Test to see whether the sides are in the same proportion.

Similar areas and volumes

You already know that if two shapes are similar their corresponding sides are in the same ratio, and their corresponding angles are equal.

Consider the following set of similar triangles.

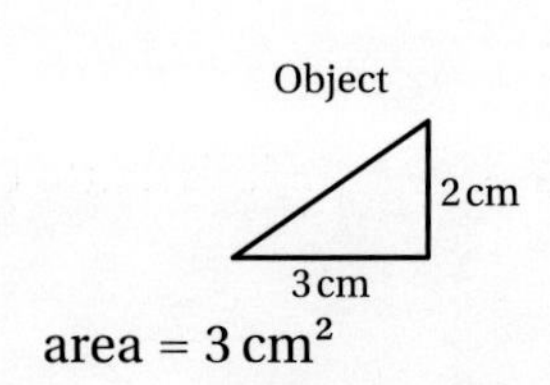

area = 3 cm^2

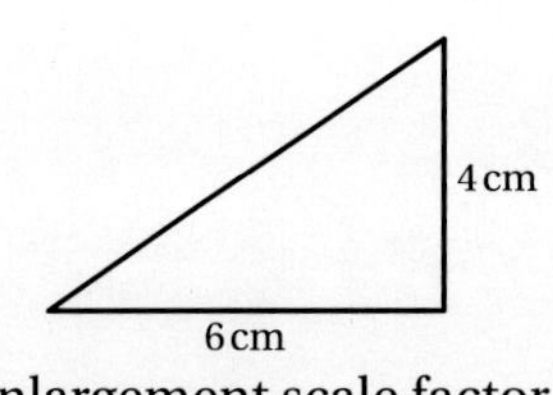

enlargement scale factor 2
area = 12 cm^2

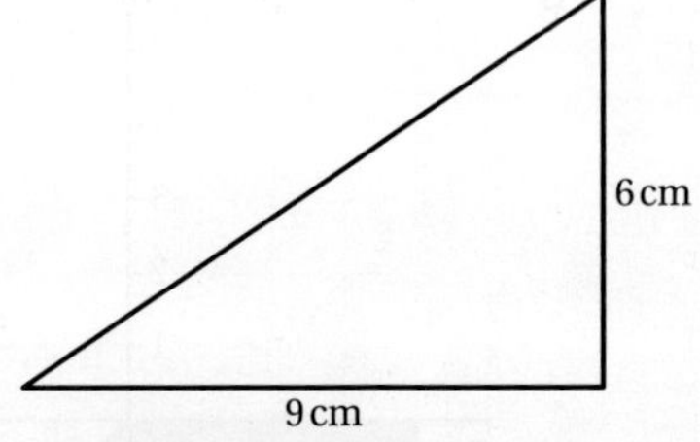

enlargement scale factor 3
area = 27 cm^2

You should notice the following:

Scale factor 2 area scale factor $4 = 2^2$

Scale factor 3 area scale factor $9 = 3^2$

In general:

Scale factor n area scale factor n^2

Consider the following set of similar cuboids.

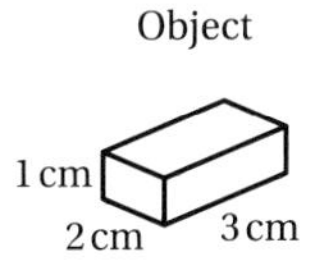

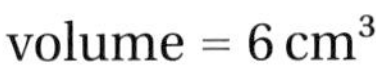

volume = 6 cm^3

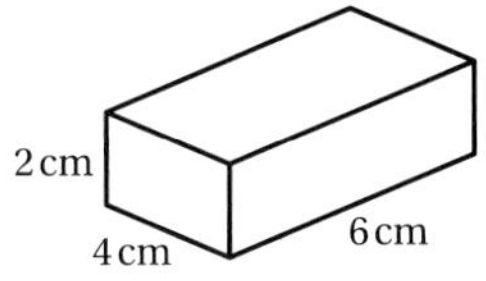

enlargement scale factor 2
volume = 48 cm^3

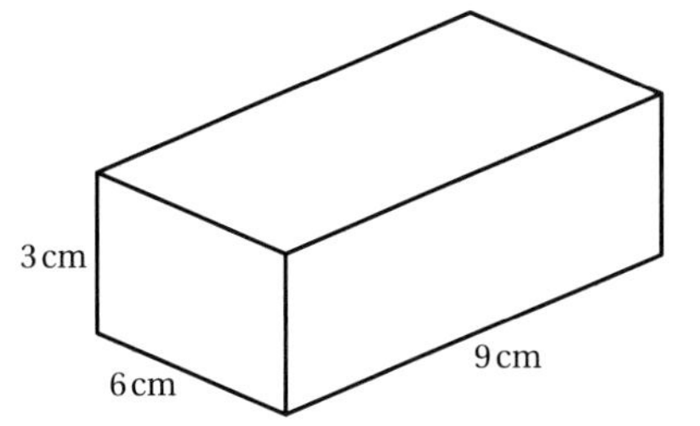

enlargement scale factor 3
volume = 162 cm^3

You should notice the following;

Scale factor 2 volume scale factor 8 = 2^3

Scale factor 3 volume scale factor 27 = 3^3

In general:

Scale factor n volume scale factor n^3

If a 2D shape is enlarged by a scale factor, a similar shape is created. All of the dimensions of the image will be enlarged by the same scale factor. If the scale factor is a, then the ratio of the area of the image to the area of the original shape will be a^2.

If a 3D object is enlarged by a scale factor of a, then the ratio of the volume of the enlarged object will be a^3.

The scale factor of enlargement for a measurement can be a whole number or a fraction. It can't be a negative number, as measurements cannot take on negative values.

WORKED EXAMPLE 7

The triangle opposite is enlarged by a factor of 3.

a What will the new dimensions and the area be?

b Calculate the ratio of the new area to the original area.

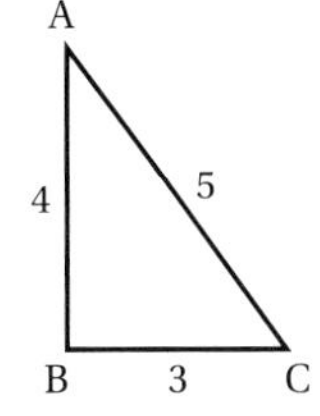

a 3AB = 12
3AC = 15
3BC = 9

Each side of the triangle will be enlarged by a factor of 3.

Area of triangle $= \frac{1}{2}$ base × height
$= \frac{1}{2} \times 9 \times 12$
$= 54$ square units

Use the new dimensions to calculate the area of the enlarged triangle.

b The original area was:

Calculate the original area.

Area of triangle $= \frac{1}{2}$ base × height
$= \frac{1}{2} \times 3 \times 4$
$= 6$ square units

The ratio of the enlargement is 54 to 6 = 9 : 1.

Notice that this is equal to the square of the scale factor: $3^2 = 9$.

WORKED EXAMPLE 8

A triangular prism with base triangle ABC from Worked example 7 has a height of 10 cm.

The whole prism is enlarged by a scale factor of 3.

a Calculate the new surface area of the triangular prism.

b What is the ratio of the new surface area to the original surface area?

c Calculate the new volume of the prism.

d What is the ratio of the new volume to the original volume?

> **Tip**
> Draw a diagram of the two triangular prisms.

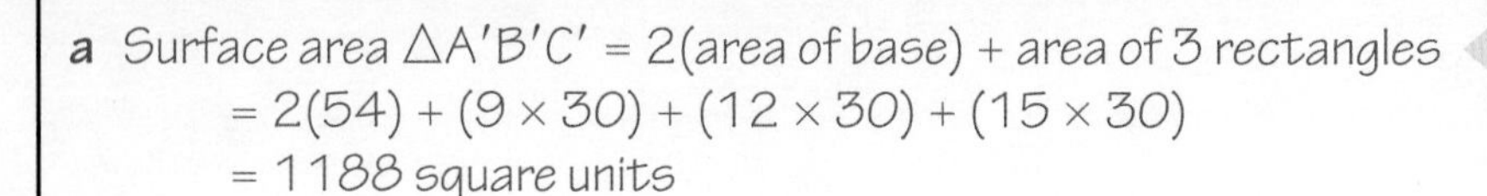

a Surface area $\triangle A'B'C'$ = 2(area of base) + area of 3 rectangles
$= 2(54) + (9 \times 30) + (12 \times 30) + (15 \times 30)$
$= 1188$ square units

Use the dimensions of $\triangle A'B'C'$ calculated in Worked example 7, to calculate the surface area of the enlarged prism. Don't forget to increase the height of the prism by a scale factor of 3.

b Original surface area = 2(area of base) + area of 3 rectangles
$= 2(6) + (3 \times 10) + (4 \times 10) + (5 \times 10)$
$= 132$ square units

Ratio is $1188 : 132 = 9 : 1$.

Calculate the original surface area, and then the ratio.

c Volume = Area of base × height
$= 54 \times$ new height
$= 54 \times 30 = 1620$ cubic units

Calculate the volume using the new dimensions.

d Original volume $= 6 \times 10$
$= 60$

Ratio is $1620 : 60 = 27 : 1$.

Calculate the volume of the original prism, and then the ratio.

Notice that the value 27 is the cube of the scale factor of the length 3: $3^3 = 27$.

EXERCISE 31F

1 Decide whether each statement below is true or false. Explain your reasoning.

a All squares are similar.

b All hexagons are similar.

c All rectangles are similar.

d All regular octagons are similar.

2 Sketch the following pairs of shapes and decide if they are similar. Explain your reasoning.

a Rectangle ABCD with AB = 5 cm and BC = 3 cm.
Rectangle EFGH with EF = 10 cm and FG = 6 cm.

b Rectangle ABCD with AB = 5 cm and BC = 3 cm.
Rectangle EFGH with AB = 10 cm and BC = 9 cm.

c Square ABCD with AB = 4 cm.
Square EFGH with EF = 6 cm.

3 These two shapes are similar. Find the missing lengths a and b.

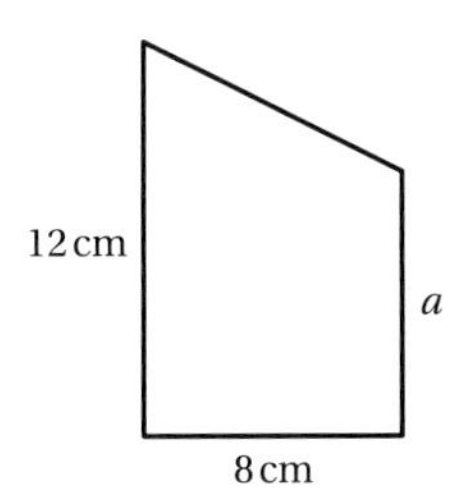

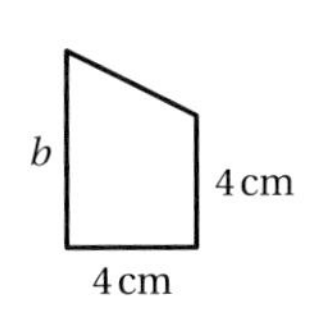

4 The two shapes on the right are similar. Find the lengths of the sides in the second shape.

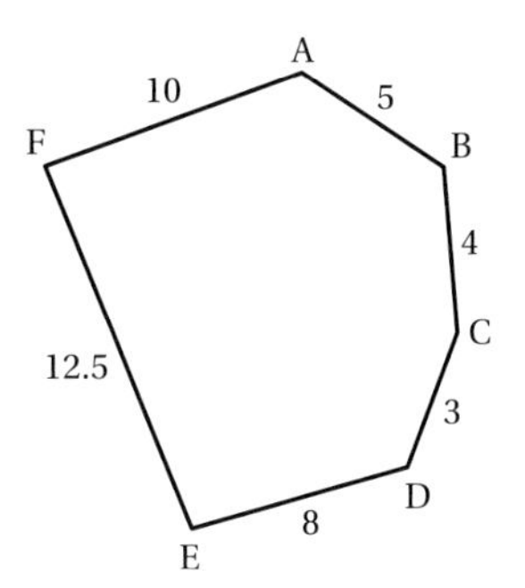

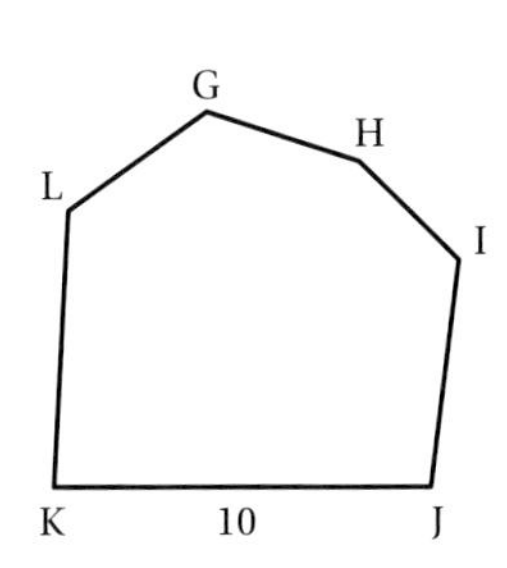

5 The first shape on the right has been enlarged by a scale factor of 1.5 to create the image GHIJKL.
AB = 5 cm and BC = 7 cm.
Find the lengths of the sides in the second shape.

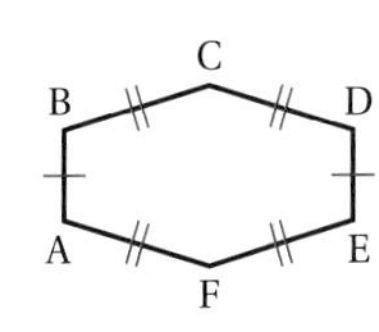

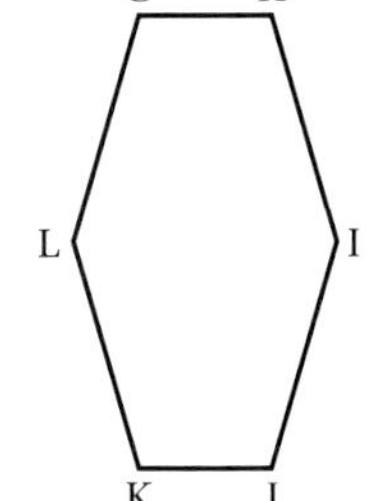

6 Two similar shapes have volumes in the ratio 1 : 512.
What is the ratio of their surface areas?

7 Triangle A and B are similar. Triangle A has an area of 48 square units. Triangle B has an area of 588 square units.
- **a** What is the scale factor of enlargement?
- **b** If Triangle A has a base of 8 units, what are the dimensions of Triangle B?

8 A canned food producer makes tins of radius 2.5 cm, and similar tins that have a radius of 7 cm.
The smaller tin has a volume of 157 ml (or 157 cm^3).
- **a** What is the height of the smaller tin?
- **b** What will the height and the volume of the larger tin be?

9 A cone has a radius of 4 cm and a height of 8 cm.
- **a** Calculate the volume and surface area of the cone.
- **b** If each measurement is increased by a factor of 3, calculate the new volume and surface area.

10 A sphere of radius 13 units is enlarged by a factor of 4.
- **a** Determine the surface area of the original as well as the enlarged sphere.
- **b** Determine the volume of the original as well as the enlarged sphere.

Checklist of learning and understanding

Similar triangles

- Two triangles are similar if all three corresponding angles are equal. Similar triangles are the same shape, and their corresponding sides are in proportion.
- The proportion between corresponding sides of similar triangles can be used to solve problems in geometry.

Find answers at: cambridge.org/ukschools/gcsemaths-studentbookanswers

Enlargements

- An enlargement is a transformation that changes the position and size of a shape.
- Enlargements are described by a scale factor and centre of enlargement.
- A fractional scale factor that is less than 1 will make the image smaller than the object, but it is still called an enlargement.

Similar shapes

- If shapes are enlarged by a scale factor, similar shapes are created with all their sides in proportion. The proportionality between lengths can be used to solve geometry problems.
- If a 2D shape is enlarged, a similar shape is created. All of the dimensions of the image will be enlarged by the same scale factor. If the scale factor is a, then the ratio of the area of the image to the area of the original shape will be a^2.
- If a 3D object is enlarged by a scale factor of a, then the ratio of the volume of the enlarged object will be a^3.

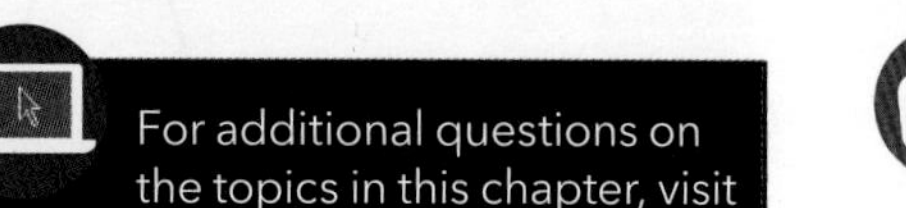

Chapter review

1 **a** Prove that triangle VWX is similar to triangle VYZ.

b Find the length of XZ.

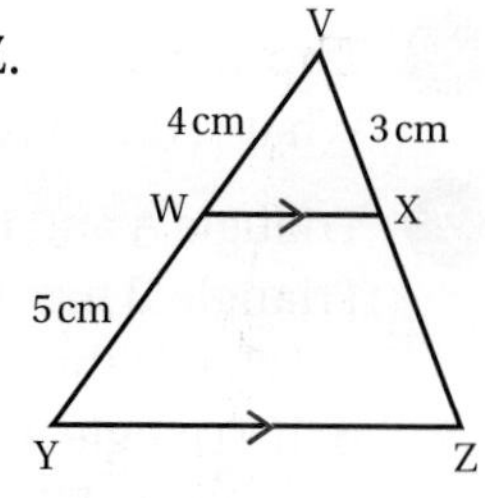

2 A tree which is 3 m high has a shadow length of 7.5 m. At the same time of day, a building casts a shadow that is 16.25 m long. Use similar triangles to calculate the height of the building.

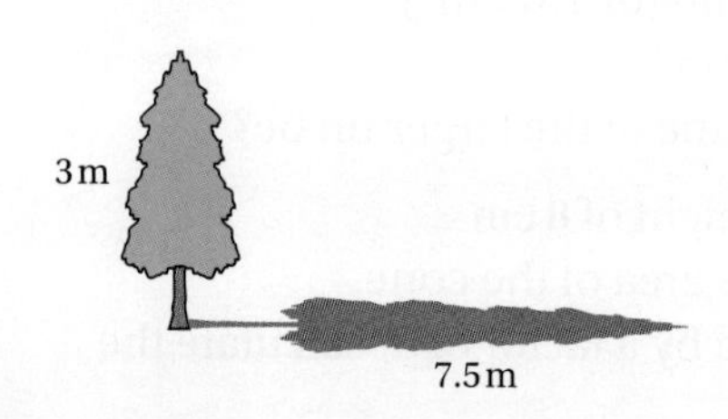

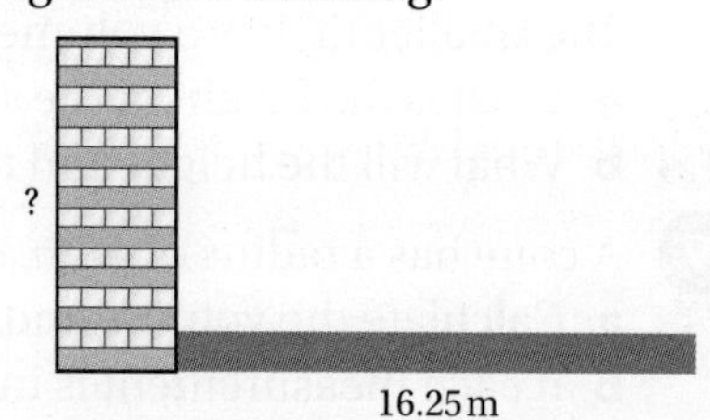

3 Draw an enlargement of ABCD by $\frac{1}{2}$, using the given point as the centre of enlargement.

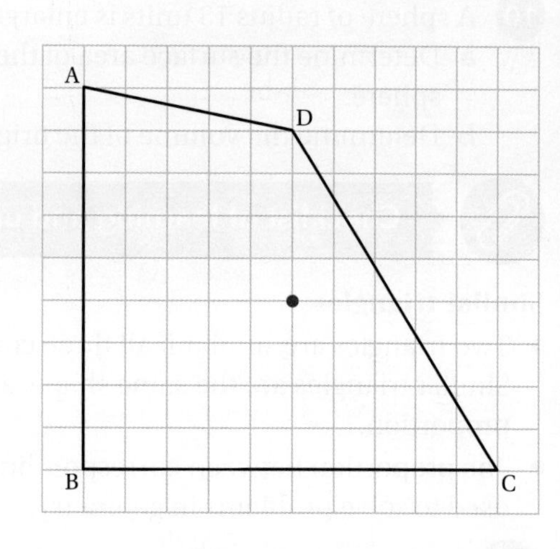

4 Draw an enlargement of this shape by 1.5, using the origin as the centre of enlargement.

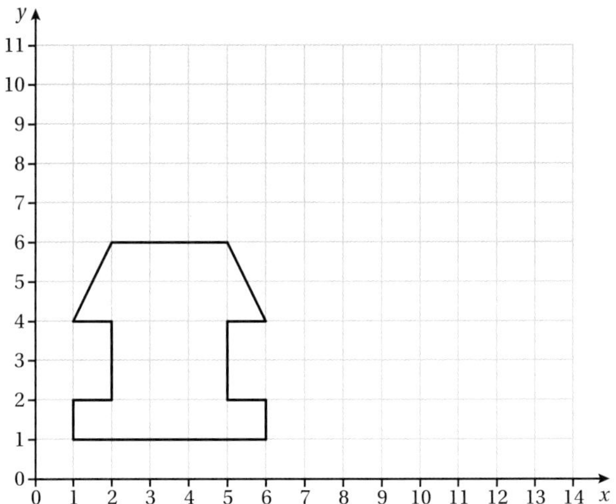

5 Are any two regular hexagons similar shapes? Explain.

6 Are any two rhombuses similar shapes? Explain.

7 These two glasses are similar.

The height of the smaller glass is 7 cm and the height of the larger glass is 14 cm.

a Ruby says: The base area of the smaller glass is half the base area of the larger glass.

Explain why she is wrong. *(1 mark)*

b The volume of the larger glass is 480 cm^3.

Calculate the volume of the smaller glass. *(2 marks)*

8 Two similar shapes have areas of 90 cm^2 and 62.5 cm^2. The perimeter of the smaller shape is 40 cm. What is the perimeter of the larger shape?

9 A square-based pyramid has a base length of 10 cm and a perpendicular height of 12 cm. If the pyramid is enlarged by a factor of $\frac{1}{2}$, calculate the new volume.

10 Describe this enlargement.
The object is labelled.

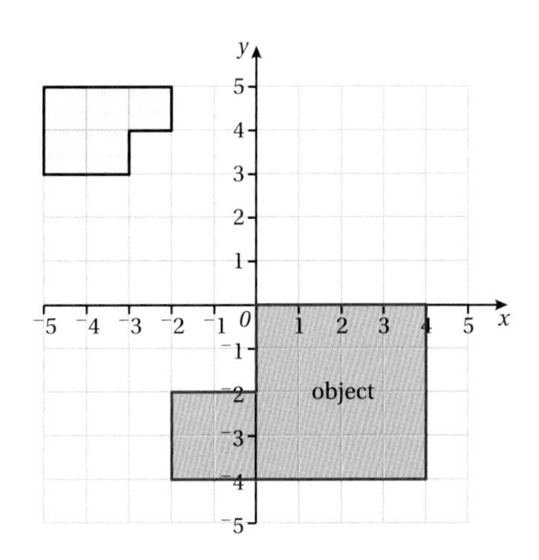

Find answers at: cambridge.org/ukschools/gcsemaths-studentbookanswers

32 Pythagoras' theorem

In this chapter you will learn how to …

- develop full knowledge and understanding of Pythagoras' theorem.
- apply in 2D and 3D problems.
- link the maths to real-life skills for industry.

For more resources relating to this chapter, visit GCSE Mathematics Online.

Using mathematics: real-life applications

Builders, carpenters, garden designers and navigators all use Pythagoras' theorem in their jobs. It is a method based on right-angled triangles which helps them to work out unknown lengths and check for right angles.

"I use Pythagoras' theorem to help me prepare floor plans and do calculations for footings and heights of buildings. Any building surveyor will have a range of tools to help them make check calculations on site." *(Building surveyor)*

Calculator tip

Make sure you know how to work with surds on your calculator and that you can round numbers to given levels of accuracy using decimal places and significant figures.

Before you start …

Ch 27	You must be able to work with exact and approximate values of surds.	**1** **a** Which are correct? **i** $\sqrt{19} = \pm 4.36$ 2dp **ii** $\sqrt{19} = \pm 361$ **iii** $\sqrt{361} = \pm 19$ **iv** $19^2 = 361$ **b** Which are correct? **i** $1.95^2 = 3.8025 = 3.80$ 2dp **ii** $\sqrt{7} = 2.645751311... = \pm 2.64$ 2dp **iii** $\sqrt{3} = 1.732050808... = 1.73$ 2dp **iv** $3.14^2 = 9.8596 = 9.86$ 2dp
Ch 9	You need to recognise and define different types of angle.	**2** Which is a right angle? Identify the other angles. **a** **b** **c** **d** **e**
Ch 5	You'll need to apply the properties of different types of triangle to solve problems.	**3** What is the area of this triangle? (4, 5, 3) **4** What do you know about angles x and y in this triangle?

Assess your starting point using the Launchpad

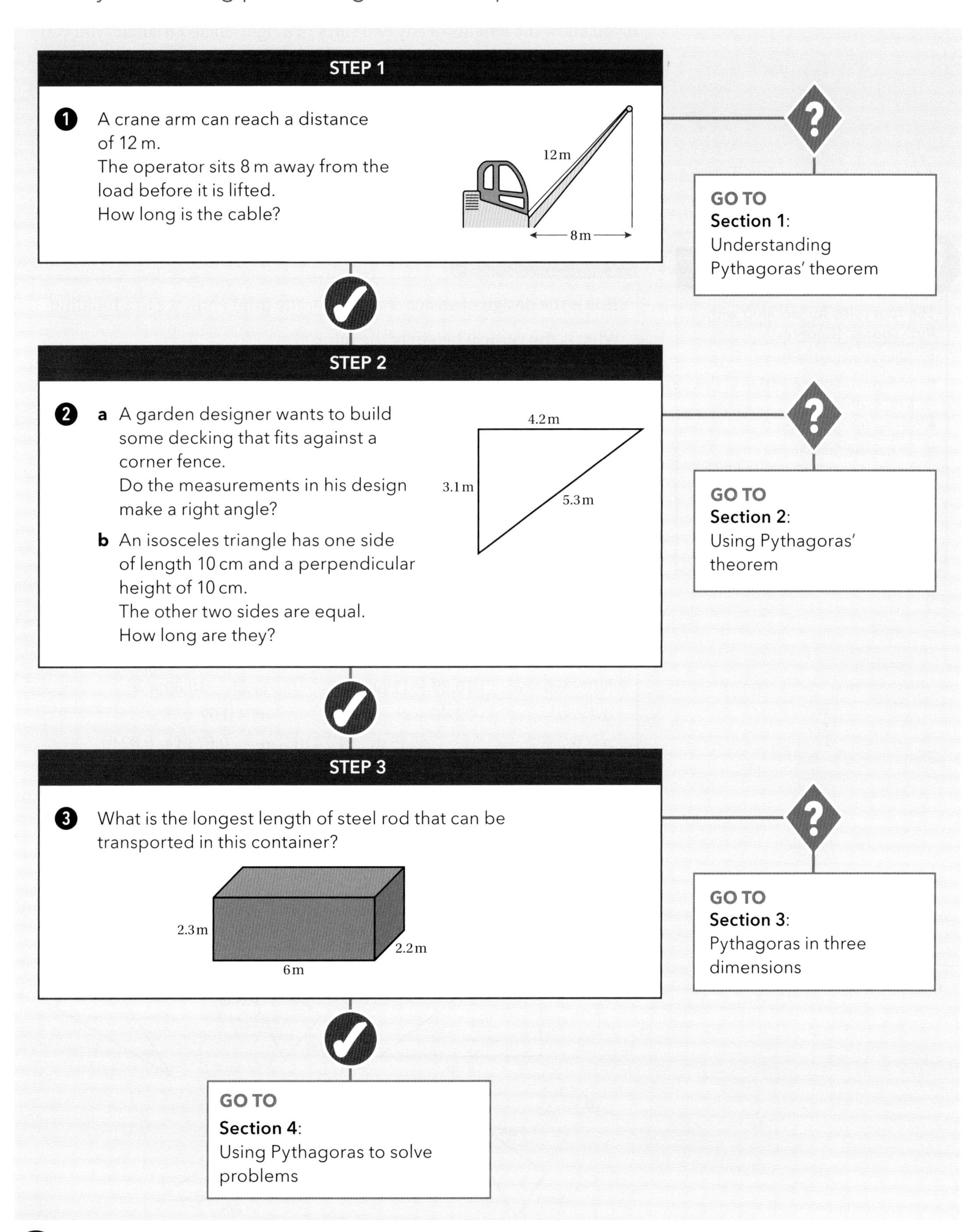

STEP 1

1. A crane arm can reach a distance of 12 m.
The operator sits 8 m away from the load before it is lifted.
How long is the cable?

?

GO TO
Section 1:
Understanding Pythagoras' theorem

✓

STEP 2

2. **a** A garden designer wants to build some decking that fits against a corner fence.
Do the measurements in his design make a right angle?

b An isosceles triangle has one side of length 10 cm and a perpendicular height of 10 cm.
The other two sides are equal.
How long are they?

?

GO TO
Section 2:
Using Pythagoras' theorem

✓

STEP 3

3. What is the longest length of steel rod that can be transported in this container?

?

GO TO
Section 3:
Pythagoras in three dimensions

✓

GO TO
Section 4:
Using Pythagoras to solve problems

Find answers at: cambridge.org/ukschools/gcsemaths-studentbookanswers

Section 1: Understanding Pythagoras' theorem

If you know the lengths of any two sides of a right-angled triangle, you can use them to find the length of the third side.

In the formula $a^2 + b^2 = c^2$, c is the hypotenuse and a and b are the two shorter sides.

The formula can be rearranged to make a or b the subject of the formula:

$$a^2 = c^2 - b^2$$

$$b^2 = c^2 - a^2$$

Tip

The four rules for working with Pythagoras' theorem:

- Always write the formula.
- Always include a sketch of the problem.
- Show full workings.
- Show the final answer to a given degree of accuracy. This could be 2 decimal places or several significant figures, depending on the detail of the problem.

WORK IT OUT 32.1

This is the design of an access ramp for the front entrance to a building.

What is the vertical height of the ramp?

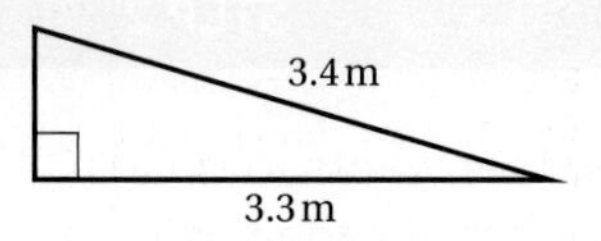

Which of these calculations is correct for this design? Why are the others wrong?

Option A	Option B	Option C
$c^2 - b^2 = a^2$	$c^2 - b^2 = a^2$	$c^2 - b^2 = a^2$
$3.4^2 - 3.3^2 = a^2$	$3.3^2 + 3.4^2 = c^2$	$3.4^2 - 3.3^2 = a^2$
$3.4 \times 2 = 6.8$	$3.3 \times 3.3 = 10.89$	$3.4 \times 3.4 = 11.56$
$3.3 \times 2 = 6.6$	$3.4 \times 3.4 = 11.56$	$3.3 \times 3.3 = 10.89$
$6.8 - 6.6 = a^2$	$10.89 + 11.56 = c^2$	$11.56 - 10.89 = a^2$
$0.2 = a^2$	$22.45 = c^2$	$a^2 = 0.67$
$a = 0.2 \div 2$	$a = \sqrt{22.45\text{ m}} = 4.74\text{ m}$	$a = \sqrt{0.67\text{ m}} = 0.82\text{ m}$
$a = 0.1$ m	(to 2 decimal places)	(to 2 decimal places)

EXERCISE 32A

1 Find the length of the hypotenuse in each of the following triangles:

a

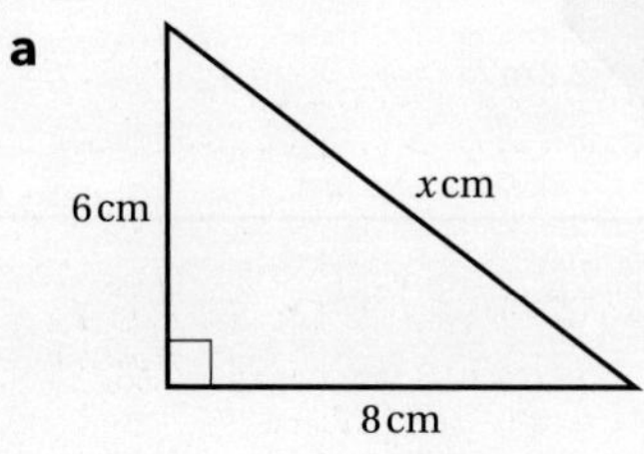

b

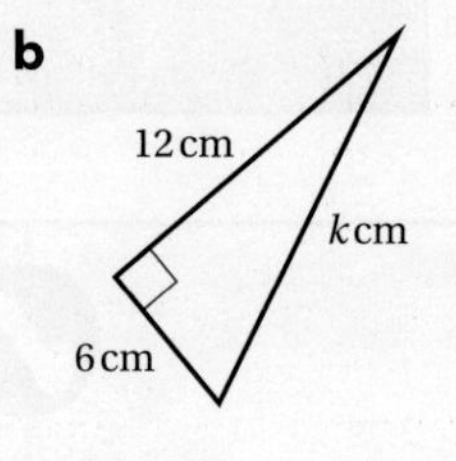

c

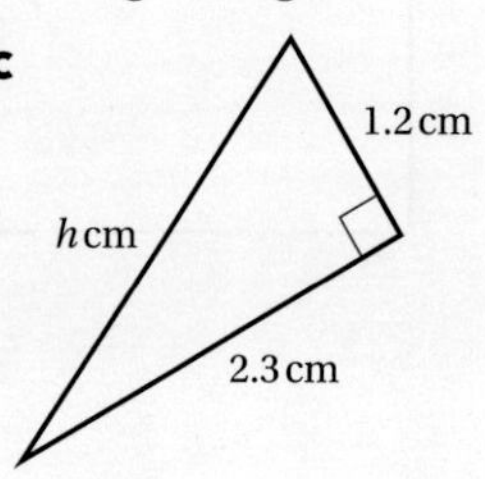

d

1.5 cm

p cm

0.6 cm

e

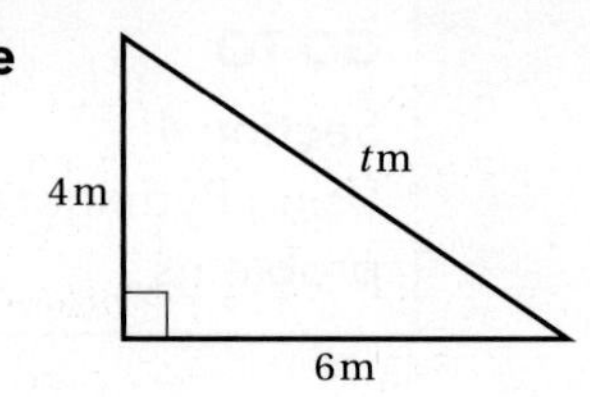

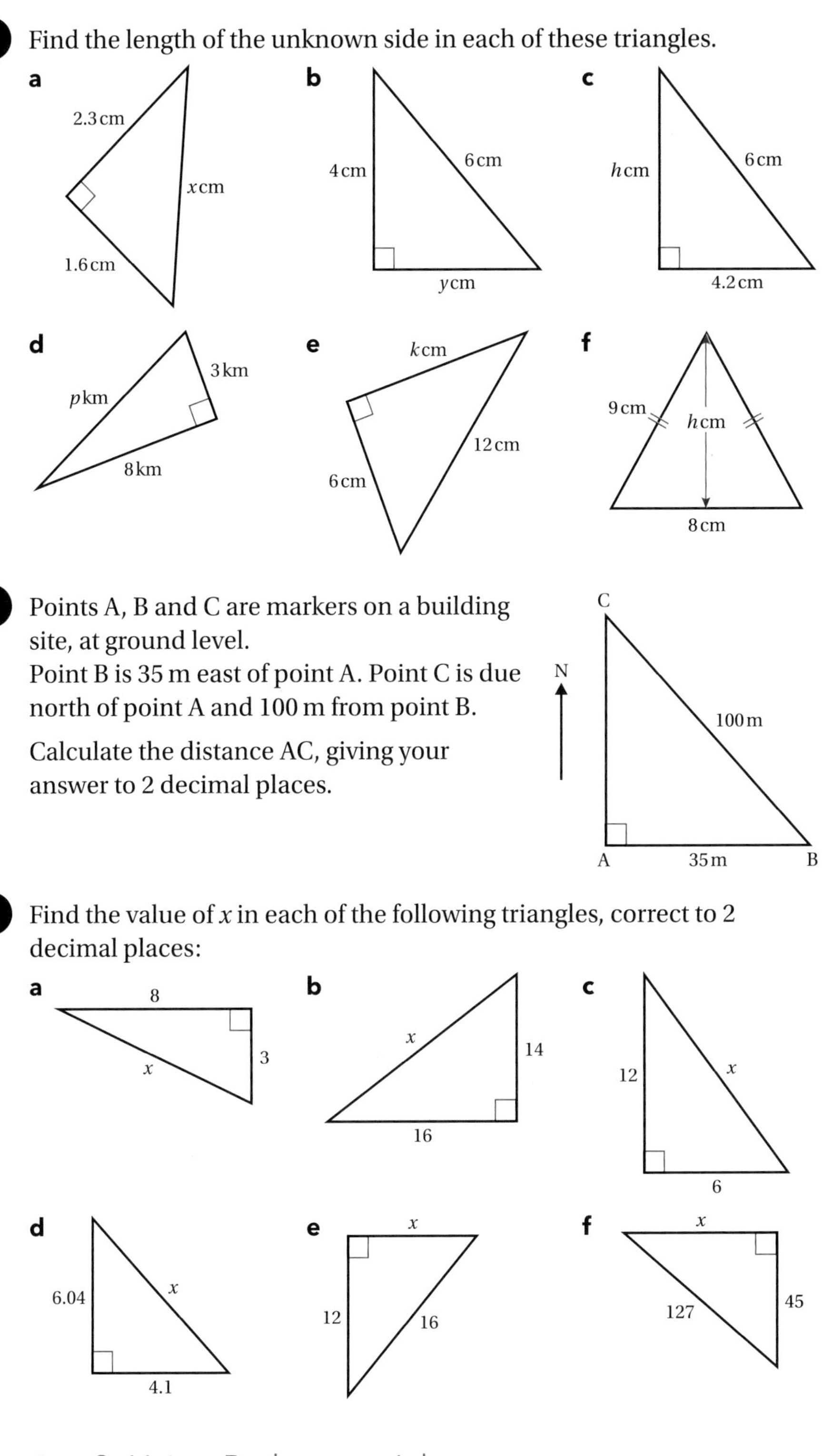

2 Find the length of the unknown side in each of these triangles.

3 Points A, B and C are markers on a building site, at ground level.
Point B is 35 m east of point A. Point C is due north of point A and 100 m from point B.

Calculate the distance AC, giving your answer to 2 decimal places.

4 Find the value of x in each of the following triangles, correct to 2 decimal places:

Section 2: Using Pythagoras' theorem

Pythagorean triples

Carpenters need to make rectangular window frames. If the frames don't have right angles at the corners, then the window won't fit. If the sides of the triangle at the corner of the frame have lengths in the ratio of 3 : 4 : 5, then the carpenter knows the angle is a right angle because $3^2 + 4^2 = 5^2$. Carpenters call this the '3, 4, 5 rule'.

Find answers at: cambridge.org/ukschools/gcsemaths-studentbookanswers

"I use 'Pythagorean triples' in my job. I measure 3 units up and 4 across the corner of a frame and check that the line that joins the end points is 5 units long. If the rule applies, then I know the frame is a right angle and that the window will fit properly." *(Window fitter)*

Key vocabulary

Pythagorean triple: three non-zero numbers (a, b, c) for which $a^2 + b^2 = c^2$.

Any set of three whole numbers that satisfy Pythagoras' theorem are called **Pythagorean triples**. One such triple is 3, 4, 5; so are any multiples of that, for example, 6, 8, 10 and 9, 12, 15. Other common triples are 5, 12, 13 and 7, 24, 25.

The converse of Pythagoras' theorem

The converse of Pythagoras' theorem states that:

> If the square on the longest side of any triangle is equal to the sum of the squares of the other two sides, then the triangle is right-angled.

WORK IT OUT 32.2

Is this triangle right-angled?

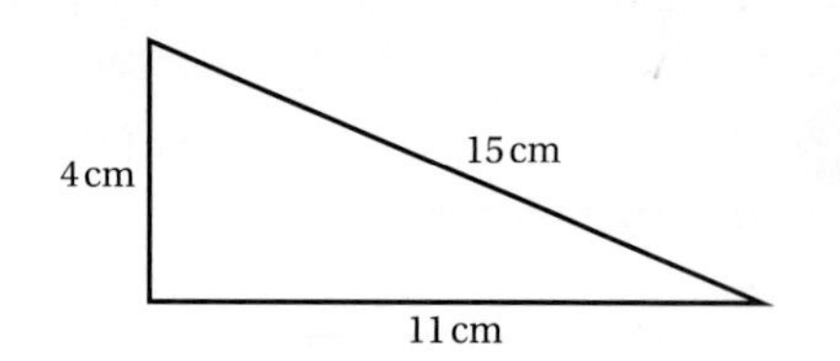

Which of these answers is correct?

Where have the others gone wrong?

Option A	Option B	Option C
$4 + 11 = 15$ So $a^2 + b^2 = c^2$ Yes, the triangle is right-angled!	$a^2 + b^2 = 4^2 + 11^2$ $= 4 \times 2 + 11 \times 2$ $= 8 + 22 = 30$ $c^2 = 15^2 = 15 \times 2 = 30$ $a^2 + b^2 = c^2$ Yes, the triangle is right-angled!	$a^2 + b^2 = 4^2 + 11^2$ $= 16 + 121 = 137$ $c^2 = 15 \times 15 = 225$ $a^2 + b^2 \neq c^2$ No, the triangle is not right-angled!

EXERCISE 32B

1. Determine which of these sets of lengths are Pythagorean triples.

   a 6, 8, 10 **b** 24, 45, 51 **c** 10, 16, 18

   d 20, 48, 52 **e** 9, 40, 41 **f** 12, 35, 37

2. Why is 3, 4, 5 the smallest possible whole number Pythagorean triple?

3. Is there a limit to the number of Pythagorean triples there are? Explain why or why not?

4 Which of the following triangles are right-angled?

a

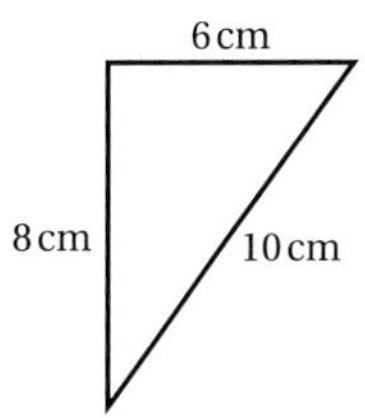

b

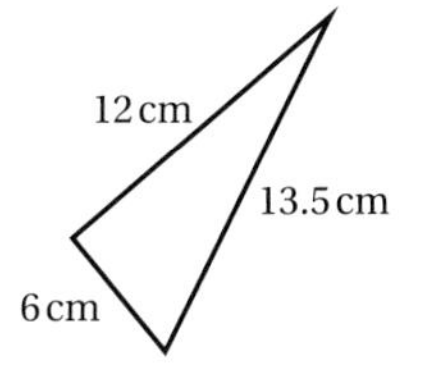

c

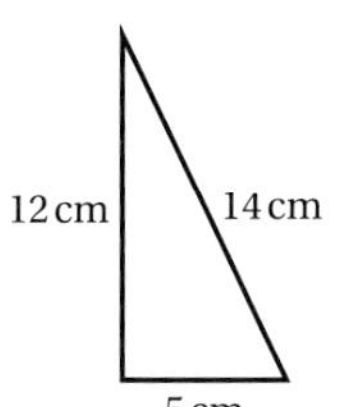

d

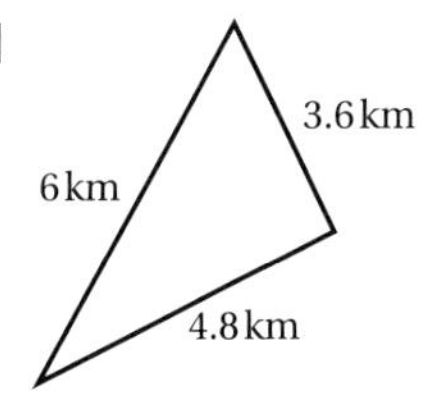

e

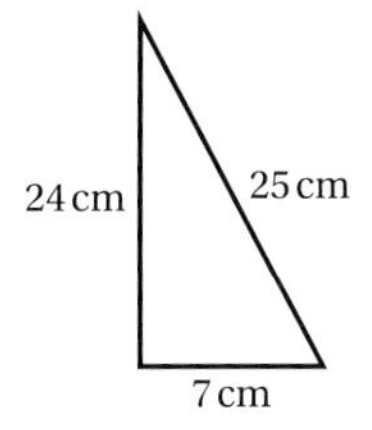

5 Determine which of the following are rectangles:

a

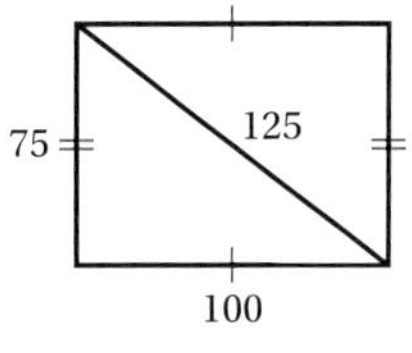

b

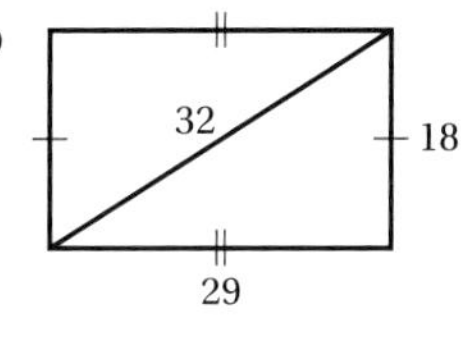

c

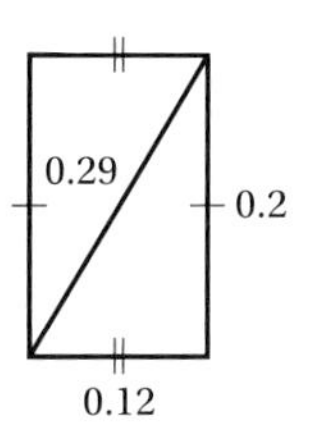

Pythagoras in polygons

You already know that polygons can be divided into triangles by drawing in their diagonals. You also know that you can decompose composite shapes into known shapes, including right-angled triangles, to work out their area and perimeter (Chapters 15 and 16).

Pythagoras' theorem is very useful for solving many geometry problems involving polygons and composite figures, particularly when you need to find the lengths of unknown sides.

WORKED EXAMPLE 1

1 Find the length of side x and then calculate the perimeter of this composite shape. Measurements are in centimetres.

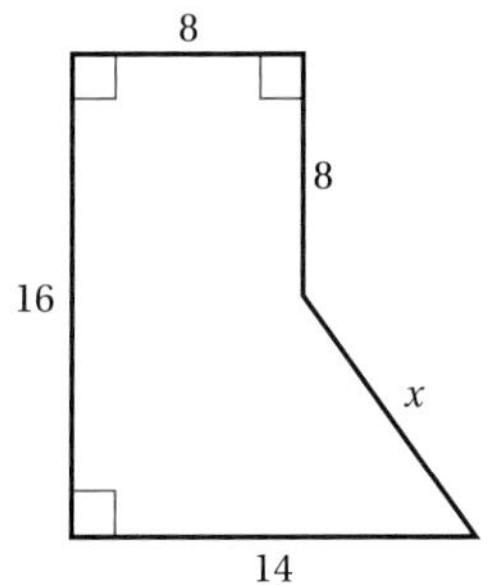

1

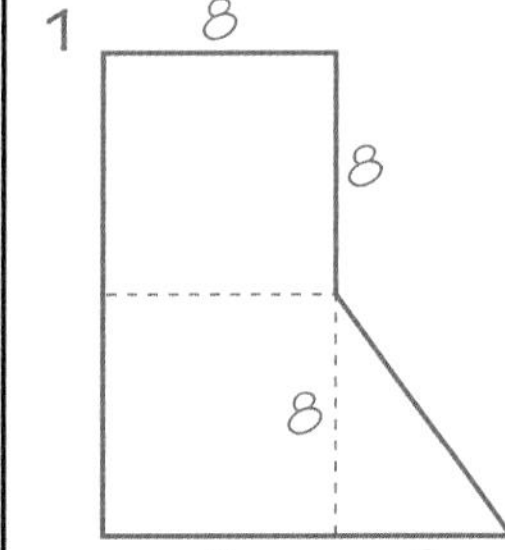

Divide the shape to make a right-angled triangle.

$8^2 + 6^2 = x^2$ (Pythagoras)

$64 + 36 = x^2$

$100 = x^2$

$x = 10$ cm

The broken lines show you that the composite shape is made up of a rectangle with dimensions 8×16, and a right-angled triangle with sides 6 and 8. The unknown, x, is the hypotenuse of the triangle. Use Pythagoras to find x.

Perimeter $= 16 + 14 + 8 + 8 + 10 = 56$ cm.

Continues on next page ...

2 Rhombus ABCD has diagonals BD = 12 cm and AC = 8 cm which intersect at E. What is the exact side length of the rhombus? Give your answer in simplified surd form.

2 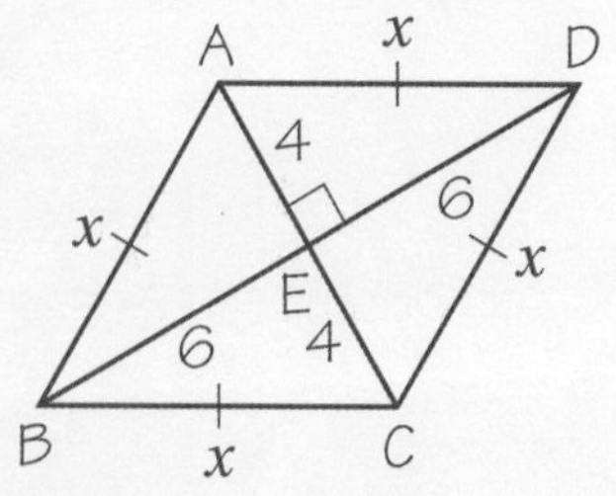

Start by doing a rough sketch and marking what you know.

AE = 4 cm and BE = 6 cm (diagonals of a rhombus bisect at right angles)

$x^2 = 4^2 + 6^2$

$x^2 = 52$

$x = \sqrt{52}$

$x = 2\sqrt{13}$

Remember to leave your answer with the surd still in place as you need to provide an **exact** answer.

EXERCISE 32C

1 What is the perpendicular height of an equilateral triangle of side length 5 cm?

2 What is the perpendicular height of an isosceles triangle of side lengths 8 cm and base length 6 cm?

3 What is the length of the sides in this quadrilateral?

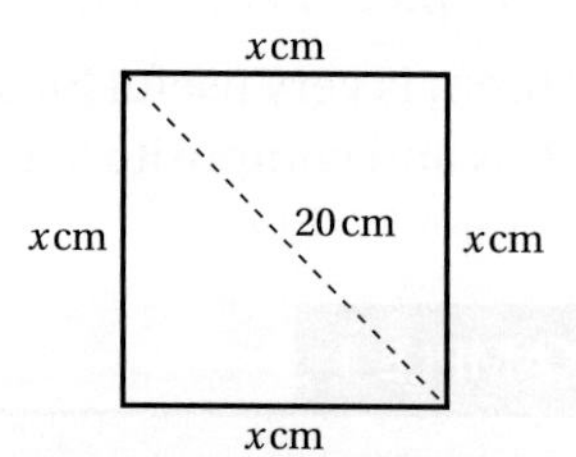

4 A scalene triangle has sides of length 14 cm, 10.5 cm and 17.5 cm. Is it a right-angled scalene triangle?

5 **a** Show that the length of the hypotenuse of this triangle is 10.0 cm to 1 decimal place.

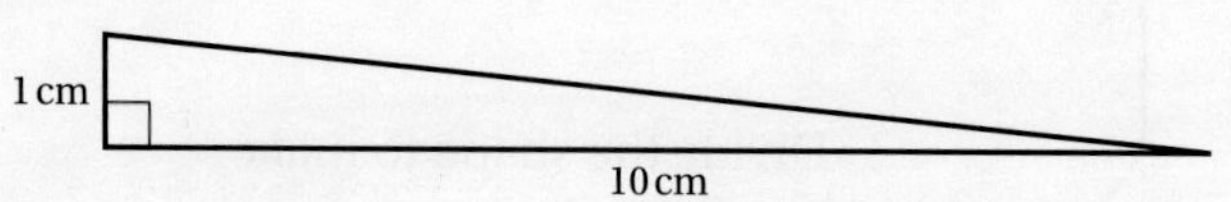

b Is it an isosceles triangle?

6 Find the length of AD in this figure.

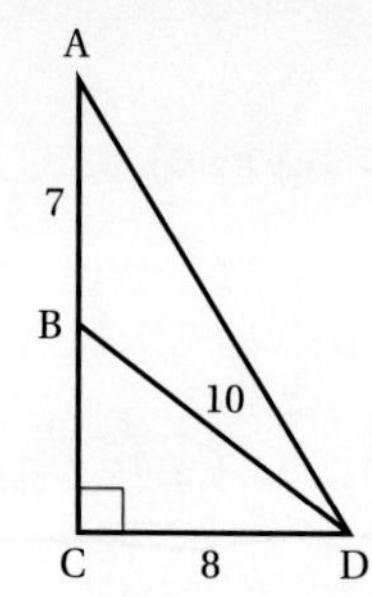

7 Calculate the length of:

a AC. **b** BC. **c** EC.

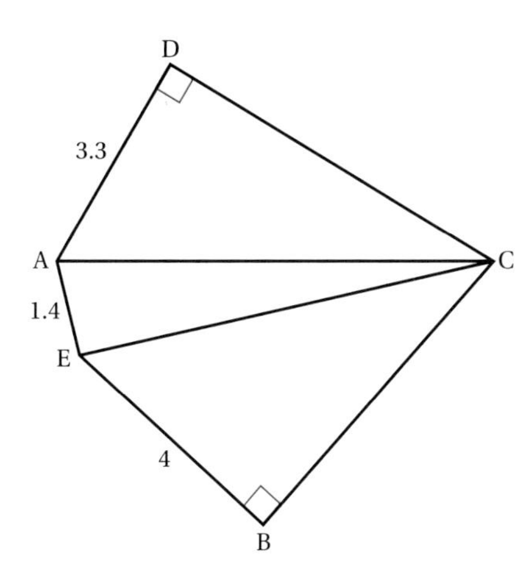

8 Find the area of trapezium ABCD.

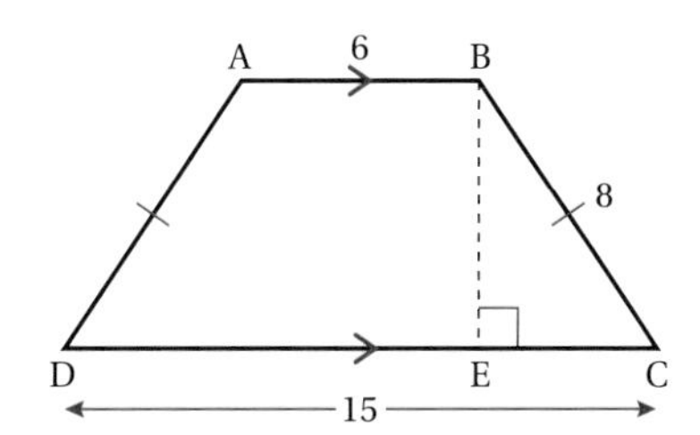

9 Find the length of side AD and calculate the perimeter of this shape.

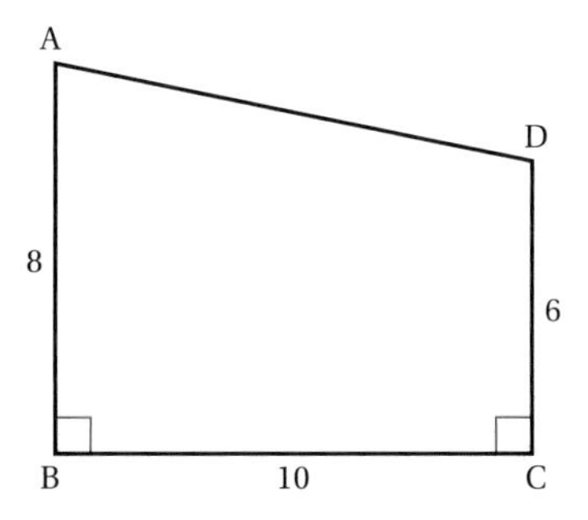

10 The area of square ACDE is 50 mm^2.
Determine the length of AB.

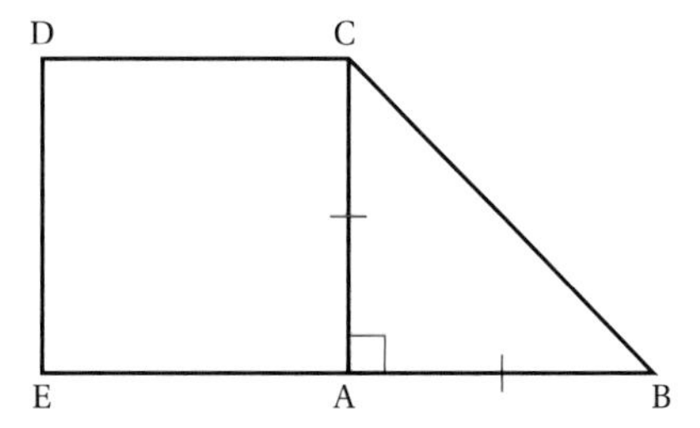

11 Find the value of x. Leave your answer as a surd in simplified form.

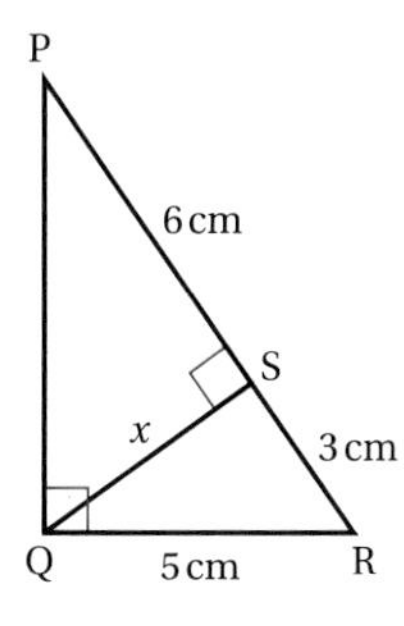

Section 3: Pythagoras in three dimensions

So far you have used Pythagoras to solve problems involving plane shapes in two dimensions. In reality, many problems involve 3D objects or spaces.

When you work with 3D shapes it is important to draw careful diagrams. In most cases, you will need to separate out triangles to work with them.

The first step in working with 3D objects is to find the right-angled triangle in the plane that has the length you need to work out. Once you've identified the correct triangle, you can draw it on its own as a 2D shape and add all the necessary information.

WORKED EXAMPLE 2

The height of a box is 7 cm, its length is 4 cm and its depth is 3 cm.

Find the length of the diagonal BS.

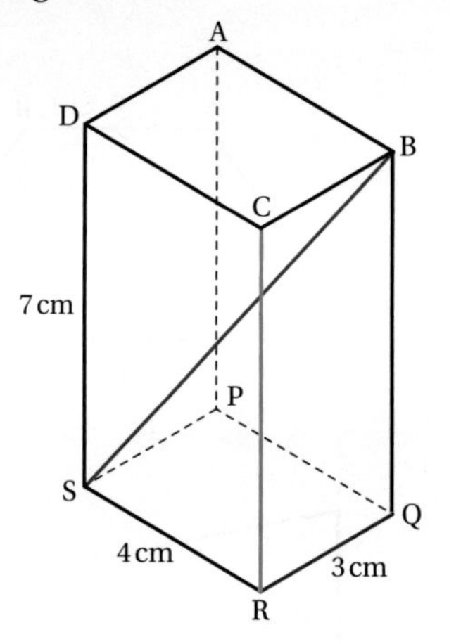

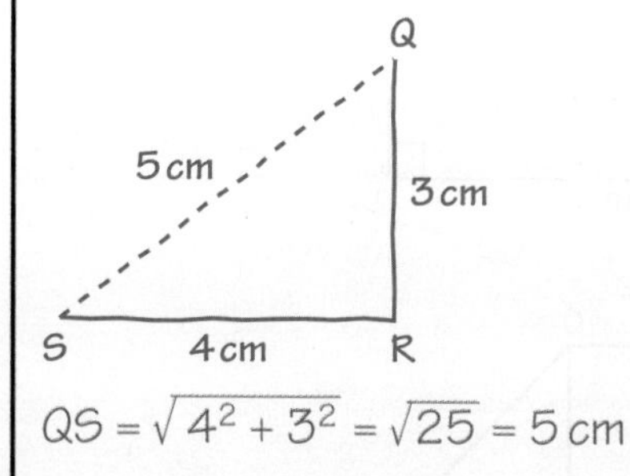

BS is the hypotenuse of the 2D triangle BQS. Find the length QS first. It is the hypotenuse of right-angled triangle QRS on the base of the box. Draw the relevant 2D triangle.

$QS = \sqrt{4^2 + 3^2} = \sqrt{25} = 5$ cm

Apply Pythagoras' theorem. (You might have spotted that this was a Pythagorean triple.)

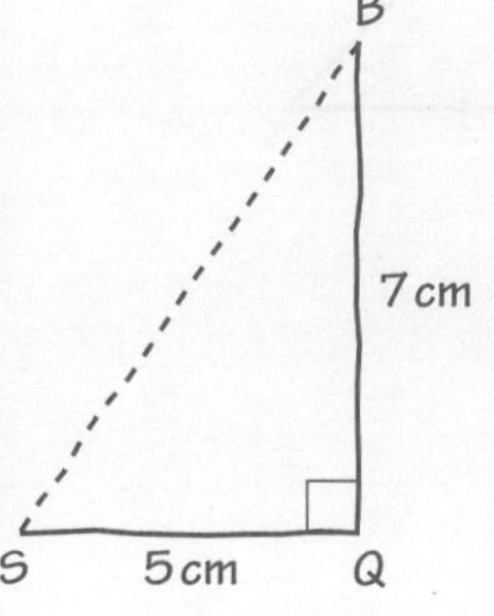

You can now draw the right-angled triangle BQS with a base of 5 cm. Use Pythagoras' theorem to find BS, the hypotenuse of this triangle.

$BS = \sqrt{5^2 + 7^2} = \sqrt{74} = 8.60$ cm (2 decimal places).

Continues on next page ...

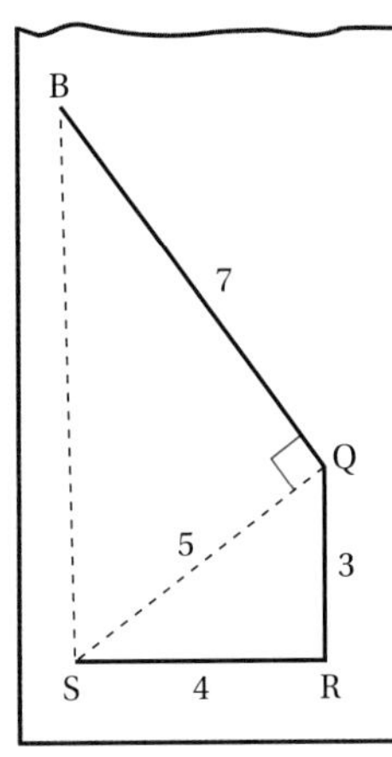

Alternatively, you could have extended your earlier diagram with an upright triangle.

EXERCISE 32D

1 This symmetrical pyramid has a rectangular base.
Find the perpendicular height of the pyramid.

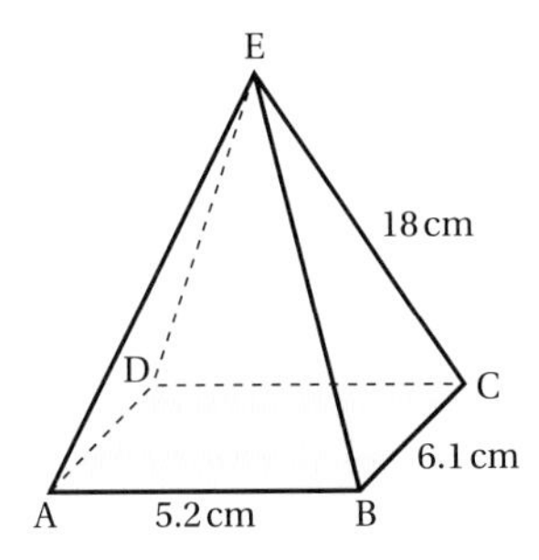

2 The diagram shows a wedge used to get trolleys through a doorway.

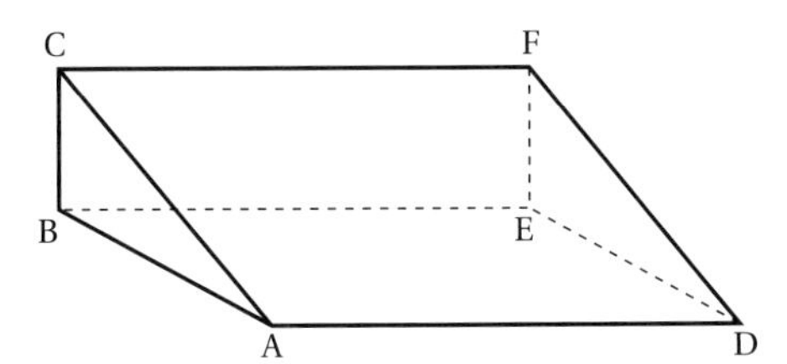

The face ABED is a rectangle and is at right angles to CBEF, which is another rectangle.

AB = 50 cm, AF = 65 cm and BE = 30 cm.

Calculate whether the wedge will fit under the lip of a door measuring 30.5 cm high.

3 An office block is the shape of a cuboid with a rectangular-based pyramid fitting exactly on the top.
How tall is the building?

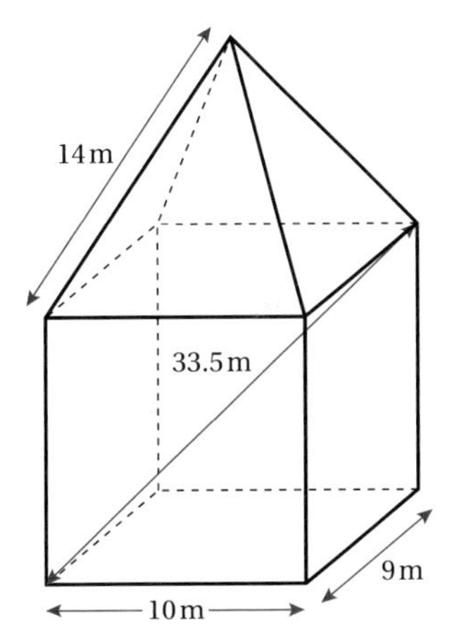

4 A spaghetti jar is the shape of a cylinder with a radius of 8 cm and a height of 0.35 m. Will dried spaghetti that is 40 cm long fit into the jar?

5 The slant height of a cone with a circular base is 100 mm and the diameter of its base is 60 mm.
Determine the perpendicular height of the cone.

Section 4: Using Pythagoras to solve problems

Pythagoras' theorem is also very useful for solving a range of problems involving real-life contexts.

Shortest and longest route problems are a good example.

Problem-solving framework

To get home, you normally cross a 100 m by 60 m rectangular football field diagonally from one corner to the other as it is the quickest route. Today there is a game playing so you have to go round the edge. Calculate how much further you have to walk.

Steps for approaching a problem-solving question	What you would do for this example
Step 1: If it is useful to have a diagram, sketch one and add the information. This may help you visualise the problem.	60 m 100 m
Step 2: Identify what you have to do.	Find the difference between the length of the diagonal and the length of the two sides added together.
Step 3: Test the problem with what you know. Can I use a ruler? What type of angle is it?	You could use a ruler, but you would have to draw a very accurate diagram to scale. As you have a right-angled triangle and side lengths you can use Pythagoras' theorem.
Step 4: What maths can I do?	Use Pythagoras' theorem to work out the length of the diagonal: $100^2 + 60^2 = c^2$ $10\,000 + 3600 = c^2$ $13\,600 = c^2$ $c = \sqrt{13\,600} = 116.6$ m (1 dp) If you walked straight across the diagonal the distance would be 116.6 m. You have to go round the outside which is 160 m (100 + 60), so you must walk 43.4 m further.
Step 5: Check your workings and that your answer is reasonable.	A diagonal pitch length of 116.6 m seems reasonable given the sides are 60 m and 100 m.
Step 6: Have you answered the question?	You were asked to find how much further you would have to walk. You have found this to be 43.4 m.

EXERCISE 32E

1 Computer gaming designers use x- and y-coordinates to place characters or objects in a game. They need to know distances between characters or how far players are apart.

If one player is at coordinate (30, 10) and the other at (15, 4), how far apart are they?

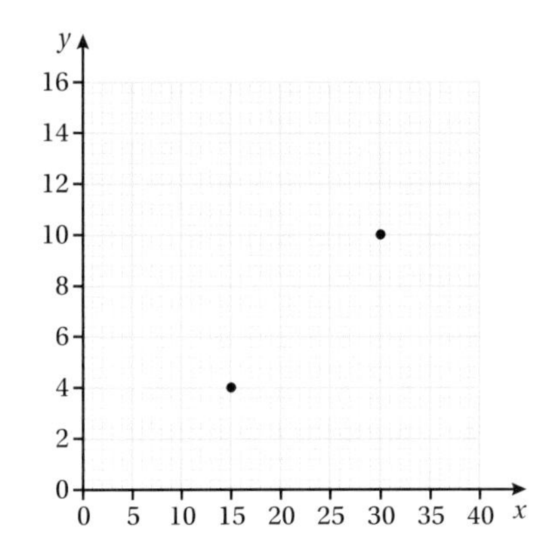

2 You are going to buy a new television. It is an 80 inch television (measured along its diagonal length).
Your current television is a 52 inch.
Both are the same height, 40 inches.

a How much wider is your new television than your current one?

b More importantly, will it fit in the 58 inch gap between the chimney and wall?

3 A rectangular shed is 6 m long by 4 m wide and has a ceiling height inside of 3 m. Will a boat mast 7 m long fit inside?

4 In which storage box can you fit the longest pole?

A

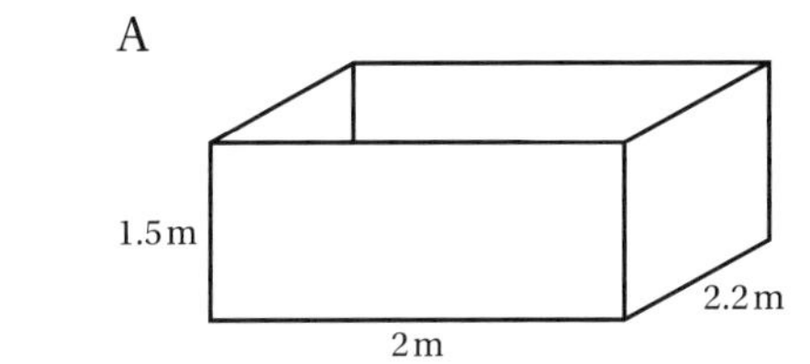

B

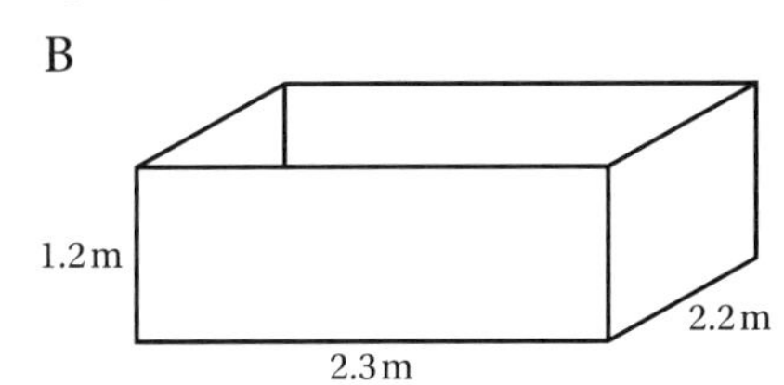

5 Find the length of the hypotenuse of the largest triangle.

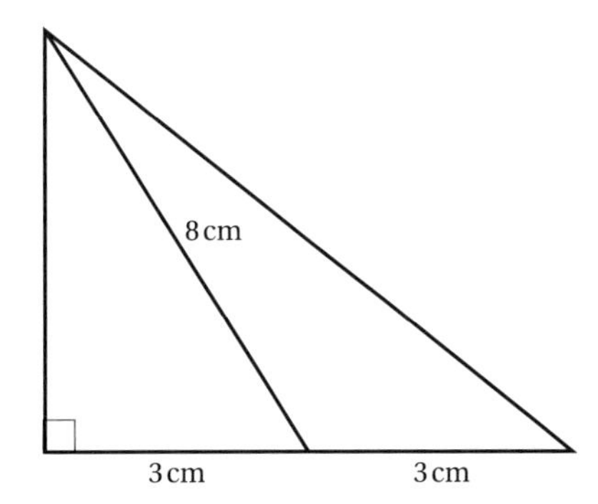

6 Here is a square shape with a brace across it. Is it constructed correctly as a perfect square, within reasonable bounds?

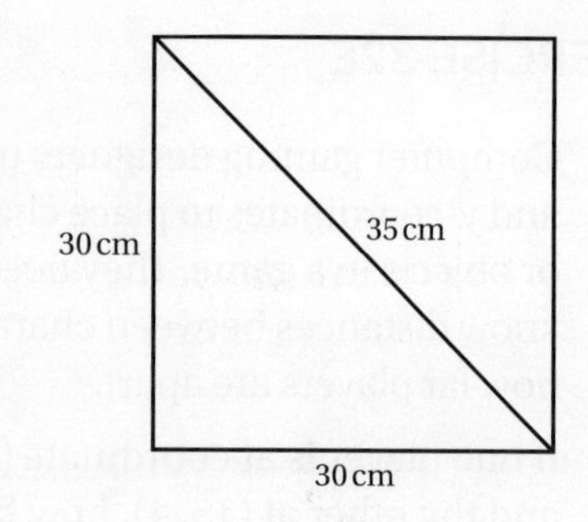

7 Calculate the length of AB in this trapezium. It is from a model of the end of a building.

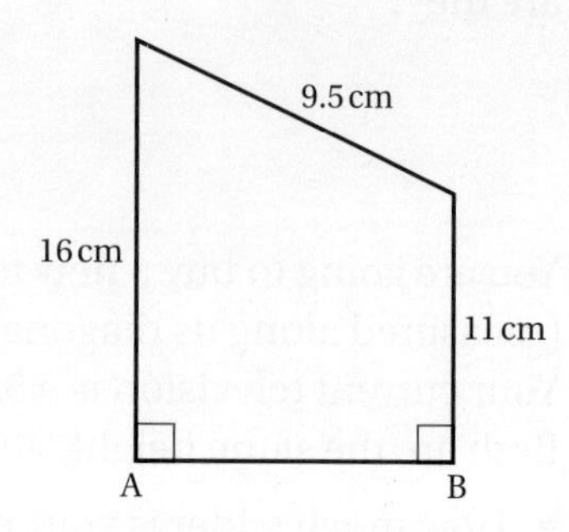

8 A plumber carries piping on the roof of his van in a cylindrical canister that is 4.6 m long.
The longest pipe the tube can carry is 4.65 m.
What is the diameter of the tube, in centimetres, to 2 decimal places?

9 A hotel has a square courtyard with sides 12 m long. The hotel owners want to construct a circular fountain with diameter 1.2 m in the middle of the courtyard.

a What is the length of each of the courtyard's diagonals?

b How many metres along each diagonal should the fountain be placed so the centre of the fountain is in the centre of the courtyard?

Tip

When facing more complex questions involving triangles remember the problem-solving steps. There are some key questions to ask yourself:

1. What information do I have? Sketch it onto a diagram if there isn't one.
2. What information do I need to find out? Is it lengths of sides?
3. What can I do? Is there a right angle somewhere or could one be created?
4. What maths can I do to find more information? Can I use Pythagoras' formula? Do I need to rearrange it?
5. Check your working. What roughly is the length I need to find? Is the result roughly what I had estimated? Is the hypotenuse the longest side of any right-angled triangles? If not, then you've definitely made a mistake!
6. Double check you've answered the question.

10 A horse paddock is the shape of a rectangle with a diagonal of 20 m and a shorter side of 10 m.

a What is the perimeter and the area of the paddock?

b How many laps of the paddock would a horse have to trot to cover a kilometre, to the nearest whole lap?

11 When tiling a bathroom a plumber uses tiles that are isosceles triangles with equal sides of length 15 cm and base 10 cm.

a What would the height of 10 rows of tiles be?

b How many rows of tiles would fit below a shelf which sits 90 cm from the floor?

12 A piece of artwork has been designed to look like this with the dimensions shown.

It is three right-angled triangles welded together. The hypotenuse of the top triangle is perpendicular to the ground.

Will it fit into a display area that is 3 m high?

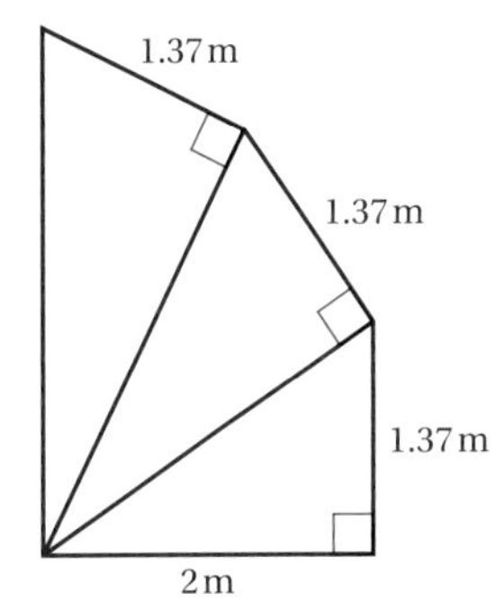

Checklist of learning and understanding

Pythagoras' theorem

- Pythagoras' theorem only applies to right-angled triangles.
- The theorem is that the square of the hypotenuse of a right-angled triangle equals the sum of the squares of the other two sides. It can be written as

  $a^2 + b^2 = c^2$
- Rearrange the formula to find either of the other two sides:

  $a^2 = c^2 - b^2$

  $b^2 = c^2 - a^2$
- You can use the theorem to find the length of the hypotenuse, the longest side. The theorem can also be used to prove there is a right angle within a triangle or to find a missing length within a right-angled triangle.
- Always draw a diagram and label it; also write out the formula you are using to fully explain what you have done.
- Pythagorean triples are three lengths that satisfy the formula and therefore prove you have a right angle; learn the common ones, such as 3, 4, 5 and 5, 12, 13.
- If you need to find an unknown length in a composite shape, try to divide the shape up so that it contains one or more right-angled triangles so that you can use Pythagoras' theorem to calculate the unknown length.

Find answers at: cambridge.org/ukschools/gcsemaths-studentbookanswers

- You can calculate unknown lengths within 3D objects by finding the right-angled triangle in the plane that has the length you need to work out. Once you've identified the correct triangle, you can draw it on its own as a 2D shape, add all the necessary information and use Pythagoras' theorem, as with 2D shapes.

For additional questions on the topics in this chapter, visit GCSE Mathematics Online.

Chapter review

1 Find the length of x and y in this figure, giving your answers correct to 2 decimal places.

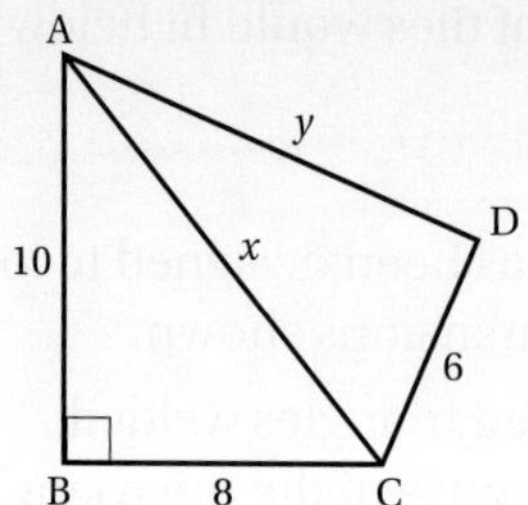

2 What is the perimeter of this kite?

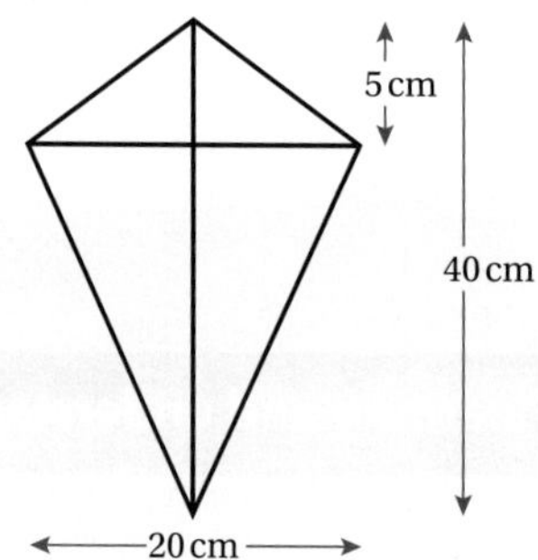

3 Find the length of FG.

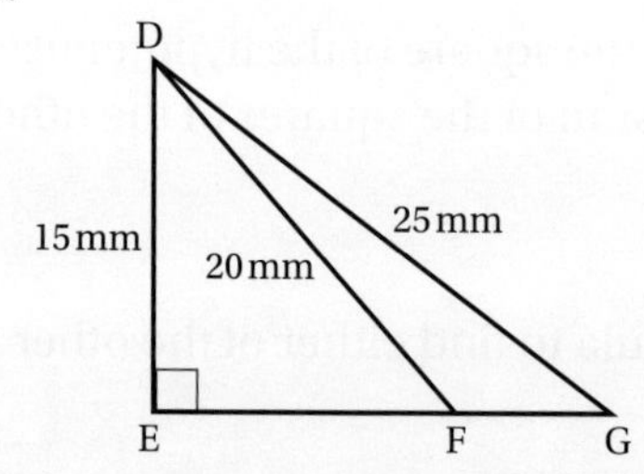

4 The base of a pyramid of height 1.6 m is a square with a diagonal of 2.4 m. Calculate the length of the sloping edge of the pyramid.

5 In the triangle, all lengths are in centimetres.

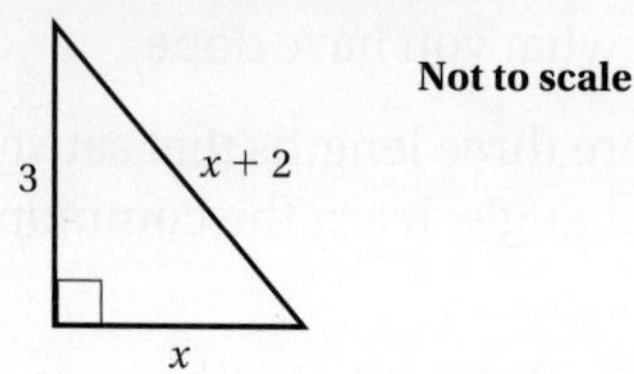

Use Pythagoras' theorem to find x. *(5 marks)*

© OCR 2013

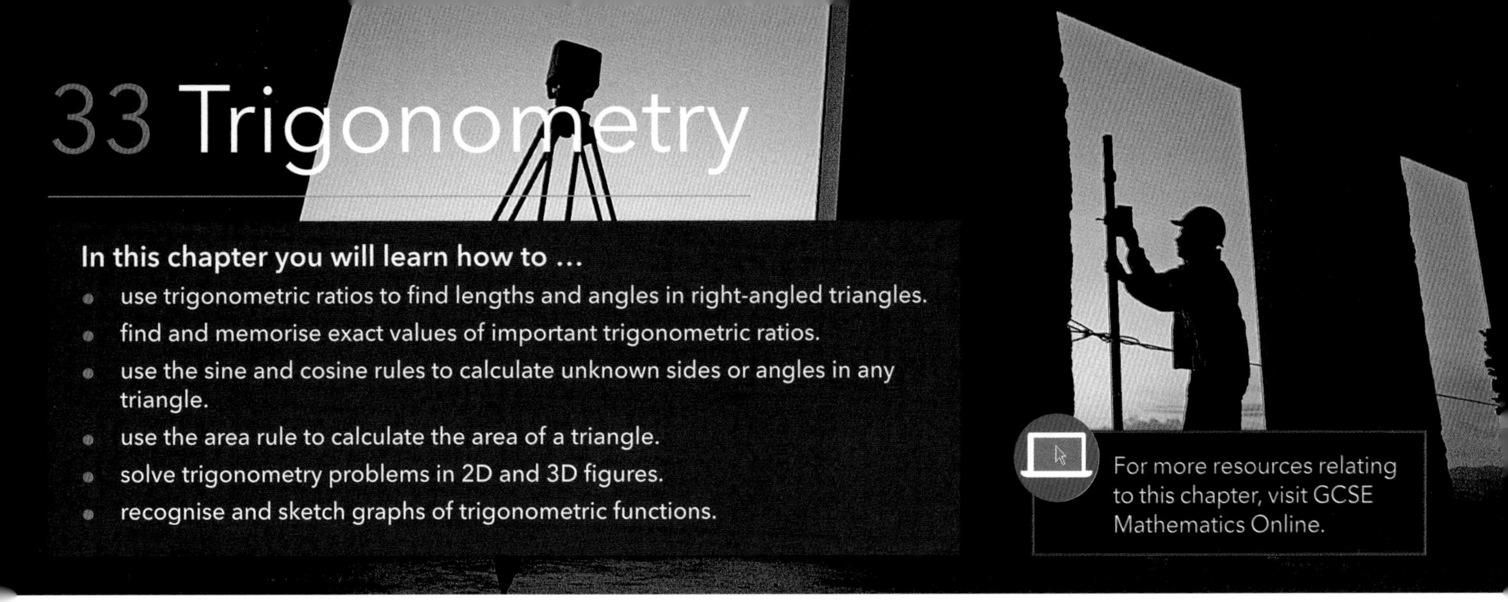

33 Trigonometry

In this chapter you will learn how to ...

- use trigonometric ratios to find lengths and angles in right-angled triangles.
- find and memorise exact values of important trigonometric ratios.
- use the sine and cosine rules to calculate unknown sides or angles in any triangle.
- use the area rule to calculate the area of a triangle.
- solve trigonometry problems in 2D and 3D figures.
- recognise and sketch graphs of trigonometric functions.

For more resources relating to this chapter, visit GCSE Mathematics Online.

Using mathematics: real-life applications

Trigonometry means 'triangle measurements' and it is very useful for finding the lengths of sides and sizes of angles. Trigonometry is used to determine lengths and angles in navigation, surveying, astronomy, engineering, construction and even in the placement of satellites and satellite receivers.

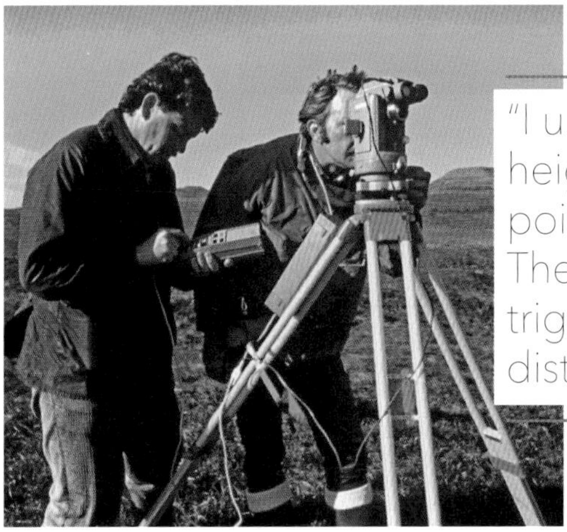

"I use a theodolite to work out the height of mountains. You basically point it at the top of the mountain. The theodolite uses the principles of trigonometry to measure angles and distances." *(Geologist)*

Before you start ...

Ch 32	You should be able to use Pythagoras' theorem to find lengths in triangles.	**1** Find the length of x in each triangle. **a** 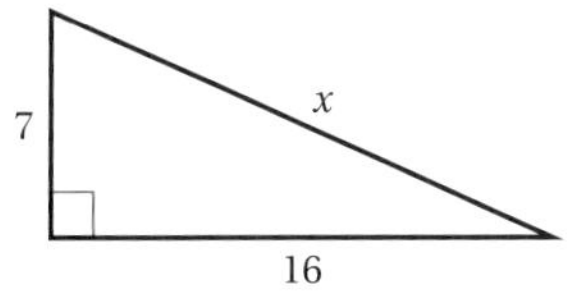 **b** 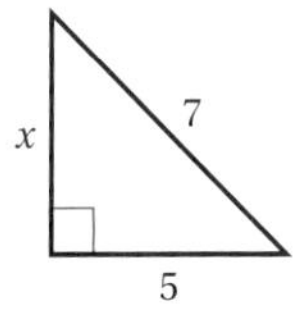
Ch 17	You must be able to work with approximate values and round to a specified number of places.	**2** What is $\sqrt{53}$ correct to 2 decimal places? **3** If $c^2 = 94.34$, what is c correct to 3 significant figures?
Ch 22, 31	You need to be able to use ratio and proportion to calculate sides in similar triangles.	**4** Find the length of AC if the ratio of sides $\frac{AB}{AC} = \frac{5}{3}$ and AB = 35 cm.

Find answers at: cambridge.org/ukschools/gcsemaths-studentbookanswers

Assess your starting point using the Launchpad

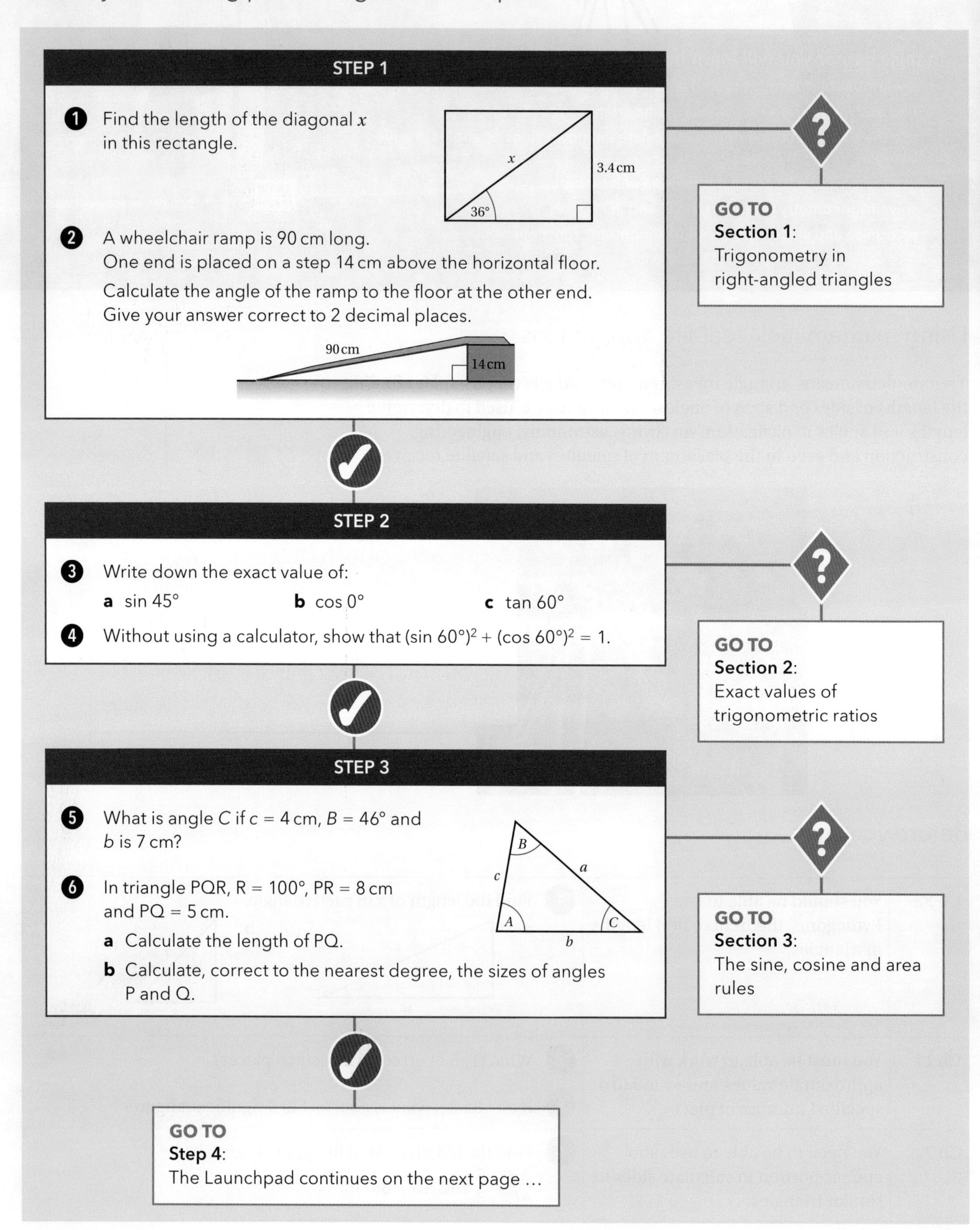

STEP 1

1 Find the length of the diagonal x in this rectangle.

2 A wheelchair ramp is 90 cm long.
One end is placed on a step 14 cm above the horizontal floor.
Calculate the angle of the ramp to the floor at the other end.
Give your answer correct to 2 decimal places.

GO TO
Section 1:
Trigonometry in right-angled triangles

STEP 2

3 Write down the exact value of:

a sin 45° **b** cos 0° **c** tan 60°

4 Without using a calculator, show that $(\sin 60°)^2 + (\cos 60°)^2 = 1$.

GO TO
Section 2:
Exact values of trigonometric ratios

STEP 3

5 What is angle C if $c = 4$ cm, $B = 46°$ and b is 7 cm?

6 In triangle PQR, R = 100°, PR = 8 cm and PQ = 5 cm.

a Calculate the length of PQ.

b Calculate, correct to the nearest degree, the sizes of angles P and Q.

GO TO
Section 3:
The sine, cosine and area rules

GO TO
Step 4:
The Launchpad continues on the next page …

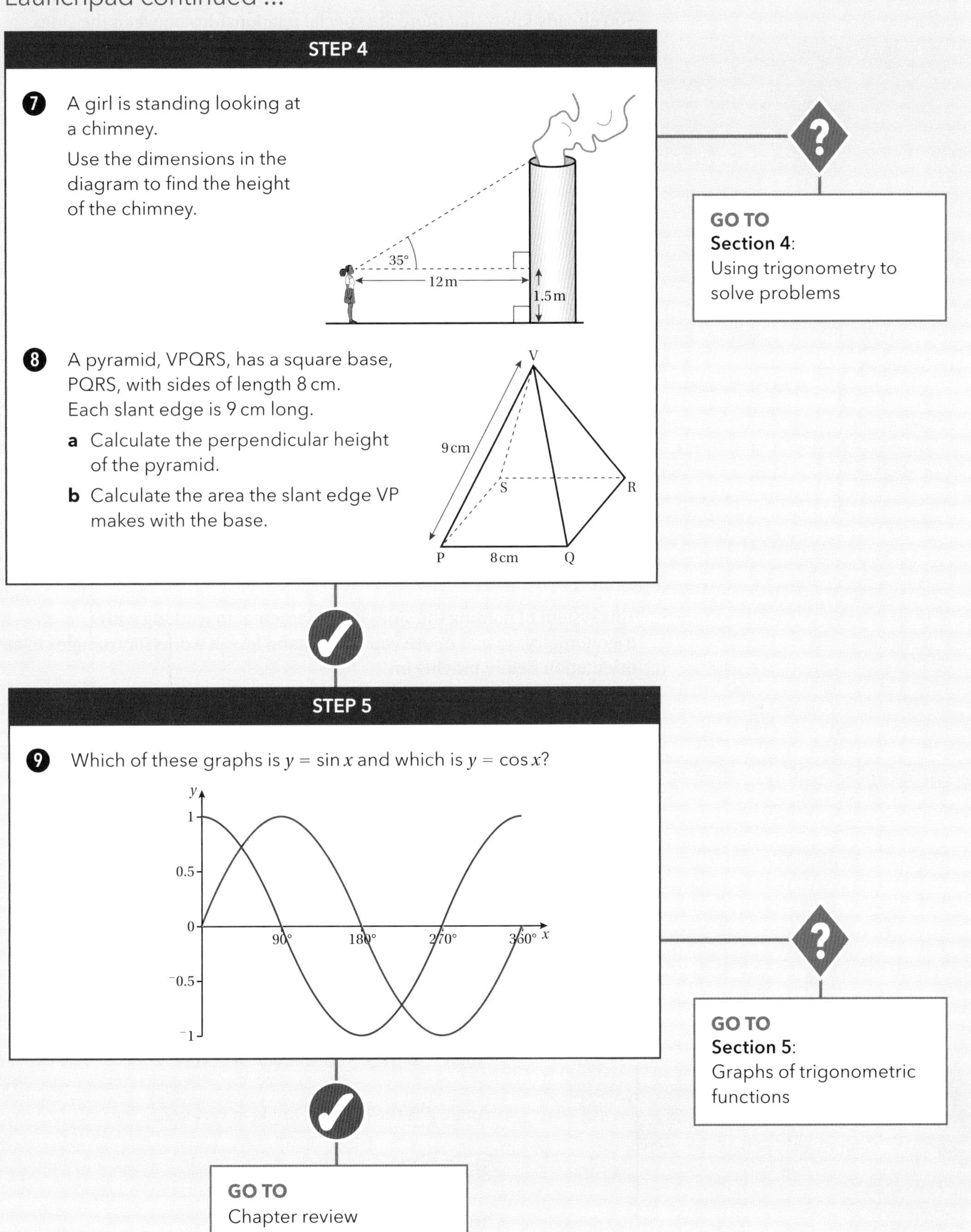

Launchpad continued ...

STEP 4

7 A girl is standing looking at a chimney.

Use the dimensions in the diagram to find the height of the chimney.

8 A pyramid, VPQRS, has a square base, PQRS, with sides of length 8 cm. Each slant edge is 9 cm long.

a Calculate the perpendicular height of the pyramid.

b Calculate the area the slant edge VP makes with the base.

GO TO
Section 4:
Using trigonometry to solve problems

STEP 5

9 Which of these graphs is $y = \sin x$ and which is $y = \cos x$?

GO TO
Section 5:
Graphs of trigonometric functions

GO TO
Chapter review

Find answers at: cambridge.org/ukschools/gcsemaths-studentbookanswers

Section 1: Trigonometry in right-angled triangles

You already know that there are special relationships between the sides of right-angled triangles and that you can use Pythagoras' theorem to find missing sides when two sides are known. You also know that the ratio of corresponding pairs of sides in similar triangles is always the same.

These facts are important for understanding and using trigonometry.

Naming the sides of right-angled triangles

The hypotenuse is the longest side of a right-angled triangle, opposite the right angle.

The other two (shorter) sides are named in relation to the acute angles in the triangle.

In these triangles, one of the acute angles is labelled θ (the Greek letter theta).

The sides can then be labelled **opposite** (that is, opposite angle θ) and **adjacent** (that is, adjacent to angle θ).

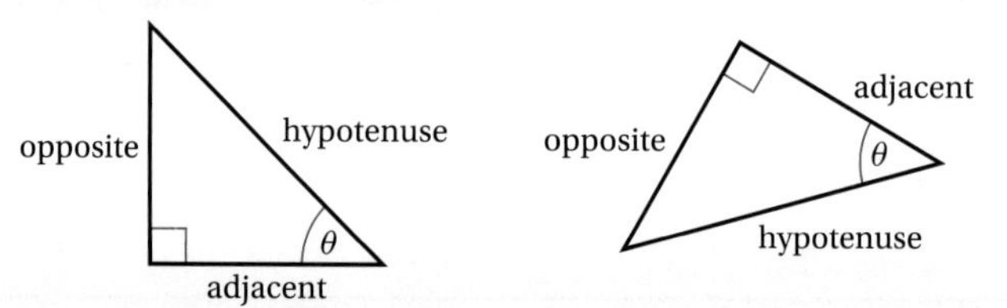

This system of naming the sides is fundamental to working with trigonometry, so make sure you understand how it works for triangles in any orientation before moving on.

WORKED EXAMPLE 1

In the right-angled triangle XYZ which side is:

a the hypotenuse?
b opposite angle Y?
c adjacent to angle Z?
d Which angle is adjacent to side XY?
e Which angle is opposite side XZ?

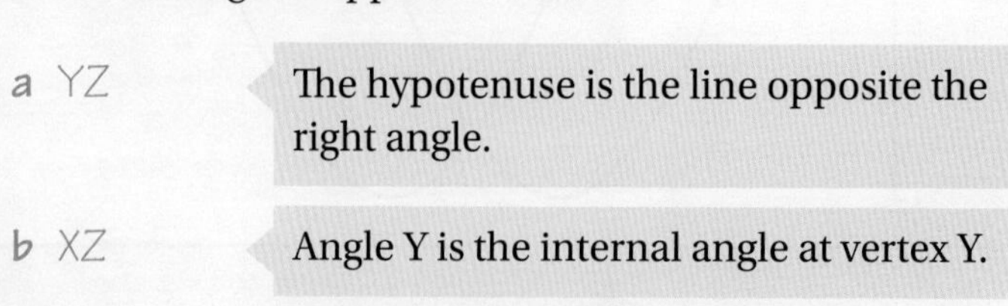

a YZ — The hypotenuse is the line opposite the right angle.

b XZ — Angle Y is the internal angle at vertex Y.

c XZ — Angle Z is the internal angle at vertex Z.

d Angle Y — The line XY is next to angle Y.

e Angle Y — The line XZ is opposite angle Y.

The ratio of sides in similar triangles

This diagram shows three similar right-angled triangles. The green sides are opposite angle θ and the red sides are adjacent to it. You can see that the ratio of $\frac{\text{opposite}}{\text{adjacent}}$ sides is $\frac{1}{2}$ for all the similar triangles.

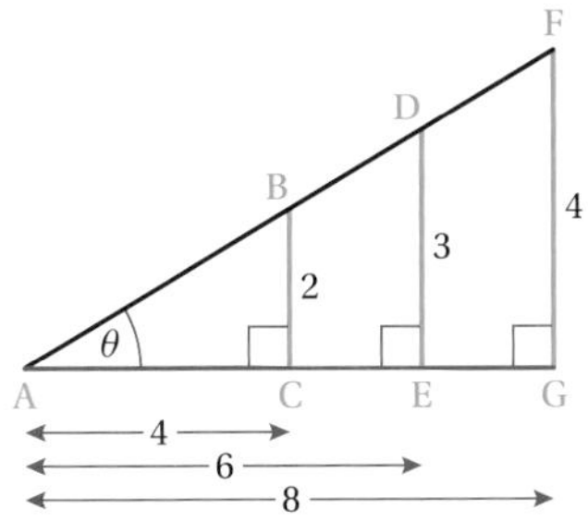

$$\frac{BC}{CA} = \frac{2}{4} = \frac{1}{2}$$

$$\frac{DE}{EA} = \frac{3}{6} = \frac{1}{2}$$

$$\frac{FG}{GA} = \frac{4}{8} = \frac{1}{2}$$

$$\therefore \frac{\text{OPPOSITE}}{\text{ADJACENT}} = \frac{1}{2}$$

Similar triangles show that for a given angle in a right-angled triangle, the ratio of corresponding sides will always be the same, regardless of the size of the triangle. Every angle therefore has a fixed value for the ratio of any two of its sides.

Depending on which sides you compare, the ratio of sides is given a special name. You will work with three ratios: sine, cosine and tangent.

The trigonometric ratios

In similar triangles, the ratio $\frac{\text{opposite}}{\text{hypotenuse}}$ is the same in each one, and the ratio $\frac{\text{adjacent}}{\text{hypotenuse}}$ is the same in each one, as is the ratio $\frac{\text{opposite}}{\text{adjacent}}$.

These ratios are called the trigonometric ratios (shortened to trig ratios) and they are named as follows:

The **sine** ratio ($\sin\theta$) is the ratio of the side opposite the angle to the hypotenuse.

The **cosine** ratio ($\cos\theta$) is the ratio of the side adjacent to the angle to the hypotenuse.

The **tangent** ratio ($\tan\theta$) is the ratio of the side opposite the angle to the side adjacent to the angle.

$$\sin\theta = \frac{a}{c} = \frac{\text{opposite}}{\text{hypotenuse}}$$

$$\cos\theta = \frac{b}{c} = \frac{\text{adjacent}}{\text{hypotenuse}}$$

$$\tan\theta = \frac{a}{b} = \frac{\text{opposite}}{\text{adjacent}}$$

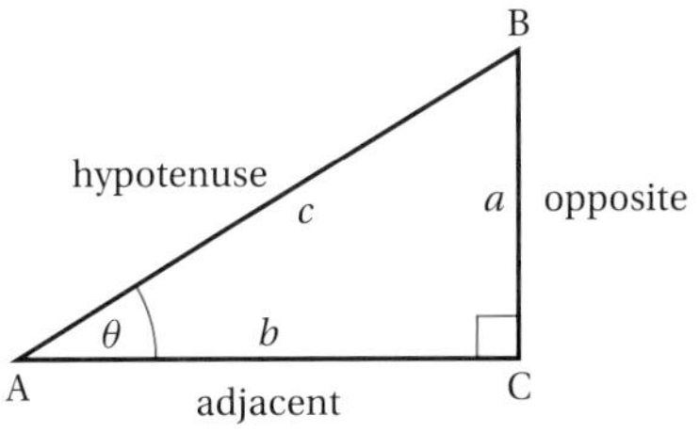

The three ratios have the same value for a particular angle, no matter how long the sides are.

For example, $\sin 30°$ is $\frac{1}{2}$ or 0.5 for **any** right-angled triangle. This means that the ratio of $\frac{\text{opposite}}{\text{hypotenuse}}$ is $\frac{1}{2}$ if the angle you are working with is 30°.

You can find the ratio for any angle using your calculator. Make sure you know how to find and use the trig function keys.

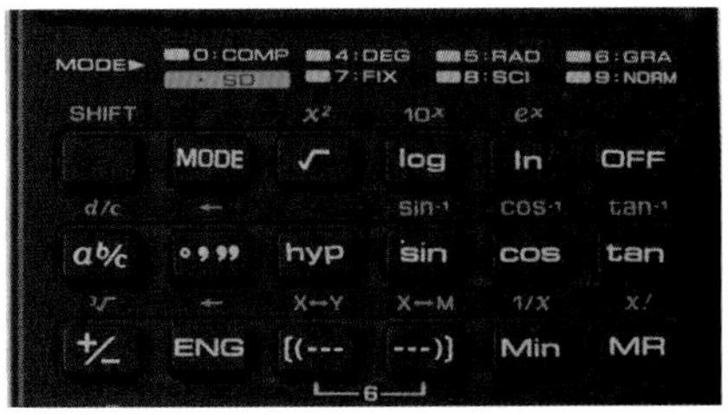

WORKED EXAMPLE 2

a For triangle ABC, find the ratio of sides used to calculate:

i sin A. **ii** cos A. **iii** tan A.

b Use your calculator to find the value of each of the following trig ratios. Give your answers to 3 significant figures where necessary.

i cos 32° **ii** sin 18° **iii** tan 80°

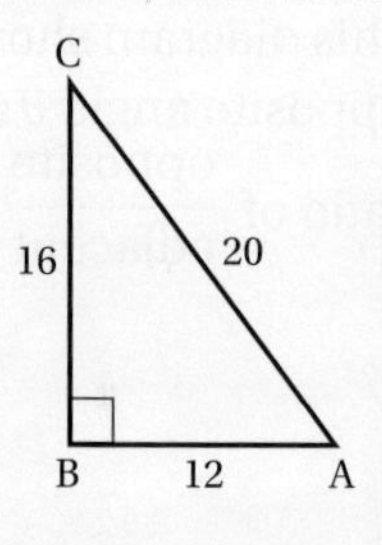

a i $\sin A = \frac{\text{opposite}}{\text{hypotenuse}} = \frac{16}{20} = \frac{4}{5}$

Substitute in the appropriate lengths for each ratio.

ii $\cos A = \frac{\text{adjacent}}{\text{hypotenuse}} = \frac{12}{20} = \frac{3}{5}$

iii $\tan A = \frac{\text{opposite}}{\text{adjacent}} = \frac{16}{12} = \frac{4}{3}$

b i cos 3 2 = 0.848

ii sin 1 8 = 0.309

iii tan 8 0 = 5.67

Press the appropriate key then enter the number for each ratio. Check your calculator to see how it works, as some calculators work differently to others.

Tip

If you have two sides of a right-angled triangle, you can find the other side using Pythagoras' theorem. If you have one side and at least one of the acute angles, you will need to use the trigonometric ratios.

Solving triangles

Finding unknown sides or angles is called solving the triangle. You can use trigonometric ratios to do this.

Finding unknown sides

If you know an angle (other than the right angle) and one side in a right-angled triangle, you can use the ratios to form **equations** that you solve to find the missing lengths.

You will not be told which ratio to use. You have to pick the right one based on the information that you have about the triangle.

You can remember the ratios using the mnemonic SOH-CAH-TOA and the formula triangles.

Tip

Circle the angle you are working with and mark the sides H, A and O to help you see which values you have and which you need. To remind yourself how to use formula triangles, see Chapter 12.

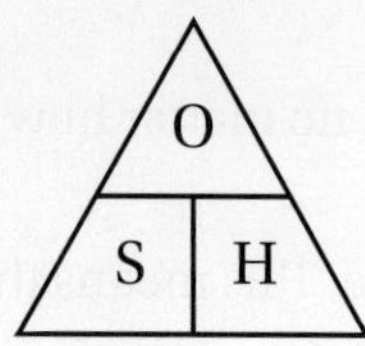

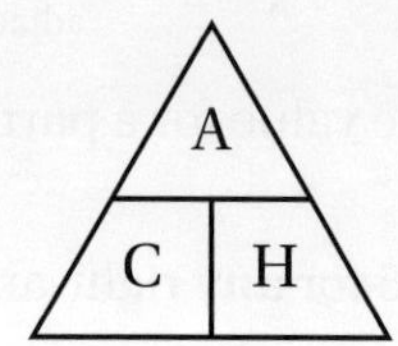

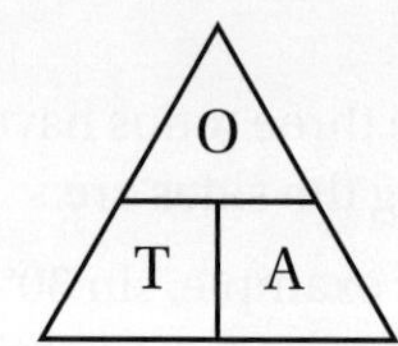

SOH: $\sin = \frac{\text{opposite}}{\text{hypotenuse}}$ CAH: $\cos = \frac{\text{adjacent}}{\text{hypotenuse}}$ TOA: $\tan = \frac{\text{opposite}}{\text{adjacent}}$

WORKED EXAMPLE 3

1 In triangle ABC, angle B = 90°, AC = 15 cm and angle C = 35°.
Calculate the length of AB correct to one decimal place.

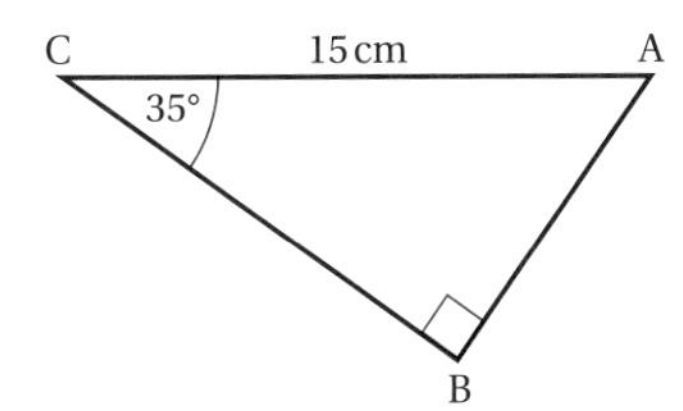

1 *Given:*
angle C = 35°
AB is opposite C
CA = 15 cm and is the hypotenuse.

Write down the information that you have been given and anything you can work out from the diagram.

$\sin 35° = \frac{\text{opposite}}{\text{hypotenuse}}$

AB is opposite the given angle (35°) and you have the length of the hypotenuse, so use the ratio with O and H, which is sin (SOH).

$\sin 35° = \frac{AB}{15}$

Substitute in the known values and solve for AB.

$AB = \sin 35° \times 15$

Multiply both sides by 15 to get AB on its own.

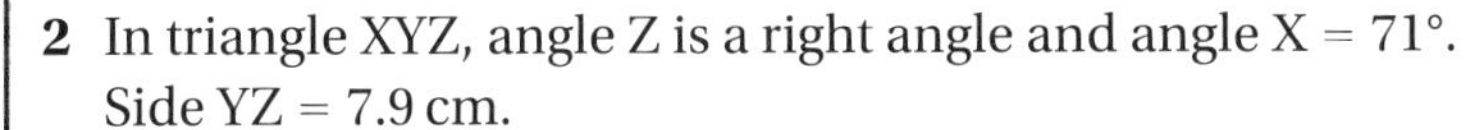

Use your calculator to find this value.

AB = 8.6 cm (1 dp)

2 In triangle XYZ, angle Z is a right angle and angle X = 71°.
Side YZ = 7.9 cm.
Calculate the length of XZ correct to 1 decimal place.

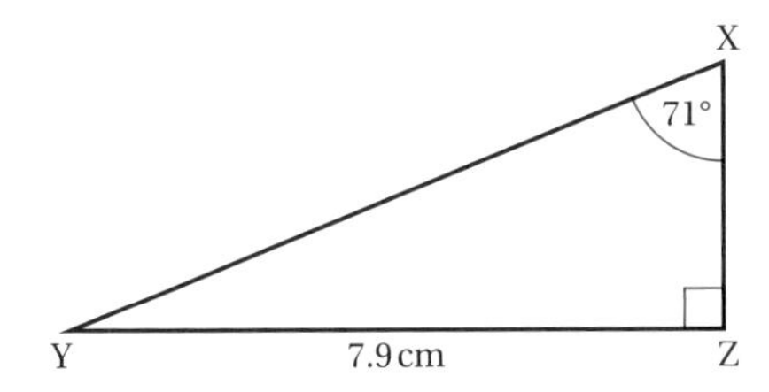

2 *Given:*
angle X = 71°
YZ = 7.9 cm = opposite side
XZ is the adjacent side

$\tan 71° = \frac{\text{opposite}}{\text{adjacent}} = \frac{7.9}{XZ}$

Use the ratio with OA which is tan (TOA). Multiply both sides by XZ to get rid of the fraction.

$7.9 = XZ \times \tan 71°$

$\frac{7.9}{\tan 71°} = XZ$

Divide by tan 71° to get XZ on its own.

Do the calculation on your calculator.

XZ = 2.7 cm (1 dp)

EXERCISE 33A

1 In each triangle, identify the opposite, adjacent and hypotenuse relative to the given angle.

a
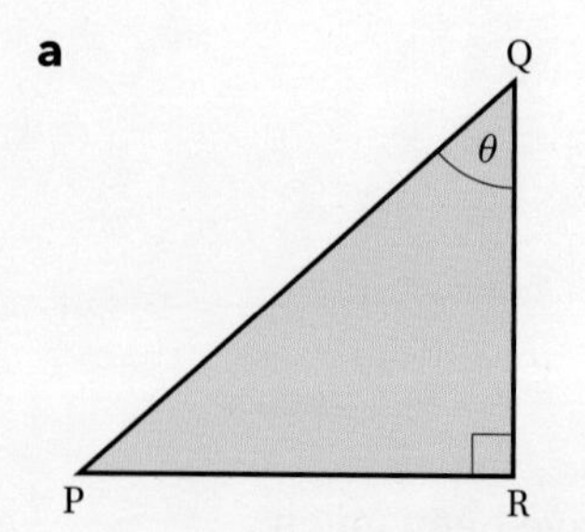

b
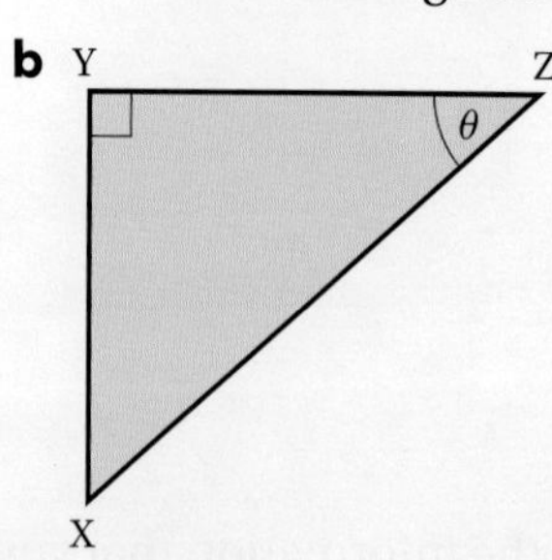

c
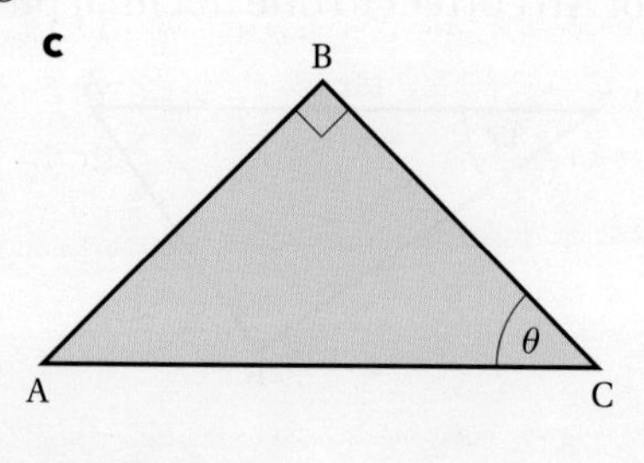

d
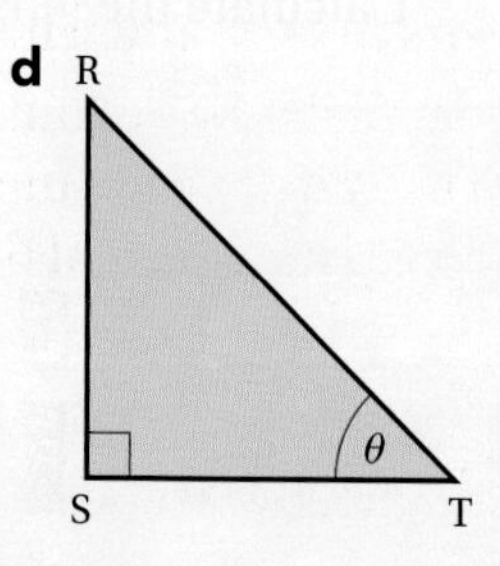

2 Use your calculator to give the value of each ratio correct to three decimal places, if necessary.

a sin 32° **b** cos 90° **c** tan 24°
d sin 30° **e** tan 87° **f** cos 49°
g sin 0° **h** cos 32° **i** tan 45°

3 Select the correct trigonometric ratio and use it to find the length of the side marked with a variable in each triangle.

a
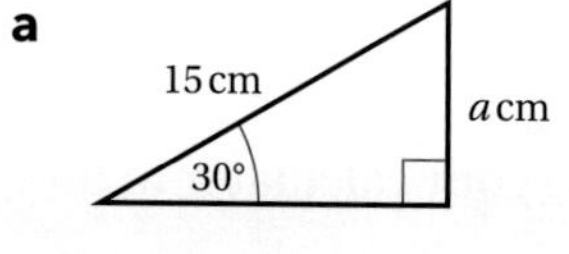

b
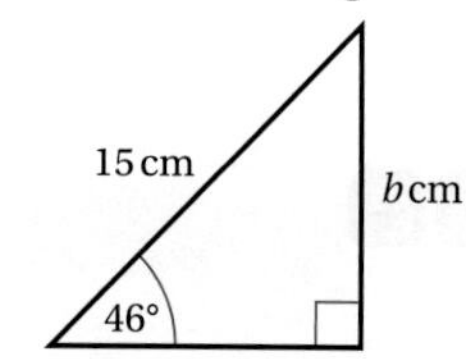

c
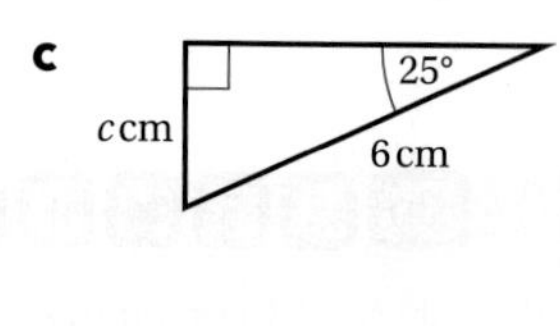

d
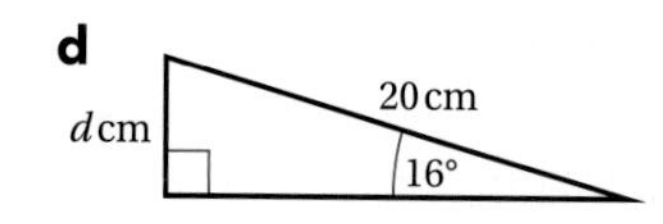

e
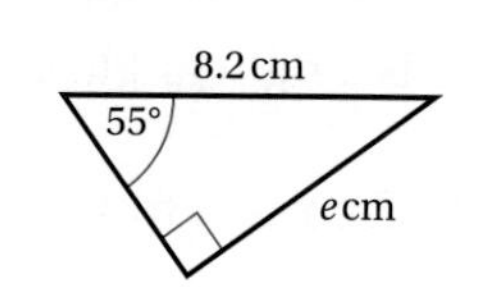

f
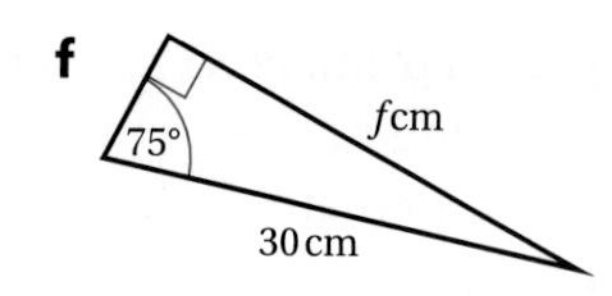

g
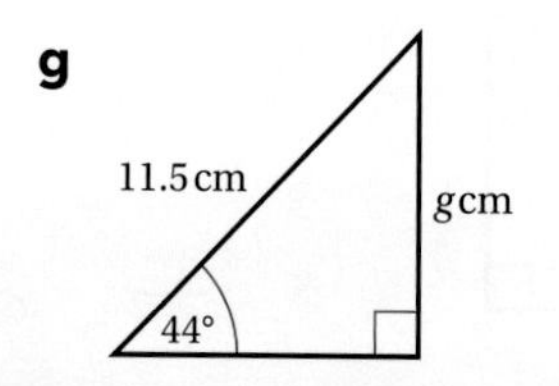

h
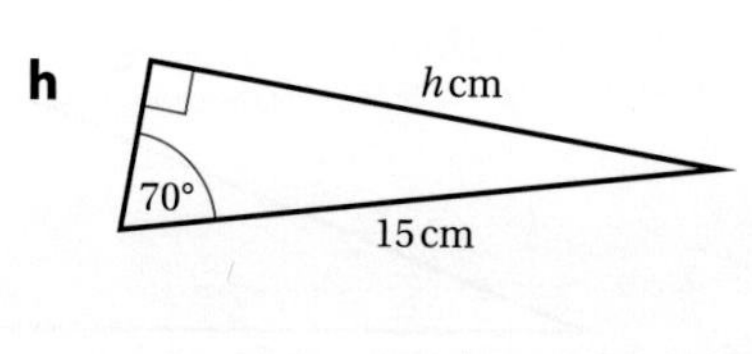

4 Determine the length of the marked side in each triangle. Give your answers correct to two decimal places.

a
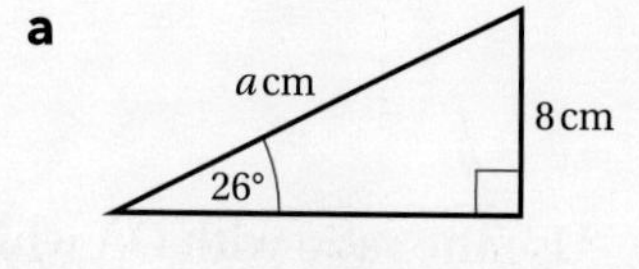

b
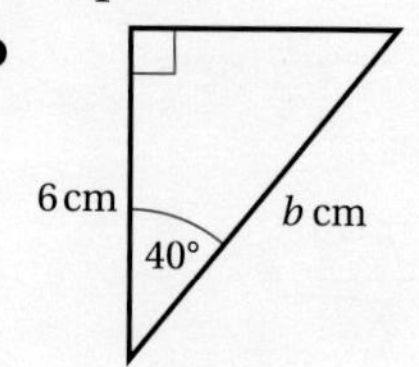

c
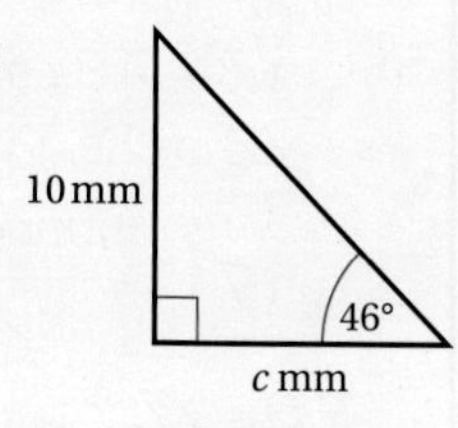

d
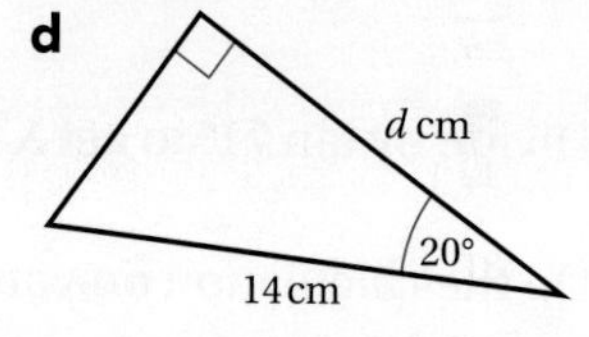

e
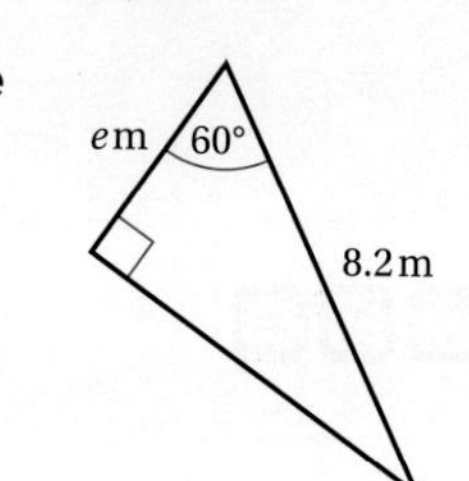

f
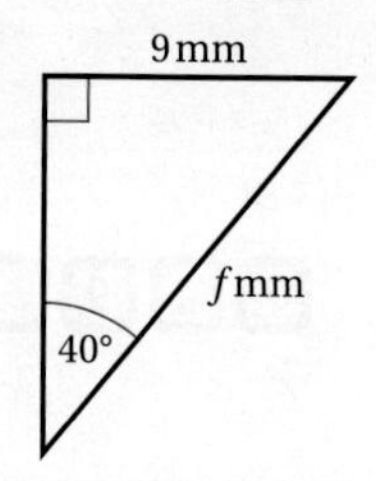

Finding unknown angles

You can use the ratio of sides to find the size of unknown angles. You find the ratio using the side lengths in the same way as before but then you need to work out the **angle** in degrees that is associated with that ratio.

To find the size of unknown angles using the trigonometric ratios you need to use the inverse function of each ratio. On most calculators these are the second functions of the sin, cos and tan buttons. They are usually marked $\sin^{-1}$, $\cos^{-1}$ and $\tan^{-1}$.

Calculator tip

Your calculator can 'work backwards' to find the size of the unknown angle associated with a particular trigonometric ratio. To find the angle if you have the ratio, key in the inverse trigonometric function on your calculator, $\sin^{-1}$, $\cos^{-1}$ or $\tan^{-1}$.

WORKED EXAMPLE 4

a Given that $\tan x = 5$, what is the size of angle x?

b Find the size of angle x in each triangle to 3 significant figures.

i

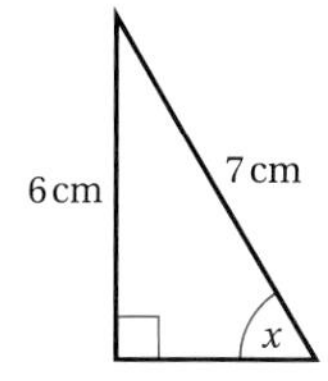

ii

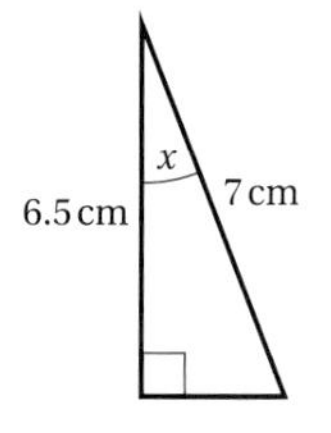

iii

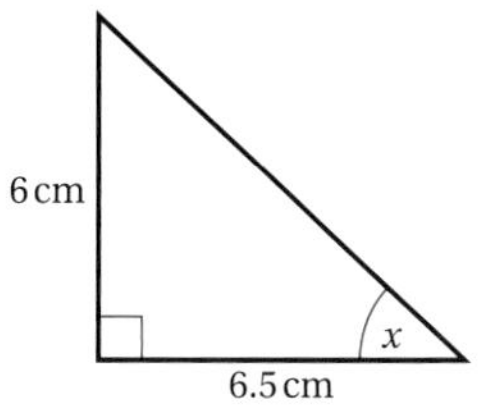

a 

Using your calculator.

Angle x is 78.7° correct to 1 dp.

b i $\sin x = \frac{\text{opposite}}{\text{hypotenuse}} = \frac{6}{7} = 0.857$

Write down the relevant information from the artwork. Find the ratio as before.

$\sin^{-1} 0.857 = x = 59.0°$

Use the inverse sin key on your calculator to find the angle in degrees.

ii $\cos x = \frac{\text{adjacent}}{\text{hypotenuse}} = \frac{6.5}{7} = 0.929$

$\cos^{-1} 0.929 = x = 21.8°$

iii $\tan x = \frac{\text{opposite}}{\text{adjacent}} = \frac{6}{6.5} = 0.923$

$\tan^{-1} 0.923 = x = 42.7°$

WORK IT OUT 33.1

For a ladder to be safe it must be inclined between 70° and 80° to the ground.

The diagram shows a ladder resting against a wall.

Is the ladder positioned safely?

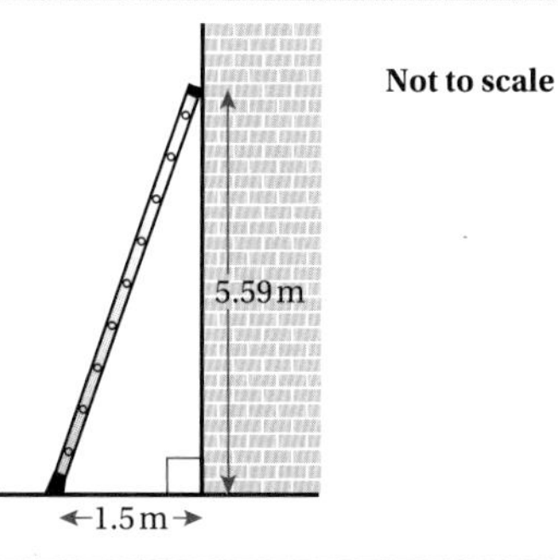

Which of these calculations gives you the answer that you need?

Option A	Option B	Option C
$\sin x = \frac{1.5}{5.59} = 0.268$	$\cos x = \frac{1.5}{5.59} = 0.268$	$\tan x = \frac{5.59}{1.5} = 3.73$
$\sin^{-1} 0.268 = 15.56°$	$\cos^{-1} 0.268 = 74.43°$	$\tan^{-1} 3.73 = 74.99°$

Find answers at: cambridge.org/ukschools/gcsemaths-studentbookanswers

Obtuse angles

Sin, cos and tan are useful for working with acute angles, that is, angles between 0° and 90°, but you can also use them for obtuse angles, that is, angles between 90° and 180°.

Angles on a straight line add up to 180° and you can use this angle rule to find the value of obtuse angles. First work out the value of the acute angle that lies on a straight line with the obtuse angle of interest, using the inverse function of the appropriate trigonometry ratio as usual. Then find the obtuse value by subtracting the acute angle from 180°.

WORKED EXAMPLE 5

The Potters have a small cupboard under the stairs that is the shape of a right-angled triangle. They have the following aerial plan of the cupboard and want to find out how far the cupboard door will swing open.

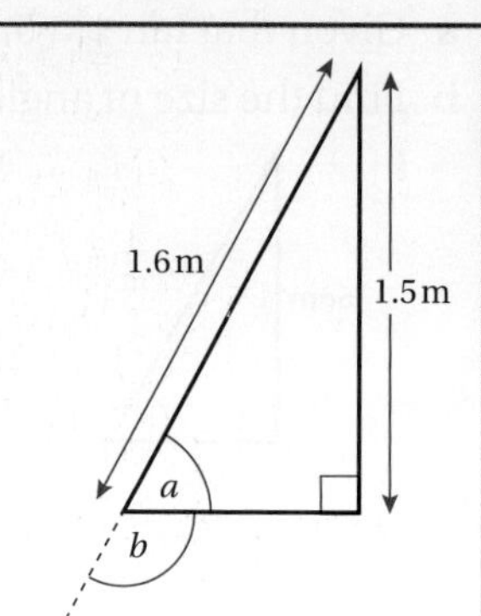

$\sin a = \frac{\text{opposite}}{\text{hypotenuse}} = \frac{1.5}{1.6} = 0.9375$

You need to calculate the angle b but you can see that it is an obtuse angle. It is also on a straight line with angle a, which is an acute angle within the triangle. Use sin (SOH).

$a = 70°$ (to the nearest degree)

Use the inverse function of sin to find the angle a.

$b = 180° - 70° = 110°$

b is 110° so the door will swing open wide enough to fit things inside.

As angle b is on a straight line with angle a, you can find b.

EXERCISE 33B

1 Given the following functions:

a $\sin^{-1} 0.7$ **b** $\cos^{-1} 0.713$ **c** $\tan^{-1} 0.1$

d $\sin^{-1} 0.732$ **e** $\cos^{-1} 0.1234$ **f** $\tan^{-1} 12$

i Determine the acute angle correct to the nearest degree.

ii Determine the corresponding obtuse angle, correct to the nearest degree.

2 Use your calculator to work out the value of each angle to the nearest degree.

a $\sin^{-1} 0.017$ **b** $\cos^{-1} 0.866$

c $\tan^{-1} 1$ **d** $\sin^{-1} 0.866$

3 **a** Find the sine, cosine and tangent of the following right-angled triangle and find the size of angle θ.

b The answers should all be the same. Why is there a small difference?

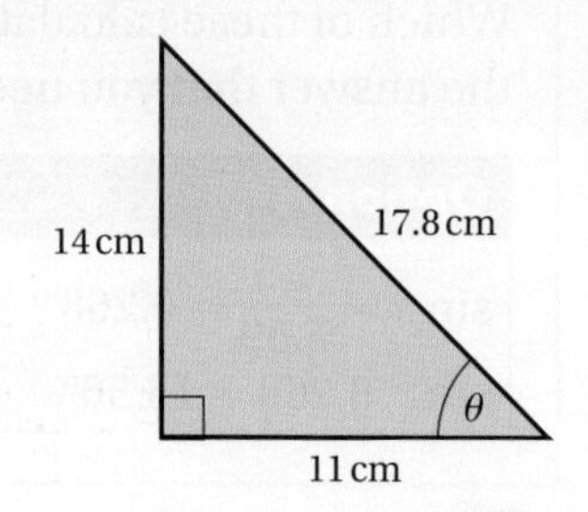

4 Calculate the value of θ for each ratio:

a $\sin \theta = 0.682$ **b** $\cos \theta = 0.891$ **c** $\tan \theta = 2.4751$

d $\sin \theta = 0.2588$ **e** $\tan \theta = 3.9469$ **f** $\cos \theta = 0.7847$

5 Find the size of each marked angle. Give your answers correct to 3 significant figures.

a

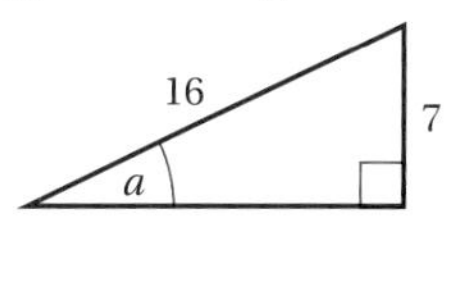

b

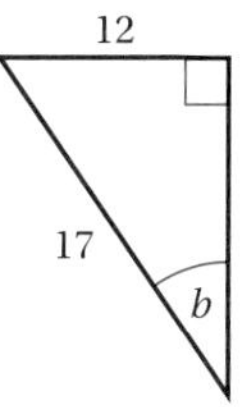

c

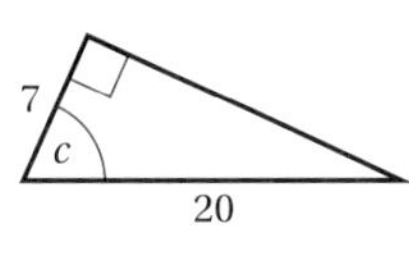

d

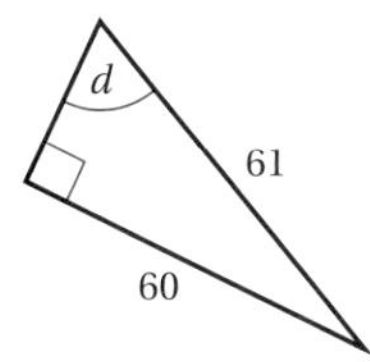

e

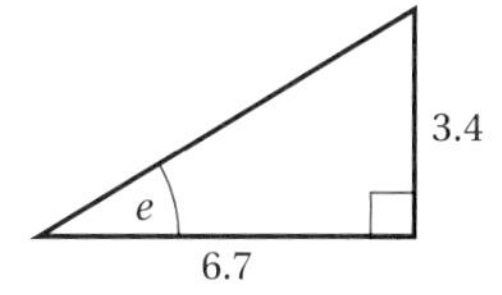

f

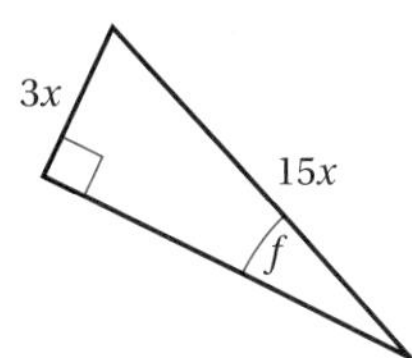

6 PQR is a right-angled triangle.

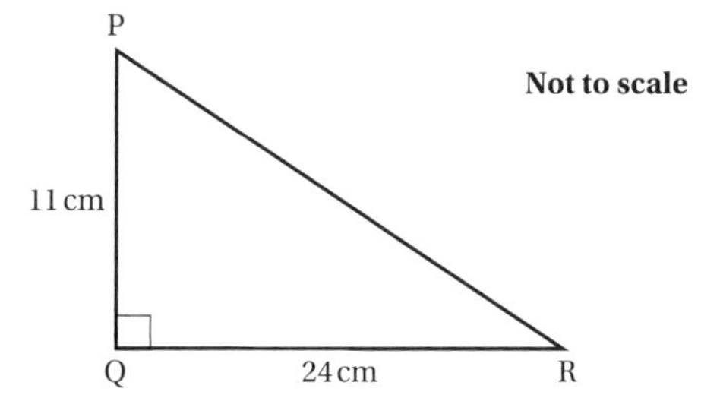

Calculate the size of angle PRQ.

7 What is the size of angle x?

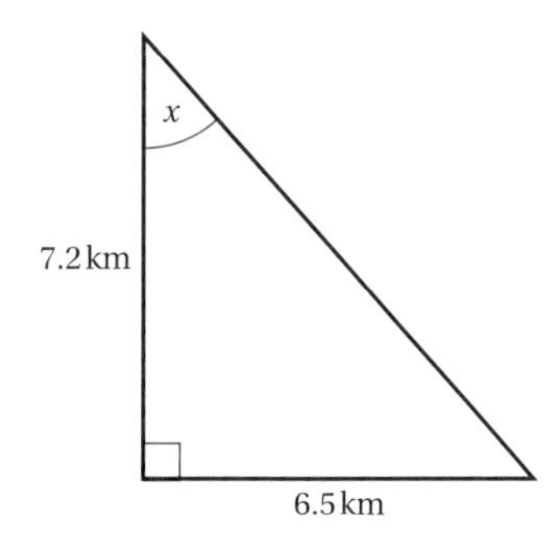

8 For each of these, sketch the triangle and calculate the required values.

a In triangle ABC, angle B = 90°, BC = 45 units and angle C = 23°. Calculate the length of AB.

b In triangle PQR, angle R = 90°, PQ = 12.2 cm and angle P = 57°. Calculate QR.

c In triangle EFG, angle G = 90°, EG = 8.7 cm and angle E = 49°. Calculate the length of FG.

d In triangle XYZ, angle Y = 90°, XZ = 36 units and angle X is 25°. Calculate the lengths of:

i XY. **ii** YZ.

Find answers at: cambridge.org/ukschools/gcsemaths-studentbookanswers

9 For each triangle, draw a sketch and then calculate the required values.

a In triangle ABC, angle C = 90°, BC = 6.7 units and AB = 9.8 units. Calculate angle A.

b In triangle DEF, angle D = 90°, DF = 13 units and EF = 17 units. Calculate angle F.

c In triangle GHI, angle I = 90°, HI = 8.2 cm and GI = 13.7 cm. Calculate the sizes of:

i angle G. **ii** angle H.

d In triangle JKL, angle J = 90°, JK = 85 mm and KL = 113 mm. Calculate the sizes of:

i angle K. **ii** angle L.

e In triangle MNO, angle N = 90°, NO = 29.8 cm and MN = 20.6 cm. Calculate:

i angle O. **ii** angle M. **iii** MO.

f In triangle PQR, angle Q = 90°, PQ = 57.3 mm and QR = 45.1 mm. Calculate:

i angle P. **ii** angle R. **iii** PR.

Section 2: Exact values of trigonometric ratios

When you work out trigonometric ratios on your calculator you often get approximate (or truncated) values because the sides of right-angled triangles are not always perfect squares.

Some values of sin, cos and tan can be calculated exactly.
You need to know the exact values of the sin, cos and tan ratios for 0°, 30°, 60°, 45° and 90 ° angles, except for tan where there is **no** ratio for 90°.

Angle θ	sin θ	cos θ	tan θ
0°	0	1	0
30°	$\frac{1}{2}$	$\frac{\sqrt{3}}{2}$	$\frac{1}{\sqrt{3}}$
45°	$\frac{1}{\sqrt{2}}$	$\frac{1}{\sqrt{2}}$	1
60°	$\frac{\sqrt{3}}{2}$	$\frac{1}{2}$	$\sqrt{3}$
90°	1	0	tan 90° is undefined

You can find the exact values of sin, cos and tan ratios for 0°, 30°, 60°, 45° and 90° using two special right-angled triangles.

Sine, cosine and tangent ratios for 30° and 60°

Take an equilateral triangle (triangle M) with side length 2 units. If this is cut in half, you get triangle N where the angles are 90°, 60° and 30°.

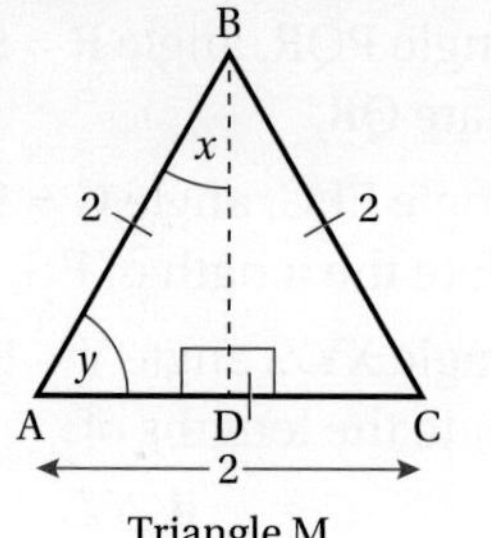

Triangle M

B
30°
2
$\sqrt{3}$
60°
A 1 D

Triangle N

Angle $y = 60°$ (angles of equilateral triangle are equal).

Angle x is bisected by BD, so it is 30°.

BD is the perpendicular bisector of the base. So, AD is 1 unit long.

Using Pythagoras' theorem, you can find the length of BD to be $\sqrt{3}$. You can use this triangle to calculate the value of the following ratios:

$\sin 30° = \frac{\text{opposite}}{\text{hypotenuse}} = \frac{1}{2}$ $\qquad$ $\sin 60° = \frac{\text{opposite}}{\text{hypotenuse}} = \frac{\sqrt{3}}{2}$

$\cos 30° = \frac{\text{adjacent}}{\text{hypotenuse}} = \frac{\sqrt{3}}{2}$ $\qquad$ $\cos 60° = \frac{\text{adjacent}}{\text{hypotenuse}} = \frac{1}{2}$

$\tan 30° = \frac{\text{opposite}}{\text{adjacent}} = \frac{1}{\sqrt{3}}$ $\qquad$ $\tan 60° = \frac{\text{opposite}}{\text{adjacent}} = \frac{\sqrt{3}}{1} = \sqrt{3}$

Sine, cosine and tangent ratios for 45°

The diagram shows a right-angled isosceles triangle with the equal sides 1 unit long.

Using Pythagoras, the hypotenuse is $\sqrt{2}$.

The base angles (x) are each 45°.

$\sin 45° = \text{opposite/hypotenuse} = \frac{1}{\sqrt{2}}$

$\cos 45° = \text{adjacent/hypotenuse} = \frac{1}{\sqrt{2}}$

$\tan 45° = \text{opposite/adjacent} = \frac{1}{1} = 1$

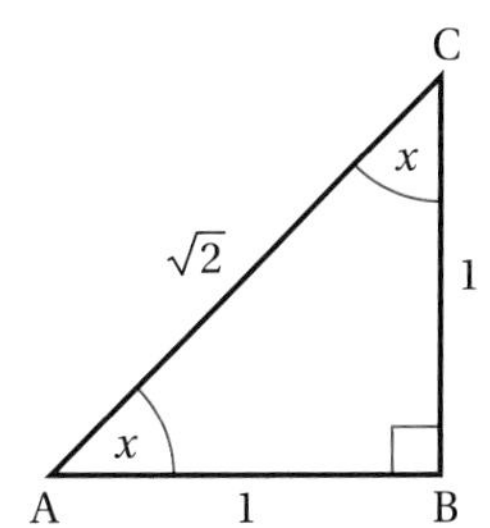

You need to know these exact values of special angles. You can always draw the triangles above to find them if you forget.

Surds

You should have noticed that some trigonometric values use surds, for example $\sqrt{3}$.

It is more accurate to leave an irrational number as a surd in a calculation or on a diagram because it gives the exact value rather than a rounded one. When you are asked to give an exact answer and you have an irrational number, you should leave your answer in surd form.

Tip

You learnt about surds in Chapter 27.

WORKED EXAMPLE 6

Without using your calculator, find the value of x. Leave your answer in exact form.

a $x = \frac{3}{\tan 45°}$ $\qquad$ **b** $x = \sin 30° \times 6$ $\qquad$ **c** $x = \tan 60° \times 2$

a $x = \frac{3}{\tan 45°} = \frac{3}{1} = 3$ — $\tan 45° = 1$ (you need to learn this).

b $x = \sin 60° \times 6 = \frac{\sqrt{3}}{2} \times 6 = 3\sqrt{3}$ — $\sin 60° = \frac{\sqrt{3}}{2}$ (you need to learn this). Use what you learnt about manipulating surds in Chapter 27 to give your answer in the simplest terms.

c $x = \tan 60° \times 2 = 2 \times \sqrt{3} = 2\sqrt{3}$ — $\tan 60° = \sqrt{3}$ (you need to learn this).

EXERCISE 33C

1 Copy and complete this table. Use it to memorise the exact values of the trigonometric ratios for the different angles.

Angle θ	sin θ	cos θ	tan θ
0°			
30°			
45°			
60°			
90°			tan 90° is undefined

Tip

Think about sin 30° and cos 60°; sin 45° and cos 45°; cos 30° and sin 60°.

2 Without using your calculator, find:

a sin 30° + cos 60° **b** sin 45° + cos 45° **c** cos 30° − sin 60°

d Explain your results with reference to complementary angles.

3 You have an equilateral triangle with side length 7. Find the perpendicular height of the triangle using each of the trigonometric ratios.

4 Find the exact value of the variable(s) in each triangle.

a

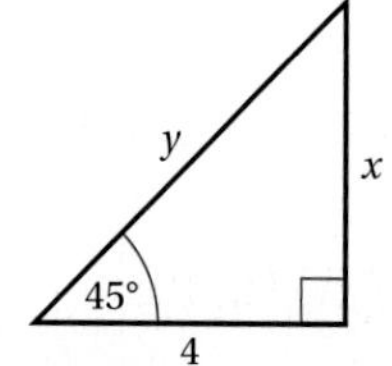

b

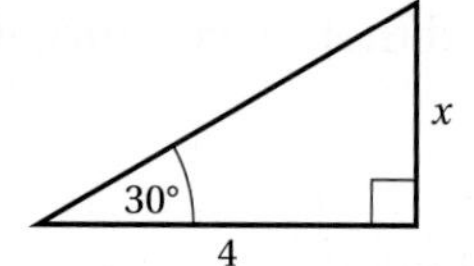

c

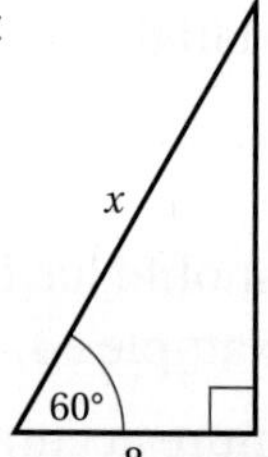

d

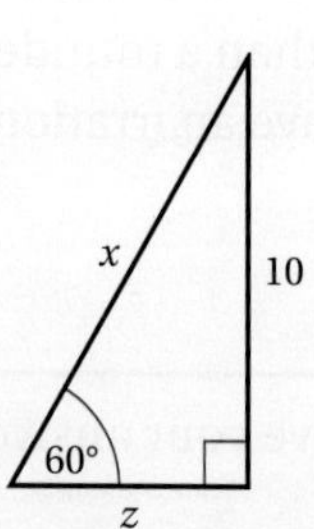

e

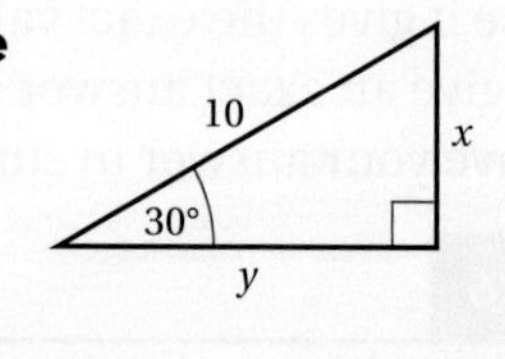

f

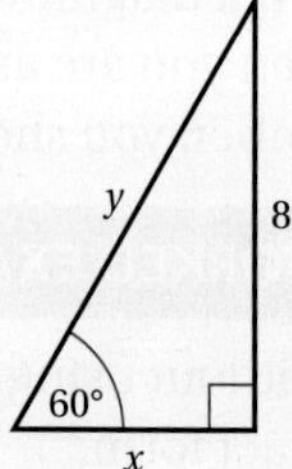

Section 3: The sine, cosine and area rules

So far you have worked only with right-angled triangles. However, trigonometry can be used to solve **any** triangle. There are two important formulae for finding lengths and angles in any triangle: the sine rule and the cosine rule.

The sine rule

In triangle CAX the sine of the base angle at A $= \frac{h}{b}$.

Rearranged, this becomes $h = b \sin A$.

In triangle CBX the sine of base angle at B $= \frac{h}{a}$.

Rearranged, this becomes $h = a \sin B$.

Therefore $b \sin A = a \sin B$.

This can be rearranged to:

$$\frac{a}{\sin A} = \frac{b}{\sin B}$$

The same ratio exists for the length c and sin C such that:

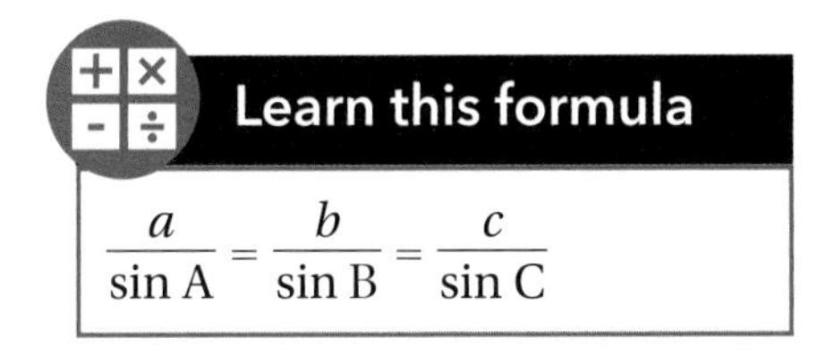

Learn this formula

$$\frac{a}{\sin A} = \frac{b}{\sin B} = \frac{c}{\sin C}$$

This is the **sine rule**.

This rule means that the ratio of any side of the triangle divided by the sine of its opposite angle is equal to the ratio of any other side divided by the sine of its opposite angle. So, the rule actually shows three possible relationships.

This version of the sine rule, with the sine ratios as the denominator, is usually used to calculate lengths.

You can invert the ratios to calculate the size of unknown angles:

$$\frac{\sin A}{a} = \frac{\sin B}{b} = \frac{\sin C}{c}$$

Tip

The sine rule is used when you are dealing with **pairs** of opposite sides and angles.

WORKED EXAMPLE 7

a Calculate length AB.

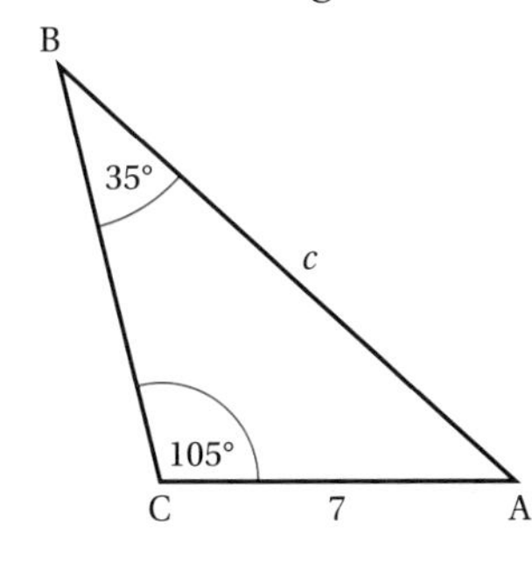

b Find the size of angle θ correct to the nearest degree.

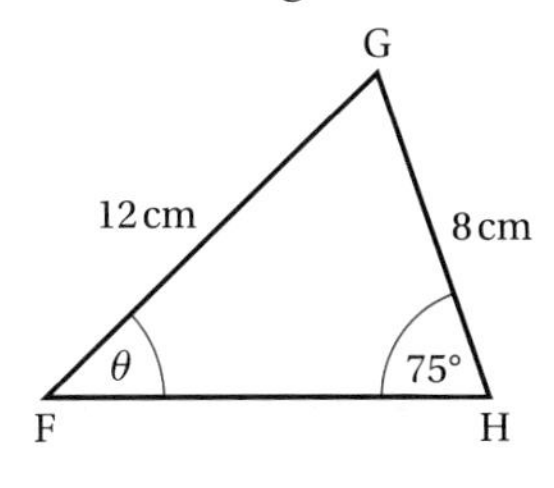

Continues on next page ...

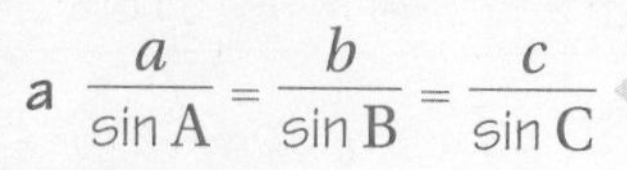

a $\dfrac{a}{\sin A} = \dfrac{b}{\sin B} = \dfrac{c}{\sin C}$

You have a length and the size of the angle opposite it, and you have the angle opposite the unknown length, so use the sine rule.

$\dfrac{7}{\sin 35°} = \dfrac{c}{\sin 105°}$

Substitute in the known values and solve for *c*.

$\dfrac{7}{0.574} = \dfrac{c}{0.96}$

$12.2 = \dfrac{c}{0.96}$

$12.2 \times 0.96 = c = 11.71$ (2 dp)

b $\dfrac{8}{\sin \theta} = \dfrac{12}{\sin 75°}$

You have the length of a side, the size of the angle opposite it and you know the length of the side opposite the unknown angle. Apply the sine rule.

$\dfrac{\sin \theta}{8} = \dfrac{\sin 75°}{12}$

Take the reciprocal of both sides to make your calculation simpler.

$\sin \theta = \dfrac{8 \sin 75°}{12} = 0.6440\ldots$

$\therefore \theta = 40°$ (to the nearest degree)

Use $\sin^{-1}$ to find the angle.

Tip

You can use the sine rule to find unknown angles as long as you know the length of one side and the size of the angle opposite it, and at least one other side length.

EXERCISE 33D

1 Find the value of x in each of the following equations.

a $\dfrac{x}{\sin 50°} = \dfrac{9}{\sin 38°}$ **b** $\dfrac{x}{\sin 235°} = \dfrac{20}{\sin 100°}$ **c** $\dfrac{20.6}{\sin 50°} = \dfrac{x}{\sin 70°}$

2 Find the length of the side marked x in each triangle.

a 50°, x, 65°, 9 cm

b 75°, 10 cm, x, 72°, 33°

c

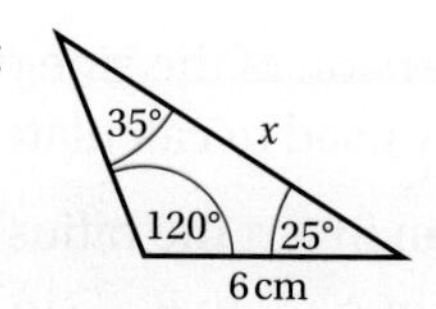

d 65°, 51°, x, 6.2 cm

e

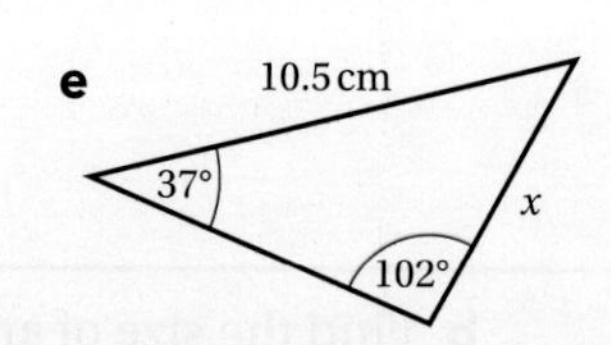

f

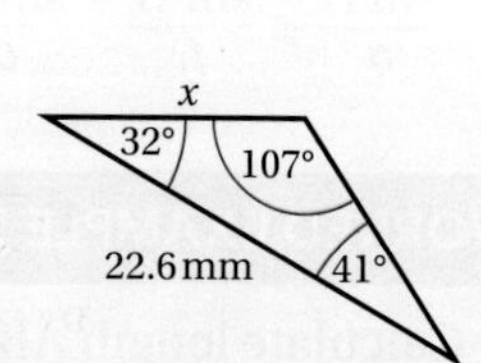

3 Find the size of the angle marked θ in each triangle. Give your answers correct to 1 decimal place.

a 9 cm, 10 cm, 65°, θ

b

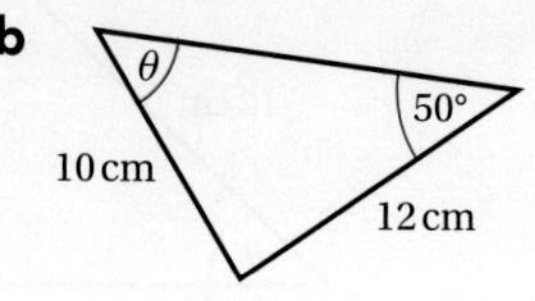

c

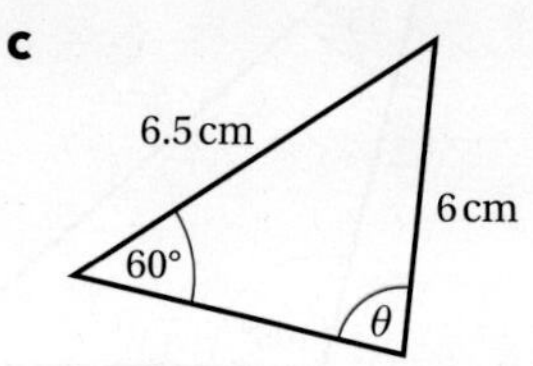

d

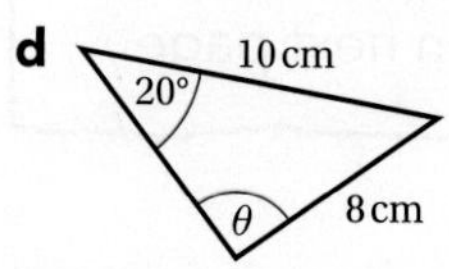

e

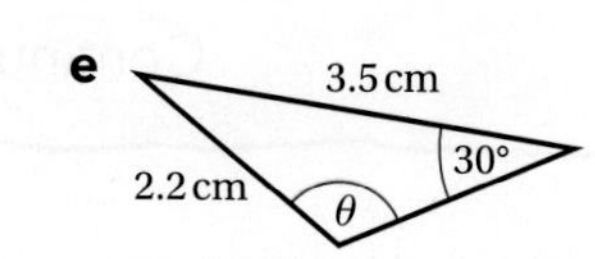

f θ, 37°, 9.2 cm, 11.8 cm

4 In triangle XYZ, angle X = 40°, side XZ = 12 cm and side YZ = 15 cm.

a Explain why angle Y must be less than 40°.

b Calculate the size of angle Y and angle Z.

5 ABCD is a parallelogram.

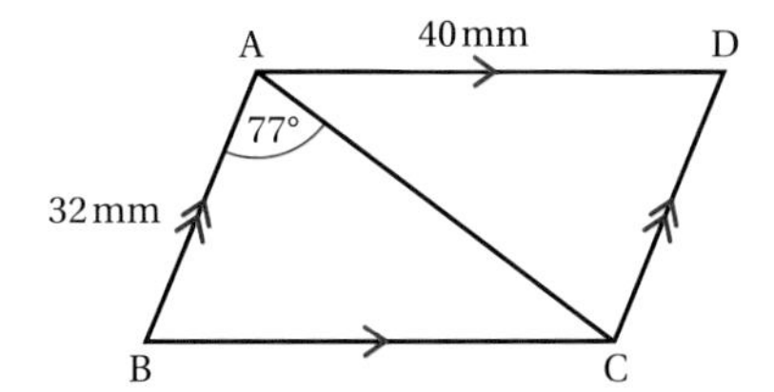

a Find the size of angle BCA to the nearest degree.

b Find the size of ABC to the nearest degree.

c What is the length of the diagonal AC in this figure?
Give your answer correct to 2 decimal places.

The cosine rule

You can use the sine rule when you know the size of an angle and the length of the side opposite the angle. If you don't have this information, you may be able to use another formula called the cosine rule.

The cosine rule is based on Pythagoras' theorem and it applies to any triangle.

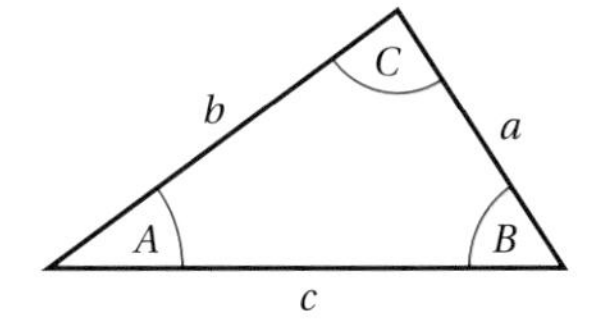

The cosine rule states that:

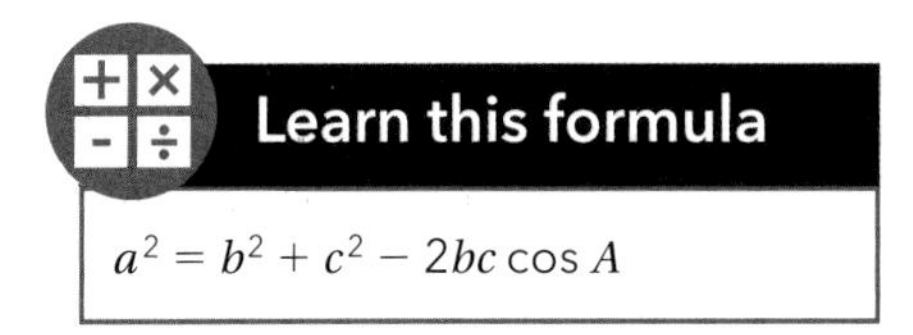

$a^2 = b^2 + c^2 - 2bc \cos A$

Notice that all three sides are used in the cosine rule but only one angle.

The side whose square is used as the subject of the formula is opposite the known angle. This form of the cosine rule is used to find unknown sides.

The formula can be rearranged to make the square of any side the subject.

$b^2 = a^2 + c^2 - 2ac \cos B$ $\qquad$ $c^2 = a^2 + b^2 - 2ab \cos C$

You can also make the cosine ratio the subject of the formula in order to calculate unknown angles.

$\cos A = \dfrac{(b^2 + c^2 - a^2)}{2bc}$ $\qquad$ $\cos B = \dfrac{(a^2 + c^2 - b^2)}{2ac}$ $\qquad$ $\cos C = \dfrac{(a^2 + b^2 - c^2)}{2ab}$

WORKED EXAMPLE 8

a In triangle ABC, angle B = 50°, AB = 9 cm and BC = 18 cm.

Find the length of AC correct to 3 significant figures.

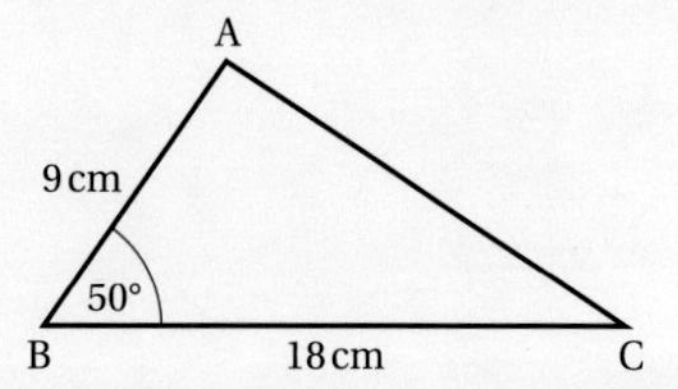

b Calculate the size of angle C.

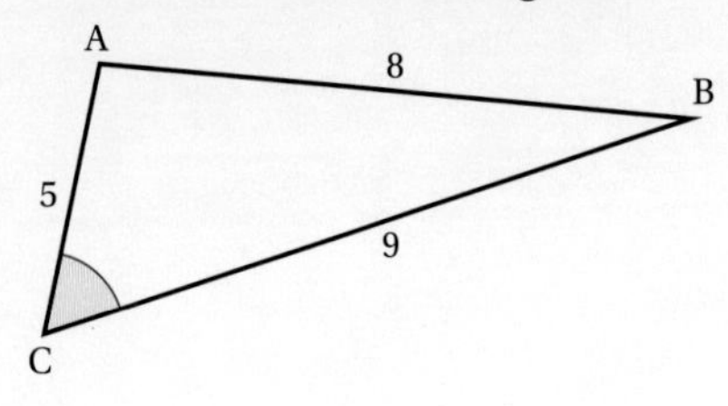

a AC = b B = 50°

$b^2 = a^2 + c^2 - 2ac \cos B$

You know two sides and an angle, so use the cosine rule. The unknown side is opposite the known angle, so you need to use a rearranged version of the rule.

$b^2 = 18^2 + 9^2 - (2 \times 18 \times 9 \times \cos 50°)$

Substitute in known values and solve for b.

$b^2 = 324 + 81 - (208.2631...)$

$b^2 = 196.7368...$

$\therefore b = \sqrt{196.7368...}$

$b = 14.0262$

AC = 14.0 cm (3 sf)

b $c^2 = a^2 + b^2 - 2ab \cos C$

$64 = 9^2 + 5^2 - (2 \times 9 \times 5 \cos C)$

You know the length of three sides and the unknown is an angle. Use the version of the rule where the side opposite the angle is the subject of the formula. Substitute in known values and solve for angle C.

$64 = 81 + 25 - 90 \cos C$

$0 = 106 - 64 - 90 \cos C$

$0 = 42 - 90 \cos C$

$42 = 90 \cos C$

$\frac{42}{90} = \cos C$

$0.467 = \cos C$

$\cos^{-1} 0.467 = C$

Angle C = 62.2°

You could also rearrange the formula to make cos C the subject of the formula and work it forward from there:

$\cos C = \frac{(a^2 + b^2 - c^2)}{2ab}$

EXERCISE 33E

1 Find the size of the side marked x in each triangle. Give answers correct to 3 significant figures.

a

b

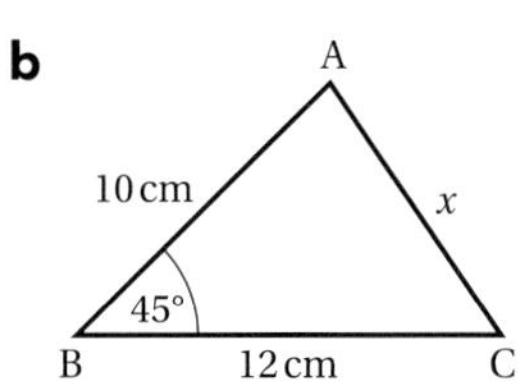

c

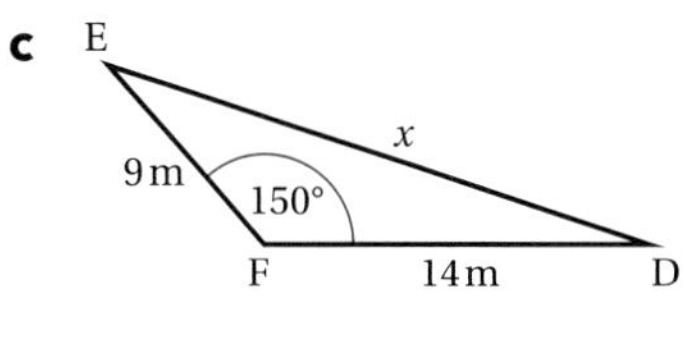

2 In triangle PQR, side PQ = 11 cm, Side QR = 9 cm and side RP = 8 cm. Find the size of angle RPQ, giving your answer correct to 3 significant figures.

3 In triangle STU, angle S = 95°, ST = 10 m and SU = 15 m.

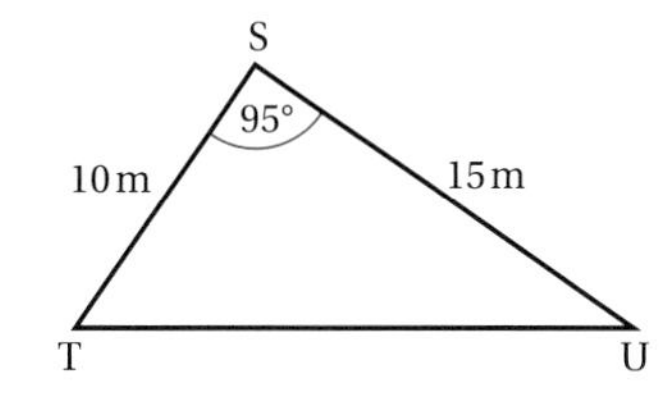

a Calculate the length of TU.

b Find the size of angles U and T.

4 Determine the size of each angle in this triangle.

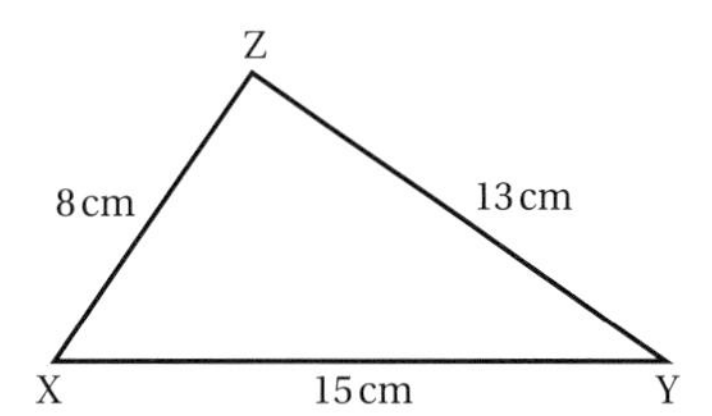

5 A boat sails in a straight line from Aardvark Island on a bearing of 060°. When the boat has sailed 8 km it reaches Beaver Island and it turns to sail on a bearing of 150°. The boat remains on this bearing until it reaches Crow Island, 12 km from Beaver Island. From Crow Island, the boat sails directly back to Aardvark Island.

Calculate:

a the length of the return journey.

b the bearing on which the boat must sail to return directly to Aardvark Island.

The area rule

You already know that the area of a triangle can be found using the formula $A = \frac{1}{2}bh$.

If you don't have the length of the base or the perpendicular height you can calculate the area of any triangle using trigonometry.

Look at triangle ABC.

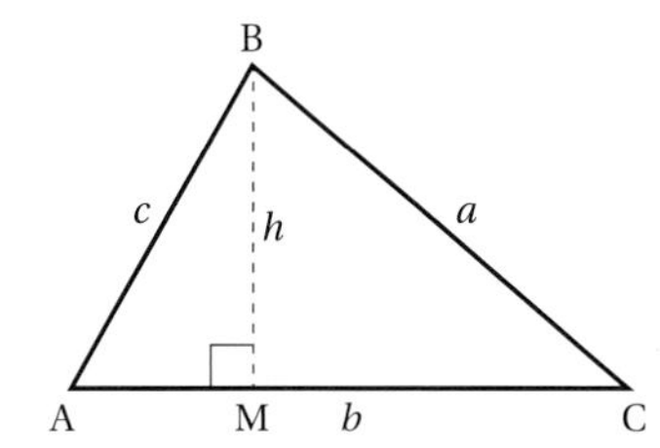

The perpendicular height (h) creates a right-angled triangle BMC.

In triangle BMC, the height (h) is the side opposite angle C and the hypotenuse is side a.

Using the sine ratio, $\sin C = \frac{h}{a}$, so $h = a \sin C$.

This means that you have a way of determining the perpendicular height (whether it is drawn in the triangle or not) and you can use it with the base length to find the area.

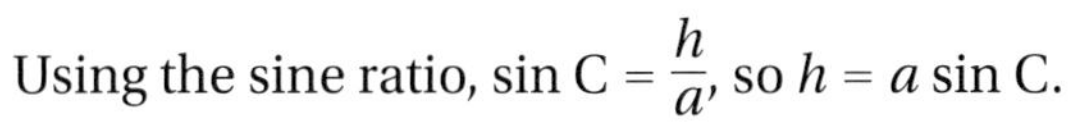

Learn this formula

Area $= \frac{1}{2}ab \sin C$

Area $= \frac{1}{2} \times$ base $\times$ height

$= \frac{1}{2}b \times a \sin C$

This gives a formula for finding the area of **any** triangle.

To use the area formula you need two sides and their included angle.

When you find the area of a triangle, you can use any side as the base. This means that the formula can be rearranged to suit the information you have.

Area $= \frac{1}{2}ac \sin B$ Area $= \frac{1}{2}bc \sin A$

WORKED EXAMPLE 9

In triangle PQR, angle P = 125°, q = 48.1 cm and r = 32.7 cm.

Determine the area of the triangle, correct to 2 decimal places.

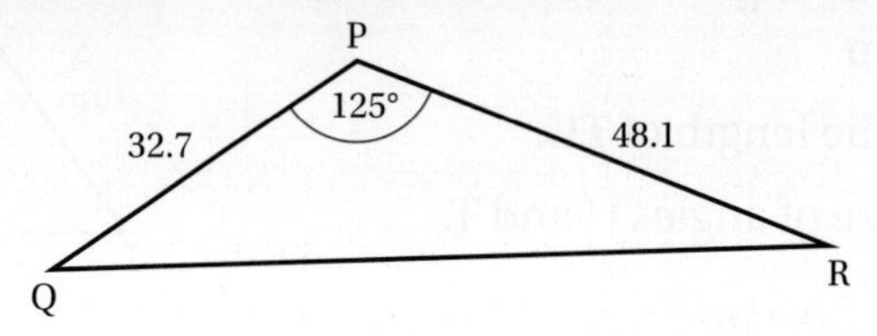

Area PQR $= \frac{1}{2}qr \sin P$

$= \frac{1}{2} \times 48.1 \times 32.7 \times \sin 125°$

$= 644.21$ cm² (to 2 dp)

(Use this version as you know the size of angle P.)

Problem-solving framework

Calculate the area to 2 decimal places for triangle ABC, where angle A = 48°, c = 5 cm and b = 7 cm.

Steps for approaching a problem-solving question	What you would do for this example
Step 1: What do you need to do and what information have you been given?	You need to calculate the area of the triangle. You have been given the measurements of two sides and the included angle, so you can use the area rule.
Step 2: Draw a diagram to show the information.	B, 5 cm, 48°, A, 7 cm, C
Step 3: What maths can you do?	Choose the formula and write it down correctly. Since you are given the size of angle A, use the formula: area of triangle ABC $= \frac{1}{2}bc \sin A$ $= \frac{1}{2} \times 7 \times 5 \sin 48°$ $= 13.01$ cm²
Step 4: Check the answer.	Redo the calculation to check. Based on an estimate of $(5 \times 7) \div 2 = 17.5$ cm, the answer seems reasonable.

EXERCISE 33F

1 Calculate the area of each triangle. Give answers correct to 2 decimal places.

a

P, Q, R; 5 cm, 4 cm, 30°

b

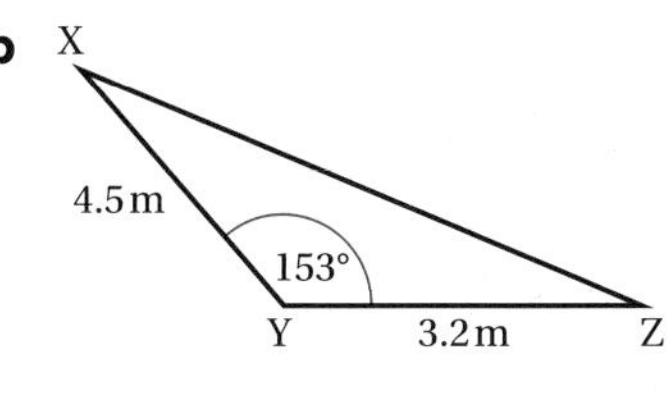

c

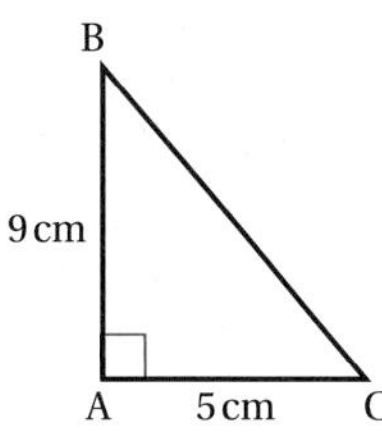

d

G, H, J; 24 mm, 28°, 53 mm

e

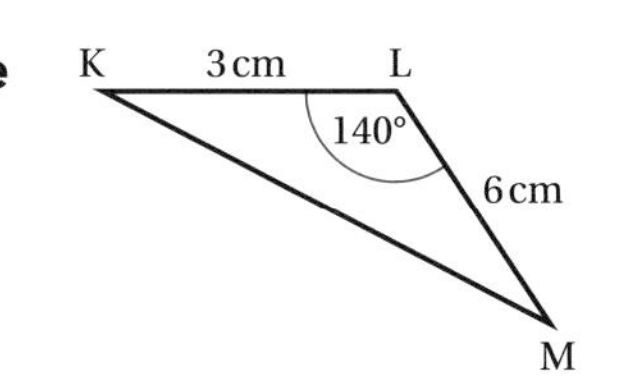

f

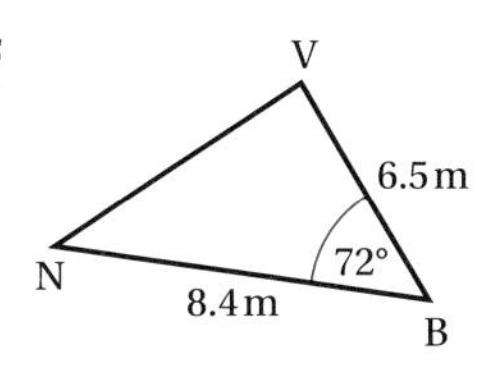

2 Calculate the area of each triangle ABC with the following measurements:

a $a = 3$ cm, $b = 6$ cm and angle C = 65°

b $b = 5.2$ cm, $c = 7.7$ cm and angle A = 105°

c $a = 6.1$ cm, $c = 5.3$ cm and angle B = 98°

d $b = 8$ cm, $c = 12$ cm and angle A = 39°.

3 Triangle XYZ has an area of 50 cm^2.

Calculate the length of XZ.

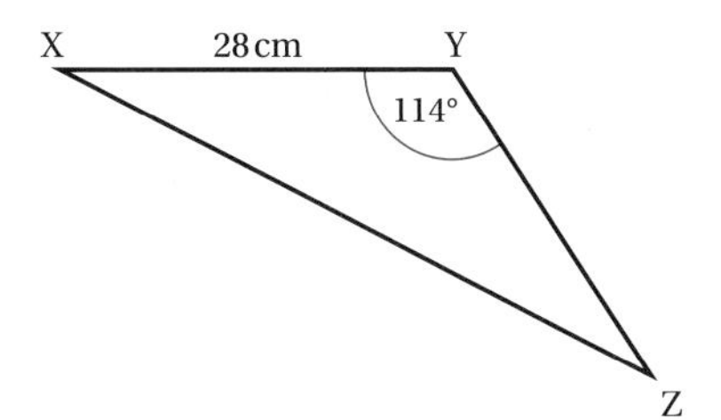

Section 4: Using trigonometry to solve problems

The trigonometric ratios together with the sine, cosine and area rules can be applied to many different types of measurement problems in both 2D and 3D figures.

If trigonometry problems do not include a sketch, it is useful to draw one. Make it large and clear and mark what you know on it. This will help you to identify the correct ratio or rule to use to solve the problem.

Angles of elevation and depression

Many trigonometry problems involve lines of sight. (In other words, the straight line from your eyes to an object you are looking at.)

Key vocabulary

angle of elevation: when looking up, the angle between the line of sight and the horizontal.

angle of depression: when looking down, the angle between the line of sight and the horizontal.

Problems that involve looking **up** to an object can be described in terms of an **angle of elevation**. This is the angle between an observer's line of sight and a horizontal line.

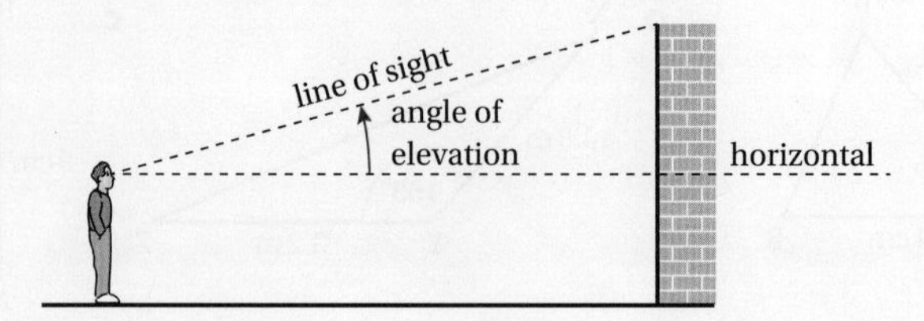

When an observer is looking **down**, the **angle of depression** is the angle between the observer's line of sight and a horizontal line.

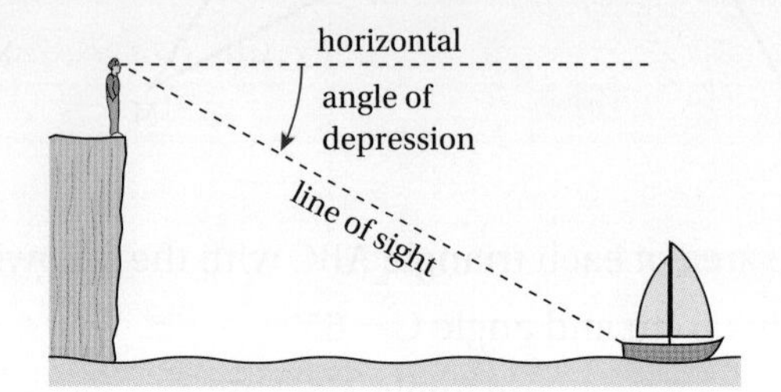

WORKED EXAMPLE 10

From the top of a lighthouse 105 m above sea level, the angle of depression of a boat is 5°.

How far is the boat from the shore? Give the answer to the nearest metre.

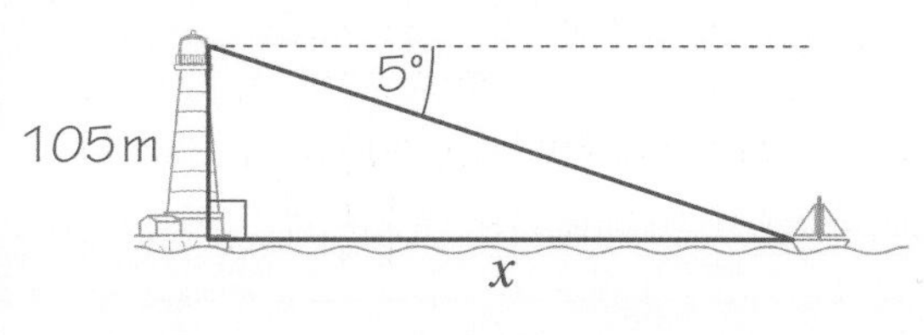

Sketch a diagram of the information you have.

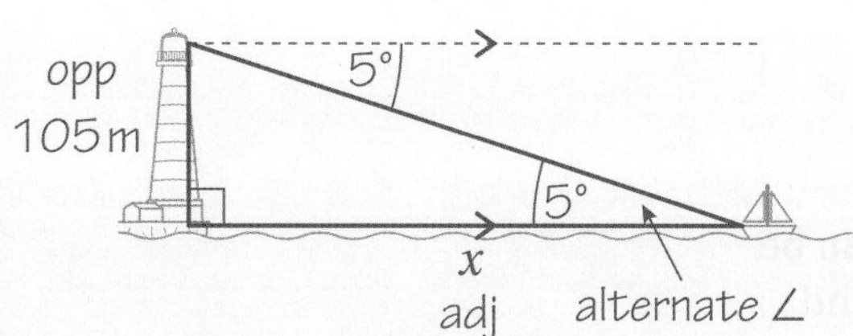

Look at the diagram carefully and mark on any additional information you can: the angle of depression is alternate to the angle at the boat (the sea level is a horizontal line parallel to the horizontal from which the angle of depression is measured). You need to find x. This is the side adjacent to 5°.

$\tan 5° = \frac{105}{x}$

The side opposite the angle is 105 m.

You need the ratio with OA, which is tan (TOA).

$\tan 5° \times x = 105$

$x = \frac{105}{\tan 5°} = 1200.155492$

$x = 1200$ m (to the nearest metre)

Problem-solving framework

A ship is laying cable along the sea bed.

The angle of the cable to the sea bed is 40° and at this angle, the length of cable between the sea bed and the sea level is 40 metres.

The ship is 50 kilometres offshore and travelling north-west.

What is the depth of the sea bed?

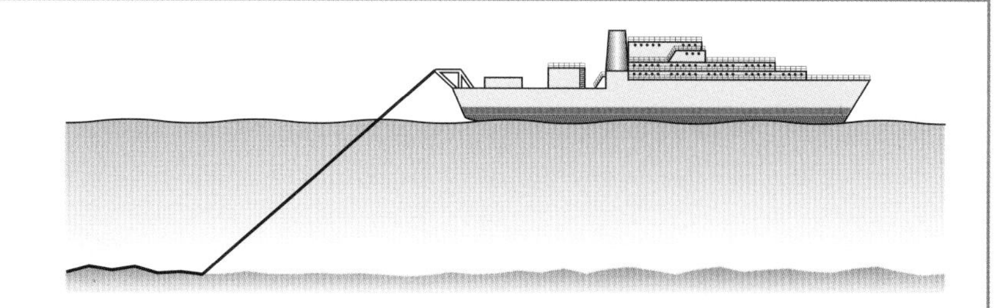

Steps for approaching a problem-solving question	What you would do for this example
Step 1: What have you got to do?	Find the depth of water.
Step 2: Is it helpful to draw a sketch?	You are given an angle and a length and asked to find another length. A sketch of the resulting triangle would be useful. sea level 40 m x 40° sea bed
Step 3: What information do you need? What information don't you need?	Angle 40° and length of cable as the hypotenuse is 40 m. That the ship is 50 kilometres offshore and the direction in which it is travelling are irrelevant.
Step 4: What maths can you do?	You have a right-angled triangle, an angle and the length of the hypotenuse. You cannot use Pythagoras as you do not know two lengths, so use trig. Decide which trigonometry function – sin, cos or tan? The required length is opposite the angle, so use sin (SOH). $\sin 40° = \frac{x}{40}$ $0.6428 \times 40 = x$ $= 25.712\text{ m}$ The depth of the seabed is 25.71 m (2 dp)
Step 5: Have you answered the question? Is your answer correct?	Yes – answer seems reasonable given that the length must be less than 40 m and as $\sin 30° = 0.5$ so half of 40 m would be 20 m.

Three-dimensional problems

To solve problems related to three-dimensional objects you need to visualise two different flat planes and how they meet at right angles. Then you think of the problem in terms of 2D triangles. One of the 2D triangles will give you the information you need to solve the other.

Tip

It often helps to sketch the 2D triangles from a 3D problem, then solve each triangle in turn.

WORKED EXAMPLE 11

The diagram shows a door wedge with a rectangular horizontal base PQRS.

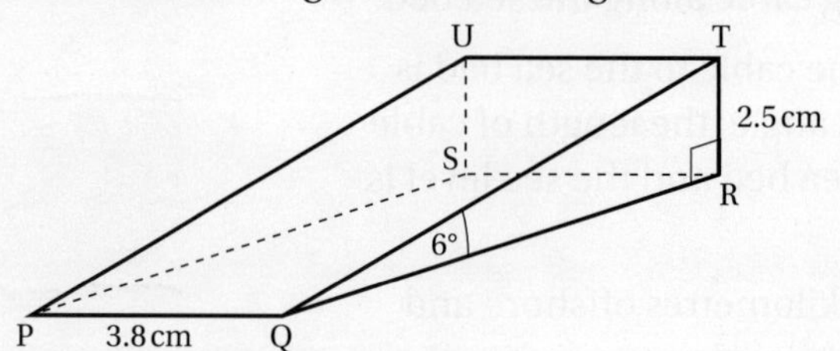

The sloping face PQTU is also rectangular. PQ = 3.8 cm and angle TQR = 6°. The height TR is 2.5 cm.

What is the length of the diagonal PT?

$\sin 6° = \frac{2.5}{QT}$

$0.1045 = \frac{2.5}{QT}$

PT is a line in the triangle PQT on the surface of the rectangle PQTU. You can find the length QT using the triangle QRT.

$QT = \frac{2.5}{0.1045} = 23.92 \text{ cm}$

$PQ^2 + QT^2 = PT^2$

$3.8^2 + 23.92^2 = PT^2$

You have the length PQ and QT, so use Pythagoras' theorem to find PT for triangle PQT.

$14.44 + 572.17 = PT^2$

$586.61 = PT^2$

$PT = \sqrt{586.61} = 24.22 \text{ cm (2 dp)}$

EXERCISE 33G

1 A slide is 4.2 m long and makes an angle of 63° with the horizontal. Calculate the height of the slide.

2 Carol is in a hot air balloon at point C in the sky.
CG is the vertical height of the hot air balloon above the ground.
David is standing on the ground at point D.
The distance between points D and G is 27 m.
The angle of elevation from David to the hot air balloon is 53°.
Calculate the length of CG.

3 Two boats are sailing out at sea a distance of 25 m apart. The angle of depression from the top of a lighthouse to one boat is 35° and to the other boat is 55°. How tall is the lighthouse?

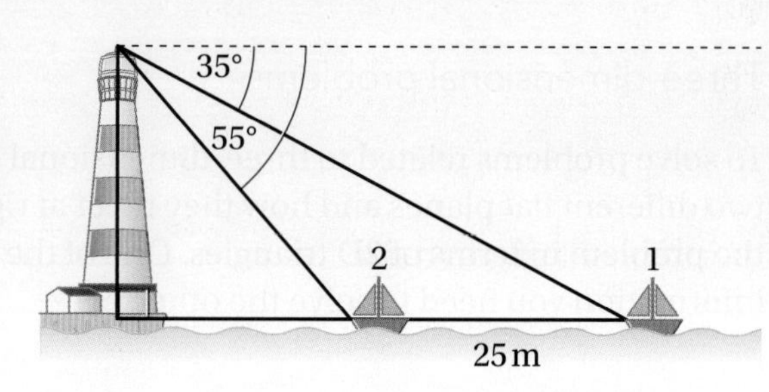

4 Mike is standing 15 m away from a flagpole. The angle of elevation from his line of sight to the top of the flagpole is 60°.

a How tall is the flagpole?

b If he moves another 10 m further away from the flagpole, how will the angle of elevation to the top of the flagpole change?

5 Two observers in different positions at A and B are watching a rare bird on a tree at C. The angle of elevation from A to C is 56° and the angle of elevation from B to C is 25°.

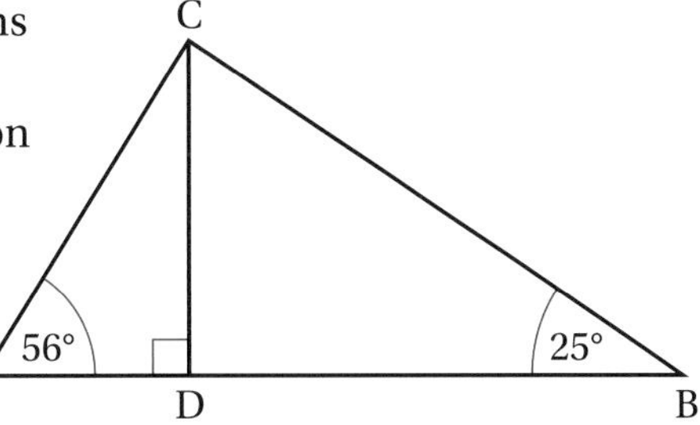

a If person B is standing 15 m from D (the base of the tree), calculate the height of the bird above the ground (the length of CD).

b Calculate the distance of person A from D.

6 A person is standing at a point P, 30 m away from a signal tower for a mobile phone operator. The angle of depression from the top of the tower to P is 56°.

Calculate the height of the tower.

7 A tree surgeon uses an instrument to measure that the angle of elevation from her point of view to the top of a tree is 20°. She is standing 10 m away from the tree.

a If she assumes that the tree is perfectly perpendicular, how would she calculate the height of the tree?

b To check her calculation, she moves another 10 m away from the tree in a straight line, and measures the angle of elevation again. What should the angle measurement be now if her first measurement and calculation were correct?

8 The sketch represents a field PQRS on level ground. The sides PQ and SR run due east.

a Determine the bearing of S from P.

b Calculate the shortest distance between SR and PQ.

c Calculate the area of the field in square metres.

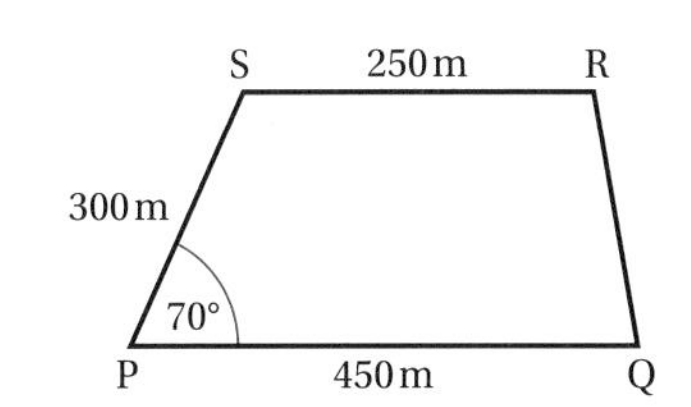

9 Find the area of a regular pentagon with sides $2a$ m long.

10 The diagram represents a room in the shape of a cuboid.

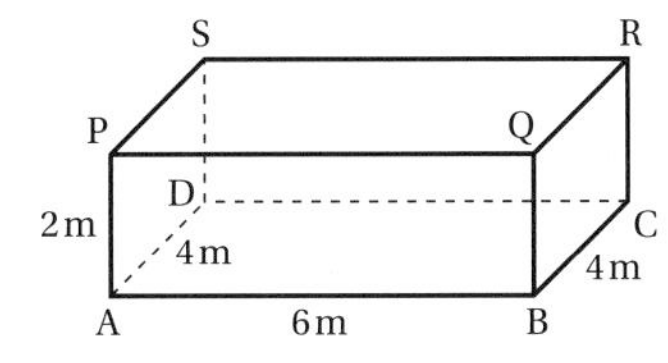

AB = 6 m, AD = 4 m and AP = 2 m.

Calculate the angle between the diagonal BS and the floor ABCD.

Find answers at: cambridge.org/ukschools/gcsemaths-studentbookanswers

ABCDE is a square-based pyramid.
AB = BC = CD = AD = 5.6 cm. N is the centre of the square ABCD. E is directly above N.

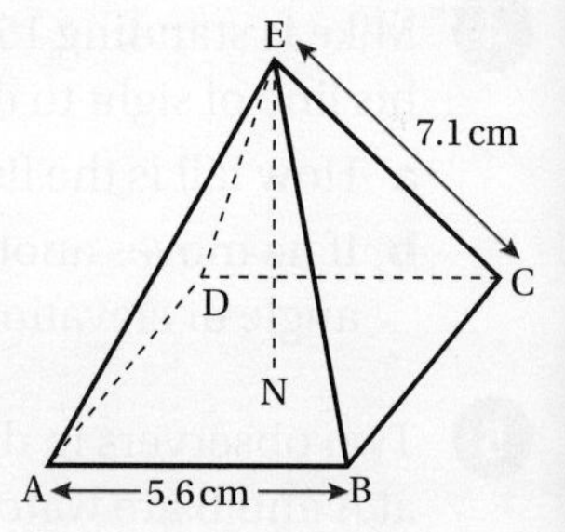

a Calculate the distance BD.

b Calculate angle EBN.

The Great Pyramid at Giza in Egypt has a square base with sides of 232.6 m long.

The distance from the top of the pyramid to each corner of the base was originally 221.2 m.

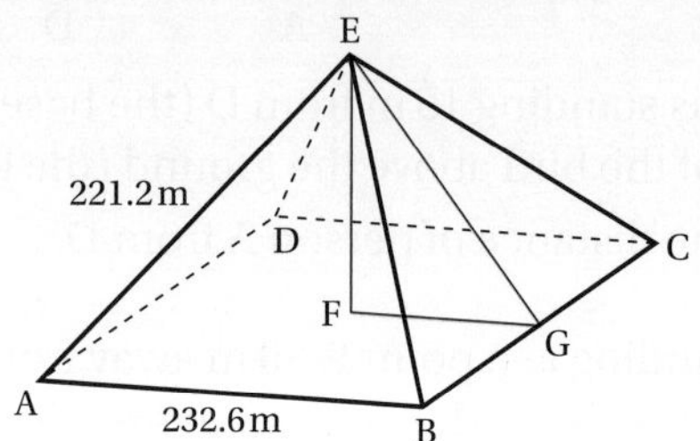

a Determine the angle each face makes with the base.

b Determine the size of the apex angle of a face of the pyramid (angle GEF).

Section 5: Graphs of trigonometric functions

Consider what happens to one capsule on the London Eye as it rotates through 360°.

As the Eye turns, the angle (bearing) of the capsule and its distance above the ground changes. Each time the capsule passes the same point, it will be at the same angle and the same height again and again.

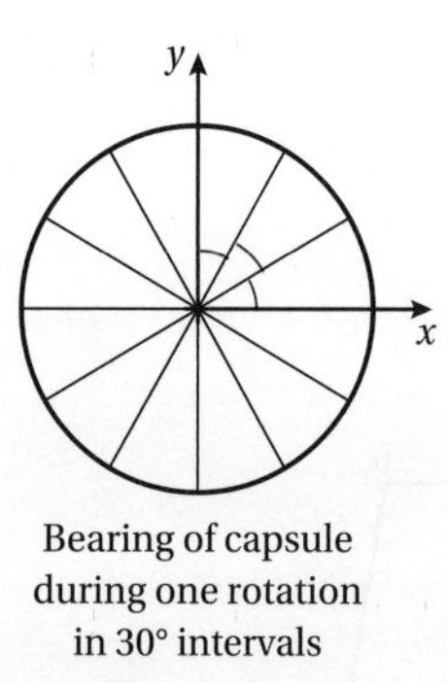

Bearing of capsule during one rotation in 30° intervals

The graph of this movement has a wave shape.

The graph of the movement over time is a **periodic graph** because the same y-values repeat at regular intervals. The graph shows that the capsule completes a rotation every 30 minutes.

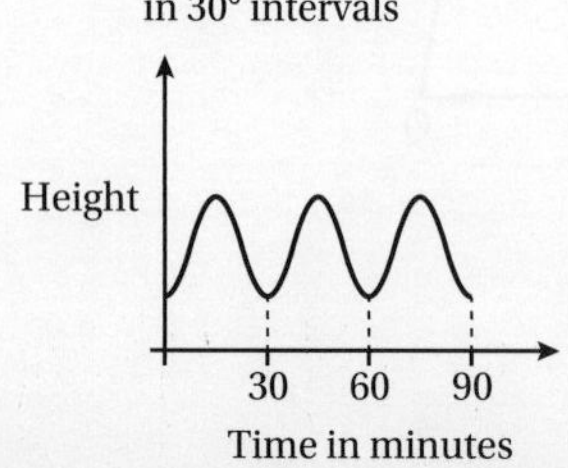

The graphs of the trigonometric functions sine, cosine and tangent are examples of periodic graphs. They are functions of an angle and so the values of x are in **degrees**.

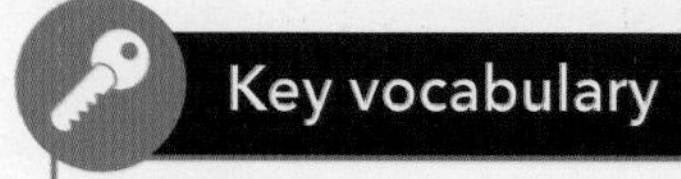

Key vocabulary

periodic graph: a graph that repeats itself in a regular way.

The sine function ($y = \sin x$)

The table below shows the positive values of $y = \sin x$ for angles from 0° to 360°.

x	0°	30°	90°	180°	270°	360°
$y = \sin x$	0	$\frac{1}{2}$	1	0	$^{-}1$	0

The x-axis for this graph needs to be labelled in degrees as the values of the x-coordinates are angles.

Plotting these values produces this graph.

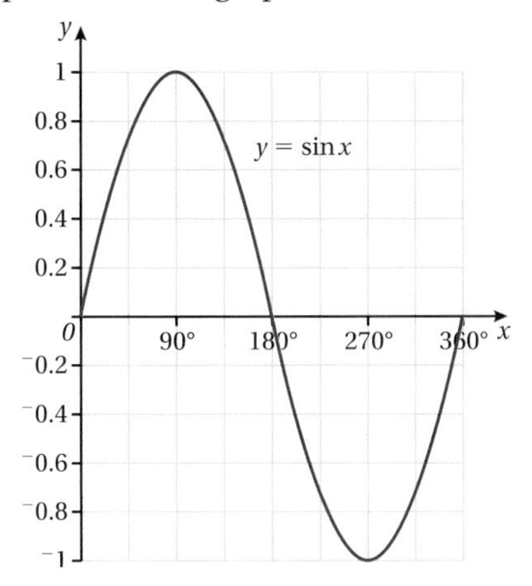

This graph only shows the values of $y = \sin x$ between 0° and 360° but the graph is not restricted to these values. Working out values from $^-270°$ to 720° would produce this graph.

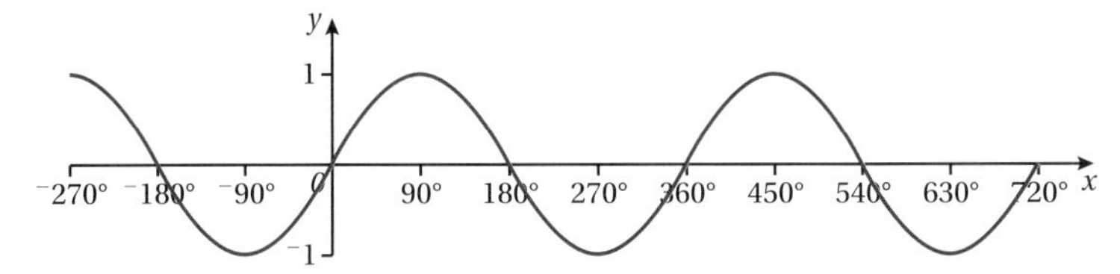

Notice that the graph repeats itself every 360° and that it intercepts the x-axis every 180°. Its y-intercept is at the origin. It has multiple, repeating, turning points with a maximum y-value of 1 and a minimum y-value of $^-1$.

The cosine function ($y = \cos x$)

The graph of $y = \cos x$ can be generated from a table of values. This is the resulting graph for $^-360° \leqslant x \leqslant 360°$.

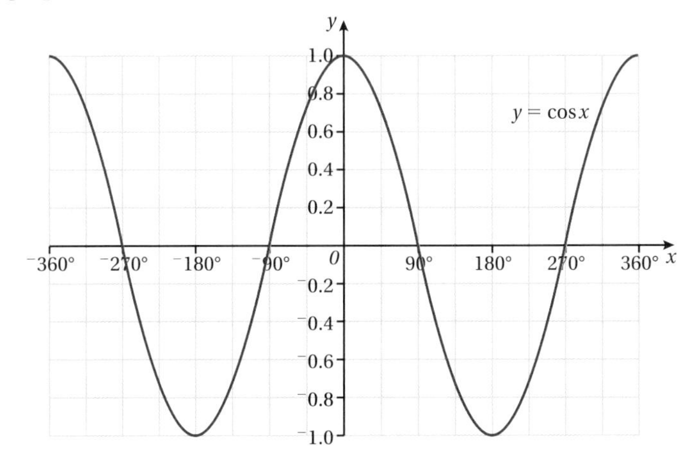

The graph of $y = \cos x$ has the same shape as the graph of $y = \sin x$ with minimum and maximum values at -1 and 1. However, it does **not** go through the origin because $\cos 0° = 1$.

The graph repeats every 360° and it intercepts the x-axis every 180° after 90°.

If you don't recognise the curve by sight, you can work out whether it is a sin or cos graph by finding values of y for different values of x. One set of values does not allow for a conclusive identification of the function, but it can help.

Consider when $x = 0°$ and $x = 90°$.

$x = 0°$	$x = 90°$
$\sin 0° = 0$	$\sin 90° = 1$
$\cos 0° = 1$	$\cos 90° = 0$

The tangent function ($y = \tan x$)

The tan function is periodic, repeating every 180°, but it is not a wave function. The graph is discontinuous and it has no minimum or maximum turning points.

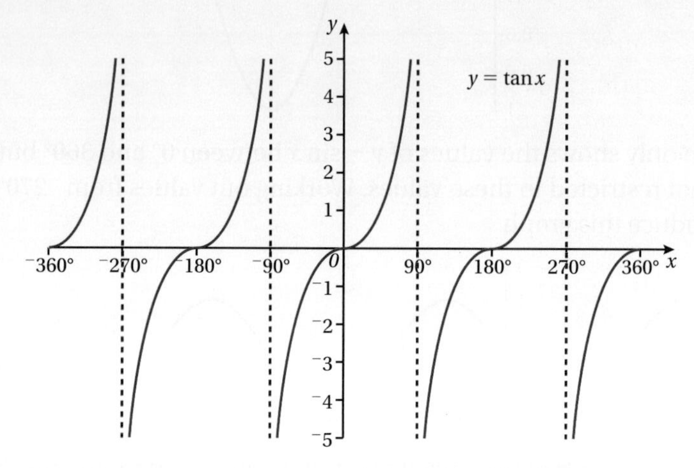

Tip

You will learn more about graphs of trigonometric functions when you deal with transformation of curved graphs in Chapter 41.

Notice also that the tan function approaches but never crosses the lines $x = {}^-270$, $x = {}^-90$, $x = 90$ and $x = 270$. The function $\tan x$ is undefined for these values.

EXERCISE 33H

1. Plot the graphs of $y = \sin x$, $y = \cos x$ and $y = \tan x$ for $-360 \leqslant x \leqslant 360$.
   - **a** Label the minimum and maximum (if it exists) on each graph and state the period over which it repeats.
   - **b** Use the graphs to determine the value of x for which $\sin x = \cos x$ between 0° and 90°.

2. Graph A shows how voltage varies over time in mains electricity. Graph B shows the electronic signal when an AC signal is added to a DC voltage source.

   Which trigonometric function does each graph most resemble? Why?

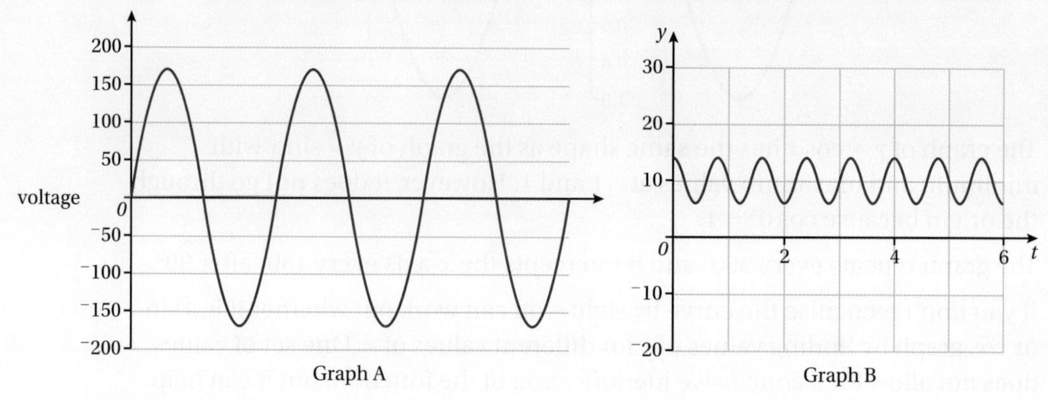

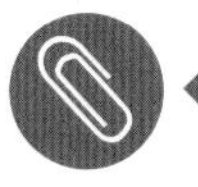

Checklist of learning and understanding

Trigonometric ratios

- In a right-angled triangle, the longest side is the hypotenuse. For a given angle θ, the other two sides can be labelled opposite (to angle θ) and adjacent (to angle θ).

- The sine, cosine and tangent ratios can be used to find unknown sides and angles in right-angled triangles.
    - $\sin \theta = \dfrac{\text{opposite}}{\text{hypotenuse}}$
    - $\cos \theta = \dfrac{\text{adjacent}}{\text{hypotenuse}}$
    - $\tan \theta = \dfrac{\text{opposite}}{\text{adjacent}}$

- You can find the value of a ratio using the sin, cos and tan buttons on your calculator.
  To find the size of an angle, use the inverse functions for each ratio.

Exact values

- You can find the exact values of sin, cos and tan for special angles. Some exact values contain square roots, known as surds.

Sine, cosine and area rules

- The sine and cosine rules can be used to calculate unknown sides and angles in triangles that are not right-angled.

- The sine rule is used for calculating an angle from another angle and two sides, or a side from another side and two known angles. The sides and angles must be arranged in opposite pairs:

$$\frac{a}{\sin \text{A}} = \frac{b}{\sin \text{B}} = \frac{c}{\sin \text{C}}$$

- The cosine rule is used for calculating an angle from three known sides, or a side from a known angle and two known sides:

$$a^2 = b^2 + c^2 - 2bc \cos \text{A}$$

- You can calculate the area of any triangle by using the area rule:

$$\text{Area} = \frac{1}{2}ab \sin \text{C}$$

Graphs of trigonometric functions

- Sin, cos and tan graphs are periodic graphs with particular shapes and characteristics.

Find answers at: cambridge.org/ukschools/gcsemaths-studentbookanswers

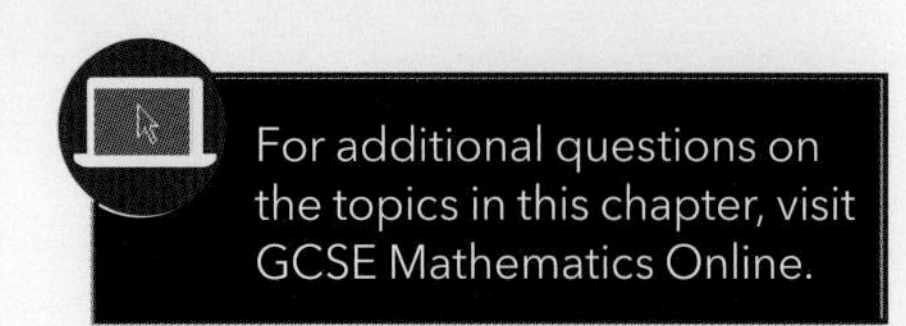

Chapter review

1 The diagram shows a triangle ABC. Angle A = 20° and angle C = 90°; AB = 32 m.

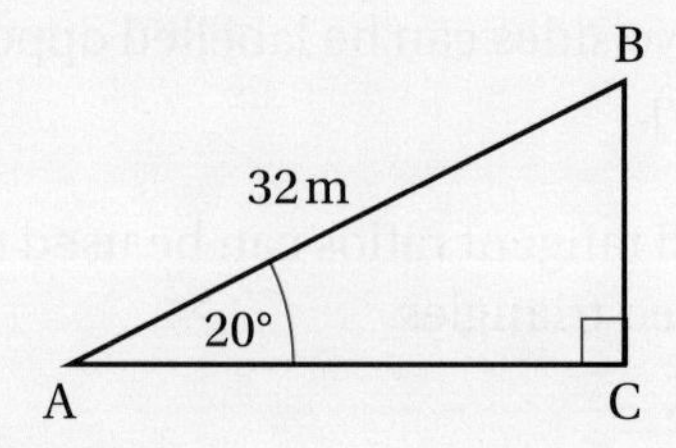

Calculate the height BC.

2 A ladder leans against the side of a house. The ladder is 4.5 m in length, and makes an angle of 74° with the ground.

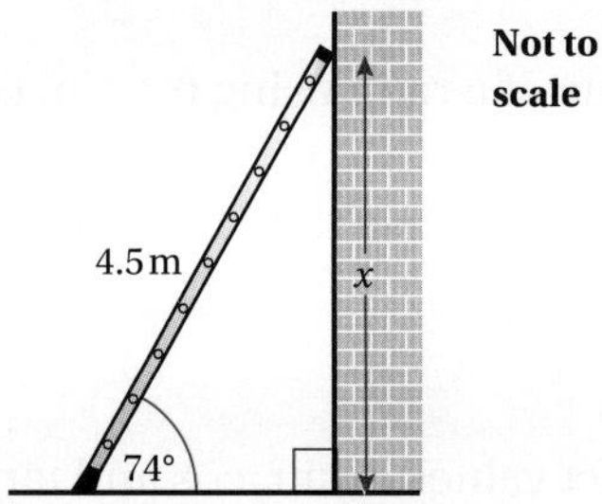

How high up the wall will it reach? (This length is marked x in the diagram.)

3 The same ladder is now placed 0.9 m away from the side of the house.

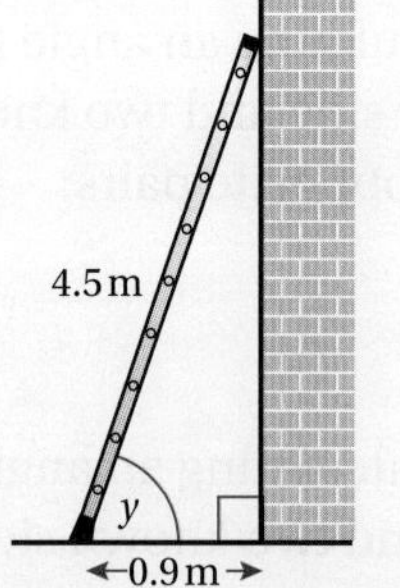

What angle does the ladder now make with the ground? (This angle is marked y in the diagram.)

4 The dimensions of the lean on the Leaning Tower of Pisa are shown below.

From the information given, what is the actual height of the tower?

5 Triangles ABC and PQR are similar. AC = 3.2 cm, AB = 4 cm and PR = 4.8 cm.

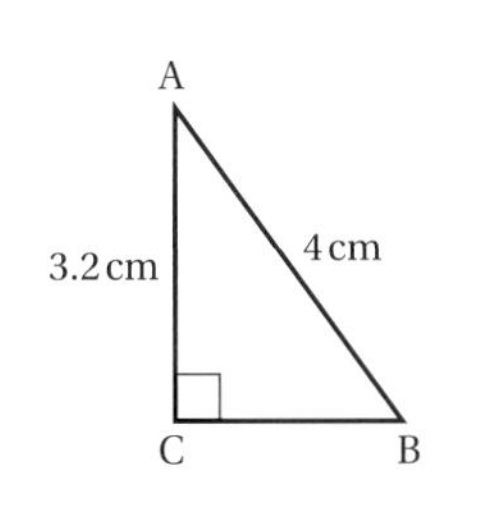

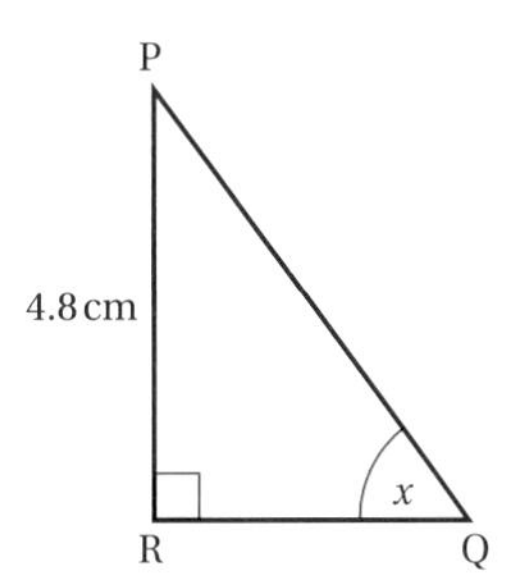

Explain why $\sin x = 0.8$.

6 What would be the round trip starting at point A all the way round, to the nearest mile?

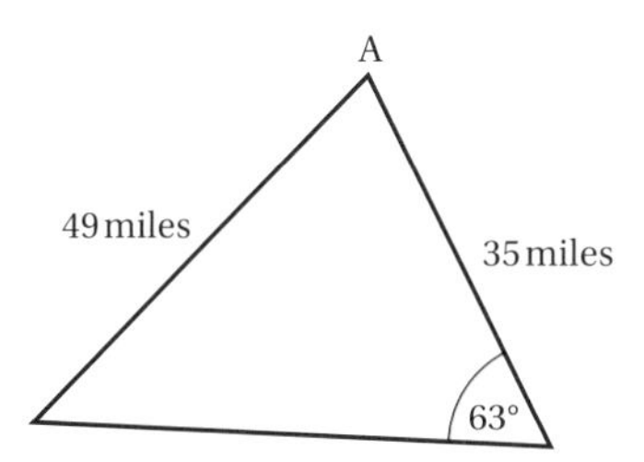

7 Determine the length, correct to one decimal place, of a side of an equilateral triangle with an area of 24 cm^2.

8 If the area of triangle FGH is 804 cm^2, with FG = 43.2 cm and GH = 38.7 cm, calculate two possible values for angle G, correct to 1 decimal place.

9 If the area of triangle ABC is 18 cm^2, with angle B = 30° and AB = 8 cm, calculate the length of BC.

Find answers at: cambridge.org/ukschools/gcsemaths-studentbookanswers

10 In triangle OAB, angle AOB = 15°, OA = 3 m and OB = 8 m. Calculate:

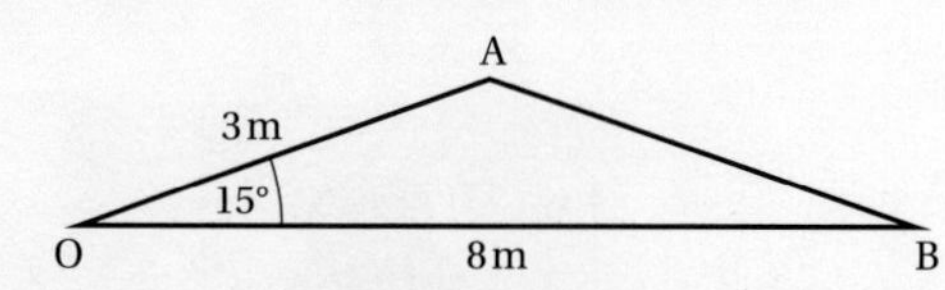

a the length of AB.

b the area of triangle OAB.

11 The diagram shows a triangle ABC.
AB = 14.7 cm, BC = 11.5 cm and AC = 19.4 cm.

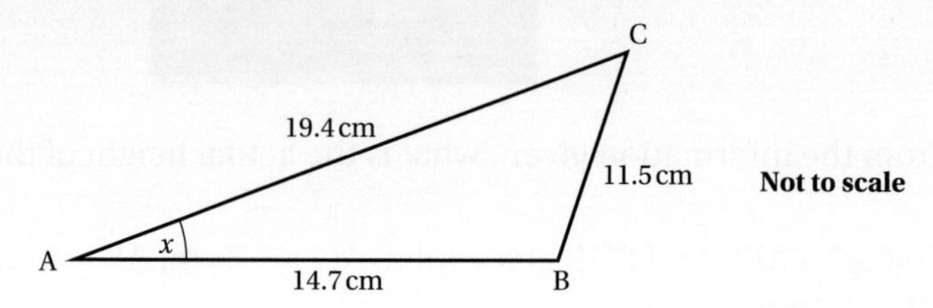

a i Show that the triangle ABC is **not** a right-angled triangle. *(3 marks)*

ii Calculate the angle x. *(3 marks)*

b Calculate the area of this triangle. *(2 marks)*

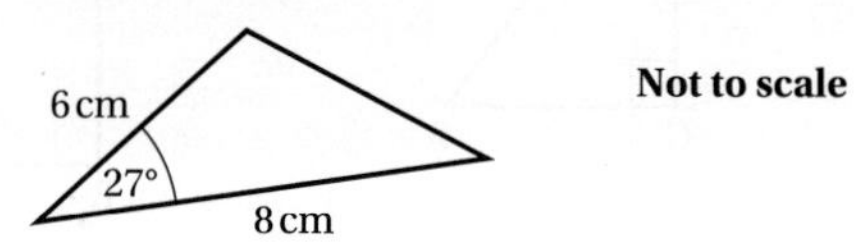

12 A cuboid is 15 cm long, 5 cm wide and 3 cm high. Calculate:

a the length of the diagonal on its base.

b the length of its longest diagonal.

c the angle between the base and the longest diagonal.

13 Pyramid VPQRS has a square base with sides of length 8 cm.
Each sloping edge is 9 cm long.

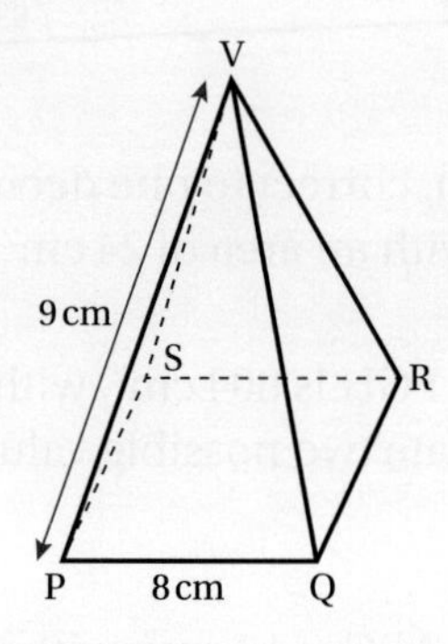

a Calculate the perpendicular height of the pyramid.

b Determine the angle that sloping edge VP makes with the base.

14 **a** Identify and give the equation of the red and blue curves on this grid.

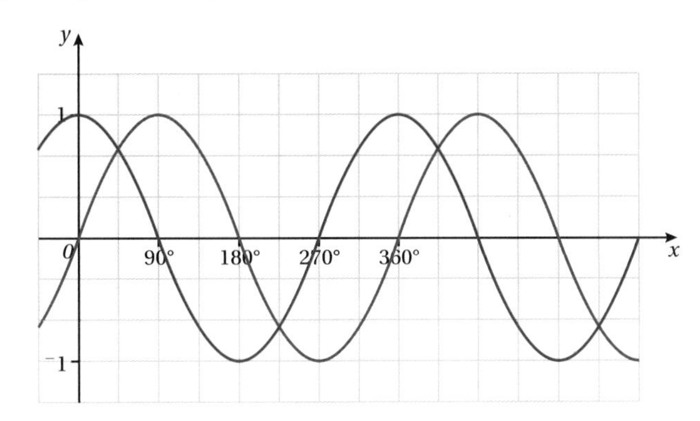

b For what values of x in the range $0° \leqslant x \leqslant 270°$ are the values of $\sin x$ and $\cos x$ equal?

34 Circle theorems

In this chapter you will learn how to ...

- use and apply circle definitions and understand their properties.
- prove and apply the standard circle theorems, using them to find related results.

For more resources relating to this chapter, visit GCSE Mathematics Online.

Using mathematics: real-life applications

Being able to find and use angles in and around circles has applications in fields such as engineering and computer-aided design. Designing cogs and camshafts, plotting navigation charts and working out angles in astronomy all use or apply circle theorems.

"Using the height of the line of sight from the ship and the radius of the Earth, I can calculate the distance of the visible horizon from the ship using circle theorems." *(Navigation officer)*

Before you start ...

Ch 5, 6, 9	You should be able to calculate the sizes of missing angles in geometry problems.	**1** Decide whether each statement is true or false. **a** $c = 120°$ **b** $a = 120°$ **c** $d = b$ **d** $b = c = 60°$
Ch 5, 15, 16	You should know how to find the circumference and area of circles and sectors of circles	**2** For each of these circle sectors, calculate the length of the arc and the area of the sector: **a** 3 cm, 72° **b** 25 mm, 45° **c** 150°, 12 cm
Ch 5, 30, 31, 32	You should be able to use given facts in geometry problems to write proofs.	**3** You need to write a proof that angle BAD = 50°. Which of these proofs is **not** correct? **a** Angle CED = 50° (the sum of angles of a triangle is 180°) Angle BAD = 50° (corresponding angle to angle CED, AB // EC) **b** Angle CEA = 100° (alternate angle to angle ECD) Angle BAD = 50° (180° − 30° − 100° = 50°) **c** Angle DBA = 100° (corresponding angle to angle ECD, AB // EC) Angle BAD = 50° (the sum of angles of a triangle is 180°)

Assess your starting point using the Launchpad

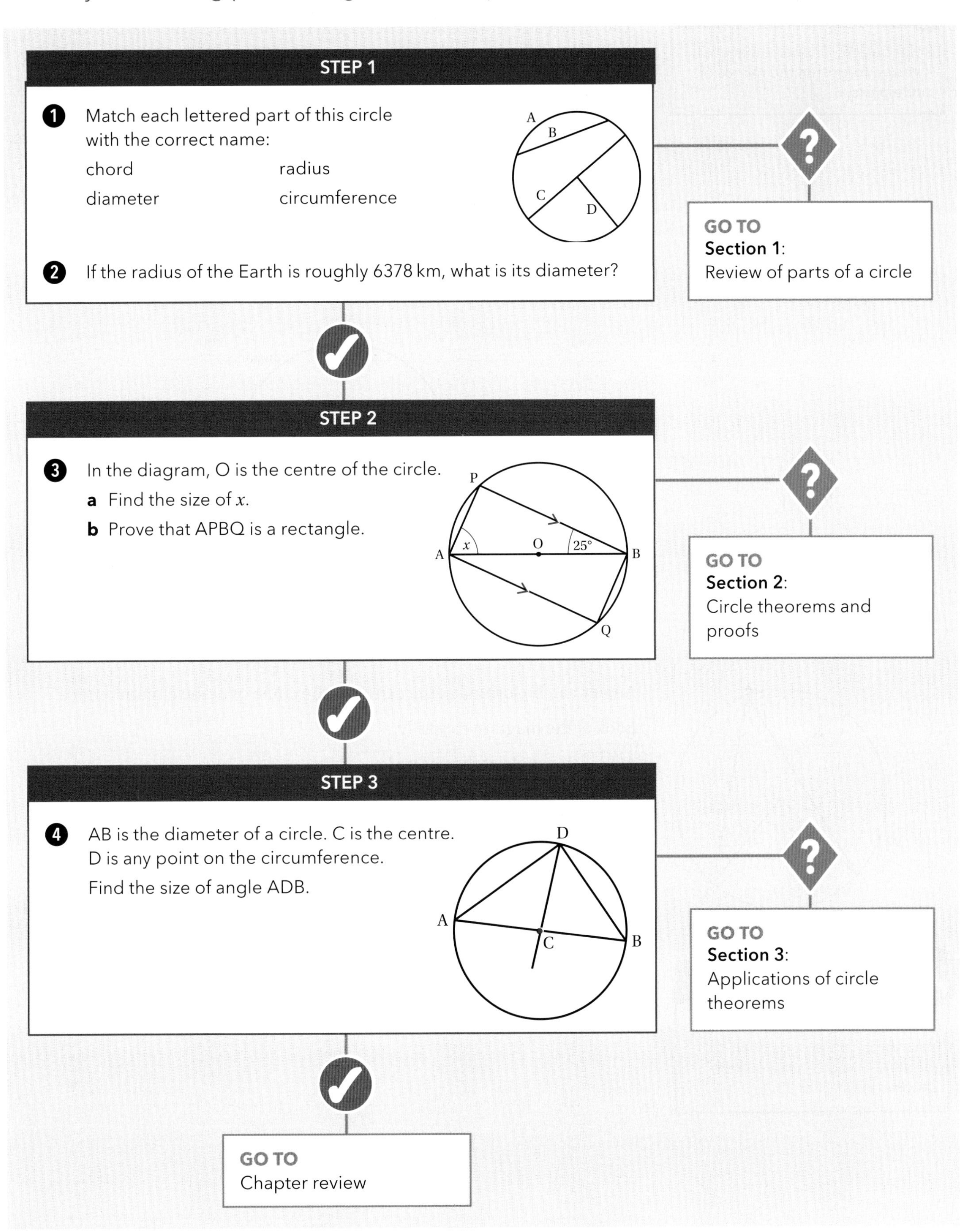

STEP 1
1 Match each lettered part of this circle with the correct name:
chord
radius
diameter
circumference
A
B
C
D
2 If the radius of the Earth is roughly 6378 km, what is its diameter?
GO TO
Section 1:
Review of parts of a circle
STEP 2
3 In the diagram, O is the centre of the circle.
a Find the size of x.
b Prove that APBQ is a rectangle.
P
A
x
O
25°
B
Q
GO TO
Section 2:
Circle theorems and proofs
STEP 3
4 AB is the diameter of a circle. C is the centre.
D is any point on the circumference.
Find the size of angle ADB.
D
A
C
B
GO TO
Section 3:
Applications of circle theorems
GO TO
Chapter review

Find answers at: cambridge.org/ukschools/gcsemaths-studentbookanswers

Tip

Refer back to Chapters 5 and 15 if you've forgotten the names of circle parts.

Section 1: Review of parts of a circle

You've already worked with circles and learned the correct names for circle parts. You will use these names as you learn to prove circle theorems, so it is worth revising them before you go on.

A circle is the locus of points that are all the same distance from a given point. That point is the centre of the circle. Any line drawn from the centre to the circumference is therefore a fixed length called a radius.

If you take any two points on the circumference and join them, you get a line called a chord. If the chord passes through the centre, then it is called the diameter and it is equivalent to two radii ($d = 2r$).

Basic terminology

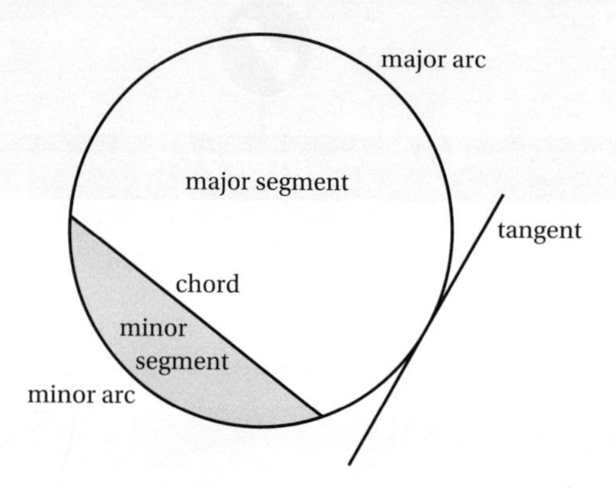

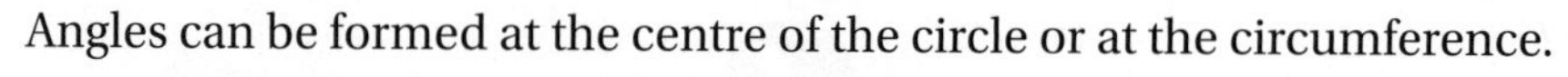

Naming angles in circles

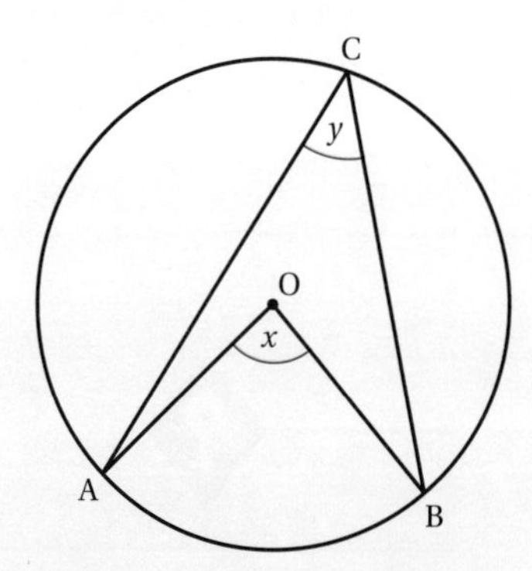

Angles can be formed at the centre of the circle or at the circumference.

Look at the diagram carefully.

AOB is the angle at the centre (x).

ACB is the angle at the circumference (y).

The arms of both these angles are 'standing' on the minor arc AB.

Angles that stand on an arc are said to be **subtended** by the arc.

Angle AOB is said to be subtended at the centre of the circle, and angle ACB is subtended at the circumference point C.

Angles can also be subtended by a chord. If you were to join points A and B by a straight line, you would create the chord AB and you can see that the angles x and y are now subtended by the chord AB and the arc AB.

Key vocabulary

subtended: an angle whose sides pass through the ends of an arc (or other curved line); you first saw this in Chapter 15.

EXERCISE 34A

1 Read the clues and identify the circle parts correctly.

a half the diameter **b** the larger part of a circle when it is divided into two parts by a chord **c** formed by two radii and an arc **d** a line outside a circle that touches the circumference at one point only **e** the locus of a point at a fixed distance from another point **f** the smaller part of the circumference when a circle is divided by a chord	A sector B tangent C minor arc D circle E radius F major segment

2 In the diagram, O is the centre of both circles. Complete the sentences using the correct mathematical terms.

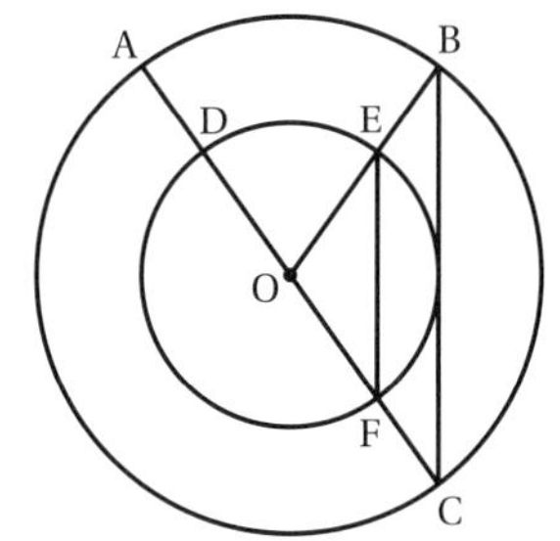

a DF is the _______ of the smaller circle.

b AO is a _______ of the larger circle.

c AC is the diameter of the _______ circle.

d ED is a _______ _______ of the smaller circle.

e Arc ACB is a _______ _______ of the larger circle.

f EF is a _______ of the smaller circle.

g Angle FOE is the angle at the _______ subtended by arc ___.

h Angle ACB is subtended by ___ at the circumference.

3 Look again at the diagram in Question **2**.

a What can you say about angles OFE and OEF? Why?

b How can you prove that angle OEF = OFE = OCB = OBC?

Section 2: Circle theorems and proofs

Angles subtended at the centre and circumference

For this theorem, we are interested in angles formed at the centre or on the circumference of a circle, although an angle can be formed at a point anywhere within or outside the circle.

Tip

Think of DE forming a base for the angle at the centre or at the circumference of the circle to stand on. Angle DOE and angle DFE stand on the arc DE.

In the diagram, the minor arc DE (or chord DE):

- subtends angle DOE at the centre O
- subtends angle F at point F on the circumference.

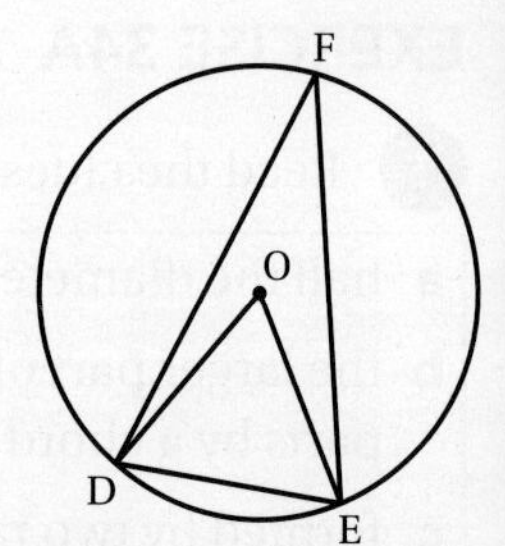

The theorem states that the angle at the centre of a circle, subtended by an arc, is twice the size of the angle at the circumference subtended by the same arc.

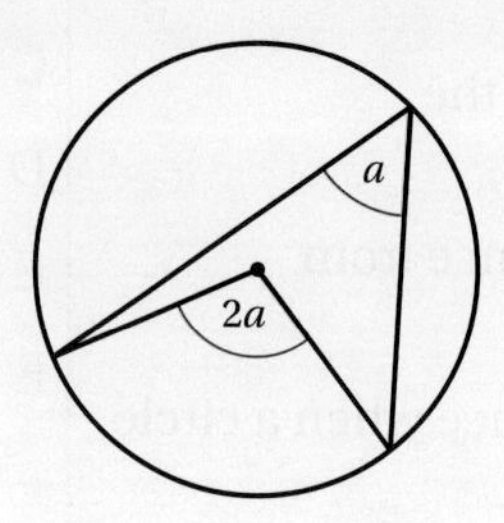

WORKED EXAMPLE 1

Use the diagram to prove that the angle at the centre of a circle, subtended by an arc, is twice the size of the angle at the circumference subtended by the same arc.

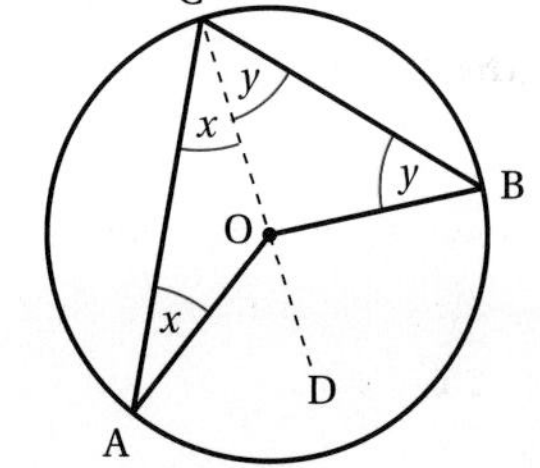

Angle AOB and angle ACB are both subtended by arc AB.

Drawing a line segment joining the centre O to C forms two isosceles triangles: triangle AOC and triangle BOC. The equal sides of each triangle are the radii of the circle.

Angle ACB = $x + y$

Angle AOC = $(180 - 2x)$

Angle BOC = $(180 - 2y)$

$360°$ = AOB + $(180 - 2x) + (180 - 2y)$

0 = AOB $- 2x - 2y$

Angle AOB = $2x + 2y$

$2x + 2y = 2(x + y)$

So angle AOB = 2 × angle ACB

Therefore the angle subtended by AB at the centre is twice the angle subtended by AB at the circumference.

Tip

When you use this theorem in a geometry problem, write '**angle at the centre theorem**' as a reason. It is also good practice to include which arc (minor or major) the angle is subtended by.

The theorem that the angle subtended at the centre is twice the angle subtended at the circumference by the same arc, will be used in the proof of other theorems. You can accept that it is true when proving other theorems.

Angles in a semicircle

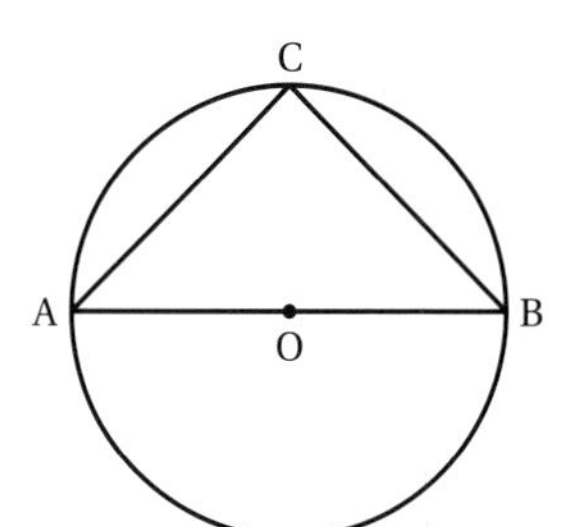

This theorem states that the angle subtended by a diameter at the circumference of a circle is a right angle. This follows from the first theorem.

In this diagram, AB is a diameter, and C is called an angle in a semicircle.

Arc AB subtends an angle of 180° at the centre (a straight line), and it subtends angle C on the circumference.

Using the first theorem, we know that the angle subtended at the centre is twice the angle subtended by the same arc at the circumference.

So angle $C = \frac{1}{2} \times 180° = 90°$

'The angle in a semicircle is always a right angle' is another fact that we can use in other proofs.

Tip

Simply write '**angle in a semicircle**' as a reason in your proofs.

Angles in the same segment

This theorem states that two angles in the same segment are equal. This theorem also follows from the first theorem.

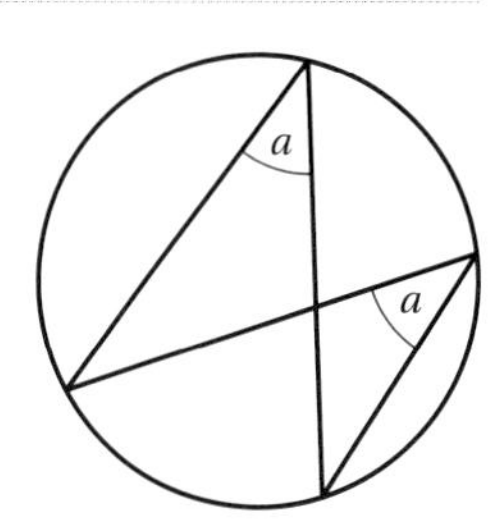

Two angles in the same segment refers to two angles at the circumference that are subtended by the same arc or chord.

WORKED EXAMPLE 2

Use the diagram to prove that two angles in the same segment are equal.

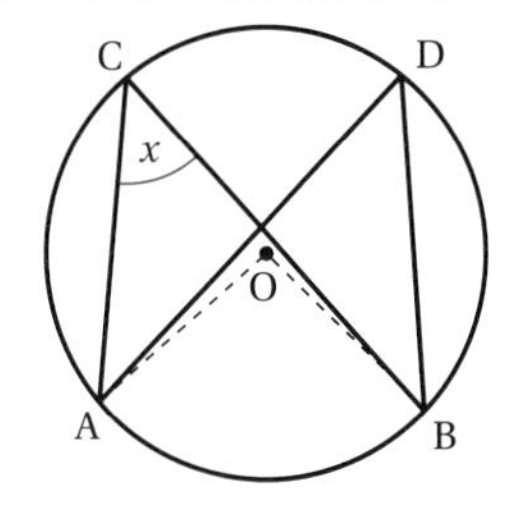

Angles C and D are in the same segment, because they are both subtended at the circumference by arc AB.

We are given that angle $C = x$.

The arc AB also subtends angle AOB at the centre. $AOB = 2x$.

This uses the result of our first theorem. We can state the reason as 'angles subtended at centre and circumference.'

Angle D is equal to x, using the same reason: angles subtended at centre and circumference.

So angle $C = \text{angle } D = x$.

'Two angles in the same segment are equal' is another fact that we can use in other proofs.

Tip

Simply write '**angles in the same segment**' as a reason in your proofs.

EXERCISE 34B

1. Prove that angle COB = 2(angle CAB).

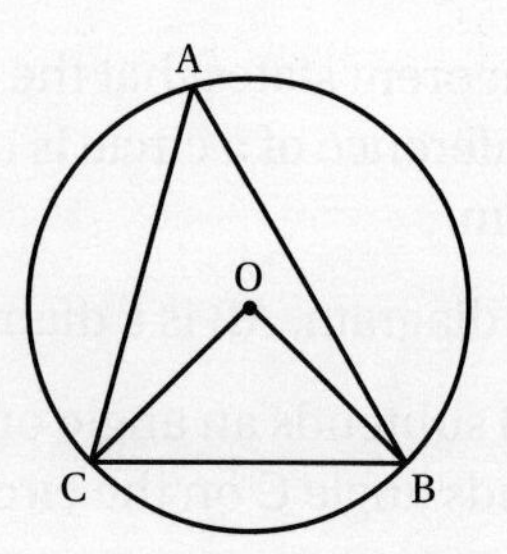

2. Prove that angle XYZ is a right angle.

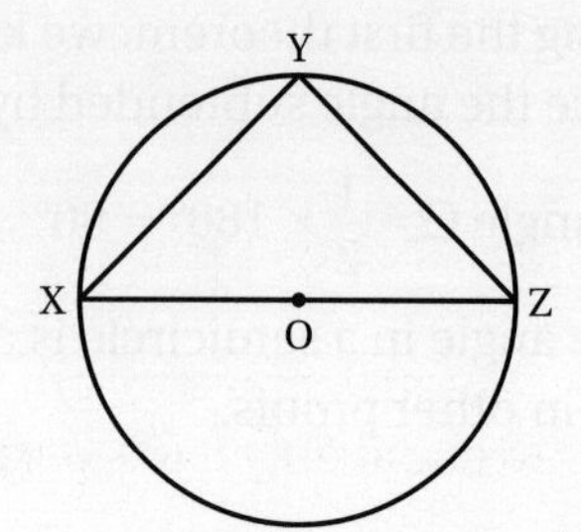

3. Prove that $y = 25°$.

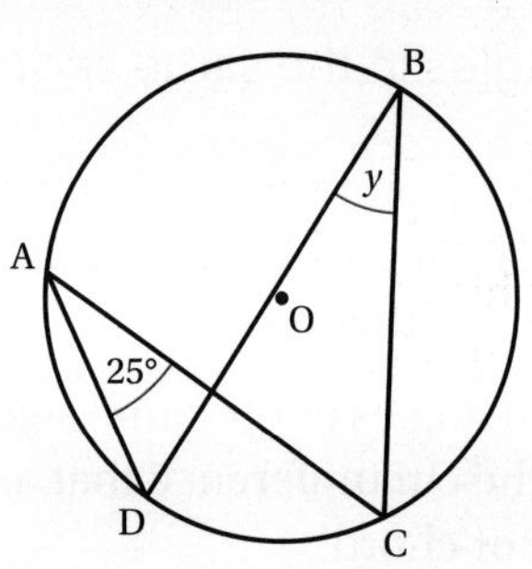

4. What size is the angle marked at POQ?

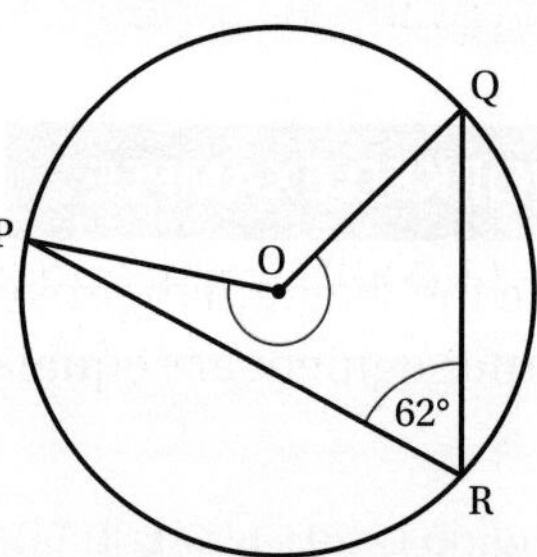

5. What is the size of angle BAC?

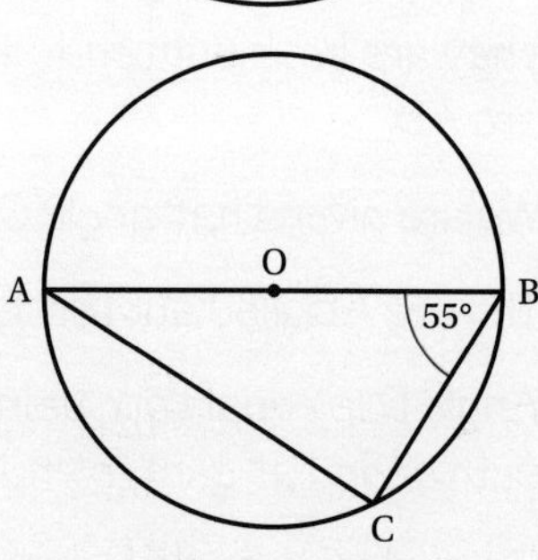

6. What is the size of angle BCD?

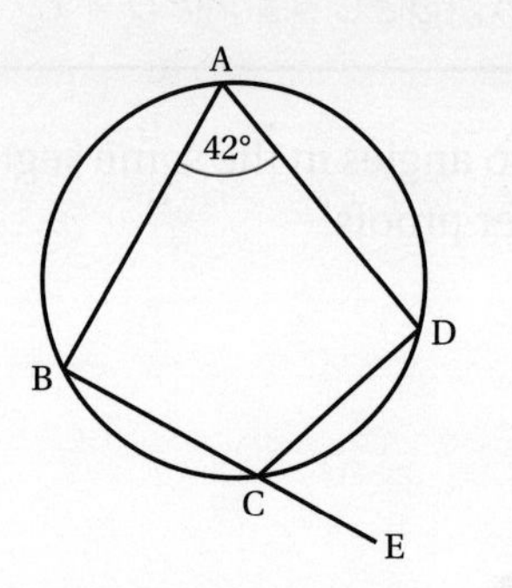

7 The angle CAB is twice the angle at CBA.

What is the size of each angle?

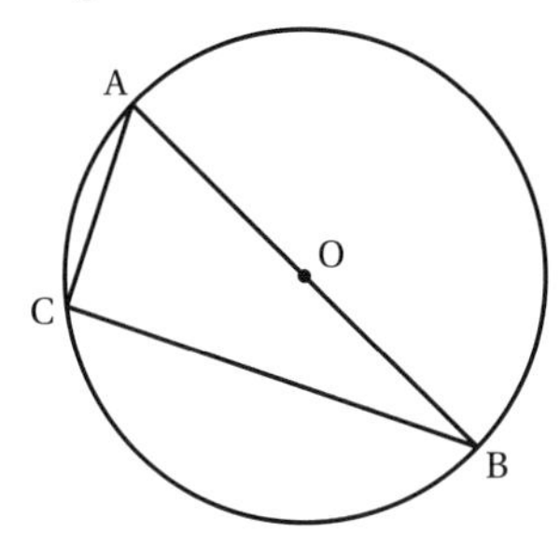

Angle between radius and chord

This theorem states that a radius or diameter bisects a chord if and only if it is perpendicular to the chord.

In other words, the theorem says that if the radius r in the diagram is perpendicular to the chord AC, then it will bisect the chord and AB = BC.

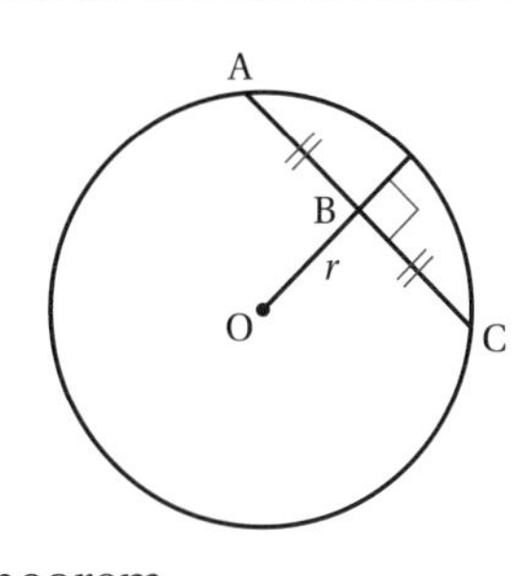

The phrase 'if and only if' means that the radius will not bisect the chord if it is not perpendicular to the chord. So we have two facts that we can use from this theorem.

WORKED EXAMPLE 3

Prove that AM = MB.

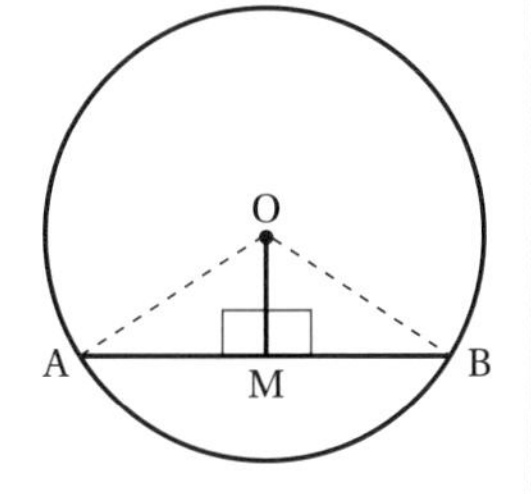

Triangle OMA and triangle OMB are congruent, because:

OA = OB (they are radii of the circle).

They have a common side OM.

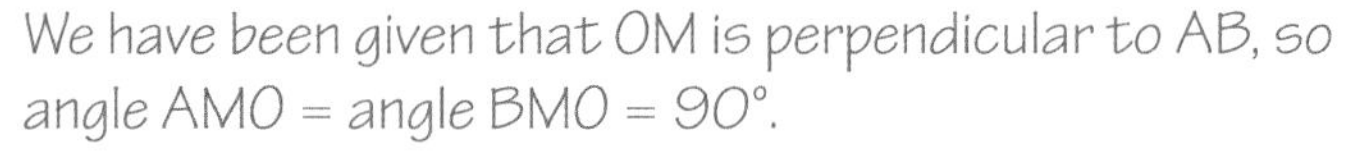

We have been given that OM is perpendicular to AB, so angle AMO = angle BMO = 90°.

Therefore AM = MB (because the triangles are congruent).

Tip

This fact is useful when solving geometry problems. State it in short as '**angle between radius and chord**'.

Angle between the radius and tangent

This theorem states that for a point P on the circumference, the radius through P is perpendicular to the tangent at P.

We know that if the line is a tangent, it touches the circle only once. If it touches the circle more than once, then it is not a tangent.

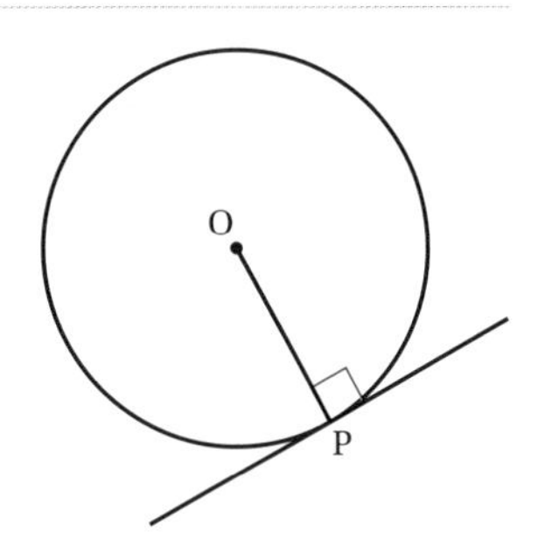

We can see that if the radius did not meet the tangent at a right angle, then the line would not be a tangent.

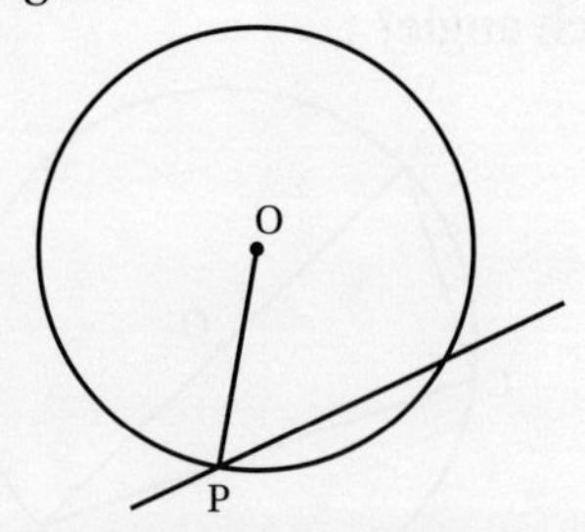

Tip

When you use this theorem in a geometry problem, write '**angle between radius and tangent**' as a reason.

This kind of proof is called a **proof by contradiction**. It shows that the only way that the line can be a tangent is if the angle at the point where it touches the circle is at a right angle to the radius drawn at that point.

Two-tangent theorem

This theorem states that two tangents to a circle, drawn from a common point outside of the circle, are equal in length.

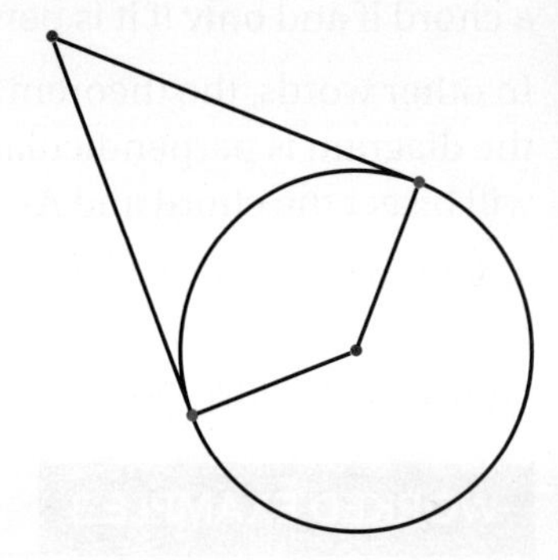

WORKED EXAMPLE 4

Prove that TA = TB.

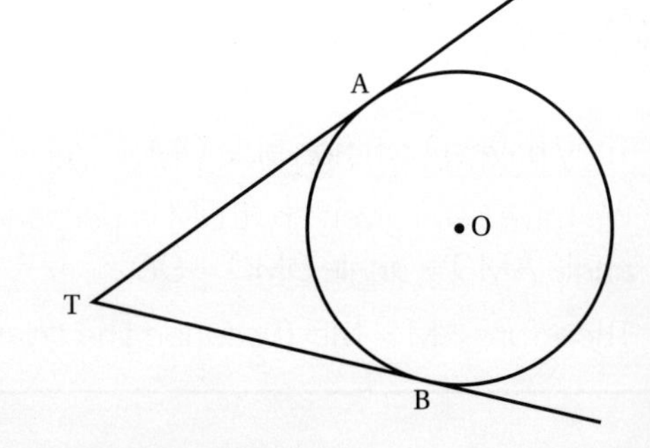

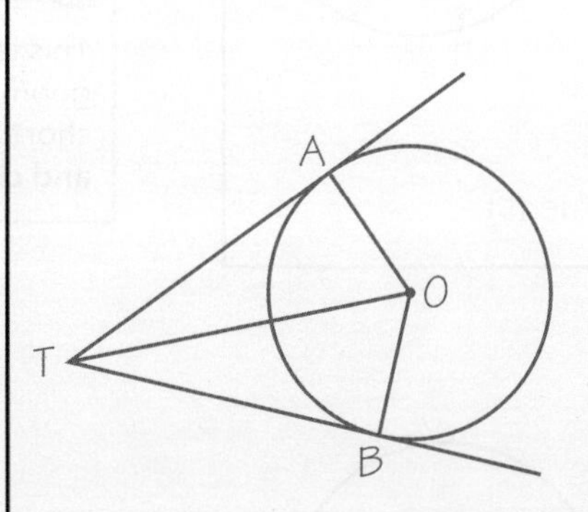

Draw radii OA and OB. Join OT.

Triangle OAT is congruent to triangle OBT for the following reasons:

- Angle TAO = angle TBO = 90° (the tangent is perpendicular to the radius).
- They both have the same side, TO.
- OA and OB are radii, so OA = OB.

So triangle OAT is congruent to triangle OBT and therefore TA = TB.

EXERCISE 34C

1. PQ is a chord of a circle centre O.
   Radius OS ⊥ PQ, and cuts PQ at R.
   Radius OQ = 25 units and PQ = 48 units.

   Calculate the length of SR.

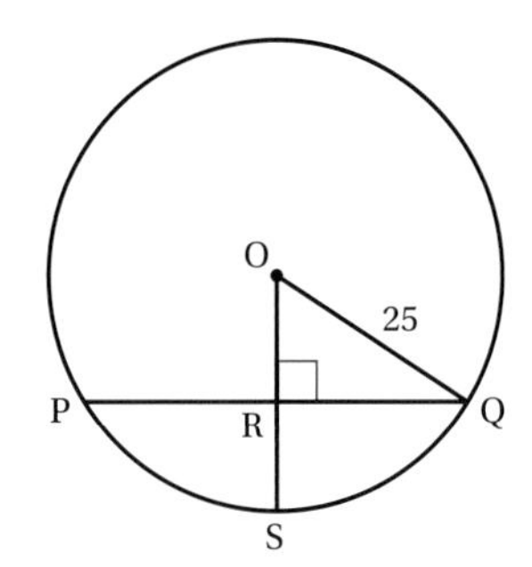

2. The circle on the right has centre O and chord PQ.
   OR ⊥ PQ. OR = 6 units and PQ = 16 units.

   Calculate the length of OQ.

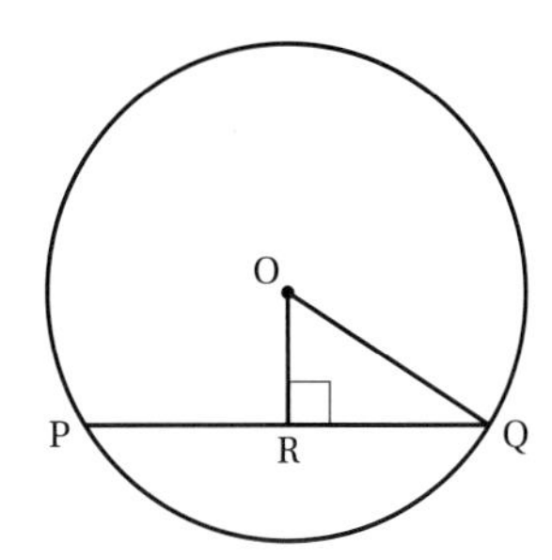

3. Identify which angles are equal.

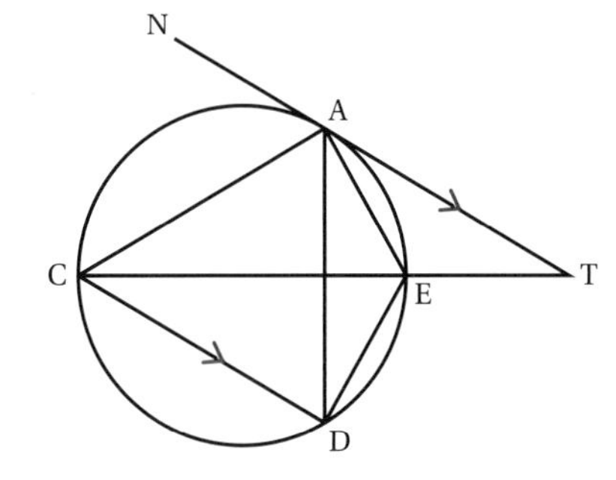

4. The diagram shows a circle with centre O and tangents LM and LP.

   Find the size of the angles marked a and b.

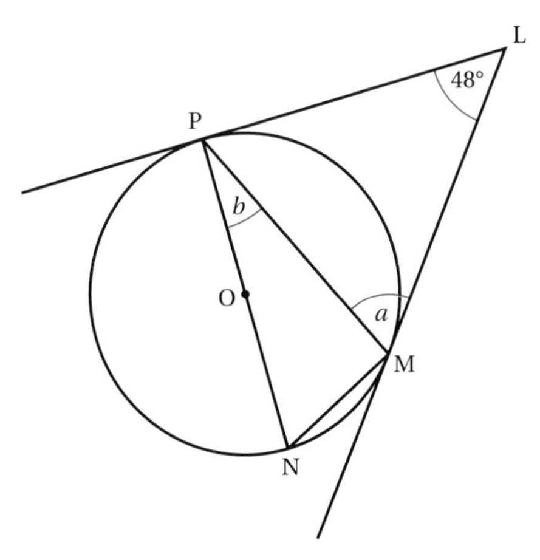

Alternate segment theorem

This theorem states that for a point P on the circumference, the angle between the tangent and a chord through P equals the angle subtended by the chord in the opposite segment.

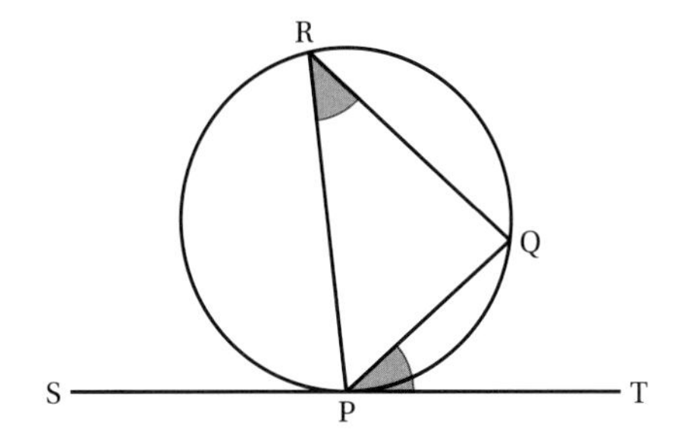

Find answers at: cambridge.org/ukschools/gcsemaths-studentbookanswers

You need to think carefully about this theorem so that you can work out when it is needed to solve a problem.

Consider what the term 'opposite segment' means. Remember from the parts of a circle that a chord divides a circle into a minor segment and a major segment. So we are looking at one angle in a minor segment and one angle in the major segment.

So in the diagram above, the chord PQ creates the angle QPT with the tangent (ST) at the point P. The angle PRQ is inside the major segment, and it is subtended at R on the circumference by the chord PQ.

WORKED EXAMPLE 5

PR is a chord with W on the minor arc and Q on the major arc.

Prove that angle SPR = PQR and also that angle TPR = PWR.

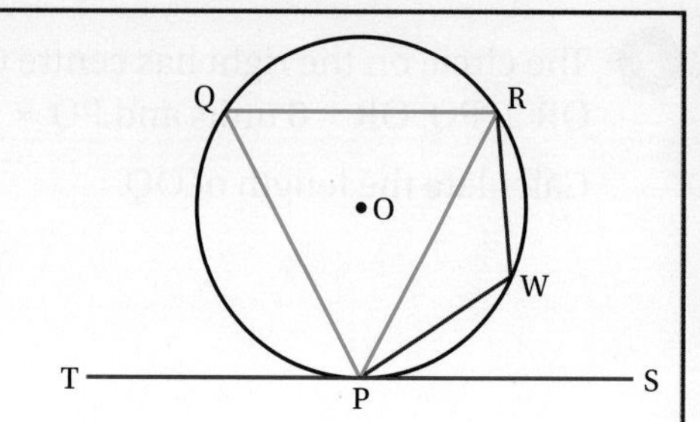

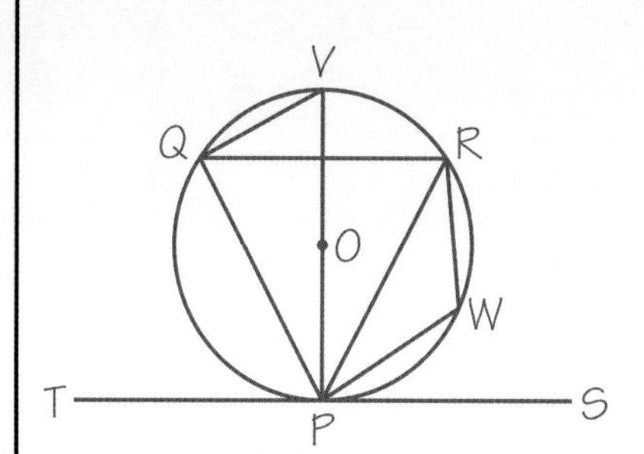

Draw diameter VP. Join points V and Q.

Angle PQV = 90° (VP is the diameter, so angle PQV is the angle in a semicircle)

Angle SPV = 90° (angle between the radius and tangent)

So Angle PQV = angle SPV

Angle RPV = angle RQV (angles in the same segment)

As angle RPV = angle RQV.

Angle SPR = angle SPV − angle RPV
= 90° − angle RPV

Angle PQR = angle PQV − angle RQV
= 90° − angle RQV

So angle SPR = angle PQR.

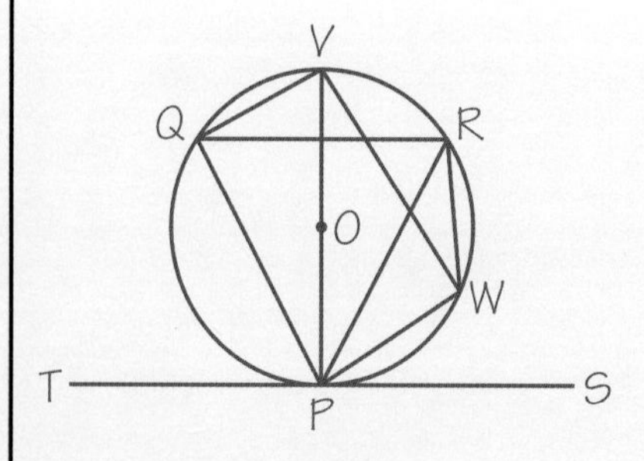

Join line segment VW.

Angle TPV = 90° (angle between radius and tangent)

Angle PWV = 90° (angle in a semicircle)

Now notice that angle RPV = RWV (angles in the same segment subtended by VR)

angle TPV + RPV = angle PWV + RWV

So angle TPR = PWR.

Angles in cyclic quadrilaterals

Key vocabulary

cyclic quadrilateral: any quadrilateral with all four vertices touching the inside of the circumference of a circle.

A **cyclic quadrilateral** is a quadrilateral which has all four vertices touching the inside of the circumference of a circle.

This theorem states that the opposite angles in a cyclic quadrilateral add up to 180°.

WORKED EXAMPLE 6

ABCD is a cyclic quadrilateral.

Prove that the opposite angles in a cyclic quadrilateral add up to 180°.

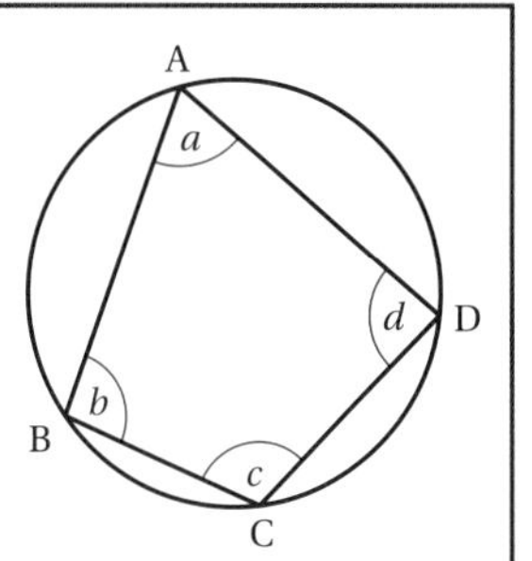

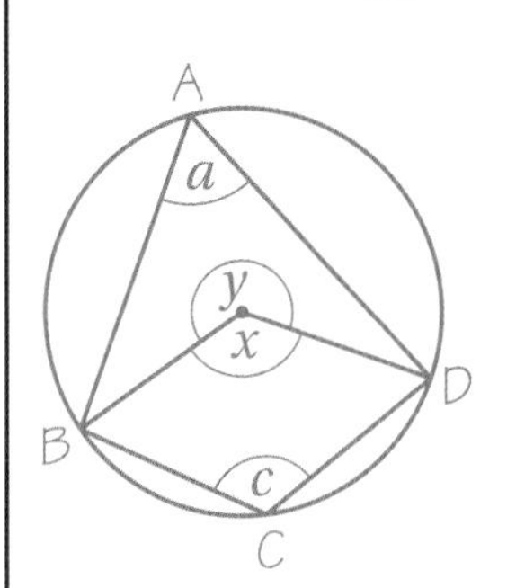

Draw radii BO and DO.

$x = 2a$ (angle at centre theorem, minor arc BD)

$y = 2c$ (angle at centre theorem, major arc BD)

$\therefore x + y = 2a + 2c$

But $x + y = 360°$ (angles at a point add to 360°)

$\therefore 2a + 2c = 360°$

$\therefore a + c = 180°$

Tip

ABCO is not a cyclic quadrilateral.

For a quadrilateral to be a cyclic quadrilateral all **four** vertices must touch the circumference. Make sure you check this carefully when you are applying the theorems.

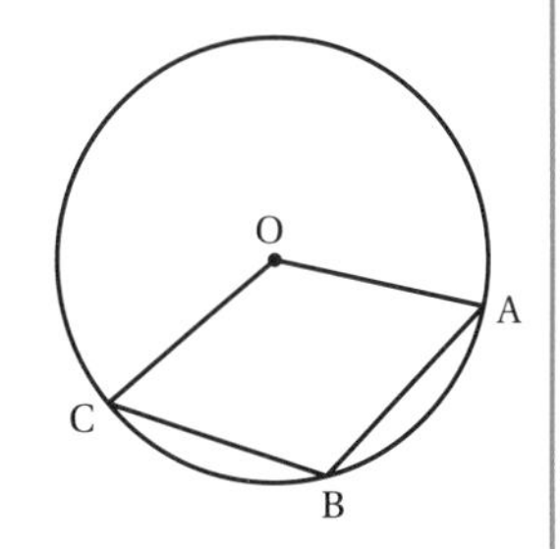

EXERCISE 34D

1 TAN is a tangent to the circle, angle TAC = angle BAN.

Prove that:

a CB is parallel to TN.

b AC = AB.

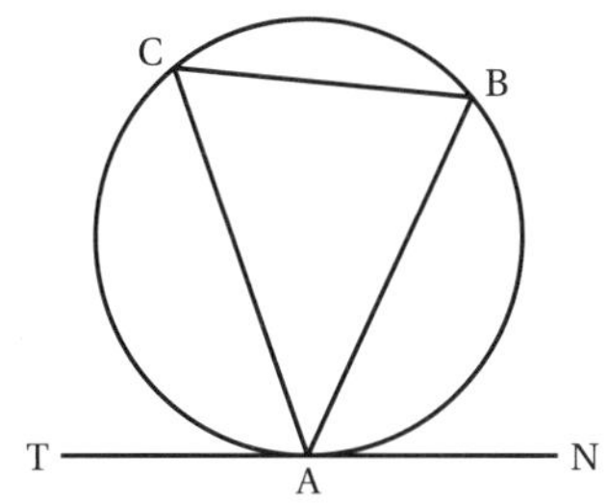

2 APB is a tangent at P to circle centre S. Angle APR = x.

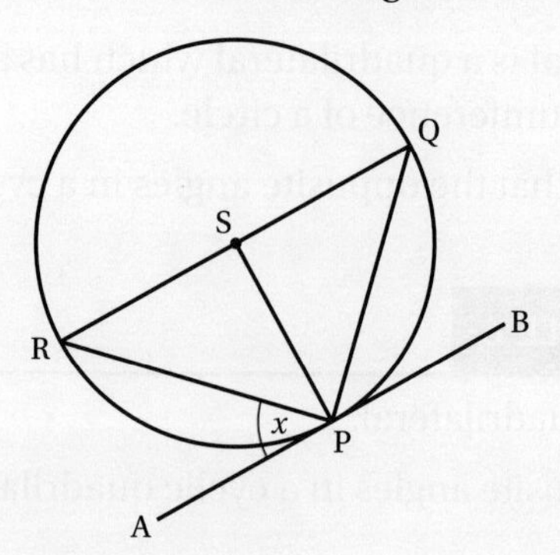

Write the following angles in terms of x:

a angle Q.

b angle QSP.

c angle RSP.

d angle RPS.

e angle QPB.

3 In each of the following examples, calculate the values of the variables. Give reasons for your statements.

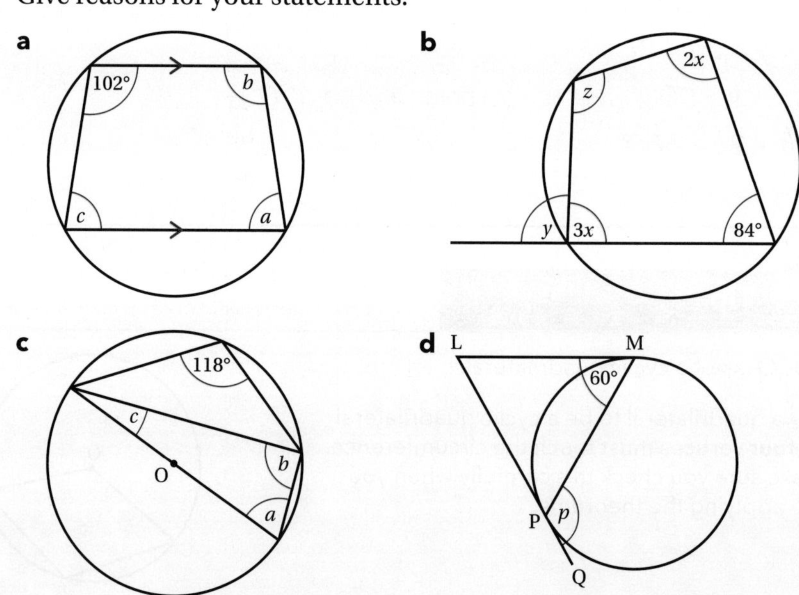

Section 3: Applications of circle theorems

To solve problems in geometry, you need to apply your knowledge correctly. You will often not see the solution immediately, but you can start to identify a pathway towards the final solution by building on the information step by step. It is useful to follow the steps provided by the problem-solving framework.

Problem-solving framework

Look at the diagram below. AOB and COD are diameters of circle with centre O. Angle ACD = 33°.

Calculate the sizes of:

a angle AOD.

b angle D.

c angle B.

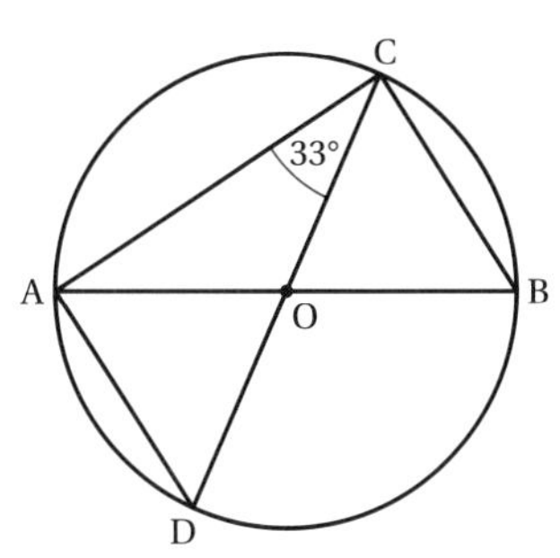

Steps for approaching a problem-solving question	What you would do for this example
Step 1: What do you have to do?	Calculate the size of various angles using one known angle and circle theorems.
Step 2: If a diagram is useful, draw one.	You have been given a diagram but it would be useful to mark on any additional information that you can work out. Angles DAC and ACB are right angles (angles in a semicircle). Lines AB = CD (both are diameters of the circle). Label angle OCB as x. Then, using angles in the same segment or isosceles triangles formed by radii, label other angles that also equal x. Angles AOD and COB = $(180° - x)$ because angles in a triangle sum to 180°, also vertically opposite angles are equal. Angle AOC = $2x$ (angles on a straight line sum to 180°). Same applies for angle BOC (also, vertically opposite angles are equal).
Step 3: What maths can you do?	Use this information to decide what method you will use. This may not be obvious immediately. You may need to spend some time studying the problem. Write out all your working and include reasoning. Even if you've already applied this reasoning to mark up your diagram, you still need to state it as part of your working. You can use the theorems as accepted facts. You can also use facts that you have proved. **a** angle ACB = 90° (angle in a semicircle) angle OCB = 90° − 33° = 57° angle DAO = angle OCB = 57° (angles in the same segment) angle ODA = angle DAO = 57° (triangle OAD is isosceles) angle AOD = 180° − 2(57°) = 66° (sum of angles in a triangle) **b** angle D = 57° (already proved in part a) **c** angle B = angle OCB = 57° (triangle OBC is isosceles)
Step 4: Have you answered the question? Do your answers seem reasonable? Check your answers.	Answers seem reasonable given the angle of 33° supplied. Checked all calculations. ✓

WORK IT OUT 34.1

Rose needs to solve the following problem.

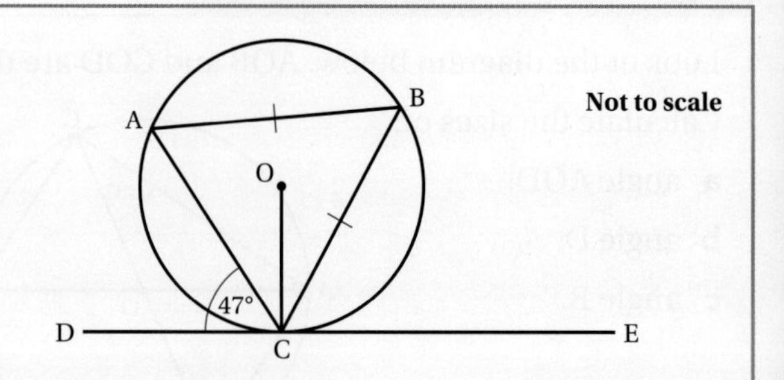

A, B and C are points on the circumference of a circle, centre O.

The line DCE is the tangent to the circle at C.

AB = BC

Angle DCA = 47°.

Calculate the size of angle OCB.

Give a reason for each stage in your working.

Which of her answers below is the correct answer?

Where have her other answers gone wrong?

Option A	Option B	Option C
Angle ABC = 47° (angles in alternate segments)	Angle ABC = 47° (angles in alternate segments)	Angle ABC = 47° (angles in alternate segments)
Angle OCB = angle ABC (radii of circle are equal, isosceles triangle)	Triangle BAC is isosceles (AB = BC) So angle BAC = angle BCA	Triangle BAC is isosceles (AB = BC)
Angle OCB = 58°	Angle BCA = 180° − 47° = 133° ÷ 2 = 66.5°	Angle BAC = angle OCB (triangle BAC is isosceles)
	Angle BAC = 66.5° so angle BCE = 66.5° (alternate segment theorem)	Angle OCB = $\frac{1}{2} \times (180° - 58°)$ (the sum of angles in a triangle is 180°)
	Angle OCE = 90° (angle between the radius and a tangent)	Angle OCB = $\frac{1}{2} \times 122°$
	Angle OCB = 90° − 66.5° = 23.5°	Angle OCB = 61°

EXERCISE 34E

1. AB is a chord of a circle centre P, with D the midpoint of AB. PA = 50 units and AB = 96 units.

   Calculate the lengths of:

   a PD

   b DE

   c DF

   d BF

   e BE

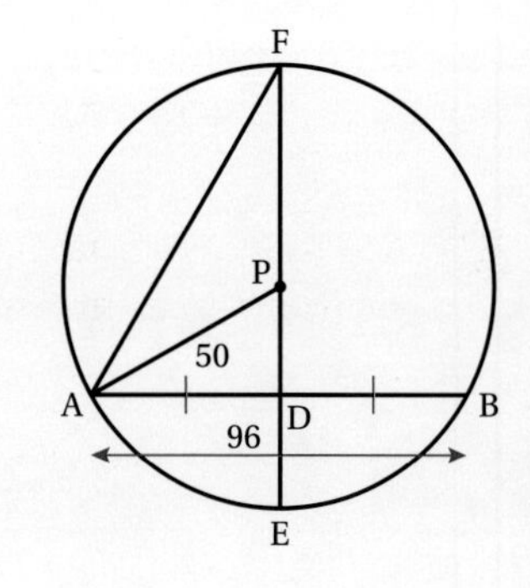

2 ABC is a tangent at B to a circle with centre O. FOB is a diameter.

GF // BD, EH = HD and angle ABG = 62°.

Write down the sizes of the following angles. Give reasons in each case.

a angle EHO **b** angle GFB

c angle GBF **d** angle FEG

e angle DBF **f** angle GEH

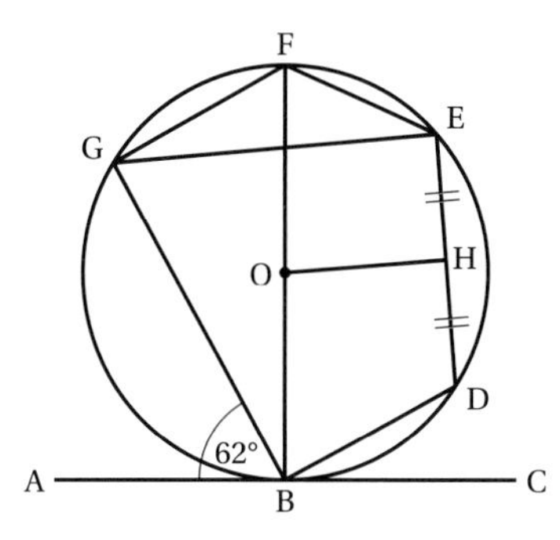

3 Find the size of angle PLM in terms of x.

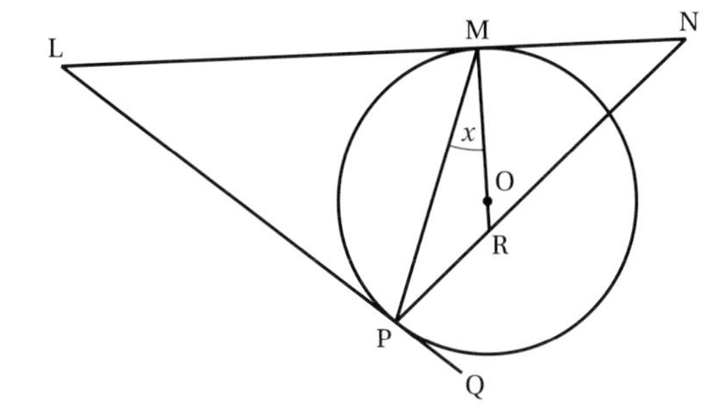

4 In the diagram, TA and TB are the tangents from T to the circle with centre O. AC is a diameter of the circle and ACB = x.

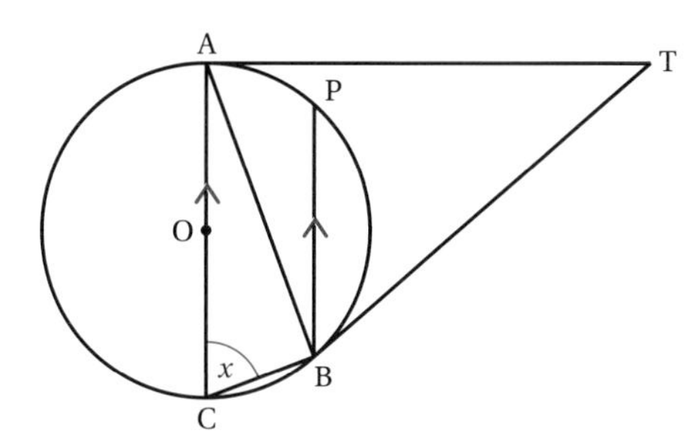

a Find CAB in terms of x.

b Find ATB in terms of x.

c The point P on the circumference of the circle is such that BP is parallel to CA. Express PBT in terms of x.

5 Calculate the diameter of each circle. Give your reasons and give your answers to 3 significant figures.

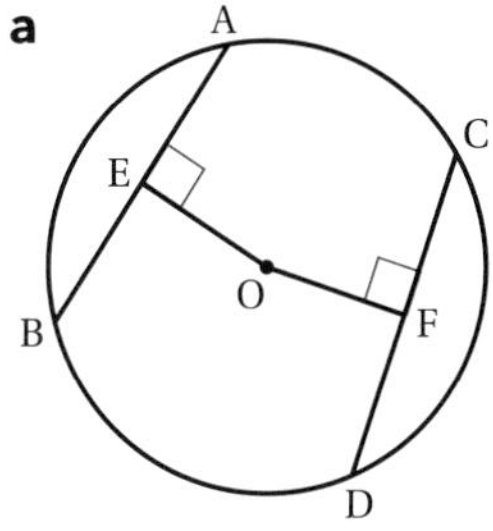

AB = CD = 11.4 cm
OF = 6.5 cm

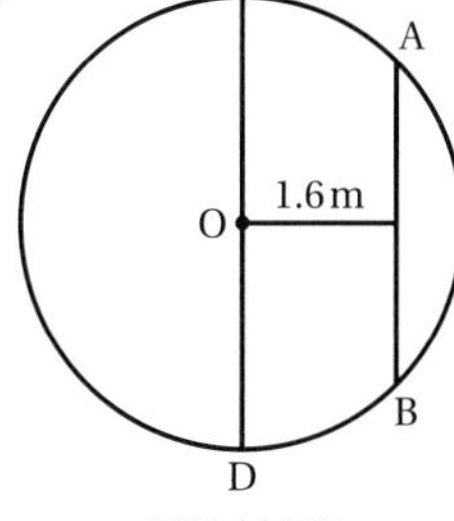

CD//AB
AB = 2.8 m

c

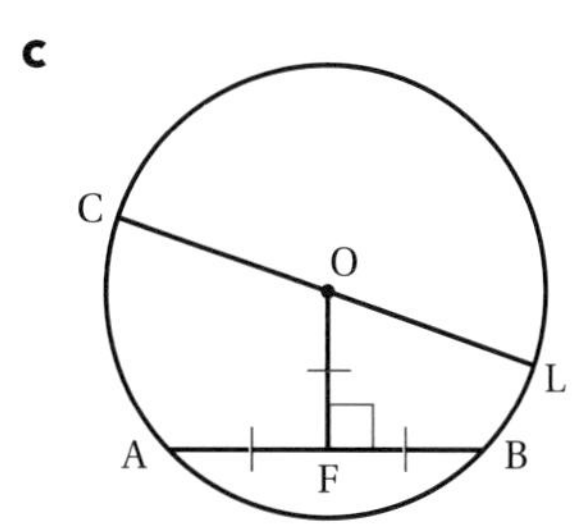

AB = 22 mm

Find answers at: cambridge.org/ukschools/gcsemaths-studentbookanswers

6 In the diagram, SAT is the tangent to the circle at point A. The points B and C lie on the circle and O is the centre.

If angle ACB = x, express, in terms of x, the size of:

a angle AOB. **b** angle OAB. **c** angle BAT.

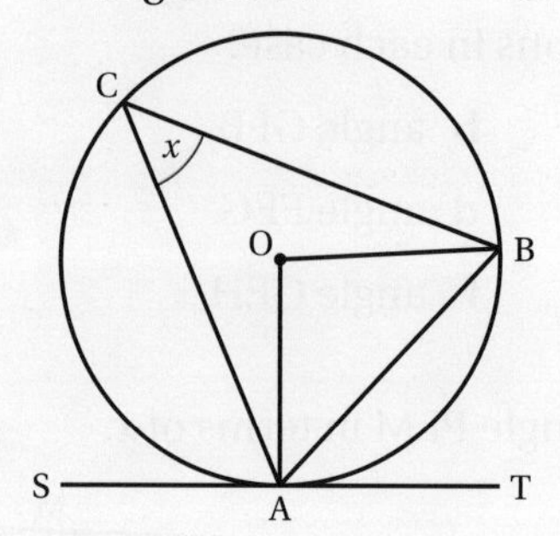

7 Find the size of each lettered angle in these sketches.

a

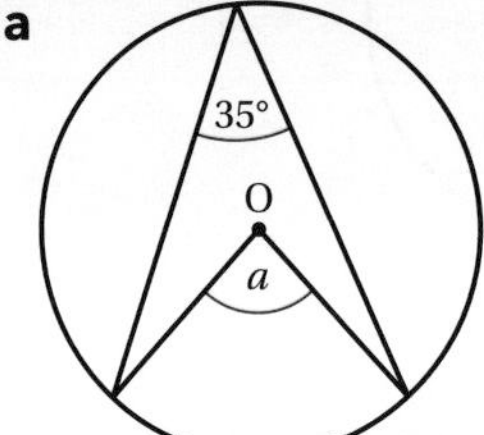

b

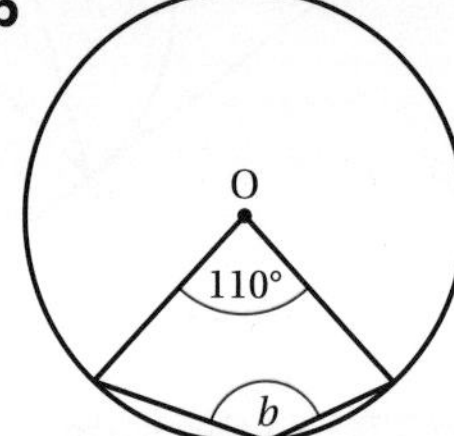

c

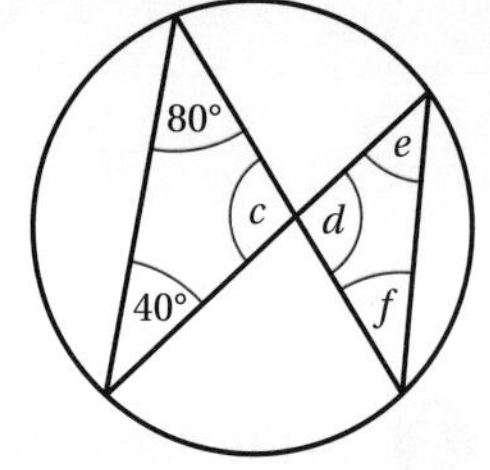

Checklist of learning and understanding

Circle theorems

- Angles at the centre theorem: the angle subtended by an arc at the centre is twice the angle subtended at the circumference.
- Angles in a semicircle: the angle on the circumference subtended by a diameter is a right angle.
- Angles in the same segment: two angles in the same segment are equal.
- Angle between radius and chord: a radius or diameter bisects a chord if and only if it is perpendicular to the chord.
- Angle between radius and tangent: for a point P on the circumference, the radius or diameter through P is perpendicular to the tangent at P.
- Two tangent theorem: two tangents from a given point outside of the circle are equal in length.
- Alternate segment theorem: for a point P on the circumference, the angle between the tangent and a chord through P equals the angle subtended by the chord in the opposite segment.
- Angles in cyclic quadrilaterals: the opposite angles of a cyclic quadrilateral are supplementary.

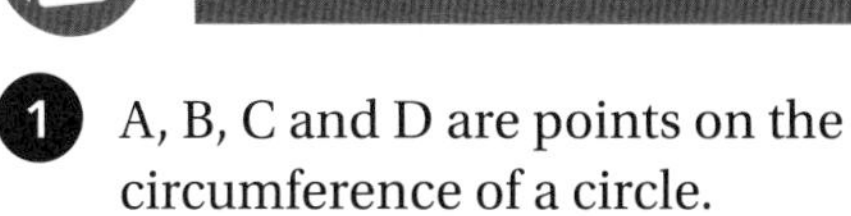

Chapter review

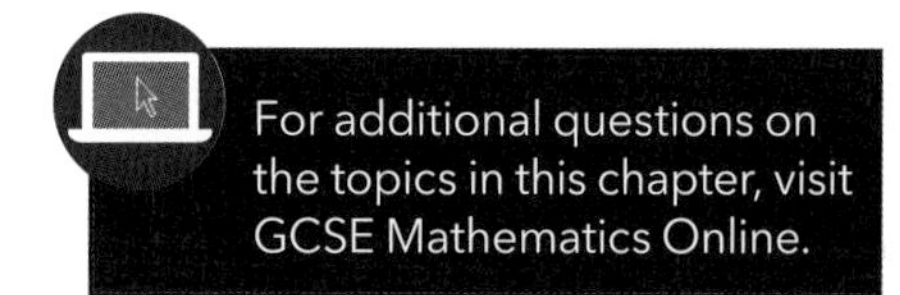

1 A, B, C and D are points on the circumference of a circle.
Angle ABD = 54°. Angle BAC = 28°.

Find the size of angle ACD.
Give a reason for your answer.

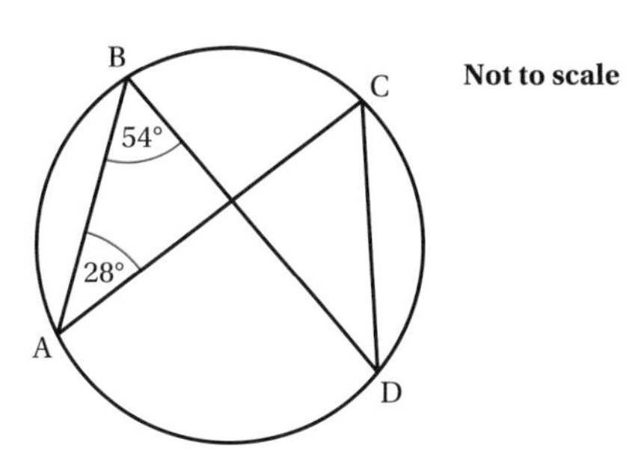

2 In the diagram, A, B, C and D are points on the circumference of a circle, centre O. Angle BAD = 70°. Angle BOD = x°. Angle BCD = y°.

a Work out the value of x. Give a reason for your answer.

b Work out the value of y.

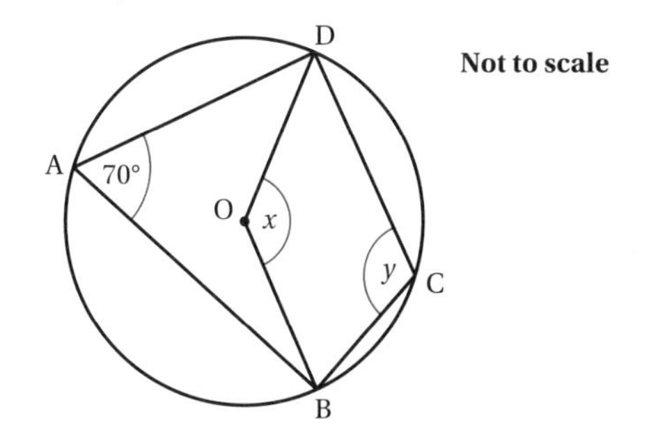

3 In the diagram B, C, D and E are points on the circumference of a circle.

AT is the tangent to the circle at B.

Angle BCE = 48° and angle BEC = 54°.

a Find angle a.
Give a reason for your answer. *(2 marks)*

b Calculate angle b.
Give a reason for each step of your working. *(3 marks)*

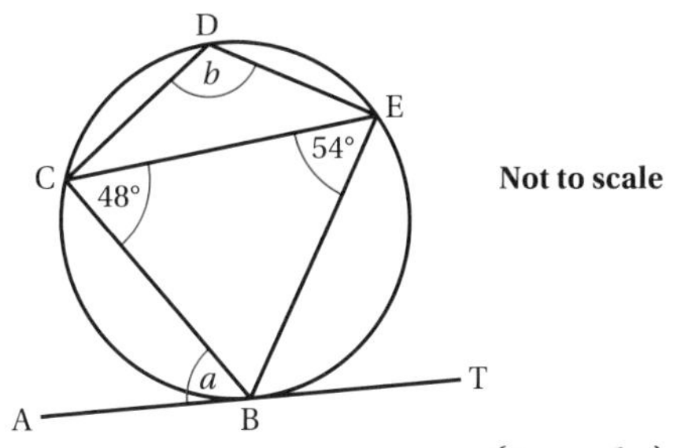

35 Discrete growth and decay

In this chapter you will learn how to ...

- set up and solve problems involving growth and decay, including simple and compound interest.
- express exponential growth or decay as a formula.

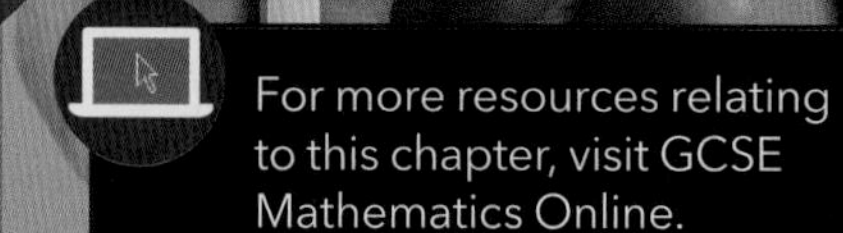

For more resources relating to this chapter, visit GCSE Mathematics Online.

Using mathematics: real-life applications

Many real-life situations involve growth (increase) or decay (decrease) as time passes. Population numbers, growth of bacteria, disease infection rates, world temperature patterns and the value of money or possessions may all increase or decrease over time.

"My computer program calculates interest on a daily basis. This means whatever is in the account gains interest, not just the initial investment." *(Investment broker)*

Before you start ...

Ch 11, 13	You must be able to convert percentages to decimals.	**1** Write each of the following as a decimal: **a** 5% **b** 190% **c** 0.4% **d** 12.5%
Ch 10	You need to be able to increase or decrease a quantity by a given percentage by multiplying by a suitable decimal.	**2** Carry out the following increases and decreases using only multiplication: **a** Increase \$44 by 22%. **b** Increase £35 by 5.5%. **c** Decrease £13 by 44%. **d** Decrease \$170 by 8%.
Ch 19	You should remember how to plot and interpret functions in the form of $y = ab^x$.	**3** The graph shows the growth rate of bacteria in cheese. **a** How many bacteria were there to start with? **b** What happens to the number of bacteria each hour? 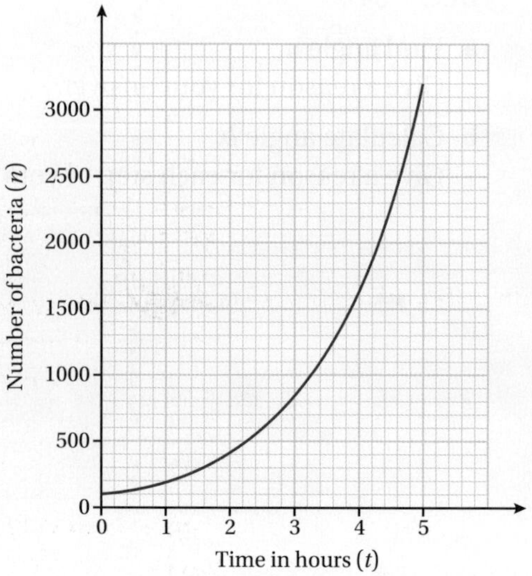 (Graph: Number of bacteria (n) 0, 500, 1000, 1500, 2000, 2500, 3000; Time in hours (t) 0, 1, 2, 3, 4, 5) **4** The function for this graph is $n = 100(2)^t$, where n is the number of bacteria and t is the time period in hours. **a** Where does the constant value of 100 come from? **b** Why is $b = 2$?

Assess your starting point using the Launchpad

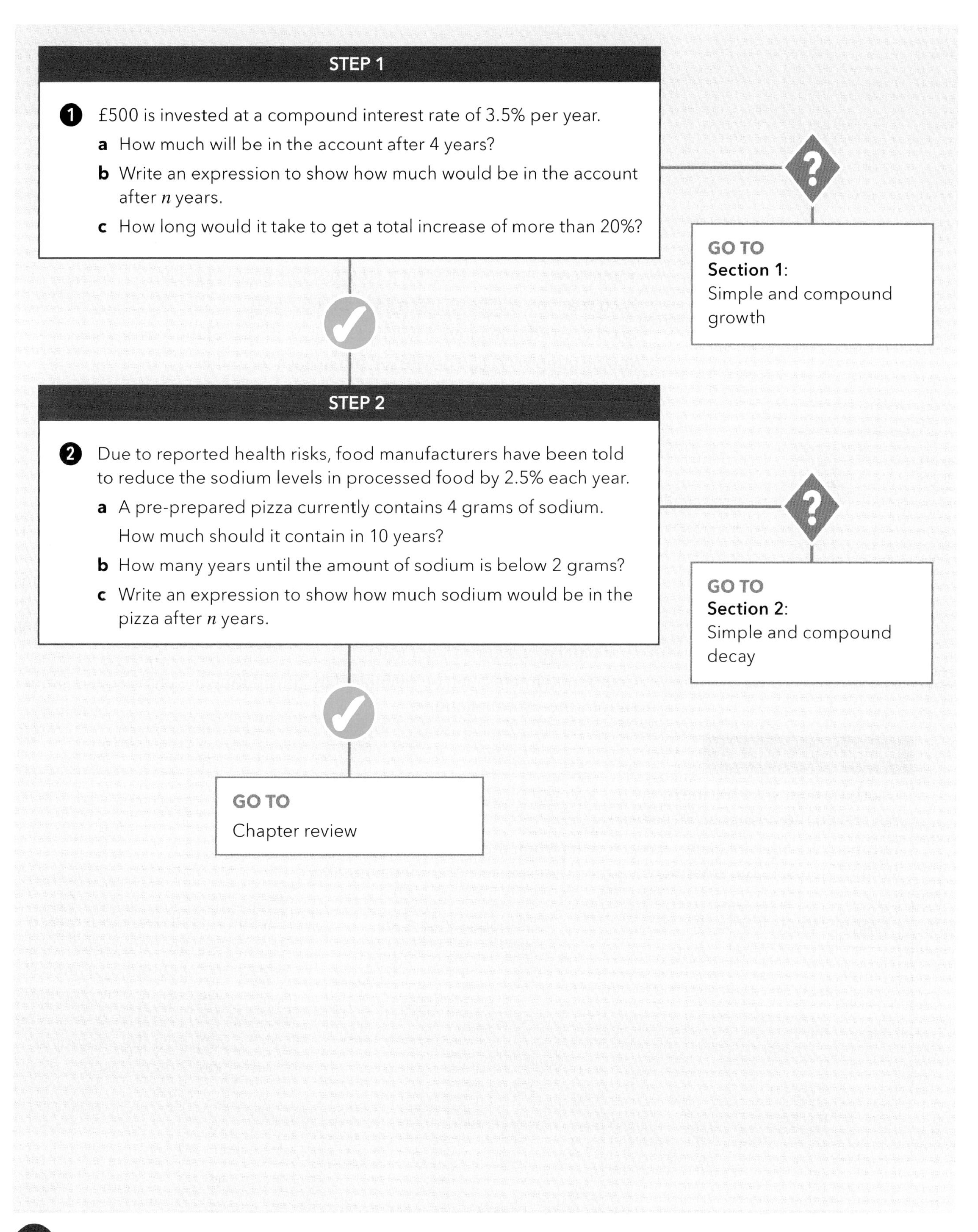

STEP 1

1. £500 is invested at a compound interest rate of 3.5% per year.
   - **a** How much will be in the account after 4 years?
   - **b** Write an expression to show how much would be in the account after n years.
   - **c** How long would it take to get a total increase of more than 20%?

GO TO
Section 1:
Simple and compound growth

STEP 2

2. Due to reported health risks, food manufacturers have been told to reduce the sodium levels in processed food by 2.5% each year.
   - **a** A pre-prepared pizza currently contains 4 grams of sodium. How much should it contain in 10 years?
   - **b** How many years until the amount of sodium is below 2 grams?
   - **c** Write an expression to show how much sodium would be in the pizza after n years.

GO TO
Section 2:
Simple and compound decay

GO TO
Chapter review

Section 1: Simple and compound growth

Simple interest

When you borrow money or buy things on credit you are normally charged inte
for the use of the money. When you invest or save money, you are paid interest
the bank or other institution in return for leaving your money with them.

The amount of money you invest or borrow is called the principal.

Simple interest is interest paid on the original principal. The same interest is pai
for each time period.

For example:

Suppose you borrow £1000 at an interest rate of 5% per year.

Each year you will be charged 5% × £1000 = £50.

In other words the interest will be £50 for each year of the loan period.

Simple interest (I) can be worked out using a formula:

$$I = PRT$$

Where P is the principal amount,

R is the rate of interest as a percentage,

T is the time period over which the interest is calculated.

Compound interest

When using compound interest, we calculate the interest on the principal amou
plus any interest that has been added. So using our previous example, after the
first loan period, the bank would work out the interest you owe based on £1050
(principal plus interest), not £1000.

Compound interest can be calculated by considering the increase as a series of
simple interest calculations.

WORKED EXAMPLE 1

Charlotte deposits £500 into a savings account for six years. The bank pays interest on the savings at 5% per year.

Compare the amount each year for 6 years when the interest is calculated using simple interest to the amount when the interest is compounded annually.

Year	Simple interest	Compound interest
1	£500 + 5% = £525	£500 + 5% = £525
2	£525 + (5% of £500) = £550	£525 + 5% = £551.25
3	£575	£578.81
4	£600	£607.75...
5	£625	£638.14...
6	£650	£670.04...

For simple interest, £25 is added every year; the interest stays constant. For compound interest, £25 is added in the first year, £26.25 (5% of £525) is added in the second year ... each year the amount of interest earned increases.

Charlotte would get a greater return on her investment if she used an account that offered compound interest.

WORK IT OUT 35.1

The population of Europe is growing at a rate of 0.2% per year. The current population is 739 million. What will the population be in 3 years' time?

Three students attempted the question. In pairs, decide who has got the correct answer. Also decide which you think is the most efficient method to find the answer.

Kayleigh	Tom	Zac
Find 0.2%: 0.2% of 739 000 000 $= 0.002 \times 739\,000\,000$ $= 1\,478\,000$ The same growth for 3 years: $3 \times 1\,478\,000 = 4\,434\,000$ Add it on: $739\,000\,000 + 4\,434\,000$ $= 743\,434\,000$	Year 1: Find 0.2% of 739 000 000 $= 0.002 \times 739\,000\,000$ $= 1\,478\,000$ Add it on: 740 478 000 Year 2: Find 0.2% of 740 478 000 $= 0.002 \times 740\,478\,000$ $= 1\,480\,956$ Add it on: 741 958 956 Year 3: Find 0.2% and add it on $1\,483\,918 + 741\,958\,956$ $= 743\,442\,874$	Increase by 0.2% means there is 100.2%, do this three times in a row. $739\,000\,000 \times 1.002 \times 1.002 \times 1.002$ $= 739\,000\,000 \times 1.002^3$ $= 743\,442\,874$

Working with compound interest and growth rates is very much like working with function machines. Each time an output is produced it goes back to becoming an input and the process is repeated. This keeps going for the allotted period of time. This kind of process is called iterative – it iterates or repeats.

Tip

Always show your working for these questions. Write down what you type into the calculator so you can check it and so that your teacher and the examiner can see how you have calculated your answer.

For example, let's suppose that the population of starlings in a park increases each year at a rate of 10%. To increase a quantity by 10% we multiply by 1.1 (100% + 10% = 110%, or 1.1).

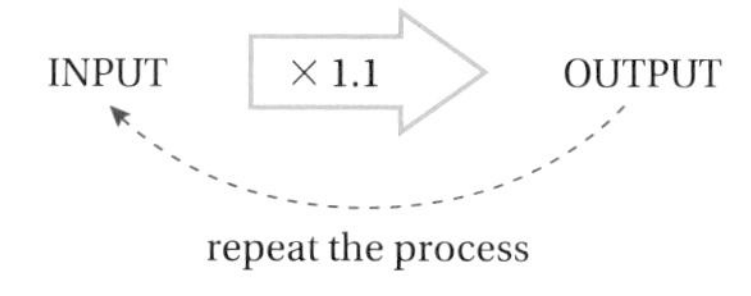

If the population started with 80 starlings and you wanted to predict how many there would be in five years' time, the first input would be 80.

$80 \times 1.1 = 88$

$88 \times 1.1 = 96.8$ (notice that if our model was to stop here we would round sensibly; however, we use the unrounded value to ensure the following year's prediction is more accurate)

$96.8 \times 1.1 = 106.48$

$106.48 \times 1.1 = 117.128$

$117.128 \times 1.1 = 128.8408$, so we would predict that there will be 129 starlings in five years' time.

This could be re-written as

$80 \times 1.1 \times 1.1 \times 1.1 \times 1.1 \times 1.1$ (you can simplify this to 80×1.1^5).

We can generalise this for n years

$$80 \times 1.1^n$$

For any model where we have an initial quantity or investment, q, a percentage growth (or interest rate) in decimal form, p, and the number of years (or period of time), n, we can represent compound interest as:

$$q \times (1 + p)^n$$

Exponential growth

When a quantity increases (grows) in a fixed proportion (normally a percentage) at regular intervals the growth is said to be **exponential**.

You should remember from Chapter 19 that increasing exponential functions produce curved graphs that slope steeply up to the right. The general function for these graphs is $y = ka^x$, where $k > 0$ and $a > 1$.

This function can be used to express exponential growth as a formula:

$$y = a(1 + r)^n$$

where a is the principal (original value),

r is the growth rate (expressed as a decimal)

and n is the number of time periods.

WORKED EXAMPLE 2

£100 is invested subject to compound interest at a rate of 8% per annum. Find the value of the investment after a period of 15 years.

Value $= a(1 + r)^n$

$= 100(1.08)^{15}$

$= 317.2169114$

Use the formula for exponential growth/compound interest, and substitute in the appropriate values.

Value of investment is £317.22 (correct to the nearest penny).

EXERCISE 35A

1 £300 is invested for three years with interest compounded annually at a rate of 2%. How much is in the account after:

a 1 year? **b** 3 years? **c** 8 years? **d** n years?

2 Copy and complete this table. The interest is compounded annually.

Investment	Interest rate	1 year	2 years	$5\frac{1}{2}$ years	n years
£250	2%				
£1500	4.5%				
	3%	£51.50			

3 £1000 is invested subject to compound interest at a rate of 3% per annum.

Plot a graph showing how much money is in the account over the first 10 years of the investment.

4 A colony of bacteria grows by 4% every hour. At first the colony has 100 bacteria.

a How many bacteria will there be after 24 hours if the colony grows at this rate?

b Write a formula to calculate the number of bacteria after n hours.

5 The population of Ireland is growing at an annual rate of 1.7%. In 2014 the population was 4.6 million.

a If this growth rate remains constant, what will the population of Ireland be in 2024?

b How many more people is this?

c Using this model, determine how many people lived in Ireland in 2012? Comment on the likely validity of your answers.

6 The Bank of England's target inflation rate is 2%. This tells you how much the cost of living, food, fuel and rent is likely to go up each year. In 1998 a month's rent was assumed to be £450.

a Assuming that the target is met, how much is rent likely to be 20 years later in 2018?

b What will rent be in n years' time?

7 Gavin is saving for a new bike. The model he wants costs £255. So far, he has saved £200. Gavin's dad has offered to pay him 8% interest for every month that Gavin does his homework, cleans his room and loads the dishwasher.

How long is he going to have to wait for the bike? Show clear working to explain your answer.

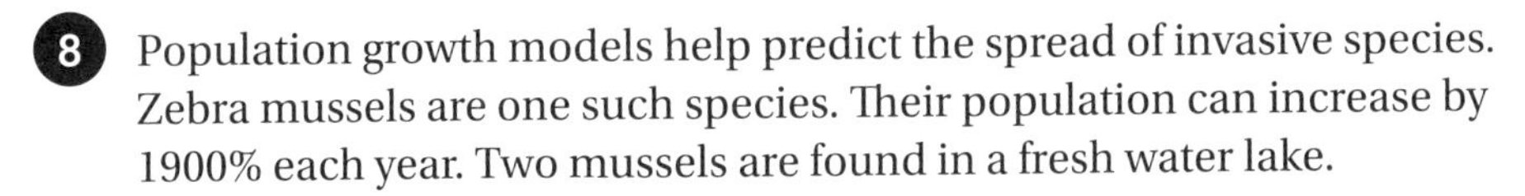

8 Population growth models help predict the spread of invasive species. Zebra mussels are one such species. Their population can increase by 1900% each year. Two mussels are found in a fresh water lake.

Should biologists be worried that this will have a significant impact over the next 10 years? Give details to explain your response.

9 Two investors are having an argument. They want to maximise their profit. They are investing for five years and have a choice. They can either have 6% simple interest or 5.5% compound.

a Which should they choose?

b Would their answer change if they were investing for four years?

10 Copy and complete this table.

Investment	Rate	1 year	2 years	3 years	*n* years
					$\$600 \times 1.015^n$
£500		£530			
\$6000			\$7260		
£750				£1296	

11 \$100 000 is invested at a rate of 5% for 10 years.

a How much more money is earned using compound interest compared with simple interest?

b What simple interest rate would be needed to achieve the same earnings?

12 House prices are rising. A two-bedroomed house cost £195 000 last year and now costs £216 450.

If the price keeps rising at the same rate, how much will this house cost in three years' time?

13 Which of the following investment models gives the highest earnings?

Model 1	Model 2	Model 3
Year 1: 5% interest	Years 1–3	Years 1–3
Year 2: 4% interest	4% compound interest	3.9% simple interest
Year 3: 3% interest		

14 A colony of bacteria grows by 5% every hour.

How long does it take for the colony to double in size?

Section 2: Simple and compound decay

Key vocabulary

depreciation: the loss in value of an object over a period of time.

When the value of something goes down we say it has depreciated. For example, a brand new car will show a **depreciation** in value of about 30% in the first year of ownership alone.

WORKED EXAMPLE 3

The value of a new computer depreciates by 30% per year. If it cost £1200 new, what will it be worth in two years' time?

Method 1

Value after 1 year = £1200 − (30% of £1200)
= £1200 − £360
= £840

Value after 2 years = £840 − (30% of £840)
= £840 − £252
= £588

Each year the value decreases by 30%. Work out the new value after one year by subtracting 30% of the cost when new. To calculate the value after two years, subtract 30% of the value after one year from the value after one year, and so on, working one year at a time.

Method 2:

Value after 1 year = 70% of £1200 = £840
Value after 2 years = 70% of £840 = £588

Decreasing an amount by 30% is the same as finding 70% of the amount.

When the number of individuals in a population declines over time, it is called decay rather than depreciation.

For example, if the population of squirrels is in decay, it means that each year there are fewer and fewer animals in the population. If the rate of decline is 10%, each year 10% of the squirrels disappear, leaving 90%. So from one year to the next the number of animals is $n \times 0.9$, where n is the total number of squirrels in the previous year.

WORK IT OUT 35.2

For every 1000 m you climb, the atmospheric pressure decreases by 12%. If the atmospheric pressure at sea level is 100 300 pascal (Pa), what would the pressure be for a skydiver at an altitude of 4000 metres?

Three students attempted the question. In pairs, decide who has got the correct answer, and also the most efficient method to find the answer.

Bob	Jenny	Ethan
12% of 100 300 $= 0.12 \times 100\,300$ $= 12\,036$ $4 \times 12\,036 = 48\,144$ $100\,300 - 48\,144$ $= 52\,156$ Pa	Decrease by 12% leaves 88% 88% of 100 300 $= 0.88 \times 100\,300$ $= 88\,264$ 88% of 88 264 $= 77\,672.32$ 88% of 77 672.32 $= 68\,351.6416$ 88% of 68 351.6416 $= 60\,149.444608$ Pa	Decrease by 12% leaves 88%, do this four times in a row. $100\,300 \times 0.88 \times 0.88 \times 0.88 \times 0.88$ $= 100\,300 \times 0.88^4$ $= 60\,149.444\,608$ Pa $= 60\,149.4$ Pa (1 dp)

Tip

A pascal is the SI unit of pressure, equal to one newton per square metre.

Exponential decay

When a quantity decreases by a fixed percentage over regular periods of time it is called **exponential decay**. The graph of exponential decay is a curve that slopes down steeply towards the right.

The general function of such decreasing exponential graphs is $y = ka^x$, where $k > 0$ and $0 < a < 1$ (in other words a is a fractional quantity).

The general formula for exponential decay is therefore

$$y = a(1 - \mathrm{r})^n$$

where a is the original value/quantity,

r is the rate of decay (as a decimal) and

n is the number of time periods.

EXERCISE 35B

1 The value of a compact car depreciates each year by 8%. A new compact car costs £11 000. How much will this be worth in:

a 1 year? **b** 3 years? **c** 8 years? **d** n years?

2 Copy and complete this table:

Initial cost	Depreciation rate	1 year	2 years	6 years	n years
£400	2%				
£2 500	15%				
£50 000	3.5%				

Find answers at: cambridge.org/ukschools/gcsemaths-studentbookanswers

3 The pesticide DDT was banned after being discovered to be dangerous. It remains in the soil for many years. It decays by being absorbed by the soil at a rate of 7% a year.

a If a farmer used 2 kg of DDT in a field in 1970, how much remains in the field in 2014?

b Will there ever be 0 g of DDT left in the field? Explain.

4 The rate at which water flows out of a tank with an opening at the bottom depends on the amount of water left in the tank. For one particular tank, the height of the water left will reduce by 15% every 5 minutes. Which of the following graphs depicts this?

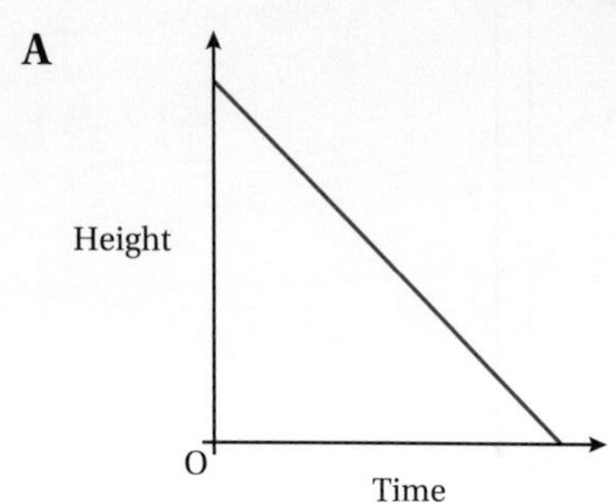

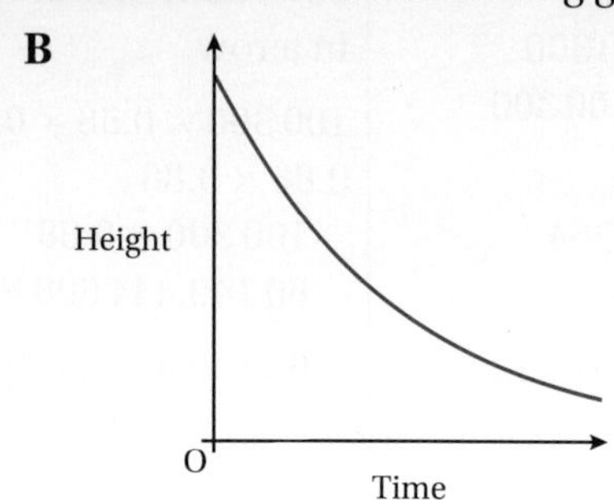

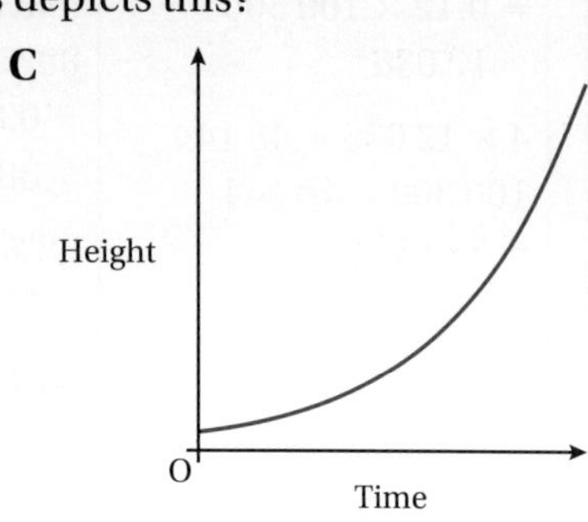

5 At the start of an experiment there are 8000 bacteria. A chemical is introduced to the population causing a reduction of 1600 in the population in an hour.

a What percentage decrease in population is this?

b Assuming the same rate of decrease, how many bacteria would you expect to be alive after 8 hours?

c How long until less than 100 bacteria are alive?

6 For every 1000 m higher you climb, the atmospheric pressure decreases by 12%.

If the sea-level atmospheric pressure is 100 300 pascal (Pa), what would the pressure be if you jumped out of a balloon flying 39 km above sea level?

7 The population of Bulgaria is decreasing at a rate of 0.6% per year. In 2014 the population was 7.4 million people.

a How many people are expected to be living in Bulgaria in 2020?

b How many years until the population dips below 7 million?

8 Copy and complete this table:

Initial cost	Depreciation rate	1 year	2 years	6 years	*n* years
					$\$7500 \times 0.925^n$
£650		£617.50			
	11%	\$30 260			
£12 000 000			£10 267 500		

9 The cost of mobile phones has been falling. Three years ago the latest model cost £400 with no contract; now the latest model costs £342.95.

If the price keeps falling, how long until the current model costs less than two-thirds of today's price?

Checklist of learning and understanding

Simple and compound growth

- Simple growth, such as simple interest, is a fixed rate of growth, calculated on the original amount.
- The formula $I = PRT$ can be used to calculate simple interest, where P = principal amount, R = the rate of interest as a percentage, and T is the time period over which the interest is calculated.
- Compound growth, such as compound interest, is calculated on the principal for the first period and then compounded by calculating it on the principal plus any interest paid or due for each previous period.
- You can work out compound growth using a multiplier for each period or by applying the formula $y = a(1 + r)^n$.

Simple and compound decay

- A drop in value of an object over time is called depreciation.
- A decline in a population is called decay.
- Simple and compound decay can be found by working out the percentage decrease using repeated subtraction, by working out the percentage remaining, or by applying the formula for compound decay: $y = a(1 - r)^n$.

Chapter review

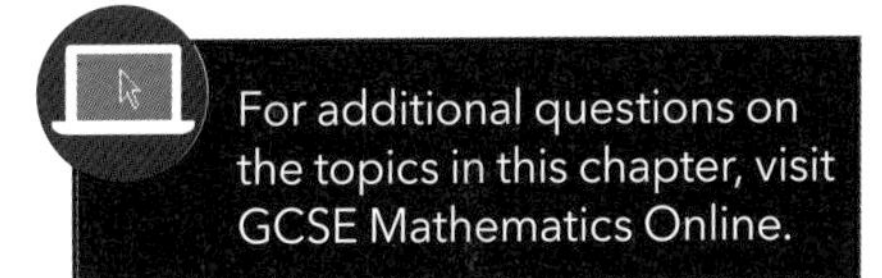

1 A camera has a cash price of £850. Nasief buys it on credit and pays a 10% deposit, with the balance to be paid over two years at a simple interest rate of 10% per year. Calculate:

a the amount of his deposit.

b the balance owing after deducting the deposit.

c the amount of interest paid in total over two years.

d the monthly payment amount for 24 equal monthly instalments.

e the difference between the cash price and what Nasief actually paid in the end.

2 Salma invests her money in an account that pays 6% interest compounded half-yearly.

If she puts £2300 in the account and leaves it there for two years, how much money will she have at the end of the period?

3 A car valued at £8500 depreciates by 30% in the first year, 20% in the second year and a further 12% in the third year.

How much is it worth after three years?

Find answers at: cambridge.org/ukschools/gcsemaths-studentbookanswers

4 Each year the education department arranges a quiz. There are 140 students in the competition to start with. During each round, half of the quiz contestants are eliminated.

How many students will still be participating after round 4?

5 The population of elephants in a national park in southern Africa increased exponentially from just 40 elephants in 1963 to 640 in 2013, as a result of conservation efforts.

Determine the growth rate r of the elephant population.

6 A population of birds decreases at a rate of 5% each year.

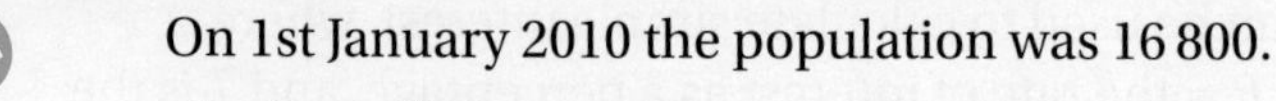

On 1st January 2010 the population was 16 800.

The formula for working out N, the size of the population, t years after 1st January 2010 is

$$N = 16\,800 \times A^t$$

a Explain why the value of A is 0.95. *(1 mark)*

b Calculate the population on 1st January 2016. *(1 mark)*

c Find the year in which the population will fall below 8000. *(3 marks)*

© OCR 2013

36 Direct and inverse proportion

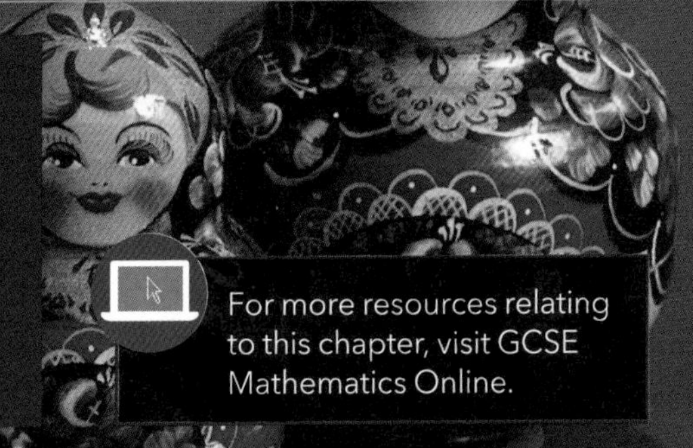

In this chapter you will learn how to …

- understand proportion and the equality of ratios.
- solve problems involving direct and inverse proportion, including graphical and algebraic representations.
- understand that x is inversely proportional to y is equivalent to x is proportional to $\frac{1}{y}$.
- interpret equations that describe direct and inverse proportion.

For more resources relating to this chapter, visit GCSE Mathematics Online.

Using mathematics: real-life applications

Proportional reasoning is very common in daily life. You use proportional reasoning when you mix ingredients for a recipe, convert between units of measurement or work out costs per unit. It is an area of maths where you can use many different methods to solve particular problems.

"I test out new dishes on my family. Then I have to scale up the recipes in proportion so that they taste just as good. Sometimes it may be for just a few people at one table in my restaurant, at other times it may be for a whole room of wedding guests." *(Chef and restaurant owner)*

Tip

Review the sections in Chapter 23 on equivalent ratios and fractions to prepare for this chapter.

Before you start …

Ch 10, 12	You need to know how many minutes there are in fractions of an hour.	**1** How many minutes are there in: **a** half an hour? **b** a quarter of an hour? **c** a third of an hour? **d** a fifth of an hour?
Ch 10, 12	You need to be able to find what fraction of an hour a given time is.	**2** What fraction of an hour is: **a** 5 minutes? **b** 24 minutes? **c** 54 minutes?
Ch 8, 14	You should know how to substitute values into formulae.	**3** $g = 3b$ **a** What is the value of g when $b = 7$? **b** What is the value of b when $g = 72$? **c** What is b when $g = 1.2$?

Find answers at: cambridge.org/ukschools/gcsemaths-studentbookanswers

Assess your starting point using the Launchpad

STEP 1

1 **a** A recipe for chocolate muffins makes 12 muffins. It uses 180 g of dark chocolate.
How much chocolate is needed to make 30 muffins?

b A car is travelling at 80 km per hour.
How far would it travel in 75 minutes?

c €1 = $1.40
How many euros is a t-shirt that costs $24?

GO TO
Section 1:
Direct proportion

STEP 2

2 The cost of carpeting a hallway is proportional to the area of the hall. One hallway measuring 15 m² costs £97.50 to carpet.

a Find a formula for the cost, c, of carpeting a hallway with area, a.

b How much would it cost to carpet an area of 32 m²?

c What area can be carpeted for £328.90?

GO TO
Section 2:
Algebraic and graphical representations

STEP 3

3 The cost of putting new soundproofing on a square dance floor is directly proportional to the square of the length of the side of the floor. A dance floor with a side length of 8 metres costs £976 to soundproof.

a Find a formula for the cost, c, of soundproofing a floor with side length s.

b How much would it cost to lay flooring on a floor with side length 7.5 m?

c To the nearest 10 cm, what side length of floor can be soundproofed for £645?

GO TO
Section 3:
Directly proportional to the square, square root and other expressions

GO TO
Step 4:
The Launchpad continues on the next page …

Launchpad continued ...

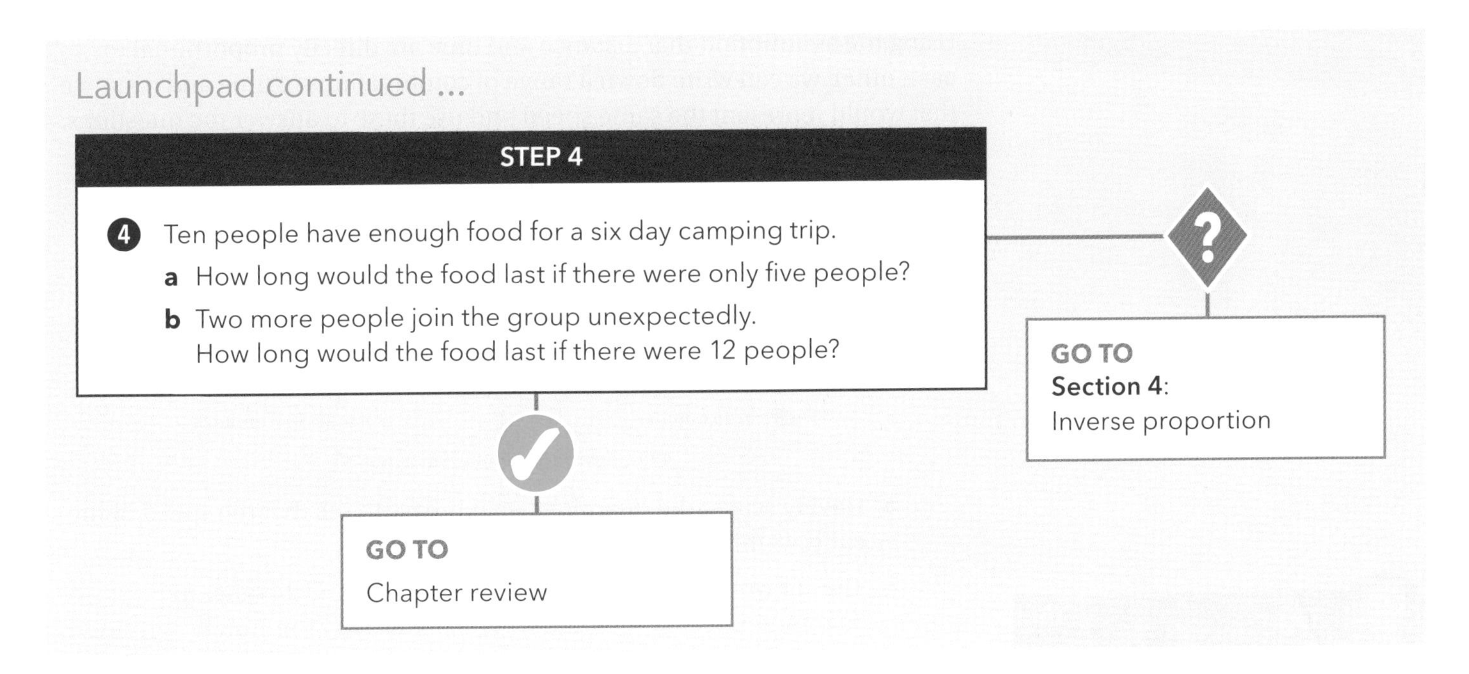

STEP 4
4 Ten people have enough food for a six day camping trip.
a How long would the food last if there were only five people?
b Two more people join the group unexpectedly.
How long would the food last if there were 12 people?
GO TO
Section 4:
Inverse proportion
GO TO
Chapter review

Section 1: Direct proportion

Key vocabulary

ratio: the comparison between two or more amounts in relation to each other; you learnt about ratio in Chapter 22.

direct proportion: two values that both increase in the same ratio.

When two quantities vary but remain in the same **ratio** they are said to be in **direct proportion**. A simple example would be the volume and price of petrol. The more petrol a driver puts into the car, the more it costs.

Scaling up recipe ingredients involves direct proportion. If you want to make double the amount of food, you need to use double the amount of ingredients. Other examples are the number of hours someone works and the amount they get paid at an hourly rate, and exchange rates between two currencies.

Decide whether each of these examples represents a directly proportional relationship:

- The time spent riding your bike and the calories you burn.
- The distance you travel and the cost of your rail ticket.
- The amount of water in a kettle and the number of cups of tea you can make.
- The cost of buying woodchips for a primary school play area and the size of the play area.
- The time it takes to walk to school and the number of friends you walk with.

In problems involving variables in direct proportion, you may be given a rate such as price per litre. If not, it can be helpful to find this rate.

Scaling up or down

Consider the following problem.

A car travels 12 miles in 15 minutes.

a At what speed is the car travelling?

b How far would the car go in 75 minutes?

c How long would it take the car to travel 80 miles?

Think how you would find the answers to these questions.

One approach is to assume that distance and time are directly proportional to each other, which allows us to develop a **mathematical model** of the problem.

Key vocabulary

mathematical model: a representation of a real-life problem; assumptions are used to simplify the situation so that it can be solved mathematically.

Using the assumption that distance and time are directly proportional to each other, we can write down a range of combinations of time and distance that would represent the same speed and use these to answer the questions. The diagram shows some combinations.

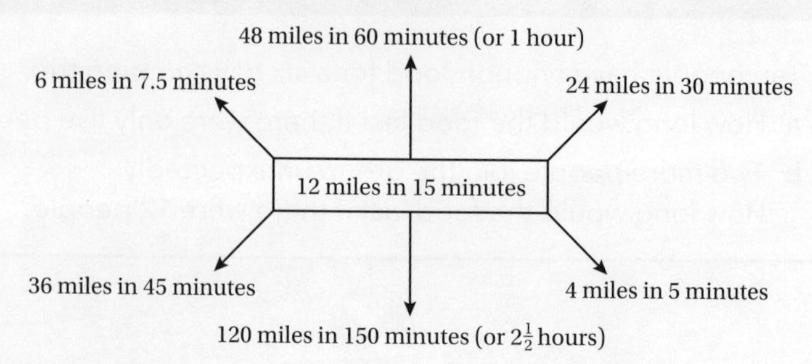

a Having scaled the quantities up it is easy to see that the speed of the car is 48 miles per hour.

b The car would travel 36 miles in 45 minutes and 24 miles in 30 minutes. Hence it would travel 60 miles in 75 minutes.

c The car would travel 4 miles in 5 minutes. Multiply both of these quantities by 20 to find that it would do 80 miles in 100 minutes (or 1 hour and 40 minutes).

These are not the only combinations that you could use to answer the original questions.

Tip

Very often when working with proportion problems it helps to write down a proportion fact you know and consider what would happen if one side is halved, doubled, tripled, multiplied by 10 and so on. These notes will often then help you to solve the original problem. For example, if you know that six eggs make two cakes then half the number of eggs, three, will make just one cake.

Unitary method

Finding the rate for one unit and using this to calculate other values, is called the **unitary method**. In the example above you are told that the car travels 12 miles in 15 minutes. If you calculate how far the car travels in one minute ($12 \div 15 = 0.8$) you find the rate per minute and you can multiply this value as required to find the distance travelled in x minutes.

Similarly, you could calculate how long it takes the car to travel one mile ($15 \div 12 = 1.25$ minutes).

Sometimes it is simpler to just scale the values up or down as necessary, and sometimes the new values won't be a neat multiple or factor and the unitary method would be easier to use. Either method works, use whichever one you prefer.

WORKED EXAMPLE 1

While in Florida, Danny bought a T-shirt. It cost \$18. He paid on a debit card. When he returned home the charge on Danny's debit card statement was £10.71. He bought a pair of jeans for \$32. Assuming the bank uses the same exchange rate, what would this charge appear as on his debit card statement?

$\$18 = £10.71$

$\$1 = £0.595$

$32 \times 0.595 = £19.04$

In this example scaling up and down is too inefficient and it is better to use the unitary method. In other words, start by finding the exchange rate of dollars to pounds. This is very much like an equivalent ratio problem in Chapter 23. Divide both sides by 18. Now use this rate to find the cost in pounds of \$32.

This example works out exactly. In reality, quite often exchange rates result in fractions of pence or cents. In these cases, exchange desks would round up or down to the nearest whole number.

EXERCISE 36A

1 A bluefin tuna fish can travel 3 km in just 20 minutes.

List some other distance–time facts about the fish assuming that it always travels at a constant rate.

2 Patrick works for four hours and gets paid £22.

What is his rate of pay per hour?

3 Jelly beans cost £1.20 for 100 g.

a How much would 50 g cost?

b How much would 300 g cost?

c How much would 1 kg cost?

d What weight of jelly beans could you buy with £4.20?

4 On holiday Ben uses his mobile to call home. A 12 minute call costs £4.20.

a How much would it cost to ring home for 18 minutes?

b Danny calls home for 20 minutes and it costs him £6.40.

Whose phone is better value, Ben's or Danny's? Why?

5 Look at this pancake recipe. It serves 8 people.

100 g plain flour 2 eggs 300 ml semi-skimmed milk

If you have 2 litres of milk, 500 g of plain flour and 9 eggs and make as much pancake mixture as possible, how many will it serve?

6 The fastest train in Europe is the French TGV from Paris to Le Mans. It travels at 320 km per hour. Assuming that the train is going at its full speed:

a How far does it travel in 2 hours?

b How far does it travel in 30 minutes?

c How far does it travel in 15 minutes?

d How far does it travel in 1 minute?

e How far does it travel in 10 seconds?

f The equator is approximately 40 000 km long.
If it were possible, how long would it take to travel around it in a TGV train?

Find answers at: cambridge.org/ukschools/gcsemaths-studentbookanswers

7 The fastest animal on land is the cheetah. It can reach speeds of up to 120 kilometres per hour, but for only a short burst of time.

How far would it travel at this speed in 15 seconds?

8 Before going to Australia, Finley exchanges £175 into Australian dollars. He gets an exchange rate of £1 = AU$1.81.

How many Australian dollars does he get?

9 When returning from her holiday, Amber exchanges her $44 back into pounds. The exchange rate is £1 = $1.68.

How many pounds does she receive?

10 Paint is sold in a variety of tins. However, the price per litre remains unchanged.

a Find the cost of each of these tins of paint.

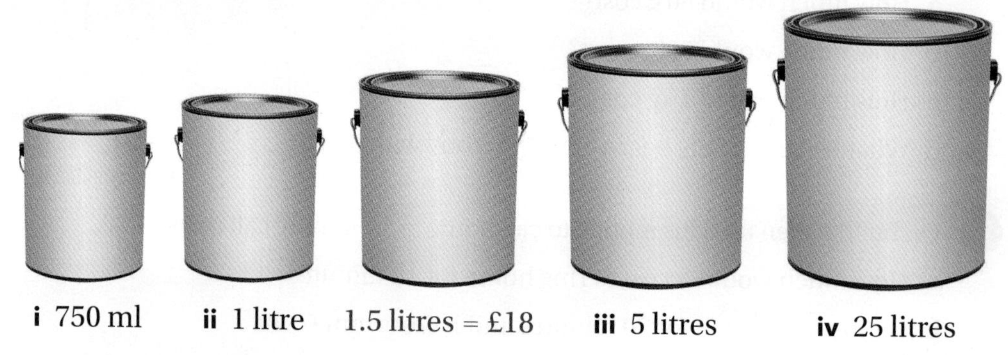

i 750 ml **ii** 1 litre 1.5 litres = £18 **iii** 5 litres **iv** 25 litres

b Is this the way items are usually priced?

11 Lucy exchanges £50 for €60.50.

a What is the exchange rate from pounds to euros?

b What is the exchange rate from euros to pounds?

12 When planning a skiing holiday in Switzerland, Ethan compares two resorts. The exchange rate from pounds to Swiss francs is £1 = CHF1.48.

Which resort is a better deal? How many pounds cheaper is it?

	Accommodation	Food	Ski rental	Flights
Bun di Scuol	£340	CHF96.20	£300	CHF102.12
Flims-Laax-Falera	CHF444.00	£100	CHF164.28	£144

13 On a recent holiday to Spain, Megan compares the price of saffron. In the UK 35 g of saffron can be bought for £2.69. In Spain it costs €11.80 for 125 g.

Which is the better deal, given that £1 = €1.21? Explain your answer clearly.

Section 2: Algebraic and graphical representations

Direct proportion problems can be represented graphically or generalised through the use of algebra. This allows you to solve problems concerning the same relationship, either by reading information off a graph or by using an algebraic formula.

WORK IT OUT 36.1

Which of these graphs show a pair of variables that are directly proportional to each other? Explain your choice.

How can you tell that the variables in the other graphs are not directly proportional to each other?

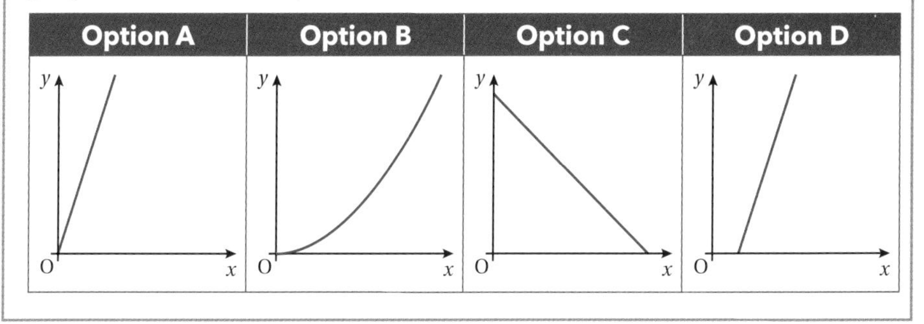

The graph of a directly proportional relationship has a fixed gradient and goes through the origin. We can express this relationship algebraically.

The mathematical symbol $\propto$ is used to indicate that two values are proportional.

$y \propto x$ means that y is proportional to x

This tells us the following relationship between x and y:

$$\frac{y_1}{y_2} = \frac{x_1}{x_2} \text{ or } \frac{y_1}{x_1} = \frac{y_2}{x_2}$$

Tip

Relate this back to the work you have done on equivalent fractions and ratios, Chapter 23.

If you pay per minute to use your mobile phone the time you speak, t, is proportional to the cost of the call, c. Hence $t \propto c$.

This means that for some fixed value k (often called the constant of proportionality), you can write a formula linking the time and cost:

$$c = kt$$

Tip

Be careful with units in questions. In this example the cost is in pence and time in minutes. To use the formula you'd need to make sure all quantities were in pence and minutes and convert any that were not.

If you pay 75p for a 15 minute call, you can calculate the value of k by substituting the known values into the formula.

$$c = kt$$
$$75 = k \times 15$$
$$5 = k$$

So the formula linking the cost in pence, c, and time in minutes, t, is $c = 5t$.

WORK IT OUT 36.2

Which of these formulae represent variables that are directly proportional to each other? Can you explain why (or why not)?

Option A	Option B	Option C	Option D
$y = 3x + 5$	$10w = h$	$\frac{s}{t} = 7$	$d^2 = 4f$

Gradients and ratios

If you plotted a graph to illustrate the car journey described earlier in the chapter where the car travels 12 miles in 15 minutes, the resulting graph would be a straight line showing distance against time. This is called a distance–time graph. The line shows direct proportion because at time 0 the car has travelled 0 miles. You know that these two variables are in direct proportion, and therefore are increasing in the same ratio.

You have seen this kind of relationship before, in the gradient of a straight-line graph:

$$\text{Gradient} = \frac{\text{vertical rise}}{\text{horizontal run}} = \frac{\text{difference in } y\text{-values}}{\text{difference in } x\text{-values}}$$

The gradient is a rate of change, it tells you how y changes as x changes. When the graph is a straight line, the gradient is constant and represents two variables in direct proportion.

If you are given a straight-line graph, you can use it to calculate the ratio between two variables by using two points on the line to calculate the gradient. Exchange rates, conversion charts and distance–time graphs are all examples of straight-line graphs that show direct proportion.

Tip

You will learn more about distance–time graphs in Chapter 39.

In a distance–time graph, time is plotted along the x-axis and distance along the y-axis. The gradient is therefore: $\frac{\text{distance}}{\text{time}}$.

In the car example above, the car travelled 12 miles in 15 minutes. At the start it would have travelled 0 miles in 0 minutes and so you have two points on the graph (0, 0) and (15, 12). You could calculate the speed using the ratio

$$\frac{\text{distance}}{\text{time}} = \frac{12}{15} = 0.8 \text{ miles/minute}$$

The gradient of a distance–time graph tells you the speed of a moving object.

EXERCISE 36B

1 The graph shows the cost of telephone calls.

Write a formula linking the cost c, in pence, and time t, in minutes, of a call.

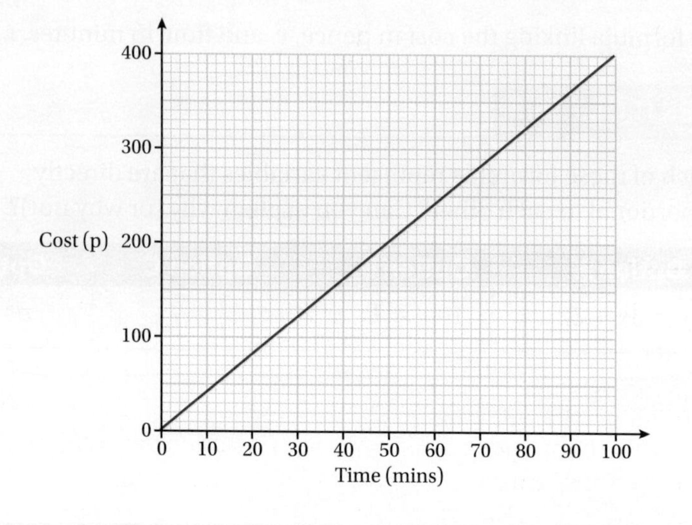

2 What information would you need to collect to compare how crowded two school playing fields are? How would you carry out the comparison?

3 This is a time–distance graph for two runners.

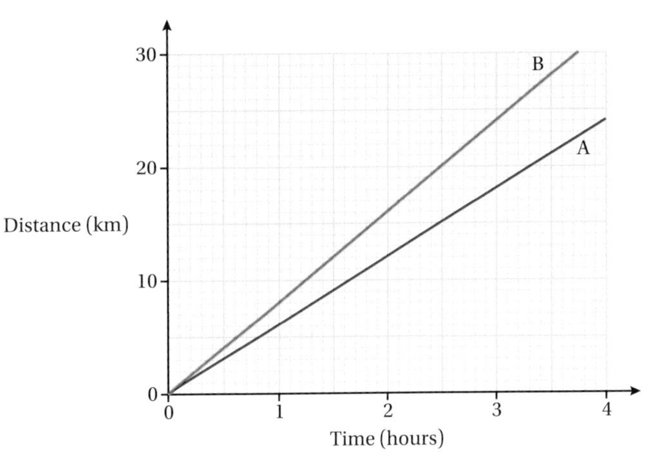

a How far had runner A travelled after 30 minutes?

b How long did it take runner B to travel 18 km?

c Which runner is going fastest?

d What is the speed of each runner?

e What assumptions have been made when drawing this graph?

4 The graph shows the cost of telephone cable.

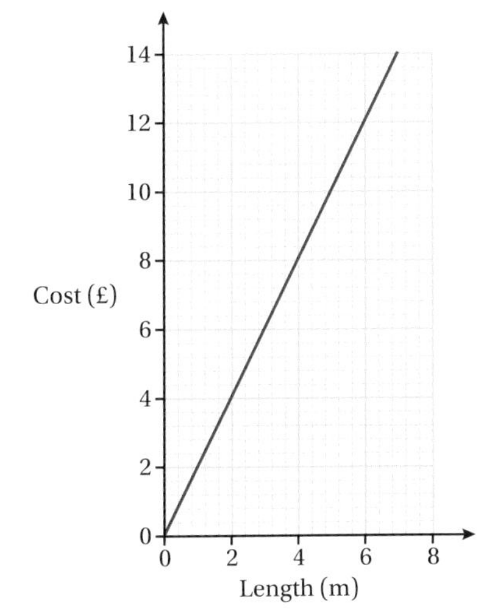

a What is the cost per metre?

b Complete this formula: cost = ______ × length.

c Write a formula linking the cost c, in pounds, and length l, in metres, of wire.

Section 3: Directly proportional to the square, square root and other expressions

In this section you'll consider directly proportional relationships where one value has to be squared, square rooted, or raised to another power before it is multiplied by the constant of proportionality.

For example, an object dropped from rest doesn't travel at a constant velocity, it accelerates; therefore the distance it travels isn't directly proportional to the time it has been falling.

In fact the distance travelled, d, by an object dropped from rest is proportional to the square of the time, t, for which it has been falling. So if you double the time, you quadruple the distance travelled; if you triple the time you multiply the distance travelled by 9, and so on.

Find answers at: cambridge.org/ukschools/gcsemaths-studentbookanswers

WORKED EXAMPLE 2

An object is dropped from rest and after 3 seconds has travelled a distance of 44.1 metres.

a How far will it have gone after 7 seconds?

b How long will it take to travel 100 metres?

a $d \propto t^2$

Hence $d = kt^2$

Write down the proportional relationship. If the time and distance were directly proportional, you could write the equation as $d = kt$. But you know that, for an object dropped from a height, the distance is proportional to the square of the time.

$44.1 = k \times 3^2$

Substitute in the given values and solve for k.

$4.9 = k$

$d = 4.9t^2$

This gives you the formula for the distance travelled in terms of the time fallen. You can use this to work out how far the object travelled after 7 seconds, and how long it will take to travel 100 metres.

$d = 4.9 \times 7^2 = 240.1$ m (1 dp)

Substitute $t = 7$.

b $100 = 4.9t^2$

This time you have the value of d and want to calculate t. Substitute in the known values and solve for t.

$\frac{100}{4.9} = t^2$

$\sqrt{\frac{100}{4.9}} = 4.52$ seconds (2 dp)

Worked example 2 is a simplified version of the calculations performed by skydivers to ensure that they open their parachutes in time to land safely.

EXERCISE 36C

1 w is directly proportional to the cube of m. When $m = 3$, $w = 108$.

a Find a formula for w in terms of m.

b Find the value of w when $m = 5$.

c Find the value of m when $w = 62.5$.

2 r is directly proportional to the square root of s. When $r = \frac{1}{2}$, $s = \frac{1}{16}$.

a Find a formula for r in terms of s.

b Find the value of r when $s = 20$.

c Find the value of s when $r = 12$.

3 The rate at which a toaster produces heat, j (joules), is proportional to the square of the current, I (amps) in the circuit (Joule's First Law). A toaster using 3.5 amps produces 857.5 joules of heat.

a Find a formula for the heat produced, j, in terms of the current, I.

b Another toaster produces 400 joules of heat. What current does this toaster draw?

4 The time it takes a pendulum to complete one full swing (from left to right and back again) is directly proportional to the square root of its length. A pendulum of length 16 cm takes 1.28 seconds to complete a full swing.

a Find a formula for the time taken, t, in terms of the length of the pendulum, l.

b How long is a pendulum that takes 2 seconds to complete one full swing?

5 The mass of a cube of gold, m, is directly proportional to the cube of its side length, s. A cube with side length 2 cm has a mass of 154.4 grams.

a Find a formula for the mass of a cube of gold in terms of its side length.

b A gold ingot has a mass of 12.4 kg. What size cube would this bar make?

Section 4: Inverse proportion

In some instances one quantity decreases as the other one increases. For example, if you increase your speed, the time it takes to travel a fixed distance is reduced. If you add more workers to a job, each working at the same rate, the overall time it takes to complete the job goes down. These types of relationship are inversely proportional.

The graph of a pair of inversely proportional variables never quite touches the x- or y-axis, but comes closer and closer to them.

WORK IT OUT 36.3

A rectangle has a fixed area of 24 cm^2. Its length, x, and height, y, can vary. Which of the graphs below represents this situation? How did you come to your decision?

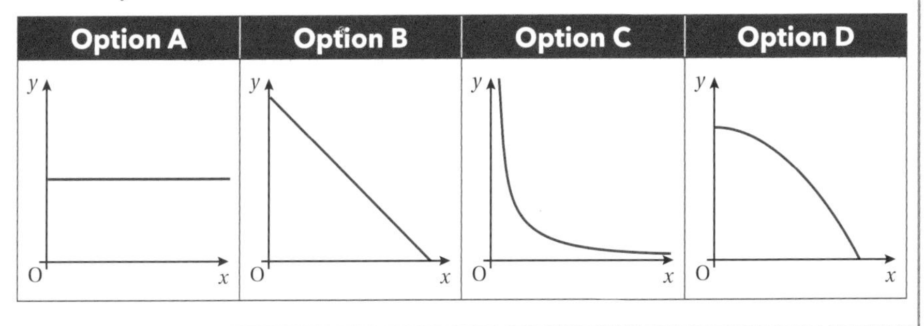

In the example of a rectangle with a fixed area (*Work it out 36.3*), the length, x, and height, h, are in **inverse proportion**. When one is multiplied the other is divided. Hence:

$$y \propto \frac{1}{x}$$

$$y = \frac{k}{x}$$

For example, let's look at the rectangle with area 24 cm^2 where it has a length of 2 cm and a height of 12 cm.

$$12 = \frac{k}{2} \text{ so } k = 24$$

$$y = \frac{24}{x}$$

Neither x nor y can equal zero.

Key vocabulary

inverse proportion: a relation between two quantities such that one increases at a rate that is equal to the rate at which the other decreases.

Find answers at: cambridge.org/ukschools/gcsemaths-studentbookanswers

WORKED EXAMPLE 3

It takes 4 people 3 days to paint the school hall.

a How long would it take 1 person to paint the school hall?

b How long would it take 2 people?

c How long would it take 6 people?

d The job needs to be completed in a day.
How many people are needed?

e What assumptions are being made?

a $4 \div 4 = 1$; $3 \times 4 = 12$

12 days

The number of people is inversely proportional to the number of days as the number of people increases the number of days it takes decreases. If you divide one value by an amount, you have to multiply the other value by the same amount to keep them inversely proportional. It will take 1 person 4 times as long to do the job as 4 people.

b $1 \times 2 = 2$, $12 \div 2 = 6$ (or $4 \div 2 = 2$, $3 \times 2 = 6$)

6 days

It would take 2 people half the time it takes one person; or twice as long as it takes 4 people.

c $1 \times 6 = 6$, $12 \div 6 = 2$

2 days

Use the values you calculated for one person to work out the time taken for 6.

d $3 \div 3 = 1$, $4 \times 3 = 12$

12 people

Alternatively, you could have used the values for one person. You can use any of the values calculated as they are all in the same proportion.

e The assumptions are that everybody works at the same rate, and that several people can paint the hall at the same.

EXERCISE 36D

1 While on holiday you budget to buy five souvenirs at \$2.40 each.

a How much money do you intend to spend?

b How many souvenirs costing \$0.80 each could you buy with your budget?

c If you need eight souvenirs of equal value, how much should you pay for each souvenir to keep within your budget?

2 Speed (s miles per hour) and travel time (t hours) are inversely proportional. The faster you travel the less time a journey takes.
A journey between Cambridge and Manchester takes 3 hours when travelling at 60 miles per hour.

a Find a formula for the time taken, t, in terms of the speed travelled, s.

b It takes 4 hours to make the journey.
What speed is this?

c How long will it take to do the journey at 75 miles per hour?

d It takes 2 hours 15 minutes to make the journey.
What speed is this?

3 A scientist is analysing the efficiency of a new antibacterial agent by exposing a large colony of bacteria to the agent and recording the size of the population over time. The graph shows the relationship between the number of bacteria in the colony and time after first exposure to the agent.

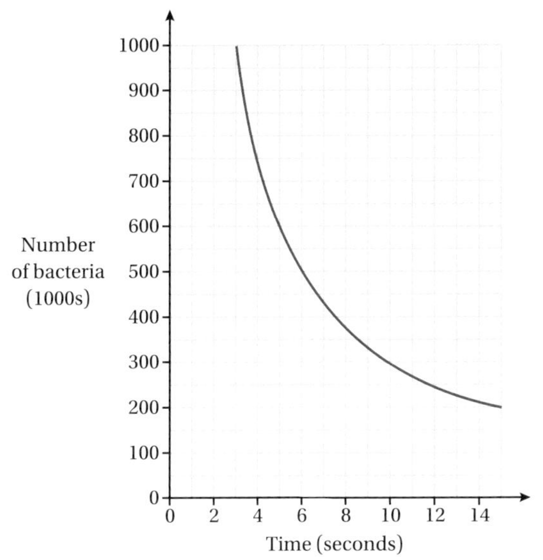

a How many bacteria are there in the colony 10 seconds after exposure?

b How many bacteria were in the colony after 1 second?

c Find a formula for the size of the colony, p, in terms of the number of seconds exposed to the agent, t.

4 A water tap is running at a constant rate (r litres per minute), filling a pond in m minutes. The greater the flow of water the less time it takes to fill the pond. Copy and complete the table and then draw a graph to represent this situation.

***m* minutes**	10	20	30	40	50	60	70	80	90	100
***r* litres per minute**						10				

Find a formula for r in terms of m.

5 The variable a is inversely proportional to the square of b. When $b = 5$, $a = 2$.

a Find a formula for a in terms of b.

b Find the value of a when $b = 2$.

c Find the value of b when $a = 0.5$.

6 The rate, r, at which a gas diffuses is inversely proportional to the square root of its molecular mass, m (Graham's Law of Diffusion). Carbon dioxide diffuses at 1 mol/s (mole per second) and has molecular mass 44 g/m (gram per mole).

Find a formula for r in terms of m.

Find answers at: cambridge.org/ukschools/gcsemaths-studentbookanswers

7 In 1963 the land speed record of 407 miles per hour was held by American Craig Breedlove. In 1997 this record was held by British driver Andy Green. His recorded speed was 340 metres per second. What was the difference, in minutes, between the times it took the two drivers to travel a mile?

Checklist of learning and understanding

Direct proportion

- If two quantities are directly proportional to each other they increase and decrease at the same rate. For example, if one is tripled so is the other, if one is halved so is the other.
- A formula for a direct proportion relationship between two variables x and y is $y = kx$, where k is the constant of proportionality and can be found by substituting in known values.
- The graph of two directly proportion variables is a straight-line graph of the form $y = mx$, where m is positive.

Inverse proportion

- If two quantities are inversely proportional to each other then as one increases the other decreases. For example, if one is tripled the other is divided by three, if one is halved the other is doubled.
- A formula for an inverse proportion relationship between two variables x and y is $y = \frac{k}{x}$, where k is the constant of proportionality and can be found by substituting in known values.
- The graph of two inversely proportional variables is of the form $y = \frac{m}{x}$, where m is positive. It looks like:

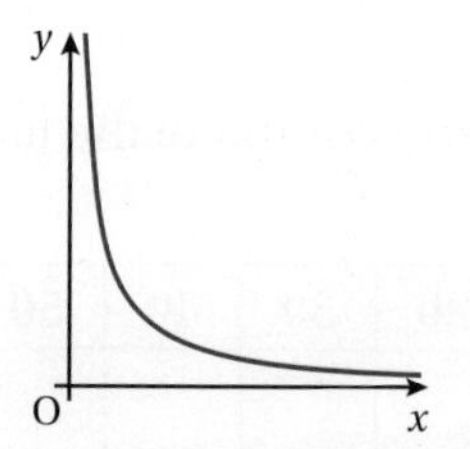

For additional questions on the topics in this chapter, visit GCSE Mathematics Online.

Chapter review

1 Wine gums cost 90p for 200 grams.

a How much would 800 grams cost?

b What mass of wine gums would you get for £2.75?

2 Look at this stir-fry recipe. It serves six people.

120 g chicken
300 g vegetables
15 tbsp of soy sauce

If you have 300 g of chicken, 500 g of vegetables and 60 tbsp of soy sauce and make the most stir-fry possible, how many will it serve?

3 The graph shows the cost of buying electrical wire.

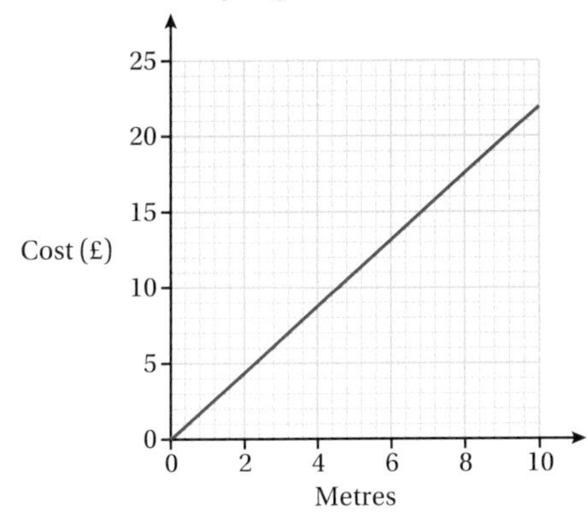

Write a formula for the cost, c, in terms of the number of metres bought, m.

4 y is directly proportional to x^2 and $y = 80$ when $x = 4$.

Write a formula for y in terms of x. *(3 marks)*

5 F is directly proportional to the cube root of g. When $g = 64$, $F = 12$.

a Find a formula for F in terms of g.

b When $g = 125$, what is F?

c When $F = 21$, what is g?

6 It takes three hairdressers an hour to style the hair of models for a fashion show.

How long would it take nine hairdressers at the same rate?

7 The rate at which water flows into a pond is inversely proportional to the time it takes to fill up. It takes 2 hours for the pond to fill when water flows in at a rate of 50 litres an hour.

Find a formula for the time in hours, t, in terms of the flow rate of the water, w.

8 The variable d is inversely proportional to the cube of e.
When $d = 3$, $e = 2$.

a Find a formula for d in terms of e.

b Find the value of d when $e = 3$.

c Find the value of e when $d = 0.5$.

37 Collecting and displaying data

In this chapter you will learn how to ...

- infer properties of populations or distributions from a sample, recognising the limitations of sampling.
- interpret and construct appropriate tables, charts and graphs for grouped and ungrouped data.
- choose the best form of representation for data and understand the appropriate use of different graphs.

For more resources relating to this chapter, visit GCSE Mathematics Online.

Using mathematics: real-life applications

We live in a very information-rich world. Knowing how to construct accurate graphs and how to interpret the graphs we see is important. Many graphs in print and other media are carefully designed to influence what we think by displaying the data in particular ways.

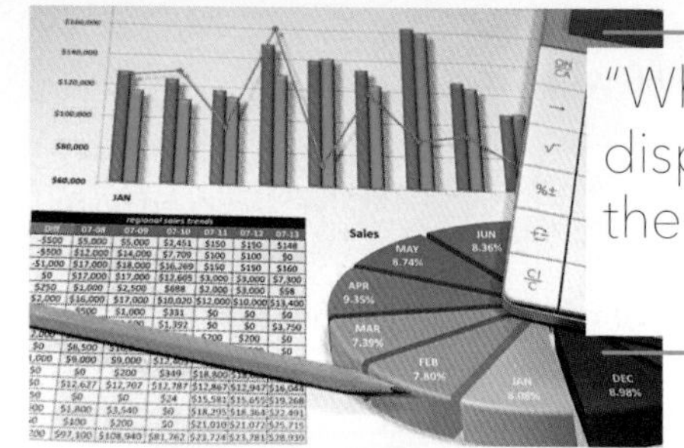

"When we have data, we need to display it so that our message has the maximum impact."

(Newspaper editor)

Tip

The key to displaying data is to choose the graph or chart that clearly shows what the data tells us without the reader having to work too hard. Getting the scale and labelling right makes a big difference when creating graphs and charts.

Before you start ...

KS3	You need to be able to sort and categorise data.	**1** What would be suitable categories for a set of adult heights ranging from 1.39 m to 1.85 m?
Ch 12, 18	You need to be able to use scales properly.	**2** **a** What is each division on this scale: 200 — 300 **b** A scale between 0 and 100 has five divisions. Which numbers should go alongside each division?
Ch 6, 9	You need to be able to measure and draw angles to create pie charts.	**3** **a** Measure these angles: **b** Draw an angle of 72° accurately.

Assess your starting point using the Launchpad

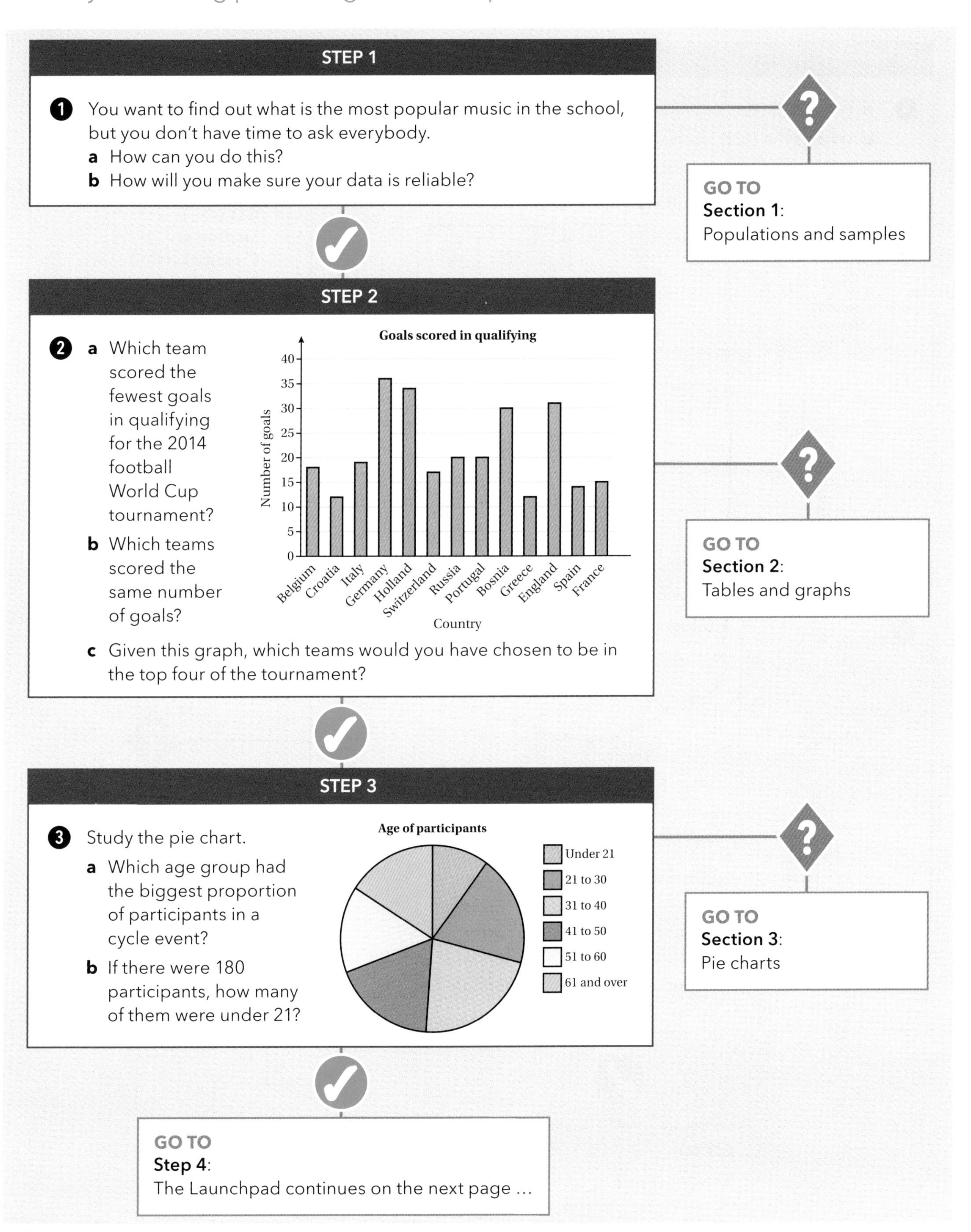
STEP 1
1 You want to find out what is the most popular music in the school, but you don't have time to ask everybody.
a How can you do this?
b How will you make sure your data is reliable?
?
GO TO
Section 1:
Populations and samples
STEP 2
2 a Which team scored the fewest goals in qualifying for the 2014 football World Cup tournament?
b Which teams scored the same number of goals?
c Given this graph, which teams would you have chosen to be in the top four of the tournament?
Goals scored in qualifying
Number of goals
0
5
10
15
20
25
30
35
40
Belgium
Croatia
Italy
Germany
Holland
Switzerland
Russia
Portugal
Bosnia
Greece
England
Spain
France
Country
?
GO TO
Section 2:
Tables and graphs
STEP 3
3 Study the pie chart.
a Which age group had the biggest proportion of participants in a cycle event?
b If there were 180 participants, how many of them were under 21?
Age of participants
Under 21
21 to 30
31 to 40
41 to 50
51 to 60
61 and over
?
GO TO
Section 3:
Pie charts
GO TO
Step 4:
The Launchpad continues on the next page …

Launchpad continued ...

STEP 4

a Why are the bars not the same width on this graph?

b What is the class interval for the yellow bar?

c How many students are between 11 and 16?

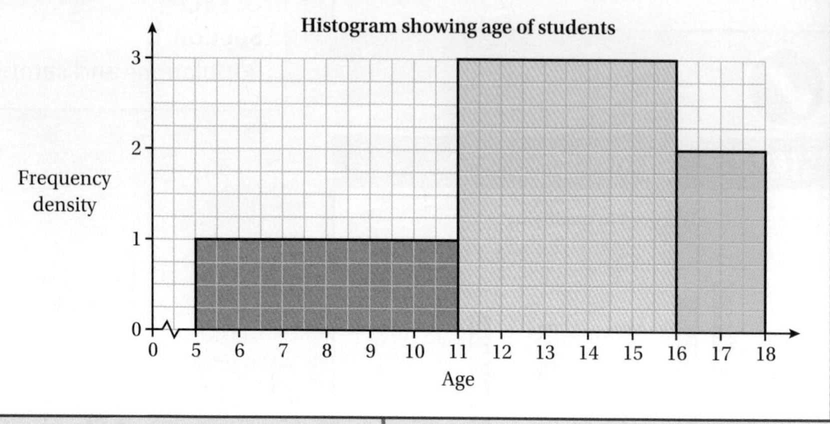

?

GO TO
Section 4:
Cumulative frequency curves and histograms

STEP 5

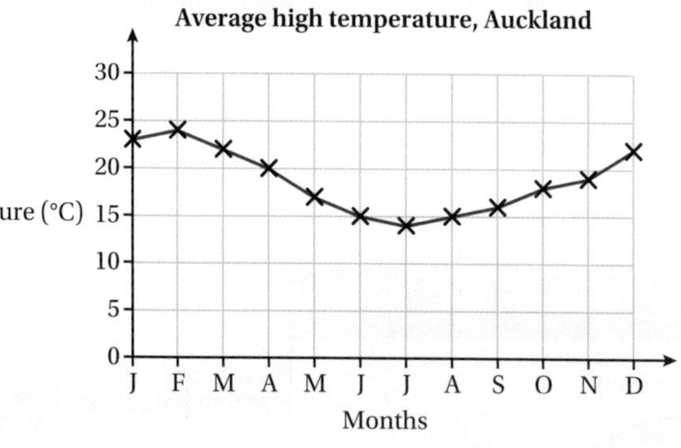

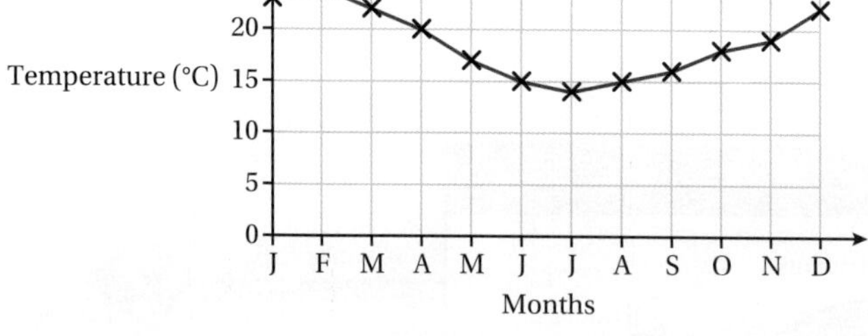

a Which month has the warmest average high temperature in Auckland?

b Between which two months does the temperature rise the fastest?

c Between which two months does the temperature fall the fastest?

GO TO
Section 5:
Line graphs for time series data

GO TO
Chapter review

Section 1: Populations and samples

Key vocabulary

population: the name given to a data set.

sample: a small set of data from a population.

representative sample: a smaller quantity of data that represents the characteristics of a larger population.

A statistical **population** is a set of individuals or objects of interest.

For example, a school might want to find the mean height of students to decide what size of equipment to buy for the gymnasium. In this example, the population would be all the students in the school.

In a large school it would be impractical to measure each student's height. It is more likely that the researcher would choose some of the students as a **sample** of the population. The sample needs to be a **representative sample** to provide useful data.

A representative sample would come from measuring a mix of male and female students from different years. It would not be a good idea to measure just the Year 7 students, or just the Year 11 students. The results from the sample can then be used to estimate the total numbers in the whole population. For example, there are 1000 students in a school and a representative sample of 50 students is chosen. Six of these students are below 150 cm tall. $1000 \div 50 = 20$, so it is reasonable to estimate that $6 \times 20 = 120$ students in the population will be below 150 cm tall.

A representative sample can be created by taking a factor that is unrelated to the property being measured, for example creating the sample by using all the students whose first name begins with a letter drawn at random. As the starting letter of your name has no effect on your height, this should give a representative sample.

Although the sample can't guarantee we will get the heights of the biggest and smallest students, it should make sure that we get a good idea of the spread of the data.

"I collect data on behalf of my company so that they can find out how likely people are to buy new products. We use quota sampling in our work. This involves choosing people with particular characteristics. For example, I may only be interested in teenage boys who play video games." *(Market researcher)*

Random sampling

In a random sample, each member of the population is equally likely to be chosen for the sample. For example, if you had a list of all the students in the school you could use a computer program to pick 10% of them at random from the list.

Other methods of random sampling include numbering all the items in the population and then choosing numbers at random using a random number generation application.

In reality, it is often difficult to choose a genuinely random sample. For example, if you are doing a survey for a school project you are mostly likely to use **convenience sampling** because you would probably survey friends and family members; that is, your sample is chosen based on being from people that it is easier for you to speak to. This method could result in a biased sample.

Key vocabulary

bias: something that affects the chance of an event occurring in favour of a desired outcome.

Care must be taken to avoid **bias** in a sample. This can happen if one section of a population is favoured. For example, choosing telephone numbers at random from a list might appear to be a sensible way to sample a population, but this excludes people who don't have a telephone, who therefore won't be represented in the sample.

In statistics, the population may be divided into groups using a particular system. **Stratified sampling** is quite common. This involves dividing the sample into groups (strata) and then choosing a random sample from each group. The size of the sample chosen from each group should be in proportion to the size of the group within the population. For example, in a school population you could use the year groups as strata. If the Year 9 students make up 30% of the school population, then 30% of the sample should come from that group.

Another method of sampling is **capture/recapture**. This method is used by biologists and ecologists to estimate a population size when counting would be impossible or impractical. A sample of the population is captured and marked then released back into the population. A new sample is then recaptured and the number of previously marked animals noted. The total population size can be estimated by multiplying the number of animals marked in the first sample by the number of animals caught in the second sample and dividing the result by the number of marked animals caught in the second sample.

WORK IT OUT 37.1

A market researcher has been asked to conduct a survey using a random sample of shoppers at a shopping mall. She suggests four options.

a Which is the only option that would produce a random sample?

b Explain why each of the other three options does not produce a random sample.

Option A	Option B	Option C	Option D
Ask all the women with children.	Ask people between 8 am and 8.30 am.	Stand outside a book shop and ask everyone who comes out.	Stop and ask every 10th person who walks by the researcher.

EXERCISE 37A

1 Which of these methods are likely to give a random sample?

a Selecting all the odd numbered houses in a street.

b Calling people on their home telephones during the day.

c Selecting everybody who is wearing trainers.

d Calling the person whose name is at the top of each page of the phone book.

e Drawing a series of names from a hat.

Give reasons for your answers.

2 A market research company wants to find out how many people are likely to buy a new baby food.

a Suggest a good place to conduct a survey of young parents.

b 35 of the 50 parents asked said they would be interested.
How many parents would you expect to be interested, in a population of 1000 parents?

3 A gym owner wants to know how many running machines to buy. She asks every member whose surname begins with an 'S' whether they will use a running machine.

a If 15 of the 28 in her sample say 'yes', what would be a sensible number of machines to buy if there are 300 members overall?

b Does she really need this many machines?

c Is there a better way of sampling her members to make sure she gets a realistic number of machines?

4 At the end of 2012 there were 28.7 million cars on the roads of Britain.

Surjay and his friends conduct a random survey of the cars passing the school and discover that of the 50 cars recorded, 3 had a sunroof, 1 had a faulty exhaust and 4 had chips on the windscreen.

Use this information to estimate how many cars in Great Britain have:

a a sunroof. **b** faulty exhausts. **c** chips in the windscreen.

5 A survey of 100 people who live on the same street in a town showed that the following radio stations are listened to for at least an hour per day:

Radio station	Number of listeners
Radio Uno	25
Ears on	12
Hip and Happening	11
Classic Numbers	6
R Town Radio	23

a Suggest one reason why the sample may not be representative.

b If the population of the town is 4550, and the sample is representative, estimate how many people listen to each of the radio stations?

6 A school wishes to consult the pupils on a change to the school uniform. They decide to use a stratified sample of 100 students.

How many students should be chosen from each year group, given that the school population is as follows?

	Year 7	Year 8	Year 9	Year 10	Year 11
Boys	98	107	184	154	145
Girls	72	121	172	162	168

Find answers at: cambridge.org/ukschools/gcsemaths-studentbookanswers

7 An ecologist is trying to estimate the population of badgers in a certain area.
She catches 65 badgers and marks them.
They are then released back into the wild and a further sample caught a week later.
In the second batch she catches 52 badgers, of which 21 were marked.
Use this information to estimate the badger population in the area.

Section 2: Tables and graphs

Using tables to organise data

Tip

You worked with frequency tables in Chapters 23 and 24 when you dealt with probability.

A frequency table is a table used to organise data and show the 'frequency' of an event, or how often it happens. For example, the number of goals scored by each of the 20 premiership teams one weekend was as follows:

5	1	3	0	1	2	4	1	1	2
0	3	1	0	0	4	0	1	3	0

In a frequency table these results would look like this:

Number of goals scored	Tally	Frequency
0	𝍸 I	6
1	𝍸 I	6
2	II	2
3	III	3
4	II	2
5	I	1

Using bar charts to display data

The data in the frequency table above can be shown on a bar chart.

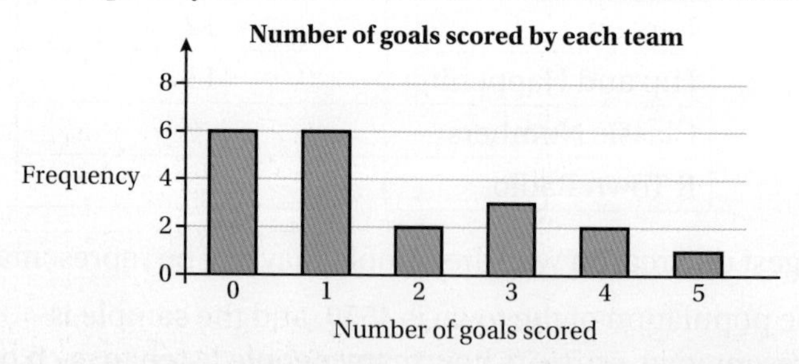

The chart has a title, a scale on the left and accurately drawn bars.

Note that there is a gap between each bar and each one is labelled.

Bar charts are used to display **discrete data**. The number of goals scored by each team is discrete data because it can only have certain values. It must be a whole number; you can't score $\frac{1}{2}$ a goal or 2.34 goals.

Sometimes it is helpful to sort data into categories. Pairs of shoes in a cupboard could be categorised into 'brown shoes', 'black shoes', etc. Each piece of data can only be in one category. This is known as **categorical data**.

A vertical line graph (or bar line chart) is very similar to a bar chart but the number of data in each category is represented by a line rather than a bar.

Key vocabulary

discrete data: data that can be counted and that only has one possible value; it is counted in integers.

categorical data: data that has been arranged in categories.

This data could also be shown using a pictogram.

In a pictogram for this data a symbol could be used to represent either each goal or a number of goals.

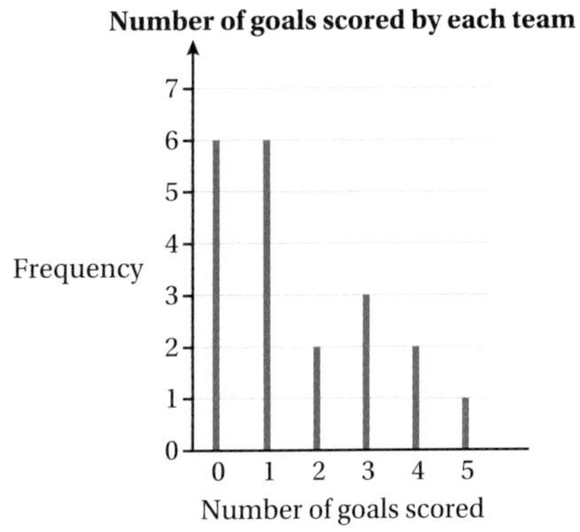

In this example, each ball represents two goals:

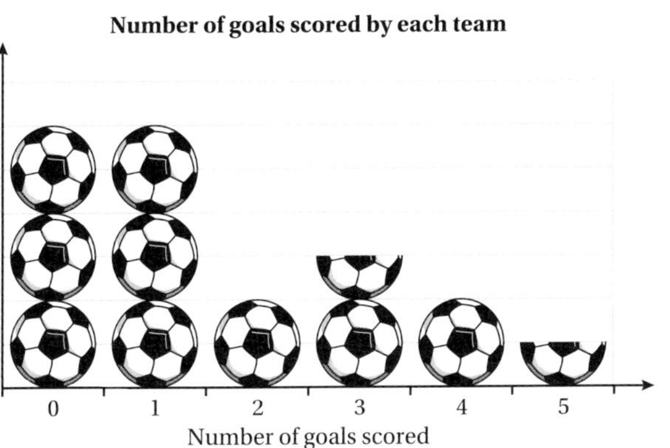

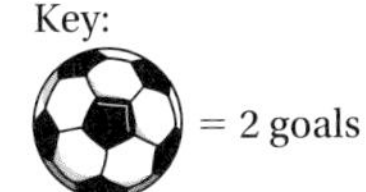

Note that a pictogram should always have a key to indicate what the symbol represents.

WORK IT OUT 37.2

Ramiz records the number of mistakes he makes in a series of maths tests.

2 3 1 3 4 2 0 3 2 6 1 1 3 2 4 2

Which graph or chart best shows this data? What is wrong with the other two?

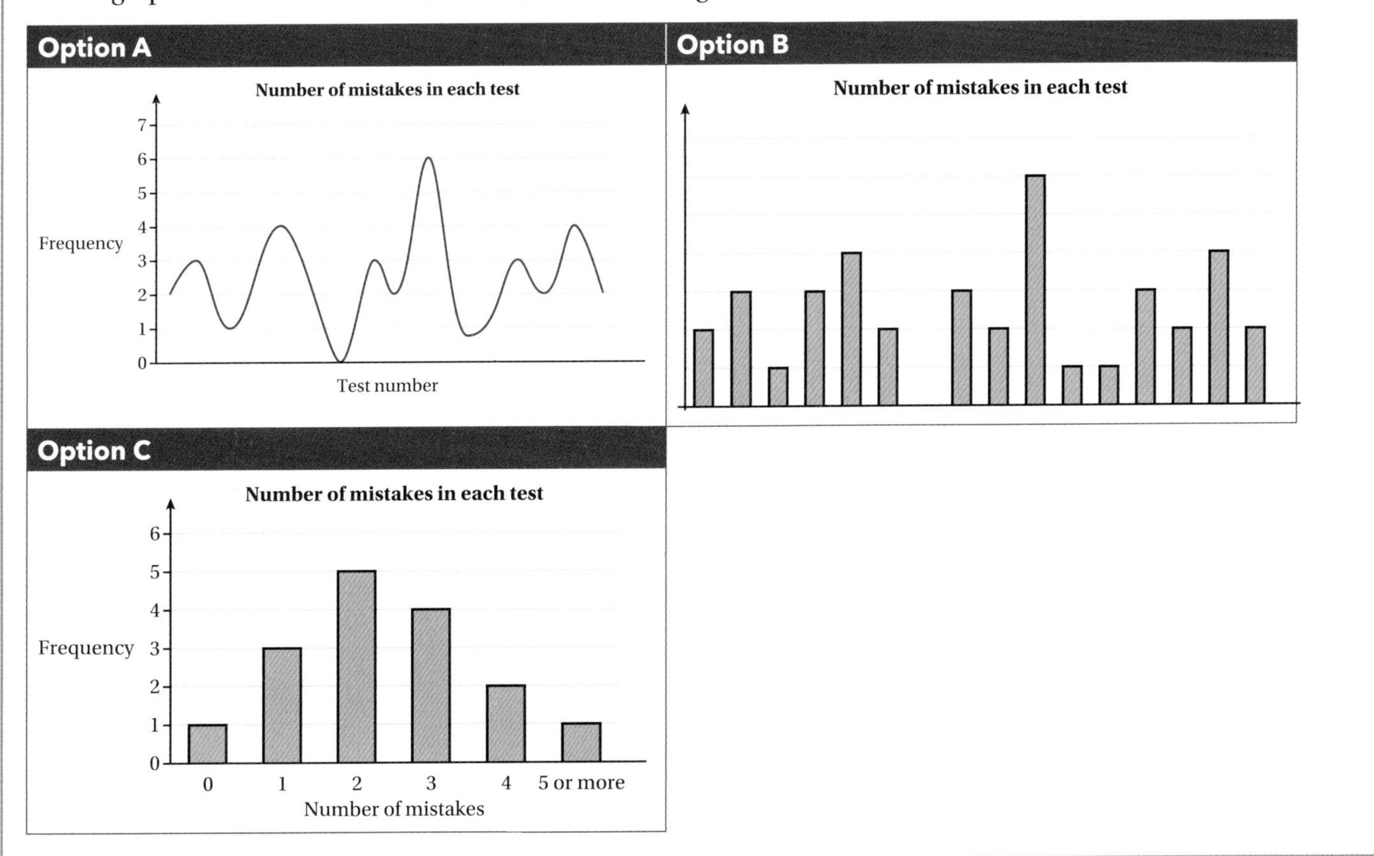

It is not always possible to say that one type of graph is better than another. The type of graph you draw depends very much on what data you have collected.

These guidelines can help you choose an appropriate graph for different kinds of data:

Tip

You might also have heard a composite bar chart called a compound bar chart; they are the same thing.

- Use bar chart or vertical line charts for discrete data that can be categorised.
- Use a pie chart or a composite bar chart if you want to compare different parts of the whole or show proportions in the data.
- Use a line graph for numerical data when you want to show trends (changes over time).
- Use scatter diagrams when you want to show relationships between different sets of data (you will deal with these in Chapter 38).
- Use histograms for continuous data with equal or unequal class intervals (you will deal with these in Section 4).

EXERCISE 37B

1 In an extended family of 30 members, 10 have blond hair, 9 have black hair, 6 have brown hair and 5 have grey hair.

Draw a vertical line graph to show this information.

2 The table below shows the percentages of people who use a particular mode of transport to work in a factory.

Mode of transport	Percentage
Car	36
Bus	27
Cycle	19
Walk	18

Show this information in a bar chart.

3 30 students were asked how many times in the last week they had visited the snack shop. Their responses were:

1 2 1 2 1 5 1 3 2 1 2 1 3 2 1 2 0 2 3 2 0 2 0 1 2 0 0 3 1 2

a Draw a frequency table for this data.

b Present this information in a suitable graph.

4 Construct a bar chart for the data in this table.

Favourite holiday destination	UK	Spain	France	USA	Greece
Frequency	9	15	17	12	8

5 A group of students were asked to choose their favourite snacks.
The results are in the table below.

Favourite snack	Number of students
Fruit	6
Crisps	8
Chocolate bar	9
Pizza slice	12
Cookie	7

Draw a pictogram to show these results.

6 The graph below shows the monthly rainfall in Lowestoft over a year.

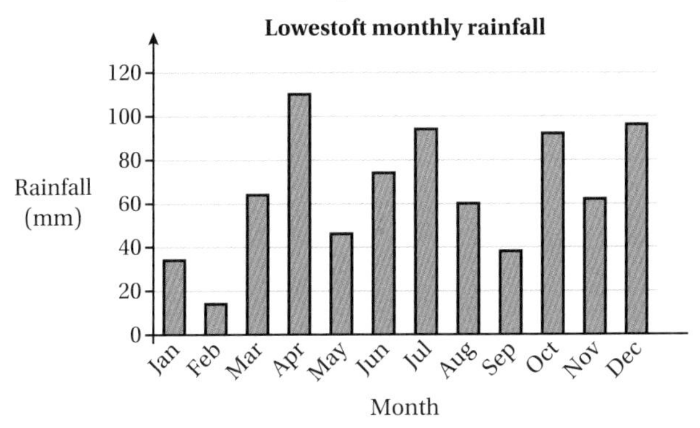

a In which month was the rainfall heaviest?

b Estimate the amount of rain that fell in April.

c Which was the driest month?

d Spring is March, April and May. Estimate how much rain fell in the spring.

e The average annual rainfall for Lowestoft is approximately 575 mm. Was this a wetter or drier year than average?

7 Jenny is carrying out a survey of what sort of snacks are bought from a shop outside her school.

She writes down the items that people buy:

Chocobar	Apple	NRG drink	Juicebar	crisps	NRG drink	Chocobar	NRG drink	Juicebar
Juicebar	crisps	Cheese puffs	Gum	Cheese puffs	Fruit chews	NRG drink	NRG drink	Chocobar
Chocobar	Juicebar	Chocobar	crisps	Chocobar	Gum	Chocobar	Cheese puffs	crisps
Cheese puffs	crisps	NRG drink	Fruit chews	NRG drink	Cheese puffs	NRG drink	Juicebar	Gum
NRG drink	Chocobar	Apple	NRG drink	Chocobar	Juicebar	crisps	Chocobar	Cheese puffs
Gum	Fruit chews	Gum	crisps	Apple	crisps	Fruit chews	Fruit chews	Fruit chews
Juicebar	crisps	Cheese puffs	Fruit chews	Gum	Cheese puffs	Fruit chews	crisps	Cheese puffs

a How could Jenny have been better organised before she started her survey?

b Use Jenny's data to create a table to show what was bought in the shop.

c Jenny will get extra credit if she can categorise her data.
Adjust your table so that the data is classified in an appropriate way.

Find answers at: cambridge.org/ukschools/gcsemaths-studentbookanswers

Tip

You might have heard a multiple bar chart called a comparative bar chart; they are the same thing.

Multiple and composite bar charts

A multiple bar chart is useful when you want to compare data for two or more groups. For example, to compare shoe sizes of male and female students in Year 9 you would show the data for male and female students in matching pairs of bars like this:

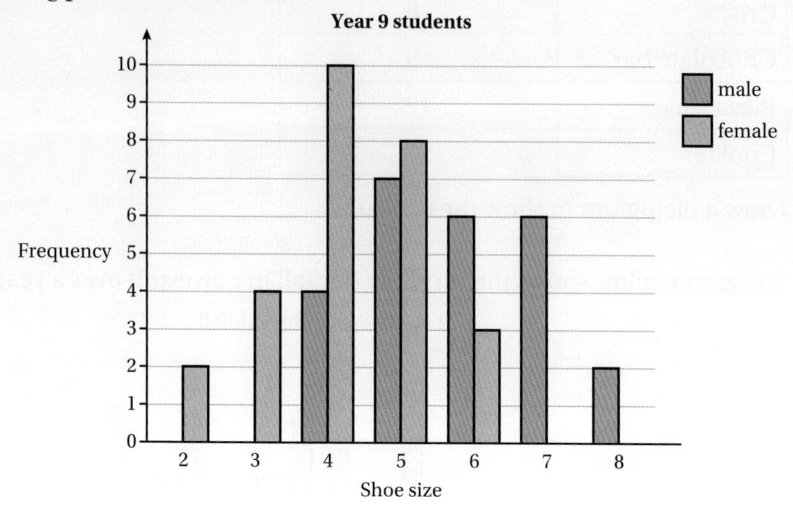

Notice that:

- The graph has a key to show what each colour bar represents.
- The two bars for male and female students who wear each size touch each other, but there is an equal space between each pair of bars.

Composite bar graphs are used to show parts of a whole. The total height of each bar represents a total amount. The height of the bar is divided into parts that show each category's share of the total amount.

To interpret a composite bar chart you need to work out what each bar represents and then do a calculation to find the fraction or percentage of the total that each part represents.

WORKED EXAMPLE 1

This composite bar chart shows the amount of water used by three different households over a four-month period.

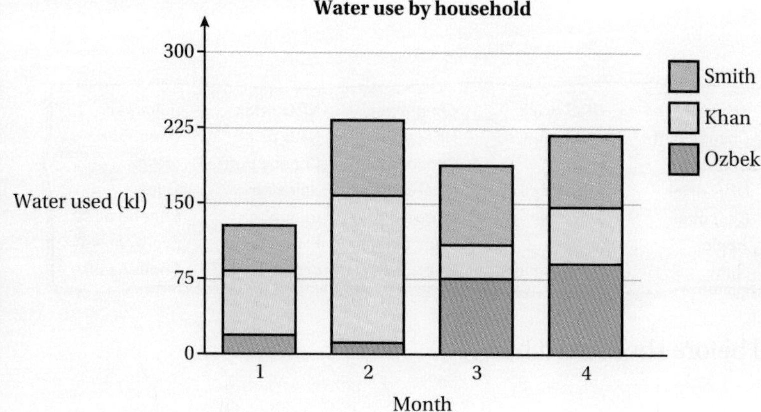

a What does each bar show?

b Which household used the greatest amount of water in month 1?

Continues on next page ...

c Describe the trend in water use for the Ozbek household over this period.

d One household had a leaking pipe in this period.
Can you work out from the graph who this was and when it happened?

a The top of each bar shows the total water used by three households in a month. Each coloured segment shows the fraction of the total used by each household.

b The Khan household.

The yellow section is bigger than the other two.

c In months 1 and 2 they used very little water. In month 3 the amount of water used increased quite dramatically and in month 4 it went up a little more.

d It is most likely the Khan household, as they had a large jump in consumption in month 2. However, it could be the Ozbeks as well. If their water pipe started leaking in month 3 and wasn't fixed, it could account for the big increase in their consumption.

EXERCISE 37C

1 Study the bar graph carefully.

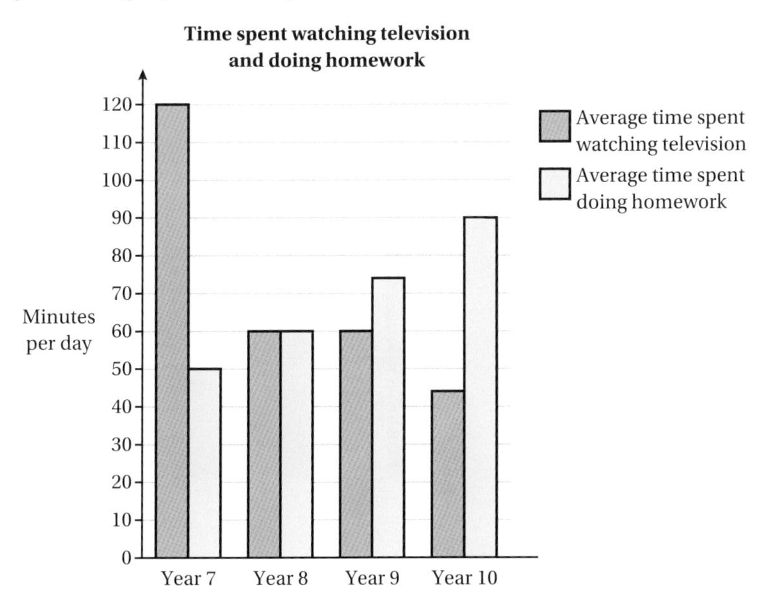

a What two sets of data are shown on this graph?

b Describe the trend in the amount of time spent watching TV as students move into higher years.

c What happens to the amount of time spent on homework as TV watching time decreases?

d How much time do Year 10 students spend on average each day on:

i homework?

ii watching TV?

Find answers at: cambridge.org/ukschools/gcsemaths-studentbookanswers

2 Patrick runs a computer company. He keeps a record of his costs and his income for four large projects in a year.

He drew this graph to compare his costs and his income for each project.

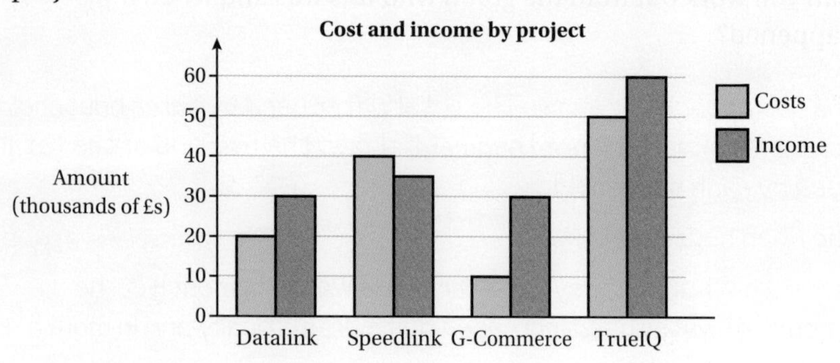

a Which project brought in most money?

b Which project brought in least money?

c Which project had the highest costs?

d Which project had the lowest costs?

e Which project gave Patrick the biggest profit?

f On which project did Patrick lose money? How can you tell?

g How much profit did Patrick make altogether?

Tip

Remember that profit = income − cost.

3 Carefully study this composite bar graph showing the proportion of total sales and how they are made for four different companies and answer the questions below.

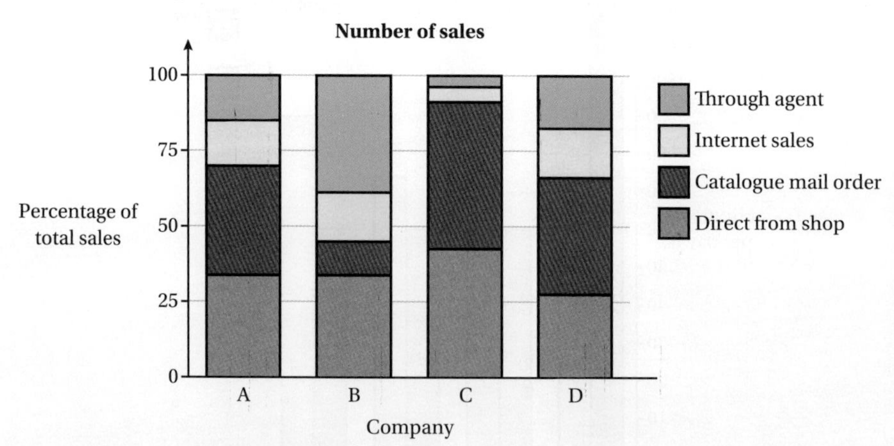

a Can you work out the value of each company's total sales from this graph? Explain your answer.

b Which company does the greatest proportion of sales direct from the shop?

c Which company makes the least of its sales through agents?

d Which company makes almost half of its sales by catalogue mail order?

e What fraction of Company A's sales are done over the Internet?

f Describe the breakdown of sales by type for Company D.

4 The chart below was drawn by a medical student writing a paper based on a World Health Organization (WHO) report on malaria.

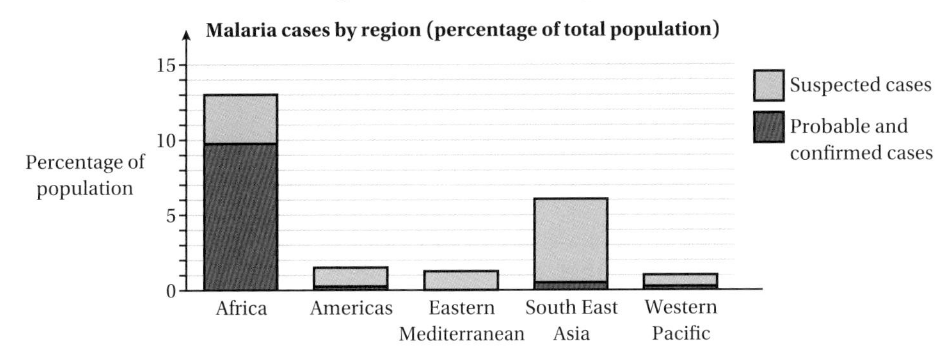

According to the chart:

a What percentage of the population of Africa had or were suspected of having malaria?

b In South East Asia what percentage of the population were suspected of having malaria?

c Were any of the suspected cases in the Eastern Mediterranean confirmed? How can you tell?

d Which region of the world has the biggest problem with malaria?

e The student used a report that was published in 2010. Do you think the student should include the date in their paper? Why? Is the data a good representation of the prevalence of malaria today?

Tip

You will learn more about misleading data in Chapter 38.

5 The table below shows exam grades for boys and girls in a Year 11 group.

Grade	Boys	Girls
A*	12	6
A	15	9
B	17	22
C	13	15
D	8	6

Construct a bar chart to show both sets of data.

Section 3: Pie charts

A pie chart is useful for displaying data when you are interested in the relative sizes or the proportions of the data.

Pie charts are always circular, so the sum of the angles at the centre must always be 360°.

When drawing pie charts that have data as percentages, each 1% will be represented by 3.6° because $360 \div 100 = 3.6$.

In this example, data has been collected that shows the area in which students in a class live:

Area	Frequency	Percentage
Reepham	12	40.0%
Whitwell	6	20.0%
Booton	3	10.0%
Cawston	2	6.7%
Salle	7	23.3%

There are 30 students altogether. Each percentage is worked out by dividing the number of students in that area by the total number of students, and multiplying by 100; for example, for Reepham: $\frac{12}{30} \times 100 = 40\%$

The pie chart shows this data:

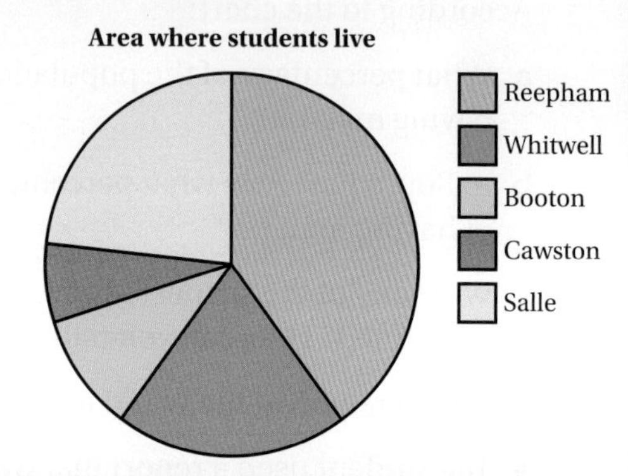

WORK IT OUT 37.3

A survey of how many minutes late a sample of 20 trains are gives the following results:

1 0 2 0 3 1 5 4 1 3 6 4 3 5 2 4 3 2 2 4

Which of the pie charts best shows this information? Explain what is wrong with the other two pie charts.

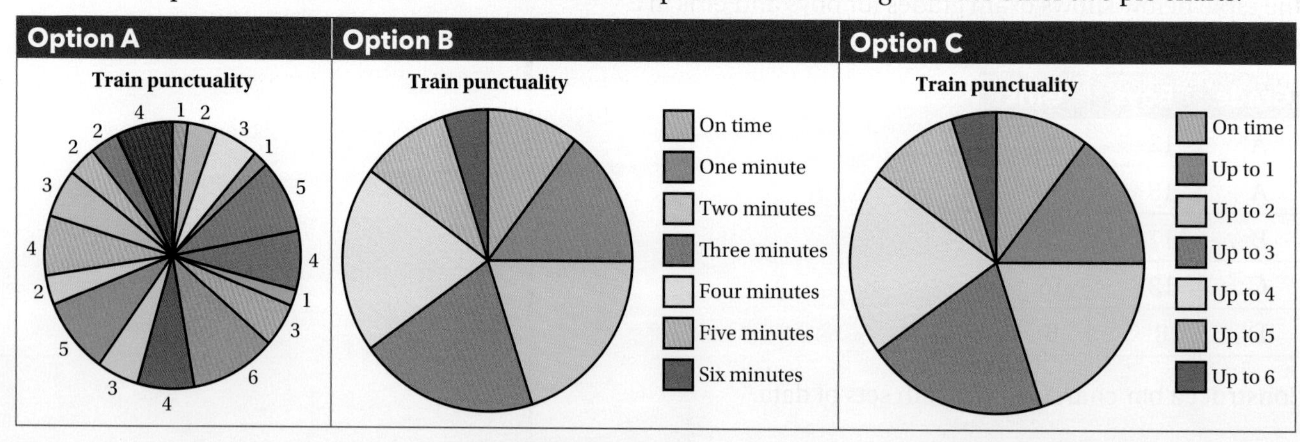

EXERCISE 37D

1 Create a pie chart to represent this data:

Electricity generation	Proportion used
Gas	28.0%
Other fuels	2.6%
Coal	39.0%
Nuclear	19.0%
Renewables	11.4%

2 The pie charts below show the population of two different countries by age.

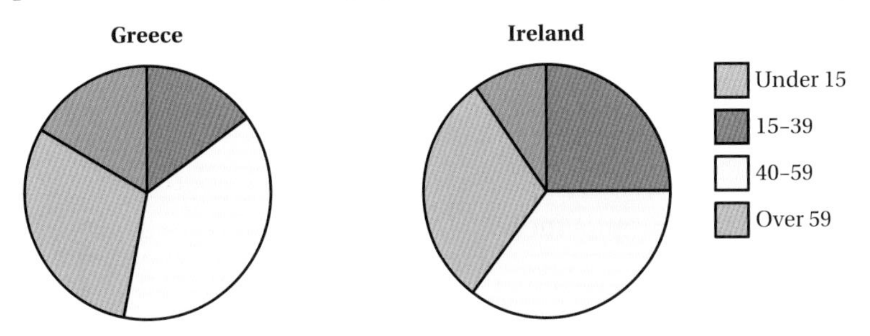

a Write down two differences between Greece and Ireland.

b Which country has the biggest proportion of over 59s?

c There are more under 15s in Ireland than Greece.
Is this statement true?

3 This pie chart shows the favourite leisure activity of 72 students.

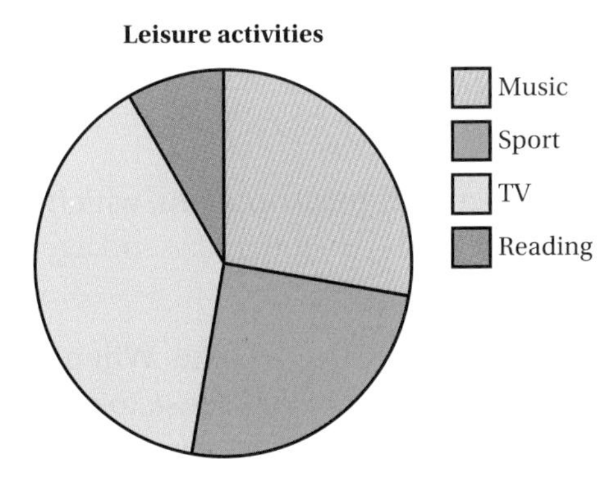

a Use a protractor to measure the sector for music. Use this measurement to work out how many students prefer music.

b Which is the most popular activity?

c How many students prefer reading?

4 The department of transport maintains data for the different types of vehicles on the road in the UK. Data for vehicles other than cars is shown for 1994 and 2013, in the pie charts below.

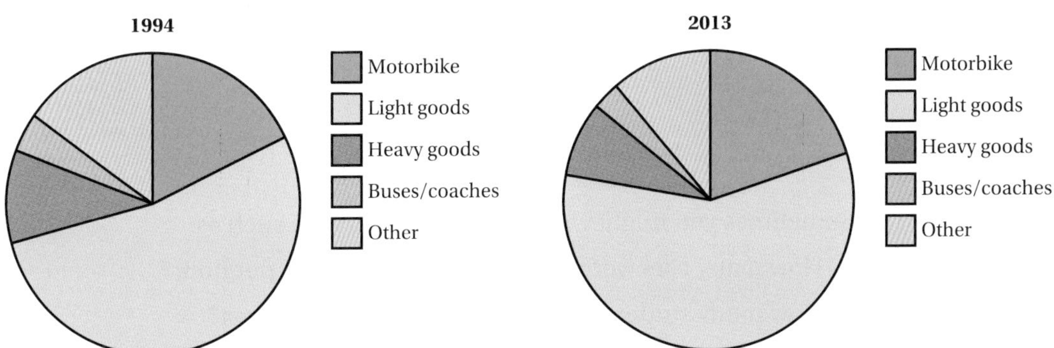

a Give two differences between the proportions for 1994 and 2013.

b If there were 6.075 million vehicles other than cars on the road in 2013, calculate the number of light goods vehicles there were.

c What percentage of vehicles other than cars were motorbikes in 2013?

Find answers at: cambridge.org/ukschools/gcsemaths-studentbookanswers

5 This data shows the destinations of students leaving a sixth form college.

Destination	College A	College B
Higher education	32	46
Further education	45	72
Employment	28	31
Gap year	12	24
Unemployment	15	22

Create two pie charts and use them to argue that one college is more successful than the other.

Section 4: Cumulative frequency curves and histograms

Grouped and continuous data

Key vocabulary

continuous data: data that can have any value within a given range.

grouped data: data that has been put into groups.

class intervals: groups into which numerical data has been grouped.

Continuous data is collected by measurement, which means the data can take on any fractional value. Height, mass and age are all examples of continuous data.

You are also going to work with **grouped data**. When you collect data with a lot of different possible values, it makes sense to group the values. For example, if you are collecting test percentages, the data values could range from 0 to 100. In this case you might group the data in tens, listing scores from 0 to 10, 11 to 20 and so on.

The groups are called **class intervals**. The top value in each interval is called the class boundary.

Generally, class intervals are equal and they do not overlap. For discrete data the intervals may be given in the form of a range such as 11–20. For continuous data the interval is often given in inequality notation. For example, heights between 5 and 10 metres might be given as $5 \leqslant h < 10$.

The size of a class interval is normally chosen so there are about 5 to 10 groups (known as 'classes') in the data set.

Cumulative frequency

Key vocabulary

cumulative frequency: the sum total of all the frequencies up to a given value.

Sometimes you might want to answer questions such as:

- How many cars were travelling above 30 miles per hour?
- How many students scored higher than 60% on test?
- How many of the strawberries in your garden weighed at least 24 grams?

Questions like these can be answered using cumulative frequencies. A **cumulative frequency** is a running total of the frequencies for each class interval. Adding the frequencies up to a particular value allows you to work out how many values were below or above that level.

The frequency table below shows the masses of strawberries picked from a garden.

Mass	Frequency	Cumulative frequency
$12 \leqslant m < 16$ g	3	3
$16 \leqslant m < 20$ g	5	$3 + 5 = 8$
$20 \leqslant m < 24$ g	9	$3 + 5 + 9 = 17$
$24 \leqslant m < 28$ g	7	24
$28 \leqslant m < 31$ g	2	26

Tip

Notice the use of inequality notation. A class interval of $12 \leqslant m < 16$ g means that all strawberries with a mass between 12 g and '15 point something' grams will be in that class. Any strawberries with a mass of 16 g or more but less than 20 g will be in the next class, and so on. You need this notation so that any given mass can only be placed in **one** class interval.

The data has been grouped into equal class intervals, and the cumulative frequency calculated as a 'running total'.

Now you can see quite quickly that 17 strawberries have a mass of less than 24 grams.

Cumulative frequencies can be plotted against the upper boundaries of each class interval to produce a curved graph.

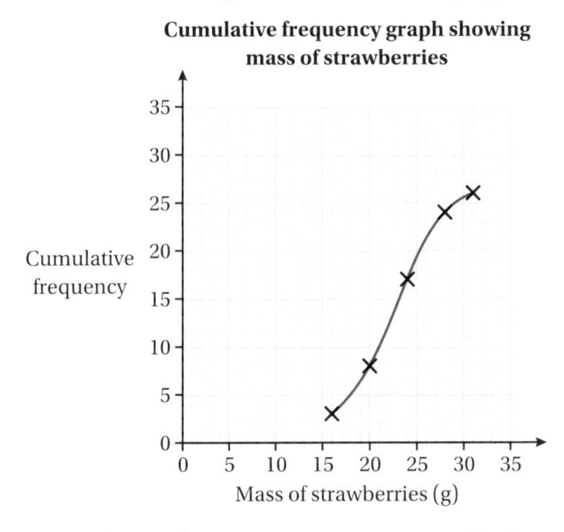

Tip

You plot the point for each cumulative frequency at the upper end of the class interval. The graphs are called cumulative frequency curves, so you must join them with a smooth curved line and not broken straight lines.

Cumulative frequency curves are called ogives (oh-jives) because they take the shape of narrow pointed architectural arches (called ogees) like the ones on the Royal Pavilion in Brighton.

Key vocabulary

histogram: a graph with bars whose area is proportional to the frequency of a variable and whose width is equal to the class interval.

Histograms

Grouped continuous data can also be plotted on a graph called a **histogram**. The histogram below shows the ages of members visiting a gym:

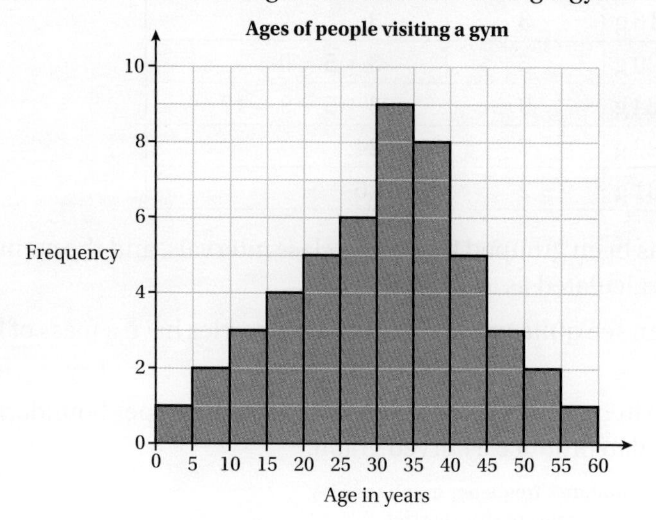

Histograms look like bar charts but there are important differences between them:

- The horizontal scale is continuous and each 'bar' is drawn above a particular class interval. For example, the first bar shows ages from 0 to 5.
- The frequency of the data is shown by the **area** of the bars.
- There are no spaces between the bars because the horizontal scale is continuous.

In this example the class intervals are equal, so the bars are the same width. When the class intervals are equal you can read the frequency off the vertical axis and you do not need to determine the area of the bars.

Histograms with unequal class intervals

When the class intervals in the data are not the same you cannot use the height of the bars to give the frequency. Instead, the vertical scale is used to give frequency density.

$$\text{Frequency density} = \frac{\text{frequency } (f)}{\text{class width}}$$

WORKED EXAMPLE 2

The ages of young people visiting a park are shown in the frequency table.

Draw a histogram to show these results.

Age (n)	Frequency
$5 \leqslant n < 10$	10
$10 \leqslant n < 16$	24
$16 \leqslant n < 18$	6
$n \geqslant 18$	0

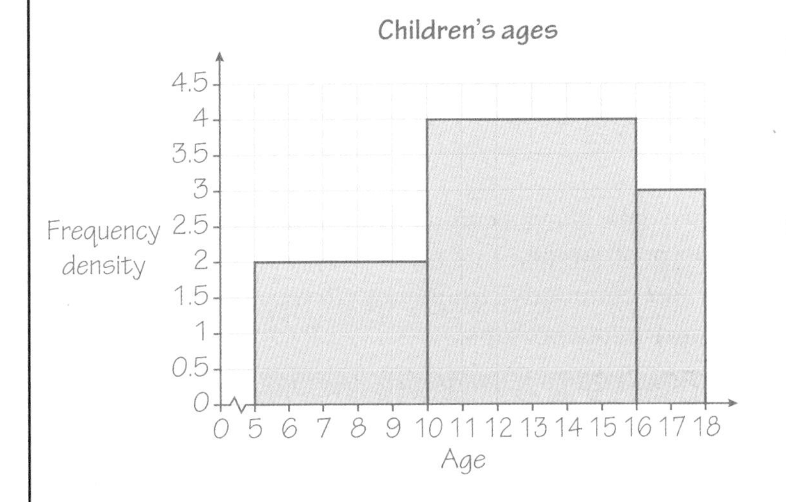

Find the class boundaries first (5, 10, 16 and 18 are the lowest values that could be in each class). Then determine the class widths: $10 - 5 = 5$, $16 - 10 = 6$, $18 - 16 = 2$.

In the histogram the **area** of the block represents the frequency, and the height of the bar represents the frequency density.

The first group has a frequency density of 2:

$$\frac{\text{frequency } (f)}{\text{class width}} = \frac{10}{5} = 2.$$

The second group as a frequency density of 4:

$$\frac{\text{frequency } (f)}{\text{class width}} = \frac{24}{6} = 4, \text{ and so on.}$$

Notice that you can rearrange the formula to find the frequency or the class interval if you have the relevant information.

Problem-solving framework

The heights of sunflowers grown by a class of children are measured and recorded.

85.4	114.9	99.6	81.6	105.0	114.7
88.0	99.2	107.0	123.8	104.5	121.4
116.7	79.4	118.5	115.8	103.8	113.6
77.5	89.9	113.6	100.7	120.3	103.5
75.9	99.0	104.1	99.0	101.8	98.7

Group the data into suitable class intervals and draw an appropriate graph to represent the data.

Continues on next page ...

<table>
<tr><th>Steps for approaching a problem-solving question</th><th>What you would do for this example</th></tr>
<tr><td>Step 1: Identify what you have to do.</td><td>You need to choose appropriate class intervals for the data and a suitable means of displaying the data and then draw it.</td></tr>
<tr><td>Step 2: What maths can you do?</td><td>You know it is best to choose a class interval that allows for between 5 and 10 classes, so start there and put the data into a frequency table.

<table>
<tr><th>Class interval</th><th>Frequency</th></tr>
<tr><td>$75 \leq h < 95$</td><td>7</td></tr>
<tr><td>$95 \leq h < 105$</td><td>11</td></tr>
<tr><td>$105 \leq h < 115$</td><td>6</td></tr>
<tr><td>$115 \leq h < 120$</td><td>3</td></tr>
<tr><td>$120 \leq h < 125$</td><td>3</td></tr>
</table>

As you have grouped data, either a cumulative frequency graph or a histogram would be appropriate.
If you chose to draw a cumulative frequency curve:
Draw up a cumulative frequency table.

<table>
<tr><th>Class interval</th><th>Frequency</th><th>Cumulative frequency</th></tr>
<tr><td>$75 \leq h < 95$</td><td>7</td><td>7</td></tr>
<tr><td>$95 \leq h < 105$</td><td>11</td><td>18</td></tr>
<tr><td>$105 \leq h < 115$</td><td>6</td><td>24</td></tr>
<tr><td>$115 \leq h < 120$</td><td>3</td><td>27</td></tr>
<tr><td>$120 \leq h < 125$</td><td>3</td><td>30</td></tr>
</table>

Construct a cumulative frequency graph remembering to use the upper boundary of each class interval.

Cumulative frequency graph showing sunflower height
</td></tr>
</table>

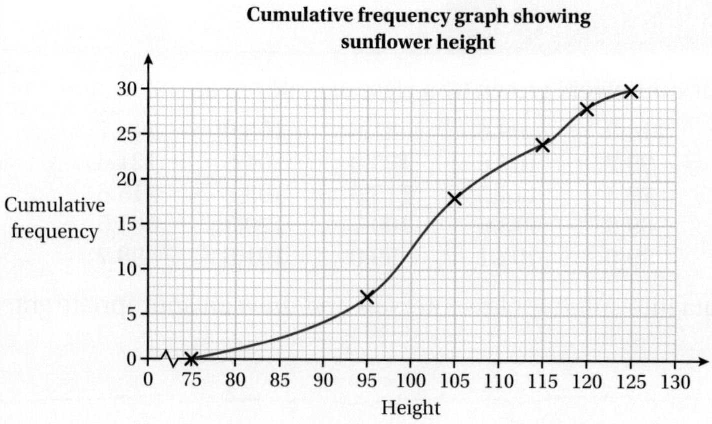

Continues on next page ...

If you chose to draw a histogram:

Calculate the frequency density of each class interval; it can help to draw up a table.

Class interval	Freq	Size of class interval	Frequency density
$75 \leqslant h < 95$	7	20	0.35
$95 \leqslant h < 105$	11	10	1.1
$105 \leqslant h < 115$	6	10	0.6
$115 \leqslant h < 120$	3	5	0.6
$120 \leqslant h < 125$	3	5	0.6

The histogram is plotted with frequency density on the y-axis and height on the x-axis.

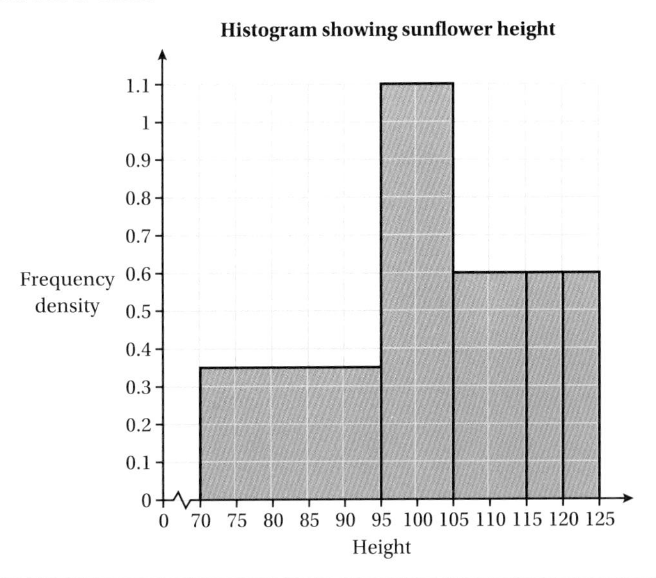

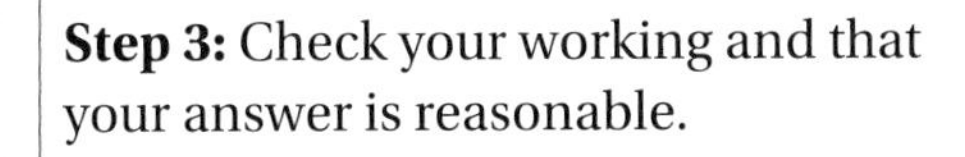

Step 3: Check your working and that your answer is reasonable.

Check the shape of the graphs.

Does the cumulative frequency graph resemble an ogive curve?

Do the areas of the rectangles match the numbers in each class interval? (Area of bar = frequency × frequency density.)

Step 4: Have you answered the question?

Yes, the data is correctly displayed.

EXERCISE 37E

Tip

Too few class intervals can oversimplify, and too many can cancel out the effect of grouping. Aim for between 5 and 10 groups.

1 A sea life centre records the mass in kilograms of small fish eaten by one of their great white sharks over a series of days.

193	210	265	203	246	216	236
343	242	208	294	229	287	266
343	223	308	235	166	231	241
255	196	230	276	247	296	266

Group the data into suitable equally-sized class intervals, and create a cumulative frequency curve.

Use your curve to estimate:

a On how many days did the shark eat less than 230 kg of fish?

b On how many days did the shark eat more than 300 kg of fish?

2 The maximum daily temperature in °C is measured every day for 25 days.

18.5	19.6	18.6	23.5	17.2
19	17.1	23	17.5	15
22.4	24.6	24.1	15.3	18.3
15.1	19.7	19.3	17.9	20.5
16	18.5	18.3	15.8	15.3

Group the data into suitable equally-sized class intervals, and create a cumulative frequency curve.

Use your graph to estimate:

a how many days the temperature was less than 20 °C.

b how many days the temperature was above 22 °C.

c how many days the temperature was between 20 °C and 22 °C.

3 The timed results from a 5-km race are given below:

20:48	18:39	26:09	23:36	21:20	26:07
23:46	23:31	20:03	21:45	23:20	22:55
24:38	25:16	22:24	18:26	21:27	21:04
17:33	22:14	20:57	24:00	19:14	24:38
19:32	26:39	17:31	23:25	22:50	21:27
25:08	22:57	23:55	20:25	25:30	24:45
21:43	19:04	18:19	17:36	22:31	25:14
22:40	21:07	24:11	21:34	25:41	23:45
25:42	24:01	20:19	26:17	20:13	25:10
20:14	24:44	26:21	23:48	22:52	24:14

Group the data into suitable equally-sized class intervals, and create a cumulative frequency curve.

Use your graph to estimate:

a how many runners completed the course in more than 25 minutes.

b how many completed the course in less than 20 minutes.

c how many runners finished in less than 19 minutes.

4 The frequency table shown gives information regarding the test results out of 50, of a group of 23 students.

Present this information in a histogram.

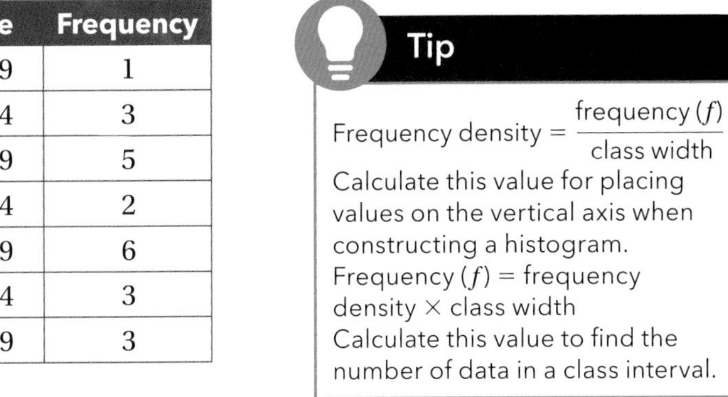

Score	Frequency
15–19	1
20–24	3
25–29	5
30–34	2
35–39	6
40–44	3
45–49	3

> **Tip**
>
> Frequency density $= \dfrac{\text{frequency } (f)}{\text{class width}}$
>
> Calculate this value for placing values on the vertical axis when constructing a histogram.
>
> Frequency (f) = frequency density × class width
>
> Calculate this value to find the number of data in a class interval.

5 In a medical test, a group of students had the distance from hip to heel measured. The measurements were made correct to the nearest centimetre. The results were as follows:

85	86	91	87	77	88	83	86	74	89	85	85	80
94	82	84	89	84	94	84	76	93	86	84	94	84

Present this information in a histogram using the classes:

a 70–79, 80–89, 90–99

b 70–74, 75–79, 80–84, 85–89, 90–94

6 The histogram on the right gives information about the results of Year 9 students in a history examination.

a How many students sat for the examination?

b If the pass mark was 60, how many students passed?

c What percentage of students obtained 90 or more?

d What percentage of students obtained less than 70?

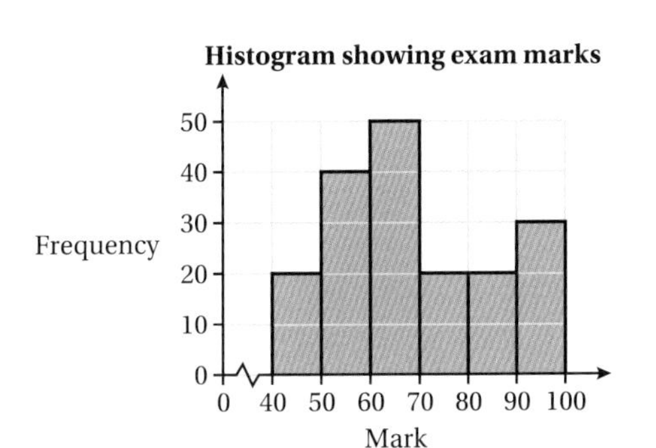

7 The histogram shows temperature measured at 42 different locations.

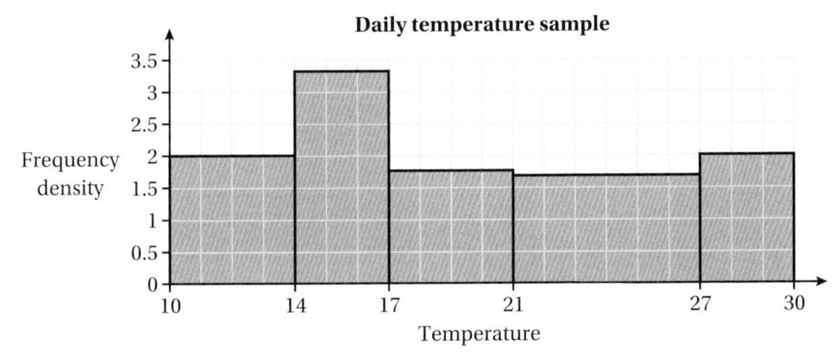

Copy and complete the grouped frequency table that was used to construct this graph.

Group	Frequency
10 ⩽ temp < 14	8
14 ⩽ temp < 17	
17 ⩽ temp < 21	
21 ⩽ temp < 27	
27 ⩽ temp < 30	

8 Data has been collected on the mass of a sample of potatoes in a crop.

The data is grouped as follows:

Group	Frequency
$100 \leqslant \text{mass} < 120\,\text{g}$	6
$120 \leqslant \text{mass} < 130\,\text{g}$	20
$130 \leqslant \text{mass} < 145\,\text{g}$	14
$145 \leqslant \text{mass} < 150\,\text{g}$	20
$150 \leqslant \text{mass} < 160\,\text{g}$	12

a What is the frequency density for the $120 \leqslant \text{mass} < 130\,\text{g}$ group?

b Which group has the highest frequency density?

c Draw a histogram to represent this data.

Section 5: Line graphs for time-series data

Some data that you collect changes with time. For example, the average temperature each month for a year, the number of cars passing through a junction each hour or the amount of minutes on your mobile phone you have left to use by the end of the month.

Line graphs are useful for showing how data changes over time. When time is one of the variables it is always plotted on the horizontal axis of the graph.

The maximum temperature at a weather station is recorded at noon every day. Data that shows change over time like this is called time-series data.

Monday	Tuesday	Wednesday	Thursday	Friday	Saturday	Sunday
15 °C	17 °C	18 °C	21 °C	16 °C	20 °C	14 °C

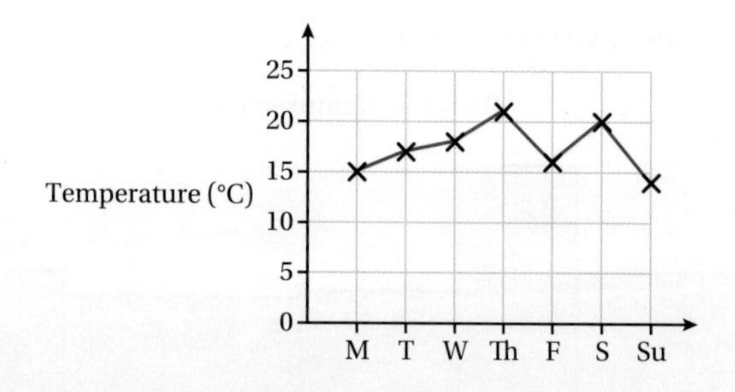

Each of the points on the line is joined by a straight line. This type of graph is used when we want to show trends over time.

EXERCISE 37F

1 **a** Construct a time-series graph for the average temperature (in °C) in a particular city, which is given in the table below.

Month	Jan	Feb	Mar	Apr	May	Jun	Jul	Aug	Sep	Oct	Nov	Dec
Average temp (°C)	15.2	16.5	17.2	19.1	19.6	20.1	22.2	24.1	21.3	19.3	17.6	16.6

b Use the time-series line graph to write a brief description of how the average temperature varies in this particular city.

2 The table below gives the annual profit (in £million) of a company over a ten-year period. Construct a time-series graph of the information.

Year	Year 1	Year 2	Year 3	Year 4	Year 5	Year 6	Year 7	Year 8	Year 9	Year 10
Profit (£million)	2.2	1.8	2.3	1.2	0.6	1.1	2.2	3.1	3.7	4.2

3 The table below gives the number of teeth extracted at a dentist's surgery each month for a year.

Month	Jan	Feb	Mar	Apr	May	Jun	Jul	Aug	Sep	Oct	Nov	Dec
Number of teeth	54	47	49	60	41	45	36	11	38	42	32	22

a Represent this information on a time-series graph.

b Briefly describe how the number of teeth extracted each month changed over the year.

c Why might the number of teeth extracted fall during August?

4 The table below gives the position of a particular five-a-side football team in a league of ten teams at the completion of each week throughout the season.

Round	1	2	3	4	5	6	7	8	9	10	11	12	13	14	15	16	17	18
Position	2	3	5	7	6	5	6	7	5	5	4	5	3	4	3	3	4	3

a Represent this information on a time-series graph.

b Describe the progress of the team throughout the season.

5 The data below shows the value of sales at a service station on a main road over a period of three years. Each quarter represents three months (a quarter) of the year. The quarters are labelled 1 to 12 in the corresponding time-series graph.

Sales quarter	Sales £thousand
Quarter 1	64
Quarter 2	82
Quarter 3	83
Quarter 4	65
Quarter 5	77
Quarter 6	89
Quarter 7	96
Quarter 8	58
Quarter 9	79
Quarter 10	92
Quarter 11	101
Quarter 12	66

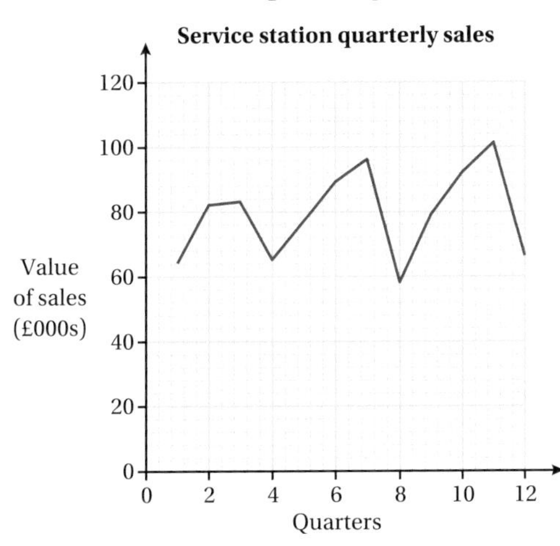

a In which quarter of each year is the value of sales highest?

b In which quarter of each year is the value of sales lowest?

c Compare the sales figures for the first quarter of each year. Are the sales figures improving from one year to the next?

Find answers at: cambridge.org/ukschools/gcsemaths-studentbookanswers

6 The table below gives the numbers of garden sheds sold each quarter during 2012–2014.

Number of sales	Q1	Q2	Q3	Q4
2012	27	32	56	41
2013	33	35	65	45
2014	38	41	72	51

a Represent this information on a time-series graph.

b Describe how the shed sales have altered over the given time period.

c Does it appear that sheds sales are seasonal?

7 This graph shows how the water level in a pond varies from month to month.

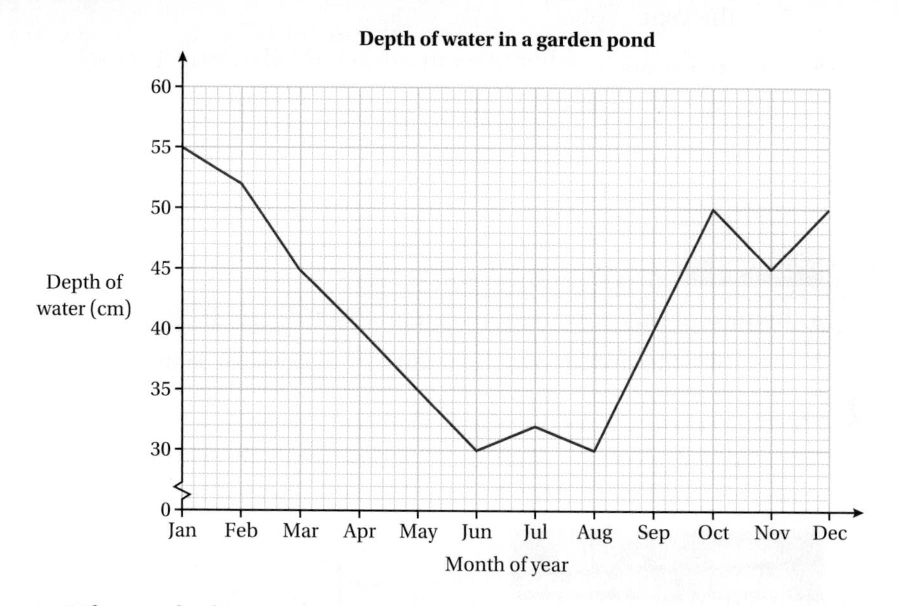

a When is the lowest depth of water?

b What do you think might have happened in July?

c When does the water level drop most rapidly?

d How much water is in the pond in May?

e What is the difference in depth between August and September?

Checklist of learning and understanding

Sampling

- A population is a set of individuals or objects of interest.
- A sample is a small set of data from a population; a **representative sample** is one that represents the characteristics of a larger population.
- In a random sample, each member of the population is equally likely to be chosen for the sample.

Tables and graphs

- A frequency table is a method of organising data by showing how often a result appears in the data.
- Data can be displayed using a number of different graphs.
- All graphs should be clearly labelled and scaled, including a title.
- Vertical line graphs and bar charts are a good way of showing discrete data, where the height of each bar or line determines the frequency.
- Multiple bar charts are a good way of comparing data for two or more groups; a composite bar chart is good for showing the component parts of a data set.
- Pictograms are an interesting visual way of displaying discrete data.
- Pie charts are used to compare categories of the same data set.

Cumulative frequency curves and histograms

- When you have continuous data, the data set can become very large so it can be helpful to group the data into class intervals. Typically, it is good to split the data set into 5 to 10 classes.
- Cumulative frequency curves are used to display grouped continuous numerical data; the cumulative frequency is a running total of frequencies for each class interval. These curves are useful for calculating how many values are above or below a given value. Cumulative frequencies can be plotted against the upper boundaries of each class interval to produce a curved graph.
- Histograms are used to display grouped continuous data; they look like bar charts but they have some important differences:
  - The horizontal scale is continuous and each bar is drawn above a particular class interval; there are no gaps between the bars unless a class interval has no data.
  - The frequency of the data is shown by the **area** of the bars.
  - The vertical axis shows the **frequency density**:

    $$\text{Frequency density} = \frac{\text{frequency } (f)}{\text{class width}}$$

    (When the class intervals are equal, the y-axis of the histogram is often taken to be the frequency.)

Line graphs

- Line graphs for time-series data are useful for showing trends and changes over time.
- Time is always along the x-axis.

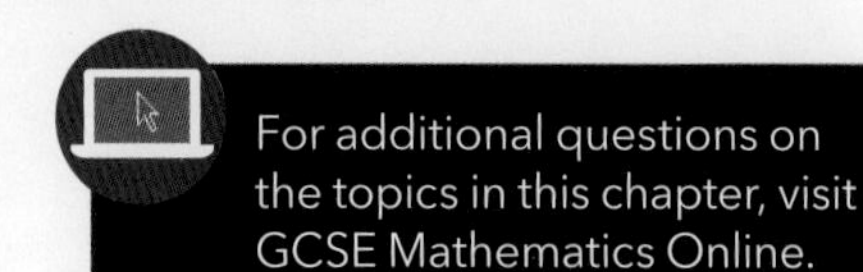
For additional questions on the topics in this chapter, visit GCSE Mathematics Online.

Chapter review

1. Bonita needs to find out how students travel to school, so she decides to take a representative sample.

   a Suggest two ways in which she could do this.

   b Bonita asks a representative sample of 50 students and gets the following results:

Car	15
Walk	17
Bus	6
Taxi	7
Bike	5

   If there are 600 students in the school, what is a sensible estimate of the number of students who walk to school?

2. Two adults are comparing how much money they spend each month.

Expenditure per month (£)	Josh	Ben
Rent	840	450
Food	250	300
Transport	350	160
Savings	250	40
Entertainment	110	250

   a Draw suitable graphs to enable you to compare the proportions of money they spend.

   b Write two sentences comparing their spending habits.

3. The profits for two companies are given for each quarter of a two-year period below:

	Company profits (£)	
	Company A	Company B
1st quarter 2013	134 820	125 912
2nd quarter 2013	138 429	189 355
3rd quarter 2013	140 721	130 969
4th quarter 2013	131 717	156 548
1st quarter 2014	103 746	219 357
2nd quarter 2014	197 028	151 296
3rd quarter 2014	187 883	249 216
4th quarter 2014	168 414	102 158

   Use the data to plot a suitable graph to compare how profits change.

   a Which company is the most successful?

   b Which is the biggest change between quarters?

4 Two machines are used in a factory to pack crisps into packets of 65 g. A random sample of the mass of 30 bags packed by each machine is given in the tables shown.

Machine A				
65.4	64.8	65.4	64.3	64.2
64.4	65.3	65.1	64.7	64.7
64.6	65.0	65.5	64.1	65.0
64.9	64.5	64.6	64.2	64.7
65.3	64.9	64.5	64.6	65.4
65.5	64.5	65.5	64.6	64.2

Machine B				
64.0	65.5	63.6	63.9	65.8
64.3	65.2	64.7	64.7	64.4
65.4	64.2	65.7	65.2	63.9
64.4	65.6	63.9	65.9	65.0
65.6	65.7	64.0	64.2	66.0
65.0	65.7	64.8	64.8	64.7

a Use class intervals of 0.5 and draw a histogram for each machine.

b Which machine is the most reliable?

5 This table summarises the ages of the members of the audience at a cinema one Saturday afternoon.

Age (*a* years)	Frequency
$15 \leq a < 20$	25
$20 \leq a < 30$	32
$30 \leq a < 40$	16
$40 \leq a < 60$	15
$60 \leq a < 90$	12

a Draw a histogram to represent this distribution. *(3 marks)*

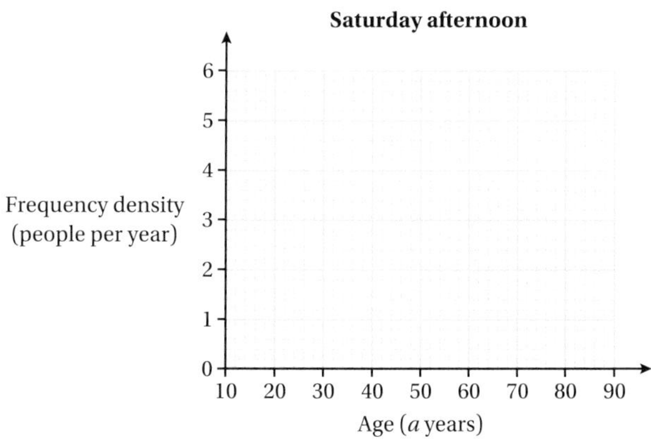

b The histogram below represents the distribution of the ages of the members of the audience at the cinema on Saturday evening.

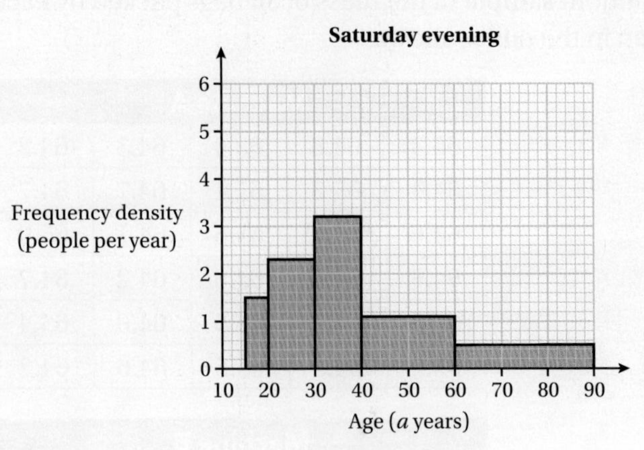

Make two comparisons between the distributions for the afternoon and the evening. *(2 marks)*

38 Analysing data

In this chapter you will learn how to ...

- calculate and compare summary statistics for ungrouped and grouped data.
- compare distributions.
- draw and interpret box plots.
- recognise when data is being misrepresented.
- plot and interpret scatter diagrams and use them to describe correlation and predict results.
- identify outliers and understand how they can indicate errors in data.

For more resources relating to this chapter, visit GCSE Mathematics Online.

Using mathematics: real-life applications

Analysing large sets of data enables financial and insurance companies to make predictions about what might happen in the future. Car insurance premiums are worked out according to typical or 'average' behaviour of large groups of people.

"We group drivers together by age and gender and use statistics to find typical driving behaviour for each group. Young drivers have more accidents, so their insurance costs more." *(Insurance broker)*

Tip

Knowing how to calculate averages and measures of spread gives us tools to compare different sets of data. Make sure you know what these are and when to use the different measures.

Before you start ...

KS3	You should remember how to find the mean, median, mode and range of a set of data.	**1** Find the mean, median, mode and range of the following sets of data. Give your answers correct to one decimal place. **a** 2, 4, 2, 7, 3, 5, 4, 2, 3, 1 **b** 40, 20, 30, 60, 50, 10
Ch 18	You should be able to plot coordinates on a set of axes.	**2** Write down the coordinates of points A, B and C on the line.
Ch 18	You should be able to recognise whether a gradient is positive or negative.	**3** Use the graph above. **a** What is the gradient of the graph? **b** What is the equation of the line?

Find answers at: cambridge.org/ukschools/gcsemaths-studentbookanswers

Assess your starting point using the Launchpad

STEP 1

1 The table shows how many cups of coffee a group of office workers drank in a particular week.

Number of cups of coffee	Frequency
0–5	16
6–10	5
11–15	5
16–20	4
21–25	5
26–30	3

a What is the modal class of the data?

b Estimate the median number of cups consumed per week.

c Estimate the mean number of cups of coffee consumed per week.

d How many people drink 10 or fewer cups of coffee per week.

2 These box plots compare the distribution of a class's marks in March and June of the same year.

a Comment on the differences in the data.

b Did the class's performance improve in June?

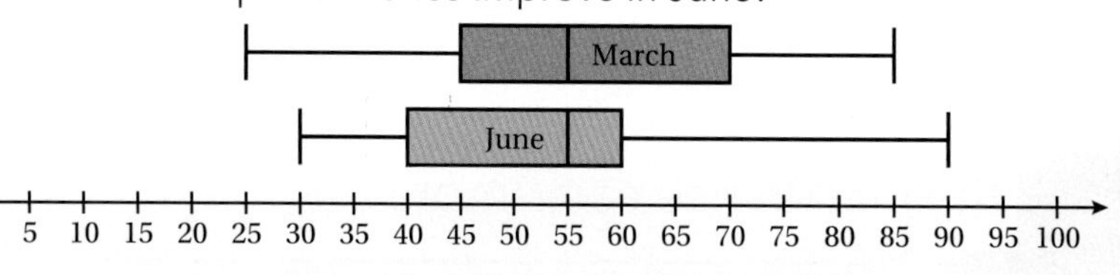

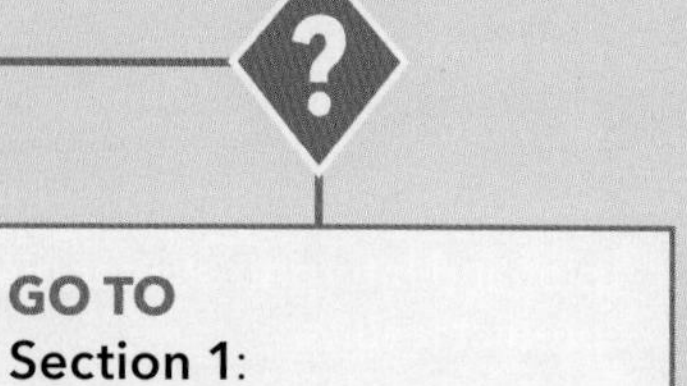

GO TO
Section 1:
Summary statistics

STEP 2

3 This graph appeared in a newspaper article.

Explain how this graph could be misleading.

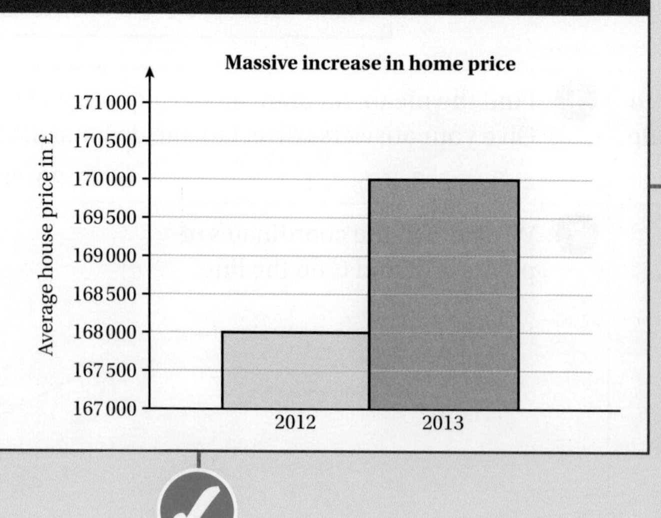

GO TO
Section 2:
Misleading graphs

GO TO
Step 3:
The Launchpad continues on the next page …

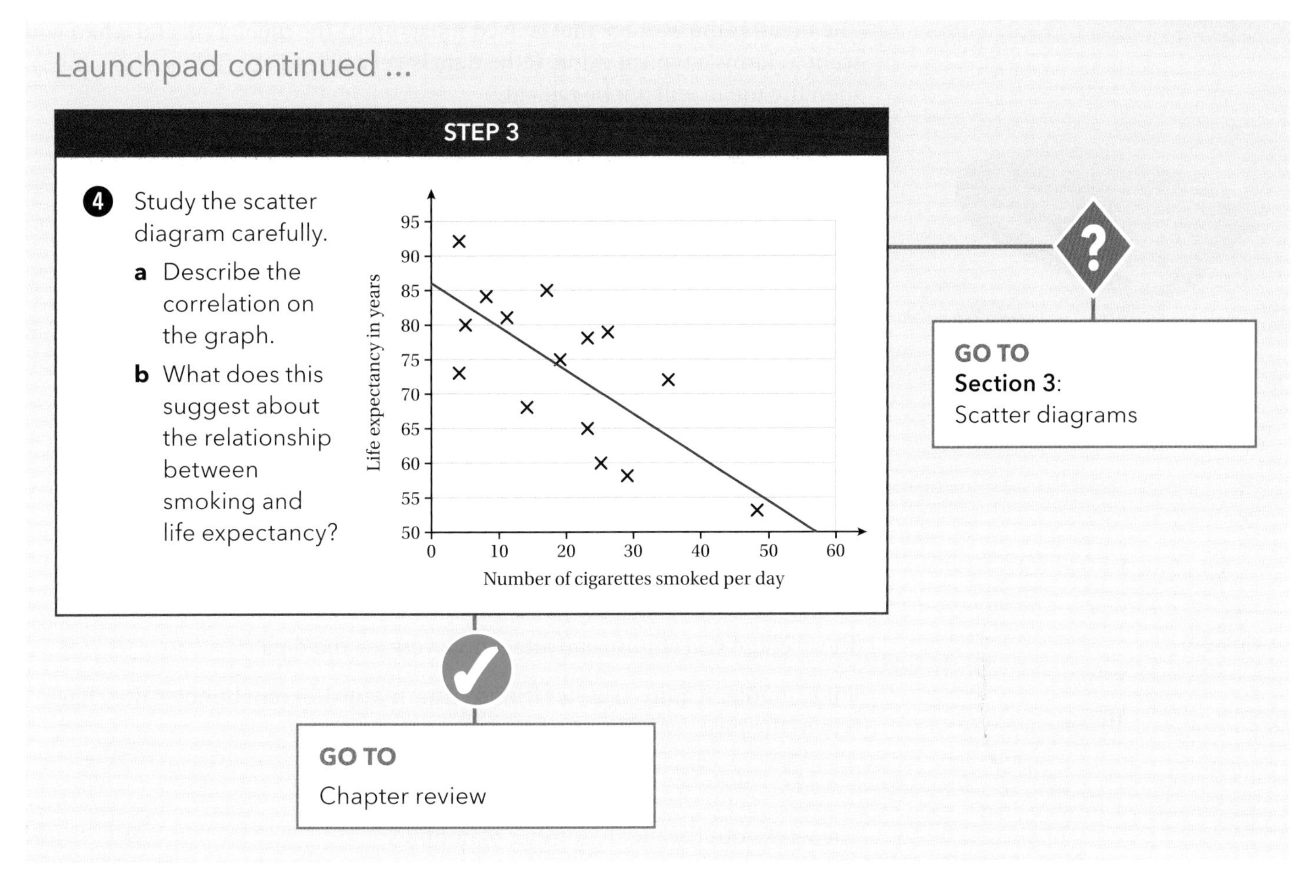

Section 1: Summary statistics

In statistics you are often asked to give a single value that summarises the data and tells you something about it. You have learnt to use four different values to summarise data:

- the mean: $\frac{\text{sum of values}}{\text{number of values}}$
- the mode: the value with the highest frequency
- the median: the middle value when the values are arranged in size order
- the range: the difference between the highest value and the lowest value.

The mean, median and mode are all types of averages, or measures of central tendency.

The range is a measure of spread or dispersion. The range is useful for determining whether the mean is distorted or not.

To describe and compare two sets of data, calculate the averages and range and write sentences to summarise what you notice.

Choosing the correct average

The type of average that you choose depends on the situation and what you want to know.

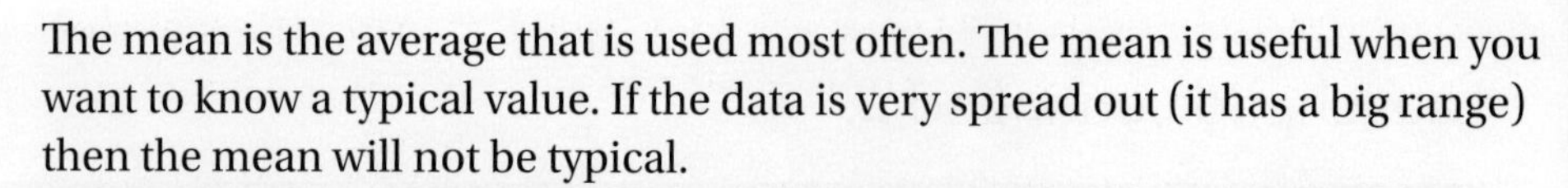

The mean is the average that is used most often. The mean is useful when you want to know a typical value. If the data is very spread out (it has a big range) then the mean will not be typical.

For example, the boss in a company earns £20 000 per month. Her nine employees earn £2000 each. This gives a mean salary of £3800, which is not typical.

In this situation, the median salary is £2000 and the modal salary is £2000. Both are more representative than the mean.

When the data is not numerical, you have to use the mode as the average.

The mode is most useful when you need to know which item is most common or most popular.

You would use the mode when you wanted to show:

- which clothing size was bought most often
- what shoe size is most common
- what brand of mobile phone is the most popular.

The average you choose can affect how you see the data.

For example, Jeanne asks her friends how many different hobbies they have had in the past year. These are their answers:

1 1 2 1 3 40 1

She works out the mean number of hobbies:

$$\frac{1+1+2+1+3+40+1}{7} = \frac{49}{7} = 7$$

Tip

A very high or a very low value in a set of data is called an **outlier**. If there is an outlier then the mean will not be a typical value.

The mean suggests that Jeanne's friends have had an average of 7 different hobbies each in the past year.

But that is not a very representative average and it does not give the typical number of hobbies. The mean number of hobbies is high because one friend had many more different hobbies than the others.

Jeanne doesn't think the mean is a good average for her set of data, so she finds the median:

1 1 1 1 2 3 40

This gives her an average of 1 hobby. That seems more typical. The mode is also 1 because most people have only had one hobby. The median and the mode are more representative and they are not affected by outliers in the data, so they are better averages in this case.

Quartiles and interquartile range

Quartiles and the interquartile range are summary statistics that give more information about the spread of data.

The median divides the data into two halves, with 50% of the data above the median, and 50% of the data below the median.

The quartiles divide an ordered data set into quarters. You work out the value of the quartiles in a similar way to the median.

The first quartile (Q_1), often called the lower quartile, is the data value one quarter of the way along the data set.

So 25% of the values are smaller than the first quartile, and three quarters lie to the right of the first quartile.

The second quartile (Q_2) is the same as the median.

The third quartile (Q_3), often called the upper quartile, is the value that lies three quarters of the way along the data set. So 75% of the values are smaller than this value.

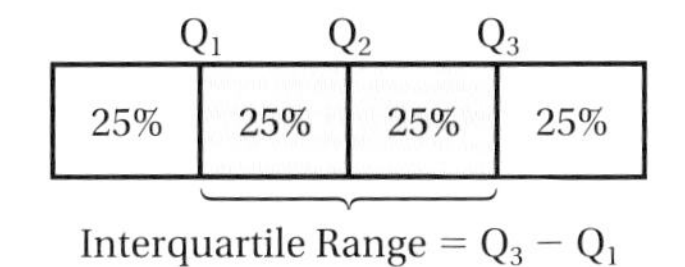

The interquartile range (IQR) is the difference between the values of the upper and lower quartiles.

IQR = third quartile − first quartile

The IQR tells us where the middle 50% of the data occurs. This is important to know, because it gives us information about whether the data is spread out far away from the median, or clustered close to the median value.

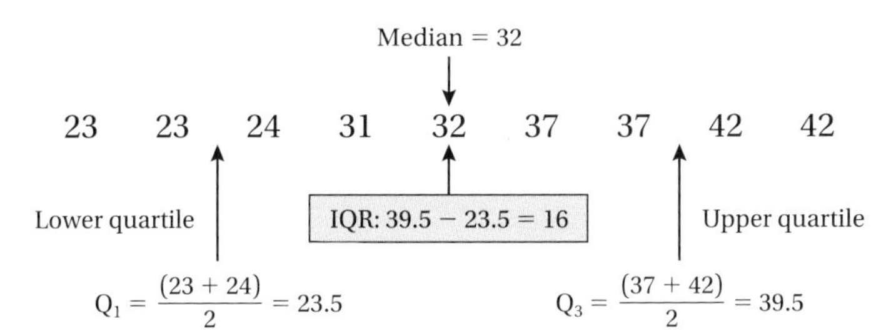

WORKED EXAMPLE 1

Josh has developed a website and is monitoring how many hits it receives per hour.

In the first two days (48 hours) it receives the following numbers of hits per hour (arranged in numerical order):

100	105	106	106	107	107	108	110	117	118
135	137	145	148	148	148	153	155	157	159
162	171	171	179	183	183	185	185	189	199
201	203	204	209	216	220	223	224	224	227
229	230	231	233	234	235	237	238		

a Find the following summary statistics:
 i the mean, median and mode. **ii** the range.

Continues on next page ...

b Josh is trying to sell advertising on his website.
Write a mathematical sentence he could use about the number of hits his site is receiving.

c Josh compares his data to a similar website run by Delia. Delia's data set has the following data values:

mean = 180 hits
median = 140 hits
mode = 135 hits
range = 200

What can Josh say to compare the two sets?

a **i** Mean $= \frac{\text{sum of values}}{\text{number of values}} = \frac{8394}{48} = 174.88$ hits (2 dp)

Median $= \frac{179 + 183}{2} = 181$ hits

Mode = 148

ii The range = 238 − 100 = 138

b There is a consistent hit rate of over 100 hits per hour with 175 hits per hour on average.

c Although the mean of the hits is a bit higher for Delia's set, the median is much lower for her set than it is for Josh's. This indicates that the mean of Delia's hits is influenced by a few high values, but usually the number of hits is lower. Delia's data shows a wider range, showing that the data is more spread out, and therefore less consistent.

Median of the data is halfway between the 24th and the 25th data values.

The value 148 is the mode because it occurs three times.

You need to decide which measure of average you are going to use.

Finding the mean from a frequency table

You can also find the mean when data is set out in a frequency table.

WORKED EXAMPLE 2

Leona is investigating the number of spots a typical ladybird has.

She observes and records the following data.

Number of spots	Frequency
1	4
2	7
3	12
4	14
5	9
6	7

Calculate the mean number of spots.

Continues on next page ...

Number of spots	Frequency	Number of spots × frequency
1	4	4
2	7	14
3	12	36
4	14	56
5	9	45
6	7	42
Total	53	197

$197 \div 53 = 3.72$

Add an extra column to the table and calculate the number of spots in total.

Divide the total number of spots by the number of ladybirds

Analysing grouped data

Data is sometimes grouped together before it is analysed. The groups are known as class intervals. Note that they **do not** overlap. For example,

Marks scored	Frequency
0–9	6
10–19	6
20–29	4
30–39	5
40–50	9
Total	30

You learnt about class intervals in Chapter 37.

When you only have the grouped data in a frequency table, it is not possible to calculate precise values for the mean, median, mode and range because you don't know the individual values.

You can identify the modal and median classes from the table.

- The modal class is the class interval that has the most elements, not the individual value that appears the most. In the table above the modal class is 40–50 marks.
- To estimate the median of grouped data, find the class interval in which the middle value occurs. It is only possible to say that the median is within that group. In the table there are 30 values, the middle value is between the 15th and 16th values, so it must fall into the class 20–29 marks.

Estimating the mean of a frequency distribution

To estimate the mean of grouped data, find the midpoint of each class interval. The midpoint is found by adding the lowest and highest possible values for each class interval and dividing by 2. Multiply each midpoint by the frequency for each class interval. Find the total of these products, and divide by the total number of values you have.

Find answers at: cambridge.org/ukschools/gcsemaths-studentbookanswers

WORKED EXAMPLE 3

Ben goes fishing and records the masses of the fish he catches in the table below:

Mass (m) in kg	Frequency	Midpoint	Midpoint × frequency
$2 \leqslant m < 4$	5		
$4 \leqslant m < 6$	8		
$6 \leqslant m < 8$	4		
$8 \leqslant m < 10$	9		
$10 \leqslant m < 12$	3		

a Complete the table.

b Find the modal class.

c Estimate the mean, median and range.

a

Mass (m) in kg	Freq	Midpoint	Midpoint × frequency
$2 \leqslant m < 4$	5	3	15
$4 \leqslant m < 6$	8	5	40
$6 \leqslant m < 8$	4	7	28
$8 \leqslant m < 10$	9	9	81
$10 \leqslant m < 12$	3	11	33

b The modal class is $8 \leqslant m < 10$

The modal class is the class with the highest frequency.

c Total of the midpoint × frequency values is 197.

Estimated mean is $197 \div 29 = 6.79$ kg (2 dp)

The total frequency is 29. The mean is only **estimated** because the data is grouped and so you cannot know its exact value.

The median class is $6 \leqslant m < 8$ kg.

Range = $12 - 2 = 10$ kg.

There are 29 values, so the middle value is value number 15. This occurs in the interval $6 \leqslant m < 8$ kg.

EXERCISE 38A

1 A large company has kept a record of how many days each employee is absent from work each year.

The results are in the table below:

Days absent (d)	Frequency	Midpoint	Midpoint × frequency
$0 \leqslant d < 5$	15		
$5 \leqslant d < 10$	23		
$10 \leqslant d < 15$	19		
$15 \leqslant d < 20$	12		
$20 \leqslant d < 25$	6		
Total			

a Complete the table.

b Use the information to find the modal class.

c Estimate the mean, median and range.

2 For a charity event, several students are throwing rubber darts at a Velcro dart board while blindfolded.

The scores they achieve are given below:

89	11	57	25	55	78
28	35	15	90	83	38
57	37	28	14	36	40
74	59	57	9	18	70
25	18	22	2	37	53
74	61	79	53	87	46
30	29	4	90	83	77

a Choose suitable class intervals and group the data.

b Estimate the mean, the median and the range of the scores.

c Is it sensible to estimate the range?

d What is the modal group?

3 A health club has measured its members' heights (in metres) before buying some new gym equipment. The data is given below:

1.68	1.68	1.58	1.72	1.58	1.75	1.89
1.84	1.55	1.65	1.66	1.84	1.55	1.81
1.47	1.55	1.58	1.66	1.55	1.61	1.68
1.57	1.57	1.69	1.65	1.75	1.55	1.73
1.64	1.85	1.53	1.65	1.77	1.66	1.75
1.75	1.59	1.88	1.82	1.62	1.69	1.67
1.63	1.66	1.84	1.77	1.52	1.84	1.53

a Use group intervals of every 5 cm, starting with the group $1.45 \leqslant h < 1.50$.
Estimate the mean and the median.

b What is the modal class?

c Use class intervals of every 10 cm, starting with the class $1.40 \leqslant h < 1.50$.
What difference does this make to your estimates of the mean and median?

4 30 runners complete a marathon race. Their times are given below (to the nearest minute):

2 hours 45 mins	3 hours 25 mins	3 hours 46 mins	4 hours 15 mins
5 hours 8 mins	4 hours 49 mins	4 hours 18 mins	3 hours 38 mins
3 hours 43 mins	3 hours 5 mins	2 hours 55 mins	4 hours 23 mins
4 hours 25 mins	3 hours 39 mins	3 hours 20 mins	4 hours 1 min
3 hours 33 mins	4 hours 6 mins	5 hours 11 mins	2 hours 51 mins
4 hours 35 mins	3 hours 19 mins	4 hours 47 mins	4 hours 28 mins
5 hours 5 mins	4 hours 19 mins	2 hours 46 mins	3 hours 18 mins
3 hours 53 mins	4 hours 35 mins		

a Group the data into suitable class intervals.

b Find the modal class.

c Estimate the mean, median and range.

Find answers at: cambridge.org/ukschools/gcsemaths-studentbookanswers

5 The mass of fruit from a farm is recorded in the table below.

Mass of fruit in kg	Frequency	Midpoint of class interval	Midpoint × frequency
$200 \leqslant m < 250$	15		
$250 \leqslant m < 300$	11		
$300 \leqslant m < 350$	13		
$350 \leqslant m < 400$	7		
$400 \leqslant m < 450$	2		
$450 \leqslant m < 500$	2		
$500 \leqslant m < 550$	2		
Total			

a Calculate an estimate of the mean mass of the fruit.

b In which interval does the median lie?

c In which interval does the third quartile lie?

Estimating values from cumulative frequency graphs

It is possible to find further summary statistics about grouped data from a cumulative frequency curve.

Remember that cumulative frequency graphs are drawn from grouped data and are plotted at the upper end of the class interval.

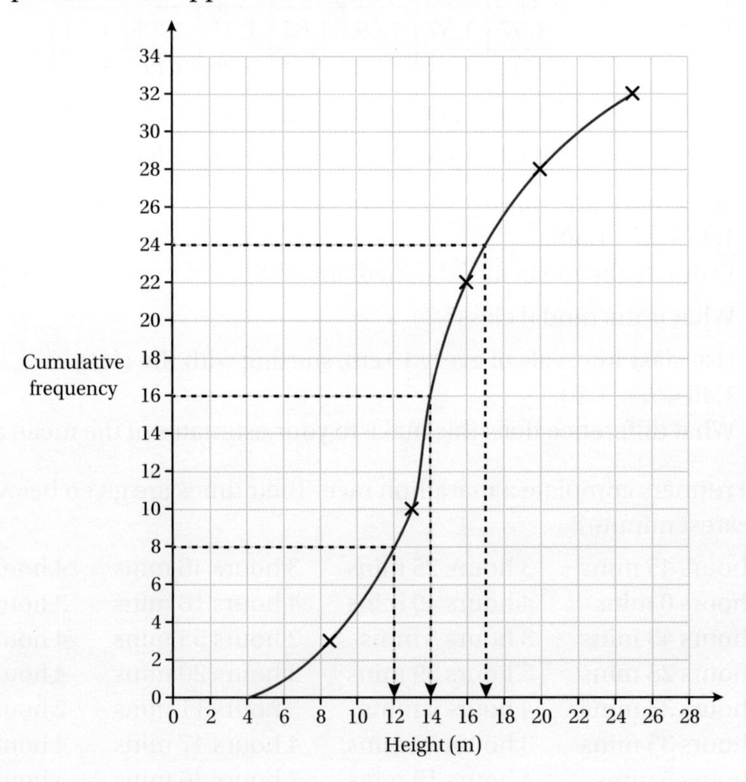

In the graph above there are 32 pieces of data, as the maximum cumulative frequency is 32.

The median of the data is therefore halfway up the vertical scale (16). Drawing a line across to the curve and down to the horizontal scale gives a value of 14. (This is an estimate of the median as the data is grouped.)

The lower quartile is the value $\frac{1}{4}$ of the total frequency. $\frac{1}{4}$ of 32 is 8, so drawing across to the curve and down to the horizontal scale gives an estimate of 12.

The upper quartile is $\frac{3}{4}$ of the total frequency, so is found from 24 ($\frac{3}{4}$ of 32) and drawing across to the curve in the same way. The estimate of the upper quartile is 17.

The interquartile range is upper quartile − lower quartile, and is a way of measuring how spread out the middle 50% of the data values are.

In this example, the interquartile range (IQR) is 17 − 12 = 5 (although this is an estimate).

WORKED EXAMPLE 4

24 dogs are weighed and their masses recorded:

46	41	44	38	49	30
27	49	43	30	37	48
30	28	46	51	43	36
47	52	37	29	38	40

a Group the data into suitable equal-sized class intervals and draw a cumulative frequency graph.

b Use the cumulative frequency graph to estimate the median and the interquartile range.

a

Class intervals	Frequency	Cumulative frequency
$25 \leq m < 30$	3	3
$30 \leq m < 35$	3	6
$35 \leq m < 40$	5	11
$40 \leq m < 45$	5	16
$45 \leq m < 50$	6	22
$50 \leq m < 55$	2	24

Remember to set the size of the classes so that there are between 5 and 10 classes.

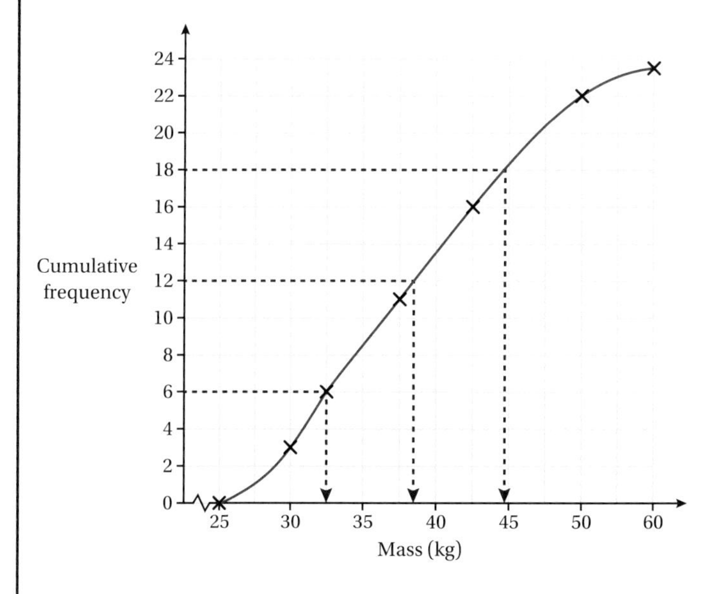

Plot the cumulative frequency against the upper boundary of each class and draw a smooth curve through the points.

b Median: approximately 38 kg

IQR: 44.5 kg − 32.5 kg = 12 kg

Add lines to your graph to help you find the median, Q_1 and Q_2. Then calculate the IQR.

EXERCISE 38B

1 The graph below shows the cumulative frequency of the ages of 30 members of a bingo club.

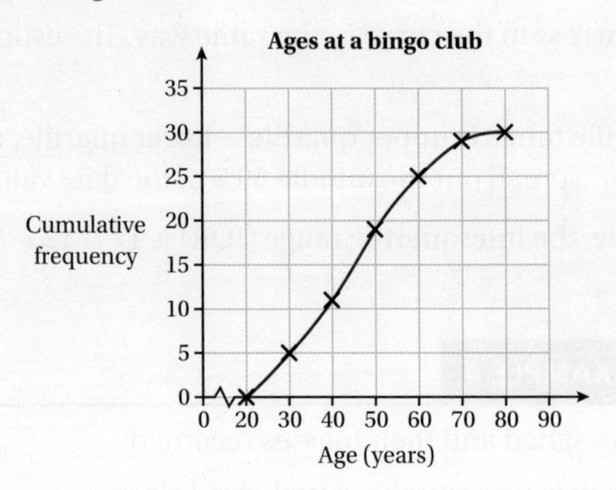

Use the graph to estimate the median and interquartile range of the ages.

2 The graph below shows the prices of 25 cars in a second-hand car dealership.

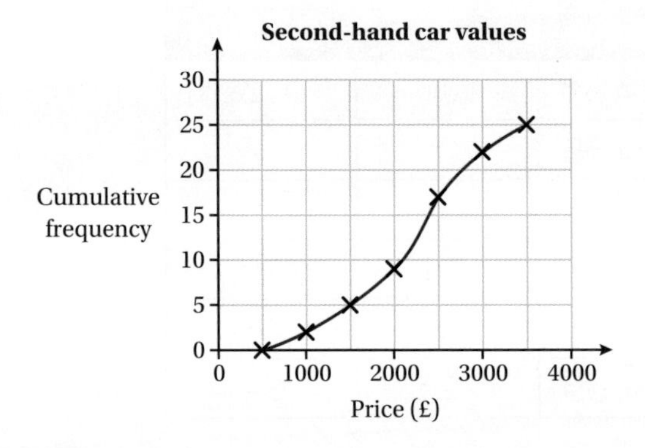

Use the graph to estimate the median and the interquartile range of the values.

3 The graph below shows the time taken for the same train journey on 32 different occasions.

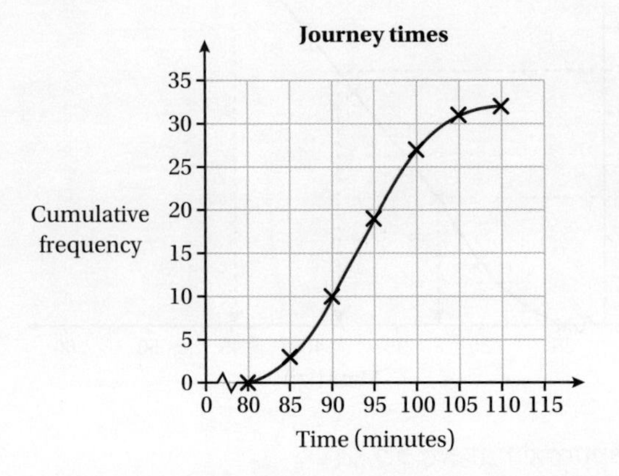

Use the graph to estimate the median and the interquartile range of the journey times.

Comparing two or more sets of data

Summary statistics allow us to compare sets of data and make decisions about them. One summary statistic on its own does not give enough information about the whole set. Think about the following:

Set A may have a similar mean value to Set B, but the median is lower than the median of Set B. This shows us that there are a few higher values in the set that have made the mean higher, but that more of the values are low.

One set may have a higher mean than the other, but the range of the data may be much wider, which shows us that many of the values are very different from the mean or median.

Using box plots to compare two sets of data

A **box plot** is a way of displaying data that also shows five summary statistics: the lowest and highest values (range), the first and third quartiles (the interquartile range) and the median. The box plot gives an instant impression of how the data set is distributed.

It is especially useful for comparing data for two different populations, by aligning the box plots on the same number line (scale).

Tip

When comparing box plots, plot them on the same scale so that you can compare the IQR, median and range.

WORKED EXAMPLE 5

The masses in kilograms of 20 students are (in order):
48, 52, 54, 55, 55, 58, 58, 61, 62, 63, 63, 64, 65, 66, 66, 67, 69, 70, 72, 79.

Draw a box plot to represent this data.

The median is 63 kg.

Start by calculating the median: there are 20 data values, so the median will be halfway between the 10th and the 11th data values; in this case they are both 63.

$Q_1 = \frac{55 + 58}{2} = 56.5$ kg.

Now find the lower quartile: consider the data values that are lower than the median. Use the same approach to find the midpoint as you used for finding the median.

$Q_3 = \frac{66 + 67}{2} = 66.7$ kg.

Repeating the process in the upper half of the data, the upper quartile is between the 15th and 16th data values.

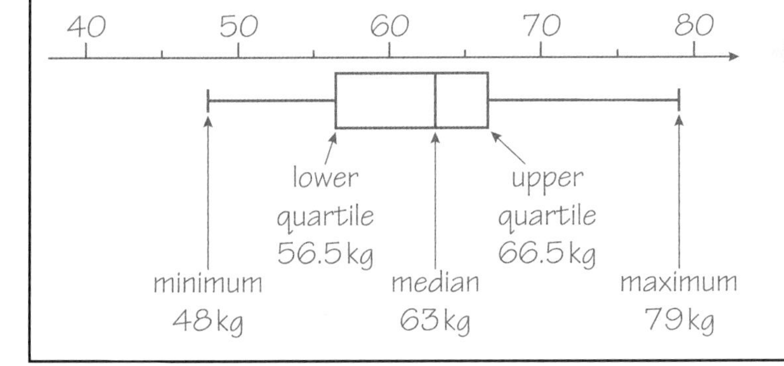

To draw a box plot, use a horizontal scale that allows the highest and lowest values. Mark the median, and the upper and lower quartiles. Create a rectangle from the upper and lower quartiles as shown. Extend a horizontal line to the highest and lowest values.

Note that the lower quartile is sometimes referred to as Q_1, and the upper quartile as Q_3.

Find answers at: cambridge.org/ukschools/gcsemaths-studentbookanswers

WORKED EXAMPLE 6

There are ten boys and ten girls in a Year 7 class.

Their heights (in cm) are:

Girls	137	133	141	137	138	134	149	144	144	131
Boys	145	142	146	139	138	148	138	147	142	146

a Use calculations of the mean, median and range to describe and compare these populations.

b Draw a box plot for both sets of data, and compare the interquartile ranges.

a Girls: 131 133 134 137 137 138 141 144 144 149

Girls: mean = 138.8 cm, median = 137.5 cm and range = 18 cm.

Boys: 138 138 139 142 142 145 146 146 147 148

Boys: mean = 143.1 cm, median = 143.5 cm and range = 10 cm.

Arrange the data in order. Then calculate the mean, median and range

b

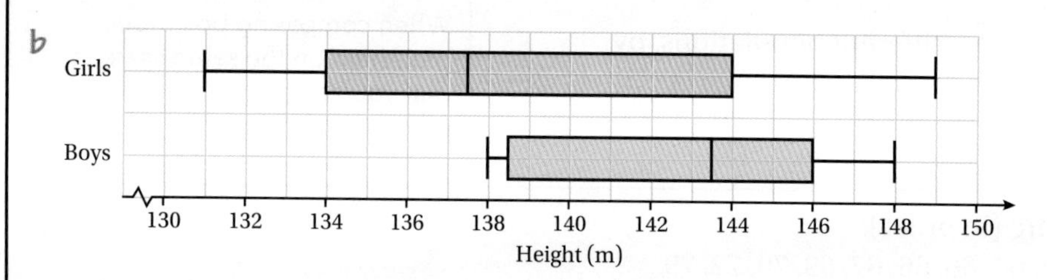

The IQR for girls (10 cm) is wider than that for boys (7.5 cm), showing that the data is more varied and spread out.

EXERCISE 38C

1 The results from two maths tests are given.

A	35	68	55	52	49	63	61	69	35	53
B	47	34	71	41	60	44	57	74	67	64

Describe and compare the results.

2 Two cricketers are having an argument about who has had the better season.

They have both batted 16 times, and the number of runs they have scored in each innings is given below.

Ahmed	27	16	36	27	55	35	51	38	44	17	41	53	7	43	48	49
Bill	2	30	44	11	26	32	13	46	40	44	0	45	15	34	14	24

a Compare and describe their records.

b Who do you think has had the better season?

3 Yusuf has recorded the time it takes to get home on two different buses.

Which bus route should he use?

Bus 127	17	17	21	23	19	20	19	18	21	22	19	22	21	20
Bus 362	23	26	20	15	15	20	26	19	18	15	16			

Explain your answer. Does it matter that he has more data about the 127 bus?

4 A factory needs to choose between two machines that are both capable of bottling soft drinks.

Both manufacturers have provided data about how many bottles each machine fills per hour.

Machine A	
Bottles (b)	**Frequency**
$200 \leqslant b < 250$	36
$250 \leqslant b < 300$	48
$300 \leqslant b < 350$	59
$350 \leqslant b < 400$	61
$400 \leqslant b < 450$	21

Machine B	
Bottles (b)	**Frequency**
$200 \leqslant b < 250$	16
$250 \leqslant b < 300$	58
$300 \leqslant b < 350$	63
$350 \leqslant b < 400$	78
$400 \leqslant b < 450$	15

Use estimates of the mean, median and the range along with the modal group to decide which machine to choose.

5 **a** According to the Office of National Statistics, the 'average' price of a house in the UK in July 2014 was £272 000.

What would be the best type of average to measure house prices?

Explain why you think so.

b Give an example of when using the mean as a measure of central tendency would be the most useful.

c Give two examples when using the mode as a measure of central tendency is the most useful.

6 The heights, measured in centimetres, of 25 students in a class are:

170 175 133 153 164 189 143 133 167 145 150 164 169
159 177 186 173 164 177 168 142 155 153 167 166

a Find Q_1, the median and Q_3.

b Find the interquartile range.

c Draw a box plot to display the data.

7 The annual incomes of 30 people, given correct to the nearest £1000, are:

54 000	67 000	92 000	78 000	54 000	87 000
102 000	112 000	132 000	45 000	256 000	89 000
78 000	98 000	34 000	75 000	65 000	100 000
34 000	68 000	79 000	81 000	82 000	103 000
21 000	345 000	98 000	67 000	105 000	98 000

a Find Q_1, the median and Q_3.

b Find the interquartile range.

c Draw a box plot to display the data.

Find answers at: cambridge.org/ukschools/gcsemaths-studentbookanswers

8 Two teams of friends have recorded their scores on a game and created a pair of box plots.

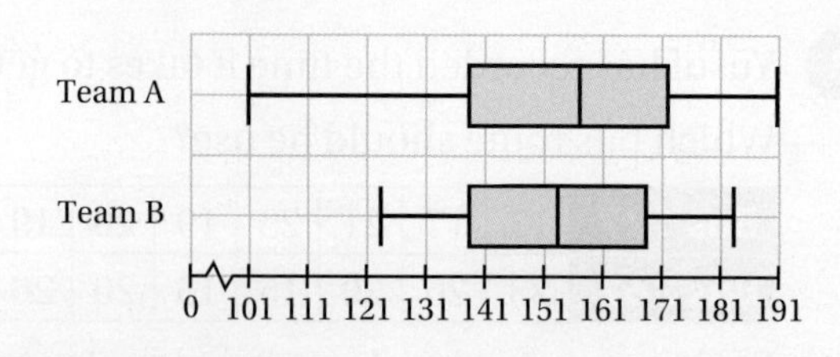

a What is the interquartile range for Team A?

b What is the interquartile range for Team B?

c Which team has the most consistent scores?

d To stay in the game you must score at least 120. Which team seems most likely to stay in?

e Which team gets the highest scores?

f Explain your reasons for part **e**.

Section 2: Misleading graphs

One of the advantages of using graphs is that they show information at a glance. But graphs can also be misleading because most people don't look at them very closely.

When you look carefully at a graph you may find that it has been drawn in a way that gives a misleading impression. Sometimes this is intentional, sometimes it is not. You need to be able to look at graphs and know if they are misleading or wrong.

When you look at a graph, you have to think about:

- the scale and whether or not it has been exaggerated in any way to give a particular impression
- whether or not the scale starts at 0; this can affect the information shown and give a misleading impression
- whether bars or pie diagrams have 3D sections which make some parts look much bigger than others
- whether the scales are labelled and whether or not the graph has a title
- whether the source of the data is given.

Here are some examples of misleading graphs.

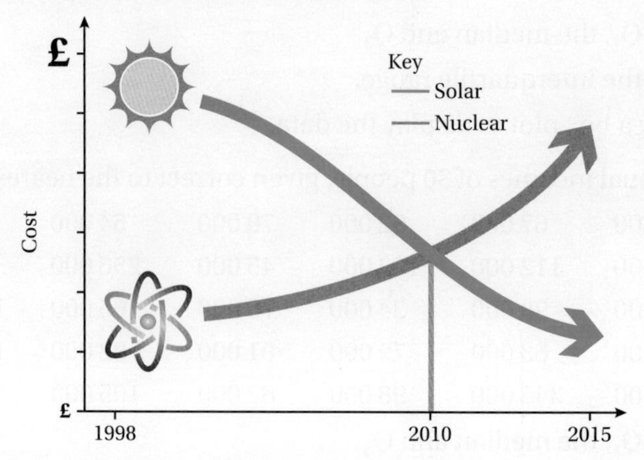

This graph seems to suggest that the price of solar energy is dropping quickly while the cost of nuclear power is increasing.

There are no values on the vertical scale, so it is not possible to decide whether that is really true and no source is given for the information. The scale could be £5 million at the bottom and £10 million at the top, in which case the graph would be very misleading.

You also don't know what costs are being compared. The graph could be comparing the cost of building a nuclear power station (very expensive) and the cost of installing 25 solar panels (much less expensive).

This graph is suggesting that recycling has increased dramatically from 1980.

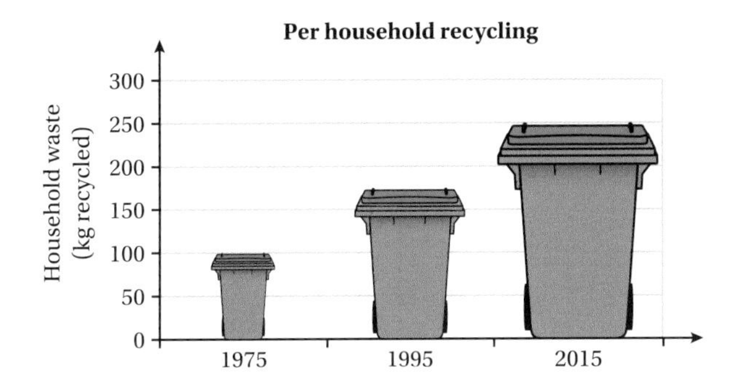

The graph uses proportion to mislead. If you look at the scale, you will see that the amount of recycled material has increased from 100 kg to 250 kg, so 2.5 times more material is recycled. The bin, however, is about six times bigger, so it looks like much more is recycled.

By drawing the pie chart in this orientation and making the sectors 3D, it looks like raisins are just as popular as peanuts and that popcorn is more popular than crisps. The real figures show that only 5% chose raisins and 11% chose peanuts, so the green sector that sticks out represents less than half of the blue sector. The other two sectors each represent 42% but they don't look the same size in this graph.

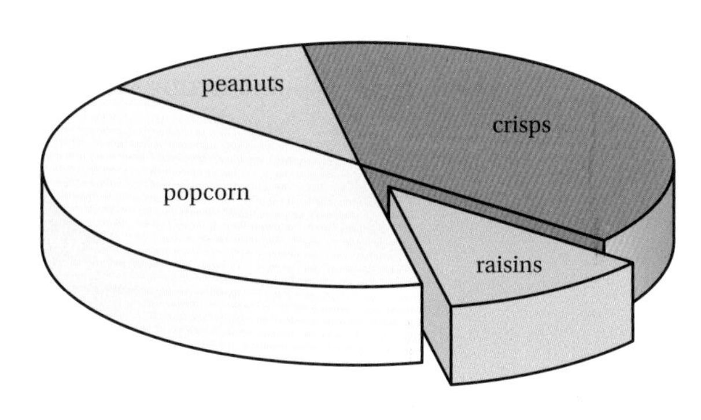

WORKED EXAMPLE 7

What is wrong with this graph?

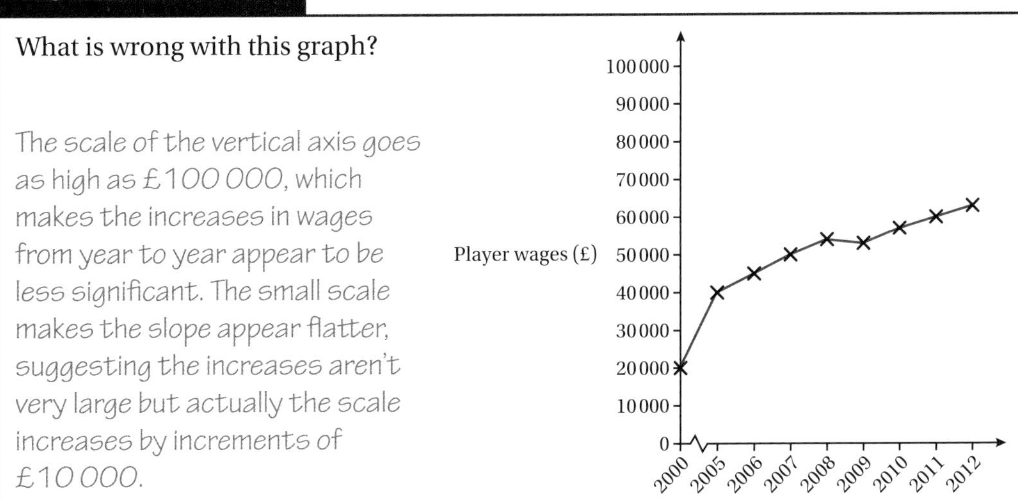

The scale of the vertical axis goes as high as £100 000, which makes the increases in wages from year to year appear to be less significant. The small scale makes the slope appear flatter, suggesting the increases aren't very large but actually the scale increases by increments of £10 000.

EXERCISE 38D

1. Identify the error in this graph.

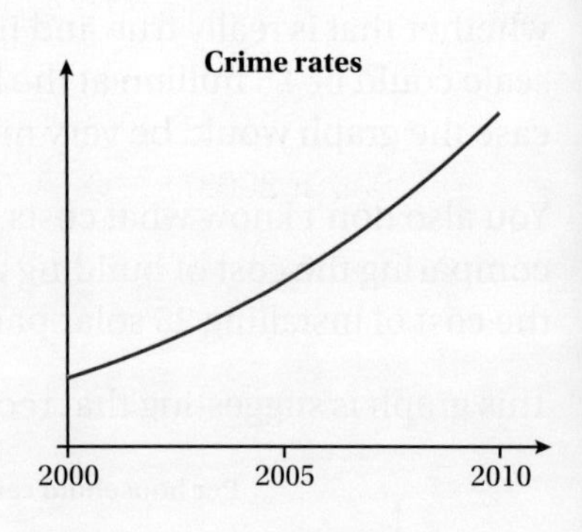

2. How is this graph misleading?

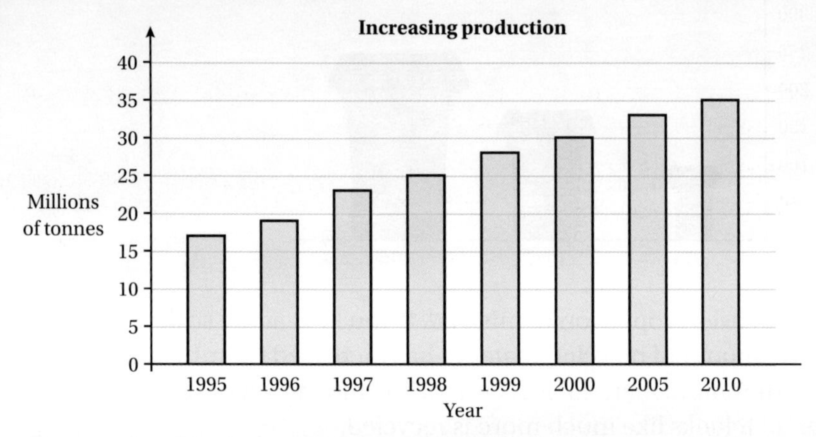

Why might someone have drawn the graph like this?

3. What is wrong with this graph?

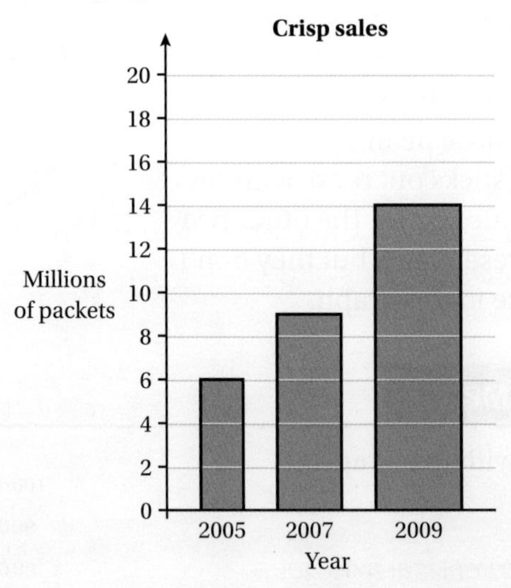

4. Look carefully at the graph.

   a What is misleading about this graph?

   b Why do you think it has been drawn this way?

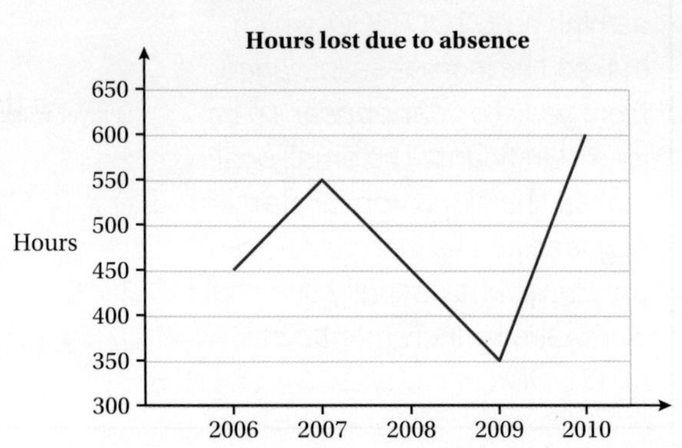

5 The same data has been presented in this 3D graph.

a Which year has the most days lost, 2006 or 2008?

b Why is it hard to tell?

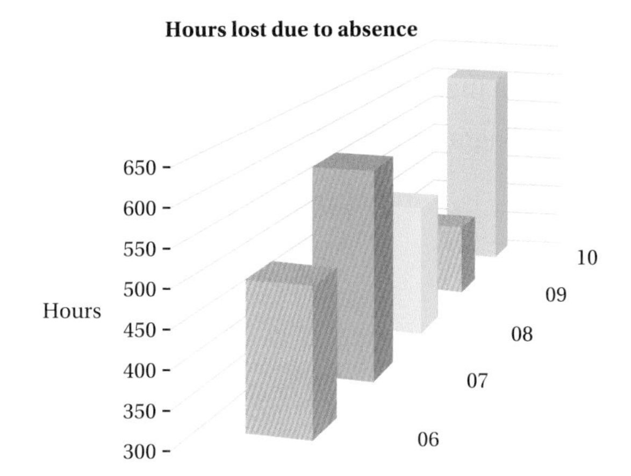

6 You are given the following data showing viewing figures for various TV programmes (in millions).

Week beginning	*Britain's Got Talent*	*The Crimson Field*	*Gogglebox*
7 April	10.03	6.89	2.75
14 April	8.45	6.31	3.37
21 April	8.63	6.25	3.48
28 April	8.45	6.01	3.47
5 May	8.58	6.33	3.54

Choose one of the TV programmes and create a graph which shows how well it has performed. You can use any type of graph, but must not change the numbers.

Section 3: Scatter diagrams

A scatter diagram is used to show whether or not there is a relationship between two sets of data collected in pairs. Data that is collected in pairs is called **bivariate data**.

Key vocabulary

bivariate data: data that is collected in pairs.

For example, you could record the number of hours different students spend studying and the results they get in a test. This would give two pieces of data for each learner: time spent studying and test results. You can think of these as a pair of number coordinates (x, y).

In bivariate data, both sets of data are numerical, so each pair of data can be plotted as a point using coordinates on a pair of axes.

Key vocabulary

correlation: a relationship or connection between data items.

Once you have plotted the data, you can look for a pattern to see whether there is a relationship or **correlation** between the two variables or not. The diagrams below show the typical patterns of correlation and what they mean.

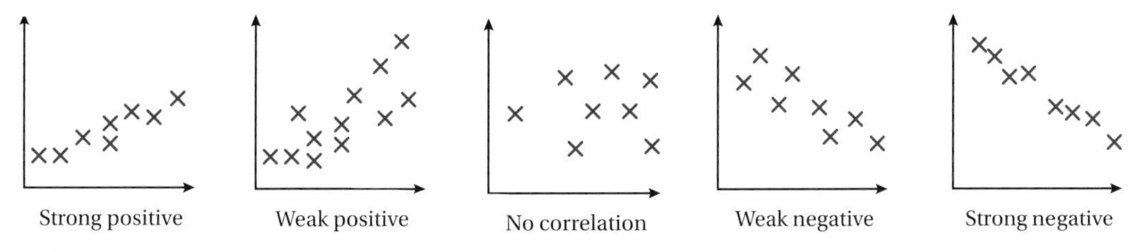

WORKED EXAMPLE 8

Nick says people who are good at maths are also good at science.

Use this data to draw a scatter diagram and comment on whether Nick is correct or not.

Maths average (%)	Science average (%)
20	22
32	30
45	39
38	40
60	60
80	70
80	72
90	90
80	25
60	65

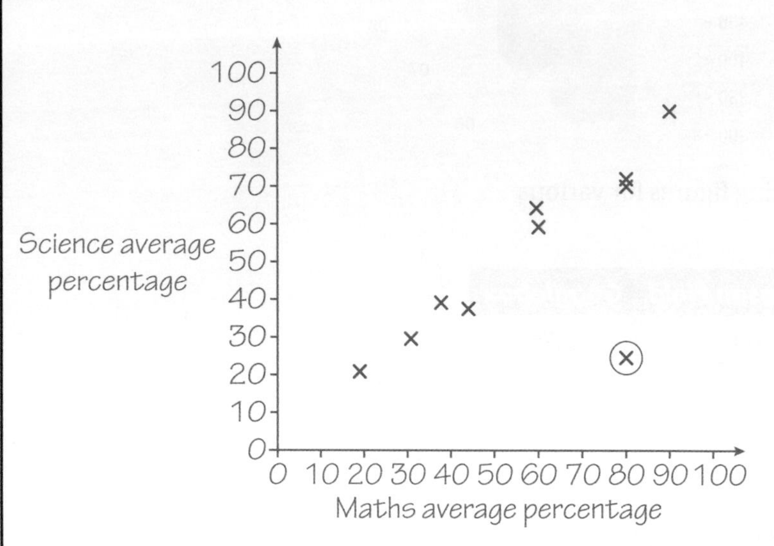

The points slope up towards the right, so it seems there is a positive correlation between maths achievement and science achievement. Nick seems to be correct.

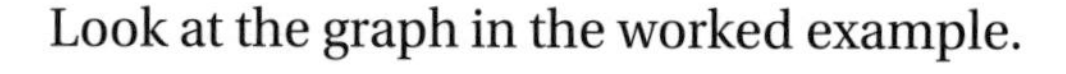

Key vocabulary

dependent variable: data that is measured in, and affected by, an experiment.

Look at the graph in the worked example.

Note that maths is on the horizontal axis and science on the vertical axis.

Science is the **dependent variable** in this case. Nick's statement is that science achievement is dependent on whether or not you are good at maths. Maths is the independent variable so it goes on the horizontal axis.

The pattern of points is used to decide whether there is a relationship. In this case, the points are grouped fairly closely and they form a thick line that slopes up to the right, so you can say there is a **positive correlation** between the scores.

If the line that formed sloped down to the left, you would say there is a **negative correlation**.

If the points are scattered such that no discernible line could be drawn, you would say there is **no correlation**.

Key vocabulary

outlier: data value that is much larger or smaller than others in the same data set.

The point that is circled is far away from the others and doesn't seem to fit the pattern. It shows a student with a high mark for maths but a low mark for science. This point is an **outlier** in this set of data.

It is important to note that correlation is **not causation**. This means that although there may be a relationship between two variables, the change in one cannot be said to be the definite reason for the change in the other; one does not necessarily cause the other.

Lines of best fit

A line of best fit is used to show a general trend on a scatter diagram.

This is a line drawn on the graph passing as close to as many points as possible.

This is the line of best fit for the scatter diagram in Worked example 8.

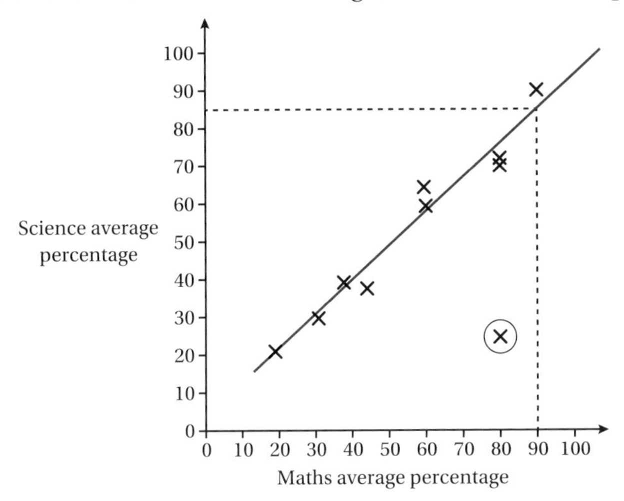

You can use the line of best fit to make predictions based on the collected data.

For example, if you wanted to predict the science results for a student who got 90% for maths, you could find this is 85% using the line. This is shown by the dotted line on the diagram.

EXERCISE 38E

1 Draw a scatter diagram for the following data and draw a line of best fit.

Homework (mins)	10	25	38	65	84	105	135	158
TV viewing (mins)	60	55	50	20	30	15	10	8

What kind of correlation is this?

2 Draw a scatter diagram to show the relationship between car engine size and fuel economy (miles per gallon).

Car miles per gallon	64	60	59	58	55	49	47	42
Car engine size (litres)	1.1	1.3	1.4	1.6	1.8	2	2.5	3

3 Mike has an ice cream stall in the local park.

He writes down the number of ice creams he sells and the maximum temperature each day for a week.

Ice creams sold	86	89	45	69	84	25	78
Maximum temperature	25°C	26°C	19°C	23°C	25°C	15°C	21°C

a Draw a scatter diagram to show the correlation between ice cream sales and the temperature.

b Can you suggest any other factors that might affect sales?

Find answers at: cambridge.org/ukschools/gcsemaths-studentbookanswers

4 During a census the number of people living in each house is recorded.

House number	1	3	5	7	9	11	13	15	17	19	21
Number of residents	1	5	1	4	2	5	6	3	5	3	6

a Draw a scatter diagram to show this data.

b What kind of correlation is this?

5 The table below shows the athlete's height and the height jumped by the last 10 men's high jump world record holders.

	Athlete height (m)	Height jumped (m)
Sotomayor	1.95	2.45
Sjoberg	2.00	2.42
Paklin	1.91	2.41
Povarnitsyn	2.01	2.40
Jianhua	1.93	2.39
Wessig	2.00	2.36
Mogenburg	2.01	2.35
Wszola	1.90	2.35
Yashchenko	1.93	2.34
Stones	1.96	2.32

a Draw a scatter diagram showing this data.

b Is there a correlation between the height of the jumper and the height he jumped?

Outliers

Outliers are data values that lie outside the normal range for a set of data. In science experiments they may be 'freak' results or the result of inaccurate measurements. It can be difficult to decide when it is reasonable to disregard an outlier, but if it is an obvious error then the value is usually just ignored.

However, if outliers are genuine results and not errors, they can't be ignored just because they spoil a pattern. Outliers will have an impact on calculating the mean and the range of a set of data, but less so when finding the median and the mode.

On a scatter diagram an outlier will be a point that is away from the main scatter of points, or might fit the line of best fit but be at an extreme value.

WORKED EXAMPLE 9

A coach records the 100 m times of her 10 athletes at the start and the end of a week of intense training. Nine of the athletes were faster than their previously recorded times, but one athlete was 2 seconds slower. The team showed a mean improvement of 0.2 seconds.

Can the coach claim to be making a significant impact on her athletes?

Continues on next page ...

Yes, the coach is making an impact on the athletes. Although the mean suggests only a small improvement, this is likely to be due to the one negative result. In terms of running times for a 100 m race, 2 seconds is a large value and could have skewed the mean. The athlete with reduced performance is an outlier.

EXERCISE 38F

1 Several students sit a maths test and their scores are given below:

54 50 47 42 54 44 36 37 45 36 55 55 52 85 39

a What is the mean score in the test?

b What is the range?

c What is the median score?

d What is the median without the outlier?

e What is the mean without the outlier?

2 Several students sit a maths and an English exam. Their scores are given below.

English	49	42	46	44	53	41	64	14	44	53	55	42
Maths	46	47	43	45	49	48	69	39	33	46	53	44

a Plot their scores on a scatter diagram.

b Draw a line of best fit on your scatter diagram.

c Are any of the points outliers?

3 The time taken to travel by train from Norwich to London in minutes is recorded for 20 journeys:

109	129	98	106	109	156	128	98	99	113
126	99	105	110	126	98	106	114	122	107

On a normal day the journey should take between 95 and 115 minutes, depending on the number of stops at stations.

a What is the mean journey time?

b The train company claim that the mean journey time is 111 minutes on a normal day.
Is this right?

Checklist of learning and understanding

Summary statistics

- The mean, median and mode are all measures of central tendency (averages). They can be found precisely for populations that are ungrouped, and estimated for grouped data.
- The range is a measure of spread. It can be found precisely for ungrouped data and estimated for grouped data.

Find answers at: cambridge.org/ukschools/gcsemaths-studentbookanswers

- Quartiles are the values that divide an ordered set of data into quarters.
- The first quartile (Q_1), sometimes called the lower quartile, is the data value one quarter of the way along the data set; 25% of the values are smaller than the first quartile, and three quarters of the data are larger.
- The second quartile (Q_2) is the same as the median.
- The third quartile (Q_3), sometimes known as the higher quartile, is the value that lies three quarters of the way along the data set; 75% of the values are smaller than this value.
- The interquartile range (IQR) is the range between the upper and lower quartiles, and is a measure of spread.
- Box plots show the range, the quartiles and the median value and are a good way of comparing data sets.

Misleading graphs

- The way that data is presented in graphs can be misleading. Watch out for uneven scales and ways of representing the data to exaggerate changes.

Scatter diagrams and correlation

- Bivariate data is data that is collected in pairs.
- Scatter diagrams can be used to look for correlations in bivariate data. A correlation is a relationship, such as one quantity increasing as another decreases. Bivariate data can have a positive, negative or no correlation.
- Correlation does not mean causation; in other words, identifying a relationship does not necessarily mean that a change in one data set is causing the change in the other.
- Outliers are pieces of data that sit outside the pattern or expected result. They can be ignored if an obvious error, but otherwise should be considered and explained.

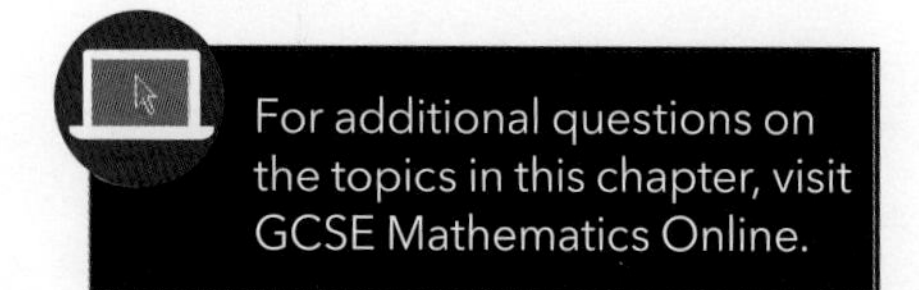

For additional questions on the topics in this chapter, visit GCSE Mathematics Online.

Chapter review

1 Windsurfers need a certain amount of wind to surf, but too much can be dangerous.
A learner would typically surf in a speed of 7 to 18 knots, but an expert would prefer to surf at above 30 knots.

Use the data on wind speed in knots measured at the same time each day for the two lakes from the table below to decide which lake is better for beginners and which for experts. Use measures of central tendency and spread to support your argument.

First lake (knots)	0	21	33	13	20	11	35	3	5	3	31
	28	19	19	22	26	40	40	4	25	21	26
Second lake (knots)	15	11	19	11	10	19	23	25	10	18	10
	16	23	15	15	20	22	10	13	11	18	18

2 The box plot shows the height of 30 Year 10 students in centimetres.

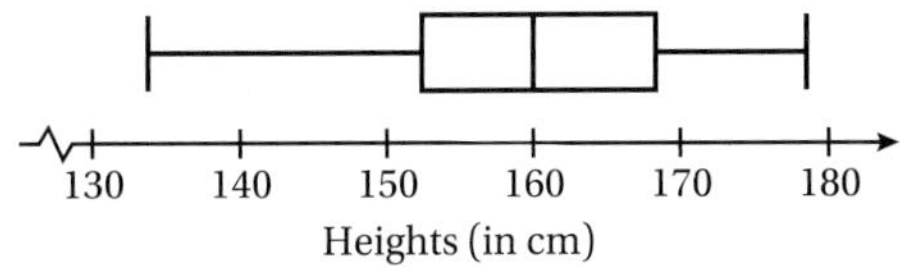

a What information is shown on the box plot?

b How does this help you make sense of the data?

3 Study the two line graphs.

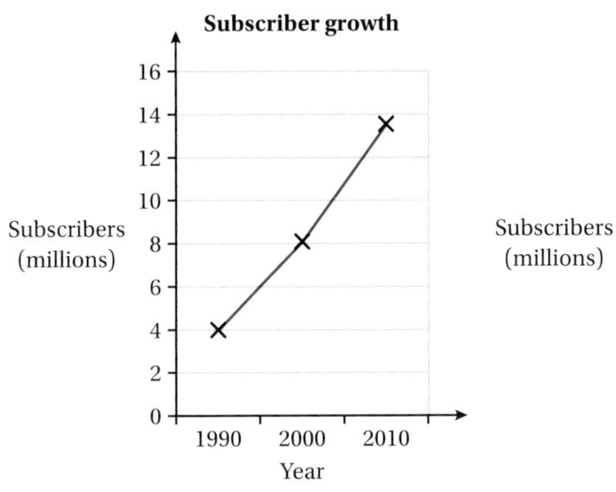

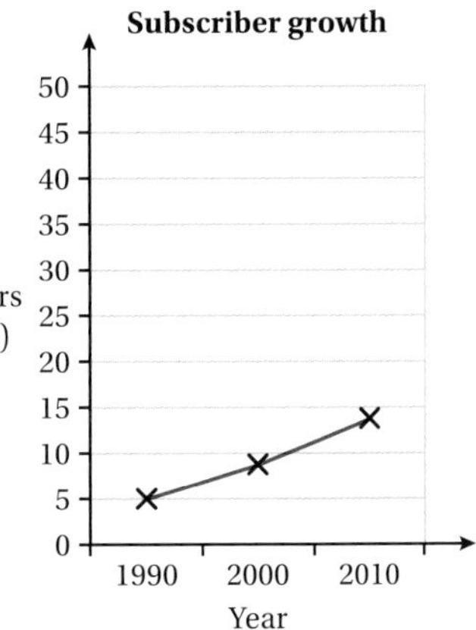

a These two graphs show the same data. Explain why they look different.

b Which graph would you use if you were a mobile phone service provider who wanted to suggest that there had been a huge increase in subscribers over this period? Why?

c Who might find the other graph useful? Why?

4 Data for the price of chocolate bars and their mass is given:

Price	45p	80p	£1.50	£3.00	£5.00	£10
Mass	35 g	80 g	175 g	320 g	540 g	1 kg

a Plot the data on a scatter diagram and draw a line of best fit.

b State what type of correlation there is.

39 Interpreting graphs

In this chapter you will learn how to …

- construct and interpret graphs in real-world contexts.
- interpret the gradient of a straight-line graph as a rate of change.
- find and interpret the gradient at a point on a curve as the instantaneous rate of change.
- plot and interpret graphs of non-standard functions in real contexts.

For more resources relating to this chapter, visit GCSE Mathematics Online.

Using mathematics: real-life applications

All sorts of information can be obtained from graphs in real-life contexts. The shape of a graph, its gradient and the area underneath it can tell us about speed, time, acceleration, prices, earnings, break-even points or the values of one currency against another, among other things.

"My car needs to perform at its optimum limits. We generate and analyse diagnostic graphs to calculate the slight changes that would increase power, acceleration and top speed."

(F1 racing driver)

Before you start …

Ch 36	You need to be able to distinguish between direct and inverse proportion.	**1** Which of these graphs shows an inverse proportion? How do you know this? 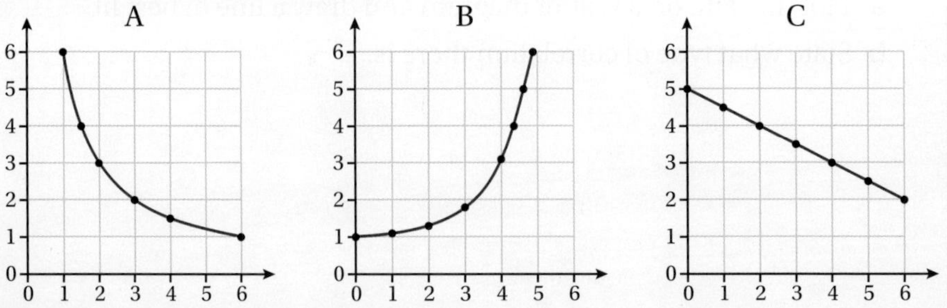
Ch 18	You need to be able to calculate the gradient of a straight line.	**2** Calculate the gradient of AB. 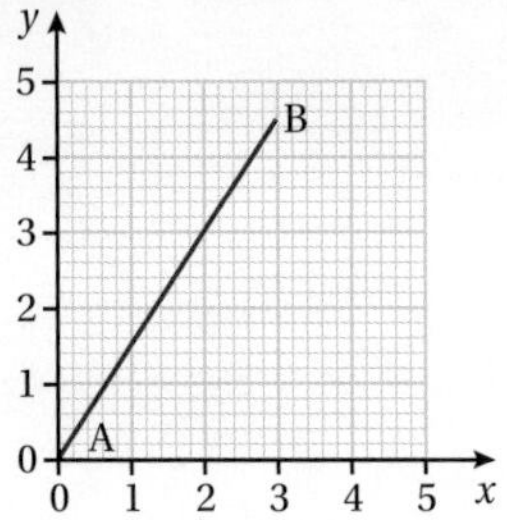

Assess your starting point using the Launchpad

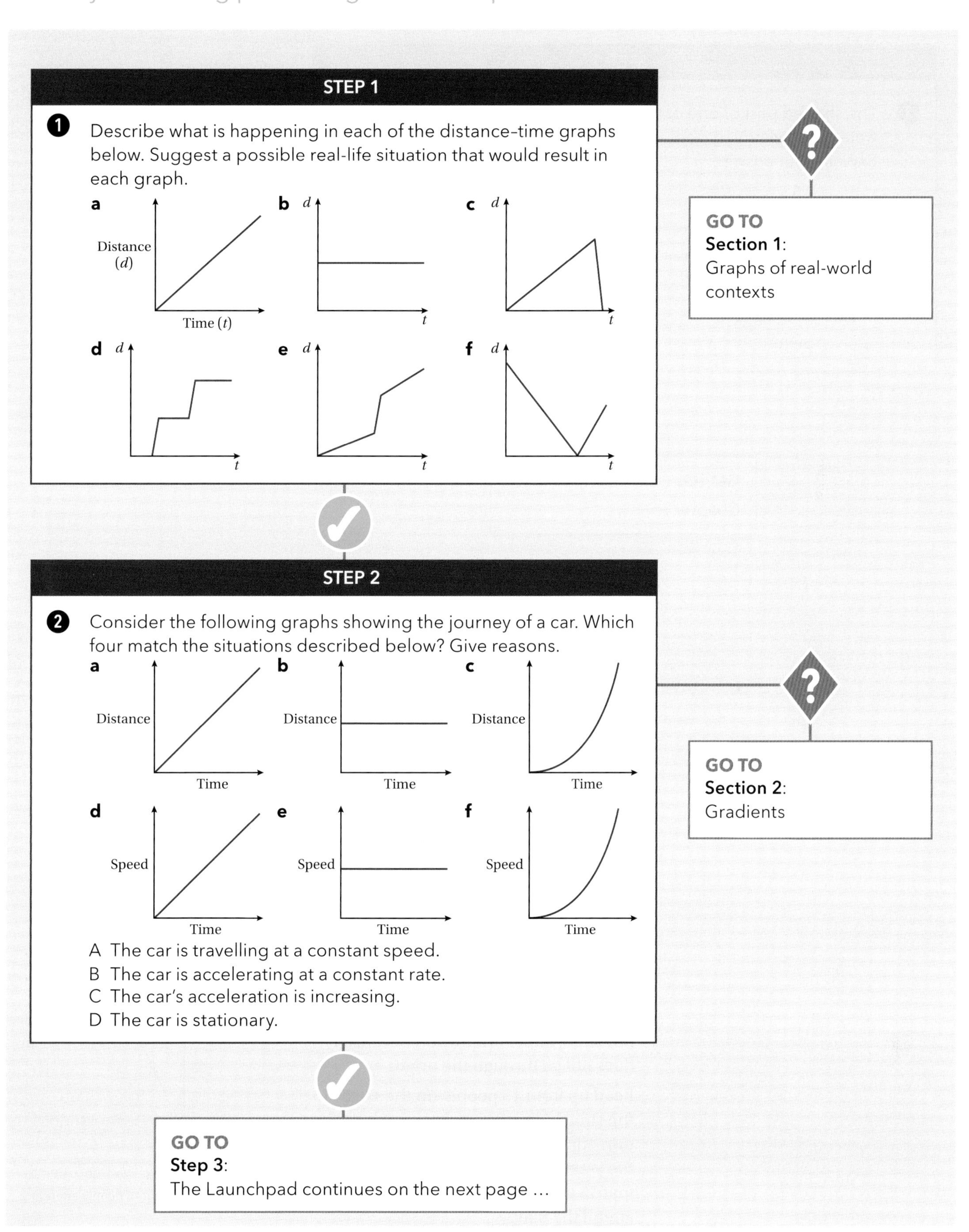

STEP 1

1 Describe what is happening in each of the distance-time graphs below. Suggest a possible real-life situation that would result in each graph.

?

GO TO
Section 1:
Graphs of real-world contexts

✓

STEP 2

2 Consider the following graphs showing the journey of a car. Which four match the situations described below? Give reasons.

A The car is travelling at a constant speed.
B The car is accelerating at a constant rate.
C The car's acceleration is increasing.
D The car is stationary.

?

GO TO
Section 2:
Gradients

✓

GO TO
Step 3:
The Launchpad continues on the next page …

Find answers at: cambridge.org/ukschools/gcsemaths-studentbookanswers

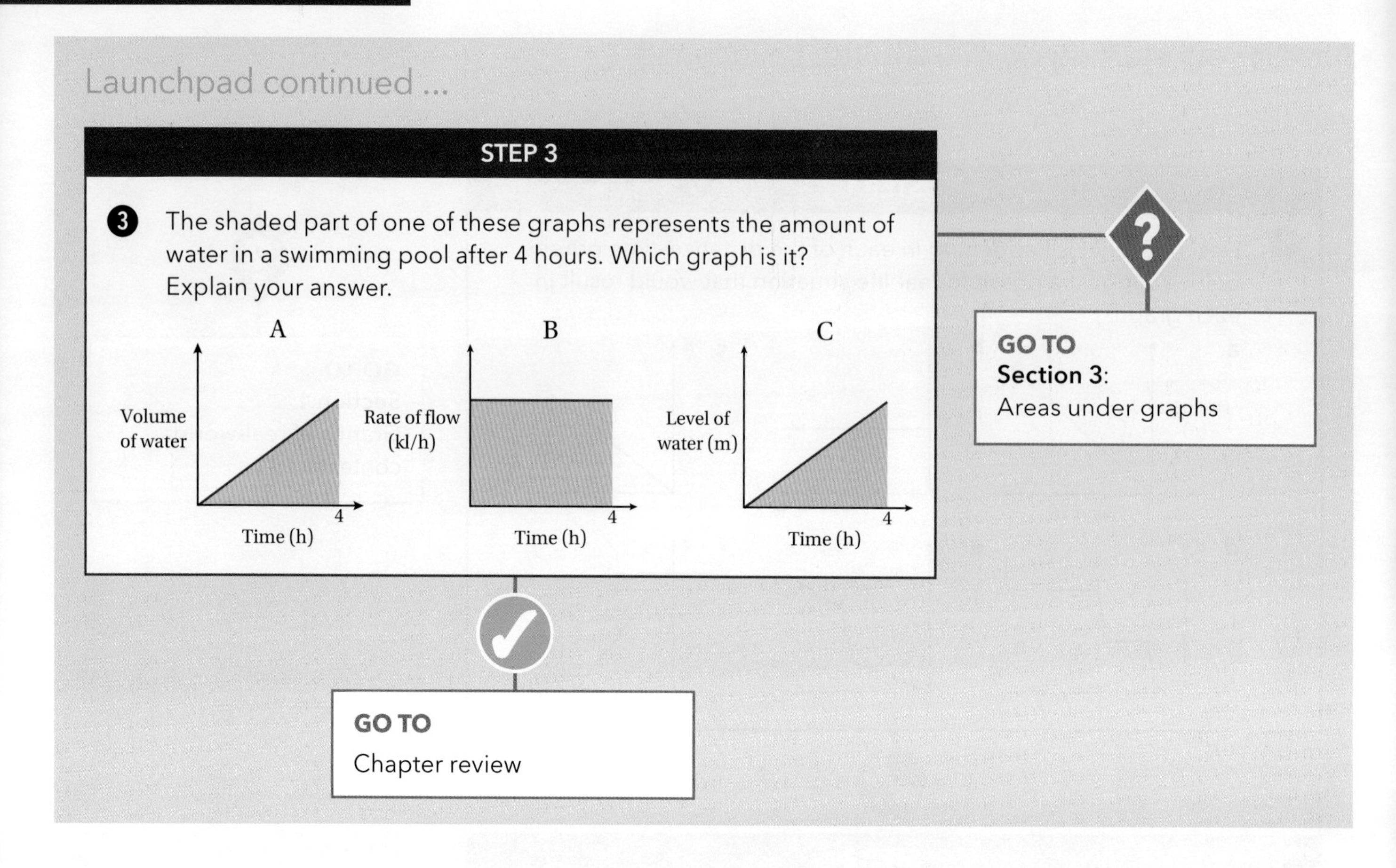

Section 1: Graphs of real-world contexts

Graphs are useful for visually representing the relationships between quantities.

For example, a group of people buy tickets to attend a play at the costs shown in the graph below. The tickets include transport costs and seats in the hall.

This graph shows lots of information.

The horizontal axis (or x-axis) shows the number of people attending. The vertical axis (or y-axis) shows the total cost.

The cost depends on the number of people attending. However, there is a cost of £10 regardless of how many people attend – this is a group charge.

There are six marked points on the graph.

This graph is a linear graph, but it does **not** show direct proportion because it does not go through the origin.

Read up from 15 people on the x-axis to the straight line. When you reach the line move across horizontally until you reach the y-axis. The cost is £40. This means that 15 people will need to pay £40 to attend the play.

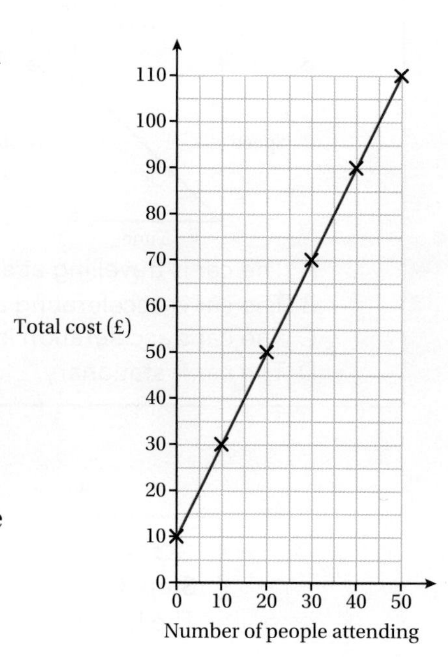

Distance-time graphs

Graphs that show the connection between the distance an object has travelled and the time taken to travel that distance are called distance–time graphs or travel graphs.

Time is normally shown along the horizontal axis and distance on the vertical.

The graphs normally start at the origin because at the beginning no time has passed and no distance has been covered.

Look at the graph. It shows the following:

- a cycle for 4 minutes from home to a bus stop 1 km away
- a 2 minute wait for the bus
- a 7 km journey on the bus that takes 10 minutes.

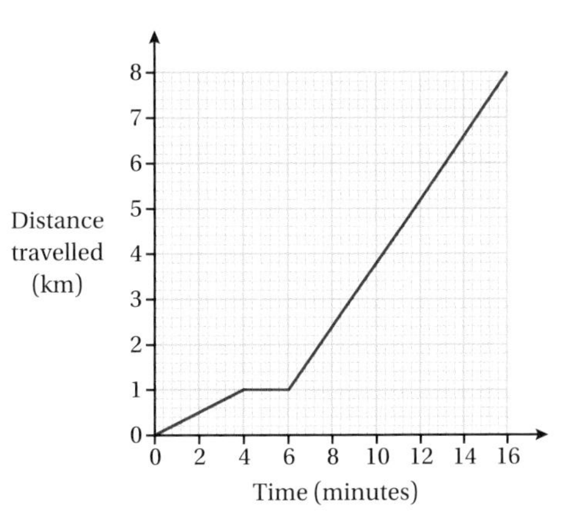

The line of the graph remains horizontal while the person is not moving (waiting for the bus) because no distance is being travelled at this time. The steeper the line, the faster the person is travelling.

WORKED EXAMPLE 1

The graph shows the relationship between the length and the width of a hall.

Find the formula for this relationship.

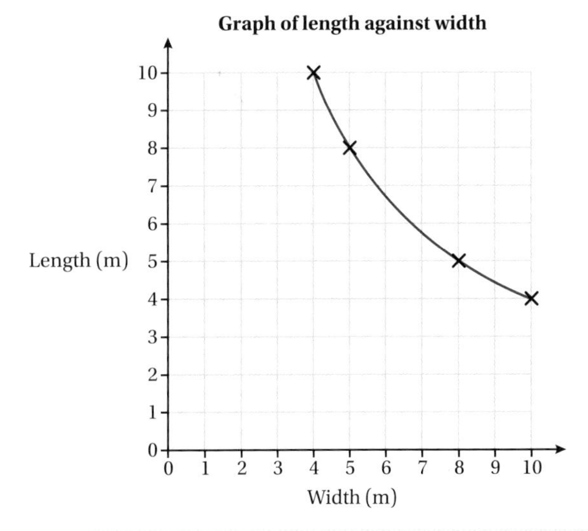

(4, 10), (5, 8), (8, 5) and (10, 4)

Write down the coordinates of some points on the line.

$4 \times 10 = 40\text{ m}^2$

The area of the hall is constant, at $= 40\text{ m}^2$.

Work out the area of the hall using one of the coordinates, for example when the hall has a length of 10 m and a width of 4 m.

$\text{length} = \dfrac{40}{\text{width}}$

From the shape of the graph, you know that it is showing inverse proportion, so as the length increases the width decreases, therefore your formula will be in the form $y = \dfrac{k}{x}$. In this case $k = 40$.

Tip

You saw conversion charts in the form of exchange rates in Chapter 36.

Because it shows a real-world context, the graph in Worked example 1 is only valid for that particular range of values.

Graphs are also useful in the real world for reading off values quickly without having to do the whole calculation. They can serve as conversion charts.

WORKED EXAMPLE 2

This graph shows the number of Indian rupees you would get for different numbers of US dollars at an exchange rate of US$1 : Rs 45. This relationship is a direct proportion.

a Use the graph to estimate the dollar value of Rs 250.

b Use the graph to estimate how many rupees you could get for US$9.

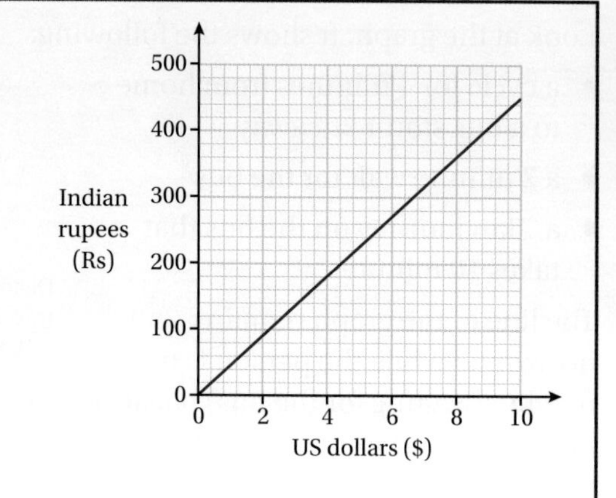

a Rs 250 is worth about $5.50.

Find 250 on the y-axis and read across and down to find the corresponding point on the x-axis.

b You could get about Rs 400 for $9.

Find 9 on the x-axis and read up and across to the find the corresponding point on the y-axis.

Speed-time graphs

Graphs that show the connection between the velocity (or speed) that an object is travelling at and the time taken are called speed-time graphs.

Time is normally shown along the horizontal axis and speed on the vertical.

The graphs normally start at the origin because at the beginning no time has elapsed (passed) and no distance has been covered so the speed is zero.

The graph below shows the car journey of someone leaving their house and heading to a dual carriageway.

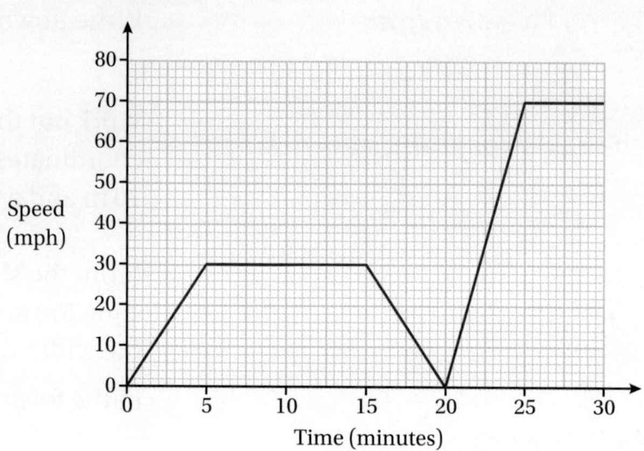

The upwards slopes indicate an increase in speed; this is known as acceleration. The horizontal line indicates a constant speed. The steeper the line, the faster the object is accelerating. A downwards slope would indicate that the object is decelerating (slowing down).

EXERCISE 39A

1 This graph shows the movement of a taxi during a four-hour period.

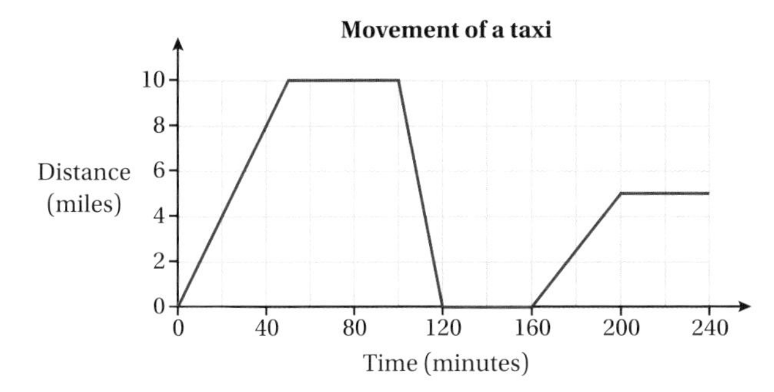

a Clearly and concisely describe the taxi's journey.

b For how many minutes was the taxi waiting for passengers in this period?
How can you tell this?

c What was the total distance travelled?

d Calculate the taxi's average speed during:

i the first 20 minutes.

ii the first hour.

iii from 160 to 210 minutes.

iv for the full period of the graph.

Tip

You saw in Chapter 14 that

$$\text{Speed} = \frac{\text{distance travelled}}{\text{time taken}}$$

2 This distance–time graph represents Monica's journey from home to a supermarket and back again.

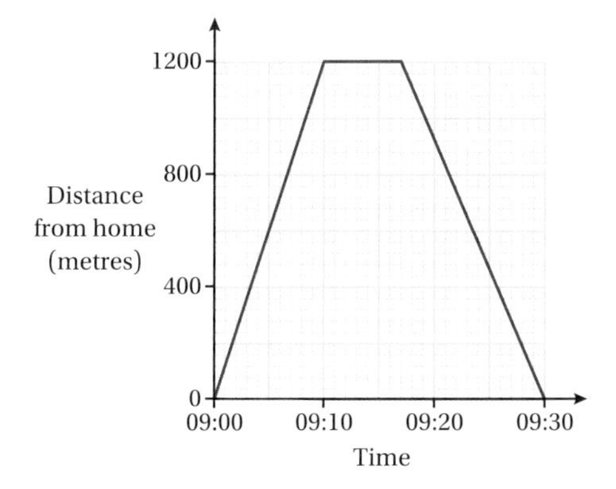

a How far was Monica from home at 09:06 hours?

b How many minutes did she spend at the supermarket?

c At what times was Monica 800 m from home?

d On which part of the journey did Monica travel faster, going to the supermarket or returning home?

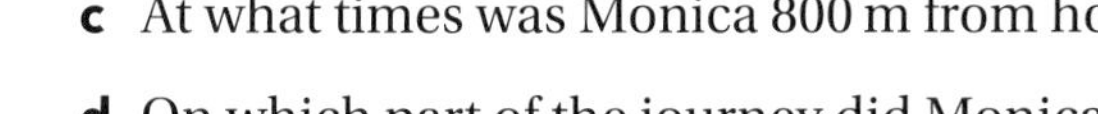
Find answers at: cambridge.org/ukschools/gcsemaths-studentbookanswers

3. A swimming pool is 25 m long. Jasmine swims from one end to the other in 20 seconds.

   She rests for 10 seconds and then swims back to the starting point. It takes her 30 seconds to swim the second length.

   a Draw a distance–time graph for Jasmine's swim.

   b How far was Jasmine from her starting point after 12 seconds?

   c How far was Jasmine from her starting point after 54 seconds?

4. A hurricane disaster centre has a certain amount of clean water. The length of time the water will last depends on the number of people who come to the centre.

   a Calculate the missing values in this table.

No. of people	120	150	200	300	400
Days the water will last	40	32			

   b Plot a graph of this relationship.

Section 2: Gradients

Speed in distance-time graphs

Tip

You learnt about kinematics in Chapter 14. Revise that section if you need to.

The steepness (slope) or gradient of a distance–time graph gives an indication of speed. A straight-line graph indicates a constant speed.

The steeper the graph, the greater the speed.

An upward slope and a downward slope represent movement in opposite directions.

The distance–time graph shown is for a person who walks, cycles and then drives for three equal periods of time.

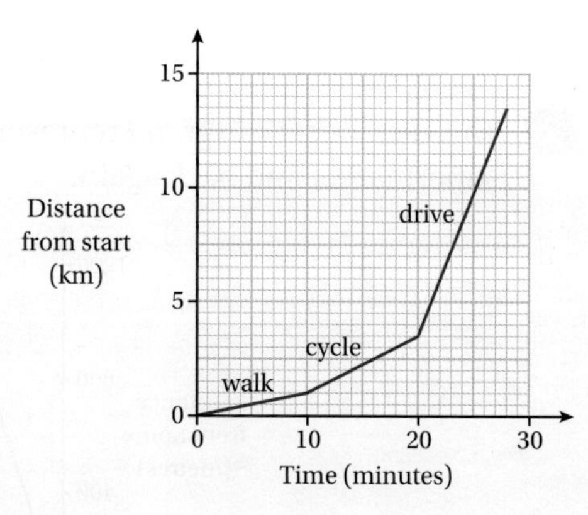

For each period, speed is given by the formula:

$$\text{Speed} = \frac{\text{distance travelled}}{\text{time taken}}$$

If a line section on a graph is horizontal, the gradient is zero and there is no speed, that is the object has stopped moving.

Acceleration in speed–time graphs

You saw in Chapter 14 that acceleration is measured units of distance and time (m/s/s or m/s^2), that is distance per time squared.

In a speed-time graph, you would get this unit by finding the ratio of speed to time:

$$\text{Acceleration (m/s}^2) = \frac{\text{speed (m/s)}}{\text{time taken (s)}}$$

So, the gradient of a speed–time graph tells you the acceleration of the object.

Using gradient triangles to interpret changing gradients

Looking at the gradient of a graph along with the axis labels gives a large amount of detail; even when there is no scale given.

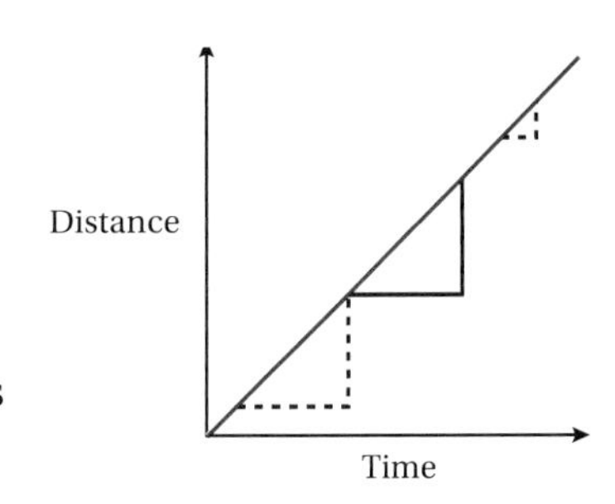

This graph shows a car journey.

Consider what happens as time moves on. In this case, as time moves on the distance covered increases. So the car is moving.

Now consider the gradient triangles drawn on the graph above. It doesn't matter where these triangles are drawn, each is similar to the other, so the sides represent the same gradient $\left(\frac{\text{rise}}{\text{run}}\right)$.

This shows that the car is moving at a constant speed.

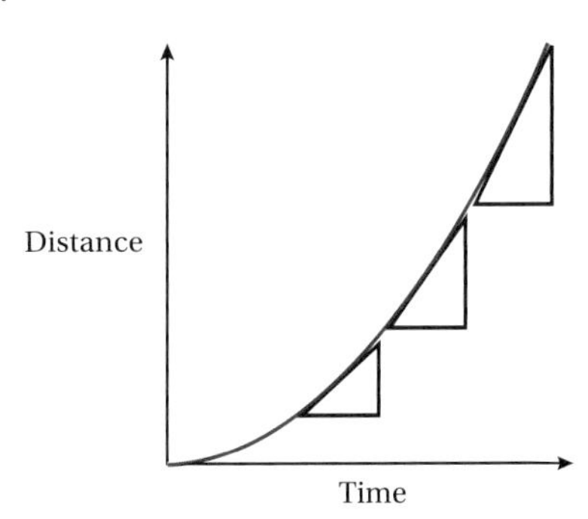

In the graph on the right, as time moves on the distance covered increases. So the car is moving.

Now consider the gradient triangles drawn on the graph; each has the same base (unit of time) but a different height.

This time the gradient triangles don't fit the graph as it is not a straight line. Instead, we've laid them against the graph at different places. The hypotenuse of each forms a tangent to the graph.

You can see by the slope of each triangle's hypotenuse that the speed is changing along the graph. Moving up the slope, the triangles' hypotenuses are getting steeper – the gradient of the graph is increasing. This shows that the car is speeding up, or **accelerating**.

EXERCISE 39B

1 The following graphs show what is happening to the level of water in a tank.

Describe what is happening in each case. Justify your answers using gradient triangles.

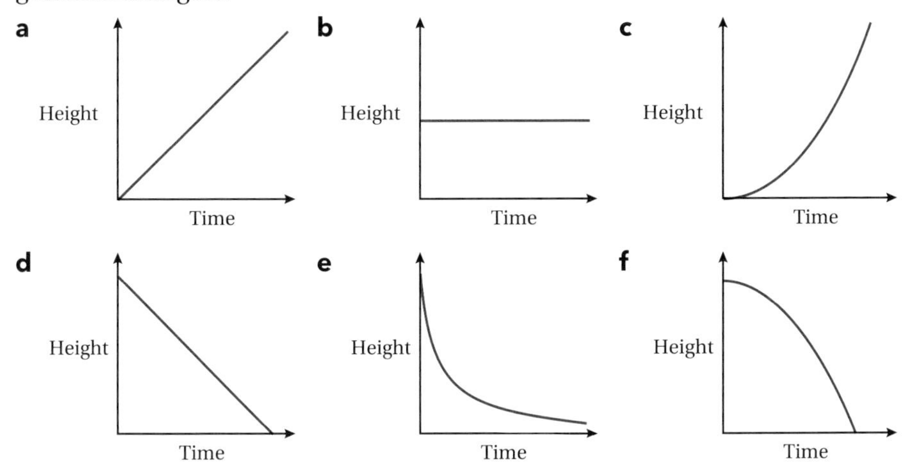

Find answers at: cambridge.org/ukschools/gcsemaths-studentbookanswers

2 The following graphs show what is happening to the price of oil.

Describe what is happening in each case, justifying your answers using gradient triangles.

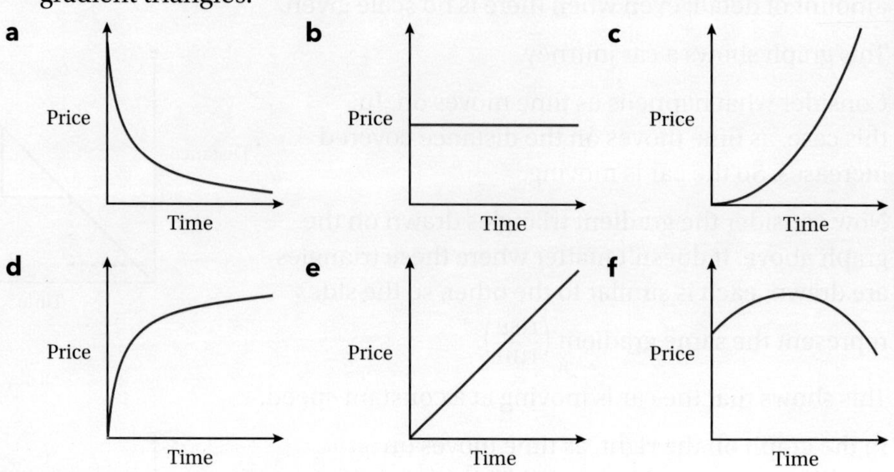

3 The following is a speed–time graph of a parachute jump.

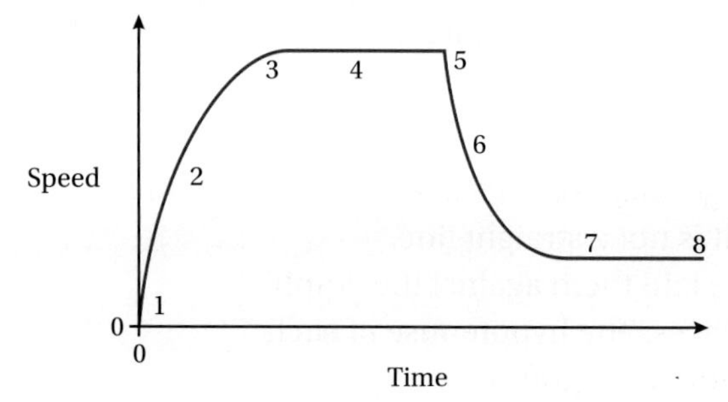

Describe what is happening to the speed and acceleration of the parachutist throughout the jump.

Finding the gradient of a curve using a tangent line

This simple graph of height against distance shows the route followed by a mountain biker on a trail.

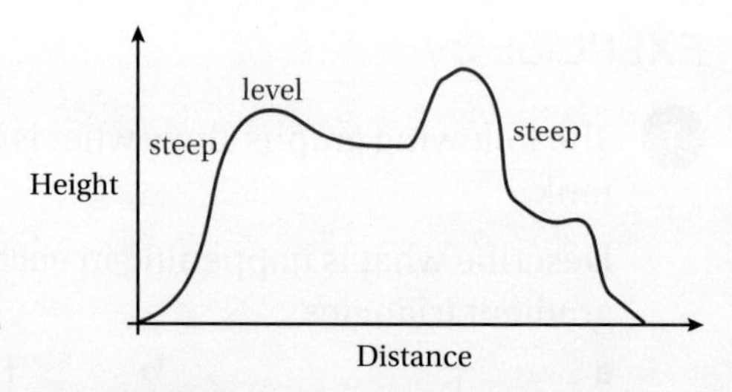

Some parts of the curve have a steep positive gradient, some have a gradual positive gradient, some parts are level and other parts have a negative gradient. It should be clear from this graph that a curved graph never has a single gradient like a straight-line graph has.

You cannot find the gradient of a whole curve but you can find the gradient at a point on the curve by drawing a tangent to it. This is known as the **instantaneous rate of change**, because it is the rate of change at that one given point.

The gradient of a curve at a point is the gradient of the tangent to the curve at that point. Once you have drawn the tangent to a curve, you can work out the gradient of the tangent just as you would for a straight line gradient:

$$\text{Gradient} = \frac{\text{change in } y\text{-values}}{\text{change in } x\text{-values}}$$

Look at the graph below to see how this works. BC is the tangent to the curve at A.

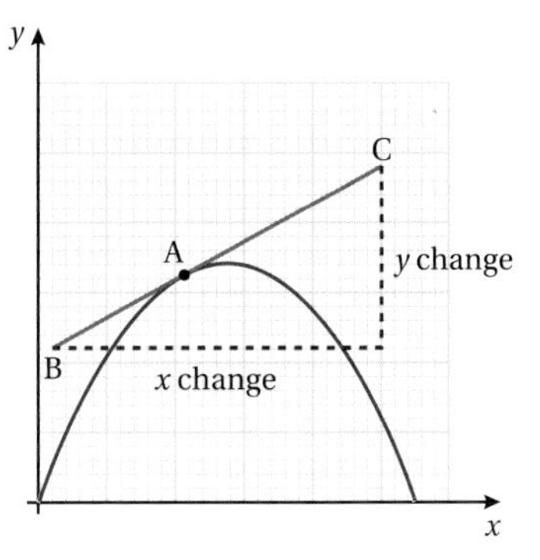

How to draw the tangent

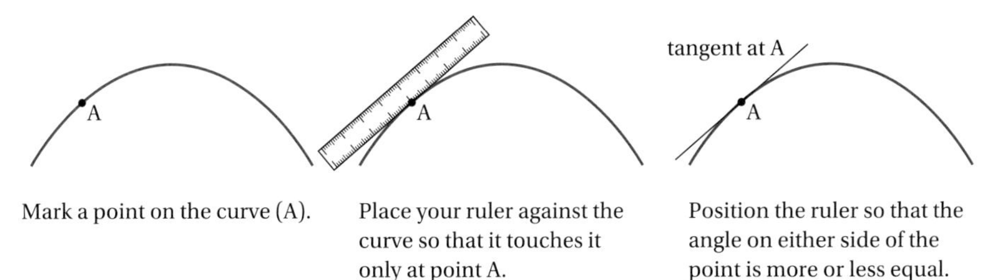

Mark a point on the curve (A).

Place your ruler against the curve so that it touches it only at point A.

Position the ruler so that the angle on either side of the point is more or less equal. Use a pencil to draw the tangent.

Calculating the gradient to a tangent

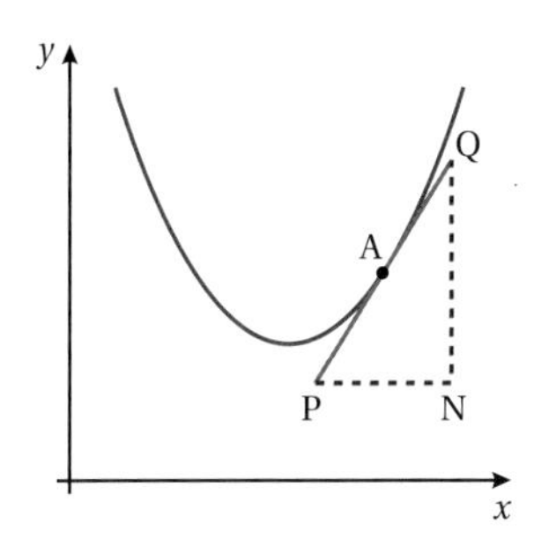

Mark two points, P and Q, on the tangent. Try to make the horizontal distance between P and Q a whole number of units (measured on the x-axis scale).

Draw a horizontal line through P and a vertical line through Q to form a right-angled triangle PNQ.

Gradient of the curve at A = gradient of the tangent PAQ

$$= \frac{\text{distance NQ (measured on the } y\text{-axis scale)}}{\text{distance PN (measured on the } x\text{-axis scale)}}$$

WORKED EXAMPLE 3

The graph shows the height of a tree (y metres) plotted against the age of the tree (x years).

Estimate the rate at which the tree was growing when it was four years old.

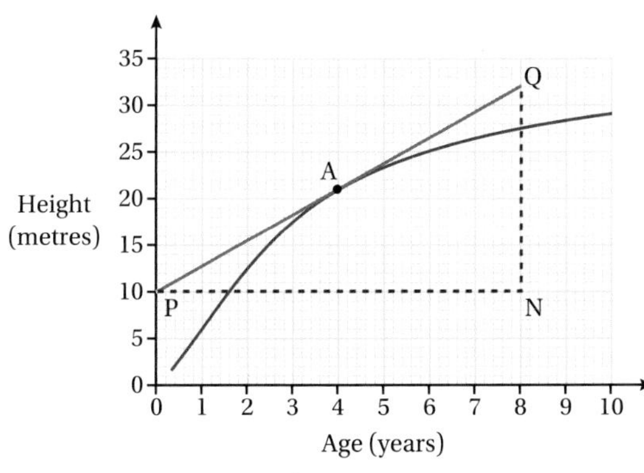

The rate at which the tree was growing when it was four years old is equal to the gradient of the curve at the point where $x = 4$. Draw the tangent at this point (A).

Gradient at A $= \frac{NQ}{PN} = \frac{22.5}{8} = 2.8$ (1 dp)

The tree was growing at a rate of 2.8 metres per year.

Find answers at: cambridge.org/ukschools/gcsemaths-studentbookanswers

Calculating the average rate of change

You know that in a straight-line graph the gradient is constant, so the rate of change is the same at every point on the graph. On a curved graph, the gradient changes, so to find the average rate of change you select two points from the curve and use these to calculate the average.

For example, take the growth of the tree in Worked example 3. To calculate the average speed of growth of the tree, choose two points on the line. At age 2 years, its height was approximately 12 m (2, 12). At age 9, its height was approximately 28 m (9, 28). Use the change in these values to calculate the average speed of growth:

$$\text{Gradient} = \frac{\text{change in } y\text{-values}}{\text{change in } x\text{-values}} = \frac{28 - 12}{9 - 4} = \frac{16}{7} = 2.3\text{m/yr (1 dp)}$$

This seems reasonable given the answer in Worked example 3.

EXERCISE 39C

1 The following graph is a distance–time graph for a drag-racing car.

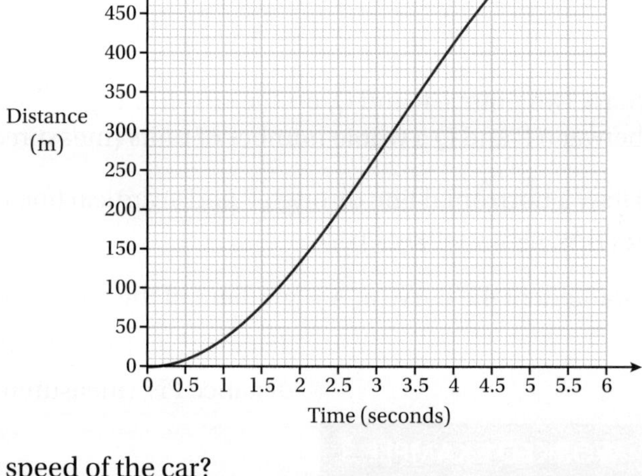

a How far had the car travelled after 2 seconds?

b How long did it take the car to travel 50 metres?

c When was the car going at its fastest speed?

d How fast was the car going after

i 0.5 seconds?

ii 3.5 seconds?

e What was the average speed of the car?

2 The following graph shows the predicted height of the tide at Milford Haven.

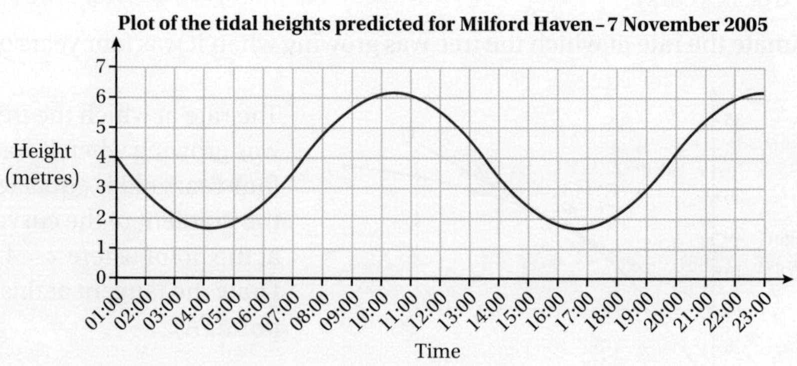

a When is the tide coming in at its fastest rate?

b When is the tide fully in?

c How fast is the tide going out at

i 4pm? **ii** 2pm?

d Why would this kind of information be useful?

3 A walker in the Lake District is training for a mountain climb.

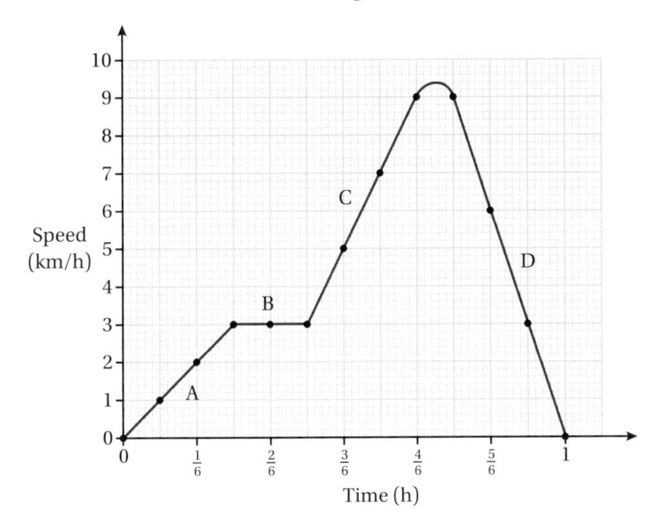

a Describe the speed changes on the route that this person is taking. Pay particular attention to the four sections of the graph.

b What is the velocity of the person after 30 minutes?

c What is happening after 43 minutes?

d What is the acceleration of the person after 10 minutes?

e At what rate is the person decelerating after 50 minutes?

4 The graph of $y = x^2$ is shown in the diagram.

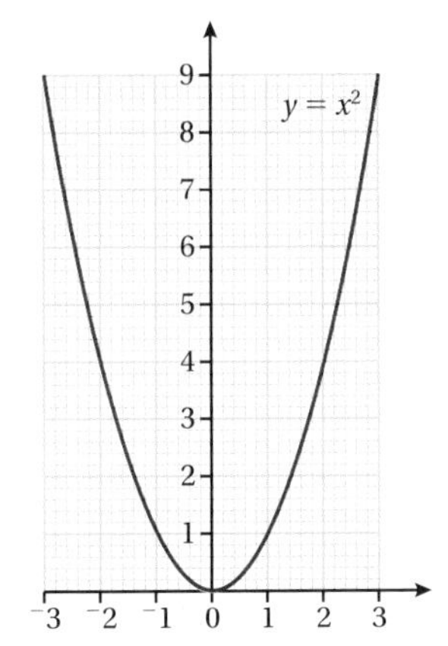

a Copy the graph using tracing paper and find the gradient of the graph at the points:

i (2, 4) **ii** ($^{-}1$, 1)

b The gradient of the graph at the point (1.5, 2.25) is 3.
Write down the coordinates of the point at which the gradient is $^{-}3$.

c If you calculated the average gradient of this curve, would it be a reasonable estimate? Why?

Find answers at: cambridge.org/ukschools/gcsemaths-studentbookanswers

Section 3: Areas under graphs

The following graph shows a car travelling at a constant speed of 40 km/h.

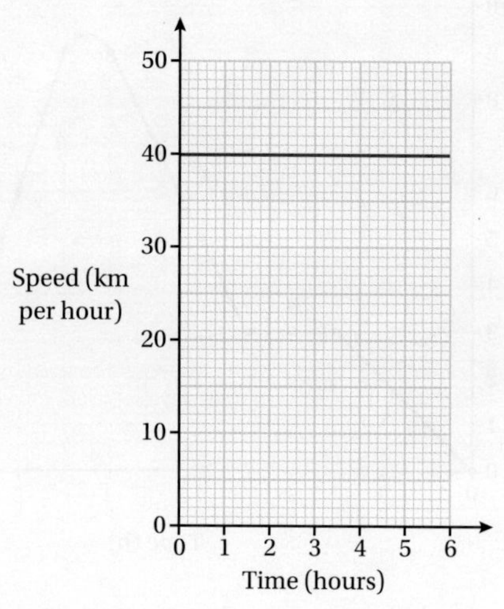

The area under the graph after 2 hours is 80; after 3 hours is 120 and after 4 hours is 160. What do you notice?

The area under the graph on a speed–time graph equals the **distance travelled**.

This makes sense if you consider the formula for calculating speed.

Speed $= \dfrac{\text{distance}}{\text{time}}$, so if we multiply both sides of the formula by time we have:

$$\text{Speed} \times \text{time} = \frac{\text{distance}}{\text{time}} \times \text{time} = \text{distance}$$

The same applies when the graph is sloping due to acceleration or deceleration, the area under the graph will tell you the distance travelled.

EXERCISE 39D

1 Find the distance travelled for each of the following vehicles in their first four hours.

a

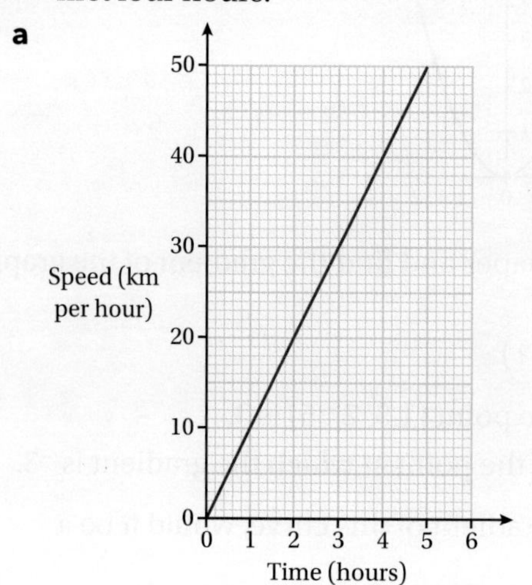

b

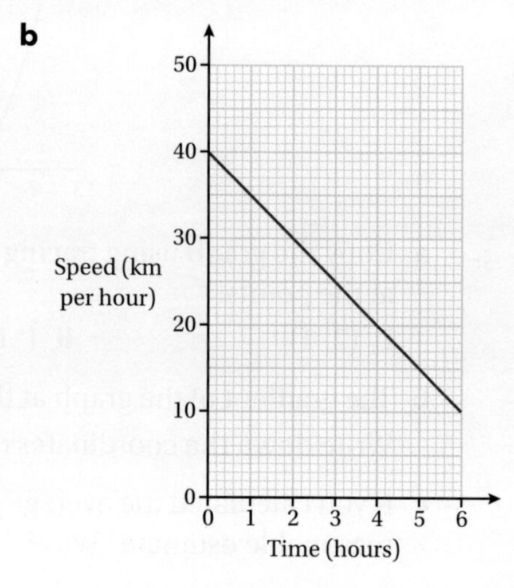

c

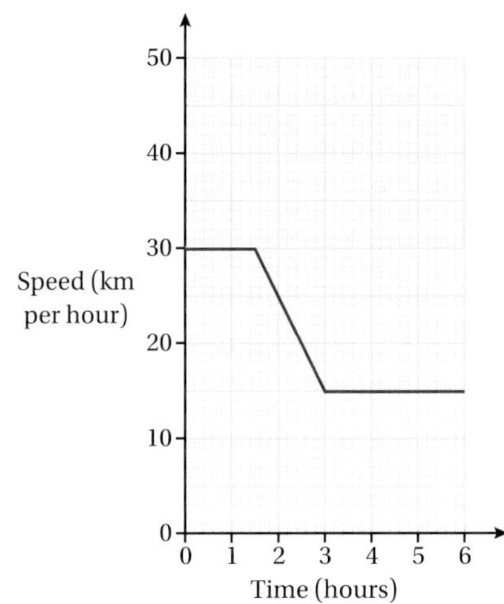

d

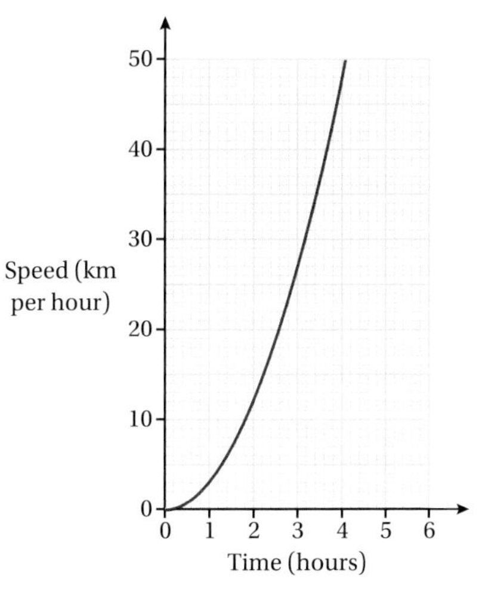

2 The following graph shows the rate of water flow in a river throughout the day.

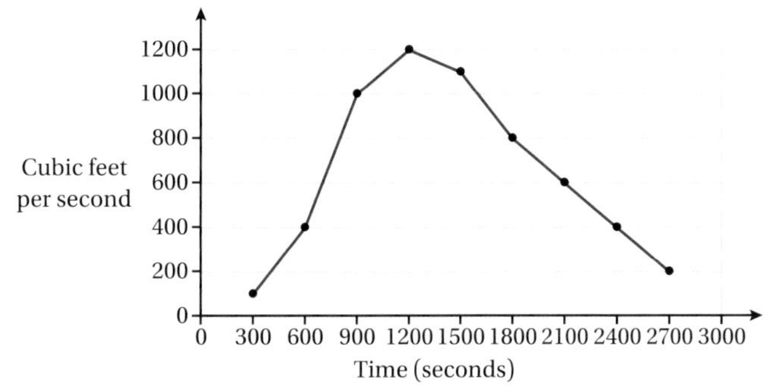

a What would the area under the graph represent?

b How much water had flowed down the river in the first 5 minutes?

c How much flowed between 10 and 15 minutes?

d How much less water flowed between 35 and 40 minutes compared to 15 and 20 minutes?

e Why might a river's flow change like this?

Checklist of learning and understanding

Graphs of real-world contexts

- Graphs are useful for visually representing the relationships between quantities.
- Graphs that show the connection between the distance an object has travelled and the time taken are called distance–time graphs.
- Graphs that show how the speed of an object changes over time are called speed–time graphs.

Find answers at: cambridge.org/ukschools/gcsemaths-studentbookanswers

Gradient

- The gradient of a distance–time graph is the speed of the object.
  - If speed is constant the gradient is constant and is represented by a straight line.
  - If the line is horizontal then the object is not moving; the gradient is zero.
- The gradient of a speed–time graph is the acceleration of the object.
  - If the line is sloping upwards it represents acceleration (increasing speed).
  - If the line is sloping downwards it represents deceleration (decreasing speed).
  - If the line is horizontal it represents a constant speed; the gradient is zero.
- Curved graphs have gradients that change along the graph continually. Gradient triangles can be used to estimate the changes in the gradient.
- The gradient at a point on a distance-time graph will give the instantaneous speed. This can be calculated by drawing a tangent at the point and finding the gradient of the tangent.
- The average speed of an object can be calculated using two points on the line.

Area under a graph

- The area under a graph can be used to calculate other values. For example, the area under a speed-time graph gives the distance travelled.

For additional questions on the topics in this chapter, visit GCSE Mathematics Online.

Chapter review

1 The graph shows how the population of a village has changed since 1930.

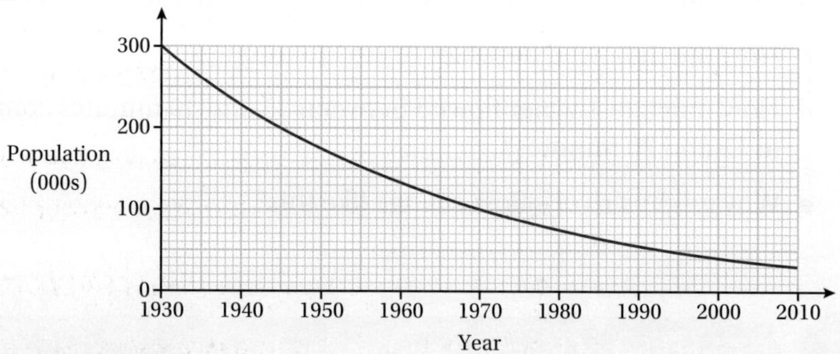

a Copy the graph using tracing paper and find the gradient of the graph at the point (1950, 170).

b What does the graph represent?

2 This graph shows Ben's journey to school.

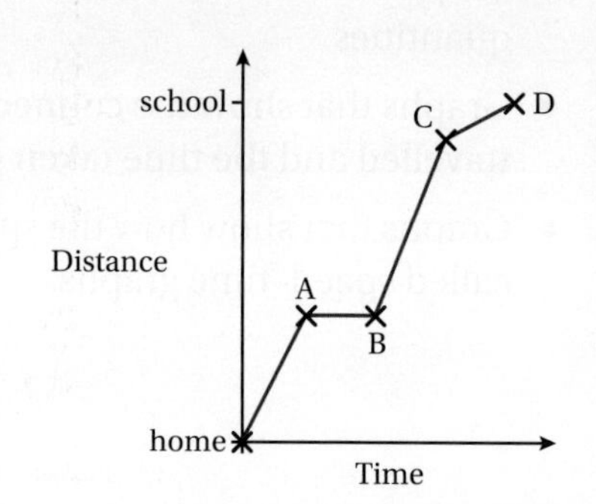

a During which part of his journey was Ben travelling fastest?

b What happened between A and B?

c Did Ben speed up or slow down at C?

3 The following graph shows the height of water in a cylindrical vase.

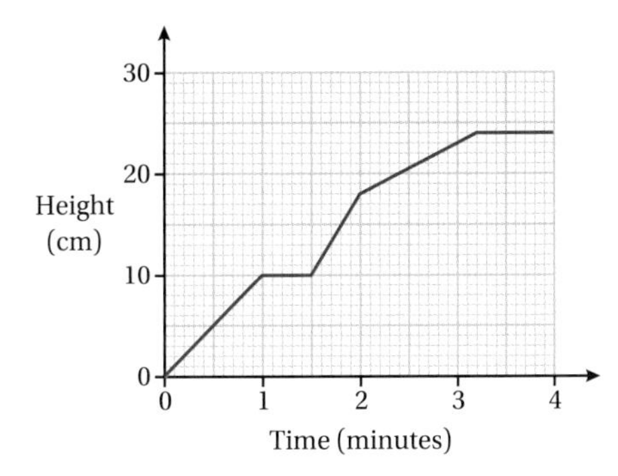

a The vase was filled to $\frac{3}{4}$ of its capacity. How tall is the vase?

b Given that the radius of the base was 3 cm, what rate was the water flowing at the beginning?

c What was the rate between 2 and 3 minutes?

4 The speed–time graph below represents the journey of a train between two stations. The train slowed down and stopped after 15 minutes because of engineering work on the railway line.

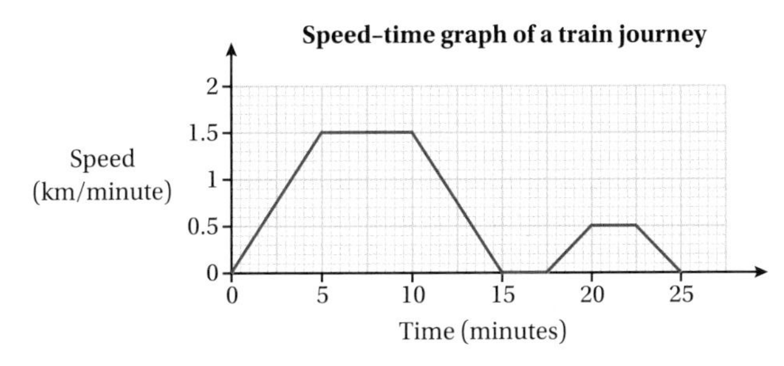

a Calculate the greatest speed, in km/h, that the train reached.

b Calculate the deceleration of the train as it approached the place where there was engineering work.

c Calculate the distance the train travelled in the first 15 minutes.

d For how long was the train stopped at the place where there was engineering work?

e What was the speed of the train after 19 minutes?

f Calculate the distance between the two stations.

40 Algebraic inequalities

In this chapter you will learn how to ...

- use the correct symbols and notation to express inequalities.
- solve linear and quadratic inequalities in one variable and represent the solution set on a number line and in set notation.
- solve (several) linear inequalities in two variables, representing the solution set on a graph.

For more resources relating to this chapter, visit GCSE Mathematics Online.

Using mathematics: real-life applications

Inequalities can be used to model and solve problems where a range of answers are possible rather than specific values, for example, all distances less than 2 m from the edge of a road. Linear programming is used in logistic and project management. It involves graphing complex constraints for a project to find a region of feasibility and identify the best solution.

"All big civil engineering projects involve a great deal of planning. We need to work with limits on time and budget. Inequalities are one way of expressing the ranges of values that have to be met and considered together." *(Civil engineer)*

Before you start ...

Ch 8	You must be able to solve linear equations.	**1** **a** If $3x + 2 = 2x + 5$, $x = ?$ **b** If $4(n + 3) = 6(n - 1)$, $n = ?$ **c** If $6(5 - 3x) = 5(2x - 5)$, $x = ?$ **d** If $\frac{3a - 2}{4} = \frac{a - 5}{2}$, $a = ?$
Ch 8	You should remember how to solve quadratic equations.	**2** $x^2 - 2x - 3 = 0$ $x = 3$ is one of the roots of this equation. What is the other root?
Ch 18	You should be confident using linear graphs.	**3** The equation of the red line is: $y = \frac{^-1}{3}x + 1$ **a** What is the gradient of the line that is perpendicular to this line? **b** What is the equation of that line if it cuts the given line at (3, 0)? 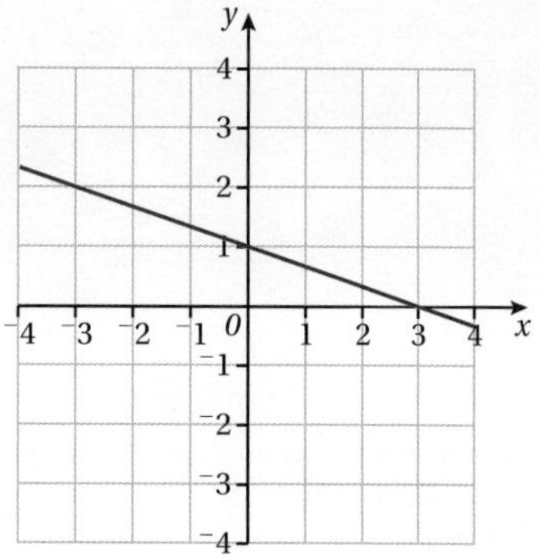

Assess your starting point using the Launchpad

STEP 1

1 Use mathematical symbols to express the following:
a p is less than 0.45
b x is greater or equal to $^-4$
c y lies between the values of 11 and 18

2 List the integers that satisfy each inequality:
a $2 < x < 5$
b $12 > x > {}^-2$
c $2 \leqslant x - 1 < 5$

?

GO TO
Section 1:
Expressing inequalities

✓

STEP 2

3 Draw a number line to represent each of the following.
a $x > {}^-1$
b $x \leqslant 2$
c $^-3 \leqslant x \leqslant 4$

4 Say whether each statement is true or false.
a On a number line, $\{x: x \leqslant {}^-3\}$ would be a line starting at and including $^-3$ and pointing in a negative direction.
b On a number line, $\{x: x > 2\frac{1}{2}\}$ would be a line starting at and including $2\frac{1}{2}$ and pointing in a positive direction.

?

GO TO
Section 2:
Number lines and set notation

✓

STEP 3

5 Solve these inequalities:
a $4x - 5 < 3$
b $3(x + 5) \geqslant 9$

?

GO TO
Section 3:
Solving linear inequalities

✓

STEP 4

6 Solve $x^2 - 4x - 5 = 0$.

7 Produce a graph of the equation $y = x^2 - 4x - 5$ and represent the solution set for the inequality $x^2 - 4x - 5 > 0$.
Use the graph to identify the range of values that would satisfy this inequality.

?

GO TO
Section 4:
Solving quadratic inequalities

✓

GO TO
Section 5:
Graphing linear inequalities

Find answers at: cambridge.org/ukschools/gcsemaths-studentbookanswers

Key vocabulary

inequality: a mathematical sentence in which the left side is not equal to the right side.

Section 1: Expressing inequalities

An **inequality** is a mathematical sentence that uses symbols such as $<$, $\leqslant$, $\neq$, $>$ or $\geqslant$ in place of an equals sign. The expressions on either side of the symbol are not equal.

The most common inequality symbols are:

$>$ greater than
$<$ less than
$\geqslant$ greater than or equal to
$\leqslant$ less than or equal to

You've already used inequality symbols to give a range of values. For example, $2 < x < 6$ means '2 is *less than* x and x is *less than* 6'. The integer values of x that satisfy this expression are 3, 4 and 5.

Another way to read this statement is to say x lies between 2 and 6.

Inequalities indicate a range of values to be considered. In $a \leqslant x \leqslant b$, x is a value that lies between the values of a and b and can be equal to a and b. This statement can also be written in the form $b \geqslant x \geqslant a$.

An inequality will have a finite number of integer solutions but an infinite number of real solutions.

Applying operations to inequalities

If you apply an operation to both sides of an inequality, then the resulting inequality is true for addition and subtraction. It is also true for multiplication and division of **positive** numbers.

When you multiply or divide both sides of an inequality by a **negative** number, then you need to **reverse** the direction of the inequality symbol in order to make the resulting inequality true.

For example, if you multiply both sides of the inequality $7 > 3$ by $^-2$, then the resulting statement is correctly written as $^-14 < {}^-6$.

If $^-x < 3$, this means that $x > {}^-3$ (multiplying both sides by $^-1$).

EXERCISE 40A

1 Complete the statements with the correct inequality symbol.

a If $7 > 3$, then $4 + 7 \square 4 + 3$ **b** If $8 < 13$, then $8 - 5 \square 13 - 5$
c If $^-5 < {}^-1$, then $^-5 + 3 \square {}^-1 + 3$ **d** If $^-4 > {}^-11$, then $^-4 - 6 \square {}^-11 - 6$

2 Complete the statements with the correct inequality symbol.

a If $7 > 3$, then $2 \times 7 \square 2 \times 3$ **b** If $8 < 13$, then $2 \times 8 \square 2 \times 13$
c If $7 > 3$, then $7 \div 2 \square 3 \div 2$ **d** If $8 < 13$, then $8 \div 2 \square 13 \div 2$

3 Complete the statements with the correct inequality symbol.

a If $7 > 3$, then $^-2 \times 7 \square {}^-2 \times 3$ **b** If $8 < 13$, then $^-2 \times 8 \square {}^-2 \times 13$
c If $7 > 3$, then $7 \div {}^-2 \square 3 \div {}^-2$ **d** If $8 < 13$, then $8 \div {}^-2 \square 13 \div {}^-2$

4 List four whole numbers that satisfy the following inequalities.

a $x > 14$ **b** $x \geqslant 6$ **c** $x \leqslant {}^-2$ **d** $x + 3 \geqslant 7$ **e** $x - 4 \leqslant 5$

5. If $x > 6$ how many values can x take?

6. If $3 < x < 8$, how many integer values can x take? How many values can x take if we include decimal values or fractions?

7. What integer values are given by $6 > x > 2$?

Section 2: Number lines and set notation

Key vocabulary

number line: a line marked with numbers in order, similar to a ruler scale.

You can use a **number line** to illustrate an inequality. When drawing and illustrating values on a number line the convention is to use an open dot (small circle) if the starting value is not included, and a solid dot if the starting point is included.

The expression $x \leqslant 11$ means numbers less than 11 including 11. So a number line representing $x \leqslant 11$ shows values starting from and including 11 with a solid dot at 11.

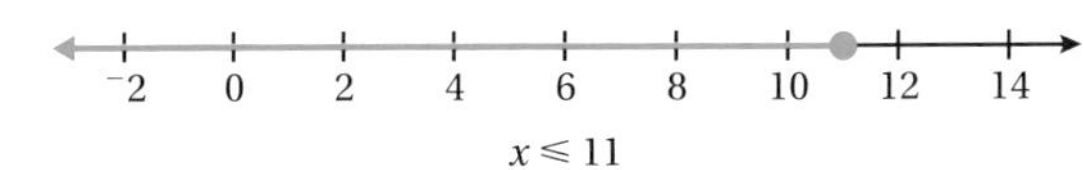

The expression $x > 11$ means numbers greater than 11. So a number line representing $x > 11$ starts at 11 but the open dot is taken to signify that 11 is not included.

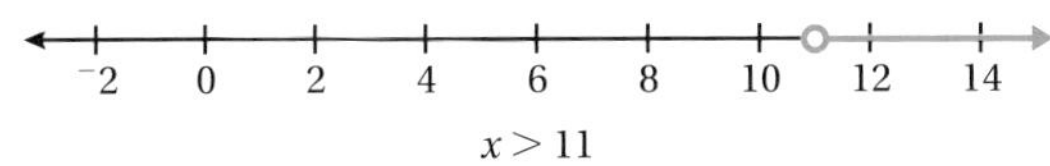

Set notation

Tip

You learnt about set notation in Chapter 24.

Another way to write statements about inequalities and the range of numbers that have been identified is to use set notation.

A set is a collection, and the elements of a set are written within curly brackets { }; these brackets are shorthand for 'the set of'. For example, {2, 4, 6, 8} is the set of the numbers 2, 4, 6, 8, which represents the even numbers between 1 and 10.

$\{x: x \leqslant 11\}$ This statement represents the set of numbers that are equal to or less than 11.

$\{x: x > 11\}$ This statement represents the set of numbers that are greater than 11.

Tip

Number lines and graphs are a good way of checking you have identified the correct range of numbers in a set. Think carefully about the use of the symbols.

You can represent some sets using number lines. For example, this number line shows the set $\{x: 3 \leqslant x < 6\}$:

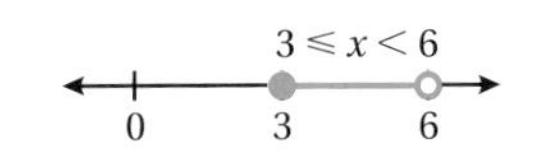

EXERCISE 40B

1. Use set notation to describe the range of values shown on each number line.

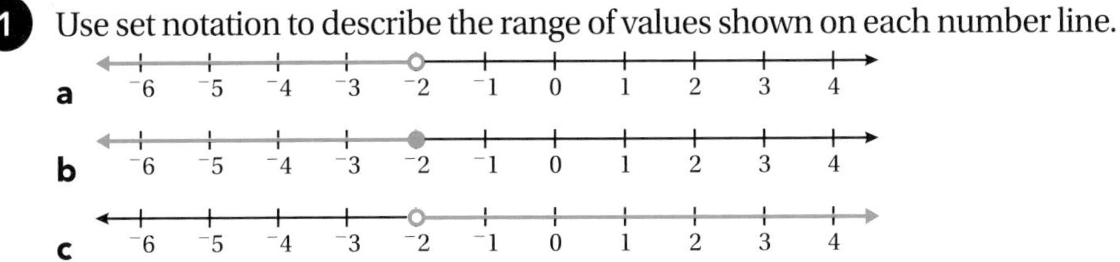

Find answers at: cambridge.org/ukschools/gcsemaths-studentbookanswers

2 Represent each set on a number line.

a $\{x: x > 8\}$ **b** $\{x: x \leqslant 0\}$ **c** $\{x: x < {}^{-}5\}$
d $\{x: x > {}^{-}1\}$ **e** $\{x: x \leqslant {}^{-}2\frac{1}{2}\}$

3 Using set notation write a statement for each of the sets identified in these number lines in two different ways.

a
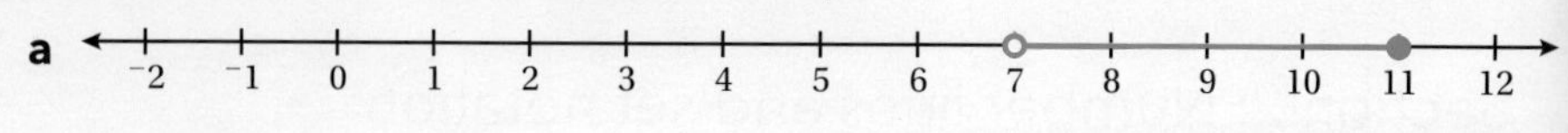

b ⁻2 ⁻1 0 1 2 3 4 5 6 7 8 9 10 11 12

c ⁻2 ⁻1 0 1 2 3 4 5 6 7 8 9 10 11 12

d ⁻2 ⁻1 0 1 2 3 4 5 6 7 8 9 10 11 12

4 Draw a number line for each of the following sets using conventional notation.

a $\{x: {}^{-}5 \leqslant x \leqslant 1\}$ **b** $\{x: 6 > x > {}^{-}1\}$

Section 3: Solving linear inequalities

Solving the linear inequality $4x - 5 < 3$ means finding all of the values for x that satisfy that inequality. You can solve inequalities using the same methods that you used for linear equations.

However, you must apply the rules that you learned in Section 1:

- If you multiply or divide an inequality by a negative value, the inequality sign must be reversed to make an equivalent inequality.
- If you swap the sides of the inequality you must reverse the signs. For example, if you have $2 < x$ and you want x on the left-hand side, you get $x > 2$. (Think of this as reading the inequality from right to left.)

WORKED EXAMPLE 1

Solve for x. Show your solutions on a number line.

a $4x - 5 < 3$ **b** $\frac{5x - 3}{2} \geqslant 11$ **c** ${}^{-}5 \leqslant 3x + 4 \leqslant 13$
d $5(x - 2) > 6$ **e** $4(7 - x) \leqslant 3$

a
$$4x - 5 < 3$$
$$4x < 8$$
$$x < 2$$

1 2 3

Add 5 to both sides, then divide both sides by 4.

b
$$\frac{5x - 3}{2} \geqslant 11$$
$$5x - 3 \geqslant 11 \times 2$$
$$5x \geqslant 22 + 3$$
$$5x \geqslant 25$$
$$x \geqslant 5$$

4 5 6 7

Multiply both sides by 2 to get rid of the denominator, add 3 to both sides.

Continues on next page ...

c $\quad {}^{-}5 \leqslant 3x + 4 \leqslant 13$

$\quad {}^{-}5 - 4 \leqslant 3x \leqslant 13 - 4$

$\quad {}^{-}9 \leqslant 3x \leqslant 9$

$\quad {}^{-}3 \leqslant x \leqslant 3$

Subtract 4 from each expression, then divide all the terms by 3.

d $\quad 2(5x - 2) > 6$

$\quad 10x - 4 > 6$

$\quad 10x > 10$

$\quad x > 1$

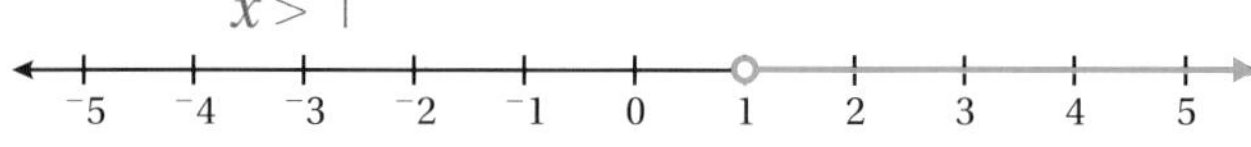

e $\quad 4(7 - x) \leqslant 3$

$\quad 28 - 4x \leqslant 3$

$\quad 25 \leqslant 4x$

$\quad 6.25 \leqslant x$

$\quad x \geqslant 6.25$

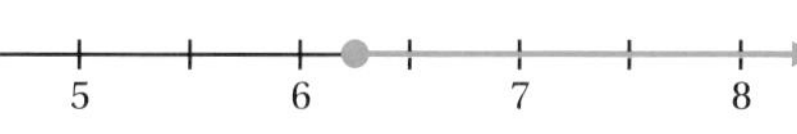

EXERCISE 40C

Tip

You can substitute your solution in the inequality to check that it is correct, in the same way you can check solutions to equations.

1 Solve these simple one step inequalities making sure you have the correct symbol in your answer.

a $4x \leqslant 20$ **b** ${}^{-}10x \geqslant 130$ **c** ${}^{-}12x > {}^{-}42$

d $\frac{{}^{-}x}{2} \leqslant 5$ **e** $\frac{{}^{-}x}{5} > 4$ **f** $3 - 2x > 5$

2 Solve each inequality. Leave fractional answers as fractions in their simplest form.

a $3(h - 4) > 5(h - 10)$ **b** $\frac{y + 6}{4} \leqslant 9$

c $\frac{1}{2}(x + 5) \leqslant 2$ **d** $3 - 7h \leqslant 6 - 5h$

e $2(y - 7) + 6 \leqslant 5(y + 3) + 21$ **f** $6(n - 4) - 2(n + 1) < 3(n + 7) + 1$

g $5(2v - 3) - 2(4v - 5) \geqslant 8(v + 1)$ **h** $\frac{z - 2}{3} - 7 > 13$

i $\frac{3k - 1}{7} - 7 > 7$ **j** $\frac{2e + 1}{9} > 7 - 6e$

3 When 5 is added to twice p, the result is greater than 17. What values can p take?

4 When 16 is subtracted from half of q, the result is less than 18. What values can q take?

5 When $2p$ is subtracted from 10, the result is greater than or equal to 4. What values can p take?

Find answers at: cambridge.org/ukschools/gcsemaths-studentbookanswers

6 The sum of $4d$ and 6 is greater than the sum of $2d$ and 18. What values can d take?

7 A number a is increased by 3 and this amount is then doubled. If the result of this is greater than a, what values can a take?

8 At a certain school, the mark out of 100 for the Term 1 exam and twice the mark out of 100 for the Term 3 exam are added together. The students must obtain at least 150 marks to achieve a satisfactory grade. A student obtains x marks in the Term 1 exam.

a Write an appropriate inequality to show the mark, y, that the student must obtain in the Term 3 exam in order to pass.

b Solve this inequality for:

i $x = 35$ **ii** $x = 49$

Section 4: Solving quadratic inequalities

A quadratic inequality contains at least one term with a squared variable and no terms with any powers higher than 2. For example:

$$x^2 - 4x > 5$$

You can solve quadratic inequalities using the methods you applied to quadratic equations.

In this example, first rewrite the inequality to make the right-hand side 0.

$x^2 - 4x - 5 > 0$ (first step)

Then factorise

$$(x - 5)(x + 1) > 0$$

The inequality 'greater than 0' means the product of the brackets is positive.

This could mean one of two things:

either both brackets are positive (in other words > 0), **or** both brackets are negative (in other words < 0).

Either $(x - 5) > 0$ and $(x + 1) > 0 \rightarrow x > 5$ and $x > {}^-1$: this results in $x > 5$

Or $(x - 5) < 0$ and $(x + 1) < 0 \rightarrow x < 5$ and $x < {}^-1$: this results in $x < {}^-1$

So the solution is the set $\{x > 5 \text{ or } x < {}^-1\}$.

Starting with a different inequality:

$$x^2 - 4x < 5 \rightarrow x^2 - 4x - 5 < 0$$
$$(x - 5)(x + 1) < 0$$

This time the product is less than zero. So:

either $(x - 5) > 0$ and $(x + 1) < 0 \rightarrow x > 5$ and $x < {}^-1$

(But this is a contradiction, as x cannot satisfy both conditions at the same time.)

or $(x - 5) < 0$ and $(x + 1) > 0 \rightarrow x < 5$ and $x > {}^-1$

This results in the solution set $\{{}^-1 < x < 5\}$.

In mathematics, graphs are usually a very effective method for deriving information about functions. Consider a graph of $y = x^2 - 4x - 5$ and you will see how easy it is to identify and verify solutions of quadratic inequalities.

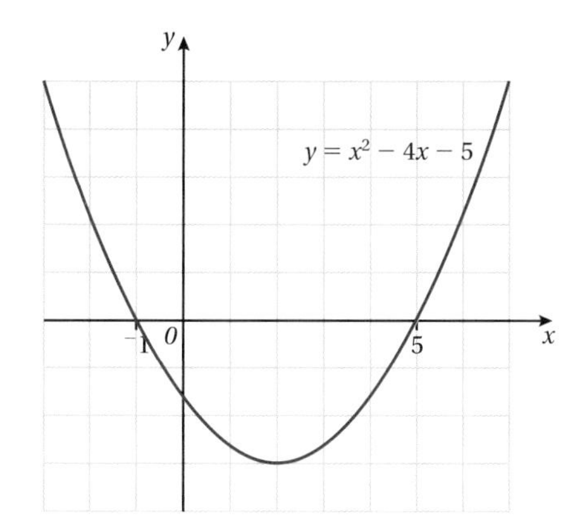

For $x^2 - 4x - 5 > 0$, this means $y > 0$. Where the graph has values above the x-axis, y is positive, so the solution to the inequality $x^2 - 4x - 5 > 0$ is the set of numbers where $x > 5$ or $x < {}^{-}1$, that is $\{x: x > 5 \text{ or } x < {}^{-}1\}$.

For $x^2 - 4x - 5 < 0$, this means $y < 0$. Where the graph has values below the x-axis, y is negative, so the solution to the inequality is the set of numbers where ${}^{-}1 < x < 5$, that is, $\{x: {}^{-}1 < x < 5\}$.

Tip

If you are solving a quadratic inequality algebraically be sure to include all the steps involved in the solution to help you apply the correct reasoning.

EXERCISE 40D

1 **a** For what values of x is $x^2 - 3x - 3 \geqslant 0$?

b For what values of x is $x^2 - 3x - 3 < 0$?

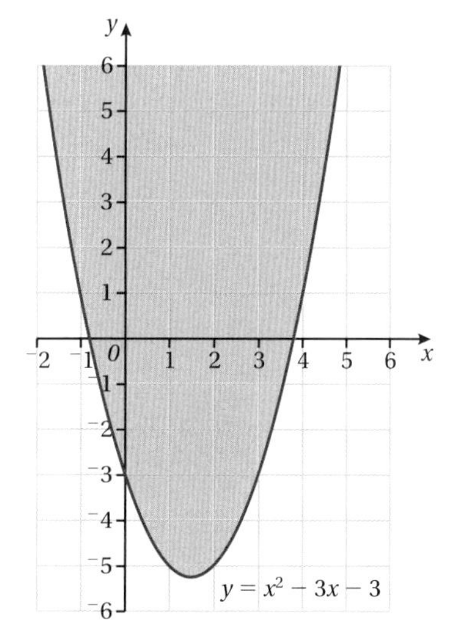

2 **a** For what values of x is ${}^{-}2x^2 + 16x - 24 \geqslant 0$?

b For what values of x is ${}^{-}2x^2 + 16x - 24 \leqslant 0$?

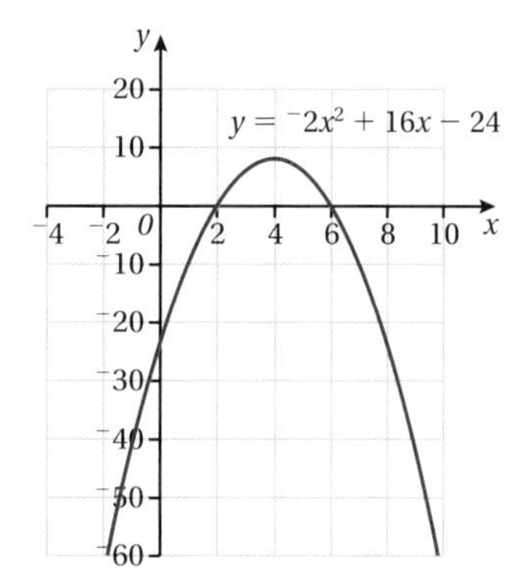

3 Sketch a graph and find all values of x such that:

a $(x - 3)(x + 2) > 0$ **b** $(x + 1)(x + 4) \leqslant 0$

c $(x - 5)(x - 2) \geqslant 0$ **d** $x(x + 3) < 0$

4 Solve for x. Sketch the graphs if you need to.

a $^-2x^2 - 5x + 12 > 0$ **b** $x^2 - 5x < 0$

c $8 + 2x - x^2 \leqslant 0$ **d** $12 - 5x - 2x^2 < 0$

5 Write the quadratic inequalities that are represented by the values on these number line graphs.

a

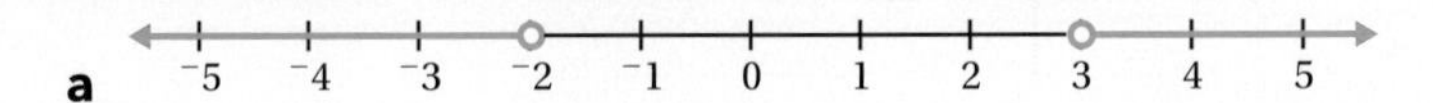

b

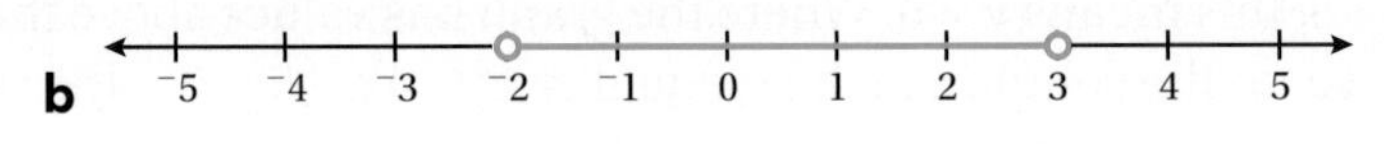

c $^-5$ $^-4$ $^-3$ $^-2$ $^-1$ 0 1 2 3 4 5

Section 5: Graphing linear inequalities

So far, you have dealt with inequalities with one variable and solutions that can be shown on a number line. Inequalities such as $y = x + 1$ have two variables connected to them (x, y). The solution to such an inequality is a region on a plane. You need to understand how to represent inequalities on a number plane so that you can use graphs to find and/or represent the solution to two or more simultaneous inequalities.

Key vocabulary

equalities: having the same amount or value.

Regions on a plane

Simple equations like $y = 3$ and $x = {^-2}$ can also be called **equalities**. y is equal to 3 and x is equal to $^-2$.

On a number plane, all the points that satisfy each equality lie on one straight line.

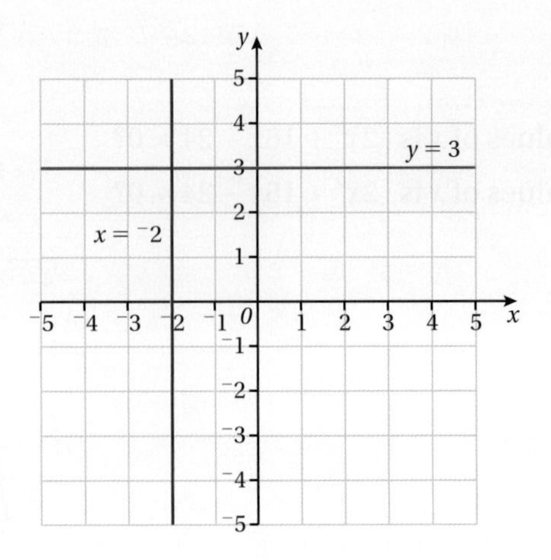

All the points that satisfy the inequality $x > 2$ lie on one side of the line $x = 2$.

The region into which these points fall is shaded on the graph below.

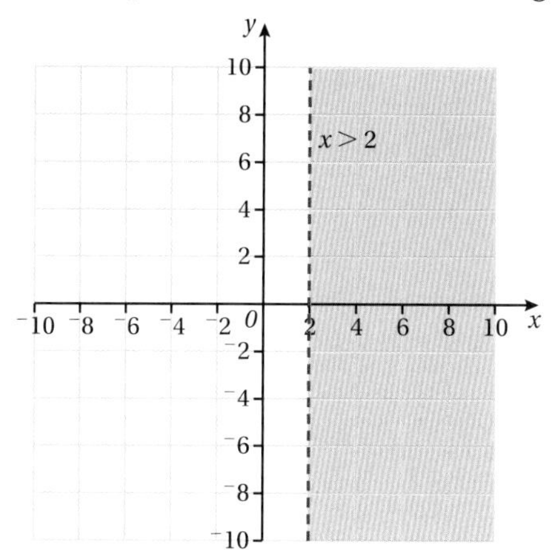

All points on the other side of the line satisfy the inequality $x < 2$.

The line itself is not included in the region $x > 2$, so it is shown as a broken line.

The diagram on the right shows the inequality $y \leqslant 2x + 1$.

In this case, the points on the line are included in the region so the line is shown as a solid line.

For inequalities linked by $<$ or $>$ symbols, the boundary line of the region is a broken line.

For inequalities linked by $\leqslant$ or $\geqslant$ symbols, the boundary line of the region is a solid line.

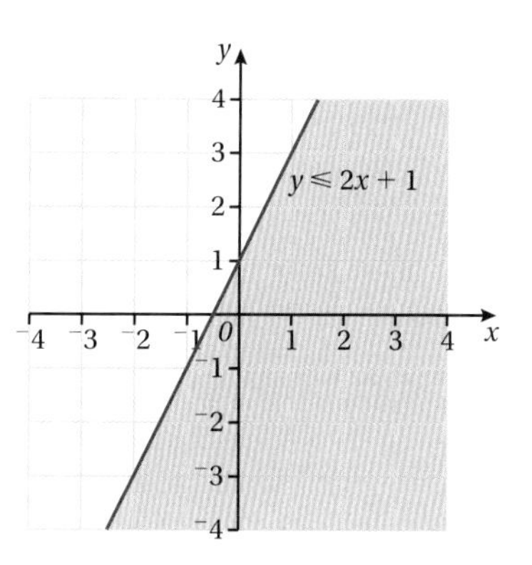

WORKED EXAMPLE 2

Draw a set of x- and y-axes from -4 to 4. Shade the region on the diagram that satisfies both statements $y > 3$ and $x < -2$. Give two points in the identified (shaded) region.

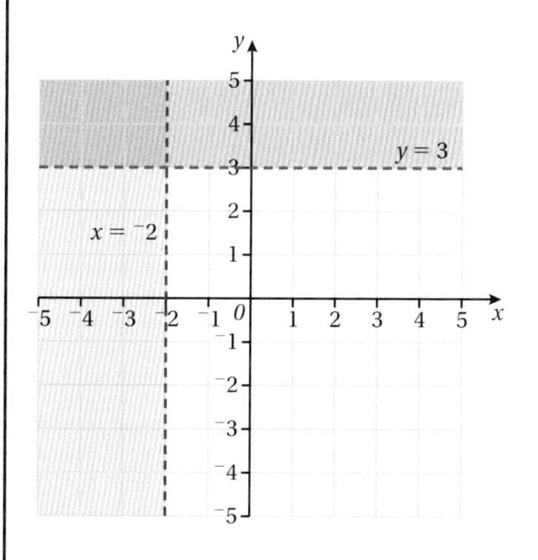

First, draw the lines $y = 3$ and $x = -2$; make sure the lines are dashed as $y = 3$ and $x = -2$ are not included in their respective inequality. Shade in the graph **above** the line $y = 3$; shade in the graph to the **left** of the line $x = -2$ in a different colour.

$(-3, 4)$ and $(-5, 5)$

You can choose any two points within the shaded region, **except** for those that contain an x-coordinate of -2 or those with a y-coordinate of 3, that is any that fall on the two dashed lines.

Verifying solutions

This is a graph of $y = {}^{-}x + 3$.

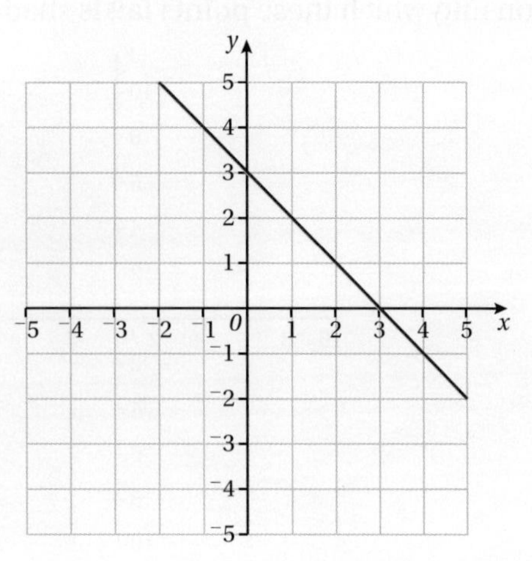

In order to identify the region that is true for $y \geqslant {}^{-}x + 3$, test a point either side of the line; for example $({}^{-}2, 3)$. Mark this point with coordinates on the graph. If $x = {}^{-}2$, then $y = {}^{-}({}^{-}2) + 3 = 5$; 3 is the y-value in the coordinate but it is **not** greater than or equal to 5 so the inequality is not true. Mark this point with coordinates on the graph.

Try (2, 2) on the other side of the line. Mark this point with coordinates on the graph.

If $x = 2$, $y = {}^{-}2 + 3 = {}^{-}1$, the value of y in the coordinate is 2, which **is** greater than ${}^{-}1$.

$2 \geqslant {}^{-}1$, so the region to the right of the line represents the inequality $y \geqslant {}^{-}x + 3$.

EXERCISE 40E

1 Sketch a graph for each of the following linear equations and on each graph shade the region defined by the inequality:

a $y = x + 1, y \geqslant x + 1$ **b** $y = {}^{-}2x + 4, y \leqslant {}^{-}2x + 4$

c $y = \frac{1}{2}x + 3,\ y \geqslant \frac{1}{2}x + 3$

2 Draw the following vertical and horizontal lines on a graph: $x = {}^{-}4$, $x = 1$, $y = 5$ and $y = {}^{-}3$.

Shade in the regions defined by the inequalities $x \geqslant 1$, $y \leqslant 5$.

State the coordinates of two points in the region where the two inequalities overlap.

3 Is the region shaded in this diagram $y \geqslant \frac{1}{3}x - 2$ or is it $y \leqslant \frac{1}{3}x - 2$? How do you know?

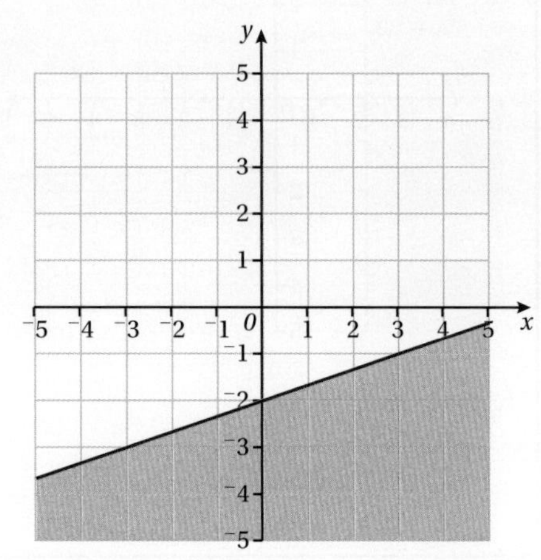

4 For each of the diagrams, find the equation of the line and write an inequality to define the shaded region.

a

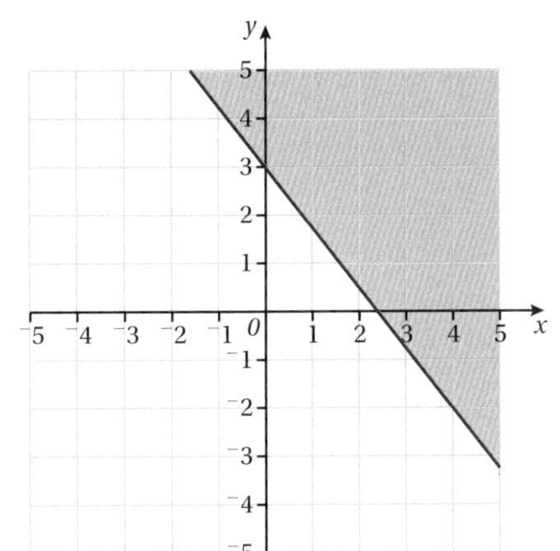

b

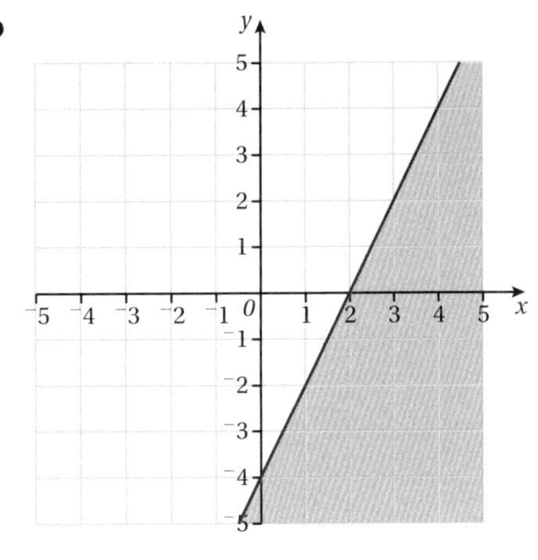

5 Check that the inequalities that define each region of the diagram are correct by substituting a point from the region.

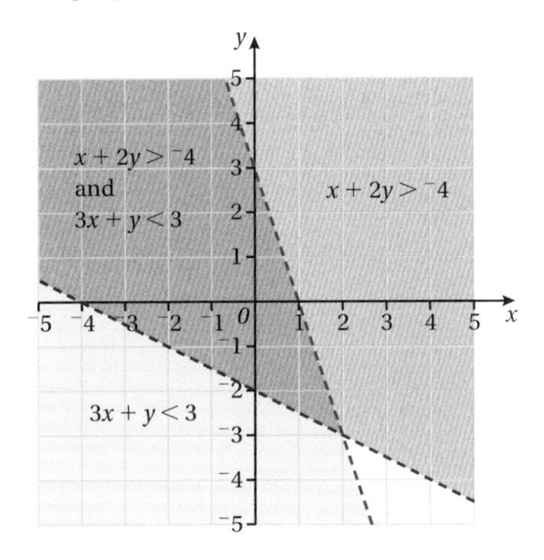

Write a pair of inequalities which define the unshaded region.

Graphing several inequalities

Many problems involving two or more inequalities can be readily solved by a graphical method.

WORKED EXAMPLE 3

A company needs two types of cupboard in its office, type A and type B. It wants as many of both kinds as will fit. The facilities manager writes an inequality based on the maximum number of each cupboard that could be in the office:

$B \leqslant \frac{^{-}3}{2}A + 12$ and $B \leqslant \frac{^{-}1}{2}A + 10$

Draw a graph to find out the maximum number of each cupboard he could have, if he has to use as many of each type as possible.

Continues on next page ...

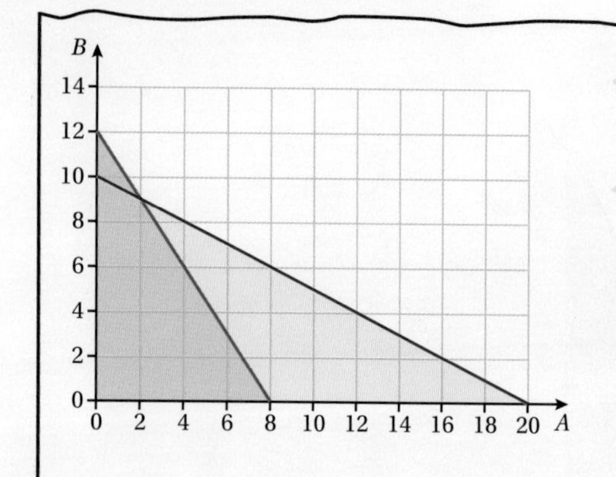

Plot $B = \frac{^-3}{2}A + 12$ and $B = \frac{^-1}{2}A + 10$; use a solid line for each.

The region in yellow satisfies $B \leqslant \frac{^-1}{2}A + 10$. The blue region satisfies only $B \leqslant \frac{^-3}{2}A + 12$. Where the two regions overlap (the green region) is where **both** inequalities are true. So the coordinates of any point here represent the number of each type of cupboard that can fit.

However, you need the **maximum** number of each type of cupboard that would fit at the same time. This can be found at the point where the two lines intersect, (2, 9). The maximum number of each type that can fit in the office is 2 of type A and 9 of type B.

Only whole number solutions are valid to solve this problem; it would not be possible to purchase a fraction of a cupboard.

EXERCISE 40F

1. Draw a sketch diagram of the two linear equations $y = {^-4}x + 8$ and $y = x + 1$.

   Identify and shade the region satisfied by the inequalities $y > x + 1$ and $y \leqslant {^-4}x + 8$.

2. Plot the following linear equations:

   $y = 2x - 3$

   $y = \frac{^-5}{4}x + \frac{5}{2}$

   $y = {^-3}$

   Identify and shade the region defined by $y \geqslant 2x - 3$, $y > {^-3}$, $y \leqslant \frac{^-5}{4}x + \frac{5}{2}$.

3. Write the equations of the two lines and identify the inequalities that represent the shaded area. Verify your answer with a point in the shaded region.

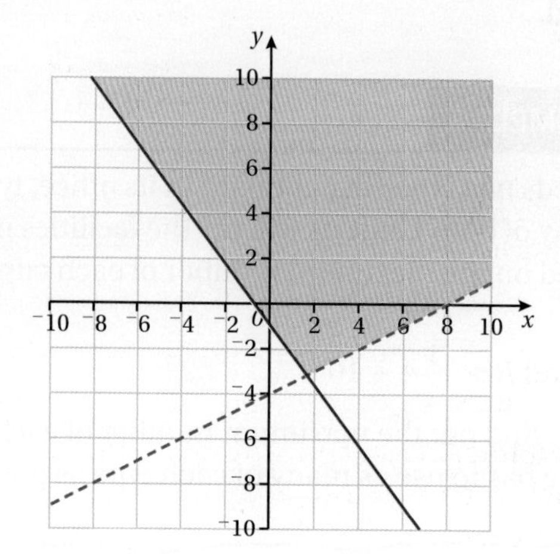

4 Write the equations of each of the lines in the graph. Write the three inequalities that identify the shaded region.

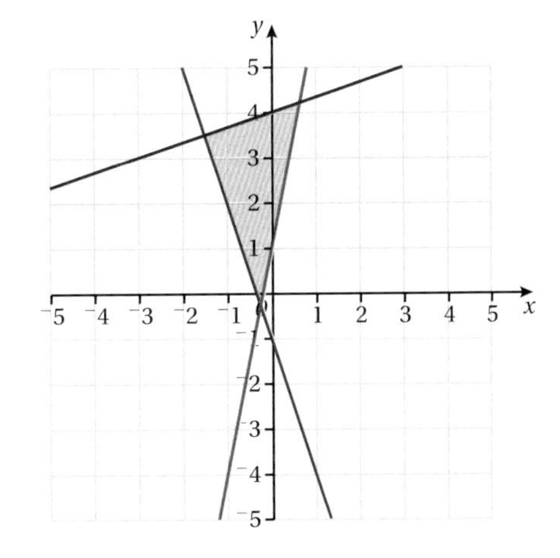

5 List the three inequalities that identify the shaded region and verify your answer with a point in the region.

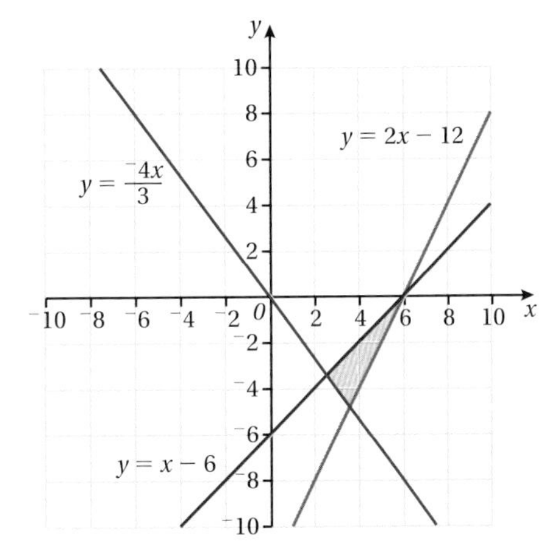

6 Solve the simultaneous equations:

$$4x - 5y = 20$$
$$7x - 2y = 14$$

Draw a sketch graph of the two lines and shade the region defined by $4x - 5y < 20$ and $7x - 2y \geqslant 14$.

Checklist of learning and understanding

Expressing inequalities

- Inequalities use the symbols $>, <, \geqslant, \leqslant, \neq$ and indicate a range of values.
- Inequalities indicate a range of values to be considered. In $a \leqslant x \leqslant b$, x is a value that lies between the values of a and b and can be equal to a and b. This statement can also be written in the form $b \geqslant x \geqslant a$.
- An inequality will have a finite number of integer solutions but an infinite number of real solutions.
- Inequalities can be illustrated on a number line:
  - A closed dot ● indicates that the starting value is included.
  - An open dot ○ indicates that the starting value is not included.
- If you add or subtract the same number to both sides of an inequality then the new inequality remains true: if $a > b$ then $a \pm c > b \pm c$; likewise, if $a < b$ then $a \pm c < b \pm c$.

Find answers at: cambridge.org/ukschools/gcsemaths-studentbookanswers

- If you multiply or divide by the same **positive** number on both sides of an inequality then the new inequality remains true: if $a > b$ and $c > 0$ then $ac > bc$ and $\frac{a}{c} > \frac{b}{c}$.
- If you multiply or divide by the same **negative** number on both sides of an inequality then you need to **reverse** the inequality sign for the new inequality to be true: if $a > b$ and $c < 0$ then $ac < bc$ and $\frac{a}{c} < \frac{b}{c}$.
  For example, if $a = 10$, $b = 4$ and $c = {}^-2$:
  $10 > 4$ but ${}^-2 \times 10 < {}^-2 \times 4$, ${}^-20 < {}^-8$ and $\frac{10}{{}^-2} < \frac{4}{{}^-2}$, ${}^-5 < {}^-2$.

Solving inequalities

- Linear inequalities in one unknown can be solved in the same way as equations but the answer includes an inequality symbol and indicates a range of values for the variable.
- Solutions can be written using set notation: $\{x: x < 3\}$ means the set of numbers x such that the value of x is less than 3. For example, integer solutions for $\{x: 2 < x \leqslant 5\} = \{5, 4, 3\}$.
- Quadratic inequalities are best solved by considering values on a graph. The shaded area in this graph represents the values of x for $y \geqslant x^2 - 1$.

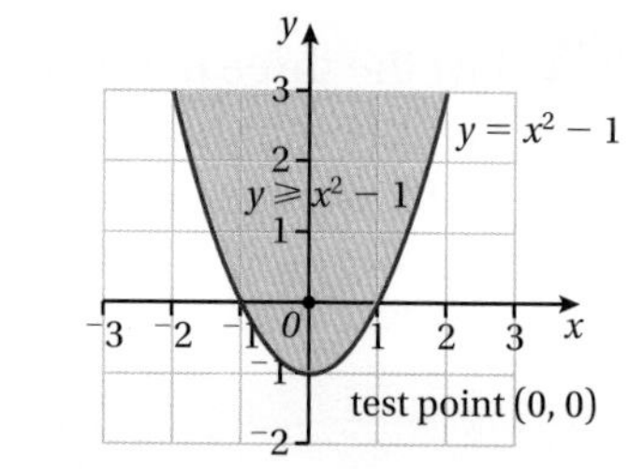

Graphing inequalities

- Problems that require more than one linear inequality can be solved by graphing and identifying the area that is the solution set.

For additional questions on the topics in this chapter, visit GCSE Mathematics Online.

Chapter review

1 Which of the following statements are true?

a $x + 11 > x - 11$

b $x \geqslant 12$ means that a number x is greater than 12.

c This number line shows the inequality $x \leqslant {}^-1$:

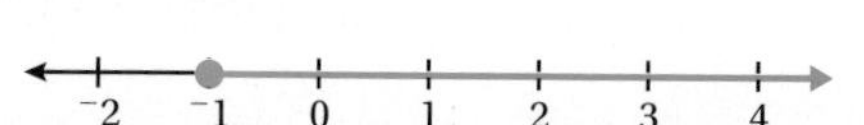

2 x is a whole number such that ${}^-3 \leqslant x < 5$ and y is a whole number such that ${}^-4 \leqslant y \leqslant 2$.

Write down the greatest possible value of:

a $x + y$ **b** $x - y$ **c** xy

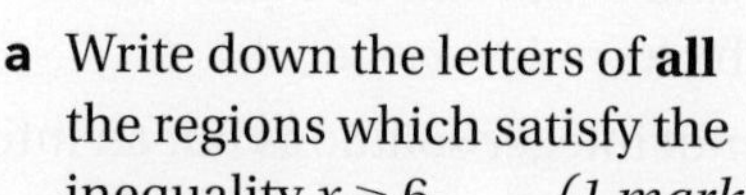

3 A, B, C, D, E, F and G are regions on a coordinate grid.

a Write down the letters of **all** the regions which satisfy the inequality $x \geqslant 6$. *(1 mark)*

b The regions D, F and G satisfy a different inequality.

Write down this inequality. *(2 marks)*

© OCR 2012

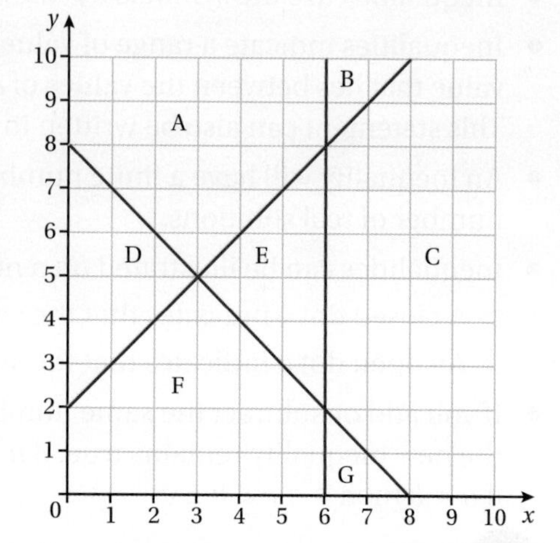

4 Frozen chickens will be sold by a major chain of supermarkets only if they weigh at least 1.2 kg and not more than 3.4 kg.

a Represent this range of values on a number line.

b Write an inequality to represent this range of values in terms of mass (m).

5 On a graph, identify the three integer values of x and y that satisfy all these four inequalities:

$4x + 3y < 12 \qquad y < 3x \qquad y > 0 \qquad x > 0$

6 On a diagram, draw straight lines and use shading to show the region R that satisfies the inequalities $x \geqslant 2$, $y \geqslant x$, $x + y \leqslant 6$.

7 Write the list of the three inequalities that identify the values in the shaded area.

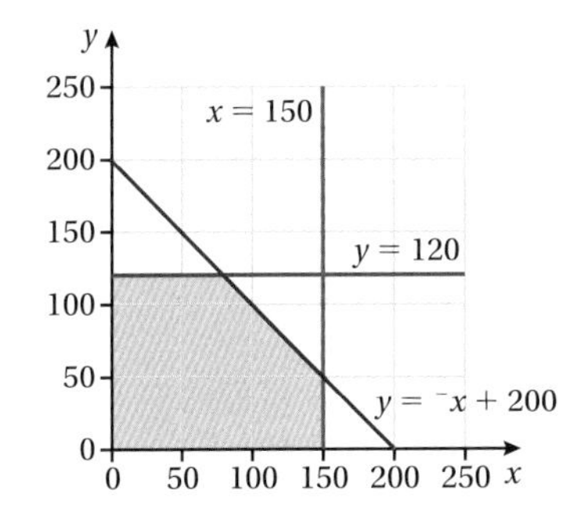

8 The shaded area in this diagram represents $x^2 - x - 12 > 0$.

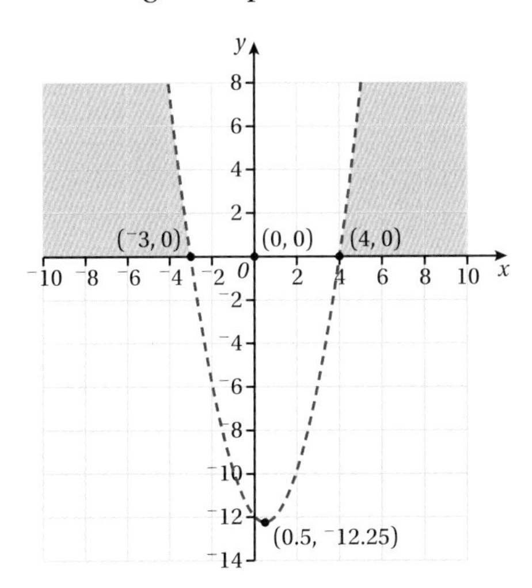

Verify that the shaded area is correct by solving the quadratic inequality $x^2 - x - 12 > 0$ and stating the solution sets for x.

41 Transformations of curves and their equations

In this chapter you will learn how to ...

- identify translations and reflections of a given graph or equation.
- sketch the graphs of these types of transformation.

For more resources relating to this chapter, visit GCSE Mathematics Online.

Using mathematics: real-life applications

Tip

Use ICT when studying this chapter. This allows you to generate graphs so you can study them quickly and experiment with changing values and comparing results.

Many people study the graphs of curves in the course of their work. Sound engineers are a good example. They mix and balance sounds by looking at curves made by sound waves.

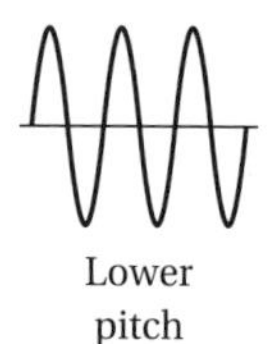

Lower pitch

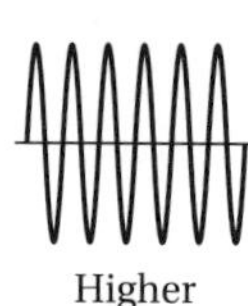

Higher pitch

"In today's music industry, we need to be talented in both the arts and the sciences." *(Sound engineer)*

Before you start ...

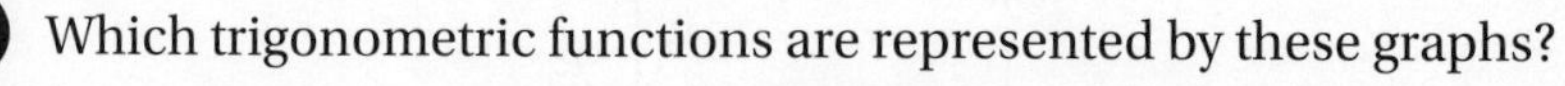

Ch 18, 19	You need to be able to recognise the graphs of standard functions: $y = mx + c$ $y = x^2$ $y = ax^2 + bx + c$ $y = \frac{1}{x}$	**1** Which of these functions would result in a curved graph and which would produce a linear graph? **a** $y = x^2 + 9$ **b** $5y + x = 10$ **c** $y = \frac{2}{x}$ **d** $y = (x + 7)^2 - 2$ Sketch each function to show its general shape.
Ch 33	You should be able to sketch the trigonometric functions: $y = \sin x$ $y = \cos x$ $y = \tan x$	**2** Which trigonometric functions are represented by these graphs? **a** 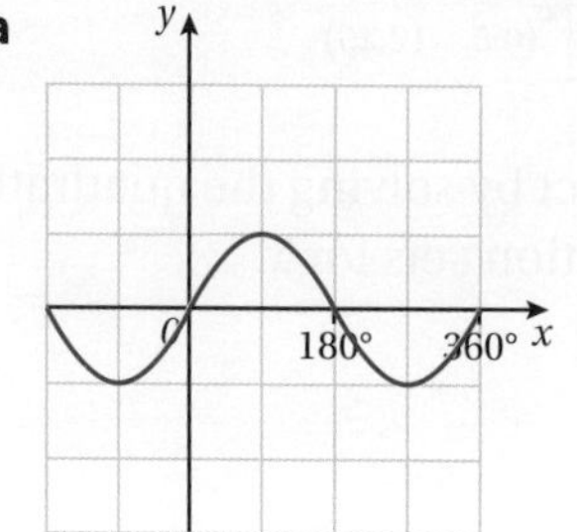 **b** 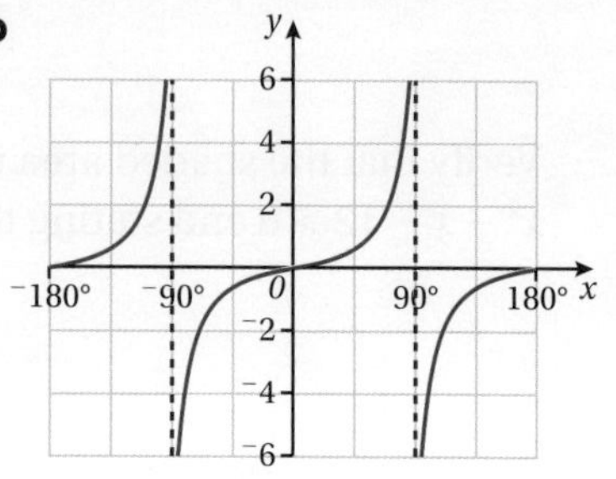 **c** Sketch and label the third trigonometric function.
Ch 8	You should be able to complete the square on a quadratic equation.	**3** Rewrite $y = 3x^2 + 6x + 7$ in the form $a(x - h)^2 + k$.

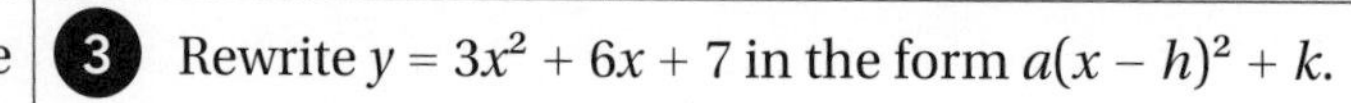

Assess your starting point using the Launchpad

STEP 1

1 This is a graph of $y = x^2$:

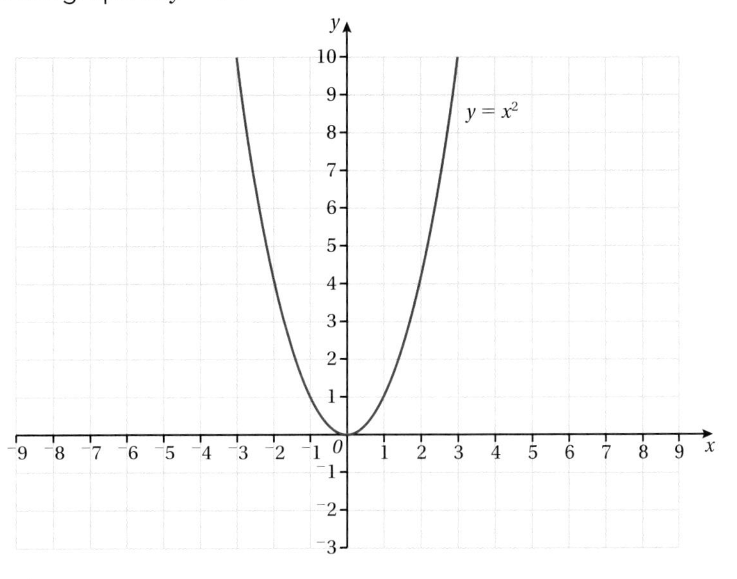

Sketch:

a $y = x^2 + 2$ **b** $y = x^2 - 2$

c Write the equation of the reflection of this graph about the x-axis.

d What happens to the graph when it is reflected about the y-axis?

GO TO
Section 1:
Quadratic functions and parabolas

STEP 2

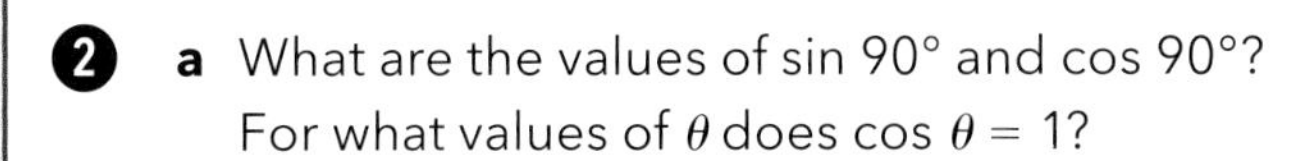

2 **a** What are the values of sin 90° and cos 90°?
For what values of θ does $\cos \theta = 1$?

b Sketch graphs of $y = \sin x + 2$ and $y = \cos(x + 90°)$.

GO TO
Section 2:
Trigonometric functions

GO TO
Step 3:
The Launchpad continues on the next page …

Launchpad continued ...

STEP 3

3 **a** What is the equation of this function, $y = x^3$ or $y = {}^-x^3$?

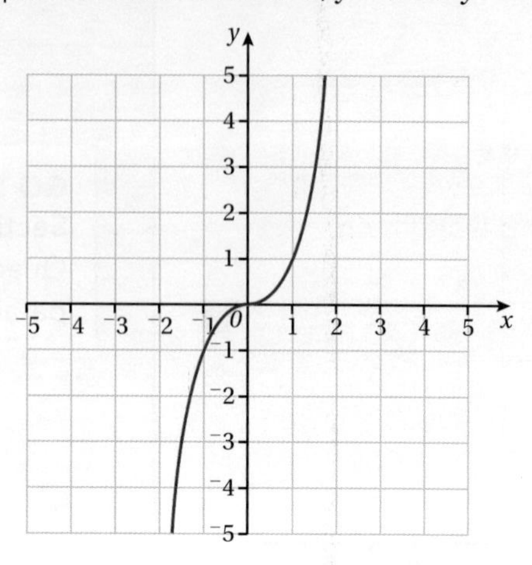

b **i** Draw its reflection about the x-axis and give the equation of the function.

ii Draw a reflection about the y-axis. What can you say about this reflection?

c Sketch the graph of $y = \frac{1}{x}$.

GO TO
Section 3:
Other functions

STEP 4

4 What is the equation of the graph of the basic hyperbola $y = \frac{1}{x}$ if it is translated 3 units to the right and reflected about the y-axis?

A $y = \frac{-1}{(x-3)}$ **B** $y = \frac{1}{(x+3)}$ **C** $y = \frac{-1}{(x+3)}$

GO TO
Section 4:
Translation and reflection problems

GO TO
Chapter review

Section 1: Quadratic functions and parabolas

The general form of a quadratic function is:

$$y = ax^2 + bx + c$$

In Chapter 8, you learnt different methods for solving quadratic equations. In Chapter 19 you learned about the features of the curved graph of a quadratic function, the parabola. In this section you are going to explore how changing the equation of a parabola results in a reflection and/or a translation of the curve.

The graph of a quadratic equation is called a parabola. These are beautiful curves that can be observed in many places in the natural and built environment.

Tip

Look back at Chapters 8 and 19 to make sure you understand the main features of quadratic equations. You should be able to solve quadratic equations and understand what is meant by the terms turning point, vertex and roots.

WORK IT OUT 41.1

Quadratic equations can have two different solutions, a single solution or no solutions. Look at these three graphs of quadratic equations.

Identify the equation that has two solutions.

What can you say about the other two equations based on the graphs?

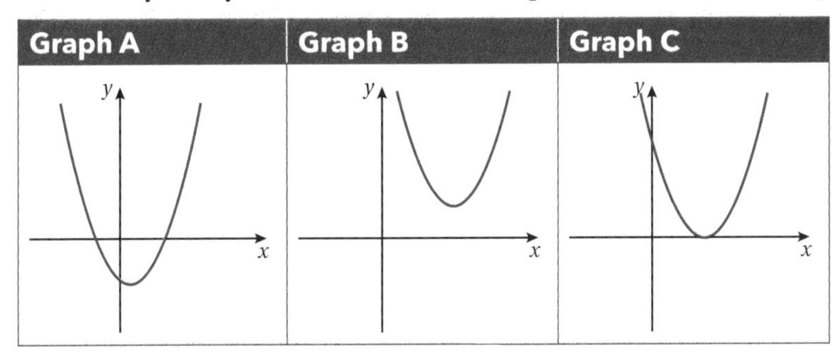

Find answers at: cambridge.org/ukschools/gcsemaths-studentbookanswers

Vertical translations

This is a graph of the function $f(x) = x^2$. The axis of symmetry is the y axis (or $x = 0$). The minimum turning point is the origin (0, 0), and this is called the vertex of the function $f(x) = x^2$.

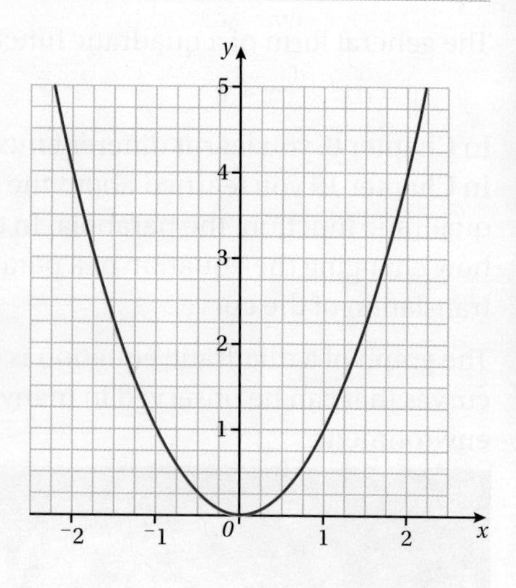

Consider what happens to the graph $y = x^2$ if we add 3 units to the function and get $y = x^2 + 3$.

For $y = x^2$ the vertex is (0, 0), also known as the turning point. But if $y = x^2 + 3$, the vertex moves up three positions and the graph cuts the y-axis at (0, 3).

$x^2 \geqslant 0$, $x^2 + 3 \geqslant 3$. Therefore the minimum value of $y = x^2 + 3$ is 3, and this is when $x = 0$. The vertex of this function is (0, 3).

Now consider the axis of symmetry. As $x^2 = (^-x)^2$, $x^2 + 3 = (^-x)^2 + 3$, so y has the same value for $\pm x$. The axis of symmetry is the y-axis.

We have shown that $y = x^2 + 3$ has a vertex (0, 3) and an axis of symmetry $x = 0$.

Adding the constant (+3) to the equation therefore translated the graph 3 units upwards (in the positive direction).

If the constant is negative, the graph is translated downwards (in the negative direction).

The diagram on the right shows the graph of $y = x^2$ in red.

The graphs of $y = x^2 + 5$, $y = x^2 + 12$ and $y = x^2 - 8$ are shown too.

Note that these graphs are all translations of $y = x^2$.

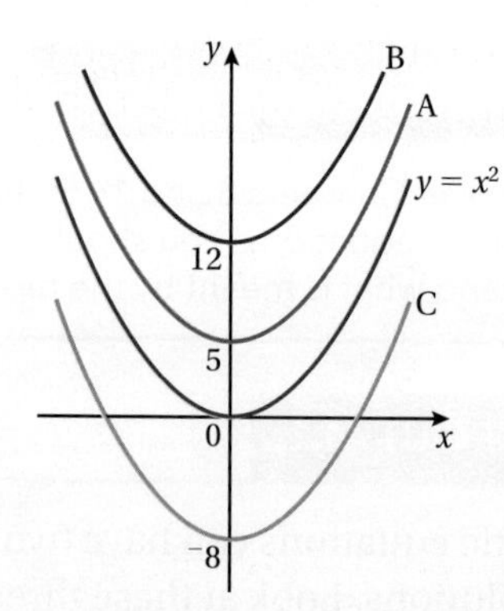

> **Tip**
>
> In mathematics we refer to **families** of lines or curves that have a main feature in common. How would you describe the family of parabolas you have just investigated?

EXERCISE 41A

1 Sketch the graphs of the following quadratic functions and state the coordinates of the vertex:

a $y = x^2 + 1$ **b** $y = x^2 - 1$ **c** $y = x^2 - 4$

d $y = x^2 + 2$ **e** $y = x^2 - 3$ **f** $y = x^2 + 3$

2 Sketch the graph of $y = {}^-x^2$. This is an image of the graph $y = x^2$.

What word can you use to describe the transformation?

3 Using your sketch for $y = {}^-x^2$, complete this sentence:

As x increases and decreases in value, y ...

4 Draw a sketch graph for each of the quadratic functions in question 1 if x^2 is now replaced by $^{-}x^2$.

What word can you use to describe the transformations to the original curves?

Horizontal translations

So far, we have considered vertical translations. The graph of $y = x^2$ was moved up and down the y-axis.

Graphs can also be translated in a horizontal direction. The result is a shift to the left or right.

Look at this graph carefully.

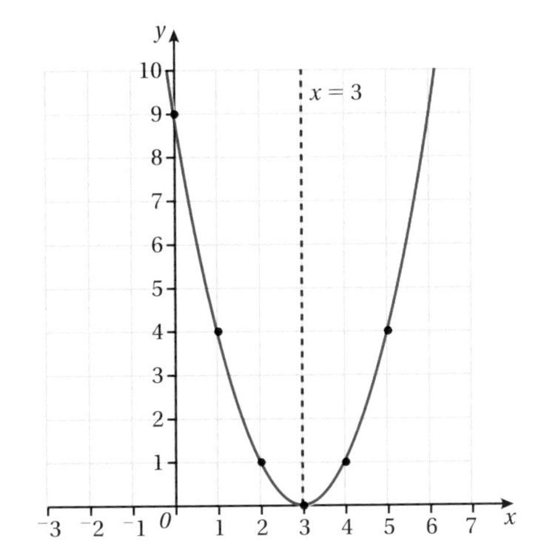

Every point on the basic parabola $y = x^2$ has coordinates (p, p^2).

If we translate the parabola three units to the right, then the vertex $(0, 0)$ goes to $(3, 0)$, and the axis of symmetry, $x = 0$, goes to $x = 3$.

For a horizontal translation of three units, the general point (p, p^2) goes to the point $(p + 3, p^2)$ as we move along the x-axis.

Doing some basic algebra:

If $x = p + 3$, then $p = x - 3$

And as $y = p^2$, substituting for p gives

$y = (x - 3)^2$

So a translation of 3 units to the right results in the function $y = (x - 3)^2$.

If we translate 3 units to the left, a similar argument would produce the result $y = (x + 3)^2$. The axis of symmetry for $y = (x + 3)^2$ is $x = {}^{-}3$.

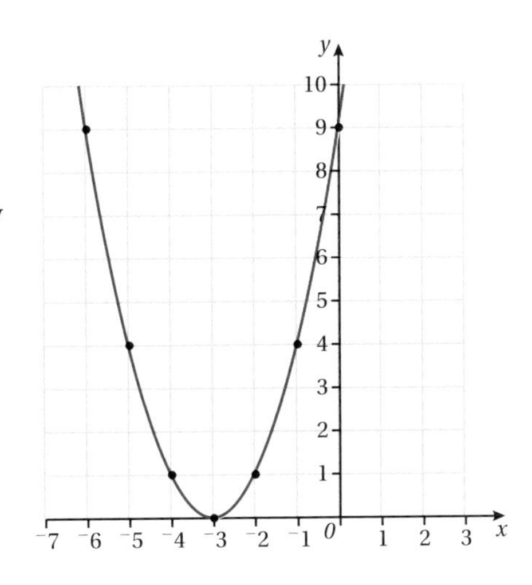

Tip

In order to draw a sketch of a graph from a translation of a quadratic function, you need to check for vertical and horizontal moves. Identify the axis of symmetry and the vertex (the turning point).

The table summarises the general case for translating a parabola.

	Equation of translated curve	Axis of symmetry	Vertex
Vertical translations	$y = x^2 + c$ $y = x^2 - c$	$x = 0$ $x = 0$	$(0, c)$ $(0, ^-c)$
Horizontal translations	$y = (x + b)^2$ $y = (x - b)^2$	$x = ^-b$ $x = b$	$(^-b, 0)$ $(b, 0)$

WORKED EXAMPLE 1

Sketch $y = (x - 2)^2 - 4$.

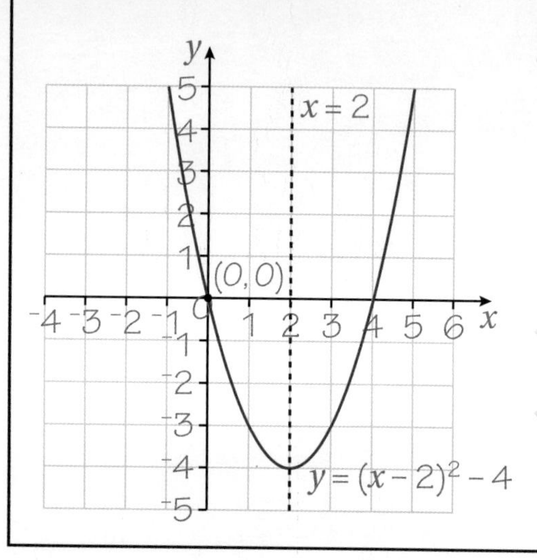

It might help to think of this as two separate translations. First deal with the vertical translation: the '-4' in $x^2 - 4$ indicates that the graph cuts the axis at $y = -4$, and this is a translation of x^2 by 4 units down. Now deal with the horizontal translation: $(x - 2)^2$ means translating x^2 two units to the right. So, overall the image is obtained by translating the graph of $y = x^2$ two units to the right and four units down.

The axis of symmetry is $x = 2$. The vertex is at $(2, ^-4)$.

When $x = 0$, $y = (^-2)^2 - 4 = 0$, so the y-intercept is $(0, 0)$, the origin.

Tip

Learning rules can be useful but it is always best if these rules are based on understanding, so that you can go back and establish the rule if you forget it.

Reflection in the x-axis

When you reflect a positive quadratic in the form $y = ax^2 + bx + c$ (where $a > 0$) over the x-axis, you will get a negative quadratic function, $y = ^-(ax^2 + bx + c)$(where $a > 0$), that is there is a reversal of signs for all the values of a, b and c. For example, $y = 2x^2 + 4x + 1$ has a reflection in the x-axis represented by the equation $y = ^-2x^2 - 4x - 1$, not $y = ^-2x^2 + 4x + 1$.

The vertex becomes a maximum turning point and not a minimum turning point (where $a < 0$) the vertex becomes a minimum turning point and not a maximum point.

The graph of the equation $y = ^-x^2$ is the image of $y = x^2$ reflected in the x-axis.

If $y = x^2$ and $y = ^-x^2$ are reflected about the y-axis the graphs remains the same because the y-axis is the axis of symmetry.

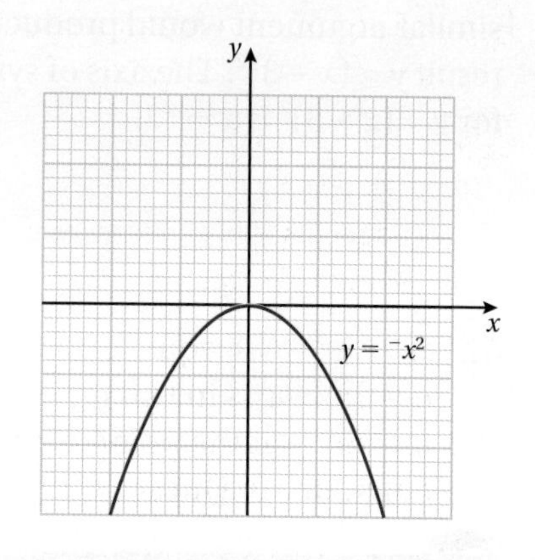

Reflection in the y-axis

When you reflect a quadratic in the form $y = ax^2 + bx + c$ about the y-axis, the line of symmetry and the x-coordinate reverse their sign.

Using the equation in Worked example 1, if $y = (x - 2)^2 - 4$ is reflected about the y-axis the axis of symmetry becomes $x = {}^-2$ and the vertex $({}^-2, 4)$.

The equation that represents this reflection is $y = (x + 2)^2 - 4$. The y-intercept remains the same, in this case $(0, 0)$.

The equation of the reflection of $y = x^2 + 4x - 5$ about the y-axis becomes $y = x^2 - 4x - 5$.

WORKED EXAMPLE 2

Sketch the graph of the equation $y = x^2 + 4x - 5$ reflected about the y-axis.

Write the equation of the resulting graph.

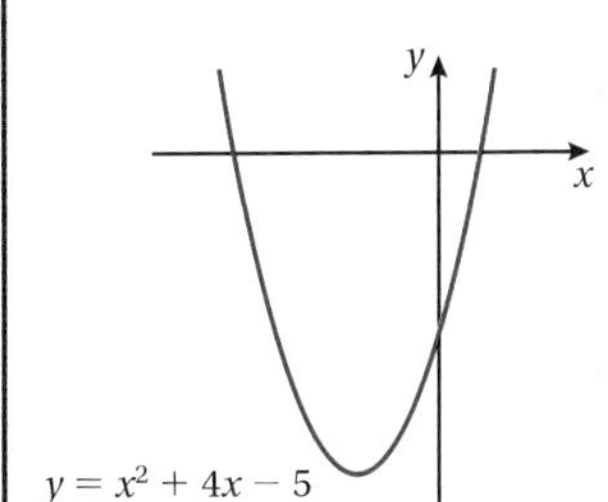

$y = x^2 + 4x - 5$

By completing the square $y = (x + 2)^2 - 9$

The axis of symmetry is $x = {}^-2$. The vertex is at $({}^-2, {}^-9)$.

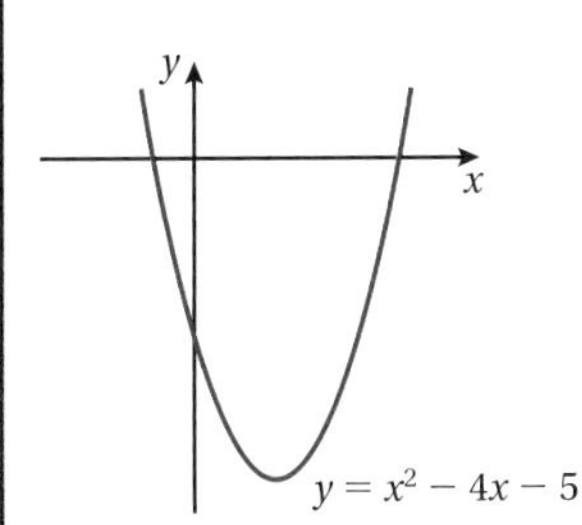

When $x = 0$, $y = {}^-5$, so the y-intercept is $(0, {}^-5)$.

If this graph is reflected about the y-axis the axis of symmetry becomes $x = 2$ and the vertex is $(2, {}^-9)$. The y-intercept remains $(0, {}^-5)$

The equation is $y = (x - 2)^2 - 9$, which gives the equation $y = x^2 - 4x - 5$.

In Worked example 2 the graph is translated 4 units to the right.

Tip

A reflection about the y-axis represents a horizontal translation.

EXERCISE 41B

1. Find the axis of symmetry, the vertex and the y-intercept of these equations.

   a $y = (x - 5)^2$ **b** $y = (x - 2)^2 + 9$
   c $y = (x + 6)^2 - 7$ **d** $y = (x - 3)^2 - 10$

2. What is the equation of the resulting curve when the graph of $y = x^2 + 4x - 1$ is reflected about the y-axis?

3. What happens to the graphs of the equations above when $(x \pm b)^2$ is replaced by ${}^-(x \pm b)^2$? Check using ICT that the graphs produced are images of the original graphs reflected in the horizontal line that passes through the vertex. Give the equation of the line of reflection in each case.

4. If a quadratic equation has real roots (solutions), the graph will either cut the x-axis in two places or touch the x-axis (the axis is a tangent to the curve at the point of contact.) State what the case is for each of the equations in Question 1.

Find answers at: cambridge.org/ukschools/gcsemaths-studentbookanswers

5 Sketch the graphs of:

a $y = (x - 5)^2$ **b** $y = (x - 1)^2 - 3$

c $y = (x + 2)^2 + 3$ **d** $y = (x - 4)^2 - 3$

e $y = (x - 1)^2 + 6$ **f** $y = (x - 4)^2 - 4$

6 **a** Find the axis of symmetry, the vertex and the y-intercept:

i for the equation $y = x^2 + 6x - 7$.

ii when the graph is reflected about the y-axis.

c Give the equation of the reflected graph.

d How many units to the right has the original graph moved?

e Draw a diagram of the two graphs on the same pair of axes.

7 **a** Find the axis of symmetry, the vertex and the y-intercept of

i for the equation $y = 2x^2 + 4x + 6$

ii when the graph in a is reflected about the y-axis.

c Give the equation of the reflected graph.

d How many units right has the original graph moved?

e Draw a diagram of the two graphs on the same pair of axes.

8 **a** Give the equation that represents a reflection about the y-axis of the graph $y = 3x^2 + 8x - 2$.

b Give the equation that represents a reflection about the x-axis of the graph $y = -x^2 + 3x + 2$.

Sketching quadratic functions by completing the square

Writing a quadratic equation in the completing the square form, $y = a(x + b)^2 \pm c$, can give you all the information you need to sketch the resulting parabola:

$(x = {}^-b)$ is the axis of symmetry

$({}^-b, \pm c)$ is the vertex

$(0, b^2 \pm c)$ is the y-intercept

WORKED EXAMPLE 3

Sketch the parabola $y = {}^-x^2 - 6x + 7$.

$y = {}^-x^2 - 6x + 7$
$= {}^-(x + 3)^2 + 16$

Rewrite this in the form $y = a(x - b)^2 + c$. (Look back at the section on completing the square in Chapter 7.)

${}^-(x + 3)^2 + 16 = 0$
${}^-(x + 3)^2 = {}^-16$
$(x + 3)^2 = 16$
$(x + 3) = \pm 4$

This equation has an axis of symmetry of $x = {}^-3$, the y-intercept is 7 and the vertex is $({}^-3, 16)$. To find the x-intercepts, solve the equation.

Continues on next page ...

So $x = 1$ or $x = ^-7$: the parabola crosses the x-axis at (1, 0) and ($^-7$, 0).

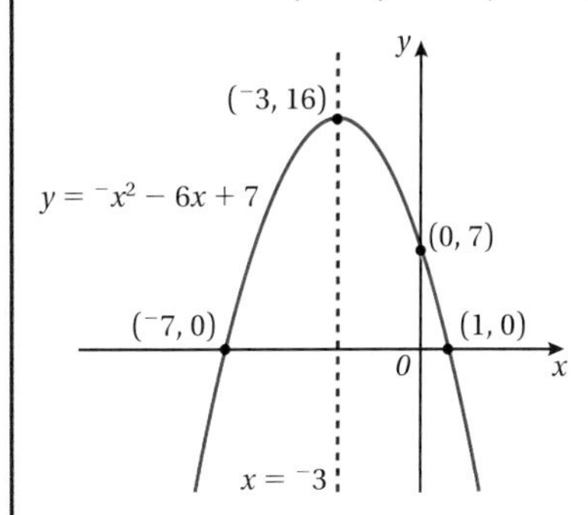

Use this information to sketch the graph.

EXERCISE 41C

Tip

Rewriting quadratic expressions by completing the square provides the information you need for a sketch. Remember that some parabolas will not cut the x-axis.

1 Complete the square to find the information you need to sketch the following graphs.

a $y = x^2 + 6x - 5$ **b** $x^2 + 8x + 4 = y$

c $y = x^2 - 4x + 2$ **d** $(x - 1)(x + 2) - 1 = y$

2 Sketch each parabola, label each of the defining features.

a $y = ^-x^2 + 3$ **b** $y = ^-x^2 - 2x$ **c** $y = ^-x^2 + 6x + 13$

d $y = ^-x^2 + 8x - 7$ **e** $y = ^-x^2 + 8x + 7$ **f** $y = ^-x^2 + x + 1$

Section 2: Trigonometric functions

Tip

Revise the section on trigonometric functions in Chapter 19 if you have forgotten the features of these graphs.

This diagram show the graphs of $y = \sin x$ and $y = \cos x$ for angles from 0° to 360°. The maximum and minimum values of the two functions are 1 and $^-1$, respectively. The graphs are called wave functions and continue in both directions repeating the same pattern at set intervals.

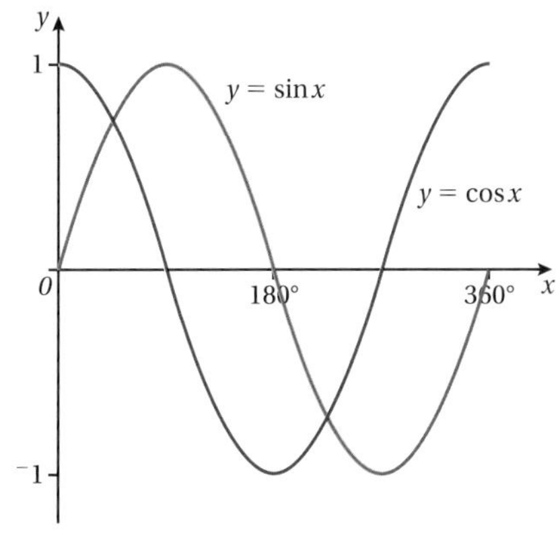

In the next exercise, you will investigate transformations of these functions so that we can produce some generalised findings. To do this, make sure you can generate these functions using ICT. It might be useful to work with another student on these questions so that you can both check your findings.

EXERCISE 41D

1 Investigate these graphs using ICT and write an explanation of what you find.

a $y = \sin x + 2$, $y = \sin x - 2$

b $y = \cos x + 2$, $y = \cos x - 2$

Find answers at: cambridge.org/ukschools/gcsemaths-studentbookanswers

2. Investigate these graphs using ICT and write an explanation of what you find.
   a $y = \sin(x + 90°)$, $y = \sin(x - 90°)$
   b $y = \cos(x + 90°)$, $y = \cos(x - 90°)$

3. Without using ICT describe the following transformations of the trigonometric functions.
   a $y = \sin x + 1$
   b $y = \sin(x + 45°)$
   c $y = \cos(x - 45°)$
   d $y = \cos x - 1$

4. What happens to the graph of $y = {}^{-}\sin x$ and what is its relationship to $y = \sin x$?

5. What happens to the graph of $y = {}^{-}\cos x$ and what is its relationship to $y = \cos x$?

Transformations of trigonometric functions

Your investigation in Exercise 41D should have helped you reach these conclusions.

Action	Transformation	Resultant image
Adding or subtracting a constant to a trigonometric function Example: $y = \sin x \pm c$	Vertical translation	The graph of the transformed function follows a parallel path to the original.
Adding or subtracting an angle to the argument of a trigonometric function Example: $y = \sin(x \pm b°)$	Horizontal translation	The graph of the transformed function moves left for an addition to the angle and right for a subtraction.
Taking the negative of a trigonometric function Example: $y = {}^{-}\sin x$	Reflection	The graph of the transformed function is a reflection of the original in the x-axis ($y = 0$).

Section 3: Other functions

You can apply what we have learnt about transformations of curves to some of the other functions you have studied in this course.

In the next exercise you can use ICT to produce graphs of the functions, but use the findings from earlier in the chapter to predict what you expect to see before you use ICT to look at the graphs of the transformations.

Again, it might be useful to work with another student to finalise your conclusions.

EXERCISE 41E

1 This is the basic cubic equation (highest power of x is 3).

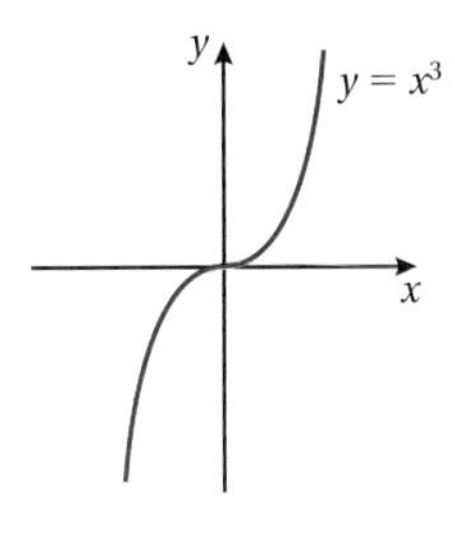

Sketch:

a $y = x^3 + 2$ and $y = x^3 - 2$

b $y = (x + 2)^3$ and $y = (x - 2)^3$

c $y = {}^-x^3$.

What is the line of reflection?

2 This is a graph of a hyperbolic function.

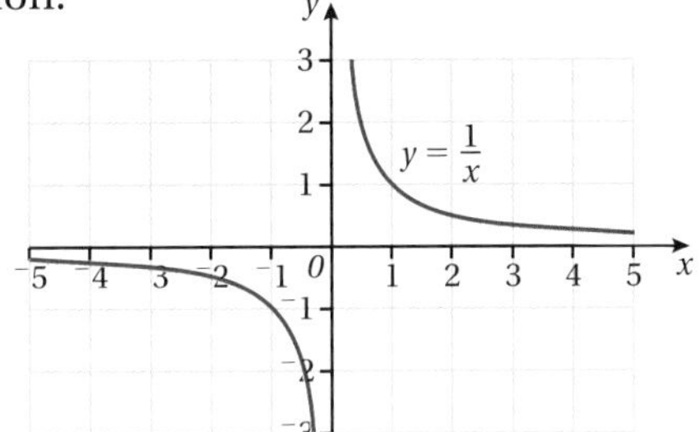

Sketch:

a $y = \frac{1}{x} + 3$

b $y = \frac{1}{x} - 3$

c $y = {}^-\left(\frac{1}{x}\right)$

3 This is a graph of an exponential function.

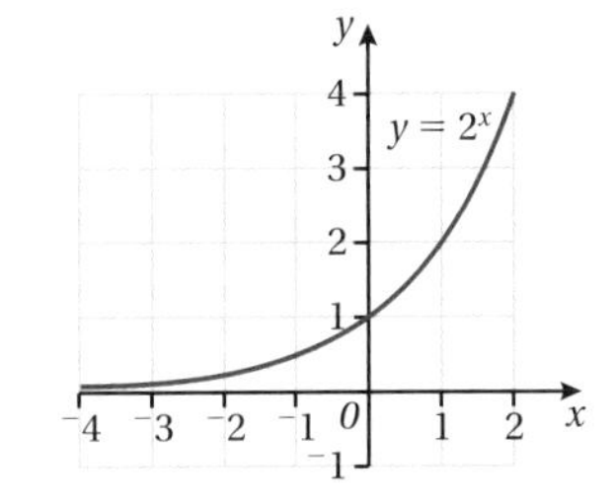

Sketch these exponential functions:

a $y = {}^-2^x$

b $y = 2^x + 3$

c $y = 2^{-x}$

Section 4: Translation and reflection problems

You have seen that changing the parameters of a function can change the position and orientation of a graph. In this section you are going to use the general principles you learned to solve problems related to transforming graphs.

EXERCISE 41F

1 Sketch the parabola $y = (x + 3)^2 - 8$. What is the equation of the resulting image if it is:

a translated 8 units up and 3 units to the right?

b translated 2 units to the left and 3 units down?

2 Consider the parabola $y = (x - 1)^2 + c$. Find the value of c if the y-intercept is:

a 1 **b** 3 **c** 0 **d** $^-7$

Sketch the graph in each case.

3 Sketch the graph of each quadratic, clearly labelling the x- and y-intercepts, the axis of symmetry and the vertex.

a $y = x^2 - 6x + 5$ **b** $y = x^2 - 4x - 12$

Find answers at: cambridge.org/ukschools/gcsemaths-studentbookanswers

Tip

Congruent means exactly the same as; remember that in geometry it means the same shape and size.

4 What is the translation of $y = \sin x$ that would result in $y = \sin x$ being congruent to $y = \cos x$?

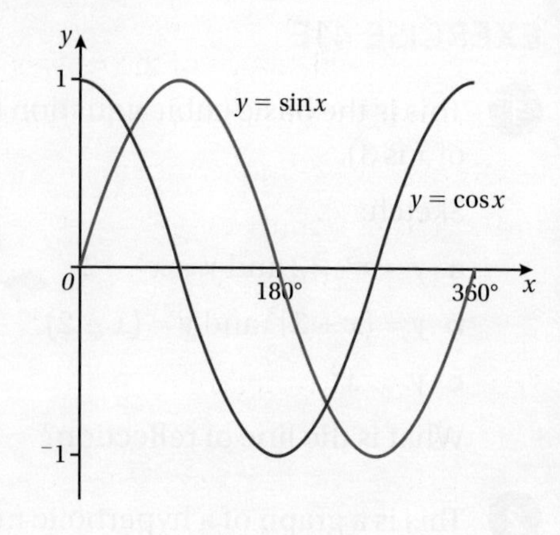

5 Draw a sketch of:

a $y = x^3 + 2$.

b the reflection of $y = x^3 + 2$ about the y-axis and about the x-axis.

6 The diagram shows the graph of $y = 2^x$ shifted horizontally to the right.

What is the equation of the image?

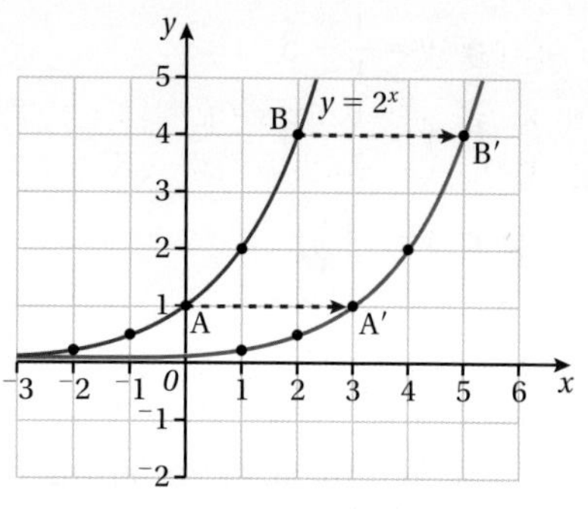

7 A basic curve has been shifted to form each of the graphs below. Decide how the graph was shifted and write the equation of the graph shown. Substitute the coordinates of the given points into the equation to check that your answers are correct.

a

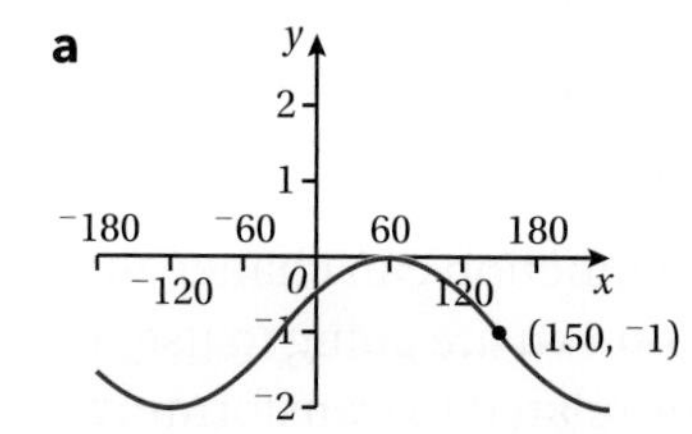

b

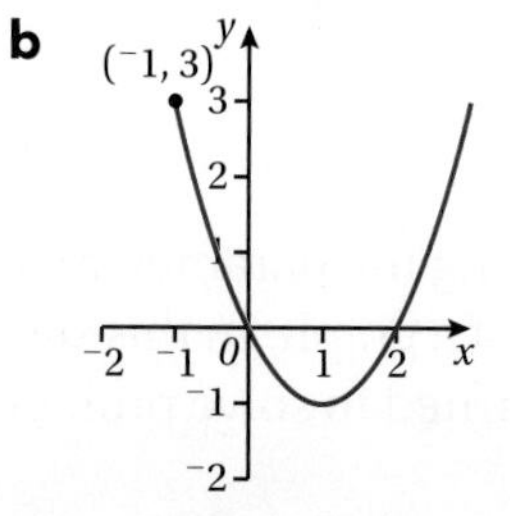

c

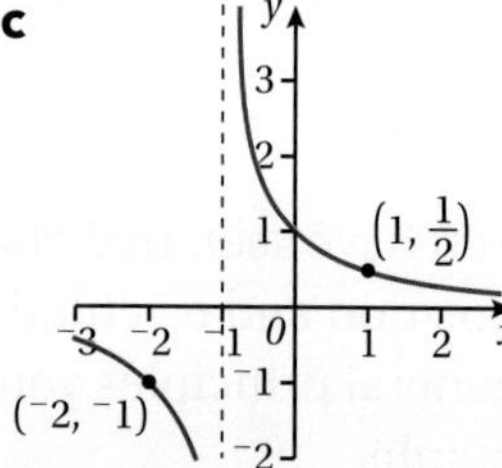

d

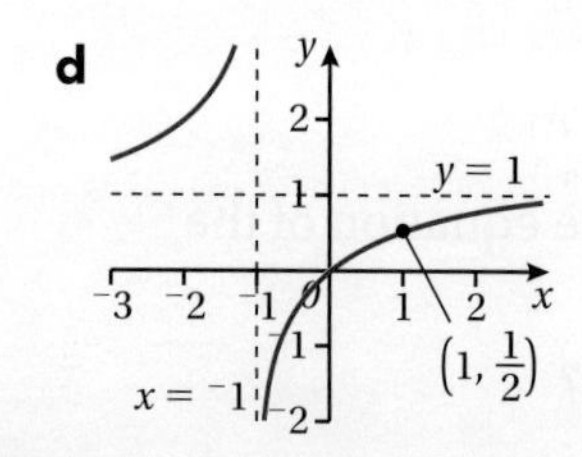

e

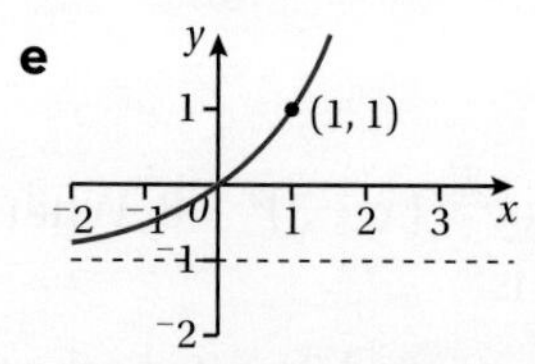

f

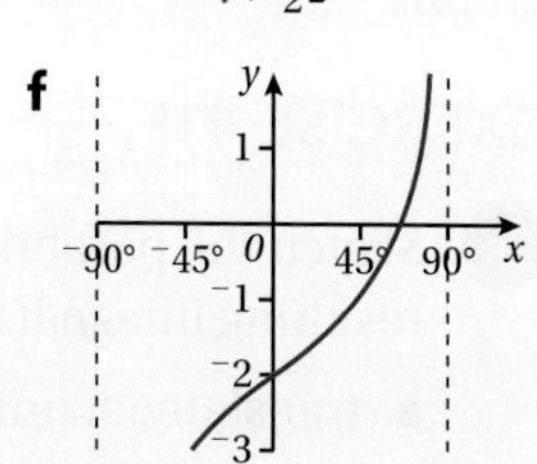

8 If you shift the graph of $y = \sin x$ to the right by 40° and down by 2 units, what will the equation of the new graph be?

Checklist of learning and understanding

Quadratic functions

- Vertical translations of the basic parabola $y = x^2$ are represented by the family of curves $y = x^2 \pm c$. The vertex is $(0, \pm c)$ and the axis of symmetry is the y-axis, $x = 0$.

- $y = {}^{-}x^2 \pm c$ represents the family of curves with a maximum turning point, which are reflections of the curves $y = x^2 \pm c$ about the x-axis.
- Horizontal translations of $y = x^2$ are represented by $y = (x \pm b)^2$.
- The axis of symmetry for $y = (x + b)^2$ is $x = {}^{-}b$ and the vertex is $({}^{-}b, 0)$.
- The axis of symmetry for $y = (x - b)^2$ is $x = b$ and the vertex is $(b, 0)$.
- Completing the square $y = a(x + b)^2 \pm c$, gives the axis of symmetry $(x = {}^{-}b)$, the vertex $({}^{-}b, \pm c)$, and the y-intercept $(0, b^2 \pm c)$.

Trigonometric functions

- $y = \sin x \pm c$ and $y = \cos x \pm c$ represent vertical translations of the functions $y = \sin x$ and $y = \cos x$.
- $y = \sin(x \pm b°)$ and $y = \cos(x \pm b°)$ represent a horizontal translation of the functions $y = \sin x$ and $y = \cos x$, $+b°$ to the left and ${}^{-}b°$ to the right.

Generally

- Generally speaking, these patterns can be applied to the graphs of other types of function as well.

Chapter review

For additional questions on the topics in this chapter, visit GCSE Mathematics Online.

1 Sketch the following graphs on the same set of axes and describe the transformation that changes graph **a** to graphs **b**, **c** and **d**.

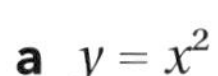

a $y = x^2$ **b** $y = x^2 - 5$ **c** $y = {}^{-}x^2$ **d** $y = (x - 5)^2$

2 This sketch shows the graph of $y = x^2$.

a On the same axes, sketch the graph of $y = 2x^2$. *(1 mark)*

b Describe the transformation that maps the graph of $y = x^2$ onto $y = x^2 - 3$. *(2 marks)*

© OCR 2013

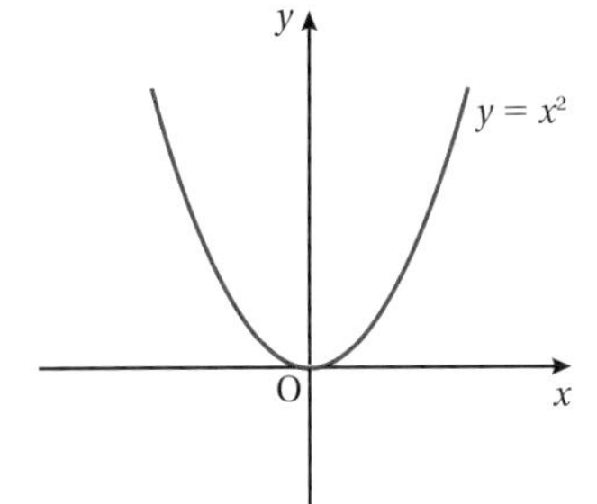

3 **a** What translation of the trigonometric function $y = \sin x$ has resulted in this shift to the right?

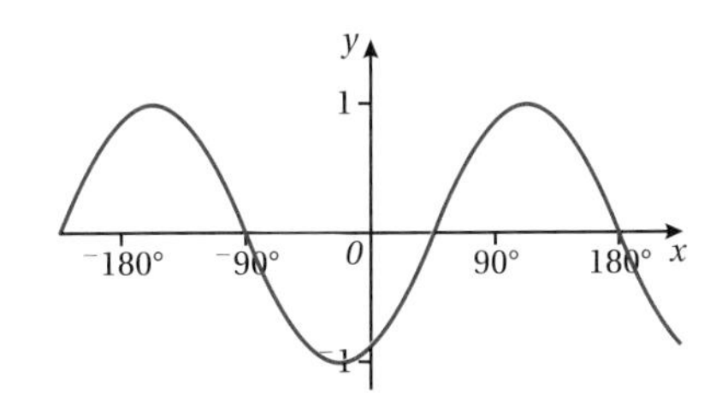

Draw a sketch of:

b **i** $y = \sin(x - 90°) + 1$

ii $y = \cos(x - 45°)$

4 Sketch the graph of:

a $y = \frac{1}{x} - 1$ **b** $y = 2^x$ reflected in the line $y = 0$

5 The reciprocal function $y = \frac{1}{x} + 2$ is changed and the new function is $y = \frac{1}{x + 1} + 2$.

What effect would this change have on the resulting graph?

Find answers at: cambridge.org/ukschools/gcsemaths-studentbookanswers

Glossary

A

Adjacent: next to each other; in shapes, sides that intersect each other.

Alternate angles: the angles on parallel lines on opposite sides of a transversal.

Angle of depression: when looking down, the angle between the line of sight and the horizontal.

Angle of elevation: when looking up, the angle between the line of sight and the horizontal.

Arc: a segment of a circle's circumference between two points.

Arc of a circle: a section of circumference between two points; a minor arc is the shorter distance between the two points, the major arc is the larger distance.

Arithmetic sequence: a sequence where the difference between each term is constant.

B

Bias: something that affects the chance of an event occurring in favour of a desired outcome.

Binomial: an expression consisting of two terms.

Binomial product: the product of two binomial expressions; for example, $(x + 2)(x + 3)$.

Bisect: to divide exactly into two halves.

Bivariate data: data that is collected in pairs.

C

Categorical data: data that has been arranged in categories.

Chord: a straight line from one point on the circumference to another that does not pass through the centre of the circle.

Circumference: distance round the boundary of a circle.

Class intervals: groups into which numerical data has been grouped.

Coefficient: the number by which a variable is multiplied.

Co-interior angles: the angles within the parallel lines on the same side of the transversal.

Common denominator: a number into which all the denominators of a set of fractions divide exactly.

Composite function: a function created by combining two or more functions.

Congruent: identical in shape and size.

Conjugate: binomial expressions with the same terms but opposite signs are said to be conjugate.

Consecutive terms: terms that follow each other in a sequence.

Constant: a number on its own.

Continuous data: data that can have any value within a given range.

Continuous variable: data that can take any numerical value within a range; it can be measured.

Conversion factor: the number that you multiply or divide by to convert one measure into another smaller or larger unit.

Coordinates: an ordered pair (x, y) identifying a position on a grid.

Correlation: a relationship or connection between data items.

Corresponding angles: angles that are created at the same point of the intersection when a transversal crosses a pair of parallel lines.

Cumulative frequency: the sum total of all the frequencies up to a given value.

Cyclic quadrilateral: any quadrilateral with all four vertices touching the inside of the circumference of a circle.

D

Decimal place (dp): place value position of a digit to the right of the decimal point.

Degree of accuracy: the number of places to which you round a number, for example to the nearest whole number, 2 decimal places, 3 significant figures.

Dependent events: events in which the outcome is affected by what happened before.

Dependent variable: data that is measured in, and affected by, an experiment.

Depreciation: the loss in value of an object over a period of time.

Diameter: a straight line from one point on the circumference to another that passes through the centre of the circle; it is twice the length of the radius.

Direct proportion: two values that both increase in the same ratio.

Discrete data: data that can be counted and that has only one possible value; it is counted in integers.

Displacement: a change in position.

E

Elevation: a view of an object from the front, side or back.

Error interval: the difference between the upper and lower bounds.

Equalities: having the same amount or value.

Equally likely: having the same probability of happening.

Equidistant: means 'the same distance from'; if all points are equidistant they are the same distance apart.

Equivalent: having the same value; two ratios or fractions are equivalent if one is a multiple of the other because they will cancel to the same simplest term.

Estimate: an approximate answer or rough calculation.

Evaluate: to find the value of, to solve.

Event: a set of possible outcomes in an experiment or situation, to which you give a probability.

Exchange rate: the value of one currency used to convert that currency to an equivalent value in another currency.

Expanding: multiplying out an expression to get rid of the brackets.

Exponent: the value of a power.

Exponential function: a function of the form $y = ak^x$, where $k > 1$.

Expression: a group of numbers and letters linked by operation signs.

Exterior angles: angles produced by extending the sides of a polygon.

F

Factorising: writing a number or expression as a product of its factors.

First difference: the result of subtracting a term from the next term.

Formula (plural **formulae**): a general rule or equation showing the relationship between unknown quantities.

Function: a set of instructions for changing one number (the input) into another number (the output).

G

Geometric sequence: a sequence where the ratio between each term is constant.

Gradient: a measure of the steepness of a line.
Gradient $= \dfrac{\text{change in } y}{\text{change in } x}$

Grouped data: data that has been put into groups.

H

Histogram: a graph with bars whose area is proportional to the frequency of a variable and whose width is equal to the class interval.

Hyperbola: the curved graph(s) formed by a reciprocal function; the curve of $y = \frac{1}{x}$ gets increasingly close to the x-axis and y-axis but never touches them.

I

Image: the new shape (once the object has been transformed).

Included angle: the angle between two lines that meet at a vertex.

Independent events: events that are not affected by what happened before.

Index: a power or exponent indicating how many times a base number is multiplied by itself.

Index notation: writing a number as a base and index, for example 2^3.

Inequality: a mathematical sentence in which the left side is not equal to the right side.

Integers: whole numbers belonging to the set {... −3, −2, −1, 0, 1, 2, 3, ...}; sometimes called directed numbers because they have a negative or positive sign.

Interior angles: angles inside a two-dimensional shape at the vertices or corners.

Inverse function: a function that reverses another function.

Inverse proportion: a relation between two quantities such that one increases at a rate that is equal to the rate at which the other decreases.

Irrational number: a number that cannot be written in the form of $\frac{a}{b}$ or as a terminating or repeating decimal.

Irregular polygon: a polygon that does not have equal sides and equal angles.

Isometric grid: special drawing paper based on an arrangement of triangles.

L

Line (or **axis**) **of symmetry:** a line that divides a plane shape into two identical halves, each the reflection of the other.

Linear equation: an equation where the highest power of the unknown is 1, for example $x + 3 = 7$. There are no fractional or negative powers and the resultant graph of the equation is a straight line.

Locus (plural **loci**): a set of points that satisfy the same rule.

Lower bound: the smallest value that a number (given to a specified accuracy) can be.

M

Mathematical model: a representation of a real-life problem; assumptions are used to simplify the situation so that it can be solved mathematically.

Midpoint: the centre of a line; the point that divides the line into two equal halves.

Mirror line: a line equidistant from all corresponding points on a shape and its reflection.

Mutually exclusive: events that cannot happen at the same time.

N

Number line: a line marked with numbers in order, similar to a ruler scale.

O

Object: the original shape (before it has been transformed).

Order of rotational symmetry: how many times a shape will fit exactly onto itself when you rotate it through 360°.

Orientation: the position of a shape relative to the grid.

Outcome: a single result of an experiment or situation.

Outlier: data value that is much larger or smaller than others in the same data set.

P

Parabola: the symmetrical curve produced by the graph of a quadratic function.

Parallel vectors: occur when one vector is a multiple of the other.

Perfect square: a binomial product of the form $(a \pm b)^2$.

Periodic graph: a graph that repeats itself in a regular way.

Perimeter: the distance around the boundaries (sides) of a shape.

Perpendicular bisector: a line perpendicular to another that also cuts it in half.

Plan view: the view of an object from directly above.

Plane shape: a flat, two-dimensional shape.

Plot: draw a graph accurately by marking points on a grid using coordinates.

Polygon: a closed plane shape with three or more straight sides.

Polyhedron: a solid object with flat faces that are polygons.

Polynomial: an expression made up of many unlike terms with positive powers for the variables.

Population: the name given to a data set.

Position-to-term rule: operations applied to the position number of a term in a sequence in order to generate that term.

Prime factor: a factor that is also a prime number.

Product: the result of multiplying numbers and/or terms together.

Proportion: a comparison of a part, or amount, to the whole; often expressed as a fraction, percentage or ratio.

Pythagorean triple: three non-zero numbers (a, b, c) for which $a^2 + b^2 = c^2$.

Q

Quadratic expression: an expression in which the highest power of x is x^2.

R

Radius (plural **radii**)**:** the distance of any point on the circumference from the centre of the circle.

Random: not predetermined.

Ratio: the comparison between two or more amounts in relation to each other.

Rational number: a number that can be expressed in the form of $\frac{a}{b}$ (or as its equivalent as a terminating or repeating decimal).

Reciprocal: the reciprocal of a number, x, is 1 divided by x, i.e. $\frac{1}{x}$. Any number multiplied by its reciprocal is 1. For a fraction $\frac{a}{b}$, the reciprocal is $\frac{b}{a}$.

Reflection: an exact image of a shape about a line of symmetry.

Regular polygon: a polygon with equal sides and equal angles.

Representative sample: a smaller quantity of data that represents the characteristics of a larger population.

Right prism: a prism with sides perpendicular to the end faces (base).

Roots: the individual values of x in a quadratic equation.

Rotational symmetry: symmetry by turning a shape around a fixed point so that it looks the same from different positions.

Rounding: writing a number with fewer non-zero digits by replacing some digits with zeroes.

S

Sample: a small set of data from a population.

Sample space: a list or diagram that shows all possible outcomes from two or more events.

Scalar: a numerical quantity (it has no direction).

Scale factor: a number that scales a quantity up or down.

Second difference: the difference between each term in the first difference.

Sector: the part of a circle enclosed by two radii and an arc.

Segment: a chord splits a circle into two segments, the smaller segment is known as the minor segment and the larger is the major segment.

Semicircle: exactly half of a circle; the diameter splits a circle into two semicircles.

Sequence: a number pattern or list of numbers following a particular order.

Set: a collection. The brackets { } are shorthand for 'the set of'. For example, {2, 4, 6, 8} is the set of the numbers 2, 4, 6, 8, which represents the even numbers.

Significant figure (sf): the most significant figure (digit) in a number is the first non-zero digit when reading the number from left to right.

Similar: shapes that have the same shape and proportions but are a different size.

Simultaneous equations: a pair of equations with two unknowns that can be solved at the same time.

Sketch: a basic graph showing the direction, gradient and y-intercept; it is not drawn by plotting a table of values.

Solution: both possible values of x in a quadratic equation.

Subject: the variable which is expressed in terms of other variables; it is the variable on its own on one side of the equals sign. In the formula $s = \frac{d}{t}$, s is the subject.

Substitute: replace letters with numbers.

Subtended: a subtended angle is one whose sides pass through the ends of an arc (or other curved line).

Supplementary angles: two angles are supplementary angles if they add up to 180°.

Surd: if $\sqrt[n]{a}$ is an irrational number, then $\sqrt[n]{a}$ is called a surd.

T

Tangent: a straight line that touches the circumference of a circle at only one point.

Term: a combination of letters and/or numbers. Each number in a sequence is called a term.

Term-to-term rule: operations applied to any number in a sequence to generate the next number in the sequence.

Transversal: a straight line that crosses a pair of parallel lines.

Truncation: cutting off all digits after a certain point without rounding.

U

Upper bound: the largest value that a number (given to a specified accuracy) can be.

V

Variable: a letter representing an unknown number.

Vector: a quantity that has both magnitude and direction.

Vertically opposite angles: angles that are opposite one another at an intersection of two lines. Vertical here means 'of the same vertex or point', not up and down.

X

***x*-intercept:** the point where a line crosses the x-axis when $y = 0$.

Y

***y*-intercept:** the point where a line crosses the y-axis when $x = 0$.

Index